TEXTBOOK OF
PRACTICAL BOTANY

TEXTBOOK
OF
PRACTICAL BOTANY

Volume - I

(Algae, Fungi, Microbiology)

By

Dr. S. Sundara Rajan

Principal

Reva Institute for Science and Technology Studies

R.T. Nagar

Bangalore – 560 024

ANMOL PUBLICATIONS PVT. LTD.

NEW DELHI - 110 002 (INDIA)

ANMOL PUBLICATIONS PVT. LTD.
4374/4B, Ansari Road, Daryaganj
New Delhi - 110 002
Ph.: 3261597, 3278000
Visit us: www.anmolbooks.com

Textbook of Practical Botany

First Edition, 2003

ISBN 81-261-1359-6 (Set)

PRINTED IN INDIA

Published by J.L. Kumar for Anmol Publications Pvt. Ltd., New Delhi and Printed at Mehra Offset Press, Delhi.

CONTENTS

PART THREE MICROBIOLOGY

Preface

Laboratory study is an integral part of the knowledge that a student acquires in the college. This is true not only for botany but for all science subjects. Theoretical study of a subject has to be supplemented with laboratory experiments to provide a comprehensive understanding.

One of the main advantages of laboratory or practical study is the first hand knowledge that a student acquires about the material of study.

A systematic, structured and disciplined method of work is necessary in the laboratory to obtain maximum benefit. In order to help the students realize their goal practical manuals indeed play an important role.

Textbook of Practical Botany aims to provide the students, a useful hand book to guide them in the proper study of the subject material. Theoretical details are kept at the minimum and more emphasis is laid on understanding the structural details to enable the student a proper identification. A large number of properly labelled original diagrams are incorporated in the book as they are integral to the understanding of the subject.

This work is organised in the following four volumes:

Volume One deals with Algae, Fungi, Bacteria, Lichens, Viruses and Microbiology.

Volume Two deals with Bryophytes, Pteridophytes and Gymnosperms.

Volume Three deals with Taxonomy and Economic Botany.

Volume Four deals with Anatomy and Embryology of Angiosperms, Ecology, Plant Physiology and Cytology.

I have taken utmost care in the presentation of matter, style of writing and choice of illustrations so that the volumes are user friendly. The examples chosen for study are such that the books should cater to the needs of all Indian Universities.

Constructive criticism and suggestions are most welcome.

I am thankful to Mr. J.L. Kumar, Managing Director of Anmol Publications Pvt. Ltd., New Delhi, for undertaking the publication of the book.

Dr. S. Sundara Rajan
Principal

Reva Institute for Science and Technology Studies
R.T. Nagar
Bangalore - 560 024

Preface

[illegible] Laboratory classes form an integral part of [illegible] a student [illegible] college. This is true not only for botany but [illegible] Commercial [illegible] supplemented with laboratory experiments to provide a comprehensive [illegible]

One of the main [illegible] is the first hand [illegible] that a student acquires [illegible]

A systematic, structured and disciplined method of [illegible] laboratory [illegible] maximum benefit [illegible] practical manuals [illegible]

[illegible]

[illegible]

I [illegible] illustrations so that [illegible] the book [illegible] all students [illegible]

[illegible] suggestions [illegible]

[illegible] Delhi, for undertaking the publication of this book.

[illegible]

CHAPTER 1

INTRODUCTION

The study of science subjects is carried out at two levels,viz., *Theoretical* and *Practical*. Theoretical study is carried out in lecture classes when the students are imparted all the details regarding a particular topic. This has to be supplemented by practical or Laboratory studies, when the students come into active contact with the objects about which they would have heard in lecture classes. Therefore it may be safely said that laboratory studies form an integral part of a comprehensive study of any topic in science subjects.

In the study of plants, laboratory assumes a special significance as the students acquire a meaningful and firsthand knowledge about the plants which they study in lecture classes. In fact as far as botany is considered any study of a plant without being supplemented by laboratory observation is futile. It is only in the laboratory study that the student can dissect, observe and analyze the various features of the plant and it is only a study of this type that makes the student acquire a meaningful knowledge about the pant. Besides observation and study in the laboratory plants are also to be studied in field to obtain an account of their habitat, habit, association and such other details.

In this chapter an attempt is made to explain some of the essentials that a student should understand before he begins a practical work.

Cleanliness: This is the most important point that a student should keep in mind while in the laboratory. The working table and the surroundings should be kept dry and spotlessly clean.. All the requirements needed by the students such as petri dishes, slides, reagent bottles, stains etc. must be properly labeled and arranged neatly. The microscope and particularly the optical parts must be cleaned with a white lint free cloth. After the completion of the assigned work, the student should take care to clean the place, remove any debris of the plant material and put it into the dust bin.

The students should work in silence and independently unless the assigned experiment is to be carried out by a group. When a particular work is assigned to a student, he must carefully plan and divide the available time for experimentation, observation and recording of the results. Whenever a doubt arises, it is always better to consult the teacher in charge, than a classmate. It is only by a proper and systematic work, that a student can derive maximum benefit from the practical classes.

Laboratory requirements

The department provides certain amenities, instruments etc to each student for the purpose of practical work. But a student also should carry some of the requirements in order to be effective in the practical class.

One cardinal principle that every student should remember in a practical class is to have one's own requirements and not to depend on anyone. Never lend or borrow any requirements. Borrowing or lending a pencil, needle or blade will disturb the work of both.

In general every student should individually possess the following requirements

1. *Two sharp pointed needles* mounted on stainless steel or plastic handle. This is essential for the purpose of separating and teasing the microscopic material.

2. *A pair of forceps:* This is essential to remove the material either from the bottle or from petri dish.

3. A sharp properly honed razor to cut sections of the plant material.

4. A new safety razor blade. This will help either to cut section or to trim the material.

5. Two fine hair paint brushes. This will be of use in lifting the sections of plant (with out damage) from the petri dish on to the slide.

6. Two sharpened pencils (H, Hb or 2b).

7. A pencil eraser.

8. A pencil sharpener.

9. Glass droppers.

10. A piece of thin, white, lint free cloth to clean the optical parts of the microscope.

11. An observation book to note down immediately the points of interest.

12. A practical record.

13. Glass slides and cover slips(if they are not provided by the department).

Scheme of Practical Work

Practical work in the laboratory will be most useful and effective if the student keeps in mind the following:

1. Maintain silence during working. All the requirements for the practical work must be absolutely clean.

2. Every student should work independently, unless the experiment involves the participation of others.

3. Before beginning the practical work, listen to the instructions of the teacher attentively and work accordingly.

4. Follow a systematic procedure in a study.

5. If the work involved is the morphological study of an organism, follow the procedure given below:

(a) Study of external characters - for example whether a plant is microscopic or macroscopic, unicellular or multicellular, filamentous or colonial etc.

(b) Study of internal structure like details of cell, tissues etc. If the material requires sectioning, sections are to be cut and internal details should be observed.

(c) Study the reproductive details like asexual and sexual reproductive organs.

6. Mount the material after properly teasing and separating the filaments, mycelia etc. Stain the material wherever necessary and cover the preparation with a cover glass.

7. Observe the slide carefully under the microscope, first under low power and then under high power.

8. Make sure that you have observed all the details as instructed by the teacher.

9. After all the observations are over, record your findings, structural details etc. in the practical records in the form of diagrams. Draw proportionate diagrams in the proper place on the drawing sheet. Diagrams should never be drawn to the very edge of the drawing sheet. Write suitable explanatory notes pertaining to the diagrams on the interleaf.

Laboratory Equipments

In addition to the individual requirements carried by the student, the department generally provides the

following equipments/instruments etc.to the students. Sometimes these equipments may be given to each student .

These are :

(1) Students compound microscope .
(2) Dissection microscope.
(3) Binocular Research microscope.
(4) Watch glasses.
(5) Cavity blocks.
(6) Slides and cover glasses.
(7) Petri dishes.
(8) Enamel tray.
(9) Polythene/glass wash bottle.
(10) Spirit lamps.
(11) Glass marking pencil.
(12) Microtome.
(13) Staining /bottles.
(14) Stone/ for honing the razor.
(15) Chemicals and stains.
(16) Culture media(to be prepared by students).

In the list of requirements mentioned above, we will discuss some in detail.

Microscope

Of all the instruments used by students in the botany laboratory, the one that is most indispensable to the students of biology is the 'Microscope'. It is not an exaggeration to say that no biological study in the laboratory is possible without the microscope.

The function of the microscope is to enlarge the small objects, so that they become visible to the human eye. In reality the magnification power of the microscope lies in its resolving capacity between one particle and another. A microscope distinguishes one particle from another which are situated very closely. A human eye cannot resolve particles that are lying closer than 0.025 mm. In its function a microscope can be compared to a human eye, but much more effective than that.

There are different types of microscopes, but all of them are based on the same principle with variations in their construction. Some of the types of microscopes used in the laboratory are:

(1) Compound microscope.
(2) Binocular Research microscope.
(3) Dissection microscope.
(4) Stereoscopic dissection microscope.
(5) Phase contrast microscope.
(6) Electron microscope etc.

Of the six types of microscopes mentioned above the compound microscope and the dissection microscope are the ones that are most frequently used by students. The Electron microscope is very costly and a very highly sophisticated instrument.

In this book the compound microscope and the dissection microscope are explained in detail as the students should be familiar with them.

Compound Microscope

The microscope has a strong 'U' shaped or 'V' shaped base. It is also called the horse shoe shaped base. Attached to the base is a bent limb. This limb can be tilted in any direction as it is attached to the base with an inclination joint.

Attached to the limb is a stage at its lower end and a body tube at its upper end. The body tube is attached to the limb with the help of a rack and pinion so that it can move up and down. To help in the movement of the tube during focusing, there are 2 knobs viz.,*Coarse adjustment* knob and *Fine adjustment knob*.

The body tube at its top end consists of the ocular or eye piece lens. The eye pieces usually have three magnifications 5x,10x,and 15x, and can be used to get a low or high magnification. Attached to the lower end of the tube is a revolving nose piece with three objective lenses low (10x) medium (40x) and high (60x). Some times there will be a fourth objective of the magnification of 100x. This is also called the oil immersion lens (while viewing through an oil immersion lens a drop of oil-liquid paraffin has to be put on the viewing slide).

Below the objectives and attached to the limb is the *stage* of the microscope where the slides are kept for viewing. The stage may be circular, square or rectangular in shape. In order to keep the slides properly placed on the stage, there are two stage clips. In some of the costly microscopes, the stage has movable sliding system operated by knobs to move the slides. This is called the mechanical stage. The mechanical stage permits the movement of the slide backwards or forwards, left or right with the turning of the knobs provided for the purpose. A hole is provided in the stage to receive the light from below and illuminate the object. Attached to the limb, below the stage is a condenser (fixed or movable up and down) lens. Below the condenser lens is an Iris diaphragm to regulate the light passing through. Iris diaphragm is a metallic disc with an aperture. The aperture can be narrowed or widened by turning the lever. The condenser which is a system of lenses condenses the light rays passing through and converges them on the object.

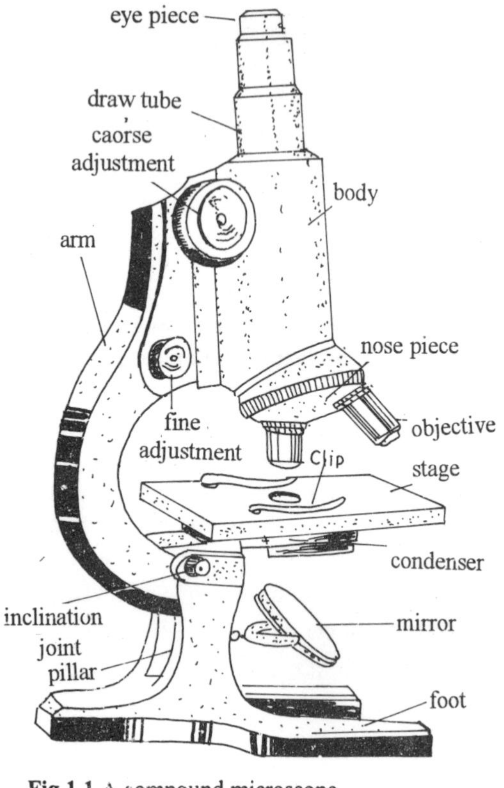

Fig.1.1 A compound microscope

The source of illumination in a student's microscope is a mirror which reflects the natural light or electric light. The mirror has two faces - plain and concave and is attached to the limb below the condenser. The mirror is adjustable and can be turned to the direction of light. When day light is the source of illumination, the concave side of the mirror is usually used to converge the rays of light. In some of the costly microscopes, instead of the mirror there is a built in illumination ,This is a tube with a lamp (9v or 12v) attached to a transformer, which can be plugged to any AC source.

Operation of the Microscope

Microscope should be placed in a well illuminated area but avoiding the direct light as it is harmful to the eyes. Generally northern light is preferable. When ' built in illumination' is employed, it is always better to use a blue or green filter to subdue the light.

The light has to be first focused on the condenser by adjusting the mirror towards the source of light. The amount of light entering the condenser may be regulated by closing or opening the iris diaphragm. The slide to be viewed is kept on the stage and fixed securely with the help of the stage clips. The object is first observed with a low power objective and then turned on to medium or high power, the specimen or section has to be covered with a cover slip. The slide should be observed with both the eyes open even while observing with monocular microscope.

Care of the Microscope

1. Clean all the parts of the microscope before and after use. The lenses and the mirror are to be cleaned with a lint free cloth or tissue paper.

2. Do not pull the microscope to the very edge of the table.

3. While lifting the microscope, use both hands. With one hand hold the base and with the other hold the arm.

4. The slide to be observed must be clean and dry (specially the under surface of the slide).

5. Do not remove the eyes piece or objectives from the microscope.

6. The microscope should be covered when not in use.

Binocular Research Microscope

This is like the ordinary compound microscope except that, it has binocular head for providing a natural stereoscopic view.

Dissection Microscope

This is a very simple microscope with only one set of lenses. Microscopes of this type will not give a very high magnification, but are very useful in dissecting flowers, plant parts etc.

The microscope has a 'V' shaped base and a limb. The stage is made up of a transparent glass plate. Below the stage is a mirror to focus the light. Attached to the limb is a mechanical arm which can be moved up and down with the help of a rack and pinion movement. The arm can also be moved horizontally. Attached to the free end of the arm is a holder to accommodate the lens (10x,15x etc.).

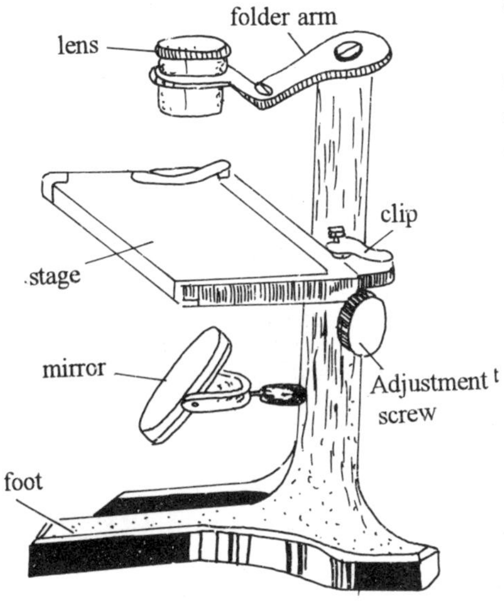

Fig.1.2 A dissecting microscope

Operation: The specimen to be observed / dissected is placed on the stage and located with the help of lens. Floral parts or others may be then dissected with the help of two needles, while still observing the specimen.

Stereoscopic Binocular Dissection Microscope

This is a combination of compound microscope and dissection microscope. While it has the higher magnification power of a compound microscope with two sets of lenses, it has retained the versatility and the manoeuvourability of a dissection microscope.

Phase contrast and Electron microscopes are not normally used by students.

Other Laboratory Equipments

Stains: These are colouring fluids. The specimens to be observed (algal or fungal material or sections of plant parts, would be generally decolourised due to preservation. These are to be coloured in order to observe them clearly. The following are the stains that are generally used in the laboratory.

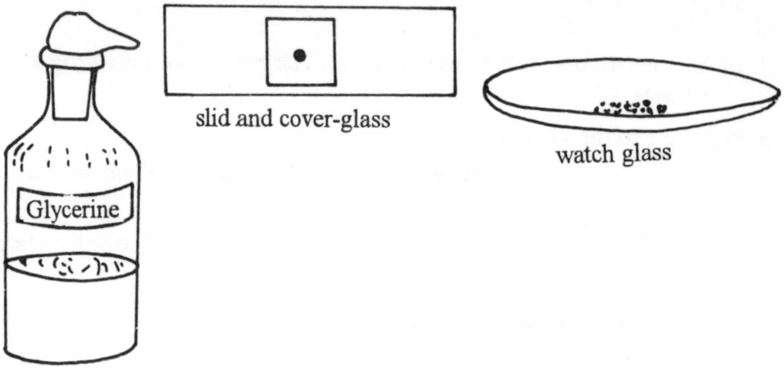

Fig.1.3 Dropping bottle

(a) Saffranin: This is the most commonly used stain. (1) It stains nucleus and lignified cell walls.
Preparation —

Saffranin —1gm.
95% (1) Ethyl alcohol –50ml.
Distilled water – 50ml.
An aqueous stain may also be prepared.
Saffranin – 18ml
Distilled water – 100ml.

(b) Fast Green

Fast green – 1gm.
90% Ethyl Alcohol – 100ml.

This stain is used to colour parenchyma cells and cytoplasm. Fast green may also be prepared in clove oil (75ml clove oil+25ml 100% alcohol).

(c) **Cotton blue.**

Cotton blue –1gm.
Lactophenol – 100ml.

(d) Haematoxylin

The dye is obtained from the heartwood of the tree *Hametoxylon campechianum,* There are two methods of preparation of the stain. These are:

1. *Heidenhain's Hametoxylin*

Dissolve 0.5 or 1.0 gm of the powder in 100 ml of distilled water. Filter and store in a dark bottle away from light. The stain is ready for use after 4-5 days of ripening. Ripening is not necessary if the stain is prepared in 70% ethyl alcohol.

2. *Delafelds Haematoxylin*

Sol.1 Aqueous solution of ferric ammonium sulphate 100ml).

Sol.2 One gm dye dissolved in 6ml of absolute alcohol.

mix solutions 1 and 2. To this mixture add 25 ml of glycerine and 25 ml absolute alcohol. The solution should be allowed to stand until it becomes dark red.

(e) Anilin blue

Also called cotton blue or China blue. It is prepared as follows:

Aniline blue –1gm.

Ethyl alcohol (90%)
or — 100ml.
Distilled water
Hydrochloric acid – 2drops.

(f) Gentian Violet (Crystal Violet)

Gentian Violet – 1gm.
90% Ethyl Alcohol 100ml.

(g) Erythrosine

Erythrosine –1gm.
90% Ethyl alcohol – 100ml.

Erythrosine may also be prepared in clove oil (Absolute alcohol 0.5 ml +95ml Clove oil).

(h) Light green

Light green –1gm.
Clove oil –95ml.
Absolute alcohol –5ml.

Light green may also be prepared without clove oil (in 100ml Absolute alcohol).

(i) Acetocarmine: The stain is prepared as follows:

Add 1 gm of stain to 100 ml of boiling acetic acid (45%)or propionic acid. Cool and decant, A few drops of saturated ferric acetate solution is added and the mixture is cooled over ice. The solution is allowed to stand for about 12 hours. Filter and store in a dark coloured bottle under refrigeration.

Fig.1.4 Brush Needle Forceps

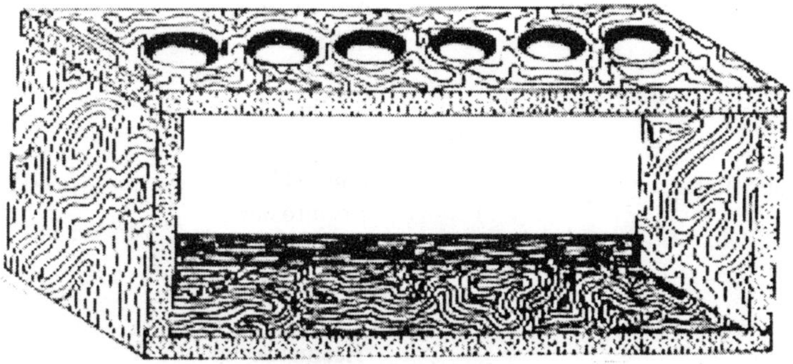

Fig.1.5 Staining rack

Specificity of Stains

Usually specific stains are employed in colouring different plant parts. The stains are used singly or in combination. The most popular stain combination are:

Saffranin - fast green.
Haematoxylin - Erythrosin.
Haematoxylin - light green.

While stain combinations are employed to study pteridophytes, gymnosperms and Angiosperms, single stains are used for Avascular cryptogams.

Green or Saffranin for Algae and Bryophytes.

Cotton blue for fungi.

The following table gives the specificity of stains to different plant parts or cell organelles.

Structure	Stain
Cellulose	Anilin blue Haematoxylin Fast green Light green
Cutinized cell wall	Gentian violet Erythrosine Saffranin
Lignified cell wall	Gentian violet Saffranin
Suberised cell wall	Saffranin
Cytopalsm	Fast green Light green Aniline blue
Chitin	Saffranin
Callose	Aniline blue
Achromatic figure	Aniline blue Erythrosine
Proteins	Saffranin
Nucleus	Heamatoxylin Saffranin Gentian violet
Chromonemata	Heamatoxylin

Chromosomes	Saffranin
Plastids	Gentian violet Haematoxylin
Mitochondria	Gentian violet
Flagella	Cotton blue

Fig.1.6 Razor

Fixing / Preserving Fluids

The plants or plant parts collected in nature are to be killed and preserved in the laboratory for the purpose of storage and study.

In killing the specimen, the various life processes are stopped at that instant, so that after preparation for microscopic study various stages may be observed.

When ever a material is collected, it is better to kill it in the field, so that there will be minimum or no distortion. Alternatively the material collected in the plastic bags or bottles should be brought to the laboratory and killed / preserved. In preservation, the material must be completely immersed in the fluid. The following are some of the killing/preservative fluids generally used:

1. *Formaline acetic alcohol (F.A.A)*

This is the most commonly used killing /preserving agent. It is popularly called standard preservative.

Ethyl alcohol (70%)	- 90ml.
Glacial acetic acid	- 5ml.
Formalin	- 5ml

2. *Formalin - Propiono - Alcohol*

Formalin	- 5ml
Propionic acid	- 5ml.
Ethyl alcohol 70%	- 90ml.

3. *Carnoy's Fluid - I*

Ethyl alcohol	- 30ml
Glacial acetic acid	- 10ml

4. *Carony's Fluid - II*

Ethyl alcohol	- 30ml.
Glacial acetic acid	- 30ml.
Chloroform	- 15ml.

Carony's fluid is widely used in killing root tips, flower buds etc. for the purpose of cytological study.

5. *Randolph's Modified Navashin Fluid*

This consists of two solutions A and B.

Solution A

Chromic acid	- 5.0 gms.
Glacial acetic acid	- 50ml
Distilled water	- 320ml.

Solution B

Commercial formalin	- 200ml.
Distilled water	- 175 ml.
Saponin	- 3gms.

Mix A and B in equal volumes before use.

Mounting media

These are chemicals used for the purpose of mounting (Temporary/Permanent) the specimens on the slide for microscopic observation. The following are some of the mounting media.

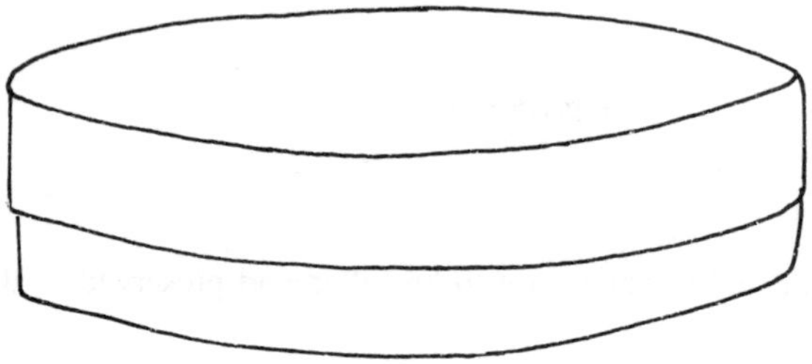

Fig.1.7 Petri dish

1. ***Acid water***

HCl – 10-15 drops

water – 100 ml.

2. ***Lactophenol***

Phenol	-100gms
Lactic acid	-100 ml.
Glycerine	-100 ml.
Distilled water	-100ml.

3. ***Glycerine***

Glycerine	- 15-20 drops
Formaline	- 1-2 drops
Distilled water	- 100ml.

4. ***Canada Balsam:*** This is a gum obtained from Gymnosperms. This fluid is readily available in the market.

5. ***Euparol:*** This is available in the ready to use form in the market.

Canada balsam and Euparol are used for permanent mounting, while the rest are used for temporary preparations.

Staining rack: This is a wooden rack used for holding the staining bottles.

Dropping bottles: This is a glass bottle having a narrow mouth fitted with a stopper. The stopper is grooved and has a beak. When the bottle is upturned the stain/fluid comes out in drops.

Slides: These are readily available in the market. The usual size 25mm x 75mm. The thickness is about 1mm. Special types of slides called cavity slides are available for specific study.

Cover slips (Cover glasses)

These are meant to cover the materials when the preparation is ready for observation. The coverslips may be round, square or rectangular.

Methods of Study of Plant Material

The following is the procedure for a scientific study in the laboratory.

Morphological study

The specimen given to study external detail should be observed either with or without a magnifying glass. Whatever details can be seen are to be recorded. If the specimen requires microscopic study, it should be suitably prepared for the purpose of microscopic observation. There are two types of microscopic preparation.

1. Whole Mount Technique

This is suitable for the purpose of external study of algal and fungal material. Take a little of the material, stain it with a suitable stain (staining procedure is explained later.) Wash the excess stain in water / alcohol. Mount in a suitable medium (temporary or permanent) and observe under the microscope.

2.Section cutting

For the purpose of anatomical study, sections are to be cut of plant parts. In algae, globule of *Chara*, stolons of *Caulerpa*., receptacles of *Sargassum* etc., can be cut. Similarly fruiting bodies of fungi, thalli of bryophytes also many be cut in a suitable plane. In higher plants cylindrical plant parts like root, stem, petiole etc. may be cut for the study of internal structure.

Planes of Cutting: In order to study the internal structure from various angles, sections are cut in various planes.

Transverse :The section is cut at right angles to the longitudinal axis.

Longitudinal : The section is cut parallel to the long axis (right angles to the transverse axis. The longitudinal section may be cut along a radius (Radial longitudinal section-RLS) or along a tangent (Tangential longitudinal section-TLS).

Cylindrical organs like roots and stems are cut transversely or longitudinally, while dorsiventral organs like thalli of bryophytes, leaves etc. are usually cut in vertical transverse plane.

Types of Section cutting

There are usually 2 types of section cutting methods employed for plant parts. These are (a)Free hand sectioning and (b) Microtome sectioning

(a) Free hand Sectioning

For the purpose of laboratory study where slides are not made permanent, usually free hand sectioning method is employed. Free hand sections are cut mostly with the help of a good sharp razor (sometimes safety razor blades also may be used).

Honing the razor : The razor used for sectioning must have a sharp, even edge. There should be no nicks. Before cutting sections, the razor edge has to be honed to maximum sharpness either on a stone or on a leather strap. Oblique strokes, with the edge forward are to be given on the sone or strap. Usually oil drops are used on the honing medium to give maximum sharpness.

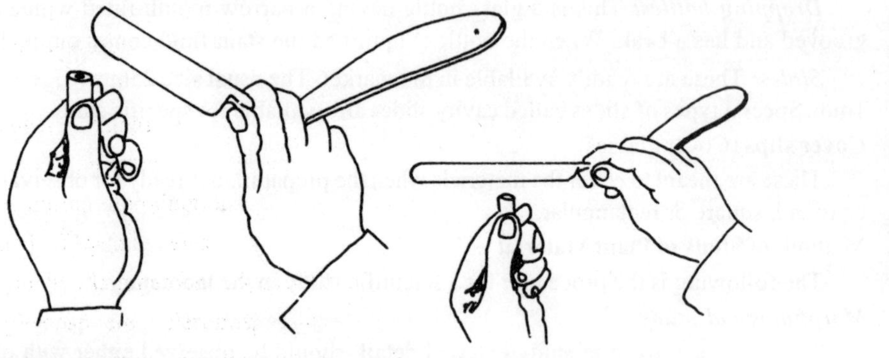

Fig.1.8 Method of holding pith Methoding of holding razor Method of section cutting

Method of cutting : Open out the razor and bend the handle backwards. Hold the razor between the index finger and thumb. The handle should be free. The remaining fingers of the hand will be touching the back edge of the razor blade.

In order to cut a transverse section of a cylindrical part like stem or root, follow the procedure given below.

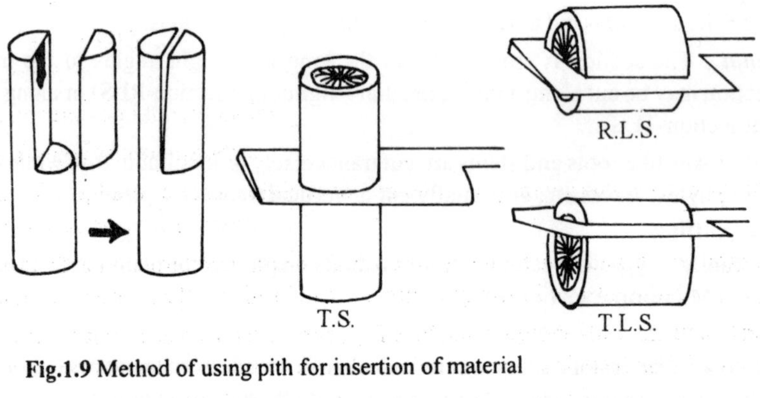

Fig.1.9 Method of using pith for insertion of material

The stem or root should be cut into a 2-3" long piece. It should be held in the hand in an erect position between the thumb and index finger. The material held in the hand should be exactly at right angles to the razor's edge. The thumb holding the specimen should be slightly at a lower level. Flood the edge of the razor with water. Now using the index finger as the platform move the razor quickly several strokes across the material. During this process sections are cut and they will be collected on the blade of the razor. The sections now may be carefully transferred with the help of a fine brush in to a petridish containing water.

For cutting sections of delicate and fragile materials like thalli or leaves, a suitable supporting medium has to be used. For this purpose a pith, carrot or potato may be used. These are to be cut into rectangular cubes and then split at one end. The material for sectioning should be inserted in the slit and sections are cut

After section cutting the razor should be wiped dry. It should be greased and encased.

Microtome Sectioning: This is generally employed very frequently in the study of bryophytes, pteridophytes and other higher plants. The microtome sectioning is a long and complicated process. It takes anywhere between 15-20 days to process a material and to obtain the sections.

Microtome sectioning involves the following steps:

1. Killing and preserving the material.
2. Dehydration and infilteration.
3. Paraffin embedding .
4. Section Cutting.

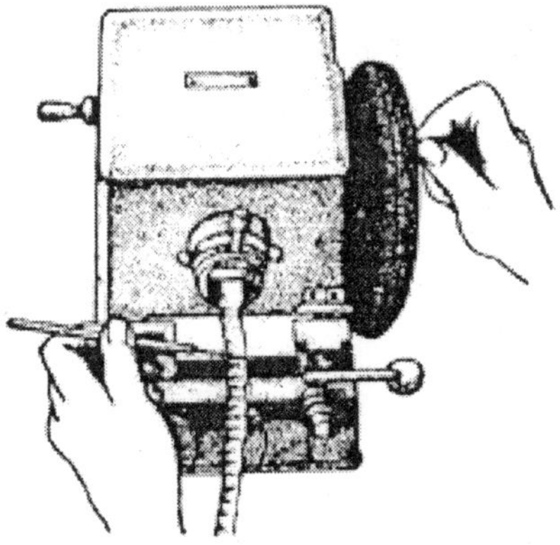

Fig.1.10 Rotary microtome

The material like stem, leaf, flower etc. is to be killed and preserved in a suitable fluid.

(a) Cut the material into bits of suitable size.

(b) Place the material in small vials containing a fixative or preservative.

(c) Label the vial giving the details of place of collection , name of the material, preservative used and date of collection.

The material should be in the killing fluid at least for six hours (preferably a day).

Dehydration: In order to dehydrate the material, remove it from the preservative and wash thoroughly in water and transfer it to a suitable tube or vial.

Dehydrate the material in alcohol series (Ethyl alcohol) as follows

Stage 1	50% Alcohol	First day
Stage 2	70% Alcohol	Second day
Stage 3	90% Alcohol	Third day
Stage 4	100% Alcohol	Fourth day
Stage 5	Absolute alcohol and xylol (50/50)	Fifth day
Stage 6	Absolute alcohol and xylol (25/75)	Sixth day
Stage 7	Xylol	Seventh day

When the material is transferred to pure xylol, there should be no fogging (i.e the solution should not turn milky). If there is fogging repeat the procedure starting from 4th stage.

The material in xylol will be completely dehydrated and looks transparent. It is now ready for infilteration by paraffin wax. The purpose of infilteration is to see that the wax enter into the material completely so that it can easily be cut.

Infilteration: Add a few pieces of paraffin wax everyday for about 4-5 days and then transfer the vial into a hot air oven, while temperature is adjusted to the melting point of wax. Everyday add a few pieces of wax. Adding of wax should be continued until all the xylol in the vial is replaced by paraffin wax. When the tube is smelt, there should be no smell of xylol. The material is now ready for embedding.

Paraffin embedding: In order to cut the material, it has to be embedded into a paraffin block. Prepare a paper boat or take a porcelain rectangular dish and smear glycerine to the inner surfaces. Pour molten paraffin into the porcelain dish and empty the contents of the vial into the dish. Before the paraffin solidifies, arrange

the material in proper rows and in such a manner as desired. Leave it for cooling for 6-8 hours. For quick cooling the dish may be transferred to a bowl containing ice water. Carefully remove the solidified paraffin block from the dish and cut it into suitable pieces. The material is now ready for sectioning.

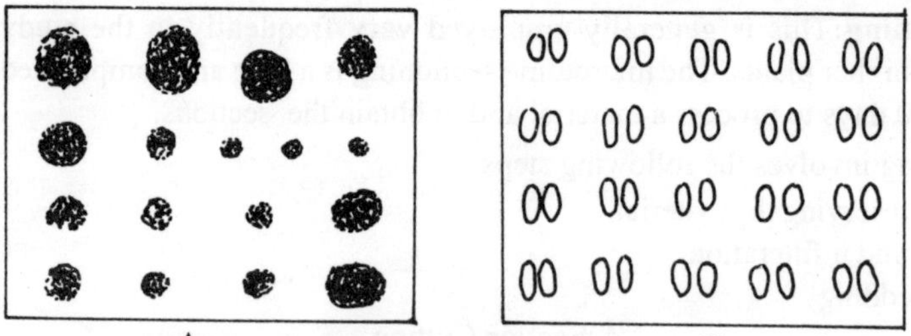

Fig.1.12 Arrangement of materials in paraffin blocks - A) Flower buds in groups B) flower buds in pairs

Sectioning: Sectioning is carried out with the help of an instrument called microtome. There are various types of microtomes like Rotary, sliding, Freezing etc. The one that is normally used is a rotary microtome. The following procedure can be adopted in the case of the rotary microtome.

1. The paraffin block is to be cut into small pieces each having the embedded material.

2. The pieces of paraffin containing the material is to be fixed to wooden 'riders'. For this purpose, heat the top of the rider and the base of the paraffin piece with a hot knife and fix. Allow it to cool. The paraffin piece is now firmly fixed to the rider. In order to make the edges of the paraffin even, cut the edges and make them parallel.

3. Fix the wooden rider into the microtome so that the paraffin piece with the material is exactly parallel to the edge of the microtome knife.

4. Adjust the knob of the microtome to give the required thickness to the sections.

5. Rotate the wheel of the microtome in a clockwise direction. The cylinder containing the rider moves up and down on the knife edge. Sections are cut in a sequence and get attached to each other forming a paraffin ribbon.

6. Keep a needle in your left hand to hold up the ribbon as it lengthens. When the ribbon is sufficiently long, cut it and transfer it on to a black cardboard. Cut the ribbons into suitable length and transfer to the slide.

7. Smear the slides with an adhesive like egg albumen or gelatin and rub it well.

8. Pour some water onto the slide with the help of dropping bottle.

9. Transfer the paraffin ribbons on to the slide and drain excess water.Retain a little water on the slide.

10. Place the slide on a slide warmer. The ribbon expands, and so also the sections of the material. Now draw out the water completely and allow the slides to cool.

11. The slides should be allowed to stand for atleast 24 hours whereupon the sections will stick to the slide permanently.

The slides are now ready for staining.

Staining procedure

The plant material (whole mount or sections) are to be stained and mounted properly before they are ready for microscopic observations.

Algal filaments, fungal mycelia, pollen grains and spores, prothalli of pteridophytes are usually stained with a single stain - saffranin or green. In some case Aniline blue is used to stain the filaments of Algae and mycelia of Fungi. The staining procedure is as follows.

Place the material either in a watch glass or cavity slide and add a few drops of stain. The period of staining varies with different materials. But usually a 5-10 minute period is sufficient. Drain out the stain and observe the material under the microscope. If the material has taken less stain, it may be kept in stain for some more period, or if it has taken excess stain it can be removed by washing either with water or 50% ethyl alcohol.

After the material is suitably stained pass it through a series of alcohol grades to dehydrate it. Usually different grades of alcohol (70%,90%,100%) taken in a filler are put on the material. After passing through absolute alcohol the material is kept in xylol and it is now ready for mounting.

Fungal mycelia can be stained with aniline blue. Usually the material taken on a clean slide is stained with a drop or two of the stain prepared in lactophenol. The material can bemounted straightway as lactophenol is also a mounting medium. If the stain is not prepared in lactophenol the material may be transferred to fresh lactophenol and mounted.

Double Staining Procedure

In the case of most of the algae and fungi there is little or no tissue differentiation and hence single staining is sufficient. In some cases of plants with relatively more tissue differentiation, a single stain is not sufficient to mark out the various tissues clearly. In such cases 2-3 stain combinations are used to stain the tissues. For example lignified and non-lignified tissues are stained with a combination of acidic and basic dyes.This type of staining is also called differential staining where different tissues take up different stains.

The following are some of the popular stain combinations.

Main stain	*Counter stain*
Haematoxylin	Saffranin
Haematoxylin	Erythrosine
Haematoxylin	Eosin
Haematoxylin	Orange G.
Saffranin	Fast green
Saffranin	Light green
Saffranin	Aniline blue
Saffranin	Crystal violet
Crystal violet	Erythrosine

Staining Procedure for free hand sections

(a) Procedure for temporary preparations

Free hand sections of root, stem etc., should be washed thoroughly in distilled water in a watch glass. After repeated washing the sections should be transferred to a watch glass containg the main stain. The sections should be kept in the stain for a period ranging from 30-40 minutes. After this period, the sections should be transfereed to a watch glass containing distilled water. The sections should be washed for 2-3 minutes whereupon excess of stain is removed. They should be observed under the microscope. If the main stain is saffranin only lignifed tissues should retain the stain, others should be destained. If the main stain is specific to non lignified tissues, only they should retain the stain. Then the sections should be washed thoroughly in water and repeatedly observed under the microscope. The sections can be mildly stroked with a fine hair brush, so that excess of stain comes out with distilled water. The sections are now ready for counter-staining. They should be transferred to a watch glass containing the counter stain (fast green etc). Counterstain acts quickly on the tissues. Hence one to two minutes of staining should be sufficient. The sections are now observed under the microscope. If there is excess of counter stain, it can be removed by

washing sections with 15-20%glycerine. After the final stianing the tissues whould be demarcated clearly with no mixing of the stains. The sections are now ready for mounting.

(b) Procedure for permanent preparaions

If the sections of plant materials are to be mounted and made into permanent slides, the procedure should be slightly altered so as to include dehydration and clearing with suitable agents. The procedure is as fallows:

1.Stain with the principle stain (Saffranin, Haemotoxylin etc).

2.Allow the sections to remain in stain for about 15-20 minutes.

3.Transfer the sections to water and wash thoroughly. The sections should be stroked mildly until no more stain comes out, and the water should remain clear. If the main stain like saffranin is produced in 50% alcohol then destaining (washing) should be done in 50% alcohol.

4.The sections should be now dehydrated in alcohol grades of 50%, 70% and 100%.

5.In each grade of alcohol keep the section for about 5 minutes.

6.Transfer the sections from one grade of alcohol to the other with the help of a fine brush. Never use needles as they injure the sections.

7.Transfer the sections to the counter stain (fast green etc). The counter stains are usually prepared in 90% alcohol.

8.Keep the section in counter stain for about a minute. Wash the sections in 90% alcohol and transfer them to 100% alcohol.

8a. In some case counter stains are prepared in clove oil. In such cases counter staining should be done after the sections are completely dehydrated with absolute alcohol.

1. Sections are transferred completely dehydrated to a solution of 50:50 Absolute alcohol and xylol followed by 25:75 Absolute alcohol and xylol and ultimately the sections are transfered to pure xylol. Now xylol should not turn milky or turbid. If it turns milky, dehydrate the sections again through combination of absolute alcohol : xylol series.

The sections are now ready for mounting.

Staining procedure for microtome sections

The slides with the paraffin ribbons adhering to them have sections of plant material.The processing and staining is as follows:

1. Immerse the slides in a coupling Jar containing xylol. Leave them for 10 minutes . The paraffin dissolves in xylol and only the sections remain on the slides.

2. The slides are now kept in 50:50 absolute alcohol and xylol.

3. Pass the slides through a down grade series of alcohol. 100%, 90% and 70%

4. Transfer the slides to pure water and wash them thoroughly to remove traces of alcohol.

5. Keep the slides in a mordant like 4% iron alum. This is necessary only when Haematoxylin is used. After about 10 minutes wash the slides in water.

6. Transfer the slides to the main stain Haematoxylin. Saffranin etc. and keep them for 15-30 minutes.

7. Remove the slides from stain and wash in water.

8. Observe the slides under microscope. Usually the sections are overstained.

9. If the main stain used in Hametoxylin, destaining can be done with 2% Iron alum or with water containing few drops of concentrated HCI. Repetaedly observe the slides under the microscope. Stop destaining when only the nuclei have retained the stain.

Caution: When destaining is done with HCI, not more than 5-6 drops should be used for 75ml of water.

Excess acid will eat away the adhesive and the sections come out of the slide.

10.Wash the slides in water overnight if the destaining agent is Iron alum, otherwise washing for 30 minutes is enough.

11.Now pass the slides through an upgrade series of alcohol (5 minutes in each grade) 70% 90% 100%

12. When the sections are completely dehydrated stain them with the counter stain (Fast green etc) for a minute or two. Wash the excess stain in clove oil and transfer the slides to pure xylol. The slides are now ready for mounting.

Mounting

(a) *Mounting Media*

Mounting media are of two categories: Temporary mounting media and permanent mounting media.

A temporary mounting medium may be just water or water mixed with glycerine. If the temporary sections are to be retained for 2-3 hours, dilute glycerine (15-20%) is recommended.

Permanent mounting media are used when the slides are to be retained indefinitely. The following are some of the permanent media.

(1) **Canada Balsam:** This is the most widely used mounting medium in botanical preparations. It is available in ready to use bottles. It is a gummy substance. If the solution is very thick add little xylol.

Canada balsam is a resinous product of a Gymnosperm, Abies balsamea. The gum has very good refractive index and as such it is highly reccommended.

(2) **Lactophenol:** Lactophenol can be prepared in the laboratorty. Mix equal quantites of phenol cystals, lactic acid glycerine and distilled water.

(3) **Glycerine gelly:B** This is prepared as follows:

Gelatine 1 part.
Water 6 parts.
Glycerine 7 parts.

Glycerine jelly is suitable even for mounting fresh specimens (even without dehyderation).

Add gelatine to water, stir thoroughly and warm. Add glycerine and cool the solution. After sometime the solution settles into a jelly.

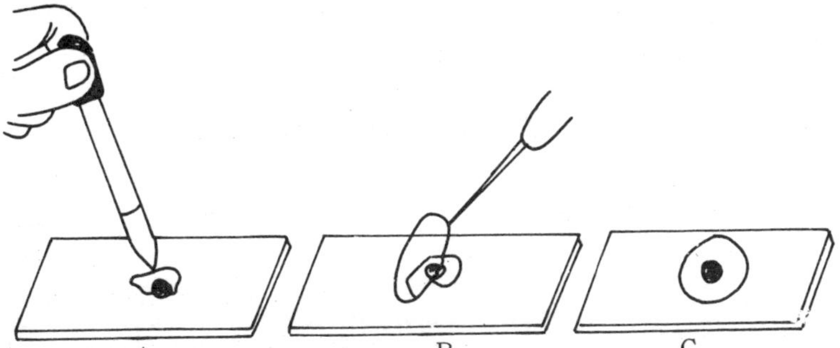

Fig.1.13 Method of mounting the material

(b) Mounting Method

The sections (free hand) are to be mounted on to clean slide in a specific mounting medium. If the sections are microtome preparations they would be already adhering to the slide. In any case the sections are to be covered with a thin coverslip or cover glass which are readily available in the market. They may be round, square or rectangular. Select the one that is suitable for your preparation.

The sections or objects should be mounted on the centre of the slide. In order to ensure placing of the object on the centre of the slide, the following procedure may be adopted. On a white paper draw a rectangle of the same size as the slide. Join the opposite corners by drawing diagonal lines. The centre of the slide lies where the two lines cross. Place the slide on the rectangle and material or objects may be kept on the central point.

Place a drop or two of the mounting medium. Hold the coverslip at the edges (see the figure) and allow it to touch the edge of the mounting medium at an angle of approximately 45°. Slowly lower the cover glass on the mounting medium with the help of a needle. Take care to see that no air bubbles enter the mounting medium. If there are air bubbles, lift the cover glass and mount again.

Excess mounting medium may be carefully wiped with the help of a clean tissue. Allow the slides to dry. If necessary the slides may be transferred to a hot air oven for drying. The slides should be labelled at one end as shown in the figure. They are now ready for microscopic observation.

Ringing of Slides

Temporary preparations may be kept for few days if the borders are sealed. This procedure is known as ringing . Ringing is done by applying wax, canada balsam or even nail polish to the edges. Ringing may be done manually or by using a ringing table having a rotary metallic disc.

Record of Practical Work

Practical work carried out in the laboratory should be recorded. The botany practicval record book has two types of sheets viz., Interleaves and drawing sheets. Interleaves are meant to write the notes pertaining to the diagrams. The drawing sheets which are thicker than the interleaves are used for diagrams. The following points may be useful in maintaining a good record.

1. Use good pencils (HHB) with sharp points and a good quality eraser.
2. Leave a minimum of one inch margin on all the sides.
3. Always write diagrams of only one plant in a page. Do not mix up diagrams of other plants.
4. On the left hand corner of the drawing sheet write the date of the experiment.
5. Give the complete systematic position of the member on the top of the drawing sheet.
6. For microscopic organisms draw first the habit sketch followed by low power and high power view/s. High power view should show the cellular details.
7. In the case of anatomical drawings, draw the ground plant first, followed by a sector enlargement to show the cellular details.
8. All parts of the diagrams must be properly labelled.
9. Stages of reproduction, if any, must be drawn after the details of vegetative structure.
10. The description corresponding to the diagrams must be written on the opposite interleaf.

Field Collection of plants

Specimens of algae, fungi and bryophytes can be collected in the field for the purpose of laboratory study. Field trip is an integral part of botanical study and all students must participate in it. It is in the field study that the students will come to understand the natural growth, habitat and environmental association of plants. During a field trip to collect specimens, the following items are to be carried:

1. Pocket lens.
2. Fixative F.A.A or Carnoy's fluid.
3. Collection bottles.
4. Blotting paper.
5. Polythene bags
6. Field note book
7. Labels.

When the specimens are collected they can be fixed and preserved for the purpose of laboratory study. Use the required preservative and immerse the specimens in it. Label the bottle showing details of

Date of collection

Location

Preservative fluid (name) used.

Name of the plant

As it is necessary for the purpose of identification, details of habitat, associations with other plants if any, colour of the fresh specimen etc. should be recorded in the field note book.

CHAPTER 2

CLASSIFICATION OF THE PLANT KINGDOM

The living world consists of plants and animals of various kinds. Evem within the plant world many kinds of different types of plants have been recognized and perhaps many more are yet to be discovered If such huge number of individuals are not properly arranged, their study will not only be difficult but also meaningless because proper comprehension is impossible. So, in order to bring an arrangement and discipline plants have been grouped into larger and smaller groups based on resemblances and differences (mostly morphological). This system of dividing plants into groups is known as "plant classification or systematics."

In general there are mainly two systems of classification, viz. (1) Artificial system and (2) Natural system. Phylogenetic.

In the artificial system of classification, as proposed by earlier botanists, importance is given to minor, superficial and frequently variable characters. For example divisions of plants based on their habit into herbs, shrubs and trees and using this as a criterion in the classification. The artificial system does not indicate the relationship existing among different plants.

The natural system of classification is based on the natural phylogenetic affinities existing between various types of plants. Natural system lays stress on the similarity in reproductive structures including anatomical character, cell structure and physiology. Such structures are more stable than the ordinary vegetative characters and are not pone to frequent alterations by the environmental influence.

According to the natural system, all the plants have been included in the plant kingdom. The plant kingdom is divided into divisions, each division into several sub-divisions each subdivision into classes, each class into orders, each order into families, each family into genera and each genus into species, and in some cases each species into several varieties.
Various natural systems of classification have been proposed time and again. It is not the idea of present attempt to give a detailed survey of all those syetems. Only a few systems have been briefly discussed below so as to give an understanding to the students the relative positions of different groups of plants in the

evolutionary sequences of plant kingdom so that they will be in a better position to understand and appreciate the different groups of non-vascular and vascular plants.

In general, two main categories can be identified in the natural systems. They are:
(1) Old or Traditional systems.
(2) Revised or Modern systems.

Traditional systems

Of all the old systems perhaps the one that is followed even today is that of Eichler (1866). He proposed that the plant kingdom can be divided into:
(A) Cryptogamae and (B) Phanerogamae

(A)**Cryptogamae:** Cryptogamae have their reproductive structures embedded in the plant body and do not bear flowers or seeds.
They include three divisions voz.,
(1) Thallophyta
(2) Bryophyta and
(3) Pteridophyta

Thallophyta: The division thallophyta is characteriosed by plants
a) With an undifferentiated flat plant body called thallus.
b) having no vascular tissue.
c) with unicellular reproductive structures.
d) without embryonal stage in the life cycle.
e) members generally aquatic (not a rule)
Thallophyta is further divided into 2 classes (1) Algae and (2) Fungi

Algae: The class Al;gae includes the members majority of which are:
(a) Aquatic in habitat.
(b) Autotrophic because of the presence of chlorophyll (green photosynthetic pigment) and other pigments.
(c) Having unicelluar or multicellular plant body.

Fungi: This includes all achloraphyllus thallophyta. As the members are pigmentless, they are not autotrophic. Their mode of nutrition is heterotropic. Fungi follow two methods of hetertrotropic nutrition viz.
(1) Parasitic and (2) Saprophytic.

In the parasitic method fungi obtain their nutrition from living organisms (plants or animals) while in the saprophytic method fungi derive nutrition from dead and decaying organisms.

Bryophyta: Plants are amphibious in their habit. They differ from thallophytes by the following respects:
(1) Possession of multicellular sex organs.
(2) Presence of an embryonal stage in the life cycle.
(3) Occurrence of heteromorphic alternation of generations in the life cycle.
(4) Sporophyte being dependent on the gametophyte.

Pteridophyta: Pteridophytes are called vascular cryptogams because, of all the cryptogams, only they have vascular tissue or conducting tissue, which helps in the transportation of food materials and also offers mechanical support to the plant body. They are seedless. Unlike in Bryophytes sporophyte is the predominant plant body but nevertheless the gametophyte is independent (with very few exceptions).

(A)**Phanerogams:** These are the plants in which the reproductive structures are not embedded. They produce seeds. Phanerogams are divided into two divisions viz.
(1) *Gymnospermae*: Seeds are naked, not covered by any envelope, no fruit formation.
(2) *Angiospermae*: Seeds enclosed by fruit wall, fruit being formed by the mature ovary which is a special envelop for the ovule (seed). Angiosperms are divided into two classes viz.,
(1) Dicotyledonae - seeds with two cotyledons.
(3) Monocotyledonae - seeds with one cotyledon.
Some botanists divide the entire plant kingdom into four major divisions viz ,
(1) Thallophyta.
(2) Bryophyta.
(3) Pteridophyta and
(4) Spermatophyta (includes Gymnosperms and Angiosperms).

Modern Systems of Classification

According to one of the modern systems there are only three major divisions in the plant kingdom. They are:
(1) Thallophyta
(2) Bryophyta
(3) Tracheophyta

The last division includes all the vascular plants, which are classified into the following sub-divisions.
(1) Psilopsida
(2) Lycopsida
(3) Sphenopsida and
(4) Pteropsida

Pteropsida includes 3 classes viz.,
(a) Filicineae
(b) Gymnospermae
(c) Angiospermae

Angiospermae has two sub classes
(1) Dicotyledonae and
(2) Monocotyledonae.

In another system, only Thallophyta and Embryophyta have been recognised as two major divisions. In Embryophyta there is the formation of an embryonal stage in the life cycle while it is absent in thallophytes. Embryophyta includes bryophytes, pteridophytes, gymnopserms and Angiosperms.

Many modern botanists consider that the plant kingdom consists of 18 major divisions.

They are:
(1) Chlorophyta
(2) Charophyta
(3) Euglenophyta
(4) Pyrrophyta
(5) Crysophyta

(6) Phaeophyta
(7) Cyanophyta
(8) Rhodophyta
(9) Schizomycophyta (Schizomycota)
(10) Myxomycophyta (Myxomycota)
(11) Eumycophyta (Eumycota)
(12) Bryophyta
(13) Psilophyta
(14) Lycophyta (Lepidophyta)
(15) Arthrophyta (Calamophyta)
(16) Filicophyta (Pterophyta)
(17) Gymnophyta
(18) Angiophyta (Anthophyta)

PART ONE
ALGAE

CHAPTER 3

GENERAL ACCOUNT AND CLASSIFICATION

The study of algae is called phycology. Algae have been known to human beings since time immemorial. though their scientific study is recent. Early Indian, Chinese, Roman and Greek literatures contain numerous references to this group of plants. In Amara kosa, a Sanskrit lexicon or the 4th or 5th century A.D., Algae are mentioned and are called '*jalanili*' or '*shivala*' meaning waterscum. Ancient Hawaiians used algae for food. At present the study of Algae is one of the important fields of Botanical science having far reaching implications from the point of view of human welfare.

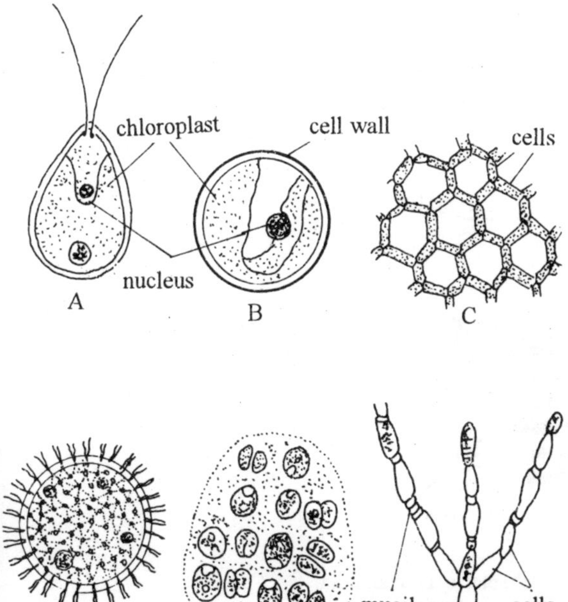

Fig.3.1 A few algae
a. Chlamydomonas; b. Chlorella; c. Hydrodictyon; d. Volvox; e. Tetraspora; f. Prasinocladus.

Introduction

According to the present trend of plant classification however algae do not constitute any compact group. The term algae refers to about eight divisions of the plant kingdom. From the point of view of convenience however the term has till been retained. All algae are chlorophyllous, thallophytes and hence autotrophic. They are non-vascular plants with predominant aquatic distribution. There are about 30,000 species of algae.

Occurrence of Habit

The members are aquatic being distributed both in fresh and salt waters. Fresh water algae abound in stagnant waters, tanks, ponds, ditches, slow running steams, rivers, etc. Some of the terrestrial forms grow abundantly wherever nature offers a damp surface. They grow on wet trunks and branches of trees or even on rocks. They may also live as epiphytes or ephizoophytes (algae growing on animals, specially molluscan shells). Fresh water algae may be found free floating or attached to rocks, stones or other higher water plants like *Elodea, Ottalia* etc.

Some of the algae are endophytic growing in company of higher plants. Some of the cyanophycean algae like *Nostoc* and *Anabena* which are known to possess the capacity of nitrogen fixation are found forming symbiotic association with plants like *Anthoceros, Isoetes, Cycas* etc.

Marine algae are commonly called seaweeds occupying a large area (sometimes several thousand miles). They may be free floating or attached. The free floating marine algae are known as "phytoplanktons". Those which are attached found growing in shallow waters form "Benthons".

Range of Vegetative Structures

The thallus organisation in algae may be classified into the following types:

(1) *Motile types:* This is the simplest type. Plant body is motile and unicellular throughout. In some cases motile forms may be colonial with little or no division of labour. Example — *Chlamydomonas, Volvox* etc.

(2) *Palmelloid type and Perdroid types:* Palmelloid or non-motile stage is found as a temporary state even in motile forms. But in some forms like *Tetraspora*, this is a permanent feature. Here the cells are surrounded by mucilage covering. Dendroid type is a slight variation of the palmelloid type. For example *Tetraspora, Prasinocladus* etc.

(3) *Coccoid forms:* In these forms the locomotory organs (flagella) are lost and the cells become rounded.

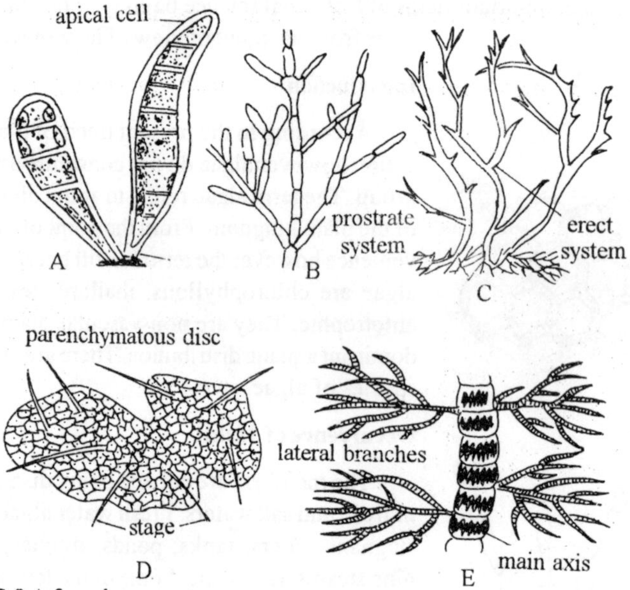

Fig.3.2 A few algae
a. Ulothrix; b. Cladophora; c. Stigeoclonium; d. Coleochaete; e. Drapanaldia.

For example *Chlorcoccum.*

(4) *Filamentous types* : The filamentous habit is believed to have originated from motile forms. Loss of motility and restriction of cell division to one plane has resulted in the filamentous types. The filaments may be free floating or attached. All the cells in the filament may be alike. One of the lower cells may be modified to help in attachement (hold fast, hapteron etc). The filaments may be branched or unbranched e.g. *Oedogonium, Cladophora.*

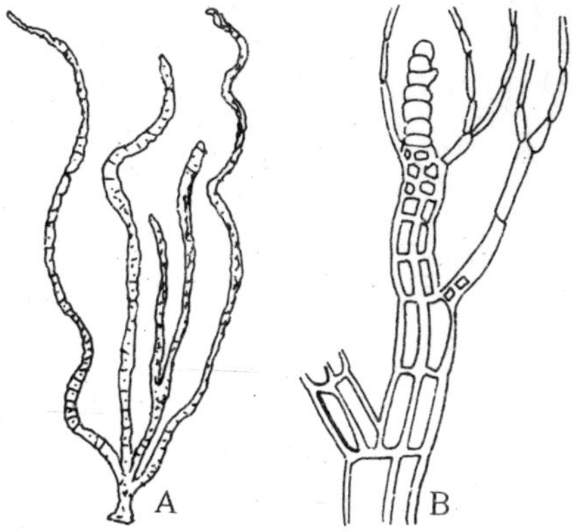

Fig.3.3 A few Algae A) Gracilaria B) Polysiphonia

(5) *Siphonous habit*: The plant body consists of tube like filaments (branched also) but no cross walls are found. Hence the entire plant body is comparable to a single cell. Cross walls appear only to demarcate the reproductive structures or to separate off the dead portions of the thallus e.g., *Vaucheria, Caulerpa* etc.

(6) *Advanced types:* They may be monosiphonous or polysiphonous filamentous types or even true parenchymatous e.g., *Batrachospermum, Polysiphonia* etc.

Reproduction

Algae reproduce by two methods viz:

(1) Asexual (includes vegetative propagation also).

(2) Sexual

Asexual Reproduction

This is brought about by a variety of methods

(1) *Cell division* : Cell division as a means of multiplication of the individuals is seen in several unicellular members e.g., *Pleurococcus* etc.

(2) *Fragmentation :* Accidental break up of the plant body results in fragments. Each fragment develops into a new individual e.g., *Spirogyra, Ulothrix* etc.

(3) *Hormogones* and *Hormocysts:* This type of vegetative multiplication is found in some cyanophyceae. Filament breaks up into 3-4 celled bits called hormogones (thick walled ones are called hormocysts) which develop into a new individual e.g., *Oscillatoria*.

(4) *Protonema:* These are called thread like bodies helping in vegetative multiplication e.g., *Chara*.

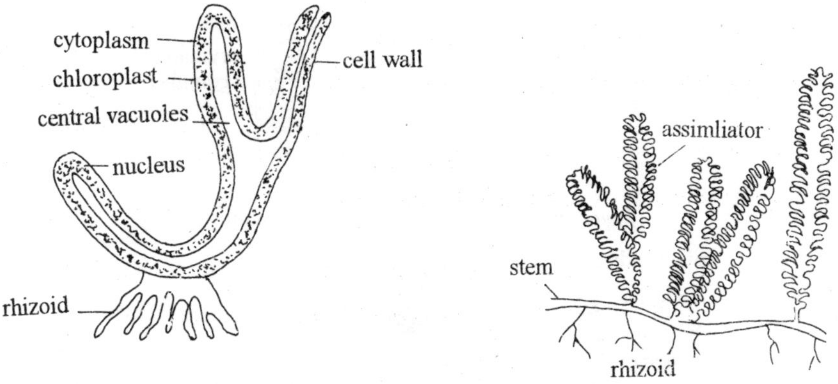

Fig.3.4 A few Algae
a. Nemalion multifidum; b. Polysiphonia flexicaulis; c. Vaucheria sp.; d. Caulerpa taxifolia.

(5) *Amylum stars:* These are starch filled stellate structures helping in asexual reproduction e.g.,*Chara*.

(6) *Akinetes:* These are non-motile oval shaped structures formed in cells usually to withstand the adverse environmental conditions. They are formed in many members of chlorophyceae.

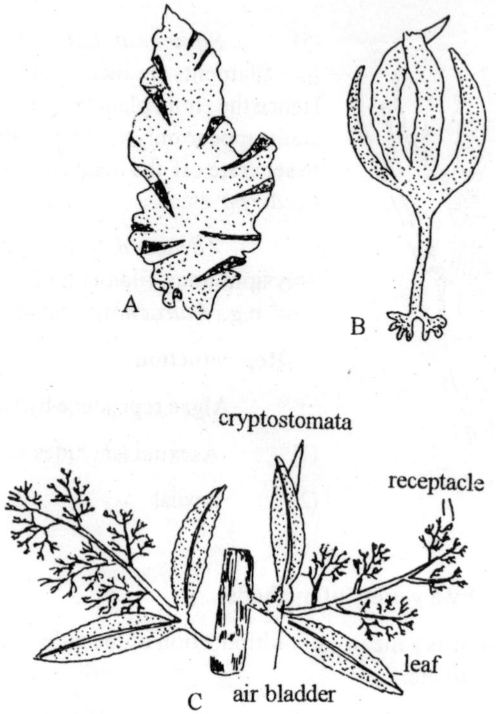

Fig.3.5 A few Algae
a. Ulva; b. Laminaria; c. Sargassum

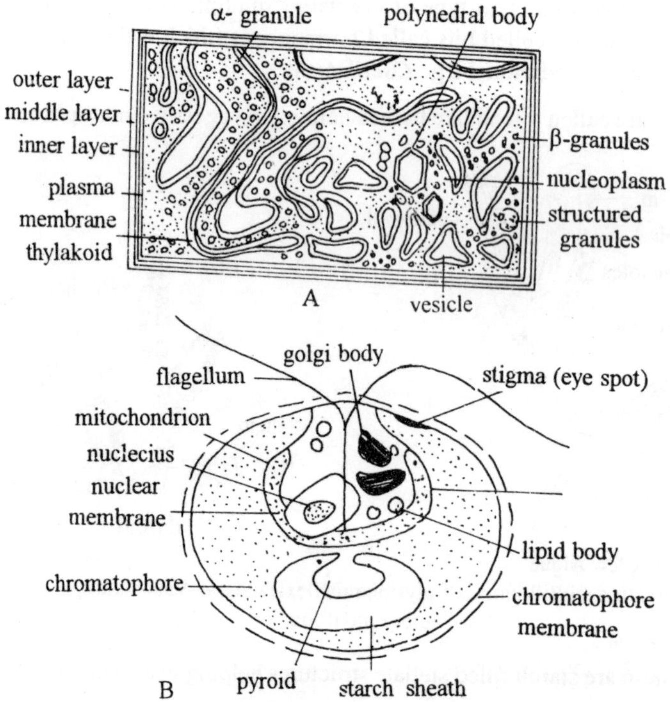

Fig.3.6 A: Diagrammatic plan of a procaryotic cell of blue-green Algae,
B: Eucaryotic cell organisation (*Heteromas':cx rotunda*).

(7) *Aplanospores* : These are non-motile spores, formed in cells. At favourable season these develop into new individuals.

(8) *Zoospores* : By far this is the commonest method of asexual increases in Algae. These are motile spores adopted to aquatic habitat. They are formed in cells (zoosporanguim) either single or in large numbers. They are biflagellate mostly, but may be quadriflagellate or multiflagellate also. Flagella may be of equal length (Iskontean) or of unequal length (stephenokontean). Zoospores are mostly pyriform in shape. They swim for some time in water and then develop directly into a new individual.

Besides the above, various other algae reproduce asexually by means of Endospores, Autospores, Auxospores, Carpospores, Monospores, Hypnospores, Statospores etc.

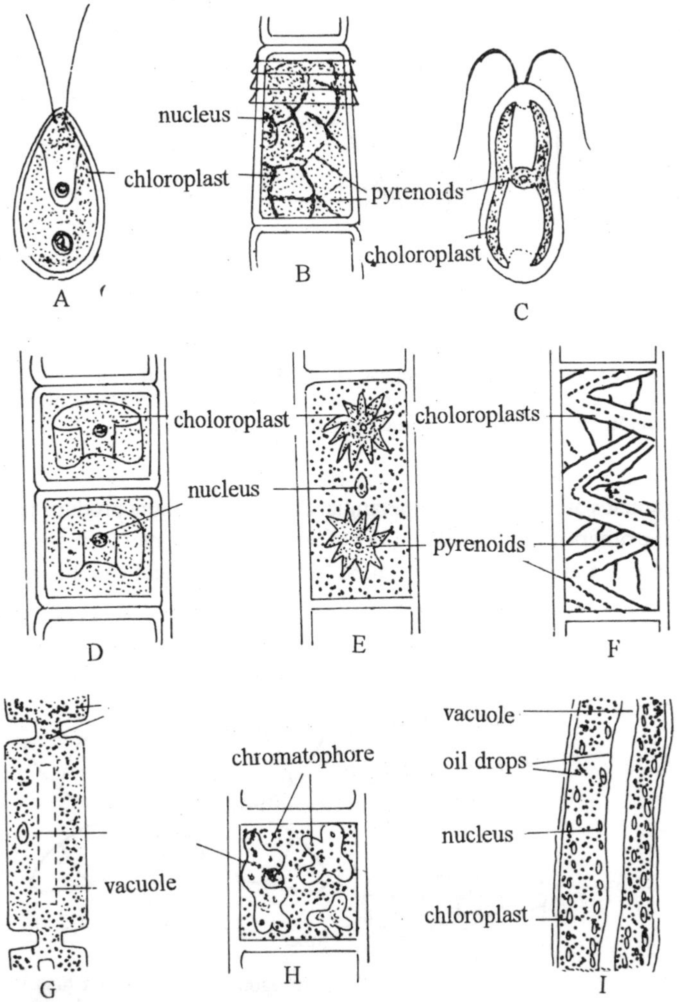

Fig.3.7 Different types of chloroplasts in Algae, A: cup shaped (*Chlamydomonas eugametos*); B. reticulate (*Oedogonium* sp); C: H-shaped (*Chlamydomonas alpina*); D: girdle shaped (*Ulothrix* sp); E. star shaped (*Zygnema* sp.); F.ribbon shaped (*Spirogyra* sp,);. G. discoid (*Polysiphonia* sp); H. irregular (*Ectocarpus* sp.,; I. discoid (*Vaucheria* sp.).

Sexual Reproduction

This is brought about by the fusion of sex cells known as gametes which are produced in gametangia.

These are of three types. viz. (1) Isogamy (2) Anisogamy and (3) Oogamy. The oogamous type of reproduction is the most advanced type.

Isogamy

In this type there is no morphological difference between the fusing gametes. The gametes are motile and come out of the gametangia. Fusion takes place outside.

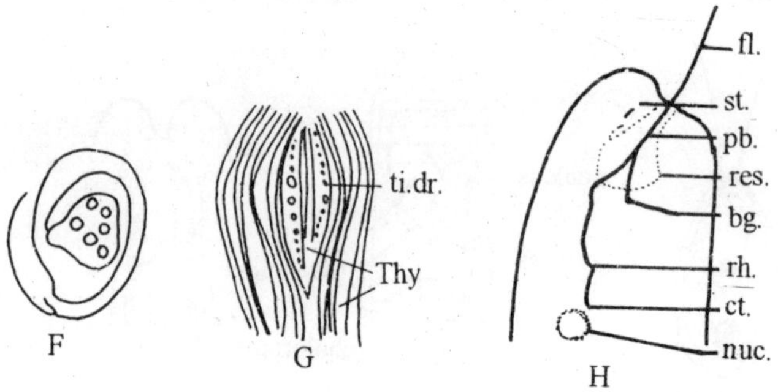

Fig.3.8 Flagella, stigma and neuromotor apparatus in Algae; A. Simple flagellum; B: Acronematic type; C: Pantonematic type; D:Stichonematic type; E: T.S. of flagellum; F: Haptonema type; G. Stigma; H: Neuromotor appratus of Euglena (bg: basal granule; ct : centriole; fl: flagellum; li dr : lipid droplets; nuc: nucleus; pl: para basal granule; rh: rhizoplast; st: stigma; thy: thylakoid).

Anisogamy

In this, two types of gametes are formed in gametangia. Some gametangia produce large number of smaller gametes (male) and some others produce less number of large sized gametes (female). Both however are liberated out. Fusion is external. Large gametes are called Macrogametes and small gametes are called microgametes.

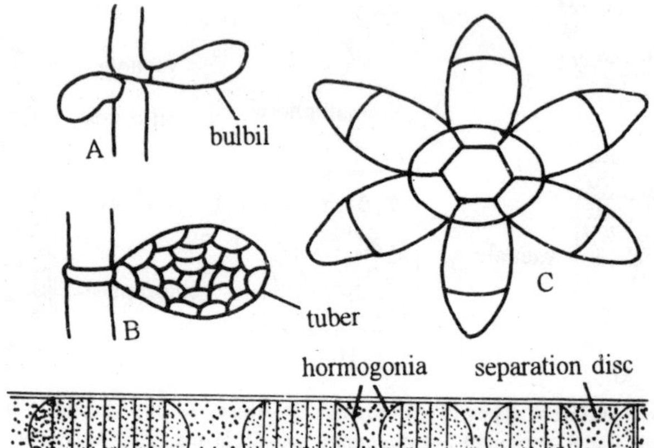

Fig.3.9 Vegetative reproduction,
A: Bulbil of Chara; B: Tuber of Chara; C: Amylum star of Chara;
D: Hormogonia of Lyngbya sp.

Oogamy

In this the male and the female gametangia are different. The former is called Antheridium (spermatangium) and the latter oogonium. Antheridium produces flagellate gametes. Oogonium produces only one non-motile female gamete (egg) which never comes out. Fertilization is internal.

Alternation of Generations

In some Algae the life history comprises of two district generations viz., sporophyte and gametophyte. The former is asexual and diploid, the latter sexual and haploid . Haploid gametophyte reproduces sexually and the fused product (zygote) directly develops into the diploid sporophyte (No alternation of generations if the zygote undergoes reduction division). This diploid sporophyte reproduces asexually by spores. During spore formation reduction division takes place, hence the spores are haploid. They develop into gametophyte. If both gametophyte and sporophyte are alike morphologically, alternation of generation is said to be isomorphic and if they are different, Heteromorphic.

Only few algae however exhibit alternation of generations.

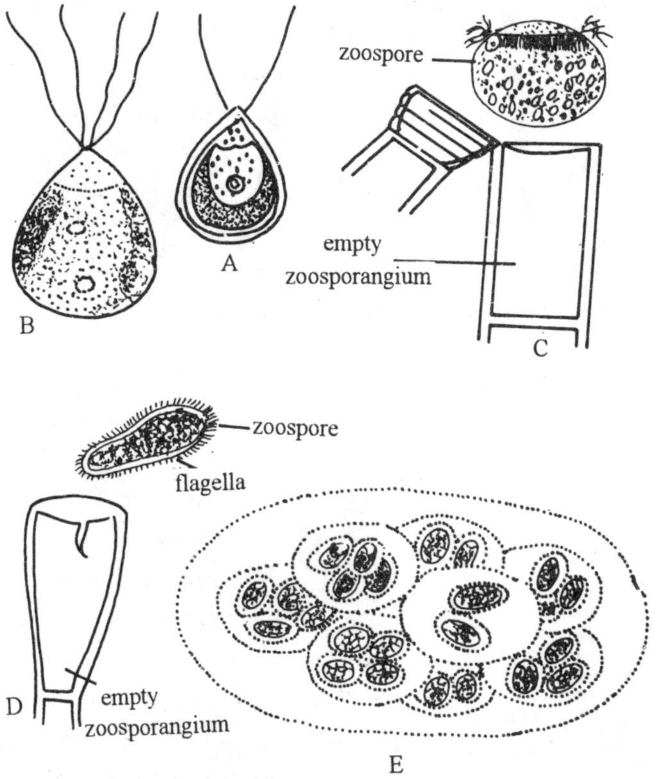

Fig.3.10 A sexual reproduction, A: Zoospore of Chlamydomonas sp, B: Zoospore of Ulothrix sp, C: Zoospore of Oedogonium sp, D: Zoospore of Vaucheria sp, E: Palmella stage in Chlamydomonas sp.

Classification

Before 20th century only four classes were recognised in Algae. They are:

(1) Chlorophyceae

(2) Phaeophyceae

(3) Rhodophyceae and

(4) Cyanophyceae.

Fritsch divided algae into eleven classes namely:

(1) Chlorophyceae

(2) Xanthophyceae

(3) Chrysophyceae

(4) Bacillariophyceae

(5) Cryptophyceae

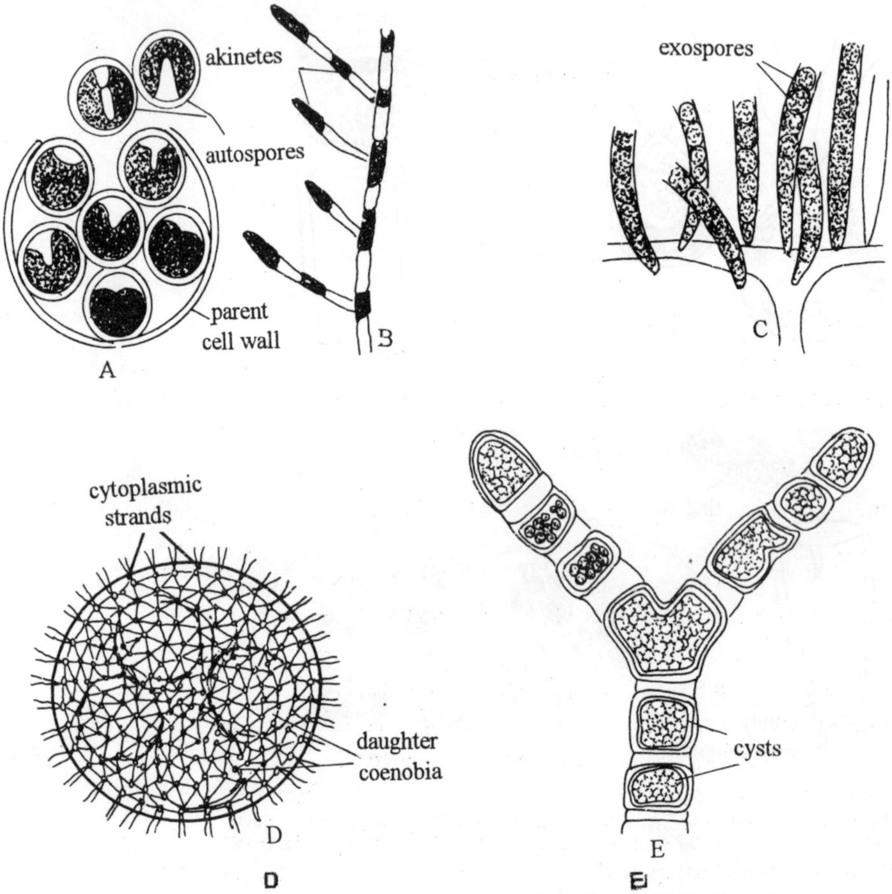

Fig.3.11 Asexual reproduction, A: Autospore formation in Chlorella vulgaris, B: Akinetes of Pithophora kewensis, C: Exopores in Chaemosiphon incurstans, D: Daughter colony of Volvox sp, E: Cysts in Vaucheria sp.

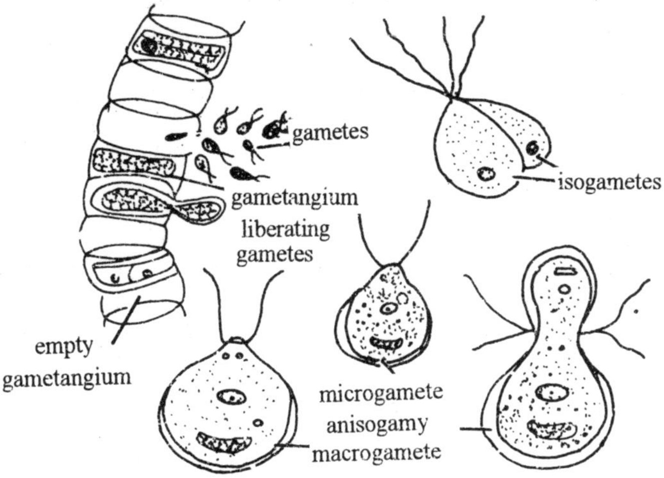

Fig.3.12 Sexual reproduction, A: Gametangia and gametes of Ulothrix sp, B: Isogamy in Ulothrix sp, C: Female or macrogamete of Chlamydomonas braunii, D: Male or microgamete of Chlamydomonas braunii, E: Anisogamy in Chlamydomonas braunii.

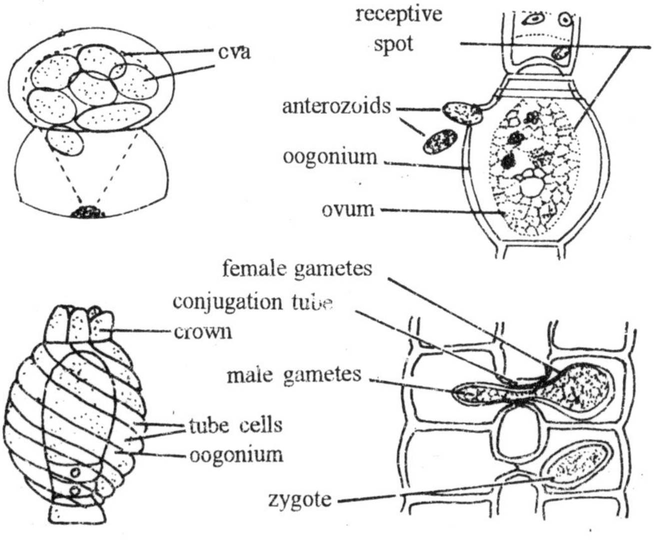

Fig.3.13 Sexual reproduction , A: Eggs or ova of Fucus sp, B: Oogamy in Oedogonium sp, C: Nucule of Chara sp, D: Apla:.ogamy in Spirogyra sp.

(6) Myxophyceae

(7) Dinophyceae

(8) Chloromonodineae

(9) Euglenophyceae

(10) Phaeophyceae

(11) Myxophyceae

(12) Rhodophyceae

In the classification proposed by Fristch importance is given to pigmentation, nature of plant body, reproductive features, metabolic products, locomotory organs etc.

Glibert M.Smith classifies algae as follows:

Division: (1) *Chlorophyta*

Class : (a) Chlorophyceae

(b) Charophyceae

(2) *Euglenophyta*

(3) *Pyrrophyta*

(a) Desmophyceae

(b) Dinophyceae

(4) *Chryosophyta*

(a) Chryosophyceae

(b) Xanthophyceae

(c) Bacillariophyceae

(5) *Phaeophyta*

(a) Isogenetate

(b) Heterogenerate

(c) Cyclosporeae

(6) *Cyanophyta*

(7) *Rhodophyta*

Rhodophyceae.

Algae of uncertain systematic position

(1) Chloromonadales (2) Cryptophyceae

Papenfuss divides algae into eight phyla. They are:

(1) Chlororophycophyta

(2) Charophycophyta

(3) Euglenophycophyta

(4) Chrysophycophyta

(5) Pyrrophycophyta

(6) Phaeophycophyta

(7) Schizophyta

(8) Rhodophycophyta

Chapman (1926) proposed a system of classification of algae based on morphological characters, pigments and biochemical differences. He divides algae into four:

(1) Euphycophyta

(2) Myxophycophyta

(3) Chrysophycophyta

(4) Pyrrophycophyta

He includes Charophyceaee, Chlorophyceae, Phaeophyceas and Rhodophyceae under Euphycophyta.

CHAPTER 4

MYXOPHYCEAE

The members of the class Myxophyceae can be identified by the following general characters.

1. Members are blue green in colour due to specific pigmentation.
2. Plastids are absent.
3. There is no motile phase in the life history.
4. Sexual recombination as understood in other higher members is absent. But genetic recombination has been reported in several members.
5. The cell interior is divided into an outer coloured chromoplasm and an inner colourless centroplasm representing the incipient nucleus. (There is no nuclear membrane).

The class consists of about 165 genera and more than 1500 species. The members are commonly called blue green algae, because the cell has bluish green pigment phycocyanin and the red pigment phycoerythrin in addition to chlorophyll, myxoxanthin, myxoxanthophyll and carotenoids.

Occurrence and Distribution

Blue green algae posses a wide range of tolerance to climatic fluctuations and are found inhabiting almost all types of ecological niche of the environment.

A good majority occur in a variety of fresh water habitats like hard water lakes, streams and wet soil, rocks damp barks etc.

Some of the members are planktonic (free floating) and marine in their habitat. Some are endophytic in higher plants, where they help in nitrogen fixation. Some myxophyceae members also enter into lichen association.

Thallus construction: Among the algae, the blue greens are the most primitive and possess a very simple plant body. The plant body is of three types, namely:

1. Unicellular forms.
2. Non-filamentous colonial forms.
3. Filamentous forms.

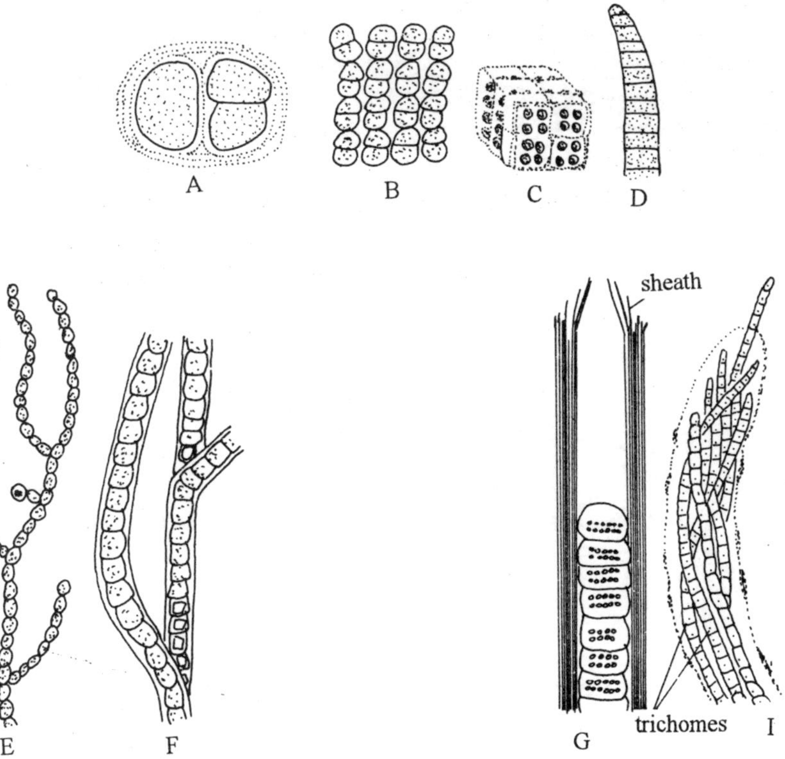

Fig.4.1 Thallus of Cyanophyceae, A-C: Nonfilamentous, and D-H filamentous; A: Chroococcus turgidus; B: Merismopedia elegans; C: Eucapsis alpina; D: Oscillatoria formosa; E: Nostochopsis lobatus; F: Tolypothrix tenuis; G: Porphyrosiphon notarisii; H: Microcoleus vaginatus.

1. *Unicellular forms:* The plant body is made up of only one cell, e.g., *Chroococcus.*
2. *Colonial forms:* The plant body is made up of aggregates of cells held together by a gelatinous matrix, e.g., *Merismopaedia.*
3. *Filamentous forms:* The plant body is multicellular. The filament is built up by the end to end joining of cells. The filament may be branched or unbranched.

Cell structure: In the blue green algae two regions are distinguishable within the protoplast. There is a peripheral region (chromatoplasm,environment.

A good majority occur in a variety of fresh water habitats like hard water lakes, streams and wet soil, rocks damp barks etc.

Some of the members are planktonic (free floating) and marine in their habitat. Some are endophytic in higher plants, where they help in nitrogen fixation. Some myxophyceae members also enter into lichen association.

Thallus construction: Among the algae, the blue greens are the most primitive and possess a very simple plant body. The plant body is of three types, namely:

1. Unicellular forms.

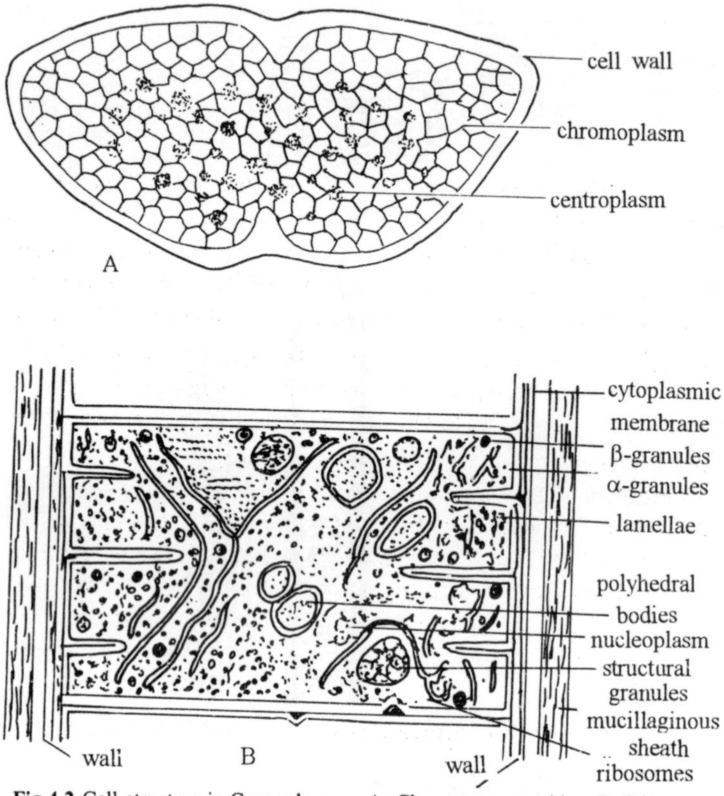

Fig.4.2 Cell structure in Cyanophyceae; A: Chroococcus turgidus; B: Diagrammatic sketch of Symphloca muscorum based on electron micrograph.

2. Non-filamentous colonial forms.
3. Filamentous forms.

1. Unicellular forms: The plant body is made up of only one cell, e.g., *Chroococcus.*

2. Colonial forms: The plant body is made up of aggregates of cells held together by a gelatinous matrix, e.g., *Merismopaedia.*

3. Filamentous forms: The plant body is multicellular. The filament is built up by the end to end joining of cells. The filament may be branched or unbranched.

Cell structure: In the blue green algae two regions are distinguishable within the protoplast. There is a peripheral region (chromatoplasm,chromoplasm) having the photosynthetic pigments and central region which is colourless. The central region is known as the centroplasm or the incipient nucleus. But the distinction between the chromoplasm and centroplasm may not be always marked as in *Chroococcus turgidus,* where the pigments are scattered throughout the protoplast.

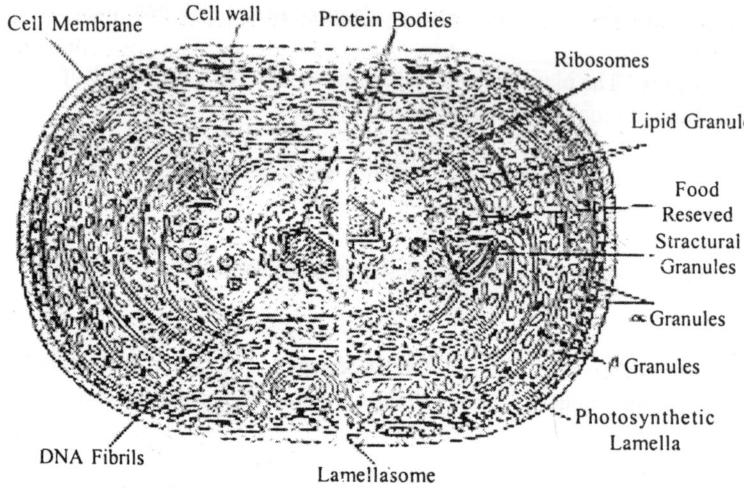

Fig.4.3 General structure of a blue-green algal cell based on electron micrograph

The cell is externally bound by a cell wall made up of cellulose and pectic compounds. The wall is surrounded by gelatinous sheath which may be stratified.

The chromoplasm is either alveolar in nature or homogenous. It has various pigments like chlorophyll, carotene, myxoxanthin, myxoxanthophyll, flavisin phycocyanin, xanthophyll and phycoerythrin.

The centroplasm is composed on DNA, RNA and proteins. There is neither any nucleolus nor a nuclear membrane.

Reproduction: It is mostly asexual. Only in a few members like *Anacystis nidulans* (Kumar H.D, 1962), *Cylindrospermum majus* (Singh R.N. and Singh R.N., 19965)etc, genetic recombination has been reported.

Asexual reproduction takes place by means of vegetative propagation and spore formation .

Vegetative Propagation:

1. *Cell division:* In unicellular members, the two daughter cells, after division mature into individuals members.
2. *Hormogones:* These are formed in filamentous species. Hormogones represent portions of filaments between two dead cells. They get separated from the filament and develop into new forms.
3. *Hormocysts:*Hormogones formed at the tip of the filament,having thick walled, nutrient rich cells are called *Hormocysts.*

Spore Formation

1. *Akinetes, Resting spores or Arthospores:* These are non-motile spores, generally formed to tide over adverse environmental conditons. Usually these spores have a thick wall and a large amount of reserve food.
2. *Endospores:* These are spherical or angular asexual cells produced in the order Chaemosiphonales.
3. *Exospores:* This is formed externally by obstriction e.g., *Chaemosiphon.*
4. *Nannospores or Nannocysts:* The young daughters cells, immediately after cell division in *Gloeocapsa* are called Nannospores
5. *Heterocysts:* These are formed in all filamentous Cyanophyceae,except Osciallatoriaceae. Heterocysts are thick walled cells and formed from vegetative cells. They may occur singly or in chains. In position heterocysts may be intercalary or terminal. At either poles, they have two polar nodules to help in maintaining cytoplasmic connection with the other cells.

Classification: Fritsch (1935) has classified the class into five orders, namely, Chroococcales, Chaemosiphonales, Pleurocapsales, Nostacales and Stigonematales.

Smith divides the class into three orders, namely, Chroococcales, Chaemosiphonales and Hormogoanles.

In this book the following members are discussed:

*Gleocapsa,Nostoc, Anabena, Lyngbya, Oscillatoria, Rivularia, Gloeotrichia, Scytonema and Stigonema.*chromoplasm)

GLOEOCAPSA
(Kuetzing 1843)

Common species: G. decorticans, G. calcarea, G. rupestris etc.

Distribution : Wet rocks or soil. Marine or fresh water.

Study: 1. External features.
2. Reproductive structures.
3. Systematic position and identification.

External Features

1. Cells are arranged in colonies of 2-8. Sometimes unicellular also.
2. Colonies are surrounded by many envelopes of mucilage. These may or may not be coloured
3. Each mucilage sheath is lamellated or stratified.
4. Cells may be spherical or oval in shape.
5. Each cell in the colony has its own mucilage sheath.
6. Cell structure typical to a cyanophycean cell and has outer chromoplasm and inner centroplasm.
7. Cell wall is two -layered -outer pectin and inner cellulose.

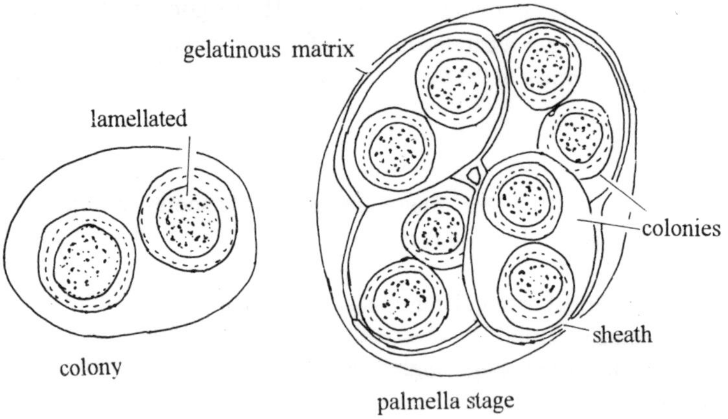

Fig.4.4 Gloeocapsa

Reproduction

1. Sexual reproduction has not been reported.
2. Asexual reproduction takes place by the following methods:
   (a) Cells divide and the daughter cells develop into new individuals.
   (b) Nannospores or Nannocytes also help in asexual reproduction.
   (c) In some cases spore formation also has been observed.

Systematic Position and Identification

A. 1. Chlorophyll present
2. Plant body a thallus
3. Cell wall is cellulosic....... *Algae*

B. 1. Blue green pigments present
2. Plastids absent
3. Protoplasm divided into chromoplasm and centroplasm
4. Sexual reproduction generally not seen....... *Myxophyceae*

C. 1. Plants unicellular or colonial
  2. Nannocysts present .....*Chroococcales.*

D. 1. Plants unicellular or colonial
  2. Filamentous plant body absent....... *Chroococaceae*

E. 1. Cell spherical or oval in shape
  2. Each cell has its own mucilage sheath
  3. Sheath stratified....... *Gloeocapsa.*

NOSTOC
(Vaucher 1803)

Common species: N. calcicala, N. columnar, N. sphaericum

Distribution: Fresh water ponds and ditches. Form mucilagenous balls. Often found on wet rocks, also found in symbiotic association with *Anthoceros, Cycas* and *Lichens.*

Study: 1. External features
  2. Reproductive structures
  3. Systematic position and identification

External features

1. Many hundreds of filaments are enclosed in the form of a mucilage ball.
2. The ball like colony has a thick mucilage envelop and may be coloured.
3. The plant body is a filament and is enveloped by a mucilage sheath.
4. The filaments are unbranched and do not have a base or apex.
5. Each filament is made up of a number of bead like cells joined end to end.
6. Here and there in the filament certain thick walled cells are seen. These are called heterocysts and may be intercalary or terminal in position.
7. Heterocysts are thick walled and pale yellow coloured. Terminal heterocyst has a single polar module, while intercalary heterocysts have two polar nodules.
8. Vegetative cells have a typical cyanophycean structure. Protoplast having centroplasm and chromoplasm. Cell wall is cellulosic.

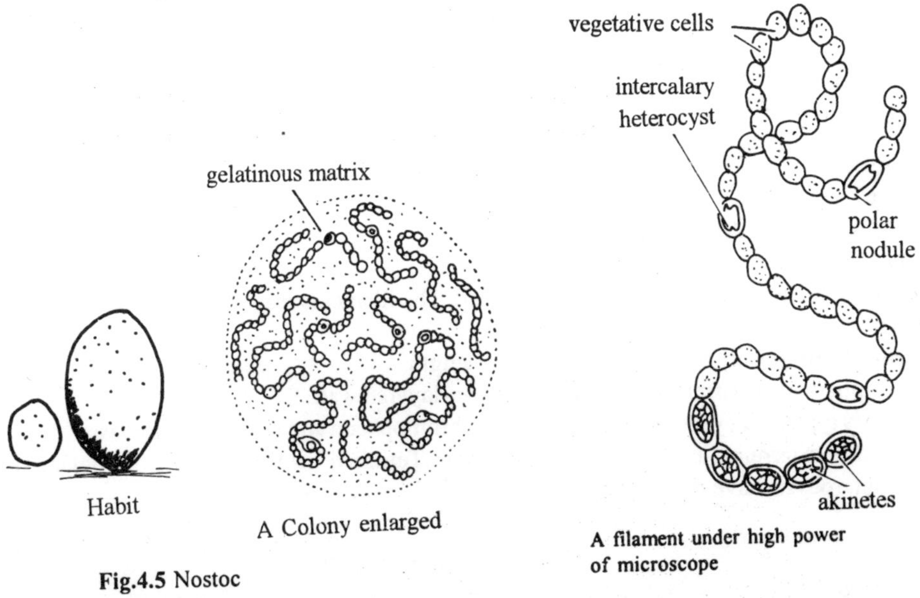

Fig.4.5 Nostoc

Reproductive features

1. Sexual reproduction has not been reported.
2. Asexual reproduction is by the formation of hormogones and akinetes.
3. Hormogones are fragments of the filament, that generally separate from the main filament near a heterocyst.
4. Hormogones develop into new individuals after separation from the main filament.
5. Akinetes are thick walled cells rich in reserve food materials. They separate from the plant and develop into a new individual.
6. Occasionally heterocysts (*N. commune*) may also germinate to form new individuals.

Systematic position and identification

A. 1. Chlorophyll present
   2. Plant body a thallus
   3. Cell wall cellulosic... *Algae*

B. 1. Blue green pigments present
   2. Plastids absent
   3. Protoplast divided into chromoplasm
   4. Sexual reproduction generally not seen... *Myxophyceae*

C. 1. Plant body is filamentous
   2. Filaments unbranched or show false branching
   3. Heterocysts present...*Nostocales.*

D. 1. Filaments to not show differentiation into base and apex.
   2. Filaments unbranched and uniseriate.
   3. Heterocysts present ...*Nostocaceae*

E. 1. Heterocysts generally intercalary
   2. Mucilage balls constitute a colony.
   3. Reproduction by Hormogones...*Nostoc.*

ANABENA
(Bory 1822)

Common species: *A. cycadacearum, A. gelatinicola, A. iyengarii* etc.

Distribution: Commonly found in wet places, ponds, rice fields etc. Some of them could be semiterrestrial. A few species are symbiotic with the roots of higher plants.

Study:

1. External features
2. Reproduction
3. Systematic position and Identification.

External features

1. Plant body is filamentous. The filament is unbranched. The filaments may or may not be in groups.
2. There is no mucilage sheath covering the filament.
3. The filaments are uniformly broad throughout their length.
4. Cells are subspherical or barrel shaped and have a two-layered wall.
5. Cell structure is typical to cyanophyceae with outer chromoplasm and middle centroplasm.
6. Filament consists of terminal and intercalary heterocysts. Terminal heterocysts have one polar module, while intercalary heterocysts have two polar nodules.

Myxophyceae

Reproduction

1. Sexual reproduction has not been reported.
2. Asexual reproduction is usually brought about by Akinetes.
3. Akinetes are thick walled cells, much larger in size than the vegetative cells. They occur either singly or in a row of 4-6. They may or may not be adjacent to a heterocyst. Akinetes germinate into a new filament.

Systematic Position and Identification

A. 1. Chlorophyll present
   2. Plant body a thallus
   3. Cell wall cellulosic...*Algae*

B. 1. Blue green pigments present
   2. Plastids absent
   3. Protoplast divided into chromoplasm and centroplasm
   4. Sexual reproduction generally not seen....*Myxophyceae*.

C. 1. Plant body filamentous
   2. Filaments unbranched or show false branching

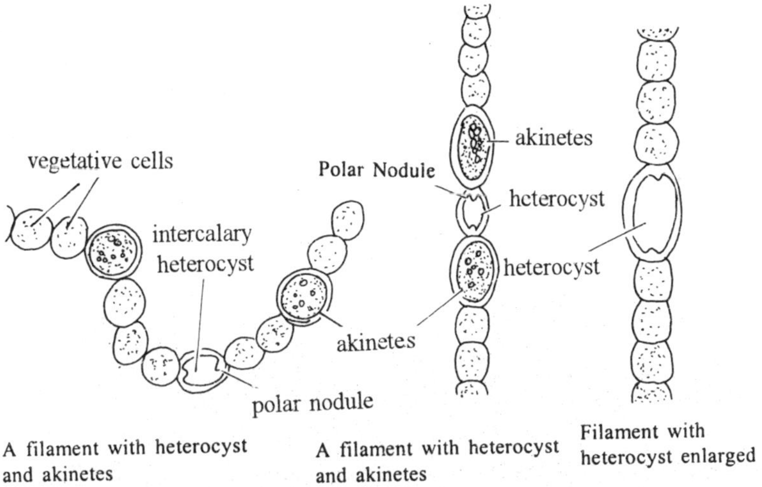

Fig.4.6 Anabena

   3. Heterocysts present...*Nostocales*

D. 1. Filaments do not show differentiation into base and apex
   2. Filaments unbranched and uniseriate
   3. Heterocyst present...*Nostocaceae*

E. 1. Mucilage sheath absent
   2. Akinetes many times larger than the vegetative cells and are usually associated with heterocysts....*Anabena*.

RIVULARIA
(AGARDH 1842)

Common species: *R.atra*

Distribution: Found in the form of shiny mucilage balls floating in water or sticking to aquatic plants such as *Elodea, Ottelia* etc. Occasionally found in salt marshes also.

Study 1. External features.
2. Reproduction
3. Systematic position and Identification

External features

1. Colonies are irregular masses, yellowish brown in colour with a greenish tinge.
2. Each colony has hundreds of filaments which radiate from the centre. The filaments are held together by the mucilage.
3. The filaments have a broad base which gradually tapers towards the apex.
4. The filaments are usually completely enveloped by the mucilage.
5. Each filament consists of a basal heterocyst. Occasionally, the heterocyst may be intercalary also.
6. Frequently, the filaments show false branching occurring beneath an intercalary heterocyst.
7. The branching is not evident as the branches separate immediately.
8. Radiating filaments of the colony result due to false branching.
9. Cell structure of vegetative cells typical to Cyanophyceae in having outer chromoplasm and inner centroplasm.

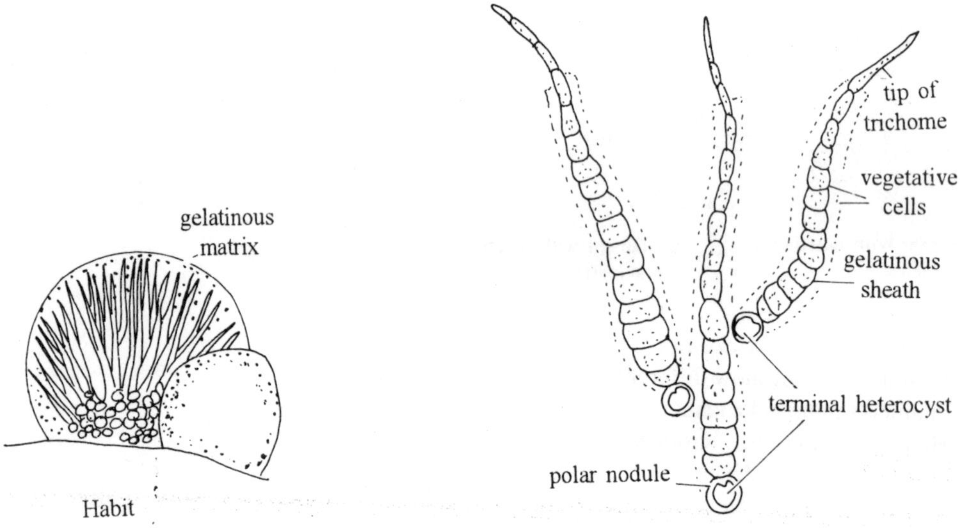

Fig.4.7 Rivularia

Reproduction

1. Sexual reproduction has not been reported.
2. Akinetes are not formed.
3. Hormogones break up the filament into bits. Each bit develops into a new individual.

Systematic position and identification

A. 1. Chlorophyll present
   2. Plant body a thallus
   3. Cell wall cellulosic...*Algae*

B. 1. Blue green pigments present
   2. Plastids absent
   3. Protoplast divided into centroplasm
   4. Sexual reproduction not seen...*Myxophyceae*

C. 1. Plant body filamentous
   2. Filaments unbranched or show false branching
   3. Heterocysts present...*Nostocales*

D. 1. Plant body filamentous and unbranched
   2. Heterocyst basal
   3. Filaments broad at the base tapering towards the apex...*Rivulariaceae.*

E 1. Akinetes absent
   2. Filement completely enclosed by the mucilage sheath
   3. Heterocyst usually basal
   4. Filaments radiate from the centre in a colony...*Rivularia*

GLOEOTRICHIA
(Agardh 18)

Common species: *G.natans, G. pisum* etc.
Distribution : Freshwater ponds, ditches immediately after rains. Found in the form of mucilagenous masses.

Study 1. External features
2. Reproduction
3. Systematic position and Identification

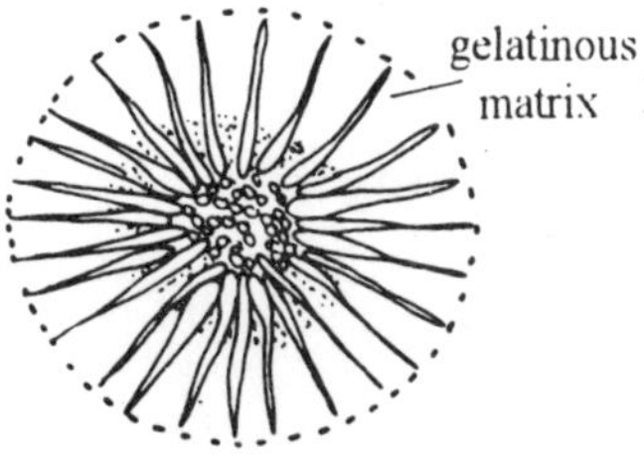

Habit

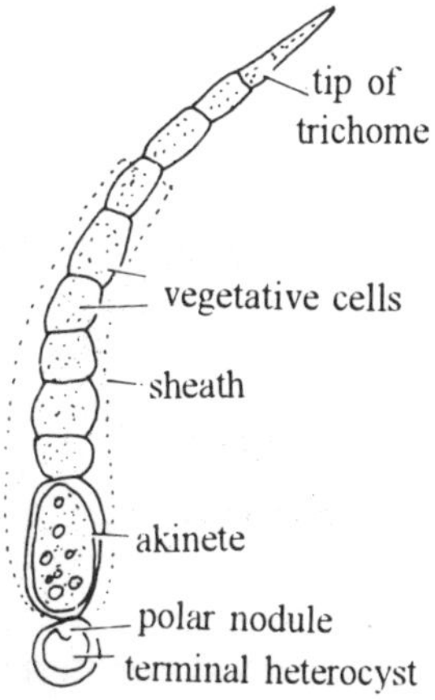

A Filament enlarged

Fig.4.8 Glocotrichia

A Filament enlarged

External features

1. Found in the form of irregular bluish green mucilagenous masses. Colonies are completely ensheathed by mucilage.
2. Many hundreds of filaments are found in each colony arranged radiately.
3. Fialments show false branching and have a mucilage sheath which incompletely envelops them.
4. Heterocyst is basal.
5. Above the heterocyst is an Akinete. Akinete is very large. 4-5 times bigger than the vegetative cells (Akinete not seen in *Rivularia*).
6. Cell structure typical to Cyanophyceae in having an outer chromoplasm and a central centroplasm.

Reproduction

1. Sexual reproduction has not been reported.
2. Asexual reproduction brought about by hormogones and akinetes.
3. Hormogones are bits of filaments which on separation develop into a new individual.
4. Akinetes separate from the plant and develop into a new individual.

Systematic position and Identification

A. 1. Chlorophyll present
   2. Plant body a thallus
   3. Cell wall cellulosic...*Algae*

B. 1. Blue green pigments present
   2. Plastids absent
   3. Protoplast divided into chromoplasm and centroplasm
   4. Sexual reproduction generally not seen ...*Myxophyceae*

C. 1. Plant body fimamentous
   2. Filaments unbranched or show false branching
   3. Heterocyst present...*Nostocales*

.D. 1. Plant body filamentous and unbranched
   2. Heterocyst basal
   3. Filaments broad at the base tapering towards the apex...*Rivulariaceae*

E 1. Mucilage sheath generally incomplete
   2. Akinete always present above the basal heterocyst...*Gloeotrichia*

LINGBYA
(Agardh 1824)

Common species : *L. spiralis, L. nigra, L. lutea* etc.

Distribution: Freshwater or salt water. Filaments usually in clusters, may be solitary also.

Study
1. External features
2. Reproduction
3. Systematic position and Identification

External features

1. Plants body is an unbranched filament. The filaments are present in clusters.
2. Filaments are flat, olive green in colour and unbranched.
3. A thick mucilage sheath envelops the filaments.

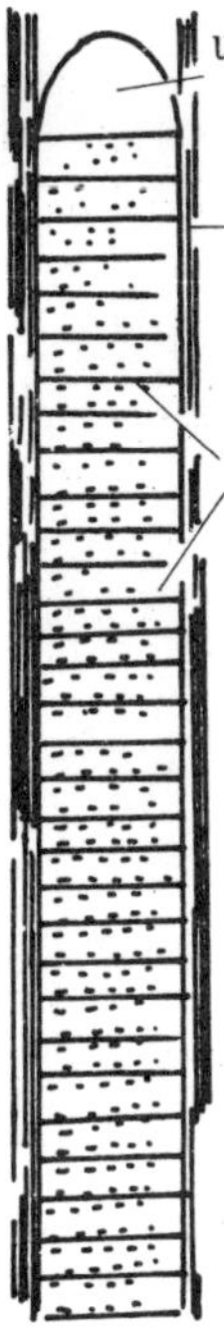

Fig.4.9 Lyngbya. External features

4. The filament is made of cells joined end to end.
5. The cells are broader than long and the terminal cell is dome shaped.
6. Cell structure typically Cyanophycean with peripheral chromoplasm and central centroplasm.
7. Cyanophycean granules and starch present.

Reproduction

1. Sexual reproduction has not been reported.
2. Asexual reproduction by means of hormogones and the separation discs. They (hormogones) develop into new individuals.

Systematic position and Identification

A. 1. Chlorophyll present
2. Plant body a thallus
3. Cell wall cellulosic...*Algae*

B. 1. Blue green pigments present
2. Plastids absent
3. Protoplast divided into chromoplasm and centroplasm
4. Sexual reproduvtion generally not seen...*Myxophyceae*

C. 1. Plant body fimamentous
2. Filaments unbranched or show false branching ...*Nostocales*

D. 1. Filaments uniseriate
2. Cells broader than long
3. Asexual reproduction by hormogones...*Oscillatoriaceae*

E. 1. Filaments enclosed by a mucilage sheath
2. Calyptra at the tip absent...*Lyngbya*

OSCILLATORIA
(Vaucher 1803)

Common species: O.martini, O. nigra, O. princepis etc.

Distribution : A very common blue green alga, found in ponds, stagnant gutter water, ditches etc. Some times the alga is found on damp soil forming dark blue green patches.

Study 1. External features
2. Reproduction
3. Systematic position and Identification

External features

1. Plant body is an unbranched filament. The filaments may be solitary or many of them interweave into knots or patches.
2. Filaments are unbranched with cells joined end to end.
3. Cells are broader than long and have a cellulosic cell wall.
4. Filaments may or may not have a mucilage envelope. When present the mucilage envelope in very thin.
5. The terminal cell of the filament is tapering. It may be hemispherical, acuminate or capitate.
6. In some species the apical wall of the terminal cell is thickened to form *calyptra.*
7. The filaments show oscillating movements hence the name *Oscillatoria*
8. Cell structure typical to Cyanophyceae and has outer pigmented chromoplasm and inner centroplasm.

Cyanophycean granules and starch present.

9. Here and there dead cells are present in the form of biconcave discs

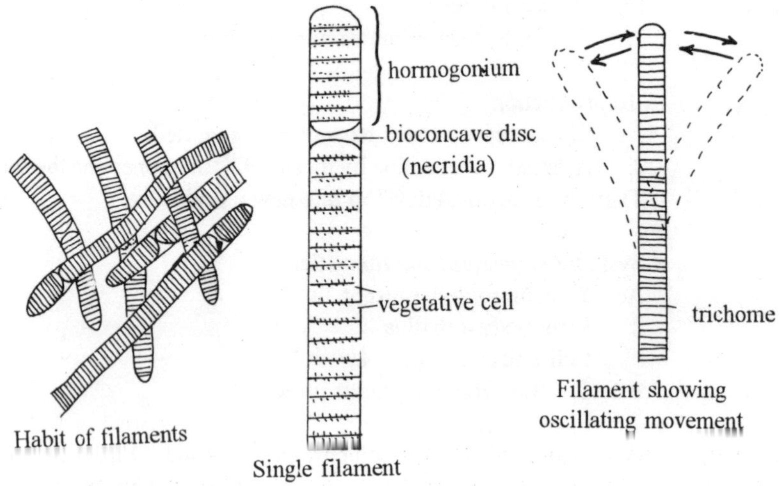

Fig.4.10 Oscillatoria

Reproduction

1. Sexual reproduction has not been reported.
2. Asexual reproduction by hormogones.
3. Hormogones are bits of filaments (5-6 cells long) and present between the two biconcave discs. The hormogones separate at these discs and grow into a new filament.

Diagrams to be drawn

1. Habit sketch
2. A few filaments enlarged.
3. Hormogone and biconcave discs.
4. Detailed cell structure

Systematic position and Identification

A. 1. Chlorophyll present
   2. Plant body a thallus
   3. Cell wall cellulosic...*Algae*

B. 1. Blue green pigments present
   2. Plastids absent
   3. Protoplast divided into chromoplasm and centroplasm
   4. Sexual reproduction generally not seen...*Myxophyceae*

C. 1. Plant body filamentous
   2. Filaments unbranched or show false branching ...*Nostocales*

D. 1. Filaments uniseriate with a terminal tapering cell
   2. Heterocysts absent
   3. Mucilage sheath thin...*Oscillatoriaceae*

E. 1. Terminal cell of the filament forms a calyptra
   2. Filaments show oscillating nature.
   3. Biconcave discs present in the filament
   4. Mucilage sheath either absent or very thin when present...*Oscillatoria*

SCYTONEMA
(Agardh 1824)

Common species: *S. rivulare*, S. *simplex*, *S. iyengarii*, *S.varuim* etc.

Aquatic or semiterrestial. Founon the bottom of ponds, pools etc. or on damp soil. Occcurs also in symbiotic association in lichens.

Study 1. External features (including branching)
2. Reproduction.
3. Systematic position and identification

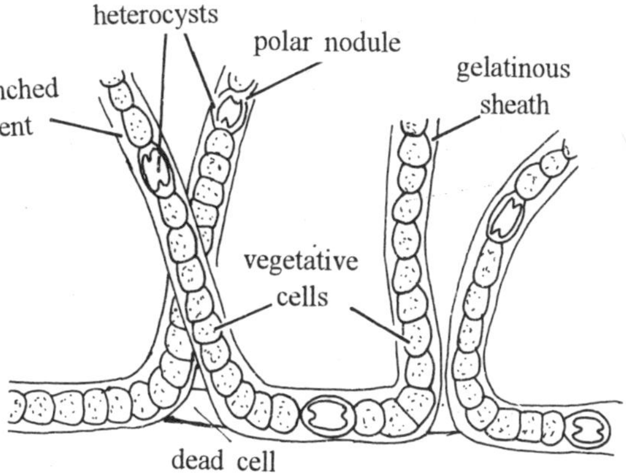

Fig.4.11 Scytonema   A portion of pseudobranched filaments

External features

5. The filamentous plant body is surrounded by a mucilage sheath. The sheath is lamellated.
6. Hetrocysts present. They are terminal as well as intercalary.
7. Filament shows false branching. One or two branches arise at a point.
8. Branching results due to the death of an intercalary cell, on which the two cells on either side resume apical growth and develop into branches.
9. Heterocysts are pale yellow in colour and have one or two polar nodules depending on their location.
10. Cell structure is typically Cyanophycean. It has a central centroplasm, surrounded by chromoplasm.

Reproduction

11. Sexual reproduction has not been reported.
12. Asexual reproduction brought about by means of hormogones, which are formed between two separation discs.
13. Akinetes have been reported in some species.

Diagram to be drawn

14. Habit sketch
15. One or two filaments enlarged
16. A portion of the filament at the region of branching
17. Cell structure

Systematic position and Identification

A. 1. Chlorophyll present
   2. Plant body a thallus
   3. Cell wall cellulosic...*Algae*

B. 1. Blue green pigments present
   2. Plastids absent
   3. Protoplast divided into chromoplasm and centroplasm
   4. Sexual reproduction generally not seen ... *Myxophyceae*

C. 1. Plant body filmentous
   2. Filaments unbranched or show false branching...*Nostacale*

D. 1. Filaments exhibit false branching
   2. Heterocysts intercalary mostly
   3. Reproduction by hormogones...*Scytonemataceae*

E. 1. False branching
   2. Mucilage sheath enveloping the filament...*Scytonema*

STIGONEMA

Common species: *S. arugineum, S. panniforme, S.ocellatum, S.dendroideum* etc.
Distribution: Mostly aquatic, could be semiterrestrial also. Sometimes found attached to aquatic weeds.

Study 1. External features
2. Reproduction
3. Systematic position and Identification

External features

1. The plant body is a filament. Many filaments accumulate to form brown mucilaginous lumps.
2. Each filament has an independent mucilaginous envelop.
3. Filaments are multiseriate at older regions but uniseriate in younger regions
4. Filaments are branched. Branching is true.
5. Cells are sub-spherical or rectangular., Sometimes broader than long.
6. Cell structure is typically cyanophycean i.e,. it has a periphereal chromoplasm surrounding a central centroplasm.
7. Growth is apical

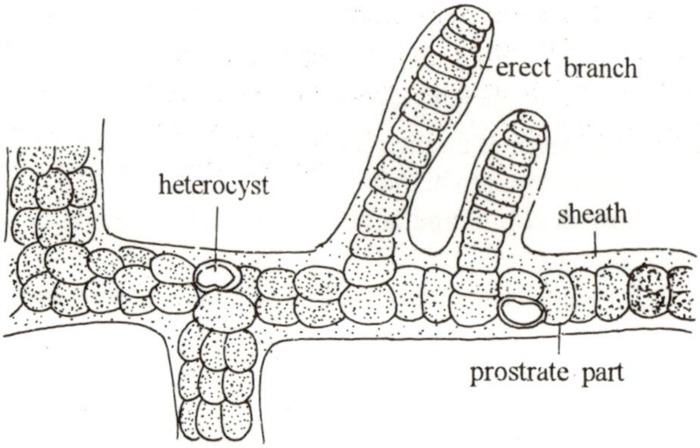

Fig.4.12 Stigonema,
Thallus of Stigonema showing true branchiug

8. Heterocysts present in some species.

Reproduction

1. Sexual reproduction has not been reproted.
2. Asexual reproduction is brought about by fragmentation, hormogones or heterocysts.
3. When the filament breaks up into bits accidentally. Each bit develops into a new individual.
4. Few celled hormogones at maturity develop into new individuals.

Diagram tc be drawn

1. Habit sketch.
2. A filament enlarged to show branching.
3. A portion of the film enlarged at the region of branching.
4. Cell structure.

Systematic Position and Identification

A. 1. Chlorophyll present
   2. Plant body a thallus
   3. Cell wall cellulosic...*Algae*

B. 1. Blue green pigments present
   2. Plastids absent.
   3. Protoplast divided into chromoplasm and centroplasm.
   4. Sexual reproduction generally not seen ...*Myxophyceae*

C. 1. Plant body is a filament
   2. Filaments show true branching...*Stigonematales.*

D. 1. Plant body is a filament with true branches.
   2. Filaments found in mucilaginous sheath.
   3. Branching of the filament is lateral...*Stigonemataceae*

E. 1. Filament exhibiting true branching.
   2. Each filament has an independent mucilaginous envelope.
   3. Older parts of the filament multiseriate while younger parts are uniseriate.
   4. Growth is apical
   5. Cells are sub-spherical or recetangular...*Stigonema.*

CHAPTER 5

CHLOROPHYCEAE

(Chlorophyta)

The class Chlorophyceae includes about 360 genera and nearly 6500 species. The members are commonly called grass green algae because of their deep green colouration. All the members have their pigments arranged in plastids (chloroplasts). They possess a definite nucleus. The members may be motile or non motile. The reproductive cells are generally flagellate. Most of the members of chlorophyceae, possess the same phtotosynthetic pigments seen in higher plants.

Occurrence and Distribution: Approximately 90% of chlorophyceae are fresh water in distribution (Smith 1966). Some orders like Ulvales and Siphonales are predominantly marine, Ulotrichales predominantly fresh water, while Oedogoiales and Conjugales exclusiverly fresh water.

Some Chlorophycean members are terrestrial. Sime like *Trentepohlia* are epiphytic growing on wet barks of trees.

Range of Thallus Construction

The members of Chlorophyceae exhibit great diversity in form and structure. In the vegetative construction, they range from unicellular motile forms up to parenchymatous viz., motile forms, palmelloid forms, coccoid forms, filamentous forms, sipihonaceous habit, uniaxial type and parenchymatous type.

Cell structure: A few primitve chlorophycean members have naked protoplasts. But in a great majority, the protoplast lies within a prominent cell wall. The cell-wall is usually 2-3 layered: the outer chitinous, middle pectic and inner cellulosic. Mucilage envelope is present in some members. Lime and silica may also be deposited over the cell wall.

The protoplast in unicellular forms consists of two small contractile vacuoles. In fialmentous forms, the cell usually has a large central vacuole traversed by a number of cytoplasmic strands. Such a cytoplasm with strands is called Primordial utricle.

In motile forms locomotion is brought about by flagella.

Chloroplasts: The members of Chlorophyceae exhibit a great range of variation in the structure and organisation of chloroplasts. Chloroplasts may be cup-sharped, discoid, stellate, spiral or reticulate, Pyrenoids (one to many) are present in chloroplsts. The pigments present in chloroplasts are chlorophyll a, chlorophyll b, carotene, and xanthophyll. Often a red coloured pigment called Haematochrome is also present.

The product of assimilation is starch and it is stored in pyrenoids.

Reproduction: Members of Chlorophyceae reproduce asexually as well sexually.

Asexual Reproduction: It is brought about by the following methods:

Cell division: In unicellular forms cell-division is a means of reproduction. The two daughter cells after a cell division develop into new individuals.

Zoospore formation: The characteristic method of axexual reproduction in majority of the forms is the formation of either bi or quardriflagellate zoospores. They are usually produced in zoosporangium which may form one to many zoospores. Usually the protoplast divides to form a number of uninuleate bits which become motile. They are set free in water. Zoospores are generally formed when there is abundance of moisture. Usually zoospores are formed during night and liberated in the morning. After a short period of swarming , the zoospores develop into new individuals.

Fragmentation: This is the usual method of multiplication among filamentous forms. If a filament, accidentally breaks, each bit will develop into a new individual.

Aplanospores and Akinetes: In some members, non-motile spores called aplanospores are produced. In some members resting spores called Aikinetes are formed.

Sexual Reproduction: It is of three types Isogamy, Anisogamy and Oogamy (for details see General account of algae).

Types of Life Cycles

Three types of Life cycles are seen in Chlorophyceae. These are Haplontic type, Isomorphic alternation of generations and Hetermorphic alternation of generations.

The haplontic type, there is no differentiation into gametophyte and soprophyte. The same plant body produces gametes and spores at different times.

In a life cycle with isomorphic alternation of generations, sporophyte and gametophyte are identical morphologically, while in hetermorphic alternation of generations, they are dissimilar morphologically.

Classification: Various types of classification have been proposed for Chlorophyceae. Of these the classification by Smith (1966) and Fritsch (1935) have been given below.

Smith's Classification: According to Smith, the division Chlorophyta is divided into two classes namely- Chlorophyceae and Charophyceae. Chlorophyceae comprises of ten orders whereas Charophyceae has only one order.

Divisions-Chlorophyta

Class **Chlorophyceae**

Orders:
1. Volvocales
2. Tetrasporales
3. Ulotrichales
4. Oedogoniales
5. Ulvales
6. Schizogoniales
7. Chlorococcales

8. Siphonales
9. Siphonocladales
10. Zygnematales

Class **Charophyceae**
Order Charales

Fritch's classification: The classification is based on the structure of vegetative and reproductive bodies. Fritsch divides the class into 9 order viz.,

1. Volvocales
2. Chlorococcales
3. Ulotrichales
4. Cladophorales
5. Chaetophorales
6. Oedogoniales
7. Conjugales
8. Siphonales
9. Charales

In this book, the classification of Fritsch (1935) is followed.

CHLAMYDOMONAS
(Ehrenberg 1833)

Common species: *C. ehrenbergii, C.nivalis, C.reticulata, C.pluristigma* etc.

Distribution : Found in stagnant water, pools, ponds and ditches.

Some species are found in saline water also.

Study: 1. Cell structure
2. Palmella stage
3. Systematic position and Identification

Cell structure

1. The plant body in unicellular. Cells may be spherical or oval in shape.

2. Outermost layer is the cell-wall, occasionally there might be mucilage covering.

3. The cell can be distinguished into a broad posterior end and a narrow anterior end.

4. At the anterior end are found two long flagella, which help in locomotion. Each flagellum arises from a basal granule or blepharoplast.

5.The two basal granules are joined by a fibril known as Paradesmos. One of the basal ganule is connected to the nucleus thro' another thread called Rhizoplast. The whole structure is called Neuromotor Apparatus.

6. Near the basal granules are two contracrtile vacuoles, whose function is osmoregulation.

7. Towards the anterior end is also found the lens shaped stigma or Eyespot, which is photosensitive.

8. Internal to the cell-wall is the living plasma membrane.

9. Towards the broader end of the cell and occupying 2/3rd of the cell interior is a single cup shaped chloroplast, with a centrally embedded Pyrenoid.

10. Filling the cup of the chloroplast is the cytoplasm.

11. The single nucleus lies suspended in cytoplasm.

12. Nutrition is autotrophic.

Reproduction: *Chlamydomonas* reproduces both asexually and sexually. However these stages cannot be easily observed in the laboratory. Only one method of Asexual reproduction viz., Palmella stage can be observed. The stage is given the name palmella stage, because of its resemblance to the alga *Palmella*.

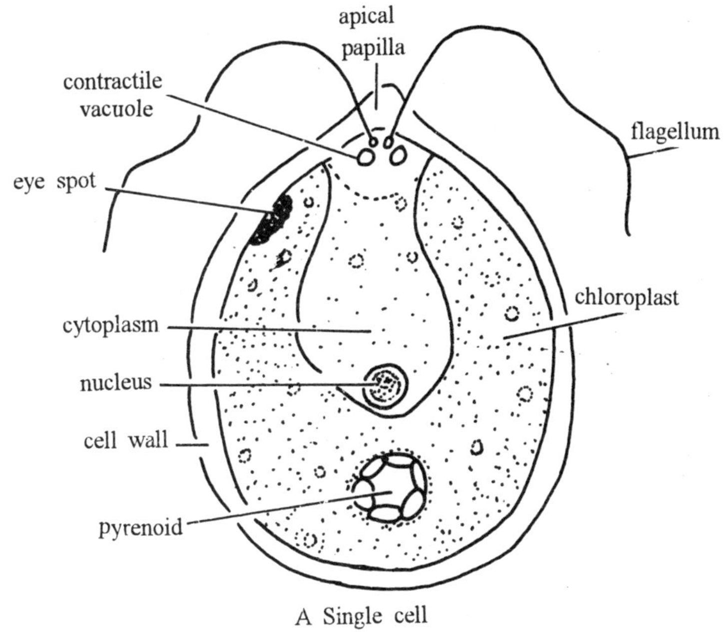

Fig.5.1 Chlamydomonas

Palmella stage:

1. Palmella stage is usually seen when the dry season will be approaching fast.

2. The cells divide, but the protoplasts do not develop flagella and remain inactive within the parent cell.

3. Many groups of cells get surrounded by a common mucilage envelope.

4. Palemlia stage consists of several generations of cells.

5. Plamella stage is a transition stage.

6. In the return of the favourable season, the mucilage envelope dissolves, and all the cells are set free. They develop flagella very soon.

Diagrams to be drawn

1. Single cell

2. Anterior part of the cell enlarged.

3. Palmella stage

Systematic position and Identification

A.

(1) Chlorophyll present
(2) Plant body is a thallus
(3) Cell wall is cellulosic *Algae*

B.

(1) Plastids present
(2) Chlorophyll is the chief pigment
(3) True nucleus present
(4) Photosynthetic product is starch

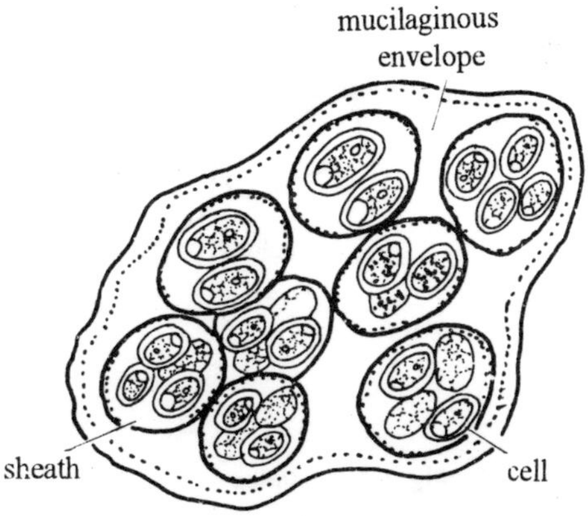

Fig. 5.2 Palmella Stage

(5) Reproductive cells motile...*Chlorophyceae*

C

(1) Plant body motile

(2) Chloroplast cup saped...*Volvocales*

D

(1) Plant body unicellular

(2) Contractile vacuoles present...*Chlamydomonadaceae*

E

(1) Cell pyriform in shape

(2) Eyespot present.

(3) Chloroplast cup-shaped with a centrally embedded stigma.

(4) Presence of Palemella stage during unfavourable conditions

(5) Cells uninucleate...*Chlamydomonas*.

PANDORINA
(Bory 1824)

Common species: *P. morum*

Distribution: Fresh water pools, ponds and ditches.

Study (1) External features of the colony

(2) Asexual Reproduction

(3) Sexual Reproduction

(4) Systematic position and Identification.

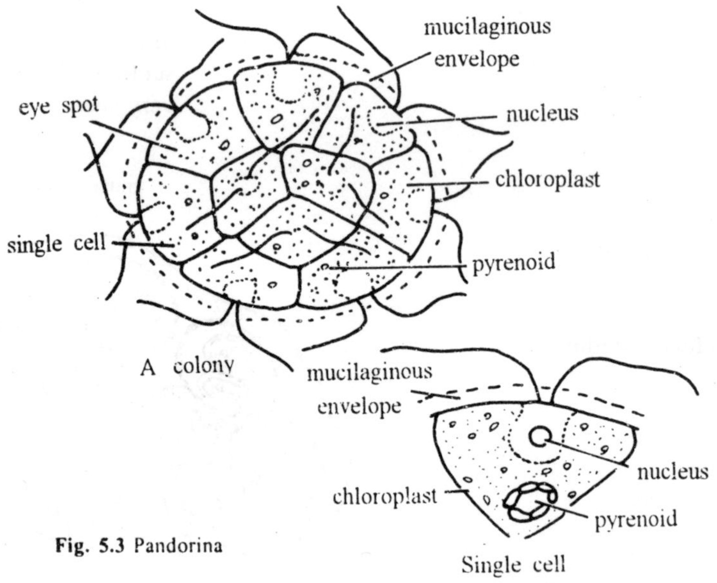

Fig. 5.3 Pandorina

External features

1. The plant body is a multicellular colony called the coenobium. It is a hollow sphere.

2. The coenobium is motile, because the cells of the colony are flagellate.

3. The number of the cells per colony range from 16-32. They are embedded in a muciliaginous matrix.

4. The cells are biflagellate and chlamydomanad (like *Chlamydomonas*) in structure.

5. The cells are arranged in tiers of four each.

6. Cells are pyrifrom having a broad posterior and narrow anterior end.

7. Cells are uninucleate with a cup shaped chloroplast.

8. There are two contractile vacuoles and an eye spot at the anterior end.

Asexual Reproduction

1. Each cell of the plant colony can divide and form daughter colonies.

2. Formation of daughter colonies takes place after the coenobia stop locomotion and settle down in a place.

3. After the formation of daughter colonies the mucilage envelop dissolves liberating them.

Sexual reproduction

1. Coenobia are dioccious
2. Reproduction is anisogamous.
3. All the cells of the colony take part in the formation of gametes.
4. Each cell of the colony divides to form a number of gametes.
5. All the cells of the colony either form macrogametes or microgamtes.
6. Fusion always takes place between a microgamete and a macrogamete, resulting in the formation of a zygote. The zygote is red in colour due to haemotachrome.
7. The zygote develops into a new coenobium.

Diagrams to be drawn

(1) A colony enlarged
(2) A cell enlarged
(3) Colony showing daughter colonies
(4) Colony showing zygotes

Systematic position and Identification

A

1. Chlorophyll present
2. Plant body is a thallus
3. Cell-wall cellulosic... *Algae*

B

1. Plastids present
2. Chlorophyll is the chief pigment
3. True nucleus present
4. Photosynthetic product is starch
5. Reproductive cells motile... *Chlorophyceae*

C

1. Plant body motile
2. Chloroplast cup shaped... *Volvocaceae*

D

1. Plant body is colonial.
2. Colonies motile
3. Cells divide in a longitudinal plane... *Volvocaceae*

E

1. Colony spherical or hollow
2. Cells arranged in tiers
3. Zygote red in colour due to the presence of haemotachrome... *Pandorina*

EUDORINA

Common species: *E.indica, E.elegans, E.illinoiensis* etc.

Distribution: Commonly found in fresh water, ponds and ditches.

Abundant as Planktonic flora in lakes.

Study (1) Structure of the colony

(2) Structure of the cell

(3) Asexual reproduction

(4) Sexual reproduction

(5) Systematic position & identification

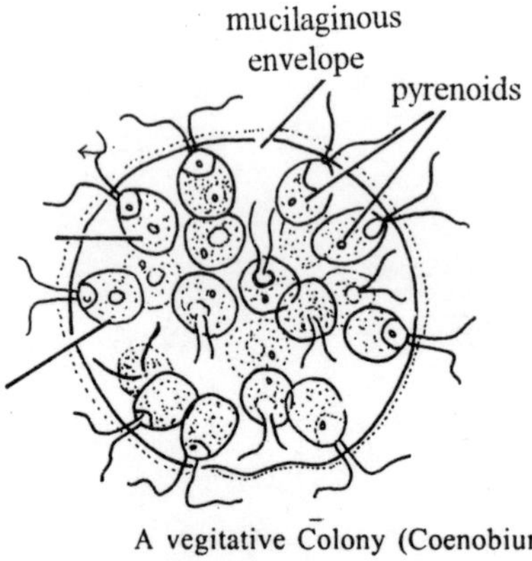

A vegitative Colony (Coenobium)

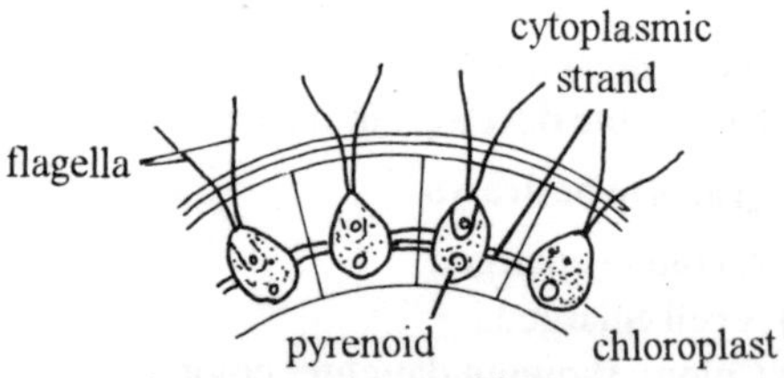

A portion of coemobium showing enlarged cells

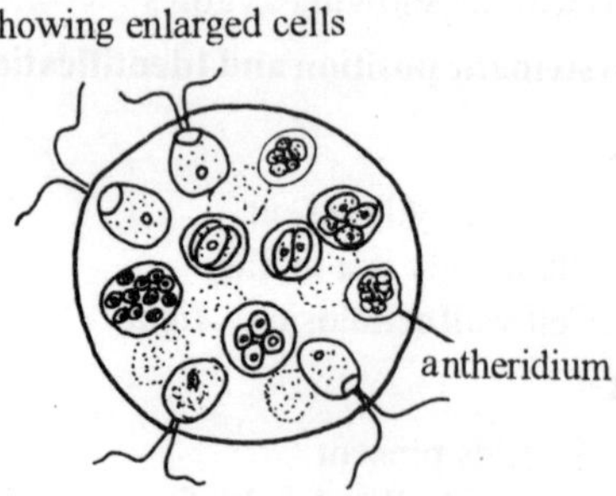

Coenobium showing reproductive cells

Fig. 5.4 Eudorina

Structure of colony

1. The plant body is a motile colony called Coenobium.
2. The colonies are subspherical or oval in shape.
3. The Coenobium is enveloped by a mucilage layer. It may be smooth or warty.

4. Cells are arranged in a single plane with the centre of the colony being hollow.

5. The number of the cells in the colony ranges from 16-64.

6. Cells are connected to one another by means of this cytoplasmic connections.

Cell structure.

1. The cell structure is Chalmydomonad.
2. Cells are uninucleate and biflagellate
3. Each cell has a single cup shaped chloroplast.
4. At the anterior end there are two contractile vacuoles
5. A single eye spot is present near the anterior region.

Reproduction

Asexual

(1) Daughter colonies are formed in each cell of the colony or only a few cells may form daughter colonies.

(2) At maturity the mucilage envelops of the parent colony ruptures releasing all the daughter colonies.

(3) Palmella stage may be formed during unfavourable conditions, when the daughter colonies may withdraw their flagella and undergo a period of rest.

Sexual

1. Colonies of *Eudorina* may be monoecious or dioecious.
2. Gametes are anisogamous.
3. Fusion takes place between a larger (female) gamete and a smaller (male) gamete.
4. Fusion of the gametes results in the formation of zygote which undergoes a period of rest.
5. Zygote has a thick and smooth wall. It is orange red coloured due to the presence of haematochrome.

Diagrams to be drawn

1. A single colony
2. A cell enlarged
3. Parent colony with daughter colonies
4. Colony with zygotes

Systematic Position and identification

A.

1. Chlorophyll present
2. Plant body a thallus
3. Cell wall cellulosic...*Algae*

B.

1. Plastids present
2. Chlorophyll is the chief pigment

3. True nucleus present

4. Photosynthetic product is starch

5. Reproductive cells motile...*Chlorophyceae*

C.

1. Plant body motile

2. Chlorophyll cup shaped...*Volvocales*

D

1. Plant body is colonial

2. Colonies are motile

3. Cells divide in a longitudinal plane...*Volovocaceae*

E

1. Colony subspherical or ovoid

2. Number of cells in a colony less than 256

3. All cells in the colony are of same size...*Eudorina.*

PLEDORINA

Common species: *P.californica, P.indica, P.illinoiensis*

Distribution: Colonies are distributed extensively in fresh water, stagnat ponds.

Study:

1. Structure of the colony

2. Cell structure

3. Asexual reproduction

4. Sexual reproduction

5. Systematic position & identification.

Structure of the colony

1. Plant body is a motile colony called coenobium.

2. The conenobia are spherical or occasionally ellipsoidal in shape.

3. Cells are arranged in a single layer at the periphery resulting in a hollow sphere. The cells in each colony ranges from 32,64, or 128.

4. The colony is enveloped by mucilage layer.

5. Cells in the colony are of two sizes: Large non flagellate reproductive cells and small flagellate vegetative cells.

6. Vegetative cells occupy about 1/3rd of the colony, the rest being occupied by reproductive cells.

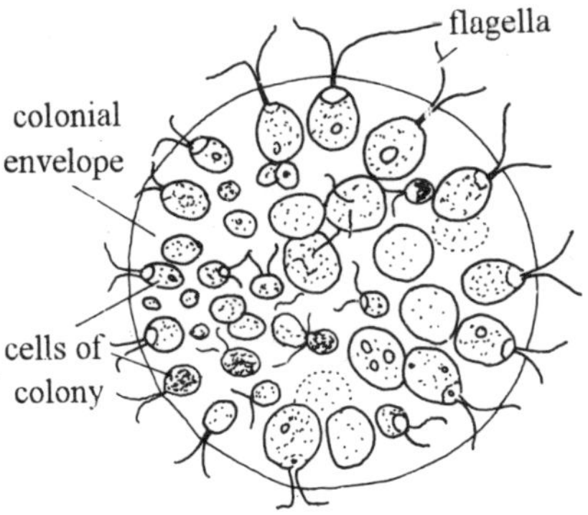

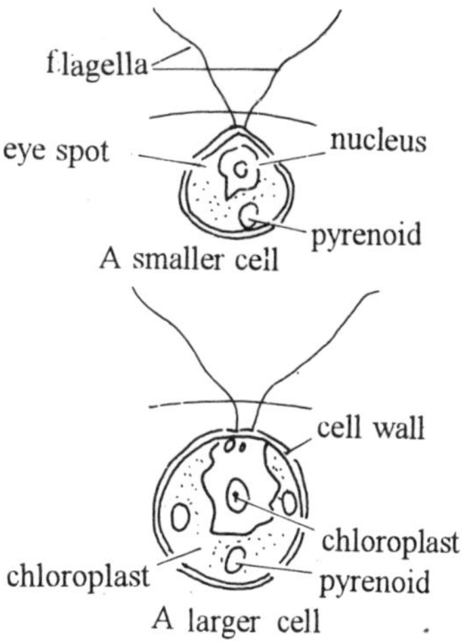

Fig. 5.5 Pleodorina

Cell structure

1. Cells are typically chlamydomonad in structure.

2. Each cell has two flagella, a cup shaped chloroplast, a stigma, and a single nucleus.

3. Flagella are absent in the reproductive cells.

Asexual reproduction

1. Daughter colonies are formed by some of the large reproductive cells.
2. The daughter colonies are trapped inside the mother colony for quite a long time.
3. After a certain period of time the daughter colonies are released by the rupture of the mother colony.

Sexual reproduction

1. Colonies reproduce sexually by means of oogonia (female) and antheridia (male). Reproduction is by oogamous method.

2. The coloneis are dioecious. Some colonies produce only antheridia and some produce only oogonia.

3. In the colony reproductive cells produce sperm plates. Each sperm plate has a number of biflagellate sperms.

4. Oogonia are formed in the female colony by the reproductive cells. Each oogonium has a single egg.

5. Fertilization results in the formation of zygote, which is red in colour and has a thick wall

Diagrams to be drawn

1. A colony enlarged

2. Cells enlarged

3. Daughter colonies

4. Colony whith oogonia

5. Colony with sperm plates

6. Colony with antheridia

7. Colony with zygotes

Systematic position and identification

A

(1) Chlorophyll present

(2) Plant body a thallus

(3) Cell-wall cellulosic...*Algae*

B.

(1) Plastids present

(2) Chlorophyll is the chief pigment

(3) True nucleus presnt

(4) Photosynthetic product is starch

(5) Reproductive cell motile... *Chlorophyceae*

C.

(1) Plant body motile

(2) Chloroplast cup-shaped...*Volvocales*

(3) Cells divide in a longitudinal plane...*Volvocaceae*

D.

(1) Plant body motile .

(2) Colonies are motile

(3) Cells divide in a longitudinal plane...*Volvocaceae*

E

(1) Colonies are ellipsoidal in shape

(2) No. of cells in a colony do not exceed 256

(3) Division of labour is seen in the cells of the colony...*Pleudorina*

VOLVOX
Linnaeus (1758)

Common species: *V.globator, V.africanus, V.prolificus* etc.

Distribution: Fresh water, ponds and ditches, very common particularly after a spell of rain.

Study

(1) External features of the colony

(2) Cell structure

(3) Asexual reproduction

(4) Sexual reproduction

(5) Systematic position and Identification.

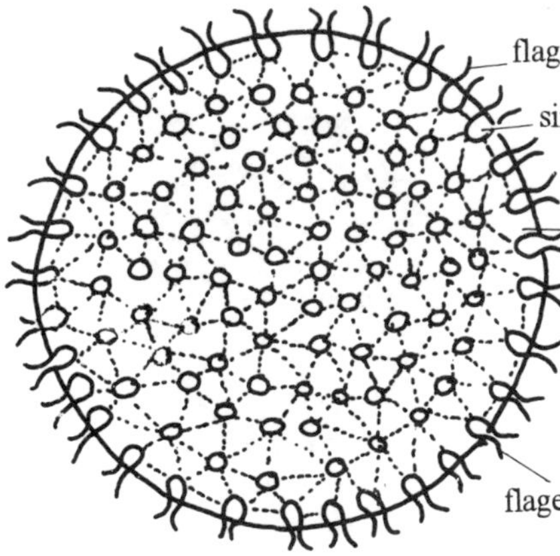

A Vegetative Coenobium

Fig. 5.6 Volvox

External features

1. The plant body is a coenobium found floating or revolving as small green balls of the size of a pin head in fresh water.

2. The colony is scarcely visible to the naked eyes and is a hollow sphere made up of cells ranging in number from 500-50,000.

3. The entire coenobium is enveloped by mucilage.

4. The colony being spherical or ovoid has no anterior or posterior end: but the end which has few or no reproductive cell is called anterior end.

5. The reproductive cells lack the flagella and the eye spot.

Asexual reproduction

1. Asexual reproduction is brought about by means of daughter colonies.

2. The cells that give rise to daughter colonies are called 'parthenogonidia' They are bigger in size than the vegetative cells, have rich cytoplasm and a prominent nucleus.

3. The gonidium divides repeatedly and forms a daughter colony. The daughter colony is soon released into water.

4. In some cases the daughter colony is held trapped in the parent colony. It matures itself and some of its

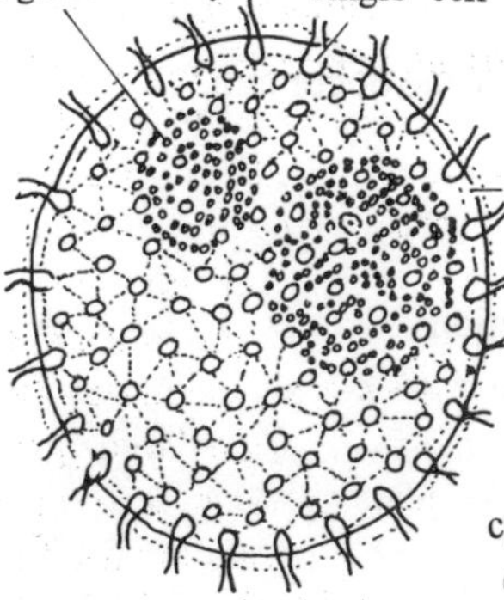

Coenobium showing daughter colonies

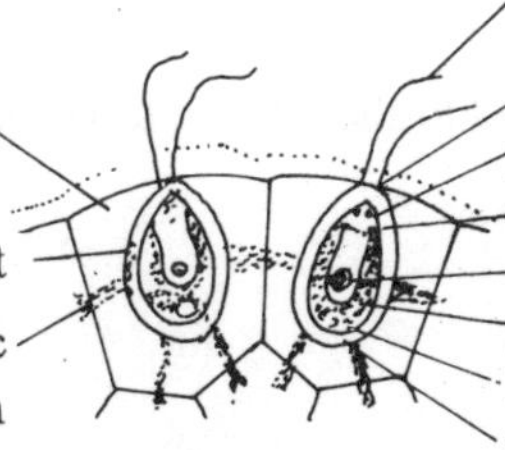

A portion of coenobium showing enlarged cells

Fig. 5.7 Volvox

cells produce asexually another young colony called grand daughter colony.

5. In some cases an old colony of Volvox may have as many as 4-5 generations of colonies.

6. At a certain stage however the cell wall ruptures and all the colonies are set free, whereupon they develop into independent colonies.

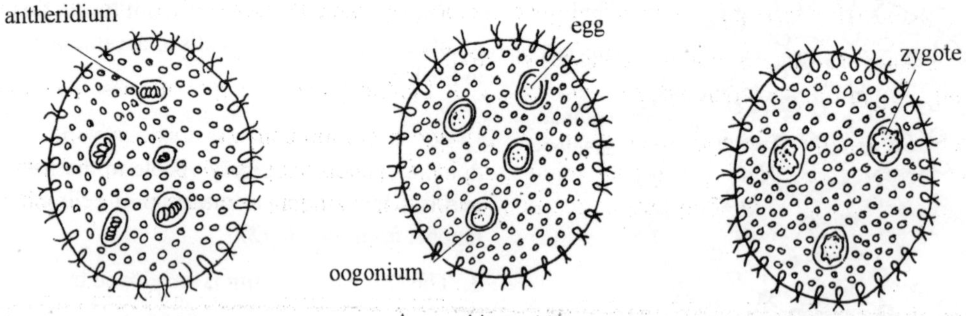

A coenobium with antheridia (anthrogonidia)

A coenobium with oogonia (gynogonidia)

A coenobium with zygote

Fig. 5.8 Volvox

A Coenobium with antheridia (androgonidia)

Sexual reproduction

1. This is of the oogamous type. The species of *Volvox* may be monoecious or dioecious.

2. Male sex organs are antheridia and female sex organs are oogonia.

3. The sex organs develop from special cells called gonidia, present at the posterior part of the colony. Antheridia develop from androgonidia while oogonia develop from gynogonidia

4. An antheridium has a number of cigar shaped biflagllate sperms arranged in a sperm plate. They are set free at maturity.

5. An oogonium has a single oosphere or egg.

6. Fertilization results in the formation of a zygote which eventually develops a thick spinescent wall. It is red in colour due to the pigment haematochrome.

7. The zygote is released from the colony by the decay of the wall by the oogonium.

Diagrams to be drawn

1. Colony showing cells (habit)

2. A few vegetative cells enlarged to show the cytoplasmic connections.

3. Colony with daughter colonies.

4. Colony with sperm plates.

5. Colony with oogonia.

6. Colony with zygotes.

Chlorophyceae

Systematic position and Identification

A.

(1) Chlorophyll present

(2) Plant body a thallus

(3) Cell wall cellulosic...*Algae*

B.

(1) Plastids present

(2) Chlorophyll is the chief pigment.

(3) True nucleus present

(4) Photosynthetic product is starch

(5) Reproductive cells motile...*Chlorophyceae*

C.

(1) Plant body motile

(2) Chloroplast cup shaped...*Volvocales*

D.

(1) Plant body is a coenobium

(2) Colonies are motile

(3) Cells divide in a longitudinal plane...*Volvocaceae*

E.

(1) Colony has more than 256 cells

(2) Division of labour seen in the cells of the colony

(3) Cells of the colony have structural and physiological differences...*Volvox*

CHLORELLA
(Beigerinok 1890)

Common spicies: *C.pyrenoidosa, C.vulgaris, C.conglomerata* etc.

Distribution: Occurs in fresh water,ponds,damp soil, sides of garden pots

Study (1) Structure or the plant body

(2) Reproduction

(3) Systematic position and Identification

Structure of plant body

1. The plant body is unicellular and green in colour.

2. The cells are non-motile, spherical or ellipsoidal in shape.

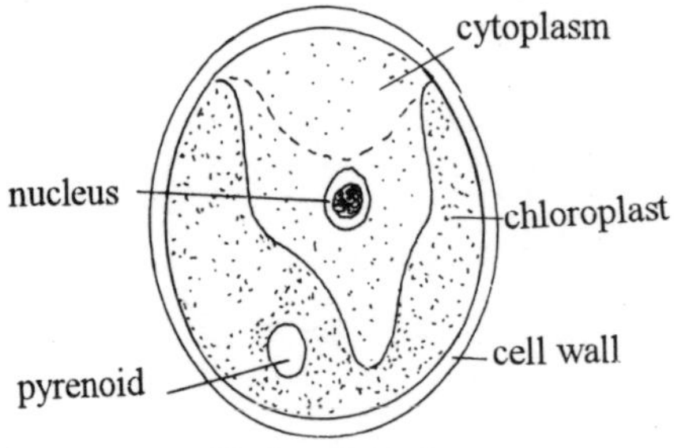

Single cell (thallus)

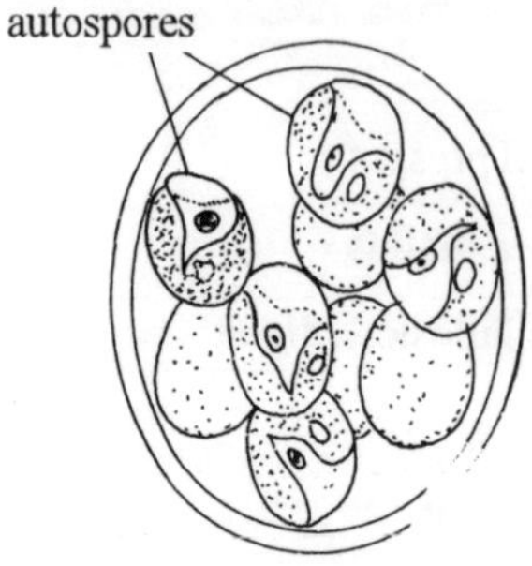

Thallus with autospore

Fig. 5.9 Chlorella

3. The cell-wall does not have a mucilaginous envelope.
4. Each cell has a single cup shaped or band shaped chloroplast: there is no pyrenoid.
5. Cells are uninucleate.

Reproduction

1. Sexual reproduction has not been reproted.
2. The cells divide and form spores called autospores.
3. The number of autospores per cell is 2-8.
4. Autospores are uninucleate and develop into a new individual.

Diagram to be drawn

1. Structure of the cell
2. Cell with autopsores

Systematic position and Identification

A.

1. Chlorophyll present
2. Plant body a thallus
3. Cell wall cellulosic...*Algae*

B.

1. Plastids present
2. Chlorophyll is the chief pigment
3. True nucleus present
4. Photosynthetic product is starch...*Chlorophyceae*

C.

1. Cells are uninucleate and non-motile.

2. Reproduction by spores which are non-motile (motile spores very rare)...*Chlorococcales*

D

1. Plant body unicellular
2. Cells with a cup shaped chloroplast...*Chlorellaceae*

E

1. Cells spherical and Uninucleate
2 Nucleus located at the centre
3. Chloroplast cup shaped ...*Chlorella*

PEDIASTRUM
(Meyer 1829)

Common species: *P.duplex, P.simplex, P.ovatum* etc.

Distribution: Found commonly in stagnant fresh water ponds and lakes.

Study

1. Structure of plant body
2. Reproduc
3. Systemat

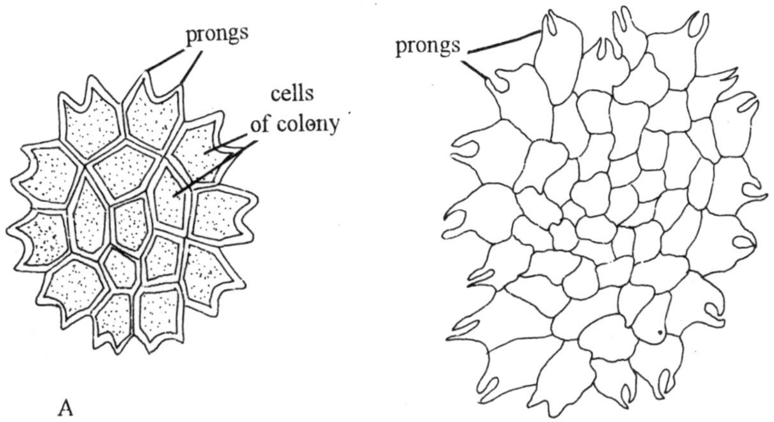

A

A vegetative colony

A vegetative colony

Fig. 5.10 Pediastrum

Structure of the Plant body

1. The plant body is a colony found freely floating on water.
2. The marginal cells of the colony have elongated projections giving the appearance of a star.
3. The cells are non motile and range in number per colony from 4-128.

4. The cells are arranged in cell plates. It is this cell plate the margin of which forms the projections giving a stellate appearance.

5. The cells have specific arrangements, which are as follows:

8 celled colony-1 in the centre, 7 in outer ring.

16 celled colony-1 in the centre, 5 in the first ring and 10 in second ring.

32 celled colony 1 in the centre, 5 in the first ring, 10 in second ring and 16 in outer most ring.

6. Cells are polyhedral and arranged closely without leaving any intercellular space.

7. Cells have a cellulosic cell-wall which may be smooth or granulate.

8. Young cells are uninucleate, but may become bi or tetra nucleate.

9. Chloroplast is parietal and has one or two pyrenoids.

Reproduction

1. Zoospores help in asexual reproduction

Diagrams to be drawn

1. A colony enlarged.

2. Few cells enlarged to show the arrangement and other details.

Systematic position and Identification

A.

1. Chlorophyll present

2. Plant body a thallus

3. Cell-wall cellulosic...*Algae*

B

1. Plastids present

2. Chlorophyll is the chief pigment

3. True nucleus present.

4. Photosynthetic product is starch...*Chlorophyceae*

C.

1. Colonies are non-motile

2. Cells uni or multinucleate...*Chlorococcales*

D

1. Colonies non-motile

2. Mature cells mutlinucleate...*Hydrodictyaceae*

3. Colony stellate in appearance

4. Cells in the colony are arranged in concentric rings...*Pediastrum*

HYDRODICTYON
(Roth 1800)

Common species: *H. indicum,.H.reticulatum* etc.

Distribution: Found floating as a net in tanks, ponds and pools

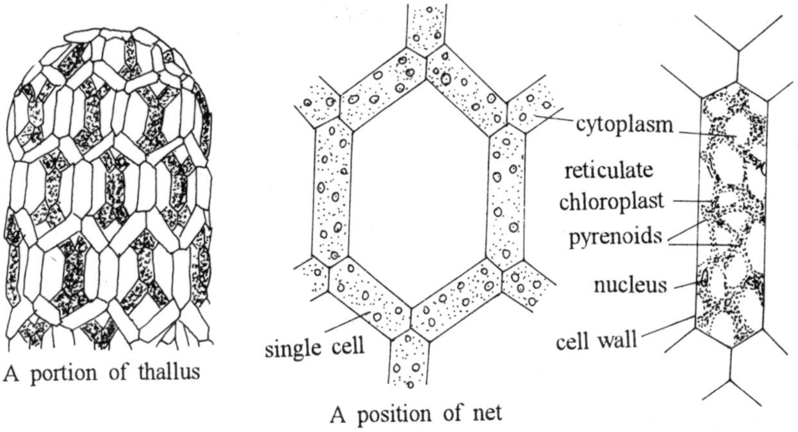

Fig.5.11 Hydrodictyon

Study

1. Structure of the colony
2. Cell sructure
3. Reproduction
4. Systematic position and Identification

Structure of the colony

1. The plant body is a reticulate colony (commonly called waternet) found floating on water.
2. The colony consists of many hundred cells attached to one another to form a reticulum.
3. At every joint usually three cells are joined, but occasionally four cells may also join at a point.
4. The meshes of the network appear like a hexagon.
5. The cells are many times longer than broad and do not hve a mucilage envelope.
6. Each cell is multinucleate and has a reticulate chloroplast with a number of pyrenoids

Reproduction

1. There is no vegetative cell division, the number of cells per colony is determined.
2. Daughter colonies are formed in the vegetative cells by the formation of zoospores which do not go out,

but arrange themselves inside the cell (to form a daughter colony).

3. Sexual reproduction is Isogamous.

Diagrams to be drawn

1. Colony (Habit sketch)
2. Few cells enlarged to show the attachment of cells.
3. A few cells enlarged
4. A cell with daughter colony.

Systematic position Identification

A

1. Chlorophyll present
2. Plant body a thallus
3. Cell wall cellulosic...*Algae*

B

1. Plastids present
2. Chlorophyll is the chief pigment
3. True nucleus present
4. Photosynthetic product is starch...*Chlorococcales.*

C

1. Colonies are non-motile
2. Young cells unicucleate...*Chlorococcales.*

D

1. There is no vegetative cell division in the colonuy
2. Adult cells multinucleate...*Hydrodictyaceae*

E

1. Colony is in the form of a reticulum
2. Cells multinucleate with a reticulate chloroplast...*Hydrodictyon.*

ULOTHRIX
(KUETZING 1833)

Common species: *U.cylindricum, U.flacca, U.tenissia, U.zonata* etc.

Distribution: Found in fresh water ponds, streams immediately after rainy season

A few species are found in salt water.

Study

1. External features

2. Reproduction

3.Systematic position & Identification

External features

1. The plant body is an unbranched filament

2. Filaments are thread like, fragile and formed by a uniserate row of cells.

3. The basal cell of the filament is modified into a hold fast to help in attachment. Hold fasts are hyaline or brownish in colour.

4. The cells are broader than long, with a two layered cell wall outer pectic and inner cellulosic.

5. Cells are uninucleate hnd have a single girdle shaped chloroplsst. There are 2-3 pyrenoids in each cell

6. There is a large vacuole in the centre of the cell.

7. The hold fast is colourless: It has no chloroplast

Reproduction-Asexual

1. Sexual reproduction is Isogamous

2. All cells except the hold fast and the terminal cell can form gametes.

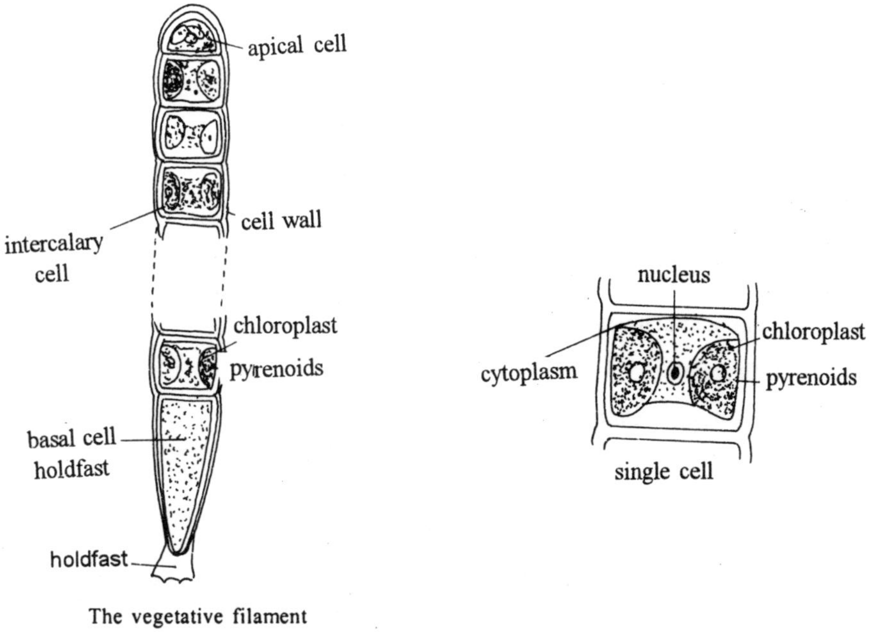

Fig.5.12 Ulothrix

Diagrams to be drawn

1. Habit sketch
2. A filament enlarged
3. A cell enlarged

Systematic position and Identification

A

1. Chlorophyll present
2. Plant body a thallus
3. Cell wall cellulosic...*Algae*

B

1. Plastids present
2. Chlorophyll is the chief pigment
3. True nucleus present
4. Photosynthetic product is starch...*Chlorophyceae*

C

1. Plant body, unbranched filament
2. Cells uninucleate with a girdle shaped chloroplast
3. Zoospore bi or quadriflagellate...*Ulotrichales*

D

1. Cells uninucleate
2. Cells are broader than long...*Ulotrichaceae*

E

1. Chloroplast Girdle shaped
2. Hold fast present, to help in attachment
3. Zoospores quadri or biflagellate
4. Sexual reproduction isogamous...*Ulothrix*

ENTEROMORPHA
(Link 1820)

Common species: *E.compressa, E.procera, E.intestinalis* etc.

Distribution: Mostly fresh water, but a few species are marine.

Study:

1. Structure of the thallus

2. Reproduction

3. Systematic position and Identification

Structure of the thallus

1. The plant body is a cylindrical or tube like thallus.

2. It is attached to the substratum with the help of a disc.

3. The thallus is one celled thick.

4. The thallus is branched, the branching being brought about by the longitudinal division of the cells.

5. The cells are many angled in shape and typically ulotrichean in structure

6. The cell wall is made up of cellulose and striated.

7. Cells are uninucleate with a parietal chloroplast. There is a single pyrenoid embedded in the chloroplast.

8. The chloroplast faces the outside of the cell, while the nucleus is inside.

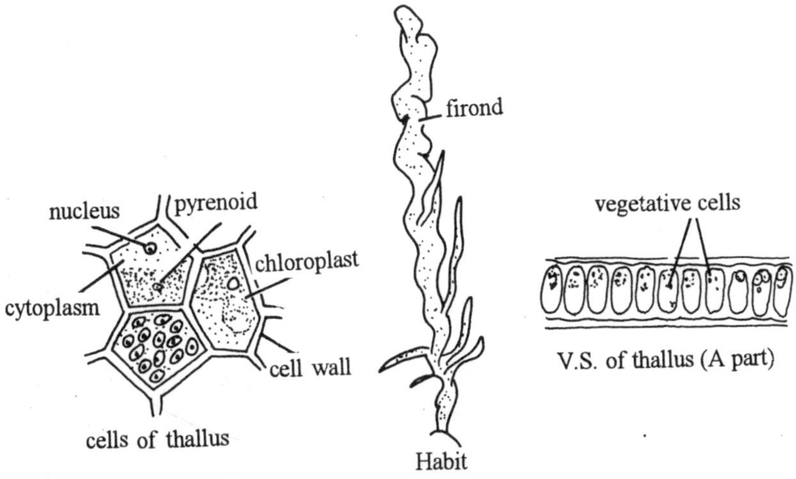

Fig.5.13 Enteromorpha

Reproduction

1. Vegetative propagation is brought about by the fragmentation of the thallus.

2. Asexual reproduction takes place by the formation of quadriflagellate zoospores.

3. Any cell of the thallus may form zoospores.

4. Sexual reproduction is iso or anisogameous.

5. The plants are dioecious, while sexual reproduction is anisogamous.

Diagrams to be drawn

1. Habit sketch

2. Section of the thallus

3. Hold fast

Systematic position and Identification

A

1. Chlorophyll present

2. Plantbody a thallus

3. Cell wall cellulosic...*Algae*

B

1. Plastids present

2. Chlorophyll is the chief pigment

3. True nucleus present

4. Photosynthetic product is starch...*Chlorophyceae*

C

1. Cells divide in more than one plane to form a sheet

2. Sexual reproduction iso or anisogamous

3. Cells uninucleate with a parietal chloroplast...*Ulotrichales*

D

1. Thallus one or two celled thick

2. Chloroplast parietal...*Ulvaceae*

E

1. Thallus is cylindrical (tubular) and branched

2. Constriction is present at the region of branching

3. Hold fast disc present in attachment...*Enteromorpha*.

ULVA-Sea Lettuce
(Agardh 1833)

Common species: *U.lactuca, U.lobata, U.reticulata, U.stenophylla* etc.

Distiribution: Most of the species are marine , found atached to rocks and other substrata in the tidal zone. A few may be fesh water also.

Study

1. External features

2. Internal features
3. Reproduction
4. Systematic position and Identification

External features

1. The thalloid plant body is a sheet two to three celled thick.

2. The plant body is anchored to the soil with the help of a hold fast.

3. The hold fast produces many rhizoidal outgrowths that grow down into the soil.

4. The rhiziods are chlorophyllous, septate, and multinucleate.

5. The cells are uninucleate, the nucleus being present towards the outerface and the chloroplast being parietal in position.

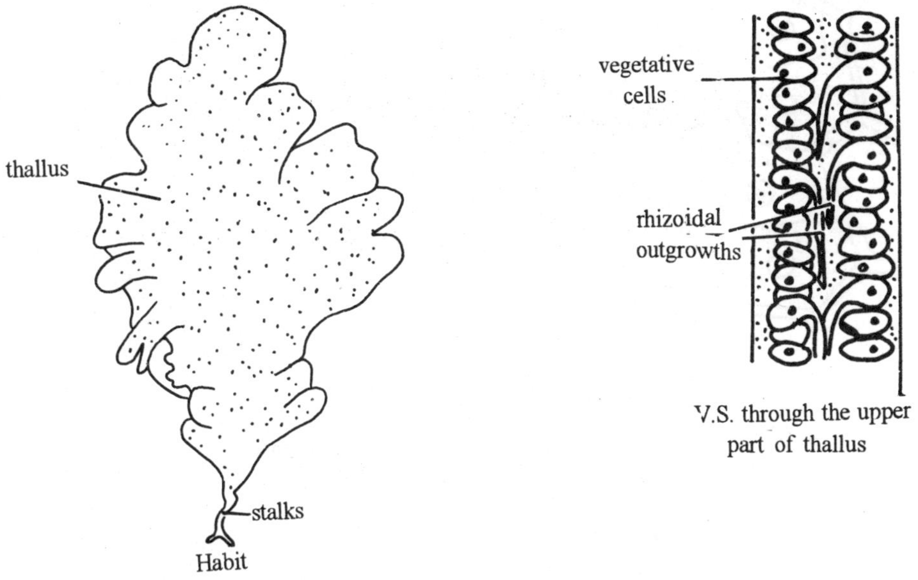

Fig.5.14 Ulva

Internal structure: A long section of the thallus shows the following features.

1. The cells of the thallus are arranged in two vertical rows.

2. Cells at the basal part of the thallus give out rhizoidal out growths, they combine together to form the hold fast.

Reproduction

1. Vegetative propagation of the thallus takes place by fragmentation.
2. Asexual reporduction takes place by the formation of quadri flagellate zoospores.
3. Any cell of the thallus can form zoospores.
4. Most of the species of *Ulva* are heterothallic.
5. Sexual reproduction is mostly by isogamy but occasionally anisogamy has also been reported.

Diagrams to be drawn

1. Habit sketch.
2. L.S. of the thallus.
3. A few cells enlarged.
4. Hold fast enlarged.

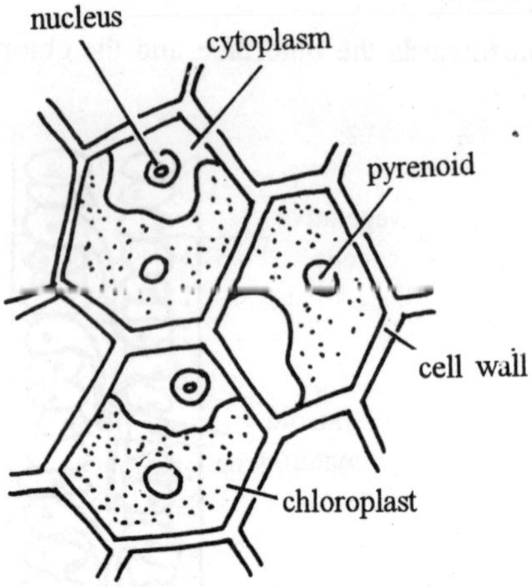

Vegetative Cells enlarged

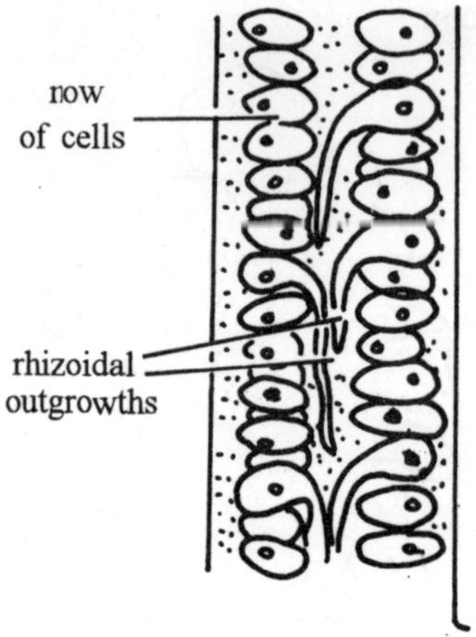

V.S. through the basal part of thallus

Fig. 5.15 Ulva

Systematic position and Identification

A.

1. Chlorophyll present
2. Plant body a thallus
3. Cell wall cellulosic...*Algae*

B

1. Plastids present
2. Chlorophyll is the chief pigment
3. True nucleus present
4. Photosynthetic product is starch...*Chlorophyceae*

C.

1. Cells divide in more than one plane to form a sheet of cells.
2. Cells uninucleate with a parietal chloroplast.
3. Sexual reproduction iso or anisogamous...*Ulvales*

D.

1. Thallus one or two celled thick.
2. Choloroplast parietal... *Ulvaceae*

E

1. Plant body flattened having two vertical rows of cells.
2. Phizoidal outgrowths combine to form a hold fast... *Ulva*

CLADOPHORA
(Kuetzing 1843)

Common species: *C. glomerata, C.crispata, C.fracata* etc.

Distribution: Found in fresh water as well as marine; seen attached to shells, stones or weeds.

Some species are aegagrophilous forming small floating balls.

Study

1. Structure of the plant body
2. Reproduction
3. Systematic position and Identification.

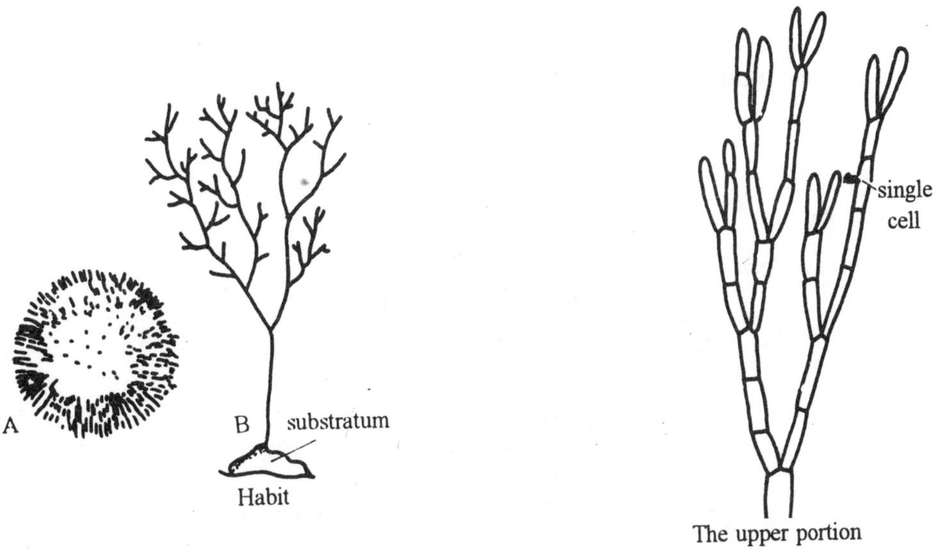

Fig. 5.16 Cladophora

Structure of plant body

1. The plant body is a branched filament found attached to certain substrata in water.
2. The filament is attached to the substratum with the help of rhizoidal outgrowths that are produced from the basal region.

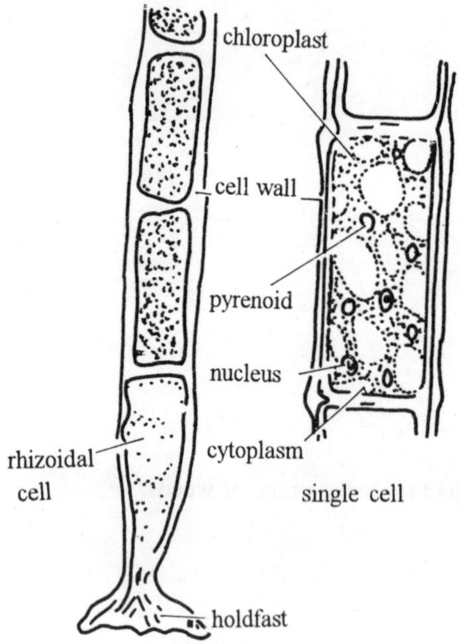

the basal part of filament

The basal part of filament

Fig. 5.17 Cladophora

3. The branching of the filament is lateral, a branch appearing just below the cross wall.

4. The branching later becomes pseudodichotomous due to the process of evection (Pushing)

5. Cells are longer than broad and have a stratified three layered cell wall.

7. The cytoplasm is periphereal, there being a large central vacuole.

Reproduction

1. Asexual reproduction is by formation of quadri flagellate zoospores.

2. Sexual reproduction is isogamous. The gametes are biflagellae.

3. Alternation of generation seen in the life cycle.

Diagrams to be drawn

1. Habit sketch

2. A filament enlarged

3. A cell enlarged

Systematic position and Identification

A

1. Chlorophyll present

2. Plant body is thallus

3. Cellwall cellulosic...*Algae*

B

1. Plastids present

2. Chlorophyll is the chief pigment

3. True nucleus present

4. Photosynthetic product is starch...*Chlorophyceae*

C

1. Cells multinucleate.

2. Filaments branched ..*Cladophorales*

D

1. Cells many times longer than borad

2. Branching is pseudodictomous...*Cladophoraceae*

E.

1. Filaments branched

2. Cells multinucleate with a reticulate chloroplast...*Cladophora*

PITHOPHORA
(Wittrock 1877)

Common species: *P.affains, P.kewensis* etc.

Distribution: Fresh water pools and ponds. Some times found attached to molluscan shells.

Study

1. External features
2. Reproduction
3. Systematic position and Identification

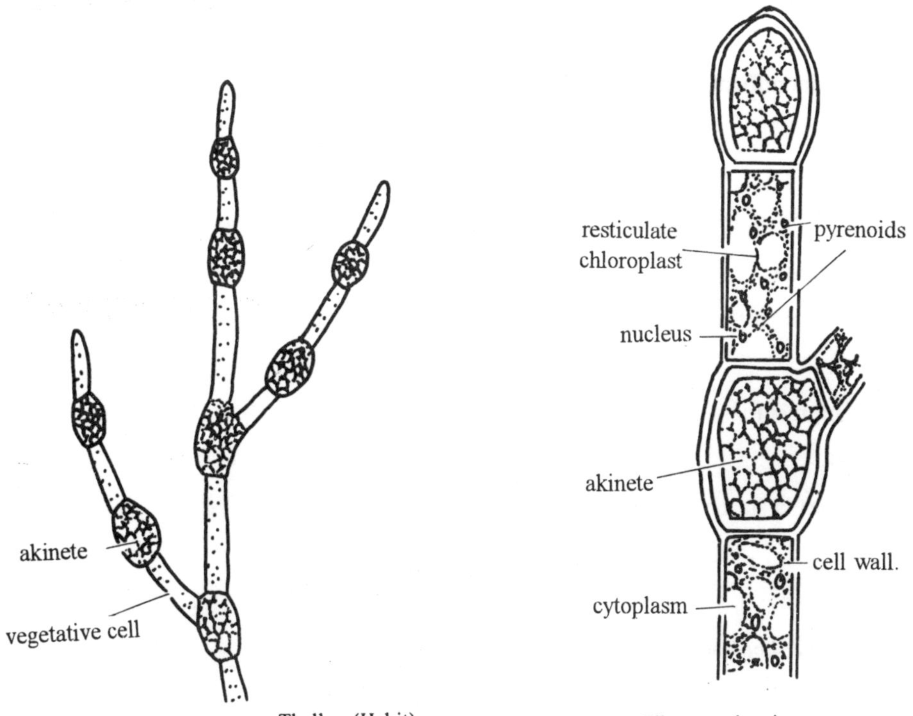

Thallus (Habit)

Filament showing akinetes

Fig.5.18 Pithophora

External features

1. The plant body is a branched filament.

2. The cells are joined end to end to form the filament.

3. The cells have a thin wall which is not lamellated.

4. The plant body is anchored to the substratum with the help of rhizoidal outgrowths.

5. Branching of the filament is same, as seen in *Cladophora*, i.e., pseudodichotomous

6. Cells are longer than broad, multinucleate with a reticulate parietal chloroplast. Embedded in the chloroplast are found a number of pyrenoids.

Reproduction

1. Fragmenation of the filament results in bits, which can develop into new individuals.
2. Akinetes are formed either in intercalary or terminal cells.
3. Akinetes are thick walled and dark brown in color and are rich in reserve food materials.
4. On germination akinetes develop into new individuals.

Diagrams to be drawn

1. Habit sketch
2. A part of the filament enlarged to show branching.
3. A cell enlarged

Systematic position and Identification

A

1. Cholorophyll present

2. Plant body a thallus

3. Cell wall cellulosic...*Algae*

B

1. Plastids present

2. Chlorophyll is chief pigment

3. True nucleus present

4. Fhotosynthetic product is starch...*Chlorophyceae*

C

1. Cells multinucleate

2. Filaments branched...*Cladophorales*

D

1. Cells many times longer than broad

2.Branching pseudodichotomous...*Cladophoraceae*

E

1. Branching a little below the cross wall
2. Cell wall not stratified.
3. Akinetes present...*Pithophora*

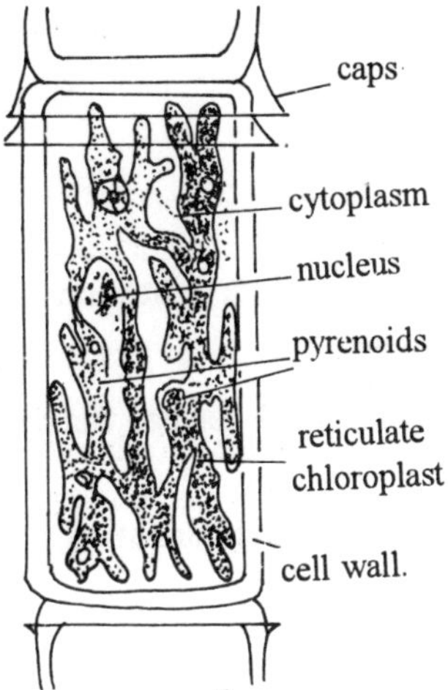

A young cell

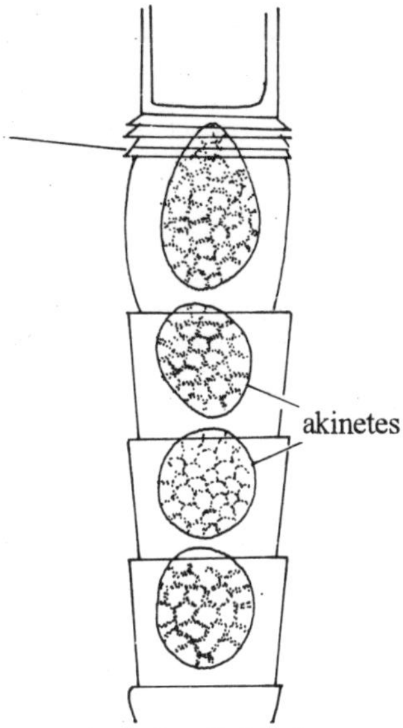

A chain of akinetes

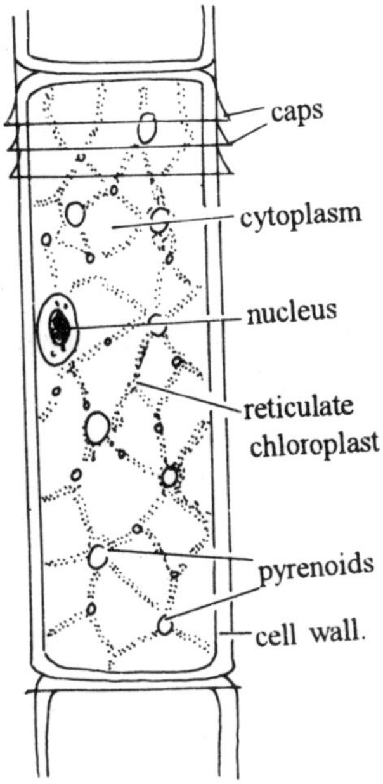

An old cell

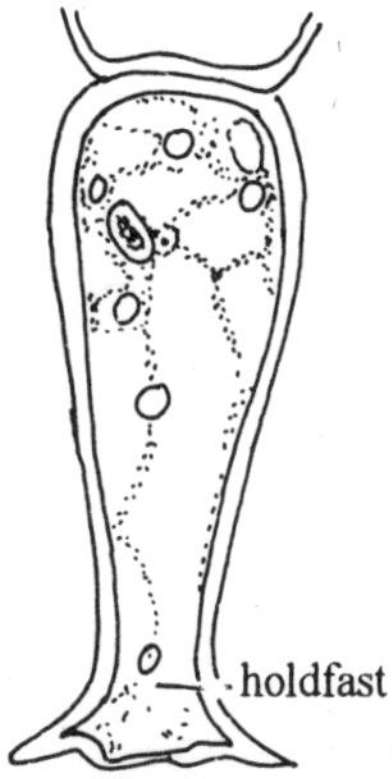

Holdfast

Fig.5.19 Oedogonium

OEDOGONIUM
(Link 1820)

Common species: *O.reliana, O.americanum, O.intermedium* etc.

Distribution: The plant is found in ponds, lakes and pools. It is usually submerged underwater attached to twigs or weeds.

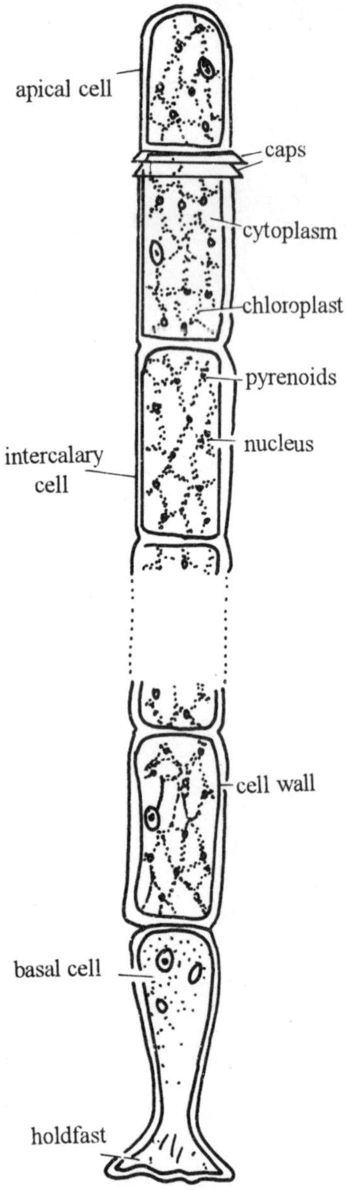

A Vegetative filament

Fig. 5.20 Oedogonium

Study

1. External morphology
2. Cell division and growth
3. Reproduction
4. Systematic position and Identification

External morphology

1. The plant body is an unbranched filament, it is deep green in colour.

2. The filament consits of cylindrical cells joined end to end.

3. The basal cell is modified into a hold fast to hlep in attachment. The hold fast consists of a poorly developed chloroplast and has its free end modified into a disc.

4. The terminal cell usually has a rounded tip.

5. The cell has a thick stratified cell wall divided into three layers, outer chitinous, middle pectic and inner cellulosic.

There is no mucilage sheath.

Cells are uninucleate with a reticulate chloroplast having a number of pyrenoids.

Cells division and growth

1. Growth is intercalary and brought about by the addition of cell in between.

2. During cell division transverse striations called 'caps' are formed as new cell are added in between. These apical caps are characteristic of *Oedogonium*.

Reproduction-Asexual

1. Some of the free floating species undergo fragmentation.

2. Zoospores are formed in all cells except the terminal cell and the hold fast cell. Each cell forms a single zoospore.

3. Zoospores are pyriform, multiceliate, the cilia attached at the anterior end.

4. The zoospore is usually released into a vesicle and then set free.

5. After a short period of free swimming, the zoospore germinates to develop into a new individual.

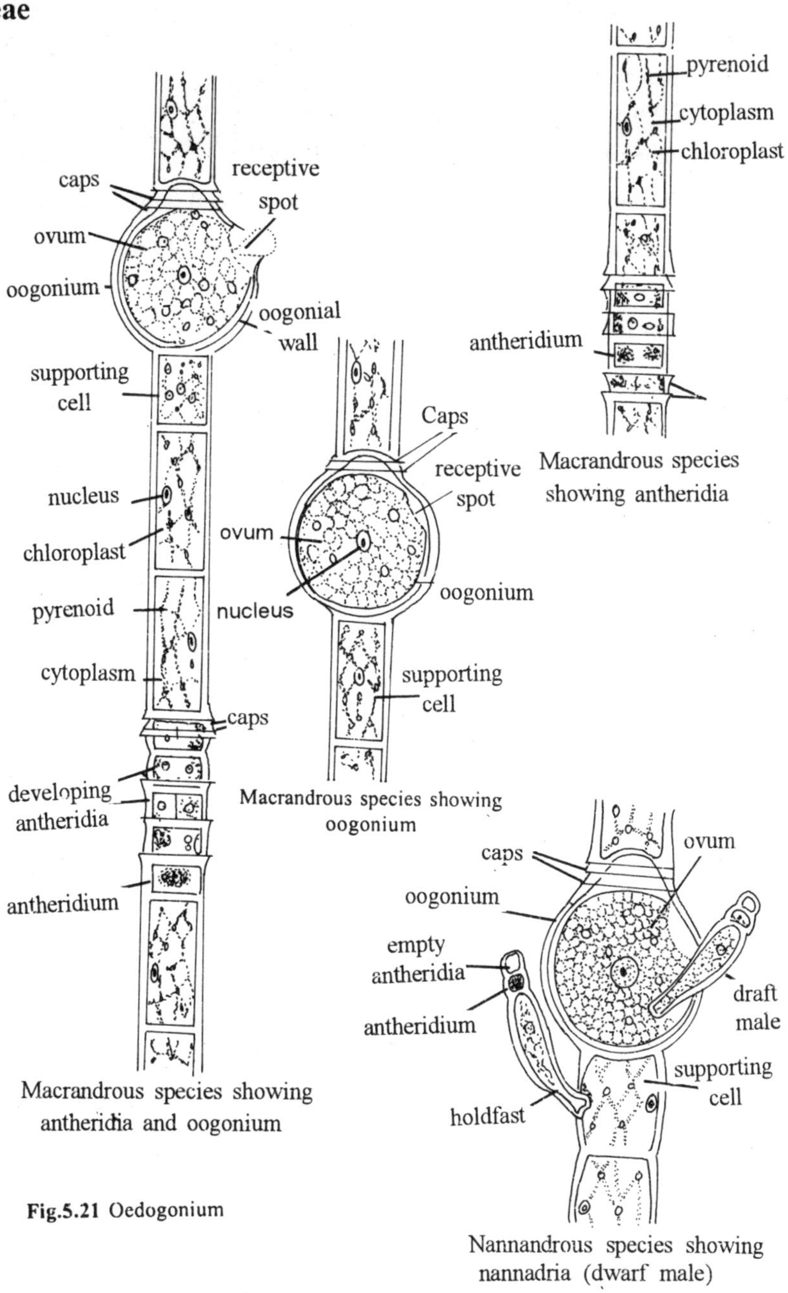

Fig.5.21 Oedogonium

Reproduction-Sexual

1. Sexual reproduction is oogamous. Sex organs are antheridia and oogonia.

2. Two categories can be identified in *Oedogonium* with reference to the development of antheridia. In macrandrous forms antheridia are formed on normal filaments while in nannandrous forms, antheridia are formed on short or dwarf male filament called Nannandria.

3. Oogonia are formed on normal filaments in both macrandrous and mannandrous forms.

4. Any intercallary cell can develop into an oogonium. The oogonium is globose, rich in cytoplasm and has a hyaline receptive sopt at one end. The oogonium is usually supported by a Suffultory cell.

5. In macrandrous forms antheridia are formed in an intercallary cell.

6. A series of transverse divisions take place in an antheridial mother cell resulting in a row of antheridia.

7. Each antheridium forms two sperms or male gametes which are exactly like the zoospore but much smaller in size.

8. In Nannandrous forms an Intercalary cell in normal filaments forms a number of androsporangia by undergoing repeated transverse divisions.

9. Antherozooids (sperms) are released into water.

10. Each androsporangium produces a single androspore. An androspore is exactly like a zoospore but intermediate in size between zoospore and antherozooid.

11. On release from the androsporangium, the androspore attaches itself to the oogonium or the suffultory cell. It germinates there and develops a short 3-4 celled filament called Nannandrium.

12. The terminal cell of the Nannandriumm forms the antheridia which produce antherozoids.

13. Antherozoids are similar in structure irrespective of the fact whether they are formed on normal filaments or nannandria.

14. On fertilisation, the egg develops into a zygote. Zygote is thick walled and somewhat reddish in colour.

15. Zygote undergoes a period of rest and divides to form four zoospores.

16. These zoospore develop into new individuals.

Diagrams to be drawn

1. Habit sketch

2. A few filaments enlarged

3. A cell enlarged to show structure and apical caps.

4. A cell with a zoospore.

5. A filament with oogonium and Nannandrium.

6. A filament with antheridia

7. A filament with androsporangia.

Systematic position and Identification

A

1. Chlorophyll present

2. Plant body is thallus

3. Cell wall cellulosic...*Algae*

B

1. Plastids present

2. Cholorophyll is the chief pigment

3. True nucleus is present

4. Photosynthetic product is starch...*Chlorophyceae*

C

1. Zoospores, antherozoids multiceliate

2. Sexual reproduction by oogamy

3. Filaments unbranched...*Oedogoniales* and *Oedoganiaceae*

D

1. Growth intercalary

2. Cap cells present

3. Nannandrium present...*Oedogonium*

BULBOCHAETAE

Common species: *B.setigera, B. intermedia* etc.

Distribution Fresh water ponds, pools etc.

Study

1. Habit and structure

2. Reproduction

3. Systematic position and Identification

Structure of the plant body

1. The plant body is a filament found usually attached with the help of the basal hold fast cell.
2. The filament is made up of cylindrical cells and is branched unlike in *Oedogonuim*.
3. The branches are profuse and usually successive branches have bulbous hairs (hence the name *Bulbochaetae*)
4. The cells are uninucleate and have an elaborate reticulate chloroplast.
5. The chloroplast has a number of pyrenoids embedded in the meshes of the reticulum.
6. The cell wall has a thin mucilage envelope.
7. Growth is intercalary and is brought about by the cell divisions of intercalary cells.
8. The details of the cell divisions and cap formation are similar to what is seen in *Oedogonium*.
9. In most species of *Bulbochaetae* however, growth is mostly basal and not more than one cap is formed per cell.
10. Each cell after its formation grows out slightly at its apex towards one side and this protuberance is cut off as a colourless cell.
11. The cell lengthens into a hair, with the ruptured wall forming an irregular sheath around the base.

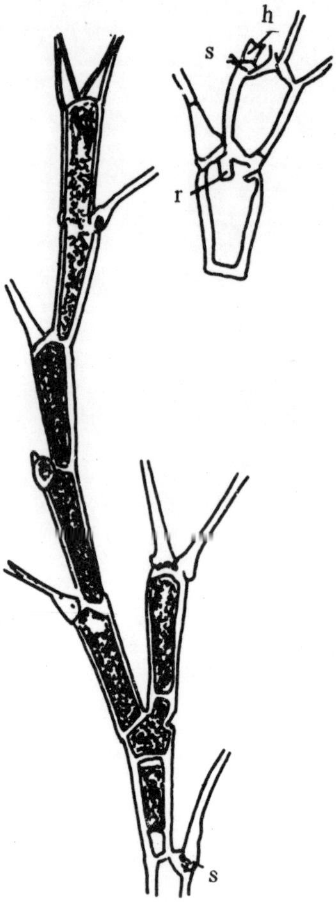

Fig.5.22 Bulbochaetae

Reproduction-Asexual

1. Vegetative propogation is brought about by mechanical fragmentation.

2. Asexual reproduction takes place by means of zoospores. Zoospores are multi-flagellate and pyriform in shape. A circlet of flagella is found at the aterior end.

3. Each cell forms a single zoospore. Rest of the details are same as in *Oedogonium*.

Reproduction-Sexual

1. Sexual reproduction does not occur quite frequently. It is of the oogamous type.

Oogonia and antheridia are found on the same filament in macrandrous species, while in nannandrous species, antheridia are formed on dwarf male filaments.

3. The nanandrous filaments arise from special types of spores called androspores produced in androsporangia.

4. Rest of the details of the formation antheridium, androsporangium and oogonium are same as in *Oedogonium*.

5. Nannandrous filaments attach themselves either to an oogonium or to a cell supporting the oogonuim.

6. The dwarf males are 3-4 celled in length and produce antheridia at their tip.

7. Fertilization is similar to what is seen in *Oedogonium*.

8. Oospore (zygote) develops a thick three layered wall and the contents assume a red colour.

9. The oospore after undergoing a period of rest develops into a new individual by undergoing reduction division and forming four zoospores.

Diagrams to be drawn

1.Habit sketch

2. A filament enlarged

3. A single cell with the hair

4. Filament with oogonia and dwarf male

5. Filament with antheridia

6. Filament with androsporangia

Systematic position and Identification

A

1. Chlorophyll present

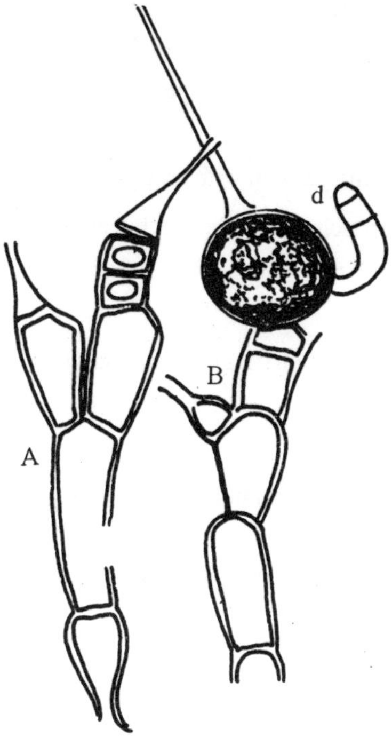

Fig.5.23 Bulbochaetae

2. Plant body a thallus

3. Cell wall cellulosic...*Algae*

B

1. Plastids present

2. Chlorophyll is the chief pigment

3. True nucleus present

4. Photosynthetic product is starch...*Chlorophyceae*

C

1. Zoospores, Antherozoids multiciliate

2. Sexual reproduction oogamous

3. Filaments branched or unbranched...*Oedogoniales* and *Oedogoniaceae*

D

1. Growth is basal or intercalary

2. Filaments branched

3. Bulbous hairs present at the tip of each terminal cell...*Bulbochaetae.*

DRAPARNALDIA

Common species : *D.glomerata, D.lillosa* etc

Distribution: Exclusively fresh-water. Found growing attahced to sand, twigs or as epiphytes.

Study:

1. Structure of the plant body

2. Reproduction

3. Systematic position and Identification

Structure of the plant body

1. The plant body is a filament. It is anchored to the substratum with the help of a hold fast. The hold fast has rhizoidal outgrowths.

2. The plant body is heterotrichous, but the prostate system is mostly absent.

3. Thin mucilage envelope surrounds the filament.

4. The plant body has a central axis made up of a uniseriate row of cells.

5. At regular intervals, the axis produces laterals or branches; these branches are in tufts.

6. The profuse development of the laterals, sometimes masks the main axis.

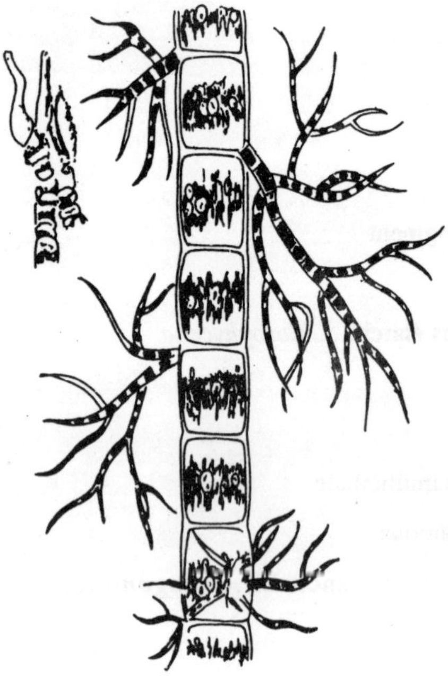

Fig.5.24 Drapanaldia

7. The apices of the branches usually terminate in hair like processes called setae.

8. The cells of the main axis are large, barrel shaped and uninucleate.

9. The chloroplast is plate like and parietal with a single pyrenoid.

10. The cells of the laterals are much smaller in size, but are similar in structure to those of the main axis.

Reproduction

1. Asexual reproduction is by quadriflagellate zoospres and hypnospores.

2. Sexual reproduction is isogamous.

3. Gametes are quadri flagellate but are much smaller in size than the zoospores.

4. The zygoe, formed by the fusion of the gametes develops directly into a new individual.

Diagrams to be drazwn

1. Habit sketch

2. A portion of the plant enlarged to show the branching .

Systematic position and Identification

A.

1. Chlorophyll present

2. Plant body a thallus

3. Cell wall cellulosic...*Algae*

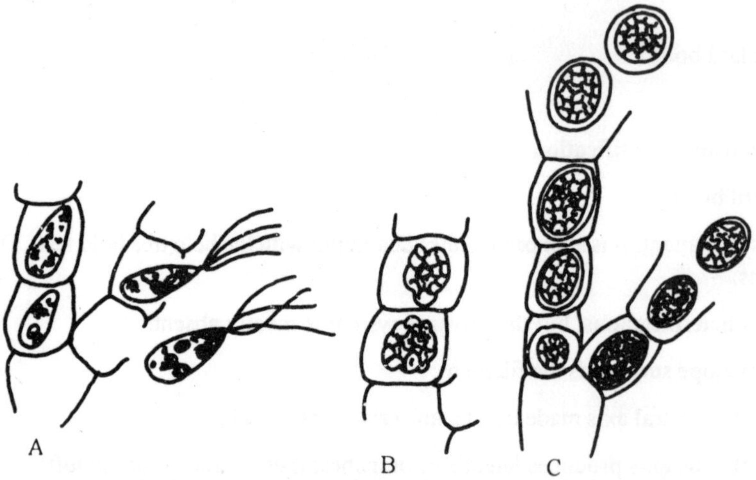

Fig. 5.25 Draparnaldia - stages in reproduction

B

1. Plastids present
2. Chlorophyll is the chief pigment
3. True nucleus present
4. Photosynthetic product is starch...*Chlorophycae*

C

1. Plant body exhibits heterotrichous habit
2. Branches end in setae...*Chaetophorales*

D

1. Cells uninucleate with the parietal chloroplast
2. Terminal cells of laterals end in setae...*Chaetophoraceae*

E

1. Prostrate system almost absent
2. Cells are barrel shaped...*Draparnoldia*

DRAPARNALDIOPSIS
(Smith)

Common species: *D.indicia, D.alpine* etc.

Distribution: Fresh water ponds, stagnant ditches. Epiphytes or free floating.

Study: 1. Habit and structure

2. Reproduction

3. Systematic position and Identification

Structure of the plant body

1. The filmaentous, multicellular plant body is anchored to the soil with the hlep of rhizoidal outgrowths.
2. The plant body is heterotrichous the prostrate system is also well developed.
3. The rhizoids are profusely branched forming a cortical covering at the basal region.
4. The erect system or the main axis consists of a uniseriate row of cells.
5. The cells in the main axis are of two types long, barrel shaped inter nodal cells and short discoid nodal cells. The two types of cells alternate regularly.
6. The nodal cells bear a system of branched laterals. The branches are of two types, I.e., the branches of unlimited growth and branches of limited growth.
7. The long laterals or branches of unlimited growth arising at nodal cells (2-4 in number) have the same structure as the main axis.

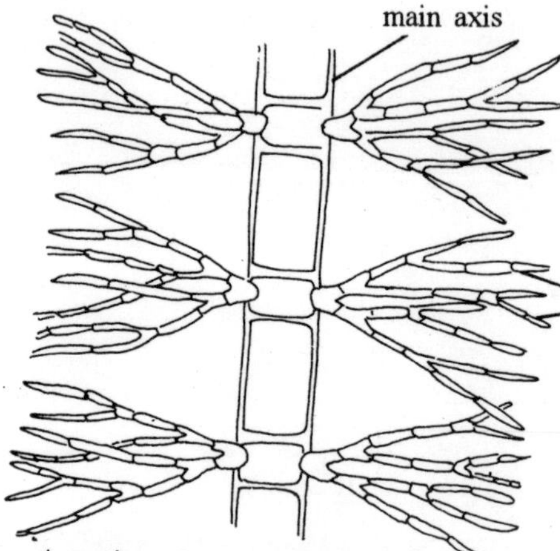

A portion of main with nodes, internodes and branches

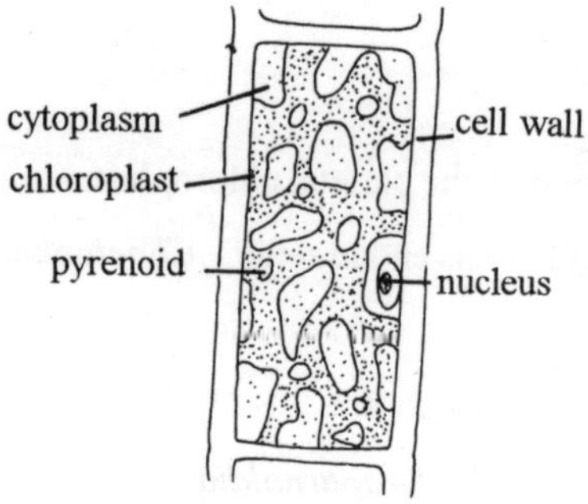

A single cell enlarged

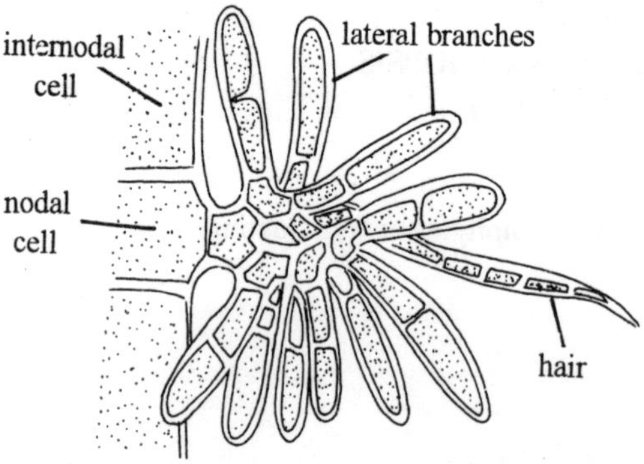

Lateral shot branches from the nodal cell

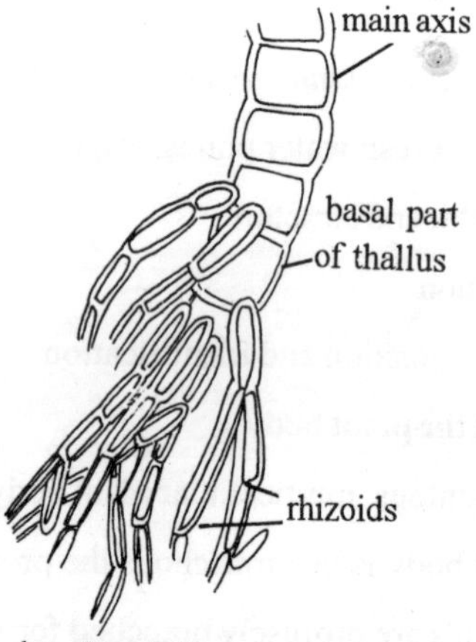

basal part of with rhizoids

Fig.5.26 Draparnaldiopsis

8. The branches of limited growth or short laterals arise either on the nodal cell of the main axis or on long laterals. They are smaller in size, profusely branched and the cells terminate in hair like processes.

9. The cells exhibit (nodal, internodal etc) the same structure in both the branches and the main axis.

10. The cells are cylindrical or barrel shaped and have a thin cell wall.

11. The chloroplast is reticulate with a number of pyrenoids.

12. The cells are uninucleate. The nucleus is located laterally in the cells of the mian axis while it is centrally placed in the cells of the branches.

Reproduction:

1. Asexual reproduction takes place by the formation of zoospores which are of three types, viz., quadriflagellate macrozoospores, quadri flagellate microspores and bi flgellate microzoospores.

2. Zoospores are formed only in the cells of the laterals.

3. Sexual reproduction is isogamous.

4. Gametes are bi flagellate and are produced in the cells of the laterals.

Diagrams to be drawn

1. Habit

2. A part enalrged to show branching

3. A cell enlarged

Systematic position and Identification

A

1. Chlorophyll present

2. Plabt body a thallus

3. Cell wall cellulosic...*Algae*

B.

1. Plastids present
2. Chlorophyll is the chief pigment
3. True nuclues present
4. Photosynthetic product is starch...*Chlorophycae*

C.

1. Plant body exhibits heterotichous habit
2. Branches end with setae...*Chaetophorales*

D.

1. Cells uninucleate with parietal chloroplast
2. Terminal cells of laterals end in setae...*Chaetophoraceae*

E

1. Prostrate system also well developed
2. Main axis consists of Nodal and Internodal cells.
3. Laterals are also of two types viz., branches of unlimited growth and branches of limited growth
4 Cells uninucleate with a parietal chloroplast.
5. Rhizoids branch profusely forming a cortical envelope...*Draparnoldiopsis*

FRITCHIELLA
(Iyengar 1932)

Common species: *F. tuberosa* is the only species recorded for the genus.

Distribution: Terrestrial alga found on damp soil, pond beds and soil rich in organic matter.

Study: 1. External structure

2. Systematic position and Identification

External structure

1. The microscopic plant body exhibits a clear heterotrichous habit.

2. Four different systems of thallus can be identified on microscopic examination. These are rhizoidal system, prostrate system, primary vertical system and secondary vertical system.

3. The rhizoidal system is sub-terranean, colourless, branched and septate. It helps in the anchorage.

4. The prostrate system consists of an accumulation of cells. The chlorophyllous cells form branched filaments. The cells of the filament occasionally may be distinguished into nodal and internodal cells. The nodal cells give rise to rhizoids from their lower face and primary vertical system from their upper face.

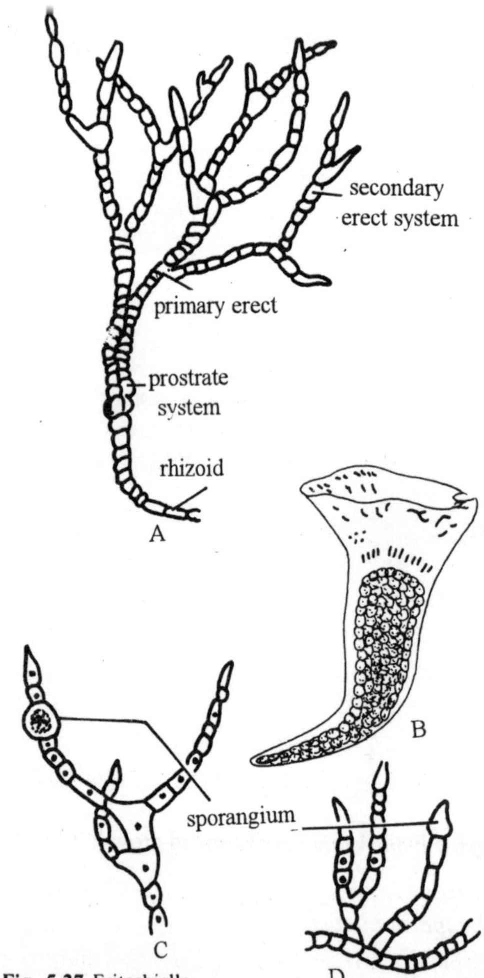

Fig. 5.27 Fritschiella

5. The prostrate system gives rise to the primary vertical system consisting of photosynthetic branched filaments. The primary vertical system is sub aerial

6. Many delicate, chlorophyllous branches arise from the primary vertical system constituting the secondary vertical system.

7. All the cells, except those in the rhizoidal system have a pareital chloroplast. The cells are uninucleate.

Diagrams to be drawn

1. Habit sketch

2. The systems of filaments (enlarged)

Systematic position and Identification

A. 1. Chlorophyll present

2. Plant body a thallus

3. Cell wall cellulosic ....*Algae*

B. 1. Pastids present

2. Chlorophyll is the cheif pigment

3. True nucleus present

4. Photosynthetic product is starch ....*Chlorophyceae*

C. 1. Plant body exhibits heterotrichous habit

2. Vertical system ends in fine filament ....*Chaetophorales*

D. 1. Cells uninucleate with parietal chloroplast

2. Terminal cells of the vertical system end in fine filaments

......*Chaetophoraceae*

E 1. Plant is terrestrial in habitat

2. Plant body having four systems of filaments ......*Fristschiella*

COLEOCHAETE

Common species : *C.scutata, C.pulvinata, C.nitellarum, C.orbricularis* etc.

Distribution : Fresh water epiphytic alga, found on seeds, ridges etc.

Study 1. External structure

2. Reproduction

3. Systematic position and identification

External structure

1. The plant exhibits a heterotrichous system but in some members, the heterotrichous system is obscure.

2. Two types of forms may be distinguished in the genus. These are discoid forms and cushioned forms.

3. In discoid forms heterotrichous systems is not evident. The branches of the thallus are laterally opposed and form a compact pseudoparenchymatous structure. Here the erect system is represented by a few setae.

4. In cushioned forms, heterotrichous system is very evident. The filaments of vertical system form a cushion which will be enveloped by mucilage.

5. Shape of the cell varies in erect and prostrate branches. In prostrate branches, cells are polygonal. In erect branches, the cells are cylindrical.

6. Cells are thin walled, uninucleate and have a parietal chloroplast with one or two pyrenoids.

7. The vegetative cells have setae or bristles. These are found only in younger cells.

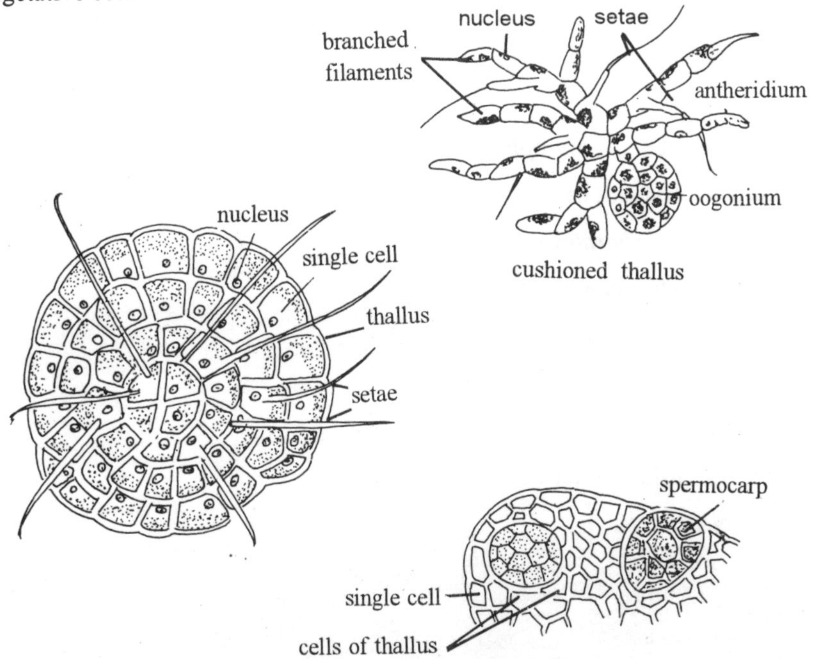

Fig.5.28 Coleochaete; Discoid thallus (habit) Discoid thallus with spermocarp

Reproduction

1. Asexual reproduction by large zoospores.

2. Sexual reproduction is oogamous.

3. Antheridia are borne on erect branches. Each antheridium produces a single biflagellate antherozoid. Antheridia are usually found in clusters.

4. Oogonia are flask shaped and are formed terminally on the lateral branches of the erect system. The oogonium is also called carpogonium because of its flask shaped structure.

5. Fertilization results in the formation of zygote which remains within the oogonium and becomes enveloped by the surrounding cells. This becomes reddish brown in colour and is called spermocarp.

6. Spermocarp has a long period of dormancy and germinates later giving rise to zoospores.

Diagrams to be drawn

1. Habit sketch

   a. Cushioned thallus

   b. Discoid thallus

2. Antheridia

3. Oogonia and spermocarp

Systematic position and identification

A. 1. Chlorophyll present

2. Plant body a thallus

3. Cell wall cellulosic ... *Algae*

B. 1. Plastids present

2. Chlorophyll is the chief pigment

3. True nucleus present

4. Photosynthetic product is starch ... *Chlorophyceae*

C. 1. Plant body exhibits heterotrichous habit

2. Branches end in setae ... *Chaetophorales*

D. 1. Cells uninucleate with parietal chloroplast

2. Terminal cells have setae

3. Filamentous branches ... *Coleochaetaceae*

E. 1. Thallus parenchymatous

2. Plant body has discoid and cushioned forms

3. Spermocarp present ...*Coleochaete*.

ZYGNEMA
(Wittr)

Common species : *Z.himalayensis, Z.tetrrestris, Z.globosum* etc.

Distribution : Widely distributed in fresh water, forming yellowish green scums.

Study : 1. External features

2. Asexual reproduction

3. Scalariform conjugation

4. Lateral conjugation

5. Systematic position and identification

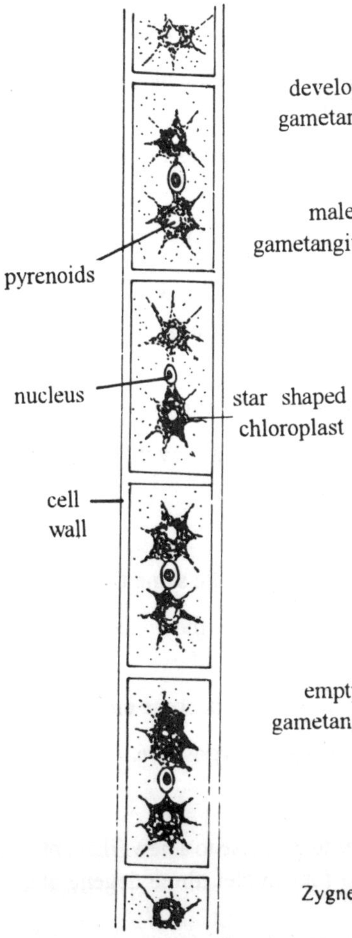

Fig.5.29 Zygnema. filament

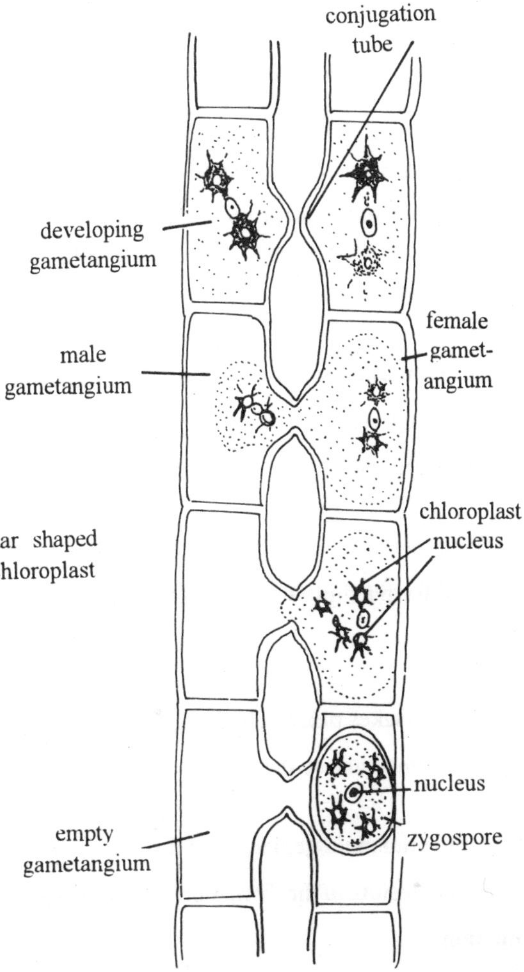

Zygnema. filaments showing scalariform conjugation

External features

1. The plant body is an unbranched filament consisting of cylindrical cells joined end to end.

2. Rarely certains cells of the filament may develop haptera to help in anchorage.

3. Cell has a two layered wall - outer pectic and inner cellulosic. There is a mucilage covering on the cell wall.

4. Cells are uninucleate and have two stellate chloroplasts having a pyrenoid each.

Asexual reproduction

1. Fragmentation of the filament results in the breaking up of the filament into bits, each bit developing into a new individual.

2. Akinetes and Aplanospores are also formed to help in asexual reproduction.

Reproduction : sexual

1. This takes place by means of conjugation

2. Conjugation is of two types - scalarifomr and lateral.

Scaliform conjugation

1. This type of conjugation takes place between the cells of two different filaments.

2. Two filaments lie parellely and small protruberances from the two filaments grow in opposite directions and meet in the centre forming a passage called the conjugation tube.

3. Conjugation tube formed in successive cells of the two filaments gives the appearance of a ladder, hence the name Scalariform.

4. The contents of the cell round off and form amoeloidal gametes. The gametes are isogamous.

5. Gamete from one cell migrates to another of the opposite filament through the conjugation tube.

6. Even though gametes are morphologically isogamous the one that migrates can be called male and the other female.

7. All the cells of a filament become empty after the migration of the gametes.

8. The gametes fuse and form a diploid zygote. The zygote develops a very thick wall and is now called the zygospore.

Lateral conjugation

1. This type of conjugation takes place between the adjacent cells of the same filament.

2. Conjugation tube is formed above the cross wall or in some cases, the cross wall between the two cells dissolves.

3. Cell contents turn into amoeboid gametes and fuse. Alternate cells become empty.

4. Fusion of gametophyte results in the formation of a thick walled zygospore.

Zygospore germination

After undergoing a period of rest, the zygospore germinates to give rise to a new filament. Before germination the diploid nucleus undergoes reduction division. Of the four nuclei, three degenerate, the surviving one forms the new filament.

Chlorophyceae

Diagrams to be drawn

1. Habit sketch
2. A filament enlarged
3. A cell enlarged
4. Scalariform conjugation
5. Lateral conjugation

Systematic position and identification

A. 1. Chlorophyll present

2. Plant body a thallus

3. Cell wall cellulosic ... *Algae*

B. 1. Plastids present

2. Chlorophyll is the chief pigment

3. True nucleus present

4. Photosynthetic product is starch ... *Chlorophyceae*

C. 1. Sexual reproduction by conjugation

2. No zoospore formation ... *Conjugales*

D. 1. Plant body is an unbranched filament

2. Gametes amoeboid ... *Zygnemataceae*

E. 1. Filamentous plant body

2. Cells have two stellate chloroplasts ... *Zygnema*

SPIROGYRA

Common species : *S.condensata, S.jogensis, S.singularis* etc.

Distribution : Found floating in fresh water ponds and ditches, commonly called pond silk or pond scum, because of its slippery nature.

Study : 1.Study of the structure of the plant body

2. Reproduction

3. Systematic position and identification

Structure of the plant body

1. The plant body is an unbranched filament. There is no differentiation into base and apex.
2. The filament has a thick mucilagenous envelope.
3. The cells are long cylindrical and joined end to end to form the filament.
4. The cell wall is three layered - outer chitinous, middle pectic and inner cellulosic. The cross wall between the cells is replicate (pectose middle lamella having cellulose on either side).

5. The cytoplasm is peripheral with a large central vacuole. Thin cytoplasnic strands traverse across the vacuole.

6. There is a single nucleus suspended in the cytoplasm.

7. The chloroplast is ribbon shaped and is a spiral with wavy edges. Embedded in the chloroplast are found a number of pyrenoids. Usually there will be only one spiral helix in the chloroplast. In some species two helices are found spirally wound in opposite direction.

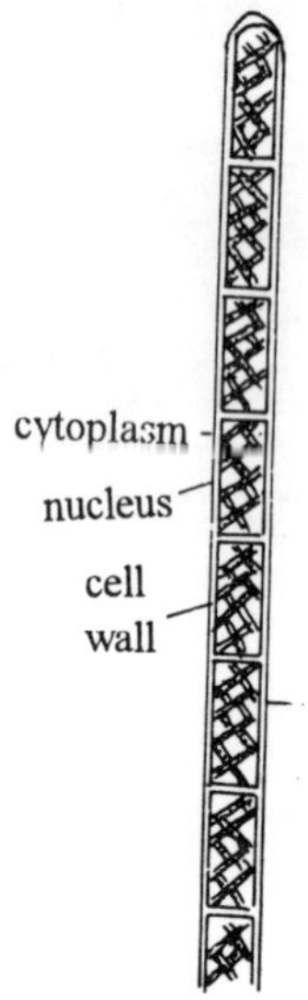

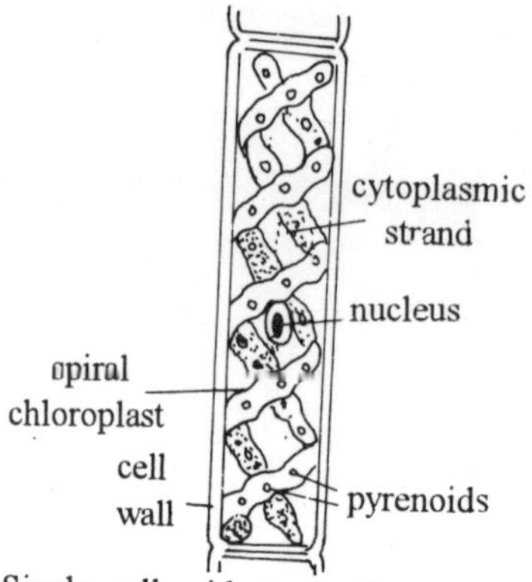

Single cell with two chloroplast

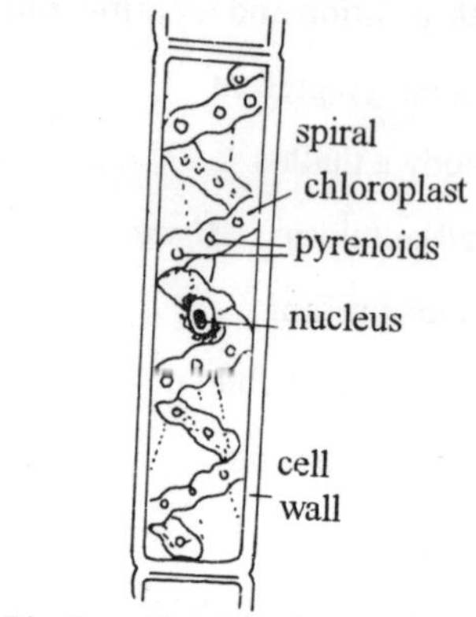

Single cell with single chloroplast

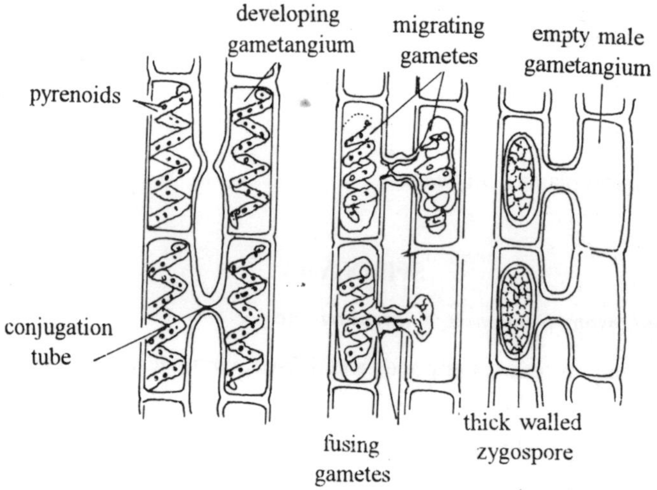

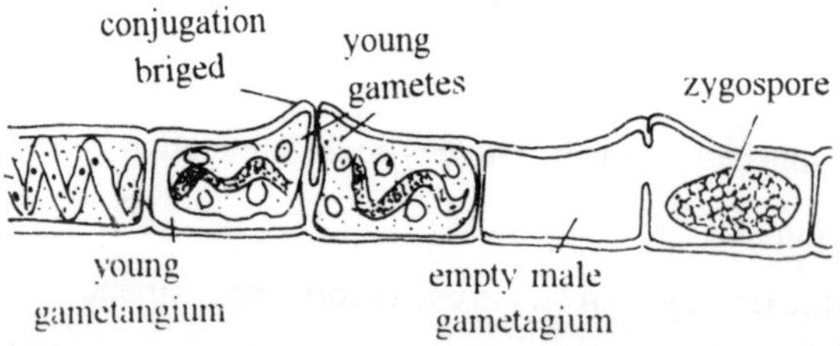

Fig.5.30 Spirogyra filament   Spirogyra filament showing lateral conjugation

Reproduction

1. Asexual reproduction by zoospores is not seen.

2. Vegetative propagation is brought about by means of fragmentation.

3. Sexual reproduction is called conjugation. The details are same as seen in *Zygnema*.

4. In *Spirogyra* also, two types of conjugation viz., scalariform and lateral are seen (for details, see *Zygnema*).

5. Structure and germination of zygospore is similar to what is seen in *Spirogyra*.

6. Zygospores are dark blue in colour and oval in shape. They have a thick sculputred wall.

Diagrams to be drawn

1. Habit sketch

2. A few filamentous enlarged.

3. A cell enlarged

4. Scalariform conjugation

5. Lateral conjugation

6. Zygospore germination

Systematic position and identification

A. 1. Chlorophyll present

2. Plant body a thallus

3. Cell wall collulose ... *Algae*

B. 1. Plastids present

2. Chlorophyll is the chief pigment

3. True nucleus is present

4. Photosynthetic product is starch ... *Chlorophyceae*

C. 1. Sexual reproduction by conjugation

2. There is no zoospore formation ... *Conjugales*

D. 1. Plant body is an unbranched filament.

2. Gametes amoeboid ... *Zygnemataceae*

E. 1. Cell wall has thick mucilage envelope.

2. Cross walls are replicate

3. Chloroplast spiral ... *Spirogyra*.

MOUGEOTIA
(Agardh 1824)

Common species : *M.capucina, M.boodlai* etc.

Distribution : Found in ponds, semipermanent pools, lakes, slow flowing streams as bright green floating masses.

Study: 1. Structure of the plant body

2. Reproduction

3. Systematic position and identification

Structure of the plant body

1. The plant body is an unbranched filament consisting of cylindrical cells joined end to end.
2. Very rarely the basal cell may develop rhizoidal outgrowths to help in anchorage.
3. Slime layer is found deposited on the cell wall. The cell wall is two layered. It has an outer pectic and inner cellulosic layer.
4. The cross wall between the cell splits into two circular discs at a very early stage.
5. The chloroplast is axile and plate like, almost filling the cell cavity. It may have 3-4 pyrenoids.
6. The cytoplasm is peripheral with a central vacuole.
7. The single nucleus is axile and plate like, almost filing the cell cavity. It may have 3-4 pyrenoids.

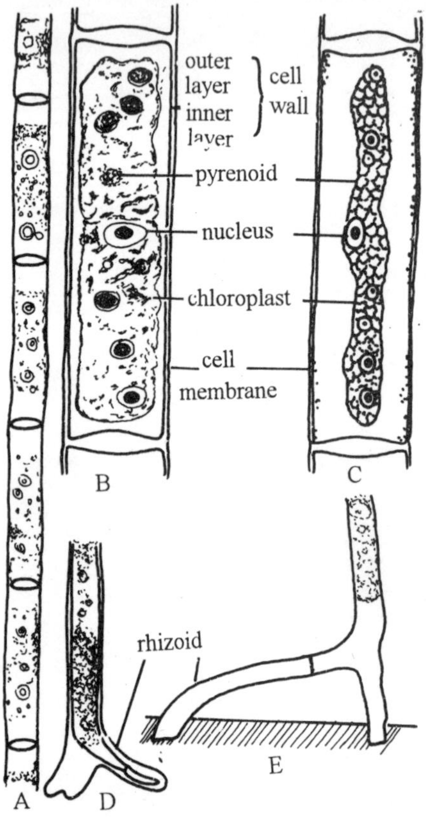

Fig.5.31 Mougeotia A, portion of a filament; B, face view of the cell showing structure, & the chloroplast in surface view; C, chloroplast in profile position; D-E, basal cell showing rhizoid formation (D-E, after Salisbury, B-C, based on Morris)

Reproduction

1. Vegetative propagation takes place by fragmentation. The filament breaks up into few celled segments. Each segment develops into a new individual.
2. Akinetes are found in some species to help in asexual reproduction. The akenites have a thick wall and rich in nutrients.
3. Occasionally aplanospores may also be fomred as in some species.
4. Sexual reproduction is by means of conjugation. It is of two types viz., scalariform and lateral.

Scalariform conjugation

1. This type of conjugation takes place between the opposite cells of two different filaments.
2. Two papillae arise from the opposite cells at corresponding points of conjugating filaments.
3. The two papillae growing in opposite directions meet in the centre establishing a conjugation tube or canal.
4. The contents of two cells form amoeboid gametes. The two gametes migrate into the conjugation tube which enlarges in diameter.
5. Gametes are isogamous. They fuse to form a zygote in the conjugation tube. The zygote secretes a thick wall around itself and will now be called zygospore.

Lateral conjugation

1. In lateral conjugation protoplasts of adjacent cells of the same filament fuse.
2. Conjugation tube is formed above the cross wall between the conjugating cells.
3. The protoplasts of the two cells migrate to the conjugation tube where they fuse to form the zygote. The zygote develops a thick wall and become the zygospore.

4. The zygospore undergoes a period of rest and then divides meiotically to form four nuclei. Out of these three degenerate. The surviving one forms the new filament.

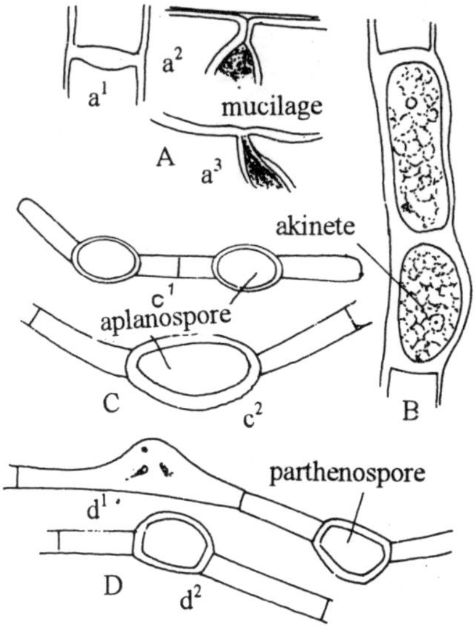

Fig.5.32 Mougeotia. Asexual reproduction, A.mechanism of cell disjunction; B.akinetes of M.capucina; C.aplanospore formation in M.boodlai; D.parthenospore formation in Gonatonema ventricosum (A.after Frisch; B.after Borge; C & D.)

Diagrams to be drawn

1. Habit sketch
2. A filament enlarged (showing Akinetes)
3. A cell enlarged
4. Stages in scalariform conjugation
5. Stages in lateral conjugation

Systematic position and identification

A. 1. Chlorophyll present

2. Plant body a thallus

3. Cell wall cellulosic ... *Algae*

B. 1. Plastids present

2. Chlorophyll is the chief pigment

3. True nucleus present ... *Chlorophyceae*

4. Photosynthetic product is starch.

C. 1. No motile reproductive cells

2. Sexual reproduction by conjugation... *Conjugales*

D. 1. Plant body an unbranched filament

2. Gametes amoeboid ... *Zygnemataceae*

E. 1. Gametes strictly isogamous

2. Fusion of gametes takes place in the conjugation tube ... *Mougeotia*

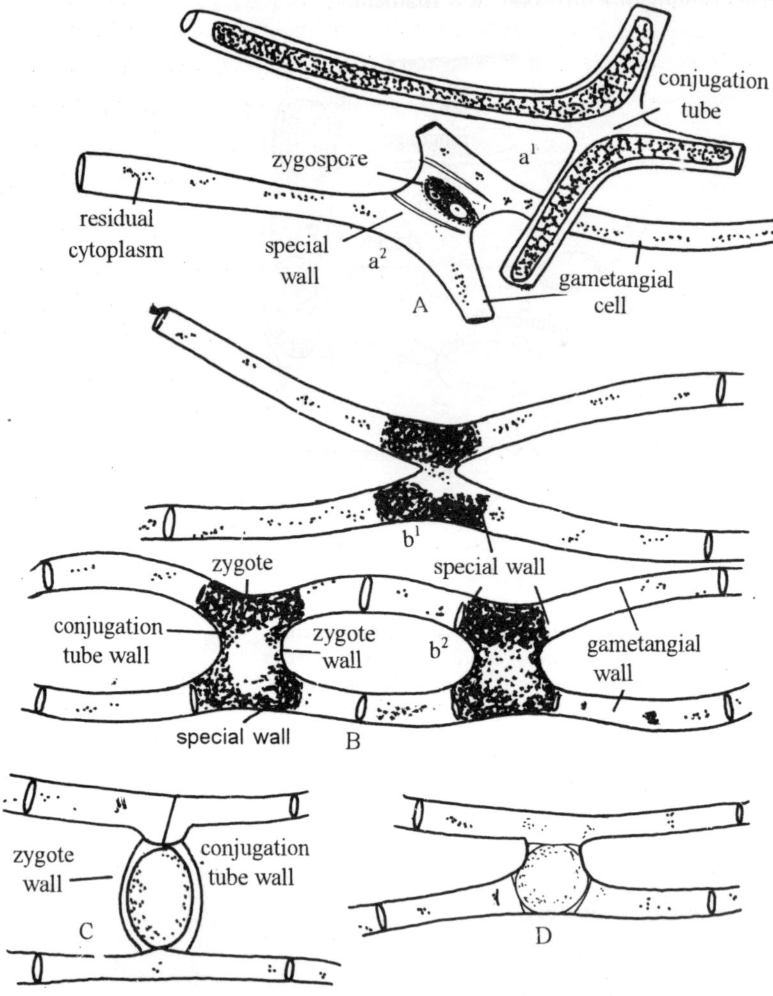

Fig.5.33 Mougeotia

COSMARIUM

Common species : *C.biretum, C.leotytis, C.turgidium* etc.

Distribution : Found in pools, ponds and stagnant waters rich in decaying organic matter.

Study : 1. Structure of the plant body

2. Asexual reproduction

3. Sexual reproduction

4. Systematic position and identification

Structure of the plant body

1. It is a fresh water free floating alga, commonly called Placoderm desmid or constricted desmid.

2. The plant body is unicellular, the cell one and half times longer than broad. Sometimes the cells join together to form chains or amorphous colonies.

3. Each cell consists of two symmetrical halves called semicells, separated by a constriction (sinus) and joined by a narrow region called Isthums.

4. Each semicell may be oval, circular or elliptical in outline.

5. The cell wall may be smooth without any architectural pattern, or the ornamental wall may have three concentric layers.

6. The outer layer is gelatinous, middle layer is cellulosic impregnated with pectic compounds and innermost is cellulosic.

7. The two inner layers of the wall are porous, through which the mucilage exudes out resulting in the jerking movement.

8. Each semicell has an axial chloroplast with radiating arms. In some each semicell may have four pariental chloroplasts.

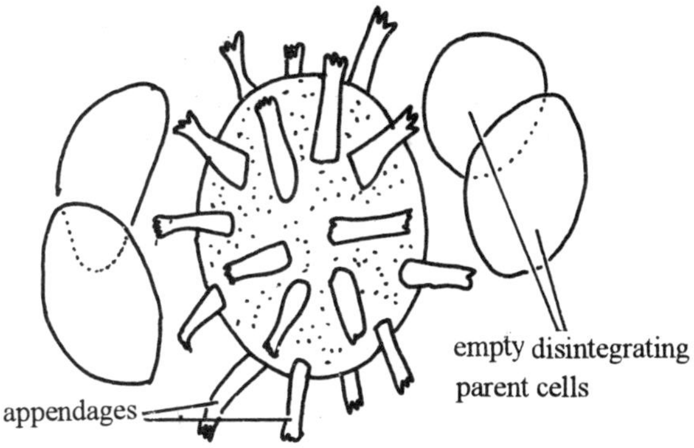

Fig. 5.34 A Zygospore of Mougeotia

9. Each chloroplast may have one or two pyrenoids.

10. The cells are uninucleate, the nucleus being situated at the isthums region.

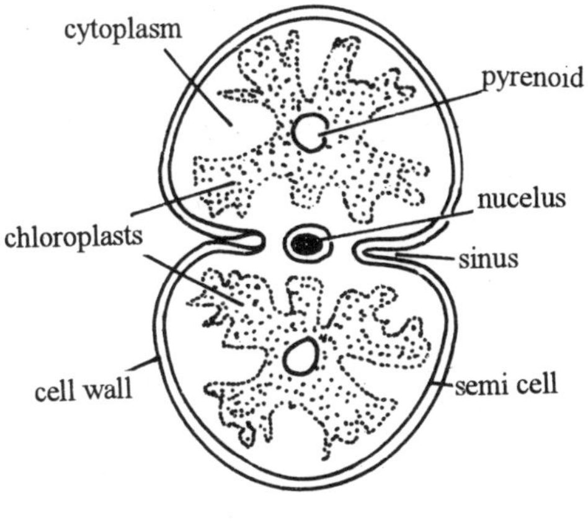

Fig.5.35 Cosmarium. A single cell

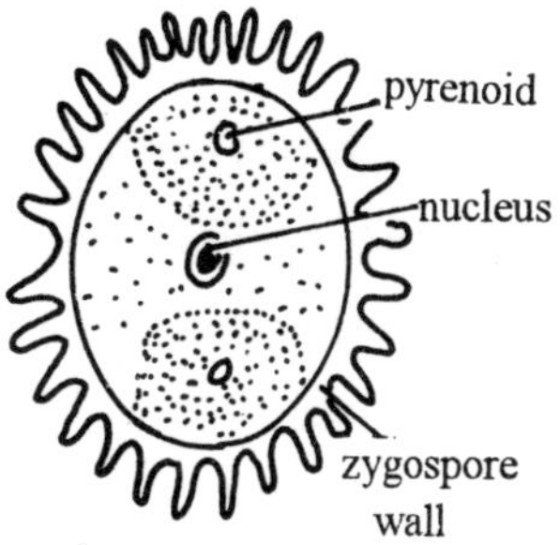

A zygospore

Asexual reproduction

1. Asexual reproduction is by cell division.

2. At the time of cell division, the nucleus mitotically divides into two. The isthums enlarges and divides into

two by means of a cross wall.

3. The two daughter cells get separted and each one develops a new semicell at the isthmus region.

4. The chloroplast also divides, but the pyrenoid may divide or a new pyrenoid may be synthesised.

5. Thus, each newly formed cell consists of one parental semicell and one new semicell.

Sexual reproduction

1. Sexual reproduction is very large and isogamous when it takes place. It is called conjugation.

2. At the time of conjugation two cells some nearer and get encased by a common mucilage envelope.

3. The cell contents of each cell get metamorphosed into an amoeboid gamete.

4. The cell wall splits at the isthmus. The gametes come out, meet at the centre freely or in the conjugation tube.

5. They fuse to form a diploid zygospore. The zygospore has a thick warty wall.

6. The remnants of the four semicells remain attached to the zygospore for some time.

7. The zygospore is brown in colour and germinates after a brief period of dormancy.

8. Nuclear fusion is effected before germination. The zygospore wall ruptures releasing two new individuals of *Cosmarium*.

Diagram to be drawn

1. Structure of the cell

2. Conjugation stages

Systematic position and identification

A. 1. Chlorophyll present

2. Plant body a thallus

3. Cell wall cellulosic ... *Algae*

B. 1. Plastids present

2. Chlorophyll is the chief pigment

3. True nucleus present

4. Photosynthetic produce is starch ... *Chlorophyceae*

C. 1. Motile reproductive cells absent

2. Sexual reproduction by conjugation ... *Conjugales*

D. 1. Plant body unicellular

2. The cell consists of two halves called semicells ... *Desmidiaceae*

2. Asexual reproduction present between the two semicells ... *Cosmarium.*

CLOSTERIUM

Common species : *C.parvulum, C.colosporum* etc.

Distribution : Found widely distributed in stagnant waters pools, ponds rich in decaying organic matter.

Study: 1. Cell structure

2. Reproduction

3. Systematic position and identification

Cell structure

1. It is commonly called Placoderm desmid or uniconstricted desmid.

2. The plant body is made up of a single cell. The cell is elongated and spindle like with tapering apices.

3. Majority of the species have a straight plant body, but some of them are crescent shaped.

4. The cell is circular in a cross section.

5. The plant body consists of two perfectly symmetrical halves called semicells.

6. A narrow protoplasmic bridge called isthmus connects the two semicells. The isthums is not constricted as in *Cosmarium*.

7. The cell is uninucleate, the nucleus being placed in the isthmus region. The nucleus has a conspicuous nucleolus.

8. At either end of the cell is a conspicuous vacuole containing calcium sulphate crystals showing brownian movement.

9. The crystals do not perform any physiological function and are probably products of excretion.

10. The cell wall is three layered (see *Cosmarium*).

11. Grooves are present on the cell wall having minute pores arranged in rows. Mucilage excretes through these pores causing movement of the plant body.

12. Each semicell has an axial curved cone like chloroplast with ribs. The chloroplast has a number of pyrenoids.

Reproduction

1. Asexual reproduction is brought about by means of cell division.

2. At the time of cell division a constriction appears in the centre.

3. The nucleus divides followed by the cell. The two semicells get separated.

4. These separate and each semicell forms a new half at the broad end.

5. Each newly formed desmid has an old semicell of the parent and a new semicell formed by it.

6. Sexual reproduction takes place by conjugation and is of the isogamous type except in *C.parvulum*.

7. During conjugation two desmid cells approach each other laterally and become encased by a common mucilage envelope.

8. After arranging them parallely, the cell wall of the desmids splits near the middle while the inner slender layers grow in opposite directions as proteuberances, meet in the centre forming a conjugation tube.

9. The protoplast of each cell gets transformed into an amoeboid gamete.

10. The gametes move into the canal and fuse forming a diploid zygospore.

11. In many species, no conjugation tube is formed. The gametes come out of the cell and fuse to form a

zygospore.

12. The wall of the zygospore is thick and smooth or ornamental.

13. The zygospore after a period of rest, germinates to give rise to two individuals of closterium.

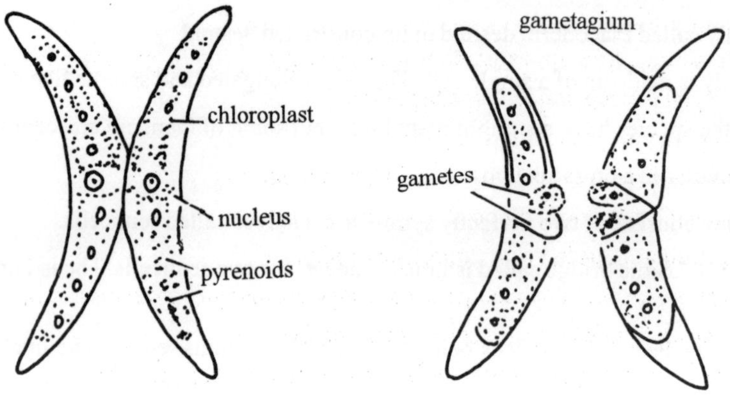

Reporductive cells showing conjugation.

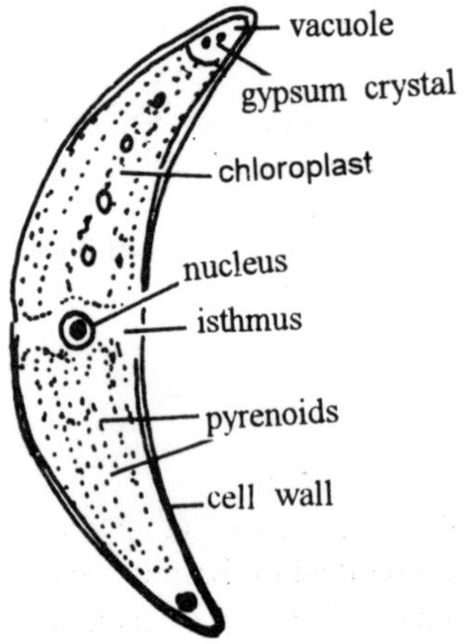

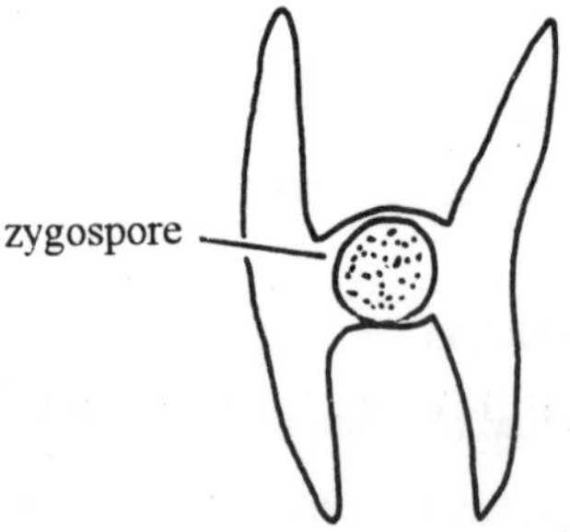

Single zygospore

Fig. 5.36 Closterium. A single cell

Diagrams to be drawn

1. Cell structure

2. Cell division

3. Stages in conjugation.

Systematic position and identification

A. 1. Chlorophyll present

2. Plant body a thallus

3. Cell wall cellulosic ... *Algae*

B. 1. Plastids present

2. Chlorophyll is the chief pigment

3. True nucleus present

4. Photosynthetic produce is starch ... *Chlorophyceae*

C. 1. Motile reproductive cells absent

2. Sexual reproduction by means of conjugation ... *Conjugales*

D. 1. Plant body unicellular

2. The cell consists of two halves called semicells ... *Desmidiaceae*

E. 1. The unicellular plant body has two halves called semicells

2. The cells elongated with no constriction at the isthmus region ... *Closterium*

CAULERPA

Common species : *C.prolifera, C.olivieri, C.fastigata, C.taxifolia, C.race- mosa scalpelliformis, C.peltata, C.cupressoides, C.chemntzia* etc.

Distribution : All the species are exclusively marine being distributed in the quite shallow waters of tropical seas. Some like *C.prolifera* and *C.ollivieri* occur in Mediterranean seas.

Study : 1. External features of the thallus

2. Transverse section of the stolon

3. Reproduction

External features of the thallus

1. The thallus is very elaborate with a great amount of external complexity resembling a vascular plant.

2. The plant body however is a coenocyte; it is made up of single cell.

3. The plant body may be distinguished into three parts viz.,

1. Horizontal stolon

2. Rhizoidal structures

3. Photosynthetic assimilators.

4. The stolon is horizontal. It is a branched rhizome like structure. It may be as long as one metre in some species.

5. The stolon on its under surface bears a number of branched rhizoids. These help in anchorage and absorption.

6. On the upper surface the stolon bears a number of green leaf like photosynthetic assimilators.

7. The photosynthetic assimilators show a great variety in their shape structure etc. In fact the species of *Caulerpa* are identified on the basis of assimilators, as the stolon and rhizone are same in all the species.

8. The species of *Caulerpa* are named after the plants and their structures which they resemble.

9. *C.scalpelliformis*, has flat assimilatiors with margins dissected (not deeply) into tooth like projections.

10. *C.peltala* has assimilators bearing a number of peltate discs.

11. *C.taxifolia* has assimilatiors much like in *C.scalpelliformis*, but dissections are very deep and the tooth like projections are narrow.

12. *C.verticellata* has assimilators in the form of cylindrical whorled structures.

13. *C.chemnitzia* has assimilators with small tubular, trumpet like lateral appendages.

14. *C.hypnoides* has moss like bushy assimilators.

15. *C.fastigata* has irregularly branched thread like assimilators.

Anatomy of the stolon

1. A cross section of the thallus is somewhat circular in outline.

2. There are no cross walls. The cell wall is made up of a mixture of pectin and cellulose.

3. Internal to the cell wall is the peripheral cytoplasm containing numerous discoid chloroplasts (no pyrenoids) and nuclei.

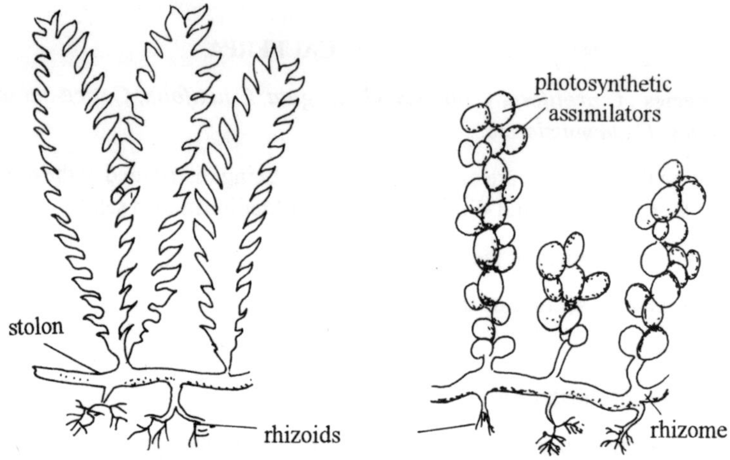

Fig. 5.37 A. Caulerpa scalpelliformis B. Caulerpa racemosa

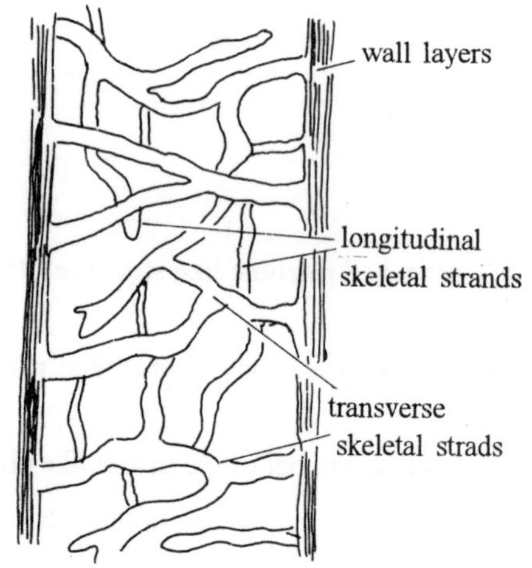

Fig. 5.38 Caulerpa - L.S. of assimilator (diagrammatic)

4. There is a large vacuole in the centre. It conists of colloids and is transversed by a number of skeletal strands called trabeculae.

5. The trabeculae are found in all parts of the plant body but they are exceptionally well developed in stolon.

6. The trabeculae develop from the wall and are right angles to the surface.

7. The trabeculae in the stolon are closely interwoven and offer mechanical support.

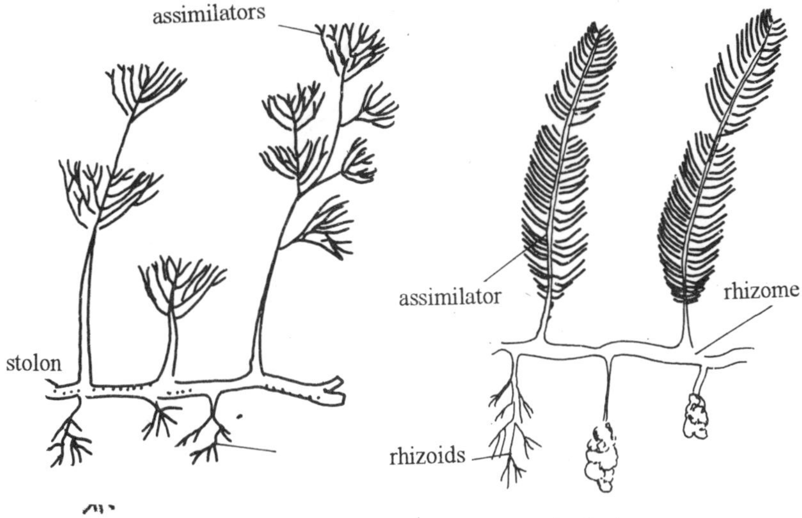

Fig. 5. 39 C.Caulerpa verticellata D. Caulerpa

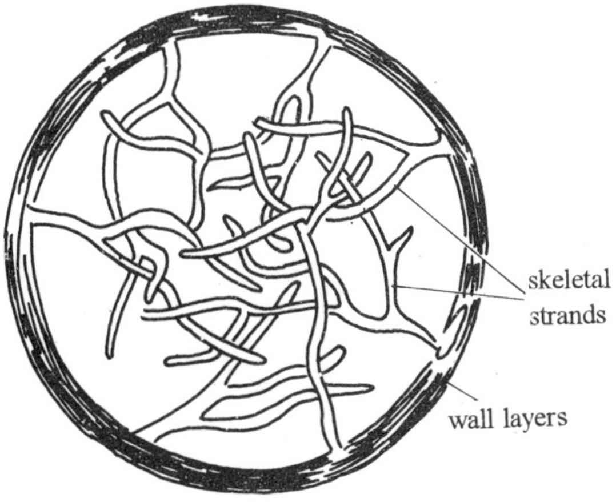

Fig.5.40 Caulerpa. T.S.of stolon (dagrammatic)

Reproduction

1. Vegetative propagation takes place by means of death and decay of the stolon. The branches get seperated and develop into individual plants.

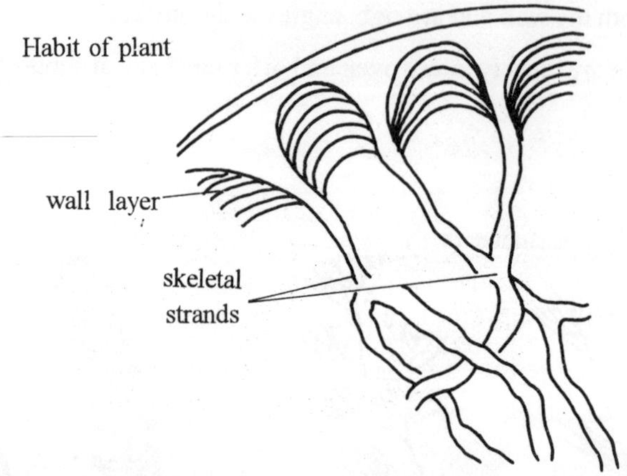

Fig. 5.41 Caulerpa A portion of T.S.of stolon (enlarged)

2. Sexual reproduction is anisogamous; species are heterothallic.

3. The assimilators act as gametangia and produce gametes which come out through the extrusion papillae (Dostal 1928).

4. The gametes fuse to form the zygote which develops into a new individual.

Diagrams to be drawn

1. Habit sketch of the species available in the laboratory.

2. Transverse section of the stolon.

3. Assimilators with extrusion papillae.

Systematic position and identification

A. 1. Chlorophyll present

2. Plant body a thallus

3. Cell wall cellulosic ... *Algae*

B. 1. Plastids present

2. Chlorophyll is the chief pigment

3. True nucleus present

4. Photosynthetic product is starch ...*Chlorophyceae*

C. 1. Thallus coenocytic, tubular with no ross walls

2. Thallus is macroscopic and looks complex

3. Chloroplasts discoid.

4. Cell has a large central vacuole...*Siphonales*

D. 1. Plants are marine in habitat.

2. Thallus has a horizontal stolon, with rhizoids below and assimilators above ...*Caulerpaceae.*

E. 1. Plant body coenocytic

2. Assimilations show great variety.

3. Extrusion papillae develop on assimilators ...*Caulerpa*

The various speices of *Caulerpa* given have to be identified on the basis of the nature of the assimilators.

CODIUM

Common spcies : *C.fragile*

Distribution : Widely distributed in warm tropical seas; grows on rocks between the tide lines.

Study : 1. Structure of the thallus

2. Internal structure

3. Reproduction

4. Systematic position and identification

Structure of the thallus

1. The plant body is a much branched, soft thallus, tubular or cylindrical in appearane.

2. The branches are interwoven into a prostrate cushion like or globose mass.

3. The prostrate system bears a number of erect cylindrical shoots which are dichotomously branched.

4. The plant body is spongy to touch and macroscopic.

5. The plant body is anchored with the help of discs or rhizoids.

Internal structure

1. A transverse section of the branch of a thallus reveals an aerial core of densely interwoven colourless filaments.

2. Lateral branchlets called utricles arise from the aerial core of filaments.

3. The utricles lie in a palisade like layer surrounding the central aerial core.

4. Each utricle has a thick peripheral layer of cytoplasm internal to the cell wall surrounding a central vacuole.

5. The cytoplasm has many nuclei and discoid chloroplasts without pyrenoids. Most of the chloroplasts are found at the tip of the utricles.

6. Filaments of the axial core develop annular ingrowths here and there especially at the bases of the utricles.

Reproduction

1. Sexual reproduction is by fragmentation of the plant body.

2. Sexual reproduction is anisogamous. The plants are heterothallic.

3. Gametangia are formed on one side of the utricles.

4. Both micro (male) and macro (female) gametes are biflagellate.

5. Male and female gametangia may be distinguished by their distinct colouration. Male gametangia are golden brown in colour while female gametngia are dark green.

6. The gametes fuse to form a zygote, which directly give rise to a new plant.

Diagrams to be drawn

1. Habit sketch

2. Cross section of the thallus

3. Gametangia arising from the utricle

Systematic position and identification

A. 1. Chlorophyll present

2. Plant body a thallus

3. Cell wall cellulosic ... *Algae*

B. 1. Plastids present

2. Chlorophyll is the chief pigment

3. True nucleus present

4. Photosynthetic product is starch ... *Chlorophyceae*

C. 1. Thallus single celled, tubular and coenocytic.

2. Thallus is a macroscopic structure

3. Chloroplasts have no pyrenoids ... *Siphonales*

D. 1. Plant body soft, spongy and tubular

2. Branches interwoven to form a macroscopic structure ... *Codiaceae.*

E. 1. Tubular plant body

2. Cross section of the thallus shows an aerial core of filaments and a palisade layer of utricles.

3. Gametangia arise from utricles ... *Codium.*

CHARA

Common species : *C.baltica, C.roxallina, C.zeylanica, C.wallichi* etc.

Distribution : Grows submerged in fresh water and upon sandy or muddy bottoms of ponds. Some species like *C.baltica* grow in salt water.

Study : 1. Structure of the plant body

2. Mount the apical cell

3. Structure of the cell

4. Anatomy of internode

5. Asexual reproduction

6. Structure of Globule (tease open and study)

7. Structure of Nucule (tease open and study)

8. Systematic position and identification

Structure of plant body

1. The plant body is commonly called stonewort or aquatic horsetail because of its superficial resemblances to *Equisetum*.

2. The plant body has a central axis 20-40 cms long and differentiated into nodes and internodes.

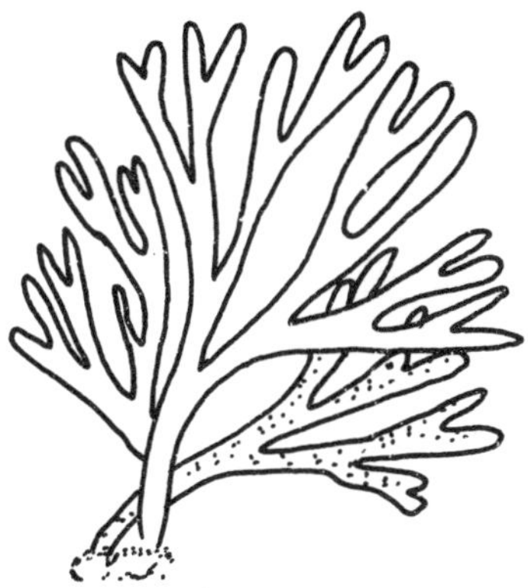

Fig.5.42 Codium. Thallus showing habit

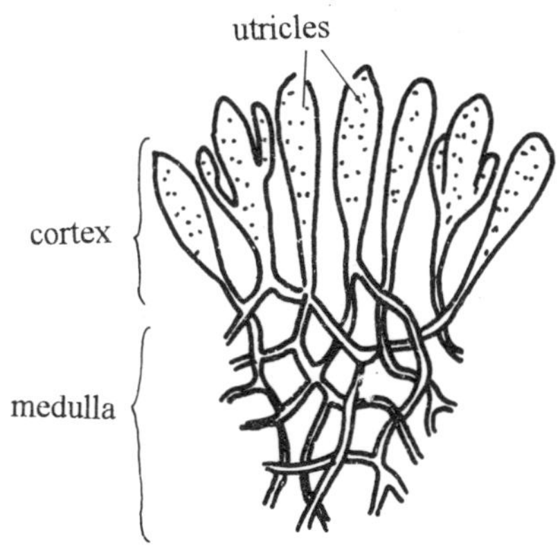

T.S. of thallus (A portion)

3. The axis is attached to the substratum with the help of rhizoids, which arise either at the base or from lower nodes.

4. The rhizoids are uniseriate and branched.

5. The main axis is branched profusely and the branches are of two types viz., branches of limited growth and branches of unlimited fruit.

6. The branches of limited growth, also called 'leaves' arise in whorls of 3-15 at the nodal region. They are small 5-15 internodes long and green in colour very often short, spiny; secondary laterals are produced from the leaves.

7. The branches of unlimited growth resemble the main axis in having nodes and internodes. They are produced at the nodes. The branches of unlimited growth at their nodal region bear the leaves.

Structure of the apical cell

1. The growth of the main axis as well as the branches takes place with the help of a single dome shaped apical cell.

2. This cuts of disc shaped segments. Each of these by a transverse division form a biconcave cell called the nodal initial and a lower biconvex internodal initial.

3. Internodal initial will not divide but simply elongates to form the internode.

4. The nodal initial divides to form nodal cells.

5. The internodal cells at later stages become covered by a sheath of cells called corticating filaments. Half of the cortication comes from the lower node and the other half from the upper node.

Anatomy of internode

1. A transverse section of the axis at the internodal region shows a central axial cell surrounded by a number of corticating cells.

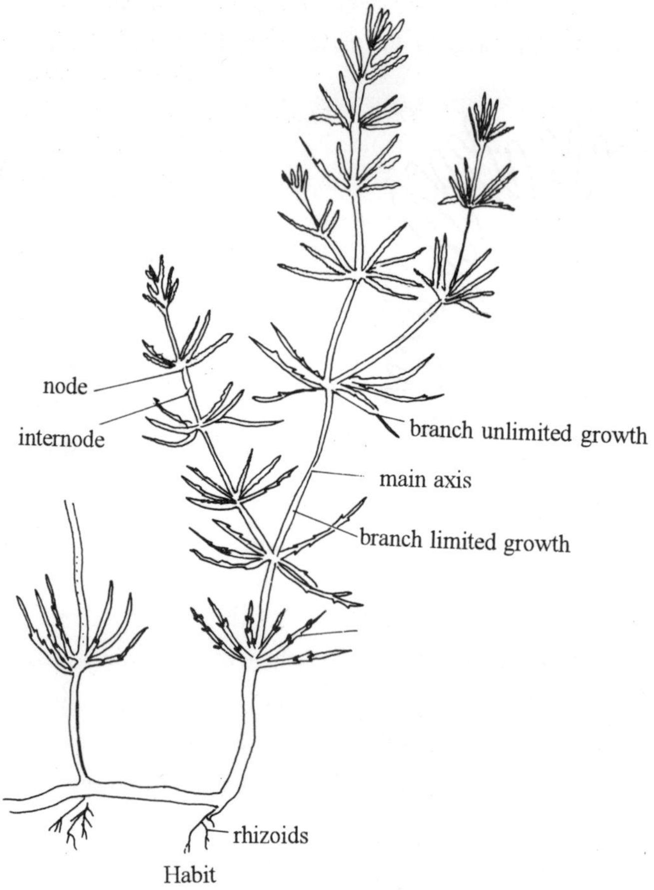

Fig. 5.43 Chara.

How to mount an apical cell ?

1. Select an apex, either of the main axis or of a branch of unlimited growth.

2. Cut it a little behind the tip. Place it on a slide and carefully go on removing the branches that surround the apex.

3. After removing sufficient number of branches (at this stage the apex will be clearly visible to the naked eye) mount the apex on a slide in glycerine.

4. After putting the coverglass, apply slight pressure either with a finger or with the back of a pencil. This will push the branches (that are still covering the apical cell) to a side. The apical cell can be clearly seen under the microscope.

Asexual reproduction

1. Asexual reproduction is brought about by Amylum stars, Bulbils and secondary protonema.

2. Amylum stars are stellate aggregate of cells rich in anylum starch. On detachment from the parent plant they grow into a new individual.

3. Bulbils are small tuberous rounded structures developed on the rhizoids. They develop into new individuals on detaching from the plant body.

4. Sometimes adventitious branches arises at nodes of aged plants. They are capable of giving rise to a new plant.

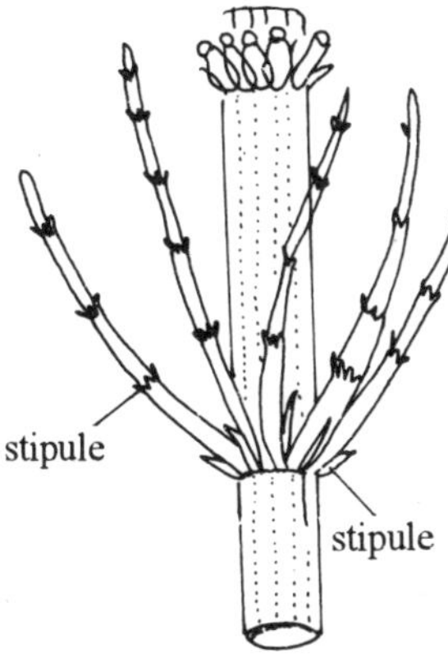

Fig. 5.44 Chara. A portion of thallus

The rhizoidal plate

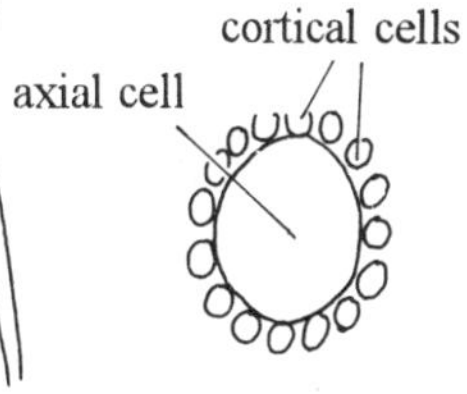

T.S. through internode

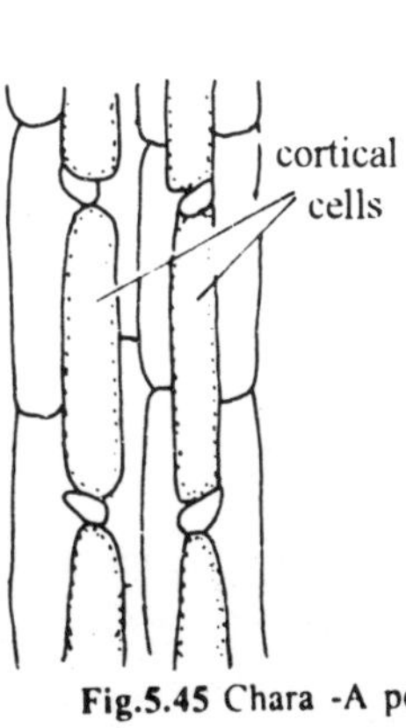

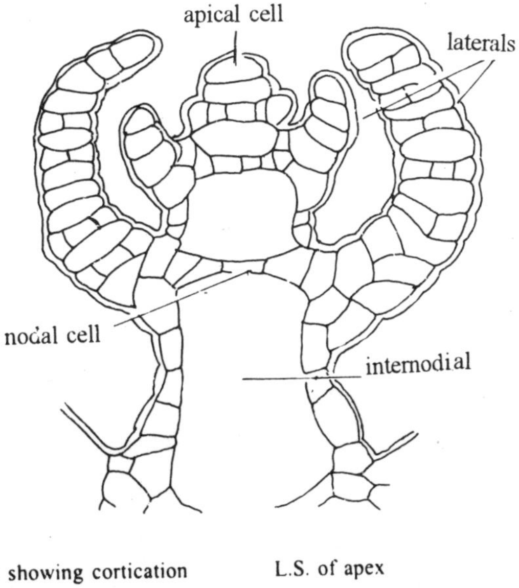

Fig.5.45 Chara -A portion of axis showing cortication

L.S. of apex

Sexual reproduction

1. Sexual reproduction is of the oogamous type. The sex organs of *Chara* are perhaps most complex in algae.
2. The male reproductive structure is antheridium, also known as the globule.
3. The female reproductive structure is oogonium, also known as the nucule.
4. Most of the species of *Chara* are monoecious. On monoecious species a pair of sex organs are produced at the nodal regions. The nucule and globule are juxtaposed with the latter below.

Structure of the globule

1. Globule is spherical in shape and attached to the node with the help of a stalk cell.
2. The wall of globule is made up of 8 shield cells. The walls of the cells are infolded to give a pseudocellular appearance.
3. At maturity of the globule, the shield cells develop organge red pigmentation.
4. The globule consists of a long pedicle cell which grows into the globular cavity.
5. Each shield cell sends inwards, a rod like cell - The manubrial cell (8 in number)
6. Each manubrial cell at its free and bears one or more capitula or head cells.
7. Each capitulum gives rise to about six secondary capitula.
8. From the secondary capitula develop 2-4 long whip like spermatogenous filaments.
9. Each spermatogneous filament consists of 25-200 cells. Each cell gives rise to a biflagellate sperm.
10. The spermatogenous filaments are colied up between the manubrial cells, filling up the cavity of the globule.

Structure of the Nucule

1. Nucule is a short stalked oval structure attached to the primary lateral one cm above the globule.
2. A mature nucule consists of a large oval egg surrounding which is an envelope of 5 tubular cells, which are spirally twisted once or twice in clockwise direction.
3. The tip of each tube cell form a small cell called the crown cell. The five crown cells together form the corona.
4. The egg has a short stalk of two cells. It has a receptive spot and rich in oil drops and starch grains.

Zygote

1. After fertilization, the zygote secretes a thick wall around and looks greenish black in colour. The wall is impregnated with calcium.

Cell structure

1. The cells at the nodes are small, have dense granular cytoplasm and are uninucleate.
2. The outer wall specially of the internodal cells is made up of a mixture of cellulose and pectin and impregnated with silica and calcium carbonate.
3. Inside the cell is a large central vacuole surrounded by a thin cytoplasmic layer.
4. The young internodal cell is usually uninucleate but as it enlarges, it becomes multinucleate.
5. Each cell has a number of ellipsoidal chloroplasts.
6. Streaming movements of protoplasm is seen inisde the internodal cells.

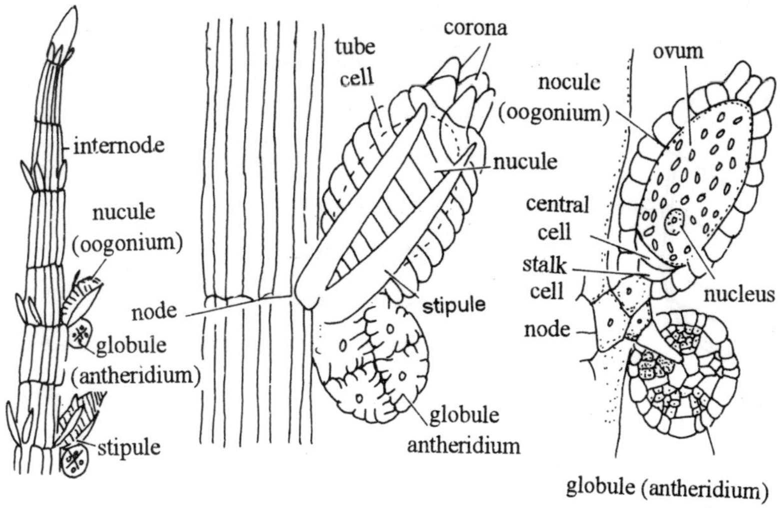

Fig. 5.46 A portion of fertile branch L.S. through sex organs to show the structure of nucule and globule Single node bearing nucule and globule

Diagrams to be drawn

1. Habit sketch
2. A part enlarged to show the branching
3. L.S apex showing the apical cell
4. T.S of nodal region
5. A nodal region enlarged showing the globule and nucule.
6. Long section of the globule
7. A nucule enlarged
8. Sectional view of a nucule.

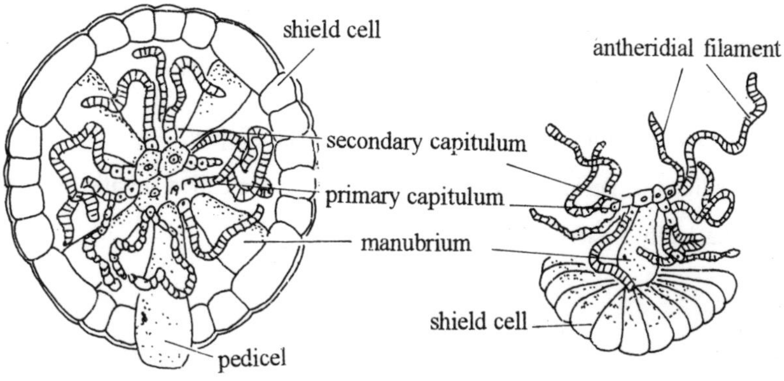

Fig. 5.47 L.S. through globule Single shield cell wall manubrium and antheridial filaments

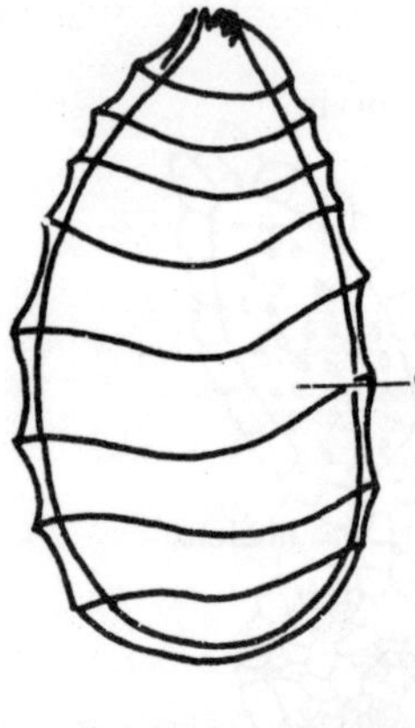

Fig. 5.48 Chara A zygote

Systematic position and identification

A. 1. Chlorophyll present

2. Plant body a thallus

3. Cell wall cellulosic ...*Algae*

B. 1. Plastids present

2. Chlorophyll is the chief pigment

3. True nuclueus present

4. Photosynthetic product is starch ...*Chlorophyceae*

C. 1. Plant body is differentiated into nodes and internodes.

2. Sex organs are unique - globule and nucule ...*Charales*

D. 1. Silica impregnated on the cell wall

2. Plant body complex morphologically ...*Characeae*

E. 1.Internodal regions have corticating filaments

2. Nucule has a corona made up of 5 cells

3. Nucule is placed above the globule ...*Chara*

NITELLA
(Agardh 1824)

Common species : *N.cornia, N.opara, N.gracilis*

Distribution : Found in fresh water pools, ponds etc. In comparatively deep waters.

Study : 1. External morphology

2. Structure of a cell

3. Structure of sex organ

4. Systematic position and identification

External morphology

1. The plant body looks morphologically complex. It is macroscopic.

2. The plant looks like the pteridophyte *Equisetum* plant hence called aquatic horse tail.

3. The plant has basically 3 parts:

a. Main axis

b. Rhizoids

c. Branches

4. The main axis is divided into nodes and internodes. Nodes are multicellular while internodes are unicellular being made up of an elongated internodal cell. There is no cortication (see *Chara*) of the internodal cell.

5. Rhizoids help in the anchorage of the plant body to the substratum. They arise from the lowest node, branched and multicellular with oblique septa. A quadrant rhizoidal plate is present, from which arise the rhizoids.

6. The branhes arise at the nodal region of the main axis. They are of two types viz., branches of limited growth and branches of unlimited growth.

7. The branches of limited growth or short laterals are 5-10 internodes long and are bonre in a whorl. At their nodal regions, the short laterals, bear the secondary short laterals, which are unicellular and hair like. The secondary short laterals are also called stipules.

8. The branches of unlimited growth, also known as long laterals, are produced at the nodal regions in pseudo-axillary positions. They resemble the main axis in all respects.

Structure of the cell

1. The cell at the internode which can be considered a typical cell has a cellulose wall.

2. Within the cell, there is a large central vacuole.

3. The cytoplasm is peripheral and encloses a single nucleus and many discoid chloroplasts. The chloroplasts do not have pyrenoids.

Reproduction

1. Sexual reproduction is oogamous. The species of *Nitella* are either monoecious or dioecious. Just as in *Chara*, here also, the antheridium is known as the globule and, the oogonium is called the nucule.

2. In monoecious species, the nucule and globule are produced at the nodal regions of secondary short laterals.

3. At every node, usually the nucule and globule are placed one above the other with the nucule below and globule above.

Structure of the globule

1. The globule is orange red in colour and is spherical in shape.

2. It is attached to the plant at the nodal region with the help of a long stalk cell or pedicel cell.

3. The wall of the globule is made up of eight shield cells.

4. Each shield cell produces into the cavity of the globule a rod shaped manubrial cell. There are totally 8 manubrial cells.

5. Each manubrial cell has 4-6 primary capitulum cells.

6. Each primary capitulum inturn bears 2-4 secondary capitula.

7. From each cell of the secondary capitulum arise 2-4 spermatogenous filaments. Each filament 25-100 cells long.

8. The cells of the spermatogenous filaments, each one of them produce a biflagellate antherozoid or sperm.

Structure of Nucule

1. The nucule is an oval structure deep green to greenish black in colour and is found in the nodal region below the globule. It has a short pedicel.

2. The nucule has a single oosphere or egg rich in granular cytoplasm. The cytoplasm also has starch and oil

as reserve food.

3. The envelope of the ovule is made up of 5 tubes which are coiled around the nucule.

4. The tip of the nucule is covered by 10 crown cells constituting the corona. The cells are arranged in two superposed tiers.

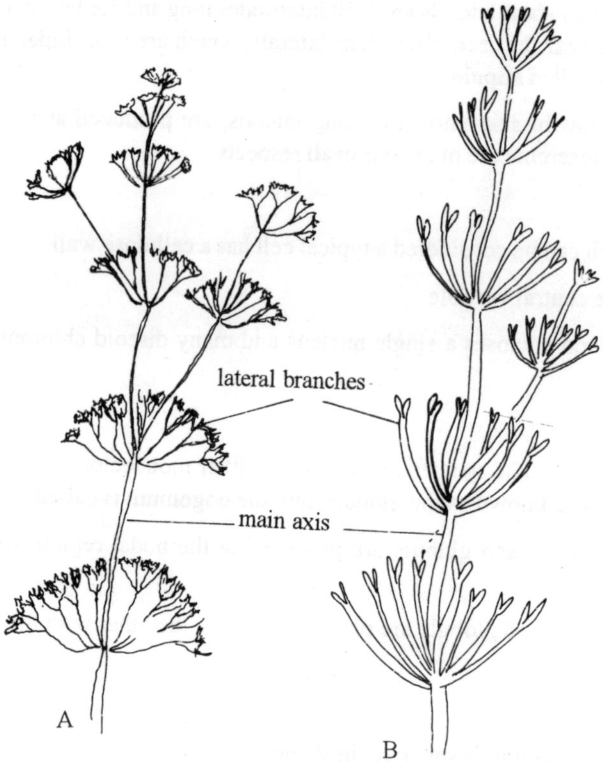

Fig.5.49 Nitella. - Habit

Fertilization

1. The cells of the corona open apart to facilitate the entry of antherozoids.
2. The zygote or oosphore develops a thick wall around it and becomes silicified.
3. The zygospore after undergoing a period of rest germinates to develop into a new individual.

Diagrams to be drawn

1. Habit sketch
2. Nodal region enlarged to show the branching
3. Nucule and globule enlarged
4. Sectional views of nucule and globule

Systematic position and identification

A. 1. Chlorophyll present

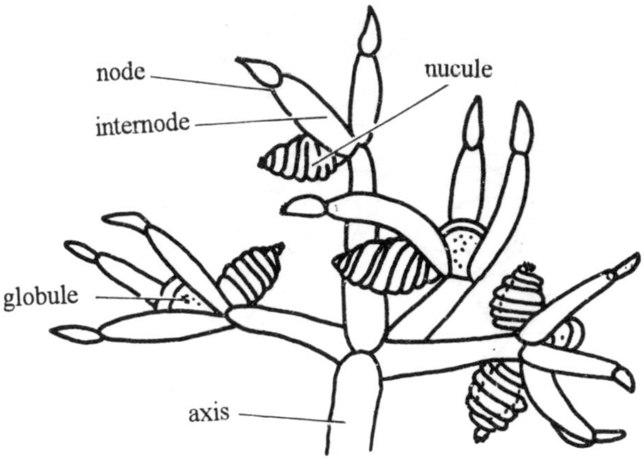

Fig.5.50 Nitella. A portion of thallus

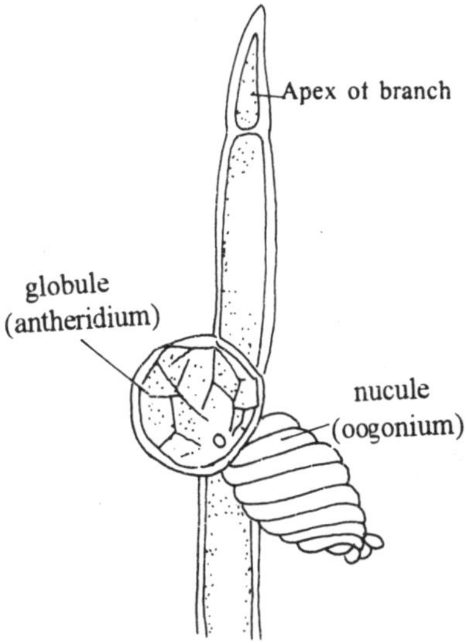

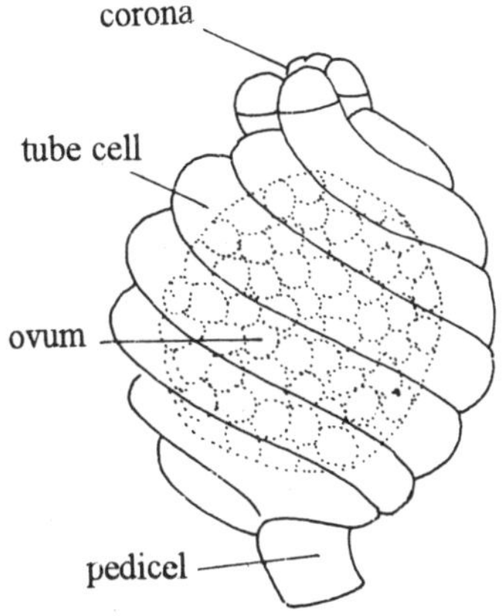

A nucule

Fig.5.51 Nitella. A node with Globule and nucule

2. Cell wall cellulosic

3. Plant body a thallus ... *Algae*

B. 1. Plastids present

2. Chlorophyll is the chief pigment

3. True nucleus present

4. Photosynthetic produce is starch ... *Chlorophyceae*

C. 1. Thallus is differentiated into nodes and internodes

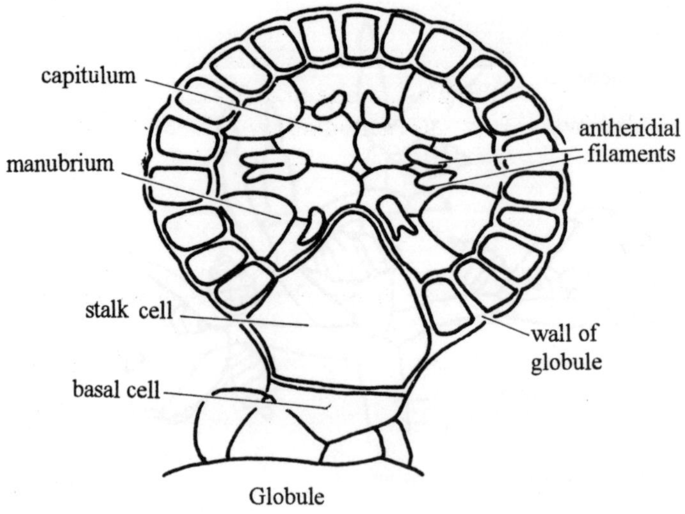

Fig.5.52 Nitella.

2. Sex organs are globule and nucule... *Charales*

D. 1. Silica deposit is found on the plant body

2. Rhizoids are multicellular... *Characeae*

E. 1. At the nodal region globule is above and nucule below

2. There is no cortication of the internode

3. The corona of the nucule has 10 cells ... *Nitella*

CHAPTER 6

XANTHOPHYCEAE

The members of the class have yellowish green chromoplasts containing chlorophyll i.e,beta carotene and xanthophyll. Food reserve is leucosin and oil. There is no formation of starch. Plant body may be unicellular or multicellular. The motile reproductive cells are biflagellate, which are unequal in length (Heterokontae)

Asexual reproduction is by zoospores or aplanspores. Sexual reproduction is mostly anisogamous and occasionally isogamous. In *Vaucheria* however sexual reproduction is oogamous.

The class includes 15 genera and 375 species (Smith) majority of which are freshwater in distribution.

Smith classifies Xanthophyceae into six orders of which Heterosiphonales is one. This order includes *Vaucheria* and *Botrydium*.

Fritsch considered Xanthophyceae as a separate class in Algae. He classified the class into 4 orders, of which heterosiphonales is one. But Fritsch classified *Vaucheria* under Siphonales of the class Chlorophyceae. In this book , the classification of Smith is followed with reference to *Vaucheria*.

VAUCHERIA
(De Condolle)

Common species : *V. sessilis, V. polysperma, V. amphibia, V. geminata, V. terrestris, etc.*

Distribution: About half a dozen species are Marine among the 40 odd species. The remainder are fresh water inhabiting well aerated streams or terrestrial. Terrestrial species grow on damp soil forming extensive green mats.

Study: 1 Structure of the plant body
2. Asexual reproduction
3. Systematic position and classification

Structure of the plant body:

1. The thallus is a sparingly branched tubular cell frequently attaining a length of several centimetrs.
2. In terrestrial forms the thallus is attached to the soil with the help of hapteron.
3. There is a centrally running continuous vacuole present throughout the length of the plant body.
4. The cell wall is thin and made up of two layers. Outer pectic and inner cellulosic.
5. The cytoplasm is peripheral. It has numerous discoid chromatophores on its outer face (towards the cell

wall) and many nuclei (coenocytic) towards the inner face.

6. The chromatophores are small, circular or elliptical in outline and have no pyrenoids.
7. Minute droplets of oil are found in cytoplasm.
8. There are no cross walls. They are formed only during the formation of reproductive structures or to separate off a dead portion.

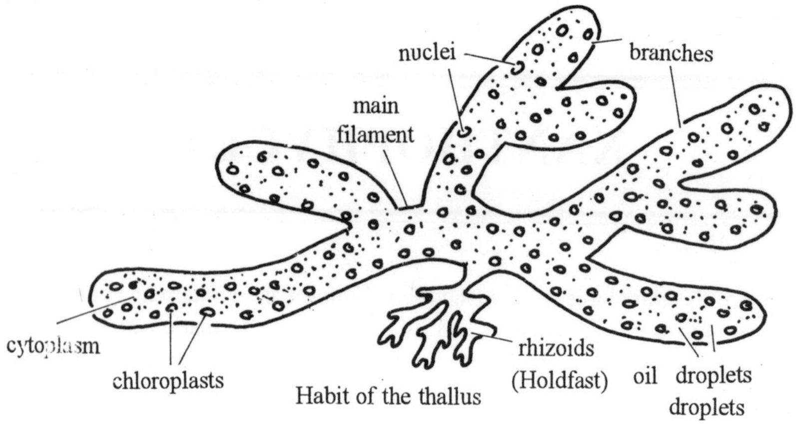

Fig.6.1 Vaucheria

Asexual reproduction

1. Asexual reproduction takes place by:
   (a) Fragmentation
   (b) Synzoospore formation
   (c) Aplanospores and Akinetes
   (d) Hypnospores
2. Any mechanical breakage will result in fragments which can develop into new individuals.
3. Zoospores are mostly formed in aquatic species.
4. Zoospores are formed singly in club shaped sporangia at the tip of the branches.
5. The sporangium is separated from the rest of the plant body my means of a cross wall.
6. At the time of formation of a zoospore, the nuclei and chromatophores reverse their position, in the zoosporangium, so that the nuclei lie outside.
7. The zoospore is multinucleate and multiflagellate. It is called a *compound zoospore or synzoospore*
8. The distal wall of the sporangium opens to form a pore through which, the zoospore squeezes itself out.
9. On being liberated out of the sporangium, the zoospore looses its flagella and germinates to give rise to a new individual.
10. Akinetes or Aplanospores or non-motile spores are usually formed in terrestrial species.
11. In each sporangium only one aplanospore or akinete is formed.
12. In some terrestrial species, the entire thallus forms a number of thick walled *hypnospores* by transverse segmentation.
13. Each hypnospore may directly develop into a new individual.
14. The contents of the hypnospores may sometimes divide to form a number of cysts.
15. The plant of *Vaucheria*, with transverse septation having a number of cysts appears like an alga by name *Gongrosira*, hence this stage is also called the *Gongrosira stage.*
16. On liberation from the plant, each cyst develops into a new individual.

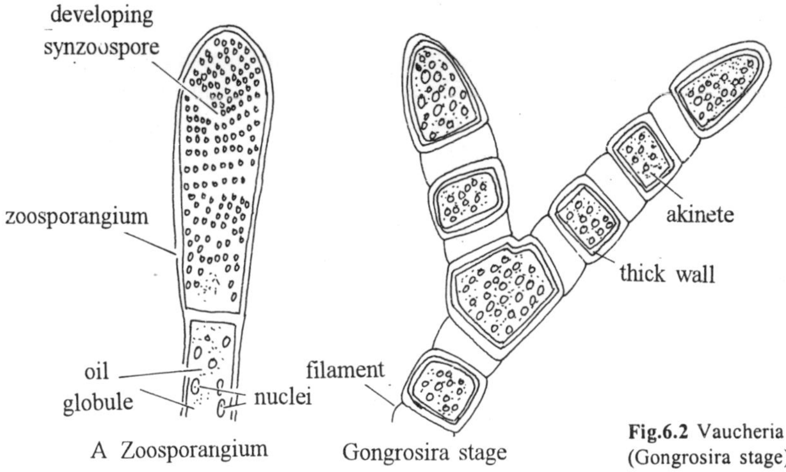

Fig.6.2 Vaucheria (Gongrosira stage)

Sexual reproduction

1. Sexual reproduction is of the oogamous type. Sex organs are antheridia and oogonia.
2. Species of vaucheria may be monoecious or dioecious (mostly monoecious).
3. Antheridia and oogonia arise as outgrowths either on the main thallus adjacent to each other or on special lateral branches.
4. In some species, a lateral branch terminates with an oogonium, with an antheridium below.
5. In certain instances, a single antheridium will be associated with many oogonia.
6. An antheridium is a tubular structure, sometimes assuming a hook shape.
7. At maturity the antheridium bends and the terminal part is cut off from the remainder by the development of a cross wall.
8. The protoplasm of the antheridium gets divided into a number of biflagelate antherozooids.
9. The mature antherozooids are liberated out of the antheridium through a terminal pore.
10. The oogonium is a globose or oval structure. It is sessile
11. The oogonium is formed by the lateral bulging of the filament.
12. Into this bulge migrates a mass of cytoplasm called wanderplasm.

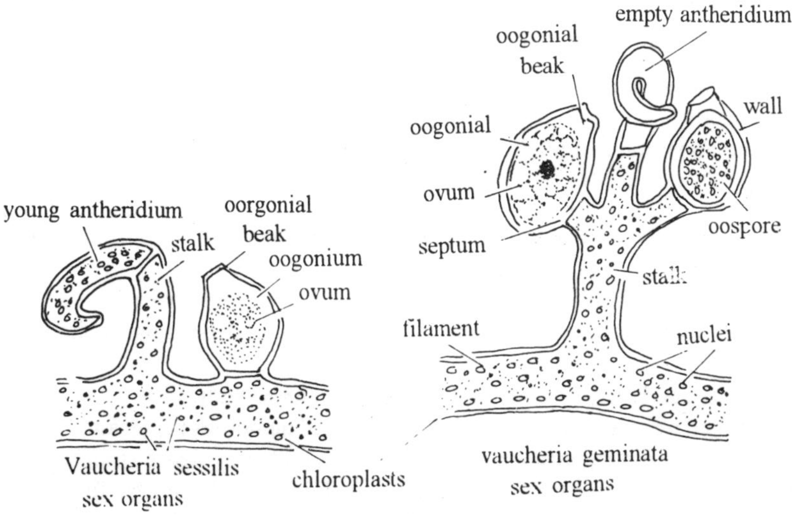

Fig.6.3 Vaucheria

13. At maturity, the oogonium gets separated from the filament by means of a cross wall.
14. An Oogonium, at the time of fertilization consists of an uninucleate egg.
15. The Oogonium develops a papilla or beak on one side to form a hyaline receptive spot.
16. It is at this hyaline spot that the sperms enter the oogonium. The fusion of the pserm and egg results in the formation of zygote. The zygote has a thick wall and develops into a new individual.

Diagrams to be drawn

1. Habit sketch
2. A part of the plant enlarged.
3. Zoosporangium with zoospore.
4. Aplanospore.
5. Antheridium and Oogonium

Systematic position and Identification

A. 1. Chlorophyll present
 2. Plant body a thallus
 3. Cell wall cellulosic...*Algae*

B. 1. Reserve food material is oil
 2. Motile reproductive cells have unequal flagella ... *Xanthophyceae*

C. 1. Plant body siphorous with no cross walls.
 2. Nuclei many in the cell ... *Heterosiphonales*

D. 1. Chloroplasts lack pyrenoids.
 2. Siphonein and Siphonoxanthin pigments absent ... *Vaucheriaceae*

E 1. Plant body a Coenocyte
 2. Thallus branched
 3. Cross walls are formed only during formation of reproductive structures ... *Vaucheria*

BOTRYDIUM
(Wall roth, 1815)

Common species: *B. granulosum, B. divisum, B. wallrothii etc.*

Distribution: Terrestrial, grows on drying up muddy banks of streams, pools, damp soil etc.

Study:
1. Structure of the plant body
2. Reproduction
3. Systematic position and phylogeny

Structure of the plant body:

1. The plant body is unicellular, vesicular and consists of a ballon shaped aerial portion and a colourless rhizoidal portion.
2. The aerial vesicular portions looks like a pin-head reaching a size 1-2 mm across.
3. The aerial vesicular portion may become cylindrical and branched, depending on the environmental conditions.
4. The aerial part of the plant body has a tough wall and is impregnated with silica.
5. The wall is internally lined by cytoplasm in which are embedded numerous nuclei and chromatophores.
6. The chromatophores are discoid and often connected to one another by dense strands of cytoplasm.
7. In young plants pyrenoids like bodies are found in chromatophores, but the reserve food is never starch.
8. Reserve food is oil or leucosin.
9. There is a large central vacuole, but the cytoplasmic strands traverse across the vacuole.

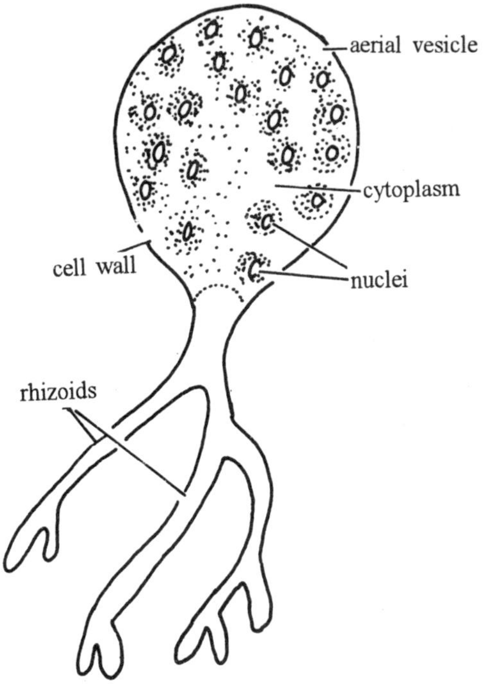

A vegetative thallus

Fig.6.4 Botrydium

10. The underground rhizoidal portion is colourless and is either sparingly or profusely branched. It has no chromatophores, but nuclie are present.

11. Rhizoidal portion helps in anchorage and absorption.

Reproduction:

1. Asexual reproduction takes place by a variety of methods.

These are : (a) buds (b) Aplanospores (c) Zoospores (d) Hypnospores or Cysts.

(a) In some species buds are formed from the aerial part. These develop into new individuals.

(b) Large number of small, uninucleate, pyriform, biflagellate zoospores are formed. These come out of the vesicle, swim for some time in water and develop into new individuals.

(c) When moisture is not sufficient to form the zoospores, the protoplast of the cell gets divided into a number of non-motile, uni or multinucleate aplanospores.

(d) The aplanospores formed during dry season called *hypnospores*. In some instances cysts are also formed. These are of 3 types.

(i) *Macrocysts:* One large cyst for the entire vesicle.

(ii) *Sporocysts:* Many cysts are formed in the vesicle.

(iii) *Rhizocysts :*

Many cysts are formed in the rhizoidal part.

2. Sexual reproduction is either isogamous or anisogamous.
3. Gamete formation is seen mostly during rainy season.
4. The species of *Botrydium* may be monoecious or dioecious.
5. During gamete formation, the protoplast in the vesicle gets divided into a number of uninucleate gametes.
6. The gametes are pyriform, have one to four chromatophores, no eye spots and have unequal flagella.
7. Gametic fusion results in the formation of zygote which develops into a new individual.

Diagrams to be drawn

1. Habit sketch
2. A plant enlarged
3. Plant showing hypnospores.
4. Plant showing rhizocysts.

Systematic position and Identification

A. 1. Chlorophyll present
2. Plant body is thallus
3. Cell wall cellulosic ...*Algae*

B. 1. Reserve food material is oil
2. Motile reproductive cells have unequal flagella ... *Chlorophyceae*

C. 1. Plant body siphonous with no cross walls.

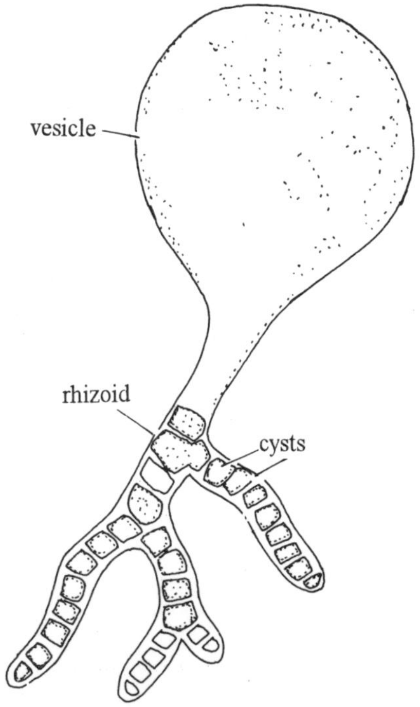

Fig.6.5 A plant body bearing rhizoidal cysts (hypnospores)

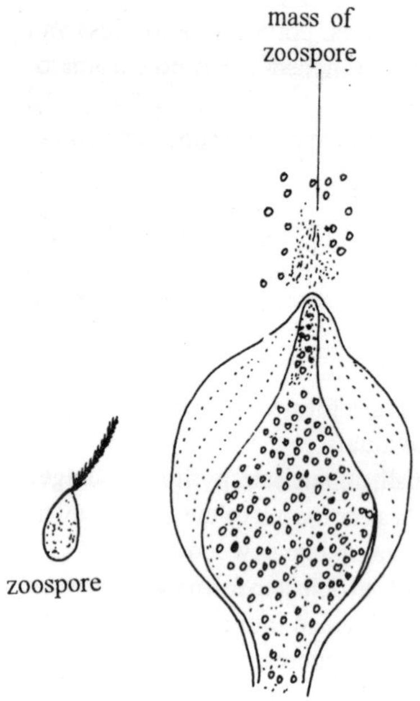

Fig.6.6 Discharge of swarmers from vesicle

2. Nuclei many in the cell...*Heterosiphonales*

D. 1. Coenocytic plant body

2. Central vacuole present

3. Plant body inhabits terrestrial habitat always ...*Botrydiaceae*

E. 1. Plant body divided into aerial vesicle and underground rhizoidal portion

2. Sexual reproduction iso or anisogamous but never Oogamous...*Botrydium*

CHAPTER 7

BACILLARIOPHYCEAE

The Bacillariophyceae are commonly called diatoms. They include a large number of unicellular and clonial algae, which vary greatly in form and size. They are most widely distributed among Algae. The primary distinguishing feature of Bacillariophyceae is that the cell wall is silicified and consists of two halves, which fit together as do the lid and box. The cell has one to many chromatophores.

Reproduction is by cell division and auxospore formation. Auxospores are usually zygotic in nature, produced by the union of gametes, but may not be always so.

There are about 170 genera, and 55,000 species majority of which are in living but some known only in fossil conditions.

Certain genera are found only in fresh water and certain others only in salt water. Many are found in both the habitats. The marine forms of diatoms constitute a predominant part of the planktonic Algae. They are so small, that as many as 35,500,000 diatoms may be present in one cubic meter of water. Many of the diatoms grow as epiphytes attached to the surface of higher plants and other Algae.

The marine planktonic diatoms together with dinoflagellates are of fundamental biological importance, as they form the first link in the food chain of all animals in the ocean.

Diatoms usually exist as single cells, but sometimes cells may be joined by mucilage to form filamentous structures. The cells have a beautiful symmetry. They may be in the form of circles, squares, boat shaped, triangles and various other types of curves.

Diatomaceous Earth: Even after the death of diatom cell, the silicified wall remains unaltered, and millions of these accumulate at the bottom of the oceans and become hardened. This fossil deposit is known as diatomaceous earth and is of great importance economically.

Daitomaceous earth is of great commercial value. It is used as mild abrasives in silver polish and tooth paste. At one time it was used as the absorbent for nitroglycerine in dynamites.

Diatomaceous earth also finds extensive use as a filter in sugar refineries, as an insulator in blast furnaces etc.

Classification: The class Bacillariophyceae is divided into two orders ,viz., Pennicae or Pennales and Centricae or Centrales.

Pennales: These are characterised by:

1. Cells are elongated or boat shaped in outline
2. They are predominantly fresh water.
3. They exhibit locomotion due to streaming movement of cytoplasm.
4. These cells are bilaterally symmetrical.

Centrales:

1. The shape of the cell is like a circular pill box.
2. They are predominantly marine.
3. There is no locomotion.
4. There is no raphe.

In this book a member of the order Pennales: *Pinnulcria* which is studied as a typical example is given.

PINNULARIA
(Ehrenberg)

Common species: *P. major, P. viridis, etc.*

Distribution: Commonly found in fresh water ponds, and ditches; forms on oily layer on top of water.

Study: 1. Cell structure
- (a) Girdle view
- (b) Valve view

2. Reproduction
3. Systematic position and Identification

Cell structure:

1. Structurally the diatom cell consists of two parts or halves which fit in like a lid and box.
2. Of the two halves, the smaller one is called *Hypotheca* and the bigger one is called *Epitheca*.
3. The cell has a size of 140 μ x 20 μ
4. Each valve or half consists of a flat top (valve face) and the margin or girdle representing the overlapping face.
5. A diatom cell can be viewed from two different angles. The view from the side is called girdle view, and the view from the top is called the valve view.

Valve view:

1. Only one valve can be seen.
2. The wall can be clearly seen and it is called frustule
3. The cell wall is made up of pectin and it is silicified.
4. On the wall are seen prominent markings called costae. The costae represent rows (striae) of small pores or cavities in the frustule.
5. Along the centre of the valve is present a long slit in the wall. This is called raphe, it is sigmoid in shape.
6. At the centre of the raphe is a swelling called the *central nodule*.
7. At either end of the raphe are also present two nodules called *Polar nodules*
8. The mucilage extends out of the wall through openings or pores and is supposed to help in locomotion.
9. In side the silicified wall is the cytoplasmic lining, surrounding vacuole.

Girdle view:

1. In the girdle or side view both the valves are seen (only one valve is seen in the valve view)
2. Costae can be seen in the frustule.
3. Inside the wall is the cytoplasm having a single nucleus.
4. One or two chromatophores may be seen distributed in the cytoplasm.
5. Oil and volutin globules are present in the cytoplasm

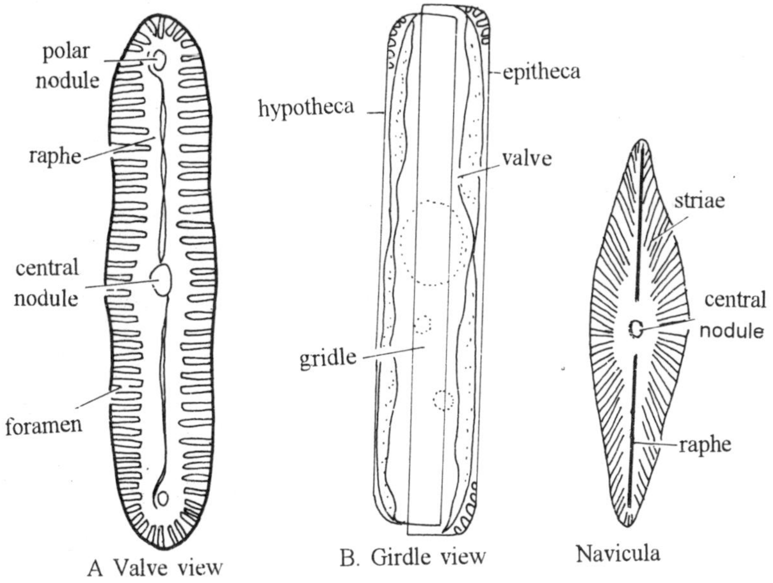

Fig.7.1 Diatoms (Value view and gridle view)

Reproduction

1. The commonest method of reproduction seen in diatoms is cell division.
2. During cell division, the two halves get separated after the nuclear division. Each daughter cell gets only one valve from the parent. It has to develop another valve by itself. The newly formed valve is always a hyoptheca. This is irrespective of the fact, whether it gets a hypotheca or epitheca from the parent cell.

Auxospores:

1. Auxospores are elongated spores formed in the diatom cell to bring about reproduction.
2. Auxosopres are both asexual as well as sexual. They are formed by any one of the following five ways.
   (i) Two cells conjugate to form a single auxospore.
   (ii) Two cells conjugate to form two auxospores
   (iii) Two cells being enveloped into a common envelop, but each giving rise to an auxospores without conjugation
   (iv) One cell gives rise to one auxospore.
   (v) One cell gives rise to two auxospores.

The first two methods given above are obviously sexual. The last three methods appear to be asexual.

Diagrams to be drawn

Systematic position and Identification

A. 1. Chlorophyll present
2. Plant body a thallus
3. Cell wall cellulosic ... *Algae*

B. 1. Plant body unicellular
2. Cell is made up of two halves
3. Silica is deposited on the cell wall ... *Bacillariophyceae*

C. 1. Cells bilaterally symmetrical
2. Striations arranged in a pennate fashion

3. Raphe present ... *Pennales*

D. 1. Raphe well developed

2. Thallus boat shaped .. *Naviculaceae*

E. 1. Cell bilaterally symmetrical

2. Two laminate chromatophores present per cell

3. Costae in many parallel rows ... *Pinnularia*

CHAPTER 8

PHAEOPHYCEAE

(Phaeophyta Brown algae)

The members of Phaeophyceae are commonly called "Brown Algae becuase of predominance of xanthophyll a yellowish brown pigment in the chromatophore. Of the six xanthophylls, only three are found in Phaeophyceae.

The class comprises of 195 genera and 1000 species. Of these, except three all are marine forms found mostly in cold waters than tropical seas. Two genera *Sargassum* and *Dictyota* however, are abundant even in warm waters. Many of them are sub-littorial in distribution found at a depth of 20-90 meters. They are normally attached to rocks and other objects by hold-fasts. In *Fucus*, the hold fast is discoid but not like in *Laminuira*. Many of them grow as epiphytes or as endophytes. In some as in *Sargassum* the plant body may get detached and float in large masses. Sometimes, they are found in such abundance in the sea, as in the sea waters between the coast of Florida and West Indies, to give the name *Sargasso sea*. Here, they form large floating masses covering an area of one quarter million square miles.

Range of vegetative Construction:

The plant body is Multicellular. It is filamentous and branched as in *Ectocarpus* or a thin flat thallus as it *Dictyota*. It may be also a tough leathery thallus as in *Fucus*. In *Ectocarpus*, the plant body exhibits heterotrichous arrangement with erect and prostrate systems. Heterotrichous organisation is absent in higher phaeophyceae.

The largest forms of brown algae are called *kelps* or *rock weeds*, and they range from one meter to several meters. One of the giant kelps *Macrocystis pyrifera* is as long as 90 meters.

Cell structure:

The outer most layer is the cell wall having an inner cellulose and an outer gelatinous portion. Calcium is reported in the middle layer of the wall, whereas the outer layer has compounds like algin, fucosin etc. Algin chemically is calcium salt of alginic acid either whole or in part. The cytoplasm has nucleus and chromatophores. Nuclear division is Mitotic.

Anatomy of thallus :

Simple Phaeophyceae like *Ectocarpus* have only one layer of cells. Unlike *Fucus, Laminaria* etc. which show differentiation into epidermis, cortex and medulla. The cells of medulla help in conduction.

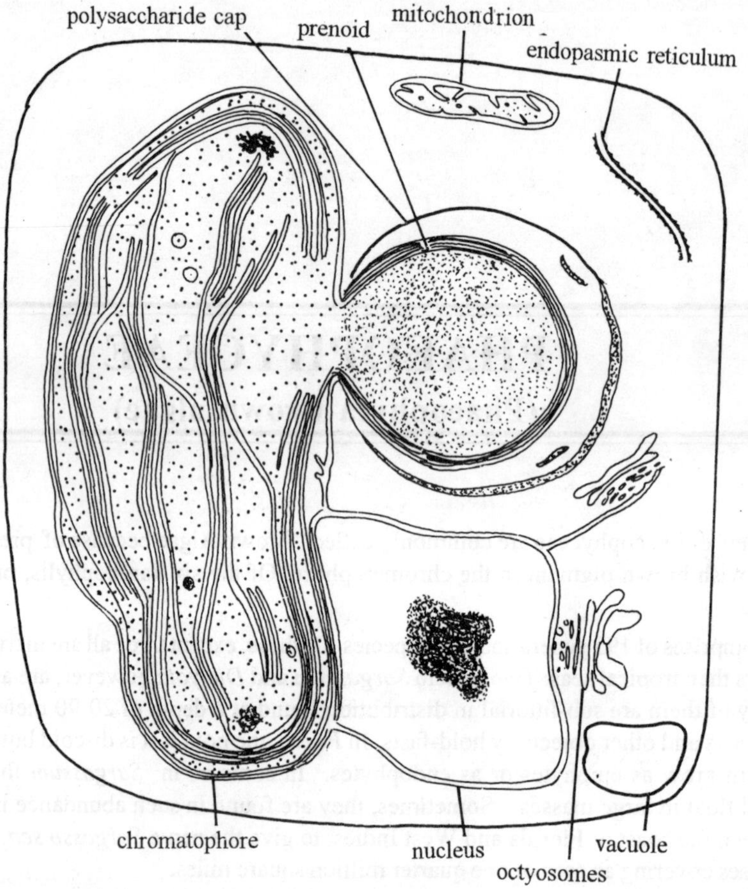

Fig.8.1 Generalised structure of Phaeophycean cell

Pigments: Chromatophores are discoid usually. They have in addition to chlorophyll (less of chlorophyll), carotene and xanthophyll. Besides these, additional brown pigments, fucoxanthin and phycophein are also present. Fucoxanthin is a carotenoid pigment (C40 H45 O6 Willstater and Pate 1914) consisting of Fucoxanthin A and B. In addition to these, others like violoxanthin, flavoxanthin, Neoxanthin, Neofucoxanthin A and B are also present. Pyrenoids are absent, though pyrenoid like bodies are reported in *Ectocarpus siliculosis*.

Product of assimilation is not starch but complex carbohydrates like Laminarin, Marnital and some fat like substance. In some like *Laminaria*, iodine is present as important reserve.

Reproduction:

Vegetative propagation is by fragmentation. Asexual reproduction is by motile or non-motile spores and they are formed in sporangia like unilocular sporangia or multilocualr sporangia. The zoospores formed in the former are haploid as they are formed after Meiosis. In the latter, zoospores are diploid. Consequently, they develop into sporophytes only. In some Phaeophyceae, like *Dictyota* non-motile spores called tetraspores are formed. Tetraspores are haploid as they are formed after reduction division.

Sexual reproduction

Is isogamous, anisogamous or oogamous.

Alteration of generations: Many Phaeophyceae exhibit isomorphic alternation of generations.

Classification: Earlier phycologists divided phaeophyceae based on structure and methods of reproduction into 2 orders.

1. *Phaeosporales*: Presence of zoospores, sexual reproduction is anisogamous or oogamous.
2. *Cyclosporales:* Zoospores absent, sexual reproduction is anisogamous or oogamous.

Kyalin (1933) divided Phaeophyceae on the basis of life cycle pattern into 3 sub classes.

(1) *Isogeneratae*: Shows isomorphic alteration of generations. It is divided into 5 orders of which order Ectocarpales is one.
(2) *Cyclosporae*: Shows heteromorphic alternation of generations. It is divided into 6 orders.
(3) *Heterogeneratae:* No alternation of generations. Has only one order – Fucales.

Fritsch divides the class Phaeophyceae into 9 orders namely:

1. Ectocarpales
2. Tilopteridales
3. Culteriales
4. Sporochnales
5. Desmarestiales
6. Laminariales
7. Sphacelariales
8. Dictyotales and
9. Fucales

In this book, *Ectocarpus, Laminaria, Dictyota, Sargassum and Fucus* are discussed.

ECTOCARPUS
(Lyngbye 1819)

Common species: *E. coniferus, E. fariculatus, E. secundus, etc.*

Distribution: Marine, distributed world wide but frequent in temperate oceans. Grows also as epiphyte on Fucales.

Study: 1. Structure of the plant
2. Reproduction
3. Systematic position and Identification

Alternation of generations:

Ecotocarpus exhibits isomorphic alternation of generations, in which the sporophyte, and gametophyte are similar morphologically. The difference between the two becomes obvious only during reproduction.

Structure of the plant (sporophyte)

1. The thallus is made up of thin, brownish, profusedly or sparingly branched filaments.
2. The thallus is differentiated into an attached prostrate system and an erect system.
3. The cells are joined end to end in an uniserriate manner.
4. The postrate system is profusedly branched.
5. The erect system is a branched tuft having the ultimate branches attenuated into an acute apex.
6. In some species, the older portion becomes corticated by a layer of descending rhizoidal branches.
7. Each filament is several centimeters long and monosiphonous. Cells are attached end to end.
8. The filaments are branched, the branching is lateral and the branches lie below the cross walls.
9. Due to evection, sometimes branching may appear dichotomous.
10. In some species branches taper and terminate in uniserriate hairs.

11. Growth in length of the prostrate system is apical.
12. In the vertical system growth is intercalary and is brought about by short meristematic vertical branches.

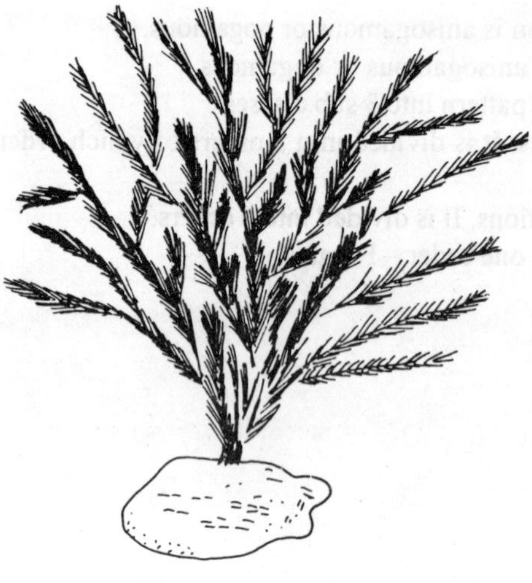

Fig.8.2 Ectocaprus, Habit

Cell structure:

1. Cell wall consists of an outer gelatinous part and an inner cellulose part. Algin is also present on the outer layer.
2. There is a central vacuole, traversed by thin cytoplasmic threads.
3. The cell has a single nucleus and many discoid chromatophores. Sometimes the chromatophores may be band shaped..

Reproduction:

1. The life cycle exhibits isomorphic alternation of generations.
2. The sporophyte reproduces asexually, while the gametophyte reproduces sexually.
3. Asexual reproduction is brought about by means of zoospores.
4. Zoospores produced in plurilocular sporangia are diploid, while zoospores produced in unilocular sporangia are haploid.

Plurilocular Sporangia:

1. These develop on the sporophyte at the end of lateral branches.
2. The sporangia are elongated cone like structures and are made up of a number of cubical cells arranged in several vertical rows.
3. Each cubical cell develops into a biflagellate zoospore. This zoospore is diploid as it develops without a reduction division on the sporophyte.
4. The zoospores develop back into sporophyte.
5. The zoospores are called neutral spores and the sporangium neutral sporangium.

Fig.8.3 Ectocaprus - A portion of thallus

Unilocular Sporangia:

1. These are also stalked, globose or pear shaped structures produced laterally on the lateral branches of the sporophyte.
2. The sporangium has many diploid nuclei, which undergo reduction division to produce a large number of haploid nuclei.
3. Each haploid nucleus develops into a biflagellate haploid zoospore.
4. The haploid zoospores develop into gametophyte

Gametophyte:

1. The gametophyte resembles the sporophyte in all respects except for the reproductive organs.

Reproduction:

1. The gametophyte reproduces sexually.

Sexual reproduction is isogamous.

2. Gametes are formed in plurilocular gametangia, which look exactly like plurilocular sporangia.
3. Plurilocular gametangia produce a number of haploid biflagellate gametes.
4. Gametes fuse to form a diploid zygote which develops into sporophyte.

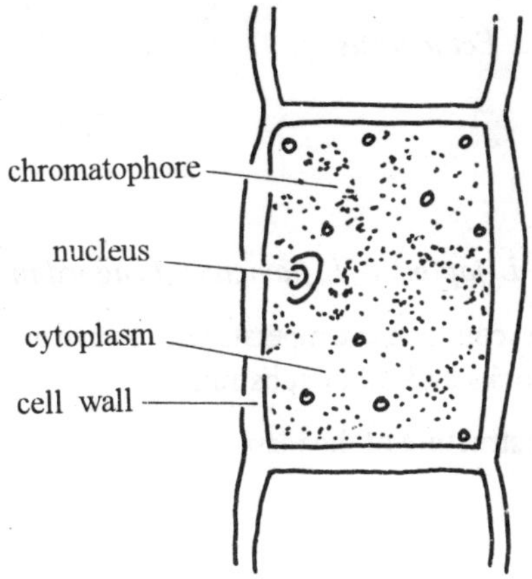

Fig.8.4 Single cell enlarged

Diagrams to be drawn:

1. Habit sketch
2. A part of the plant to show branching
3. Plurilocular sporangia
4. Unilocular sporangia
5. Plurilocular gametangia

Systematic position and Identification

A. 1. Chlorophyll present
2. Plant body a thallus
3. Cell wall cellulosic ...*Algae*

B. 1. Chromatophores brown in colour
2. Reserve food material is mannitol and lamenarin
3. Flagella attached on the sides in reproductive cells ... *Phaeophyceae*

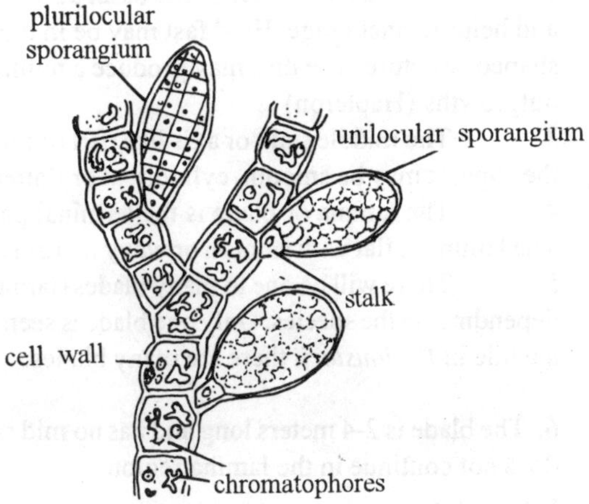

A portion of thallus showing unilocular and plurilocular Sporangia

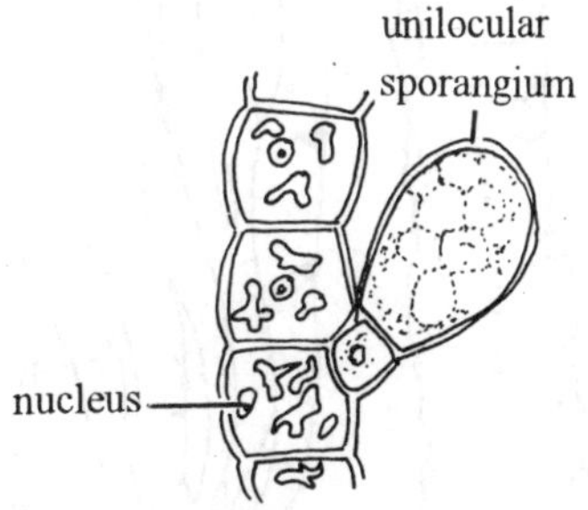

A filament with plurilocular sporangium

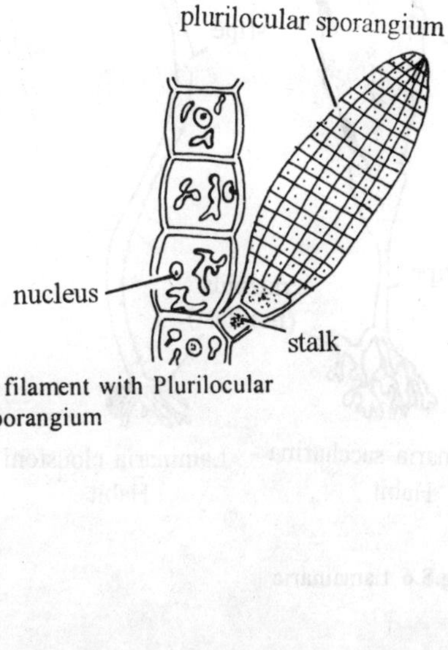

A filament with Plurilocular Sporangium

Fig.8.5 Ectocarpus

C. 1. Plant body filamentous
2. Growth trichothallic

3. Life cycle exhibits isomorphic alternation of generations... *Ectocarpales*

D. 1. Thallus monosiphonous

2. Habit heterophores...*Ectocarpaceae*

E. 1. Chromatophores are discoid or band shaped

2. Sporangia plurilocular or unilocular

3. Pyrenoid like structures present in the chromatophores ... *Ectocarpus*

LAMINARIA
(Lamouroux, 1813)

Common species: *L. japonica, L. flavicans, L. agardhii, etc.*

Distribution: Marine, common in temperate seas. Found in arctic zones, attached to rocks and other substrata.

Study: 1. External structure of the plant

2. Internal structure

3. Reproduction

4. Systematic position and Identification

External structure:

1. The sporophytic plant body is distinguishable into three parts, namely, hold fast, stipe and blade. Altogether the plant reaches a length of 12-15 meters.

2. Hold fast is present at the basal part of the plant body and helps in anchorage. Hold fast may be in the form of a disc shaped structure. The disc may produce a number of rhizoidal outgrowths (Hapteron)

3. The middle stalk or axis like part of the plant is called the stipe. It may be smooth, cylindrical or flattened.

4. The lamina or blade is the terminal part of the plant. The lamina is flat with smooth or wavy margins

5. There will be one to many blades (lamina) on the stipe depending on the species. Only one blade is seen in *L.saccharin* a while in *L. cloustorii* there are many blades.

6. The blade is 2-4 meters long and has no mid rib i.e., the stipe does not continue in the lamina region.

Internal structure:

The internal structure of the stipe can be studied by having a cross as well as a long section.

Long section of the stipe:

1. A long section shows 3 regions viz, epidermis, cortex, and central medulla.

2. Epidermis has barrel shaped cells with chromatophores. These cells are meristematic and hence epidermis is also known as *Meristoderm*. A cuticle is found bordering the epidermis.

3. The cortex has vertically elongated cells. It can be distin-

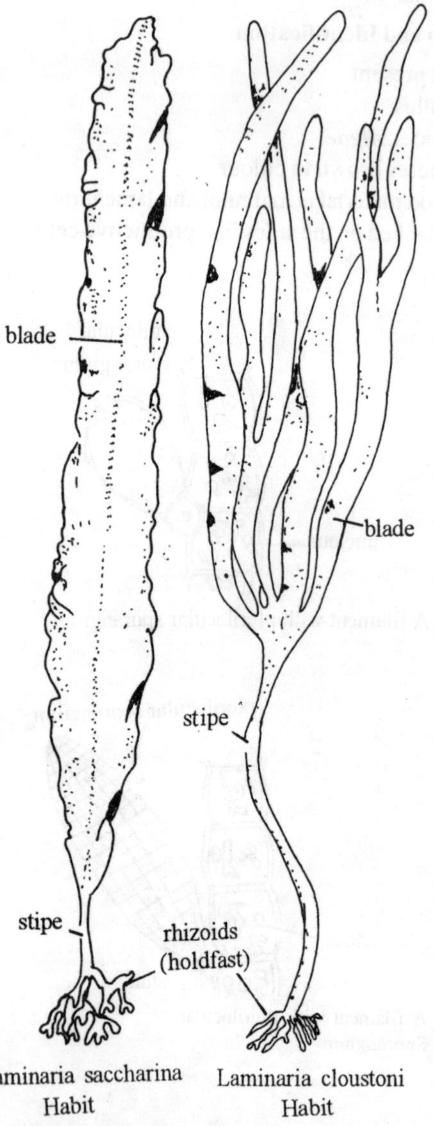

Fig.8.6 Lamainaria

guished into outer cortex and inner cortex.

4. In the medulla regions are found loosely arranged cells.

Cross section of the stipe:

1. In the cross section also, three regions viz., epidermis, cortex and medulla are seen.
2. Epidermis is similar to what is seen in long section . Cells are small and cubical.
3. Internal to the epidermis is the cortex. This is divided into outer cortex and inner cortex.
4. Outer cortex has elongated cells, pale yellow in colour having pointed ends.
5. The cells of the inner cortex are comparatively broad and have thick walls. The cells of the inner cortex develop certain tubular outgrowths towards the medulla region.
6. Mucilage canals are found in the cortex. They are lined by secretory cells.
7. The central region called medulla, is composed of loosely arranged elongated cells or *hyphae*.
8. These cells are called *trumpet hyphae* owing to their expanded end (looking like a trumpet). The trumpet hyphae have a perforated plate much like the seive plates of vascular plants.

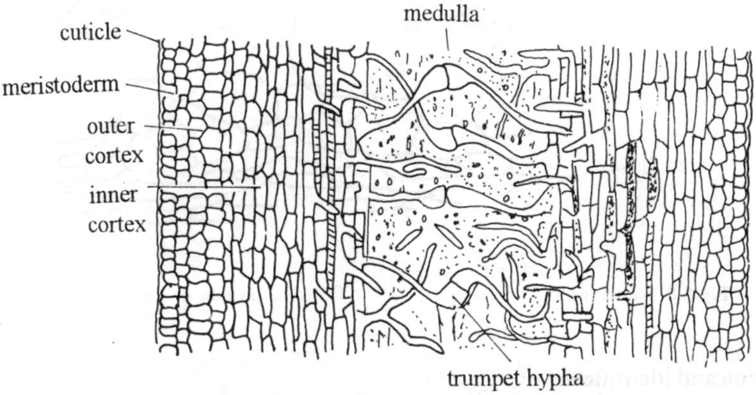

T.S. of a portion of stipe

Fig.8.7 Lamainaria

Lamina

1. Anatomically the lamina resembles the blade. There is an epidermis, cortex and central medulla.
2. Epidermis is the outermost layer. Cells are small and have many chromatophores. The epidermis is covered by a gelatinous layer.
3. The cells of the cortex are isodiametric in shape and those present near the epidermis posses chromatophores. Many mucilage cavities are found in the cortex.
4. Medulla is composed of vertical hyphae like cells arranged in the form of a reticulum.

Reproduction:

1. The sporophyte reproduces asexually by the development of unilocular sporangia. The unilocualr sporangia produce zoospores after reduction division.
2. The unilocular sporangia are arranged in clusters called sori on either surface of the lamina. Between the sporangia are present many hair like processes called paraphyris.
3. The cells in sporangia undergo reduction division to produce a large number of haploid zoospores. These zoospores develop into gametophytes.

Sexual Reproduction:

1. The gametophytes are small and filamentous. They are branched. The male and female gametophytes are

different.

2. Sexual reproduction is Oogamous.
3. Antheridia are produced at the tips of lateral branches.
4. Oogonia are small globose structures.
5. The male and female gametophytes are highly reduced structures.

Diagrams to be drawn:

1. Habit sketch
2. T.S. and L.S. of stipe
3. T.S. of lamina

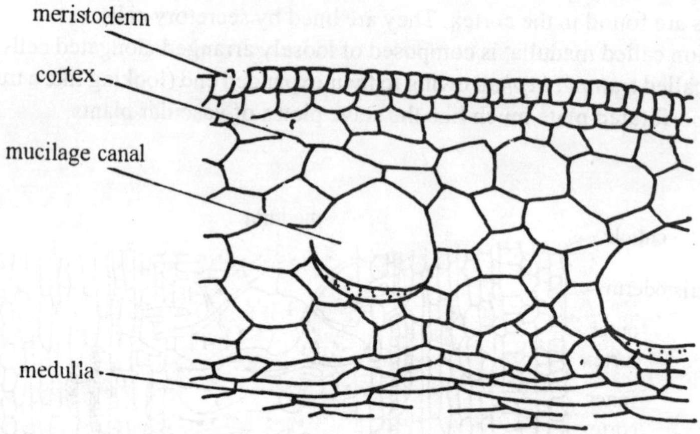

Fig.8.8 Lamainaria T.S. of a portion of blade

Systematic position and identification:

A. 1. Chlorophyll present
   2. Plant body is a thallus

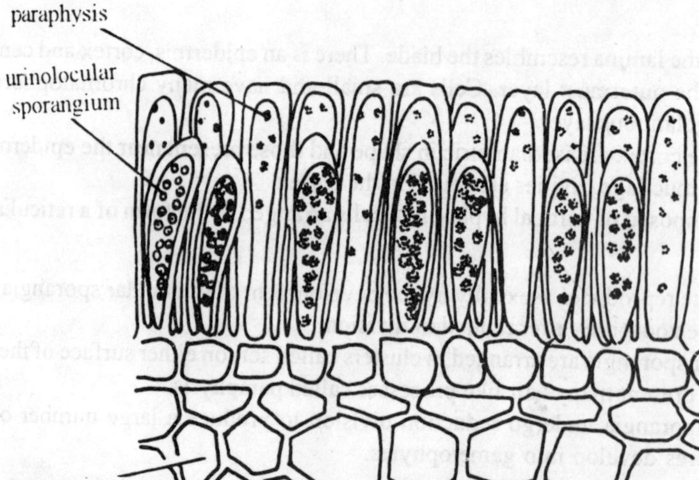

Fig.8.9 Lamainaria T.S. through a part of corus

3. Cell wall cellulosic ... *Algae*

B. 1. Chromatophores brown in colour.
2. Reserve food material is mannitol and lamenarin
3. Flagella lateral in position ... *Phaeophyceae*

C. 1. Alternation of generations heteromorphic
2. Gametophytic stage is telescoped
3. Sporangia aggregate into sori ...*Laminariales*

D. 1. Plant body divided into lamina, stipe and hlodfast.
2. Paraphyses present in between sporangia
3. Sori present on either surface of lamina ... *Laminariaceae*

E. 1. Blade without longitudinal ribs ...... *Laminaria.*

DICTYOTA
(Lamouroux, 1809)

Common species: *D. dichotoma.*

Distribution: Marine alga, found widely distributed in tropical seas, growing attached to rocks in the intertidal zones.

Study: 1. External morphology
2. Internal structure
3. Reproduction
4. Systematic position and identification

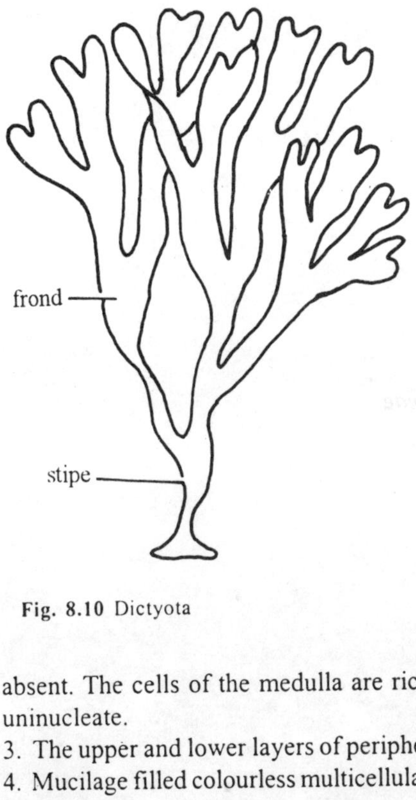

Fig. 8.10 Dictyota

External morphology:

1. The plant body is a flat ribbon like thallus, found attached to rocks or other substrata with the help of a disc like hold-fast.
2. The thallus is dichotomously branched
3. The hold-fast that helps in anchorage is disc shaped. Occasionally however it may be divided irregularly.
4. The thallus consists of a cylindrical stalk and an extensively branched top portion
5. The growth of the plant is brought about by a dome shaped apical cell. At the time of branching, the apical cell vertically divided into two and each daughter cell gives rise to the branches.
6. The cells of the thallus found below the apical cell are arranged in a radial manner.

Internal structure of the thallus :

1. A cross section of the thallus reveals three layers of cells viz., two layers of small cells and a middle layer of large cells.
2. The central layer of large cells also called medulla, has thin walled cells. The cells have a large vacuole. They have but few chromatophores. Sometimes the chromatophores may be entirely absent. The cells of the medulla are rich in fucosan vesicles and many refractive granules. The cells are uninucleate.
3. The upper and lower layers of peripheral cells are smaller in size and have many chromatophores.
4. Mucilage filled colourless multicellular hairs are present on both theupper and lower faces of the epidermis.

Reproduction:

1. Asexual reproduction takes place by means of tetrasporangia.
2. The tetrasporangia are spherical cells present on either surface of the thallus. They may occur singly or in clusters called sori.
3. The tetrasporangia are diploid as they are formed on the sporophytic plant. They undergo reduction division to form four tetrahedrally arranged tetraspores.
4. The tetraspores develop into gametophytes.

Sexual reproduction:

1. Sexual reproduction is oogamous. The gametophytes are dioecious. The sex organs namely antheridia and oogonia are borne on both the surfaces of the thallus.
2. Both antheridia and oogonia occur in clusters called sori.

Antheridia :

1. The antheridial sori are produced on either surface of the thallus. Each sorus has 100-300 antheridia.
2. The sorus on all sides is covered by long sterile hair like cells forming a protective sheath.
3. Each antheridium is an elongated sac and has a short stalk made up of a single cell.
4. The antheridium is multinucleate. Each nucleus surrounds itself with a little cytoplasm and develops into an uniflagellate antherozoid.

Oogonium :

1. The oogonium are also clustered into sori. Each sorus has a group of 25-50 oogonia.
2. There are no sterile cells in the oogonial sorus to form an envelope.
3. Oogonia also have a single celled stalk, sac like in appearance and have a single egg each.
4. Fertilization results in the formation of egg. Egg and antherozoids meet outside. Fusion is external.

Diagrams to be drawn

1. Habit sketch
2. T.S of the thallus
3. Surface view of tetrasporangia
4. T.S of thallus showing tetrasporangia
5. T.S. of thallus through an antheridial sorus
6. T.S. of thallus through an oogonial sorus

Systematic position and identification

A. 1. Chlorophyll present
2. Plant body is a thallus
3. Cell wall cellulosic ... *Algae*

B. 1. Chromatophores brown in colour.
2. Reserve food material is mannitol and Lamenarin ... *Phaeophyceae*

C. 1. Thallus flat and dichotomously branched.
2. Asexual reproduction by non-motile spores.
3. Growth is apical ... *Dictyotales*

D. 1. Sex organs aggregated into sori.
2. Sperms flagellate with non motile egg.
3. Fertilization external ... *Dictyotaceae.*

E. 1. Thallus dichotomously branched.
2. No midrib in the thallus.
3. Central medulla is single layered ... *Dictyota.*

SARGASSUM
(Agardh, 1821)

Common species : *S.cinerium*

Distribution : All the species are confined to tropical seas. Found attached to rocks in intertidal zones. Sometimes may from large floating masses extending over several miles.

Study : 1. Morphology of the thallus.
2. Anatomy of stem
3. Reproduction
4. Systematic position and identification

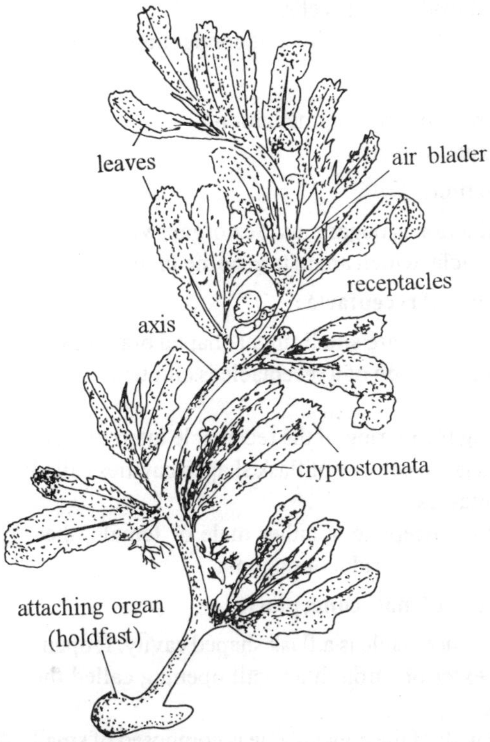

Fig. 8.11 Sargassum. Habit

Morphology of the thallus :

1. The plant body is macroscopic more of less bushy resembling an angiosperm plant.
2. The plant body is mostly attached to rocks and other substrata, sometimes found free floating also.
3. The plant has a central axia or stem which is attached in the young condition to the substratum with the help of hold fast.
4. The axis is differentiated into nodes and internodes.
5. The axis is branched. It bears richly branched primary laterals or long shoots of unlimited growth.
6. The main axis as well as long shoots bear a number of flat leaf like photosynthetic laterals (Branch of I order).
7. Leaves are narrow with a terete margin. They may or may not have a midrib. They help in photosynthesis.
8. The leaves are sterile and have numerous dots called *cryptostomata*. These lead internally into empty chambers called *cryptoblasts*.
9. The axis also bears a system of secondary branches (Branches of II order) which are modified into air bladders to help in buoyancy.
10. In the axils of leaves there is a tertiary branching (Branches of III order) consisting of numerous finger shaped projections. These are called receptacles and are fertile.
11. The receptacles internally have conceptacles bearing reproductive organs.

Anatomy of stem

A cross section of the thallus shows three regions viz., epidermis or meristoderm, cortex and medulla.

1. Epidermis is the outermost layer. This is often invested with a mucilage covering.
2. The cells of the epidermis are columnar and have no intercellular spaces.
3. The cells are meristematic, and have large member of plastids and fucosan granules.
4. Cortex is multilayered consisting of thick walled polygonal cells.
5. The cells of the cortex help in storage, besides offering mechanical support.
6. The innermost cortical cells gradually merge with the cells of medulla and have gelatinised walls.
7. Medulla is the central zone and it has loosely arranged cells.
8. The outermost region of medulla has thick walled cells but the inner region has thin walled cells.
9. The cells of meduall help in conduction.

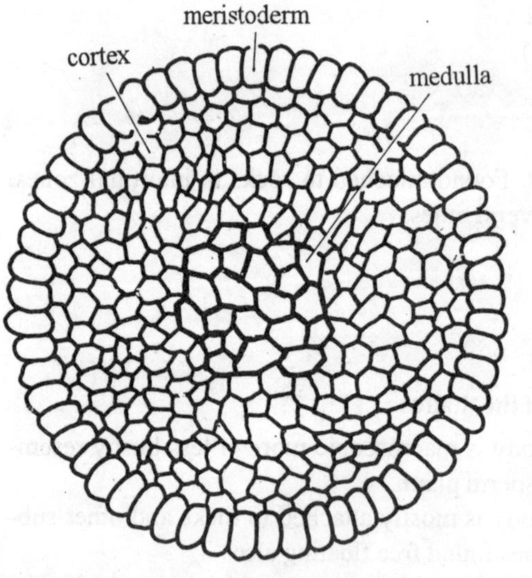

Fir. 8.12 Sargassum. T.S. of axis

Anatomy of leaf

1. The leaf in its internal structure resembles the stem.
2. The outline is elliptical and has a number of pores leading to cryptostomata.
3. Outermost layer is epidermis, made up columnar cells which are closely packed. There is a mucilage envelope for the epidermis.
4. Internal to the epidermis is a layer of cells rich in chromatophores. This is the chief photosynthetic zone of the leaf. This is the assimilating tissue.
5. Next to the assimilatory tissue is the storage zone consisting of thin walled isodiametric cells.
6. In the middle of the leaf is a zone of conducting tissue.

Asexual reproduction

1. This is brought about by fragmentation.
2. No zoospores are formed.

Sexual reproduction

1. Sexual reproduction is oogamous. Sex organs viz oogonia and antheridia are borne in flask shaped structures called conceptacle, which are in turn borne in receptacles.

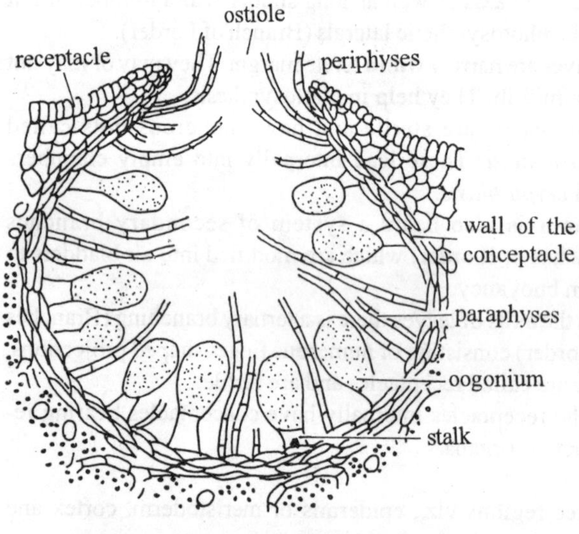

Fig. 8.13 Sargassum, Female conceptacle enlarged.

Structure of receptacle :

1. Receptacles are small finger shaped branches.
2. A cross section of a receptacle is elliptical in outline.
3. Lining the maring of the receptacle internally are found a large number of flask shaped cavities called conceptacles.
4. Each conceptale is either male or female, they bear either antheridia or oogonia.

Structure of male conceptacle :

1. The conceptacle is a flask shaped cavity. It opens to the exterior through a small opening called the ostiole.
2. The wall of the conceptacle is composed of small flat cells having many chromatophores.
3. From the floor of the conceptacle arise many hair like structures called paraphyses.
4. Near the ostiole also, hair like structures are found. These are called periphyses.
5. The antheridia are borne on some of the branched paraphyses.
6. The antheridia are small globose structures having a two layered wall.
7. About 64 antherozoids are produced inside the antheridium. The antherozoids are biflagellate.

Structure of female conceptacle :

1. Female conceptacles are similar in structure to the male conceptacles, except that they posses oogonia.
2. Oogonia do not develop on the paraphyses. They arise directly on the wall of the conceptacle.
3. The oogonium has a single celled stalk which is embedded in the wall of the conceptacle.

4. The oogonium has a 3 layered wall.
5. A young oogonium has eight nuclei, but seven of them abort, so that a mature oogonium has a single egg.

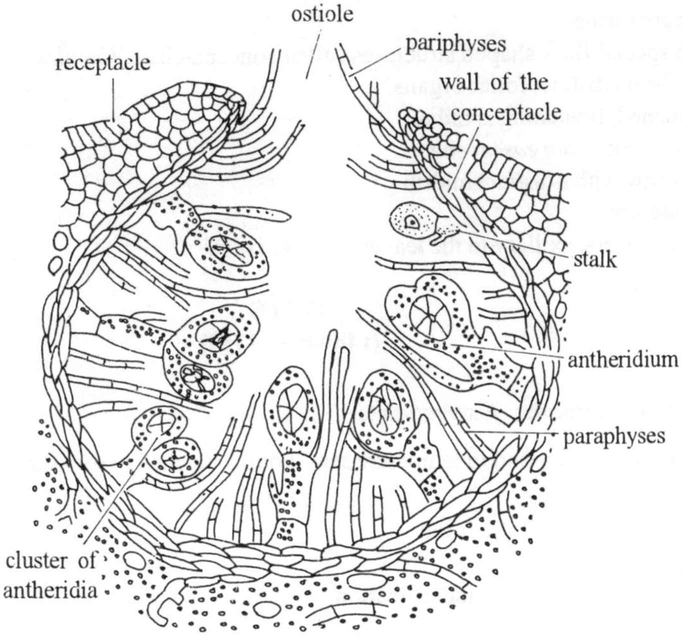

Fig. 8.14 Sargassum, Male conceptacle enlarged.

Structure of a sterile conceptacle:

1. There are conceptacles that do not bear any sex organs.
2. Numerous paraphyses arise from the floor of the conceptacle.
3. Near the ostiole are hairs called periphyses whose cells are rich in chromatophores.

Fertilization:

1. The fusion of the gametes results in the formation of zygote which directly develops into a new plant of *Saragssum*.

Diagrams to be drawn

1. Habit sketch
2. T.S. of stem
3. T. S. of leaf
4. Ground plan of T.S.of Receptacle
5. A sterile conceptacle enlarged
6. A male conceptacle enlarged
7. An antheridial branch enlarged
8. A female conceptacle enlarged
9. An oogonium enlarged

Systematic position and identification

A. 1. Chlorophyll is present
2. Plant body is a thallus
3. Cell wall cellulosic ... *Algae*

B. 1. Chromatophores brown in colour.
2. Reserve food material is mannitol and Laminarin ... *Phaeophyceae*
C. 1. Plants have a complex morphological structure.
2. They show a high degree of anatomical differentiation.
3. No zoospores are formed.
4. Sex organs in special flask shaped structures called conceptacles... *Fucales*
D. 1. Plant body bears distinct foliar organs.
2. Thallus is branched, branching is radial.
3. Air bladders present ... *Sargassaceae.*
E. 1. 'Leaves' narrow with a distinct midrib
2. Air bladders present
3. Receptacles in clusters, axillary to the leaves ... *Sargassum.*

FUCUS
(Linnaeus, 1753)

Common species : *F.vesiculosus, F.serratus* etc.

Distribution : Marine alga, found in temperate seas. Found attached to rocks in intertidal zones.

Study : 1. External morphology
2. Internal structure of the frond
3. Reproduction
4. Systematic position and identification

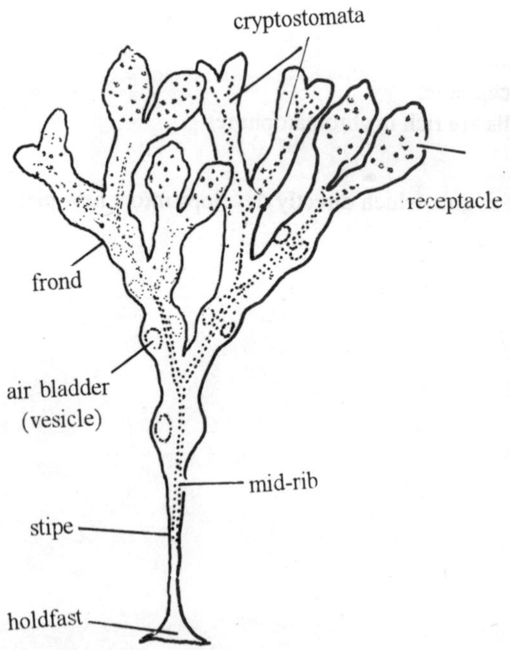

Fig. 8.15 Fucus External features

External morphology

1. The plant body is a flattened dichotomously brnached ribbon like thallus.
2. The dark brown thallus is leathery in texture and is anchored to the substratum with the help of hold fast.
3. The plant body is divided into three parts viz., hold-fast, stipe and frond.
4. The stipe is a stalk like structure arising from hold fast. It continues in the region of the frond in the form of a midrib.
5. The terminal flattened ribbon like portion of the plant body is called the frond.
6. The frond is dichotomously branched and the surface is slippery due to the presence of gelatinous material.
7. The frond has a midrib, along the line of which in some speices are found air bladders. These help in buoyancy.
8. The margin of the frond may be smooth or serrate.
9. The growth is brought about by an apical cell situated in a furrow.
10. Along the length of the frond on the surface are found numerous, small dot like structures called Cryptostomata or sterile conceptacles.
11. Mature fronds, at their tips bear swollen fertile structures called receptacles. These receptacles contain many

flask shaped cavities called conceptacles.

Anatomy of frond :

1. A cross section of the frond shows three regions viz., meristoderm, cortex and medula.
2. Meristoderm or epidermis consists of meristematic cells. It is made up of a single layer of closely packed cells.
3. Often the meristoderm is enveloped by a cuticle.
4. The cells of the meristoderm have numerous chromatophores.
5. Cortex consists of large, thin walled cells. The cells become broader towards the centre of the frond. They have but few chromatophores.
6. The central region called medulla, consists of loosely arranged cells with intercellular spaces filled by mucilage.
7. The cells of medulla help in conduction.

Anatomy of midrib (Stipe)

1. In its internal structure, the stipe resembles the frond. It has an outer meristoderm, middle cortex and inner most medulla.
2. Meristoderm and cortex are similar in structure to that of the frond.
3. The medulla consists of long irregular cells called hyphae which offer mechanical support.

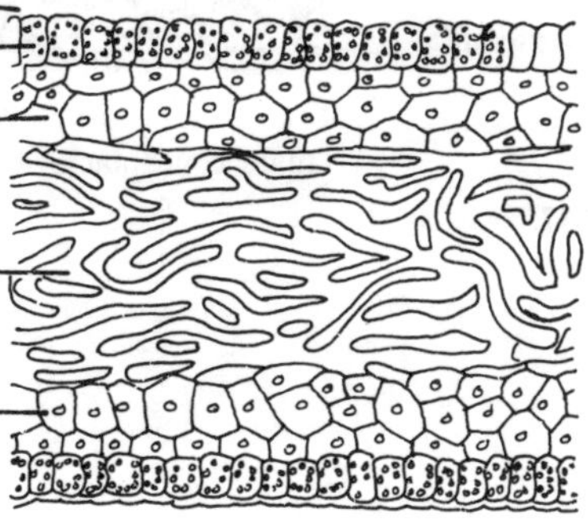

Fig. 8.16 Fucus T.S. of wing (a Portion)

Reproduction

1. Vegetative propagation is brought about by means of fragmentation.
2. Sexual reproduction is oogamous brought about by oogonia and antheridia.
3. The sex organs are bone in flask shaped cavities called conceptacles which in turn are borne in the terminal receptacles.
4. Each receptacle in a transverse section shows a number of conceptacles lining the maring internally.
5. Species of *Fucus* may be monoecious or dioecious.
6. Conceptacles also may be bisexual (bear both sex organs) or unisexual.

Structure of the male conceptacle

1. The conceptacle is a flask shaped structure. It has an inner basal layer of fertile cells.
2. Many paraphyses (sterile hairs) arise from the floor of the conceptacle. These are multicellular and may be branched or unbranched.
3. The conceptacle opens to the exterior through a small opening called the ostiole.
4. The antheridia arise internally on short or long filaments which terminate in a paraphysis.
5. The antheridia are small, pyriform and stalked.
6. The antheridia have a two layered wall viz the outer Exochite and inner Endochite.
7. Each antheridium produces many biflagellate uninucleate antherozoids.

Structure of the female conceptacle

1. The conceptacle is similar in structure to the male except that it bears oogonia.
2. Many oogonia are found in each conceptacle arising directly from the wall of the conceptacle.
3. The female conceptacle has paraphyses, ostiole etc. Just like the male conceptacle.
4. Oogonia have a short stalk. They arise singly and are globose in shape.
5. The wall of the oogonium is very thick. It has eight functional eggs or oospheres.
6. At maturity, the eggs come out of the conceptacle through the ostiole. However they remian attached to the

conceptacle base.

7. Fusion of the gametes results in a zygote which develops into a new individual.

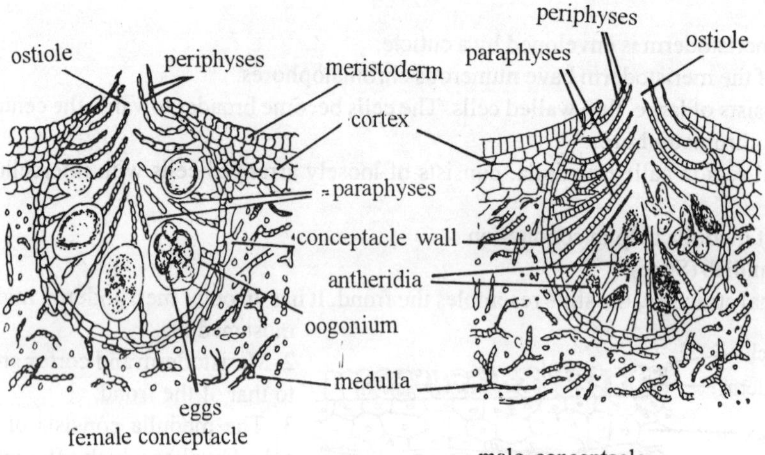

Fig. 8.17 Fucus

s

Diagrams to be drawn :

1. Habit sketch
2. Anatomy of frond
3. T.S. of receptacle (ground plan)
4. Male and female conceptacles enlarged.

Systematic position and identification

A. 1. Chlorophyll present
2. Plant body is a thallus
3. Cell wall cellulosic ... *Algae*

B. 1. Chromatophores brown in colour.
2. Reserve food is mannitol and Lmainaria... *Phaeophyceae*

C. 1. Plants have a complex morphological structure.
2. They show a high degree of internal tissue variety.
3. No zoosopres are formed.
4. Sex organs in conceptacles ... *Fucales*

D. 1. Frond has a distinct midrib
2. Air bladder intercalary (when present)
3. Oogonia have 8 functional eggs ....*Fucaceae*

E. 1. Plants have a definite hold fast
2. Frond leathery dark brown with prominent midrib.
3. Branching is dichotomous
4. Conceptacles unisexual or bisexual ... *Fucus*

CHAPTER 9

RHODOPHYTA – Red algae

The red Algae are primarily distinguished from other Algae, by their sexual reproduction in which, aflagellated male gametes (spermatia) are passively transported to the female sex organ, carpogonium. Plastids of red algae contain several pigments among which *r* phycoerythrin is usually present in large quantities to mask pigments and give red colour.

Occurrence

There are about 400 genera and 2,500 species. About 50 species belonging to about a dozen or more genera are fresh water, where they are restricted to rapidly flowing streams and falls.

A majority of red Algae are marine in distribution, the members may be attached or occasionally free floating. They are widely distributed in all global oceans including Arctic and Antarctic. They grow attached to rocks and other substrate. Some may be epiphytic.

Morphology of the thallus

Except for two genera, all others are multicellular. The thallus is fundamentally filamentous. The thalli are macroscopic reaching a size of 30 cms. The filaments are branched, may be monosiphonous or polysiphonous.

Pigments: The red algae owe their name to the red colour. But, some fresh water forms like *Batrachospermum* are green. They possess phycoerythrin (red) and phycocyanin (blue-green) in addition to chlorophyll. Phycoerythrin is in abundance so as to mask chlorophyll. Phycoerythrin is absent in fresh water forms like *Batrachospermum*.

Cell structure

Generally cells are uninucleate but some are multinucleate having 3000 to 4000 nuclei in a single cell. Except in a few there is a central vacuole in the majority of species with cytoplasm forming a thin layer internal to cell Wall. The cell wall is made up of cellulose and various other pectic compounds (Kylin 1943). Cells of primitive red algae have a single stellate chromatophore having a pyrenoid. These however lack the starch sheath found in chlorophyceae. In advanced members, chromatophores are discoid. Chromatophores have carotenes, xanthophylls, red pigments, blue green pigments in addition to chlorophyll. The product of assimilation is called Floridean starch. It is considered to be related to amylopectin. Many red algae accumulate soluble sugar Floridoside, a galactoside of glycerol. Cells are connected to one another, through plasmodesmata.

Fig.9.1 Batrachospermum External features (habit)

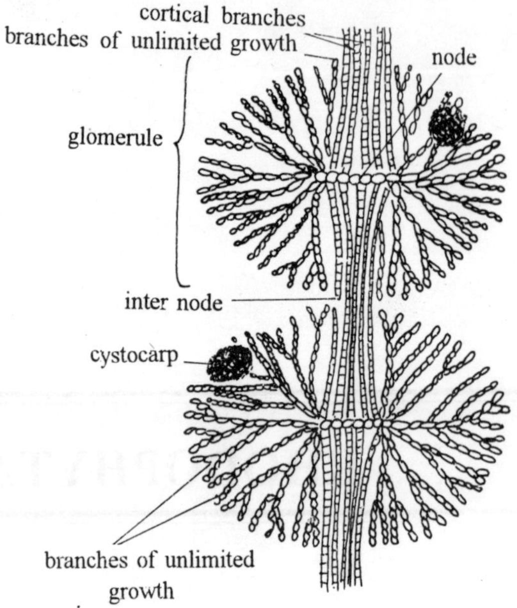

Fig.9.2 Batrachospermum. A portion of thallus showing whorl of laterals (glomerules)

Reproduction

Motile reproductive cells are entirely wanting. Asexual reproduction by a variety of spores like monospores, carpospores, tetraspores etc.

Sexual Reproduction

The male sex organ is spermatangium and the female, the flask shaped carpogonium having a long neck called trichogyne. Spermatia liberated from spermatangia fuse with the egg of the carpogonium to form zygote. The zygote may or may not undergo reduction division, and forms carpospores. In the former case, the carpospores are haploid and develop into a diploid generation in which there is reduction division producing diploid generation through tetraspores.

Majority of red algae show alternation of generations which may be biphasic or triphasic.

Classification: The class Rhodophyceae is divided into two subclasses, viz., *Bangioideae* and *Florideae*. In Bangioideae forms are simple and carpospores are haploid. There is only one order Bangiales. In Florideae, the plant body is highly complex and carpospores are diploid. Members are both fresh water and marine. This sub class includes six orders, viz., Nemalionales, Gelidiales, Cryptomeniales, Gigartinales, Rhodomenales, and Ceramiales. Rhodomenales, Gigartinales and Ceramiales are discussed with an example for each order.

BATRACHOSPERMUM
(Roth 1797)

Common species: *B. moniliforme, B. densum, B. vagum, etc.*

Distribution: Found in fresh water streams and rivulets. Usually found attached to substrata like stones and twigs.

Study: 1. Structure of the plant body
2. Reproduction

3. Systematic position and Identification

Structure of the plant body:

1. The plant body is a greenish or pink bushy filamentous structure.
2. It consists of a central axis divided into nodes and internodes.
3. The main axis is uniseriate (monosiphornous). It is anchored to the substratum with the help of old shoots forming a prostrate system.
4. The main axis is corticated and at the nodal region it bears branches.
5. Branches are lateral and are of 2 types, viz., branches of unlimited growth and branches of limited growth.
6. The branches of limited growth resemble the main axis in all respects and in turn are divided into nodes and internodes.
7. The branches of limited growth are also known as *glomerules* are short and 10-15 cells long. They are made up of bead like cells.
8. The branches of limited growth are found both on the main axis as well as on the branches of unlimited growth at the nodal regions.
9. The glomerules arise in large numbers at the nodal regions giving a bushy appearance to the plant. Occasionally however they may be also found sparsely along the internodal region.
10. The cells are rectangular in shape, uninucleate and have a wall made up of cellulose.
11. There are many discoid chromatophores in each cell. Each chromatophore has a pyrenoid.

Reproduction:

1. Vegetative propagation is brought about by means of fragmentation. Each fragment can develop into a new individual.
2. Sexual reproduction is oogamous. Sex organs are *spermatangia* and *carpogonia.*
3. Species of *Batrachospermum* may be *monoecious* or *dioecious.*
4. The sex organs are usually borne at the tips of short laterals (branches of limited growth).
5. The terminal cell of a short lateral divides and forms 204 spermatangia or antheridia.
6. Each spermatangium produces an uninucleate non-motile *Spermatium* or *male cell.*
7. At maturity the spermatia are liberated from the plant. They are dispersed by water current. But they do not have any flagella.
8. The female reproductive structure, viz., the carpogonium is also produced at the tip of a short lateral.
9. The carpogonium is a unicellular flask shaped structure. It has a bulbous basal region and an elongated neck like portion called the trichogyne.
10. There is a single egg at the basal region of the carpogonium.
11. Fertilization of the egg by spermatia results in the formation of a zygote which develops into *Cystocarp* or *Carposporophyte.*

Structure of cystocarp

1. This is a dark compact structure, formed in between the glomerules.
2. The cystocarp develops from the carpogonial base and is surrounded by many vegetative branches or filaments.
3. The cystocarp is made up of a large number of filamentous structures called '*gonimoblast filaments*'. It is the accumulation of these filaments that constitutes the cystocarp.
4. The terminal cell of each gonimoblast filament develops into a carposporangium.
5. Each carposoprangium produces a single dark coloured carposore. It is an account of this, that the cystocarp is called a *Carposporophyte.*
6. The carpospores are released on maturity.
7. Each carpspore develops into a filamentous structure called *Chantransia* or Juvenile stage.

Chantranisa stage:

1. The Chantransia is produced by the generation of a carpospore.

2. Each Chantransia has a prostrate as well as a vertical system of filaments. It exhibits heterotrichous habit.
3. The Chantransia stage resembles an alga by the name *Chantransia*, hence the name.
4. Chantransia reproduces asexually by the formation of monospores.
5. Eventually some of the branches of Chantransia assume a bushy habit and develop into the typical plant body of *Batrachospermum*.

Diagrams to be drawn:

1. Habit sketch
2. A part enlarged to show branching
3. Spermatangia and carpogonia
4. Cystocarp
5. Chantransia

Systematic position and Identification:

A. 1. Chlorophyll present
   2. Plant body is a thallus
   3. Cell wall cellulosic ... *Algae*

B. 1. Motile reproductive cells absent
   2. Chromatophores red pigmented ... *Rhodophyceae*

C. 1. Plant body has monosiphonous axis.
   2. Chromatophores discoid .
   3. Juvenile stage present in the life cycle... *Nemalionales*

D. 1. Plant body is found in fresh water.
   2. Life cycle is diphasic and Haplobiontic ...*Batrachorpermaceae*

E. 1. Plant body bushy
   2. Short laterals have bead like cells
   3. Juvenile stage present in life cycle ... *Batrachospermum*

GRACILARIA
(Greville, 1830)

Common species: *G. armata, G. compressa etc.*

Distribution: Marine in distribution occuring also as a lithophyte at the intertidal zones in tropical seas.

Study:
1. Morphology of the thallus.
2. Anatomy of the thallus
3. Reproduction
4. Systematic position and classification

Morphology of the thallus:

1. The thallus is a branched, cylindrical or compressed structure made up of a number of filamentous structures.
2. The thallus also called frond is attached to the substratum with the help of a discoid hold fast. The hold fast consists of a number of closely packed thread like structures.
3. The fronds are 25-40 cms, long, dark red in colour and dichotomously branched.
4. Some of the branches of the frond have attenuated apices, terminating in hair like structures.
5. The fronds are monosiphonous, have uniserriately arranged cells. The fronds have a gelatinous envelope on their surface.

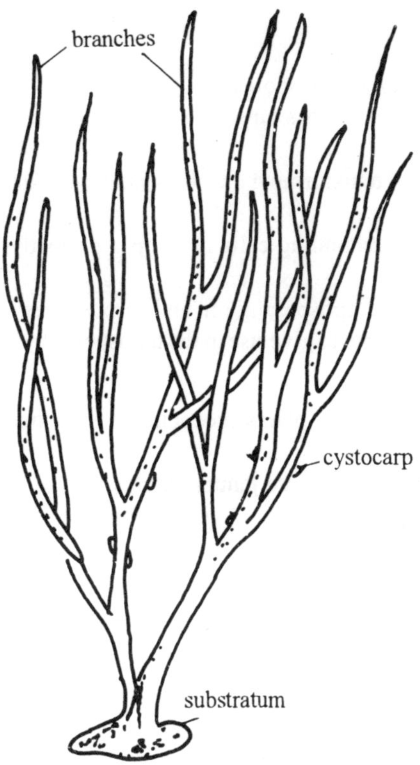

Fig.9.3 Gracilaria. Thallus bearing cystocarps.

Anatomy of the thallus:

1. A transverse section of the thallus reveals three regions, namely, peripheral layer, cortex and the medulla.
2. The cells of the peripheral layer are small uninucleate and rich in cytoplasmic contents.
3. The mature cells of the peripheral layer may be multinucleate.
4. Unicellular hairs are produced from some of the peripheral cells.
5. Internal to the peripheral layer is the cortex. It is made up of many small cells. The cells have many chromatophores.

Reproduction:

1. All the species of *Gracilaria* are dioecious.
2. The life cycle exhibits alternations of generations.
3. There are three types plants, male plants, female plants and sporophytes. All of them are similar morphologically.
4. In the male plant, the reproductive structures are spermatangia.
5. The spermatangia develop from the cortical cells. Each spermatangium produces a spermatium. The spermatia are non-motile structures. They are dispersed by water current.
6. In the female plant, the female reproductive organs, namely, the carpogonia develop from the cortical cells.
7. Some of the cortical cells distined to produce carpogonia, develop into supporting cells. Each supporting cell produces a hypobasal cell and a carpogonium proper.
8. The carpogonium is made up of a single flask shaped cell. It has a swollen base in which is situated the egg and an elongated trichogyne.
9. The fusion of the egg and spermatium results in the formation of the zygote which develops into the cystocarp or the carposporophyte.

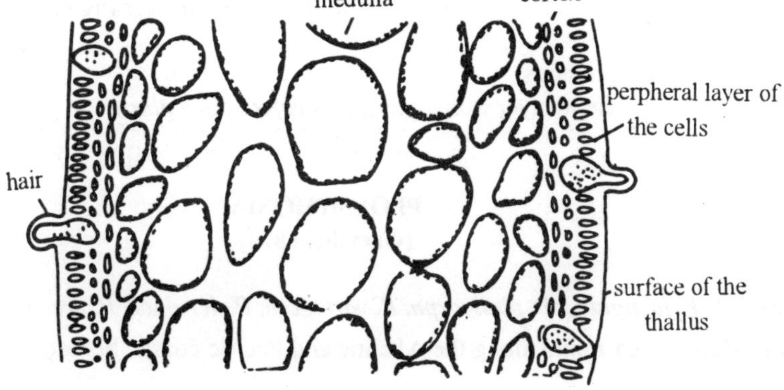

Fig.9.4 Gracilaria. T.S. thallus

Structure of the cystocarp:

1. The cystocarp is a globose structure. It has a short stalk.
2. The cystocarp has an opening called the ostiole, for the liberation of the spores.
3. At the base of the cystocarp is a placenta which gives rise to filamentous structures called the gonimoblast filaments.
4. Each gonimoblast filament is 3-4 celled long with the terminal cell transforming into a *carposporangium*.
5. Each carposporangium produces a diploid carpospore.
6. When the cystocarp is producing gonimoblast filaments, the surrounding cells develop a protective sheath.
7. The cystocarp has a wall, a placental cell, gonimoblast filaments, carposporangium and carpospores.
8. At maturity, the carpospores are liberated from the plant and germinate to give rise to the tetrasporophyte.

Structure of the tetrasporophyte:

1. Tetrasporophytes are exactly similar in morphology to the gametophytic plants except that it is asexual.
2. The sporophytes produce tetrasporangia from the cortical cells.
3. Each tetrasporangium produces four haploid tetraspores which develop into gametophytes.

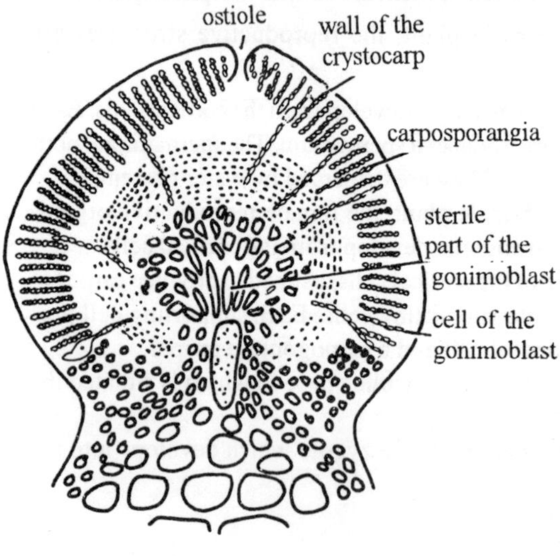

Fig.9.5 Gracilaria L.S. cystocarp

Diagrams to be drawn:

1. Habit sketch
2. Anatomy of the thallus
3. Spermatangium
4. Carpogonium
5. Tetrasporophyte

Systematic Position and Identification

A. 1. Chlorophyll present
2. Plant body is thallus
3. Cell wall cellulosic ... *Algae*

B. 1. Motile reproductive cells absent
2. Chromatophores red pigmented
3. Cystocarp present ... *Rhodophyceae*

C. 1. Thallus dichotomously branched
2. Male and female sex organs develop from cortical cells...*Gigartinales*

D. 1. Carpogonia develop from hypobasal cells.
2. Tetraspores arranged in the form of a cross ...*Gracilariaceae*

E. 1. Thallus is monosiphonous
2. Epidermis or peripheral layer with a gelatinous layer...*Gracilaria*

PLOYSIPHONIA
(Greville, 1824)

Common species: *P. fastigata, P. platicarpa, P. variegata, P. derulace, P. urceolata, etc.*

Distribution: Marine alga found along the Altantic and Pacific coasts. Sometimes grows as epiphytes on Fucales.

Study: 1. Structure of the plant body
2. Cell structure

3. Reproduction
4. Systematic position and identification

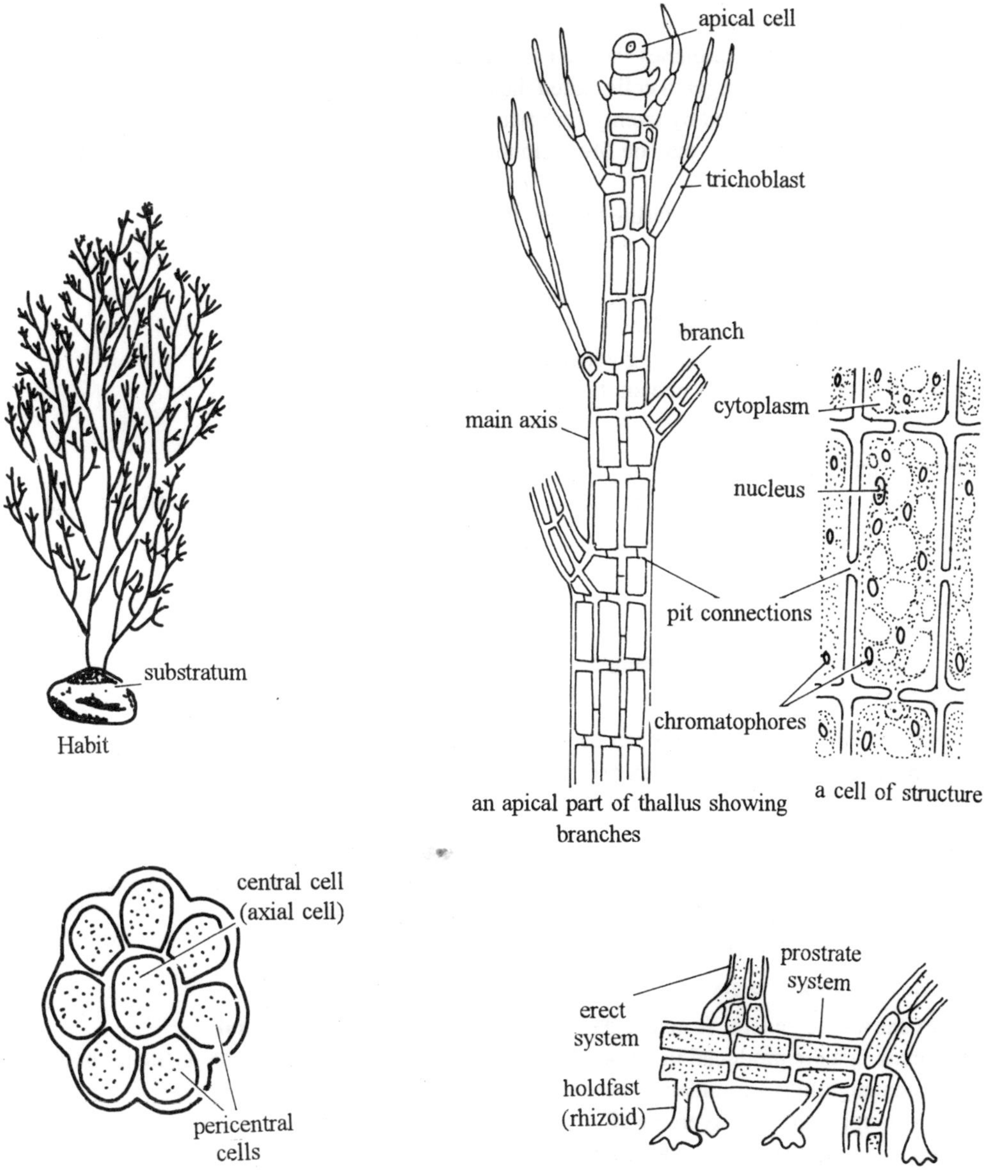

Fig.9.6 Polysiphonia. T.S. of axis   A part of thallus with rhizoids

Structure of the plant body:

1. The plant body is a complex, branched, filamentous thallus.
2. It forms, dense, tufted, bushy filaments of purple red colour.
3. The thallus shows distinction into two types of filaments, viz., (a) Basal prostrate or decumbent filaments

and (b) vertical filaments that arise from the prostrate filaments.

4. The prostrate filament is attached to the substratum with the help of aseptate rhizoids.
5. The free ends of the rhizoids enlarge to from disc like haptera.
6. The filaments show dichotomous or lateral branching.
7. The main axis as well as the branches show a polysiphonous (multiaxial arrangement)
8. There is a central row of elongated cells called the *axial siphon.*
9. The axial siphon is surrounded by twenty four pericentral siphons.
10. Some of the ultimate branches may be Monosiphonous (uniaxial) terminating into hair like structures called trichoblasts.
11. The trichoblasts may be sterile or fertile bearing the sex organs.

Cell structure:

1. Cells are uninucleate with a number of discoid chromatophores.
2. The chromatophores produce floridean starch.
3. There is a central vacuole surrounding which is a thin layer of cytoplasm.
4. Adjacent cells are connected to one another by means of cytoplasmic strands.

Life cycle:

1. The life cycle exhibits triphasic, isomorphic alternation of generations.
2. Three types of individuals are distinguished in the life cycle. All these are identical except for the reproductive structures they bear.
3. The three types of individuals are:

(a) Female plant bearing carpogonia.
(b) Male plant bearing spermatogonia.
(c) Sporophytic plant bearing tetrasporangia

Reproduction in the gametophytes:

1. Individuals of *Polysiphonia* are dioecious.
2. Sexual reproduction is oogamous, sex organs are spermatangia (male) and carpogonia (female).

Female plant bearing carpogonia:

1. The plant shows the general structure described under structure of plant body.
2. The carpogonia are flask shaped structures having a swollen basal region and an elongated neck called the trichogyne. There is a single egg at the base of carpogonium.
3. Carpogonia are produced either on carpogonial branches or on trichoblasts.

Male plant bearing Spermatangia:

1. The plant shows the general structure described under structure of plant body.
2. The spermatangia are produced in dense clusters near the apex of the plant on fertile trichoblasts.
3. Each spermatangium is a rounded or oblong structure and uninucleate.
4. The spermatangium has a 3 layered wall.
5. Each spermatangium produces a non motile uninucleate spermatium.
6. The spermatium gets dispersed by water current.
7. The fusion of the gametes results in the formation of the zygote.
8. The zygote develops into cystocarp or carposporophyte.

Structure of the Cystocarp:

1. The cystocarp looks like a Grecian urn.
2. It has an envelope made up of compactly arranged vegetative cells.
3. The cystocarp has an opening celled the Ostiole to facilitate the liberation of spores.
4. At the basal region, inside the cystocrap is seen a placental cell.

5. From the placental cell arise a number of gonimoblast filaments.
6. The tip of each gonimoblast filament has a carposporangium. Each carposporangium produces a diploid carpospore which on germination develops into the tetrasporophyte.

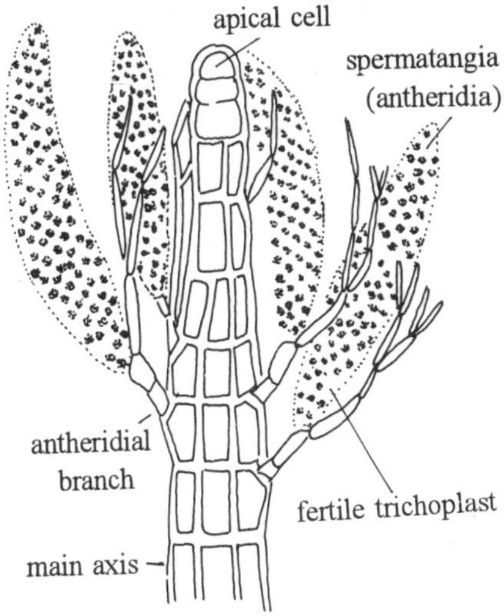

thallus bearing spermatangia

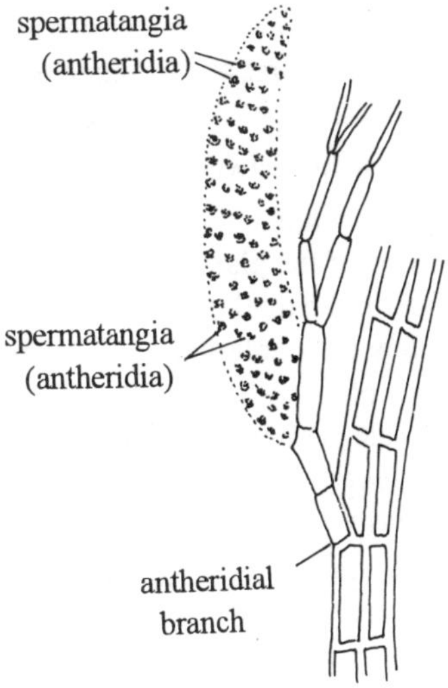

A fertile branch bearing spermatangium

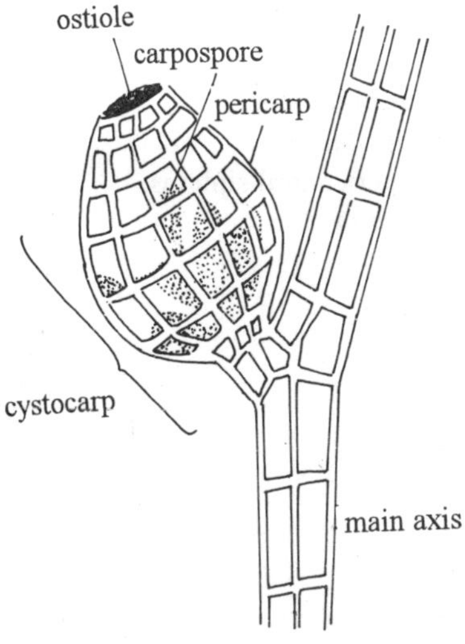

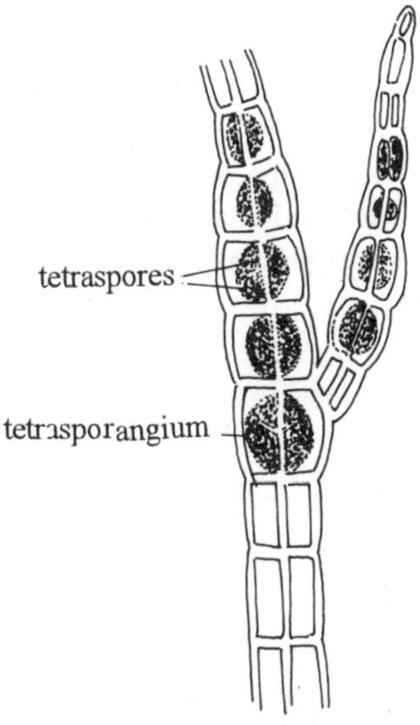

Fig.9.7 Polysiphonia. A filament with cystocarp

A tetrasporophyte bearing tetrasporangia

Tetrasporophyte:

1. The tetrasporophyte in its general structure is same as what is described under structure of the plant body.
2. Tetrasporophyte reproduces asexually by means of tetrasporangia.
3. The tetrasporangia are spherical structures borne on one celled stalks.
4. The terasporangia arise laterally from the pericentral cells of the plant.
5. The tetrasporangium divides meiotically to produce four haploid tetraspores.
6. The tetraspores are arranged tetrahedrally. On germination they develop into gametophytes either male or female.

Diagrams to be drawn:

1. Habit sketch
2. A part of the plant enlarged to show branching.
3. T. S. of the thallus.
4. A few cells enlarged to show cytoplasmic connections
5. Male plant with spermatangium.
6. Female plant with carpogonium
7. Cystocarp
8. Tetrasporophyte

Systematic position and Identification:

A. 1. Chlorophyll present
   2. Plant body is a thallus
   3. Cell wall is cellulosic... *Algae*

B. 1. Motile reproductive cells absent
   2. Chromatophores red pigmented
   3. Cystocarp present... *Rhodophyceae*

C. 1. Plant body polysiphonous
   2. Life cycle is triphasic ... *Ceramiales.*

D. 1. Cells uninucleate with discoid chromatophores
   2. Trichoblast bearing spermatangia shows dichotomous branching
   3. Carpogonia on unbranched trichoblasts ....*Rhodomelaceae*

E 1. Plant body has a central siphon surrounded by many pericentral siphons.
   2. Tetrasporangia borne singly on the thallus... *Polysiphonia*

PART TWO
FUNGI

CHAPTER 10

INTRODUCTION AND GENERAL CHARACTERS

Fungin represent one of the most important groups of plant kingdom. The study of mushrooms (Gr. *mykes* = mushroom; *logos* = study). Though the name mycology was originally meant to include only the mushrooms, now a days it has a broad meaning. It encompasses all the fungi. The reason for mushrooms being the first to be studied among the fungi is that they are one of the largest among fungi and hence attracted the attention of naturalists even before the invention of microscope.

With Leunwenhock inventing the microscope in the seventeenth century, the systematic sudy of fungi began. Pier Antonic Micheli- the Italian Botonist did a great lot to found the science of mycology and with Justice he deserves to be called the founder. In 1729 the published Nova plantarum genera in which he included his researches of Fungi.

What are Fungi?

It is really rather difficult 1/ to define the term fungi giving enough justice to its scope. Without entering into any air spilitting argument we can genrally accept the defination given by Alexopoulois (1925). He defines fungi as "nucleated spore bearing acholorophyllours orgainsms which generally reproduice sexuallyn and asexually whose filamentous, branched somatic structre are typically surrounded by cell walls containing cellulose or Chitin or both."

Presebnt day fungi include both common plants like mushrooms, puff balls, toadstools, morels rusts, smust and host of other members.

Fungi & Mankind

Of what importance is the study of fungi to mankind? This question can be very easily answered by listing, both the good and bad done by fungi to man.

Fungi are friends as well as foes to mankind. Tholugh the systematic study if fungi is only two and a half centuries old, the manifestations of this great fgrou; of plants on human beings, has been felt since time immemorial. Alexopolous in a klighter vein mentions "ever since the first toast was proposed over a shell full of wine. And the first loaf of bread was baked fungi are known to mankind." It is perhaps impossible to imagine how the world would have been without these fungi. Without fungi of course we would not have had such deadly plant diseases like rusts and smits which take such a heavy toll of one wheat crop . Ourpotatoes would not have had the notorious late blight area plams would have been safe without kole roqes, grapes

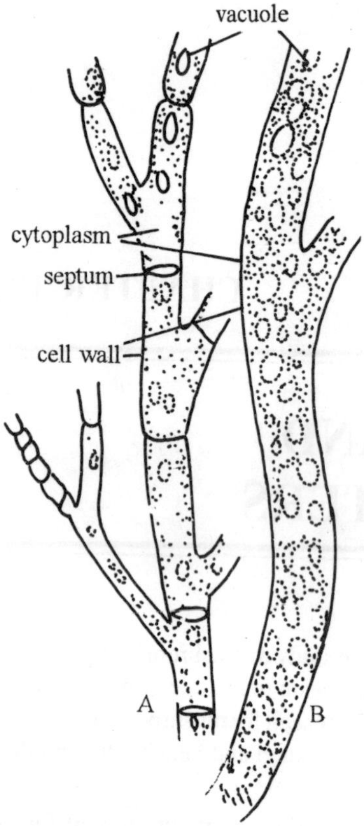

Fig. 10.1 Somatic structures of fungi. A-septate hyphae, B-Aseptate and coenocytic hypa

would not have suffered with mildew disease, and we need not have bothered about food articles which now get the attack of moulds; now much of care would have needed to maintain aspectic conditions in the laboratory. Our clothes, shoes would have been from the greeen moulds and we would not woulld have been a real paradkise to live in without these ugly creatures. But at the same time let us view the other side of the coin. Mankind would not have heard anything about alcoholic beverages the famous campaign would be non existent, there wouod be nno bakeries and perhaps there would be dead bodies of orgainisms all around for there would not bne any saprophytic fungi to decompose the, there would not be that great potent drug penicillin the morels , well these are a few of the favorite things we would have missed. And we really cannot afford to miss these. To provide both good and bad fungi are there and it is in his own interest that man is striving to have an intimate knowledge of these organisms. So that he can improve the races in which he is interested, and try to check the growth of others and that are harmful to him. The above given explanaation is just a representative of what the fungi are capable.

General characters

Fungi comprise more than 3900 genera and 80,000 species.

Occurrence

Fungi have a great potentially for quick adaptation, hence they have colonised all possible types of habitats. Primitive fungi are aquatic in their habitat. Generally higher and higher fungi amphibions and terrestrial habitats. Majority of the terrestrial forms are saprophytic in the soil or on animal and plant remains.

The fungi constitute a group of living orgainsms which do not have chlorophjyll, roots, stems or leaves that are familiarly met with common plants. They are usually filamentous and multicellular (could be unicellular also). With few exceptions they exhibit little division of labour.

Nutrition and Growth

As fungi are achlorophyllous, they obtain nutrition by hjeterotropic mode.

This is of 2 types.

1. Parasutuc Method: [G.Parasites eating beside another] Parasitic fungi live on other orgainisms (plants, animals including human beings and obtain their nutritiohn from a living host. In the process they may cause diseases to the host. Parasites may be ectoparasites, or endoparasites depending on their lication in then body of the host. Ectoparasites live on the body of the host(e,g zysiphjales), while endoparasistes live inside the tissues of the host.

Parasites may also be classifed into obligate parasites and faculatative parasites. Obligate parasits are tose which can liver only on a living protoplasm, while facultative parasites depending on the circumstances may live on dead organic matter.

2. Saprophytic method; (Gr. - sapros-roten + bios - life). Fungi which obtain nutrition from dead and decaying organic remains are called apsprophytes are also 2 types (1) Obligate saprophytes and (2) Facultative saprophjytes. Fungi which live on dead organic matter and are incapable of infecting living organisms

are called obligate saprobes, those capable of causing disease on a living host or living on dead organic matter depending on the situation are called faculative saprobes.

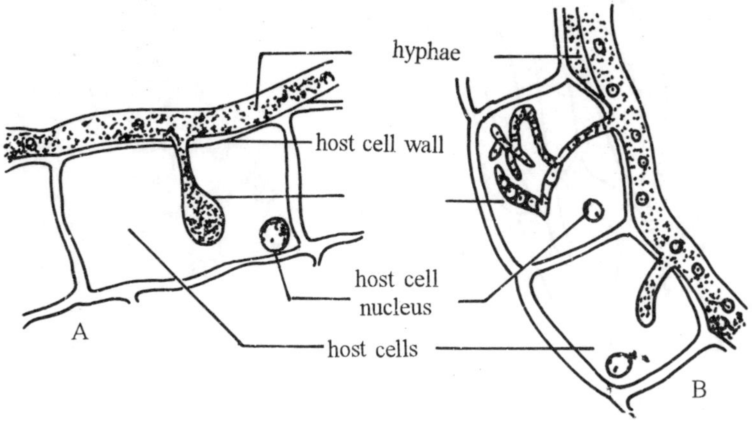

Fig.10.2 Types of Haustoria in fungi. A-Saccate haustorium, B-Branched haustoria (In some cases, the hyphae are intercellular)

Mycorrhiza

Some fungi, enter into an association with the roots of higher plants and it is known as mycorrhiza (Gr-Mykes mushroom rhiza-roots). In many cases the association is mutually beneficial. Though much been discovered (Kelley-1950) with regard to mycorrhiza the exact relationship is doubtful.

Fungi can synthesise all their nutritional requirements if they are initially supplied with an organic food such as carbohydrates. Laboratory-studies have indicated that fungi require all the essential and trace mineral elements for their growth. They have synthesise their vitamin requirements genreally. While some fo the fungi are omnivorous and can grow on any moist organic matter, others are specific about nutrition and grow only on particular species.

Most fungi are capable of growing in the temeprature range between 0 and 35 c, the optimum temperature between 20-30c , But they can withstand at least for few hours such low temperature as 195 c,. As a contrast to bacteroa fimgo [refer am acodoc ,edoi, fpr grpwtj wotj a PH of 6 being most suitable.

While light is not very much necessary for the growth it is needed at least at the time of spore function.

Somatic Structure

The plant body of fungus generally consist of thread or tube like cylindrical filament structure called Hypha.(plant-Hypac Gr-Hypho-web). A hypha consists of thin wall and has inner lining of cytoplasm. A hypha may or may not have cross wall. The crosswalls are called Septa. (Sing -Septum. La-Septum -hedge or partitions). A hypha with septa is called septate, without it is called aspetate.

Cell wall and Composition

The chemical composition varies in different fungi. Whuile cellulose is the chief componinet in some, in most of the higher fungi the wall is mainly composed of chitin, Besides these, hemicellulose, lignin, like substance have also been found in fungal cell-wall. However, none of these occurs in isolation, the fungal cell wall being very complex. The competition of the wall may vary in young and lod hyphae in fifferent PH etc.

All fungi have a definite nucleus. During nuclear divcision definite chromosomes are orgainsed. Some mycologists believe that nuclear division in fuingi is not directly comparable to mitosis.

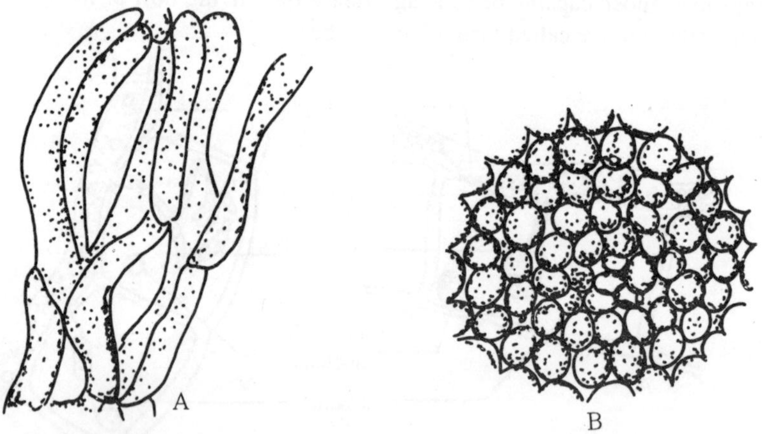

Fig. 10.3 Type of gungal ti-sues
A-Plectenchyma, B-pseudoparenchyma (a cross sectional view)

The aspetate hyphae usually have number of nuclei distributed in them, the condition being termed dcoenocytie (Gr -Koinois-commonn+Kytos a hallow vessel). In septate hyphae, the cell may have uninucleate, binucleate or multinucleate. Vacuoles, oil drops and other inclusions are generally present in the cytoplasm.

The fungal plant body usually consist of a mass of-hyptace called mycelium (plants-mycelia). In formiung a plant body the mycelia may be loose or form thick stands called Rhizomorphs (Gr -rhizo-rootm Morphat-shape), In this mycelial strands are closely united to form root like structures exhibiting division of labour. Rhizomorphs are resistant to adverse environment conditions.

In parasitic fungi, the hyphae produce special absorbing branches to draw nutrition from the host. These are called haustoria (sing-gaystriuyn, L-haustosdrinker). Haustoria are produced in mycelia which are intercellular they may be knob like elongated or branched like a miniatrue root system. Hustoria are not produced in artificual culture.

Saprophytic fungi do not producer any haustoria as they absorb the food by direct diffusion causing disintegration of he organic matter.

Fungal hyphae have a great longivity. As Alexopoulous states fungal colonies in natu8re, of more than 400 years old are known and it is possible that some mycelia are thousands of years old.

A mycelium starts its life as a short germ tubne produced by the germination of a spore (Gr spore-seed). Spores are asexual propogating units of fungi.

In higher fungi, the mycelium has tendency to organisc into compact or loosely interoven tissues to form compact macroscopic strutcyures. The general term plectecnchyma (or Plekoweave + enchyna-influsion, I,e..a woven tissue) is used to designate all types of fungal tissues there are 2 general types of plecyenchyma prosenchyma (or Pros towards+ enchymo-infusion which the hyphae can be identified. Tjhey are arraged more or less parallel to to one another. Pseudoparenchyma (gr. Pseudo-false +Parenchyma-a type of palnt tissue), consists of closely packed more or less isodiametric or oval cells resembling the parenchyma tissue of higher plants. In this type of tissue the hyphae has their identity and are more distingusihable.

Prosenchyma and pseudo parenchyma tissues compose many somatic structure in higher fungi. Two such structures are storma (plants. Stromata, Gr stroma =matters)and the sclerotium. (Plants-sclerotia, Grskeleros =hareds). A stromata, is a flat compact mattress like structure in which frucification are formed. A sclerotium is a hard resting body which is resistant to adverse environmental conditions. It remains dormant and germinates during favourable condition.

Reproduction: Fungi reproduces bvoth asexually and sexually. In some fungi, during the formation of reprodctive structures (asexual or sexual) the entire thallus is used up. Such a case is common in lower fungi and they ae called Aolocarpic (Holos-whole +karpos-fruit), In higher fungi however only vegetative and reproductive phases can occur toghether, these fungi are called Eucarpic (Gr Eu good, karpos fruit) Holocarpic nature is regarded as Perimitive.

Asexual Reproduction: Asexual reproduction is of frequeent occurrende in fungi then sexual and its is that which accounts for the rapid distribution of fungi, Asexual reproduction takes p;lace by a variety of methods. They are (1) Fragtmetation (2) Fission (3) Budding (4) Spore formation.

Fragmentation:

This is of 2 types : (a)Accidental and (b) purposive. In the first type when they mycelium accidentally breaks up into several fragments each one will develop into a new individual. In the second type, fragmentation of the myucelium, takes place as normal means of propagation . The hyphae breakes up individual cells. Eacu such isolatyed cells in called Oidium. (plants-oidia, Gr iodion = small egg) or Arthospore (Gr Arthos- joint spore- seed). Each arthosopre develops into a new individual .... During unfavourable conditions arthospores surround themselves with tensile wall. Suchj spore are called Cjhlamydospores (Gr chalmyds= mantle) they can withstand the inflavourable conditions.

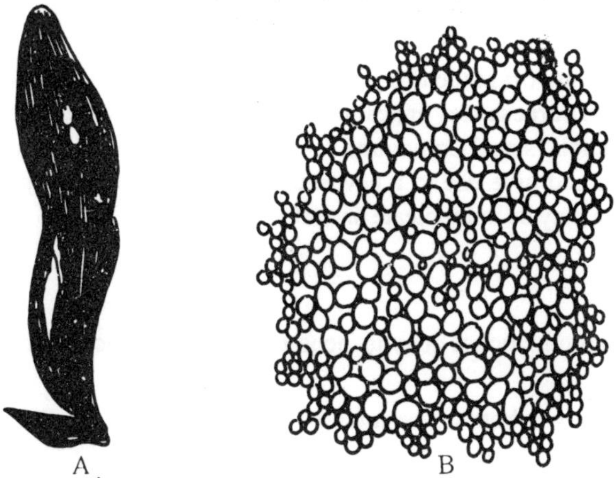

Fig. 10.4 Modifications of mycelium into a sclerotium. A- entire sclerotium, B- cross sectional view of a sclerotium.

Fission is the simple splitting of the cell into the daugheter cells. First constriction of the nucleus of the cell takes place followed by the division of the nucleus and the formation of cell-wall Fission as a means of reproduction is seen in unicellualr forms like yeats and bacteria.

Budding: the production of a new cell by the otgrowth (bud) of a cell is called budding. As the bud is being formed, the nucleus in the parental cell divides with one of the daughter nuclei entering the bud. The bud enventually breaks from the parental cell and develops into a new individual.

Sporeformation: By far the commonest mehtod of asexual reproduction is by spore formation spore vary in colour from ligaline through green yellow, orange, red, brown to black, in size from minute to large, in shape from globose through oval, oblojng, needle shaped to helical, in number of cell from one to many, in the arrangement of cells and in the way in which spores themselves are borne (Alexopoulos 1952), As Alexopoulos states it is this infininte variety of apores that makes the st6udy fascinationg. While some fungi produce only one type others producer many types of spores.

Spore formation may usually be categorised into 2 types- Internal and External.

Internal spore formation:

The spores are formed internally in some parts of the surface. The parts of the hyphae which are set apart for spore production are called sporangia (sing-sporangium: Gr -angion-vessel). The spores themselves are called sporongi spores. A spiragium may be globase or oval or scarcely distinguishable from the rest of then hyphae. Generally it is a sac like structrure whose contents are transformed into many spores. At the time of sporangial formation the type of a hypjhase usually swells into which cytoplasem nuclei migrate. Later this is cut off from the reminder of the plant by a cross wall. Protoplamic cleavage takes place resulting in the formation of spores. Sporangiospores may be mitole. Usually lower fungi produce motile. Motile spores are called planospores or zoospores (Gr zooanimal) In higher fungi the spore are motile and are called aplamospores. (Gr a= not + plant wander). The zoospopres have locomotory organs called flagella (singular-flagellum and flagellum = whip) which helps swimming. Zoospores may have one (uniflagellate) or two (geflagellate) flagella. The flagella found in fungi are of two types _ whiplash and tinsel. The whiplash flagellum is unbranched and has b basal rigid portion much longer than the terminal flexible portion The tinsel flagellum is a branched feathery sturcture consisting of a rachis with lateral hair like projections.

Flagellar Apparatus:

This is a complex structure consisting of (a) flagellum (b) belpharo9plast and Rhizoplast. Blephjaroplast (Gr -blepharis = cy-clash) which is the base of the flagellum is inside the cell. Rhizoplast is thread like sturc5ure throught which the flagellar a base is connected to the nucleus. Electron microgrqphs have revealed that the flagellum consists of eleven fibres of which nine are peripheral and two central. This 9+ 2 arrangement is seen in all motile cells except in bacteria.

The zoospores are usually pyriform in shape. In some casesthey are reniform also the flagella are terminal in pyriform and lateral in reniform zoospores. On liberation from the sporangium zoospores. On liberation fom the sporangium zoospores directly germinate to give rise to new individual.

External spore Formation

In some cases, the spores are formed by the cutting off at the tips of hyphae in baseptalous succession. There are specialised branches of the hyphae for this purpose they are called conidiophores (Gr-konis= dust, phore = stalk). The spores themselves are called Conidia (singal Conidium). Conidia are non-motile. They are wind disperesed.m Either they directly germinate and give rise to a new indivudal or produce zoospores which in turn develop into new individuals Canidia may be uninucleate or multinucleate.

Sexual Reproduction:

As in other organisms the ultimate aim of sexual reproduction is genetic recombination which accompanied by the fusion of compatible nuclei.

Fungi may be homeothathic or heterothalic. In the former, both sex organs are formed on the same thallus and in the latter they are formed on different thalli.

The sex-cells are called Gametes. They are produced in sex-organs. Sex 0rgans are called gametangia, if they produce gametes internally, or gametosphores if they produce gametes externally. Gametaphores however are found only in case of male (Gr, gametes=husband & angion-vessel). If gametangia produce only one type of gametes. They are called Isogametangia and the gametes themselves are called isogaments. If the gametes are of two different types such gametes are called anisogametes or heterogameetes and the gametamgia are called Heterogametangia. Heterogametangia are of two types, male and female. The male gametangium is called the antheridium and the female loogonium (Gr. Antheros-flowery=egg+gonos=pffspring). The female sex organs is Ascomyucetes is ascogonium.

Having thus explained the nature of sex-organs, we may now learn the different types of sexual reproduction met within fungi. They are:

(1) **Planogametic Copulation:** This type is mostly seen in lower fungi, this involves the fusion of two motile gemetes (planogametes). The gametes may be isogametes or anisogametes. In the cse, fusion is outside the gametangia. This is seen in some aquatic fungi belonging to lower phyconycetes.

(2) **Gometangial contact:** In a large majority of fungi, the male and the female gametes to undifferntiated uninucleate protoplasts. Here the sex-organs are antheridia and oogonia. In this method, the two sex-organs come in contact. Usually to effect the contact, a conjugation tyube or fertilization tube is formed from the antheridium which penetrates the oogonium. This forms a passage for the migration of male nuclei into the oogonium. After one nuclei migrate the antheridium disintegrates.

(3) **Gametangial copulation:** In this method, the entire contents of the contacting gametangia fuse. At the time of reproduction the two gametangia come nearer and become closely expressed the walls. The two dissolve and the antheridical contents flow into the oogonium.

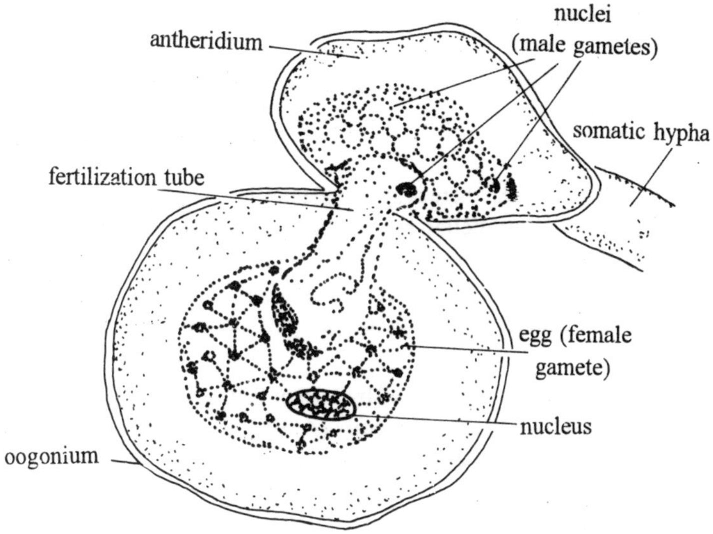

Fig. 10.5 Sex organs of fungi.
Antheridium and oogonium showing gametangial contact

(4) **Spermatization:** In some higher fungi, special erect branches of the hyphae (spermatiophore) are produced which cut off uninucleate male reproductive cells called spermatia (Gr-spermatin-little seed). The spermatia are formed much the same way in which the conidia are formed. The spermatia are carried by wind or insects to the female reproductive structure the receptive hypha the cell of which is fertile, the spermatial nucleus migrate into the receptive hypha.

Stages in Sexual Reproduction

Generally in lower fungi like phycomycetes, sexual reproduction is accomplished in three phases. In these formsm haploid phase in predominant with a short diploid phase restricted to the zygote, the three phases are:

1. *Plasmogamy:* (Gr-plasma-amolded boject+ gamos - mariage union). In this the two protoplasts and male and female) fuse bringing the neclei closer.

2. *Karyogamy:* (Gr-karyo-nut, nuycleus). In the phase, the two nuclei fuse.

3. *Meiosis*: (Gr-memosis-reduction). This immediately follows karyogamy and restores the haploid complement of the chromosomes. In short, plasmagamy fuses the two protoplasts Karyogamy brings about the fusion of two nuclei resulting in the formation of the diploid nucleus, meiosis restores the haploid condition.

In higher fungi (Ascomycetes and Basidimomycetes, however after plasmogamy, karyogamy is delayed for a very long time.

In the intervening phase the male and female nuclei just remain toghether without fusing this condition is called Dikaryotic phase. Only at the end of the life cycle, kayogamy takes place which is soon followed by meiosis.

As a result of dikaryotic phase in the life history, three types of mycelia (with reference to nuclear complement) alternate. They are:

(1) *Primary mycelium:* This is monokaryotic and haploid eproducing sexually after plasmogamy this produces.

(2) *Secondary mycelium:*which is dikaryotic, this occupies much of the life cycle. At the end of the life cycle, in some parts of the secondary mycelium karyogamy takes place. Tjhere are

(3) Teriary mnycelia which are haploid karyogamy is soon followed by meiosis resulting ibn the formation of haploid cells which gives rise to primary mycleium.

Sexuality: As we go from lower to higher fungi there is agradual elimination of sexorgans and in higher fungi, sex-organs are entirely wanting. Sexual reprodiction here takes place only by somatogamy. But this should not be taken as a loss of sexuality. Because in somatogamy there is no random fusion of any two hyphae. The compatible hyphae are carefully selected, the compatibility being determined by genetic factors. Many of the higher fungi though lack sex-organs are heterothallic I,e, they are self incompatible. Heterothallism is of tow types.

(1) **Bipolar heterothallism** in which sexualtiy is governed by a set of factors (AB) and (2) Tetrapolar heterothallism in which sexuality is detemined by two sets of factors (AaBb). In these two cases only thalli which carry the oppsite factors (A and B in bipolar, AB& ab in tetrapolar) conmate.

This clearly indicates that there is no loss in sexuality in higher fungi. In fact it is the most sophisticated tuype of sexual differentiated.

Classification: Fungi belong to the sub-diveision thallophyta. Eichler (1886) divided thallophyta into Algae and Fungi. Fungi included bacteria time moulds and lichens.

Modern mycologists have placed fuingi under 3 different classes, namely:

(1)Schizomycetes (2) Myxomycetes (3) Eumycetes

Fitiz patrick (1930) removed myoxmcyetes from fungi and divded thallophyta into

1. Myoxthallopjh7ta (slime moulds)
2. Euthallopjhyta (included algae, fungi, bacteria and lichens) Engler clssified fungi as follows;

Eumycetes (true fungi),

Section A-Lowerfungi

Class 1: Phycomycetes

SubclassI-Oomycetes

subclassii-Zygomycetes

Section B-Higher fungi

Class 2: Ascomycetes

subclass I-Hemiasci

Subclass ii - Euasci

Class 3: Basidimycetes

Subclassi-Hemibassidii

Subclassiieubasidii

subclasii-Eibasidii

Section c-Deuteromycetes or fungi imperfectii

Clement & shear (1931) divided fungi into 5 divisions (1) Phycomycetes (20 Ascomycetes (3) Basidiomycetes (4) Promycetes (5) Deuteromuycetes.

Gwynne Vaughan and Barnes (1926) classified fungi into 5-classes, namely Phycomycetes, Archimycetes, Ascomycetes Basidiomycetes and Deuteromycetes.

They divided phycomycetes into oomycetes basidiomycetes into protobasidiomycetes and Autobasidiomycetes.

Wolf and Wolf (1948) classifed fungi into Phycomycetes, Ascomycetes, Gasidiomycetes, Myxomycetes and Deuteromycetes.

Gaumann and Dodge (1928) followed the classification of wynne Vaughan in the main divisions.

Tippu classified fungi into

1.Slchizomycophyta Eq. Bacteria

2. Myxomycophyta Eq, Slime moulds.

3. Eumycophuyta-This includes four classes.

4. Phycomycetes. 2. Ascomycetes 3. Basidiomycetes 4. Deuteromycetes.

Alexopoulos (1956) places all fungi in the divisions mycota. He however keeps away the order Acrqsiales and labryintherlales form fungi as their affinitas are uncertain

The division mycota is divided into two sub-divisions (1) Myxomycotina and (2) Eumycotina

Myxomycotina has only one class-Myxomycetes

Eumycotina has the following classes

(a) Chytridiomycetes (b) Hypochytridiomycetes © Oomycets

(d) Plasmodiopjhonomycetses (e) Zygomycetes (f) Trichmycetes (g) Ascomycetes (h) Basidiomycetes (I) Deuteromycetes.

CHAPTER 11

MYOXMYCETES

The members of the class myxomuycetes are commonly mosit, shady and called Slime moulds. They are not true fungi, in that they do not possess all the charcters that we asociate with plants and animals. Anton de bary (1887) in fact considered them to be very close to animals and named them My cetozoa (Gr Mykes-fungus, zoon-animal) Bessey (19500 another famous mycologist also followed de Bary.

The term myxomycetes, for the slime moulds was first coined by Macbride (1899) According to Mabrisdol myxomycetes originated from a protozoan like ancestor.

Occurrence: Myxomycetes are found quite frequently inamycool places. Decaying wood humus rich soil, large fungi and decomposing debris of plant material are the favorite places to look for slime moulds. A few of the myxomycetes are found on the lawn of glass more than 450 speicies of slime moulds have been discovered so far.

The slime moulds are universal in distribution, but a few of them are confined totropics.

The vergative stages of myxomycetes do not exhibit any plant charcaters at all. Tjhere is no cell wall. The cell wall makes its appearance only during the formation of spores.

The vegetative plant bodynof most of the myxomycetes is naked, irregualr mass of protaplasm called plsmodium. The plasmodium is amoeboids in shape and multinucleate. The protoplasm can be divided into an outer. Ectoplasm and an inner endoplasm. Food vacuoles are present in endoplam. Nutrition is holozoic. The plamodium moves with the help of pseudopodia.

The free living plamocium becomes stationary during reproduction. Sporangia of various shape and colours areproduced. The spores have a cell wall and the first indication of the plant nature of myxomycetes.

Classification: The class myxomycetes (equivalent to division myxomycota of some mycologists) is divided into three orders, namely plamodiophorales, stemonitales and Acrasiales.

In this book Plasmodiophora and stemonites are discussed.

Plasmodiophoira

Common Species: P.brassicae

Distribtion: The fungus causes disease to cruciferous plants mainoy to cabbage. The disease is know as Club root disease figer & toe disease of clubbing disease. Occurrence: The fungus is an obligate parasite occupying the underground portion of the cryciferous plants.

Study 1. Wisease symptoms

2. Structure of the Pathogen
3. Reprodction

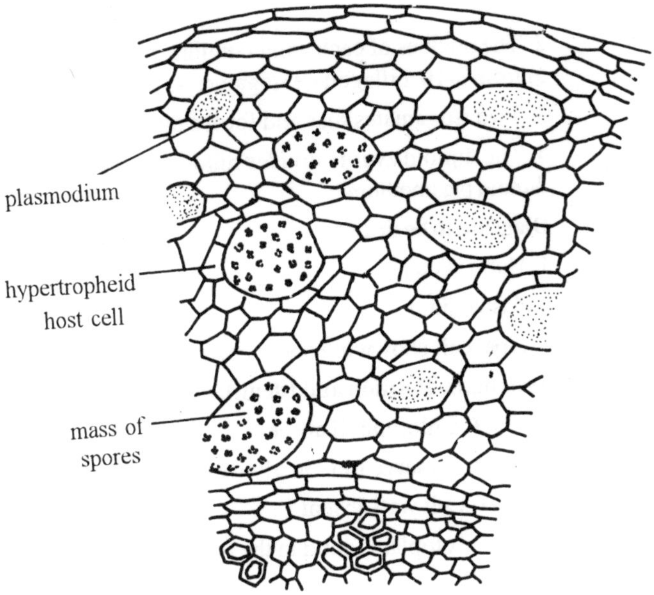

Fig. 11.1 T.S of host root showing plasmodium and spores of Plasmodiophora

4. Systematic position and Identification.

Disease symptoms:

(1) The symptoms of infection are visible in the unnderground parts and mainly in the roots.

(2) Irregular growth or hypertorphy resultls in the root, due to the infection by the fungus

(3) The hypertrophoid roots get deformed appearing like club or knotted fingers and toes, hence the name of the disease.

Structure of Pathogen

1. They vegetative plant body of the fungus is a naked mass of protoplasm called Piasmodium

2. Multiple fission in the plasmodium results in the formation of a number of daughter plamosida which develop into individual plants.

Asexual Reproduction

1. The nuclei in the plasmodium are diploid. They undergo reduction division when the plasmodia are in the cells of the root of host plant

2.Afger reduction, division, many individuals, spjherical spores are formed. The spores are non-motile and have achitinous wall.

3. The death and decay of the root cells of the host sets of the sporesfree.

4. These spores (x) metamorphose themselves into a biglagellate zoospores

5. The biflagellate zoospores infect the healthy plants, where it forms an irregularly shaped muyxmoeba.

6. On gaining entry into the host plant, they repeatedly ivide to form a haploid plasmodium.

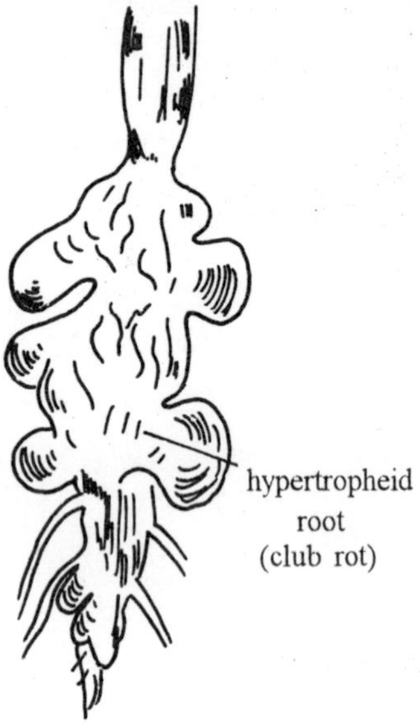

Fig.11.2 Club rot of cauliflower caused by Plasmodiophora

Sexual reproduction:

1. Each nucleus of the haploid plasmodium, gets isolated surrounds itself by a little cytoplasm and forms a gametangium.

2. The nucleus of each gametangium divides into mitotically to form 8-10, biflagellate isogametes.

3. Two isogamates fuse and form a diploid zuygote.

4. The diploid zygote divides mitotically and develops into diploid plasmodium.

Diagram to be drawn

1. Habit sketch

2.Plasmodium

3. Host cell showwing spores or zygotes.

Systematic Position and Identification:

A. 1. Chlorophyll absent

2. Reserve food in glyucogen...Fungi

B. 1. Cell wall absent

2.protoplam found in the form of plasmodium ...Myxomycetes

C 1. No frutification

2. Fungus is an endoparasite

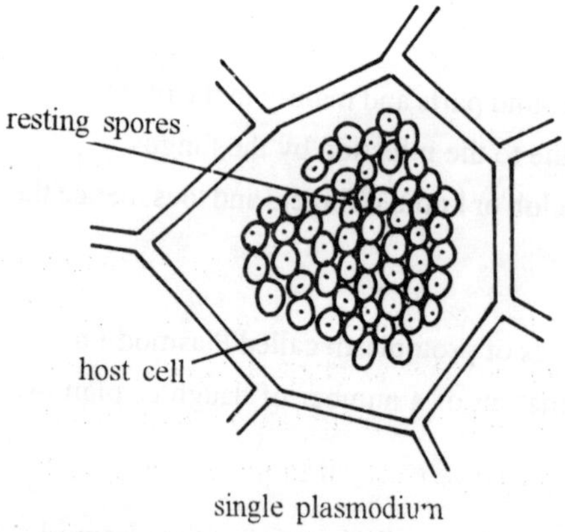

Fig 11.3 single cell

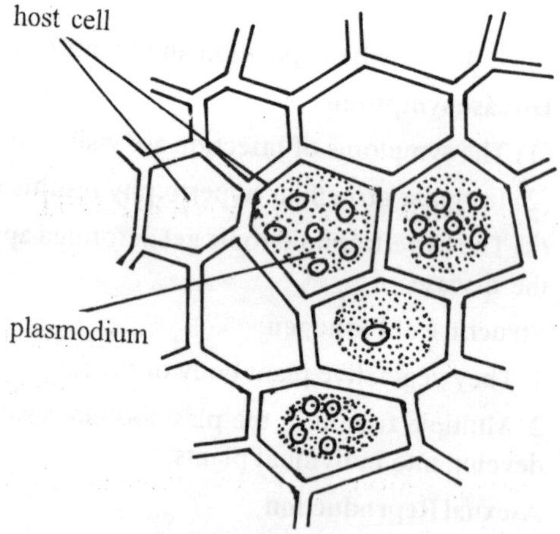

Fig. 11.4 Cells with spores masses

3. Vegetative development iside the host plant...Plasmodiophorals

D 1. Gametes terminally flagellate

2. Myxomocebae formed..Plasmodiophoraceae

E. 1.Plant body is a plamodium

2. Both diploid and haploid plasmodia are found

3. Infection courses club-shaped roots in the ost..Plasmodiophors

STEMONTES

Common species : S. herbatica

Distribution: The saprophutic fungus grows on dead and decaying wood, leaves decomposed vegetable matter etc.

Study: 1. Structure of Plant body

2. Reproduction

3. Systematic position and Identification

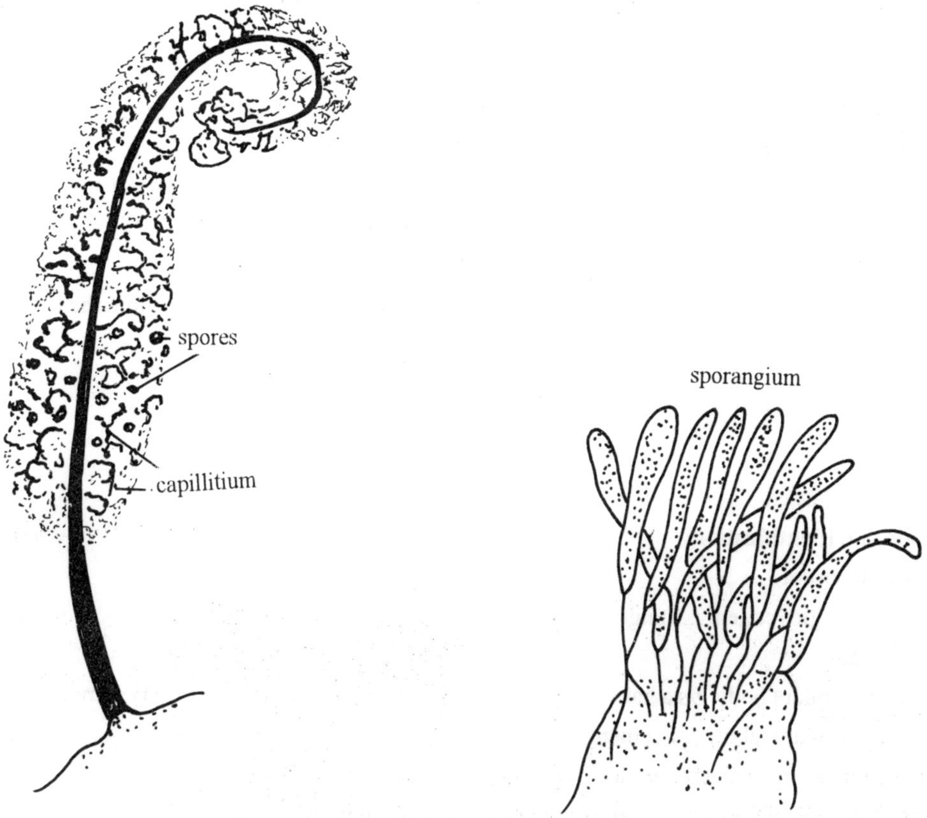

Fig. 11.4 Stemonites,

Structure of plant body:

1. The plant body is vegetative of stemonites is a plasmodium.

2. It is a naked mass of protoplasm without any cell wall. The protoplasm has 2 aone an outer ectoplams and inner endoplasm.

3. The plamodium is multnucleate and also has many food vacuoles.

4. The plasmodium is amoeboid in whape and moves with the help of pseudopodia, Nutrition is holophyte
5. After a certain period of vegetative phase, the plasmodium becomes stationary on some substatum.
6. The entire contents of the plasmodium transform themselves into a sporangium

(It is is this sporangium growing on leaves or wood that is usually given in the laboratory as the specimen of Stemonites).

Reproduction:

1. Sexual reproduction is byu spore formation-spores are produced in sporangia.
2. From the underlying plasmodium a sporanmgium develops
3. The sporangium has two part the stalk and the sporangium proper.
4. The stalk is called the stipe and it is an elongated structre
5. The sporangium proper has a central columella, which is the contninuation of the stipe.
6. The columella is surrounded by fine threads of cytoplasm which form a reticulum. The cytoplamic thread are called the capillitial threads.
7. The surface (margin) layer of the capillitium develops a compact wll like layer called the Peridiium.
8. Within the meshes of trhe capillitial network are found manyu nuclei which ndergo reduction division to form haploid spores.
9. The formation of the spores is the fist stage in the life cycle indicating the plant charcters of stmonites
10. The spores on liberation develop into amoboid which are falgellate
11. The swarm cells divide by means of fission to multipluy their number
12. The swarm cells fuse to form the zygote, whicvh by repeated mitotic divisions develops into the Plasmodium

Diagrams to be drawn:

1. habit skektch
2. Sporangiumenlarged
3. A part of capillitum enlarged

systematic position and Identification:

A. 1.. Chlorophyll absent
   2. Reserve food is gloycogen....Fungi

B. 1. Cell wall absent
   2. The vegetative pllant body i9s a naked mass of protoplasm called plasmodium ..Myxomycetes

C. 1. Plasmodiujm shwos amoeboid movement
   2. Fungus is a saprophyte ....Stemonitales

D. 1. The contents of the plamodium get transformed into a sporangium
   2. Plasmodium is always diploid....Stmonitaceae

E. 1. Sporangium consists of capillial network with a central coliomella
   2. Spores are haploid and uninucleate....Stemonites.

CHAPTER 12

PHYCOMYCETES

The class phycomycetes represents one of the lower classes of true fungi (Gr.Phycos-seaweed, myces-mushroom). They include about 254 genera and 1200 species. They are very common in occurrence. The bread mould (mucor), the water mould (saprolegnia), the white rust of mustard- all are phycomycetes.

The plant bod6y in phycomycetyes is a conenocytic, asepate much branched mycelium, septa appearing either to demarcate the dead portions or at the time of formation of reprodcutive structures. Some of the lower forms are holocarpi9c, the rest Eucarpic.

Many primitive phycoycetes are predominantly acquatic in their distribution. Every higher forms (except zygomycetes) show dependence on moisture. They may be parasites or saprophytes parasites grow on other fuhngi, algae, seed plants, fio,ns and animals also. Saprophuyticphycomycetyes occur in abundance in damp soil, dung and remains of plants &n animals.

Reproduction: Phycomycetes reproduce by both asexual and sexual means. Asexual reproduction is brought about by the following w3ays (1) Fragmentation (2) Spore formation.

Phycomycetes produce both zoospores as well as condidia. Saprolegnia, phytopthora, phythium & others produce zoospores. Albugo produces conidia. In mucorales, asexual reproduiction is by aplanospores. Chlamydospores are also found in some members.

Sexual reproduction:

Phycomycetes reproduce sexually by the following methods:

1. Planogametic copulation
2. Gametangial copulation
3. Gametangial contact.

In acquatic phycomycetes, sexual reproduction is by Planogametic copulation and in higher members it is by other two methods.

Stages in sexual reproduction:

There are only 3 stages, namely :

(a) Plasmogamy

(b) Karyogamy

(c) Meiosis

In some cases as Mucor and others, faiulure of fusion will result in the formation of parthenospores of Azygospores.

Classification

The filamentous phycomycetes were divided according to their mode of reproduction into two orders, namely: (1) Oomycetes (2) Zygomycetes. In oonycetes reproduction is oogamous and in zygomycetes it is isogamous (Martin 1950). The non-filaments phycomycete4s having round or lobed mycelical thallus were place in Archimycetes (gaumann 1949). But inclusion of all oogamous under omycetes and all isogamous forms is zygomycetes seems rather artifical. Baumann (1952) divided the class into 5 orders based on mainly on the features of the thallus, flagellar type and reproduction. These are chytridiales Blastocladiales, monoblephoriales, Oomycetes and zygomycetes.

Alexopoulos (1956) divides phcomycetes into 7 orders. Chytridiales, Monoblepariales, Plasmodiphorales Saproleniales, pernosporales, Mucorales and Entompohthorales.

Sparrow (1943) classifies phycomycetes into 3 series based on flagellar charcters.

Series 1-Uniflagellatae (flagellum one)

Series 2-Biflagellatae (flagella two)

Series 3-Aplanatae (no flagella)

Mc clean and cook (1956) have divided phycomycetes into five orders, namely Chryidiales, Arcylistorles, Saproligniales, Blastocladiales, Monoblepharidales.

The following examples are discussed here,

Synchitrium, Achlyam Saprolegnia, Pythium, Phytophtora, Albugo, Sclerospora, Peronospora, Rhizopius, Mucur, Pilobolus.

ACHLYA

Occurrence: Most of the species of the fungus are acquatic sparophuytes found in ponds, pools and stagnant waters. They are also common on decaying vbegetables lmatter, dead flies etc.

Study: 1. Culture in the laboratory.

2. Structure of the plant body.
3. Reproduction
4. Systematic position and identification

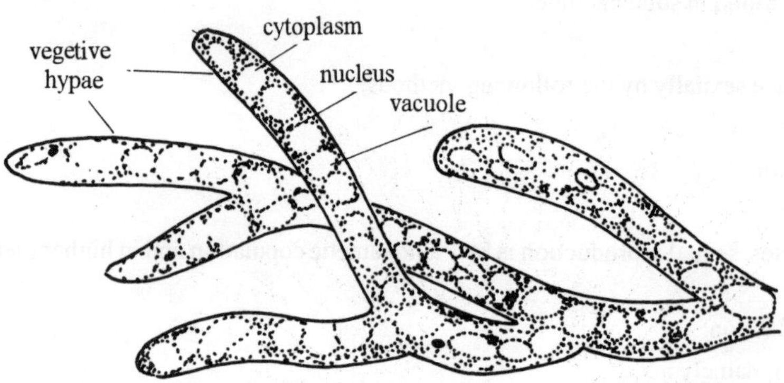

Fig. 12.1 Achylya. Vegetative mycelium

Method of culture in the laboratory:

The fungus can be cultured in the laboratory in the same way as Saprolegnia

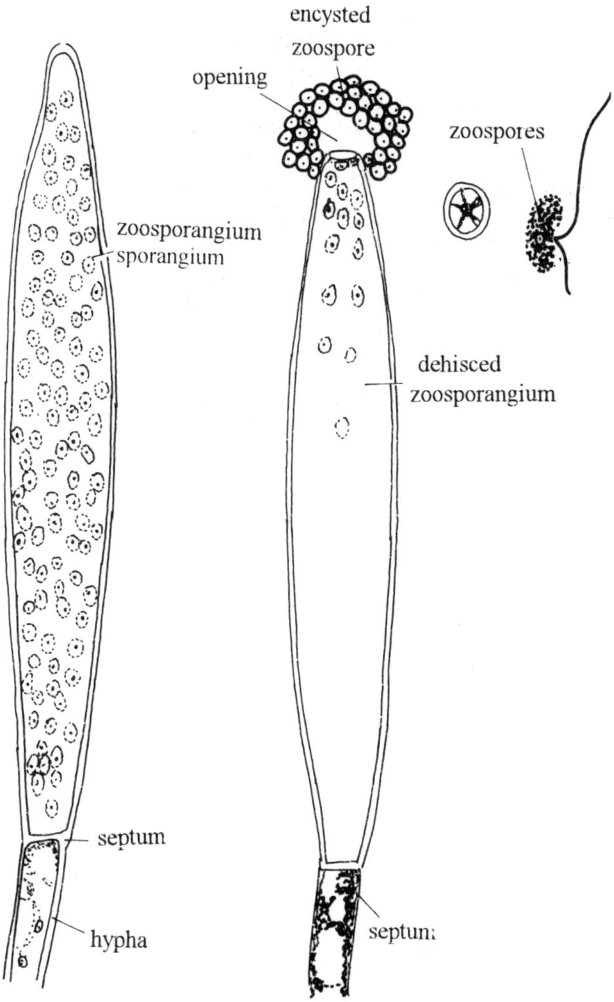

Fig. 12.2 Achykya. Single zoosporangium   Empty zoosporangium with encysted zoospores at the apex

Structure of plant body:

1. The plant body is mycelium consisting of numorous cotton7y white thread like structures.

2. The myceklium is aseptate, coenocytic and much brnached.

3. The mycelium produces certain erect h7yphae which grow vertyicalluy down into the substratum and help in the absorption of food material.

4. The hyphae are tubular and have a central continuous vacuole.

5. The cytop,asm is peripheral and has many nuclei. The cytoplasm also has oil globules and glycogen.

Reproduction :

1. Vegettive propagation is brought about by means of fragmentation. Each fragement develops into a new individual.

2. Asexual reproduction takes place by zoospores formed in sporangia.

3. Sporangia are elongated cylindrical struc tyures developed at the tips of hyphae.

4. The sporangia are separated from the rest of the hpjhuyase by cross wall.

5. each sporangium produces a large number of primaruy zoospores which are biflagellate and pyriform in shpae.

6. The primary zoospores undergo encystment and then give rise tokidney shaped secondary zoospores, which have laterally attached flagelia.

7. The secondary zoospores give rise to territory zoospores, which are similar on structure to the former.

8. The phenomenon of production more than two motile phases in the life cycle is calle polylanetism.

9. The zoospores germinate to form a mycelium.

Sexual reproduction :

1. Sexual reproduction is oogamous. Species may be homothallic or heterothallic. The sex-organs are antheridia and oogonia.

2. Antheridia develop at the tip of hypha. They are multinucleate.

3. Oogonia are globose and multinucleate. They have lmany functional eggs.

4. Antheridia and oogonia are separated from the rest of the hypha by a cross wall.

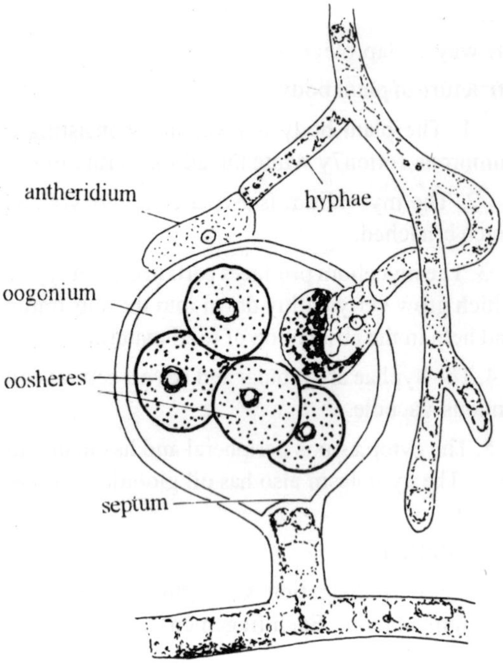

Fig. 12.3 Achylya. Sex organs

5. Sexual reproduction is by gametangial contact. One or more antheridia apply themselves to oogoniuum, produce fertilization tubes.

6The fertiliztion tubes release their contents inside the oogonium.

7. The fertilization tubes release their contents inside the oogonium. The zygote of oospore is thick walled and develops into a new individual .

Systematic position and Identification:

A. 1. Chlorophyll absent

2. Reserve food is glycogen ..Fungi

B. 1. Mycelium aseptate and Cloencoyctic

2. Septa are formed only during reproduction..Phycomycetes

C. 1. The plant is an acquatic saprophyte

2. Sexual reproduction is oogamus..Saprolegnales

D. 1. Zoosporangia cylindrical

2. Oogonium has many eggs...Saprolegniaceae

E. 1. Rhizoidal hyphjae present

2. Asexual reproduction exhibits polyplanetism...Achlya

SAPROLEGNIA

Coomon species: S.parasitica, S.terax etc.

Occurrence: Saprolengia is an acquatic saprophyte. Majority of the species grow in dead animals and inside in water. S. parasitic is a paraditic species.

Study: 1. Culture method

2. Vegetative Structure

3. Reproduction

4. Systematic position and Identification

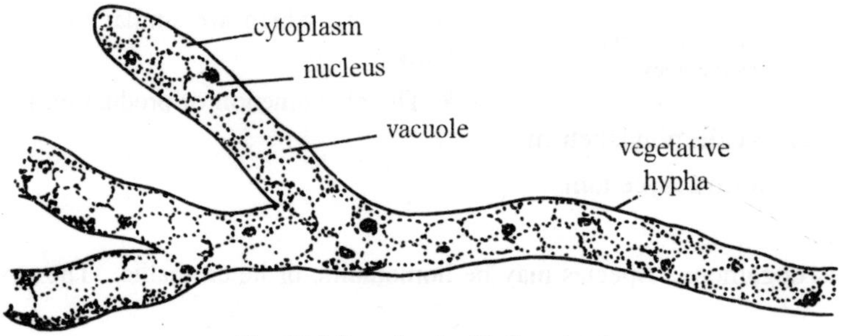

Fig. 12.4 Saprolegnia. Thallus structure

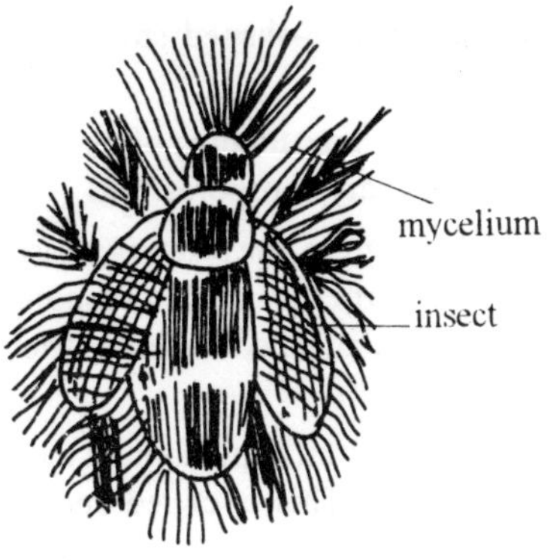

Fig. 12.5 Saprolegnia. Habit

Methods of cultruing saprolelgnia in the laboratory:

1. Select a clean petridish.

2. Place on or tow dead flies, small pieces of meat and some egg albumen

3. Add a little pond water

4. Close the petridish

5. Within a few days, the body of the housefly will be full of thread like mycelia of saprolegnia.

Vegetative structrue:

1. The plant body is a mycelium. It is a soft whitish thread strey.

2. The mycelium is coenocytic and profuself branched forming a white mould like growth on the substratum.

3. Hyphae are aseptate and speta are formed only when the plants enters the reproductive phase.

4. The cytoplasm has many nuclei, oil globules and glycogen.

Reproduction:

1. Saprolegnia reproduces asexually as well as sexually,

2. Asexual reproduction is broughjt about by means of zoosporangia.

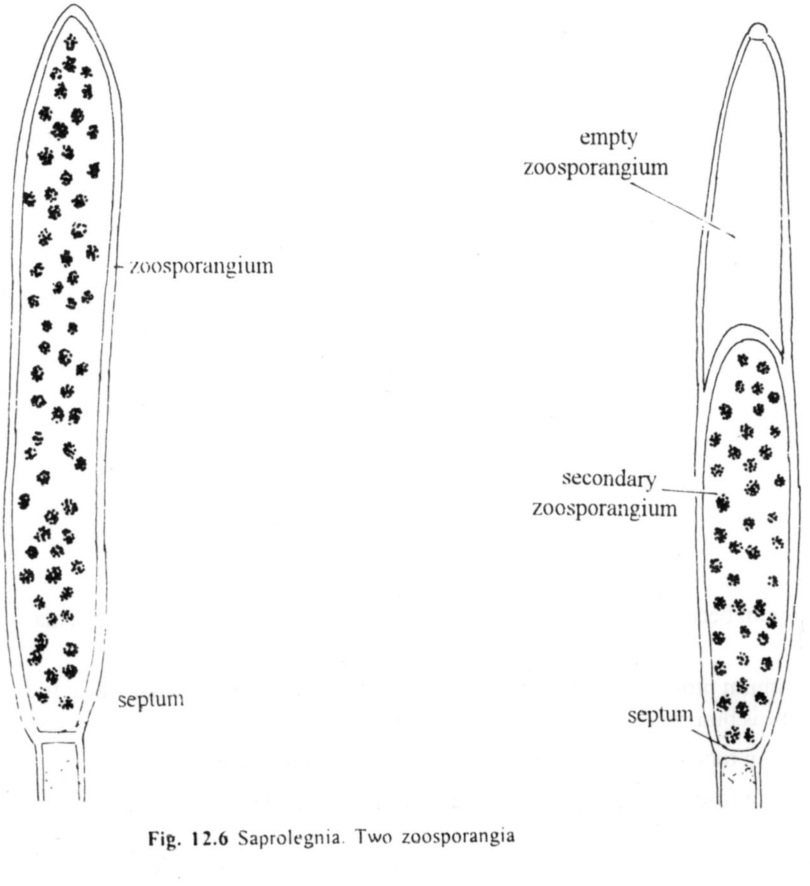

Fig. 12.6 Saprolegnia. Two zoosporangia

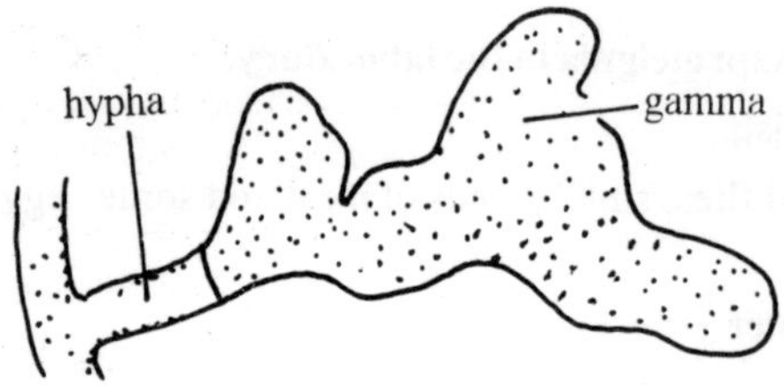

Fig. 12.7 Saprolegnia. Hypha with gemma

3. Zoosporangia are elongated club-shaped structures that develop at the tips of hyphae.

4. Each zoosporangium produce large number of plriform bi-flagellate zoospores called lprimary zoospores.

5. On liberation from the sporangium, the primary zoospores undergo encystment and then form secondary zoospores are reniform in shape with lateralluy attached flagella. They develop into new individuals.

7. After the liberation of contents from a sporangium the cross wall beneath forms another sporanmgium. Within the old sporangial walls. This is called sporangial prolieferation.

Sometimes as many as 6-7 sporangia may be formed within an old sporangial wall

Sexual Reproduction:

1. It is oogamous. Sex organsare oognis and antheridia. They are formed at the tips of the hyphae.
2. Species of saprolegnia are homothallic
3. The oogonium is a globose structure. Atmaturity has 6-8 eggs or oospheres. All of them are functional.
4. The antheridium is a club shaped structure and has many male nuclei or or sperms
5. Sexual reproduction is effected by means of gametangial contact.

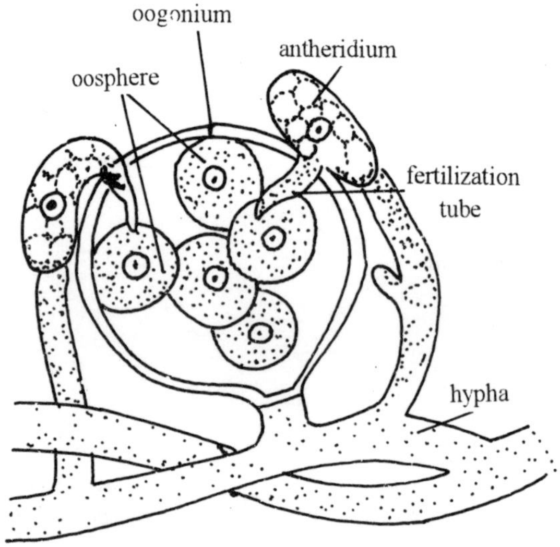

Fig. 12.8 Saprolegnia. Mycelium with sex

6. The antheridium becomes clsoely applied to the oogonium one or more antheridia may get attached to the oogonium.

7. Each antheridium produces a fertilization tube which piercds the oogonial wall. In some instances, the fertilization tube is branched.

8. The fertilization tuber bursts opening the oogonium releasing the male nuclei.

9. The male nuclei fuse with the eggs, resulting in formation of zygotes.

10. Lying in the oogonial cavity are found many zygotes or oospores.

11. the zygote produces a ge:mtube and develops into a new individual.

Diagrams to be drawn:

1. Habit sketch
2. A part of the mycelium enlarged
3. Zoosporangium
4. Antheridia and Oogonia
5. Zygotes in the oogonial cavity

Systematic position and identification :

A. 1.Chlorophyll absent

2. Reserve food glycogen...Fungi

B 1. Mycelium aseptate and Clenocytic

2. Septa are formed only during reproduction....Phycomycetes

C. 1. The plans is an acquatic saprophyte

2. Sexual reproduction is oogamous ...Saprolegniales

D 1. Zoosproangia cylindrical

2. Oogonium has many eggs...saprolegniaceae

E 1. Sporangial proflieration is seen

2. Dipolanetism is exhibited....Saprolegnia.

SYNCHITRIUM

Common species: S. endbioticm, S.trichosanthiodes, S.rhytzi, S. sesami etc.

Occurrence: The fungus is parasitic and a pathogen causing several polant diseases. Many of the economically important plants are infected by the fungus

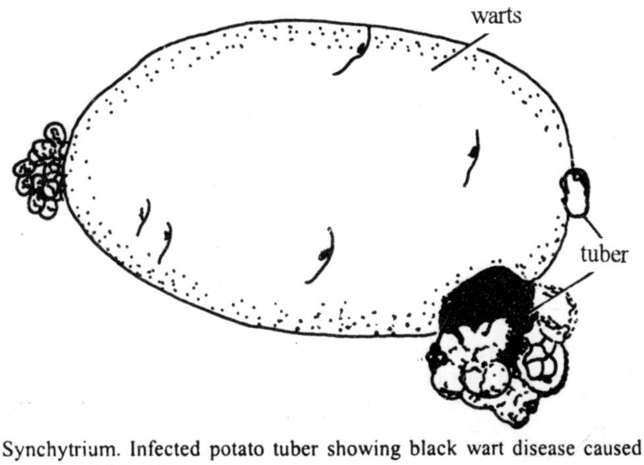

Fig. 12.9 Synchytrium. Infected potato tuber showing black wart disease caused by Synchytrium endobioticium

Study: (1) Diseases caused by the fuugus.

(2) Diseases symptoms

(3) Structure of the fungus

(4) Systematic position and Identification

Diseases caused by fungus:

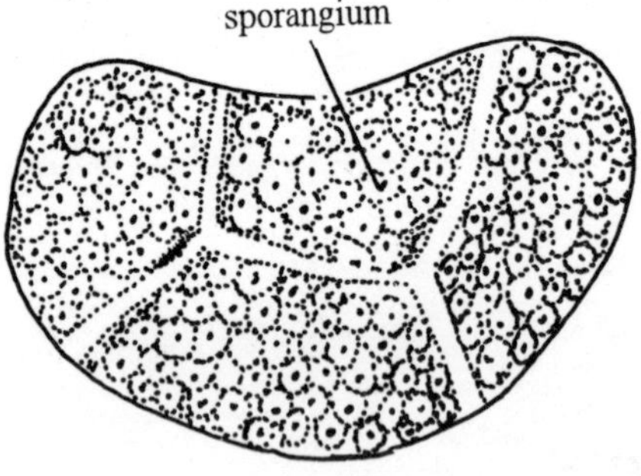

Fig 12.10. Synchitrium. A Sorus

1. Many flowering plants get the infection by the fungus and the disase is called wart disease.
2. Potato tubers are infected by S.endobioticum causing potato warts.
3. S.trizchosanthoides causes the wart disease of cucurbits speically snake guards.
4. Perystrophye bicyculata, a member of the family A canthaceae is infected by S.rhytzii.
5. Sesamum indicum of the family Pedaliaceae is infected by S.sesami

Disease symptoms

1. The characteristic symptoms a of the disease are the development of small warts on the infzected parts
2. In potato the tubers shwo the charceterstic warts
3. Many time hypertrophy and decay of host cells wall also results

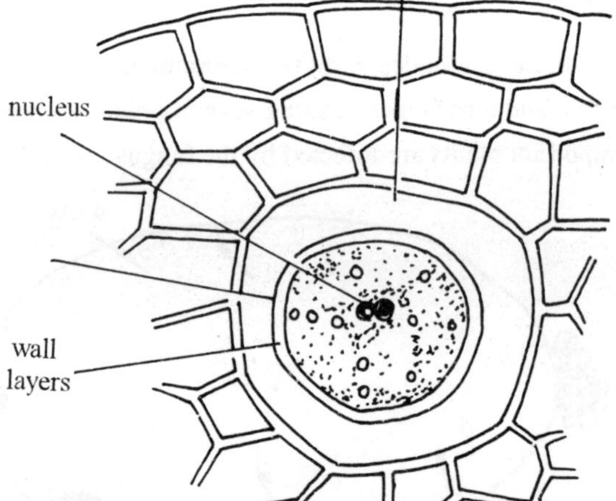

Fig. 12.12 Synchytrium. Winter sporangium within infected host cell

Structure of the fungus:

(1) The plant body of synchitirium is unicellular.

(20 The cell remains within the host cell, i.e. intercellular., This condition is also said to be endobiotic.

(3) The thallus is globose and has a thick or sometimes think chitenous wall surrounding it..

The thick walled cell is also called Prosorus.

Asexual Reproduction;

1. The plant is holocarpic I,e., the whole of the plant body gets converted into a reproductive structure.

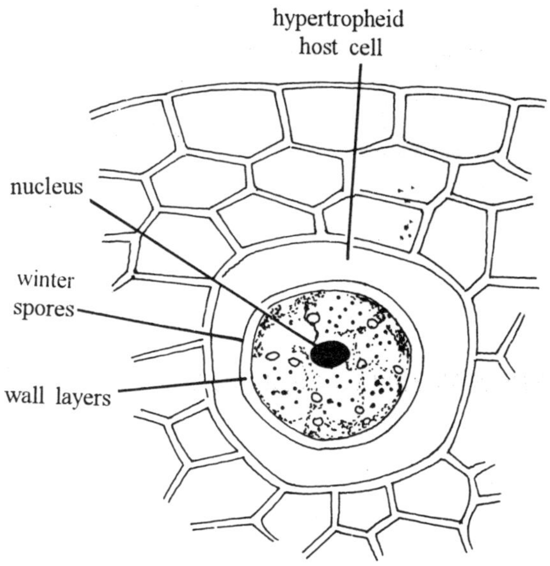

Fig. 12.12 Synchytrium. Winter sporangium within infected host cell

2. Asexual reproduction is brought about the help of zoospores.

3. The zoospores can bring about a fresh infection.

Sexual Reproduction.

1. Sexual reproduction is resorted to only during unfavourable conditions. Otherwise, the usual method of multiplication by zoospores.

2. The persons (the plant body) producers many gametes which are isogamous

3. The gametes fuse to form a diploid zygote which will be motile for sometimes.

4. After a short free swimming period the zygote penetrates the host cell and undergoes a period of rest there. This is called the Resting spore.

5. at the return of favourable conditions, the zygote undergoes reduction division and produces a number of uninucleate haploid zoospores.

Diagram to be drawn:

1. Habit sketch of infected plant
2. Section of the infected part to show hypertrophy
3. Host cell shwoing the parasite

Systematic Position and Identification:

A 1. Chlorophyll absetn
  2. Reserve food is glycogen .. ..Fungi

B. 1. Plant body aseptate
  2. Cells (plant body) uni or multinucleate...Phycomycetes

C 1. Asexual reproduction by uniflalgellate zoospores... Chytridiales

D. 1. The plant body forms a posrus
  2. The cell is holocarpic....Synchitriaceae

E. 1. Warts are produced on the host
  2. Resting spore (zygote) is formed .....Synchitium

PYTHIUM

Common Species : P. debaryanum, P.graminicolum, P.myniotylum

Occurrence: Many speices are amphibious. They mostly occur in moist humus soil and infect the hpocotyle region of seedling of plants like ginger, tobacco, mustard, wheat etc. After entering the host, the fungus acts as a parasite and causes disease like damping off, soft rot foot root etc.

Study: 1. Vegetative structure of Pathogen

2. Reproduction

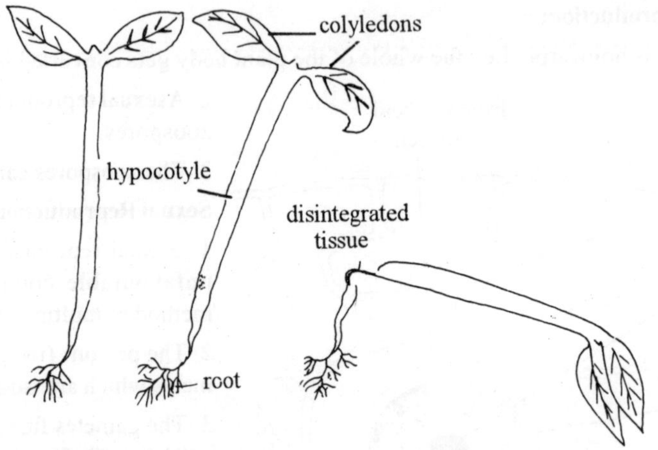

Fig. 12.13 Pythium. Host seedings showing damping off disease caused by

3. Disease symptoms
4. Systematic position and Identification

Diseases caused by the genus:

1. P.debaryanum-Damping off tobacco seedlings.

Fig. 12.14 Pythium. T.S. of host seeding showing zoosporangia

2. P.aphandiermatum-Stem rot of papaya
3. P.graminicola-foot rot of what seedlings
4. P. hindicum-Fruit rot of Lady's finger

Vegetative structures:

1. The plant body of the fungus is a mycelium
2. The mycelium is aseptate, coenocytic and much branched.
3. Haphae have a vacuolated cytoplasm in which a large nuber of nuclei are embedded.
4. The mycelia superficially resemble the filaments of vaucheria
5. Reserve food is in the form of glycogen
6. No haustoria are produced.

Asexual Reproduction:

1. Asexual reproduction takes place by the formation of sporangia. Sporangia may be terminal or intercalary.
2. Sporangia are small, golobular or sac like and are of two types viz., globose or spheroidal and elongate or filamentous.
3. Globose sporangia are globe like and terminal in position
4. Filamentous sporangia are indistingusihable from the hyphae.
5. The sporangia whether globase of filametnous are indistinguishable from the hyphae by a cross wall.
6. The contents of the sporangia form numberous zoospores which are reniform and bifflagellate.
7. Each sporangium has a beak !ike structure which enlarges to from a vesicel. It is into this vesicle that the

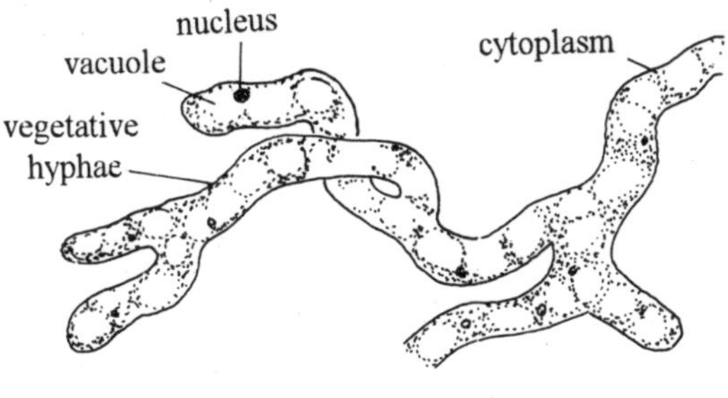

A portion of mycelium

Fig 12.15 Phythium .Apromotion of mycelium

zoospores are released first. Later the veiscle ruptures setting the zoospores free.

8. On germination each zoospore developinto a new individual

9. In some speices of pyuthium, the sporangia produce onon-motile conidia which germinate to give rise to new individuals.

Sexual Reproduction:

1. Sexual reproduction is oogamous Sex-organs are antheriedia and oogonia. They are usually formed towards the end of the growing season.

2. Pythium is homothallic

3. Antheridium is small and clubshaped, it usually terminates an antheridial branch. It is cut off from the rest of the hypha by a cross wall.

4. The antheridium is multinucleate in the beginning. Later all the nuclei except one degenerate, levaving only one functional male nucleus.

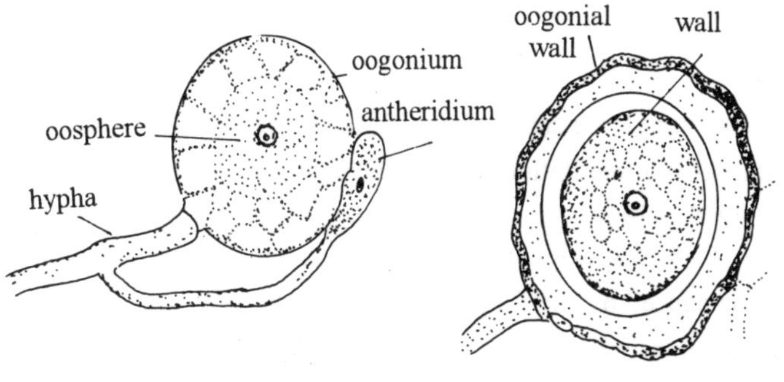

Fig. 12.16 Pythium. Sex organs An oospore

5. The oogonium is a globose structure terminating a female branch of hypha

6. the inside of the oogonium has two zones of cytoplasm viz., outer periplsm and central ooplasm.

7. The ooplasm has a single egg.

8. Sexual reproduction is effected by means of gametangial contact.

9. A fertilization tube develops from the antheridium. It penetrates the oogonium relasing the contyents inside.

10. The fusion of the male and female gamete results in the formation of a zygote or oospore.

11. The oosore after undergoing a resting period develops into a new individual

Disease symptoms

The symptoms of the disease are noticeable only in the seedling stage after they have just come out of the soil. The seedlings become plale green. The stem at its base, i.e., just above the ground level exhibit decay of he tissues. The tissues from brown. As aresult of the decay at the base of the seedling collapses.

In the fruit rot and root rot diseases, the tissues rot, and get decomposed resulting in the death of the plant.

Diagrams to be drawn:

(1) Habit of damping off of seedlings

(2) Zoosporangium

(3) Ooganium and antheridum

Systematic Position and Identification:

A. 1. Chlorophyll absent
   2. Reserve food is glycogen ...Fungi

B. 1. Mycelium aseptate and coenocytic
   2. Septa are formed only during reproduction
   3. Oospore is thick walled...Phycomycetes

C. 1. Sexual reproduction not by Planogametic copulation
   2. Oogonium has a single oosphere ...Peronosporales

D. 1. Sporangiophores not different from somatic hyphae...Pythiaceae

E. 1.Pathogen causes damping off of the seedlings.
   2. Host stem has brown decay at the base
   3. Zoospores are released into a vesicle
   4. Facultative parasite...Pythium

PHYTOPHITHORA

Common species: P.infestans, P.arecae,P.palmivora, P.cola casiae

Occurrence: The fungus is a facultative parasite, occurs in soil as a saprophyte also gaining entry into the host lant at the opportune time.

The genus has more than 75 species and nearly all of them may be cultured artificially on the medium.

Study: 1. Disease caused by the fungus
2. Disease symptoms
3. Vegetative structure
4. reproduction
5. Systematic position and Identification

Diseases caused by the fungus and symptoms:

Various species of phytophthor a cause disease to economically important crops resulting in heavy damges. The following are some of the diseases

1. Late blight of Potato: This is caused by P:infestans and is of great significance as the crop is destroyed heavily. The disease appears periodically in the plains and hilly regions of Northern India.

The diseases symptoms are first noticeable on the leaves. The infected regions show small and pale coloured spots which eventually turn brownish or purplish discolouration become softened and the tuber decays.

As the disease symptoms appears late in the growing season. It is called late blight.

2. Kole roga of Arecapalms: The disease is caused by p:arecae and is very common in Shimoga and other districts of Karnataka, where area palms are cultivated extensively. The disease usually spreads rapidly during rainy season.

The symptoms of the disease are first manifested on the nuts which show a water soaked area towards the base of which it gradually spreads. As a result of this , the nuts lose their green colour and young nuts drop down causint heavy loss. The disease also causes withering of leaves and drying up of the tops of trees.

3. Bud rot of palms: This is caused by p:oalmivora. Trees like coconut (coconucifera) and palmyra palm (Borassus flabelliformis) are infected by the fungus.

4. Colacasia blight: The disease is caused by P.Colocasiae. Leaves and corm of colocasia plant are destroyed. Broad patches appearing on the leaves are the characteristic symptoms of the disease.

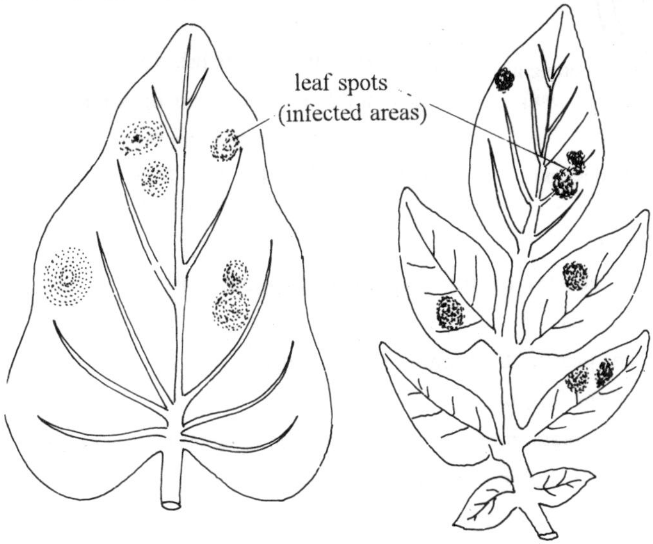

Fig. 12.17 Phytophthora. Colocasia leaf showing symptoms of late blight caused by Phytophthora

Potato leaf showing symptoms late blight caused by Phytophthora

5. Seedling blight: The disease caused by P.oarasitica destroys the seedling of castor (Ricinus communis) and many other plants.

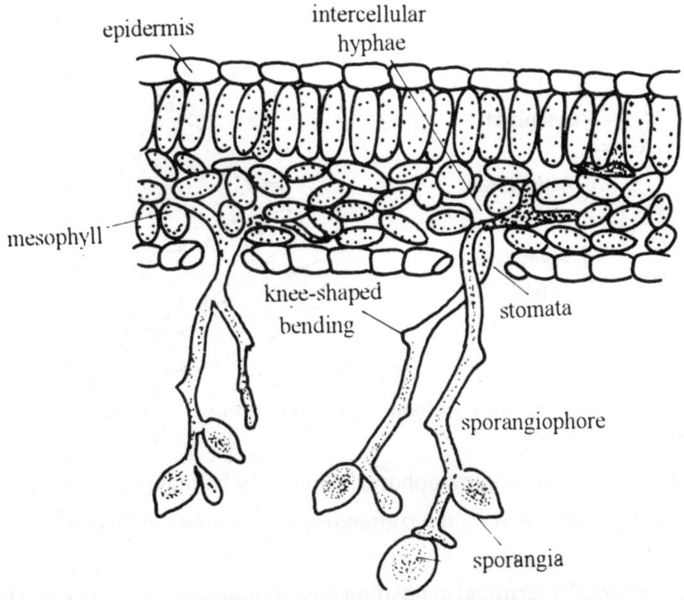

Fig. 12.18 T.S. of host leaf showing sporangiophores and sporangium of Phytophthora

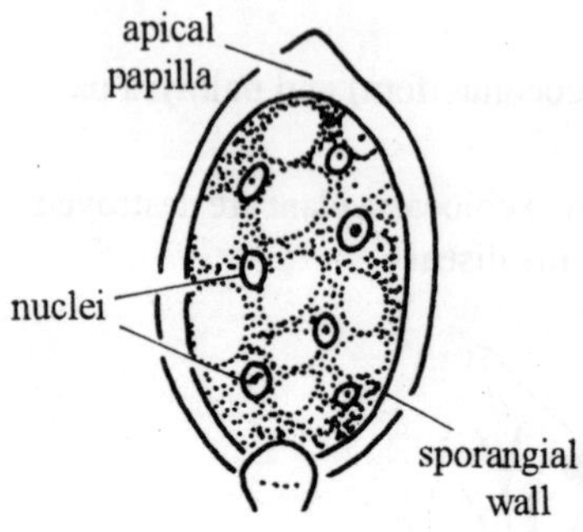

Fig. 12.19 Phtophthora. Single spo-

6. Bud rot of cocoa: P.faberi causes the decay of the cocoa plant in the young condition.

7. Fruit drop: This disease is brought about by P.ohaseoli, which causes young fruits to drop down.

8. *Bud rot of tomato* (Lycopersicum esculentus). Betil leaf (piper betel) etc.

Specimens given in the laboratory

1. Infected leaves or tubers of potato.

Observe the patches on the leaves, cut a section of the leaf to see the reproductive parts. Scrape an infected patch of the tuber and observe under the microscope.

2. Infected nuts of Areca

(Scrape the patches and observe the mycelia under the microscope)

3. Blight of colocasia leaves

(cut a section of the leaf passing through infected patch)

Vegetative structure of the fungus :

1. The plant body is an aseptate, much branched, hyaline coenocytic mycelium.

2. The hyphae may be intercellular, but sometimes intra cellular also.

3. Septa are formed only during reproduction.

4. The mycelium sends out haustoria into the host cells to absorb nutrition. Haustoria may or may not be branched.

5. Oil globules and glycogen present in cytoplasm as reserve food materials.

Asexual reproduction

1. Asexual reproduction is brought about by sporangia.

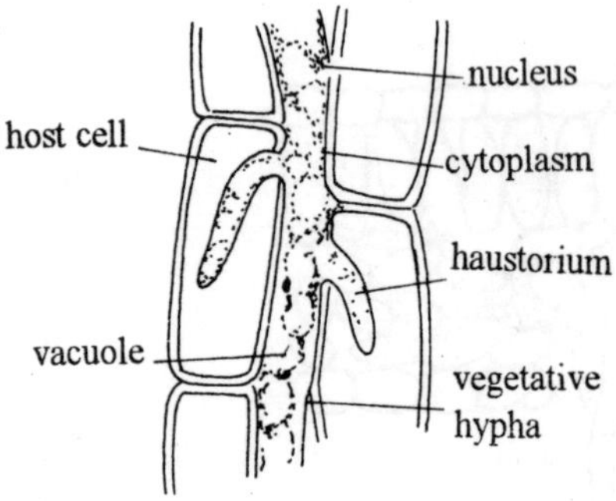

Fig. 12.20 Phytophthora. A portion of mycelium showing haustoria of Phytophthora

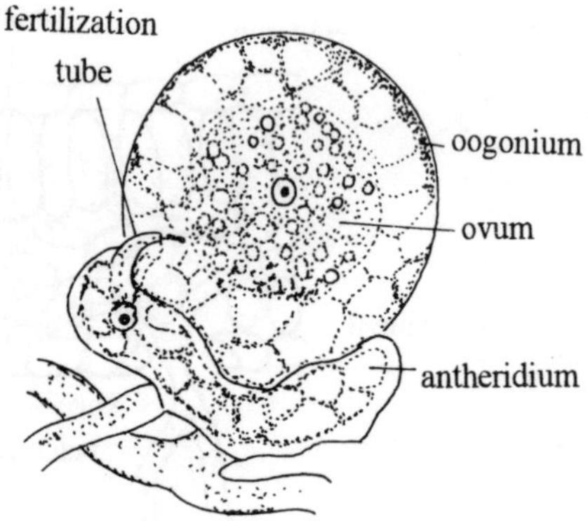

Sex organs

2. The sporangia are formed on sporangiophores which may be branched or unbranched.

3. In the case of infected potato leaves, the sporangiophores come out through the stomata of lower epidermis.

4. The sporangia are originally terminal in position on a sporangiophore. Later however the sporangia lateral

due to the process of evaction.

5. Sporangia are lemon shaped and have a papilla or break.

6. The sporangia reproduction multinucleate and deciduous ie., they are detached from the sporangiophores and are dispersed by wind.

7. On the availability of a suitable host plant a sporangium germinates directly into a new plant. In such a case of sporangium is called a conidium.

8. When moist conditions are prevailing in the atmosphere, the sporangium produces a number of zoospores, and each zoospore develops into a new individual.

Sexual reproduction

1. Sexual reproduction is oogamous. Sex organs are antheridia and oogonia.
2. Species of phytopthora are both homothallic and heterothallic.
3. In homothallic species sex-organs are arranged as follows:

a. Paragynous : Anthcridia and oogonia developed on different hyphae. They are placed side by side.

b. Amiphigynous : Antheridium decelops first. The oogonial hypha pierces thro' the antheridium. Antheridium is present like a collar at the base of the oogonium.

1. Antheridium is club shaped and multinucleate.
2. Oogonium is globose or oval and has a single egg.
3. Sexual reproduction is effected by means of gametangial contact. A slender fertilization tube develops.
4. Zygote or oospore is thick walled. It germinates to form a new mycelium.

Diagrams to be drawn :

1. Habit of infected potato, colocasia or arecanut.
2. Section thro' infected leaf (except for arecanut). To show mycelia and reproductive structures specially the sporangiphores.
3. Sex-organs to show paragynous and amphigynous arrangement.

Systematic position and identification

A. 1. Chlorophyll absent
   2. Reserve food is glycogen ... Fungi

B. 1. Mycelium aseptate and coenocytic
   2. Septa are formed only during reproduction
   3. Oospore thick walled ... Phycomycetes

C. 1. Sexual reproduction not by planogametic copulation
   2. Oogonium has a single oosphere ... Peronosporales

D. 1. Sporangiophores similar to somatic hyphae and have indefinite growth Pythiaceae

E 1. Sporangiophores branched sympodially
   2. Sporangia have an apical papilla and deciduous
   3. Zoospores not released into a vesicle ... Phytophthora

ALBUGO

Common species : *A.candida, A.portulaceae, A.evoles uli, A.tragoponis* etc.

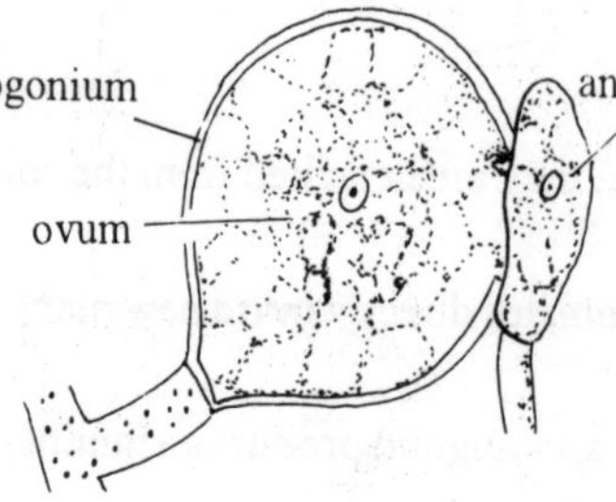

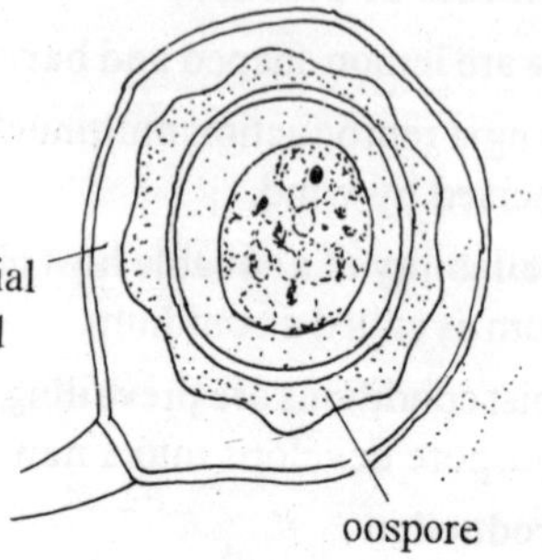

Fig. 12.21 Albugo. Sex organs An oospore

Occurrence : Albugo is an obligate parasite attacking mostly the carciferous plants. The fungus causea a disease called White rust disease. Members f Amaranthaceae, Portulacaceae are also infected by this genus.

Study : 1. Disease caused by the fungus.

2. Disease symptoms.

3. Vegetative structure

4. Reproduction

5. Systematic position and identification

Diseases caused by the fungus.

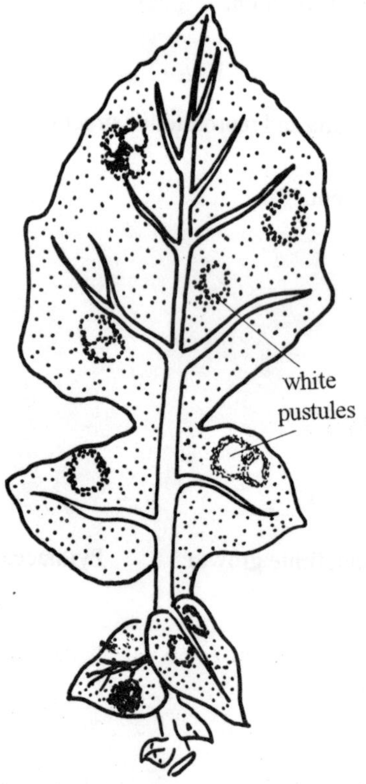

Fig. 12.22 Brassica leaf showing symptoms of white rust disease caused by Albugo

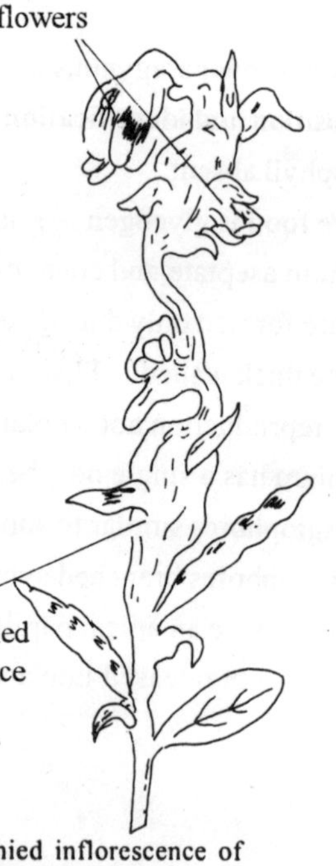

Hypertrophied inflorescence of Brassica infected by Albugo (Cystopus)

Various species of Albugo cause diseases to many flowering plants. The following is a list of host and pathogenic species.

1. A.candia - white rust of cruciferous plants.
2. A.evoluvli - white rust of Evoluvulus alsinoides
3. A.ipomoeae pandoranae white rust of Ipomoea
4. A.bliti - white rust of Amaranthaceae
5. A.portulaceae - white rust of portulaca.

Disease symptoms : The young diseased plants show marked swelling blisters, pustules and twisting in their infected parts which may be leaves, stems, branches flowers of fruits except the roots. In extreme cases, the floral axis gets deformed, may show gall formation and even swell up to 12 - 16 times their nomral size. This is called phyllody. The leaves become succulent and distorted, while the floral organs show marked hypertrophy in as much they become fleshy and turn violet in colour. At lateral stages, the disease is easily recognisable by the white, smooth and mealy pathces that appear on the undersurface of the leaves. The pustules water powdery on account of the spores produced.

Vegetative structure

1. The plant body is an aseptate, coenocytic, much branched mycelium.
2. The mycelium is intercellular growing in between the host cells.

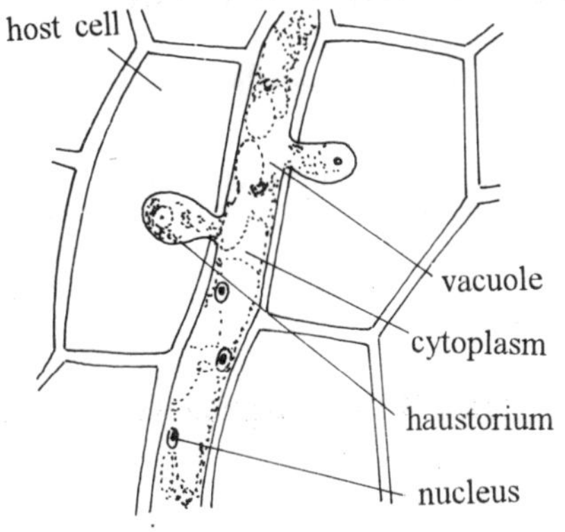

Fig. 12.23 Albugo. Aportion of mycelium showing haustoria

3. Hyphal protoplasm is granular. It contains numerous nuclei, oil globules and glycogen.
4. The mycelium produce short knob like haustoria into the host cells to absorb nutrition.

Asexual reproduction

1. Asexual reproduction is brought about by means of conidia.
2. A section of the infected leaf passing thro' a white pustule will show the asexual reproductiive structure.
3. The mycelium produces a number of erect, unbranched and clushaped, hyphae. These are called conidiophores.
4. The conidiophores are multinucleate and have thickened side walls. They are not separated from the underlying mycelium by any cross septum.
5. The conidiophores at their tip oroduce a chain of conidia in basipetalous succession (older conidia at the top younger ones below).
6. In between, the conidia in the conidial chain is present a gelatinous disc or pad, which helps in the disjoining of the chain.
7. The conidia reproduction multinucleate and non-motile
8. As a result of the production of large number of conidia, a pressure is exerted on the epidermis which rupture releasing them (conidia) outside.
9. The conidia are wind dispersed on the availability of a suitable substratum, they germinate directly into a new individual.
10. If the atmosphere is having enough moisture, the conidia function like sporangia and produce many biflagellate zoospores. Each zoospore develops into a new individual.

Sexual reproduction :

1. Sexual reproduction is oogomous. Sex-organs are antheridia and oogonia.
2. Sex-organs are generally seen in hypertrophied parts.
3. Antheridium is a small club shaped structure terminating a hypha. It has many male nuclei.
4. Oogonium is globose. A mature oogonium has one or many eggs. The oogonium with many eggs is called a Compound Gospture.

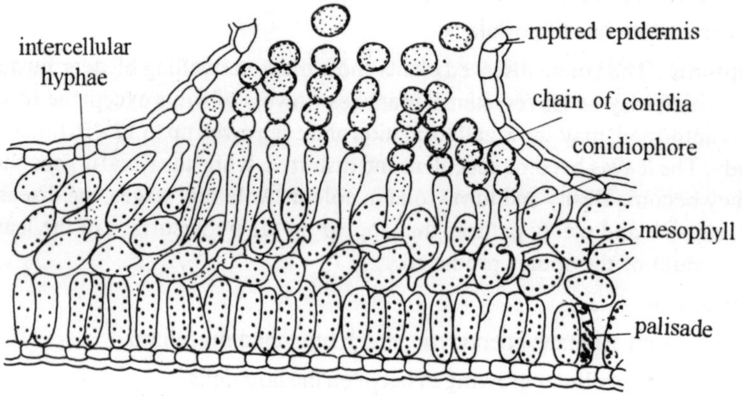

Fig. 12.24 T.S. of host leaf showing conidiophores and conidia of

5. Sexual reproduction is effected by gametangial contact. Fertilization tube developed from the antheridium helps in the transfer of male nuclei to the oogonium.
6. Zygote or oospore has a thick sculptured wall. The wall sculpturing is used as a taxonimic creiterion.

Specimens given in the laboratory

1. Infected leaves showing white blisters.
2. Hypertrophied stem, inflorescence etc.

Diagrams to be drawn

1. Habit sketch of infected plant parts.
2. Section of leaf showing conidiophores.
3. Section of a hypertrophied part to show sex-organs oospores etc.

Systematic position and identification

A. 1. Chlorophyll absent
   2. Reserve food is glycogen ... Fungi

B. 1. Mycelium aseptate and coenocytic
   2. Septa are formed only reproduction
   3. Oospore thick walled ... Phycomycetes

C. 1. Sexual reproduction not by palnogametic copulation
   2. Oogonium has a single oosphore (rarely many) ...Peronosporales

D. 1. Obligate parasites
   2. Conidiophores unbranched
   3. Oosphore has a sculptured wall...Albuginceae

E. 1. White blisters on the leaves
2. Hypertrophy of inflorescence
3. Haustoria button shaped
4. Conidia in basipetalous succession ... Albugo

SCLEROSPORA

Common species : *S. graminicola, S.sorghi, S.philippinesis* etc.

Occurrence : The species of Sclerospora are obligate parasites infecting many families of flowering plants. Graminaceous members are mostly infected by the fungus causing severe reduction in yield.

Study : 1. Diseases caused
2. Disease symptoms
3. Vegetative structure
4. Reproduction
5. Systematic position and identification

Diseases caused : The species of sclerospora infect the members of the family gramineae like jowar, bajra etc. The following is a list of some of the diseases:

1. Greenear disease or downy mildew of bajra

The disease is caused by S.grominicola

The disease symptoms are of two types - green ear symptoms and downy mildew symptoms. The symptoms of the disease depend on the time of infection. If the infection takes place at the seedling stage, downy mildew symptoms are produced and if the infection is quite late ie., at the flowering stage, greenar symptoms are produced.

Downy mild symptoms :

1. Leaves are infected and proliferation of leaves takes place.

2. Leaves get etiolated as chlorosis sets in mainly on the upper surface.

3. Under surface of the leaf gets invested with soft whitish feathery mycelium. The pustules are very soft to touch like the down feathers hence the name downy mild.

4. The plant appears stuned and busty.

Green ear symptoms :

1. Leaves form a number of vegetative buds.

2. Inflorescence exhibits the green ear symptoms.

3. The entire inflorescence appears like a cluster of leafy appendages.

4. Stamens, awans, pistil etc become hypertrophied and distorted, complete sterility sets in and no grains are formed.

Fig. 12.25 Sclerospora Proliferated flower

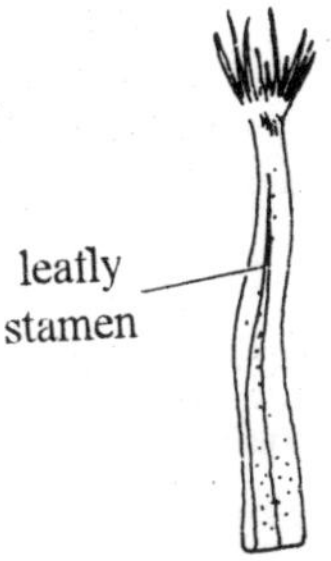

Infected stamen

Downy mildew of jawar : This is caused by S.sorghi. the disease is prevalent in the coastal regions of South India.

Disease symptoms are mainly visible as patches or pustules of downy growth of the leaves.

Downy mildew of maize : The disease is caused by S.philippinensis. the disease symptoms are almost similar to those of mildew of jowar.

Vegetative structure of fungus :

1. The plant body is a coenocytic aseptate, much branched mycelium.
2. The mycelium is intercellular and produces vesicular haustoria to absorb nutrition from the host cells.
3. Septa appear only at the time of development of reproductive structures.

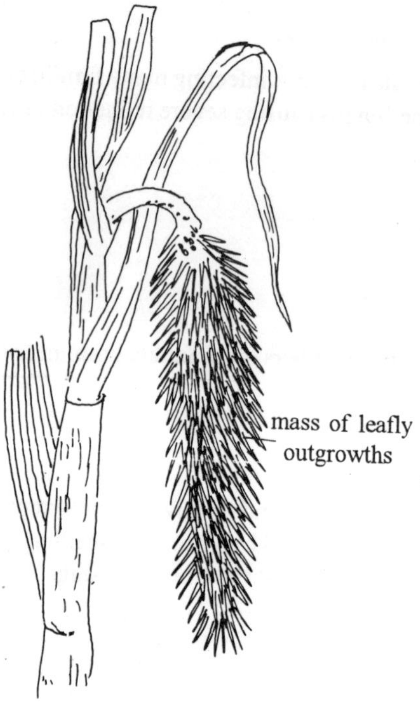

Fig. 12.26 Sclerospora. An infected ear of bajra showing green ear disease caused by Sclerospora

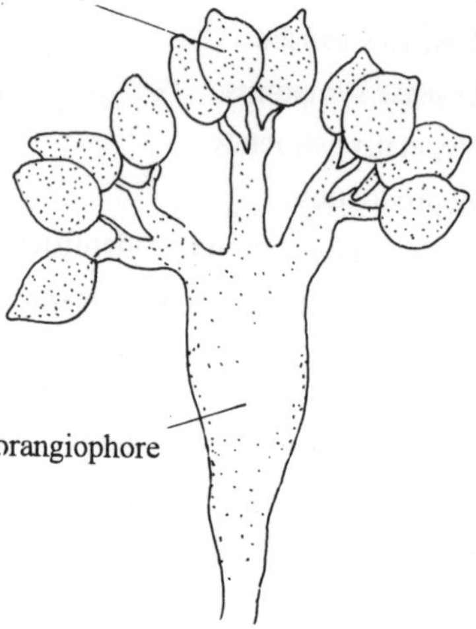

Conidiophore with conidia of Sclerospora

Reproduction

1. Asexual reproduction takes place by the formation of spores or conidia on sporangiphores (conidiophores).
2. The sporangiosphores are erect, broad and branched hyphae bearing a number of sporangia.
3. Sporangiophores may be solitary or produced in groups of 3-4.
4. Sporangia are elliptical or oval and have a papillate outgrowth at the tip.
5. Sporangia vary in size 13-34 metre (long) and 12-23 metre (broad).
6. The sporangia develop during night and are shed during day time.
7. The sporangia are deciduous and germinate by producing reinform zoospores.

Sexual reproduction

1. Sexual reproduction takes place by means of oogamy. Sex-organs are antheridia and oogonia.
2. Fertilization results in the development of zygote. Zygotes are usually formed at the end of the growing season of host. They can be observed at brownish areas on the hast plant.
3. Oospore (zygote) has an outer wall which is unevenly thick. On germination the oospore develops into a new individual.

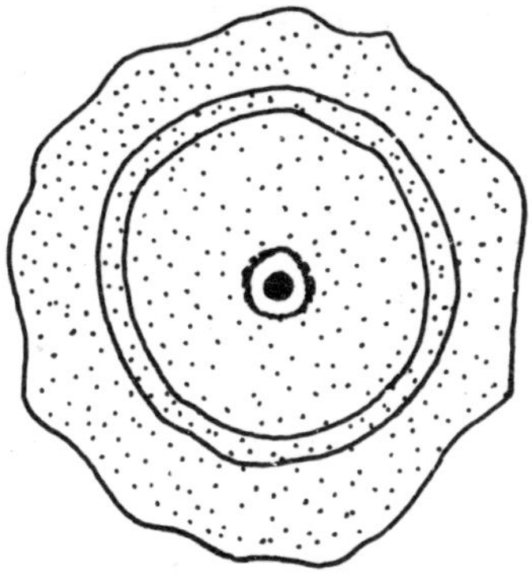

Fig. 12.27 Sclerospora. An oospore

Specimens given in the laboratory :

1. Leaves having downy pustules.
2. Green ear of Bajra.

(Cut the sections of leaves and tease open the green ear to study the malformations).

Diagrams to be drawn:

1. Habit sketch of downy mildew symptoms
2. Habit sketch of green ear symptoms
3. Deformed flower (enlarged)
4. Sporangiophores

Systematic position and identification

A. 1. Chlorophyll present
   2. Reserve food is glycogen ... Fungi

B. 1. Mycelium aseptate and coenocytic
   2. Septa are formed only during reproduction
   3. Oospore thick walled ... Phycomycetes

C. 1. Sexual reproduction not by planogametic copulation
   2. Sexual reproduction oogamous
   3. Oogonium has a single oosphere ... Peronosporales

D. 1. Sporangiophores exhibit dichotomous branching
   2. Sporangia borne singly on the sporangiophore ... Pernosporaceae

E. 1. Sporangiopores caducous and stout
   2. Oospore wall united with that of the Oogonium ... Sclerospora

PERONOSPORA
(Ainsworth & Bisby 1954)

Common species : *P.pisi, P.parasitica, P.distructor* etc.

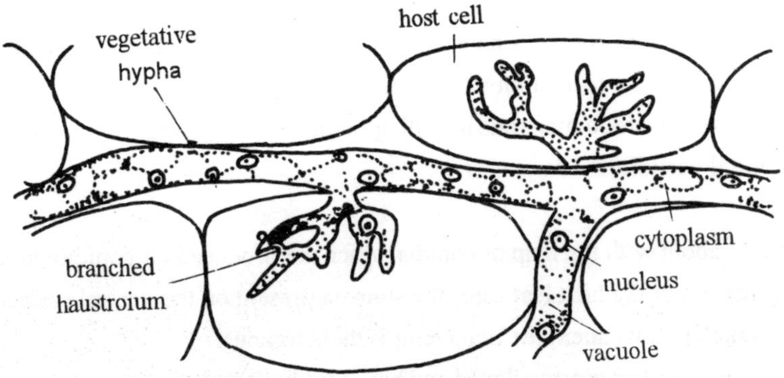

Fig.12.28 Infected host cells showing haustoria of Peronospora

Occurrence : The genus peronospora is an obligate parasite infecting many families of flowering plants. The disease caused is popularly called downy mildew because of the soft downy growth of the hyphae.

Study : 1. Diseases caused by the fungus

2. Diseases symptoms

3. Vegetative structure

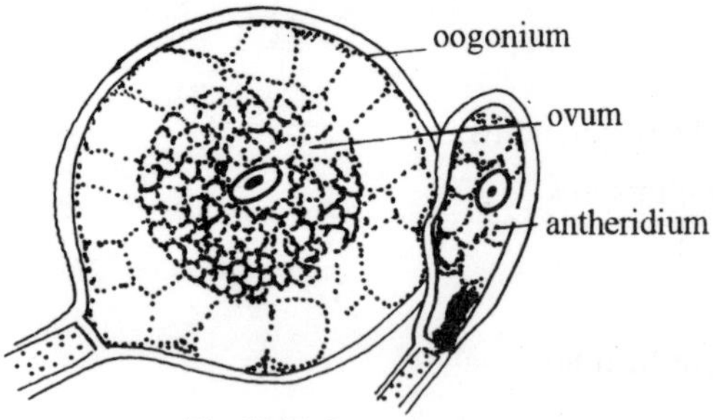

Fig 12.29 Peronospora Sex

4. Reproduction

5. Systematic position and identification

Diseases caused by the fungus :

The genus pronespora has about 75sp. Most of which are pathogenic causing diseases to many economically important plants. Some diseases caused by them are below :

1. Downy mildew of crucifers. This disease is caused by P.parasitica. cruciferous plants like raddish (Raphanus sativus), Cabbage (Brassica oberaceavar capitata) Cauliflower (Brassica oleracea var botrytis) mustard (Brassica compestis) etc.

2. Downy mildew of peas. P.pisi caused the downy mildew peas (pisum sativus).

3. Downy mildew of onion. P.destructor caused the downy mildew of onion (Allicumcepa).

This is caused by P.cherianthi. The host plant is cherianthus cherii.

Diseases symptoms :

1. Downy mildew symptoms appear quite early in the development of the host plant.

2. When the plants are about 10-12 cms height, yellow brown patches of irregular shape appear on the upper surface of the leaves.

3. The lower surface of the leaves at these regions become yellowish invested with soft downy growth of the mycelia.

4. The down growth of the fungus consists of conidiophores.

5. The disease symptoms may appear also on the fruits resulting in the decay of the pericarp and reduction in the size of the seeds.

The appearance of brownish yellow patch on the uppersurface of the leaf and downy growth of mycelia on the lower surface is charactristic of all the species of peronospora.

Vegetative structure :

1. The plant body is coenocytic, much branched, aseptate mycelium.

2. The mycelium is intercellular and produces knob like haustoria which absorb nutrition from the host cells.

3. Septa appear in the hyphae only during the formation of reproductive structures.

4. The granular cytoplasm is rich in reserve food materials like oil, globules, glycogen etc.

5. The tubular hyphae have a central vacuole.

Reproduction :

1. Asexual reproduction is brought about with the help of conidia which are produced on conidiophores.

2. The conidiophores usually come out of the host leaf thro' the stomata present on the ventral surface.

3. The conidiophores give out branches at its apex; the branching is dichotomous.

4. The cellemate branches also called sterigmata are reflexed and bear conidia at their tips.

5. The conidia may be oval or elliptical in shape, greenish yellow or purple tinged in colour. The conidia are sensitive to dessication.

6. Conidia are wind dispersed and on germination develop into a new individual.

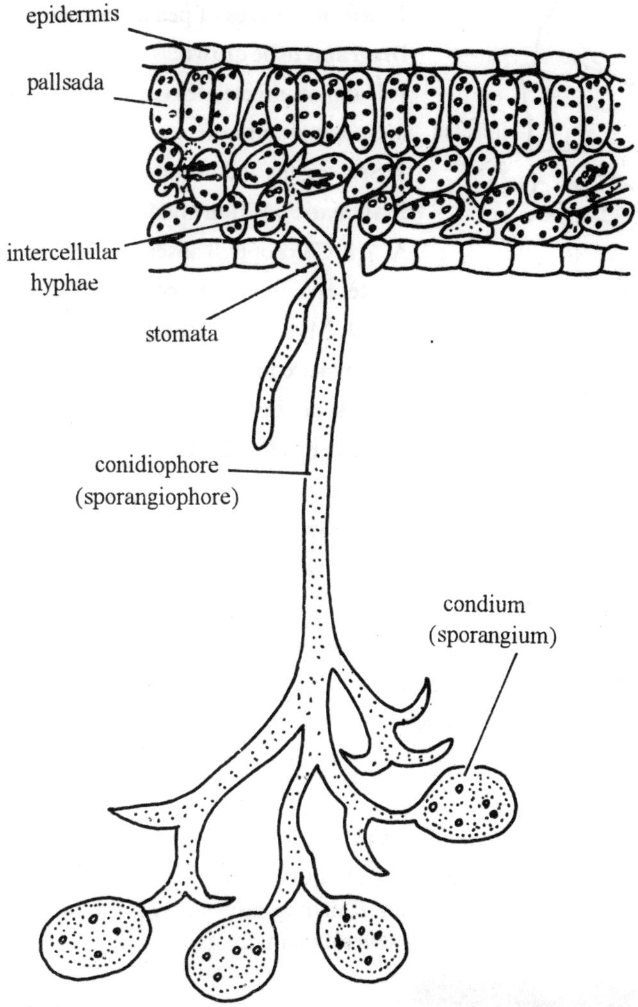

Fig. 12.30 T.S. of host leaf showing sporangiophore and sporangia of Perenospora

Sexual reproduction:

1. Sexual reproduction is oogamous. Sex-organs are antheridia and oogonia. The sex-organs are arranged paragynously (see-phytophthora).

2. Antheridium is a small tubular structure. A mature antheridium has a single functional male nucleus.

3. Oogonium is a globose structure. It has a central ooplasm with an oosphere and a peripheral ooplasm.

4. Sexual reproduction is effected by means of gametangial contact. A slender fertilization tube develops from the antheridium. It pierces through the oogonial wall and released its contents inside.

5. The zygote or oospore which is a result of fusion is light yellow in colour and has a sculptured wall called epispore. Oospores can be seen in the host tissues.

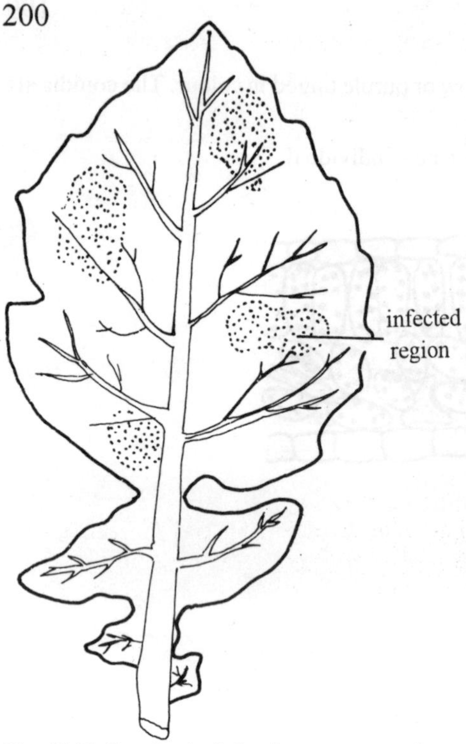

Fig. 12.31 Brassica leaf showing symptoms of

6. After a period of rest the oospore develops into a new individual by producing a germ tube.

Specimens given in the laboratory:

1. Infected leaves of pea or raphanus.

Diagrams to be drawn:

1. Habit of infected leaves
2. Sporangiophores
3. Sex-organs and oospore

Systematic position and identification

A 1. Chlorophyll absent
2. Reserve food is glycogen ... Fungi

B 1. Mycelium aseptate and coenocytic
2. Septa are formed only during reproduction
3. Oospore thick walled ... Phycomycetes

C 1. Sexual reproduction oogamous
2. Oogonium has a single oosphore ... Preonosporales

D 1. Sporangiophore exhibit dichotomous branching
2. Sporangia borne singly on the sporangiophore ... Peronosporaceae

E 1. Ultimate branches of the sporangiophores are recurved
2. Conidia germinate by a germ tube
3. Oospore germination direct ... Peronospora

RHIZOPUS

Cómmon species : *R.nigricans, R.stolonifer, R.arrhizus, R.artocarpi* etc.

Occurrence : Species of the genus are the most ubliquitous of the fungi. The members are saprophytes growing on any dead and decaying organic matter. Rhizopus grows very commonly on stale bread hence it is popularly called the bread mold. It also causes several plant diseases.

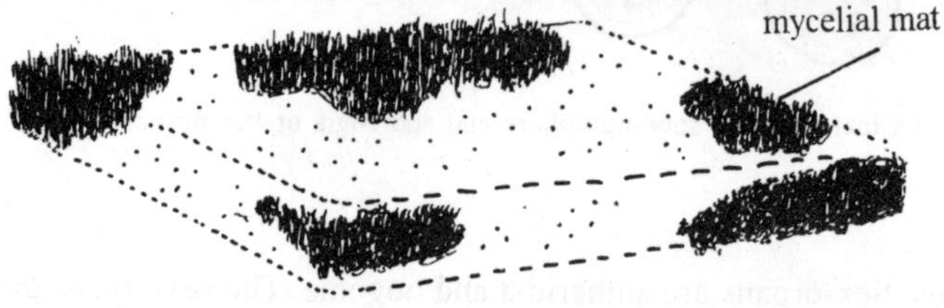

Fig. 12.32 Rhizopus. A slice of bread showing

Study

1. Diseases caused and disease symptoms
2. Method of culturing
3. Vegetative structure

4. Reproduction
5. Systematic position and identification

Diseases caused : Some species of Rhizopus causes the fruit rot diseases in some flowering plants. The following are some of the diseases.

1. Apple rot .... *R.arrhizus*
2. Fruit rot of jack fruit .... *R.artocarpi*
3. Soft rot of strawberries .... *R.stolonifer*
4. Decay of bread .... *R.nigricans*

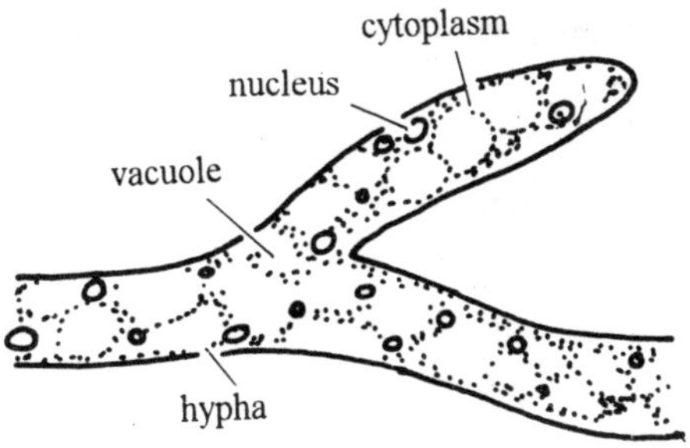

Fig. 12.33 Rhizopus. A portion of mycelium

Disease symptoms of the fruit rot

1. Young fruits drop down prematurely
2. Infected peduncles get decolorised ultimately turning black
3. Mature fruits begin to rot and get decayed

Method of culturing in the laboratory

1. Take a few slices of bread. Add a little water to moisten them.
2. Keep the slices of bread in a petridish and cover it.
3. After few days a soft cottony growth of mycelium may be observed. This is the vegetative phase.
4. A little later small while pin head like structures appear from the mycelium. These are the young sporongia.
5. After few days the white pinheads turn black. This represents the mature sporangia liberating the dark colored spores.
6. It is generally different to observe the sexual stages.

Vegetative structure

1. The plant body of Rhizopus is a much branched, coenocytic, aseptate mycelium.

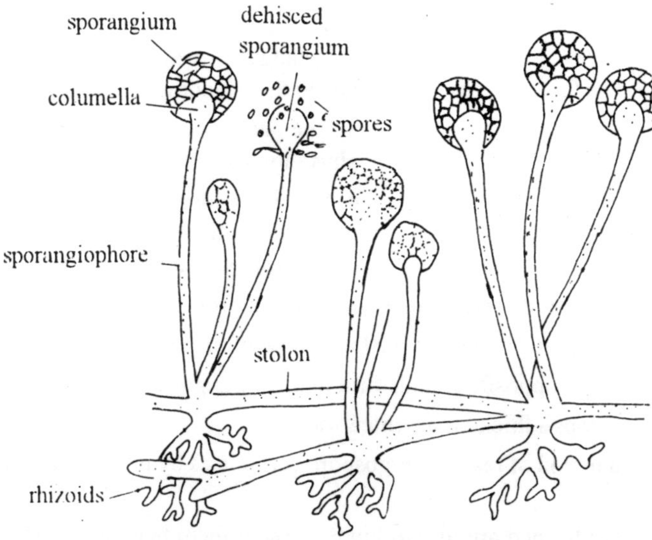

Fig. 12.34 Rhizopus. A portion of mycelium

2. The hyphae are whitish and tubular. There is a centrally running continuous vacuole cell wall in chitinous.

3. The cytoplasm is peripheral. It has numerous nuclei and has oil and glycogen as reserve food material.

4. Cross walls appear only at the time of formation of reproductive structures.

5. The mycelium which is fairly old, has three types of hyphae. These are -

a. Stolon or distributive hyphae : These are horizontal tubular hyphae that grow above the substratum. There and there swollen portions are found on the stolon.

b. Rhizoidal hyphae : From the swollen points of the stolon arise short branched hyphae called rhizoids. These grow vertically down into the substratum helping in anchorage and absorption.

c. Sporangiophores : These are erect, tubular, unbranched hyphae arising at the same points, where rhizoids arise. Sporangiophores at their apex bear a globular or oval sporangia.

Reproduction - Asexual

1. Asexual reproduction takes place by non motile spores produced in sporangia.

2. Sporangia, terminate the unbranched sporangiphores.

3. Occupying the central region of the sporangium is found a dome shaped columella, while protoplast is in continuation with that of sporangiophore.

4. In between the columella and the wall of the sporangium a large number of spores are produced.

5. The spores are non motile and have a pigmented wall due to which they appear dark brown.

6. Spores are spherical, non motile and multinucleate. They are wind dispersed and on the availability of a suitable substratum develop into new individuals.

7. Occasionally thick walled chlamydospores may be formed as a means of asexual reproduction.

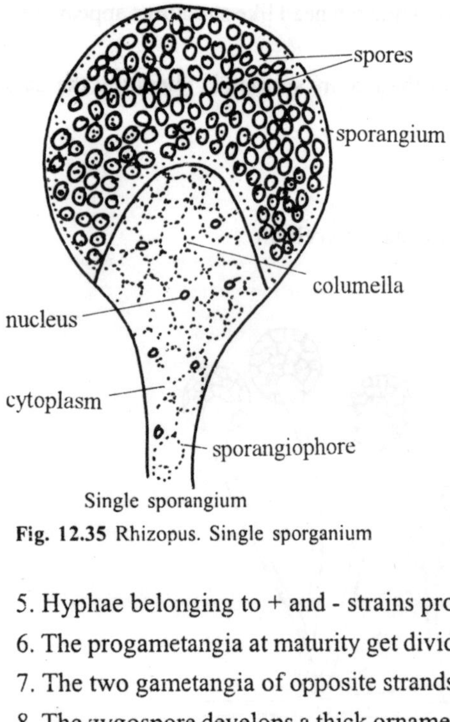

Single sporangium

Fig. 12.35 Rhizopus. Single sporganium

Reproduction - Sexual

1. Sexual reproduction in Rhizopus takes place by means of gametangial copulation.

2. The gametangia reproduction isogametangia, but the gametangia formed on the same mycelium will not fuse. This is due to the phenomenon of heterothallism.

3. The gametangia that fuse are morphologically similar but physiologically different.

4. Since the compatible gametangia cannot be distinguished into male or female, they are designated as + and - strains in order to show that they are different.

5. Hyphae belonging to + and - strains produce short outgrowths called progametangia.

6. The progametangia at maturity get divided into, a basal suspensor and a terminal gametangium.

7. The two gametangia of opposite strands fuse to form a zygospore.

8. The zygospore develops a thick ornamented wall.

9. After undergoing a period of rest, the zygospore germinates by producing an erect hypha, whose free end forms a sporangium.

10. The spores that are formed are of two kinds; 50% of them belong to + strain and the rest - strain. But morphologically the spores are thick.

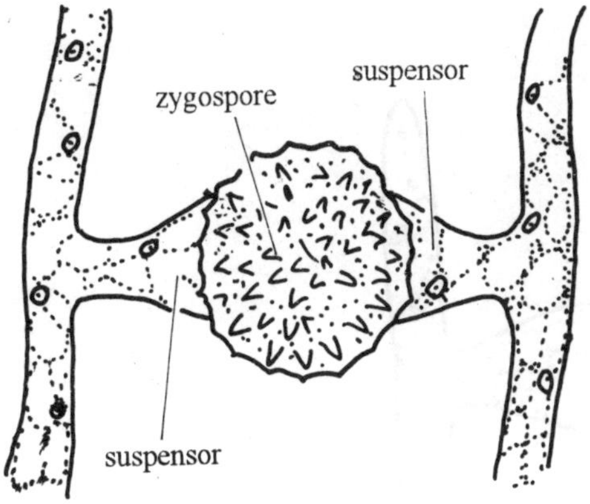

Fig 12.36 Rhizopus. A

Diagrams to be drawn

1. Habit sketch
2. Plant body enlarged
3. Sporangium enlarged
4. Stages in sexual reproduction.

Systematic position and identification

A. 1. Chlorophyll absent
   2. Reserve food is glucogen ....**Fungi**

B. 1. Plant body aseptate and coenocytic
   2. Septa formed only during reproduction ....
Phycomycetes

C. 1. Asexual reproduction by non motile spores
   2. Nutrition saprophytic ....**Mucorales**

D. 1. Conidia not present
   2. Columella present inside the sporangium......................... **Mucoraceae**

E. 1. Rhizoidal hyphae present beneath the sporangiophores
   2. Stolon will not absorb nutrition from the substratum ....**Rhizopus**

MUCOR

Common species : *M.mucedo, M.javanicus, M.priani* etc.

Occurrence : Same as Rhizopus. It is essentially a saprophyte.

Study : 1. Habit
2. Vegetative structure
3. Asexual reproduction
4. Stages in sexual reproduction
5. Systematic position and identification

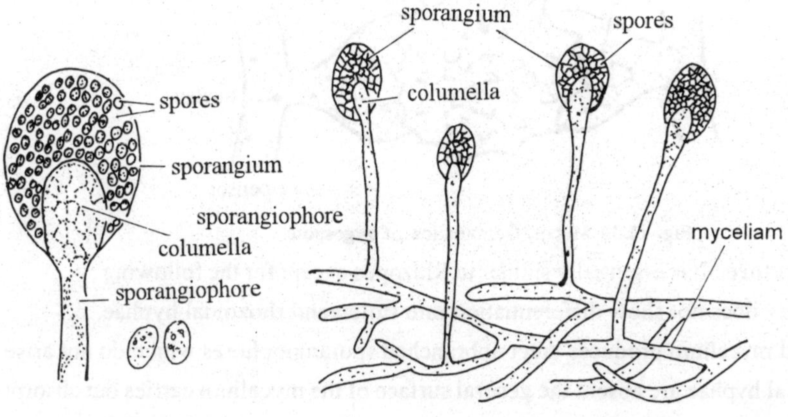

Fig. 12.37 Mucor. A portion of mycelium showing sporangia Single sporangium and spores of Mucor

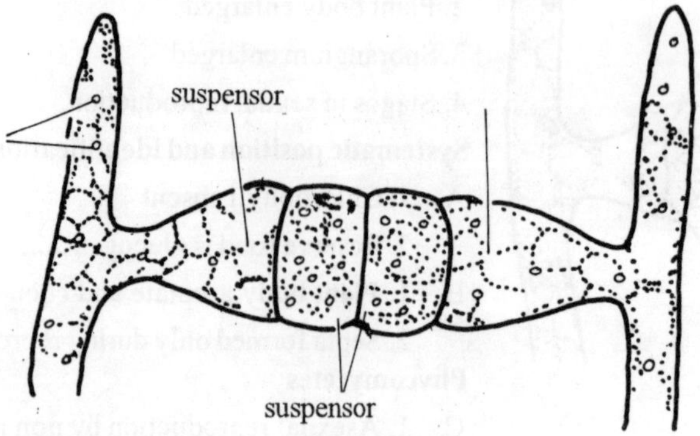

Fig. 12.38 Mucor. Gametangia (sex organs)

Habit : Same as Rhizopus

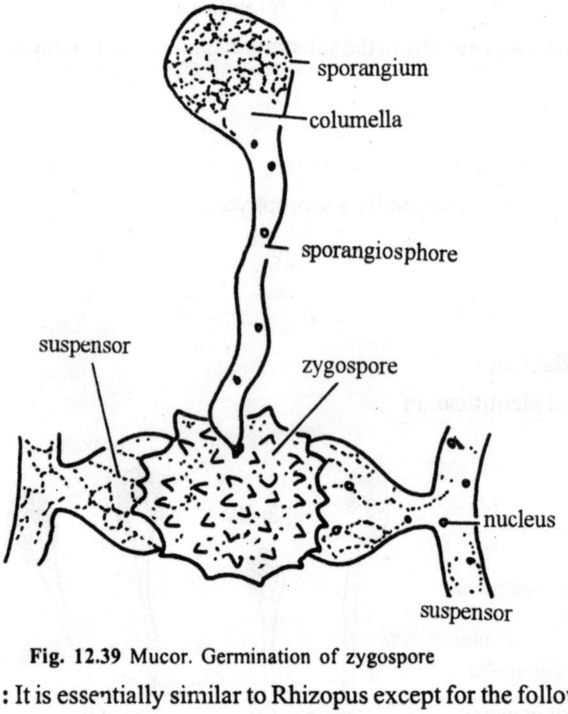

Fig. 12.39 Mucor. Germination of zygospore

Vegetative structure : It is essentially similar to Rhizopus except for the following :

1. The plant body does not show differentiation into stolon and rhozoidal hyphae.
2. The branched mycelium produces erect unbranched sporangiophores which do not arise in groups.
3. Since rhizoidal hyphae are absent the general surface of the mycelium carries out absorption.

Asexual reproduction : Same as Rhizopus.

Sexual reproduction : Same as in Rhizopus.

Diagrams to be drawn : Same as in Rhizopus.

1. Plant body does not show differentiation into stolon and rhizoidal hyphae.

2. Sporangia arise singly ... Mucor

PILOBOLUS

Common species : *P.crystallinus*

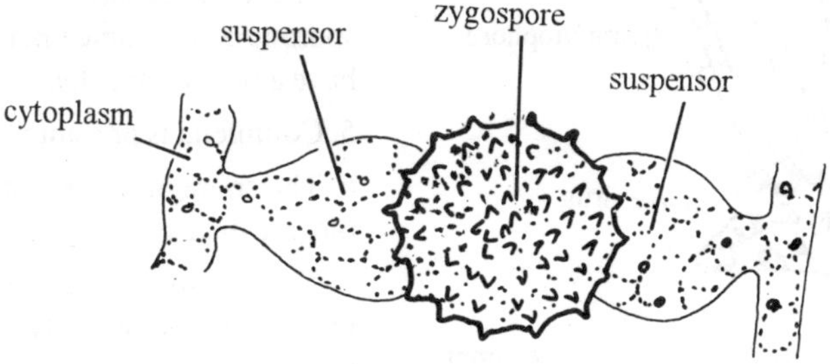

Occurrence : The species of Pilobolus and Coprophilous (dung loving).

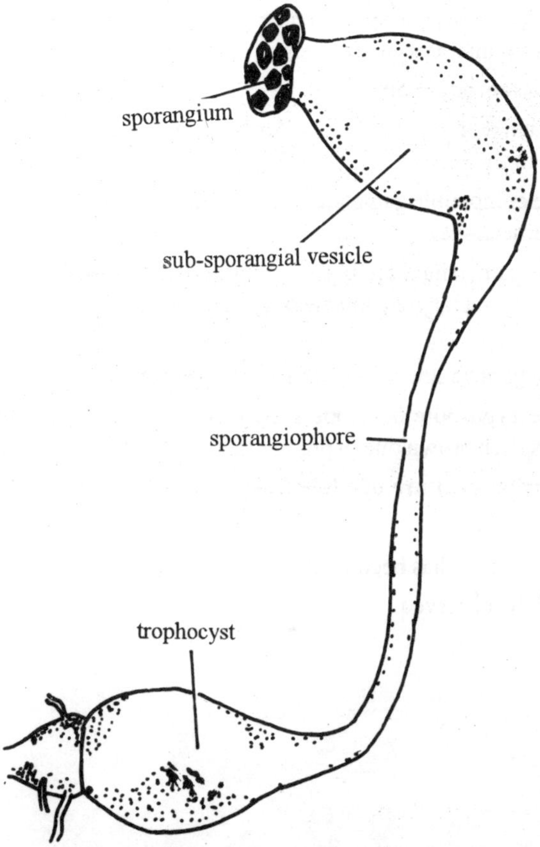

Fig. 12.41 Pilobolus. Single sporangiophore with sporangium

All of them are saprophytic growing on cowdung, horse dung etc.

FIg12.40 mucor A zygospore

Study : 1. Vegetative structure

2. Reproduction

3. Systematic position and identification

Vegetative structure

1. The plant body is an aseptate, coenocytic, much branched mycelium.

2. The mycelium is usually just above the substratum. It produces rhizoids below and sporangiophores above.

3. Rhizoids help in anchorage and absorption.

4. There is a central vacuole in the tubular hyphae. The peripheral cytoplasm has many nuclei, oil globules and glycogen.

Asexual reproduction

1. Asexual reproduction is brought about by spores which are non-motile.

2. Sporangiophores arise from the horizontal mycelium. They are erect and unbranched.

3. The sporangiophores have four parts. They are :

a. Trophocyst

b. Sporangiophore stalk

c. Sub-sporangial vesicle

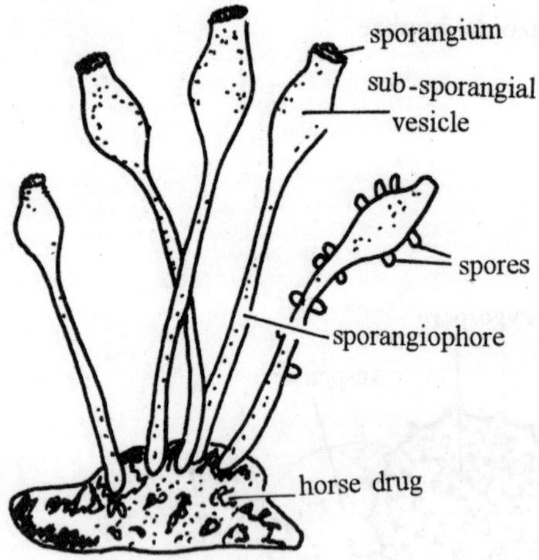

Fig. 12.42 Pilobolus. Discharge of spores from sporangium

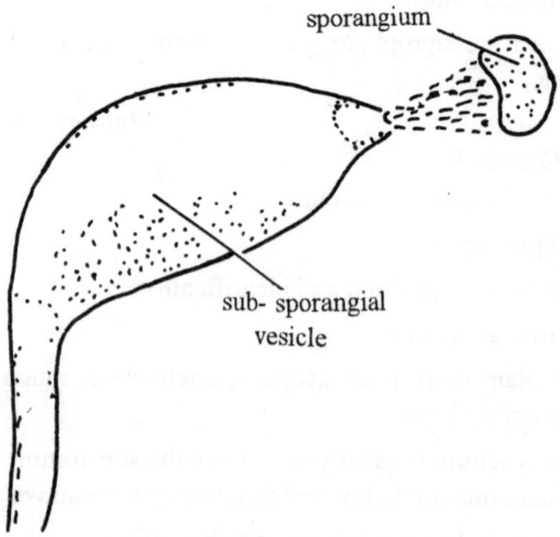

Fig 12.43 Piobous Fungus growing on dung

d. Sporangium

1. The swollen base of the sporangiophore is called Trophocyst (Gr.trophe = food + Kysis = bladder).

2. The stalk of the sporangiophore is tubular and narrow.

3. The tip of the sporangiophore is swollen to forma vesicle called sub-sporangial vesicle.

4. The terminal part consists of the sporangium proper. The sporangia are somewhat longitudinally compressed and have a rigid dark coloured terminal wall.

5. Columella is present inside the sporangium.

6. The sporagnium is deciduous. It is discharged with explosive dehiscence mechanism.

7. The sub-sporangial vesicle ruptures in a vertical slit due to osmotic pressure. This causes a jet of sap to force out the sporangium.

8. The sporangia stick to grasses and other plants and when these are consumed by herbivorous animals, they (sporangia) are dissolved in the alimentary canal setting of the spores free. The spores come out thro' the fecal matter and grow into new individuals.

Sexual reproduction :

1. Sexual reproduction takes place by means of gametangial copulation.

2. The compatible gametangia are similar morphologically (Isogametangia).

3. The gametangia are borne on curved suspensors, when the two gametangia come nearer, they appear like a pair of tongs.

4. The gametangia fuse and form a zygospore.

5. The zygospore develops a thick and smooth wall. The wall is dark brown due to pigmentation.

Laboratory culture of pilobolus :

1. Collect fresh dung of herbivorus animals, preferably of horse.

2. Keep it enclosed in a petridish. Add a little water, if the dung has become dry.

3. After few days, profuse growth of the mycelium may be observed.

4. A little later sporangiophores also will develop.

Diagrams to be drawn :

1. Habit

2. Plant body enlarged

3. Sporangophore enlarged

4. Sex-organs and zygospore

Systematic position and identification

A 1. Chlorophyll present

2. Reserve food is glycogen ... Fungi

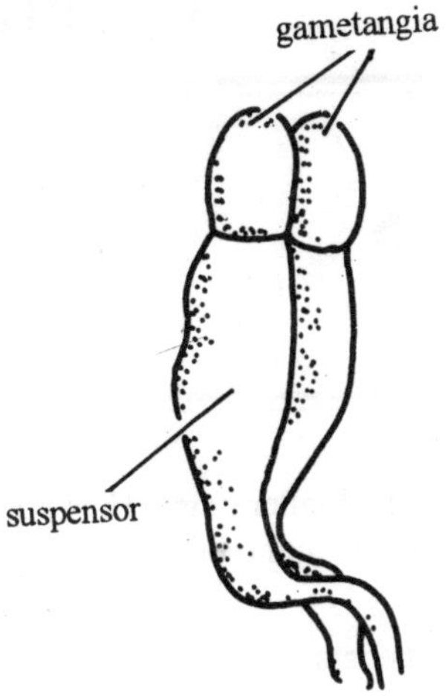

Fig. 12.44 Pilobolus. Young gametangia (sex organs)

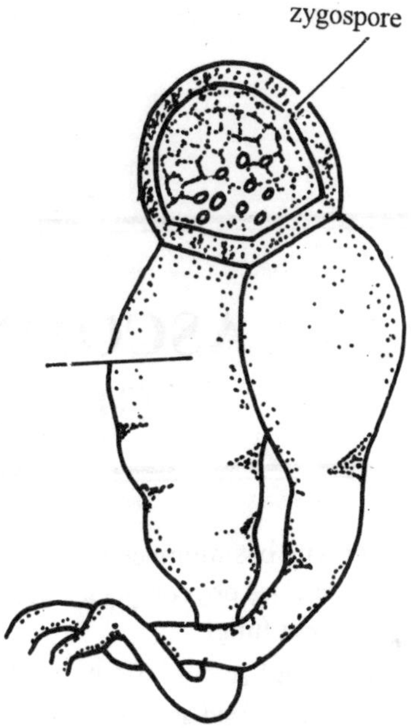

A zygospore

B 1. Plant body, a septate, coenocytic much branched mycelium.

2. Sexual reproduction produces zygospores ... Phycomycetes

C 1. Members mostly saprophytes

2. Asexual reproduction by aplanospores ... Mucorales

D 1. Sporangia have small columella

2. Terminal wall of the sporangium is rigid and remarkable ... Piloblaceae

E 1. Habitat is coprophilous

2. Sub-sporangial vesicle present

3. Sporangia are discharged by explosion mechanism ... Pilobolus

CHAPTER 13

ASCOMYCETES
(The Sac Fungi)

Ascomycetes together with basidiomycetes are often referred to as the higher fungi. They are considerably more advanced than the class phycomycetes in their complex structure and reproductive features, and are probably derived from lower fungi.

According to Aisnworth and Bisby (1950), there are about 1820 genera and 15,500 species of Ascomycetes. The Ascomycetes are called so, because they produce their spores in a special sac like structure called Ascus. In great majority of Ascomycetes as a rule each ascus produces eight ascospores. But this may occasionally vary.

The yeasts, some of the green and black moulds, the ergot fungus, the powdery mildnews, the cup fungi, the delicious morels, the buffeles and Neurospora- the Cinderella of geneticists are all Ascomycetes.

Occurrence: Ascomycetes are found in variety of habitats practically in all seasons of the year. The members may be parasitic or saprophytic. As parasites they are generally endoparasites; but some are ecto parasites (crysiphales). The saprophytic Ascomycetes usually live on the soil or on decaying logs of wood and leaves and produce large frutifications. A few forms are hypogean (gr-hypo-under geo-earth) developing and remaining underground. Many members are coprophilous (gr-koprosdung + phillen to love) growing only on the dung of certain animals. Some of the Ascomycetes are also marine in distribution. Marine Ascomycetes may be saprobic on various types of organic matter sub-merged or floating on parasitic or maine algae and higher plants.

Ascomycetes and Human Welfare

Ascomycetes are important in human welfare in multifarious ways. Like other fungi, they also do both good and bad to man.

Chactonium destroys the fabrics containing cellulose. In destroying crops, crop plants as well as timber and ornamental trees Ascomycetes are our worst fungous enemies causing such disease as apple sacb, brown rot of stone fruits, powdery mildew, food rot of cereals, ear rot of corn etc. (Alexopoulos 1956). Ascomycetes have not spared human beings also. Aspergillosis (lung infection) and Ringworm disease are the common infections by Ascomycetes. Claviceps purpurea producing the ergot of rye is poisonous if the infected grains with sclerotic are consumed. On the other hand some Ascomycetes are highly beneficial. Who can forget the fermenting activity of yeast in bakeries and berweries, the drug penicillin obtained by penicillium, the useful activities of Aspergillis in industry and fine delicacies such as morels and turffels.

Somatic structure

Except for members such as yeasts, the plant body of Ascomycetes is a septate much branched mycleium. Cell wall contains a large proportion of chitin. The cells of the hyphae may be uninucleate, binucleate or multinucleate. The cross septum dividing the cell is porous permitting the streaming of cytoplasm from one cell to another.

Ascomycetes mycelium is often organised into fungal tissues. They may be prosenchymatous or pseudo parehcnymatous. The tissues are chiefly associated with the fruitifications.

Asexual reproduction

Asexual reproduction is brought aboyt by a variety of methods such as fragmentation, fission, budding, orthospores formation chlamydospores formation and conidial formation.

Fission and budding are the usual modes of reproduction in unicellular forms such as yeasts. Conidia may be produced in isolated conidiosphores or conidiosphores may aggregate into flat materials like hyphal masses called acervuli (sing.acervulus).

Sexual reproduction

As in other classes of fungi even in Ascomycetes sexual reproduction aims at the fusion of two compatible nuclei. This takes place by a variety of methods. Sex organs may or may not be formed in Ascomycetes. When formed the male reproductive structure is a club-sjaped antheridium and the female reproductive structure is a flask shaped ascogonium. It has a swollen base and terminal elongated portion called Trichogyne. Both the antheridia and ascogoma are multinucleate. Sexual reproduction as in yeast may be by the fusion of isogametangia.

In general, Ascomycetes sexually reproduce by the following methods.

1. Gametangial copulation, 2. Gametangial contact, 3. Spermatization, 4. Somatogamy.

Stages in sexual reproduction

Unlike in Phycomycetes, there are four stages in Ascomycetes at which sexual reproduction is accomplished. They are plasmogamy dikaryotic stage, karyogamy and reduction division. As a result of this there is regular alternation between a monorayotic mycleium, a dikaryotic mycelium and a short lived diploid mycellium. Sexual reproducing monokaryotic mycelium gives ise to secondary mycelium after plsmogamy. In this dikaryotic mycelium karyogamy takes place at a late stage producing a diploid generation which immediately unergoes reduction division to produce haploid cells (ascospores). These germinate to develop into the monokaryotic mycelium.

Life cycle pattern (the process of Ascus formation)

In all Ascomycetes, the life cycle pattern is uniform with marginal variations here and there.

At the time of reproduction, the two sex organs come nearer and the tip of the antheridium comes into contact with the tip of trichogyne. The separating walls dissolve and the antheridial contents migrate into the ascogonium. Plasmogamy takes place first. The male and female nuclei are irregularly distributed. Later they pair and the paris of nuclei are gradually migrate to the periphery of the ascogonial base. At this stage, small protrusions develop from this ascogonial base which eventually elongate to form hyphae. These are called ascogenous hyphae as they produce ascus later. The pairs of nuclei (one male and female) migrate into the ascogenous hyphae. One pair into ascogenous hypha. Herer the nuclei start dividing and produce many nuclei. But even the daughter nuclei remain in pairs and are not scattered. Always a male nuclei and female nuclei will remain in pairs even among daughter nuclei. This type of division where there is mutual attraction between the nuclei of different genotype is called Conjugate division.

After the division, large number of nuclei (in pairs) are formed. The ascogenous hypha now gets divided into many cells. When cell walls are laid down, care is taken to see that a cell always encloses one male and one female nucleus.

Further development now proceeds usually in the terminal binucleate cell produces in hook like structures known as crozier towards a side. Now the two nuclei divide with a specific spindle orientation. In one of the nuclei the spindle is parallel to the long axis of the ascogenous hypha, in the other it is oblique so that one of the daughter nuclei enters the crozier. After the division four nuclei are produced, the cell itself gets divided into three cells : (a) antepenultimae cell having a single nucleus. (b) A penultimate cell having two nuclei (c) an ultimate cell (crozier) having a single nucleus. Penultimate cell is now called the ascus mother cell; as it is the one which produces the ascus. A protrusion develops from the ascus mother cells which elongates to form a sac like structure called Ascus. The diploid nucleus migrates into the ascus where it undergoes meiosis, followed by a mitosis to produce eight haploid nuclei. Each of the nuclei ultimately forms a spore- the Ascospor. After the spore formation the ascus breaks open releasing the ascospores which germinate and form the monokaryotic mycelium.

The above mentioned account gives a general picture of the process of ascus formation in Ascomycetes. While only the penultimate cells of the ascogenus hyphae forms asci, other cells may also form them. Further, after the formation of ascus, the crozier and antepenultimate cell may fuse to form another terminal cell and the entire process is repeated.

In some forms, the antheridium is not formed or even when formed does not function. In such cases the ascogonial nucleu themselves pair and the process goes on as usual.

Structure and development of identificaiton

While the asci are formed naked and the ascospores are libeated as soon as they are formed, in a great majority of Ascomycetes the asci are enclosed in a compact protective structure clled fruiting body or frutification. Usually when the asci as ascogenous hyphae are being formed, the vegetative hyphae grow round the and formal protectie envelop or the sex organs are formed in the innerlining of the already formed frutification. The main function is to afford production to these delicate reproductive structures. The general name given to a frutification in Ascomycetes is Ascocarp.

[Gr.Askos = sac + karpos = fruit]. Ascocarpus vary in structure and compostion. On the basis of this they may be classified into following four categories [Alexopoulos, 1956].

1. Cleistothecium. Peiclosto thecia. Gr.kleistas = closed + theke case. This is a completely closed structure, may be lgobose or ovoid. The wall is many layered and is called the peridium. Within this, the asci lie scattered. The ascospores will be liberated only on the decay of the peridium. In this type, there is no scope for the regular liberation of ascospores.

2. Perithecium. Plperithcia. Gr.Pere - around. These are flask shaped structures possessing a small opening called ostiole to facilitate the liberation of ascospores. Lining the inner wall of the perithecium at both are numerous asci parallely arranged. Intermingled with the asci, sterile hair like processes are found called Paraphyses (Sing.paraphysis).

3. Apothecium. (Plapothecia. Gr.Apothika store house). This is a open type of frutification and cup like with a broad opening sometimes saucer shaped also. It has multilayered walled called the peridium. Lining the inner margin of the cup is the lymenial region (fertile) containing asci and paraphysis.

Of the thrre typed of above named frutifications, apothecium is supposed to be the most advanced and deistothecium most primitive.

4. Ascotroma. (Plascostromata). In some, the asci are formed direcctly in a cavity in the stroma, so that stroma itself forms the wall of ascocarp.

In addition to the above, various other modifications of the frutifications occur.

Classification

The frutification is an important criterion in the classification of Ascomycetes. Their formation of non-formation and type of frutification when formed are all of great help in the classification.

Gwynnevaughan and Barnes (1937) divided Ascomycetes into three sub-classes namely.

Sub class : 1. Plectomycetes

(Ascocarps if present have no opening, asci are irregularly arranged)

Sub-class : 2. Pyrenomycetes

(Ascocarps flask saped when ripe asci in parallel series)

Sub-calss 3. Discomycetes

(Ascocarp and wide open when ripe, asci in parallel series)

But many modern mycologists classify Ascomycetes as fllows:

Class - Ascomycetes

Sub-class 1. Hemi Ascomycetes

(Protoqsomycetes) (no-frutification)

Sub-class 2. (Euascomycetes) (true Ascomycetes)

(frutification present)

Based on the nature of the frutification, the sub-class Euascomescetes is divided into three series namely

Plectomycetes (Cleistothecium)

Pyrenomycetes (Perithecium) and

Discomycetes (Apothecium)

SACCHAROMYCES (YEAST)

Common species : *S.cerveisiae, S.ludiwigii, Schizosaccharomyces octosporus*

Occurrence : The members are commonly called yeasts. They are widely distributed in nature. Yeasts are saprophytic and are commonly found in substrata rich in sugars like nectar of flowers, surface of fruits like grapes apples etc. Milk and other food undergo fermentation due to yeasts.

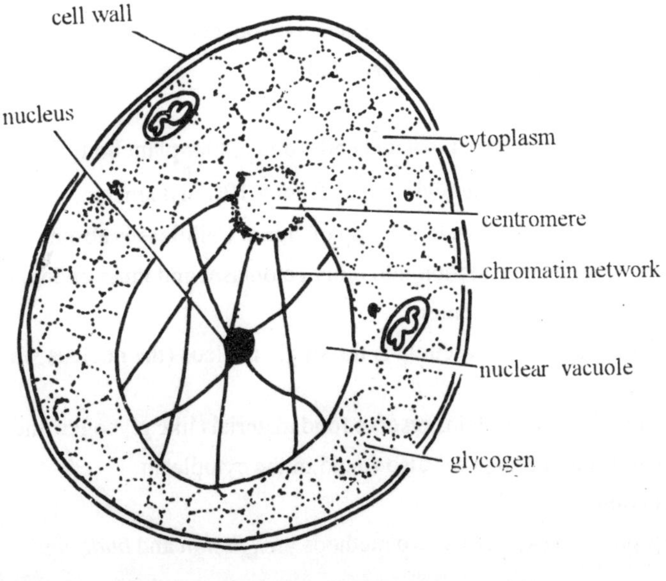

Fig. 13.1 Saccharomyces. A highly magnified cell of Saccharomyces

Study:

1. Method of culturing yeasts
2. Vegetative structure
3. Reproduction
4. Systematic position and identification

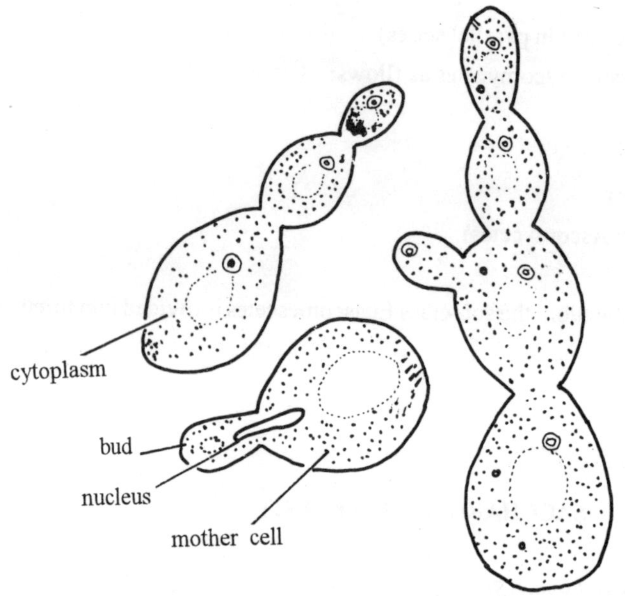

Fig. 13.2 Saccharomyces. Yeast cells showing budding

Culture of yeasts

1. Place small pieces of yeast cake in sugar solution
2. Yeast cells multiply and a dense cluster of cells is produced
3. After few days the yeast cells can be harvested

Vegetative structure

1. The thallus of yeast is made up of isolated cells (unicellular plant body)
2. The cells are usually oval or spherical ranging from 5-10μ in diameter
3. Each cell has an outermost cell wall made up of chitin. Cellulose is absent
4. The cytoplasm shows differentiation into outer ectoplasm and inner endoplasm
5. There is a large vacuole in the center
6. Placed immediately above the vacuole is the single nucleus (the nucleus was mistaken previously as the centrosome)
7. Cytoplasm is granular and contains reserve food materials like glycogen and oil
8. Mitochondira and starch grains are also found in the cytoplasm

Asexual reproduction

1. Yeast cells reproduce asexually by two methods viz., *fission* and *budding*.

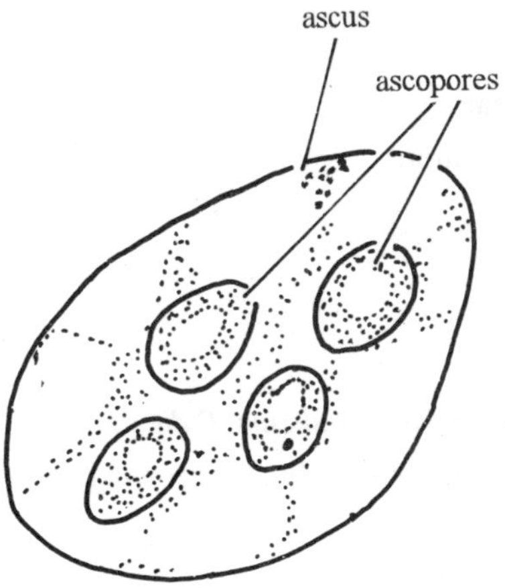

Fig. 13.3 Single ascus of Saccharomyces.

2. Fission yeasts are called *schizosaccharomyces* while budding yeasts are called *saccharomyces*.

3. During fission, the single nucleus gets elongated.

4. A constriction appears in the center of the nucleus and it becomes dumb-bell shaped.

5. With the deepening of the constriction, the nucleus gets divided into two bits.

6. The two daughter nucleu migrate to opposite poles. A cross septum appears in the center of the cell dividing it into two.

7. The two daughter cells formed thus grow into adults cells.

8. Budding is another method of reproduction. During budding a small bulge arises from one side of the cell. The bulge increases in size and is now called a bud .

9. Meanwhile, the nucleus of the mother cell divides into two daughter nuclei; one of those migrates into the bud.

10. A small constriction appears at the region of attachment of bud and the parent cell.

11. The constriction deepens and the bud gets separated. It grows into a new individual.

12. In some cases a bud does not get separated from the mother cell but, produces another bud. In this way a chain of buds is formed. This chain may be branched or unbranched. The chain looks like a mycelium hence it is called a Rhizomycelium or a Pseudomycelium.

13. Eventually the chain breaks and the individual cells get separated and grow into new cells.

Sexual reproduction

1. Sexual reproduction takes place by means of gametangial copulation, with two cells functioning like gametangia.

2. The two gametangia fuse forming a diploid zygote.

3. The zygote which is diploid functions like an ascus.

4. The diploid nucleus is ascus undergoes reduction division followed by a mitosis to form eight ascospores. In some cases only four ascospores are formed.

5. Ascospores on liberation develop into new individuals.

Diagrams to be drawn

1. Yeast cell magnified.

2. Yeast cells showing fission and budding.

3. Ascus with ascospores.

Systematic position and identification :

A. 1. Chlorophyll absent

2. Reserve food is glycogen ....**Fungi.**

B. 1. Mycelium is septate or plant body unicellular

2. Characteristic spores - ascospores are produced inside an ascus.

3. Ascospores usually light, rarely otherwise ....**Ascomycetes.**

C. 1. No fruiting body.

2. Saprophytic nutrition

3. Asci not in clusters ....**Endomycetales**

D. 1. Asexual reproduction by budding or fission

2. Plant body unicellular ....**Saccharomycetaceae**

E 1. Members produce enzyme zymase and bring about fermentation.

2. Plant body unicellular ....**Saccharomyces.**

PENICILLIUM

Common species : *P.notatum, P.chrysogenum, P.camembertii, P.roquefortii, P.italicum, P.digitatum* etc.

Occurrence : Commonly called green or blue mold, *Penicillum* is a saprophytic fungus growing on rotten fruits decayed vegetable, meat etc. Some of them grow on moist leather also.

Study

1. Diseases caused and symptoms
2. Method of culture in the laboratory
3. Vegetative structure
4. Reproduction
5. Systematic position and identification

Diseases caused

1. P.expansum, causes decay of apples, pears and grapes in storage.
2. P.italicium and P.digitatum cause decay of citrus fruits like orange, sweet lime etc.

Disease symptoms are visible in the form of greenish or slightly bluish leisons on the fruit wall. Those are the places where reproductive structures are found.

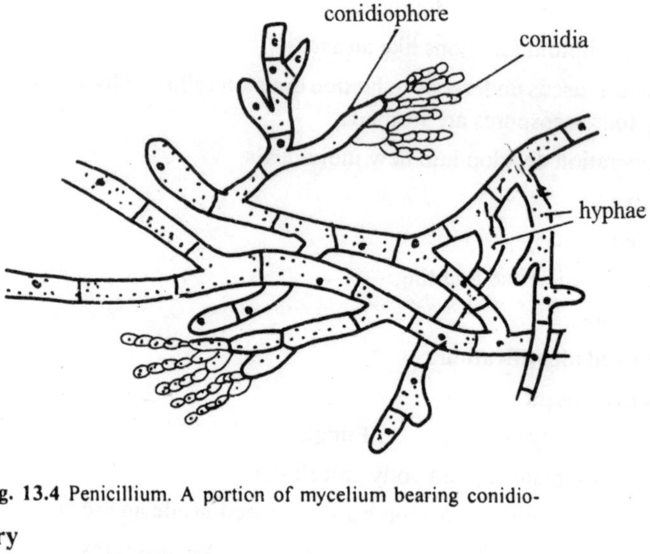

Fig. 13.4 Penicillium. A portion of mycelium bearing conidio-

Culture in the laboratory

1. Keep a stale orange in moist condition.
2. Sprinkle water to keep it wet.
3. Keep the orange enclosed.

4. After few days greenish blue coating appears on the rind.

5. Scrape the coating on to a microslide and observe under the microscope.

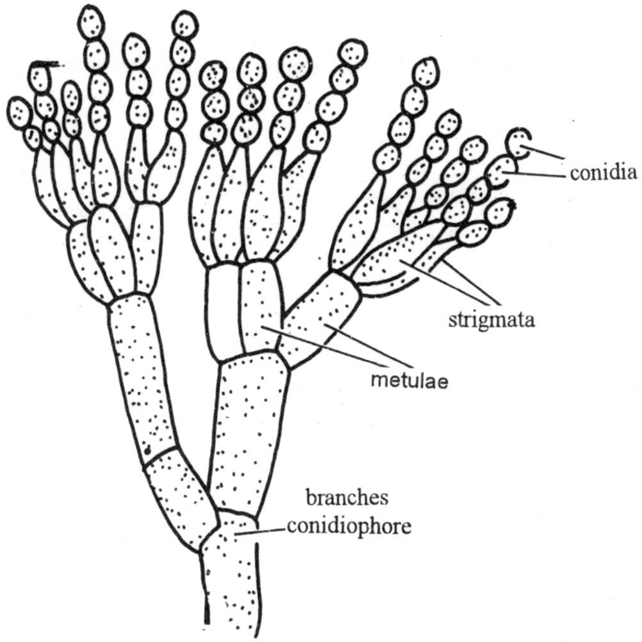

Fig. 13.5 Penicillium. Single conidiophore

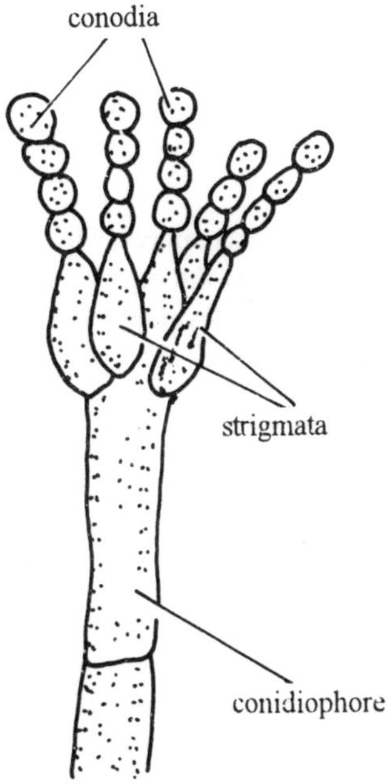

Fig 13.6Single conidiophore (monoverticillate)

Vegetative structure

1. The plant body is a mycelium. It is slender composed of profusely branched septate hyphae.

2. The mycelium is divided into cells. Cells are either monokaryotic or dikaryotic.

3. The cells wall between the cells has an opening, through which cytoplasm streams from cell to cell.

Asexual reproduction

1. Vegetative propagation is brought about by means of fragmentation of hyphae. Each fragment develops into a new individual.

2. Asexual reproduction by conidia is the most common method of multiplication in penicillium.

3. Vegetative mycelium produces conidiophores.

4. Conidiophores are erect and branched once or twice.

5. Each conidiophore or its branches bear at their tips a whorl of bottle shaped structures called Sterigmata or Phialides.

6. The branches of the conidiophores which bear phialides are called metulae.

7. The sterigmata at their tips produce a chain of conidia. Conidia are produced in basipetalous succession.

8. The entire structure consisting of conidiophores, metulae sterigmata and conidia resembles a brush or broom, hence the name Penicillium (Gr.Penicillus = brush)

9. Conidia are unicellular. They may be uni or multinucleate.

10. Upon falling on a suitable substratum, conidia germinate into a new mycelium by producing a germ tube.

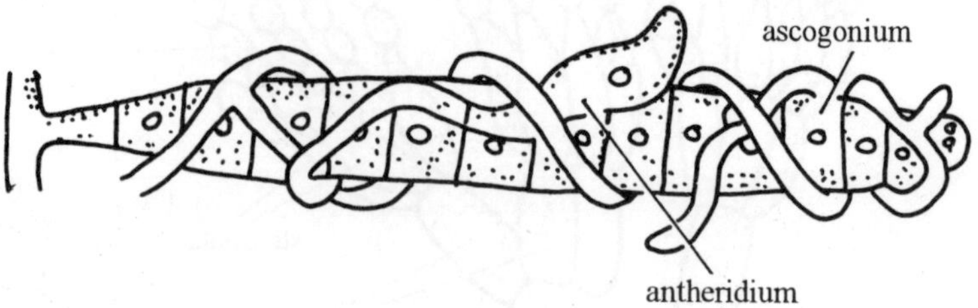

Fig.13.7 Penicillium. Sex organsconidiophores

Sexual reproduction

1. Sexual reproduction is by gametangial contact. Sex organs are antheridia and ascogonia.

2. Sexual reproduction is very rare.

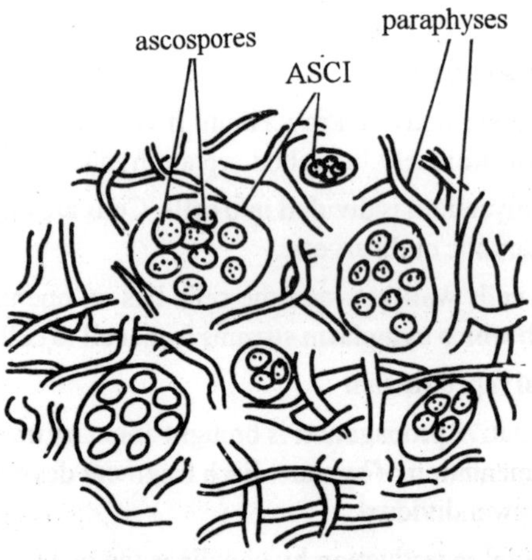

Fig 13.8 Cleistothecium in sectionphores

3. The fruiting body produced by penicillium is a cleistothecium.

4. The clesitothecium is a completely closed structure. It has a mutlilayered wall called peridium.

5. Lying scattered in the cavity of cleistothecium are many asci. Between the asci, sterile hyphae called paraphyses are found.

6. Each ascus has eight ascospores. The ascospores come out of the cleistothecium by the rupturing of the wall of the latter. On germination they develop into a new mycelium.

Diagrams to be drawn

1. Habit of infected host
2. Mycelium
3. Condiophore
4. Cleistothecium

Systematic position and identification

A. 1. Chlorophyll absent

2. Reserve food is glucogen ....**Fungi.**

B. 1. Mycelium is septate

2. Ascospore produced in ascus ....**Ascomycetes**

C. 1. Conidia are chief mode of multiplication

2. Asci scattered inside the cleisothecium ....**Aspergillales**

D. 1. Fructifications not stalked

2. Peridium made up of pseudoparenchyma ....**Aspergillaceae**

E. 1. Conidiophores branched, but no vesicle

2. Asexual structures resemble a broom ....**Penicillium**

ASPERGILLUS

Common species : *A.niger, A.fumigatus, A.flavus* etc.

Occurrence : The species of Aspergillus are generally saprophytic growing on stale bread, decaying vegetable meet etc. It also grows on jams, jellys, footwear, clother etc.

A few species of aspergillus are parasitic causing lung diseases.

Study

1. Diseases caused and symptoms
2. Laboratory culture
3. Vegetative structure
4. Reproduction
5. Systematic position and identification

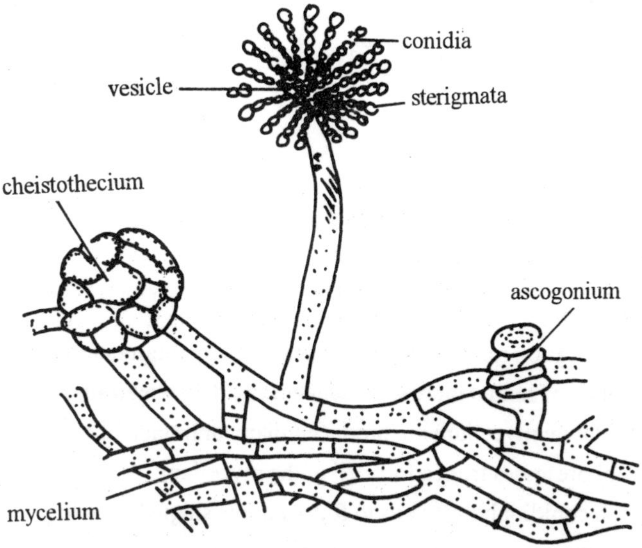

Fig.13.9 A portion of mycelium showing conidiophore cleistothecium and sex organs of Aspergillus

Diseases caused and symptoms

1. Rot disease : This is caused by A.niger. the hosts that are usually attached are pomegranate (Punica granatum), Dates (Phoenix dactylifera), Figs (Ficus carica) etc.

2. Aspergillosis : This is an allergic type of lung disease caused by the spores of A.niger, A.fumigatus and A.flavus, cattle and other animals and human beings also get the infection.

3. Otomycosis : A type of ear disease in human beings caused by several species of Aspergillus.

Disease symptoms

1. Saprophytic species produce symptoms in the form of greyish green or bluish green growth of the mycelium.

2. Fruits get rotten and decomposed due to the entry of fungus through wounds.

3. The symptoms of Aspergillosis resemble those of tuberculosis.

Laboratory culture

1. Place a piece of bread or citrus fruit under a bell jar.

2. Keep it moist

3. After few days greenish blue or yellowish mycelia grow profusely.

Vegetative structure

1. The plant body of the fungus is a much branched, septate closely interwoven mass of mycelium.

2. Hyphae are divided into cells. Cells are multinucleate suspended in a granular cytoplasm.

3. Oil globules are present in the form of reserve food.

4. While most of the myeclium is above the substratum, certain hyphae grow into the substratum, possibly to help in anchorage and absorption.

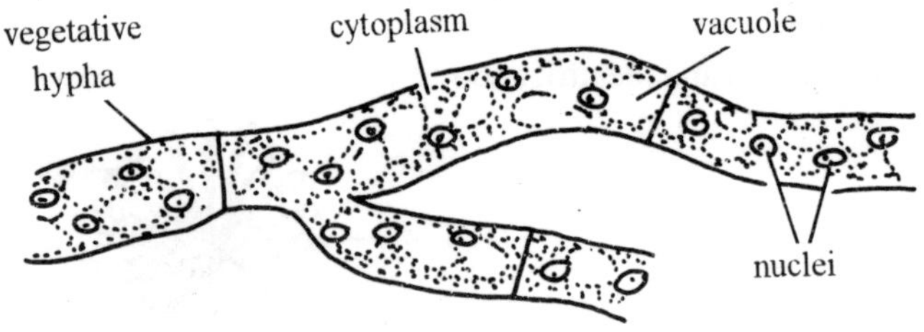

Fig 13.10 Hypha enlarged

Asexual reproduction

1. Asexual multiplication may take place vegetatively by fragmentation of the hyphae.

2. The chief method of sexual increase however is by the formation of conidia.

3. From the vegetative mycelium are produced many unbranched erect hyphae called conidiophores.

4. Each conidiophore arises from a foot cell of the mycelium.

5. The tip of each conidiophore swells to form a dome shaped vesicle.

6. The vesicle from its surface gives rise to a number of sterigmata or phialides, which are bottle shaped structures. The sterigma are multinucleate.

7. Sometimes a second layer of sterigmata may arise from the first layer in which case, the first layer is called primary sterigmata and the second layer is called the second sterigmata.

8. The sterigmata of the second layer (first layer, if there are no secondary sterigmata) produce a chain of conidia. The conidia are produced in basipetalous succession.

9. The conidia reproduction small, rounded unicellular structures. They may be uni or multinucleate. The conidial wall is pigmented in hues of black, green and blue.

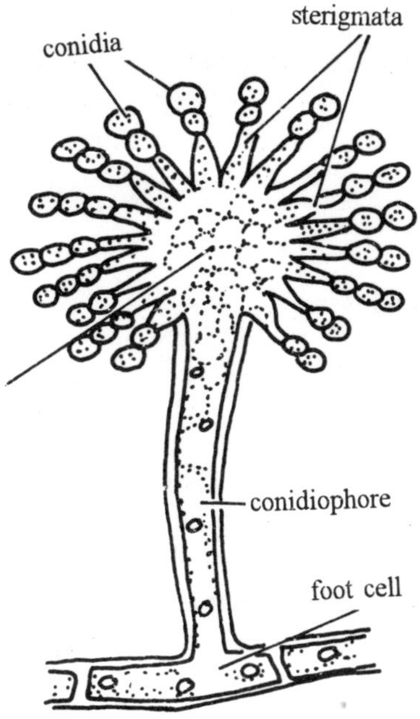

Fig.13.11 Aspergillus. A mature conidio-

10. The conidial wall is double layered. The outer epispore is spiny, while the inner endospore is smooth. On germination the conidium develops into a new individual.

Sexual reproduction

1. Sexual reproduction is by means of gametangial contact. Sex organs are antheridia and ascogonia.

2. Antheridium is small elongated and multinucleate.

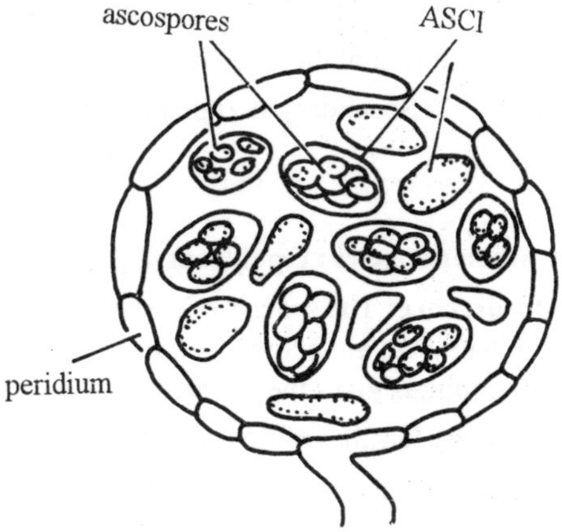

Fig. 13.12a Cross-section of

3. Ascogonium is a flask shaped structure with a bulbous base and a narrow neck called trichogyne.

4. Fruiting body is a cleistothecium. Many asci lie scattered in the cavity of the fruiting body. Each ascus has eight ascospores.

5. On liberation from the cleistothecium, the ascospores germinate to give rise to new individuals.

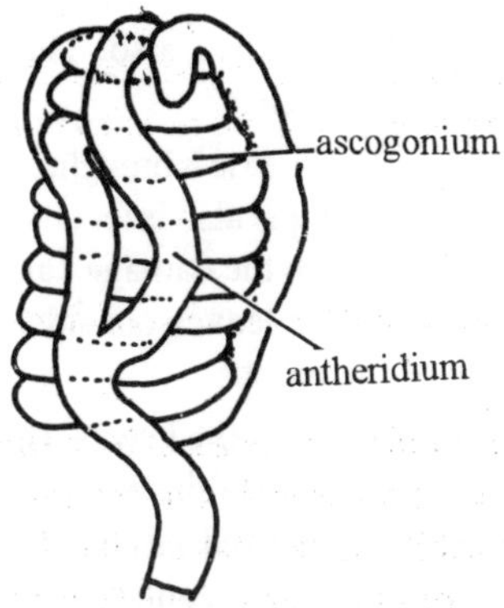

Fig. 13.13A Aspergillus. Sex organs

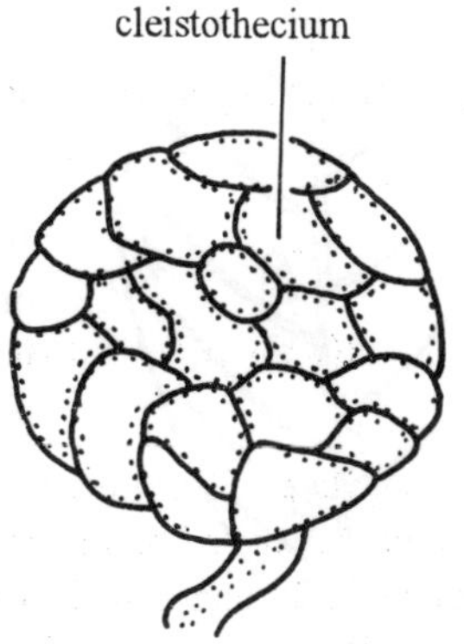

Cleistothecium

Diagrams to be drawn

1. Habit of infected material
2. Mycelium with condiophores

3. A part of the condiophore enlarged

4. Cleistothecium.

Systematic position and identification

From A to D identification. See penicillium.

E 1. Conidiophores unbranched

2. Tip of the conidiophore swells to form a vesicle

3. Secondary sterigmata present sometimes.......... Aspergillus

ERYSIPHAE

Common species : E.polygoni, E.graminis, E.ci choriacearum etc.

Occurrence: The species of eryisphae cause a disease which is commonly called powdery mildew disease. The members are ectoparasites sending only jaustoria into the host. The superficial growth of the cottony white fungus gives the appearance of sprinkled powder hence the name powdery mildew.

Many families of flowreing plants like Gramineae, papilinaceae, caricaceae etc are infected.

Study : 1. Diseases caused and disease symptoms.

2. Vegetative structure

3. Reproduction

4. Systematic position and identification

Diseases caused and disease symptoms

Various species of enisphae cause diseases to economically important. The following is a brief description of some of the diseases and their symptoms.

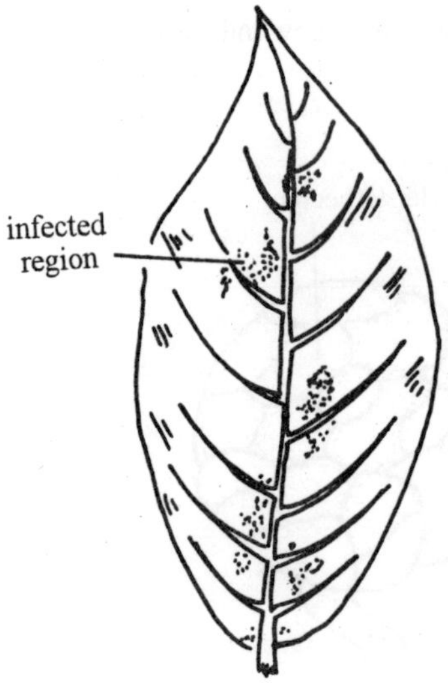

Fig.13.14 Erysiphe. Host leaf infected with powdery mildew caused by Erysiphe

1. **Powdery mildew of peas :** The disease is caused by E.polygont. It makes its appearance when the plants are fruiting during the month of Jan-Feb. The disease becomes rampant during day weather.

Disease symptoms are visible first on the leaf. Later they spread to the other parts. Powdery patches appear on either surface of the leaf. Gradually the patches spread to tendrils also. The mycelium of the fungus is seen at the region of patches.

2. **Powdery mildew of barley**. The disease is brought aboty by E.graminis, Varhordei. Besides barley, the fungus also infects wheat, rye, oats and other graminaceous plants. Though the damage caused by the fungus is not very extensive, death of the leaves considerably, reduces the photosynthetic efficiency of the plant.

Disease symptoms are first visible on the leaves. Both surfaces of the leaves, specially the upper surface exhibits powdery patches. Gradually the disease spreads to the stem and inflorescence also. Infected plants show increased rate of respiration and transpiration. Photosynthesis gets retarded.

3. **Powdery mildew of cucurbits**. Many cucurbits specially pumpkin, nibbed gourd are infected by the fungus E.cichoracearum. This pathogen also infects plants like tobacco, potato, lettuce, sun-flower, mango, castor etc.

Disease symptoms become prevalent when the atmosphere is moist. Small spots appear on leaves and stems, gradually wider giving the characteristic powdery appearance. Dark fruiting bodies of the fungal are visible on the leaves at later stage. Plants get depoliated and fruits drop down prematurely due to infection.

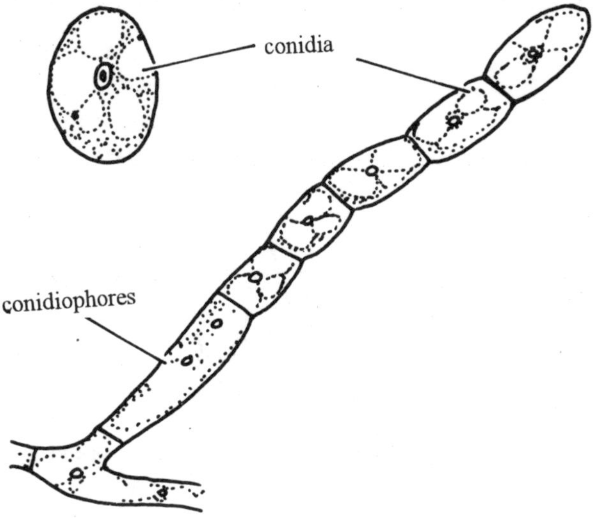

Fig. 13.15 Erysiphe. Single conidiophore with chain of

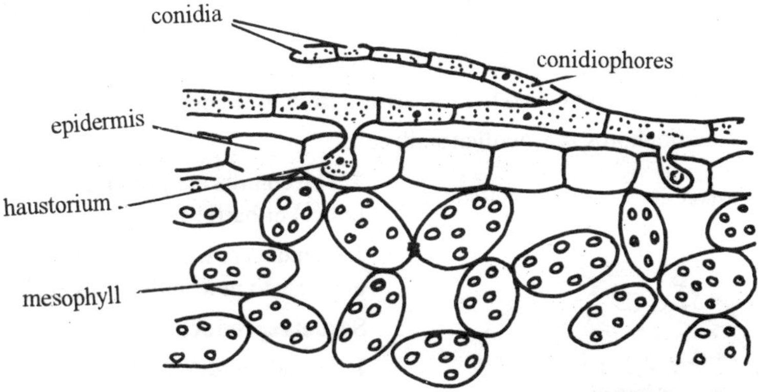

Fig. 13.16 Ectotrophic mycelium and haustoria

Specimens given in the laboratory

1. Leaves of pea, pumpkin or papaya etc.
2. Scraping of the powdery patches will reveal the mycelium.
3. Free hand section of the leaf will show the haustoria.

Vegetative structure

1. Plant body of the fungus is a mcuh branched, septate mycelium.
2. The fungus is an ectoparasite; the mycelium growing superficially on the body surface of the host plant.
3. The superficial mycelium however sends erect, unbranched saccate hyphae called haustoria into the cells to absorb the nutrition from the host.

4. The haustroia may be saccate (E.cichoracearum) or branched (E.polygoni).
5. The cells are uninucleate and have granular cytoplasm.

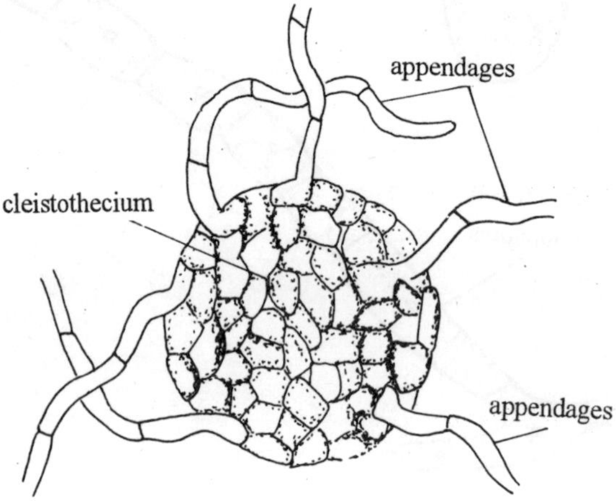

Fig. 13.17 Erysiphe. Single cleistothecium

Asexual reproduction

1. Asexual reproduction is brought about by conidia produced on conidiophores.
2. The conidiophores are erect unbranched hyphae arising from the underlying mycelium.
3. At the base of the conidiophores is a swollen stalk cell.

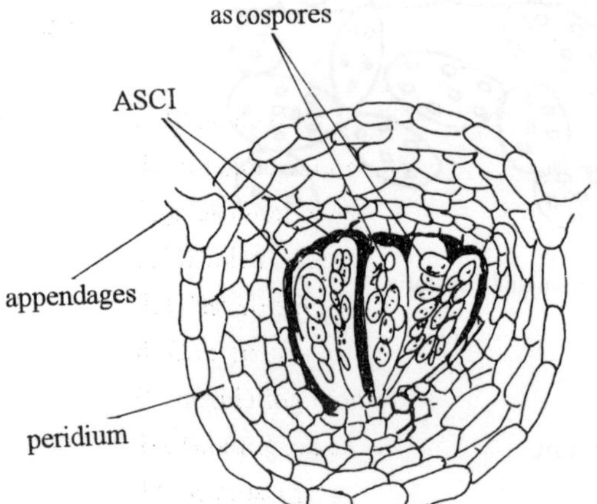

Fig 13.18V.S. of cleistothecium

4. The conidia are produced inncyains. They are elliptical or barrel shaped and uninucleate.
5. Conidia are wind dessiminated; they germinate to give rise to new individuals erysiphae.

Sexual reproduction

1. Sexual reproduction is effected by antheridia and ascogonia.
2. Sex-organs are produced late in the growing season of the host plant.
3. Cleistothecia are the characteristic frutifications produced as a result of sexual reproduction.
4. Cleistothecium is a globase structure. It has a wall peridium made up of 6-7 layers of cells.
5. The superficial cells of the peridium give out appendages. These appendages are of various kinds and shapes. They are used as criteria in the categorization of genera and species.
6. Lying in the cavity of the clustothecium are found a number of asci. Each ascus usually has eight ascospores.

7. The ascospores are unicellular, uninucleate, haploid and on coming out of the fruiting body (the peridium breaks open irregularly) germinate to form a new individual.

Diagrams to be drawn:

1. Habit of infected leaves.
2. Mycelium with conidia.
3. Cleistothecia with appendages.

Systematic position and identification

A. 1. Chlorophyll absent

2. Reserve food is glycogen ... fungi

B. 1. Mycelium is septate

2. Ascospores produced in Ascus ... Ascomycetes

C. 1. Members are ectoparasites

2. Powdery patches appear on the host ... Erysiphales

D. 1. Mycelium tubular and hyaline

2. Condidia produced in large numbers ... Erysiphaceae

E. 1. Ascospores unicellular

2. Many asci in cleistothecia

3. Appendages on cleistothecia myceloid ... Erysiphae

PHYLLACTINIA

Common species : P.dalbergiae.

Occurrence : The fungus is a semi-endo parasite. It groups commonly on the leaves of Dalbergia, cassia, morus acacia, betula etc.

Powdery patches appear on the leaves generally on the lower surface.

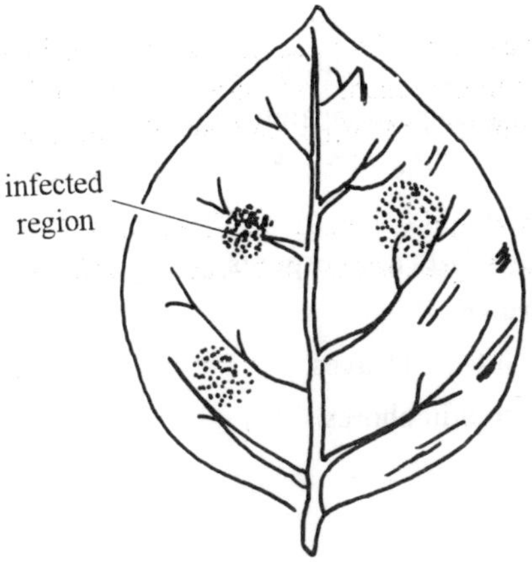

Fig. 13.19 Dalberga leaf infected with powdery mildew caused by Phyllactinia

Study : 1. Disease symptoms
2. Vegetative structure
3. Reproduction
4. Systematic position and identification

Disease symptoms :

1. Powdery pathces of white hyphae seen on the leaves.
2. In the older stages of the leaf organe coloured cleistothecia are seen. Mature cleistothecia are dark coloured.

Vegetative structure

1. The plant body is a septate, much branded mycelium.
2. The hyphae of the mycelium are not completely superficial, they are semi endoparasites.
3. The hyphae produce haustoria, which penetrate the mesophyll cells of the host and absorb nutrition.
4. Hyphae have uninucleate cells with granular cytoplasm.

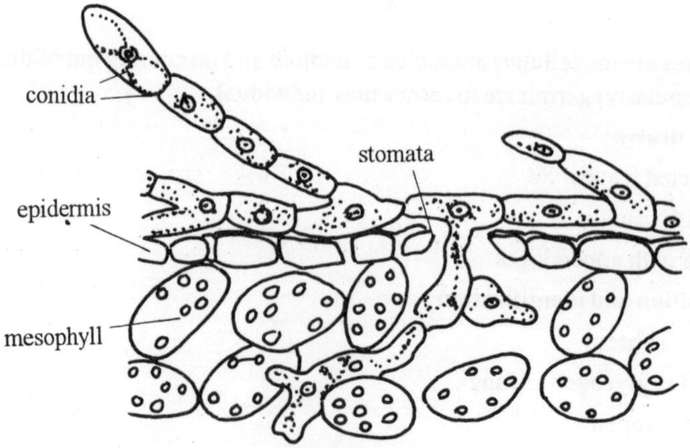

Fig 13.20 Ectotrophic mycelium

Asexual reproduction

1. Asexual reproduction is brought about by the fcrmation of conidia developed on conidiophores.
2. Conidiophores arise from the underlying mycelium. They are unbranched and 3-4 cells long.
3. Conidia are unicellular and uninucleate. They are disseminated through the agency of wind.

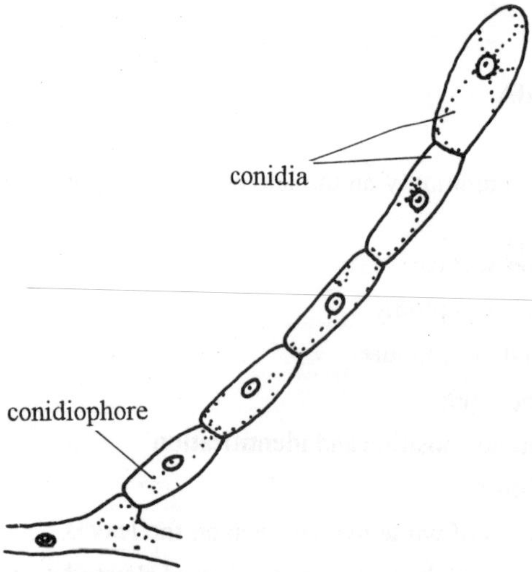

Fig. 13.21 Phyllactinia. Single conidiophore with chain of conidia

Sexual reproduction

1. Sexual reproduction is brought about by means of gametangial contact.
2. Fruiting bodies are called cleistotheia which are globular in shape.
3. Cleistothecia are dark coloured and have characteristic appendages numbering 5-10.
4. Each appendage has a bullbous base and a pointed tip.
5. Some sap oozing hyphae are also present on the cleistothecium. These help in the dispersal.
6. Cleisothecium has a hpseudoparenchymatous peridium.
7. Lying in the cavity of the cleistothecium are found a number of asci.
8. Each ascus has only two ascospores.

Diagrams to be drawn

1. Habit sketch of infected leaves.
2. Mycelium with conidiophores.
3. Cleistothecium showing appendages.

Systematic position and identification

A. 1. Chlorophyll absent
2. Reserve food is glycogen ... Fungi

B. 1. Mycelium septate
2. Ascospores produced in ascus... Ascomycetes

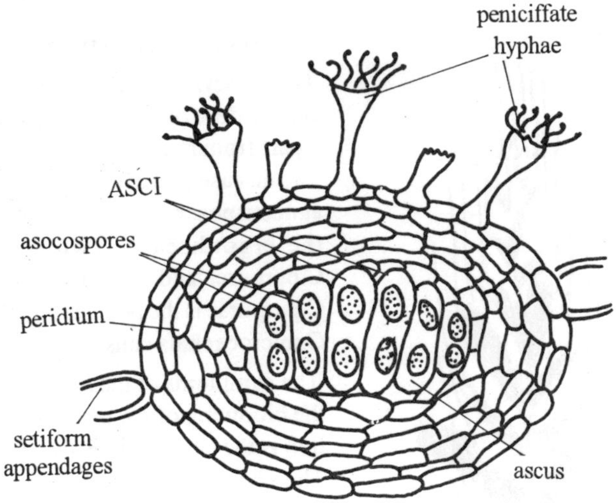

Fig 13.22V.S. of cleistothecium

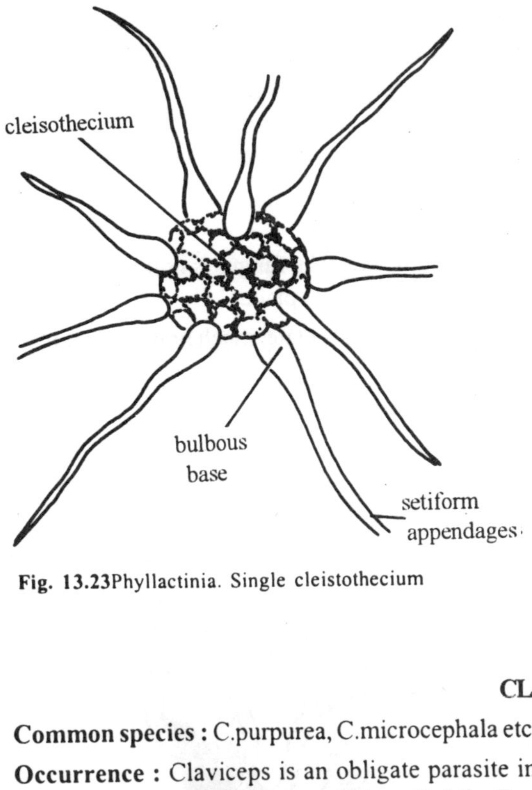

Fig. 13.23Phyllactinia. Single cleistothecium

C. 1. Mycelium tubular and hyaline

2. Powdery patches appear on the host ... Erysiphals

and. 1. Mycelium tubular and hyaline

2. Conidia produced is large numbers

3. Asci found in cleistotheia ... Erysiphaceae

E. 1. Plant body semiendoparasitic

2. Frutification is a cleistothecium

3. Appendages on the cleistothecium have a bulbous base and a pointed tip.

4. Only two ascospores produced is the ascus ... Phyllactinia.

CLAVICEPS

Common species : C.purpurea, C.microcephala etc.

Occurrence : Claviceps is an obligate parasite infecting the members of gramineae, especially rye, oats, barley, etc. The disease is popularly called the Ergot ids.

Study : 1. Diseases caused and symptoms.

2. Vegetative structure

3. Reproduction

4. Systematic position and identification

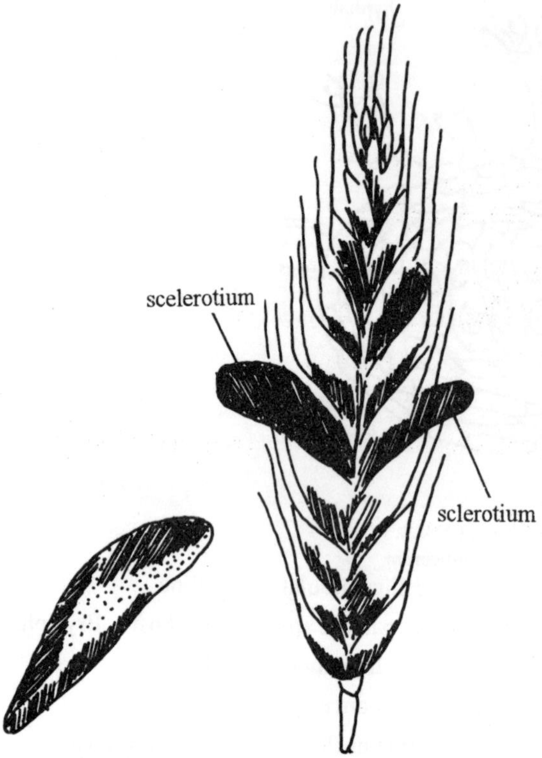

Fig. 13.24 Claviceps. Single sclerotium

Infected inflorescence of rye showing sclerotia of Claviceps

Diseases caused and symptoms :

Tribes, oryzae, phalarideae, agrostidae, avene, festuceae and hordeae of the family graminae get the ergot disease. Apart from rye, oats, barley, peral millet, sugarcane etc. The disease also affects grasses oike agropyron, elymus, agnostispoa, bromus, cynodon etc. The disease is called the ergotdisease. Two species of claviceps viz., C.purpurea and C.microcephala are the most important C.purpurae is the popular ergot causing erfot of rye, while C.microcephala causes ergot of bajra.

The symptoms of the disease will be visible only during the blooming of the plant. The first sign of visible infection is the exudation of sweet sap from the infected ears. The characteristic symptom of the disease however in the appearance of dark known coloured hard masses of fungal tissue called sclerotia. It is these sclerotia which are called ergots that have given the name to the disease.

Consumption of ergots mixed with the grains results in a disease called ergotism. Ergotism affects cattle as well as human beings. In cattle loss of horn, hooves, tall etc takes place, while in human beings loss of hair, nails or even gangrene may set in.

Vegetative structure

1. The plant body of the fungus is noticeable in the ovarian tissues of the host plant.

2. The mycelium is septate, much branched and the cells are multinucleate.

3. The mycelium develops as a result of the germination as ascospores.

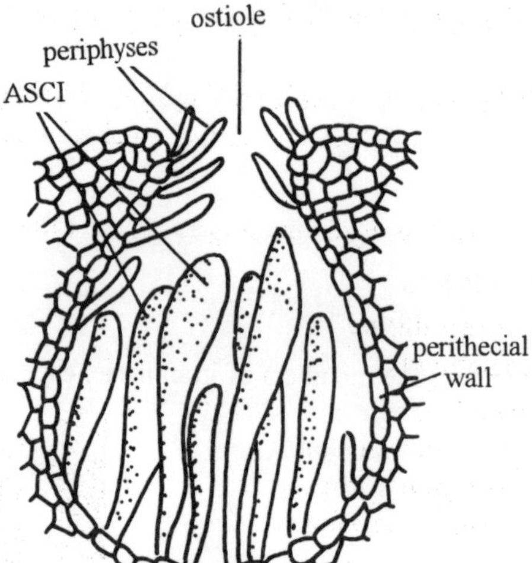

Fig.13.25 Claviceps. V.S. of ascocarp

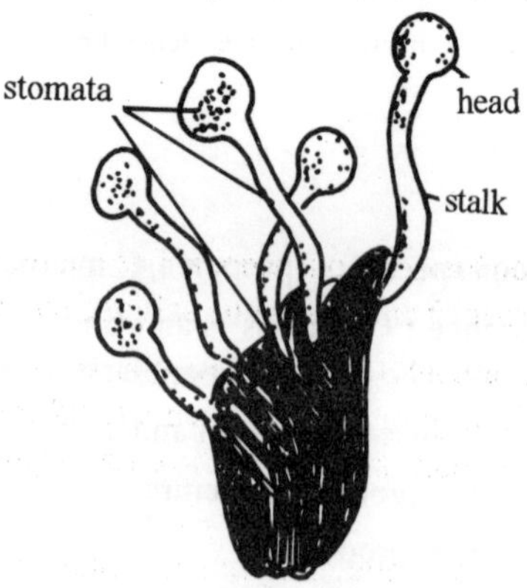

Germinating sclerotium of Claviceps

Asexual reproduction

1. The fungus reproduces asexually by the formation of conidia.

2. Conidiopores are produced inside the ovary of the host plant. They are aggregated into clusters called ascerculus.

3. Conidia are uninucleate and unicellular on liberation from the ovary (usually by insects as there is nectar) they infect other healthy plants.

4. The asexual stage is called the sphacelia stage or honeydew stage. The former name is a misnomer originally this stage was identified as a separate individual and named sphacelia.

Because of the production of nectar the stage is called the honey dew stage.

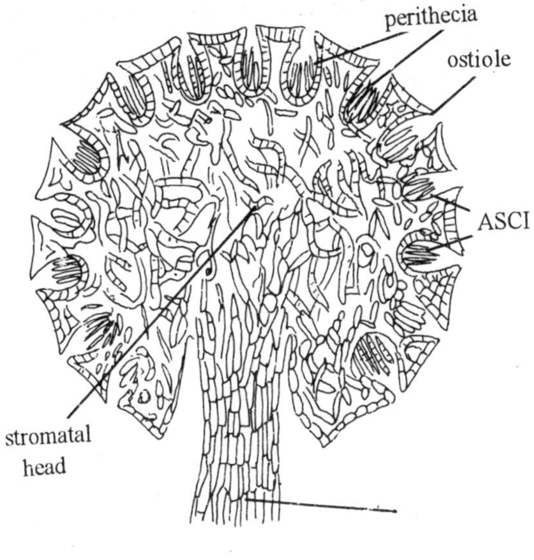

Fig. 13.26 Claviceps. V.S. of stromata

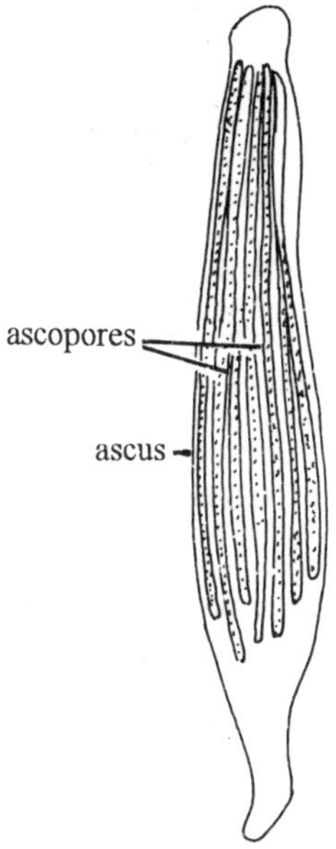

An ascus containing ascosppores

Sclerotium stage: This is the characteristic stage of the fungus also known as the ergot, the sclerotium is a hard brown mass of fungal tissue. Slightly larger than the grain which it replaces. The ergots will be peeping out of the inflorescence. They are rich in alkaloids.

2. The sclerotium is pink or purple in colour sometimes.

3. The sclerotiums stage constitutes the resting stage of the fungus. Usually sclerotia reach the soil during the threshing of grains.

Ascostroma stage

1. After overwintering the sclerotium germinated by putting forth small pin head like structure called ascostroma.

2. Each ascostroma is about an inch long. It has a stalk and a rounded head.

3. In a mature ascostroma, small cavities appear which later develop into perithecia.

4. Each perithecium is a flask shaped structure. It has a small opening called the ostiole.

5. Lining the base of the perithecium inside are a number of asci and sterile hairs (paraphysis).

6. Each ascus has light ascospores.

7. Ascospores are long, thread like and acicular (needle like). On liberation from the perithecium, they are dispersed by wind. They germinate on the stigmatic surface of a suitable host plant.

Specimens given in the laboratory

1. Infected ears of host plant

2. Sclerotia

3. Sclerotia with ascostroma

Diagrams to be drawn

1. Habit of infected ear

2. Section showing the conidiophores, conidia etc

3. Sclerotium (ergot) sectional view

4. Sectional view of ascostroma.

Systematic position and identification

A. 1. Chlorophyll absent

2. Reserve food is glycogen ... Fungi

B. 1. Mycelium septate

2. Ascus and ascospores produced ... Ascomycetes

C. 1. Ascospores acicular

2. Perithecia found in ascostroma ... Clavicipitales

D. 1. Ascus has a perforated lid to facilitate spore dispersal ... Clavicipitacea

E. 1. Ergots dark brown in colour and hard

2. Ascospores acicular and arranged parallely in the ascus ... Claviceps.

XYLARIA

Common species : X.hypoxylon, X.mate, X.polymorpha etc

Occurrence : Xylaria is cosmopolitan in distribution. It usually grows as saprophyte on decaying wood. X.hypoxylan causes root rot of apples while X.mali causes black rot of apple trees.

Study : 1. Structure of the plant body

2. Reproduction

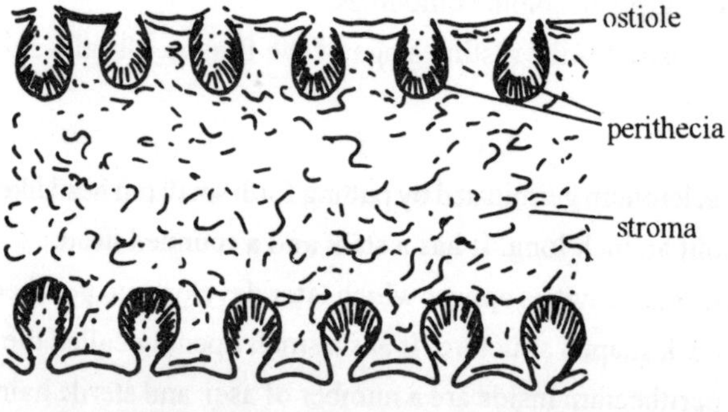

Fig. 13.27 Xylaria. Stromata in cross-section

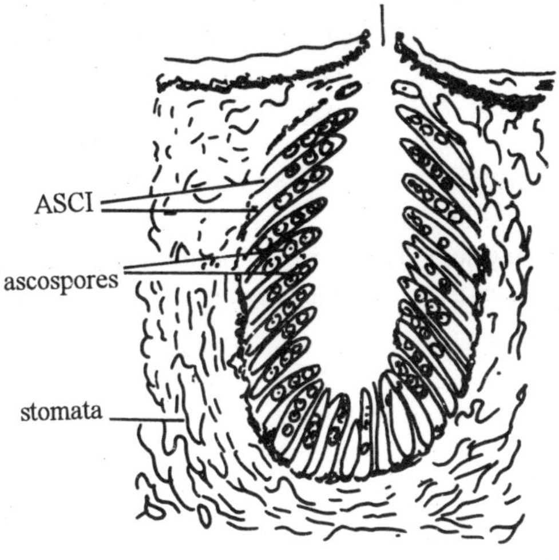

Fig 13.28 V.S. of perithecium

3. Systematic position and identification

Structure of the plant body

1. The mycelium is well branched and at intervals froms thick strands.

2. These strands are heterotropic and grow towards light.

The hyphae are septate and branched. Cells are uninucleate.

Reproduction :

1. The strands of mycelia develop into dark coloured stroma.

2. These are erect stout and cylindrical structures.

3. These may be branched or unbranched.

4. The stoma develop numerous perithecia which are arranged at right angles to the surface of the stroma.

5. During early stages stroma is covered by conidiophores.

6. A transverse section of the stroma reveals a circular outline and a wall mode up of closely packed hyphae.

7. Near the peripheral region are found a number of small flask shaped cavities called perithecia.

8. Each perithecium opens to the exterior, with the help of a small opening called the ostiole.

9. Lining the base of the perithecial cavity are found a number of asci. Intermingled with the asci are found the paraphyses.

10. Each ascus has eight cylindrical ascospores on liberation from the ascus, they germinate to give rise to new individuals.

Specimens given in the laboratory

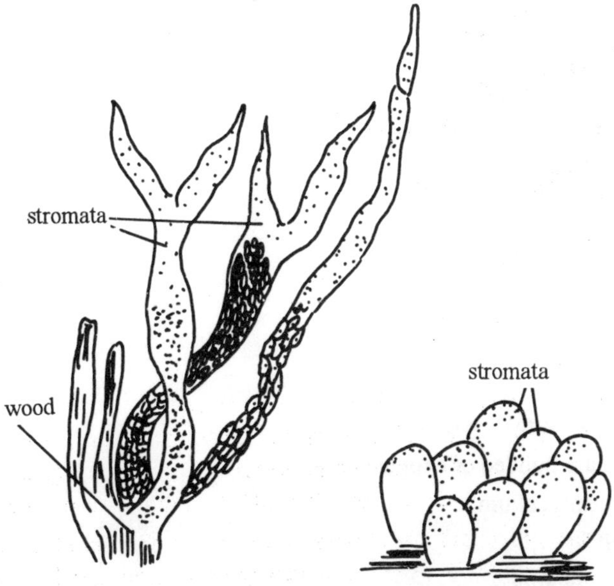

Fig. 13.29 Xylaria. Stromata growing on rotten wood

1. Stroma of xylaria (observe the external details and study the intenal structure by taking a cross section).

Diagrams to be drawn

1. Habit of stroma

2. Ground plan of transverse section

3. A perithecium enlarged

Systematic position and identification

A. 1. Chlorophyll present

2. Reserve food is glycogen ... Fungi

B. 1. Mycelium septate

2. Ascus and ascospores are formed ... Ascomycetes

C. 1. Perithecia dark coloured and pyriform.

2. Sunken perithecia ... Sphareiales

D. 1. Dark coloured stroma of fungal tissue

2. Ascospores dark coloured ... Xylariaceae

E. 1. Stroma club shaped and sometimes branched

2. Ascospore cylindrical ... Xylaria.

PEZIZA

Common species : P.pustutata, P.catinus, P.vesiculosa etc.

Occurrence : This saprophytic fungs occurs on dung, rotting logs of wood etc. Commonly called cup fungus, peziza has 13 species reported from India. Some species grow on moist leaves, heavily mannured soils etc.

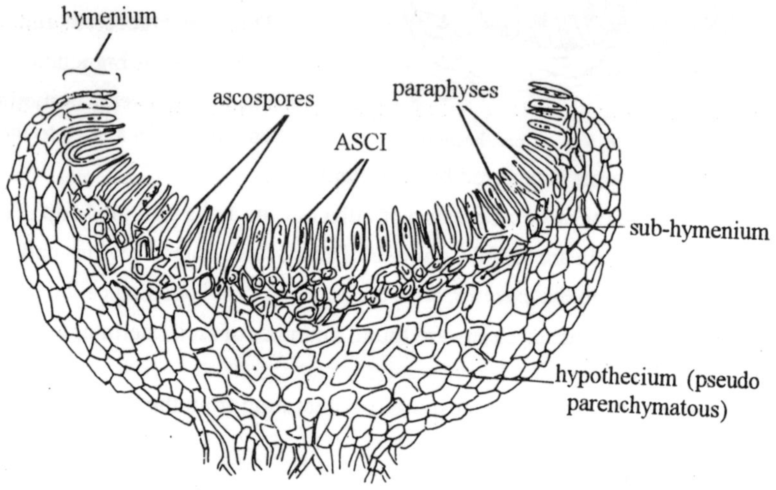

Fig. 13.30 Peziza. V.S. of apothecium

Study : 1. External features of plant body.

2. Reproduction and structure of fructification

3. Systematic position and identification.

External features :

1. The plant body is a mycelium. It is extensively branched
2. The hyphae entangle among themselves to form a network
3. The mycelium is septate and each cell is multinucleate.

Asexual reproduction

1. This is brought about by conidia and thick walled chlamydosphores
2. Conidia are produced on conidiophores and they are budded of exogenously.
3. Chlamydospores are produced from the cells of the mycelium

Sexual reproduction and structure of fructification

1. Sexual reproduction takes place by means of somatogamy
2. The asci are arranged in ascocarp called Apothecia

3. Apthoecium is a small cup shaped structure and it is dull white is colour.
4. A vertical long section of the apothecium will reveal the following structure
5. The saucer or disc like fruiting body has a short stalk. Occasionally it may be sterile also.
6. The wall of the apothecium is made up of closely racked fungal hyphae. The wall is called Peridium
7. Internal to the peridium is the medulla made up of loosely interwoven hyphae
8. The cavity of the cup is made up of hymenal layer
9. Below the hymenium is the hypothecium or the subhymenium
10. The hymenium is the fertile layer. It conists of parallely arranged asci and paraphyses (sterile hairs)
11. Each ascus has eight ascospores arranged uniseriately in an oblique manner
12. Ascospores on germination will develop into a new myalium.

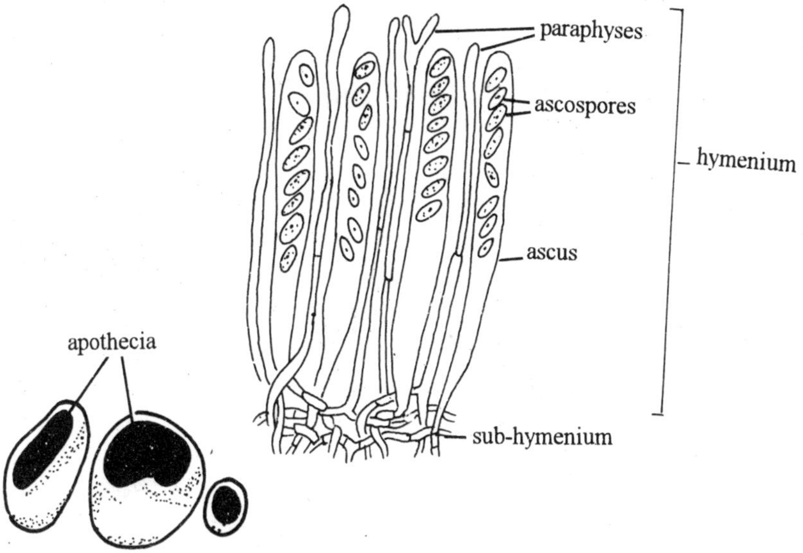

Fig. 13.31 Peziza. Cups (ascocarps) A portion of hymenium enlarged

Specimens given in the laboratory

1. Apothecia
2. Long section of apothecium (slide)

Diagrams to be drawn

1. Habit of apothecium
2. Long section of apothecium (slide)

Diagrams to be drawn

1. Habit of apothecium
2. Long section of apothecium
3. A few asci enlarged

Systematic position and identification

A. 1. Chlorophyll absent

2. Reserve food is glycogen ... fungi

B. 1. Mycelium septae

2. Asci and ascospores produced ... Ascomycetes

C. 1. Fruitifications not found in stroma

2. Apothecia fleshy ... Pezizales

D. Fruitification not differentiated into stalk and terminal pileus.

2. Fruitification disc or saucer shape ... Pezizaceae.

E. 1. Apothecia stalked or subsessile

2. Hyphae substerranean

3. Apotheica not large (usually 1-5 cms in diameter... Peziza.

PYRONEMA

Common species : P.confluens.

Occurrence : A saprophytic fungus usually growing on humans soil rich in mineral matter occasionally it also grows on dead and decaying logs.

Study : 1. Structure of the plant body

2. Reproduction

3. Systematic position and identification

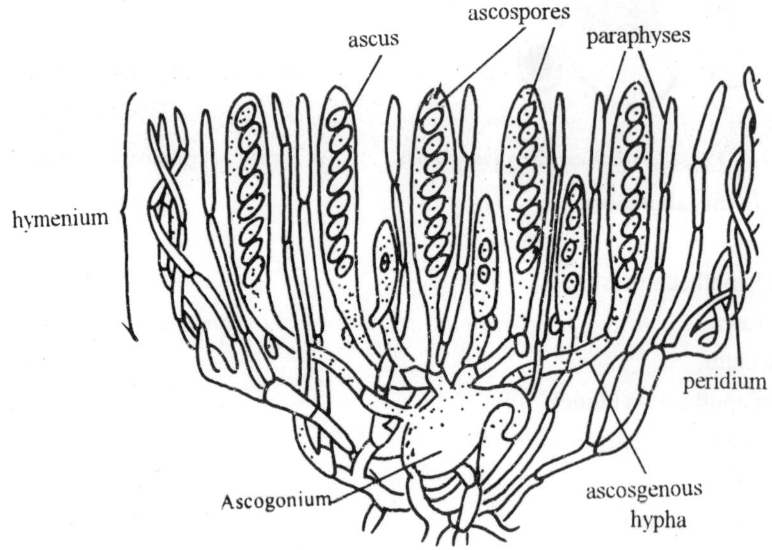

Fig. 13.32a L.S. of apothecium

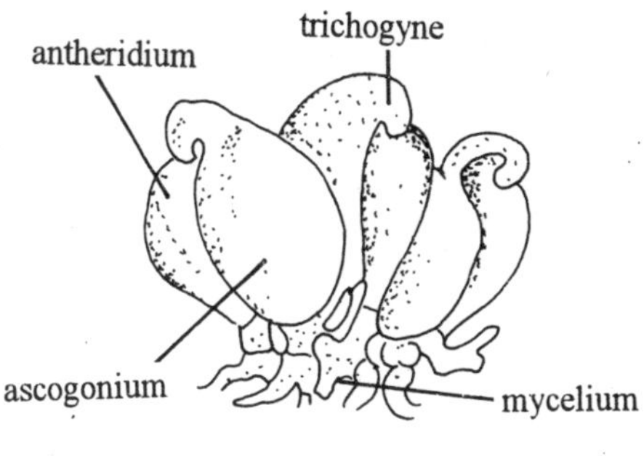

Fig. 13.33 Pyronema. A portion of mycelium with sex organs of Pyronema

Structure of the plant body

1. The plant body of the fungus is a mycelium.

2. The mycelium grows on the substratum. It is septate and much branched. The cells are multinucleate with 6-10 nuclei in each cell.

Reproduction

1. Asexual reproduction takes place by means of fragmentation

2. Sexual reproduction is of the oogamous type. It is effected by means of gametangial contact.

3. The sex-organs are antheridia (male) and ascogonia (female). Species of pyronema are homothallic.

4. Antheridia are club-shaped structures developing on short branches of the hyphae. They have many male nuclei.

5. Ascogonium is a bit elliptical or sub globoase in shape. It has a short curved appendage at its apex known as trichogyne.

6. Ascogonium has a number of functional nuclei at its base.

7. Sexual reproduction leads to the formation of the fruiting body called the apothecium.

8. The apothecia are small disc shaped structure surrouned by a well called peridium.

9. Lining the apothecial cup internally is the hymenial layer.

10. The hymenial layer consists of parallely arranged asci intermingled with sterile hairs called paraphyses.

11. Each ascus has eight ascospores arranged in a uniseriate manner. Ascospores are uninucleate.

Specimens given in the laboratory

1. Apothecia

2. Sex-organs (slide)

3. Section of apothecium (slide)

Diagrams to be drawn:

1. Apothecium (habit sketch)

2. Sex-organs

3. Long section of apothecium

Systematic position and identification

A. 1. Chlorophyll absent

2. Reserve food is glycogen ... Fungi

B. 1. Mycelium septate

2. Ascus and ascospores produced ... Ascomycetes

C. 1. Frutifications not found in stroma

2. Apothecium fleshy or leathery ... Pezizales

D. 1. Frutification not differentiated into stalk and pileus.

2. Frutification disc or saucer shaped... Pezizaceae

E. 1. Mycelium superterranena

2. Sex-organs prominent, apothecium very small ... Pyronema.

MORCHELLA

Common species : M.esculenta, M.conica, M.deliciosa etc.

Occurrence : Popularly known as Morel, morchella, has a delicious and highly priced frutification (M.esculenta). the species of morchella are found growing in decidious forests on decaying wood or on soil with plenty of leaf litter. Attempts to cultivate the edible species have been methods with only partial success.

Study : 1. Structure of the plant body

2. Structure of frutification

3. Ascus and ascospores

4. Systematic position and identification

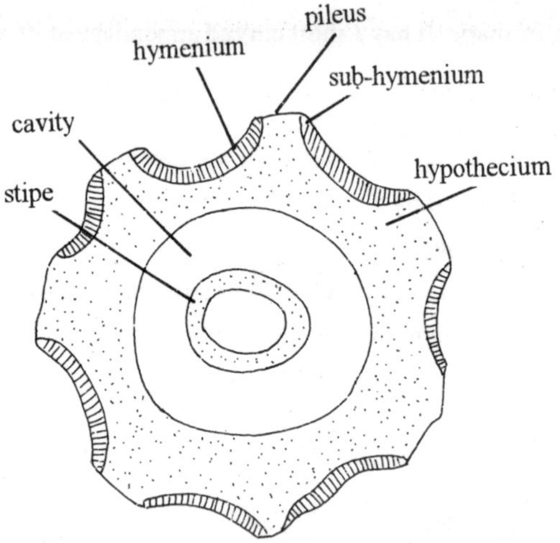

Fig. 13.34 Morchella. Cross-section of pileus

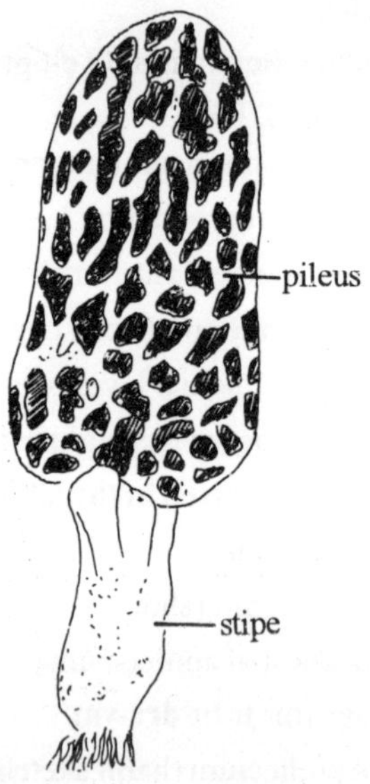

An ascocarp

Structure of the plant body

1. The plant body of the fungus is a subtteranean mycelium growing a few inches beneath the soil surface

2. The mycelium is much branched, with septate hyphae; cells are multinucleate.

Structure of the frutification

1. The underground mycelium produces a compact, macroscopic pseudoparenchymatous ascocarp (frutification)

2. The ascocarp is aerial ie., it comes above the ground and about 2-5 inches in height.

3. The ascocarp has hues of grey or dark brown which depends on species and stage of growth.

4. The ascocarp may be distinguished into two region, viz the stipe and pilcus, which are of equal length.

5. The stipe is the stalk, made up of pseudo parenchymatous tissue

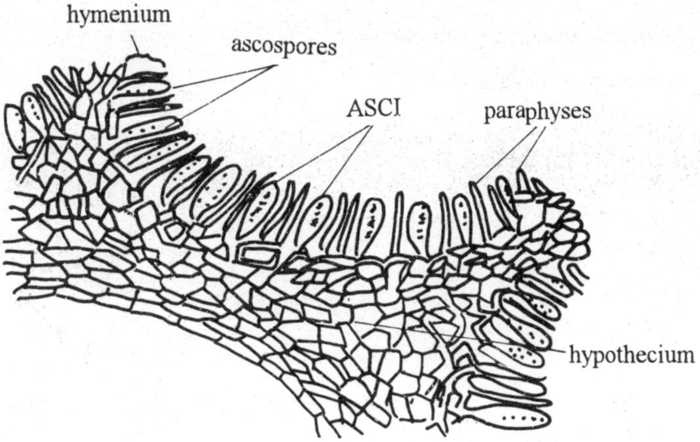

Fig.13 .35 Morchella. A cross-section of a portion of pileus enlarged

6. In the centre of the stipe is found a hollow cavity
7. The pileus is a conical structure and represents the fertile portion of the asocarp.
8. The surface of pilecus is smooth at young stages, while in a mature pileus it consists of many cavities (depressions) and ridges which look like a broad meshed network.
9. The cavities represent the fertile regions (asci) while the ridges are sterile.
10. A section of the pileus passing through the ridges and depressions shows the arrangement of asci.
11. Extending into the cavities from the margin are found a number of asci and paraphyses. This region is the hymenial region.
12. At the base of the hymenium is the subhymenium, where the hyphae are closely interwoven.
13. The arrangement of the asci is at right angles to the ridges.
14. Asci are arranged parallely in a single layer and have eight uniseriately arranged ascospores.

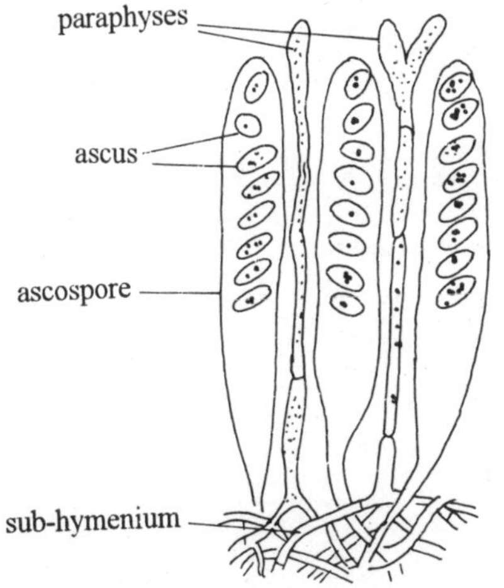

Fig 13.36 A portion of hymenium enlarged

Specimens given in the lba

1. Ascocarp
2. Section of the ascocarp (slide)

Diagrams to be drawn

1. Habit of ascocarp
2. Sectional view of pileus
3. A few cavities enlarged
4. Ascus with ascospores anᴊ paraphyses

Systematic position and identification

A. 1. Chlorophyll absent

2. Reserve food is glycogen ... Fungi

B. 1. Mycleia septate

2. Asci and ascospores produced ... Ascomycetes

C. 1. Frutifications not found in stroma.

2. Apothecium fleshy or leathery ...Pezizales

D. 1. Ascocarp distinguishable into stipe and pileus

2. Apothecia not saucer shaped ... Helvellaceae

E. 1. Pileus has cavities and ridges

2. Stipe is sterile and hollow ... Morchella

CHAPTER 14

BASIDIOMYCETES

Basdiomycetes resmelbes one of the largest classes of fungi with nearly 550 genera and 20,000 species. They differ from all other fungi in that they produce spores called Basidiospores on the outside of the spore producing club shaped hypha called basidium. Basidiospores are uninucleate and are very much comparable to arcospores.

Occurrence and economic importance

Basdiomycetes are both parasites and saprophytes. Some of them are coprophilous also. Parasitic members of Basdiomycetesare most dangerous from the point of view of food crops. The custs and smits are so motorious Basdiomycetes cause great damage to tinular and rail road sleepers in that their mycelia gradually eat away into the tissues.

However some of the higher Basdiomycetes are fine delicacies. Notable among these is the filed mushroom. Agaricus compestris bisparous, which is cultivated large quantities to meet the heavy demand of mushrooms livers the world over.

Somatic structure

The plant body is a septate, much branched mycelium, that laucrours into the substratum in reach of nutrition. The mycelium is usually white, bright, yellow or orange coloured. In some forms, mycleia fuse to form thick strands called rhizomorphs.

The mycelium of Basdiomycetes passes through three stages mainly primary mycelium, secondary mycelium and tetriary mycelium before the life cycle is completed. The primary mycelium is usually formed from the germination of a hap basidiospore and it is minokaryotic, and the cells are uninucleate. During initial stages however the primary mycelium may be uninucleat. But this is a short lives stage. The secondary mycelium originates from the primary mycelium as a result of sexual reproduction, only plasmogamy takin place. Much of the life cycle is spent in the dikaryotic secondary mycelium stage. At the completion of life cycle tertiary mycelium is produced as a result of karyogamy. Soon reduction division takes place in this, producing haploid cells (basidiospores) which develops into monokaryotic mycelium.

Clamp connections

This is an interesting structure frequently observed in the hyphae of secondary mycelium. This is seen particularly in narrow hyphae at the time of cell division. The function of clamp, connections is probably to act as bypass and to allow the easy migration of nuclei from one end to the other when the narrow limen of the hyphae may obstruct their movement. In blood hyphae where nuclear migration is easy, no clamp connections have been found.

Knip (1915) first discovered, clamp connections. Their formation is as follows.

When a lunucleate cell is about to divide, its two nuclei lie a short distance from each other. A hook like clamp appears between the two nuclei. One of the nuclei enters the clamp while the other remains in the amin cell. The two undergo division with a specific spidle orientation. In one nucleus it is parallel and in the other it is oblique. After the division four nuclei are formed. Meanwhile clamp gradually bends over the cell and the cell wall dissolves at the point of contact making a passage. After the nuclear division one nucleus will be lfet in the clamp and it migrated through it to the other end of the cell. In the other nucleus that divides in the cell of the daughter nuclei gradually migrates to the other end of the cell. After this migration one can see two groups of two nuclei (both of course belonging to different sexes). A cross wall is laid down between the two groups of nuclei dividing the cell into two.

Electron microscope studies of the cross septum of leasidiomycetes has shown it to be a parenthesis [()] like structure appearing like a barrel with open ends. This type of a septum is known are dolipore septum or parenthsome. It is difficult to evaluate the significance of the septal apparatus.

Some mycologists compare the clamp connections to the crozier formation in Ascomycetes.

Asexual reproduction

Asexual reproduction takes place by a variety of methods such as fragmentation., budding, conida, arthrospores (oidia) etc. Here, only such methods of asexual reproduction are given above help in the increase of the primary mycelium.

Sexual reproduction

Takes place by two methods

1. Spermatization
2. Somatogamy

Stages of sexual reproduction

There are four stages namely

1. Plasmogamy
2. Dikaryotization
3. Karyogamy
4. Reduction division.

Life cycle pattern. Formation of basidiospores: The primary mycelium starts its life with germination of a basidiospore. This gives rise to secondary mycelium. In this at definite points special club shaped hyphae develop where karyogamy takes place. These are called basidia. Generally two types of basidia are methods with in Basdiomycetes.

1. Phragmobasidium
2. Holobasidium.

1. Structure and development of a phragmodbasidium. This is found in primitive Basdiomycetes like usedinales and ustilaginales. Here the basidium is septate and develops from a spore called telentospores (chlamydospores). The spore puts forth a tube into which the diploid nucleus migrates. The spore is formed on secondary mycelium and will be bimucleate in the begenning. Later karyogamy takes place, this tube is the basidium. The diploid nucleus undergoes reduction division and produces four nuclei. The basidium now divides into four uninucleate cells. From each one of the cells a short stalk is produced called sterigma (plural:sterigmata). The nucleus in each cell migrates to the tip of the stigma which swells to form a spore into which the nucleus migrates. This is the basidiospore. Each basidium produces four basidiospores are liberated from the strigmata. They germinate and develop into a new individual.

2. Structure and development of a holbasidium. Holbasidium is an unseptate basidium and it is formed in higher Basdiomycetes. The terminal cell of a dikaryotic hypha develop into a club shaped structure called the basidium. Karyogamy takes place in this basidium and a diploid nuclei. In the meantime the basidium develops four stigmata at the distal end into the tips of which the nuclei migrate. The tip of each sterigmata swells to forma basidiospore.

The basidiospores show a great variety in structure and form. Typically they are uninuclate and unicellular. They may be globose, and elongated or saurage shaped, colourless or pigmented. In many cases the pigments are green, yellow, orange, ochre, pink, brown or black coloured.

Liberation of spores. Basidiospores are discharged with a force from the sterigma when the spores mature, a droplet of water begins to form at its basal and which gradually enlarges. When it reaches a certain volume the basidisopres is shot off the sterigma, taking the droplet with it. The sterigma may or may not collapse after the spores is discharged.

Fruiting body. In primitive Basdiomycetes the basidia are naked and the spores are discharged as soon as they are formed. There is no protective covering for basidia. In bodies of various types, their frutifications corresponding to ascocarp of the highr Ascomycetes are termed basidiocaprs (Gr.basidium = small + base + karps = fruits). Basidiocarps may be thin and crust like gelatinous, cartilagenous, papery, fleshy, spongy, corky, woody or of any texture. In size they vary from microscopic to 3 meters. Frutifications of Basdiomycetes are the most familiar among fungi, mushrooms, puffballs, earthstars, stinkhornes are all examples of basidiocarps of the fungi that bear them.

Basidiocarps may be open from the very first exposing the basidia (Hynenomycetes) or they may open at a late stage at maturiry releasing all the basidiospores (Gastromycetes).

Inside the basidiocarp, the basidia are formed in definite layers called hymenal layers which are comparable to the layers of Asci in Ascomycetes. Intermingled with the basidia, sterile hairs (paraphyses) are formed. Some mycologists regard them as immature basidia. In some however, paraphyses are larger than basidia and are known as cystidia (Sing.cystidium, Gr.kystis = bladder).

Sexual compatibility. About 90% of Basdiomycetes are heterothallic in that sexuality is differentiated at the genetic level though they may be morphologically alike. This is particularly true in some members which sexually reproduce by somatogamy. Somatogamous reproduction wherein two vegetative hyphae fuse apparently seems to be a situation where sexuality is logt. But this is not the case as evidenced by heterothallium. In about 37% of heterothallic species sexual differentiation is bipolar that is governed by a set of factors (Aa). In the remaining 63% it is tetrapolar ie., sexuality is governed by two sets of factors (AaBb).

Classification

This nature of basidium and frutification are of great aid in the classification of Basdiomycetes.

Gwynnevaughan and Barnes (1926) clasisfy Basdiomycetes into 3 sub classes namely 1. Aemibasidiomycetes, 2. Protobasidiomycetes, 3. Autobasidiomycetes.

In hemibasidiomycetes member of basidiospore are indefinite. It includes an order ustilaginales (smuts).

In protobasidiomycetes basidiospores are definite in number, usually four, and basidium is a phragmobasidium. This includes 3 orders.

Nredaneles (smuts)

Auricularies and

There mellales.

In Autobasidiomycetes basidiospore are four in number, basidium is a holobasidium. This includes two orders. Hymenomycetales and Gasteromycetales.

Majority of modern mycologists classify Basdiomycetes into 2 sub classes:

Subclass 1. - Heterobasidiomycetideae (Hemibasidiomycetedeae)

Subcaclass 2 - Homobasidiomycetideae (Eubasidiomycetedeae)

Subclass - heterobasidomycetideae is primtive. Here no basidiocapr is formed and the basidium is septate. This include 3 orders.

Ustilaginales, Usedinales and Teremellales.

Subclass - Homobasidiomycetideae is advanced. Here basidiocarp is formed and basidium is unseptate. Depending on the nature of the basidiocarp this is divided into two series.

1. Hymenomycetes and 2. Gasteromycetes.

In Hymenomycetes the basidiocarp is a closed structure. It breaks open only at maturity releasing the basidiospores. This include four orders.

a. Hymenogastrales, b. Nidulariales, c. Lycoperdales, d. Sclerodermatels.

USTILAGO

Common species : U.maydis, U.sutamenea, U.hordei, U.muda, U.kolleri, U.arenae etc.

Occurrence : Majority of the species of ustilage are parasites causing diseases known as smuts. The grains get destroyed and a sooty mass of spores are produced.

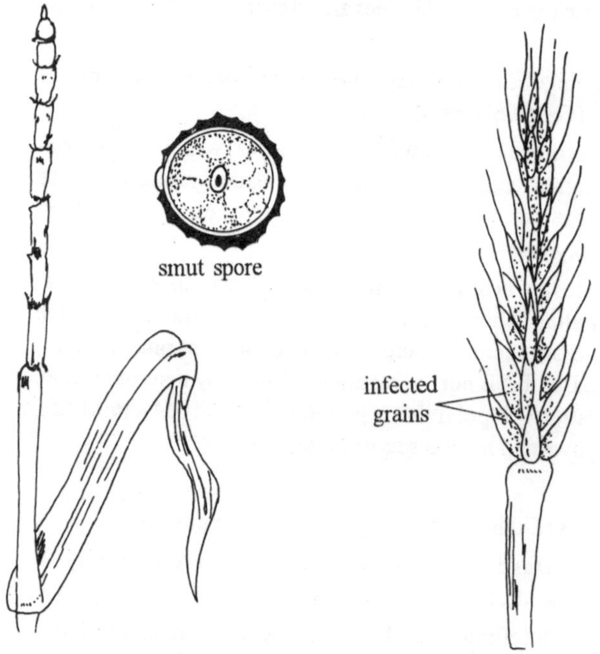

Fig. 14.1 Ustilago. Wheat ear covered showing loose smut caused by

Barley eay showing smut caused by Ustilago

Study : 1. Diseases caused and symptoms of the disease

2. Vegetative structure

3. Reproduction

4. Systematic position and identification

Diseases caused and symptoms of the disease

1. Loose smut of wheat. This disease is caused by U.tritici. The disease is prevalent throughout India. The symptoms are as follows:

a. The disease symptoms appear at the onset of ear development

b. The avarian small breaks open releasing the black chlamydospores

c. All the avaries of the spikelet except the awns get converted into smut spores

d. The smut sorus is covered by a thin membrane in the beginning, but it soon ruptures releasing the dark chlamydospores. The spores are wind dispensed

e. As the spores are dispensed and there is no membrane covering the spores for a long time, the disease is said to be the loose smut

f. After the release of spores, the barren rachis remain exposed

g. The spores have smooth endosporium and a thick echimulate exosporium.

2. Covered smut of barley. This disease is caused by U.hordei and is cosmpolitan in distribution. The disease symptoms are:

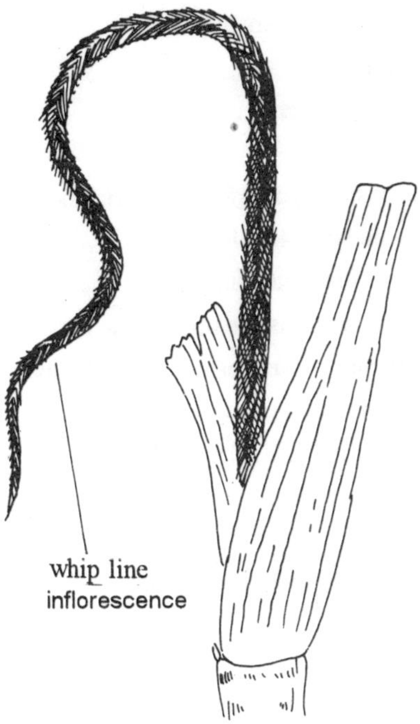

Fig. 14.2 Ustilago. Whipsmut of sugarcane caused by Ustilago scitaminea

a. Disease manifestation is at the time when the ears are coming out of the boot

b. No grain remains free of the disease in an air; all of them turn black. Each grain gets converted into a smut sorus

c. Awns of the flower remaind disease free. Initially the sorus has a membranous covering, but it soon ruptures to release the spores

d. In the chlamydospores, both the exosporium and endosporium are smooth. The endosporium however is thin.

3. Loose smut of barley: This is caused by U.nuda

a. The grains turn black due to accumulation of spores

b. Spores are discharged when the ear is still attached to the plant.

4. Whip smut of sugarcane. This is caused by U.sutaminea. due to the disease the stem tip gets converted into 3-4 feet long dark coloured whip like structure. The main shoot tip as wll as laterals also get converted into smut whips.

5. Coversed smut of oats. The disease is caused by U.holleri. In this the spores are not exposed they are covered and come out only during the threshing of the grains.

Vegetative structure

1. The plant body is a mycelium. It is septate and much branched.

2. There are two types of mycelia - primary mycelium and secondary mycelium.

3. The cells of the primary mycelium are uninucleate while those of secondary mycelium are dikaryotic ie., (they have two, one male and one female nuclei).

Reproduction

1. Reproductive parts, ie., spores are visible only in the ears of the host plants. These are chlamydospores (also called smut spores, brand spores and teleutospores).

2. The characteristic spores chlamydospores are produced on the secondary mycelium as a result of sexual reproduction.

3. The chlamydospores have a two layered wall. The outer wall called exosporium is thick, pigmented eithe smooth or ornamental. The inner endosporium is smooth may be thin or thick.

Specimen given in the laboratory

1. Infected ears of the host plants like wheat, maize, oats etc. Mount the spores by crushing the smut sorus. Study the symptoms.

Diagrams to be drawn:

1. Habit sketch of infected ears of the host.

2. A few chlamydospores.

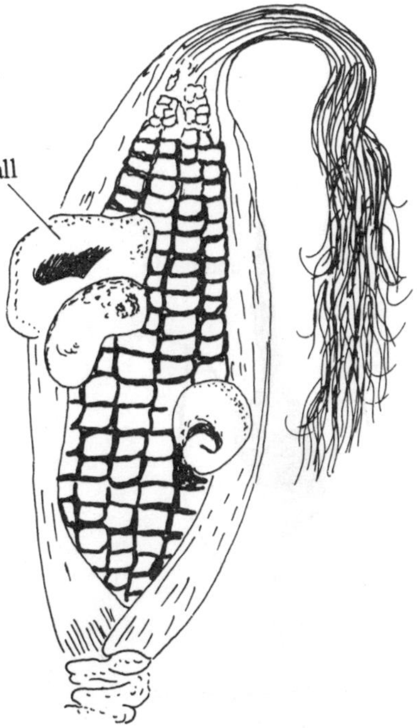

Fig. 14.3a Maize smut caused by Ustilago

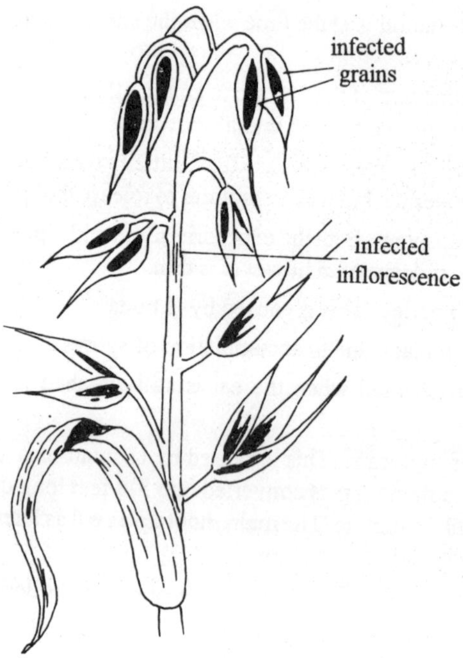

Fig. 14.4 Ustilago. Covered smut of oat caused by Ustilago kolleri

Systematic position and identification

A. 1. Chlorophyll present

2. Reserve food is glycogen ... Fungi

B. 1. Mycelia septate

2. Basidium and basidiospores formed ... Basdiomycetes

C. 1. Basidia divided into 4 cells

2. No basidiocarp...Protobasidiomycetes

D. 1. Disease symptoms visible only in the ears.

2. Characteristic smut spores formed ... ustilaginales

E. 1. Basidiospores sessile

2. Basidium fragmented ... Ustilaginaceae

F. 1. Chlamydospores black, and separate into single cells (not in pairs)

2. Sorus dusty when mature ... Ustilago.

SPHACEOLTHECA

Common species : S.sorghi, S.cruenta, S.reliana etc.

Occurrence : Ther germs sphacelotheca is one of the most notorious among smuts and infects a number of hosts like saghum, barley, maize etc., causing a great loss to the crop.

Study : 1. Disease caused and symptoms.

2. Vegetative structure

3. Reproduction

4. Systematic position and identification

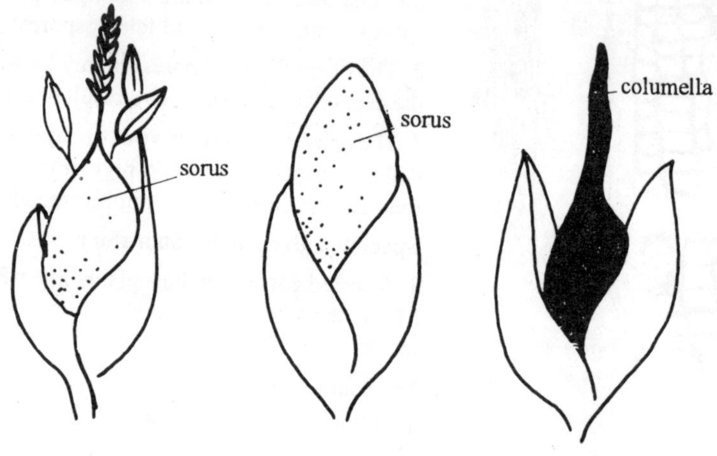

Fig. 14.5 Sphacelotheca. An infected flower — A grain containing sorus — A sorus showing columella

Diseases caused and symptoms :

A. Grain smut of sorghum : The disease is caused by S.sorghii and is the most destructive disease of sorghum in India.

The disease symptoms are as follows:

1. The disease manifests itself when the ears are emerging out of the boot.
2. Smut sorus is formed in the grain and each sorus has a central columella made up of a hard tissue.
3. Surrounding each columella are found a number of spores.
4. The sori measure in length from 1/6" to 1/2" in length.
5. Within the spore sac sterile cells are also found intermixed with the spores.
6. Spores are globose to oval, dark brown in colour, smooth walled and 5-9 in diameter.
7. The smut is externally seed borne.

B. Loose smut of sorghum: This disease is not as common as the grain smut in our country.

The disease symptoms are as follows : The disease is caused by S.cruenta.

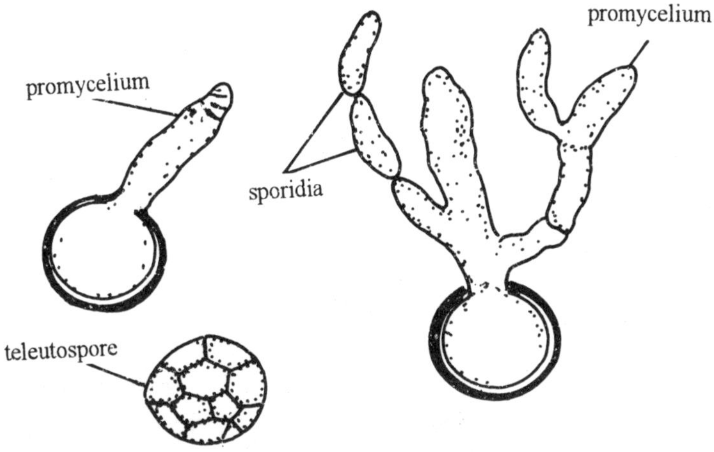

Fig. 14.6a Germinating teleutospores showing promycelium

1. Infected plants are shorter with thin stalks.
2. The ears are produced earlier than in the healthy plants.
3. All the spikelets of the ear get the disease.
4. The covering membrane of the sorus ruptures early releasing the powdery mass of dark spores.
5. The columella in the spore sac perists even after the dissemination of the spores.
6. The wall of the sorus is made up of loosely arranged cells.
7. The spores are round or elliptical darker brown than in grain smut.

C. Head smut of sorghum and maize : The disease is caused by S.reliance. The disease is mostly found in Kashmir, but present in other states also. The disease symptoms are :

1. In the infected plant the ear is completely damaged.
2. In maize large sori of the smut replace the tassel.

Fig.14.7 Ear of Sorghum infected by Sphacelotheca sorghi

3. In such cases floral bracts grow out into leafy bracts sometimes into small shoots.

4. In sorghum, the entire inflorescence is converted into a big sorus about 4" long and 2" wide.

5. At the early stages a white membrane encloses the round.

6. A network of black fibres traverses the spore mass and remains adhering even after the liberation of the spores.

7. Spores are reddish brown to black, finely echinate and irregular to spherical in shape.

Vegetative structure

1. The plant body of sphacelotheca is a septate branched mycelium.

2. The crosswall is porous and allows the movement of cytoplasm from one cell to another.

3. The mycelium is intercellular and hyaline.

4. Haustoria developed from the mycelium help in the absorption of food from the host tissues.

5. The primary mycelium is dikaryotic and produced from the primary mycelium as a result of sexual reproduction.

Reproduction

1. The cells of the secondary mycelium is the ovaries isolate themselves acquire thick walls and transform themselves into chlamydospores.

2. The chlamydospores also called teleutospores or brand sproes are dark coloured and dikaryotic.

3. Due to the infection, the grain will turn into a compact mass of fungal tissue called sorus. The rosi may be long or short.

4. The tough wall of the sorus may entirely be a fungal tissue or may also have the host tissue.

5. Inside the sorus is a central hard columella extending from the base to apex.

6. The columella represents the host tissue and is composed of parenchyma and vascular elements as in S.sorghii.

7. The spores are round or oval on liberation they germinate to give rise to basidia and basidiospores.

Specimens given in the laboratory

1. Infected ears of maize, sorghum etc.

Diagrams to be drawn:

1. Infected ears

2. Sours with columella

3. Chlamydospores

Systematic position and identification

A. 1. Chlorophyll absent

2. Reserve food is glycogen ... Fungi

B. 1. Mycelium septate

2. Basidia and basidiospores produced ... Basdiomycetes

C. 1. Basidium is a phargmolbasidium

2. Basidiocarp is abent ... Protobasidiomycetes

D. 1. Disease symptoms muscle only on the ears

2. Characteristic smut spores are formed ... Ustilaginales

E. 1. Basidiospores sessile

2. Basidium fragmented ... ustilagenacae

F. 1. Chlamydospores entangle in a mass due to the presence of black fibres.

2. Spores are reddish brown to black finely ecnirate ... Sphadcelotheca.

PUCCINIA

Common species : P.graminia, P.recondita, P.stiforms

Occurrence : The genus Puccinia of which 700 species have been reported is an obligate parasite infecting the members of the family gramineae like wheat, barley, oats, rye etc. Puccinia causes these types of rust diseases on wheat namely.

1. P.graminis Black or stem rust
2. P. stiformis (= P.gimarum) yellow or striped rust.
3. P.tritieina Brown or orange leaf rust.

The appearance of red, brown or deep red patches on the leaves are the characteristic symptoms :

Study : 1. Life history

2. Disease symptoms of major types of rusts

3. Systematic position and identification

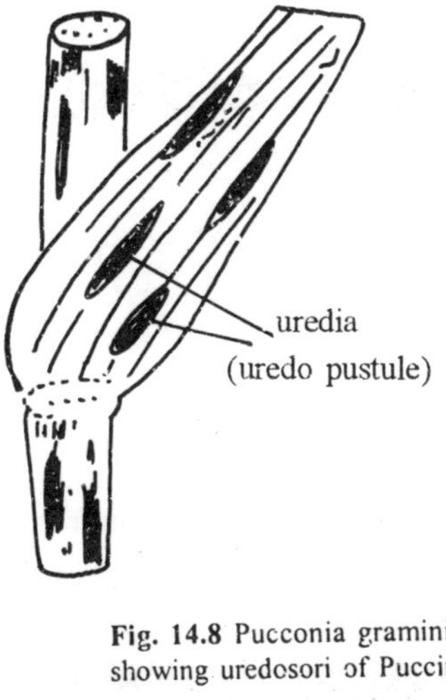

Fig. 14.8 Pucconia graminis. Wheat shoot showing uredosori of Puccinia

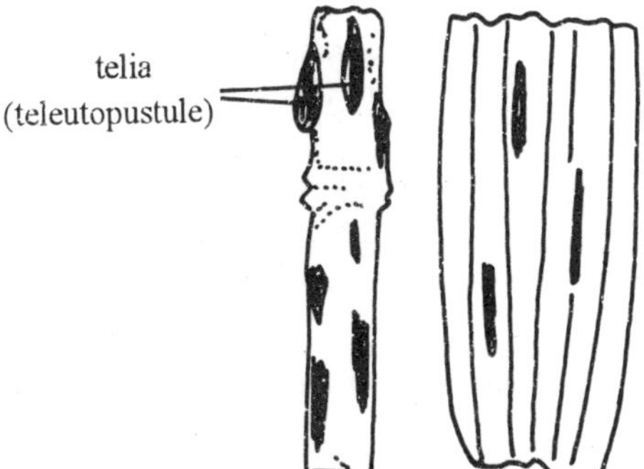

Wheat stem and leaf showing teleutosori

Specimens given in the laboratory

1. Infected leaves of bar berry
2. Infected leaves of wheat
3. Pyenial stage (slide)
4. Aerial stage (slide)
5. Teleutisours stage (slide)

Life history :

1. Puccinia granunis is a heteroecious and macrocyoic rust
2. It requires two hosts to complete the life cycle hence known as heteroecious.
3. The two hosts are primary host or permanent host wheat - Triticum and secondary host or alternate host Barberry - Berbahis.
4. The rust is said to be macrocylic because it produces more than one kind of binucleate spores on its secondary mycelium.
5. The life history can be studied in two distinct phases based on the host.

Stages of life cycle on Barberry :

1. The barberry leaves are fist infected by basidiospores of the rust.
2. The germinating basidiospore produces the primary mycelium in between the mesophyll cells of the host leaf.
3. The primary mycelium is septate, much branched with uninucleate celis.

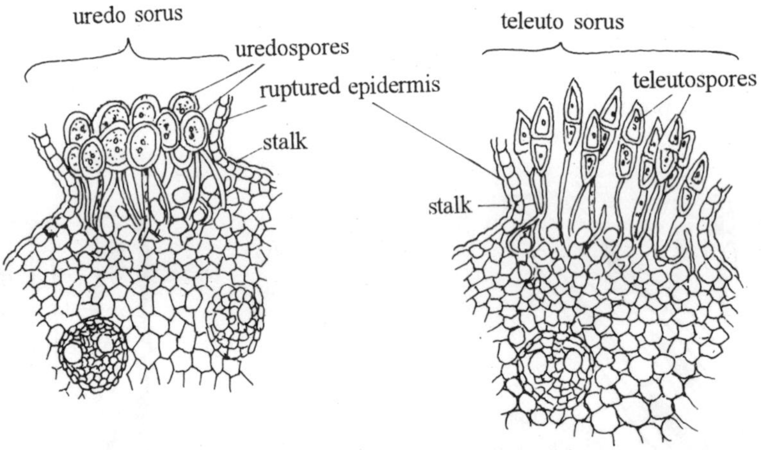

Fig.14.9 Puccinia graminis. T.S. of wheat Stem through uredosorus

T.S. of wheat stem through teleutosorus of Puccinia graminis

4. The cross walls are swollen (Dolipore septum) and allow the streaming of cytoplasm from cell to cell.
5. The primary mycelium after a certain period of growth reproduces sexually by forming sex organs in cup like structures called spermagonial cups or pyenial cups.
6. The formation of these structures is externally manifested by the appearance of black patches on the upper surface of the barberry leaves.
7. Across section of the leaf passing through an infection patch will show the details of reproductiive

structures.

8. Below the upper epidermis of the barberry leaf small flask shaped structures are seen. These are the spermagonial cupsor pyenia.

9. Each spermagonial cup has an opening called ostide, through which male sex cells are liberated.

10. Lining the basal region of the spermagonial cup are found many erect, unbranched hyphae called spermatiohpores. Each spermitopore cuts off large number of unicelllar, uninucleate spermatia from this tips.

11. Spermatia are the male gametes. They are man flagellate.

12. Some elongated branched hyphae also arise in the spermagonial cups. These are the female sex organs called receptive hyphae. The receptive hyphae come out of the spermagonial cup through the ostiole.

13. The basal cell of the receptive hyphae is fertile.

14. The spermagonial cup or the infected host cells reproduce a sweet juice. Insects visiting the leaves to get the nectar will help to transfer spermatic from one spermagonial cup to the other.

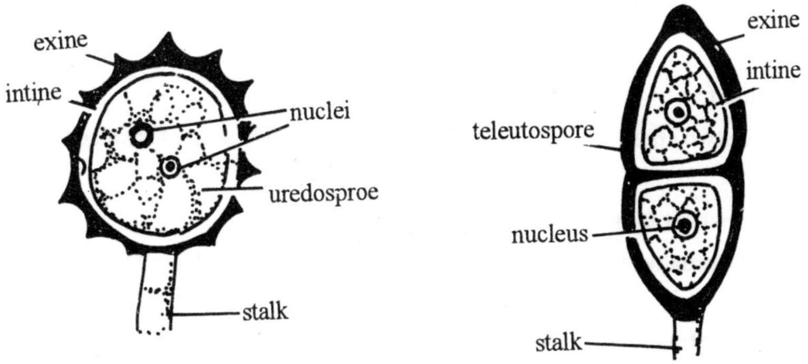

Fig. 14.10 Puccinia graminis. Single uredospore Single teteutospore

15. As a result of sexual reproduction which is by spermatization, secondary mycelium is produced. This reproduces by means of Aecia or Aecidia.

(Note the name Pyenia for the spermagonia is an old name given when the sexual nature was not known similarly spermatic were called Pyenospores).

Aerial stage

1. This is produced above the lower epidermis of the barberry leaf.
2. The accia are invested cup-like structures and have a firm wall called peridium.
3. Inside the serium many chains cells are found. All cells are dikaryotic (binucleate)
4. In each chain there are smaller cells and larger cells, which alternate regularly.
5. The smaller, cells are called disjunctor cells or intercalary cells while the larger cells are the aeciospores.
6. Aeciospores are produced in large numbers and erect a pressure on the lower epidermis of the barberry leaf.
7. The lower epidermis of the barberry leaf ruptures and the aeciospores are released.
8. The aeciospores carry the infection to the primary host ivz., the wheat plant.

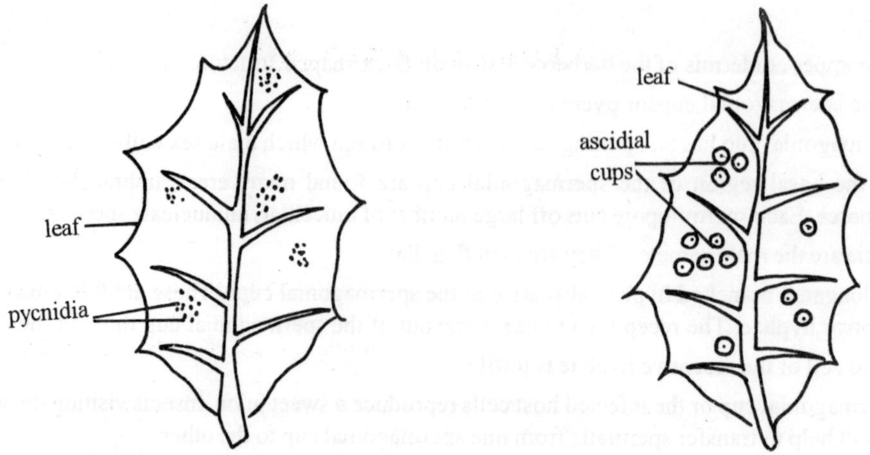

Fig. 14.11 Puccinia graminis. Braberry leaf showing pycnidia on upper surface

Barberry leaf showing aceidial cups on lower surface

Stages of life cycle on wheat :

1. The aeciospores produce two characteristic sori on the wheat plant. These are Usedosorus and teleutosours.

Uredosorus stage :

1. This is produced on the wheat by the germination of aeciospores.
2. The formation of chedosorus on the leaves and stem is externally manifested by the appearance of red patches.
3. The details of the usedosours can be studied by taking a cross section of the leaf passing through a red patch.
4. A little below the epidermis, in between the mesophyll cells is found the secondary mycelium.
5. The secondary mycelium is much branched, septate with dikaryotic cells.
6. The usedospores arise in aggregates from the secondary mycelia.
7. Each usedospore has a long stalk and a single ovarl or eliptical cell. The cell is dikaryotic.
8. The cell wall is thick and two layered. The outerwall is echihulate. The wall has red pigments hence the appearance of red patches on the leaves of wheat
9. The usedospores are called summer spores as they are produced in the beginning of summer. They are also called repeating spores because a usedospore can infect another wheat plant and produce a usedosorus.

Note : While the initial infection to a wheat plant may come from an aciospore, the rapid srpeading of the disease is due to usedospores.

10. At the end of summer the rust produces the next stage ie., the teleutosours in dried leaves, stubble etc.
11. The presence of the teleutosorus is manifested by the appearance of dark known patches on the leaves and stem.
12. Each teleutosorus represents the aggregation of teleutospores. The teleutospores are also called winter spores as they are produced at the beginnig of winter.
13. Each teleutospore has a long stalk and two terminal cells. Each cell is dikaryotic. The cell wall is thick, brown pigmented, but does not have ornamentaion.

14. Teleutospores on germination produce basidia and basidiospores.

Disease symptoms of some major rust disease caused by Puccinia :

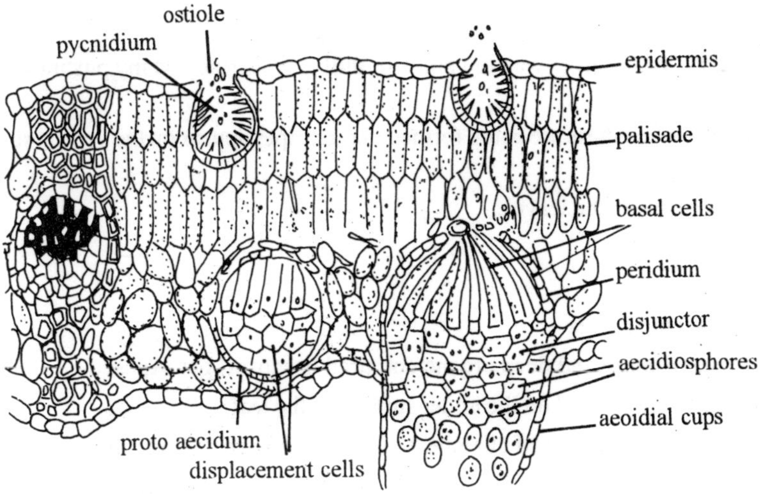

Fig. 14.12 Puccinia graminis. V.S. of barberry leaf showing pycnidia and aecidial

Black rust of wheat :

1. The disease symptoms manifest themselves at the time of maturity of grains.
2. Long, brown coloured pustules on leaf, leaf sheath and stem indicate the presence of infection.
3. The pustules are more in number on the stem. Tl. ..stules have usedosori internally.
4. The pustules are 1/4 long and on maturity put out a mass of usedospores.
5. Teleutosori are darker than usedosori, they are also produced in the stems and leaves.
6. The disease causing agent in P.graminis var tritici.

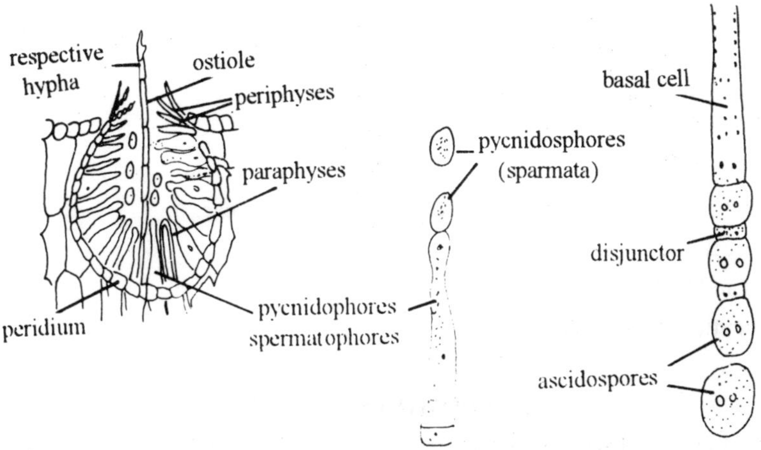

Fig. 14.13 Puccinia graminis. V.S. of pycnidium Single spermatiophore Chain of aecidiospores

Stripe rust of wheat :

1. In this disease leaves and inflorescence are completely destroyed.

2. The disease makes its appearance before the grains are mature.

3. When the disease is not intense usedosori.

4. The deep green colour of the leaves gives way to yellow or dull green stripes. Many pustules are found in each stripe.

5. Teleutosori appear little late; when these are being formed black stripes appear on the ventral surface of the leaves.

6. The geleutosori form dark rusts on the leaves.

7. The disease is caused by P.striforms.

Sex or orange rust wheat :

1. The dosori appear mostly on the leaves

2. The epidermis on the leaves rupture and deep organe coloured usedosori get exposed.

3. Teleutosori appear quite later are formed on the under surface of the leaf.

4. The teleutosorus is divided into a number of chambers due to the presence of paralyses.

Diagrams to be drawn :

1. Habit of infected barberry leaf.

2. Section of the barberry leaf to show sporogonia and aecia.

3. Habit of infected wheat leaf.

4. Usedosours.

5. Teleutosorus.

Systematic position and identification

A. 1. Chlorophyll absent

2. Reserve food material si glycogen ... fungi

B. 1. Mycelia septate

2. Basidia and Basidiospores formed ...Basidiomycetes

C. 1. Basidiocarp absent

2. Basidia septate ...Protobasidiomycetes

D. 1. Rusty appearance of pustules

2. Obligate parasites ... Uredinales

E. 1. Basidiospores formed laterally

2. Teleutospore loosely fasciledons or more than one celled ... Pucciniaceae.

F. 1. Fungus is heteroecious.

2. Hosts are wheat and barley

3. Fungus is marrocytic ... Punccinia.

UROMYCES

Common species : U.jasmini, U.fabae, U.pisi etc.

Occurrence : The species of Uromyces may be heteroecious or autoecious. U.pisi is heteroecious while

U.fabae is autoecious. Lentils, peas, jasmins and infected by the various epcies of uromyces.

Study : 1. Disease symptoms and disease caused.

2. Study of life history

3. Systematic position and identification

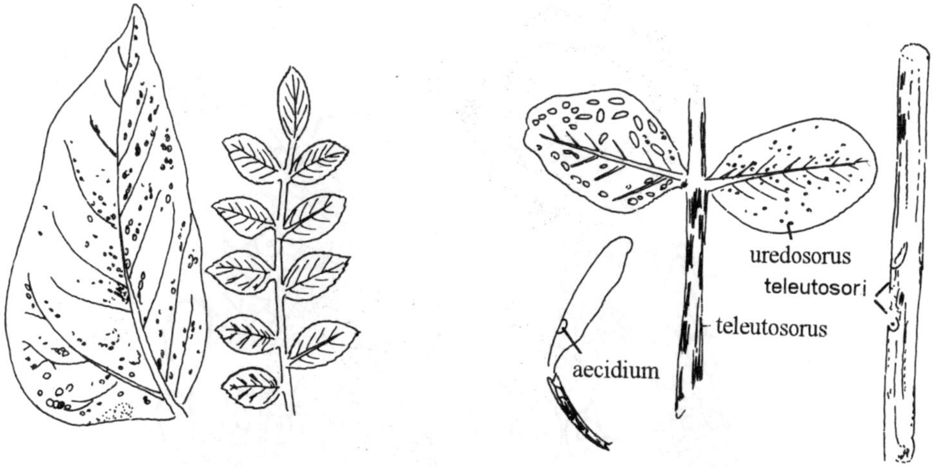

Fig. 14.14 Uromyces A. U.appendiculatus on Vigna leaf. B. U.cicer-arientini on gram leaf. C. U.fabae uredosorus and teleutosorus on leaf and stem, and aecidium on pod. D. U.decoratus (on sunnhemp); teleutosori onstem.

Diseases and symptoms :

1. Rust of Bengal gram :

a. The disease is caused by U.cuer arietini

b. Disease symptoms appear when the plants are four months old

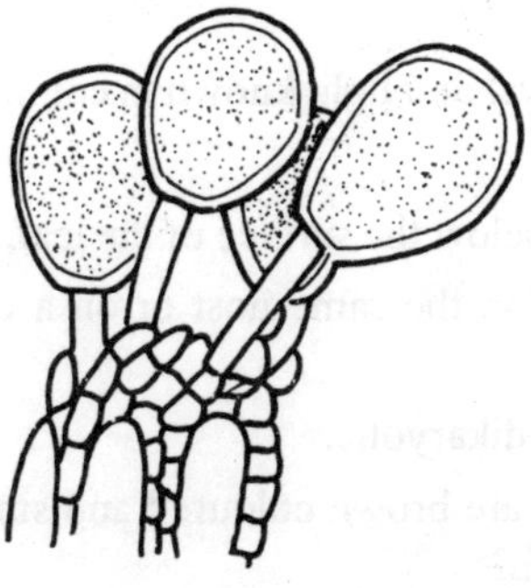

Fig. 14.15 Uromyces. A few uredospores of U.appendiculatus

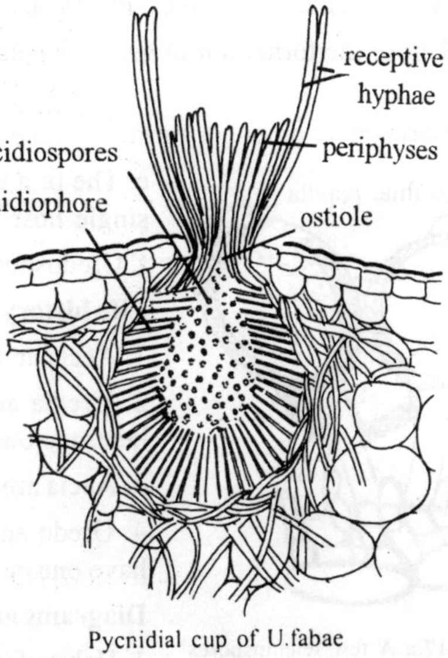

Pycnidial cup of U.fabae

c. Leaves aggregate with the appearance of known patches on them.

The patches which are small in the beginning coalse and become larger.

Disease symptoms may appear on stem and either surface of the leaves.

The spermagonial and aerial stages of this fungus are not known. Redospores are rounded with enumulate wall.

Teleutospores are deep known.

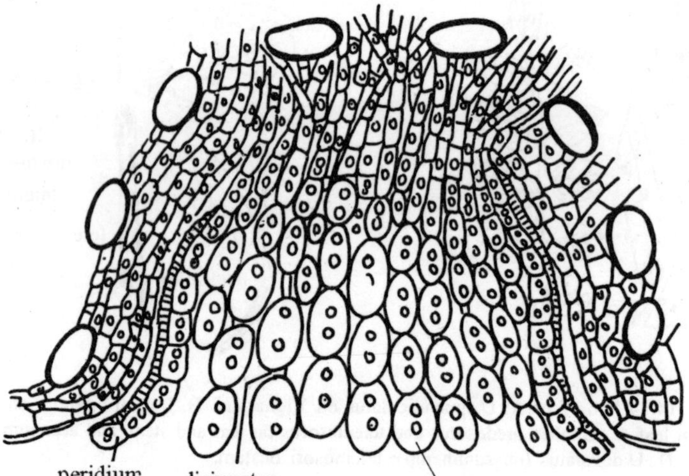

Fig. 14.16 Uromyces. Aecidial cup of U.erythronii

Rust of peas and lentils.

1. The disease is caused by U.fabae.

2. Aecia appear fist on the leaves. Pyenia are rare. Sori appear on all green parts of the plant.

3. Even before the formation of aecia, leaves, petioles, tenduts, stem and fruits turn yellow first and later brown.

d. Uredosori appear on either surfaces of the leaves

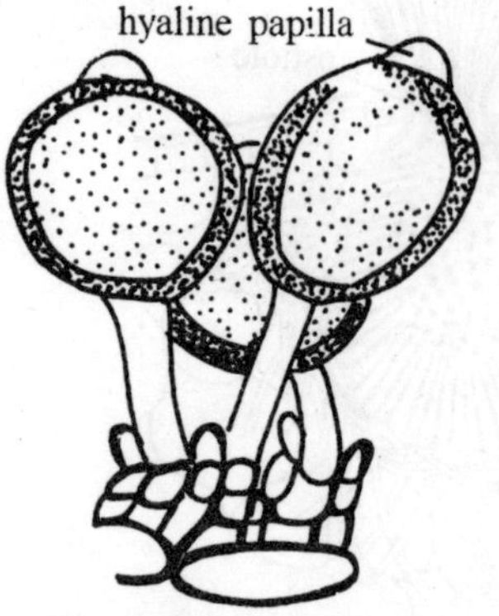

Fig. 14.17.a A few teleutospores of U.appendiculatus

e. The fungus is autoecious, completing all the stages of the life cycle in a single host

f. Uredospores rounded or oval and light known in colour.

Life history :

1. Aecia and pyenia appear below the surface of the leaf.

2. Uredia and Telia appear on the same host or on a different host (in heteroecious).

3. Aecia are cup-shaped and dikaryotic.

4. Uredo and Teleutospores are brown coloured and single celled. They have one or more germpores.

Diagrams to be drawn:

1. Habit of infected leaves.

2. Sectional view of Aecia, pyenia, uredenia and telia.

Systematic position and identification

A. 1. Chlorophyll present

2. Reserve food is glycogen ... Fungi

B. 1. Mycelium septate

2. Basidia and basidiospores formed ... Basdiomycetes

C. 1. Frutifications absent

2. Basidia septate... Protobasidiomycetes

D. 1. Rusty appearance of pustules

2. Oldigate parasites ... Uredinales

E. 1. Basidiospores formed laterally

2. Teleutospores one or two celled free or loosely fasided ... Pucuniaceae

F. 1. Teleutospores one celled

2. Teleutosori do not have a preiodium ... Usomyces.

EXOBASIDIUM

Common species : E.vexans, E.vaunii etc.

Occurrence : The members are parasite on the families Ericaceae, Theaceae etc and cause diseases like blistee rust, gill disease etc.

Study : 1. Disease caused and symptoms

2. Vegetative structure

3. Reproduction

4. Systematic position and identification

Disease caused and symptoms

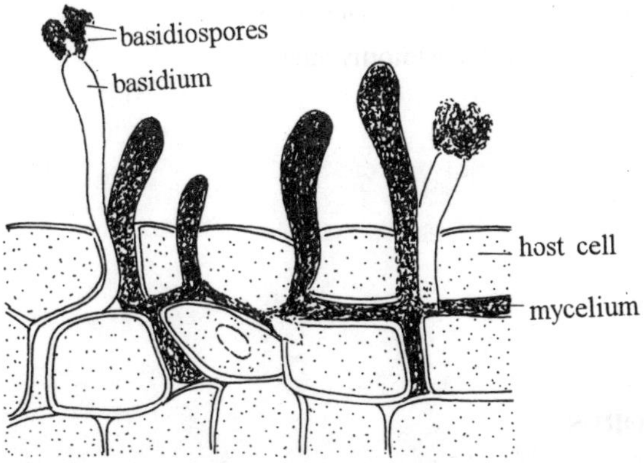

Fig. 14.18 Exobasidium

1. Blister blight of tea :

a. This disease is caused by Exobasidium vexans

b. The disease becomes wide spread during rainy season

c. Disease symptoms appear on the leaves in the form of purple coloured dots

d. The lower surface of the dots have dowry growth of the fungus

e. The dots appear depressed above and swollen below hence the name blister blight

f. The leaves drop down and the plant collapses due to the disease

2. Gall disease :

a. The disease is caused by E.vaccinii

b. Members of Ericaceae like cranberries and backleberries get the disease

c. Both leaves and twigs get the infection.

Vegetative structure

1. The plant body of the fungus is an intercellular much branched, septate mycelium consisting of tubular hyphae.

2. The hyphae produce haustoria to absorb nutrition from the host cells.

3. The mycelium is subepidermal or deep in the mesophyll tissue or cells.

Reproduction :

1. The fungus reproduces sexually by means of conidia. The conidiospores come out of the host leaf through the epidermis.

2. The conidia are single celled oval in shape unimucleate and germinate to give rise to a new individual.

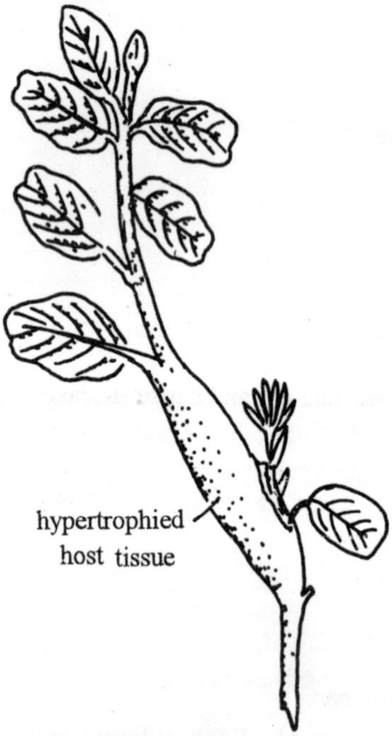

Fig. 14.19a Exobasidium

Sexual reproduction

1. Sexual reproduction results in the formation of basidia which are aseptate.

2. The basidia come out through the epidermis of the host leaf and bear 4-8 basidiospores at their tip.

3. The basidiospores may be unicellular or may have 2-3 cells by the development of septa.

4. The basidiospores may undergo budding to increase their number or many directly form a new individual.

Diagrams to be drawn:

1. Habit of infected leaves.

2. Sectional view of leaf to show conidia.

3. Sectional view of leaf to show basidia.

Systematic position and identification

A. 1. Chlorophyll absent

2. Reserve food is glycogen ... Fungi

B. 1. Mycelium septate

2. Basidia and basidiospores formed ... Basdiomycetes

C. 1. Basidium aseptate ... Holobasidiomycetes

D. 1. No basidiocarp

2. Basidia superficial on the host ... Exobasidiales

E. 1. Basidiospores become multicellular due to septation

2. Basidia club shaped ... Exobasdiaceae

F. 1. Basidia comes out in between the epidermal cells.

2. Paralyses absent ... Exobasidium.

POLYPORUS

Common species : 1. P.sulphurses, P.versicolar, P.betulinus etc.

Occurrence : The species of polyporus are parasite on a variety of forest trees causing decay and destruction. Oak, sissoo, albizzia etc get the attack of the fungus.

Study : 1. Disease caused and symptoms

2. Vegetative structure
3. Reproduction
4. Systematic position and identification

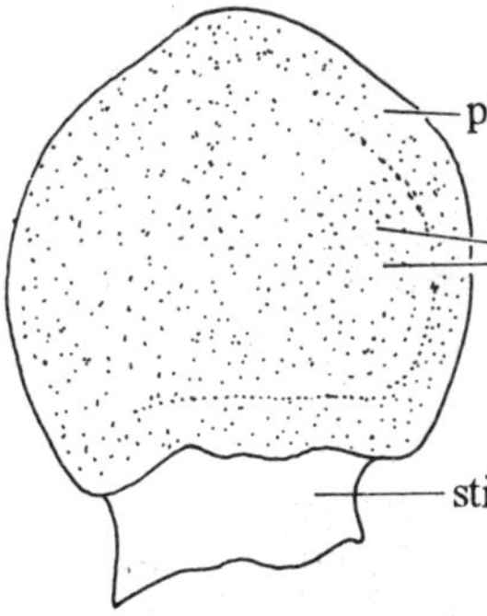

Fig. 14.20 Polyporus. fruiting body (the basidiocarp)

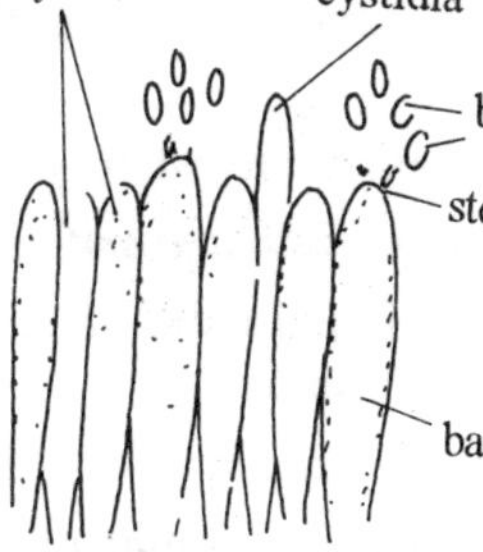

Hymenium showing basidia and basidiospores

Disease caused and symptoms

1. Wood rot of oaks : This disease is caused by P.sulphurous wherein the wood gets decayed. P.unnabearinus also caused the wood rot oaks.

2. Heart rot of elms : The heart wood and sapwood get the attack of the fungus. P.squamosas elm, many other trees also get the attack.

3. Wood rot: Various tree species are attacked by P.versicolar.

4. Rot of birch : Species of Betula get infected by P.betulinus.

Vegetative structure

1. The plant body of the fungus is a mycelium. The mycelium is septate and much branched.

2. The mycelium has uninucleate cells (primary mycelium) in the beginning later the cell becomes dikaryotic (secondary mycelium). In many hyphae of secondary mycelium clamp connections are seen.

In cases of severe infection by the fungus the hyphae materials may even penetrate the heart wood.

Fig. 14.21 Polyporus. Basidiocarps growing on tree trunk

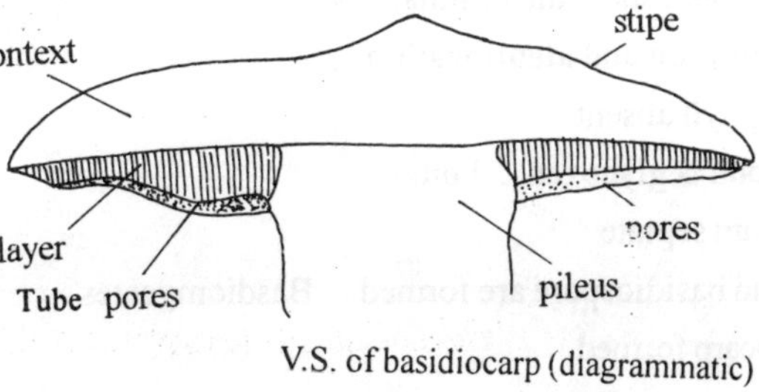

V.S. of basidiocarp (diagrammatic)

Reproduction

1. Sexual reproduction in the primary mycelium leads to the development of the fruiting body, the basidiocarp.
2. It is this basidiocarp which is seen above the substratum. The basidiocarp is built up lay the secondary mycelium.
3. The basidiocarp has a short stalk and an expanded sheath and disc shaped structure viz., the pileus.
4. The mature basidiocarp commonly called the bracket fungus is woody, cocky or leathery and becomes hard at it matures.
5. The stalk or stipe is 2-4" long, somewhat cylindrical and deep or dark brown in endour.
6. The pilus is also light brown coloured and its upper surface is smooth and slightly wavy.
7. Many concentric rings are found at the margin of the pileus. The basidiocarp its perennial and the concentric rings epreonst the growth rings.
8. The lower surface of the pileus is smooth or flat and has numerous fine pores.
9. Each pore actually represent the opening of a tube, the inner surface of which has basidia.
10. In a sectional view the basidiocarp shows the following regions viz., context, trama, pores and hymenium.
11. The context represents the outer fibrous part of the basidiocarp made up of compact thick walled hyphae.
12. The trams has loosely arranged branched hyphae.
13. Extending vertically down from below the context are several tubes which open out on the central surface in the form of pores.
14. The hymenium constitutes the inner lining of the pores. It consists of basidia and cystidia.
15. Cystidia are regard as sterial basidia. They do not bear basidiospores.
16. The basidia which project into the cavity are aseptate (holobasva).
17. At the tip of each basidium are 4 sterigmata bearing a basidiocarp each.
18. The basidiospore are released into the pores from where they get dispersed.
19. The spore output is enormous, millions of spores per basidicoarp are produced.
20. The basidiospore are oval, and on germination develop into the primary mycelium.

Specimens given in the laboratory

1. Basidiocarp
2. Section of the pileus (slide)

Diagrams to be drawn

1. Habit of basidiocarp
2. Section of the pileus
3. A pore enlarged show the basidia.

Systematic position and identification

A. 1. Chlorophyll absent

2. Reserve food is glycogen ... Fungi

B. 1. Mycelium septate

2. Basidia and basidiospore are formed ... Basdiomycetes

C. 1. Basidiocarp formed

2. Basidia aseptate ... Eubasidiomycetes

D. 1. Basidiospores released in instalments they are not completely covered till maturity ... Hymenomycetes.

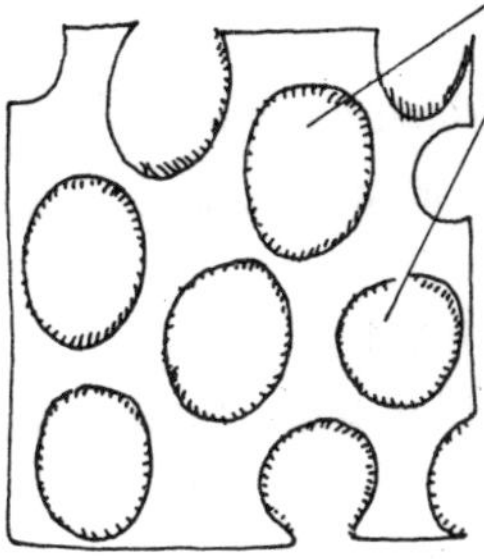

Fig. 14.22 Polyporus. Surface view of pileus showing pores

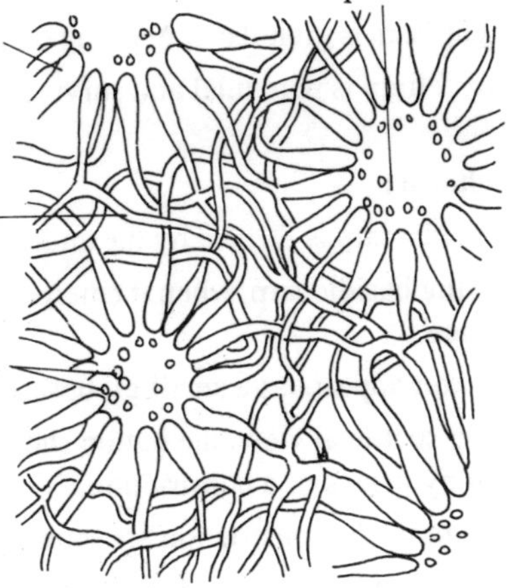

Cross-section of a portion of pileus

E. 1. Frutification macroscopic

2. Hymenial layers protected ... Agaricales

F. 1. Hymenial layer opens out into tubes

2. Numerous pores present on the undersurface of the pileus ... Polyporaceae

G. 1. Ventral surface of the pileus porous and different in colour from the upper surface.

2. Mature basidiocarp hard and woody ... Polyporus

AGARICUS

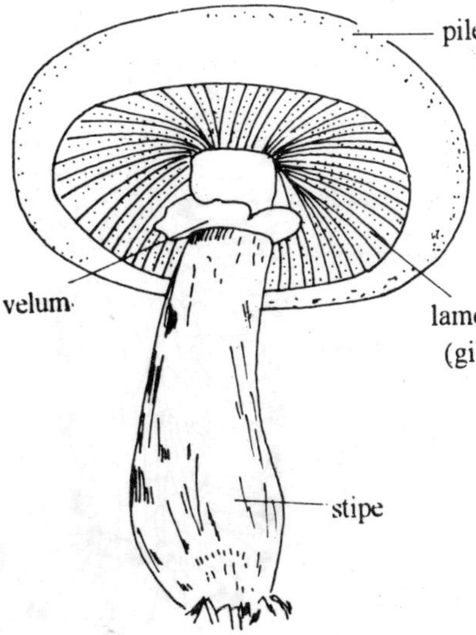

Fig. 14.23 Agaricus. A basidiocarp

Common species : A.compestris bisoprous

Occurrence : Commonly called mushroom, the genus Agaricus is a widely distributed saprophytic, fungus growing on dead logs, timeber, on soil with plenty of leaf little.

Study : 1. The vegetative mycelium

2. Basidiocarp
3. Structure of the gill
4. Systematic position and identification

Vegetative mycelium

1. The actual plant body of Agaricus is a subterranean mycelium

2. The mycelium is extensively branched with anastomosing hyphae. The network of hyphae are called spawns.

3. The hyphae are septate tubular and hyaline.

4. The cyliplasm is granular containing oil globular and other serve food materials.

5. The mycelium is perennial and survives for several years producing basidiocarps which come above the ground.

The basidiocarp

1. The balsidiocarp is formed by the underlying mat of the hyphae after dikaryotoization.

2. The young basidiocarp, when it emerges out of the ground will be button shaped, hence it is known as the button stage.

3. The button appears as a small globose whitish structure above the ground.

4. In a long sectional view, the button shows a slightly expanded top portion separated from the lower ones or portion by means of a constriction.

5. A little above the constriction are found two chambers - the lamellar cavities.

6. The margin of the pileus is attached to the stipe with the help of a thin membrane called veil or velum.

Mature basidiocarp:

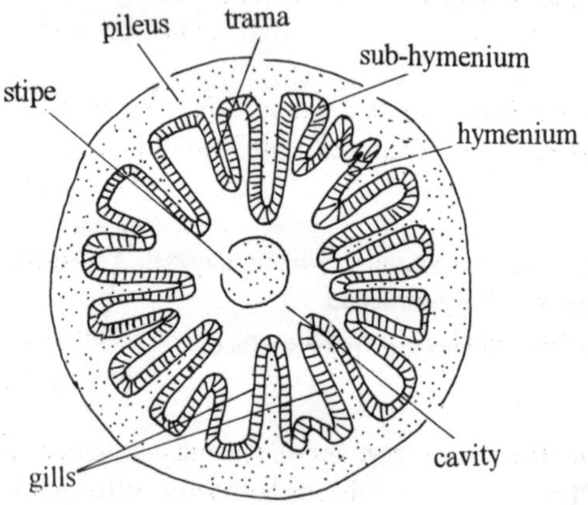

Fig. 14.24 Agaricus. Cross-section of pileus

1. The mature basidiocarp of Agaricus is a small umbrella structure.

2. It has a long, dull white coloured, cylindrical stalk called the stipe.

3. The stipe bears at its tip an expanded portion called the stipe.

4. Midway on the stipe is found the annulus representing the represents of the velum, which gets torn as the pileus expands.

5. The upper surface of the pileus is smooth or undulated.

6. The under surface of the pileus has got a number of plates of lamellae radiating from the centre. These lamellae are called gills.

7. The gills that extend right from the centre to the margin of the pileus are called primary gills while those that start midway and reach the margin are called secondary gills.

Structure of a gill :

A vertical transverse section of the gill shows the following features :

1. At the central region are present loosely arranged fungal hyphae constituting the trama

2. On either side of the trama are found two to three layers of compactly arranged hyphae which looks like rounded cells. This is called the subhymenium.

3. External to the sub-hymenium and arranged at right angles to the flat surface of the gill are found basidia constituting the hymenium.

4. The hymenium also has cystidia, the sterile hyphae inter-

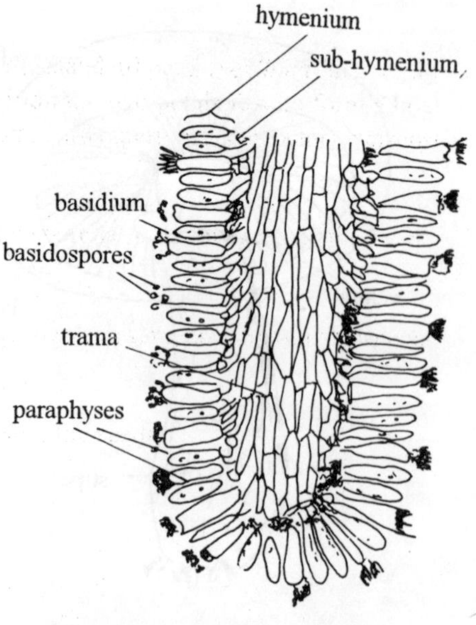

Fig 14.25 V.S. of a gill magnified

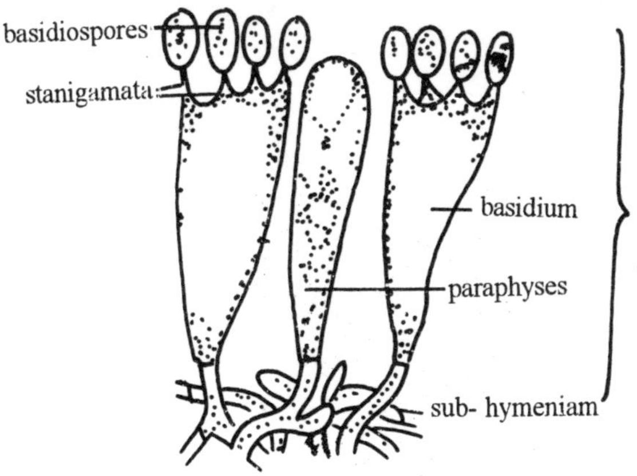

Fig 14.26 A portion of hyme-

mingled with the basidia.

5. The basidia are club shaped and aseptate. They are called holobasidia.

6. The basidiospores are small, and oval in shape. They are wind dispersed and on the availability of a suitable substratum, they germinate to give rise to the monokaryotic primary mycelium of a new individual.

Specimens given in the laboratory

1. Basidiocarp

2. Section of a gill (slide)

Diagrams to be drawn :

1. Basidiocarp

2. Section of a gill

3. A few basidia enlarged

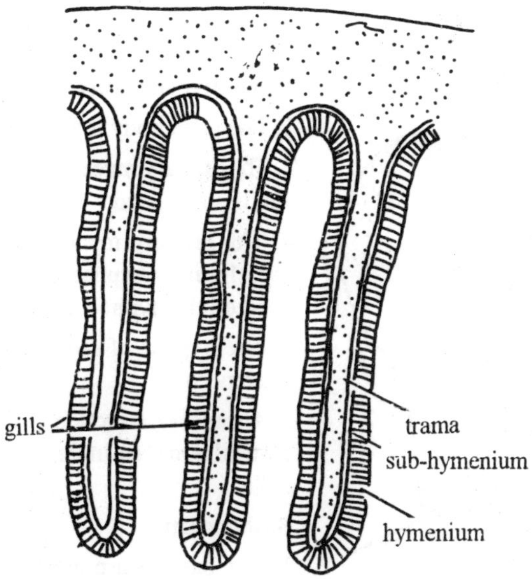

Fig. 14.27 Agaricus. V.S. of a portion of pileus

Systematic position and identification

A 1. Chlorophyll present

2. Reserve food is glycogen ... Fungi

B. 1. Mycelium is septate

2. Basidia and basidiosporec produced ... Basdiomycetes

C. 1. Basidiocarp produced

2. Basidium aseptate ... Eubasidiomycetes

D. 1. Basidiocarp exposed from the beginning

2. Basidiospores released in instalments ... Hymenomycetes
E. 1. Gills present
2. Basidiocarp macroscopic ... Agaricales
F. 1. Basidiocarp soft and fleshy
2. Gills narrow and flattened ... Agaricaceae
G 1. Gills radiating from the centre of the pileus.
2. Annulus present
3. Primary and secondary gills present
4. Stipe easily seperates from the pileus... Agaricus

LYCOPERDON

Common species :

Occurrence : Commonly called puff balls, this saprophytic fungus usually grows on decaying wood, litter filled soi, etc.

Study : 1. Structure of vegetative mycelium.
2. Basidiocarp
3. Systematic position and identification

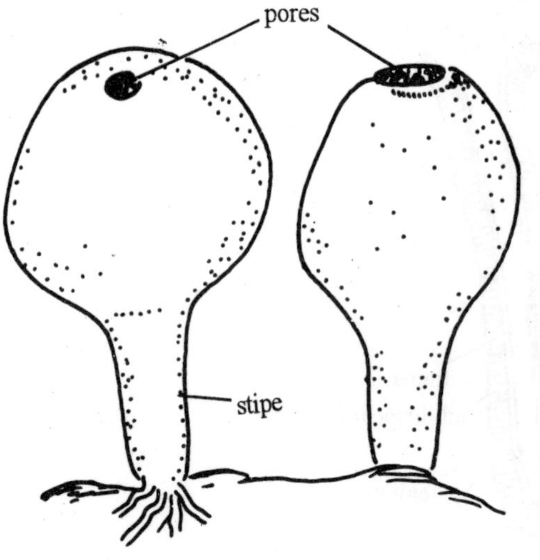

Fig. 13.28a L.S. of young basidiocarp of Lycoperdon

Vegetative structure

1. The plant body is a subterranean mycelium.

2. The primary mycelium consists of sliort uninucleate cells and is short lived.

3. The primary mycelium soon develops into dikaryotic mycelium by hyphae fusions. Clamp connections are noticed on the hyphae of the secondary mycelium.

4. The hyphae of secondary mycelium form knots called rhizomorphs.

5. A transverse section of the rhizomorph reveals an outer cortical zone, an inner subcortical zone and a central medulla.

Basidiocarp

1. The basidiocarp arises like a tiny out growth from the central region of the rhizomorph.

2. During early stages, the young basidiocarp is homogenous, ie., hyphae are uniformly distributed.

3. A fairly mature basidiocarp is found or pyriform and is attached either laterally or terminally to the rhizomorph.

4. The basidiocarp may range in rise from 1-6".

5. The basal portion of the basidiocarp remains sterile.

6. In the upper portion of the basidiocarp, a little later differentiation takes place into outer periderm and inner gleba.

7. The periderm has two layers viz., the other peridium and the inner peridium.

8. The inner region has many small cavities called global cavities.

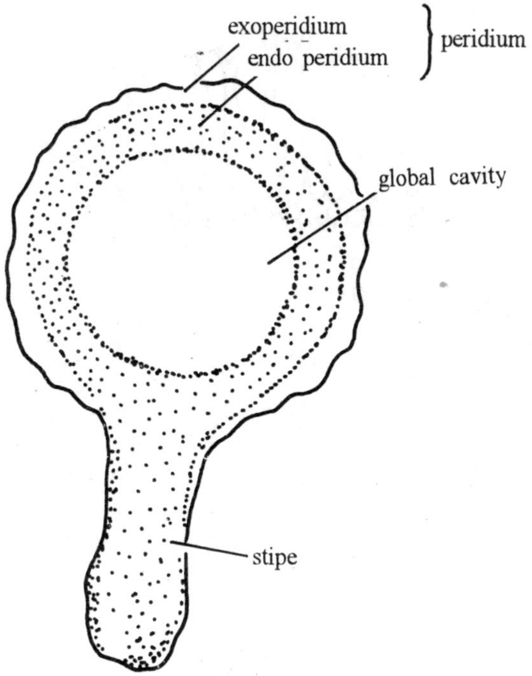

Fig. 14.29 Lycoperdon. The

9. The inner lining of the cavities has many loosely interwoven hyphae with broken ends.

10. The hyphae are made up of binucleate cells and eventually they form the basidia.

11. In a clearly mature baisidum, the baisdia project into the cavities.

12. The basidia reproduction holobasidia and each are produces four basidiospores.

13. All the basidiospores in a basidisocarp discharged simultaneously in a suffy mass, whenever the frustification breakes open.

14. The basidiospores germinate to give rise to a new individual, on falling upon a suitable substratum.

Specimens given in the laboratory

1. Basidiocarp

2. Section showing global cavities (slides)

Diagrams to be drawn :

1. Habit of lycoperdon (basidiocarp)

2. Section of the basidiocarp showing global cavities.

3. A.glebal cavity enlarged to show the basidia.

Systematic position and indentifcation.

A. 1. Chlorophyll absent

2. Reserve food is glycogen ... fungi.

B. 1. Mycelium septate

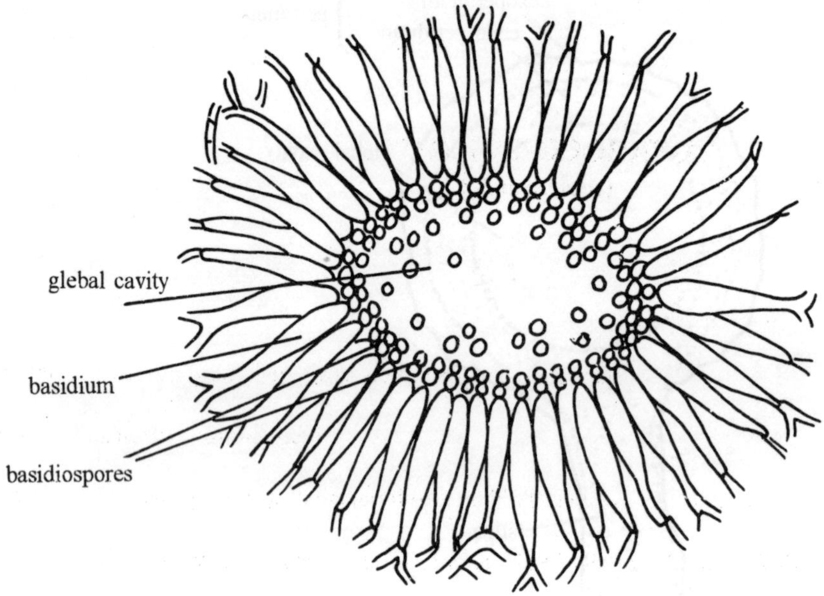

Fig. 14.30 Lycoperdon. Glebal cavity showing basidia and basidiospores

2. Basidia and basidiosproes produced ... Basidiomycetes.

C. 1. Basidiocarp formed.

2. Basidia aseptate ... Eubasiliomycetes.

D. 1. Basidiocarp completely closed.

2. Basidia exposed only at maturity... Gasteromycetes

E. 1. Gleleal cavities do not separate from the peridium.

2. Glelea powdery at maturity with light coloured spores ... Lycoperdales.

F. 1. Peridium of basidiocarp is two layered.2. Lapilitia (hyphae) present among spores... Lycoperdalac

G. 1. Glelea lacunose

2. Globose or pyriform basidiocarp.

3. Peridium breaks open at maturity... Lycoperdon.

CHAPTER 15

DEUTEROMYCETES

Deuteromycetes are called imperfect fungi for they are known only in their asexual stage. The sexual stage or the perfect stage is lacking in them or has not been discovered. These fungi possess septate hyphae that reproduce asexually by means of conidia. Many of these are saprophytes but a great majority of them are parasites causing disease to plants and animals and human beings and hence are of great seonomic importance.

According to Alexopoulos, the conidical stages of Deuteromycetes resemble those of Ascomycetes and can be regarded as ascomycetes whose assigerous stage has been dropped during the course of evolution. These have been instances in cases, if ascigerous stage is found they can be placed in Ascomycetes. In some cases, the perfect stage discovered have placed these fungi in Basidiomycetes. Hence the class Deuteromycetes is an artificial one to include genera in which the life cycles are not fully known to classify them.

Alexopoulos dyings fungi imperfects as "when may consider the fungi imperfect, therefore as conidiae stages of Ascomycetes or more readily Basidiomycetes whose sexual stages have not been discovered or no longer exist". The parasexual cycle discovered in some imperfect fungi indicates that some of them at least never had a perfect stage. There are some fungi which have never produced any spores at any stage.

Somatic structure

The thallus of fungi impe.fects consists of well developed septate branched mycelia. The cells are usually multinucleate. The septa are perforated permitting the streaming of cytoplasm.

Deuteromycetes asexually reproduce by condidia which are borne on conidiophores. The conidiophores themselves may be simple isolated structures or are produced in special structures or are produced in special structures like acervalus, cynnema, pyridium etc. In addition to this, some deuteromycetes specially animal and human pathogens produce other types of spores called microconidia, balstospores or arthospores. Microconidia as the very name indicates are small conidia, while blastospores are asexual cells produced as a result of budding or directly from ahypha. Arthrospores are produced by the disjoining and isolated of cells. They are also known as oidia.

Some of the Deuteromycetes have parasexuality. This type of sexuality conferred only some advantages of sexual reproduction. Parasexual cycle may be defined as a cycle in which plasmogamy, karyogamy and haplodization (meiosis) takes place but not at specified time or at specified points in the life cycle of an organism.

Classification. Deuteromycetes do not fit into the system of classification followed for other fungi which are mainly based on the characters of the sexual stage. Hence an artificial, but a practical system of classifying

Deutromycetes has been followed in which these are many form order, form families, form genea etc. The idea of including the prefix form is to indicate the members so grouped need not be related. In few morphological characters agree, the members are grouped together. The chief characters used in the classification of Deuteromycetes are the type of frutifications, and the shape, colour and septation of conidia. To date, between 15,000 to 24,000 species belonging to nearly 1200 genera have been included in Deutromycetes.

The form class Deutromycetes is divided into the following orders.

1. Sphaeropsidales. This order includes 500 genera and 12,500 species. Conidia reproduction borne in Pynidial cavities.

2. Melaneonials. This order comprises 80 genera and 12,000 species which are mainly parasite conidia are borne in acervuti, which are sub spidermal or sub cuticular in the host.

3. Monitates. The order consists of nearly 10,000 speices. Many of the monilates are of immense importance to human beings. It is this group that most of the human pathogens belong. Conidiophores may be simple or branched.

4. Mycelia sterilia. This is a conglomerate assemblage of about 20 genera and 350 species. They do not produce any spores.

In this book the genera :

Collectrodium, Pyricularia, Alternaria, Fusioarium and Lercospora are discussed.

COLLECTORICHUM

Common species : C.faleatum, C.gossypi, C.uscinass etc.

Occurrence : The genus is a faculative parasite attacking a variety of flowering plants. It causes many diseases known as anthracrose, red rot etc. The plants infected are sugarcane, mango, chillies etc.

Study : 1. The disease caused and symptoms

2. Vegetative structure

3. Reproduction

4. Systematic position and identification

Diseases caused and symptoms:

A. Red rot of sugarcane. The disease is caused by C.faleatum. The perfect stage of the fungus has been included under Ascomycetes.

1. Disease infects all parts of the plant. Disease symptoms appear when the sugar synthesis is about to begin in the stem.

2. The leaves at the apex get decolourised and drop down. This is followed by the collapses of the entire shoot apex.

3. Anatomically the diseased stems reveal long stripes of red colour with whites across.

4. Red patches also appear on the leaves.

B. Fruit rot and die back of chillies. The disease is caused by colletotrichum capsici.

1. The disease symptoms are of two types- fruit rot and die back.

2. Fruit rot is noticed in ripe red coloured fruits

3. Black spot first appears on the fruit and then enlarges to form a stripe.

4. Diseased fruits lose their healthy red colour and turn light yellow. Ultimately the fruit decays.

5. Rotten fruits show conidiospores and sclerotic.

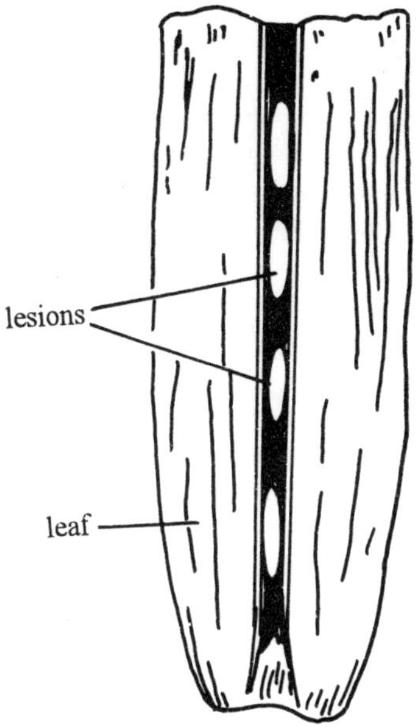

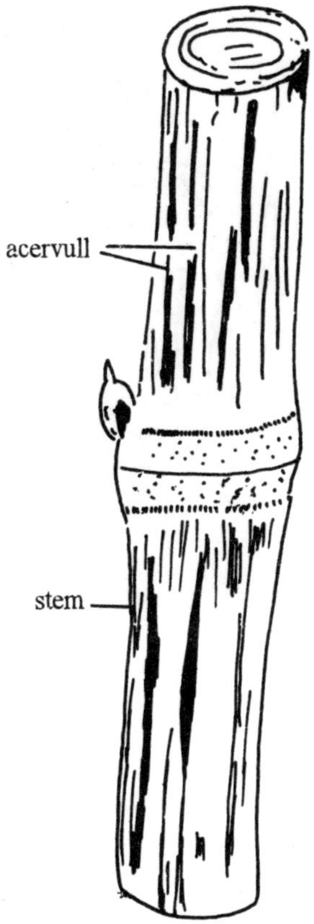

Fig. 15.1 Colletotrichum. A portion of sugarcane leaf showing red rot disease

A portion of sugarcane stem showing red rot

In die black disease, the shoot apex starts decaying and it proceeds down. As a result of this the branch or entire plant collapses.

Diseased branches have water rockea reasons. A chervil of the fungus are found in these areas.

C. Black spot disease of mango. The disease is caused by C.gleosporioides. The perfect stage of this fungus has been included in Ascomycetes under the name glomeralla singulata.

1. The disease symptoms appear on the skin of the ripening fruits, in the form of slightly depressed patches of different shapes.

2. The skin may crack open and the pulp gets decayed.

3. Disease spots may also appear on the leaves and branches.

4. Tissues of the infected, region decay and die. Tips a branches also drop down due to the diseases.

D. Anthracnose of cotton. The disease causing agent is C.gossypi and C.lindeminthianum.

E. Smudge of Ominos. The pattrogen is C.ureinans.

Vegetative structure.

1. The plant body of the fungus is a septate much branched mycelium.

2. The hyphae are endophytic and inter or intercellular. They are branched and have oil globules as reserve food.

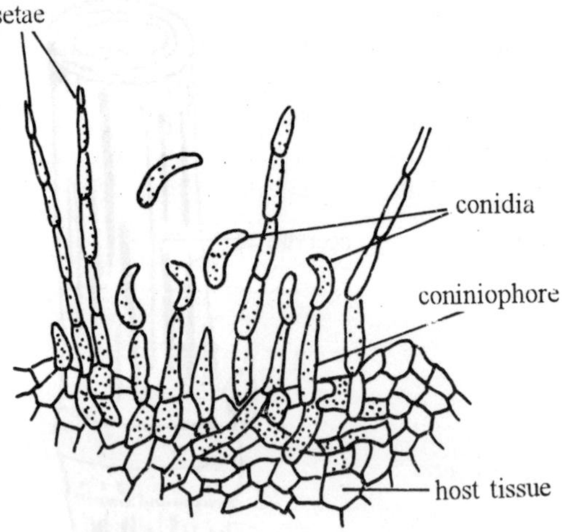

Fig. 15.2 Colletotrichum. V.S. of infected leaf through acervulus

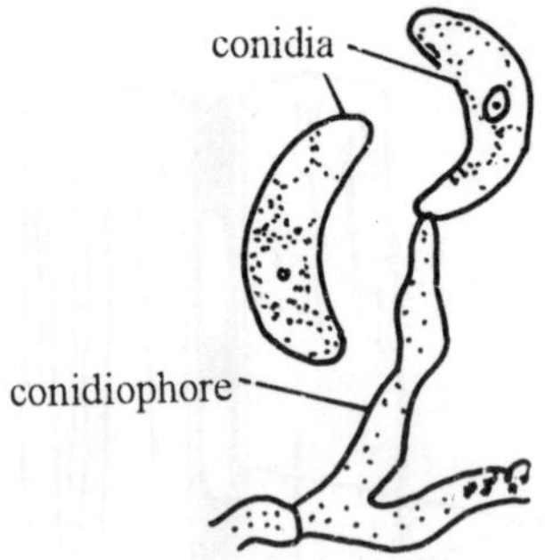

Conidiophores and conidia

Reproduction :

1. Sexual reproduction (perfect stage) has not been reported.
2. Asexual reproduction takes place by the formation of conidia.
3. The conidiospores aggregate into clusters called acervuli, which may be regarded as equivalent to fructifications (no sexual stage is seen).
4. The acervuli or materials shaped or disc or saucer shaped.
5. The acervuli are found in the form of crusts on the best.
6. Within the acervulus, the conidophores are arranged in a palisade like manner.
7. The conidiophores are erect and unbranched. At their tips, they cut of conidia.
8. Each conidium is unicellular, faleate with round ends.
9. In between the conidiophores, sterile hairs called setae are present.

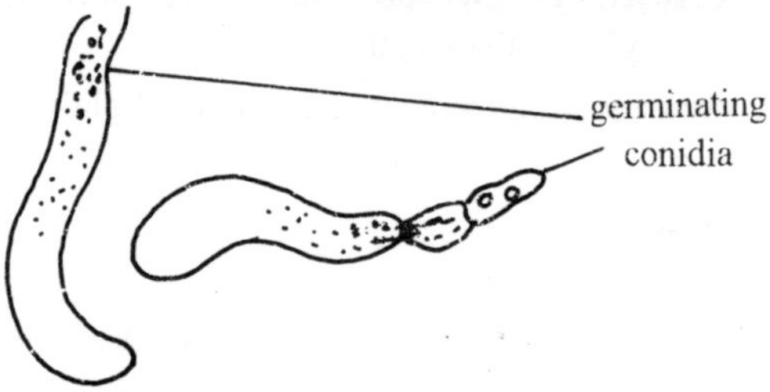

Fig. 15.3 Colletotrichum. Conidia

Diagrams to be drawn:

1. Habit sketch of infected host.
2. Section showing Acervalus.
3. Conidia enlarged.

Systematic position and identification

A. 1. Chlorophyll absent

2. Reserve food is glycogen ... Fungi

B. 1. Perfect stage absent

2. Only asexual stages (conidial) are seen ... Deutromycetes

C 1. Conidia found in Acervulus ... Melanconiales

D 1. Acervulus has conidiophores and seta ... Melanconiaceae

E 1. Red stripes on the host

2. Seta found in acervulus ... Colletrotrichim.

PYRICULARIA

Common species : P.oryzae

Occurrence : The fungus is a parasite on many flowering plants and specially on paddy. The disease is extensive in cold regions.

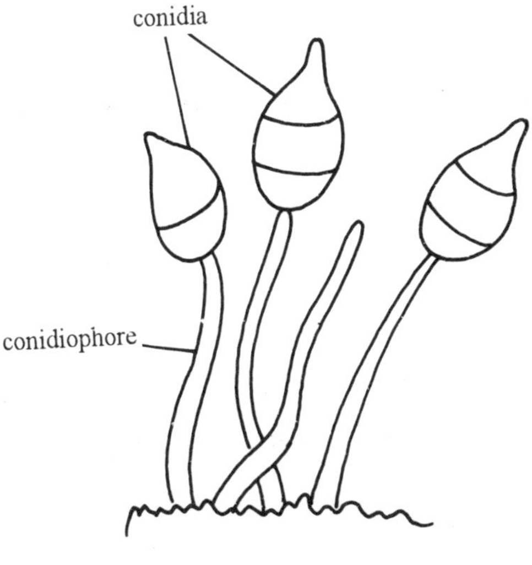

Fig. 15.4a Conidiophores and conidia of

Study : 1. Diseases caused and symptoms

2. Vegetative structure

3. Reproduction

4. Systematic position and identification

Disease caused and symptoms

A. Blast or rotten neck disease or Rice. The disease is caused by P.oryzae causing loss in yield as much as 75%.

1. The disease symptoms are visible mostly on the lamina. However leaf sheath, petiole, stem and calyx may also exhibit symptoms of the disease.

2. Small bluish blisters appear on the leaves. These may be rounded or elongated. The centre of the lesions are grayish green in colour.

3. The blisters have water soaked appearance with brown margin.

4. The characteristic symptoms of the disease is the rotting of the peduncle just below the flowers. Due to this, the ear collapses, hence the name rotten neck disease.

Vegetative structure :

1. The plant body of the fungus is a septate and much branched mycelium.

2. The hyphae are olive oil coloured and have a number of cross walls.

3. The mycelium may be intercellular or intercellular.

Reproduction :

1. Perfect stage has not been reported.

2. Asexual reproduction is by conidia borne on conidiophores.

3. Conida are small oval in shape and made up of three cells.

4. Conidiophores are simple, unbranched and septate with 2-4 cross walls; colour grayish yellow.

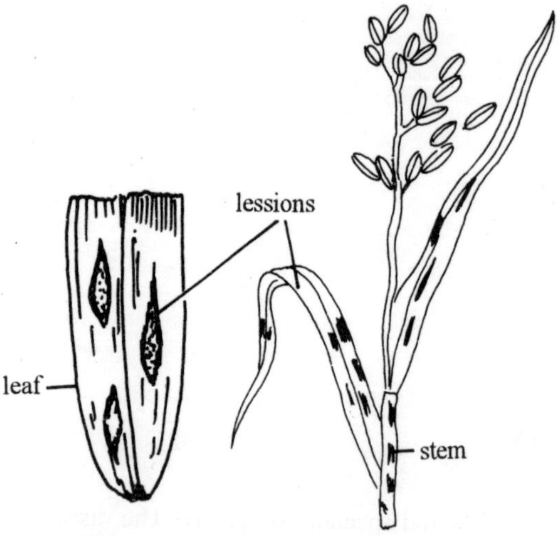

Fig. 15.5 Pyricularia. Rice leaf lesions

Rice inflorescence showing lesions

Diagrams to be drawn:

1. Habit sketch of infected leaves and inflorescence.
2. Conidiophores and conidia enlarged.

Systematic position and identification

A. 1. Chlorophyll absent

2. Reserve food is glycogen... fungi

B. 1. Perfect stage absent

2. Only conidial stages is seen ... Deutromycetes

C. 1. Conidiophores free

2. Acervulus, synnema etc absent ... Monilales.

D.1. Conidiophores and conidia olive oil coloured ... Moniliaceae

E 1. Conidia 2-4 celled and oval or pyriform.

2. Conidiospores septate

3. Conidia borne singly on the conidiophores ... Pyricularia.

ALTERNARIA

Common species : A.solani, A.brassicae etc.

Occurrence : Majority of the members of the genus are parasites but a few could be saprophytes also.

Study : 1. Disease caused and symptoms.

2. Vegetative structure

3. Reproduction

4. Systematic position and identification

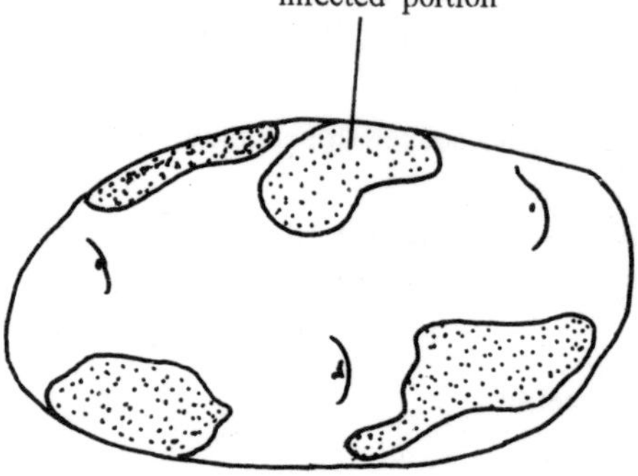

Fig 15.6 Infected potato tuber

Diseases caused and symptoms :

1. *Early blight of potato :* This disease is caused by A.solani and is widespread in areas wherever potato is cultivated.

a. Light brown patched appear first on the leaves.

b. Leaves on the lower part of the plant get the attack of the disease first.

c. Infected tissues exhibit irregular lines, with the leaf appearing like atarget board.

d. Occasionally the disease may spread to the tubers also.

2. *Alternaria leaf spot of cabbage* : Hencifous plants like cabbage, mustard, cauliflower, knol knol, raddish etc get the attack of leaf spot. The there species of Alternaria viz A.brassicae, A.brassicola and A.raphani cause the leaf spot disease. A.raphani is specific to raddish only.

a. The disease symptoms appear in the form of black dots. These are circular with a light blue coloured central region.

b. Many times the spots have concentric lines at the centre.

3. *Leaf blight of wheat :* The disease is caused by A.tritrcola.

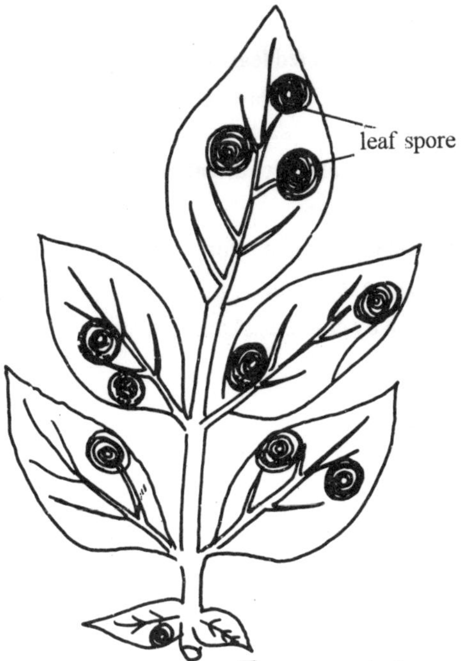

Fig. 15.7 Alternaria. Infected potato leaf showing leaf spots

a. Disease symptoms first apear on the leaves in the form of scattered oval patches.

b. Gradually, the patches occupy the entire leaf area.

c. The grayish known patches destroy the chlorophyll completely.

Vegetative structure :

1. The plant body is a mycelium. The mycelium is much branched with septate hyphae.

2. The hyphae are light yellow, hyaline and semitransparent in the young condition.

3. Mature hyphae are of the colour of olive oil.

Reproduction :

1. Perfect stage of the fungus is not seen.

2. Asexual reproduction takes place by means of conidia.

3. Cɔndiophores are simple, unbranched and septate, they come out of the host leaf through the stomata.

4. Conidia are multicellular and have 5-10 cross walls. In some cases the ends of conidia have break like projections. They are dictyosporous, ie., have both transverse and longitudinal septa.

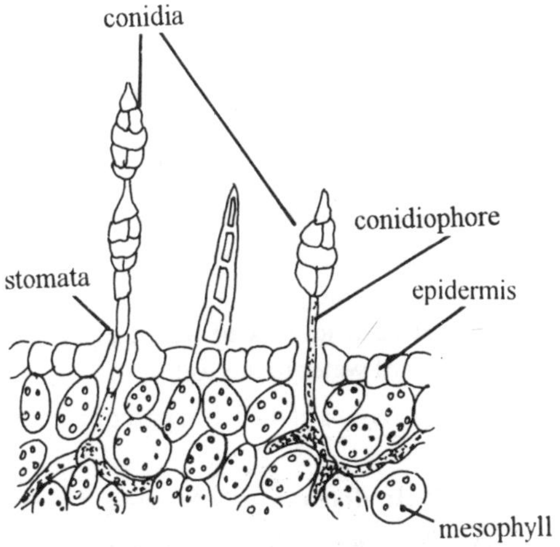

Fig. 15.8 Alternaria. T.S. of host leaf showing conidiophores and conidia

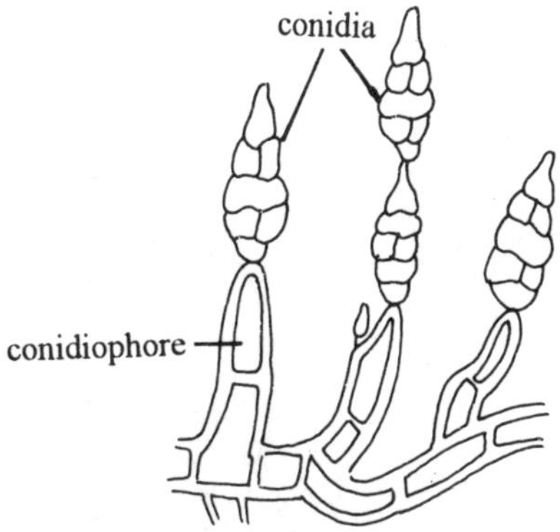

Conidiophores and conidia

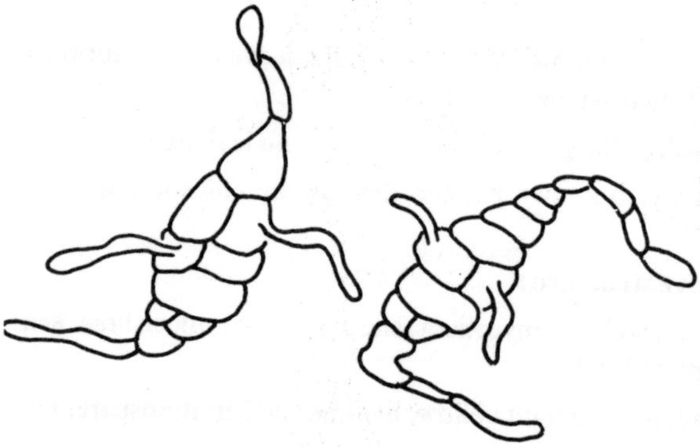

Fig. 15.9 Alternaria. Germinating conidia

Diagrams to be drawn:

1. Habit sketch of infected leaves.
2. Section of the leaf to show conidiophores and conidia.

Systematic position and identification

A. 1. Chlorophyll absent

2. Reserve food is glycogen... fungi

B. 1. Perfect stage is absent

2. Reproduction by conidia ... Deutromycetes

C. 1. Conidiophores free.

2. Acervulus, symnema etc absent ... Moniliales

D.1. Hyphae and reproductive organs semitransparent or olive in colour... Dematiaceae.

E 1. Conidiophores single and erect.

2. Conidia are dictyosporous ... Alternaria.

FUSARIUM

Common species : *F.solani, F.lini, F.oxysporum, F.vasinfectum* etc.

Occurrence : The species of Fusarium are parasitic and generally cause a disease called *Wilt.*

Study

1. Diseases caused and symptoms.
2. Vegetative structure
3. Reproduction
4. Systematic position and identification.

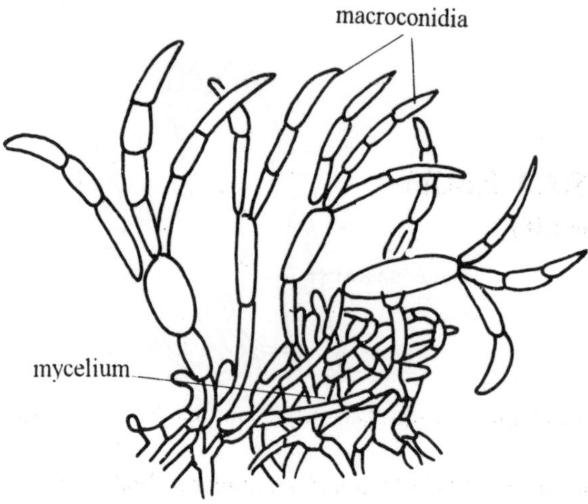

Fig. 15.10 Fusarium. A sporodochium

Diseases caused and symptoms

1. Wilt disease of red gram : The disease causing agent is *F.oxysporum*. The plants of red gram are invariably attacked, wherever they are cultivated.

a. The characteristic symptom of the disease is the wilting of the plant. The plants look as if they are suffering from lack of water, even though there is plenty of water in the soil.

b. The tissues at the basal part of the stem turn dark in color.

2. Wilt disease of cotton : The pathogen is F.vasinfectum. The symptoms of the disease are similar to what is seen in the wilt of red gram.

3. Wilt of linseed : *Fusarium lini* is the fungus which causes the disease.

a. Disease symptoms may appear on the host at all stages of growth.

b. Infected seedlings drop down and the cotyledons wither away.

c. In older plants, deep green or brown patches appear on the leaves.

d. Fungus is present in the roots as well as the basal part of the stem.

4. Wilt of banana : The disease is caused by *F.oxysporum, F.cubens.*

a. Disease symptoms are visible on the seedlings 3-5 months old.

b. The first symptom of the disease is the appearance of yellow stripe on the lowermost leaf.

c. Subsequently the leaves turn yellow and fall off.

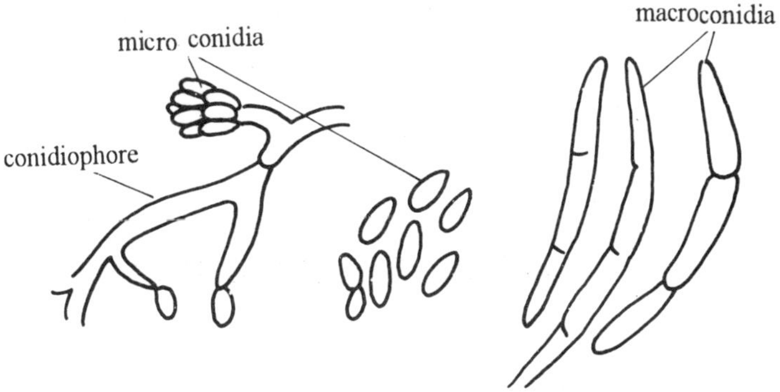

Fig 15. 11 Mycelium with microconidia and macroconidia of Fusarium

d. Leaf sheaths also crack in a linear fashion.

e. When the disease is intense, all the leaves fall down and only the 'Pseudostem' remains.

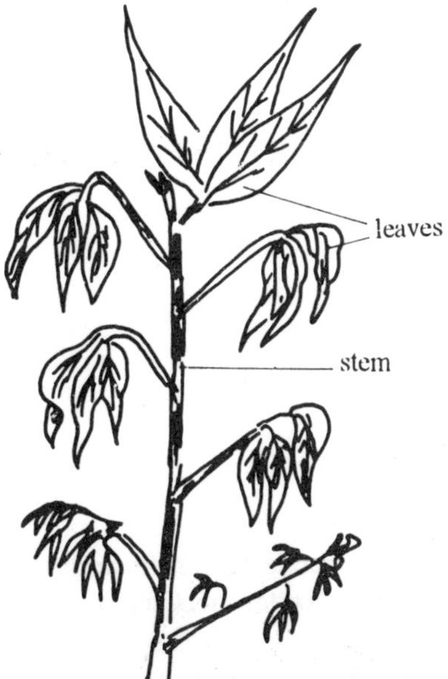

Fig. 15.12 Fusarium. Wilt of pigeon pea (Arhar) caused by Fusarium

Vegetative structure

1. The plant body is a septate, much branched mycelium.

Reproduction

1. Prefect stage is not seen.

2. Asexual reproduction is brought about by a variety of methods viz., conidia, (microconidia and macroconidia), chlamydospores, sclerotia etc.

3. Microconidia are small rounded or oval in shape; sometimes crescent shaped also. They are present in small groups and are produced at the tips of branched or unbranched conidiophores.

4. Macroconidia are produced in a fructification called *Sporodochia* which are cushion like. They (conidia) are long, sickle or crescent shaped. The conidia are septate.

5. Chlamydospores are formed by the cells of the hyphae which round off and develop into a thick walled chlamydospore.

6. In some cases, the hyphae aggregate and form small, black compact *Sclerotia*.

Specimens given in the laboratory

1. Infected hosts

2. Conidia and conidiophores (slides)

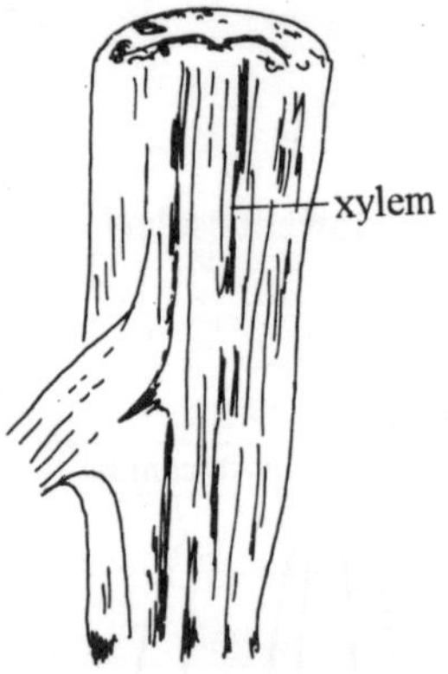

Fig. 15.13 Fuxarium. T.S. of infected root (diagrammatic)

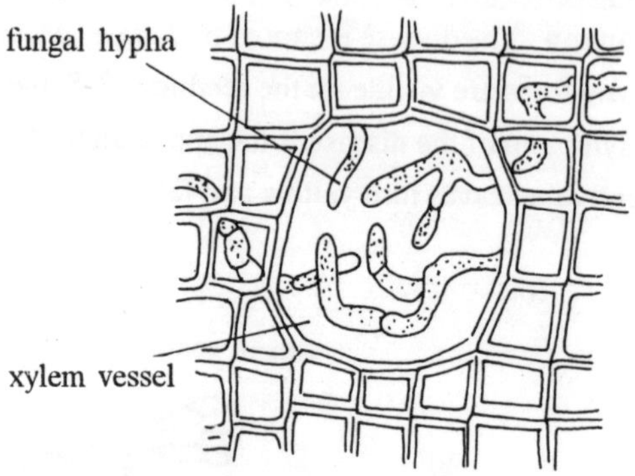

Mycelium in vascular tissue

Diagrams to be drawn

1. Habit sketch of infected host
2. Conidiophores and conidia
3. Sclerotia
4. Chlamydospores

Systematic position and identification

A. 1. Chlorophyll absent
   2. Reserve food is glycogen ....**Fungi**

B. 1. Perfect stage absent
   2. Asexual reproduction by conidia ....**Deuteromycetes**

C. 1. Conidiopores generally free
   2. A cervulus, synnema etc absent ....**Moniliales**

D. 1. Mycelium hyaline
   2. Sporodochia present ....**Tuberculariaceae**

E. 1. Conidia of two types
   2. Hyphae in the vessels of the host ....**Fusarium**

CERCOSPORA

Common species : C.personatia, C.arachnidicola, C.beticola, C.apii etc.

Occurrence : The various species of cercospora are parasitic pathogens infecting various flowering plants and cause leaf spot diseases. The members attack include groundnut, chillies, cotton etc. On groundnut, the genus causes a disease called Tikka disease.

Study : 1. Diseases caused and symptoms.
2. Vegetative structure
3. Reproduction
4. Systematci position and identification

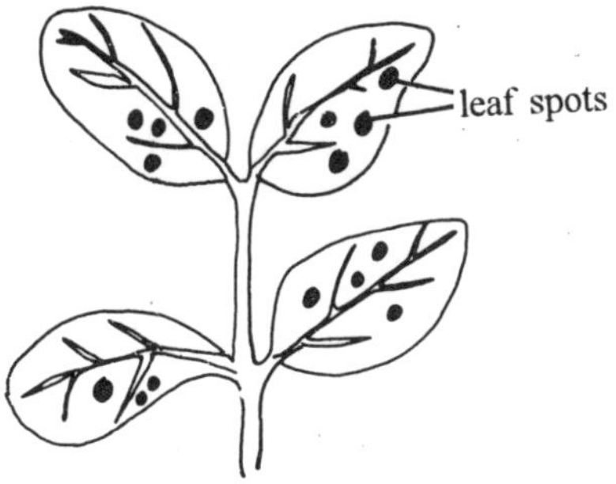

Fig. 15.14 Cerospora. Ground nut leaf showing leaf spots caused by Cercospora

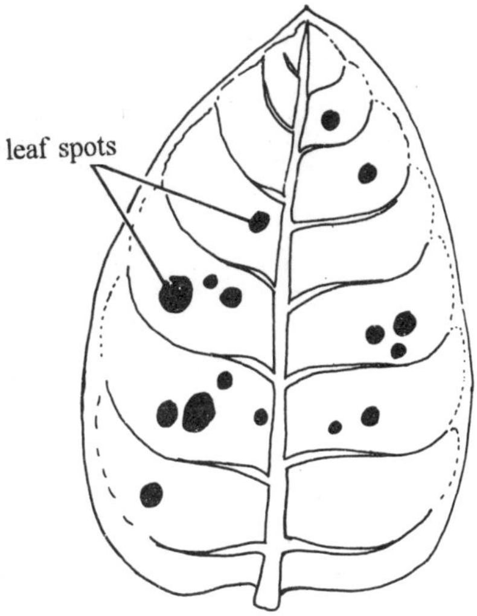

Calotropis leaf showing leaf spots caused by Cercospora

Diseases caused and symptoms :

1. Tikka disease of ground nuts : The disease is caused by two species of cercospora viz., C.personata and C.arachnidicola.

2. All the aerial parts of the host plant get the infection.

3. As the first symptoms of the disease, light green patches appear on the leaves. The epidermal cells of the leaf lose contact with the mesophyll.

4. In case of infection by C.arachnidicola, the patches are somewhat circular with a diameter of 1-10 mm.

5. The patches have a yellow halo right from the beginning.

The patches which are discrete in the beginning later coalsce.

The case of infection by C.peronata, the patches are strictly circular with a diameter of 1-6 mm.

There is no yellow halo in case of infection by C.personate

Other diseases caused by cercospora :

a. Leaf spot of Dhal - C.indica

b. Leaf spot of Cotton - C.gossypium

c. Leaf spot of Rice - C.oryzae.

Vegetative structure :

1. The plant body is a septate, much branched mycelium.

2. The hyphae are brownish in colour, slender and grow in between the cells of the host.

3. The hyphae produce haustoria as seen in C.personata. in C.arachnidicola, however the haustoria are absent as the mycelium is intercelluar. C.arachnidicola etc.

4. In C.personate the hyphae form stroma.

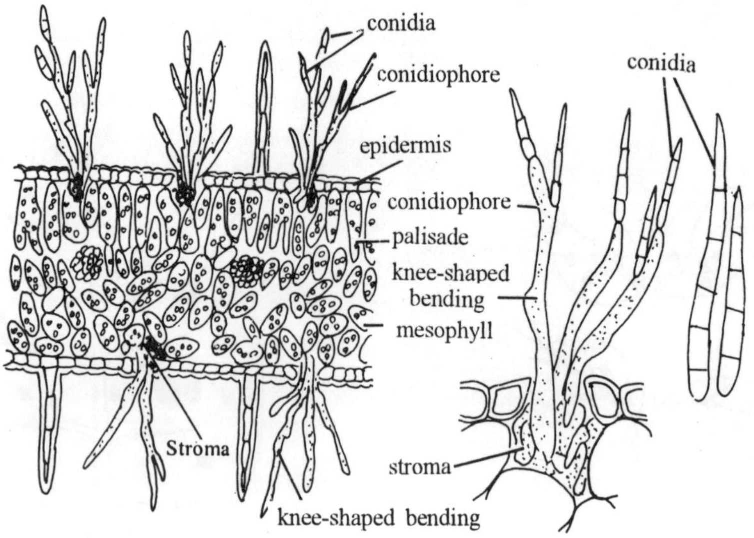

Fig. 15.15 Cercospora. T.S. of host leaf Conidiophores and conidia

Reproduction :

1. The perfect (sexual) stage is absent.

2. Asexual reproduction is brought about by means of Conidia.

3. The conidiospores are known coloured, erect, reptale and come out of the host leaf by piering through the epidermis.

4. Conidia reproduction light yellow in colour, septate with 4-12 cross walls, they range in length from 0.38 - 108 ll.

Diagrams to be drawn :

1. Habit sketch of infected leaves.

2. Section of the leaf showing conidiophores, conidia etc.

Systematic position and identification :

A. 1. Chlorophyll absent

2. Reserve food is glycogen ... fungi.

B. 1. Perfect stage is absent

2. Asexual reproduction by conidia... Deyteromycetes.

C. 1. Conidiophores generally free

2. Acrervulus, synnema absent ...Moniliales

D. 1. Fruiting body absent.

2. Mycelium and conidia brown coloured ...Dematiaceae

E. 1. Conidia septate only transversely.

2. Conidia leave a sear at the point of contact with the conidiophore after liberation ... Cercaspora.

CHAPTER 16

LICHENS

The light green threads that hang down from the branches of trees, the thick grayish green lathery patches that cover the rocks and clad the trunks of trees, the heavy hard incrustations that form an unispereable part of the exposed rocks - these are the lichens. Lichens represent one of the ingenious creations of nature in colouring and occupying and hostile environment. They form a world wide vegetation with 400 genera and nearly 16,ppp species.

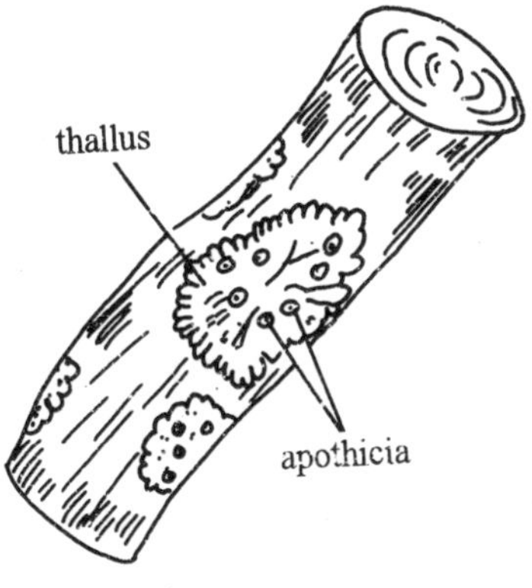

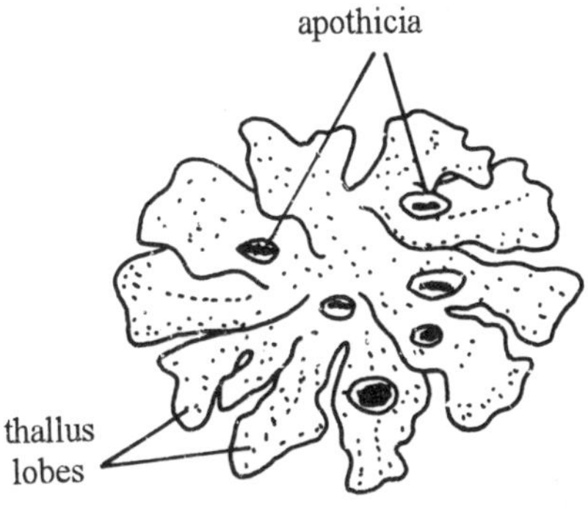

Fig. 16.1 Lichens. Crustose lichen growing on wood

Parmelia. A foliose lichen

Occurrence and habit

Lichen thalli have a wide range of tolerance to climatic fluctuations and as such they colonies and settle permanently even on such habitats which are inhospitable to other vegetation. They thrive and multiply in areas where no other vegetation is possible. They can enduce and withstand long periods of drought and come to active and normal life at the return of favourable conditions. They are cosmopolitian in distribution.

Nature and composition :

Lichen is not an individual plant, though the plant body looks like that of an individual organism. Lichens represent an association of two plants of the group Thallophyta. The duality of lichen thallus was first proposed by Schvendnec (1867). He identified the licher thallus as an association of an alga and fungus. In the licher association the algal partner is called the phycobiont and the fungal partner is called the mycoloeiont.

The mycoliont always belongs to the class ascomyetes or basidiomycetes of the 400 genera of lichens in about only four, the mycoliont belongs of Basidiomycetes.

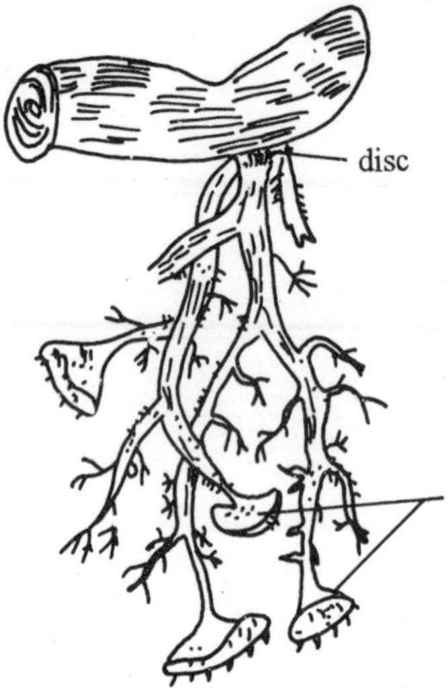

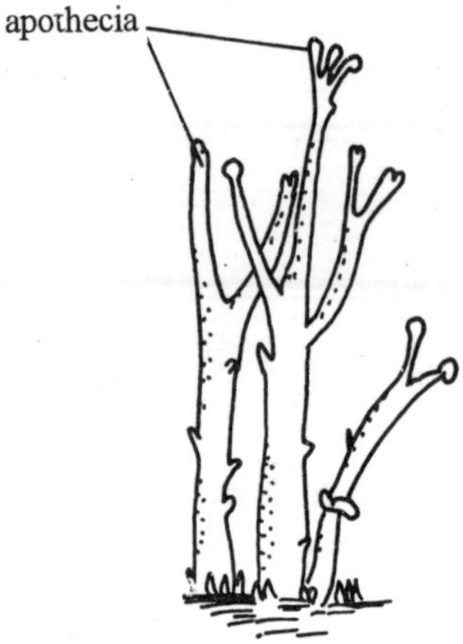

Fig. 16.2 Lichens. Usnea. A fruticose lichen   An erect fructicose lichen

The plycoliont belongs to the class cyanophyceae or chloroplyceae unlike in the case of mycoliant, phycolionts can be identified up to genus and species for they also exist independently other than the lichens association. The cyanophycean members htat have been found as phycolaionts are Nostoc, Gloeccapsa, Rivularia etc. The chorophycean algae found in lichens are protococeus, cystococeus, palmella, Trentepohlia etc.

Classification :

While the phycoliont of the lichens thallus is found to exist independently, nomycoliont can have any other made of existece other than lichens association. It is for this reason that the mycoiont is used as a criterion in the classification.

Lichens have been classified into two groups based on the mycoleiont.

Ascolichens - mycolionts belongs to Ascomycetes.

Basidiolichens - mycolionts belong to Basidiomycetes.

There is also a third group called Myxolichens in which the mycoleiont belongs to mycomycetes. Myxohichens however do not exist in nature. They have been synthesised in the laboratory.

Study : 1. Different types of Ascolichens

2. Vegetative structure on the thallus
3. Anatomy of the thallus
4. Structure of the Apothecium
5. Structure of other reproductive organs.
6. Structure of Basidiolichens.

Ascolichens : These are classified into three types based on the nature of the thallus. These are :

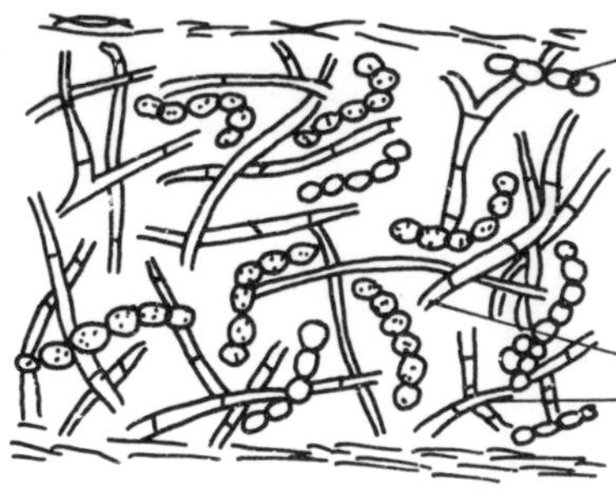

Fig. 16.3 Lichens. V.S. of homoiomerous foliose lichen

Custose Lichens :

1. They usually grow on rock or on barks of trees.
2. They form thick or thin incrustations.
3. The thallus is small and dorsiventral
4. Lower surface secretes a gumy substance which sticks the thallus to the substratum and makes it inseperable.
5. Small apothecia are found on the upper surface.

Foliose Lichens :

1. The thallus is leafy and flattened.
2. It is broder than in curtose lichens
3. They usually grow on barks of trees and occasionally or rocks also.
4. The thalli attach themselves to the substratum with the help of rhizoid like outgrowths called rhizens. Rhizines develop from the ventral surface.
5. The free end a rhizines is expanded into a disc like structure to help in attachment.
6. On the dorsal surface of the thallus are visible many opotheica.

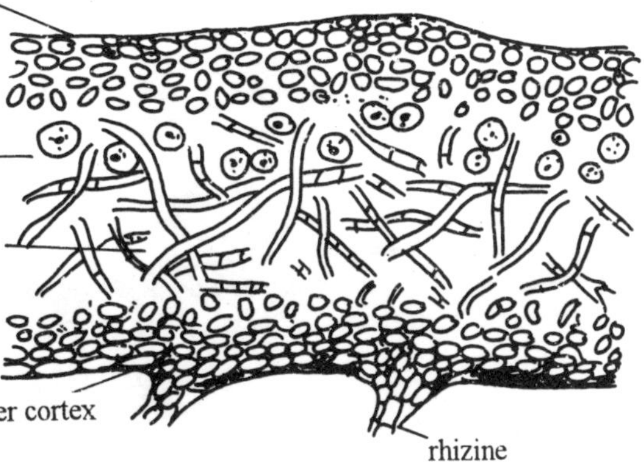

Fig. 16.4a V.S. of heteromerous foliose lichen

Fruticose lichens :

1. These have cylindrical branched thalli; that grow upright or long down from branches of trees.
2. The plant body is attached to the substratum by dense clusters of hyphae.
3. In some fruiticose lichens the basal portion is foliage like with upright cylindrical branches.
4. Apothecia are borne on the upright branches.

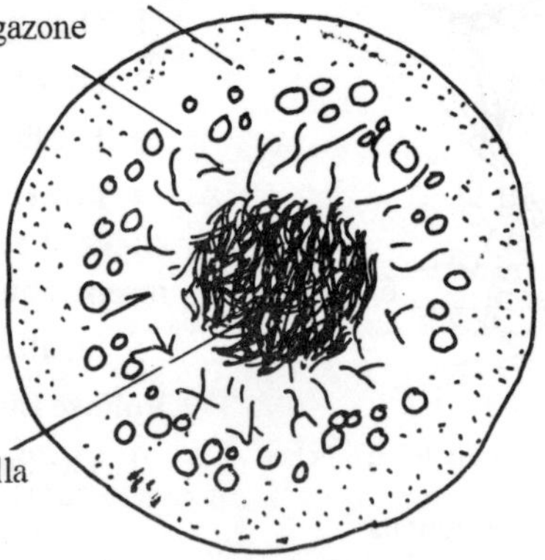

Fig. 16.5 Lichens. T.S. of fructicose lichen

Vegetative structure of the thallus :

The thallus bears a number of out growths or appendages performing various functions. Some of these are :

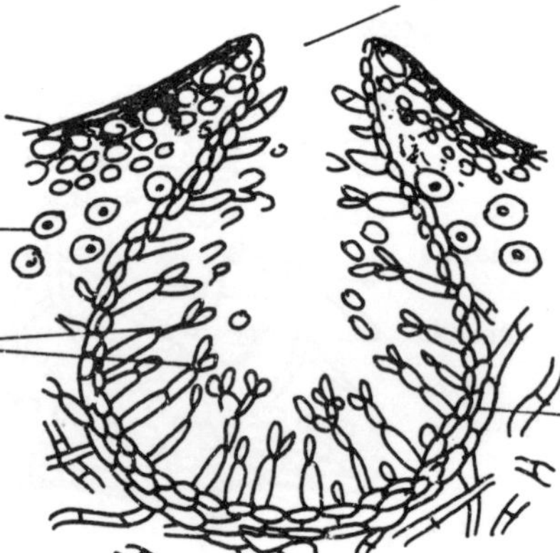

Fig. 16.6a V.S. of pycnidium

Breathing pores :

1. These are breaks or openings found in the upper cortex.
2. These pores help in respiration, acts as arrating organs.

Cyphellae :

1. Cyphellae are also arrating organs.
2. They are present as circular or oval depressions in the lower cortex.

Isidia :

1. These are small protuberances developing from the dorsal surface of the thallus.
2. The algal partner present in the Isidia is range as in the main thallus.
3. Isidia are supposed to increase the photosynthetic area.

Cephalodia :

1. These are dark coloured outgrowths present on the dorsal surface of the thallus.
2. Each cephalodium has a few algal cells surrounded by fungal hyphae.
3. The algal partner present in cephalodium is different from the one present in the main thallus.
4. Many Lichenologists regard cepholodium as an independent epiphyllous lichen.

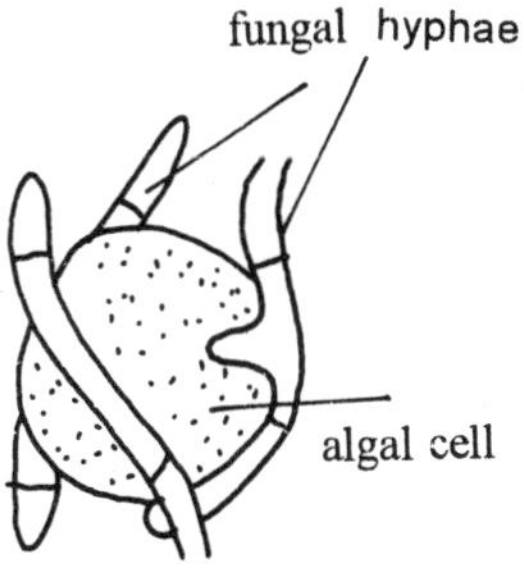

Fig. 16.7a Soredia

Soredia :

1. These are small regular buds like out growth present on the thallus.
2. Soredia may be scattered or present in aggregates called *soralia*.
3. Each soredium has fungal hyphae entangling a few algal cells (same as in the thallus).
4. At maturity the soredium detaches itself from the main thallus and develops into a new licher.

Anatomy of the thallus :

1. Anatomically the crustose and foliose lichens are identical.
2. Fruticose lichens are slightly different.
3. A vertical transverse section of either crustose or foliose lichen shows the following features.
4. On the upper and lower, surface are found the upper and lower cortex respectively.
5. They are made up of closely interwoven fungal hyphae.
6. Below the upper cortex is the algal zone.

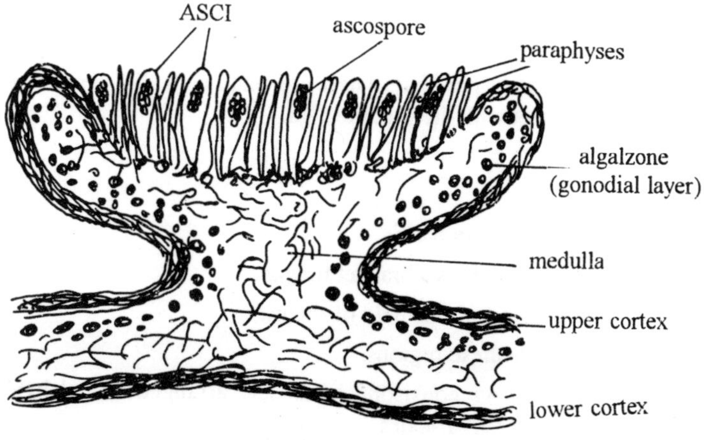

Fig. 16.8 Lichens. V.S. of apothecium

7. This was formerly called the gonidial layer when the dual nature of the licher thallus was not known.

8. In the algal zone are found algal cells dispersed among the loosely arranged fungal hyphae.

9. In the centre (above the lower cortex) in the medulla, where fungal hyphae are found body interwoven.

10. The lower cortex gives rise to rhizoids which help in anchorage.

11. The frutiose lichens the outline of a cross section will be circular.

12. There is an outer cortex, internal to which is the algal zone. In the centre is the medulla.

13. In some lichens the algal cells do not form a definite layer, but are uniformly distributed all over. They are called *Homoimerous,* while those with a restricted algal zone are called *Hoteromerous.*

Structure of Apothecium :

1. Sexual reproduction leads to the development of frutifications like apothecium.

2. Apothecium producing lichens are called *Gymnocarpic.*

3. Some lichens produce *Peritherica*, they are called *Pyrenocarpic.*

4. Apothica are very much similar to the one found in piziza in structure.

5. They are small cup like structures ranging in midth from 10 - 25 mm.

6. The inner lining consists of the hymenium having asci and paraphyses.

7. The wall of the apothecium is called the peridium and is made up of closely joined fungal hyphae.

8. Internal to the wall is the algal zone. (In Lecanorine type).

9. In some apothecia (lecidine type) algal cells are completely absent.

10. Each ascus - consists of eight ascospores, which are liberated by means of a terminal pore in the ascus.

Structure of other reproductive organs :

1. *Perithecia* : These are found in some aquatic lichens. The perithecia are flask shaped structures. They have asci and paraplyses. Lichens which produce perithecia are called *Pyrenocarpic.*

2. *Soredia* :

3. *Pynia* : These are small cup shaped structures found embedded in the thallus.

4. Each perithecium open to the exterior through a small opening called the ostiole.

5. Lining the inner region of the perithercium are many hyphae producing pyeniospores.

6. Many people however apine, that pycnia are in reality spermagonial cups and the so called pyuriospores are spermatia.

Oidia : Some lichens produce *oidia* by their phyphae. These are small cells that germinate to give rise to a new licher, provided they come in contact with the suitable alga.

Basidiolichens **:**

1. They are very rare in nature.

2. A typical example of a basidiolichen is *corapavonia.*

3. The much lobed thalloid plant body resembles a bracket fungus.

4. The thallus is attached to the free trunks with the help of rhizines.

5. A vertical section of the thallus reveals these regions.

6. The uppermost part is the *superior layer* consisting of loosely arranged fungal hyphae. The fungus belongs to the class Basidiomycetes, order Agaricales.

7. Beneath the superior layer is found that the algal layer having *chroococcus* or *sytonema.* The algae are intagled between the hyphae.

8. The *inferior layer* found on the lower face represents a dense materials of hypae.
9. The inferior layers the sub hymenium and the hymenium. The hymenium has a number of holobasidia arranged like a palisade layer.
10. Each baisidium has four sterigmata, terminating which are found in the basidiospores.
11. Occasionally paraphyses are found in between the basidia. They could be immature basidia also.

CHAPTER 17

BACTERIA

Bacteria are one of the smallest among the micro-organisms. Though small in size, they are by no means so in their significance. Bacteria are lightly botent organisms. From the point of view of human welfare, they are of greatest significanced. It causing such deadly diseases like cholera, influenza, tetanus, leprosy, tuberculosis, typhoid etc, to name only a few, they are mans worst enemies. They are mans best friends also in maintaining N2 balance in nature, in helping him in various industries (preparation of alcohols and acids, curing of coffee, tea and tobacco) and in various other aspects.

Credit for discovering bacteria goes to Anton Van Leeuwenhock (1976) a dutch scientist. The name of Louis pasteur should be mentioned, for it was he who discovered the fermenting capacity of bacteria. Robert Koch, a German bacteriologist discovered that tuberculosis is caused by leacteror T.J. Burill (1879) an American Scientist discovered that bacteria are responsible for many plant diseases also.

Distribution. Bacteria are omnipresent. They are present in air, milk, water, soil, arctic regions, hot springs, in plants, animals, human beings etc.

Morphology. Bacteria are unicellular organisms. Based on the cell shapes four morphological forms have been identified in bacteria. These are :

1. *Bacillus.* The cells are rod shaped and elongated. They are the second largest among bacteria. Ex. Mycobacterium.

2. *Rocus.* The cells are spherical. These are the smallest among the bacteria. Ex. Streptocomus Staphyloccus etc.

3. *Spirillum.* The cell will be in the form of a loose spiral. These forms are the largest among bacteria. Ex. Rhodospirillum.

4. *Comma.* The cell will be in the form of punctuation mark, comma. Ex. Vibriochobrae.

Coccus forms exist in various ways:

Micrococcus - single isolated cells.

Diplocossus - cells present in pairs.

Streptoccus - cells exist in chains.

Staphylococus - cells exist in sheets.

Locomotion. Except cocci, all the other forms of bacteria may possess organs of locomotion called flagellae. On the baiss of number and position of flagella bacterial forms are classified as follows:

Atrichous - no flagella.

Monotridcous - many flagella present at one end.

Sephotridious - many flagella present at one pole.

Amphitrichous - two tusts of flagella attached at either end of the cell.

Peritrichous - flagella present around the cell.

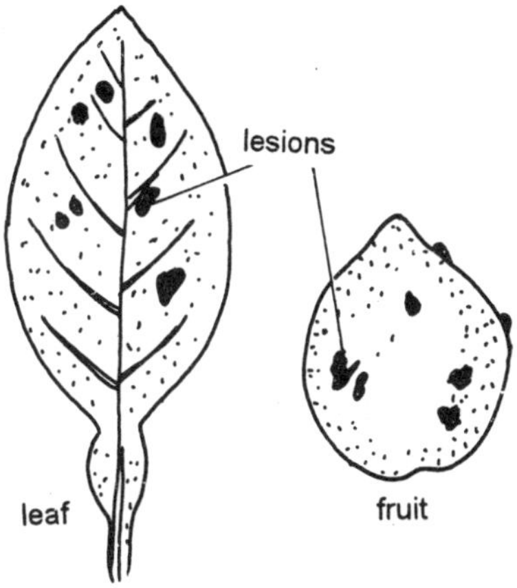

Fig. 17.1 Bacteria. Citrus canker

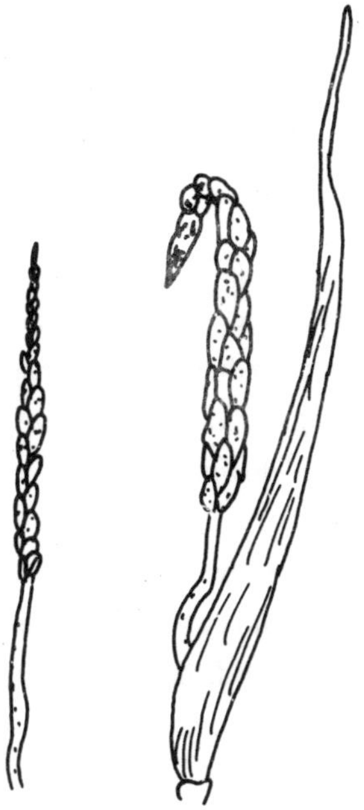

Yellowing rot of wheat

Cell structure.

1. The outermost layer is the cell wall. It is rigid and mainly consists of polysacharides.

2. External to the cell wall, in some bacteria there is a gelatinous covering known as the capsule. This is composed of polysacharides and polypeptides.

3. Next to the cell wall, internally is the plasma membrane.

4. The cytoplasm of bacteria may not always have organdles like mitochondria and endoplasmic reticulum.

There is a nucleus, but it lacks a nuclear membrane (prokaryotic).

Riloosomus present in cytoplasm help in protein synthesis.

Reproduction. Bacteria reproduce asexually as well as sexually. Sexual reproduction takes place by genetic recombination and difficult to observe in the laboratory.

Asexual reproduction takes place by means of fission, budding and undospore formation.

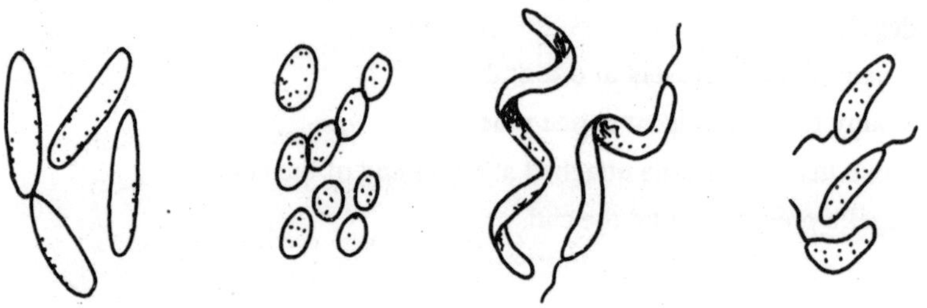

Fig. 17.2 Bacteria. Various shapes of bacterial

Fission :

1. Usually bacteria divide by means of leinary fission in which the cell divides into two halves.
2. The division is perpendicular to the long axis of the cell.
3. The nucleus elongates and divides into two.
4. The cell wall is laid down in the centre of the cell.

Budding :

1. A protrusion develops from a cell into which a little cytoplasm migrates.
2. The nucleus splits into two and one bit enters the protrusion.
3. The protrusion cuts off from the cell by means of a cross wall. This is the bud.
4. At maturity, the bud separates itself from the parental cell and develops into a new individual.

Endospore :

1. The endospore formation in one of the unique features of bacteria.
2. This is formed to withstand, the adverse environmental conditions.
3. Only one cell is formed per cell. It may like in the centre or towards one side of the cell.

Culture of bacteria.

For the purpose of study, bacteria, can be cultured in the laboratory using several culture media. Two of them which are simple and easy to prepare are given below:

PDA medium. The medium is prepared as given below. The ingredients are potato, dextrose sugar, agar and water.

Skinned potato - 250 gms

Glucose - 20 gms.

Agar - 15 gms.

Water - 1 litre

(PH - 6 to 6.5)

1. But the peeled potato into small pieces.
2. Boil it in 500 ml of water and filter through a Muslim cloth.
3. Add glucose to this solution. Stir briskly.
4. In the remaining 500 ml water dissolve agar.

5. Mix both the solutions and make up the volum to 1 litre.

6. Sterlise the solution in an autoclave.

7. To do plate culturing, pour solution into as many petriplates as required and then sturtise it.

8. It slant culturing is to be done, pour the required quantity of solutions into as many best tubes as needed and sterilize cover the mouth of the test tube with cotton and cover it with paper.

Czapeck's Agar. Add the following ingredients and sterilize in an autoclave.

Sucrose - 30 gms

Sodium nitrate - 20 gms

Dipotassium phosphate - 1gm

Magnesium sulphate - 0.5 gm

Potassium chloride - 0.5 gm

Ferrous sulphate - 0.01 gm

Agar - 15.0 gms

Distilled water - 1 litre

The medium can be poured into either petriplates or testubes as required before sterilizing.

Sources for bacteria for culture:

1. Obtained from diseased plants. For this purpose crush the diseased part of the plant tissue and macrate it. Dissolve this in a little water. This will be the cover of bacteria.

2. Bacteria in tap water.

3. Bacteria at the tip of nails.

4. Bacteria in garden soil. Dissolve a little amount of garden soil in w a. Allow it to settle. Use the superintend.

Plating the bacteria :

1. Keep a bunsen burnere or spirit lamp burng on the table.

2. Wash the table surface with formalin.

3. Keep the petriplates close to the spirit lamp.

4. Take a platinum loop, hold it to the flame, withdraw and dip it in the source of the bacteria.

5. Carefully open the petriplate or the test tube, touch the medium with the platinum loop containing bacteria and close the petriplates or the test tubes immediately.

6. In case you want to culture bacteria present at your nail tip, open the petriplate having the medium, touch it and close the petriplates immediately.

7. In case you want to culture bacteria present in air, open the petriplates containing the sterilized medium for about a second and close it immediately.

8. If you want to culture bacteria present in your breath, open the petriplates, breath on it for a second and close the petriplates immediately.

Culturing:

1. After the required bacteria are palted on the medium keep the petriplates or test tubes either at room temperature or in a hot air oven witn 28oC.

2. Within a few days bacterial colonies appear on the medium. These can be stained and identified.

Staining of bacteria:

1. One of the simplest staining procedure is the staining by cruptal violet also known as Grams stain.

2. The staining procedure is as follows:

a. Take a clean microslide. Heat it a little on the open flame

b. Take a little of the bacterial colonies and smear it and spread evenly

c. Allow it to dry for a few minutes

d. Add a few drops of the Grain's stain. Leave it for 2 minutes

e. Wash the stain with water and add a few drops of Iodine solution (potassium iodide). Allow it to settle for 2 minutes

f. Wash it thoroughly with alcohol followed by water and observe

g. Alcohol is a decolouring agent and it will remove the stain from some bacteria

h. Bacteria may appear to have pink colour on their cell walls. Such bacteria are called Gram positive bacteria

i Bacteria that have not retained the stain after the application of the decolouring agent are called Gram negative bacteria

j. The capacity of the bacteria to retain or not to retain the Gram's stain depends on the chemical composition of the cell wall, which is of importance in the classification

k. In medical bacteriolog, grams staining technique is employed to identify pathogenic bacteria

Plant disease caused by bacteria :

Bacteria plant diseases may be broadly classified into three categories namely:

1. Parenchyma diseases: In this type of disease, the soft tissues decay patches or pustules appear on various parts of the plant. These are leaf spot disease, canker etc.

2. Vascular diseases or Bacteriotracheoses. In infected plants bacteria multiply in large number and accumulate in the human of the vascular elements. As a result of the blockade of the human translocation, ascent of sap etc will sto; and the plant wilts.

3. Gallas or tumours. The infection by bacteria learning about hyptertropy on the roots or stems resulting in the development of tumours and galls.

There are about 40 and odd bacteria plant diseases occuring in India. Plants like Rice, wheat, maize, litrus, apple tomato, onion, pea etc infected. Symptoms of a few plant diseases are listed below:

Litrus canker :

1. The disease attacks litrus plants like lime, sweet lime, orange etc.
2. The pathogen is pseudomonas citri.
3. The fruits have brown or dark coloured cruptions are pustules.
4. The pustules are corky and hard.
5. The pathogen is an aexoleic, monotrichous bacillum.
6. The disease symptoms appear in the form of known sales resulting in the yellowing of leaves.
7. The disease symptoms appear on all parts of the plant except the roots.
8. The seafley impture enlarge to form rocky pustules.
9. Occasionally, the disease may form a girdle around branches. The portion of the branch above the girdle dies.
10. The rocky outgrowths on the rind do not spoil the pulp of the fruit.

Blight disease of Rice:

1. The disease infects rice plants (ocyza sativa).
2. The disease symptoms may appear on all the aerial parts, but specifically present on the leaves.

3. Lamina and sheathing leaf leases exhibit yellow starks 5-10 mm long.

4. As the disease increases in intersity the streams enlarge to form patches.

5. In some instances, sap containing bacteria may exceed from the plant.

Brown of potato:

1. The host plant is solanum tubersum (potato)

2. The pathogen is pseudomous radanacearum

3. Leaves lose their green colour and turn yellow.

4. The bacteria infect the vascular tissues which turn known when a cut in maede sap containing bacteria exudes out of the plant.

5. The plants generally collapse before they can produce tubers. Even if tubers are produced, they get discoloured and rot with a brown colour. Sap also exudes out of the tubers.

Leaf spot of Mango:

1. The host plant in Mangifera indica (Mango).

2. The pathogen is pseudomonas mangiferae indicae.

3. Water soaked areas appear at leaf tips, which turn known at later stages.

4. Death of leaftissue occurs as the lesions enlarge causing mecrosis.

5. The disease may infect the fruitsales, in which case, they crack open and drop down in a permative condition.

CHAPTER 18

VIRUSES

Fig.18.1 Viruses. Leaaf roll of potato

Viruses are the smallest leological entities, that are on the broder link between living and non living. They do not possess all the characters of the living beings. They are acellular, but have the capacity to reproduce like the living beings. The most surprising thing about viruse is that they are not destroyed when their organisation is disturbed. They can be cuptalized and kept in a bottle like another chemical, and yet when allowe to enter a living host, they start reproducing. This has led to their being described as "chemicals in a test tube but living beings in a host".

The name virus (= poison) was first coined by Louis pasteur while studying rabies disease he named the cause as virus. Four names figure prominently in the discovery of viruses.

1. A dolf meyer (1886) set to find out the reasonf for tobacco mooaic disease came to the conclusion, that no fungus or bacterium causes the disease. But he observed that the sap of the diseased plant was capable of transmitting the disease to healthy plants. He claimed that the sap filtered through a bacterial filter lose the infectivity.

2. Iwanowsky (1892) of Russia proved that even filtered sap retains the capacity of infection. But he did not find out the exact cause of the disease.

3. Beijernick (1896), a Dutch scientist demonstrated that there is a disease causing principle (definitely not a bacterium) present in the sap which beings about the mosaic disease. He called the infective principle Contaguim vivum fluidum.

4. Stanley (1935) an American biochemist, creptallised the tobacco mosaic viruse and showed that the cruptalls when introduced into tobacco plants would being about the disease.

Types of viruses:

1. *Plant viruses :* These infect the plants and being about the diseases. RNA is the genetic material in them.
2. *Animal viruses:* These infect animals and human beings brining about the disease. DNA is the genetic

material in them.

3. *Bactriophages.* These infect the bacteria and kill them. DNA is the genetic material.

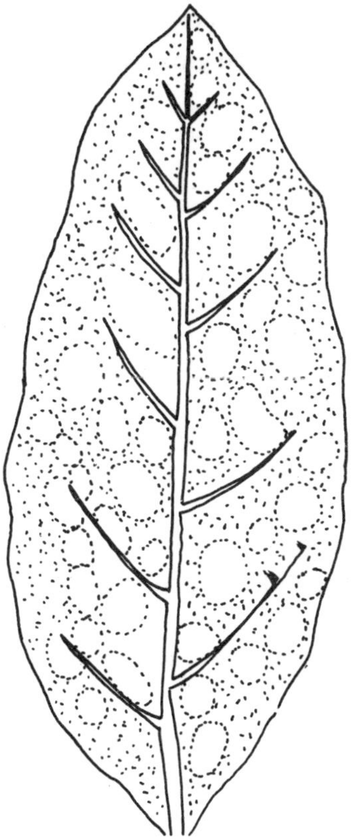

Fig.18.2 Viruses. Tobacco mosaic

Structure of a vision:

1. Virion is the techniacl term for viruses.
2. It consists of a shell a coat of protein and enclose the mucleiio.
3. The nucleoid may be DNA or RNA, but never both.
4. The protein roat is called capsid. It consist of many cub-called capsomeres.

Properties of viruses:

1. They can resist high temperature even after being subjected to 80oC, they can infect plants.
2. All viruses are obligate parasites. They multiply only within a living cell.
3. They can be piltered through filter papers.
4. They another highly resistant to auids, alkalies and salts.
5. They can be transmitted from plant to plant either through mechanical or biological means.
6. They have a high degree of host specialization.

Transmission : Viruses are transmitted from plant to plant through

1. Transmission by vegetative propagation
2. Seed transmission
3. Soil transmission
4. Transmission by contact
5. Transmission by insect vectors

Symptoms of viral plant diseases:

Viruses produce a number of external symptoms, by which a virus plant can be easily detected.

1. **Chlorosis.** This refers to the yellowing of the leaves. This is due to the disappearance of the chlorophyll at places leaving yellow spots. These yellows spots appear like a mosaic pattern, hence the disease may also be called mosaic disease.

2. **Neurosis.** This refers to the death of the cells. It is usually restricted to small areas on the leaves, sometimes it extents over large areas causing complete death of the plants. Neurostic spots are also called lesions.

3. **Distortions.** This is a common symptoms of viral diseases. It may be in the form of cushing or ciniling of leaves caused by the slow growth of the veins, sometmes there may be the production of a cluster of thin wiry shoots called Witches becomes.

4. **Vein branding and vein clearing.** Vein branding refers to the disappearance of chlorophyll and banding surrounding the veins, where as in vein clearing, chlorophyll disappears along the veins of the leaves.

5. **Yellowing :** When the chlorophyll completely disappears from the infected parts, which consequently turn yellow, the disease is known as yellowing.

Plant disease causal by viruses :

Viruses cause a number of plant diseases. A few of the symptoms are described below:

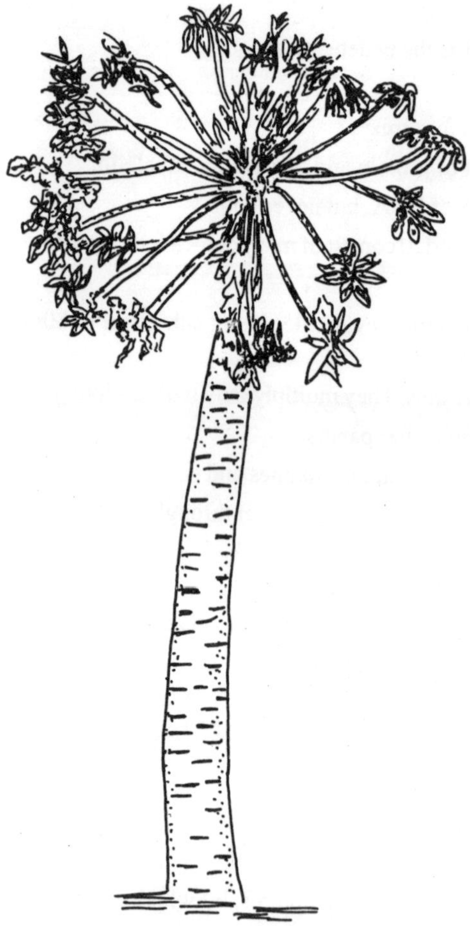

Fig.18.3 Viruses. Leaf curl of papaya

Tobacco mosaic :

1. The disease is brought about by tobacco mosaic virus (TMV).

2. TMV infects tobacco and many other flowreing plants.

3. Typical mosaic pattern of yellow and green areas appear on the leaf.

4. On the young tobacco leaves, the veins may show clearing followed by molting.

5. Finally dark green spots appear, which develop into blister like areas.

6. The plant may become stunted.

Leaf curl of potato:

1. Young plants with few leaves exhibit disease symptoms prominently.

2. Seed borne infections usually result in the seedling becoming infected. The viruses are syptermic (present inside the plant system).

3. Leaf curl virus causes the disease.

4. Pathces of yellow appear on the leaves appear in a scattered fashion.

5. As the growth proceeds, the margins of the leaves curl invards giving a crinkled look.

6. Yield gets considerably reduced, as the tubers are not of the normal size.

Leaf curl of papaya :

1. The leaf and virus causes the disease.

2. The symptoms of the disease appear conspiousolly on the leaves.

3. The infected plants slows leaves of reduced size. Sometimes malformation also appears on the leaves.

4. At advanced stages of the disease, the leaf margins curl inwards.

5. If the infection appears quite early, the plants may not produce flowers at all.

6. The leaves drop down and ultimately fall down.

7. The plant may survive for a long time even after infection, but it will produce little or no fruits.

Bean mosaic :

1. The host plants infected are beans, double beans belonging to the germs phaseolus.

2. The virus is bean mosaic virus.

3. The disease appears on the leaves characteristically producing patches of green and yellow.

4. All the leaves of the plant get the disease.

5. Eventhough the plant may not die due to the infection, the yield gets considerably reduced due ot lack of photosynthesing pigments.

Leaf curl of Tobacco:

1. This is caused by leaf curl virus.
2. Insect vertons transmit the disease.
3. Leaf gets badly curled and crumpled.
4. Heavily infected leaves drop down.

PART THREE
MICROBIOLOGY

CHAPTER19

MICROSCOPY

Experiment no. 1 ***Study of parts of a compound microscope*** **(Bright Field (Light) Microscope)**

The bright field microscope or light microscope or the compound microscope is the most common and the most indispensable instrument in the microbiology laboratory.

Basically the microscope has two types of parts - mechanical parts and optical parts. The mechanical parts provide the structural frame work of the microscope and support the optical parts. The optical parts are involved in the magnification of the object. The following is a detailed description of the various parts of a light microscope.

Base, Pillar and Inclination joint

The base and the attached pillar serve to support the entire microscope. The base is generally 'U' (horse shoe) shaped or sometimes 'V' shaped. Usually the pillar and the base are made heavy in order to minimize vibration. The inclination joint joins the arm of the microscope to the pillar. The joint permits the tilting of the arm of the microscope to any degree as desired by the observer.

Arm and body tube

The arm is a slightly bent solid piece of iron which is attached at one end to the pillar by the inclination joint. The other end of the arm has the body tube to which the principal optical systems are attached. The body tube is a cylindrical structure and at the point of attachment to the arm there is a graduated rack and pinion which helps in the movement of the body tube up and down.

Draw tube, revolving nose piece

Fitting properly inside the upper end of the body tube is the draw tube, which may be drawn up. The ocular lens fits into the draw tube. The purpose of the draw tube is to adjust the tube length so that the objective lens and the ocular are at a specified distance.

The revolving nose piece is a circular rotating disc attached to the body tube. Objective lenses are attached to this.

Adjustment Knobs

The body tube can move up and down on the rack and pinion and the movement is regulated by two sets of knobs - the coarse adjustment knob and the fine adjustment knob. The coarse adjustment knob brings in a greater degree of movement while the fine adjustment knob is used for slight and fine adjustment of the body tube. The adjustment knobs are meant to bring in the object into proper focus. While the range of the coarse adjustment knob is greater, that of the fine adjustment knob is very limited.

Stage : This is a small platform attached to the arm of the microscope. This is meant to keep the objects (on slides) for the purpose of observation. It has a hole in the center exactly in line with the body tube and the condensor below to allow the light to pass through. In most of the modern microscopes, the stage is fitted with a mechanical device called mechanical stage to help in the vertical and horizontal movement of the slide.

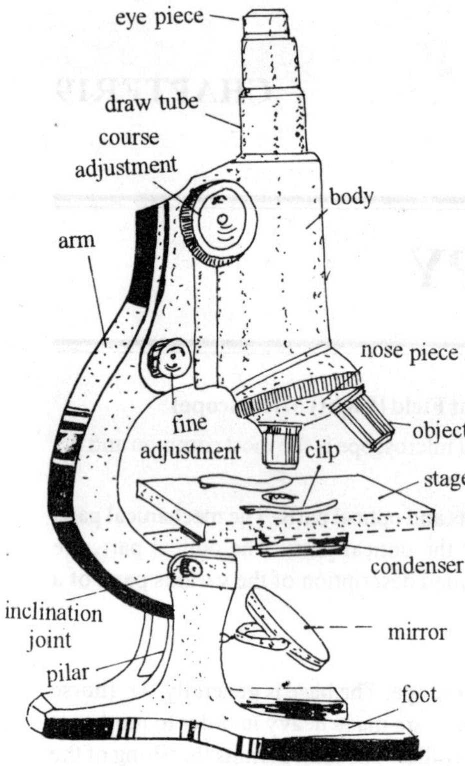

Fig.19.1 A compound microscope

Illumination : In order to collect and reflect the light beams there are two kinds of device attached to the base of the microscope. These are - mirror or an artificial illumination (built in illumination).

The mirror collects the light either from a natural source (sun light) or an artificial source (electric light) and reflects the rays into the microscope. The mirror has both sides reflective - one side is plain, the other in concave. Usually concave side is used when natural light is focused and plain side is used while focusing artificial light.

In some of the compound microscopes there is provision of built in illumination (instead of a mirror) which has a devise to use 8 or 12 volt bulb and provide direct illumination on to the stage.

Sub stage Condensor; Iris diaphragm

The light either reflected by the mirror or by a built in source will be diffuse and does not have sufficient density. In order to condense the light and focus it on to the object, there is a sub stage condenser. This consists of a condensing lens. The condensing lens may be fixed or in some microscopes it can also move up and down for suitable adjustment on a rack and pinion. The condenser helps to focus all the light on the object. Beneath the condenser there is an is diaphragm which regulates the quantity of light entering into the condenser. The iris diaphragm may be closed and opened with the help of a lever to regulate the entry of light.

Optical parts

There are two sets of optics are lenses in a compound microscope. These are eye pieces (ocular) and objective.

Eye pieces (Ocular). The ocular or eye piece is a short tube with two lenses which fits into the upper part of the draw tube. The main function of the eye piece is to magnify the image of the object formed by the objective. Eye pieces are maked 5X, 10X, 15X etc.

The eye piece commonly used is the 10X Huygenian eye piece with monocular tube. These eye pieces are also referred to as negative oculars. The positive ocular or Ramsden eye pieces are constructed with convex surfaces of both lenses facing inwards.

Objective lenses : These are attached to the revolving nose piece and are very important as they affect the quality of the image seen by the observer. Based on their type of construction objectives are classified into achromatic, apochromatic and fluorite. The first one is simple in construction and less expensive, while the other two are complicated are more expensive and used in costly microscopes. These two are also corrected for most of the common aberrations that occur in the lenses.

The objectives are of the following four types based on their magnification - 10X, (Low) 40X, (Medium) 60X, (High) and 100X (Oil immersion).

The main functions of the objective lenses are - 1. Concentration of light rays coming from the object, 2. Forming the image of the specimen and 3. Magnifying the image.

Care of the Microscope

1. Clean all the parts of the microscope before and after use. The lenses and the mirror are to be cleaned with a lint free clothe or tissue paper.
2. Do not pull the microscope to the very edge of the table.
3. While lifting the microscope use both hand and with one hand hold the base.
4. The slide to be observed must be clean and dry (specially the under surface of the slide).
5. Do not remove the eyes piece or objectives from the microscope.
6. The microscope should be covered when not in use.

Experiment no. 2 ***Use of mechanical stage***

In many of the research microscopes the stage is fitted with a mechanical device called the mechanical stage. This device is used to fix the slide on the stage and help in its adjustment during observation. To observe different portions of the slide it will be necessary to move the slide vertically as well as horizontally. In ordinary microscopes slides are adjusted manually. But while observing under high magnifications such as high power and oil immersion any manual movement of the slide would bring about shifting of the focus and observation will be difficult. Besides since the objective lens under high power and oil immersion will be very close to the slide manual movement not only disturbs the specimen under observation it might also damage the objective lens. Mechanical stage avoids all these problems and helps in accurate movement of the slides horizontally as well as vertically.

The mechanical stage has two bent clips and two adjustment knobs. These are fitted to a horizontal graduated bar. The mechanical stage is fitted on to the stage of the microscope with the help of screws. The slide for observation is fixed to the mechanical stage using the clips. After bringing into the focus the slide under observation using the adjustment knobs of the mechanical stage the slide can be moved in any required direction without disturbing the focus.

Mechanical stage has an advantage over the manual movement of the slides particularly while viewing under higher magnification.

Experiment no. 3 ***Use of oil immersion lens***

The oil immersion lens provides the highest possible magnification in the compound microscope. As the objects are viewed through a drop of oil the objective lens is called the oil immersion lens. The working principal of the oil immersion lens is as follows -

When the objective lens (less than 100x) other than oil immersion is used it is referred to as a dry objective because in the space between the specimen and the front of objective there is only air. Air has a refractive index of 1.0. Rays whose angular divergence from the centre of the optical axis is greater than half the angular aperture can not enter the objective lens and are lost. If a drop of oil (Cedar wood oil) is placed in between the specimen and the objective lens, many rays whose angle of divergence is more, also can enter the objective lens as the refractive index of oil is about 1.5. Thus NA of the objective is increased which results in better resolution and higher magnification.

Experiment no. 4 ***Use of dissecting microscope***

The dissecting microscope is a very useful tool in dissecting out plant parts particularly those which are very minute. This microscope combines the easy manouverability of a dissecting instrument and the advantage of higher magnification of a compound microscope. This microscope can also be used to observe specimens that are too big for a compound microscope. Even live specimens growing on substrates in petridishes can also be observed.

The dissecting microscope is binocular and provides a stereoscopic view (three dimensional) of the image. The dissecting microscope consists of the same parts like a compound microscope. The stage is a transparent thick glass plate and attached to this stage is a substage mirror or a built in source of illumination. The microscope has a fine adjustment knob. The optical system consists of ocular and objective lenses.

Procedure for the use of microscope

1. Clean the optical parts of the microscope with the help of a tissue paper.
2. Adjust the illumination in such a way that light is concentrated on the stage. If the microscope does not have a built in illumination use the mirror provided to focus the light on to the stage.
3. Adjust the objective lenses to proper position by turning the revolving nose piece.
4. Adjust the ocular lenses in such a way so that only one image is seen.
5. Place the object of study on the stage and observe it.
6. If the material is to be dissected or separated it can be done so during observation itself using 2 mounted needles.

Experiment no. 5 *Micrometry*

Micrometry or microscopic measurements refer to measurements of microbes (length, breadth) seen under the light microscope. Many a time, the size of the microbes is an essential criterion besides their morphological characters. The following are necessary to carry out microscopic measurements.

1. A good microscope with 10X eye piece and 10X, 40X and 100X objectives.
2. Ocular micrometer
3. Stage micrometer

1. A good microscope (with straight tube): This is necessary to carry out the measurements.

The eye piece may be 10X or 15X with objectives of 10, 40 and 100. It should be noted that when measurements are to be carried out at each magnification, the original calibration (see later) conducted should be at the same magnification.

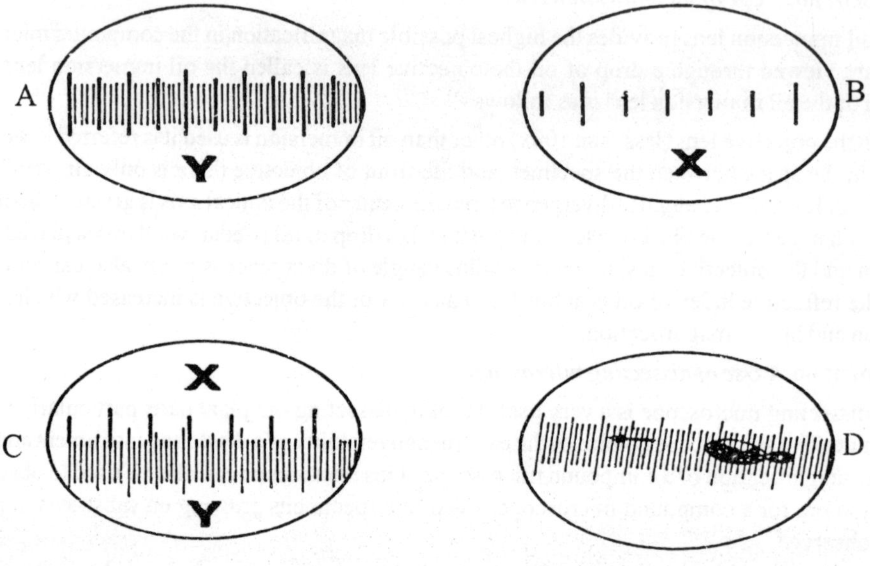

Fig. 19.2 Calibration of ocular micrometer

A. Graduations of ocular micrometer; spacing is arbitrary, B. Graduations of stage micrometer; lines are 10micrometers apart, C. Overlapping of ocular micrometer graduations (Y) over stage micrometers graduations (X), D. Measurement of a yeast cell.

2. Ocular micrometer : This is a small circular disc of glass which has graduations engraved on one surface. The distance between the lines is fixed, but the lines are arbitrarily engraved in the sense it is not a standard measurement of mm, microns etc. In other words the lines on the ocular micrometer do not indicate any measurement of the length.

3. Stage micrometer: Also known as objective micrometer the stage micrometer is a glass slide that has inscribed lines, which are exactly 0.01 mm(10 micrometer) apart. The lines can be easily seen when the slide is placed on the stage and viewed through microscope.

Procedure for measurement

The first step in the measurement is the calibration of lines of ocular with that of stage micrometer.

What is calibration ? This refers to comparison or super imposition of scales of ocular as well as stage micrometer. This is necessary to determine how many graduations of ocular coincide with one graduation of the stage micrometer. It has already been said that the lines of the ocular are arbitrary while the lines of stage micrometer are exactly 10 micrometers apart. So while using only the ocular micrometer for measurement it becomes necessary to know as to what is the distance between graduations.

At this stage it can be said, that instead of the cumbersome procedure of calibration why not use only the stage micrometer which has the measurements?

This is difficult because the microbes then have to be mounted directly on the stage micrometer which is always not possible. Hence calibration of ocular with the stage micrometer and removing the stage micrometer.

Procedure for calibration

The first step is the installation of ocular micrometer into the ocular lens. This can be done by unscrewing the top lens in the ocular, placing the micrometer inside the tube and screwing back the lens. In some of the latest microscopes, the ocular micrometer comes with a ring and it can be easily inserted into the bottom of the ocular tube.

Next step is placing the stage micrometer on the stage of the microscope and focussing it properly. After focussing, the stage micrometer has to be adjusted so that its graduations align with the graduations of the ocular properly. After the graduations are aligned, count how many ocular divisions equal one division (0.01mm) of the stage micrometer. If seven ocular divisions equal one division of the stage micrometer, the value for each division of ocular is 0.01/7 or 0.00143 mm. In other words each ocular division is about 1.43 mm part. Having known this, the stage micrometer may be replaced with the slide containing microbes for measurement. To determine the size of an organism all that one has to do is to count the number of graduations, the microbe measures and to multiply this number with 1.43 mm to give the actual size.

Note : Calibration has to be made for each pair of ocular and objective lenses as the magnifications vary. Ideally, it is better to use the same microscope used for calibration, also for measurements.

Experiment no 6. Calibration of an ocular micrometer for different objectives of the microscope

Materials :

1. Compound microscope
2. Ocular micrometer
3. Stage micrometer
4. Objectives of different power - low, high and oil immersion

Procedure :

1. The eye piece lens from the microscope should be removed and an ocular micrometer disc is inserted into the eye piece by carefully unscrewing the lenses. Care should be taken to see that the engraved side of the micrometer should be downwards. The eye piece should be inserted back into the microscope. When

observed the divisions of the ocular micrometer will be seen in sharp focus. If properly fitted there will be no difference in the lines under different objectives.

2. The stage micrometer should be kept on the microscope stage and by adjusting the fine adjustment knob, the engraved scale should be brought into proper focus.
3. The lines of the ocular should be brought parallel with those of the micrometer by carefully turning the eye piece. In this way the lines of the ocular micrometer coincide with the lines of the stage micrometer.
4. After the above mentioned adjustment count the number of divisions in both the ocular and stage micrometer and find out how many lines of the ocular micrometer coincide with the lines of the stage micrometer.
5. Several readings (alteast five) should be taken randomly by repeating the above steps.
6. Repeat the procedure above for all the three objectives ie., - low, high and oil immersion (100 X).

Results :

The observations should be recorded in a tabular form. In order to obtain an accurate measurement the average of the five readings taken from the ocular micrometer coinciding with the stage micrometer with different objectives must be taken into consideration. Thereafter the value of calibration for different objectives can be made.

1. Low - power objective (10 x 10)

Assume that 20 ocular micrometer divisions = 30 stage micrometer divisions.

Therefore one division of ocular micrometer = 30/20 x 10 = 15 mm.

2. High - power objective (10 x 40)

Assume 24 ocular divisions = 9 stage micrometer divisions. Therefore one division of ocular micrometer = 9/24 x 10=3.75 mm.

3. Oil immersion lens (10 x 100)

Assume 20 ocular divisions = 3 stage micrometer divisions. Therefore one division of ocular micrometer = 3/20 x 10=1.5 mm.

Experiment no. 7 *Measurement of cell concentration (counting of cells) of bacteria, fungal spores etc using the counting chamber (Haemocytometer).*

In many experiments concerning microbes the concentration of the cells, spores etc has to be calculated in liquid suspension. For this purpose several methods are employed. One such method is the use of haemocytometer to estimate the number of cells/spores per ml of a given sample.

The principle of haemocytometric estimation is based on the cell counts in a specific volume of suspension contained in a microchamber with known capacity.

Construction of the haemocytometer

The haemocytometer is also called Petroff-Housser chamber and it consists of a special microscopic slide. The following are the details of the structure of the haemocytometer.

The haemocytometer is a thick glass slide which has a depression in the center having a grove on both sides. The centre of the depression consists of a specific area divided into 9 squares measuring 1sq mm. The centre square is further divided into 25 smaller squares each with a side of 0.2 mm. Each of these squares is again divided into 16 squares covering an area of 0.0025 sq mm. Counting is done in this small square under the microscope. The procedure for counting is as follows-

1. Clean both the slide and coverslip to make it free of all dust particles.
2. Place a known quantity of dilute bacterial suspension with the help of a pipette syringe.
3. Place the coverslip firmly on the suspension.

4. Allow the suspension to settle. Place the slide on the stage of the microscope.
5. Count the number of cells in the small squares. Take an average of 50 squares.

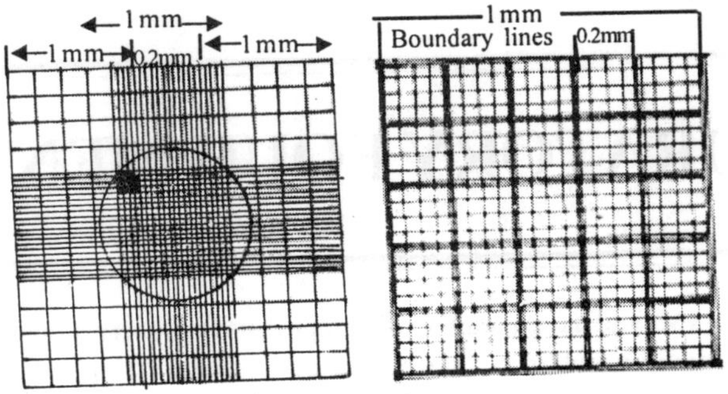

Fig. 19.3 Haemocytometer

Calculation of cell number

1. The area of the square is 0.05 mm x 0.05 mm = 0.0025 mm^2
2. The volume of each square is 0.0025 sq mm x 0.1 mm = 0.00025mm^2
3. Average count per square (say 100)
4. The sample is diluted 10^4 (10,000 times)

The number of cells in the original sample may be calculated as follows

100 1/0.00025 x 10,000 cells = 6 x 10^4 cells.

CHAPTER 20

LABORATORY EQUIPMENTS

The microbiological laboratory should have a number of equipments for observations of microbes, staining of microbes, for sterilization of medium and for qualitative and quantitative estimation of microbes and their metabolic products. The working principle and the method of using of a number of microbiological instruments are described in this chapter.

Experiment no. 1 ***Study of construction and uses of Autoclave***

Autoclave

An autoclave is basically a double walled metallic vessel made of thick stainless steel or copper, one end of which has an opening fitted with a tightly closing door. The opening is meant for keeping in the materials to be sterilized. The lid or door is provided with a pressure gauze to measure the steam pressure. In some autoclaves there are two gauzes- one to show the pressure in the outer chamber and another to show the pressure in the inner chamber. At the top there will be safety valve to permit the escape of steam from the chamber should it get too high. This is necessary to prevent explosion of the chamber. There are valves to hold the steam and another to send it into the sterilizing chamber. There is also an exhaust valve below to let the steam escape from the bottom of the inner chamber. A little away (in some autolcaves) there is a thermometer attached just beyond the exhaust valve to show the temperature of the steam as it comes from the chamber.

The articles to be sterilized are kept loosely in a basket provided with holes all around for the easy circulation of the steam. In order to increase the area of materials to be kept, perforated shelves may be used.

Autolcave is the most desired apparatus used for sterilizing solid and liquid media for microbial cultures, heat stable liquids, heat resistant equipments and instruments, glassware, rubber products and surgical instruments.

In an autoclave, steam pressure is built up to a 15lbs per square inch at 121°C for about 20-30 minutes. Sterilizing in an autoclave is done primarily by high temperature and not by the pressure. A very high amount of heat energy is transferred from the steam to the object when saturated steam settles on the surface of the objects. It has been estimated that saturated steam can heat an object 2500 times more efficiently than hot air at the same temperature.

Operation of the Autoclave

1. Load the autoclave with the materials to be sterilised such as culture media in tubes, flasks etc.
2. Close and lock the door securely.
3. Slowly open the air outlet valve to allow the air in the chamber to be displaced by the incoming steam.

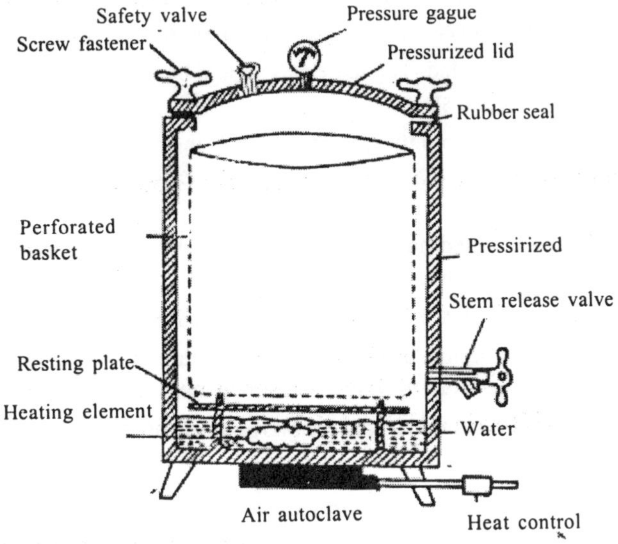

Fig. 20.1 Structure of an autoclave commonly used for sterilization of glasswareand / or media

4. Gently open the steam inlet valve to allow the steam into the sterilising chamber.
5. Observe the increase in temperature in thermometer.
6. Close the air outlet valve as the temperature approaches 100°C. At this stage all the air inside the sterilization chamber would have been displaced by the steam.
7. Watch the thermometer to see that the temperature should not exceed the required temperature range. In most of the modern autoclaves however there is a temperature regulation mechanism which does not allow the temperature to go beyond the required limit.
8. For routine sterilization of culture media 15 pounds pressure is sufficient and this should be maintained for 15 minutes at a temperature of 121°C.

Note : The thermometer measures the temperature of the steam in the discharge line. Hence eventhough the pressure gauge shows 15 pounds pressure many a time the temperature may not reach 121°C if there is still some air inside. Therefore it is necessary to count the sterilization time of 15 minutes after the thermometer shows the temperature of 121°C.

9. At the end of the period shut off the autoclave by closing the steam inlet valve. Allow the pressure to drop to 0. Never open the autoclave door until the pressure drops to 0. If necessary the operating valve can be opened to release the pressure rapidly.
10. After you are assured that there is no pressure inside the sterilization chamber open the door and remove the materials.

Checking the efficiency of autoclave : There are two methods to check whether the autoclave is sterilizing properly, ie., whether the required temperature is reached or not. These methods are -

Chemical - A chemical indicator which changes colour at the required temperature may be kept in the sterilizing chamber. Note whether required change occurs.

Biological - A preparation of dried bacterial spores may be kept in the sterilization chamber. After sterilization, the viability of the spores may be checked.

For proper and efficient sterilization certain precautions are to be taken.

1. Autoclave should not be overloaded. The materials should be arranged in such a way that every part of the material should be exposed to steam.
2. Effective steam pressure is built up if there is no air inside the chamber. This can be ensured by keeping the outlet valve open until a jet of continuous steam comes out of it; only then the valve should be closed.
3. Sterilization time must be counted not from the time it is switched on but from the time the required pressure (ie., 15 lbs per square inch) is built up.
4. Ensure there is sufficient water in the autoclave before switching on. The lid should be closed securely.
5. At the end of the sterilization period, allow the steam pressure drop to zero and only then open the lid. Opening the lid before pressure drops to zero will result in the lid being blown off due to inbuilt pressure.
6. During sterilization the recommended steam pressure should not be exceeded on any account.

Experiment no. 2 ***Study of construction of Hot air oven and its uses***

In all the microbiological laboratories one of the common pieces of apparatus is hot air oven. Hot air oven is a sterilizer using dry heat.

Hot air oven consists of an insulated double walled cabinet heated by electricity and is constructed to withstand a high temperature. The walls of the oven are made up of stainless steel or aluminum and are designed to prevent heat conduction from inside the chamber. There is a motor and fan fitted either at the bottom or back to circulate hot air inside the chamber. This hot air increases the temperature inside the chamber thus sterilizing it. A thermostat regulates the temperature at the desired level and a thermometer is fitted for recording the temperature. The shelves within the hot air oven are perforated to allow proper air circulation.

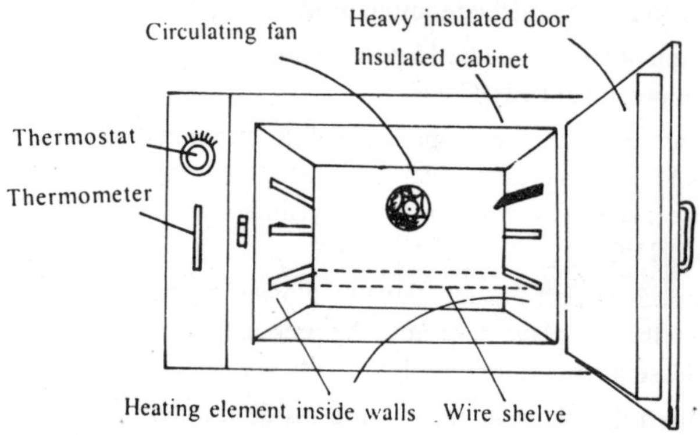

Fig. 20.2 A hot air oven

Hot air oven is primarily used for sterilization of test tubes, conical flasks, petridishes, pipettes etc. The glasswares should be empty. It (hot air oven) may also be utilized for sterilization of instruments (surgical), hypodermic needles and for some non volatile fluids like melted petroleum jelly. It cannot be used for sterilizing culture media, alcoholic or other volatile solutions.

Procedure for using hot air oven

Before sterilization, all the glassware should be thoroughly washed and wiped dry and then neatly wrapped in craft paper. Only then they should be kept inside the chamber. Conical flasks or test tubes should be plugged with cotton wool before sterilization.

After the glasswares and other articles to be sterilized are loaded on to the hot air oven, it should be switched on and the temperature should be slowly raised to 160°C (in some cases 170°C) and it should be maintained at this level. An exposure of atleast one hour (in some cases two hours) is necessary to effect proper sterilization of the glassware. This is necessary to allow the relatively slow penetration of the dry heat (which is slower than wet heat-steam). The temperature should not exceed in any case 180°C for cotton plug and wrapping paper would be burnt. But some bacteriologists opine that a temperature which slightly chars the wrapping paper is desirable. In addition, a higher temperature may also result in the cracking of glassware. After the period of heating, the door of the oven must not be opened until the temperature has fallen to less than 100°C because sudden rush of cool air from outside into the oven may crack the glassware. The following table gives the temperature range and the relative time required for sterilization in a hot air oven.

Temp. (°C)	**Time** (Minutes)
120	480
140	180
150	150
160	120
170	60
180	20

Experiment no. 3 ***Study of construction and uses of Laminar air flow apparatus***

In microbiological laboratories, the microbial population in the air will be generally high due to activities such as opening the culture plates, observing them under microscope, transfer and inoculation of culture, etc. Air in such rooms can not always be made microbe free by other sterilization methods. Hence it is the air in the laboratory that needs to be filtered and sterilized.

The laminar air flow system is used for reducing the danger of infection while working with pathogenic microbes. The laminar air flow system works on the principle of application of fibrous filters in air filtration. In this system, air of a closed cabinet or room is made to pass through high efficiency particulate air filters pack (HEPA). The HEPA filters the air and does not allow any suspended particle above 0.3 mm dimension to go out and as such the air is free of all suspended particles (above 0.3 mm size). The laminar air flow apparatus sucks the air in the room continuously and blows out the air through a pack of filters. The air is blown out in uniform velocity and in parallel flow line.

The laminar air flow apparatus consists of a platform or a work table covered from all sides except the front portion. The work table or the platform is usually fixed to a stand which has wheels to facilitate the movement of the apparatus. The back of the laminar air flow apparatus consists of HEPA air filter through which the air is blown. The top of the apparatus has a number of control switches to switch on / off UV light, tubelight and to turn on the air flow. There is a mercury level indicator which indicates the room pressure.

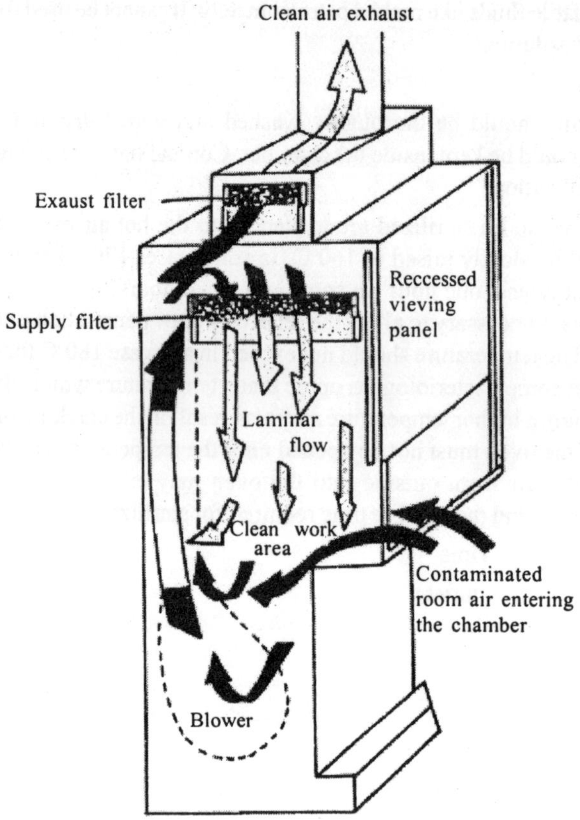

Fig 20.3. The Laminar hood

Procedure for use

1. The platform of the apparatus must be cleaned with a liquid disinfectant and wiped clean.
2. Switch on the UV light for about 15 minutes before starting the experiment. This assures the elimination of all microorganisms that may be present on the work bench.
3. Switch on the air flow and start the experiment such as inoculation etc and let it continue till the experiment is over.
4. Switch off all the control switches at the end of the experiment.

Experiment no. 4 ***Study of construction and uses of Incubator***

The culture of microorganisms in the laboratory is an extremely delicate one as the microbes require various physical, chemical and nutritional parameters. Among the physical parameters a suitable temperature range is perhaps the most important factor governing the development of colonies on a culture medium. Many a time the room temperature may not be suitable for the culture of certain microbes. For this purpose an instrument which maintains a constant temperature throughout is used to culture the microbes. This instrument is called incubator.

Construction

An incubator is basically similar to hot air oven and is made up of an insulated metal cabinet. The opening is controlled by two doors- an inner glass door and an outer asbestos insulated metal door. There is an indicator which turns on as soon as the incubator is switched on. There is another indicator which turns on in the beginning but goes off as soon as the required temperature is reached. This is regulated by a thermostat switch.

The main difference between hot air oven and incubator is the limit of the temperature in the later. An incubator operates between room temperature and 50°C.

Procedure for use

Switch on the incubator and fix the thermostat switch for the required temperature. Open the doors of the incubator and place the culture glassware (tubes, flasks, petriplates etc) on the racks and close both the doors. At the end of the required period remove the culture material and observe the growth.

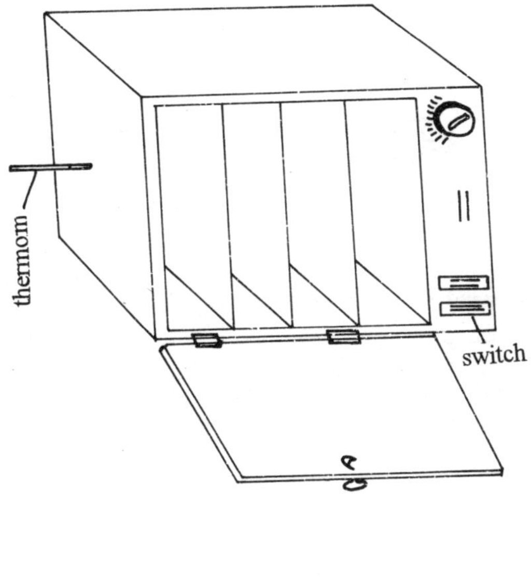

Fig. 20.4 Incubator

Some precautions

1. Unnecessarily donot open the doors of the incubator during incubation period. If necessary only the outer door should be opened and the culture material can be observed through the inner glass door.

2. Donot overcrowd the racks with the culture material. Allow sufficient space between them for hot air to pass through.

3. Never increase the temperature beyond the required limit.

Experiment no. 5 ***Study of construction and uses of Ultra violet lamp (UV lamp)***

One of the methods of sterilization of equipments and other microbiological materials is sterilization by radiation. Both ionizing as well as non ionizing radiation can be used for sterilization. Normally however ionizing radiation is not used in the laboratory as it is not only expensive but also very hazardous and requires very careful handling. Non-ionizing radiation however is less hazardous, specific and easy to handle. UV radiation is the most popular non-ionizing radiation used in the microbiological laboratory. For this purpose table top UV lamps are available. Using this, room, chamber, air, tools etc an be sterilized.

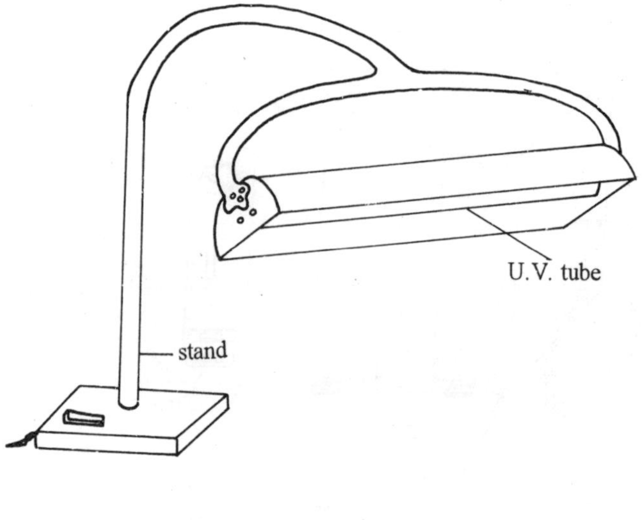

Fig. 20.5 Ultra violet lamp

A UV lamp is almost like a table lamp and has a special filter over the source of elimination to emit UV light. Articles that need to be sterilized are kept in the path of light and lamp is switched on for the required period. At the end of this period the UV light is switched off and the materials taken out.

Care should be taken to see that bare hands and skin are not exposed to direct UV light as it may be injurious in the long run.

Experiment no. 6 ***Study of construction and uses of Inoculation chamber***

Many a time in the microbiological laboratory several experiments like transfering cultures from one medium to another for subcultuing and isolation is necessary. For all these purposes maintenance of aseptic condition is necessary. An equipment called inoculation chamber permits these operations under aseptic conditions.

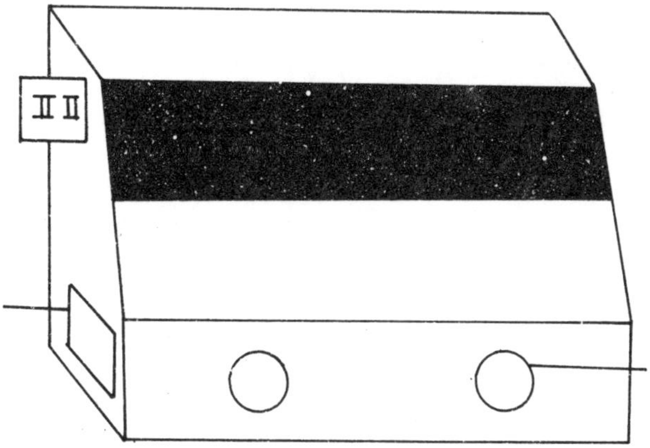

Fig. 20.6 Inoculation chamber

Construction of inoculation chamber

Inoculation chamber is a cabinet and has four doors. Of these two doors are on the opposite sides and they can be used to place the cultures etc into the chamber. The front panel has two circular doors which allows the insertion of the hands of the operator into the chamber to conduct inoculation, transfer of culture etc. The cabinet on top is provided with an UV lamp as well as an ordinary tube light. The UV lamp is used for sterilizing the internal environment while the tube light provides illumination if and when necessary.

Procedure for use

1. Clean the interiors of inoculation chamber using a suitable disinfectant such as 70% alcohol.
2. Through the doors present on the opposite sides of the chamber keep the materials required for experimentation inside the chamber.
3. Switch on the UV light for about 15 minutes to sterilize the interior of the chamber.
4. Transfer of cultures inoculation etc can be carried out after this period.
5. It is preferable to keep a spirit lamp which can be used for flaming the loops before and after inoculation.

Experiment no. 7 ***Study of construction and uses of Inoculation loops and needles***

In the laboratory one of the most frequently conducted operations are inoculation of the medium and the transfer of the culture from one source to the other. Many a time it will also be necessary to separate the colonies of bacteria, actinomycetes etc to stain and study them properly. For this purpose inoculation loops and needles are very essential. Inoculating loops consist of a handle made of glass on to which is fitted a specific length of platinum wire, nichrom or eureka wire. The free end of the platinum loop is bent in the form of a small loop. Usually the platinum wire along with the loop measures about 8 cms in length. Usually two types of loops are used in the laboratory. These are - the smaller loop and the larger loop. The smaller loop has a diameter of 3mm and the loop is on the same axis as that of the platinum wire. The larger loop however has a 5 mm diameter and is bent at right angles to the main axis of the wire. These loops can hold a thin film of culture.

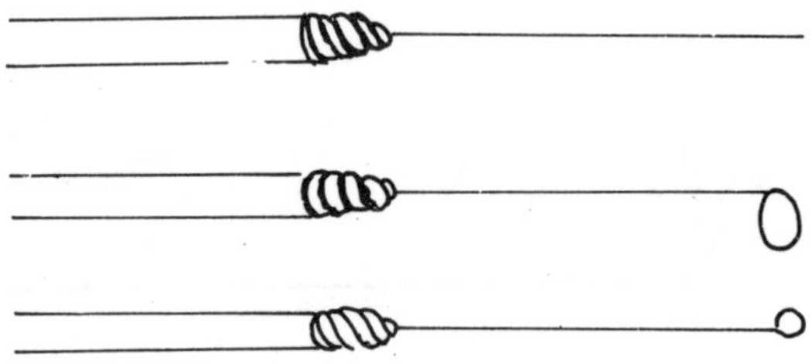

Fig. 20.7 Inoculation loops; small above, large below and needle

Use of the loops

Small loops are used to transfer extremely small quantities of the culture either from liquid or solid medium. The large loops are used to transfer larger quantities like entire colonies from solid media.

Needles

These are usually mounted straight pointed needles. The free end is very sharp. It is used either for teasing and separating the hyphae or filaments in a preparation. It can also be used for inoculating a stab culture.

Experiment no. 8 ***Study of construction and uses of Seitz filter (Asbestos filter)***

In many of the sterilization processes the materials to be sterilised are delicate and thermolabile. They may also be subjected to chemical change if treated with chemical disinfectants. For the sterilization of these materials filteration is the most suitable method. For this purpose a variety of filteration apparatuses and procedures are available. One such filter is the Seitz filter (Asbestos filter).

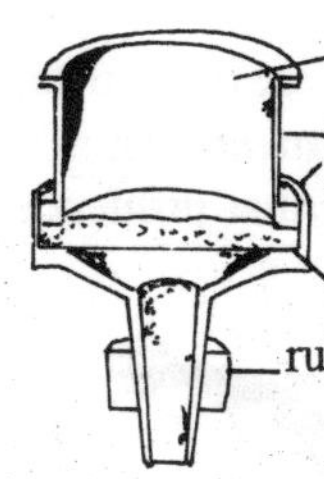

Fig. 20.8 Seitz filter assembly

Construction of a Seitz filter

A Seitz filter consists of a compressed filter sheet of asbestos fibres. The fibers are 2-6 mm in thickness. The asbestos sheets contain washed asbestos fibers, cotton liner and other filter materials. The thickness of the fibres and gap between them determines the efficiency of the filter.

Seitz filters (asbestos) are available in ready to use form either as small discs or squares of different sizes. Filters available in the market can be used only once and then discarded. Seitz filters are very soft and can be easily damaged by even slightly rough handling.

Filter sheets of different pore sizes are available and can be chosen depending on the requirement. The filter sheets or discs are always fitted to a filter assembly for filteration purposes. The filter disc is fitted to a funnel like structure through which the liquid to be filtered is sucked through vacuum or pressure. After completion of filteration, the filter should be discarded and the entire unit sterilised.

Procedure for use of the filteration apparatus

1. The filter available in the market must be sterilised first before use. For this purpose the filter should be wrapped in brown paper and sterilised in an autoclave.
2. Dry the filters for the required time in the hot air oven.
3. Disassemble the filter apparatus and using a sterilised forceps gently transfer the filter pad on to the apparatus.
4. Reassemble the apparatus and connect the suction tube to a vacuum pump.
5. Pour the solution to be sterilised through the top of the metallic funnel and switch on the vacuum pump.
6. Collect the filterate which is sterilised.

Experiment no. 9 ***Study of construction and uses of Centrifuge***

A centrifuge is an instrument which can produce centrifugal force because of accelerated rotation around an axis. This instrument is used in the microbiology laboratory mainly to separate out particles or heavier or lighter liquids that are mixed with the original liquid. Essentially, the centrifuge uses the principal of centrifugal force to separate components of varying densities. Apart from the biology laboratories, centrifuge finds applications in many industries like food processing industry to separate the unwanted particles from food. The washing machine used for washing clothes also makes use of centrifugal force in the spin dry cycle.

Principle of working

It has been mentioned above that centrifuge makes use of centrifugal force and brings about sedimentation (separation) of particles based on their density. Sedimentation may be defined as settling of the particles depending on their size and density. Another factor that influences sedimentation is the viscosity of the medium.

Fig. 20.9 Placing tubes in a clincal centrifuge

Centrifugal force may be defined as the force with which the body going in a circle or curve wants to fly away from the centre of the circle. Centrifugal force is usually measured in dynes and acts along the radius of the path in which the mass is being carried at a given speed. This speed or force can be calcuated as follows.

$CF = \omega^2 \times r$

where CF = Centrifugal force

ω = angular velocity and

r = radius

Angular velocity is related to rotations per minute (rpm) as per the following formula

$$\omega = \frac{2\pi \times rpm}{60} \text{ radius per second}$$

The centrifugal filed is usually expressed as relative centrifugal field in *g* units. This is calculated as follows.

$$RCF = \frac{4\,\pi^2\,(rpm)^2\,r}{3600 \times 980} = g \text{ units}$$

(RPM refers to the revolutions per minute of any object, while *g* units refer to the extra gravitational force (over and above the normal *g* force of 980 cm/sec 2) that the object is subjected to due to centrifugal force.

Parts of a centrifuge

A common laboratory centrifuge usually has a limited speed upto 8 or 10,000 rpm. It usually does not have refrigeration. It includes a metallic chamber which encloses the internal parts and a cover with a latch. Inside the centrifuge head is the rotor, the shaft which turns the rotor and a motor drive assembly. Most of the centrifuges have a power switch, a break assembly, speed control and automatic timer with switch off facility.

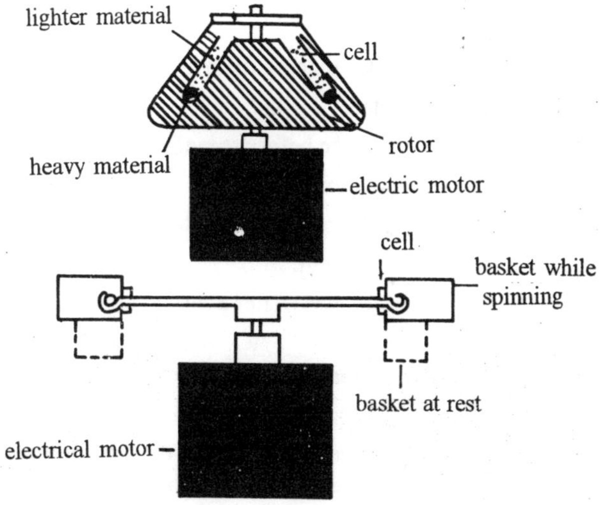

Fig. 20.10 Centrifuges with rotors and with baskets

Centrifuge motors are generally DC motors which turn faster with the increase in voltage. Occasionally AC motors are also used and speed control is achieved through stepwise reduction of number of poles in the magnetic field. The motors employ a direct drive system to the head through the shaft.

The rotors are attached to the shaft of the motor and are meant to keep the samples for spinning. The rotors are generally either of the angle type or swinging bucket type. In the angle rotor sample tubes are inserted into cavities in the rotor that are constructed at an angle to the axis of rotation. The rotor head is a solid block of metal usually made of alluminium. In the swinging bucket rotor the sample tubes are placed in holders which is attached to the rotor body by a pair of yokes. The metal holders are allowed to swing. While at rest the container tubes are parallel to the rotor axis, at operational speeds they will be perpendicular to the rotor axis.

Procedure of use

1. Carefully transfer required amount of material to be centrifuged to the tubes.
2. Place the centrifuge tubes in the rotor head.
3. The materials (liquid) should be filled into the centrifuge tubes only upto a specific level.
4. Centrifuge tubes placed diagonally opposite must be perfectly balanced.
5. In case only one centrifuge tube or odd number of centrifuge tubes are to be used for spinning, distilled water should be filled in another centrifuge tube and placed diagonally opposite the sample tube to acheive balancing.
6. After placing the centrifuge tubes close the lid and perform any other operations such as adjusting the temperature, speed etc.
7. Switch on the power and spin at the required speed and time.
8. At the end of the spinning switch off the apparatus, open the lid and take out the material.

Care of centrifuge

1. Balance the containers with sample before placing them in the opposite cavities of the rotor.
2. The sample tubes should not be overfilled.
3. The rotor should be checked for any damage before spinning.
4. The speed given for a rotor should not be exceeded.
5. The unit should be switched off (power) should any irregularity noticed during operation.
6. Rotors and chambers must be absolutely clean.
7. Rotors shold be chekced for any metal fatigue. After a certain amount of use, the rotors should be used at less than the allowed speed.

Experiment no. 10 *Study of construction and uses of pH meter*

In many microbiological experiments, the culture medium must be of a specific type with reference to alkalinity or acidity. This is necessary to culture specific microbes. Similarily during enzymatic studies, the enzymes are to be harvested and studied in a particular medium with a specific pH, otherwise it (enzyme) will be inactive. When buffer solutions are prepared care is taken to see that they have a specific pH and the pH is checked with the help of an instrument called pH meter. We will first understand what is pH and its importance before studying the working of a pH meter.

The hydrogen ion concentration of a solution is called the pH. All aqueous solutions and water normally contain some concentration of H^+ ions. Pure water is (1/10,000,000 N) with respect to H^+ ions at -22°C. In other words, the hydrogen ion concentration of pure water is 10^7. If we add some acid to water, the hydrogen ion concentration increases. For example the concentration may increase to 10^{-2}. In such a situation the OH^- concentration gets reduced to 10^{-12}.

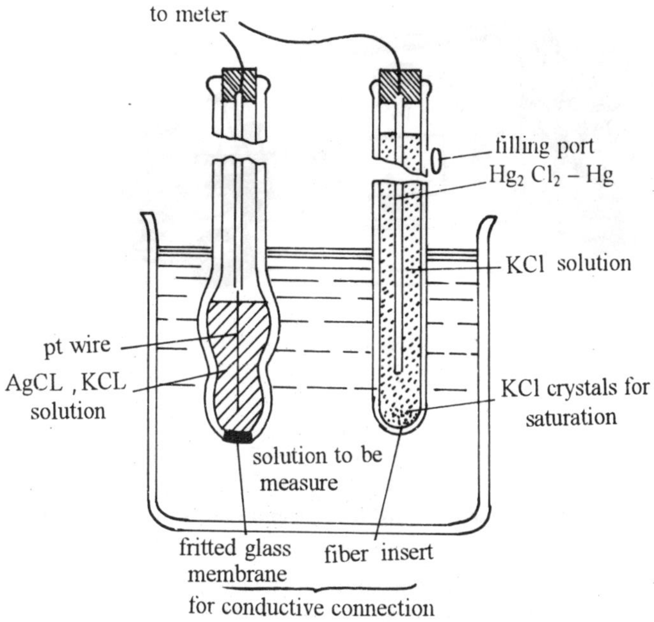

Fig. 20.11 pH meter electrode system. The electrodes measure the acidity or alkalinity of a solution through a range of 1 to 14.

The hydrogen ion concentration is generally expressed as 10^{-} Sorensen suggested that the term pH can replace 10^{-}. The term pH stands for potential of hydrogen. pH is defined as the negative logarithm of the hydrogen ion concentration. The pH scale covers a range of values from 0 -14. A litre of pure water has a hydrogen ion concentration of 0.0000001 or 10^{-7}. Since pH is equal to the negative logarithm of the hydrogen ion concentration, then

$$pH = -\log 10^{-7}$$

$$pH = -\log \frac{1}{10^{-7}} = 7$$

Pure water then has a pH of 7 and is considered neutral. Values of pH below 7 are acidic and above 7 are basic.

Construction of pH meter

A pH meter essentially consists of a pair of electrodes connected to a meter scale indicating the pH value. The pH value is determined by measuring the direct current voltage produced between these electrodes when immersed in a solution. The electrode pair is made up of a reference electrode (*calomel electrode*), that produces a constant potential, independent of the pH of the solution and a measuring electrode (*glass electrode*) in which the potential varies with the pH of the solution.

Fig.20.12 pH meter

The calomel electrode consists of a glass tube containing liquid mercury and mercurous chloride (calomel) in a reservoir filled with a saturated solution of potassium chloride. The glass electrode contains a silver chloride coated platinum wire immersed in a potassium chloride and silver chloride solution. Electrical contact is made by each of the electrode combinations with the solution at the tube end.

As the electrode pairs are in contact through the solution (whose pH is to be measured) the electrical output of the complete unit is the algebraic sum of the two electrode potentials. While the potential of the calomel electrode is constant that of the glass electrode varies depending on the solution. This potential changes 59 millivolts per pH unit of the unknown solution. Thus when the voltage flow is directed to a suitable multivoltmeter, the pH value is directly indicated.

Electrodes are of different sizes depending on the quantity of the sample to be checked. While standard pH meter might have electrodes for use in 10 - 15 ml of solution, some micro electrodes are very tiny and they can check samples as little as 1 to 0.5 ml.

In some cases both the calomel and glass electrodes are combined within a single unit. Such electrodes are refered to as combined electrodes. In a combined electrode the platinum wires from both the electrodes are connected to a potentiometer which detects the difference in the electrical potential of the sample.

pH meters have two different types of display. In one case the display will be on a scale and it has to be read by the oscillation of a needle. In another case there is a direct digital display.

Calibration of pH meter for using

1. The instrument has to be switched on atleast about 20 minutes before use.
2. The electrode is washed with distilled water and then wiped clean with tissue paper.
3. The temperature should be adjusted to the room temperature.
4. Rinse the electrode with a small amount of buffer solution (at pH 4).
5. Take a buffer solution in a beaker and immerse the electrode in it gently.
6. Find out the pH of the buffer by using the selector knob. If it doesnot show pH 4 which is the concentration of the buffers adjust the knob in such a way that it shows pH 4.
7. Allow the instrument to be in this position until the value remains constant.
8. The pH selector knob should be turned back to the standby position.

9. Remove electrode from the buffer and rinse in distilled water.
10. Take a buffer with pH 9.2 and repeat the procedure used for pH 4 buffer.
11. Adjust the selector knob so that it shows required pH.
12. The pH meter now is properly standardised to take the readings from the sample.

Procedure for determining the pH of an unknown sample

1. Wash and clean the electrodes properly and wipe them dry.
2. Adjust the temperature to the prevailing room temperature.
3. Immerse the electrode in the sample solution.
4. Turn on the selector knob and read the pH value.
5. Remove the electrode from the sample, rinse with distilled water, wipe it and again, keep it immersed in beaker containing distilled water.

Care of the pH water

1. Glass electrodes when used for the first time should be dipped in distilled water or 0.1 mol/HCl solution for several hours before use.
2. The solution should be thoroughly stirred before measuring the pH.
3. The temperature of the solution should be kept constant as pH is heat sensitive.
4. After every measurement, the glass electrode must be thoroughly washed with distilled water.
5. The electrodes should not be allowed tc dry; they must always be immersed in distilled water.
6. Before use, the instrument should be calibrated with a standard solution (a solution of known pH).

Experiment no. 11 ***Study of construction and uses of Nephelometer***

Nephelometer is an instrument used for analysing the concentration of a substance in a liquid, based on the varying intensity of the light as it passes through the liquid.

The working of Nephelometer is based on the measurement of the intensity of the scattered light (light rays that get deflected at right angles to the direct rays as they pass through a medium) as a function of the concentration of the dispersed phase.

Small particles in the medium undergo scattering to produce a symmetrical pattern of secondary rays with a maximum intensity at 90° ie., at right angles to the direct beam (which does not undergo deflection). Since nephelometry is involved in the analysis of deflected rays, the measurements are always done at 90° (secondary rays).

The instrument essentially consists of a light source, a sample tube to hold the turbid suspension and a photo cell.

The light source is a tungsten lamp, the light rays from which are focussed on to the sample tube holding the suspension to be analysed. As the light passes through the suspension, two kinds of light rays are produced - direct rays and deflected or scattered rays. Direct rays are absorbed by a light shield as they are not measured by nephelometer. The scattered rays which undergo scattering (by the suspension) to the extent of 90° are sensed by a phototube which is kept at right angles to the sample tube. The amount of scattered light that comes out of the sample suspension is directly proportional to the concentration of the sample compound.

Measurements of turbidity

The following reagents are required for measurement of turbidity.

1. Turbidity free water; Highly purified distilled water can be obtained by passing distilled water through not a membrane filter with pore sizes out exceeding 0.2 micrometer. The water must have turbidity not more than 0.02 NTU (Nephelo turbidity unit)
2. Standard turbidity suspension (stock solution)

The stock solution is prepared with formazin particles with a strength of 400 NTU. The stock solution is prepared as follows.

(I) Solution A : Dissolve one gram of hydrazine sulphate [(NH_2). $H_2S°4$] in distilled water and make the volume to 100 ml in a volumetric flask.

(II) Solution B : Dissolve 10 gms of hexamethylene tetramine [$(CH_2)6 N_4$] in distilled water and make the volume to 100 ml in a volumetric flask.

Standard turbidity stock solution is prepared as follows.

Mix 5 ml of solution A and 5 ml of solution B in a 100 ml volumetric flask and allow it to stand for, 48 hours at room temperature. Dilute the solution with distilled water up to 100 ml (up to the mark in the flask). This suspension will have a turbidity of 400 NTU and it can be used for 4-6 months.

3. Standard turbidity suspension can be prepared by taking 25 ml of stock and diluting it with distilled water to make the volume to 100 ml.

This suspension will have a strength of 100 NTU. Turbidity readings are taken as follows.

1. Switch on the instrument and allow a warm up period of 10-15 mintues.
2. Select the turbidity range (ie., 0-100 NTU) using the selection knob.
3. Insert the sample tube with turbidity free distilled water into the holder and cover with light shield.
4. With the set zero control, adjust the meter so that it reads '0'.
5. Remove the tube and replace it with standard turbidity suspension. Adjust the control so that the meter indicates 100 NTU, the actual strength of the suspension.
6. Replace the standard suspension with the unknown sample and read the turbidity directly from the meter reading in NTU scale.

In case the turbidity of sample is more than 100, use suitable dilutions with distilled water and calculate NTU as follows.

$$NTU = \frac{A \times (B+C)}{C}$$

Where A = NTU found in diluted sample

B = Volume of distilled water (in ml) added to the sample

C = Volume of sample used for dilution

For example -

A = 30

B = 80

C = 20

$$NTU = \frac{30 \times (80+20)}{20}$$

= 300 NTU

Experiment no. 12 ***Study of construction and uses of Colourimeter / Spectrophotometer***

Colourimeter and Spectrophotometer are used to measure the quantity or concentration of a substance in a liquid sample by analysing the intensity of light rays that come out through the sample. The amount of light absorbed is a measure of the concentration of the substance (note in nephelometry only the scattered light is analysed). When light rays strike a substance some amount is absorbed and some amount is transmitted. In both colorimeter and spectrophotometer, the amount of light absorbed is calculated as optical density (OD) and this is an indicator of the concentration of the absorbing substance.

Spectrophotometer and colourimeter both work on the same principles of light absorbance/transmission, but the former is a more versatile and sophisticated device than the latter.

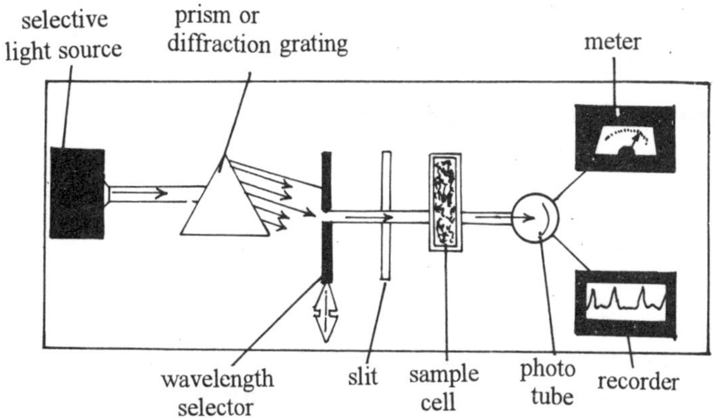

Fig. 20.13 Fundamental system of a spectrophotometer

Photometry

This refers to the measurement of light transmitting power of a solution in order to determine the concentration of substances present in it. In photometric measurements always a monochromatic light is employed. In order to obtain monochromatic light either a prism or a diffraction grating is used. The diffraction grating consists of thousands of engraved parallel slits which can disperse the beam of light into its individual rays.

The absorption of monochromatic light by substances and its bearing on concentration has been explained on the basis of Beer's law and Lambert's law. Beer's law states that when the monochromatic light passes through an absorbing medium the intensity of the light that comes out is inversely proportional to the concentration of light absorbing substance in the medium.

Colorimeter

It is simpler in construction when compared to a spectrophotometer and has a limited application. It consists of a light source, a filter to produce monochromatic light (blue, green, red etc) and a photocell with detector to measure the light intensities. The tube (cuvette) containing the solvent is placed on the light path.

The solvent chosen will have 100% transmittance of light. It is then replaced by the sample tube and the light that is transmitted is analysed in a multivoltmeter which shows how much amount of light is transmitted. The meter consits of a scale which shows percentage light absorbed and percentage light transmitted. The amount of light absorbed is measured in units of optical density (OD) and is a direct indication of concentration (Note : A colorimeter has only filters and everytime they have to be changed if wave length of light is to be changed).

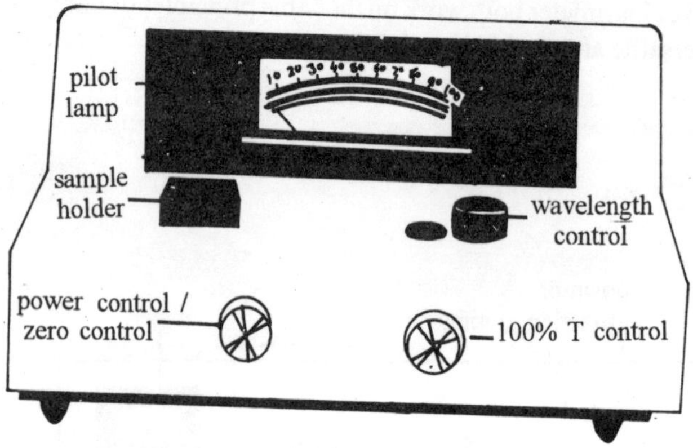

Fig. 20.14 A Spectronic 20 Spectrophotometer

Spectrophotometer

This is constructed on the same principle as the colorimeter, but may be said to be a modified version. The major difference between the two is the ability to select a wide range of wavelength within the visible spectrum. A spectrophotometer uses a grating instead of filters. A knob is provided to manipulate the grating to provide the monochromatic light of the desired wavelength; the range that can be selected is as low as 1-2 nm.

In a spectrophotometer, the source of light is a tungsten filament incandescent bulb for visible range, and a hydrogen gas discharge tube for light in the UV region. Glowing rods of silicon carbide or zirconium oxide produce wave lengths in the infrared region. Since glass is not suitable for UV region, optical system in spectrophotometers is usually made of quartz. The sample tubes (Cuvettes) are also made up of quartz.

The monochromatic beam is selected using the selection knob. The rays then pass through the sample tube and then are focussed on to the photocell which converts the light intensity into electrical impulses and shows it on a multivoltmeter which functions in the same way as in a colorimeter. In some modern spectrophotometers the optical density is displayed digitally.

Quantification of the compound

After obtaining the OD of a compound, quantification can be done by comparing it with a standard curve. A standard curve is prepared by taking series of concentration of a standard compound and noting down the ODs. The ODs and concentration are then plotted on a graph sheet to give a straight line (standard curve). OD of a test compound can then be compared with the standard curve and the concentration calculated.

Experiment no. 13 ***Study of construction and uses of Colony Counter***

In many microbiological experiments it will be necessary to count the number of viable spores, cells or colonies in a culture. This is required to find out whether a microbial colony can grow on multiply on a particular type of medium. In order to count the colonies on a petri plate an instrument called colony counter is used.

By counting the number of viable colonies one can find out the number of viable spores or cells in a particular culture since each cell or spore that is viable forms a colony.

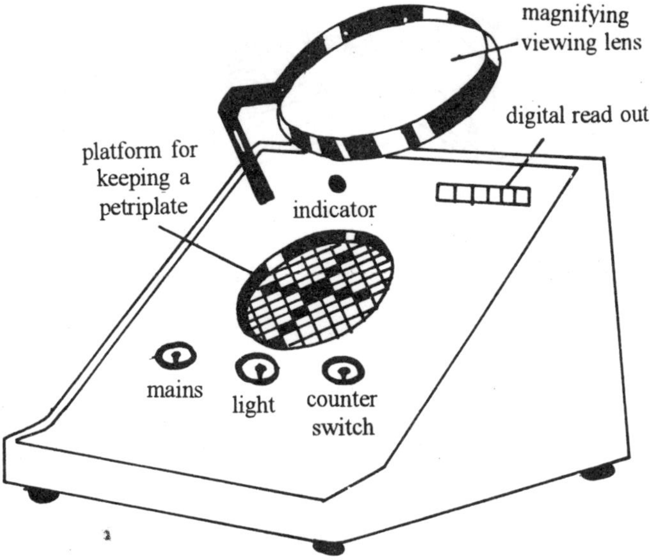

Fig. 20.15 A Quebec colony counter

Construction of colony counter

Colony counters are of two types manual colony countes and electrical colony counters. The manual colony counter is made up of a small cabinet having a circular grid in the center. There is a provision to illuminate the grid from below by a built in light source. Above the grid there is a magnifying glass which can magnify upto 1.5 times. This facilitates better counting.

Electrically operated colony counters are easy to use and avoid the possible mistakes in a manual counter. In most of the electrical operated colony counters there is a pencil shaped electrode and an LCD display which immediately registers the number of colonies counted. There is a reset button which brings the display back to zero.

Procedure for using colony counter

1. Petriplate containing the colonies is placed on the circular grid of the colony counter.
2. Power is switched on to illuminate the grid.
3. Colonies are then counted manually in a manual counter.
4. In the electrically operated counters the pencil shaped electrode is removed from its position and each colony on the petriplate is touched.

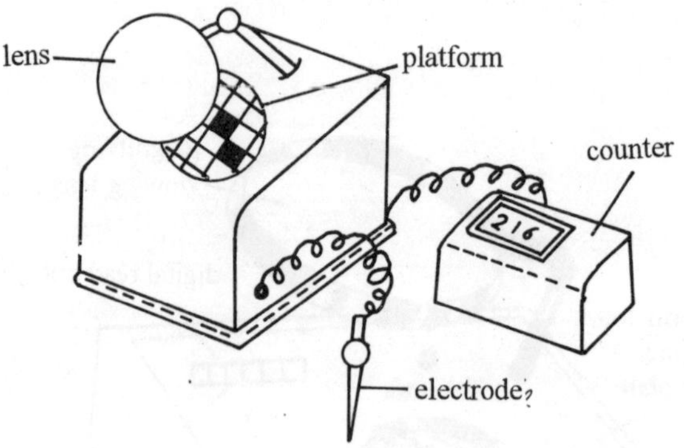

Fig. 20.16 An Electric colony counter

5. As each colony is touched by the pencil electrode it is automatically registered and the number is displayed digitally.

In a special type of electrical colony counter called Qubec colony counter there is a counter switch which when depresed immediately displays the number of colonies on the petriplate on a digital read out.

CHAPTER 21

CLEANING OF GLASSWARE AND STERILIZATION

In the microbiological laboratory cleanliness is very important as it is very necessary to maintain aseptic conditions. For this purpose all the glasswares such as flasks, culture tubes, petridishes, pipettes, beakers etc must be completely clean and sterilised. Only then they should be used for the purpose of conducting any experiment.

Besides the glassware even the work areas such as table tops and many other instruments which are exposed to air will certainly contain microbes. In addition to this if any experiment has been conducted previously on the work table it should be thoroughly disinfected before conducting any other experiment. Even the unused media must be either sterilised again or discarded.

Preparation of chromic acid (cleaning solution)

The glasswares used for any experiment must first be thoroughly cleaned before sterilising them in an autoclave. An ideal cleaning solution generally used in the microbiological laboratory is the chromic acid solution This is prepared as follows -

Dissolve 63 gms of potassium dichromate (even sodium dichromate can be used) in 35 ml of water by heating. Cool the mixture and after the mixture is at the room temperature add concentrated sulphuric acid to make up the total volume to 1 litre. This is the stock solution of chromic acid.

Method of cleaning the glassware

The glassware to be cleaned should be filled with the solution or the glasswares may be immersed in the solution taken in a glass trough. Allow the glassware to stand in the solution for one to two days. Remove the glassware from the solution wash it with tap water several times followed by distilled water. The glassware can then be sterilised suitably and used for experiments.

The cleaning solution can be used several times. Handling of this solution must be done extremely carefully as sulphuric acid is extremely corrosive and might burn the hands. If splashed on the hands accidentally immediately wash with a solution of sodium carbonate (alkali solution).

Glasswares containing unwanted culture medium should first be treated with 3% lysol for 24 hours. This treatment will kill all the microorganisms in the culture. Subsequently the glassware can be immersed in boiling detergent (5% solution) for 15 minutes and then autoclaved for 15 minutes.

The glasswares should be completely cleaned in distilled water and all the traces of media and any other debris attached must be completely cleaned. The cleaned glassware must be thoroughly dried before using them for further experiments.

The glassware should be air dried first before sterilising them. Instead of air drying the glassware may also be dried in drying chambers. Care must be taken to remove all traces of water before placing them inside a dried chamber. After washing it is preferable to keep the glassware in a stand for sometime to allow for the drying of water and only then they should be transferred to the drying chamber.

Sterilisation of glassware

A number of methods are available for sterilisation of glassware and other materials used for microbiological experiments. These methods may be broadly classified into physical methods, radiation methods, ultrasonic methods and chemical methods.

Physical methods include sterilisation by dry heat (in hot air woven) or moist heat (the Arnold sterilizer) or by steam under pressure (autoclave). All these methods physically eliminate the microbes. For details of these see chapter 3 - Laboratory equipments.

Filtration is another physical method of sterilization. For this purpose different types of filters are available.

1. Seitz filter (Asbestos filter)
2. Porcelain filter (Chamberland filter)
3. Fritted glass filter (Sintered glass filter, Morton filter)
4. Diatomaceous earth filter (Berkefeld and Mandler filter)
5. Membrane filters (Ultra filters)

(For the detailed description of the above mentioned filters see College Microbiology Vol I by the same author).

Radiation methods include subjecting the instruments media etc to the effect of ionizing and non-ionizing radiation. Normally ionizing radiations are not used in the laboratory as they are non specific, expensive and also hazardous. Non-ionizing radiation such as UV rays are commonly used in the microbiological laboratory for the purpose of sterilization.

Chemical method of sterilisation is mostly employed where physical methods like heating and boiling cannot be used. A variety chemicals such as alcohols, halogens, phenolic compounds, aldehydes and detergents are available for this purpose.

Plugging of glassware

Culture equipments such as tubes and conical flasks are to be plugged with cotton wool before sterilization. This is necessary to allow the air into the medium after filtering it of all microbes and other particles. The cotton used for the plug must be extremely clean. The plug should neither be too tight nor too loose. The best method of preparing a cotton plug is to roll a suitable length of cotton and turning the edge while rolling. The rolled plug should then be used for plugging the mouth of flask, tube etc. A clean glass rod can be used to push the plug into the mouth of the tube. Care should be taken to see that the glassware should be completely dry before putting the cotton plug. Both absorbent as well as non absorbent cotton can be used for plugging the glassware. If absorbent cotton is used for plugging it is better to cover the glassware around the neck and mouth with a piece of brown craft paper. This will prevent the cotton plug from getting wet during sterilisation.

The plug should not be removed until it is required to pour the culture medium or for the purpose of inoculating the media. The plug should be removed under aseptic conditions and that to only for a very short period.

Petridishes individually or in groups of 4-5 should be wrapped in clean craft paper and tied with a string before subjecting them to sterilisation.

CHAPTER22

CULTURE MEDIA

Microbial study involves various methodologies. Microbes present in natural conditions grow, multiply and die. In the process they affect over life in various ways both for good as well as bad. Naturally then we have to understand their life processes (microbes) as to how they grow, reproduce and multiply; then we should know what are their nutritional requirements and how they are affected if certain requirements are not available. In natural conditions, microbes are rarely present in pure condition; there will always be mixed populations. Under such circumstances, it will be difficult to study them. In order to study the various life processes of microbes, they must be cultured on artificial media under laboratory conditions; isolated in pure form and only then a meaningful study is possible.

A culture may be defined as growth of microbes under contamination free condition in the laboratory on a suitable medium.

The term culture is usually employed for an induced growth of microbes under *in vitro* (Latin; *vitro* - glass) conditions. Microorganisms as the very name indicates are extremely minute in size. It is difficult to obtain any information by the study of a single individual. Hence microbes are always cultured to obtain populations of cells. Microbial populations are usually refereed to as colony (pl. Colonies). A colony develops from the multiplication of a single cell and hence the population of cells are always homogenous. When a colony consists of only one type of cell it is called pure culture. In a mixed culture there will be cells of different types.

CULTUREMEDIUM

Microorganisms can be cultured in the laboratory under aseptic conditions using sterilized glassware and other appliances. In order to cultivate the microbes, they should be provided nutrition on a medium in nearly as natural a way as possible. Preparation of suitable nutrient substances is the single most important factor in the cultivation of microbes in the laboratory. Nutrient substance prepared in a suitable manner (so that the microbes can grow on them) is refereed to as the culture medium (Pl.media). Different microorganisms require different nutrient materials. Thus there will be different types of media. Basically media may be classified into the following four types-

1. Natural (empirical) media
2. Semisynthetic media
3. Synthetic media and
4. Living media

Experiment no. 1 ***Preparation of Broth culture media***

The broth culture medium is also known as liquid culture medium. It may be defined as a sterile liquid nutrient solution which does not have any solidifying medium.

Broths are of different types. The choice depends on the requirement of microbe to be cultured. In most of the cultures two types of broth are used - nutrient broth and glucose broth. Broth cultures are ideal where growth performance of microbes are to be studied by techniques such as turbidemetric method. Turbidity or cloudiness in the nutrient solution indicates concentrated growth of microbes which are evenly dispersed in the medium. Turbidity may be measured in a spectrophotometer or colorimeter, where optical density indicates the measure of growth rate.

Broth cultures are also suitable for studying pellicle formation wherein fungi cultured in liquid medium form large mats floating on the surface. Fungal nutritional studies also can be carried out using a broth clulture. Study of rate of sedimentation of bacterial cells is also possible using a broth culture.

Nutrient broth and glucose broth are two of the general liquid media used for the cultivation of a variety of bacteria. The following is the method used for the preparation of broth culture.

Requirements

1. Nutrient broth

a. Peptone 5.0 g

b. Beef extract 3.0 g (pH 7.0)

c. Distilled water 1000 ml

2. Glucose broth

a. Peptone 10.0 g

b. Glucose 5.0 g

c. Sodium chloride 5.0 g (pH 7.3)

d. Distilled water 1000 ml

In addition to the above, the following are also required 1 NHCl, 1 NaOh, pH meter, distilled water, hot plate, autoclave, culture tubes, flasks etc.

Procedure

Nutrient broth. Put the weighed amount of peptone (5g) and beef extract (3g) in 500 ml of distilled water. Heat the mixture and agitate with glass rod to dissolve the constituents. Add distilled water to make the volume to one litre. Adjust the pH of the medium to 7.0 either adding acid or alkali depending on the necessity. Take test tubes and pour 10 ml each. Conical flasks also may be used instead of test tubes in which case pour 25 ml to each flask. Apply cotton plugs. Autoclave at 121°C under 15 lb pressure for 15 minutes. Allow the autoclave to cool. Remove the broth tubes (flasks) and store at room temperature for use.

Glucose broth may be prepared in the same fashion as the nutrient broth.

Experiment no. 2 ***Preparation of Potato dextrose agar (PDA) medium for growing fungi***

Materials : Potato, glucose, peptone, agar, beakers, wire gauge, flask, distilled water, funnel, cheese cloth etc.

Procedure

1. Slice and peel the potato tubers.

2. Boil 200 gms of potato tubers in approximately 500 ml of water for about 1/2 an hour.
3. Decant the supernatant and filter it using a cheese cloth.
4. Collect the filtrate in a beaker. This constitutes the potato extract.
5. Boil 20 gms of agar in a separate flask (1 litre capacity) containing 500 ml of distilled water until it becomes a thick brownish solution.
6. Mix the potato extract with the agar solution.
7. Add 20 gms of glucose and 2 gms of peptone into this mixture.
8. Stir the mixture till it becomes homogenous. If necessary boil the mixture.
9. Add sufficient quantity of distilled water to make up the volume to 1 litre. This constitutes the PDA medium.
10. The pH of the medium should be 5.6. The medium should be adjusted to this pH by adding either acid (1/10 normal HCl) or alkali (1/10 normal NaOH).
11. The medium should be kept warm until it is poured into suitable culture tubes, flasks or petridishes.

Experiment no. 3 ***Preparation of Beef extract agar medium for growing bacteria***

Materials : Beaf extract, peptone, agar-agar, distilled water, beaker, 1 litre volumetric flask.

Procedure

1. Dissolve 3 gms of beef extract and 5 gms of peptone in 500 ml of distilled water taken in a beaker.
2. Heat the solution to about 60° so as to dissolve all the ingredients.
3. Boil 15 gms of agar in 500 ml of distilled water taken in a separate flask. Continue boiling until it becomes a thick brownish solution.
4. Mix the beaf extract peptone solution into agar-agar solution and stir to bring about homogenisation.
5. Add required quantity of distilled water to make up the volume to one litre.
6. The pH of the medium should be adjusted to between 6.8 - 7 by adding either acid (1/10 normal HCl) or alkali (1/10 normal NaOH).
7. Keep the medium warm until required for use.

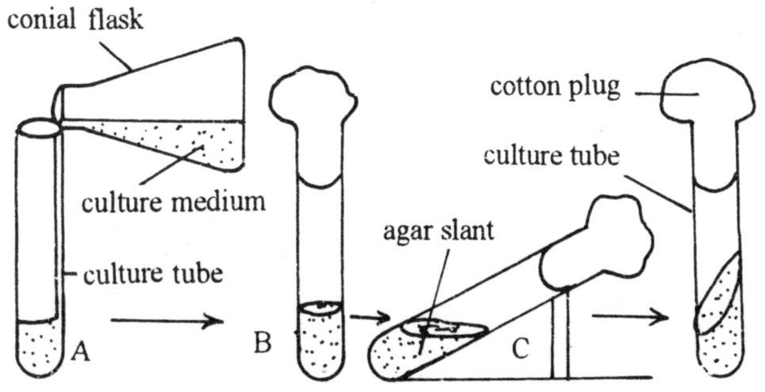

Fig.22.1 Preparation of agar slant and stab

Experiment no. 4 ***Preparation of Agar slants and stabs in a culture tube***

Materials : Culture tubes, agar medium, cotton, paper, rubber band, trays etc.

Procedure

1. Heat the agar medium to about 60° C to bring it to liquid form.
2. Take thoroughly cleaned and sterilised test tubes.
3. Pour approximately 10 ml of agar medium into culture tubes.
4. Plug the mouth of the culture tube with cotton plug. Cover the plug with the help of craft paper and tie it with rubber band.
5. Keep the tubes in a test tube stand to get agar stabs.
6. Keep the tubes in a slanting position in a tray.
7. Allow the medium to dry. You will now have agar stabs and agar slants.
8. This can be used for culturing of bacteria, fungi etc.

CHAPTER 23

PREPARATION OF STAINS

Chemical substances used to stain microbes are commonly called dyes. There are two kinds of dyes - Natural and synthetic.

Natural dyes are obtained from living sources and are used mainly for histological purposes while synthetic dyes are chemically prepared. It is the synthetic dyes that are mainly used for staining the preparations of microbes.

Microbial synthetic dyes are chemically coal tar (aniline) preparations. Chemically a dye may be defined as an organic compound containing a benzene ring plus a chromophore and an auxochrome group.

Dyes belong to three groups- Acidic, basic and neutral. These terms however do not refer to their reactions either as acidic or basic. The terms refer to the type of charge (positive or negative) which the dye molecule has.

The acidic dyes (picric acid, acid fuchsin, eosin etc) are anionic (negative) and ionize to impart a negative charge on the dye molecule. They are used to stain the cytoplasmic components which are more alkaline in nature.

Basic dyes (methylene blue, crystal violet, safranin etc) are cationic and ionize to provide the dye molecules a positive charge.

For instance methylene blue is actually a salt - methylene blue chloride which dissociates into

MBC - methylene blue^{+} + chloride^{-}

The colour of the dye is in the positively charged methylene blue. These dyes react with those components of the cell that are acidic in nature (nucleic acids). Basic dyes usually belong to chloride, sulphate or acetate group.

Neutral stains are formed by the combination of acidic and basic dyes in aqueous form. The colouring matter in neutral dyes is present in both anionic and cationic groups.

Types of stain (classification)

A wide range of dyes are available either singly or in combination for the purpose of staining. These basically fall into the following four categories :

1. Simple stains
2. Differential stains
3. Structural stains and
4. Stains to test cell reserve products

Experiment no. 1 ***Preparation of different types of stains for staining of microbes***

1. Carbol Fuchsin:

Basic fuchsin	1 gm
Absolute alcohol	10 gm

Add 100 ml of 5% solution of phenol in distilled water. Shake well. Filter before use.

2. Eosine :

Water soluble eosine	1 gm
Distilled water	100 ml

Dissolve and then add 5 ml of 95% alcohol.

3. Giemsa stain :

Giemsa stain	3.8 gm
Glycerol	125 gm
Methyl alcohol (solvent)	100 ml

4. Haematoxylin :

Glacial acetic acid	10 ml
Haematoxylin	2 gm
Potassiumalum	in excess
Distilled water	100 ml
Absolute alcohol	100 ml
Glycerol	100 ml

5. Methylene blue (stain) :

Saturated alcoholic solution	60 ml
KOH (0.01%)	200 ml

6. Nigrosine :

Nigrosine	10 gm
Formaline	10 drops
Distilled water	100 ml

7. Safranin :

Saturated alcoholic solution of safranin	10 ml
Distilled water	100 ml

8. Acid fast stain :

Carbol fuchsin (Ziehl's)

Solution A

Basic fuchsin (90% dye content)	0.3 g
Ethyl alcohol (95%)	10.0 ml

Solution B

Phenol crystals (C.P)	5.0 g
Distilled water	95.0 ml

Dissolve the solutes of each solution in the respective solvents mentioned. Mix solutions A and B and allow it to stand overnight. Filter the final preparation and store in a glass-stoppered bottle.

9. Acid-alcohol

Ethyl alcohol (95%)	97.0 ml
Hydrochloric acid (37%)	3.0 ml

Add the acid to the alcohol slowly. Store the solution in a properly labelled, glass-stoppered bottle.

10. Capsule stain :

Crystal violet (1%)

Crystal violet (85% dye content)	1.0 g
Distilled water	100.0 ml

Copper sulphate solution (20%)

Copper sulphate ($CuSO_4.5G_2O$)	20.0 g
Distilled water	80.0 ml

11. Cell wall stain :

Mordant

5% aqueous solution of tannic acid

Stain

0.2% aqueous solution of crystal violet

12. Cytoplasmic membrane stain

Plasmolysis

10% aqueous solution of potassium nitrate

13. Flagella stain :

Loeffler's flagella mordant

20% aqueous tannic acid	100.0 ml
Ferrous sulphate	20.0 g
10% basic fuchsin in alcohol	10.0 ml
Distilled water	40.0 ml

Dissolve ferrous sulphate crystals in distilled water by gently warming and add the remaining ingredients.

Loeffler's flagella stain

1% basic fuchsin in alcohol	20.0 ml
3% aniline water	80.0 ml

14. Fungal stains

Water-iodine solution

Gram's iodine (as in Gram's stain)	10.0 ml
Distilled water	30.0 ml

Lactophenol cotton blue (stain)

Lactic acid	20.0 ml
Phenol crystals	20.0 g
Glycerol	40.0 ml

Distilled water	20.0 ml
Cotton blue (1% aqueous)	2.0 ml

Add lactic acid and glycerol/glycerine to the distilled water and mix thoroughly. Add phenol crystals and heat gently in hot water with frequent stirring until the crystals completely dissolve. Add the dye and mix thoroughly. Store the stain in a brown bottle.

15. Gram stain :

Crystal violet (Hucker's)

Solution A

Crystal violet (90% dye content)	20.0 g
Ethyl alcohol (95%)	20.0 ml

Solution B

Ammonium oxalate	0.8 g
Distilled water	80.0 ml

Dissolve crystal violet in ethyl alcohol and the ammonium oxalate in distilled water, mix solutions A and B.

Gram's Iodine

Iodine	1.0 g
Potassium iodide	2.0 g
Distilled water	300.0 ml

Dissolve iodine and potassium iodide in distilled water.

Ethyl alcohol (95%)

Ethyl alcohol (100%)	95.0 ml
Distilled water	5.0 ml

Safranin

Safranin (2.5% solution in 95% ethyl alcohol)	10.0 ml
Distilled water	100.0 ml

CHAPTER24

STAINING OF MICROBES

Microorganisms are not visible to the naked eye. Their presence may be inferred by two methods - either by direct visual observation or by analysing the specific chemical products they produce.

As the microorganisms are very minute their direct visual observation is possible through the use of specific microscopes. Microscopic studies of microorganisms such as bacteria, molds etc., can be made by observing the living unstained organisms under phase contrast microscope. But use of phase contrast microscope is not always possible and many a time other kinds of microscopes like ordinary light microscope (bright field microscope) has to be used. Living microorganisms are generally transparent, colourless and usually merge with the surroundings particularly when placed in an aqueous medium. Hence in order to increase the visibility of the microorganisms and contrast them from their surroundings they (microorganisms) are stained (coloured). Stained microorganisms are dead and their observation becomes easy as they become fixed to a position under the microscope.

The purpose of staining microorganisms may be summarized as follows :

1. Provides a contrast to the organisms from their surroundings.
2. Proper observation of details of shape, size and other morphological details becomes possible.
3. Observation of internal structures like the nature of cell wall, cell inclusions, details of flagella etc., becomes easy.
4. Enables us to study microbes under higher magnifications by use of oil immersion lens.
5. Taxonomic identification of microbes will be easier in stained preparations (for example gram positive and gram negative bacteria).

Although microorganisms do not appear vastly different from their surroundings they differ markedely to chemically. It is this chemical difference that enables us the distinguish microbes by staining, the stain or dye generally reacting with the microbe but not with the background.

Staining procedure :

Staining solutions are prepared by dissolving a particular dye in either distilled water or alcohol. As concentrated alcohols destain the cells, only dilute alcohol is used (usually only about 1% concentration). Usually dilute solutions of stain applied for a longer period to the sample yield better results than stronger solutions applied for a shorter period. This however may not be true always.

In some instances some of the stains do not have chemical affinity to the cell components which they are to colour. In such cases a mordant is used. Mordant is not a dye but it has affinity to the cell component and when combined with the stain serves to adhere the stain to the cell component. Some of the commonly used mordants are - Tannic acid, salts of ammonium, tin, zinc, iron, copper, chromium etc. In any staining procedure requiring the use of mordant, the microbial smear is applied first, the mordant and then the stain.

Preparation of slide for staining

In preparations for structure of the microbes, they have to be smeared and fixed to the glass slide on which staining can be done. If the cell preparation is not fixed to the slide it will be washed out during staining. The method described below is a generalized one and applicable to any microbial preparation.

1. Take a glass slide, clean it thoroughly with alcohol, wipe it clean and expose it to flame to kill any adhering contaminants and also to eliminate any fatty material sticking to the slide.
2. Hold the edges of the slide and keep it on a clean table with the flamed side facing up.
3. With the help of a wire loop (flame it before use) transfer a little of the microbial culture on the slide. If the microbes are in a solid medium, put a drop of double distilled water on the slide and then transfer the microbial culture.
4. Spread the culture evenly with the help of a wire loop so that it forms a thin film on the slide.
5. Allow the slide to dry in the air by holding it high above a bunsen burner.
6. When the film is dry, pass the slide (with the film side up) three times quickly through the flame. Proper care should be taken at this stage as over heating will distort the shape of the microbes. The slide should be warm and not hot when placed against the back of your palm.

The purpose of fixation is to kill the microbe, coagulate the protoplasm and make it stick to the slide. An ideal fixing agent will retain the structure of the cell its original form without causing the formation of artefacts. Gentle heating is one of the fixing agents, others being alcohol and various other chemicals.

Experiment no. 1 *Staining of bacteria - Gram's stain*

One of the most widely applied differential stains, Gram's stain was first developed by Hans Christian Gram (1884) a Danish physician. By the application of gram staining it is possible to classify all bacteria into gram negative and gram positive. In this staining procedure, the bacterial smear is subjected to a stain (crystal violet or gentian violet) and a decolorizing agent followed by a counter stain (safranin). The bacteria which retain the violet stain appear dark blue (or violet) and are called gram positive and those that get decolourized due to the action of the decolourizing agent are called gram negative.

The differential response of bacteria to Gram's stain is related to the chemical composition and architecture of the surface layers of cell wall in both the types of bacteria. The gram negative cell wall is thin, complex and multilayered with relatively more lipid contents. On being subjected to alcohol treatment the lipids readily dissolve resulting in decolourization. In contrast to this, gram positive bacterial cell walls are thick, less complex and have less lipids. When treated with alcohol, dehydration results closing the openings in the cell wall thereby not allowing the leakage of stain and the bacteria do not get decolourized.

The Gram's stain requires four different solutions - a basic dye, a mordant, a decolourizing agent and a counter stain.

1. Basic dye : Crystal violet is the preferred basic dye. It reacts with the acidic components of the cell. It takes about 10 seconds to react with the cell.

2. Mordant : A mordant increases the affinity of the dye to the cell by chemically combining with it. Staining is deeper and can not be washed out easily if a mordant is used. Iodine solution is the mordant used in gram's staining.

3. Decolourizing agent : As the name indicates it is an agent that removes the dye from a stained cell. Some stained cells decolourize more easily than others (depending on their chemical composition).

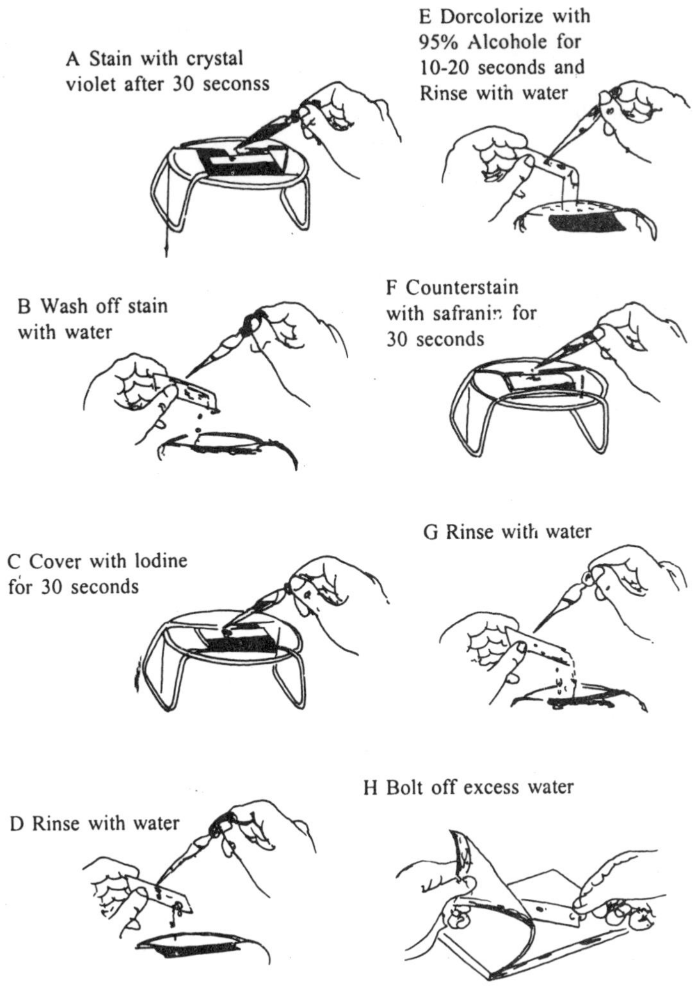

Fig. 24.1 Gram staining procedure

4. Counter stain : This is also a basic dye different in colour than the initial (primary) one. The purpose of the counter stain is to give the decolourized cells a colour different from the first one. The microbes which are not decolourized by alcohol retain the first stain while the decolourized ones take up the counter stain (second stain).

Materials :

1. Bacterial culture - *Staphylococcus* and *Escherichia coli*
2. Gram's staining reagents
   a. Iodine solution
   b. Ethyl alcohol (95%)
   c. Safranin
3. Staining tray, wash bottle, droppers

4. Distilled water
5. Glass slides
6. Inoculating loops
7. Bunsen burner / spirit lamp
8. Microscope

Procedure :

1. Prepare separate smears of *E.coli* and *S.aureus* on two different slides. Fix these preparations after allowing them to dry. Heating may be done lightly.
2. Stain the smears with crystal violet stain for about 30 seconds.
3. Rinse with water for a few seconds.
4. Add few drops iodine solution to both the smears. Allow the solution the react for 30 seconds.
5. Rinse with distilled water.
6. Add few drops of 95% ethyl alcohol to the smears. Add the alcohol drop by drop and stop adding alcohol when no more colour flows out from the smears. For a thin smear 10-20 seconds time may be enough for complete decolourization of gram negative bacteria.
7. Counter stain both the smears with safranin for about 20-30 seconds.
8. Wash with distilled water and blot dry using an absorbent paper.
9. Air dry the slides.
10. Add a drop of Canada balsam (permanent preparations) or glycerin (temporary mounts), put a cover glass each over the preparations and observe under the oil immersion objective of the microscope.

Observations:

1. Indentify which smear has taken the violet stain and which has taken only the counter stain.
2. Make sketches and classify them based on gram reaction (positive or negative).

Conclusions : Bacteria that appear blue are referred to as gram positive and those appearing pink are called gram negative.

Staphylococcus aureus bacteria appear purple or blue in colour hence they are classified as gram positive while E.coli bacterial rods appear pink hence they are classified as gram negative.

Experiment no. 2 *Staining of bacteria - Negative staining* (*Indirect staining*) :

In negative staining, the simple stain used is an acidic dye (Nigrosine, India ink etc) and it does not stain the microbe (bacteria) but stains only the background. Upon examination after staining, the microbes appear transparent in a dark background. Negative staining has two advantages - (i) cells will not shrink or distort because no heat fixing is necessary and (ii) capsulated bacteria which are difficult to stain can be observed by this method.

Purpose : To observe the size and shape of the microbial cell.

Materials :

1. 24 hours old culture of bacteria (*Staphylococcus aureus*)
2. Nigrosine solution
3. Clear glass slides
4. Staining tray

5. Inoculating loop
6. Bunsen burner / spirit lamp
7. Lens paper
8. Microscope

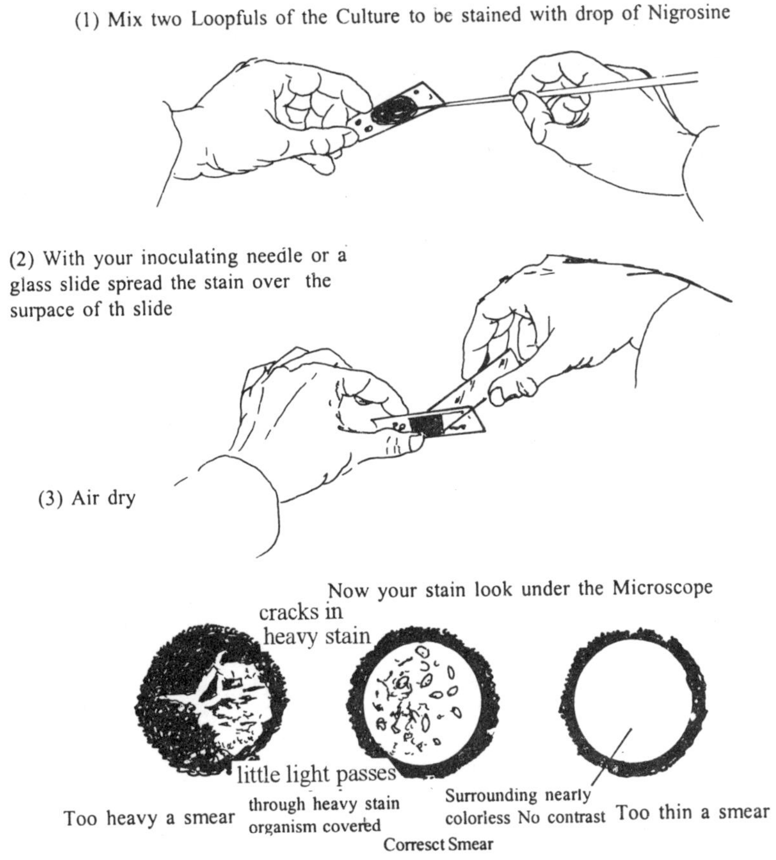

Fig. 24.2 The negative stain

Procedure :

1. Place a drop of nigrosine stain at one end of the slide.
2. Transfer a loopful of the microbial culture into the drop of stain. Allow it to mix thoroughly; stir it gently with the loop.
3. Take another clean glass slide and place its edge on the suspended microbe (in stain) at an angle of 30° (as shown in the figure) and spread it like a thin layer on the slide.
4. Allow the smear to air dry.

(**Note**: The smear should not form a thick layer. If it does, microscopic light will not pass through or on drying the stain may crack. In both cases observation is not possible. On the other hand too thin a smear will not allow sufficient contrast to observe the bacteria).

Observations : on observation under oil immersion lens of the microscope, spherical cells occurring singly as well as in clusters can be seen as transparent objects against a deep blue background.

STRUCTURAL STAINING

Many structural components of bacteria such as flagella, capsule, wall, spore etc cannot be distinguished by the staining methods explained above. Special staining procedures are necessary to visualize them and these are called structural stains. The principle of structural staining is same as that of differential staining. The structure (of bacteria) in question has a different degree of affinity to the stain so that it contrasts with other structures and can be easily identified based on colour reaction. In some cases the cell is treated with a mordant in order to increase the affinity of the structure to the dye. In case of wall staining the mordant is positively charged and it increases the adhering capacity of the negatively charged dye. In capsular staining either only the capsule is stained with the cell unstained or both capsule and cell are stained with different dyes with which they have affinity.

Structural stains are of the following types. Capsule stains, wall stains, cell membrane stains, flagella stains, endospore stains and nuclear stains.

Experiment no. 3 ***Staining of bacteria - Capsule staining***

Some bacteria have a gelatinous or mucilaginous covering external to the cell wall. This envelope is of different thickness depending on the bacterial species. This structure is called a capsule when round or oval in shape and a slime layer when it is irregularly shaped, of uneven thickness and loosely bound to the bacterial cell.

Capsule is partly synthesized by the cytoplasm and is composed of polysaccharides, glycoproteins or a polypeptide. Capsules play an important role in determining the pathogenicity of the species. Consequently capsulated forms are called virulent and non capsulated forms are called avirulent. Capsules are also important from immunological stand point. Capsule staining assists in the identification of pathogens in medical microbiology.

Purpose : To stain and identify the presence (or absence) shape, and thickness of the capsule.

Materials :

1. Culture of *Enterobacter, Klebsiella, Clostridium* etc (any one)
2. Congo red
3. Acid alcohol
4. Acid fuchsin
5. Staining tray
6. Inoculating loop
7. Glass slide
8. Blotting paper
9. Microscope

Procedure :

1. Transfer bacterial culture on to a clean glass slide with the help of inoculating loop.
2. Prepare a smear, allow it to air dry.
3. Apply congo red stain.
4. Wash with water

5. Add acid alcohol for 15 seconds
6. Wash with distilled water
7. Flood the smear with drops of acid fuchsin for about 60 seconds
8. Wash with distilled water
9. Dry the slides with a blotter

Observations : Microscopic examination reveals that the capsules are colourless surrounding the red cells against a dark blue background.

Experiment no. 4 *Staining of bacteria - Endospore staining*

In some of the bacteria, specially those like *Clostridium* and *Bacillus* certain special structures called endospores are produced. These are extremely resistant to heat, radiation, treatment of chemicals etc., that destroy other non spore forming bacteria. This resistance is primarily due to a thick wall - the spore coat. The spore coat does not take the stain easily. Considerable amount of heating is required in order to make the stain penetrate the spore coat. Once stained, the spores do not get decolorized easily. There are two staining procedures widely used by bacteriologists to stain the endospores. These are the Schaeffer-Fulton method and Dorner method.

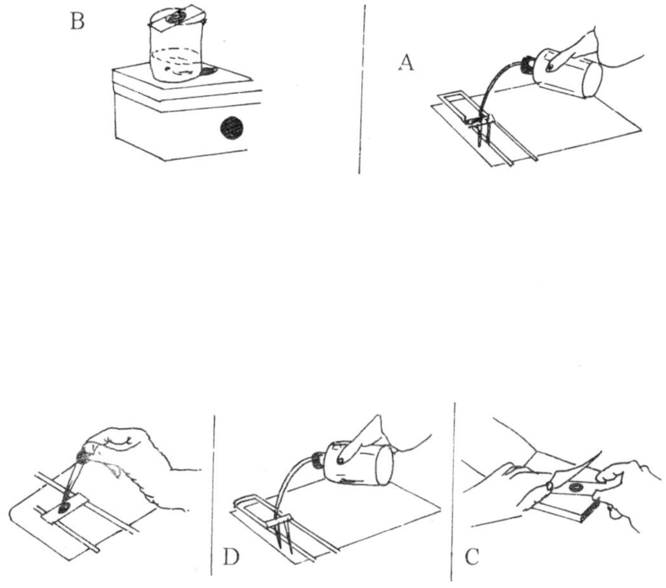

Fig. 24.3 The Schaeffer-Fulton spore stain method

A. Cover the smear with malachite green. Steam over boiling water for 5 minutes. Add additional stain if stain boils off.

B. After the slide has cooled sufficiently rinse with water for 30 seconds.

C. Counterstain with safranin for about 20 seconds.

D. Rinse briefly with water to remove safranin.

E. Blot dry with blotting paper and examine slide under oil immersion.

Purpose : To stain and observe the shape, size and contour of the spore by differential staining technique.

Schaeffer - Fulton method

Materials :

1. Cultures of bacteria - *Bacillus subtilis*
2. Malachite green (5% aqueous solution)
3. Safranin (5% aqueous solution)
4. Staining tray
5. Glass slides
6. Inoculation loop
7. Water bath / Spirit lamp
8. Microscope

Procedure :

1. Prepare the bacterial smear on a clean glass slide
2. Air dry the slides and heat fix the smears on to the slide by gently warming them
3. Cover the smears with the solution of malachite green
4. Keep the slide on a suitable stand and steam it from below for 5 minutes. If the stain dries up, add more stain to the smear from time to time.
5. Wash the slides gently under tap water.
6. Counter stain with safranin for about 30 seconds.
7. Wash the smear with distilled water.
8. Dry the slides with a blotter.

Observations : Under the oil immersion lens of the microscope observe the preparation. The spores (in Bacillus subtilis) stain green and the cells stain red. Each cell will be having one spore. The size, shape and the position of spore within the cell are the points to be observed carefully.

Dorner method

This method produces red coloured spore within a colourless sporangium (cell). Nigrosine is used to provide a dark background for contrast.

Materials :

1. Culture of bacteria (*Bacillus subtilis* or *Clostridium* spp)
2. Electric hot plate (Bunsen burner)
3. Carbol fuchsin stain
4. Nigrosine solution
5. Inoculating loop
6. Test tubes, beakers etc.

Procedure :

1. Prepare a heavy suspension of bacteria by taking several loopfuls in few drops of sterile water taken in test tube.
2. Add 5 drops of carbol fuchsin to the suspension.
3. Boil in waterbath the entire suspension (taken in test tube) for about 10 minutes. Cool the test tubes to room temperature.

4. Collect one or two loopfuls of stained bacterial suspension on to a clean glass slide and add a drop of nigrosine solution.
5. Spread the mixture on the slide with the edge of another clean slide (as done for negative staining).
6. Air dry the smear (do not heat)
7. Put a cover glass (in a drop of glycerine) and observe under the microscope.

Observations : Under the oil immersion lens of the microscope, the spores look red on taking the carbol fuchsin stain, the sporangium (cell) is transparent while the background (as nigrosine is a negative stain) is dark.

Experiment no. 5 ***Staining of bacteria - Flagella staining***

The motility of bacteria is brought about by long (upto 70mm) thread like structures called flagella. The flagella are thin proteinaceous structures that originate from cytoplasm and project out of the cell wall. In spite of their long length (which should be sufficient for a light microscopic observation) they are extremely thin (20nm or 0.01mm in diameter) and can be seen only under electron microscope. However with some special staining techniques flagella can be stained and made visible under light microscope. In order to do this a mordant is coated over the entire length of flagella to increase their apparent diameter and stained with a specific dye (with the advent of electron microscope more details of flagella can be observed than is possible by staining technique. However since use of electron microscope is not possible always this technique can be followed).

Flagellar staining is not an easy technique and requires careful handling of the material, controlled staining and suitable bacterial culture. There are three techniques of flagellar staining - Leifson's method, Lee's method and Rhode's method.

Liefson's method :

Materials

1. Bacterial culture (eg., *Proteus vulgaris*)
2. Liefson's stain
3. Clean glass slides
4. One ml distilled water blank
5. 95% alcohol
6. Inoculation loop

Procedure :

1. Take a clean glass slide and pass it through the blue portion of the flame of Bunsen burner several times.
2. When the slide is still hot, mark with a wax (China) marking pencil a thick line along the edge of the slide (see fig) leaving about an inch on one side for holding the slide.
3. Cool the slide to room temperature.
4. Take a loopful of bacterial culture. Add a drop of water.
5. Slowly tilt the slide to allow the bacterial culture to flow down and spread. Do not allow the culture to cross the wax boundary lines.
6. Allow the smear to air dry. Do not apply heat.
7. Cover the smear with few drops of Liefson's stain (the stain contains mordant also) and allow it to react for sometime (about 10 minutes).

8. Wash gently with distilled water to remove traces of excess stain.
9. Air dry the slide, and observe under the microscope.

Observations : On examination under the oil immersion lens of the microscope, the bacterial cells appear like pink rods with pink stained flagella. In *Proteus vulgaris*, cells are peritrichous (ie., flagella present all round the cells).

Lee's method:

Materials :

1. Culture of *Spirillum* or *Bacillus*
2. Osmic acid
3. Lee's stain (A and B solutions to be mixed at the time of staining)

Solution A	Magnesium sulphate	3.5gms
	Tannic acid	4.0gms
	Distilled water	55.0ml
	5% acetic acid	20.0ml

(Dissolve sulphate and tannic acid in warm distilled water. Add acetic acid a little later)

Solution B Basic fuchsin 1.5gm dissolved in 100ml of 95% ethyl alcohol.

4. Water bottles
5. Glass slides, cover slips
6. Canada balsam
7. Microscope

Procedure :

1. Place a loopful of bacteria on a very clean glass slide. Spread the drop into a smear.
2. Kill the bacteria with fumes of 2% osmic acid for about a minute. For this purpose invert the slide containing smear on a petridish having a few drops of osmic acid (Note : Osmic acid fumes are injurious to eye).
3. Air dry the slides.
4. Rinse the slides with water for about a minute, drain the water, rinse again and dry over a mild flame.
5. Flood the slides with stain (freshly mixed solutions A and B). Allow the stain to react for about a minute.
6. Drain the excess stain and air dry.
7. Wash in water and dry over a mild flame.
8. Mount the smear in canada balsam and observe.

Observation : On observation under oil immersion objective, the flagella can be clearly seen projecting out of the cell.

Rhode's method :

This method is mainly used for Spirochetes.

Materials :

1. Culture of Spirochetes
2. Rhode's mordant
3. Ammoniacal silver nitrate

4. Inoculating loop
5. Glass slides
6. Wash bottles, test tubes, pippètes etc.

Procedure :

1. Take a 24hrs old culture of Spirochetes and pippete out 2-3 ml of sterile distilled water into it.
2. Allow the bacteria in the agar culture to diffuse into distilled water.
3. Transfer the turbid suspension into another clean test tube containing 2-3 ml of 1% formalin. Hold the test tube against light and see whether it is opalescent. Add few more drops of formalin if necessary.
4. Wash the suspension in water, centrifuge and resuspend the deposit in 1% formalin.
5. Take a clean slide and place a loopful of the suspension at one end of it.
6. Tilt the slide so that the suspension flows down and forms a thin film.
7. Air dry the film (smear).
8. Add Rhodes mordant to the bacterial smear. Allow it to react for 3-5 minutes.
9. Wash thoroughly with water.
10. Heat the ammoniacal silver nitrate solution to boiling point and apply on the bacterial smear. Leave it for 3-5 minutes.
11. Wash gently with water.
12. Drain off water and blot dry.

Observation : On observation under the oil immersion lens of the microscope, the flagella can be clearly seen projecting out of the cells.

Experiment no. 6 *Staining of bacteria - Acid fast staining (Ziehl-Neelsen method):*

The acid fast stain is a differential stain that measures the resistance of a microbial cell to decolourizing acidic agents. Developed first by Paul Ehrlich in 1882 and later modified by Ziehl-Neelsen this method allows to measure the property of acid fastness by bacteria. Widely used for the identification of *Mycobacterium tuberculosis* and *M.leprae*, in this staining procedure bacteria are classified into acid fast if they retain the primary stain (carbol fuchsin) after washing with string acid and non acid fast if they get decolorized after washing with the acid. Non acid fast bacteria retain the counter stain methylene blue.

The property of acid fastness is correlated with a high lipid (mycellic acid) content in the cell wall. In acid fast staining, bacteria are first stained with the primary stain (carbol fuchsin), decolourized with acid alcohol and counter stained with methylen blue. Used primarily for the identification of tuberculosis and leprosy bacilli, acid fast stain serves also as a good identification tool for a number of harmless saprophytic bacteria.

Purpose : To classify bacteria into acid fast and non acid fast based on their affinity to carbol fuchsin dye.

Materials :

1. Culture of *Mycobacterium* (better to select a non pathogenic species *M.smegmatis*) and *Staphylococcus aureus* or *E.coli*.
2. Acid fast stain reagents
   a. Carbol fuchsin
   b. Acid-alcohol (3% HCl+2.5HNO_3+95% alcohol)
   c. Mehylene blude

3. Staining rack
4. Hot plate/water bath
5. Wash bottles
6. Distilled water
7. Glass slides
8. Bunsen burner
9. Inoculating loop
10. Blotting paper
11. Microscope

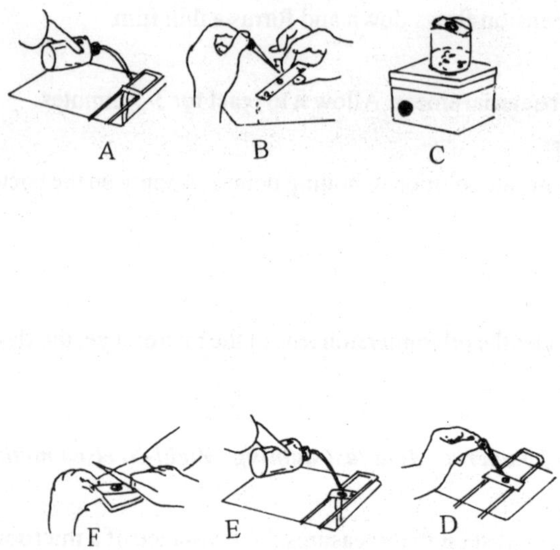

Fig. 24.4 Ziehl-Neelsen acid-fast staining procedure

Procedure :

1. Prepare separate smears of *M.smegmatis* and *S.aureus* on two slides.
2. Allow the smears to dry in air and fix with heat.
3. Heat the slides with a jet of steam coming from water bath for about 5-10 minutes.
4. Stain the smears with drops of carbol fuchsin. Take care not to dry the smear. Keep adding stain to prevent drying.
5. Cool the slides and wash with distilled water.
6. Decolourize the smears in acid alcohol for about 10-30 seconds until they are faint pink coloured.
7. Rinse with distilled water.
8. Counter stain with methylene blue for about 30-90 seconds. No heating is required.
9. Wash with distilled water.
10. Blot dry with blotting paper carefully. Do not wipe the slide.

Observation : Cells of *M.smegmatis* appear red coloured and are recognised as acid fast as they have retained the acid fast stain while cells of *S.aureus* appear blue and are recognized as non acid fast as they have retained only the counter stain and not the primary stain.

Experiment no. 7 ***Staining of bacteria - Acid fast staining (Fluorescence method):***

Flurochrome acid fast staining (fluorescence method) is used in conjunction with Ziehl-Neelsen method when a large number of samples are to be tested. Fluorescent technique has an advantage over the Ziehl-Neelsen method in the sense that the slides can be examined under a lower magnification (60x or less), while a Ziehl-Neelsen prepared sample has to be observed under oil immersion. This technique can be used for rapid screening of acid fastness because microbes appear iridescent. For confirmation of positive samples, detailed observation may be made using Ziehl-Neelsen technique.

In flurochrome staining, the smears are stained with auramine-rhodamine for 20 minutes and subsequently decolourized with acid alcohol. Counter staining is done using potassium permanganate. As soon as the slides are dry, they are to be examined under a fluorescence microscope. Acid fast bacteria appear as yellow orange fluorescent rods in a dark field. Areas that are fluorescent can be examined more carefully under oil immersion objective.

Purpose : To distinguish acid fast and non-acid fast bacteria using flurochrome acid fast staining.

Materials

1. Culture of *Mycobacterium phlei* and *Staphylococcus aureus*.
2. Auramine-rhodamine stain
3. Acid alcohol (2.5 Hcl in 70% ethanol)
4. Potassium permangamate (0.5% solution)
5. Glass slides
6. Inoculating loop
7. Wash bottles
8. Bunsen burner
9. Fluroescent microscope

Procedure :

1. Prepare smears (separately or mixed) of *Staphylococcus aureus* and *Mycobacterium phlei*. If preparing a mixed smear on the same slide, the inoculum should be vigorously agitated to disperse both the organisms.
2. Allow the smear to air dry before heat fixing.
3. Flame fix the slide over Bunsen burner. Avoid overheating the smear. A slide warmer also may be used.
4. Cover the smear with auramine-rhodamine stain for about 20 minutes (at room temperature).
5. Rinse with water to remove excess stain.
6. Decolorize with acid alcohol for about 7-10 seconds.
7. Rinse thoroughly with water.
8. Cover the smear with potassium permanganate solution and allow it to react for 3 minutes. This solution actually eliminates background fluorescence even though it is called counter staining. Excessive counter staining is to be avoided as otherwise complete fluorescence is eliminated.
9. Rinse with water and dry the slide with a blotter (or air dry).

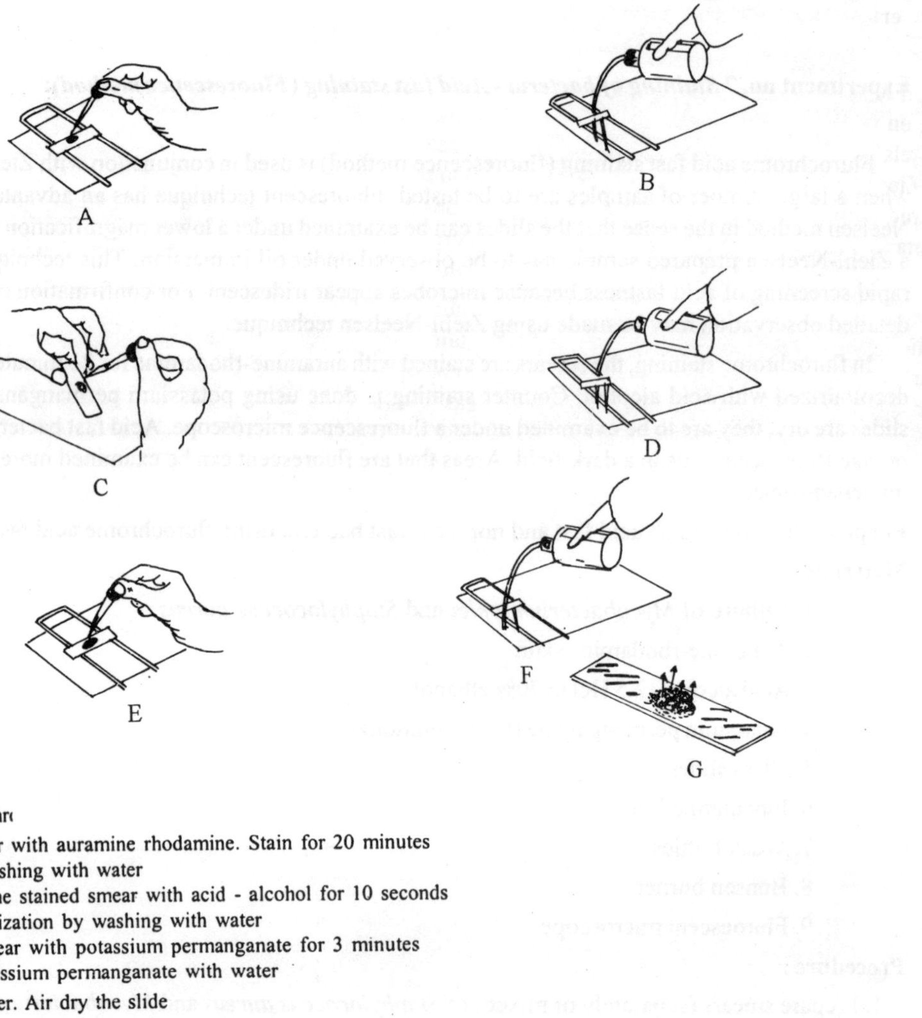

Fig. 24.5 Fluochr

A. Cover a smear with auramine rhodamine. Stain for 20 minutes
B. Destain by washing with water
C. Decolourize the stained smear with acid - alcohol for 10 seconds
D. Stop decolourization by washing with water
E. Cover the smear with potassium permanganate for 3 minutes
F. Wash off potassium permanganate with water
G. Shake off water. Air dry the slide

Observation :

Using a fluorescent microscope the slides should be scanned with low/high power (use a cover glass if necessary) objective. When fluroescent areas are located, change to oil immersion for a detailed observation. *M.phlei* appear as fluorescent rods while *S.aureus* cocci are not fluorescent. *M.phlei* may be recognised as acid fast.

Staining of fungi

For the purpose of microscopic observation fungal mycelia have to be stained with suitable dyes. There are two staining techniques for the study of fungi. In the lactophenol dye staining the cytoplasm of the mycelia are stained. While in the Giemsa dye staining the nucleus of the fungal mycelia are stained. Both these techniques are described below.

Experiment no. 8 ***Staining of fungi - Lactophenol cotton blue staining***

Lactophenol cotton blue is a commonly used stain for making temporary preparations of fungal cultures. The dye stains the cytoplasm giving a light blue tint against which the hyphal walls are clearly demarcated.

The lactophenol cotton blue stain consists of the following four constituents.

1. Phenol - a fungicide.
2. Lactic acid - functions as a cleaning agent.
3. Cotton blue - stains the cytoplasm.
4. Glycerin - serves as a medium for temporary preparation.

Materials :

a. A fresh culture of a suitable fungus
b. Lactophenol cotton blue stain
c. Needles
d. Glass slides and cover slips
e. Spirit lamp or bunsen burner
f. 70% alcohol
g. Nail polish
h. Compound microscope

Procedure :

1. Take a clean glass slide and put a drop of the stain in the center.
2. Transfer a small part of the fungal culture consisting of mycelia, spore bearing organs etc on to the stain with the help of a needle.
3. Using two needles very gently tease the material so that the mycelia and hyphae get separated.
4. Gently mix the stain with the fungal mycelia.
5. Carefully cover the preparation with the help of a clean cover glass. Avoid air bubbles.
6. This preparation may be temporarily sealed so that it can be kept for observation for a very long time. This sealing is done with the help of nail polish as follows.

Gently heat the slide on a low flame; avoid too much heating, remove air bubbles if any. With the help of a cheese cloth remove any excess stain on medium from around the cover slip. If necessary use a blotter. Apply a thin layer of nail polish uniformly all around the edge of the cover slip. Allow the preparation for drying over night. Apply a second coat of nail polish over the first coat. Allow it to dry, the slide is now ready for microscopic observation.

Experiment no. 9 *Staining of fungi - Nuclear staining*

A number of stains which selectively react with the nucleic acids are used to stain the nuclei of fungi. Some of the common stains used for this purpose are Haematoxylin, Giemsa, Feulgen and Carmine. In this experiment the method of nuclear staining using Giemsa is explained below.

Materials

1. Suitable stain, phosphate buffer (6.9pH),
2. Alcohol (70%, 95%)
3. One molar HCL
4. Acetic acid

5. Lactic acid
6. Euparol
7. Distilled water
8. Slides and cover glass
9. Petriplates
10. Wash bottles
11. Needles
12. Bulsen burner
13. Water bath
14. Microscope
15. Suitable fungal culture

Procedure :

1. Culture the fungi on a thin agar film placed on clean glass slides. Keep the glass slides in a petri dish containing moist filter paper. Fix the spores and mycelia on to the slide in a 3:1 mixture of alcohol and acetic acid or in a mixture 3:1:1 alcohol, acetic acid and lactic acid.
2. Wash the preparation gently in 95% ethyl alcohol.
3. Transfer the preparation on to a medium (petri dish) containing 70% alcohol.
4. Remove the slides from the alcohol and gently wash them in distilled water.
5. Dip the slides for 5 minutes in 1M HCl at room temperature.
6. Transfer the slide on to 1m HCl on to a water bath (60°C) and keep it there for 7 minutes.
7. Remove the slide from water bath, gently wash in water and then in phosphate buffer (repeated washes are necessary).
8. Transfer the slide on to the Giemsa stain kept in a suitable container. (Stain should be prepared in phosphate buffer at the rate of 1:15)
9. Keep the slide in the stain for 2 hours.
10. Remove the slides and rinse in phosphate buffer followed by distilled water.
11. Air dry the slide at room temperature.
12. Add one drop of euparol on to the stained culture and cover it with cover glass.

Observation :

1. Observe the slides under the microscope using high power as well as oil immersion objective. The nuclei are stained rose or pink surrounded by colorless or light blue colored cytoplasm.

Experiment no. 10 ***Staining of Actinomycetes***

Actinomycetes are gram positive microbes and they are characterised by the development of a branched filamentous body. Their cellular characters are similar to bacteria but the nature of the body is like that of fungi. Hence they are often refered to as mold like bacteria.

Actinomycetes can be cultured in the laboratory and observed after staining for the morphological details.

Materials :

1. Culture of a suitable actinomycete (*Streptomyces*)
2. Bismark brown stain
3. Toluidine blue stain
4. Ammonium sulphate
5. Sterile cellophane
6. Mounting medium (Canada Balsam)
7. Clean glass slides
8. Distilled water
9. Inoculating needles
10. Bunsen burner
11. Microscope etc.

Procedure :

1. Obtain the culture of *Streptomyces* on the surface of sterile cellophane placed on a suitable culture medium like nutrient agar.
2. Incubate the culture at 30°C until the development of spore producing structures.
3. Carefully take out the cellophane containing the microbial culture from the agar surface.
4. Carefully transfer a portion of the culture on to a clean glass slide and proceed for staining.
5. Mix Bismark brown and Toluidine blue along with ammonium sulphate in the following proportion
   a. 2 parts of Bismark brown solution (0.1% W/V)
   b. 2 parts of Toluidine blue solution (0.1% W/V)
   c. One part of saturated ammonium sulphate solution
6. Stain the microbial culture with the above mixture of stain for about 2-3 minutes. Allow the stain to react with the culture.
7. Gently rinse the slide in distilled water.
8. Remove excess water and allow it to air dry.
9. Mount the stained culture in Canada balsam.
10. Carefully put a cover glass over the preparation.
11. Observe under different magnifications of the microscope.

Observations

The preparation may be observed under high power objective and later under oil immersion objective. Note the colour of the mycelium as well as the spore producing structures.

Results

The vegetative mycelium of the *Streptomyces* takes a light yellow colour and the spores take a blue colour. Also present in the cytoplasm are red coloured granules in the aerial hyphae. The colony of *Streptomyces* appears rather rough and granular. The hyphae of the microbe are coenocytic and branched.

Experiment no. 11 *Staining of Spirochetes*

Spirochetes are helical bacteria and possess a periplasmic flagella. There is no cell wall. Spirochetes have variety of sizes and shapes. They can be obtained from natural sources like the teeth of human beings where they are harmless. In the laboratory spirochetes may be stained and observed following Fontana's method.

Materials

1. Fontana's fixative fluid
2. Ammonical silver nitrate
3. Fontana's mordant
4. Ethyl alcohol
5. Canada Balsam
6. Inoculating loop
7. Bunsen burner
8. Microscope etc.

Procedure :

1. With the help of a sterile inoculating needle gently scrape a little material from the teeth.
2. Prepare a smear of the material on a clean glass slide and allow the smear to dry.
3. Pour Fontana's fixative gently on the smear and allow it to react for about 30 seconds. Remove the fixative and apply the fixative again for the next 30 seconds.
4. Repeat step 3 for the third time.
5. Remove the fixative and treat the slide with absolute alcohol for 3 minutes. At the end of this period remove the alcohol.
6. Dry the slide gently on the bunsen burner.
7. Treat the slide with Fontana's mordant, heat the slide and allow the warm mordant to react with the smear for about 30 seconds.
8. Wash the slides with distilled water.
9. Air dry the slide.
10. Cover the preparation with ammonical silver nitrate solution.
11. Steam the slide for 30 seconds on a boiling water bath.
12. After steaming the smear gets a brown colourization.
13. Wash the preparation gently with distilled water and dry it with the help of blotting paper.
14. Place a drop of Canada balsam on the smear and cover it with cover glass.
15. Observe the preparation on the oil immersion object of the microscope.

Results

Brown coloured helical cells can be seen against a light yellow background.

Experiment no. 12 *Staining of Mycoplasmas (PPLO)*

Technically called mollicutes the mycoplasma were first discovered during the study of cattle suffering from Bovine pleuroneumonia. Mycoplasma are wall less similar to the L forms of bacteria. They can be stained and observed in the laboratory according to the following procedure.

Materials

1. Culture of PPLO

2. Methylene blue 2.5 gms
3. Azurell 1.25 gms
4. Maltose 10 gms
5. Sodium carbonate 0.25 gms
6. Distilled water 100 ml
7. Slides
8. Microscope etc

Procedure

1. PPLO stain can be prepared by mixing methylene blue, azurell, maltose and sodium carbonate (proportion as given above) in 100 ml of distilled water. Allow the stain to settle.
2. Place a drop of the stain on a clean cover slip and allow it to dry.
3. Carefully cut a small block of agar medium with PPLO colonies on it. (Use a sterile scalpel for this purpose)
4. Place the agar block with the colonies at the centre of a clean glass slide. Let the colony side of the medium be facing up.
5. Carefully put the cover glass over the PPLO colonies.
6. Seal the preparation with paraffin.
7. Observe under the microscope.

Results

Mycoplasma colonies have a dense blue center and a light blue peripheral region.

CHAPTER 25

TYPE STUDIES I - BACTERIA

Microscopic observation of microorganisms like bacteria and others is of great help in understanding their structural details. Microbes are observed under the microscope both in the living condition as well as in the dead condition. The living cells are observed to find out their natural size and shape, their ability of locomotion and response to the various environmental factors. In the other technique the cells are killed, fixed and stained and observed into the microscope.

Live specimens of microbes can be observed by the following two methods -

Wet mount method and hanging drop method. Both these methods are useful for determining the various parameters of the living cells.

Experiment no. 1 ***Wet mount technique for observation of living microbes***

Materials

1. Culture of suitable microbes such as bacteria/microalgae/protozoa etc.
2. Slides and cover slips.
3. Pipette

Procedure

1. Carefully transfer a little quantity of the microbe on to the slide using the pipette.
2. Separate pipettes are to be used for each microbial culture.
3. Cover the microbial preparation with a thin cover glass.
4. Press the cover slip gently with the blunt end of needle or pencil.
5. Observe under the microscope.

Results

1. Many microbes which are pigmented can be easily seen and even their motility may be observed. Record your observations.

Experiment no. 2 *Hanging drop technique for demonstrating bacterial motility*

The hanging drop technique is a very useful method for demonstrating the structural and other features of non stained preparations of bacteria.

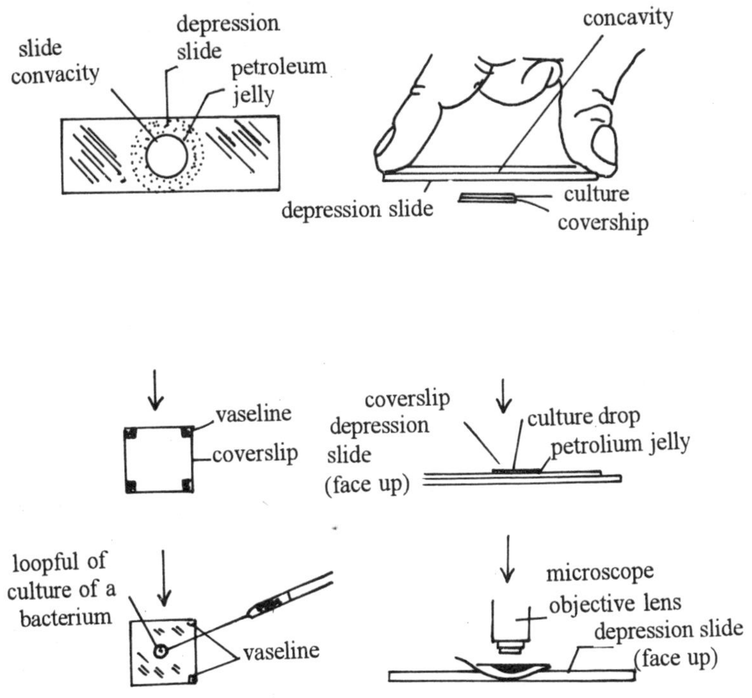

Fig.25.1 Hanging drop technique for absorbing bacterail motility

Materials

1. 12 hour old broth culture of suitable bacteria (*Proteus vulgarise*)
2. Cavity slide and cover glass.
3. Vaseline or petroleum jelly.
4. Match sticks.
5. Glass marking pencil.
6. A good microscope.

Procedure

1. Take a cavity slide (hanging drop slide), thoroughly clean it and flame it on either spirit lamp or bunsen burner.
2. Keep the slide on the table with the cavity facing up.
3. Gently smear a little vaseline or petroleum jelly around the cavity of the slide.
4. Take a clean cover slip and apply vaseline on each of the four corners using a match stick.
5. Keep the cover glass on a clean paper with the vaseline smeared slide up.
6. Gently transfer a little quantity of the culture on to the center of the cover slip. Donot take too much of the culture as the excess quantity would fill the cavity of the slide.

7. Take the cavity slide and keep it on the bacterial culture on the cover slip in such a way that the culture would be in the center of the depression of the cavity slide.
8. Press the slide gently to seal it with the cover slip.
9. Lift the preparation and turn it in the reverse direction in such a way that the bacterial culture would form a hanging drop from the cover slip into the cavity of the slide.
10. The bacterial culture forms a suspended drop (hanging drop) into the cavity of the slide.
11. Observe the preparation under low power view of the microscope.
12. Observe also under high power and oil immersion objectives.

Observation and results

Bacterial cells of different sizes and shapes can be observed. Flagellated bacteria can be seen moving to and fro across the field of the microscope.

Experiment no. 3 ***Study of characters of Staphylococcus***

Staphylococci are gram positive, they are spherical, ovoid, non motile organisms arranged usually in groups. On a nutrient agar medium they form colonies which vary in color from white, yellow to golden yellow.

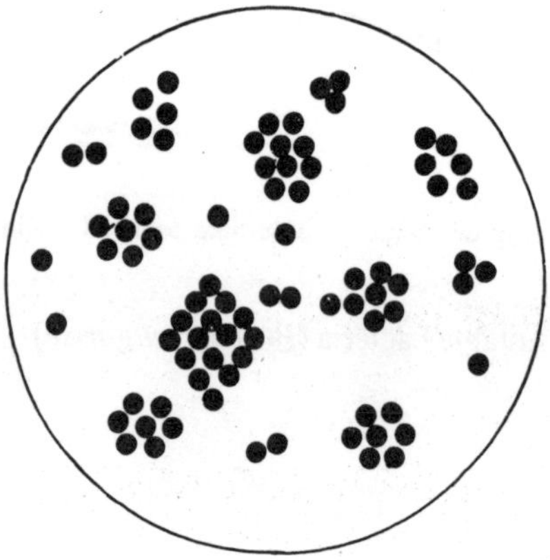

Fig. 25.2 Staphylococcus

On the basis of pigmentation three types of *Staphylococci* have been identified these are -

1. *Staphylocccus aureus* - produces golden yellow colonies and pathogenic.
2. *Staphyloccus albus* - produces white colonies and non pathogenic.
3. *Staphyloccus citrus* - produces lemon yellow colonies and non pathogenic.

S.aureus cells are non motile and they do not form spores. They are gram positive and have a size of around 0.72 - 1.2 mm. The cells form characteristic clusters looking like a bunch of grapes. This appearance is due to incomplete wall formation after successive cell divisions. Cell divisions here take place in perpendicular planes. Some of the strains of *S.aureus* possess capsule which can be demonstrated by India ink staining.

The colonies of *S.aureus* are sharply demarcated, round, sometimes convex and have a diameter of 4mm. The surface of the colony is smooth and glistening. On nutrient agar the colonies look like oily drops. Biochemically *S.aureus* is capable of producing catalase.

Another species - *S.albus* can be cultured on nutrient agar and the colonies are white in color.

Staphyloccus citrus is a saprophyte and forms lemon yellow colored colonies on nutrient agar. It does not produce any toxin or coagulase.

Experiment no. 4 ***Study of characters of Streptococcus***

The Streptococci are gram positive cells, spherical in shape and are arranged in chains. The cells are non motile and non sporulating. Streptococci are important human pathogens and can be cultured on media enriched with blood or serum.

Streptococci are classified on various criteria such as morphology, cultural characters, biochemical reactions and antigenic structure.

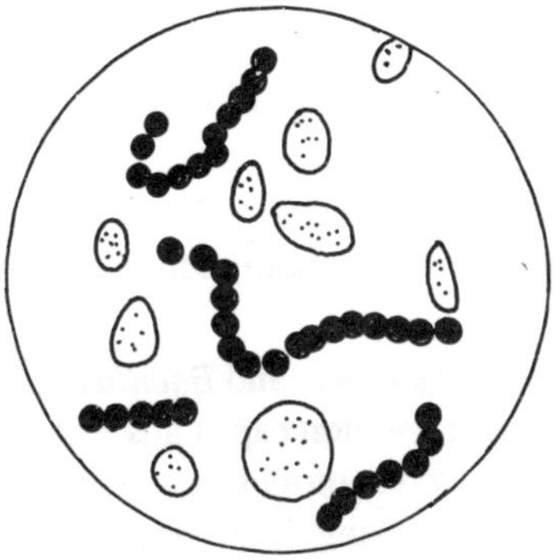

Fig. 25.3 Streptococcus

Streptococcus pyogenes is one of the most important human pathogens. The cells are arranged in chains and each cell is 0.5 - 1mm in diameter. Chains are formed due to cell division in a single plane and the failure of daughter cells to separate after division. The length of the chain depends on the type of medium on which the microbe is cultured. The cells are capsulated.

Culturally *Streptococcus pyogenes* is a facultative anaerobe and grows luxuriantly at a temperature of 37°C. The cells grow very well on enriched media such as serum broth, blood agar etc.

Biochemically *Streptococcus pyogenes* can ferment lactose, glucose, maltose etc producing acid but no gas. It is catalase negative. It does not liquify gelatine, the cells can be killed by heat at 56°C for about 80 minutes.

Streptococcus pyogenes causes a number of diseases in human beings like respiratory infection, skin infection, infection of the genital tract etc.

Experiment no. 5 ***Study of features of Bacillus***

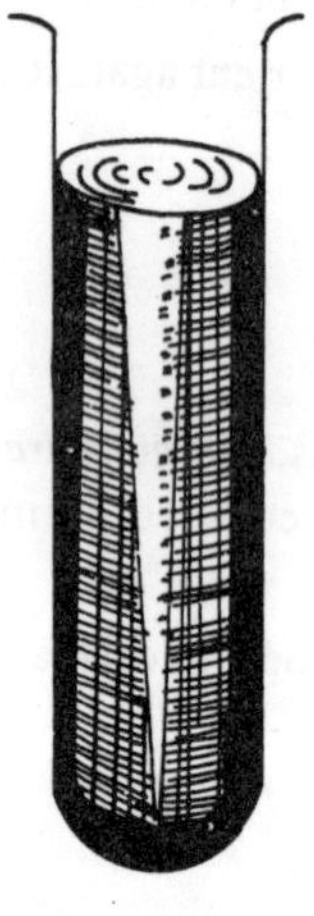

Fig 25.4 Gelatin stab showing

Bacilli are rod shaped, produce endospores and are gram positive. There are two types of bacilli - aerobic bacillus and anaerobic Clostridia.

Among the aerobic bacillus the Anthrax bacillus (*Bacillus anthracis*) is the most important. This bacillus is a parasite inhabiting cattle and sheep. Infection in man is very rare. It is mostly accidental due to contact with infected animals.

The cell is non motile, non acid fast and gram positive. The size of the cell is 1 x 3-4 mm. The cells may occur singly or sometimes grouped into short chains. The entire chain is surrounded by a polypeptide capsule. Sporulation is seen only when the bacillus is in the soil. Oxygen is necessary for sporulation.

Bacillus anthracis can be cultured either on nutrient broth or on agar plate. It may be also cultured on blood agar, gelatine stab and on some selective media consisting of EDTA. An optimum temperature of 37°C is suitable for vegetative growth. Sporulation occurs between 25°C - 30°C.

On agar plate the colony is irregular, round, 2-3 mm in diameter. Other colony characters include raised dull opaque appearance with greyish white color. On higher magnification the colonies appear like a tangled mass of long hairs looking like a cural (Barrister wig). On blood agar the colony does not cause haemolysis.

The vegetative cells can be killed at 56°C for 30 minutes. The spores are viable for many years in a dry state and can survive in 5% phenol for weeks. Spores are usually central in position in the cell but occasionally may be sub terminal also.

Other species of bacillus include *Bacillus subtilus* and *Bacillus cereus. B.subtilus* is a gram positive rod occurs either singly or in chains. It is motile and non capsulative. It can be cultured on blood agar and has haemolytic properties. It may be also cultured on nutrient broth and agar. *B.cereus* causes food poisoning. It can grow in stale food producing entrotoxin causing diarrhoea.

Experiment no. 6 ***Study of features of Clostridium***

Clostridium belongs to the family Bacillaceae and is non aerobic. The cells are gram positive, sporulating and generally spindle shaped but the cells are highly pleomorphic. The cells are motile with Peritrichus flagella. There are three important species of *Clostridium.* These are *Clostridium tetani, Clostridium perfringens, Clostridium botulinum.*

Morphologically the cells have a size of 4.8 x 0.5 μ. They show considerable variation in length. The spores are spherical, terminal and bulging. This gives the appearance of a drumstick to the cells. The cells are non capsulated and motile.

Clostridium can be cultured on the medium in the absence of oxygen. There are no Cytochrome pigments and as such oxygen cannot be used as an nitrogen acceptor. *Clostridium* tetani can be cultured on a variety of media such as cooked meat medium, nutrient agar medium, blood agar medium and lactose egg yolk agar medium. *Clostridium tetani* cannot ferment any sugar while *Clostridium botulinum* can ferment glucose and maltose.

The genus *Clostridium* is a highly pathogenic one. *Clostridium tetani* causes tetanus. While *Clostridium botulinum* causes food poisoning. *Clostridium perfingens* causes gas gangrene.

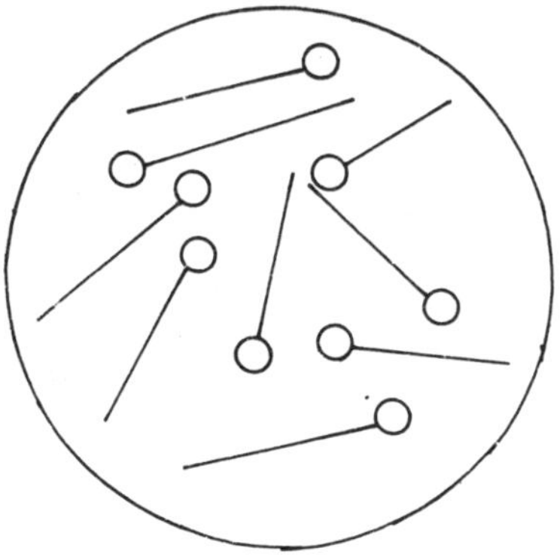

Fig. 25.5 Clostridium tetani

Experiment no. 7 ***Study of features of Lactobacillus***

Lactobacillus is a gram positive rod. The cells are mostly non motile. They are elongated and sometimes appear bent. The cell walls of *Lactobacilli* consists of peptido glycans and polysaccharides. In some cases even teichoic acid and lipids are also present.

Lactobacilli are normally present in the mouth. Some species like *L.acidophilus* may be present in the intestine where it helps to produce vitamin B_{12}. Lactobacilli can be cultured on a medium containing inorganic salts, inorganic nitrogen, amminoacids and organic carbon.

Lactobacilli find importance in dairy industry where they help in the manufacture of cheese and other milk products.

Lactobacilli can be obtained from butter milk and curds for laboratory studies.

Experiment no. 8 ***Study of features of Escherichia coli***

Escherichia coli is a parasite living in the intestine of human beings and other animals.

Morphologically the cells are gram negative, non capsulated. The cells are short and plumpy with a size range of 2 - 4 μ x 0.4 - 0.7 μ. Escherichia coli do not form spores and they are also non motile. The cells are aerobic but occasionally may be facultative anaerobes growing on simple medium. The colony grows profusely at an optimum temperature of 37°C. *Escherichia coli* can be cultured on liquid broth, nutrient agar, blood agar etc. The colonies appear 12-18 hours after incubation on a solid medium. The colonies are circular 1-3 mm in diameter, smooth and colorless. The edge of the colony is entire. On blood agar some strains may show beta haemolysis.

Escherichia coli can ferment a variety of sugars such as lactose, glucose, sucrose, maltose etc producing acid and gas. The cells are highly resistant and survive for months in soil and water. They cannot withstand a temperature of 60°C for more than 20 minutes. The cells are sensitive to antibiotic such as streptomycin, tetracycline etc.

Escherichia coli may sometimes become pathogenic causing diarrhea.

CHAPTER 26

TYPE STUDIES II- CYANOBACTERIA

The members of the class Myxophycae can be identified by the following general characters.

1. Members are blue green in colour due to specific pigmentation.
2. Plastids are absent.
3. There is no motile phase in the life history.
4. Sexual recombination as understood in other higher members is absent. But genetic recombination has been reported in several members.
5. The cell interior is divided into an outer coloured chromoplasm and an inner colourless centroplasm representing the incipient nucleus. (There is no nuclear membrane).

Experiment no. 1 ***Study of external features of Nostoc***

CYANOBACTERIA
NOSTOCALES
NOSTOCACEAE
NOSTOC

Common species: *N. calcicala, N. vommunr, N. sphaericum*

Distribution: Fresh water ponds and ditches. Form mucilagenous balls. Often found on wet rocks, also found in symbiotic association with *Anthoceros, Cycas* and *Lichens.*

Study: 1. External features
2. Reproductive structures
3. Systematic position and identification

External features

1. Many hundreds of filaments are enclosed in the form of a mucilage ball.
2. The ball like colony has a thick mucilage envelop and may be coloured.
3. The plant body is a filament and is enveloped by a mucilage sheath.
4. The filaments are unbranched and do not have a base or apex.
5. Each filament is made up of a number of bead like cells joined end to end.

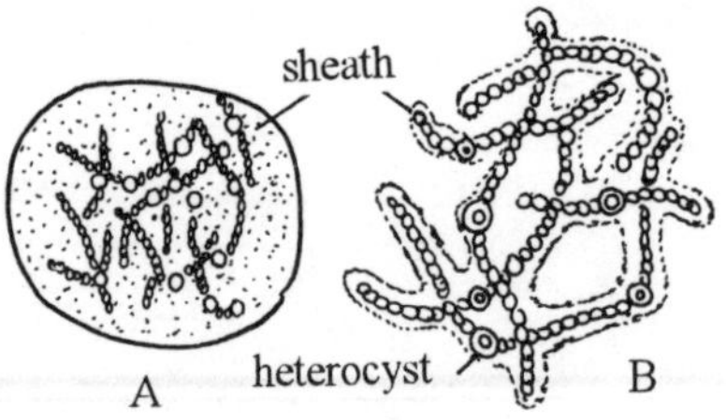

Fig. 26.1 Nostoc A. Colony under low power, B. Slightly enlarged colony

6. Here and there in the filament certain thick walled cells are seen. These are called heterocysts and may be intercalary or terminal in position.
7. Heterocysts are thick walled and pale yellow coloured. Terminal heterocyst has a single polar nodule, while intercalary heterocysts have two polar nodules.
8. Vegetative cells have a typical cyanophycean structure. Protoplast having centroplasm and chromoplasm. Cell wall is cellulosic.

Reproductive features

1. Sexual reproduction has not been reported.
2. Asexual reproduction is by the formation of hormogones and akinetes.
3. Hormogones are fragments of the filament, that generally separate from the main filament near a heterocyst.
4. Hormogones develop into new individuals after separation from the main filament.

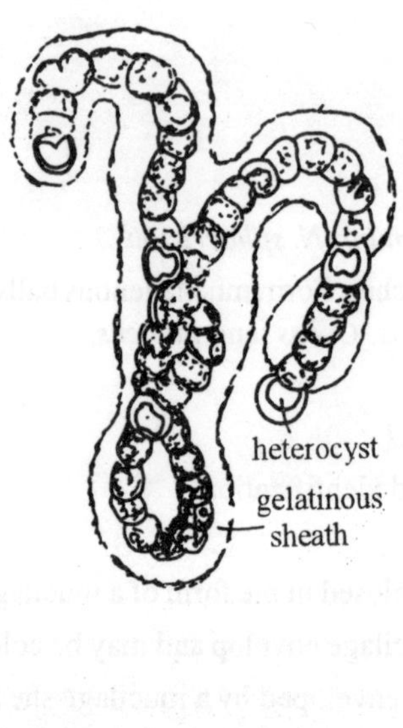

Fig. 26.2 Nostoc
Filaments showing the heterocysts and thick gelantinous sheath

5. Akinetes are thick walled cells rich in reserve food materials. They separate from the plant and develop into a new individual.
6. Occasionally heterocysts (*N. commune*) may also germinate to form new individuals.

Experiment no. 2 *Study of external features of Anabaena*

CYANOBACTERIA
NOSTOCALES
NOSTOCACEAE
ANABAENA

Common species: *A. cycadacearum, A. gelatinicola, A. iyengarii* etc.

Distribution: Commonly found in wet places, ponds, rice fields etc. Some of them could be semiterrestrial. A few species are symbiotic with the roots of higher plants.

Study: 1. External features
2. Reproduction
3. Systematic position and Identification.

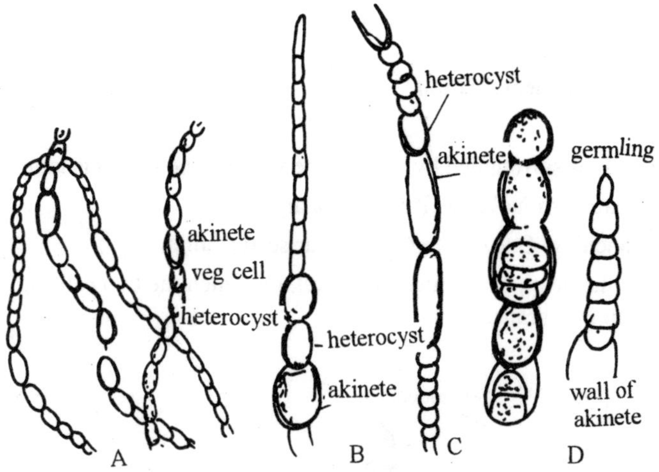

Fig. 26.3 Anabena, A,B,C. Filaments with heterocysts and akinetes, D. Germination of akinetes.

External features

1. Plant body is filamentous. The filament is unbranched. The filaments may or may not be in groups.
2. There is no mucilage sheath covering the filament.
3. The filaments are uniformly broad throughout their length.
4. Cells are subspherical or barrel shaped and have a two-layered wall.
5. Cell structure is typical to bacteria with outer chromoplasm and middle centroplasm.
6. Filament consists of terminal and intercalary heterocysts. Terminal heterocysts have one polar module, while intercalary heterocysts have two polar nodules.

Reproduction

1. Sexual reproduction has not been reported.
2. Asexual reproduction is usually brought about by Akinetes.
3. Akinetes are thick walled cells, much larger in size than the vegetative cells. They occur either singly or in a row of 4 - 6. They may or may not be adjacent to a heterocyst. Akinetes germinate into a new filament.

Experiment no. 3 ***Study of external features of Scytonema***

CYANOBACTERIA
HORMOGONALES
SCYTONEMATACEAE
SCYTONEMA

Common species: *S. vivulave, S. simplex,, S. iyengarii, S.varuim* etc.

Distribution : Aquatic or semiterrestial. Found on the bottom of ponds, pools etc. or on damp soil. Occurs also in symbiotic association in lichens.

Study : 1. External features (including branching)
2. Reproduction.
3. Systematic position and identification

External features

1. The filamentous plant body is surrounded by a mucilage sheath. The sheath is lamellated.
2. Hetrocysts present. They are terminal as well as intercalary.
3. Filament shows false branching. One or two branches arise at a point.
4. Branching results due to the death of an intercalary cell, on which the two cells on either side resume apical growth and develop into branches.

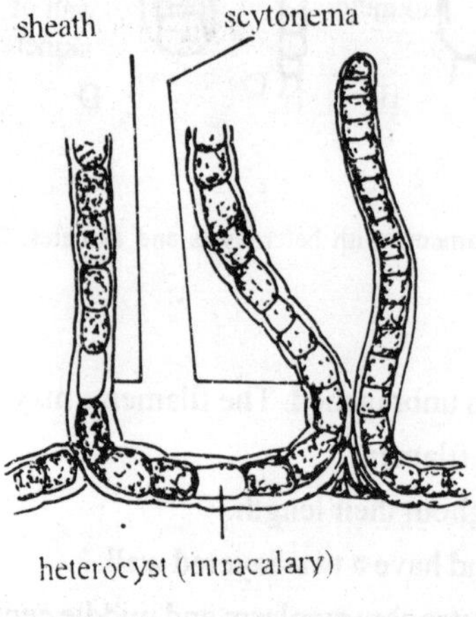

Fig.26.4 Scytonema. Habit.

5. Heterocysts are pale yellow in colour and have one or two polar nodules depending on their location.
6. Cell structure is typically cyanophycean. It has a central centroplasm, surrounded by chromoplasm.

Reproduction

1. Sexual reproduction has not been reported.
2. Asexual reproduction brought about by means of hormogones, which are formed between two separation discs.
3. Akinetes have been reported in some species.

Diagrams to be drawn

1. Habit sketch
2. One or two filaments enlarged
3. A portion of the filament at the region of branching
4. Cell structure

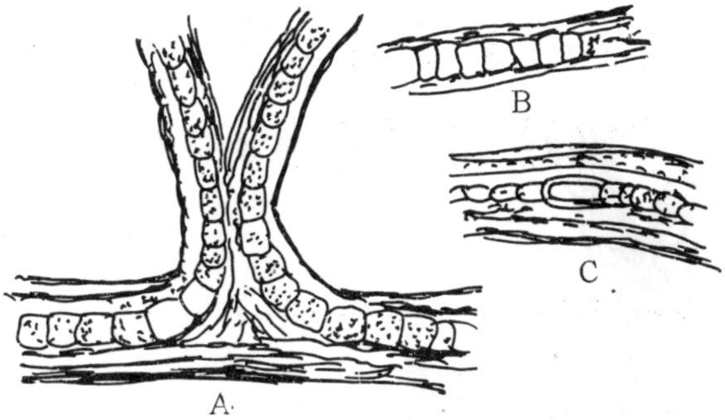

Fig. 26.5 Scytonema, A- A portion of the plant showing false branching, B- A few cells covered with mucilage, C- A part of the branch showing heterocyst

Experiment no. 4 ***Study of external features of Microcystis***

CYANOBACTERIA (CYANOPHYCEAE)
CHROOCOCCALES
CHROCOCCACEAE
MICROCYSTIS

Common species : *Microcystis flos-aquae* and *M.aeruginosa.*

Distribution : Found in ponds, lakes, tanks, rivers and stagnant rain water pools. Occasionally found in salt lakes.

Study : 1. External features.

2. Reproductive structures.

External features :

1. The plant body is unicellular.
2. Members aggregate themselves into colonies.
3. The colonies may be shapeless, spherical, ellipsodial or elongated.
4. The cells are spherical in shape occasionally may be elongated also.
5. A common mucilaginous matrix packs all the cells together.
6. The mucilage sheath is hyaline and can be demonstrated only with the help of special stains.
7. The cell structure is typical to the cyanobacterial cell.
8. Numerous gas vacuoles are found in the cell.
9. Gas vacuoles are irregular in shape.

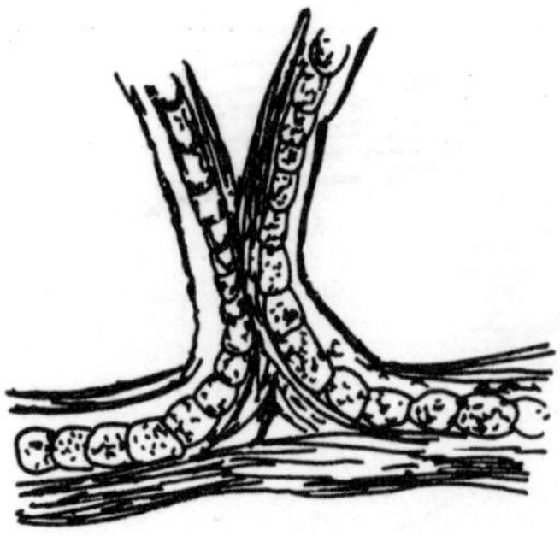

Fig. 26.6 Scytonema, Origin of false branching

Reproduction :

1. Sexual reproduction has not been reported.
2. Cell division is the only means of reproduction.
3. Cells divide in all the planes.
4. In elongated cells the plane of division is transverse and a part of the colony separates to produce new colonies.

Fig. 26.7 Microcystis pseudofilaments

Experiment no. 5 ***Study of external features of Spirulina***

CYANOBACTERIA (CYANOPHYCEAE)
NOSTOCALES
OSCILLATORIACEAE
SPIRULINA

Fig. 26.8 A diagrammatic view of Spirulina platensis.

Common species : *Spirulina subtilisima*, *S.major* and *S.subsala.*

Distribution : Found in fresh water ponds and ditches.

Study : 1. External features.

2. Reproductive structures.

External features :

1. The plant body is a trichom.
2. It is formed by a single long spirally twisted cell.
3. Occasionally cross walls may be found making the trichome multicellular.
4. Some times the cross walls are incomplete.
5. The spiral twisting of the cell in the trichome may be loose or close.
6. The cells are cylindrical in shape and have no mucilage sheath.
7. The terminal cell is rounded and has no calyptra.
8. The cells exhibit active rotational movement.
9. Cell wall is multilayered.
10. There are no chloroplast in the cell. The pigments are present in thylakoids.
11. Gas vacuoles are present in the cell.
12. The nucleus is prokaryotic.

Reproduction :

1. Sexual reproduction has not been reported.
2. Asexual reproduction takes place by means of fragmentation.

CHAPTER 27

TYPE STUDIES - III ALGAE

Experiment no. 1 ***Study of external features of Chlorella***

Common species: *C.Pyrenoidose C vuigaris, C.conglomerata* etc.

Distribution: Occurs in fresh water,ponds,damp soil, sides of garden pots.

Study

(1) Structure or the plant body

(2) Reproduction

(3) Systematic position and Identification

Structure of plant body

1. The plant body is unicellular and green in colour.
2. The cells are non-motile, spherical or ellipsoidal in shape.
3. The cell-wall does not have a mucilaginous envelope.
4. Each cell has a single cup shaped or band shaped chloroplast: there is no pyrenoid.
5. Cells are uninucleate.

Reproduction

1. Sexual reproduction has not been reported.
2. The cells divide and form spores called autospores.
3. The number of autospores per cell is 2-8.
4. Autospores are uninucleate and develop into a new individual.

Diagram to be drawn

1. Structure of the cell
2. Cell with autopsores

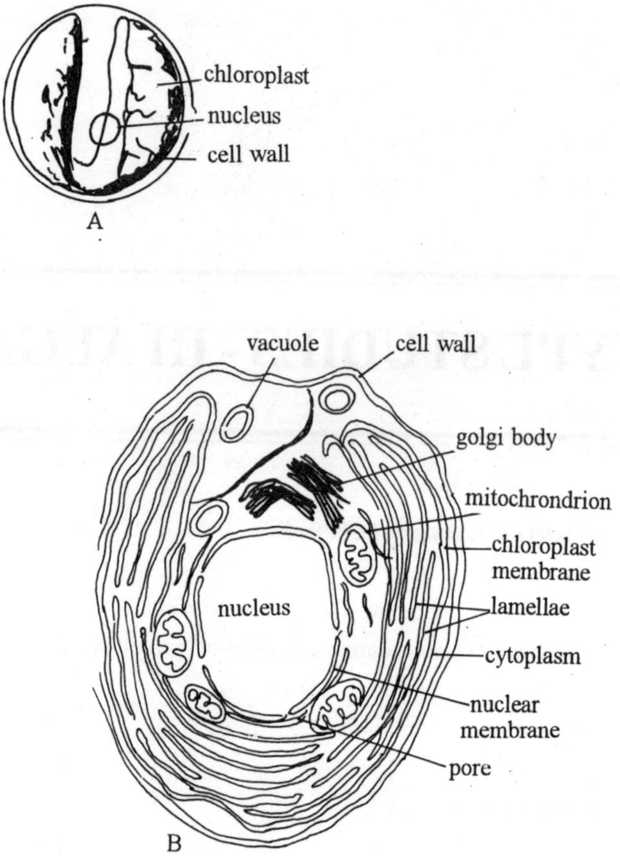

Fig. 27.1 A,B. Chlorella. A,A cell under light microscope; B, A cell under electron microscope (diagrammatic)

Systematic position & Identification

A. 1. Chlorophyll present
   2. Plant body a thallus
   3. Cell wall cellulosic...*Algae*

B. 1. Plastids present
   2. Chlorophyll is the chief pigment
   3. True nucleus present
   4. Photosynthetic product is starch...*Chlorophyceae*

C. 1. Cells are uninucleate and non-motile.
   2. Reproduction by spores which are non-motile (motile pores very rare)...*Chlorococcales*

D. 1. Plant body unicellular
2. Cells with a cup shaped chloroplast...*Chlorellaceae*

E. 1. Cells spherical & Uninucleate
2. Nucleus located at the centre
3. Chloroplast cup shaped ...*Chlorella*

Bacillariophyceae

The Bacillariophyceae are commonly called diatoms. They include a large number of unicellular and colonial algae, which very greatly in form and size. They are most widely distributed among algae. The primary distinguishing feature of Bacillariophyceae is that the cell wall is silicified and consists of two halves, which fit together as do the lid and box. The cell has one to many chromatophores.

Reproduction is by cell division and auxospore formation. Auxospores are usually zygotic in nature, produced by the union of gametes, but may not be always so.

Experiment no. 2 *Study of external features of Diatoms* (Pinnularia)

Common species: *P. major, P. viridis, etc.*

Distribution: Commonly found in fresh water ponds, and ditches; forms an oily layer on top of water.

Study

1. Cell structure
   (a) Girdle view
   (b) Valve view
2. Reproduction
3. Systematic position and Identification

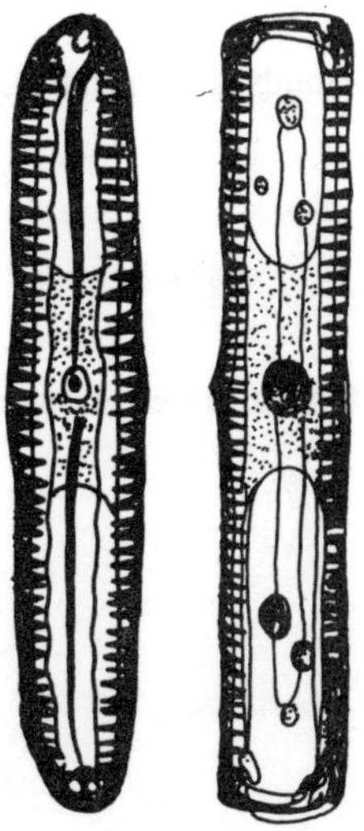

Fig 27.2 Pinnularia
A. Valve view or top view, B Girdle view

Cell structure:

1. Structurally the diatom cell consists of two parts or halves which fit in like a lid and box.
2. Of the two halves, the smaller one is called *hypotheca* and the bigger one is called *Epitheca*.
3. The cell has a size of 140 μ x20 μ
4. Each valve or half consists of a flat top (valve face) and the margin or girdle representing the overlapping face.
5. A diatom cell can be viewed from two different angles. The view from the side is called girdle view, and the view from the top is called the valve view.

Valve view:

1. Only one valve can be seen.
2. The wall can be clearly seen and it is called frustule
3. The cell wall is made up of pectin and it is silicified.
4. On the wall are seen prominent markings called costae. The costae represent rows (striae) of small pores or cavities in the frustule.

5. Along the centre of the valve is present a long slit in the wall. This is called raphe, it is sigmoid in shape.
6. At the centre of the raphe is a swelling called the *central nodule.*
7. At either end of the raphe are also present two nodules called *Polar nodules*
8. The mucilage extends out of the wall through openings or pores and is supposed to help in locomotion.
9. Inside the silicified wall is the cytoplasmic lining, surrounding a vacuole.

Girdle view:

1. In the girdle or side view both the valves are seen (only one valve is seen in the valve view)
2. Costae can be seen in the frustule.
3. Inside the wall is the cytoplasm having a single nucleus.
4. One or two chromatophores may be seen distributed in the cytoplasm.
5. Oil and volutin globules are present in the cytoplasm

Reproduction

1. The commonest method of reproduction seen in diatoms is cell division.
2. During cell division, the two halves get separated after the nuclear division. Each daughter cell gets only one valve from the parent. It has io develop another valve by itself. The newly formed valve is always a hyoptheca. This is irrespective of the fact, whether it gets a hypotheca or epitheca from the parent cell.

Auxospores:

1. Auxospores are elongated spores formed in the diatom cell to bring about reproduction.
2. Auxosopres are both asexual as well as sexual. They are formed by any one of the following five ways.
   (i) Two ceils conjugate to form a single auxospore.
   (ii) Two cells conjugate to form two auxospores.
   (iii) Two cells being enveloped into a common envelop, but each giving rise to an auxospore without conjugation.
   (iv) One cell gives rise to one auxospore.
   (v) One cell gives rise to two auxospores.

The first two methods given above are obviously sexual. The last three methods appear to be asexual.

Experiment no. 3 ***Study of external features of Gracilaria***

Common species: *G. armata, G. compressa etc.*

Distribution: Marine in distribution occurring also as a lithophyte at the intertidal zones in tropical seas.

Study: 1. Morphology of the thallus.
2. Anatomy of the thallus
3. Reproduction
4. Systematic position and classification

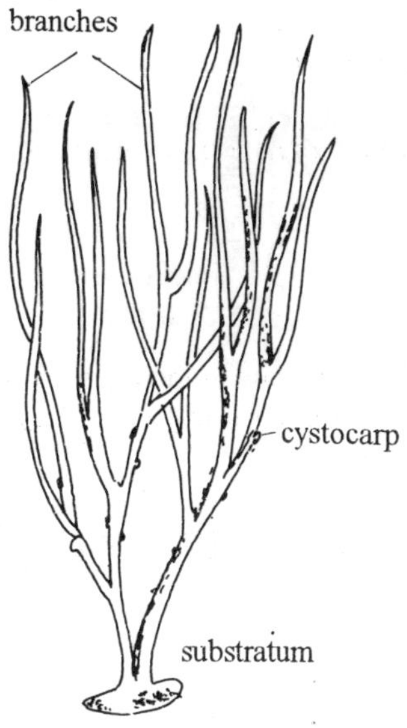

Fig. 27.3 Gracilaria. Thallus bearing cystocarps

Morphology of the thallus:

1. The thallus is a branched, cylindrical or compressed structure made up of a number of filamentous structures.
2. The thallus also called frond is attached to the substratum with the help of a discoid hold fast. The hold fast consists of a number of closely packed thread like structures.
3. The fronds are 25-40 cms, long, dark red in colour and dichotomously branched.
4. Some of the branches of the frond have attenuated apices, terminating in hair like structures.
5. The fronds are monosiphonous, have uniseriately arranged cells. The fronds have a gelatinous envelope on their surface.

Anatomy of the thallus:

1. A transverse section of the thallus reveals three regions, namely, peripheral layer, cortex and the medulla.
2. The cells of the peripheral layer are small, uninucleate and rich in cytoplasmic contents.
3. The mature cells of the peripheral layer may be multinucleate.
4. Unicellular hairs are produced from some of the peripheral cells.
5. Internal to the peripheral layer is the cortex. It is made up of many small cells. The cells have many chromatophores.

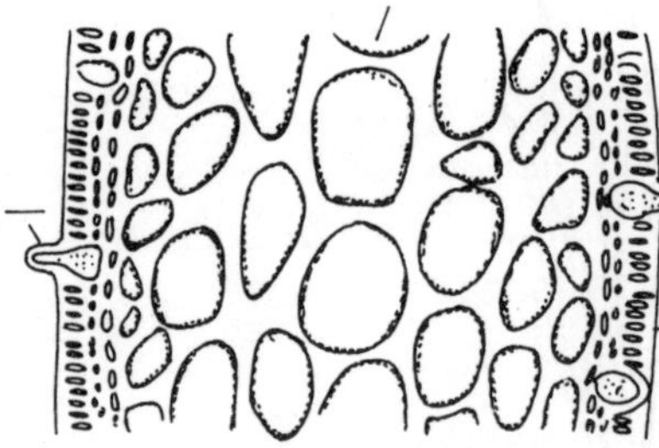

Fig. 27.4 **Gracilaria. T.S. thallus**

Reproduction:

1. All the species of *Gracilaria* are dioecious.
2. The life cycle exhibits alternation of generations.
3. There are three types plants, male plants, female plants and sporophytes. All of them are similar morphologically.
4. In the male plant, the reproductive structures are spermatangia.
5. The spermatangia develop from the cortical cells. Each spermatangium produces a spermatium. The spermatia are non-motile structures. They are dispersed by water current.
6. In the female plant, the female reproductive organs, namely, the carpogonia develop from the cortical cells.
7. Some of the cortical cells destined to produce carpogonia, develop into supporting cells. Each supporting cells produces a hypobasal cell and a carpogonium proper.
8. The carpogonium is made up of a single flask shaped cell. It has a swollen base in which is situated the egg and an elongated trichogyne.
9. The fusion of the egg and spermatium results in the formation of the zygote which develops into the cystocarp or the carposporophyte.

Structure of the cystocarp:

1. The cystocarp is a globose structure. It has a short stalk.
2. The cystocarp has an opening called the ostiole, for the liberation of the spores.
3. At the base of the cystocarp is a placenta which gives rise to filamentous structures called the gonimoblast filaments.
4. Each gonimoblast filament is 3-4 celled long with the terminal cell transforming into a *carposporangium*.
5. Each carposporangium produces a diploid carpospore.
6. When the cystocarp is producing gonimoblast filaments, the surrounding cells develop a protective sheath.

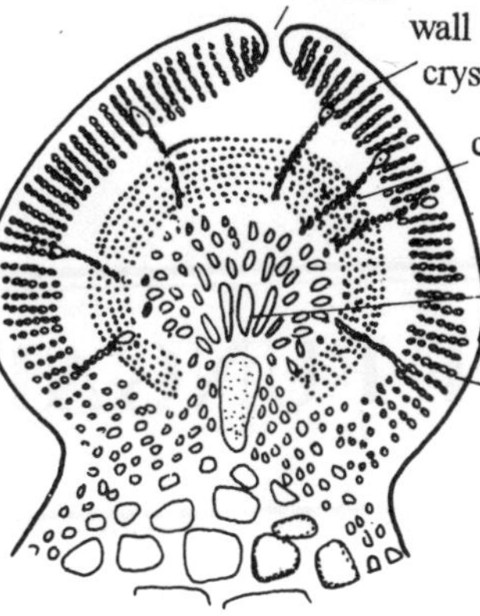

Fig. 27.5 Gracilaria. L.S. Cystocar

7. The cystocarp has a wall, a placenta cell, gonimoblast filaments, carposporangium and carpospores.
8. At maturity, the carpospores are liberated from the plant and germinate to give rise to the tetrasporophyte.

Structure of the tetrasporophyte:

1. Tetrasporophytes are exactly similar in morphology to the gametophytic plants except that it is asexual.
2. The sporophytes produce tetrasporangia from the cortical cells.
3. Each tetrasporangium produces four haploid tetraspores which develop into gametophytes.

Diagrams to be drawn:

1. Habit sketch
2. Anatomy of the thallus
3. Spermatangium
4. Carpogonium
5. Tetrasporophyte

Systematic Position and Identification

A. 1. Celorophyll present
2. Plant body is thallus
3. Cell wall cellulosic ... *Algae*

B. 1. Motile reproductive cells absent
2. Chromatophores red pigmented
3. Cystocarp present ... *Rhodophyceae*

C. 1. Thallus dichotomously branched
2. Male and female sex organs develop from cortical cells...*Gigartinales*

D. 1. Carpogonia develop from hypobasal cells.
2. Tetraspores arranged in the form of a cross ...*Gracilariaceae*

E. 1. Thallus is monosiphonous
2. Epidermis or peripheral layeres with a gelatinous layer...*Gracilaria*

CHAPTER 28

STUDY OF FUNGAL TYPES

PYTHIUM

Common species: *P.debaryanum, P.aphanidermatum, P.graminicolum, P.myriotylum* etc.

Occurrence : Many species are amphibious. They mostly occur in moist humus soil and infect the hypocotyl region of seedlings of plants like ginger, tobacco, mustard, wheat etc. After entering the host, the fungus acts as a parasite and causes diseases like damping off, soft rot, foot rot etc.

Study

1. Vegetative structure of pathogen
2. Reproduction
3. Diseases symptoms
4. Systematic position and identification

Diseases caused by the genus

1. *P.debaryanum*- damping off of tobacco seedlings.
2. *P.aphanidermatum* - stem rot of papaya.
3. *P.myriotylum* - stem rot of ginger.
4. *P.graminicola* - foot rot of wheat seedlings.
5. *P.indicum* - fruit rot of lady's finger.

Vegetative structure

1. The plant body of the fungus is a mycelium.
2. The mycelium is aseptate, coenocytic and much branched.
3. Septa appear only during the formation of reproductive structures.
4. Hyphae have a vacuolated cytoplasm in which a large number of nuclei are embedded.
5. The mycelia superficially resemble the filaments of *Vaucheria.*
6. Reserve food is in the form of glycogen.
7. No haustoria are produced.

Asexual reproduction

1. Asexual reproduction takes place by the formation of sporangia. Sporangia may be terminal or intercalary.

Fig. 28.1 Pythium
Damping off of seedling (Note the rooted stem at the base)

2. Sporangia are small, globular or sac like and are of two types viz., globose or spheroidal and elongate or filamentous.
3. Globose sporangia are globe like and terminal is position.
4. Filamentous spornagia are indistinguishable from the hyphae.
5. The sporangia whether globose or filamentous are cut off from the hyphae by a cross wall.
6. The contents of the sporangia form numerous zoospores which are reniform and biflagellate.
7. Each sporangium has a beak like out growth which enlarges to form a vesicle. It is into this vesicle that the zoospores are released first. Later the vesicle ruptures setting the zoospores free.
8. On germination each zoospore develops into a new individual.
9. In some species of *Pythium*, the sporangia produce non motile Conidia which germinate to give rise to new individuals.

Sexual reproduction

1. Sexual reproduction is oogamous. Sex organs are antheridia and oogonia. They are usually formed towards the end of the growing season.
2. Pythium is homothallic.
3. Antheridium is small and club shaped, it usually terminates an antheridial branch. It is cut off from the rest of the hypha by a cross wall.
4. The antheridium is multinucleate in the beginning. Later all the nuclei except one degenerate, leaving only one functional male nucleus.
5. The oogonium is a globose structure terminating a female branch of the hypha.
6. The inside of the oogonium has two zones of cytoplasm viz outer Periplasm and central ooplasm.

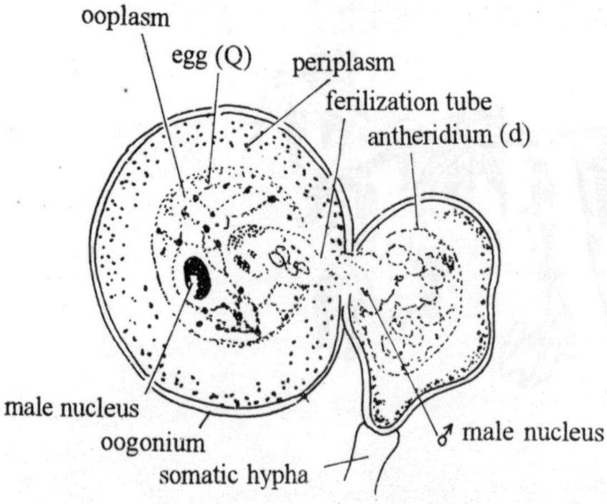

Fig. 28.2 Pythium
Sex organs showing gametangial contact

7. The ooplasm has a single egg.
8. Sexual reproduction is effected by means of gametangial contact.
9. A fertilization tube develops from the antheridium. In penetrates the oogonium releasing the contents inside.
10. The fusion of the male and female gamete results in the formation of a zygote or oospore.
11. The oospore after undergoing a resting period develops into a new individual.

Disease symptoms

The symptoms of the disease are noticeable only in the seedling stage after they have just come out of the soil. The seedlings become pale green. The stem at its base, ie., just above the ground level exhibit decay of the tissues. The tissues turn brown. As a result of the decay at the base the seedling collapses.

In the fruit rot and root rot diseases, the tissues rot and get decomposed resulting in the death of the plant.

Diagrams to be drawn

1. Habit of damping off of seedlings.
2. Zoosporangium.
3. Oognoium and antheridium.

Systematic position and phylogeny

A. 1. Chlorophyll absent
2. Reserve food is glycogen ....**Fungi**

B. 1. Mycelium aseptate and coenocytic
2. Septa reproduction formed only during reproduction
3. Oospore is thick walled.................**Phycomycetes**

C. 1. Sexual reproduction not by planogametic copulation
2. Oogonium has a single oosphere .... **Peronosporales**

D. 1. Sporangiopores not different from somatic hyphae .... **Pythiaceae.**

E 1. Pathogen causes damping off of seedlings.

2. Host stem has brown decay at the base

3. Zoospores are released into a vesicle

4. Facultative parasite ....**Pythium.**

RHIZOPUS

Common species : *R.nigricans, R.stolonifer, R.arrhizus, R.artocarpi* etc.

Occurrence : Species of the genus are the most ubiquitous of the fungi. The members are saprophytes growing on any dead and decaying organic matter. *Rhizopus* grows very commonly on stale bread hence it is popularly called the bread mold. It also causes several plant diseases.

Study

1. Diseases caused and disease symptoms
2. Method of culturing
3. Vegetative structure
4. Reproduction
5. Systematic position and identification

Diseases caused : Some species of *Rhizopus* causes the fruit rot diseases in some flowering plants. The following are some of the diseases.

1. Apple rot .... *R.arrhizus*
2. Fruit rot of jack fruit .... *R.artocarpi*
3. Soft rot of strawberries .... *R.stolonifer*
4. Decay of bread .... *R.nigricans*

Disease symptoms of the fruit rot

1. Young fruits drop down prematurely
2. Infected peduncles get decolorised ultimately turning black
3. Mature fruits begin to rot and get decayed

Method of culturing in the laboratory

1. Take a few slices of bread. Add a little water to moisten them.
2. Keep the slices of bread in a petridish and cover it.
3. After few days, a soft cottony growth of mycelium may be observed. This is the vegetative phase.
4. A little later small white pin head like structures appear from the mycelium. These are the young sporongia.
5. After few days the white pinheads turn black. This represents the mature sporangia liberating the dark colored spores.
6. It is generally difficult to observe the sexual stages.

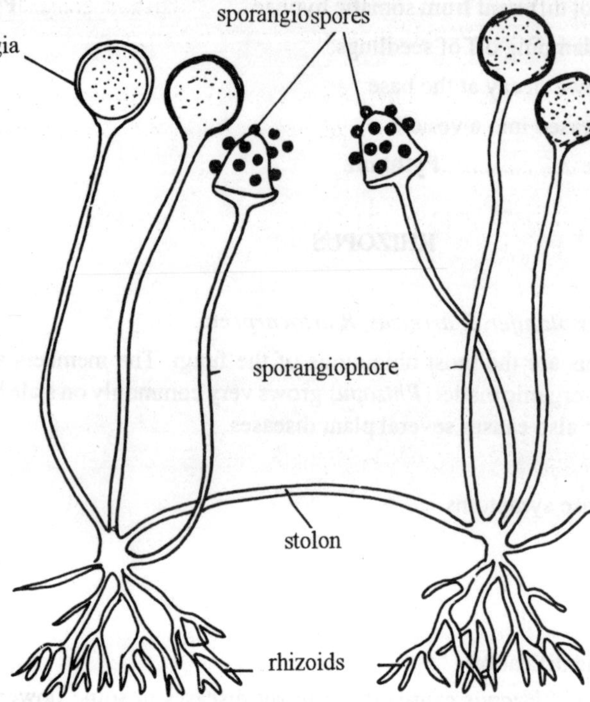

Fig. 28.3 Rhizopus Vegetative and asexual structure

Vegetative structure

1. The plant body of *Rhizopus* is a much branched, coenocytic, aseptate mycelium.
2. The hyphae are whitish and tubular. There is a centrally running continuous vacuole; cell wall in chitinous.
3. The cytoplasm is peripheral. It has numerous nuclei and has oil and glycogen as reserve food material.
4. Cross walls appear only at the time of formation of reproductive structures.
5. The mycelium which is fairly old, has three types of hyphae. These are -

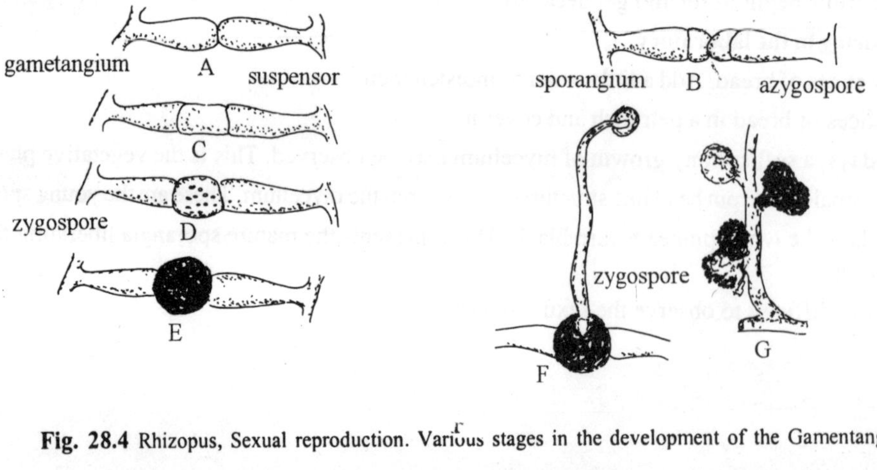

Fig. 28.4 Rhizopus, Sexual reproduction. Various stages in the development of the Gamentangia and Zygospore

a. Stolon or distributive hyphae : These are horizontal tubular hyphae that grow above the substratum. Here and there swollen portions are found on the stolon.

b. Rhizoidal hyphae : From the swollen points of the stolon arise short branched hyphae called rhizoids. These grow vertically down into the substratum helping in anchorage and absorption.

c. Sporangiophores : These are erect, tubular, unbranched hyphae arising at the same points, where rhizoids arise. Sporangiophores at their apex bear a globular or oval sporangia.

Reproduction - Asexual

1. Asexual reproduction takes place by non motile spores produced in sporangia.
2. Sporangia, terminate the unbranched sporangiphores.
3. Occupying the central region of the sporangium is found a dome shaped columella, while protoplast is in continuation with that of sporangiophore.
4. In between the columella and the wall of the sporangium a large number of spores are produced.
5. The spores are non motile and have a pigmented wall due to which they appear dark brown.
6. Spores are spherical, non motile and multinucleate. They are wind dispersed and on the availability of a suitable substratum develop into new individuals.
7. Occasionally thick walled chlamydospores may be formed as a means of asexual reproduction.

Reproduction - Sexual

1. Sexual reproduction in *Rhizopus* takes place by means of gametangial copulation.
2. The gametangia are isogametangia, but the gametangia formed on the same mycelium will not fuse. This is due to the phenomenon of heterothallism.
3. The gametangia that fuse are morphologically similar but physiologically different.
4. Since the compatible gametangia cannot be distinguished into male or female, they are designated as + and - strains in order to show that they are different.
5. Hyphae belonging to + and - strains produce short outgrowths called progametangia.
6. The progametangia at maturity get divided into, a basal suspensor and a terminal gametangium.
7. The two gametangia of opposite strains fuse to form a zygospore.
8. The zygospore develops a thick ornamented wall.
9. After undergoing a period of rest, the zygospore germinates by producing an erect hypha, whose free end forms a sporangium.
10. The spores that are formed are of two kinds; 50% of them belong to + strain and the rest - strain. But morphologically the spores are thick.

Diagrams to be drawn

1. Habit sketch
2. Plant body enlarged
3. Sporangium enlarged
4. Stages in sexual reproduction.

Systematic position and identification

A. 1. Chlorophyll absent
   2. Reserve food is glycogen ....**Fungi**

B. 1. Plant body aseptate and coenocytic

2. Septa formed only during reproduction .... **Phycomycetes**

C. 1. Asexual reproduction by non motile spores

2. Nutrition saprophytic ....**Mucorales**

D. 1. Conidia not present

2. Columella present inside the sporangium......................... **Mucoraceae**

E. 1. Rhizoidal hyphae present beneath the sporangiophores

2. Stolon will not absorb nutrition from the substratum .... **Rhizopus**

SACCHAROMYCES (YEAST)

Common species : *S.cerveisiae, S.ludiwigii, Schizosaccharomyces octosporus*

Occurrence : The members are commonly called yeasts. They are widely distributed in nature. Yeasts are saprophytic and are commonly found in substrata rich in sugars like nectar of flowers, surface of fruits like grapes apples etc. Milk and other food undergo fermentation due to yeasts.

Study :

1. Method of culturing yeasts
2. Vegetative structure
3. Reproduction
4. Systematic position and identification

Culture of yeasts

1. Place small pieces of yeast cake in sugar solution
2. Yeast cells multiply and a dense cluster of cells is produced
3. After few days the yeast cells can be harvested

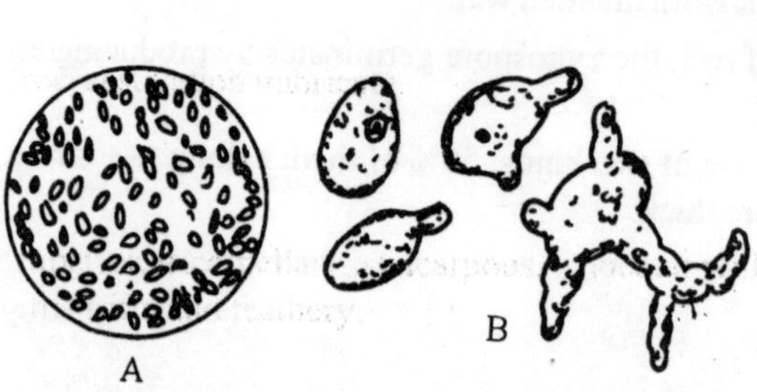

Fig. 28.5 Yeasts

A.Yeast cells under microscopic (Medium power view), B. Enlarged yeast cells

Vegetative structure

1. The thallus of yeast is made up of isolated cells (unicellular plant body)
2. The cells are usually oval or spherical ranging from 5-10m in diameter
3. Each cell has an outermost cell wall made up of chitin. Cellulose is absent.
4. The cytoplasm shows differentiation into outer ectoplasm and inner endoplasm
5. There is a large vacuole in the center
6. Placed immediately above the vacuole is the single nucleus (the nucleus was mistaken previously as the centrosome)
7. Cytoplasm is granular and contains reserve food materials like glycogen and oil
8. Mitochondira and starch grains are also found in the cytoplasm

Asexual reproduction

1. Yeast cells reproduce asexually by two methods viz., *fission* and *budding*.
2. Fission yeasts are called *Schizosaccharomyces* while budding yeasts are called *Saccharomyces*.

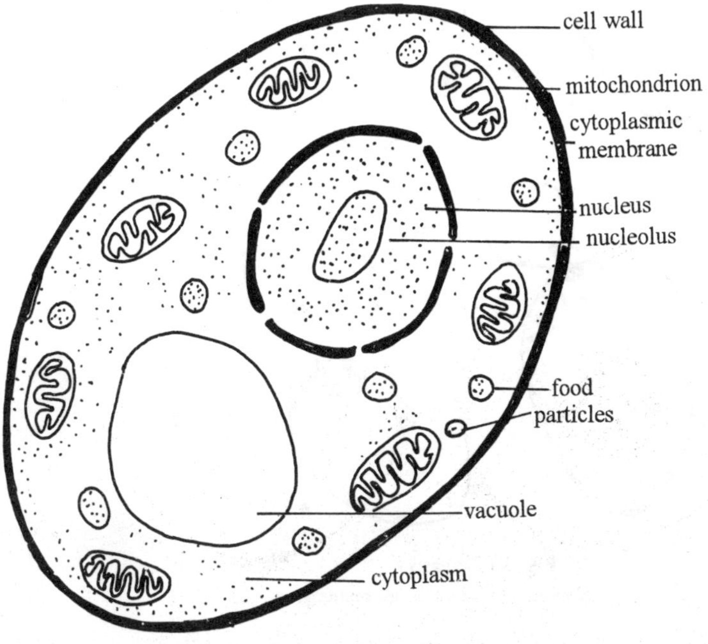

Fig. 28.6 Yeasts
Detailed cell structure (Electron microscopic view)

3. During fission, the single nucleus gets elongated.
4. A constriction appears in the center of the nucleus and it becomes dumb-bell shaped.
5. With the deepening of the constriction, the nucleus gets divided into two bits.
6. The two daughter nuclei migrate to opposite poles. A cross septum appears in the center of the cell dividing it into two.

7. The two daughter cells formed thus grow into adults cells.
8. Budding is another method of reproduction. During budding a small bulge arises from one side of the cell. The bulge increases in size and is now called a bud.
9. Meanwhile, the nucleus of the mother cell divides into two daughter nuclei; one of those migrates into the bud.
10. A small constriction appears at the region of attachment of bud and the parent cell.
11. The constriction deepens and the bud gets separated. It grows into a new individual.
12. In some cases a bud does not get separated from the mother cell but, produces another bud. In this way a chain of buds is formed. This chain may be branched or unbranched. The chain looks like a mycelium hence it is called a *Rhizomycelium* or a *Pseudomycelium*.
13. Eventually the chain breaks and the individual cells get separated and grow into new cells.

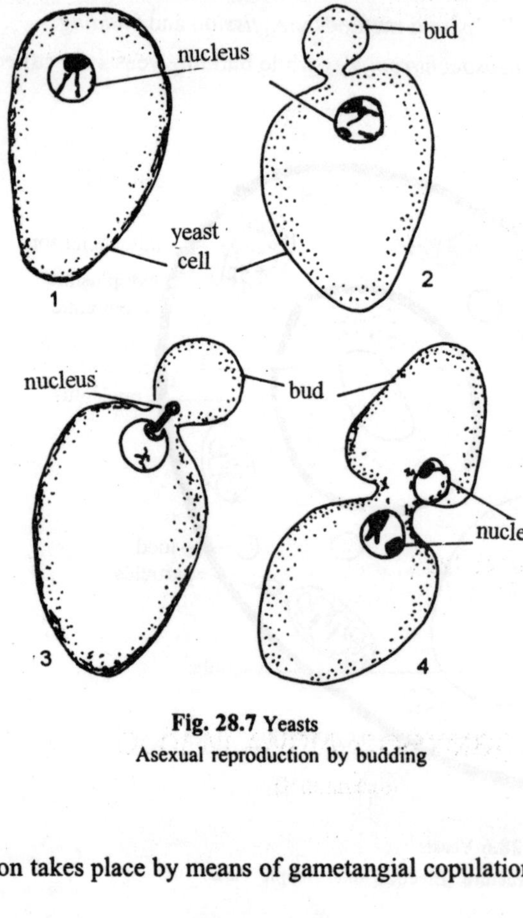

Fig. 28.7 Yeasts
Asexual reproduction by budding

Sexual reproduction

1. Sexual reproduction takes place by means of gametangial copulation, with two cells functioning like gametangia.
2. The two gametangia fuse forming a diploid zygote.
3. The zygote which is diploid functions like an ascus.
4. The diploid nucleus is ascus undergoes reduction division followed by a mitosis to form eight ascospores. In some cases only four ascospores are formed.
5. Ascospores on liberation develop into new individuals.

Diagrams to be drawn

1. Yeast cell magnified.
2. Yeast cells showing fission and budding.
3. Ascus with ascospores.

Systematic position and identification :

A. 1. Chlorophyll absent
   2. Reserve food is glycogen ....**Fungi.**

B. 1. Mycelium is septate or plant body unicellular
   2. Characteristic spores - ascospores are produced inside an ascus.
   3. Ascospores usually light, rarely otherwise .... **Ascomycetes.**

C. 1. No fruiting body.
   2. Saprophytic nutrition
   3. Asci not in clusters ....**Endomycetales**

D. 1. Asexual reproduction by budding or fission
   2. Plant body unicellular .... **Saccharomycetaceae**

E. 1. Members produce enzyme zymase and bring about fermentation.
   2. Plant body unicellular ....**Saccharomyces.**

PENICILLIUM

Common species : *P.notatum, P.chrysogenum, P.camembertii, P.roquefortii, P.italicum, P.digitatum* etc.

Occurrence : Commonly called green or blue mold, *Penicillum* is a saprophytic fungus growing on rotten fruits, decayed vegetable, meat etc. Some of them grow on moist leather also.

Study

1. Diseases caused and symptoms
2. Method of culture in the laboratory
3. Vegetative structure
4. Reproduction
5. Systematic position and identification

Diseases caused

1. *P.expansum*, causes decay of apples, pears and grapes in storage.
2. *P.italicium* and *P.digitatum* cause decay of citrus fruits like orange, sweet lime etc.

Disease symptoms are visible in the form of greenish or slightly bluish leisons on the fruit wall. Those are the places where reproductive structures are found.

Culture in the laboratory

1. Keep a stale orange in moist condition.
2. Sprinkle water to keep it wet.
3. Keep the orange enclosed.
4. After few days greenish blue coating appears on the rind.
5. Scrape the coating on to a microslide and observe under the microscope.

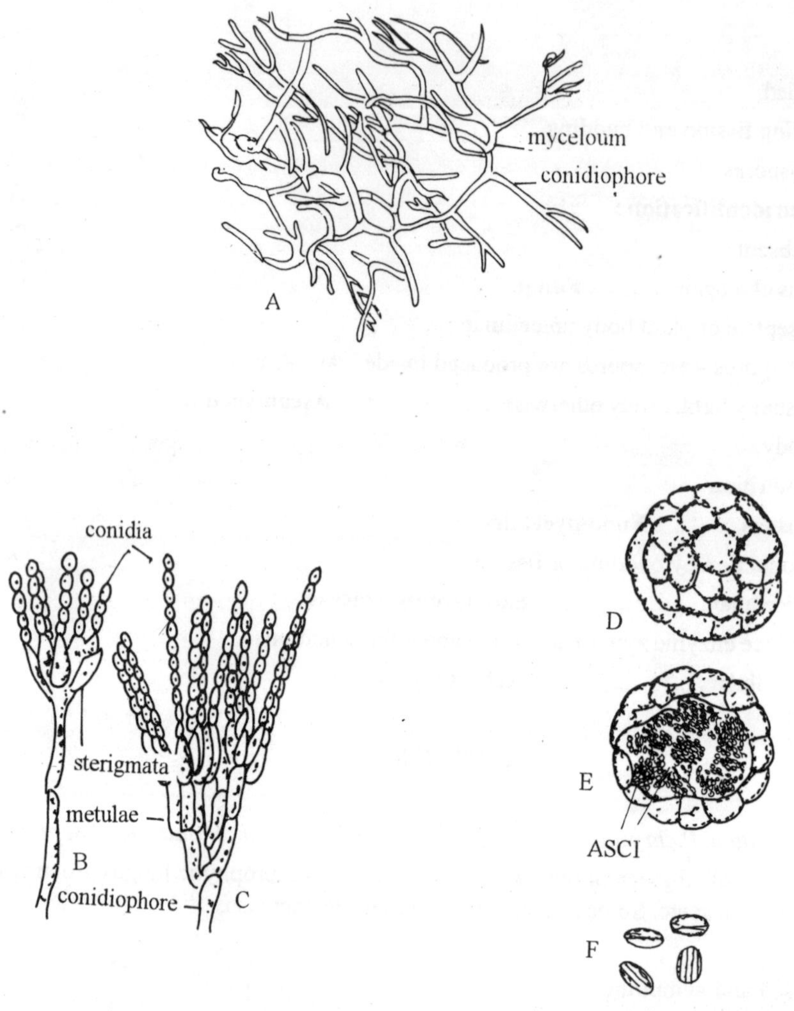

Fig. 28.8 Penicillium

A. Somatic structure, B-C. Enlarged condiophores, D. Cleishothecium, E. Cleistothecium showing Asci, F. Ascospores

Vegetative structure

1. The plant body is a mycelium. It is slender composed of profusely branched septate hyphae.
2. The mycelium is divided into cells. Cells are either monokaryotic or dikaryotic.
3. The cells wall between the cells has an opening, through which cytoplasm streams from cell to cell.

Asexual reproduction

1. Vegetative propagation is brought about by means of fragmentation of hyphae. Each fragment develops into a new individual.
2. Asexual reproduction by conidia is the most common method of multiplication in *Penicillium*.
3. Vegetative mycelium produces conidiophores.
4. Conidiophores are erect and branched once or twice.
5. Each conidiophore or its branches bear at their tips a whorl of bottle shaped structures called Sterigmata or Phialides.

6. The branches of the conidiophores which bear phialides are called metulae.
7. The sterigmata at their tips produce a chain of conidia. Conidia are produced in basipetalous succession.
8. The entire structure consisting of conidiophores, metulae, sterigmata and conidia resembles a brush or broom, hence the name *Penicillium* (Gr.*Penicillus* = brush)
9. Conidia are unicellular. They may be uni or multinucleate.
10. Upon falling on a suitable substratum, conidia germinate into a new mycelium by producing a germ tube.

Sexual reproduction

1. Sexual reproduction is by gametangial contact. Sex organs are antheridia and ascogonia.
2. Sexual reproduction is very rare.
3. The fruiting body produced by *Penicillium* is a cleistothecium.
4. The clesitothecium is a completely closed structure. It has a mutlilayered wall called peridium.
5. Lying scattered in the cavity of cleistothecium are many asci. Between the asci, sterile hyphae called paraphyses are found.
6. Each ascus has eight ascospores. The ascospores come out of the cleistothecium by the rupturing of the wall of the latter. On germination they develop into a new mycelium.

Diagrams to be drawn

1. Habit of infected host
2. Mycelium
3. Condiophore
4. Cleistothecium

Systematic position and identification

A. 1. Chlorophyll absent
   2. Reserve food is glucogen ....**Fungi.**

B. 1. Mycelium is septate
   2. Ascospore produced in ascus ....**Ascomycetes**

C. 1. Conidia are chief mode of multiplication
   2. Asci scattered inside the cleisothecium ....**Aspergillales**

D. 1. Fructifications not stalked
   2. Peridium made up of pseudoparenchyma .... **Aspergillaceae**

E. 1. Conidiophores branched, but no vesicle
   2. Asexual structures resemble a broom ....**Penicillium**

ASPERGILLUS

Common species : *A.niger, A.fumigatus, A.flavus* etc.

Occurrence : The species of Aspergillus are generally saprophytic growing on stale bread, decaying vegetable meet etc. It also grows on jams, jellys, footwear, clother etc.

A few species of *Aspergillus* are parasitic causing lung diseases.

Study

1. Diseases caused and symptoms
2. Laboratory culture
3. Vegetative structure
4. Reproduction
5. Systematic position and identification

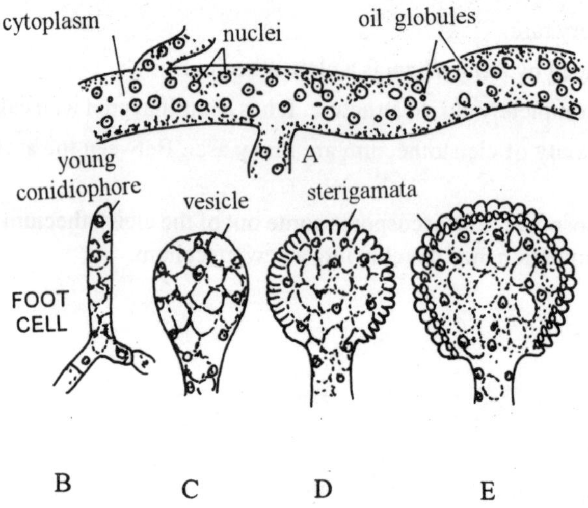

Fig. 28.9 Aspergillus. A- A portion of hypha; B-E. Development of vesicle

Diseases caused and symptoms

1. **Rot disease :** This is caused by *A.niger*. the hosts that are usually attached are pomegranate (*Punica granatum*), Dates (*Phoenix dactylifera*), Figs (*Ficus carica*) etc.
2. **Aspergillosis :** This is an allergic type of lung disease caused by the spores of *A.niger, A.fumigatus* and *A.flavus*, cattle and other animals and human beings also get the infection.
3. **Otomycosis :** A type of ear disease in human beings caused by several species of *Aspergillus*.

Disease symptoms

1. Saprophytic species produce symptoms in the form of greyish green or bluish green growth of the mycelium.
2. Fruits get rotten and decomposed due to the entry of fungus through wounds.
3. The symptoms of Aspergillosis resemble those of tuberculosis.

Laboratory culture

1. Place a piece of bread or citrus fruit under a bell jar.
2. Keep it moist
3. After few days greenish blue or yellowish mycelia grow profusely.

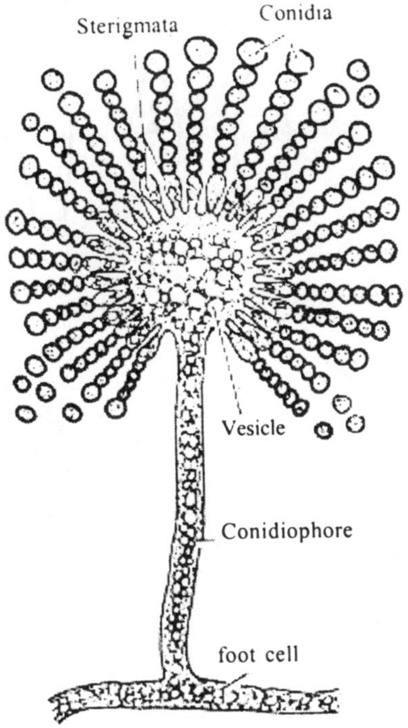

Fig. 28.10 A mature condiophore with conidia

Vegetative structure

1. The plant body of the fungus is a much branched, septate closely interwoven mass of mycelium.
2. Hyphae are divided into cells. Cells are multinucleate suspended in a granular cytoplasm.
3. Oil globules are present in the form of reserve food.
4. While most of the myeclium is above the substratum, certain hyphae grow into the substratum, possibly to help in anchorage and absorption.

Asexual reproduction

1. Asexual multiplication may take place vegetatively by fragmentation of the hyphae.
2. The chief method of sexual increase however is by the formation of conidia.
3. From the vegetative mycelium are produced many unbranched erect hyphae called conidiophores.
4. Each conidiophore arises from a foot cell of the mycelium.
5. The tip of each conidiophore swells to form a dome shaped vesicle.
6. The vesicle from its surface gives rise to a number of sterigmata or phialides, which are bottle shaped structures. The sterigma are multinucleate.
7. Sometimes a second layer of sterigmata may arise from the first layer in which case, the first layer is called primary sterigmata and the second layer is called the second sterigmata.

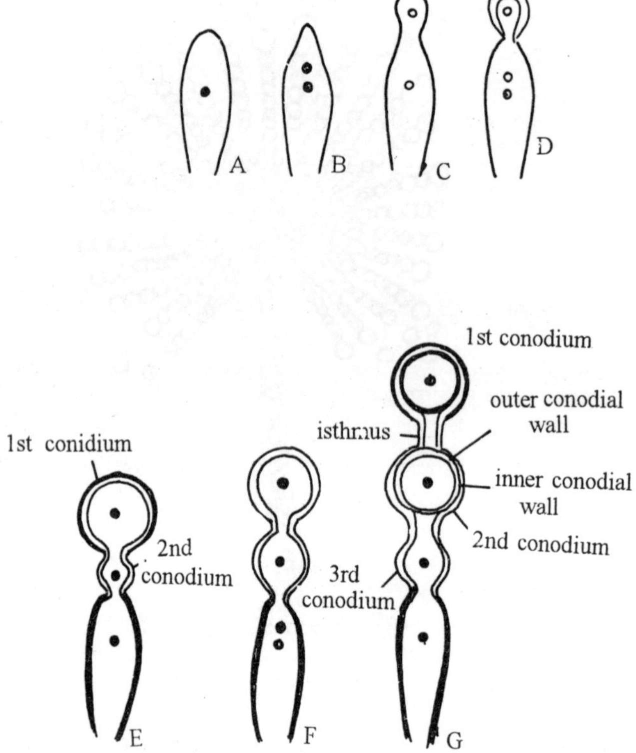

Fig. 28.11 A-G. Aspergillus niger, showing development of conidia from a phalide

8. The sterigmata of the second layer (first layer, if there are no secondary sterigmata) produce a chain of conidia. The conidia are produced in basipetalous succession.
9. The conidia are small, rounded unicellular structures. They may be uni or multinucleate. The conidial wall is pigmented in hues of black, green and blue.
10. The conidial wall is double layered. The outer epispore is spiny, while the inner endospore is smooth. On germination the conidium develops into a new individual.

Sexual reproduction

1. Sexual reproduction is by means of gametangial contact. Sex organs are antheridia and ascogonia.
2. Antheridium is small elongated and multinucleate.
3. Ascogonium is a flask shaped structure with a bulbous base and a narrow neck called trichogyne.
4. Fruiting body is a cleistothecium. Many asci lie scattered in the cavity of the fruiting body. Each ascus has eight ascospores.

5. On liberation from the cleistothecium, the ascospores germinate to give rise to new individuals.

Diagrams to be drawn

1. Habit of infected material
2. Mycelium with condiophores
3. A part of the condiophore enlarged
4. Cleistothecium.

Systematic position and identification

From A to D identification. See *Penicillium*.

E. 1. Conidiophores unbranched
2. Tip of the conidiophore swells to form a vesicle
3. Secondary sterigmata present sometimes.......... *Aspergillus*

FUSARIUM

Common species : *F.solani, F.lini, F.oxysporum, F.vasinfectum* etc.

Occurrence : The species of *Fusarium* are parasitic and generally cause a disease called *Wilt*.

Study

1. Diseases caused and symptoms.
2. Vegetative structure
3. Reproduction
4. Systematic position and identification.

Diseases caused and symptoms

1. **Wilt disease of red gram :** The disease causing agent is *F.oxysporum*. The plants of red gram are invariably attacked, wherever they are cultivated.
   a. The characteristic symptom of the disease is the wilting of the plant. The plants look as if they are suffering from lack of water, even though there is plenty of water in the soil.
   b. The tissues at the basal part of the stem turn dark in color.
2. **Wilt disease of cotton :** The pathogen is *F.vasinfectum*. The symptoms of the disease are similar to what is seen in the wilt of red gram.
3. **Wilt of linseed :** *Fusarium lini* is the fungus which causes the disease.
   a. Disease symptoms may appear on the host at all stages of growth.
   b. Infected seedlings drop down and the cotyledons wither away.
   c. In older plants, deep green or brown patches appear on the leaves.
   d. Fungus is present in the roots as well as the basal part of the stem.
4. **Wilt of banana :** The disease is caused by *F.oxysporum, F.cubens*.
   a. Disease symptoms are visible on the seedlings 3-5 months old.
   b. The first symptom of the disease is the appearance of yellow stripe on the lowermost leaf.
   c. Subsequently the leaves turn yellow and fall off.
   d. Leaf sheaths also crack in a linear fashion.
   e. When the disease is intense, all the leaves fall down and only the 'Pseudostem' remains.

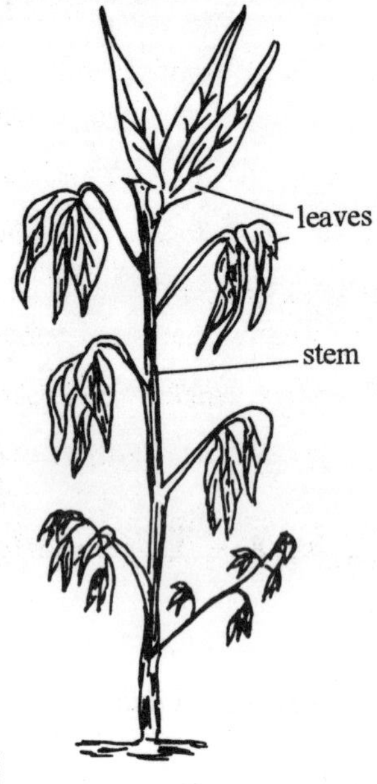

Wilt of pigeon pea (Arhar) caused by *Fausarium*

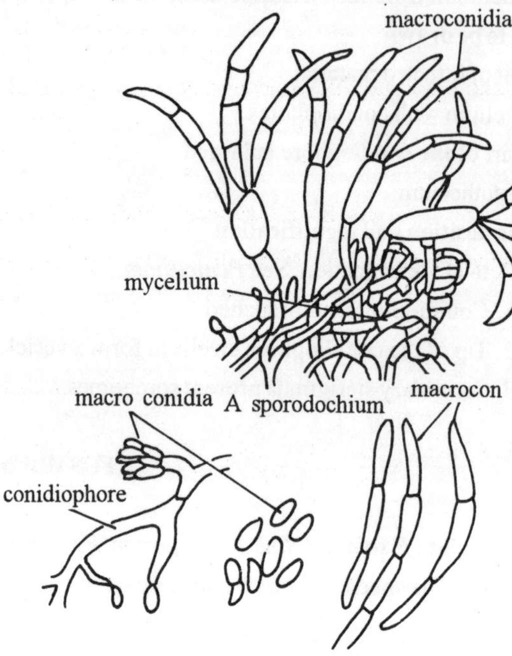

Mycelium with microconidia and macroconidia

Fig. 28.12 Fusarium

Vegetative structure

1. The plant body is a septate, much branched mycelium.

Reproduction

1. Prefect stage is not seen.
2. Asexual reproduction is brought about by a variety of methods viz., conidia, (microconidia and macroconidia), chlamydospores, sclerotia etc.
3. Microconidia are small rounded or oval in shape; sometimes crescent shaped also. They are present in small groups and are produced at the tips of branched or unbranched conidiophores.
4. Macroconidia are produced in a fructification called *Sporodochia* which are cushion like. They (conidia) are long, sickle or crescent shaped. The conidia are septate.
5. Chlamydospores are formed by the cells of the hyphae which round off and develop into a thick walled chlamydospore.
6. In some cases, the hyphae aggregate and form small, black compact *Sclerotia.*

Specimens given in the laboratory

1. Infected hosts

2. Conidia and conidiophores (slides)

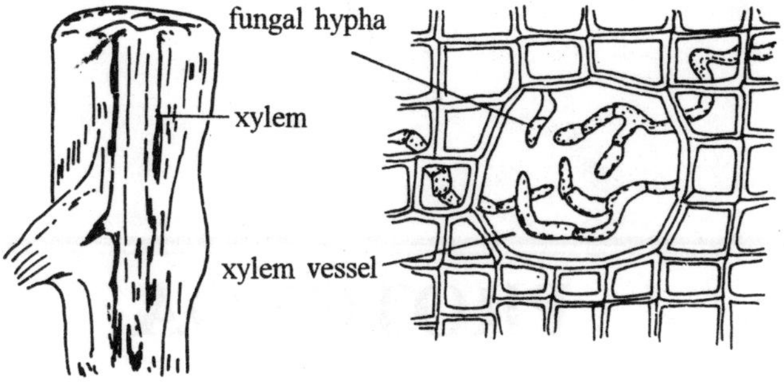

Fig. 28.13 Fusarium

Diagrams to be drawn

1. Habit sketch of infected host
2. Conidiophores and conidia
3. Sclerotia
4. Chlamydospores

Systematic position and identification

A. 1. Chlorophyll absent
   2. Reserve food is glycogen ....**Fungi**

B. 1. Perfect stage absent
   2. Asexual reproduction by conidia ....**Deuteromycetes**

C. 1. Conidiopores generally free
   2. A cervulus, synnema etc absent ....**Moniliales**

D. 1. Mycelium hyaline
   2. Sporodochia present ....**Tuberculariaceae**

E. 1. Conidia of two types
   2. Hyphae in the vessels of the host ....**Fusarium**

CHAPTER 29

PROTOZOA

Experiment no. 1 ***Study of external features of Trypanosoma***

Phylum Protozoa	Sub Kingdom Protozoa
Class Flagellata	Phylum Sarcomastigophora
Order Kinetoplastida	Class Zoomastigophora
	Order Kinetoplastida

Trypanosomes are flagellated parasites. They occur as parasites in all classes of vertebrates and are pathogenic to human beings and domesticated animals. The chief human pathogenic trypanosomes are *T.gambiense* and *T.rhodesiense* causing the characteristic "African sleeping sickness". Trypansomes are mostly prevalent in tropical countries.

Structure : The body of this protozoan is typically spindle like with both ends pointed. A firm pellicle (membrane complex) surrounds the body and maintains the shape. The pellicle encloses cytoplasm embedded in which is the nucleus. The nucleus is large and vesicular. There is a single flagellum arising from the posterior end of the body. The flagellum is a long whip like structure arising from a darkly staining granule- the blepharoplast. In addition to this, another granule the parabasal body may also be present. The two structures (blepharoplast and parabasal body) together constitute the kinetoplast. The flagellum runs along the entire length of the body on the edge of an undulating membrane. The undulating membrane is an extension of cytoplasm which connects the flagellum with the body. The flagellum helps in the swimming movement of the parasite in the blood stream.

Nutrition : Food materials are absorbed from the blood plasma by the general surface of the body. The pellicle also helps in gaseous exchange and also diffusion of nitrogenous waste material from the body. There is no contractile vacuole in the cytoplasm.

Reproduction : Trypanosomes reproduce asexually by longitudinal binary fission. The first to initiate the division is the kinetoplast followed by the nucleus and then the cytoplasm. The flagellum may also split, but in most cases one half of the cell takes the old flagellum while the other half develops its own new one.

Experiment no. 2 ***Study of external features of Trichomonas***

Phylum protozoa	Sub kingdom protozoa

Class flagella	Phylum Sarcomastigophora
Order Kinetoplasta	Class Zoomastigophora;
	Order Kinetoplasta

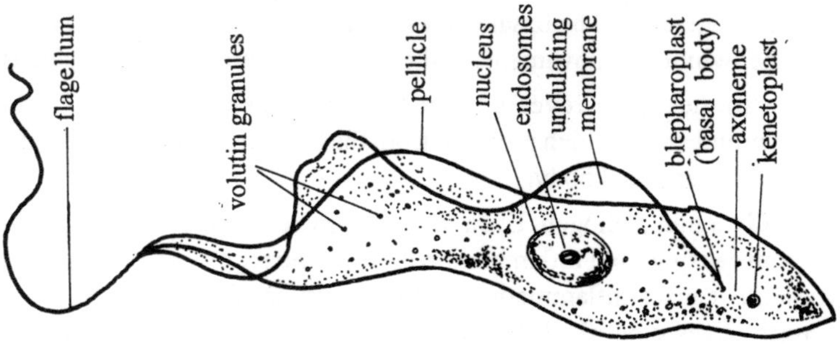

Fig. 29.1 Protozoa, Structure of Trypansoma

Trichomonas is a common parasite found in many vertebrates including man. Besides mammals it has also been reported from birds, reptiles, amphibia, molluscs and termites. Three species are found in man.

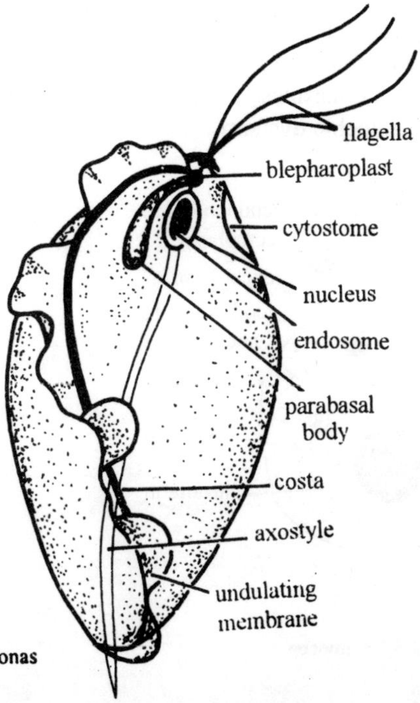

Fig. 29.2 Trichomonas

T.hominis	-	Colon
T.tenax	-	mouth
T.vaginalis	-	urinogenital tract

Structure : The body of *Trichomonas* is ovoid in shape and somewhat narrower at the hind end. In size, the body ranges from 10-30μ in length and 5-15μ in breadth. The parasite has four free anterior flagella and one recurved flagellum at the posterior end. The undulating membrane extends halfway along the length of the body to which flagella are attached. The basal bodies are found at the anterior end and these give rise to flagella. Flagella help in the swimming movements of the parasite. There is a single nucleus in the cell.

Electron microscopic studies have revealed that the flagellates (including Trichomonas) posses longitudinally arranged microtubercles which protrude from the back of the body. Other additional structures are a cytosomal groove, and a parabasal body.

Trichomonas is generally a non pathogen. Even in human males, where its presence has been recorded from the urethra, apart from causing mild inflammation there are no reports of any complications. In human females however *T.vaginalis* causes inflammation and irritation besides leading to the discharge of foul smelling liquid. The parasite lives superifically and there is no record of its invading the tissues. Researches have indicated that hormonal changes result in reduced acidic conditions which might encourage the activity of the parasite. Transmission of pathogen may be via male or due to infected material contaminating the vaginal orifice.

Experiment no. 3 ***Study of external features of Entamoeba***

Phylum - Protozoa	Sub kingdom - Protozoa
Class - Sarcodina	Phylum-Sarcomastigophora
	Class Rhizopoda.

Most of the amoebae are free living, but some have become endoparasites and have adapted themselves to live in the intestines of many invertebrates and vertebrates. *Entamoeba* is one such endoparasite, many species of which are pathogenic, some harmless also. *E.histolytica* is the most important among all the members and causes the notorious amoebic dystenty.

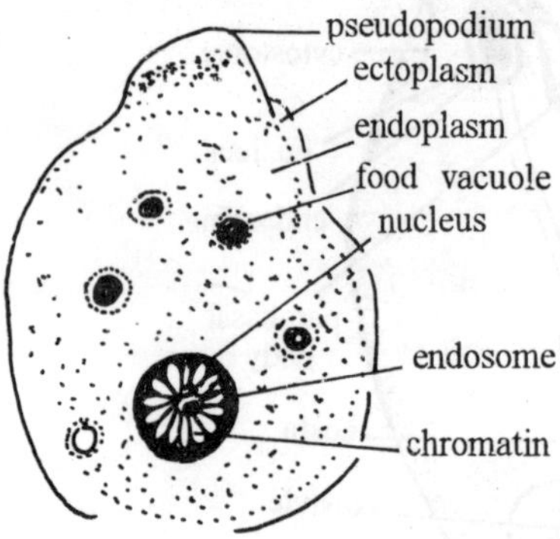

Fig. 29.3 Entamoeba

Historically *E.histolytica* was first discovered in 1859 by Lable, while its pathogenic nature was confirmed by Losch (1814). Schavdin (1903) gave the name *E.histolytica*.

Distribution : The pathogen is widely distributed being found in tropical, subtropical and temperate regions. Amoebic infection is quite high in India, China and South American countries. According to an estimate about 10% of human population has amoebic infection (Craig, 1962).

Habit and habitat : *E.histolytica* occurs in a large variety of mammals like dogs, cats, rats, monkey, baboon, chimpanzee, gorilla etc causing the intestinal infection. In man it occurs in the mucous and submucous layers of the colon. The toxic substance secreted by the parasite dissolves the mucous lining leading to bleeding. From the intestine, the parasite is carried along the blood stream to various organs like liver, kidney, lungs and even the brain. The parasite feeds on the tissues, bacteria, blood cells etc. The disease which is characterized by loose, blood smeared stools is called amoebic dysentery which is often fatal if untreated.

Cell structure : The parasite has an active stage called trophozite apart from the encysted condition.

The trophozoite resembles amoeba in structural details and ranges in size from 15-25μ. The cell is externally bound by a thin, elastic, selectively permeable membrane (plasmalemma). The membrane encloses the cytoplasm which is divided into an outer ectoplasm and an inner endoplasm. Ectoplasm is thin and hyaline while endoplasm is thick and granular. Entamoeba forms a single pseudopodium.

Cytoplasm (endoplasm) encloses nucleus and other organelles. There is a single nucleus which is vesicular and rounded. The chromatin material is concentrated along the margin of the nuclear membrane. The karyosome is small and is located at the center of the nucleus. Often it is surrounded by a clear halo. The nucleus is about 4 - 6μ in diametre. Entamoeba has six chromosomes.

Food vacuoles, present in the cytoplasm have the remnants of the RBC, WBC, tissues and bacteria. There are no contractivel vacuoles.

Experiment no. 4 ***Study of external features of Plasmodium***

Phylum Protozoa	Sub kingdom Protozoa
Class Sporozoa	Phylum Apicomplexa
	Class Sporozoa

Distribution

Plasmodium vivax is an intracellular parasite seen in the RBCs. It is distributed both in temperate and tropical countries. It requires two hosts to complete the life cycle. These are

1. Man or other vertebrates - primary host or definitive host.
2. Female Anopheles mosquito or other blood sucking insects - secondary host or intermediate host.

Asexual cycle or schizogony takes place in the human beings, while sexual cycle or gamogony and sporogony take place in the female Anopheles mosquito. The species of Anopheles which act, as vectors for the transmission of the malarial parasite in India are - A.stephensi, A.maculatus, A.fluvialitis and A.culcifancis.

Morphology: The adult parasite is called the trophozoite. It lives in the RBC of man. It is amoeboid in shape, uninuleate and has a number of granules and vacuoles in cytoplasm.

Electron microscopic studies have shown that the cell is enveloped by a unit membrane the plasmalemma. Endoplasmic reticulum, ribosomes (both attached and free), mitochondria and golgi apparatus are the other cell organelles present in the cell. The nuclues is eukaryotic and has a doulbe membrane envelop. It also has a nucleolus. The trophozoites feed on the RBCs with the residual product called haemozoin being scattered in the cytoplasm.

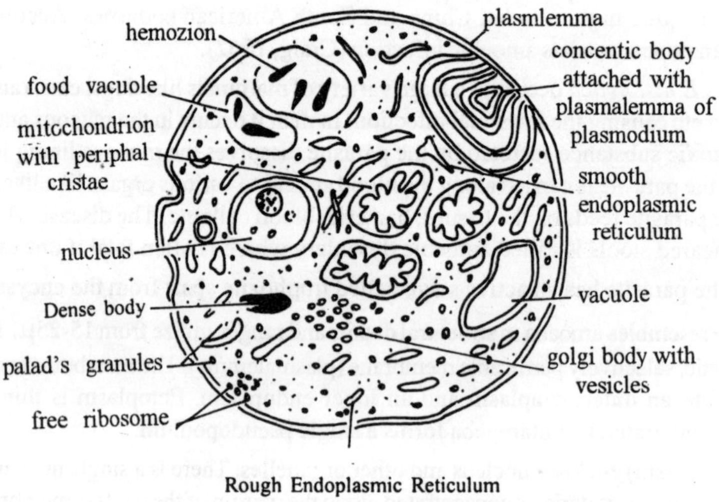

Fig. 29.4 Sporozoite of Plasmodium as seen in electron microscope

Life cycle

The life cycle of *Plasmodium vivax* is completed in two hosts - man and mosquito. The life cycle may be said to comprise of three stages.

1. Schizogony (Asexual) in man
2. Gamogony (sexual) completed in mosquito and
3. Sporogony (Asexual) in mosquito.

Experiment no. 5 *Study of external features of Balantidium*

Phylum Protozoa	Sub kingdom Protozoa
Class Ciliophora	Phylum Ciliophora
	Class Kinetofragminophorea

The class ciliata has a number of free living members with only *Balantidium coli* as a human pathogenic parasite. It is commonly found in the intestine of man, pig, sheep, camels, cockroaches etc. It is also found in the rectal contents of frog.

B.coli is oval in shape and may have a size of 200 mm and is the largest parasitic protozoa found in the intestine of man. The parasites are located mainly in the lumen of the intestine and obtain their food by feeding on bacteria. Occasionally diarrhoea mixed with blood may result due to *Balantidium* infection.

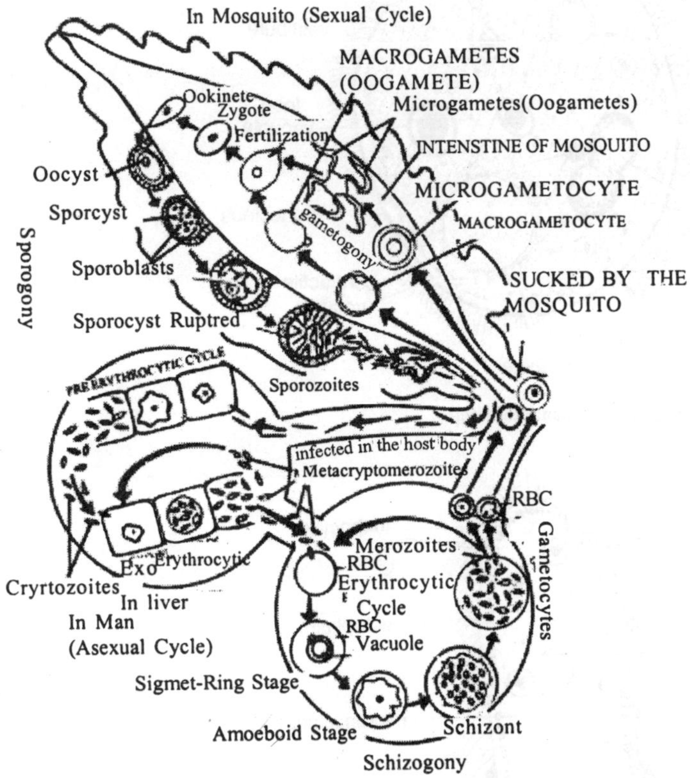

Fig. 29.5 Life-cycle of malarial parasite

The trophozoites are oval in shape and the entire body is covered with longitudinal and slightly spiral rows of cilia. There is a large preistomial depression or vestibule leading to cytosome at the anterior end. The broad posterior end bears a cytopyge. The cytoplasm may be distinguished into outer ectoplasm and inner endoplasm. The endoplasm has food vacuoles, two contractile vacuoles and two nuclei- a macronucleus and a micronucleus. The micronucleus controls sexual reproduction while the macronucleus regulates metabolic activities. Nutrition is holozoid with food consisting of tissue fragments, bacteria erythrocytes etc.

Reproduction in *Balantidium* is both asexual as well as sexual. Asexual reproduction is by binary fission while sexual reproduction is by conjugation.

Experiment no. 6 ***Study of external features of Paramecium***

Phylum Protozoa	Sub kindgom Protozoa
Class Cliphora	Phylum Ciliophora
	Class Kineto frangminoporea

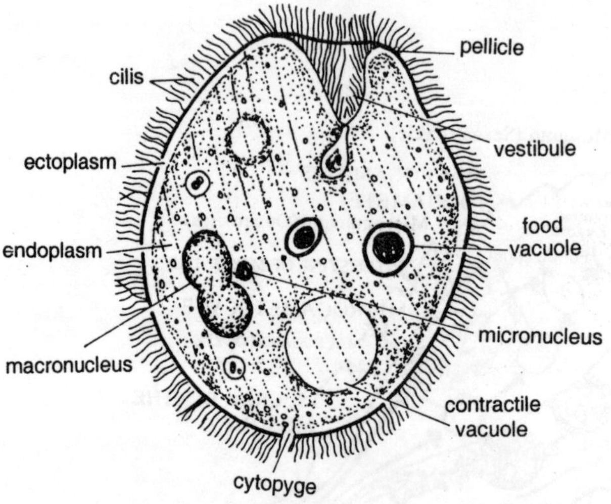

Fig. 29.6 Balantidium cell

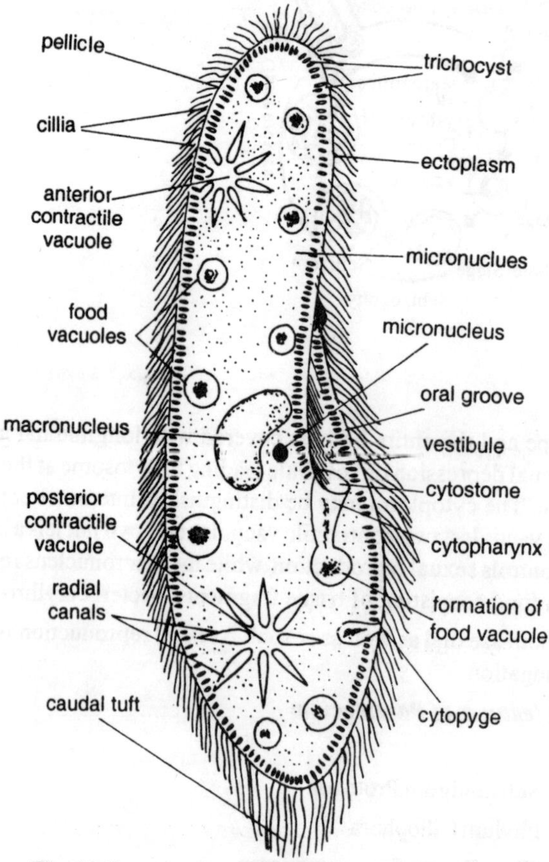

Fig. 29.7 Paramecium, from life.

Paramecium is popularly called 'slipper animalcule' because of the shape of the body. It occurs abundantly in fresh water ponds which are rich in decaying animal and vegetable remains. A drop of pond water freshly collected will show under the microscope these animlacules moving to and fro in the filed (of the microscope).

The body of Paramecium is made up of a single cell about 0.3mm in length and is shaped like the sole of a slipper hence the name slipper animalcule. One end which is somewhat narrow is called the anterior end and the broad end is called the posterior end. The anterior end is in front during locomotion. One of the flat surfaces which always faces the substratum and on which the animal glides is called the oral or ventral surface while the opposite end is known as the aboral or dorsal surface. Present in the oral surface near the anterior end is an oral groove. The oral groove leads to the mouth (of the animal) called cytosome. The mouth opens into a gullet or cytopharynx which passes obliquely downwards pointing to the posterior end. Usually food vacuole is always seen towards the end of the gullet.

Externally the body is covered by a cell membrane to which are attached a number of small hair like structures called cilia. These are the locomotory organs. The animal moves with the help of cilia and in addition, the cilia also help in gathering food particles due to their lashing movement. The cilia are arranged in longitudinal rows which run lengthwise in the hinder part of the body but follow the spiral twist on the oral surface. Two to three rows of cilia lining the gullet are somewhat larger than those present on the general surface.

Internal to the membrane (pellicle) is the cytoplasm divided into two regions-ecotplasm and endoplasm. Ectoplasm is the outer clear non granular region while endoplasm is the inner region made up of a granular mass.

Two contractile vacuoles are present in ectoplasm. Each contractile vacuole has a central vesicle and a number of radiating canals. The endoplasm consists of a macronucleus and micronucleus. In some paramecia a number of wedge shaped particles called kappa particles are present.

CHAPTER 30

TYPE STUDIES VI - HELMINTHES

The group helminthes meaning worm (in Greek) constitutes a large asemblage of simple and diverse types of parasitic and free living acoelomate and pseudocoelomate worms. The term worm however applies not just to a single phylum but several, like flat worms, round worms, earth worms etc which belong to different phyla.

In the present day usage the term helminthes is restricted to two phyla of invertebrates. These are

1. Platyhelminthes or flat worms
2. Aschelmixnthes or round worms

Experiment no. 1 ***Study of external features of Plannaria***

Plannaria belongs to the class Turbellaria. The name turbellaria is derived from latin word turbella means stirring. The animals produce small turbulance in water due to their ciliary action. There are two kinds of Plannarians - fresh water (eg. *Dugesia*) and terrestrial (eg. *Bipalium*).

The body of plannarians is broad, flat and leaf like in shape. They are free living in fresh water (sometimes saltwater also), or moist soil. Very rarely some may be commensals. There are two ends in the body. The anterior end is blunt while the posterior end is tapering. Anterior end is squarish or triangular with the apex pointing forward. This is the head. There are two projections called auricles at the base of the head. These however are not sensory. On the dorsal side of the head there are two black eyespots. On the ventral side, near the middle region is found the mount through which protrudes the pharynx. The muscular walls of the pharynx help in trapping the food. In mature worms, behind the mouth is situated a small genital pore. There are very minute excretory openings on the dorsal surface. The body wall is covered with ciliated epidermis.

Experiment no. 2 ***Study of external features of Liver fluke*** (*Fasciola hepatica*)

Liver fluke is an endoparasite in sheep, cattle, goats etc. It lives in the bile duct of sheep. Rarely it may be even found in the liver of man also. The body of liver fluke is flat and unsegmented about 25mm long and 12-13mm broad. The body is reddish brown in colour and bilaterally symmetrical. The anterior region is broad while the posterior region is pointed. The anterior region has the head lobe with an oral sucker enclosing the mouth. On the ventral side is a ventral sucker (acetabulum) helping the attachment of the parasite to the host. Anterior to the ventral sucker is found the gonopore through which eggs pass out.

The body wall of liver fluke is triploblastic and is made up of five layers - cuticle, basement membrane, circular layer of muscles, longitudinal layer of muscles and a layer of diagonal muscles.

Liver flukes are hermaphrodites. The male reproductive system consists of a pair of testes from which arise two tubes (vas deferens) which join to form a seminal vesicle. This opens to the genital atrium through the ejaculatory duct. The female reproductive system has a single, tubular, branched ovary found on the right side of the body in front of the testes. The branches of the ovary join to form an oviduct which enlarges to from the uterus.

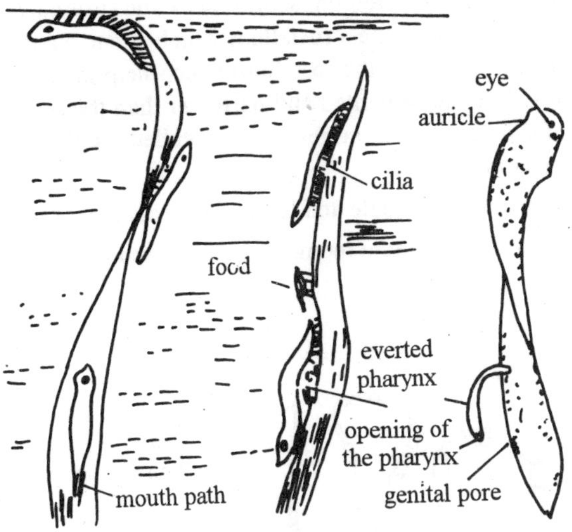

Fig. 30.1 A - Planarians as seen in nautre; and B- Planaria, side view

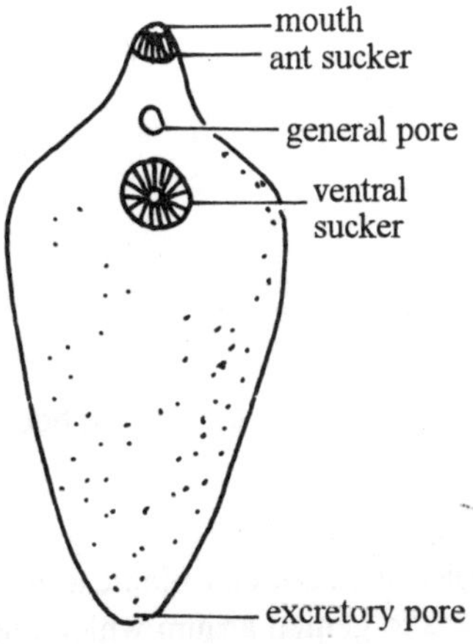

Fig 30.2 External features of Fasciola

The life history of liver fluke is complicated and involves two hosts - pond snail (intermediate host) and sheep (primary host). It also comprises the following larval stages - miracidium, sporocyst, redia, circaria and metacercaria. The miracidium larva is hatched out of the fertilized egg from sheep. It is a free swimming larva, lives for about 8 hours and gains entry into the pond snail. It turns into sporocyst within the liver of the pond snail where another larval stage the redia larva is formed. Finally the cercaria comes out of the body of the pond snail and swims freely in water for about 2-3 hours and settles on the leaf of any water plant. It encloses itself by a gelatinous wall and is now known as the metacercaia. The metacercaria are swallowed by sheep and the life cycle continues.

The infection by liver fluke causes liver rot in sheep and also bilary colic in man.

Experiment no. 3 ***Study of external features of Tapeworm*** **(*Taenia solium*)**

Taenia solium or pork tape worm is commonly found wherever pork is eaten. It is found as an endoparasite in the gut of vertebrates including man.

The body of tapeworm is elongate and flattened dorsiventrally like a ribbon. The length may be as much as 3-5 metres. The body is divided into three regions scolex, neck and strobila. Scolex or the head region is a pear shaped structure and has a terminal cone - rostellum which is surrounded by two rows of chitinous hooks. There are four suckers a little below the rostellum which help in attachment. The neck is situated immediately behind the scolex. It is unsegmented and narrower than the scolex and strobila. The strobila constitutes the main body and is made up of a number of units called proglottids. Each tapeworm may have as many as 800-1000 linear repeating units.

A mature proglottid consists of a set of male and female reproductive organs.

The body wall consists of cuticle, basement membrane and musculature.

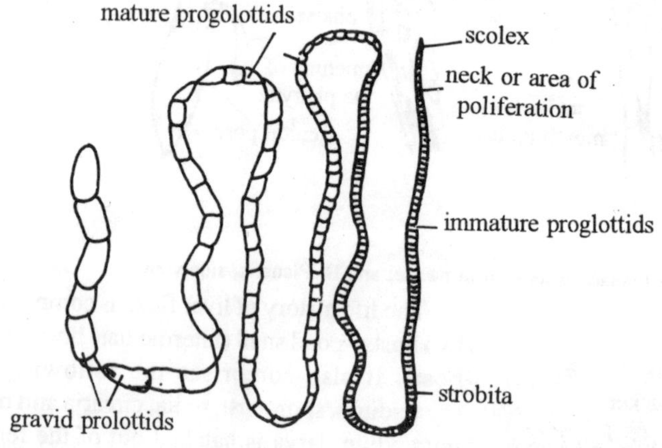

Fig. 30.3 Taenia solium. Entire

There are no respiratory organs in tapeworm; respiration taking place through the general body surface. Excretory system is same as found in trematodes - the protonephric type.

The nervous system has two ganglia situated in the scolex region.

Tape worm is hermaphrodite. The male reproductive system consists of numerous minute testes, efferent ducts, vas deferens (a large tube) and cirrus sac. The cirrus sac opens to genital atrium which open to the exterior through the gonopore.

The female reproductive system has a bilobed ovary which leads to oviduct. Uterus arises from the oviduct. The oviduct develops into an enlarged ootype where it joins the vitelline duct. Uterus arises from the oviduct. The vagina is a narrow tube arising from the junction of oviduct with the vitelline duct and opens to exterior through the gonopore.

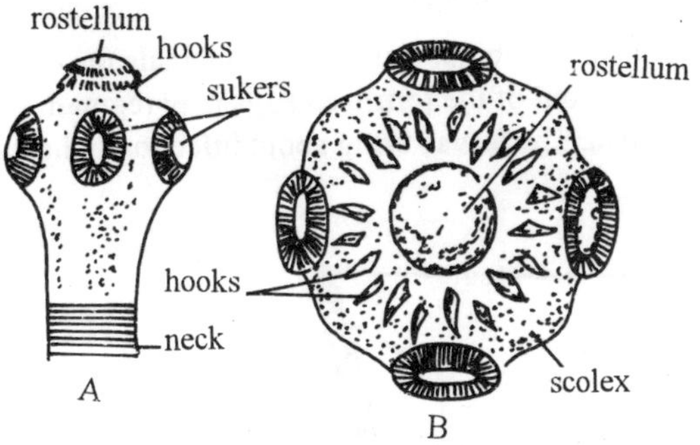

Fig. 30.4 Taenia solium. A- Scolex in magnified, B- Frontal

Experiment no. 4 ***Study of external features of common round worm*** (*Ascaris lumbricoides)*

Round worm is the most common endoparasite found in the intestine of man. It is worldwide in its distribution. In human beings *Ascaris* is more common among children than in the adults because of their propensity to contact contaminated material. The round worms usually inhabit the intestine, but in some cases they may even enter the bile and pancreatic ducts. They may also perforate the gut walls causing peritonitis.

Sexes are separate in *Ascaris*. Both male and females are elongated, cylindrical and have pointed ends. Males are 15-31 cm long and 2-4 mm in diameter while the females are 20-35 cms long and 4-6 mm in diameter. In addition, the males have a curved posterior end with two copulatory spicules protruding from the slit like opening. The females have an uncurved posterior end.

There are four longitudinal lines or streaks running throughout the length of the body. A triangular mouth is present at the anterior end of the body. It is surrounded by three denticulated (teeth like) lips. A circlet of four sensory papillae are found on the lips. There is an excretory pore situated midventrally a little behind the mouth. In females, a genital pore is present midventrally at about 1/3 distance from the anterior end. The anus is also present only in female.

The body wall is made up of a multi layered cuticle, epidermis and muscle layers. The body cavity is a pseudocoelom filled with a fluid.

Experiment no. 5 ***Study of external features of hook worm*** (*Ancylostoma duodenale)*

Ancylostoma is the common hook worm infecting man in tropical, subtropical and temperate regions of Asia, Africa, Pacific islands and South America. Apart from *A.duodenale, A.caninum* and *A.braziliense* infect cats and dogs.

The hook worm is found in the small intestine of man causing several minor discomforts like itching, bronchial problems, abdominal discomfort etc, hookworm does not have any hooks. Sexes are separate, the females are longer (than males) and measure about 10-13 mm while males are about 8-11 mm. The anterior end of both males and females are similar and curved dorsally. A large cup shaped buccal capsule is present to help in attachment to the intestinal wall. The posterior end of female is cone like and terminates in a minute spine. The posterior end of male is broad and has an umbrella like structure called bursa.

The male reproductive system consits of tubular testis while in female the ovaries are in the form of two convoluted tubes. The larve which come out of the fertilized eggs may regmain alive in moist soil for more than four weeks after coming out of man. The larvae bring about infection in healthy human beings.

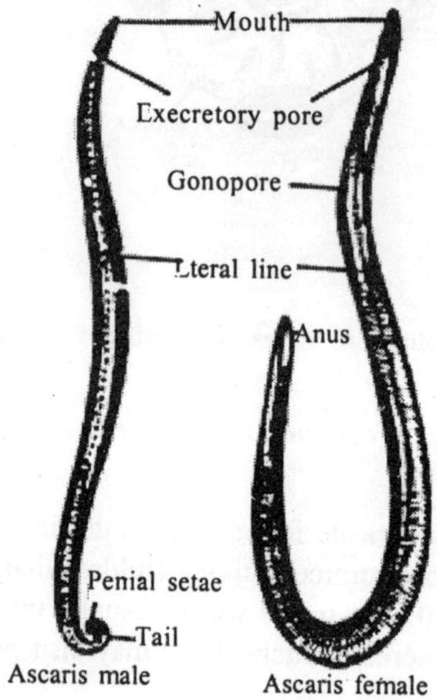

Fig. 30.5 Ascaris lumbricoides

Experiment no. 6 ***Study of external features of Filaria worm*** **(*Wucherria bancrofti*))**

Wuchereria or Filaria is the most harmful human parasite widely distributed in tropical and subtropical regions. The worm inhabits the coelomic membrane, lymph vessels and other tissues of man. In our country Filaria infection in prevalent in the west coast. Infection causes elephantiasis characterized by huge swelling of legs and affected parts. Adult worms are viviparous and produce a larva called microfilaria. The life history involves mosquito and human beings as two hosts. Microfilaria are transferred from affected human beings to healthy ones through mosquitoes. The disease is caused by the adult Filaria worm which brings about inflammation of the lymphatic system blocking the passage resulting in huge swelling.

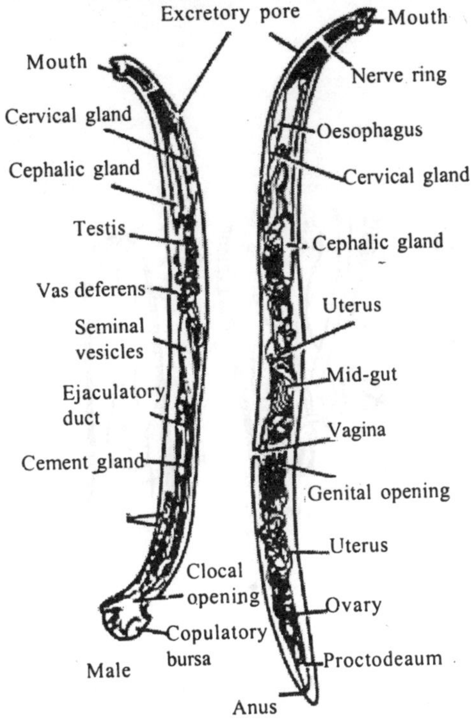

Fig. 30.6 Ancylostoma duodenale.

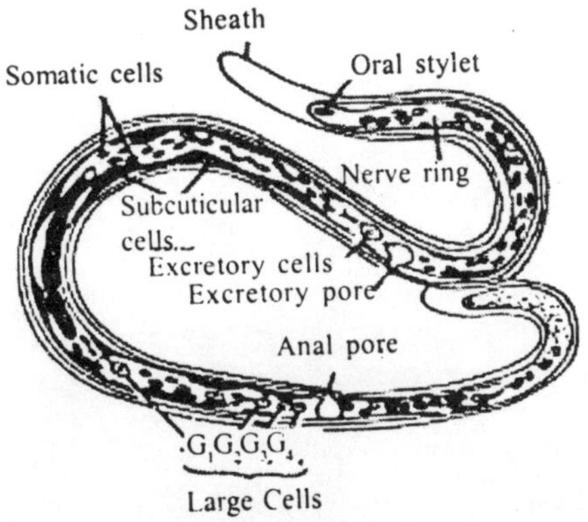

Fig30.7 Wuchereria. Microfilaria

Filariasis can be treated with arsenic preparations, Diethyl carbamazines etc. Prevention involves destruction of mosquitoes, either by insecticides or by checking their breeding, use of mosquito nets repellants etc.

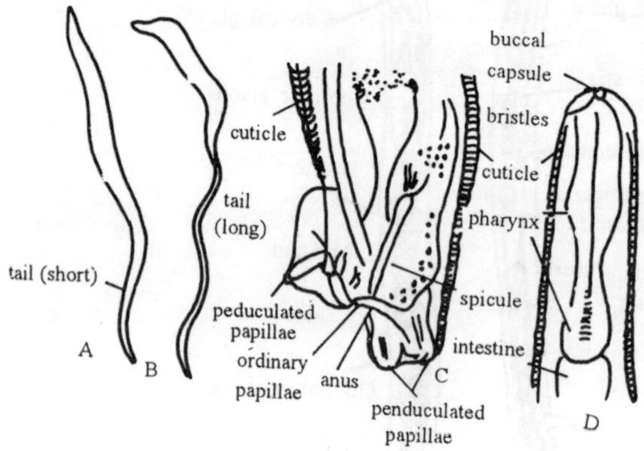

Fig. 30.8 Oxyuris equi. A and B - Short and long tailed femlaes; C- Poster of male, side view; and D - Anterior end of a female, dorsal view.

Experiment no. 7 ***Study of external features of Pinworm (Oxyuris)***

Oxyuris, commonly called pinworm is the most common pathogenic worm found in the terminal part of the colon in children. The females are longer than the males. The mature female creeps out of the anal aperture and lays eggs between the buttocks. This causes itching and while scratching, the worms may get transferred to mouth if hands are not washed properly. Use of antiseptic soap, trimming finger nails and such other hygienic methods will prevent infection.

TEXTBOOK OF
PRACTICAL BOTANY

TEXTBOOK
OF
PRACTICAL BOTANY

Volume - II

(Bryophyta, Pteridophytes, Gymnosperms)

By

Dr. S. Sundara Rajan

Principal

Reva Institute for Science and Technology Studies

R.T. Nagar

Bangalore – 560 024

ANMOL PUBLICATIONS PVT. LTD.

NEW DELHI - 110 002 (INDIA)

ANMOL PUBLICATIONS PVT. LTD.
4374/4B, Ansari Road, Daryaganj
New Delhi - 110 002
Ph.: 3261597, 3278000
Visit us: www.anmolbooks.com

Textbook of Practical Botany

First Edition, 2003

ISBN 81-261-1359-6 (Set)

PRINTED IN INDIA

Published by J.L. Kumar for Anmol Publications Pvt. Ltd., New Delhi and Printed at Mehra Offset Press, Delhi.

CONTENTS

PART THREE
GYMNOSPERMS

Preface

Laboratory study is an integral part of the knowledge that a student acquires in the college. This is true not only for botany but for all science subjects. Theoretical study of a subject has to be supplemented with laboratory experiments to provide a comprehensive understanding.

One of the main advantages of laboratory or practical study is the first hand knowledge that a student acquires about the material of study.

A systematic, structured and disciplined method of work is necessary in the laboratory to obtain maximum benefit. In order to help the students realize their goal practical manuals indeed play an important role.

Textbook of Practical Botany aims to provide the students, a useful hand book to guide them in the proper study of the subject material. Theoretical details are kept at the minimum and more emphasis is laid on understanding the structural details to enable the student a proper identification. A large number of properly labelled original diagrams are incorporated in the book as they are integral to the understanding of the subject.

This work is organised in the following four volumes:

Volume One deals with Algae, Fungi, Bacteria, Lichens, Viruses and Microbiology.

Volume Two deals with Bryophytes, Pteridophytes and Gymnosperms.

Volume Three deals with Taxonomy and Economic Botany.

Volume Four deals with Anatomy and Embryology of Angiosperms, Ecology, Plant Physiology and Cytology.

I have taken utmost care in the presentation of matter, style of writing and choice of illustrations so that the volumes are user friendly. The examples chosen for study are such that the books should cater to the needs of all Indian Universities.

Constructive criticism and suggestions are most welcome.

I am thankful to Mr. J.L. Kumar, Managing Director of Anmol Publications Pvt. Ltd., New Delhi, for undertaking the publication of the book.

Dr. S. Sundara Rajan
Principal

Reva Institute for Science and Technology Studies
R.T. Nagar
Bangalore - 560 024

CHAPTER 1

INTRODUCTION

The study of science subjects is carried out at two levels,viz., *Theoretical* and *Practical*. Theoretical study is carried out in lecture classes when the students are imparted all the details regarding a particular topic. This has to be supplemented by practical or Laboratory studies, when the students come into active contact with the objects about which they would have heard in lecture classes. Therefore it may be safely said that laboratory studies form an integral part of a comprehensive study of any topic in science subjects.

In the study of plants, laboratory assumes a special significance as the students acquire a meaningful and firsthand knowledge about the plants which they study in lecture classes. In fact as far as botany is considered any study of a plant without being supplemented by laboratory observation is futile. It is only in the laboratory study that the student can dissect, observe and analyze the various features of the plant and it is only a study of this type that makes the student acquire a meaningful knowledge about the pant. Besides observation and study in the laboratory plants are also to be studied in field to obtain an account of their habitat, habit, association and such other details.

In this chapter an attempt is made to explain some of the essentials that a student should understand before he begins a practical work.

Cleanliness: This is the most important point that a student should keep in mind while in the laboratory. The working table and the surroundings should be kept dry and spotlessly clean.. All the requirements needed by the students such as petri dishes, slides, reagent bottles, stains etc. must be properly labeled and arranged neatly. The microscope and particularly the optical parts must be cleaned with a white lint free cloth. After the completion of the assigned work, the student should take care to clean the place, remove any debris of the plant material and put it into the dust bin.

The students should work in silence and independently unless the assigned experiment is to be carried out by a group. When a particular work is assigned to a student, he must carefully plan and divide the available time for experimentation, observation and recording of the results. Whenever a doubt arises, it is always better to consult the teacher in charge, than a classmate. It is only by a proper and systematic work, that a student can derive maximum benefit from the practical classes.

Laboratory requirements

The department provides certain amenities, instruments etc to each student for the purpose of practical work. But a student also should carry some of the requirements in order to be effective in the practical class.

One cardinal principle that every student should remember in a practical class is to have one's own requirements and not to depend on anyone. Never lend or borrow any requirements. Borrowing or lending a pencil, needle or blade will disturb the work of both.

In general every student should individually possess the following requirements

1. *Two sharp pointed needles* mounted on stainless steel or plastic handle. This is essential for the purpose of separating and teasing the microscopic material.

2. *A pair of forceps:* This is essential to remove the material either from the bottle or from petri dish.

3. A sharp properly honed razor to cut sections of the plant material.

4. A new safety razor blade. This will help either to cut section or to trim the material.

5. Two fine hair paint brushes. This will be of use in lifting the sections of plant (with out damage) from the petri dish on to the slide.

6. Two sharpened pencils (H, Hb or 2b).

7. A pencil eraser.

8. A pencil sharpener.

9. Glass droppers.

10. A piece of thin, white, lint free cloth to clean the optical parts of the microscope.

11. An observation book to note down immediately the points of interest.

12. A practical record.

13. Glass slides and cover slips(if they are not provided by the department).

Scheme of Practical Work

Practical work in the laboratory will be most useful and effective if the student keeps in mind the following:

1. Maintain silence during working. All the requirements for the practical work must be absolutely clean.

2. Every student should work independently, unless the experiment involves the participation of others.

3. Before beginning the practical work, listen to the instructions of the teacher attentively and work accordingly.

4. Follow a systematic procedure in a study.

5. If the work involved is the morphological study of an organism, follow the procedure given below:

(a) Study of external characters - for example whether a plant is microscopic or macroscopic, unicellular or multicellular, filamentous or colonial etc.

(b) Study of internal structure like details of cell, tissues etc. If the material requires sectioning, sections are to be cut and internal details should be observed.

(c) Study the reproductive details like asexual and sexual reproductive organs.

6. Mount the material after properly teasing and separating the filaments, mycelia etc. Stain the material wherever necessary and cover the preparation with a cover glass.

7. Observe the slide carefully under the microscope, first under low power and then under high power.

8. Make sure that you have observed all the details as instructed by the teacher.

9. After all the observations are over, record your findings, structural details etc. in the practical records in the form of diagrams. Draw proportionate diagrams in the proper place on the drawing sheet. Diagrams should never be drawn to the very edge of the drawing sheet. Write suitable explanatory notes pertaining to the diagrams on the interleaf.

Laboratory Equipments

In addition to the individual requirements carried by the student, the department generally provides the

following equipments/instruments etc.to the students. Sometimes these equipments may be given to each student .

These are :

(1) Students compound microscope .
(2) Dissection microscope.
(3) Binocular Research microscope.
(4) Watch glasses.
(5) Cavity blocks.
(6) Slides and cover glasses.
(7) Petri dishes.
(8) Enamel tray.
(9) Polythene/glass wash bottle.
(10) Spirit lamps.
(11) Glass marking pencil.
(12) Microtome.
(13) Staining /bottles.
(14) Stone/ for honing the razor.
(15) Chemicals and stains.
(16) Culture media(to be prepared by students).

In the list of requirements mentioned above, we will discuss some in detail.

Microscope

Of all the instruments used by students in the botany laboratory, the one that is most indispensable to the students of biology is the 'Microscope'. It is not an exaggeration to say that no biological study in the laboratory is possible without the microscope.

The function of the microscope is to enlarge the small objects, so that they become visible to the human eye. In reality the magnification power of the microscope lies in its resolving capacity between one particle and another. A microscope distinguishes one particle from another which are situated very closely. A human eye cannot resolve particles that are lying closer than 0.025 mm. In its function a microscope can be compared to a human eye, but much more effective than that.

There are different types of microscopes, but all of them are based on the same principle with variations in their construction. Some of the types of microscopes used in the laboratory are:

(1) Compound microscope.
(2) Binocular Research microscope.
(3) Dissection microscope.
(4) Stereoscopic dissection microscope.
(5) Phase contrast microscope.
(6) Electron microscope etc.

Of the six types of microscopes mentioned above the compound microscope and the dissection microscope are the ones that are most frequently used by students. The Electron microscope is very costly and a very highly sophisticated instrument.

In this book the compound microscope and the dissection microscope are explained in detail as the students should be familiar with them.

Compound Microscope

The microscope has a strong 'U' shaped or 'V' shaped base. It is also called the horse shoe shaped base. Attached to the base is a bent limb. This limb can be tilted in any direction as it is attached to the base with an inclination joint.

Attached to the limb is a stage at its lower end and a body tube at its upper end. The body tube is attached to the limb with the help of a rack and pinion so that it can move up and down. To help in the movement of the tube during focusing, there are 2 knobs viz.,*Coarse adjustment* knob and *Fine adjustment knob*.

The body tube at its top end consists of the ocular or eye piece lens. The eye pieces usually have three magnifications 5x,10x,and 15x, and can be used to get a low or high magnification. Attached to the lower end of the tube is a revolving nose piece with three objective lenses low (10x) medium (40x) and high (60x). Some times there will be a fourth objective of the magnification of 100x. This is also called the oil immersion lens (while viewing through an oil immersion lens a drop of oil-liquid paraffin has to be put on the viewing slide).

Below the objectives and attached to the limb is the *stage* of the microscope where the slides are kept for viewing. The stage may be circular, square or rectangular in shape. In order to keep the slides properly placed on the stage, there are two stage clips. In some of the costly microscopes, the stage has movable sliding system operated by knobs to move the slides. This is called the mechanical stage. The mechanical stage permits the movement of the slide backwards or forwards, left or right with the turning of the knobs provided for the purpose. A hole is provided in the stage to receive the light from below and illuminate the object. Attached to the limb, below the stage is a condenser (fixed or movable up and down) lens. Below the condenser lens is an Iris diaphragm to regulate the light passing through. Iris diaphragm is a metallic disc with an aperture. The aperture can be narrowed or widened by turning the lever. The condenser which is a system of lenses condenses the light rays passing through and converges them on the object.

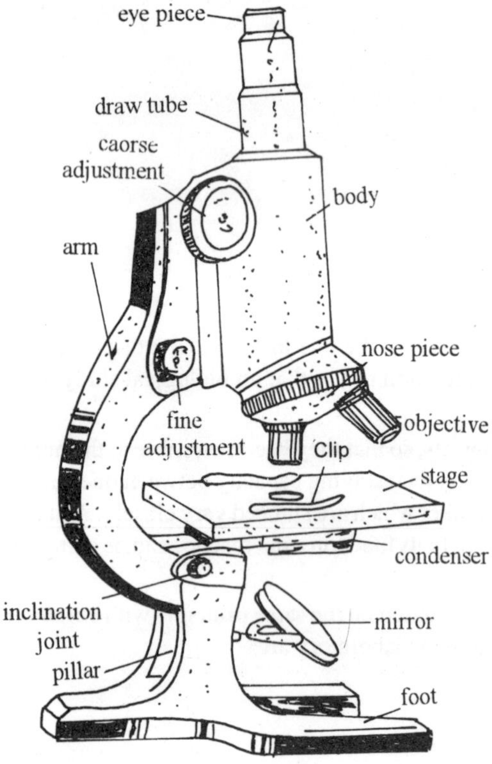

Fig.1.1 A compound microscope

The source of illumination in a student's microscope is a mirror which reflects the natural light or electric light. The mirror has two faces - plain and concave and is attached to the limb below the condenser. The mirror is adjustable and can be turned to the direction of light. When day light is the source of illumination, the concave side of the mirror is usually used to converge the rays of light. In some of the costly microscopes, instead of the mirror there is a built in illumination ,This is a tube with a lamp (9v or 12v) attached to a transformer, which can be plugged to any AC source.

Operation of the Microscope

Microscope should be placed in a well illuminated area but avoiding the direct light as it is harmful to the eyes. Generally northern light is preferable. When ' built in illumination' is employed, it is always better to use a blue or green filter to subdue the light.

The light has to be first focused on the condenser by adjusting the mirror towards the source of light. The amount of light entering the condenser may be regulated by closing or opening the iris diaphragm. The slide to be viewed is kept on the stage and fixed securely with the help of the stage clips. The object is first observed with a low power objective and then turned on to medium or high power, the specimen or section has to be covered with a cover slip. The slide should be observed with both the eyes open even while observing with monocular microscope.

Care of the Microscope

1. Clean all the parts of the microscope before and after use. The lenses and the mirror are to be cleaned with a lint free cloth or tissue paper.

2. Do not pull the microscope to the very edge of the table.

3. While lifting the microscope, use both hands. With one hand hold the base and with the other hold the arm.

4. The slide to be observed must be clean and dry (specially the under surface of the slide).

5. Do not remove the eyes piece or objectives from the microscope.

6. The microscope should be covered when not in use.

Binocular Research Microscope

This is like the ordinary compound microscope except that, it has binocular head for providing a natural stereoscopic view.

Dissection Microscope

This is a very simple microscope with only one set of lenses. Microscopes of this type will not give a very high magnification, but are very useful in dissecting flowers, plant parts etc.

The microscope has a 'V' shaped base and a limb. The stage is made up of a transparent glass plate. Below the stage is a mirror to focus the light. Attached to the limb is a mechanical arm which can be moved up and down with the help of a rack and pinion movement. The arm can also be moved horizontally. Attached to the free end of the arm is a holder to accommodate the lens (10x,15x etc.).

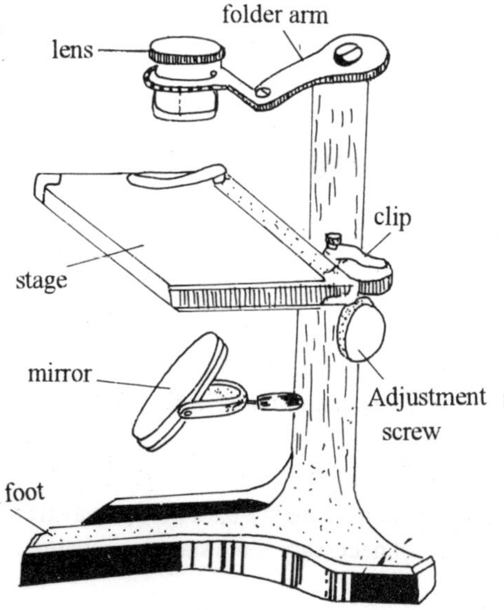

Fig.1.2 A dissecting microscope

Operation: The specimen to be observed / dissected is placed on the stage and located with the help of lens. Floral parts or others may be then dissected with the help of two needles, while still observing the specimen.

Stereoscopic Binocular Dissection Microscope

This is a combination of compound microscope and dissection microscope. While it has the higher magnification power of a compound microscope with two sets of lenses, it has retained the versatility and the manoeuvourability of a dissection microscope.

Phase contrast and Electron microscopes are not normally used by students.

Other Laboratory Equipments

Stains: These are colouring fluids. The specimens to be observed (algal or fungal material or sections of plant parts, would be generally decolourised due to preservation. These are to be coloured in order to observe them clearly. The following are the stains that are generally used in the laboratory.

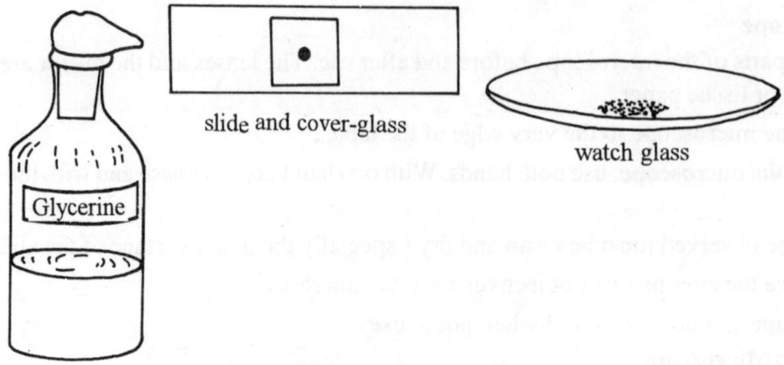

Fig.1.3 Dropping bottle

(a) Saffranin: This is the most commonly used stain. (1) It stains nucleus and lignified cell walls.

Preparation —

Saffranin —1gm.

95% (1) Ethyl alcohol –50ml.

Distilled water – 50ml.

An aqueous stain may also be prepared.

Saffranin – 18ml

Distilled water – 100ml.

(b) Fast Green

Fast green – 1gm.

90% Ethyl Alcohol – 100ml.

This stain is used to colour parenchyma cells and cytoplasm. Fast green may also be prepared in clove oil (75ml clove oil+25ml 100% alcohol).

(c) Cotton blue.

Cotton blue –1gm.

Lactophenol – 100ml.

(d) Haematoxylin

The dye is obtained from the heartwood of the tree *Hametoxylon campechianum,* There are two methods of preparation of the stain. These are:

1. *Heidenhain's Hametoxylin*

Dissolve 0.5 or 1.0 gm of the powder in 100 ml of distilled water. Filter and store in a dark bottle away from light. The stain is ready for use after 4-5 days of ripening. Ripening is not necessary if the stain is prepared in 70% ethyl alcohol.

2. *Delafelds Haematoxylin*

Sol.1 Aqueous solution of ferric ammonium sulphate 100ml).

Sol.2 One gm dye dissolved in 6ml of absolute alcohol.

mix solutions 1 and 2. To this mixture add 25 ml of glycerine and 25 ml absolute alcohol. The solution should be allowed to stand until it becomes dark red.

(e) Anilin blue

Also called cotton blue or China blue. It is prepared as follows:

Aniline blue –1gm.

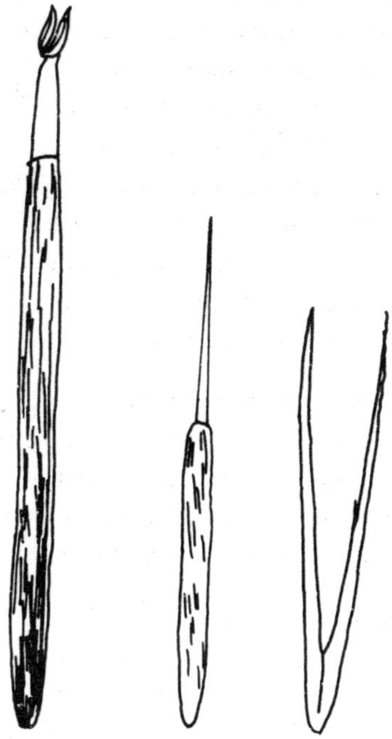

Fig.1.4 Brush Needle Forceps

Ethyl alcohol (90%)
or — 100ml.
Distilled water

Hydrochloric acid – 2drops.

(f) Gentian Violet (Crystal Violet)

Gentian Violet – 1gm.

90% Ethyl Alcohol 100ml.

(g) Erythrosine

Erythrosine –1gm.

90% Ethyl alcohol – 100ml.

Erythrosine may also be prepared in clove oil (Absolute alcohol 0.5 ml +95ml Clove oil).

(h) Light green

Light green –1gm.

Clove oil –95ml.

Absolute alcohol –5ml.

Light green may also be prepared without clove oil (in 100ml Absolute alcohol).

(i) Acetocarmine: The stain is prepared as follows:

Add 1 gm of stain to 100 ml of boiling acetic acid (45%)or propionic acid. Cool and decant, A few drops of saturated ferric acetate solution is added and the mixture is cooled over ice. The solution is allowed to stand for about 12 hours. Filter and store in a dark coloured bottle under refrigeration.

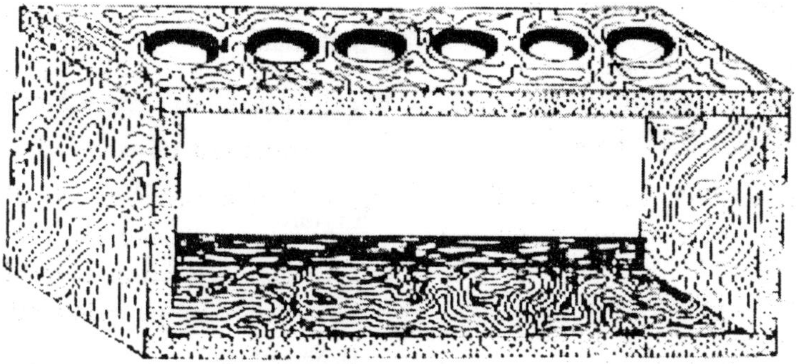

Fig.1.5 Staining rack

Specificity of Stains

Usually specific stains are employed in colouring different plant parts. The stains are used singly or in combination. The most popular stain combination are:

Saffranin - fast green.
Haematoxylin - Erythrosin.
Haematoxylin - light green.

While stain combinations are employed to study pteridophytes, gymnosperms and Angiosperms, single stains are used for Avascular cryptogams.

Green or Saffranin for Algae and Bryophytes.

Cotton blue for fungi.

The following table gives the specificity of stains to different plant parts or cell organelles.

Structure	**Stain**
Cellulose	Anilin blue Haematoxylin Fast green Light green
Cutinized cell wall	Gentian violet Erythrosine Saffranin
Lignified cell wall	Gentian violet Saffranin
Suberised cell wall	Saffranin
Cytopalsm	Fast green Light green Aniline blue
Chitin	Saffranin
Callose	Aniline blue
Achromatic figure	Aniline blue Erythrosine
Proteins	Saffranin
Nucleus	Heamatoxylin Saffranin Gentian violet
Chromonemata	Heamatoxylin

Chromosomes	Saffranin
Plastids	Gentian violet Haematoxylin
Mitochondria	Gentian violet
Flagella	Cotton blue

Fig.1.6 Razor

Fixing / Preserving Fluids

The plants or plant parts collected in nature are to be killed and preserved in the laboratory for the purpose of storage and study.

In killing the specimen, the various life processes are stopped at that instant, so that after preparation for microscopic study various stages may be observed.

When ever a material is collected, it is better to kill it in the field, so that there will be minimum or no distortion. Alternatively the material collected in the plastic bags or bottles should be brought to the laboratory and killed / preserved. In preservation, the material must be completely immersed in the fluid. The following are some of the killing/preservative fluids generally used:

1. *Formaline acetic alcohol (F.A.A)*

This is the most commonly used killing /preserving agent. It is popularly called standard preservative.

Ethyl alcohol (70%)	- 90ml.
Glacial acetic acid	- 5ml.
Formalin	- 5ml

2. *Formalin - Propiono - Alcohol*

Formalin	- 5ml
Propionic acid	- 5ml.
Ethyl alcohol 70%	- 90ml.

3. *Carnoy's Fluid - I*

Ethyl alcohol	- 30ml
Glacial acetic acid	- 10ml

4. *Carony's Fluid - II*

Ethyl alcohol	- 30ml.
Glacial acetic acid	- 30ml.
Chloroform	- 15ml.

Carony's fluid is widely used in killing root tips, flower buds etc. for the purpose of cytological study.

5. *Randolph's Modified Navashin Fluid*

This consists of two solutions A and B.

Solution A

Chromic acid	- 5.0 gms.
Glacial acetic acid	- 50ml
Distilled water	- 320ml.

Solution B

Commercial formalin	- 200ml.
Distilled water	- 175 ml.
Saponin	- 3gms.

Mix A and B in equal volumes before use.

Mounting media

These are chemicals used for the purpose of mounting (Temporary/Permanent) the specimens on the slide for microscopic observation. The following are some of the mounting media.

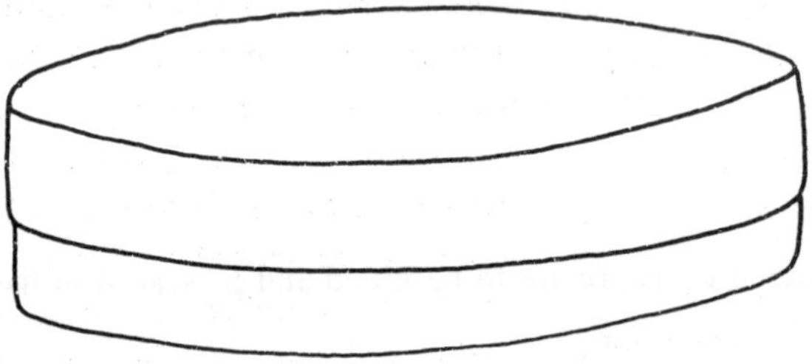

Fig.1.7 Petri dish

1. ***Acid water***

HCl – 10-15 drops
water – 100 ml.

2. ***Lactophenol***

Phenol	-100gms
Lactic acid	-100 ml.
Glycerine	-100 ml.
Distilled water	-100ml.

3. ***Glycerine***

Glycerine	- 15-20 drops
Formaline	- 1-2 drops
Distilled water	- 100ml.

4. ***Canada Balsam:*** This is a gum obtained from Gymnosperms. This fluid is readily available in the market.

5. ***Euparol:*** This is available in the ready to use form in the market.

Canada balsam and Euparol are used for permanent mounting, while the rest are used for temporary preparations.

Staining rack: This is a wooden rack used for holding the staining bottles.

Dropping bottles: This is a glass bottle having a narrow mouth fitted with a stopper. The stopper is grooved and has a beak. When the bottle is upturned the stain/fluid comes out in drops.

Slides: These are readily available in the market. The usual size 25mm x 75mm. The thickness is about 1mm. Special types of slides called cavity slides are available for specific study.

Cover slips (Cover glasses)

These are meant to cover the materials when the preparation is ready for observation. The coverslips may be round, square or rectangular.

Methods of Study of Plant Material

The following is the procedure for a scientific study in the laboratory.

Morphological study

The specimen given to study external detail should be observed either with or without a magnifying glass. Whatever details can be seen are to be recorded. If the specimen requires microscopic study, it should be suitably prepared for the purpose of microscopic observation. There are two types of microscopic preparation.

1. Whole Mount Technique

This is suitable for the purpose of external study of algal and fungal material. Take a little of the material, stain it with a suitable stain (staining procedure is explained later.) Wash the excess stain in water / alcohol. Mount in a suitable medium (temporary or permanent) and observe under the microscope.

2.Section cutting

For the purpose of anatomical study, sections are to be cut of plant parts. In algae, globule of *Chara*, stolons of *Caulerpa*., receptacles of *Sargassum* etc., can be cut. Similarly fruiting bodies of fungi, thalli of bryophytes also many be cut in a suitable plane. In higher plants cylindrical plant parts like root, stem, petiole etc. may be cut for the study of internal structure.

Planes of Cutting: In order to study the internal structure from various angles, sections are cut in various planes.

*Transverse :*The section is cut at right angles to the longitudinal axis.

Longitudinal : The section is cut parallel to the long axis (right angles to the transverse axis. The longitudinal section may be cut along a radius (Radial longitudinal section-RLS) or along a tangent (Tangential longitudinal section-TLS).

Cylindrical organs like roots and stems are cut transversely or longitudinally, while dorsiventral organs like thalli of bryophytes, leaves etc. are usually cut in vertical transverse plane.

Types of Section cutting

There are usually 2 types of section cutting methods employed for plant parts. These are (a)Free hand sectioning and (b) Microtome sectioning .

(a) Free hand Sectioning

For the purpose of laboratory study where slides are not made permanent, usually free hand sectioning method is employed. Free hand sections are cut mostly with the help of a good sharp razor (sometimes safety razor blades also may be used).

Honing the razor : The razor used for sectioning must have a sharp, even edge. There should be no nicks. Before cutting sections, the razor edge has to be honed to maximum sharpness either on a stone or on a leather strap. Oblique strokes, with the edge forward are to be given on the sone or strap. Usually oil drops are used on the honing medium to give maximum sharpness.

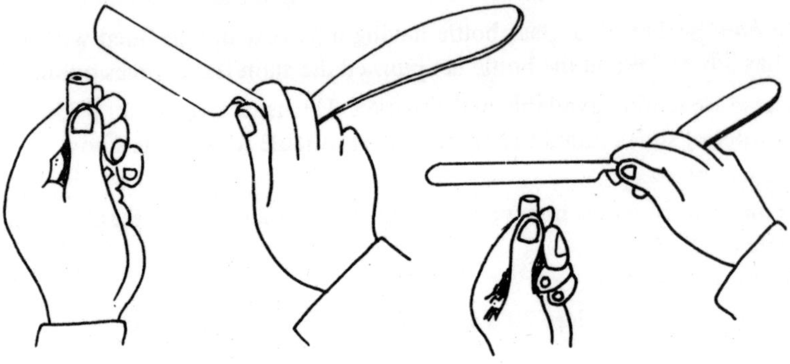

Fig.1.8 Method of holding pith | Methoding of holding razor | Method of section cutting

Method of cutting : Open out the razor and bend the handle backwards. Hold the razor between the index finger and thumb. The handle should be free. The remaining fingers of the hand will be touching the back edge of the razor blade.

In order to cut a transverse section of a cylindrical part like stem or root, follow the procedure given below.

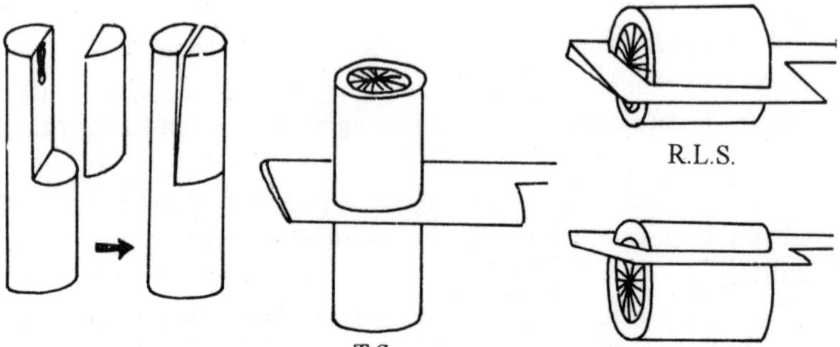

Fig.1.9 Method of using pith for insertion of material

The stem or root should be cut into a 2-3" long piece. It should be held in the hand in an erect position between the thumb and index finger. The material held in the hand should be exactly at right angles to the razor's edge. The thumb holding the specimen should be slightly at a lower level. Flood the edge of the razor with water. Now using the index finger as the platform move the razor quickly several strokes across the material. During this process sections are cut and they will be collected on the blade of the razor. The sections now may be carefully transferred with the help of a fine brush in to a petridish containing water.

For cutting sections of delicate and fragile materials like thalli or leaves, a suitable supporting medium has to be used. For this purpose a pith, carrot or potato may be used. These are to be cut into rectangular cubes and then split at one end. The material for sectioning should be inserted in the slit and sections are cut

After section cutting the razor should be wiped dry. It should be greased and encased.

Microtome Sectioning: This is generally employed very frequently in the study of bryophytes, pteridophytes and other higher plants. The microtome sectioning is a long and complicated process. It takes anywhere between 15-20 days to process a material and to obtain the sections.

Microtome sectioning involves the following steps:

1. Killing and preserving the material.
2. Dehydration and infilteration.
3. Paraffin embedding .
4. Section Cutting.

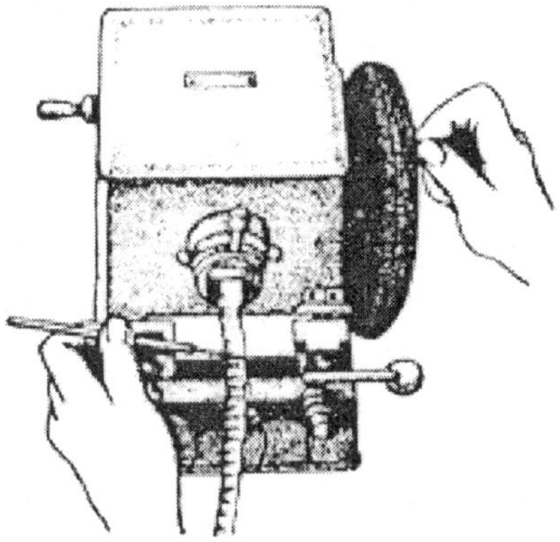

Fig.1.10 Rotary microtome

The material like stem, leaf, flower etc. is to be killed and preserved in a suitable fluid.

(a) Cut the material into bits of suitable size.

(b) Place the material in small vials containing a fixative or preservative.

(c) Label the vial giving the details of place of collection , name of the material, preservative used and date of collection.

The material should be in the killing fluid at least for six hours (preferably a day).

Dehydration: In order to dehydrate the material, remove it from the preservative and wash thoroughly in water and transfer it to a suitable tube or vial.

Dehydrate the material in alcohol series (Ethyl alcohol) as follows

Stage 1	50% Alcohol	First day
Stage 2	70% Alcohol	Second day
Stage 3	90% Alcohol	Third day
Stage 4	100% Alcohol	Fourth day
Stage 5	Absolute alcohol and xylol (50/50)	Fifth day
Stage 6	Absolute alcohol and xylol (25/75)	Sixth day
Stage 7	Xylol	Seventh day

When the material is transferred to pure xylol, there should be no fogging (i.e the solution should not turn milky). If there is fogging repeat the procedure starting from 4th stage.

The material in xylol will be completely dehydrated and looks transparent. It is now ready for infilteration by paraffin wax. The purpose of infilteration is to see that the wax enter into the material completely so that it can easily be cut.

Infilteration: Add a few pieces of paraffin wax everyday for about 4-5 days and then transfer the vial into a hot air oven, while temperature is adjusted to the melting point of wax. Everyday add a few pieces of wax. Adding of wax should be continued until all the xylol in the vial is replaced by paraffin wax. When the tube is smelt, there should be no smell of xylol. The material is now ready for embedding.

Paraffin embedding: In order to cut the material, it has to be embedded into a paraffin block. Prepare a paper boat or take a porcelain rectangular dish and smear glycerine to the inner surfaces. Pour molten paraffin into the porcelain dish and empty the contents of the vial into the dish. Before the paraffin solidifies, arrange

the material in proper rows and in such a manner as desired. Leave it for cooling for 6-8 hours. For quick cooling the dish may be transferred to a bowl containing ice water. Carefully remove the solidified paraffin block from the dish and cut it into suitable pieces. The material is now ready for sectioning.

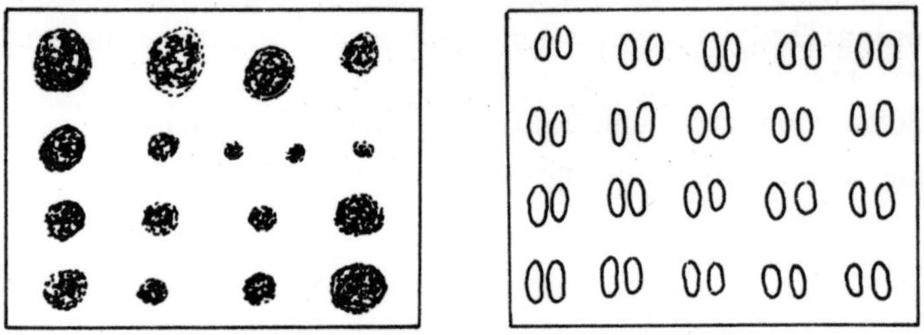

A B

Fig.1.12 Arrangement of materials in paraffin blocks - A) Flower buds in groups B) flower buds in pairs

Sectioning: Sectioning is carried out with the help of an instrument called microtome. There are various types of microtomes like Rotary, sliding, Freezing etc. The one that is normally used is a rotary microtome. The following procedure can be adopted in the case of the rotary microtome.

1. The paraffin block is to be cut into small pieces each having the embedded material.

2. The pieces of paraffin containing the material is to be fixed to wooden 'riders'. For this purpose, heat the top of the rider and the base of the paraffin piece with a hot knife and fix. Allow it to cool. The paraffin piece is now firmly fixed to the rider. In order to make the edges of the paraffin even, cut the edges and make them parallel.

3. Fix the wooden rider into the microtome so that the paraffin piece with the material is exactly parallel to the edge of the microtome knife.

4. Adjust the knob of the microtome to give the required thickness to the sections.

5. Rotate the wheel of the microtome in a clockwise direction. The cylinder containing the rider moves up and down on the knife edge. Sections are cut in a sequence and get attached to each other forming a paraffin ribbon.

6. Keep a needle in your left hand to hold up the ribbon as it lengthens. When the ribbon is sufficiently long, cut it and transfer it on to a black cardboard. Cut the ribbons into suitable length and transfer to the slide.

7. Smear the slides with an adhesive like egg albumen or gelatin and rub it well.

8. Pour some water onto the slide with the help of dropping bottle.

9. Transfer the paraffin ribbons on to the slide and drain excess water.Retain a little water on the slide.

10. Place the slide on a slide warmer. The ribbon expands, and so also the sections of the material. Now draw out the water completely and allow the slides to cool.

11. The slides should be allowed to stand for atleast 24 hours whereupon the sections will stick to the slide permanently.

The slides are now ready for staining.

Staining procedure

The plant material (whole mount or sections) are to be stained and mounted properly before they are ready for microscopic observations.

Algal filaments, fungal mycelia, pollen grains and spores, prothalli of pteridophytes are usually stained with a single stain - saffranin or green. In some case Aniline blue is used to stain the filaments of Algae and mycelia of Fungi. The staining procedure is as follows.

Place the material either in a watch glass or cavity slide and add a few drops of stain. The period of staining varies with different materials. But usually a 5-10 minute period is sufficient. Drain out the stain and observe the material under the microscope. If the material has taken less stain, it may be kept in stain for some more period, or if it has taken excess stain it can be removed by washing either with water or 50% ethyl alcohol.

After the material is suitably stained pass it through a series of alcohol grades to dehydrate it. Usually different grades of alcohol (70%,90%,100%) taken in a filler are put on the material. After passing through absolute alcohol the material is kept in xyloi and it is now ready for mounting.

Fungal mycelia can be stained with aniline blue. Usually the material taken on a clean slide is stained with a drop or two of the stain prepared in lactophenol. The material can bemounted straightway as lactophenol is also a mounting medium. If the stain is not prepared in lactophenol the material may be transferred to fresh lactophenol and mounted.

Double Staining Procedure

In the case of most of the algae and fungi there is little or no tissue differentiation and hence single staining is sufficient. In some cases of plants with relatively more tissue differentiation, a single stain is not sufficient to mark out the various tissues clearly. In such cases 2-3 stain combinations are used to stain the tissues. For example lignified and non-lignified tissues are stained with a combination of acidic and basic dyes.This type of staining is also called differential staining where different tissues take up different stains.

The following are some of the popular stain combinations.

Main stain	*Counter stain*
Haematoxylin	Saffranin
Haematoxylin	Erythrosine
Haematoxylin	Eosin
Haematoxylin	Orange G.
Saffranin	Fast green
Saffranin	Light green
Saffranin	Aniline blue
Saffranin	Crystal violet
Crystal violet	Erythrosine

Staining Procedure for free hand sections

(a) Procedure for temporary preparations

Free hand sections of root, stem etc., should be washed thoroughly in distilled water in a watch glass. After repeated washing the sections should be transferred to a watch glass containg the main stain. The sections should be kept in the stain for a period ranging from 30-40 minutes. After this period, the sections should be transfereed to a watch glass containing distilled water. The sections should be washed for 2-3 minutes whereupon excess of stain is removed. They should be observed under the microscope. If the main stain is saffranin only lignifed tissues should retain the stain, others should be destained. If the main stain is specific to non lignified tissues, only they should retain the stain. Then the sections should be washed thoroughly in water and repeatedly observed under the microscope. The sections can be mildly stroked with a fine hair brush, so that excess of stain comes out with distilled water. The sections are now ready for counter-staining. They should be transferred to a watch glass containing the counter stain (fast green etc). Counterstain acts quickly on the tissues. Hence one to two minutes of staining should be sufficient. The sections are now observed under the microscope. If there is excess of counter stain, it can be removed by

washing sections with 15-20%glycerine. After the final stianing the tissues whould be demarcated clearly with no mixing of the stains. The sections are now ready for mounting.

(b) Procedure for permanent preparaions

If the sections of plant materials are to be mounted and made into permanent slides, the procedure should be slightly altered so as to include dehydration and clearing with suitable agents. The procedure is as fallows:

1.Stain with the principle stain (Saffranin, Haemotoxylin etc).

2.Allow the sections to remain in stain for about 15-20 minutes.

3.Transfer the sections to water and wash thoroughly. The sections should be stroked mildly until no more stain comes out, and the water should remain clear. If the main stain like saffranin is produced in 50% alcohol then destaining (washing) should be done in 50% alcohol.

4.The sections should be now dehydrated in alcohol grades of 50%, 70% and 100%.

5.In each grade of alcohol keep the section for about 5 minutes.

6.Transfer the sections from one grade of alcohol to the other with the help of a fine brush. Never use needles as they injure the sections.

7.Transfer the sections to the counter stain (fast green etc). The counter stains are usually prepared in 90% alcohol.

8.Keep the section in counter stain for about a minute. Wash the sections in 90% alcohol and transfer them to 100% alcohol.

8a. In some case counter stains are prepared in clove oil. In such cases counter staining should be done after the sections are completely dehydrated with absolute alcohol.

1. Sections are transferred completely dehydrated to a solution of 50:50 Absolute alcohol and xylol followed by 25:75 Absolute alcohol and xylol and ultimately the sections are transfered to pure xylol. Now xylol should not turn milky or turbid. If it turns milky, dehydrate the sections again through combination of absolute alcohol : xylol series.

The sections are now ready for mounting.

Staining procedure for microtome sections

The slides with the paraffin ribbons adhering to them have sections of plant material.The processing and staining is as follows:

1. Immerse the slides in a coupling Jar containing xylol. Leave them for 10 minutes . The paraffin dissolves in xylol and only the sections remain on the slides.

2. The slides are now kept in 50:50 absolute alcohol and xylol.

3. Pass the slides through a down grade series of alcohol. 100%, 90% and 70%

4. Transfer the slides to pure water and wash them thoroughly to remove traces of alcohol.

5. Keep the slides in a mordant like 4% iron alum. This is necessary only when Haematoxylin is used. After about 10 minutes wash the slides in water.

6. Transfer the slides to the main stain Haematoxylin. Saffranin etc. and keep them for 15-30 minutes.

7. Remove the slides from stain and wash in water.

8. Observe the slides under microscope. Usually the sections are overstained.

9. If the main stain used in Hametoxylin, destaining can be done with 2% Iron alum or with water containing few drops of concentrated HCI. Repetaedly observe the slides under the microscope. Stop destaining when only the nuclei have retained the stain.

Caution: When destaining is done with HCI, not more than 5-6 drops should be used for 75ml of water.

Excess acid will eat away the adhesive and the sections come out of the slide.

10.Wash the slides in water overnight if the destaining agent is Iron alum, otherwise washing for 30 minutes is enough.

11.Now pass the slides through an upgrade series of alcohol (5 minutes in each grade) 70% 90% 100%

12. When the sections are completely dehydrated stain them with the counter stain (Fast green etc) for a minute or two. Wash the excess stain in clove oil and transfer the slides to pure xylol. The slides are now ready for mounting.

Mounting

(a) *Mounting Media*

Mounting media are of two categories: Temporary mounting media and permanent mounting media.

A temporary mounting medium may be just water or water mixed with glycerine. If the temporary sections are to be retained for 2-3 hours, dilute glycerine (15-20%) is recommended.

Permanent mounting media are used when the slides are to be retained indefinitely. The following are some of the permanent media.

(1) **Canada Balsam:** This is the most widely used mounting medium in botanical preparations. It is available in ready to use bottles. It is a gummy substance. If the solution is very thick add little xylol.

Canada balsam is a resinous product of a Gymnosperm, Abies balsamea. The gum has very good refractive index and as such it is highly reccommended.

(2) **Lactophenol:** Lactophenol can be prepared in the laboratorty. Mix equal quantites of phenol cystals, lactic acid glycerine and distilled water.

(3) **Glycerine gelly:B** This is prepared as follows:

Gelatine 1 part.
Water 6 parts.
Glycerine 7 parts.

Glycerine jelly is suitable even for mounting fresh specimens (even without dehyderation).

Add gelatine to water, stir thoroughly and warm. Add glycerine and cool the solution. After sometime the solution settles into a jelly.

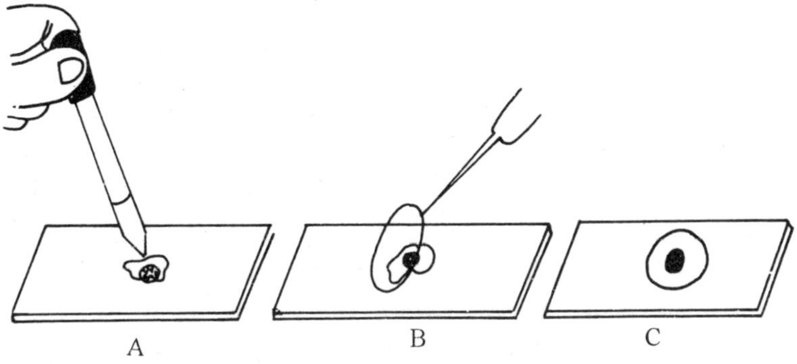

Fig.1.13 Method of mounting the material

(b) Mounting Method

The sections (free hand) are to be mounted on to clean slide in a specific mounting medium. If the sections are microtome preparations they would be already adhering to the slide. In any case the sections are to be covered with a thin coverslip or cover glass which are readily available in the market. They may be round, square or rectangular. Select the one that is suitable for your preparation.

The sections or objects should be mounted on the centre of the slide. In order to ensure placing of the object on the centre of the slide, the following procedure may be adopted. On a white paper draw a rectangle of the same size as the slide. Join the opposite corners by drawing diagonal lines. The centre of the slide lies where the two lines cross. Place the slide on the rectangle and material or objects may be kept on the central point.

Place a drop or two of the mounting medium. Hold the coverslip at the edges (see the figure) and allow it to touch the edge of the mounting medium at an angle of approximately 45°. Slowly lower the cover glass on the mounting medium with the help of a needle. Take care to see that no air bubbles enter the mounting medium. If there are air bubbles, lift the cover glass and mount again.

Excess mounting medium may be carefully wiped with the help of a clean tissue. Allow the slides to dry. If necessary the slides may be transferred to a hot air oven for drying. The slides should be labelled at one end as shown in the figure. They are now ready for microscopic observation.

Ringing of Slides

Temporary preparations may be kept for few days if the borders are sealed. This procedure is known as ringing . Ringing is done by applying wax, canada balsam or even nail polish to the edges. Ringing may be done manually or by using a ringing table having a rotary metallic disc.

Record of Practical Work

Practical work carried out in the laboratory should be recorded. The botany practicval record book has two types of sheets viz., Interleaves and drawing sheets. Interleaves are meant to write the notes pertaining to the diagrams. The drawing sheets which are thicker than the interleaves are used for diagrams. The following points may be useful in maintaining a good record.

1. Use good pencils (HHB) with sharp points and a good quality eraser.
2. Leave a minimum of one inch margin on all the sides.
3. Always write diagrams of only one plant in a page. Do not mix up diagrams of other plants.
4. On the left hand corner of the drawing sheet write the date of the experiment.
5. Give the complete systematic position of the member on the top of the drawing sheet.
6. For microscopic organisms draw first the habit sketch followed by low power and high power view/s. High power view should show the cellular details.
7. In the case of anatomical drawings, draw the ground plant first, followed by a sector enlargement to show the cellular details.
8. All parts of the diagrams must be properly labelled.
9. Stages of reproduction, if any, must be drawn after the details of vegetative structure.
10. The description corresponding to the diagrams must be written on the opposite interleaf.

Field Collection of plants

Specimens of algae, fungi and bryophytes can be collected in the field for the purpose of laboratory study. Field trip is an integral part of botanical study and all students must participate in it. It is in the field study that the students will come to understand the natural growth, habitat and environmental association of plants. During a field trip to collect specimens, the following items are to be carried:

1. Pocket lens.
2. Fixative F.A.A or Carnoy's fluid.
3. Collection bottles.
4. Blotting paper.
5. Polythene bags
6. Field note book
7. Labels.

When the specimens are collected they can be fixed and preserved for the purpose of laboratory study. Use the required preservative and immerse the specimens in it. Label the bottle showing details of

Date of collection

Location

Preservative fluid (name) used.

Name of the plant

As it is necessary for the purpose of identification, details of habitat, associations with other plants if any, colour of the fresh specimen etc. should be recorded in the field note book.

PART ONE
BRYOPHYTA

CHAPTER 2

GENERAL CHARACTERS AND CLASSIFICATION

Bryophytes include nearly 960 genera and 24,000 sporophyte. They represent a unique group of plants intermediating between aquatic thallophytes and terrestrial pteridophytes. They occupy and inhabit a zone which is neither completely land nor completely water. It is for this reason Bryophytes are often regarded as the amphibians of the plant kingdom.

Bryophytes are intermediate between thallophytes and pteridophytes not only in their habitat but even in the structure and organisation of the plant body also. As such one can see the blending of algal and pteridophyte characters in Bryophytes.

Bryophytes abound on habitats which offer moisture. They grow in large numbers on wet soil, damp cliffs of rocks and even on the branches of trees as epiphytes (porella).

Life cycle pattern

The life cycle of Bryophytes consists of a regular hetermorphic alternation of generations. The gametophyte is the predominant plant body in the life cycle whereas the sporophyte is short lived generation totally or partially dependent on the gametophyte. In fact whenever we refer to the plant body in Bryophytes, it is the gametophyte that wehave in mind.

Structure of Gametophyte

The gametophytes of Bryophytes can be broadly categorised into 2 types namely (1) Thalloid forms (2) Leafy forms.

Thalloid forms. In this type, the plant body consists of a dorseventrially flattened structures called the thallus which groups prostrate on the substration. The thallus is usually dichotomously branched with or without a mid-rib on the dorsal surface. The thallus is green and nutrition is photosynthetic. The thallus is anchored to the soil with the help of root like structures called rhizoids. The rhizoids are unicellular except in musci. In Hepaticae two types of rhizoids are found namely smooth or simple rhizoids pegged or tuberculate rhizoids. In pegged rhizoids the inner layer of the wall is unfolded into small projections which is supposed to increase the absorpture area of rhizoids. No such infolding of the wall is seen in smooth rhizoids basides rhizoids multicellular wedge shaped structures are also found in some heptaciae. These are called scales. Scales are protective in function.

The thalloid gametophytes anatomically show only parenchyma tissues, while there is no tissue differentiation in some members (Anthocros) in Riccia, Marchantea and others. The internal tissues are distinguishable into upper photosynthetic region and the lower storage region. The cells of the former have a number of chloroplasts while those of the latter are rich in starch regions.

2. Leafy forms. In leafy forms the gametophyte consists of a central axis on which are arranged the "leaves". The axis is comparable to the stem. It may be branched or unbranched. It may grow prostate (porella) or erect (funaria). It is anchored to the soil with the help of rhizoids. The rhizoids in musci are multicellular, the leaves usually arise spirally on the stem. The leaves are usually one cell in thickness but towards the centre may be more than one cell in thickness.

Reproduction

The gametophyte reproduces by 2 methods namely 1. Vegetative propagation, 2. Sexual reproduction

1. Vegetative propagation : Vegetative reproduction is an accessory means of reproduction to bring about a rapid increase in the number of gametophytes. In fact it is the vegetative propagation which is mainly responsible for the rapid spreading of the gametophyte is an area. Vegetative propagation is brought about by a variety of methods.

2. Sexual reproduction. This is the usual method of reproduction of the gametophyte by which the sporophyte is formed. Sexual reproduction is of oogamous type. Sex organs are the antheridium (male) the archegonium (female). Gametophytes may be homothallic or heterothallic. Sex organs may be produced directly on the thallus or on special branches set apart for that purpose as in Marchantia.

Structure of the Antheridium

Antheridium is usually a globose structure (Hepaline) or elongated club shaped (musci) as in some cases. It usually develops from an initial cell called the antheridial initial. A mature consists of a stalk and the terminal swollen region in which the antherozoid mother cells which give rise to antherozoids (sperms). The antherozoids of tryophytes are spirally coiled biflagellate. At maturity the wall breaks open and the antherozoids are liberated.

Structure of the Archegonium

Archegonium is a flask shaped structure. It also develops from an initial cell called the archegonial initial. It may be stalked or sessile. The basal fulbous region is called the neck. The neck region has a single layered jacket, while the venter is surrounded by jacket more than one cell in thickness. In the venter region is found the egg above which is found a small cell the venter canal cell. In the neck region there is a raw of central cells calle neck canal cells varying in number from four (Hepatica) to sixteen masses. The tip of the neck is covered by 4 neck cells or cover cells.

Fertilization

Fertilization is dependent on external moisture. At the time of fertilization the contents of the archegonium except the egg, dissolve forming a mucilagenous mass. Due to the pressure erected by this, the neck cell open apart and the mucilage oozes out.

The antherozoids liberated from the antheridium are said to be attracted chemotactically by the mucilage towards the archegonium (chemical analysis of the mucilage shows the presence of malic acid besides other constituents). Many antherozoids enter the archegonium. Ultimately one of the antherozoids succeeds in fusing with the egg forming the zygote.

Sporophyte

Sporophytic generation in tryophytes is not predominant in that it is dependent on the gametophyte either completely or partially. It is short lived and perishes soon after the liberation of the spores. The zygotes first develops into embryo which matures later into the sporophytes.

The mature sporophyte may be an undifferentiated sac containing spores as in Riccia or may show differentiation to a lesser or higher degree. Usually the sporophyte consists three parts namely foot, setae and capsule. Foot is the basal portion and is embedded in the gametophyte drawing nutrition from it. Seta is an elongated stalk like structure serving to draw the capsule away from the gametophyte. In the capsule region which is swollen, haploid spores are fomred from spore mother cells. All the spore mother cell may or may not develop into spores. Many of them become bodies known as elaters that serve to dessimate the spores (Marchantia porella).

The spores on dispersal develop into the gametophyte either directly (Hepaticae) or through the protonemal stage (Mosses).

Apogamy and Apospory

Both these protonema are of rare occurrence in Bryophytes. Apogamy is the production of a sporophyte without involving fusion of gametes. Any part of the gametophyte develops into a sporophyte. Such sporophytes are haploid and will not survive for long.

Apospory is the production of a gametophyte from a sporophyte without involving spore production ie., circumventing reduction division. Aposporous gametophytes will be diploid. Among Bryophytes, apospory has been recorded in musci and anthocerotae (Bornhagen 1926, Rink 1935).

Classification

The name Bryophyta was first coined by Braun in 1864. But he included algae, fungi, lichend and mosses in his group Dry ophyla. It was Schimper (1879) who first recognised Bryophyta as a division in the sense as it is used today.

Eichler (1833) classified Bryophytes into two groups namely Hepatica and meisci. Till recently this classification was followed by majority of botanists. Engler (1892) followed Eichler in the classification divides Marchantia Jungermanniales and Anthcerotales. Many eimnent botanists like Bessey (1911), Fritsch (1929), Wetlsteir (1935), Verdoon and Evan (1989) follow the system of Engler.

Underwood (1894) and Baryex Janezeroslei (1872) however recognised the special features of anthocerotales and opined that it cannot be included under Hepatica.

Consequently Gayet (1897) and other biologists emphasised the need to separate anthocesolde from Hepaticae. Majority of botanists including Campbell (1940), Smith (1955) and Takhtajan (1953) opine that Bryophyta should be divided into 3 classes namely Hepaticae, Anthocerotae and Meisei.

Rothmaler (1951) changed the endings of the class names as follows :

Hepaticae to hepaticopsida.

Anthrocerotae to Anthoceropsida.

Mesci to Bryopsida.

This change is accordance with the international code of Botanical nomenclature. Prokauer (1957) changed the name Anthroceropside to Anthrocrotopsida.

CHAPTER 3

HEPATICOPSIDA

The members of Hepaticopsida are commonly called liver worts (Hepatica = liver) because of the boded nature of the thallus which resembles the liver in outline. This class includes approximately 225 genera and 8500 species (Smith 1955).

Hepaticopsida of temperate climate are mostly restricted to shady terrestrial habitats. Such habitats include severmps and bags. In tropical climate also their distribution is limited to cool, moist shady areas like the floor of forests.

There is great diversity in the external make up of gametophyte. But essentially the plant body is a dorsiventrally flattened thallus except in some members like porella. The thalloid gametophyte is dichotomously branched with or without a prominent midrib. In leafy gametophytes the leaves are arranged in two or three rows on the axis.

Anatomically, the thalloid gametophyte may or may not show internal differentiation into storage and photosynthetic regions.

Sex organs are formed on the dorsal surface of the thallus developed from a superficial cell. The sporogonium may or may not show differentiation. When differentiated the sporophyte consists of foot seta and capsule. All the spore mother cells are not fertile. Some of them become sterile and form either nurse cells or claters, the latter helping in spore dispersal.

The class is divided into 4 orders namely Marchantiales, Spherocarpalles, Lalabeyales and Jungranannniales.

In this book Ricca and plagiochasma marchantia, porella and pellia of Hepaticopsida are discussed.

RICCIA

Common species : R.tumalyensis, R.curciata, R.pearsoni, R.fluitens etc.

Occurrence : This species of Riccia occur on damp soil moist rocks etc. They grow very well in cool and shady places. R.fluitens is an aquatic member.

Study : 1. External features of the gametophyte

2. Internal structure of the thallus

3. Methods of vegetative propagation

4. Structure of the sex organs

5. Structure of the sporogonium

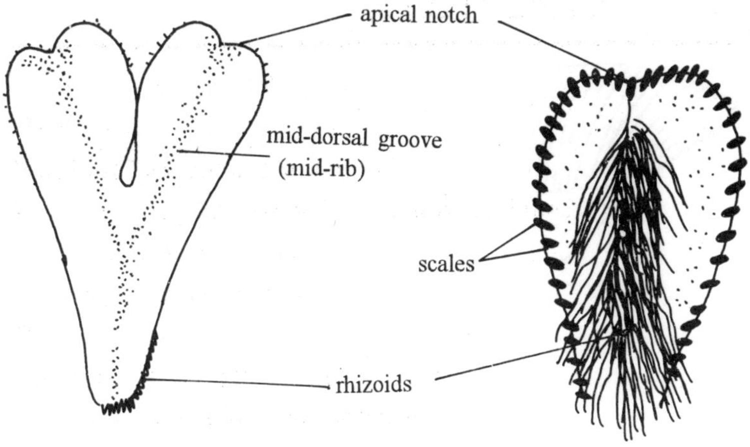

Fig. 3.1 Riccia. Dorsal surface of thallus — Ventral surface of thallus

Practical work to be done

1. Cut the vertical transverse section of thallus
2. Mount a few rhizoids and scales
3. Observe the thallus with a hand lens.

External features of the thallus

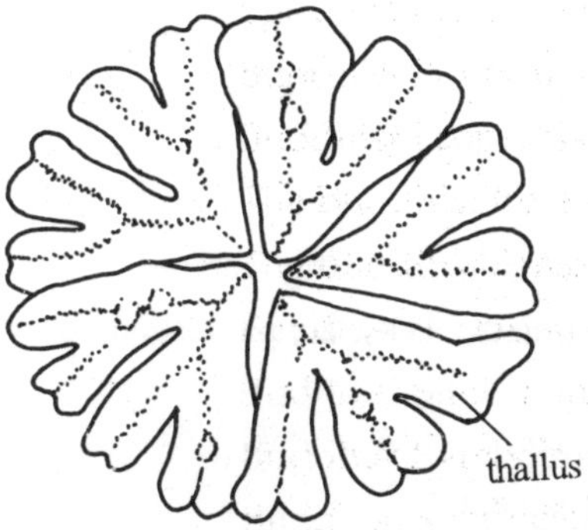

Fig. 3.2 Riccia. Rosette

1. The plant body of Riccia is a dorsiventrally flattened, deep green coloured thallus.
2. The thalli grow in groups called Rossets. In this, many thalli radiate from a central point to give a circular outline.
3. The thallus is small, thick towards the centre, gradually thinning out towards the margin.
4. The thallus show typical dichotomous branching.
5. There is an apical notch in which is situated the growing point (apial cell).
6. The thallus can be clearly distinguished into dorsal and ventral surfaces.

Dorsal surface :

1. This is the upper part facing the sunlight
2. A mid rib is present and it is represented by a median furrow.
3. Here and there small dots are present, embedded in the thallus. They are the sex organs. Sometimes the dots are dark coloured in which case they represent the sporogonium.

Ventral surface

1. The ventral surface of the thallus is somewhat dark coloured and it bears a number of rhizoids and scales.
2. Rhizoids are long tubular extensions of the epidermal cells. They are unicellular and help in absorption and fixation.

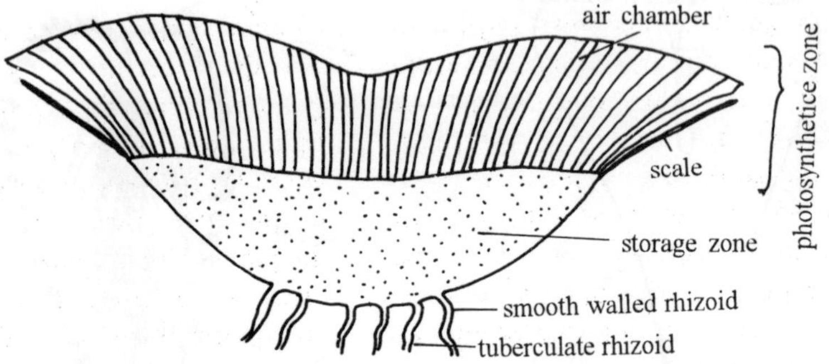

Fig. 3.3a V.T.S of tallus (diagrammatic)

3. The 2 types of rhizoids are seen in Riccia, they are smooth or simple rhizoids and pegged or tuberculate rhizoids.

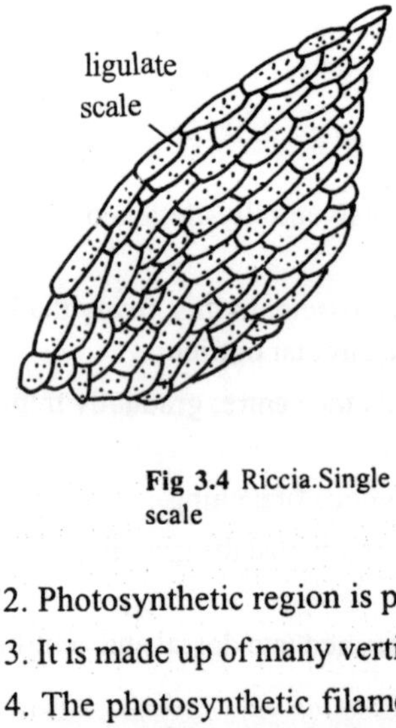

Fig 3.4 Riccia.Single scale

4. In smooth rhizoids the inner wall is even with the outerwall.

5. In pegged rhizoids the inner wall here and there is folded into small projections extending into the lumen. The projections are called pegs.

6. Scales are multinuclear, but one cell thick, wedge shaped structures. They are violet in colour.

7. They are present in 2 rows along the margin of the thallus, exotically near the growing apices.

8. The scales protect the young growing points.

Internal structure of the thallus

A transverse section of the thallus shows the following features.

1. Two distinct regions are noticed internally. These are photosynthetic region and storage region.

2. Photosynthetic region is present above ie., towards the dorsal surface.

3. It is made up of many vertical rows of cells called photosynthetic filaments.

4. The photosynthetic filaments are unbranched being made up of cells joined end to end. The cells are chlorophyllous, constituting the chief photosynthetic tissue of the plant.

5. The photosynthetic filaments are arranged parallel with air spaces between them.

6. There is no layer of cells that can be called upper epidermis.

7. The uppermost cell of each photosynthetic filament functions as an epidermal cell. Hence one can say the upper epidermis is discontinuous with air spaces between cells.

8. The storage region consists of irregular closely packed parenchyma cells.

9. The cells generally lack chloroplasts as their main function is storge.

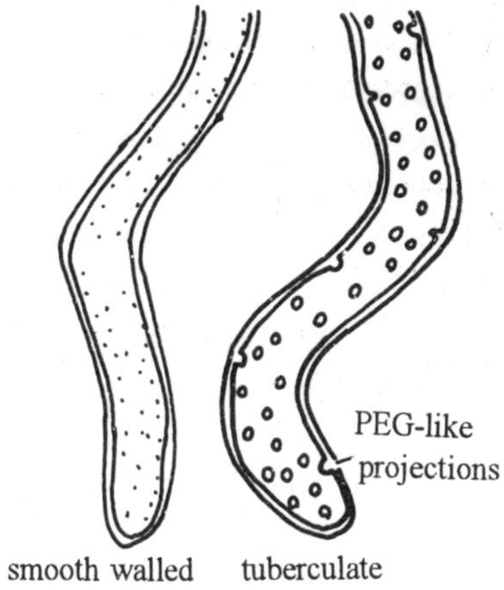

Fig. 3.5 Riccia. Rhizoids

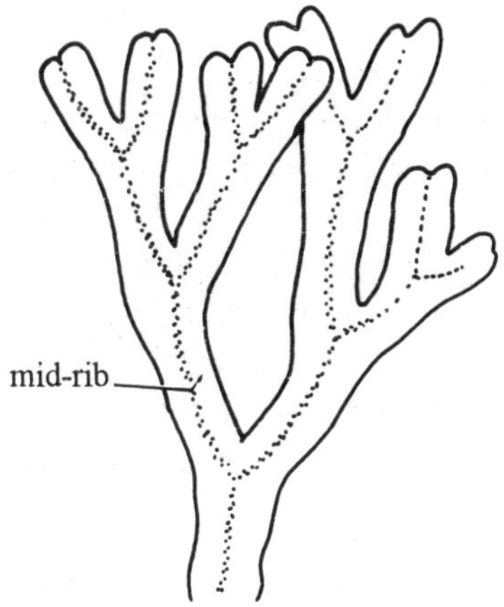

Riccia flutians. Thallus

Vegetative propagation

1. By progressive death and decay. Older parts of the thalli die, leading to the separation of dichotomies which develop into new individuals.

2. Persistent apices. During unfavorable conditions all the parts of the thallus, except the apex die. The apex survives and develops into a new individual at the return of the favorable condition.

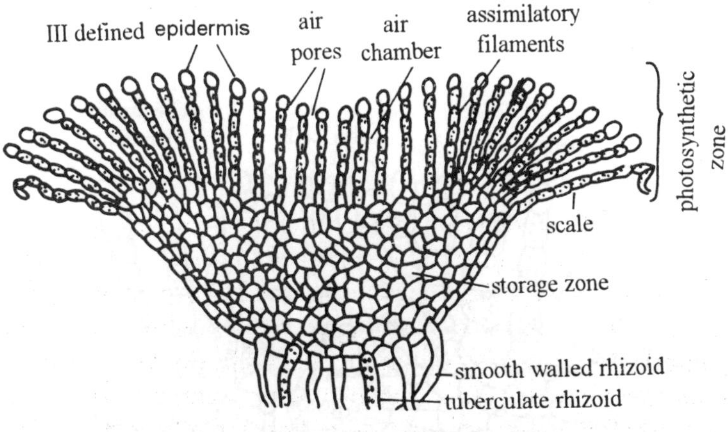

Fig. 3.6 Riccia. V.T.S. of thallus

Sexual reproduction

1. Sexual reproduction in the gametophyte is of the oogamous type.

2. Sex organs are antheridia (male) and archegonia (female)

3. Species of Riccia may be homothallic or heterothallic.

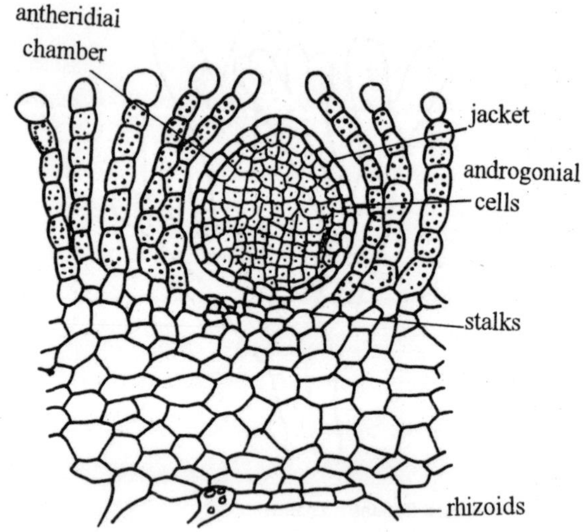

Fig. 3.7 Riccia. V.T.S of thallus passing through antheridium (A part)

4. Sex organs are found embedded in the gametophyte tissue on the dorsal surface near the median groove.

5. Sex organs develop in acropetalous succession, younger ones towards the apex.

6. In homothallic species, antheridia mature early (protandious).

Structure of the antheridium

1. A transverse section of the thallus will reveal a cavity called the antheridial chamber.

2. Lying in the cavity of the antheridial chamber will be a single antheridium.

3. Each antheridium is a sub globose or pear shaped structure having a multicellular stalk.

4. The upper sub-globose part of the antheridium is called the capsule.

5. The capsule has a single layered wall called the jacket layer.

6. Inside the wall is a mass of cubical cells called androcyte mother cells.

7. Each androcyte mother cell divides diagonally to form two triangular andiocytes.

8. Each androcyte metamorphoses itself into a biflagellate coiled antherozoid or sperm.

9. At maturity the antheridium absorbs water and breaks open to release the antherozoids which come out in a mass.

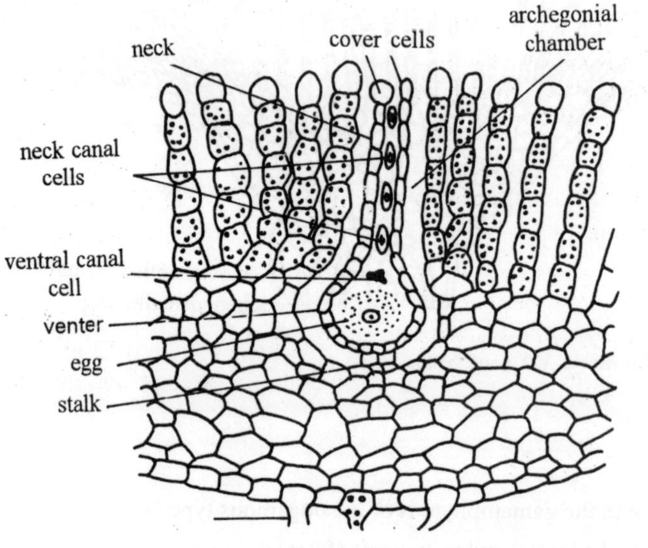

Fig. 3.8a V.T.S. of thallus passing through archegonium (A part)

Structure of the archegonium

1. A transverse section of the thallus reveals an archegonial cavity.

2. Each archegonial cavity encloses a single archegonium.

3. The archegonium is a flask shaped structure having a bulbous base called the neck.

4. The wall of the archegonium is one celled thick. It has six rows of cells constituting the wall.

5. At the venter region inside the archegonium is an egg.

6. Above the egg cell is a single venter canal cell.

7. Inside the neck are found four to six neck canal cells.

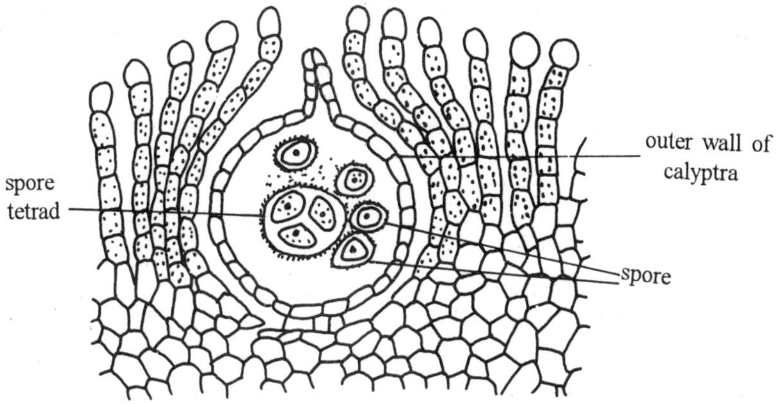

Fig. 3.9 Riccia. V.T.S. of thallus passing through sporogonium (A part)

8. The tip of the neck is covered by four cover cells or lid cells (in a long section however only 2 cells are seen).

9. At the maturity of the archegonium, all the cells inside, except the egg dissolve, which comes out in the form of a mucilaginous mass.

10. The lid cells open apart and the antherozoids enter inside.

11. Fertilization of the zygote egg results in the formation of a zygote which eventually develops into the sporophyte (sporogonium).

Structure of the sporogonium

1. The sporogonium is a globose structure and it is completely embedded inside the thallus.

2. It has a small peg like projections at the tip representing the degenerated neck of the archegonium

3. There is no differentiation of the sporophyte into foot, seta and capsule.

4. The sporogonium has a single layered wall called the calyptra.

5. Within the sporogonium are found a number of spore mother cells which undergo reduction division to form tetrads of spores. Tetrads are tetrahedral.

6. Some of the spore mother cells degenerate and form nurse cells, which provide nutrition to the surviving spore mother cells.

7. There are no elasters.

8. All the spores are of the same kind chmosporous.

9. The spore will be pyrimidal in shape, after separating from the tetrad, later it becomes round.

10. The spore is surrounded by a thick sculptured wall.

11. The wall has 3 layers viz., outer thick cultenized exospore, middle thin mesospores and inner thin endospore.

12. Spores are uninucleate and haploid. They are liberated at the decay of the sporogonium and germinate to develop into new individual gametophytes.

Systematic position and identification

A. 1. Plant body is a parenchymatous gametophyte.

2. It is leafy or thalloid

3. Vasculture and roots absent

4. Sex organs surrounded by a sterile jacket layer

5. Life cycle exhibits heteromorphic alternation of generations with a predominant gametophyte
... Bryophyta

B. 1. Thallis are lobed and dorsiventrally flattened or leafy

2. Chloroplasts do not have pyrenoids

3. Scales and two types of rhizoids present

4. No columella in the sporophyte ... Hepaticopsida

C. 1. Plant body is thalloid and dichotomously branched

2. Scales are present on the ventral surface

3. Thallus differentiated into photosynthetic and storage regions.

4. Air chambers present ... Marchantiales

D. 1. Gametophytes flat and thalloid

2. Photosynthetic filaments unbranched.

3. Sporogonium does not have foot and seta ... Ricuaceae

E. 1. Thalli grow in a rosette with individual radiating from the centre.

2. Two rows of scales present at the margin

3. Upper epidermis indistinct and discontinuous

4. Sporogonium has only spore mother cells and nurse cells.

5. Elaters are absent

6. Spores have a sculptured, three layered wall... Riccia

PLAGIOCHASMA

Common species : P.articulatum, P.simlenses, P.cordatum, P.intermedium, P.appendiculatum etc.

Occurrence : The gametophytic plant occurs in moist, cool and shady places. Majority of the species are frequently in mountainous regions. They grow on moist soil, rocks, trunk of trees etc.

Study : 1. External morphology

2. Internal structure

3. Structure of sex organs

4. Structure of sporophyte

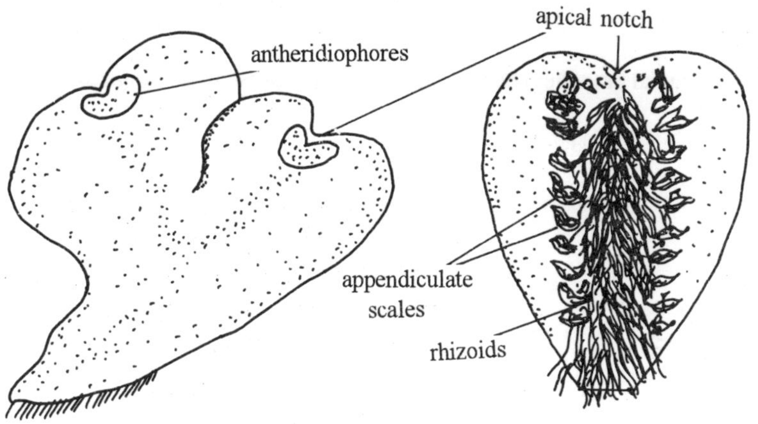

Fig. 3.10 Plagiochasma. Dorsal surface of thallus

Practical work to be done

1. Cut a cross section of thallus
2. Section the thallus to show the sex organs
3. Section the thallus to study the sporophyte

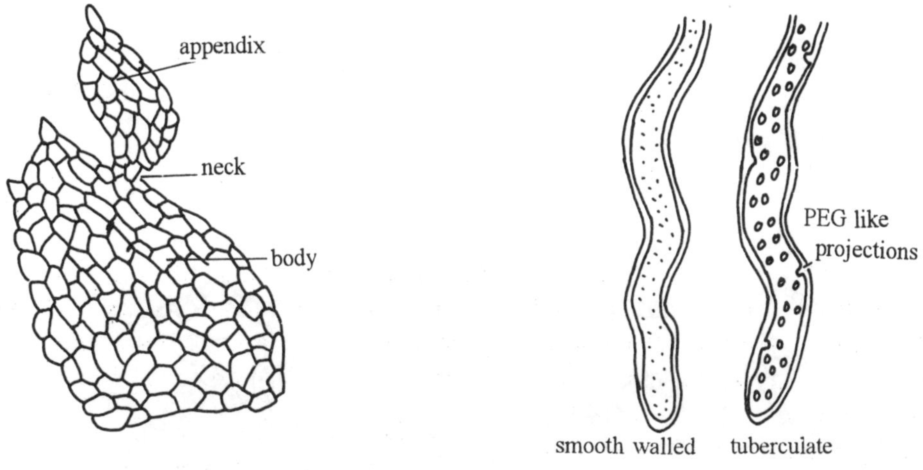

Fig. 3.11 Plagiochasma. An appendiculate Scale Rhizoids

External morphology

1. The gametophytic plant body is a dorsiventrally flattened thallus
2. The thallus is dichotomously branched and deep green in colour
3. The dorsal surface of the thallus is greener than the ventral surface and has a number of air pores

4. The air pores internally lead to air chambers and each pore is surrounded by 2-4 concentric rings of cells. Each ring has 6-8 cells.

5. Male and female receptacles (special branches) bearing sex organs are present on the dorsal surface of the thallus.

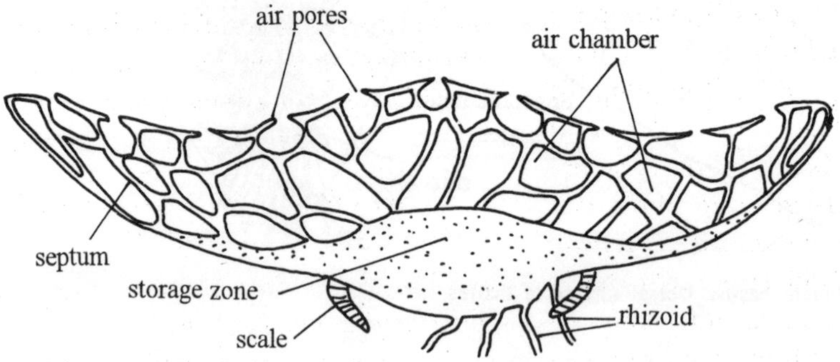

Fig. 3.12 Plagiochasma. V.T.S. of thallus (diagrammatic)

6. There is no distinct mid-ribbon the dorsal surface, but the central region is thicker than the margins.

7. The ventral surface is pale green in colour and tears scales and rhizoids. The margins of the thalli on the ventral surface is purple.

8. Two rows of scales are present along the

9. The scales are multicellular and wedge shaped. They have one or two appendages.

10. Rhizoids are present as extensions of the lower epidermis helping in anchorage and absorption.

11. Two types of rhizoids are present viz., simple and pegged (for details see Riccia).

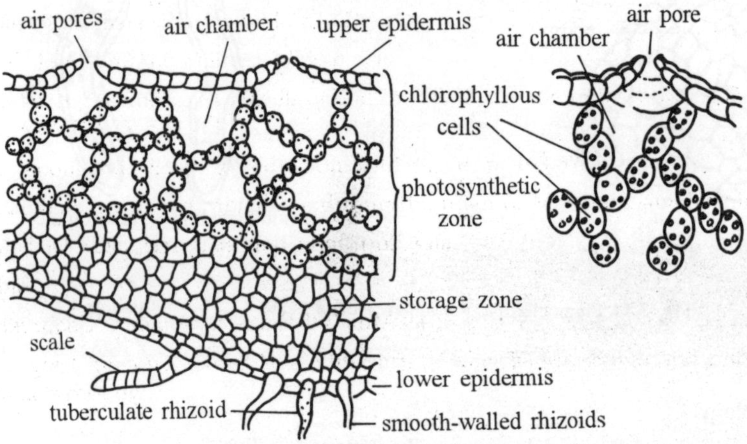

Fig. 3.13 Plagiochasma. V.T.S of thallus (A part)

A portion of V.S. of thallus showing air pore

Anatomy of the thallus

1. A cross section of the thallus shows the two distinct regions placed in between the two epidermal layers.
2. The epidermis is a distinct, more or less continuos layer.
3. It has air-pores here and there.
4. The air pores are slightly raised above the level of epidermal cells.
5. The air pores internally lead to an air-chamber.
6. Below the epidermis is the photosynthetic zone.
7. This zone consists of many chronehcyma cells forming a loose reticulum with large meshes. Each mesh represents an air chamber.
8. Below the photosynthetic region is found the storage region.
9. The storage region consists of closely packed parenchyma cells.
10. The lower epidermis is a single continuos layer of cells. It gives rise to rhizoids and scales.

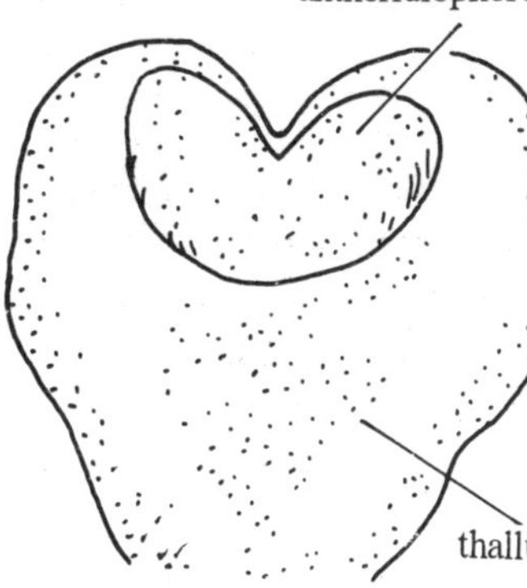

Fig. 3.14 Plagiochasma. Thallus showing male receptacle

Vegetative propagation

1. Special adventitious branches of the thalli develop at the apex. These separate from the parent plant and envelop into a new individual.
2. By the progressive death and decay of older parts, the dichotomies get separated and develop into new individuals.

Sexual reproduction

1. It is oogamous, sex organs are antheridia and archegonia. Majority of the species are homothallic.
2. The sex organs are found in special structures called receptacles. Consequently there are male and female receptacles.

Male receptacle

1. The male receptacle are borne on the dorsal surface of the thallus. They are horse shoe shaped structures and are sessile.
2. Antheridia develop on the male receptacle in acropetalous succession.
3. The male receptacles consists by many chambers called the antheridial chambers. Each chamber encloses a single antheridium
4. Each antheridial chamber opens to the exterior through an antheridial pore.
5. The antheridium has a short multicellular stalk and a globose capsular region.
6. The jacket of the antheridium is single layered and it encloses a large number of androgonial cells.
7. A mature antheridium will have a large number of antherozoids.

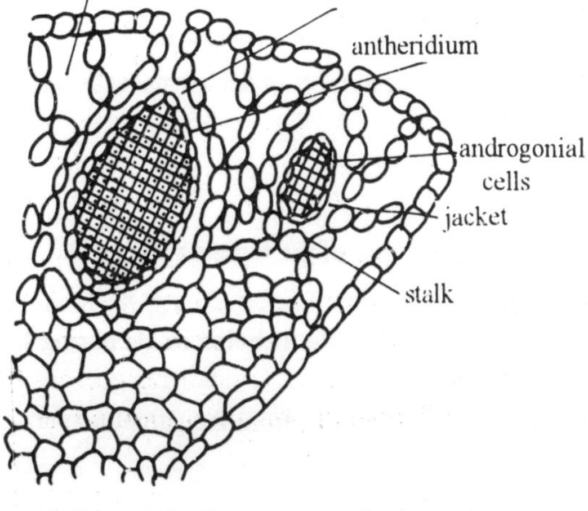

Fig. 3.15a L.S. of male receptacle

Female receptacle

1. The female receptacles are borne on the dorsal sur-

face. They are terminal in P.articulatum.

2. Each female receptacle also known as carpocephalum has a short stalk and a terminal disc.

3. Stalk does not have rhizoidal grove and in young condition, the stalk may be absent.

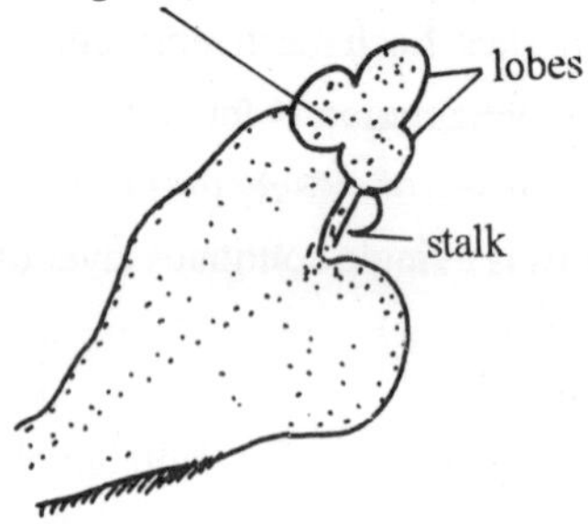

Fig. 3.16 Plagiochasma. Thallus showing female receptacle

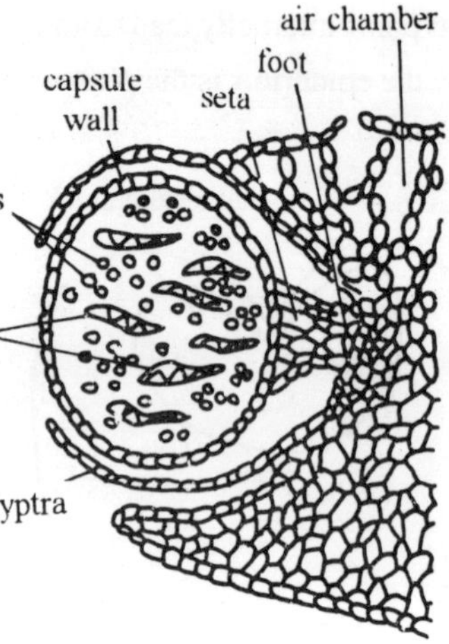

L.S. of female receptacle passing through sporogonium

4. The disc is flat, broad and has 3-9 lobes.

5. Each lobe of the disc encloses an archegonium

6. The archegonia are borne on the dorsal surface of the disc, when young. But, later they come to lie towards the ventral surface due to the centripetal growth of the disease.

7. In a mature disc, the archegonia will be having their neck pointing down.

8. The overgrowth of the disc surrounding the archegonium will form an involucure.

9. The archegonium is a flask shaped structure having a bulbous basal venter and a narrow neck.

10. The archegonium is similar to that of Riccia (see Riccia for details). The number of neck canal cells ranges from 6-9.

11. Fertilization is same as in Riccia.

Structure of the sporophyte

1. The sporophyte is seen hanging down from the ventral surface of the disc of the female receptacle.

2. The sporophyte is differentiated into three parts viz fobt, seta and capsule.

3. Foot is somewhat bulbous and is made up of polygonal parenchyma cells. It helps in ahcnoring the sporophyte to the gametophytic tissue.

4. The seta is a stalk like structure and it seves to carry the capsule slightly away from the gametophyte.

5. The capsule is the terminal part of the sporophyte. It is sowewhat globose in shape.

6. The wall of the capsule is made up of a single layer of cells except at the apeex, where it is double layered to fform an operculum or lid.

7. The lid is thrown off at maturiry to facilitate the liberation of spores.

8. Within the sporophyte are found the dipllid spore mother cells.

9. Some of the spore mother cells become sterile, elongate and form the elaters.

10. Elaters are spindle shaped structures and have one or two spiral band like thickenings which are hygro-scopic elaters help in spore dispersal.

11. The remaining spore mother cells undergo reduction division and form tetrahedral trytrads of haploid spore mother cells.

12. Spores are sub globose, uninucleate and have a two layered wall. The outer wall exine is thick and ornamented, whkle the inner wall intine is thin.

13. The spores on liberation and on the availibality of a suitble sub-stratum develop into new individual of gametophyte.

Diagrams to be drawn

1. Habit sketch

a. Dorsal surface

b. Ventral surface

2. Rhizoids and scales (enlarge)

3. Transverse section of the thallus

4. Male and female receptacle

5. Sections through the receptacle to show antheridia and archegonia

6. Section of the female receptacle to show sporophyte

7. A few elates and sproes (enalrged)

Systematic position and identification

A. 1. Plant body is a parenchymatous gametophyte.

2. It is leafy or thalloid

3. Vasculature and roots absent

4. Sex organs surrounded by a jacket layer

5. Life cycle exhibits a hereromorphic alternation of genration with a predominant gametophyte. Bryophytes

B. 1. Thalli are lobed and dorsiventrally flattened or leafy

2. Chloroplasts do not have pyrenoids

3. Scales and 2 types of rhizoids present ... Hepaticopsida

C. 1. Plant body is thalloid and dichotomously branched

2. Scales arepresent on the ventral sruface

3. Thallus differnaitaed into photosynthetic and storage regions.

4. Air chambers present ... Marchangiacales

D. 1. Gametrophytes flat and thalloid

2. Photosynthetic filaments absent

3. Scales have appendages

4. Air pores simple

5. Capsule has terminal operculum... Rebauliaceae

E. 1. Male receptacle horse shoe shaped

2. Capsules have a two lipped envelope of the involuere

3. Ventral margins of the thallus purple in colour ... Plagiochasma.

MARCHANTIA

Common species : M.polymorpha, M.palmata, M.nepalensis etc

Occurrence : The thallus of Marchanita grows luxuriantly in cool and moist climates. The species of Marchantia are abundant in hilly regions. In South India, the Nilgiri hill ranges show luxuriant growth of Marchantia.

Study : 1. External features of the thallus

2. Internal structure

3. Vegetative propagation

4. Sexual reproduction

5. Structure of the sporophyte

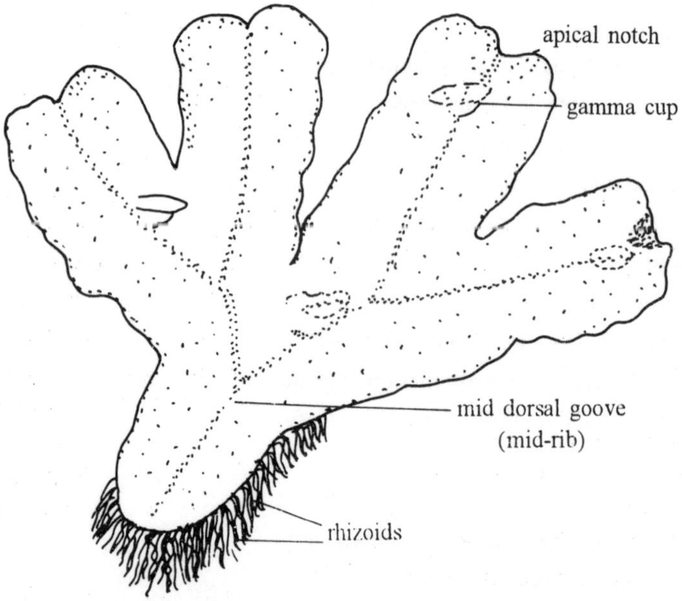

Fig. 3.17 Marchantia. Dorsal surface of thallus

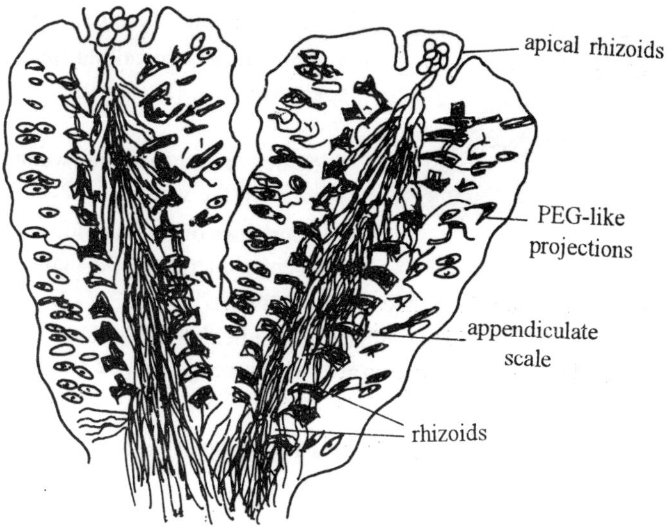

Fig. 3.18 Marchantia. Ventral surface of thallus

I. External features of the thallus :

1. The plant body of Marachantia is a dorsiventrally flattened dichotomously branched thallus.

2. The thallus is grass green in colour with an undualted margin.

3. There is a prominent median mid-rib on the dorsal surface represented by a groove on the ventral surface, the mid-rib is represented by a ridge.

4. If the thallus is observed with a hand lens, many pologynoal areas may be seen on the dorsal surface. These are the areoles and demarcate a photosynthetic chamber within.

5. In the centre of each areole is a dot like structure. This represents the air pore or the chimnery pore.

6. The growing point of the thallus is situated in the apical notch.

Reproductive structures present on the dorsal surface:

1. Along the mid-ribs here and there are found small cup shaped structures. These are called the gemmae cups. Inside the gemma cup are found a large number of gemmae.

2. The thalli of Marchantia also have special branches to bear the sex organs.

3. The thalli are dioecious. The male thalli bears the antheridiophore and the female thalli bears the archegoniophore.

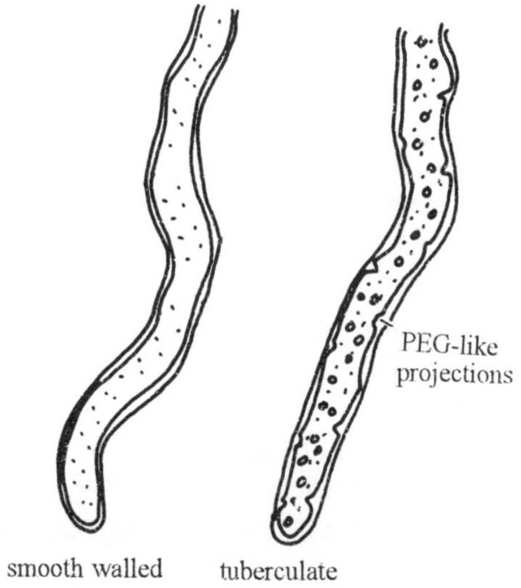

Fig. 3.19a Rhizoids

Ventral surface of the thallus

1. The ventral surface of thallus is pale green in colour.

2. The mid-rib on the ventral surface is represented by a ridge.

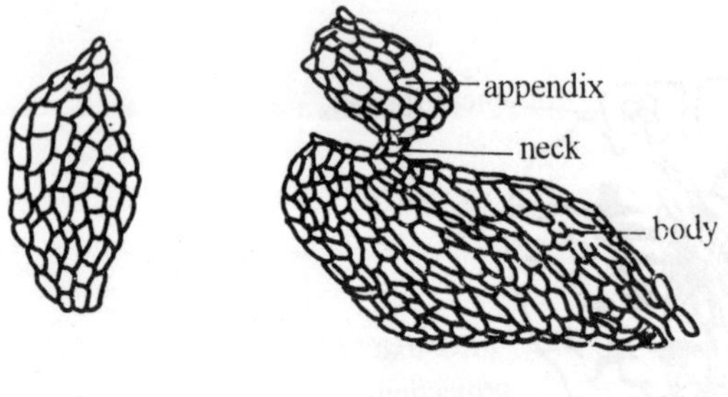

Fig. 3.20 Marchantia. Appendiculate scale

3. Along the mid-rib region on the ventral surface are found scales and rhizoids.

4. Rhizoids are unicellular and develop from the lower epidermal cells. They are of 2 types viz., simple and pegged rhizoids.

5. Scales are multicelluar wedge shaped structures also present on the ventral surface.

6. Four rows of violet coloured scales are present along the ventral margin of the thallus, specially near the grow ing points.

7. Scales are of 2 types - Appendiculate and ligulate. Appendiculate scales have an appendage arising out of a wedge shaped structures, while legulate scales are simple wedge shaped structures.

Internal structures of the thallus

1. A vertical transverse section of the thallus shows two distinct regions viz., photosyntheci region and storage region placed in between the two epidermal layers.

2. The upper epidermis is made up of a single layer of barrel shaped cells.

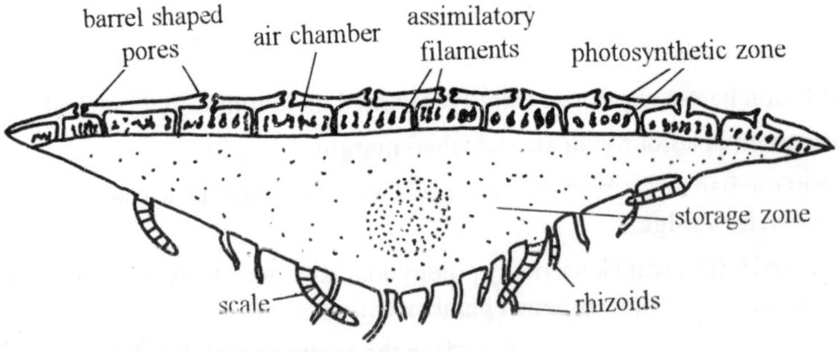

Fig. 3.21 Marchantia. V.T.S. of thallus (diagrammatic)

3. The continuity of the epidermal layer is broken by the presence of air pores. These pores have a rim or border made up of a few concentric rings of cells. As a result, the pores are slightly elevated, looking like a chimney, hence they are also called chimney pores.

4. There will be one chimney pore at the centre of each aereole.

5. Internal to the upper epidermis is the photosynthetic region divided into a number of chambers by means of parenchymatous partitions.

6. It is these chambers that constitute the aereoles superficially.

7. Arising from the floor of each photosynthetic chamber are a number of branched filaments. The cells of these filaments are chlorophyllous and constitute the cheif photosynthetic tissue of the plant.

8. Next to the photosynthetic region is the storage region, made up of closely packed polygonal cells.

9. The cells are rich in starch grains.

10. The storage tissue also consists of oil and mucilage cells.

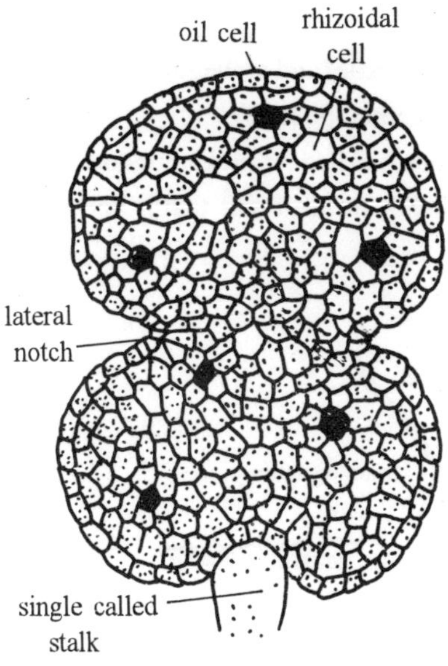

Fig. 3.22a Single gemma

11. The mid-rib region consists of thick walled cells.

12. Below the storage region is the lower epidermis made up of a single layer of cells.

13. Scales and rhizoids arise from the cells of lower epidermis.

Vegetative propagation

1. Progressive death and decay of the older parts and the consequent separation of dichotomies brings about vegetative propagation.

2. The commonest method of vegetative propagation is by means of gemmae produced in gemina cups.

3. A transverse section of the thallus passing through a gemma cup reveals the following details.

4. The gemma cup has fringed margin and is made up of parenchyma cells.

5. Rising from the floor of the gemma cup are found a number of gemmae at different stages of growth.

6. Each gemma has a single celled short stalk.

7. The gemma is a dumble shaped structure and is made up of gemma parenchyma cells.

8. The cells of gemma have chloroplasts.

9. The gemma is many celled thick at the centre, but one celled thick towards the margin.

10. There are 2 notches in the gemma situated in the lateral margins. It is at these notches that the growing points are situated.

11. Oil cells and rhizoidal initials are also found in each gemma.

12. A mature gemma separates from the gemma cup and develops into two new individuals. Each growing point of the gemma gives rise to a new thallus.

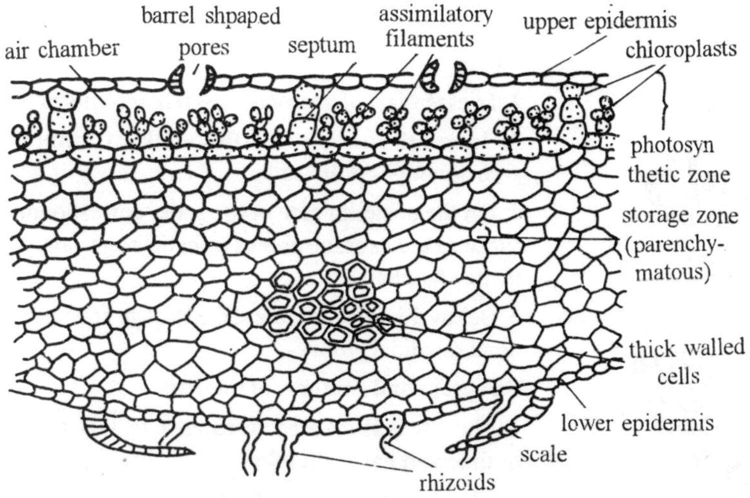

Fig. 3.23 Marchantia. V.T.S. of thallus (A part)

Sexual reproduction

1. Sexual reproduction is of the oogamous type- sex organs are antheridia (male) and archegon (female).

2. The sex organs are fomred on special branches of the thallus called "Gametophores"/

3. The male gametophore is called Antheudiophore and the gametophore is called Archegoniophore.

4. The antheudiophore and archegoniophores arise at the apical notches on different thalli. Species of Marchantia are strictly dioecious.

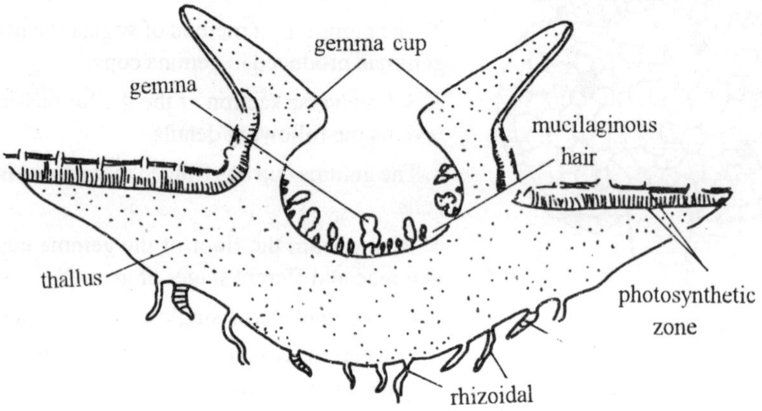

Fig. 3.24 Marchantia. V.T.S. of thallus passing through gemma cup (diagrammatic)

Structure of an Antheridiophore

1. Each antheridiophore has a cylindrical stalk and a terminal peltate disc.

2. The stalk has two longitudinal grooves called rhizoidal grooves. All along the groove the epidermal cells produce rhizoids.

Anatomy of the stalk - (transverse section)

1. The outline of the section is somewhat circular.

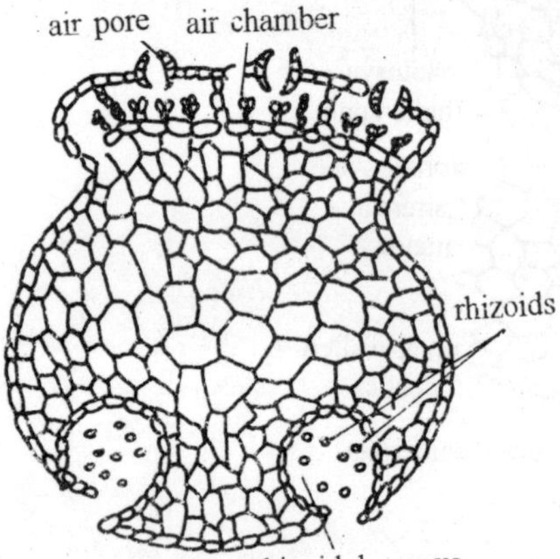

Fig 3.25 Marchantia - T.S Antheridiophore

2. In essentials of anatomy the stalk resembles the thallus.

3. Internal structure of the stalk of antheridiophore and archegoniophore are same.

4. On the posterior side of the stalk, corresponding to the dorsal surface, are present photosynthetic zone, air pores, photosynthetic chambers etc.

5. Below this zone is present the storage zone having two longitudinal rhizoidal grooves.

Structure and anatomy of the disc

1. The disc is also called the receptacle. It is irregularly circular in outline and has eight lobes.

2. The dorsal surface of the disc is slightly convex.

3. The antheridia are found embedded in the dorsal surface of the disc.

4. A vertical median longitudinal section of the disc will show the details of anatomy.

5. Anatomically the disc resembles the thallus in having an upper photosynthetic zone and a lower storage zone.

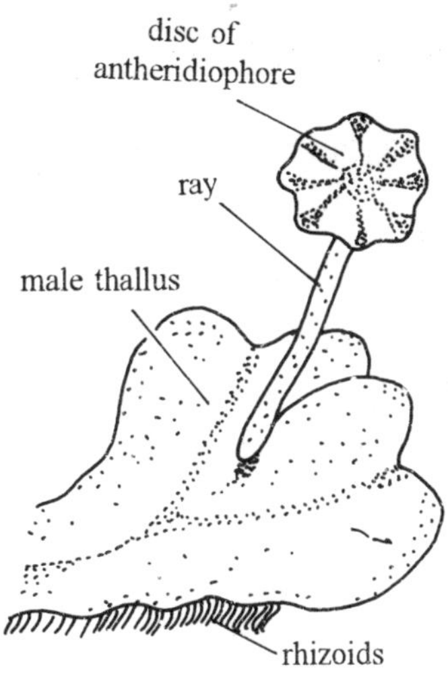

Fig. 3.26 Marchantia. Thallus bearing antheridiophore

6. There is an upper epidermis, whose continuity is interrupted by chimney pores.

7. The photosynthetic zone consists of photosynthetic chambers.

8. Unlike in the thallus, the photosynthetic chambers alternate with cavities called antherideal chambers.

9. Each antheudial chamber is a flask shaped cavity opening on the dorsal surface of the disc through an opening called the ostiole.

10. Chimney pores and ostioles alternate on the disc.

11. Each antheridial cavity encloses a single antheridium.

Structure of the antheridium

1. The antheridium is a sub globose or oval structure arising from the floor of the antherudeal chamber.

2. The antheridium has a short milticellular stalk.

3. The wall of the antheridium is made up of single layered jacket enclosing the antherozoid mother cells.

4. The antherozoid mother cells eventually form the coiled, biflagellate antherozoids.

5. The antheridia develop in acropetalous succession ie., oldest antheridium at the centre and the youngest near the margin.

Archegoniophore or the female receptacle

1. The archegoniophore also arises on the dorsal surface of the thallus near the growing point.

2. It has a stalk and a peltate disc.

3. The stalk is similar to that of the antheridiophore in structure and anatomy.

4. The disc is smaller than that of antherideophore and it has got 6-8 finger like projections called rays.

5. The dorsal surface of the disc in convex.

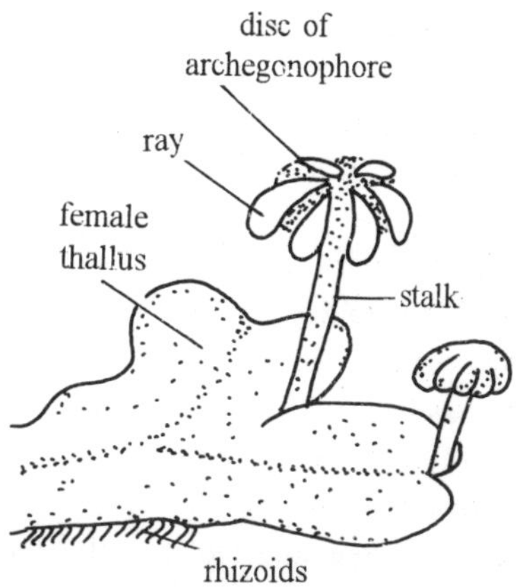

Fig. 3.27 Machantia. Thallus bearing archegoniophore

6. Archegonia are found on the dorsal surface of the disc in young conditions, while after fertilization they become inverted and pushed downwards due to the centrifugal growth of the disc. Ulimately the archegonia will be hanging down with their neck pointing downwards.

Long Section of a Young Archegoniophore

1. The disc in its anatomy resembles the thallus in having photosynthetic zone above and storage zone below.

2. Archegonia are found on the dorsal surface of the disc in erect position.

3. They develop in acropetalous succession ie., older archegonia towards the centre and younger archegonia towards the margin.

4. Each lobe of the disc corresponding to a ray has a row of archegonia.

Structure of the archegonium

1. Each archegonium is a flask shaped structure.
2. It has a bullous base called venter and a narrow elongated part called the neck.
3. The venter region encloses a venter canal cell and an egg cell.
4. The wall of the neck is made up of six rows of cells.
5. The neck has 4-6 neck canal cells. The tip of the neck is covered by means of four cover cells or lid cells.

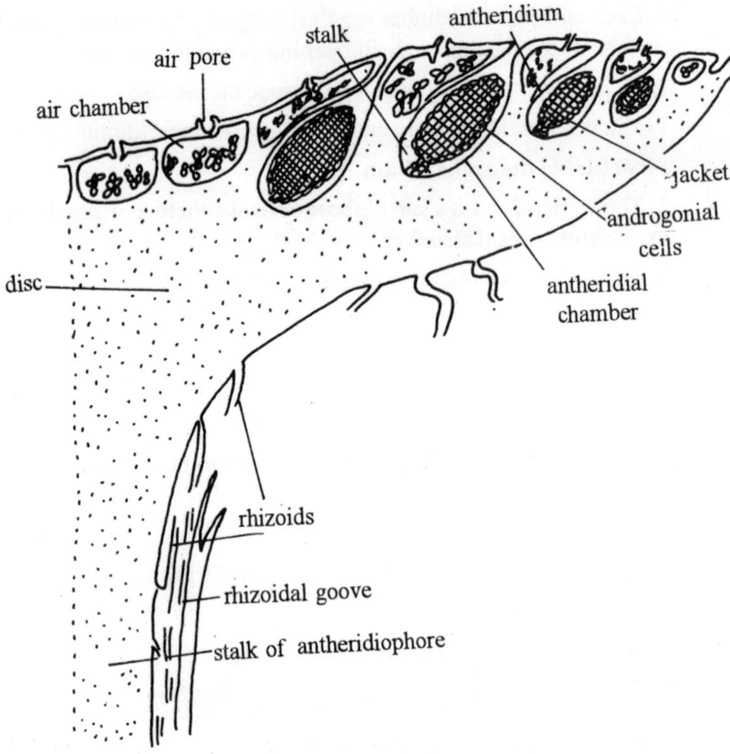

Fig. 3.28 Marchantia. L.S. of antheriophore (diagrammatic)

Structure of a mature archegoniophore

1. Along section of a mature archegoniophore will show the following details.
2. The disc as usual shows the upper photosynthetic zone and a lower storage zone.
3. The archegonia containing sporophytes at different stages of growth are seen in an inverted condition on the ventral surface of the disc.
4. The older archegonea are found towards the periphery, while the younger ones are found towards the centre ie., towards the stalk.
5. All the archegonia are enclosed in a common envelope called perichatium.
6. The ventral surface of the disc produces another envelope in between the archegonia. This is called the perrgynoium. In addition to this, the venter also provides an envelope to the sporophyte, this is called Calyptra.

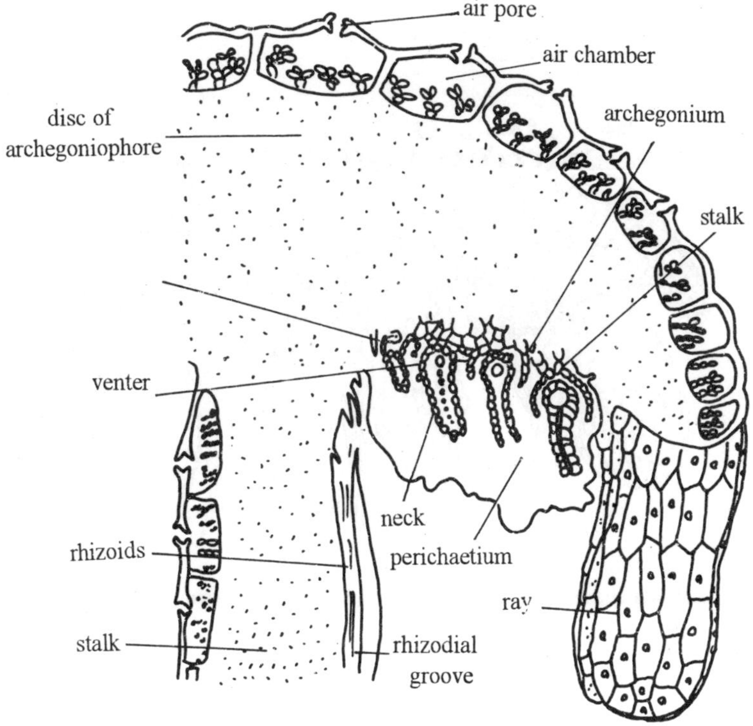

Fig. 3.29 Marchantia. L.S. of archegoniophore (diagrammatic)

Structure of a mature sporophyte

1. The mature sporophyte of *Marchanita* will be in an inverted position hainging down from the ventral surface of the archegonial disc.

2. The sporophyte has 3 distinct regions viz., foot, seta and capsule.

3. The foot is made up of parenchymatous cell. It is Sagittate (arrow head like) and helps to anchor the sporophyte in the gametophytic tissue.

4. The seta is an elongated stalk. The cells are vertically extended.

5. The terminal part of the sporophyte is the globose capsule.

6. The wall of the capsule is made up of a single layer of cells.

7. Within the capsule, in the young condition there will be a number of spore mother cells. Some of them become sterilized to form the elaters, while the rest undergo reduction division to form tetrads of haploid spores.

8. Elaters are spindle shaped structures having special thickening. They are hygroscope and help in the dispersal of the spores.

9. The spores are uninulceate and have a double layered wall. The outer layer exine is highly ornamented, while the inner intine is thin. The spores on liberation develop into new individuals.

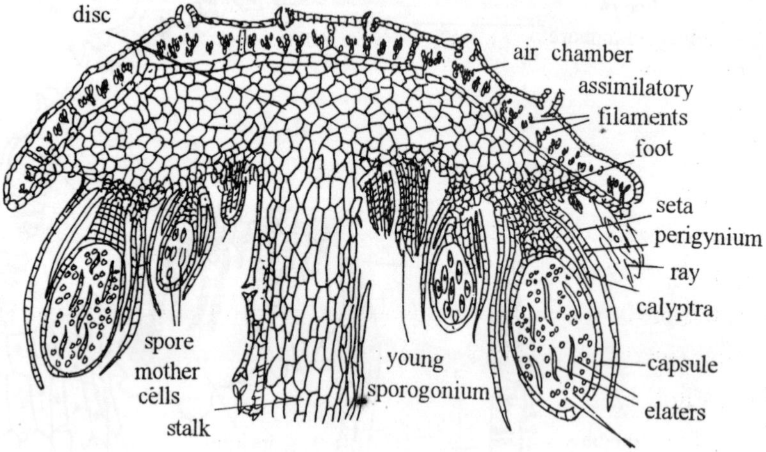

Fig. 3.30 Marchantia. V.L.S. of archegoniophore showing young sporogonia

Diagram to be drawn

1. Habit sketch - Dorsal and Ventral views
2. Dorsal view enlarged to show gemma cups antheridiophore and archegoniophore (on dif thalli)
3. T.S. of the thallus
4. T.S. of the thallus through a gemma cup (ground plan)
5. T.S. of stalk of antheridiophore
6. Vertical section of antheridiophore.
7. Vertical section of archegoniophore
8. A sporophyte enlarged.

Systematic position and identification

A. 1. Plant body is a parenchymatous gametophyte.

2. It is leafy or thalloid.

3. Vasculature and roots absent.

4. Sex organs surrounded by a sterile jacket layer.

5. Life cycle exhibits heteromoprhic alternation of generations with a predominant sporophyte ...Bryophyta.

B. 1. Thalli are lobed and dorsiventrally flattened or leafy.

2. Chloroplasts do not have pyrenoids

3. Scales and 2 types of rhizoids present

4. No columella in the sporophyte ...Hepaticopsida.

C. 1. Plant body thalloid and dichotomously branched

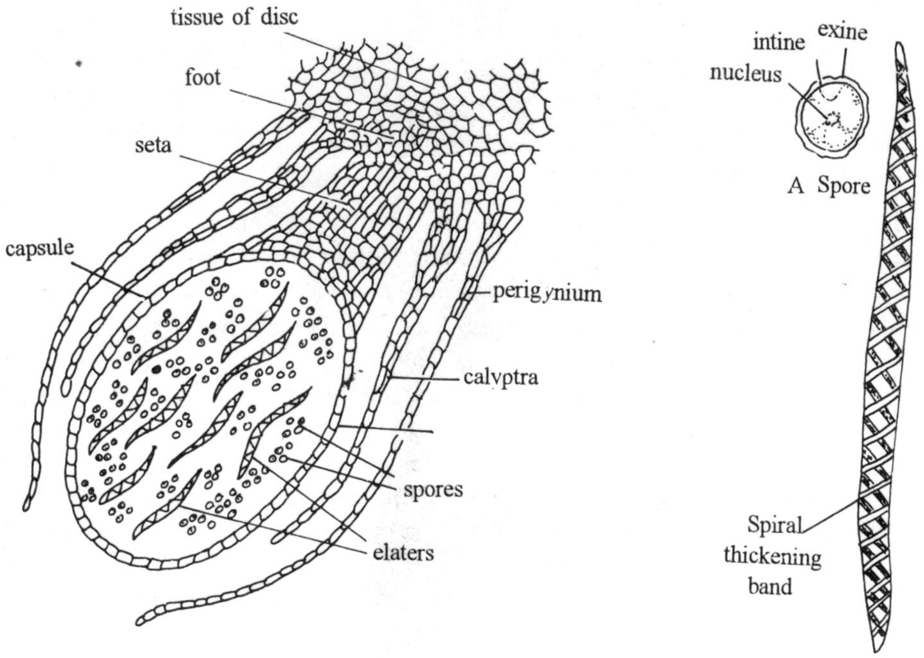

Fig. 3.31 Marchantia. L.S. of sporophyte and elater

2. Scales are present on the ventral surface.
3. Thallus differentiated into photosynthetic and storage regions
4. Air chambers present ...Marchantiales.

D. 1. Sex organs borne on gametophores.
2. Air pores are like chimney.
3. Sterile cells form elaters...Marchantiaceae.

E. 1. Four rows of scales presents
2. Scales are of 2 types.
3. Aereoles present on the dorsal surface.
4. Photosynthetic filaments branched.
5. Gemma cup with fringed margins, it is not crescent shaped.
6. Mucilage cells present in the storage region of the thallus ...Marchantia.

PORELLA

Common species : *P.platyphylla, P.rotundifolia etc.*

Occurrence : The genus Porella also known as Madotheca is a typical and widely distributed member of the order. Jungermanniales. The genus has 180 species of which about 34 are recorded from India. Porella is usually found in moist shady places growing on logs, trunks of trees and water dripping rocks.

Study : 1. External morphology

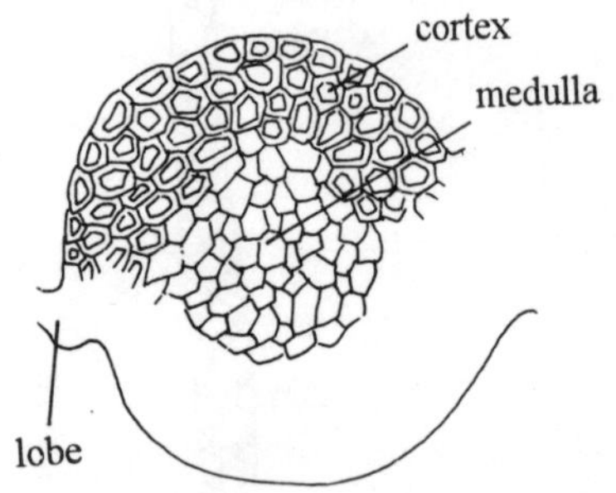

Fig. 3.32Porella. T.S. of axis

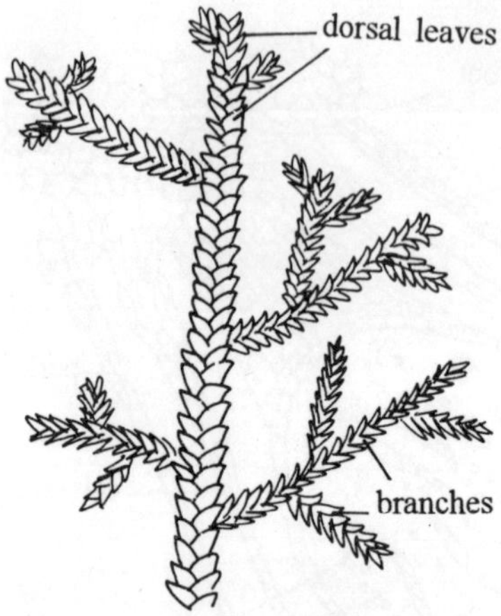

Habit of plant body

2. Anatomy of axis
3. Reproduction
4. Structure of the sporophyte
5. Systematic position and identification

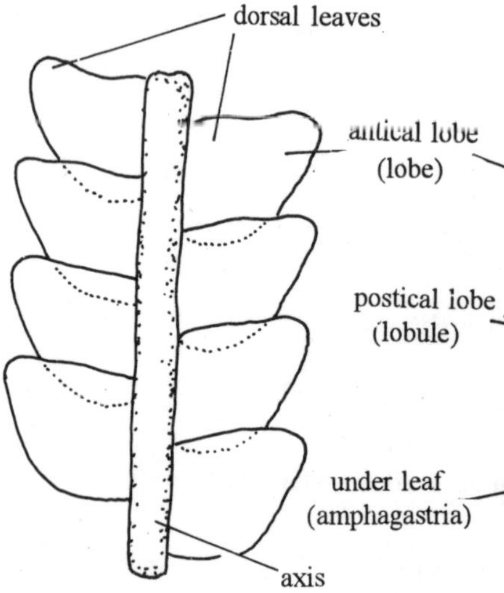

Fig. 3.33 Porella. Dorsal view of a part of leafy shoot

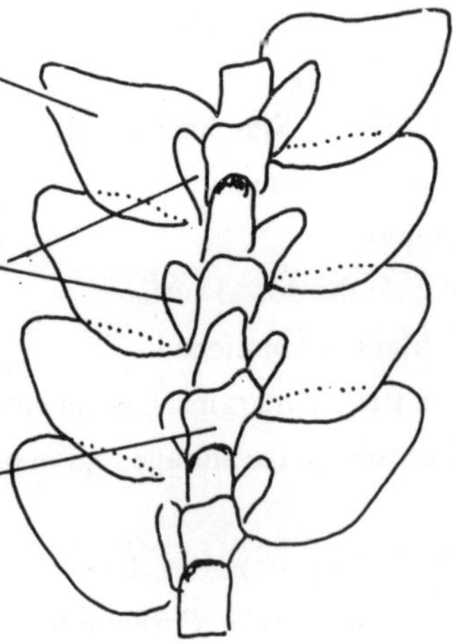

Ventral view of a part of leafy shoot

External morphology

1. The Gametophyte of porella is greenish and leafy. When mature grows to a length of 15 cms.

2. It grows flat on the surface of substratum.

3. The plant is dorsiventrally differentiated.

4. The gametophyte consists of a central axis or stem and is covered by leaf like expansion.

5. The stem shows monopodial branching.

6. The stem and branches bear 2 types of leaves viz., dorsal leaves and ventral leaves.

7. The dorsal leaves also called paragastica are found on the dorsal surface arranged in 2 rows.

8. The dorsal leaves completely cover the stem and on the dorsal surface thier arrangement is said to be incubous ie., the anterior (upper) margin of each leaf overlaps the posterior (lower) margin of next leaf found above.

9. On the ventral surface the arrangement of dorsal leaves is said to be Succubous ie., the lower margin of each leaf overlaps the upper margin of leaf below.

10. Such an arrangment of leaves enables the easy flow of water.

11. Each dorsal leaf is bibobed, it has a large Postical lobe and a small antical lobe.

12. Arranged on the ventral surface and covering them is found a row of leaves called ventral leaves or amphigastria.

13. Totally on the axis there are 3 rows of leaves (2 rows of dorsal leaves and 1 row of ventral leaf).

14. In the leaves of Porella as in other foliose liver worts, the mid-rib is absent.

15. Each leaf consists of a single layer of cells.

16. From the undersurface of gametophyte arise a few simple rhizoids which help mainly in fixation and rarely in absorption.

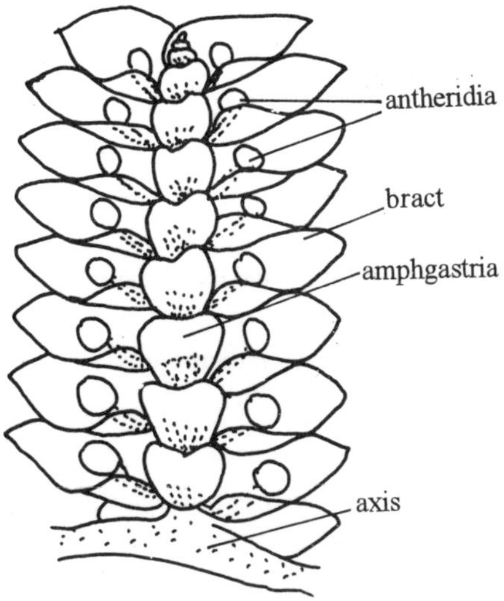

Fig. 3.34 Porella. Antheridial branch

Anatomy of the Axis

1. The stem at younger regions has only parenchymatous tissue.

2. The cells are thin walled and filled with chloroplasts.

3. The epidermal layer is not well differentiated.

4. In the older portions of the stem, there is differentiation into outer cortical region and inner outer cortical region and inner medullary region.

5. The cortex is made up of a smaller cells with slightly thickened walls.

6. The medulla consists of layer cells with thinner walls.

Vegetative propagation

1. Progressive death and decay of older parts up to a branching will result in the separation of a branch which will develop into a new individual.

Sexual reproduction

1. Sexual reproduction is of the oogamous type. Sex organs are antheridia (male) and archegonia (female).

2. Parella is dioecious, male plants are smaller than the female plants.

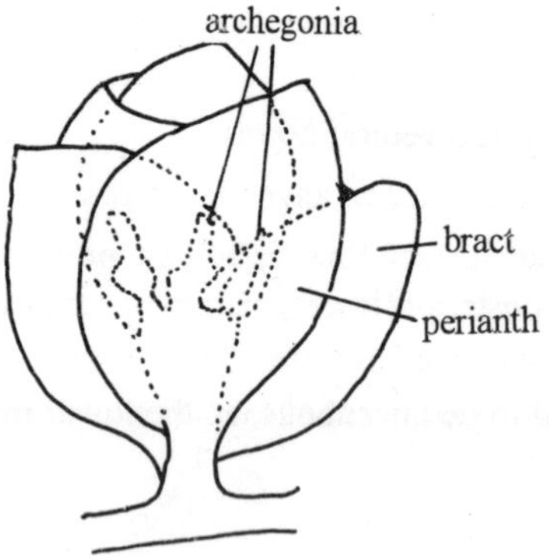

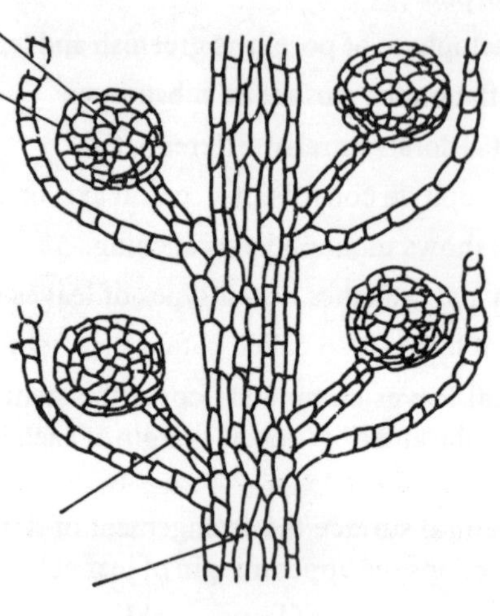

Fig. 3.35a Archegonial branch L.S. of antheridial branch

Male plant and antheridium

1. The plant produces a number of short lateral antheridial branches, which project out more or less perpendicular to the main axis.

2. Every antheridial branch is composed of a number of compactly arranged small leaves called perigonial leaves

3. In the axil of each perigonial leaf arises an antheridium.

4. A well developed antheridium has a long stalk and the spherical antheridium proper.

5. The stalk is two celled in breadth.

6. The jacket of the antheridium is single layered at the apex, two layered at the sides and three layered at the tax.

7. Within the antheridium are found antherizoid mother cells, which later give rise to antherozoids.

Female plant and archegonium

1. The female plant produces archegonia on special branches called archegonial branches.

2. Each archegonial branch is almost like a vegetative branch, but ends in a few larger leaves called the Perichactial leaves or bracts.

3. The archegonia appear in terminal clusters.

4. A well developed archegonium is a flask shaped structure with a basal bulged venter and an upper narrow neck.

5. The tip of the neck is covered by four cover cells and within the neck are present 6-8 neck canal cells.

6. The venter wall is 2-3 layered thick and it has an egg cell below and a venter and cell above.

7. The neck is covered by five vertical rows of cells forming the wall (six rows are found in Marchantiales).

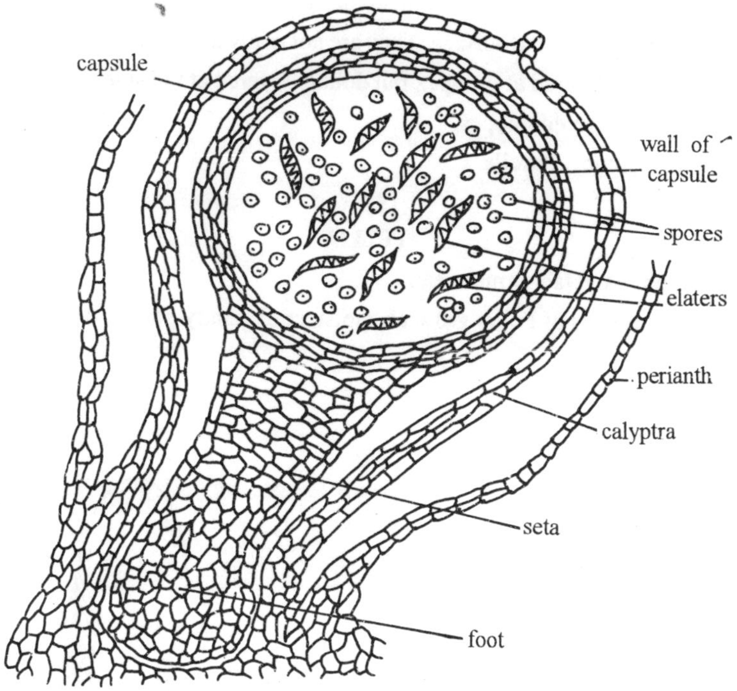

Fig. 3.36 Porella. L.S. of sporogonium

Systematic position and identification

A. 1. Plant body is a parenchymatous gametophyte.

2. It is leafy or thalloid.

3. Vasculature and roots absent.

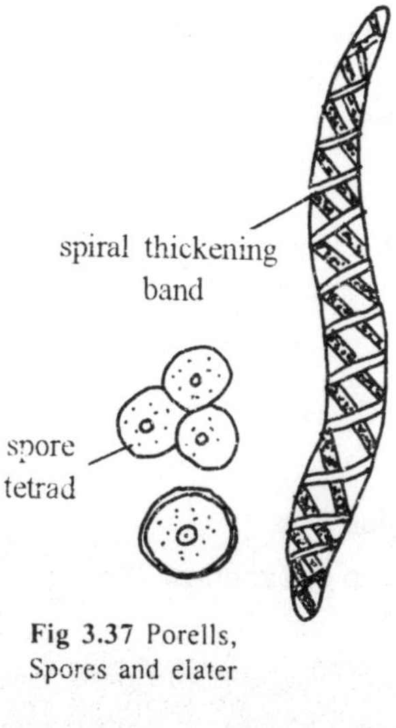

Fig 3.37 Porells, Spores and elater

4. Sex organs surrounded by a sterile jacket layer.

5. Life cycle exhibits a heteromorphic alternation of generations with a predominant gametophyte ... Bryophyta.

B. 1. Thalli are lobed and dorsiventrally flattened or leafy.

2. Chloroplasts do not have pyrenoids.

3. No columella in the sporophyte ... Hepaticopsida.

C. 1. Internal tissue differentiation absent in the gametophyte.

2. Scales are absent.

3. Rhizoids only one type (simple).

4. Archegonial neck has 5 rows of cells ...Jungermanniales.

D. 1.Gametophyte is leafhy

2. 3 rows of leaves present on the axis.

3. Archegonia produced at the tips of archegonial branches ...Jungermanniae.

E. 1. Dorsal leaves have 2 lobes and arranged in incubous manner.

2. Simple rhizoids arise from the ventral surface. They are not many in number.

F. 1. Antheridial jacket single layered at the apex but double layered at the sides.

2. Sporophyte has two protective envelopes.

3. No elaterophore in the capsule ...Porella.

PELLIA

Common species : P.epiphylla, P.fabborniana etc.

Occurrence : The gametophytes of Pellia grow on moist, cool, shady places, on trunks of trees or on wet rocks forming extensive mats.

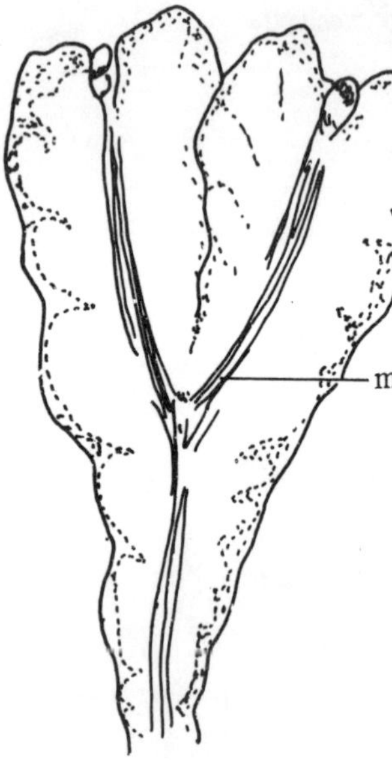

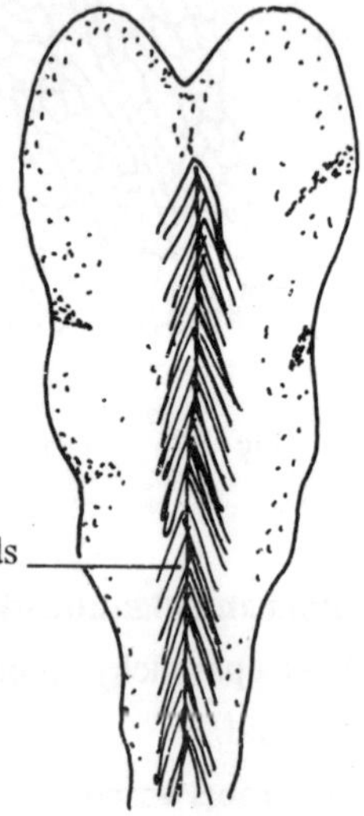

Fig. 3.38 Pellia. Dorsal surface of thallus   Ventral surface of thallus

Study : 1. External morphology of thallus
2. Internal structure
3. Reproduction
4. Structure of the sporophyte
5. Systematic position and identification

External morphology

1. The plant body is a dorsiventrally flattened dichotomously branched thallus.

2. The margins of the thallus are undulate and sinuous, often the lobes overlap one another.

3. There is a midrib in the centre, which if often purple in colour.

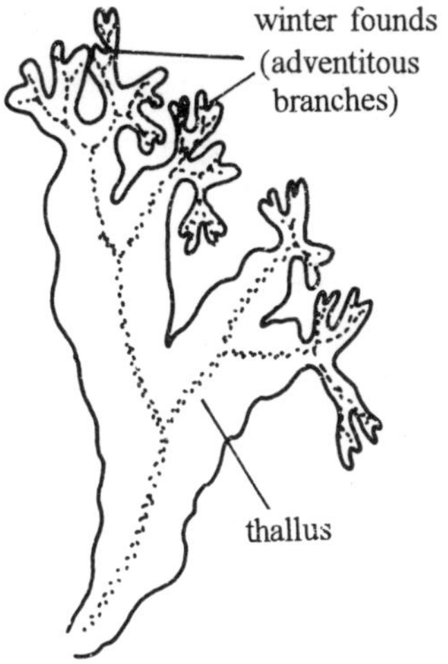

Fig. 3.39 Pellia. Thallus showing winter frongs

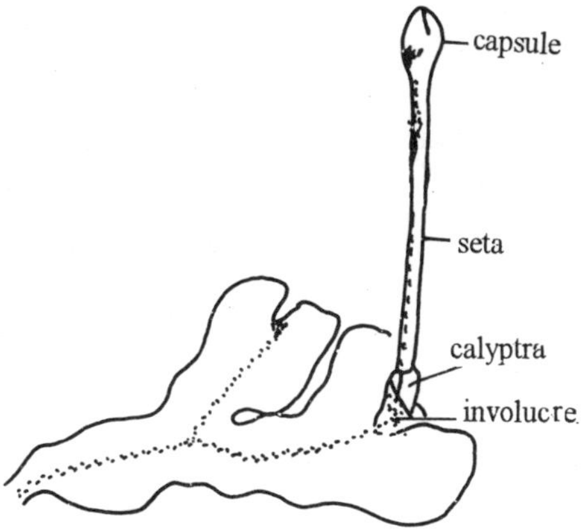

Thallus with sporogonium

4. The dorsal surface of the thallus is much greener in colour than the ventral surface.
5. The thallus on its ventral surface bears only smooth rhizoids.
6. Pegged rhizoids and scales are absent
7. The growing points of the thallus are situated in the apical nothces present at the tips of lobes.
8. The presence or absence of moisture decides the morphological appearance of the thallus.

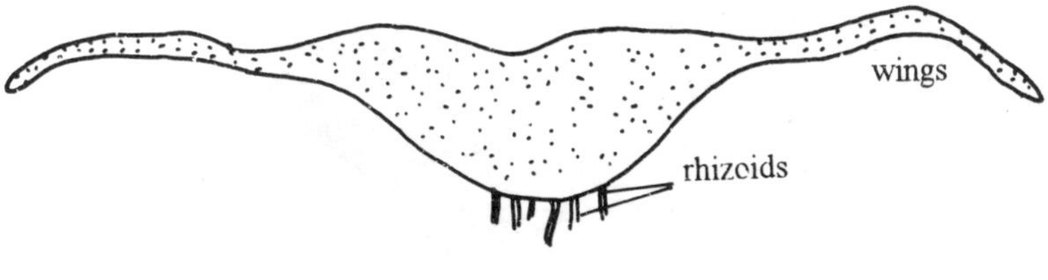

Fig. 3.40 Pellia. V.T.S. of thallus (diagrammatic)

9. Thalli growing near water are delicate, narrow and ribbon like, with a distinct mid-rib, while those growing on a dry soil are short thick and deep green without prominent mid-rib.

Anatomy of the thallus

1. A transverse section of the thallus shows on undifferentiated inner region.
2. The upper epidermis or lower epidermis are made up of a singel layer of compact cells rich in chloroplasts.
3. All the cells in between the epidermal layers are alike and there is no differentiation into storage and photosynthetic zones.

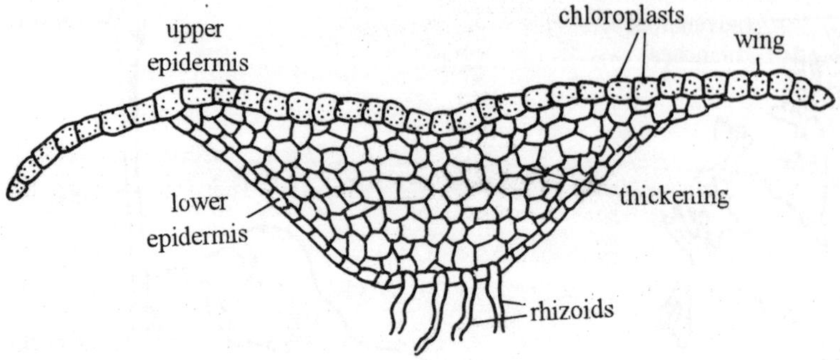

Fig. 3.41 Pellia. V.T.S. of thallus (cellular)

4. All the cells have chloroplasts (not many) and starch grains. Some cells also have oil contents.
5. The mid-rib region consistscf some thick walled cells which extend throughout the length of the thallus.
6. The mid-rib region is 8-14 cells thick while the margins are thin.
7. Smooth rhizoids arise from the lower epidermis in the midrib region.

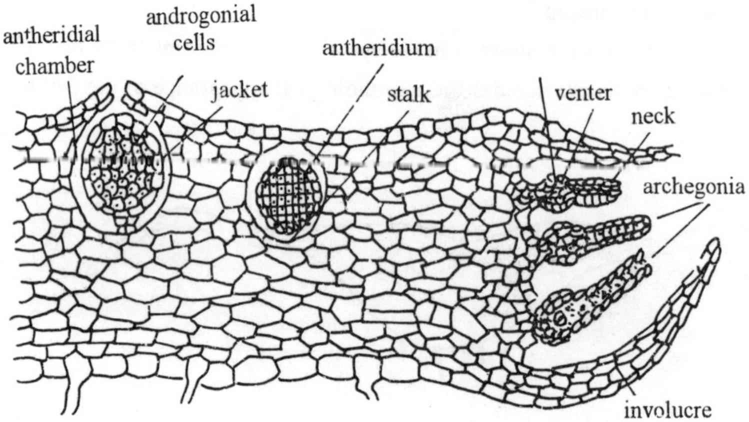

Fig. 3.42 pellia. V.L.S. of thallus passing through sex organs

Vegetative propagation

1. Adventitious branches arising on the dorsal surface, separate off and develop into new individuals.

Sexual reproduction

1. Sexual reproduction is oogamous. Sex organs arc antheridia (male) and archegonia (female).
2. Species of porella may be monoecious as in P.epiphylla or dioecious as in P.nessiana.

3. Monoecious species are usually protandrous.
4. Sex organs are found embedded on the dorsal surface.

Antheridium

1. Male thalli in dioecious species are smaller than the female thalli.
2. Presence of small knob like outgrowths on the dorsal surface indicate the presence of antheridia.
3. Antheridia are found embedded in small cavities.
4. Each cavity encloses a single antheridium and opens out by a small opening.
5. Each antheridium is a globose structure having a short multicellular stalk.
6. The wall of the antheridium is made up of a single layered jacket.

The antherozoid mother cells present inside the antheridium give rise to biflagellate coiled antherozoids.

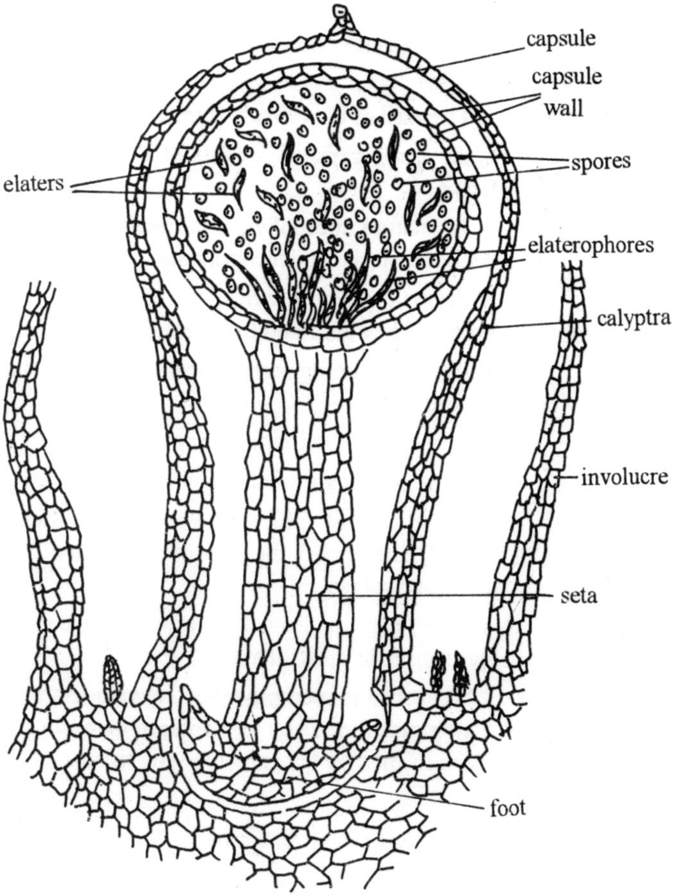

Fig. 3.43 Pellia. V.L.S. of sporogonium

Archegonium

1. The archegonia are found in groups of 6-12 on the dorsal surface, a little behind the apical region.
2. The archegonia are produced on a slightly elevated part called the receptacle.
3. All the archegonia are covered by means of a flap like envelope called the involucre.
4. The involucre arises from the thallus as an outgrowth.
5. Each archegonium is a flask shaped structure, and has a short stalk.
6. The archegonium is similar in structure to what is seen in Porella.
7. The wall of the neck is made up of five vertical rows of cells typical to Jungermanniales.

Structure of the sporophyte

1. The sporophyte is divided into foot, seta and capsule
2. The foot is well developed, somewhat bulbous and forms a collar like sheat surrounding the basal portion of the seta.
3. The seta has vertically elongated cells. It is short in the young sporophyte, but becomes long at maturity.
4. The capsule occupies the terminal part of seta and is globose in shape.
5. The wall of the capsule is made up of three layers of cells.
6. A young capsule consists of a uniform, mass of archesporial tissue.
7. In mature sporophytes, the archesporium at the basal region, becomes sterile and forms a hum f tis called elaterphore.
8. From the dlaterophore arise a number of elaters with spiral thickenings. They are fixed to the elaterophore.
9. In the remaining part of capsule are found haploid spores as well as free elaters.
10. Free elaters are similar in structure to fixed elaters.
11. The spores are small, haploid and uninucleate.
12. The wall of the capsule consists of thin walled cells at four places. These mark the lines of dehiscence. The capsule opens out into four valves.
13. On liberation from the sporophyte, the spores develop into new gametophyte individuals.
14. The mature sporophyte has 2 envelopes viz calyptra and involure. Calyptra is formed by the venter of the archegonium; while involucre is formed by the gametophytic tissue.

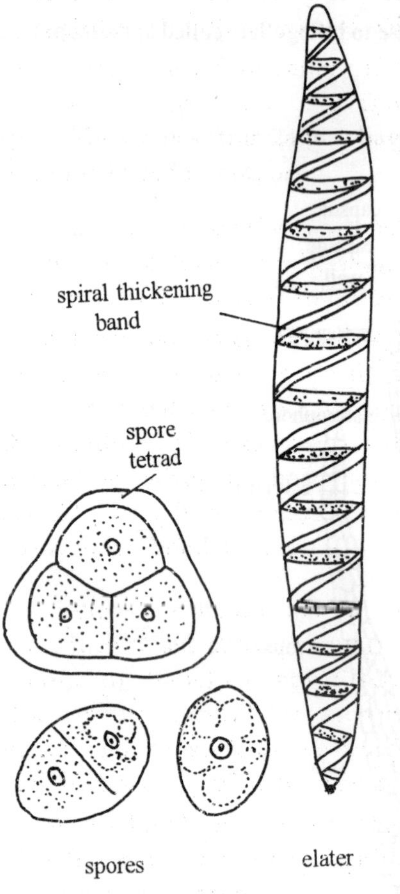

Fig 3.44 Porella

Diagrams to be drawn

1. Habit sketch - dorsal and ventral views.
2. Cross section of the thallus
3. Cross section of the thallus to show antheridia
4. Section of the thallus to show archegonia
5. Long section of the sporophyte

6. Elaterophore, elaters and spores (enlarged)

Systematic position and identification

A. 1. Plant body is a parenchymatous gametophyte

2. It is leafy or thalloid

3. Vasculature and roots absent

4. Sex organs surrounded by a sterile jacket layer

5. Life cycle exhibits, heteromorphic alterntion of generations with a predominant gametophyte Bryophyta.

B. 1. Thalli are lobed and dorsiventraly flattened or leafy.

2. Chloroplasts do not have pyrenoids

3. No columella in the sporophyte ... Hepaticopsida

C. 1. Internal tissue differentiation absent in the gametophyte.

2. Scales are absent

3. Rhizoids only of the simple type

4. Archegonial neck has 5 rows of cells in the wall... Jungermanniales.

D. 1. Apical cell does not get transformed into an archegonial initial

2. Plant body is thalloid ... Melzerinae.

E. 1. Antheridia develop singly in cavities.

2. Archegonia found in clusters

3. Elaterophore present ... Pelliaceae

F. 1. Thallus has sinores overlapping lobes

2. Archegonia enveloped by involucre

3. Knob like outgrowths present on the thallus to indicate the underlying antheridia

4. Elaterophore present at the base of the capsule

5. Capsule splits into 4 valves ... Pellia

CHAPTER 4

ANTHOCEROTOPSIDA

The class Anthocerotopsida is a small but unique group among Bryophytes. It is characterised by the following features :

1. The gametophytic plant body is a dorsiventrally flattened thallus, with no internal tissue differentiation.
2. Rhizoids are present on the ventral surface; they are of the simple (smooth) type.
3. Pegged rhizoids and scales are absent.
4. The thallus consists of mucilagenious cavities inhabited by cyanophycean algae.
5. Each cell of the gametophyte has a single chloroplast with a pyrenoid
6. The sporophyte is capable of indefinite growth
7. The central region of the sporophyte is made up of columella derived from the endothecium.

The class antherocerotopsida includes a singel order Anthocerotales with a single family Anthocerotaceae. The family includes about 5-6 genera with nearly 300 species all of which are distributed widely but have a special preference to tropical and warm temperate regions. The 6 genera in the family are Anthoceros, Phaeoceros, Aspiromitus, Notothylos, Dendroceros and Megaceros. Of all the genera, Anthoceros is the largest with about 200 species.

ANTHOCEROS

Common species : *A.hallii, A.lavis, A.erectus, A.fusiformis, A.punctatus* etc.

Occurrence : The genus with its nearly 200 species is cosmopolitan in its distribution, but having a preference to temperate and tropical regions. About 25 species found in India, growing on moist shady places or on the side bunds or rice fields.

Study : 1. External morphology

2. Internal structure
3. Vegetative propagation
4. Sexual reproduction
5. Structure of the sporophyte
6. Systematic position and identification

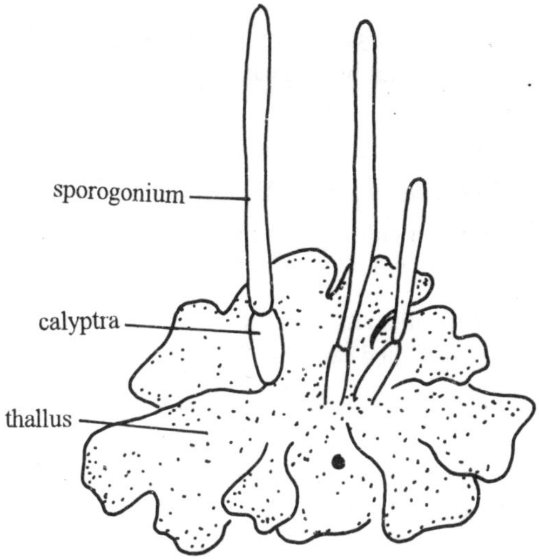

Fig. 4.1 Anthceros. Thallus bearing sporophytes

External morphology

1. The gametophytic plant body is a dorsiventrally flattened, lobed horizontally gorwing thallus.

2. The thallus is thicker at the centre than at the margins, but there is not distinct midrib.

3. The dorsal surface of the thallus may be smooth, warty or rough with ridges.

4. On the dorsal surface are found many stalk or hornlike structures (hence the name Hornwort) representing the sporophytes.

5. Arising from the ventral surface are a number of smooth rhizoids.

6. Pegged rhizoids and scales are absent.

7. Small dark green patches are found on the ventral surface (this can be seen with a hand lens) representing the position of algal colonies inside.

8. The branching of the thallus is not strictly dichotomous.

Internal structure of the thallus

1. A vertical transverse section will show the following features.

2. The internal tissue does not show any differentiation into storage and assimilatory zone.

3. The entire inner region consists of isodiametric parenchyma cells.

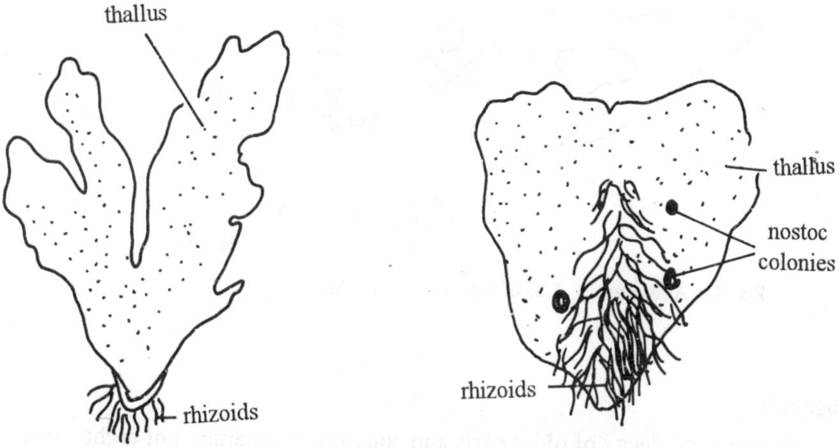

Fig. 4.2 Anthceros. Ventral surface of thallus showing Nostoc colonies

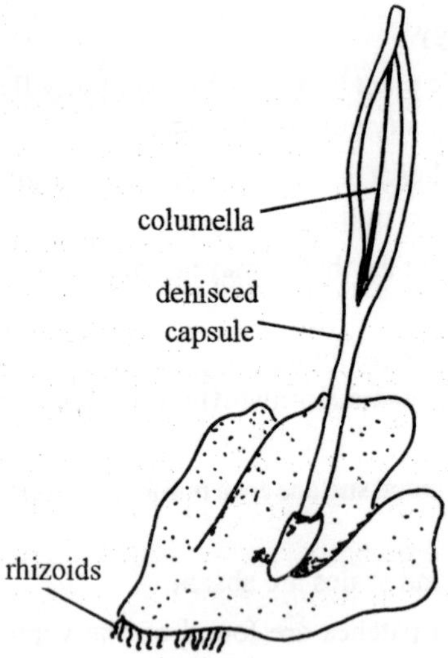

Fig 4.3 Anthoceros
Thallus bearing dehisced

4. However, on either surface (dorsal and ventral) the superficial layers are slightly different and can be esignated as upper and lower epidermis.

5. Epidemal cells are smaller in size and more regular in arrangement.

6. All the cells hae a single lens shaped chloroplast with a pyrenoid.

7. Occasionally the number of chloroplasts may be more than one.

8. The cells are uninucleate.

9. There are no airchambers or cavities, but on the ventral surface are found certain small mucilage filled cavities.

10. These are intercellular cavities opening through small pores on the ventral surface.

11. These cavities are often inhabited by endophytic algae such as Nostoc, the Nostol filaments entering the thallus through the pore.

12. The relationship between the alga and Anthoceros is believed to be symbiotic.

13. Nostoc helps in nitrogen fixation, while Anthoceros provides shelter for the alga.

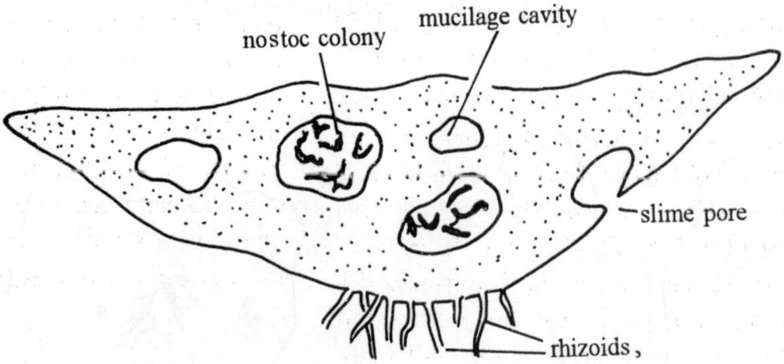

Fig. 4.4 Anthceros. V.T.S. of thallus (diagrammatic)

Vegetative propagation

1. By the progressive death and decay of older parts and consequent separation of dichotomies.

2. Small tubers are formed on the thallus which develop into new individual.

3. During unfavourable conditions, the apex remains persistent, while the other parts die. The persistent apex develops into a new individual.

Sexual reproduction

1. It is of the oogamous type. Sex organs are antheridia (male) and archegoma (female).

2. Species of Anthoceros may be monoecious (eg. A.fusiformis, A.punctatus etc) or dioecious (eg. A.erectus, A.hallii etc)

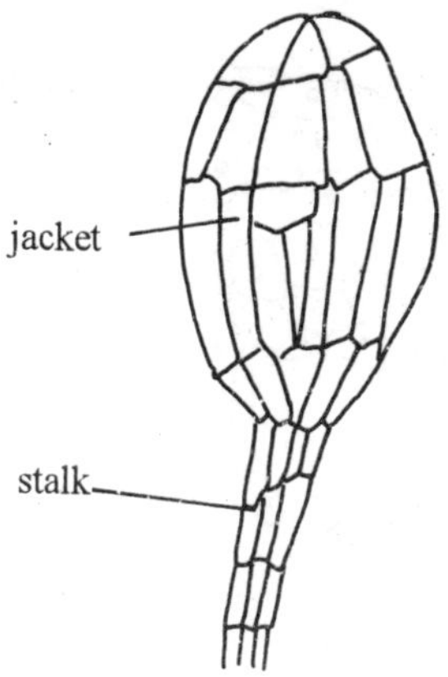

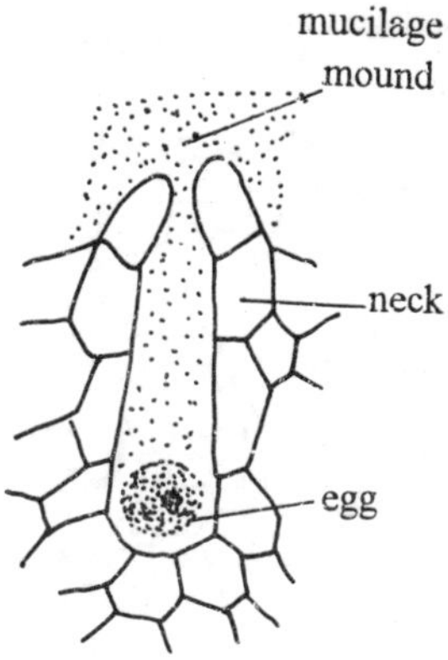

Fig. 4.5 Anthceros. Single antheridium (external features)   Mature archegonium

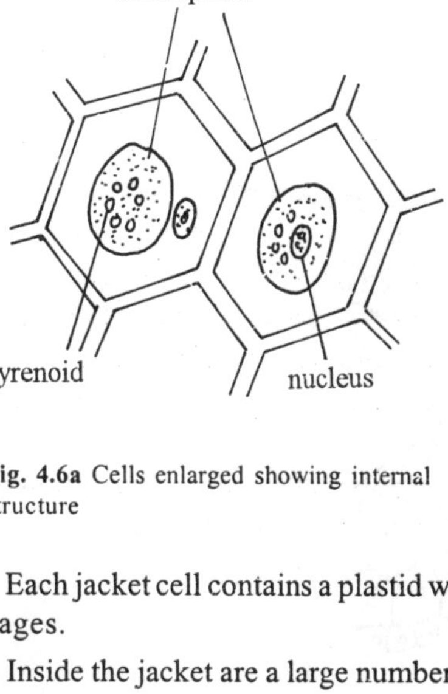

Fig. 4.6a Cells enlarged showing internal structure

3. Sex organs develop very close to the apical cell.

4. Antheridia and archegonia are found on the dorsal surface of the thallus.

Antheridium

1. Antheridia are found embedded on the dorsal surface of the thallus, in small cavities called antheridial cavities.

2. Each antheridial cavity or chamber has a covering or roof made up of two layers of cells.

3. Each antheridial chamber may enclose one to many (16-20) antheridia.

4. A mature antheridium shows a clubshaped or pouch like body borne on a slender or massive stalk.

5. The jacket of the antheridium is composed of a single layer of cells.

6. Each jacket cell contains a plastid which is colourless in the uoung antheridium, but becomes green at later stages.

7. Inside the jacket are a large number of androcytes, each of which forms an antherocoid.

8. At maturity, the roof of the antheridial chamber ruptures, so also the jacket of the antheridium and the antherocoids are released.

Archegonia

1. The archegonia is a flask shaped structure and is embedded on the dorsal surface of the thallus.

2. The archegonium is histologically continuous with the thallus and is scarcely distinguishable.

3. Only the cover cells project above the level of the thallus.

4. The wall of the neck is difficult to distinguish from the gametophytic tissue .

5. Within the neck are 4-6 neck canal cells.

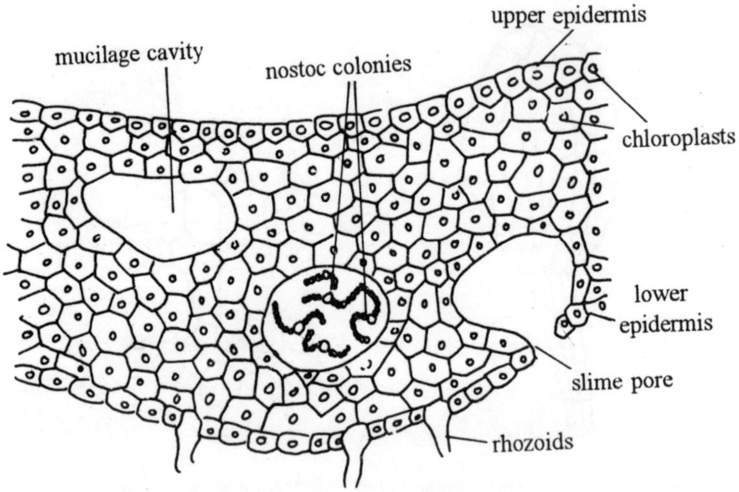

Fig. 4.7 Anthceros. A portion of V.S. of thallus (cellular)

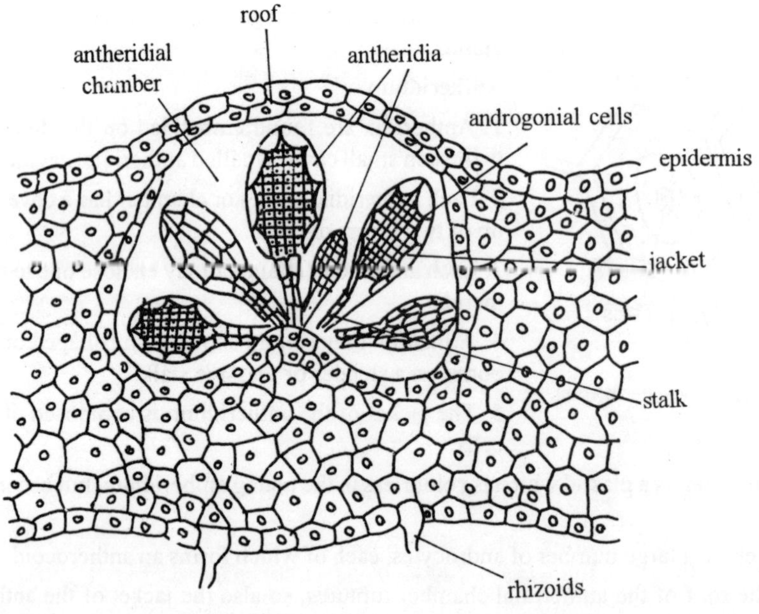

Fig. 4.8 Anthceros. A portion of V.S. of thallus showing antheridia

6. The venter region has a venter canal cell and an egg cell.
7. This is not much difference is sinze between neck and venter. The neck is as broad as the venter.
8. Neck canal cells and venter canal cells degenerate to facilitate the entry of antherocoids.
9. The antherocoids fertilize the egg and it becomes a zygote.

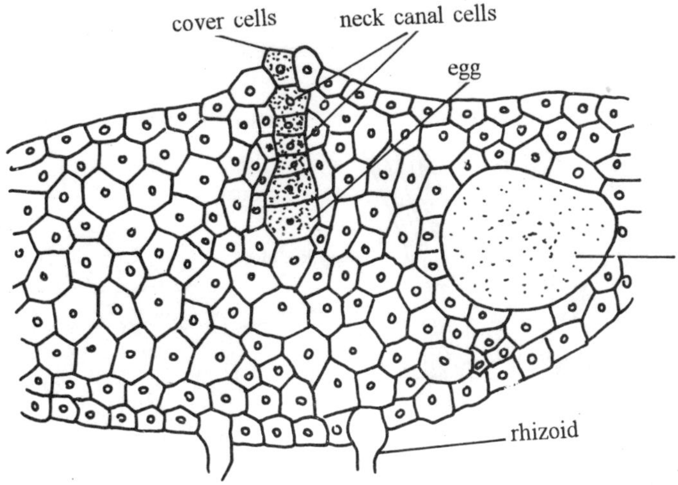

Fig. 4.9 Anthoceros. V.S. of thallus passing through archegonia (A part)

Structure of the sporophyte

1. The sporophyte has 3 parts viz., foot, staining and capsule.

2. The foot is bulbous and wholly parenchymatous; the superficial cells are palisade like.

3. Foot is embedded in the thallus and absorbs nutrition from the surrounding cells.

4. The seta is intermediate in position. It is short and consists of actively dividing cells.

5. The capsule of the sporophyte at maturity comes out of the thallus in the form of small horny structures which may be 5-15 cms long.

6. The central region of the capsule is made up of columella, which extends nearly to the tip of the capsule.

7. The cells of the columella are thickened and vertically elongated. They are derived from endothecium of the embryo.

8. Around the columella is a cylinder of sporogenous tissue, which shows the whole sporogenous series from one layered archesporium at the base, to the mature spores and pseudoelats at the tip.

9. Pseudoclaters are modified sterile spore mother cells and they lack the spiral thickenings.

10. The wall of the capsule is made up of 4-6 layers of parenchyma cells which have chloroplasts.

11. Epidermis is the outermost layer and it has stomata.

12. The presence of chloroplasts and stomata indicates the mitritanal independence of the sporophyte.

13. Spores are spherical or tetrahedral in shape and thick walled.

14. The wall has an outer thick opaque exospore and an inner endospore. Spores have oil droplets.

15. The spores on germination develop into new individuals.

Diagrams to be drawn

1. Habit sketch dorsal and ventral views.

2. Transverse section of thallus showing antheridia.

3. Transverse section of thallus showing archegonia.
4. L.S. of sporophyte.

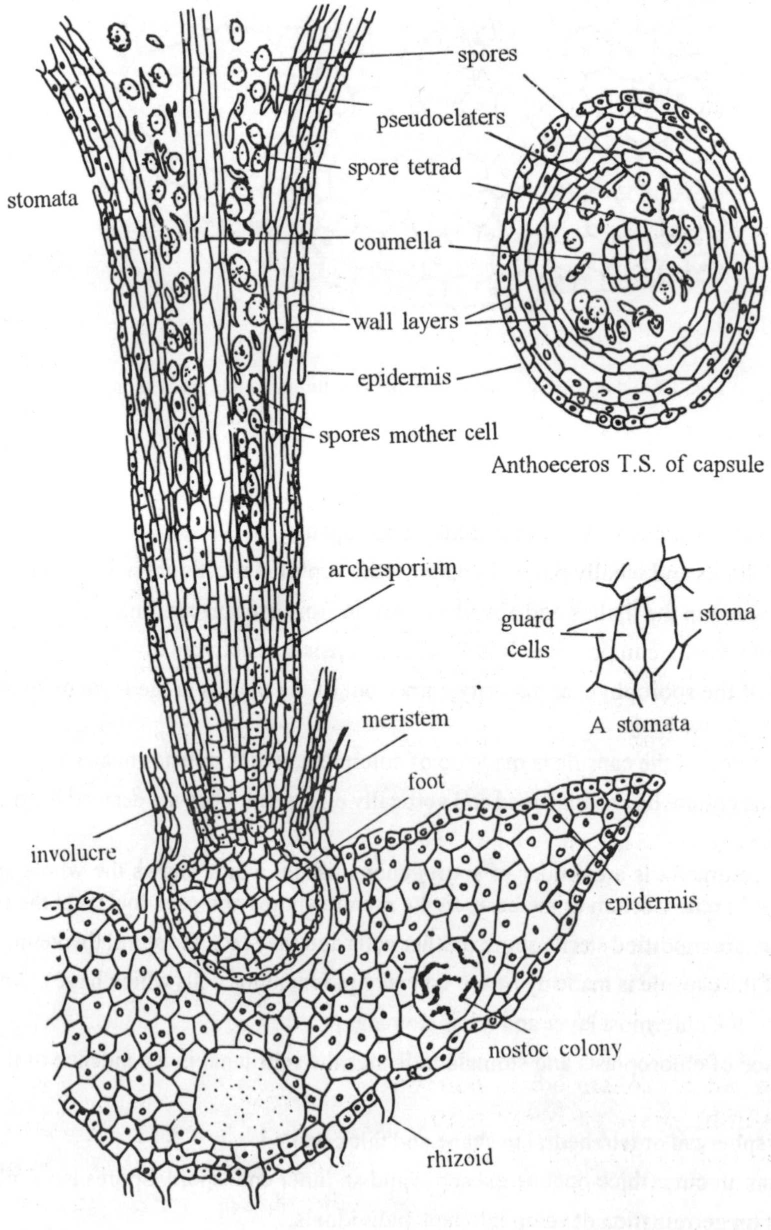

Fig. 4.10 Anthoceros. L.S. of sporogonium

Systematic position and identification

A. 1. Plant body is a parenchymatous gametophyte.

2. He is leafy or thalloid; vasculature and root absent.

3. Sex organs surrounded by a sterile jacket layer.

4. Life cycle exhibits heteromorphic alternation of generations with a dominant gametophyte... Bryophytes.

B. 1. Gametophyte thalloid.

2. Internally the gametophyte does not exhibit any tissue differentiation.

3. Cells have a single chloroplast with a pyrenoid...Anthocerotopsida.

C. 1. The thallus has uniform isodiametric cells.

2. Rhizoids smooth only, pegged rhizoids and scales are absent ...Authocerotales.

D. 1. Sporophyte has indeterminate growth.

2. Seta is short and meristematic.

3. Columella present in the capsule...Anthocerotaceae.

E. 1. Nostoc colonies present in mucilage cavities of the thallus.

2. Wall of the capsule chlorophyllous.

3. Involucre covering present only at the base of the sporophyte ...Anthoceros.

CHAPTER 5

BYROPSIDA

The class Bryopsida is the third largest the class of bryophyta including 660 genera and 14,500 species. The characteristic features of Bryopsida are as follows:

1. The gametophyte is differentiated into 2 stages, viz., (a) prostrate, thalloid, branched protonema and (b) erect leafy gametophore.
2. The gametophore has a central axis on which are spirally arranged the leaves bearing six organs.
3. Rhizoids are multicellular and have oblique cross walls.
4. Internally, the axis has a peripheral region and the central region; the latter performing conduction.
5. Sex organs are formed from superficial cells of the gametophore and show a definite apical growth in the early stage.
6. The sporophyte consists of foot, seta and capsule.
7. The capsule wall is multilayered with an epidermis having functional or non-functional stomata.
8. the archespenium is either derived from endothecium or from amphithecium usually enclosing a central columella.
9. Archsporium gives rise to only spores. Elaters are absent.
10. Spores on germination first develop into a filamentous structure called protonema. Later protenema gives rise to the adult gametophyte.

Classification

The class is divided into 5 subclasses. They are:

1. Sphagnidae – sphagnum
2. Andreaidae
3. Bryidae – Funana
4. buxabaumidae
5. Polytrichidae – Polytrichum.

SPHAGNUM

Common species : *S. Tenellum, S. molluscum, etc.*

Occurrence: Sphagmum is the monotypic genus of the family sphagnaceae. It is commonly called the Peat moss or bog moss. The genus has 336 species of which 17 are recorded from India. Majority of the species grow in aquatic areas, where the water is soft and acidic. Some species are terrestrial growing on damp soil or dripping rocks.

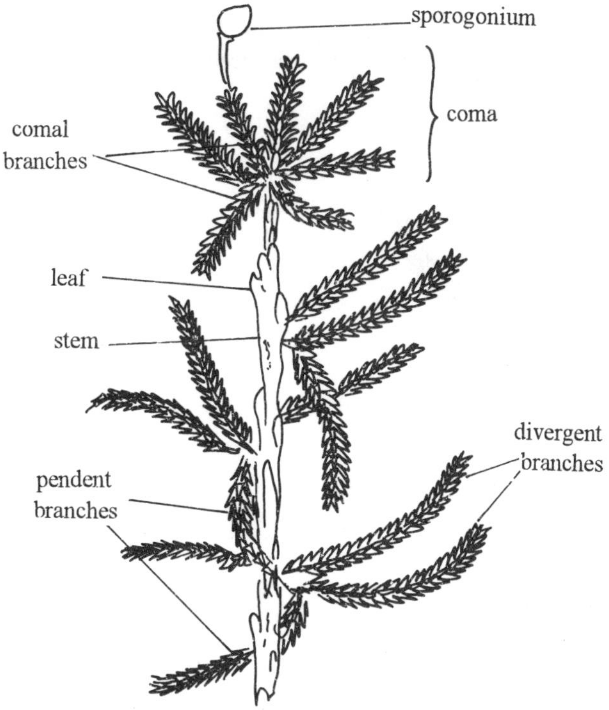

Fig. 5.1 Sphagnum. Plant body (Gametophore)

Study:
1. External morphology
2. Anatomy of the axis.
3. Vegetative propagation
4. Sexual reproduction.
5. Structure of the sporophyte
6. Structure of the protonema.
7. Structure position and identification

External Morphology:

1. The plant body of Sphagnum is the gametophyte often called the gametophore.
2. The colour of the gametophyte differs from species to species. It may be pale or dark green, orange, brown, rose, pink etc.
3. The gametophyte has a thick central axis called the stem. The stem shows lateral branching. The branches arise in tufts of 6-8.
4. The stem is anchored to the soil with the help of numerous rhizoids rising from the base.
5. Rhizoids are multicellular with oblique septa.
6. In aquatic forms, rhizoids usually disappear, especially in adult gametophores.
7. The amin axis as well as the branches are densely covered with small sessile 'leaves'.
8. The 'leaves' are spirally arranged.

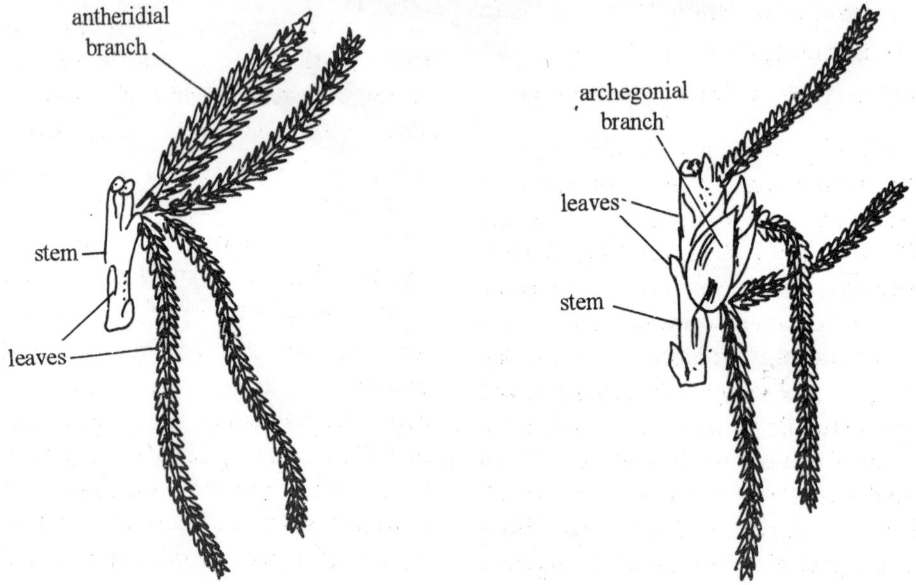

Fig. 5.2 Sphagnum. A portion of stem showing male branch

A portion of stem showing female branch

9. At the proximal end of the stem the short branches crowd themselves to form a head or coma.
10. The branches are of only one type in submerged species.
11. In floating species, the branches are of two types, viz., diverging branches or branches of limited growth and pendulans branches or branches of unlimited growth.
12. Diverging branches are short erect and grow up, while pendulous branches are long and droping downwards. pendulous branches are also known as Flagelliform branches.
13. the pendulous branches form a loose envelope around the stem, helping in the conduction of water.
14. This branch is exactly similar to the main stem in structure.
15. The leaves are spirally arranged, small, sessile and scaly. They have no midrib.
16. the leaves are one cell thick and have two types of cells, viz., chlorophyllous cells and hyaline cells.

17. The two types of cells are alternately arranged giving a net like appearance.
18. The chlorophyllous cells also called assimilatory cells are narrow and filled with chloroplasts. They help in photosynthesis. They are uninucleate.
19. The hyaline cells are rhamboidal in shape without any protoplast in them. They are dead cells.
20. These cells are colourless and filled with water.
21. The surfaces of the walls of hyaline cells have small circular or oral pores.
22. The inner surface of these cells are also traversed by annular and spiral thickenings.
23. these thickenings give mechanical support and present the cells from being folded.

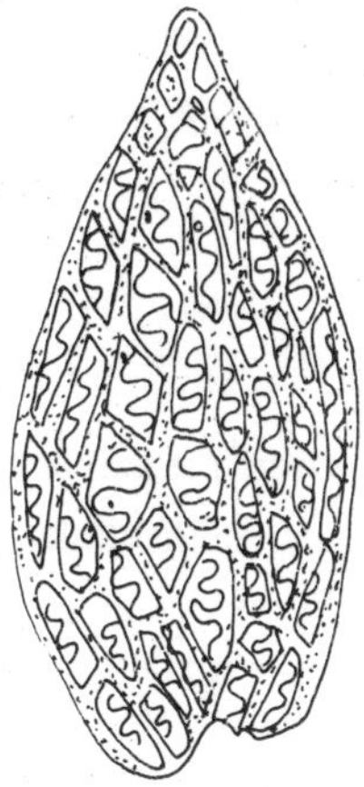

Fig. 5.3 Sphagnum. Single leaf

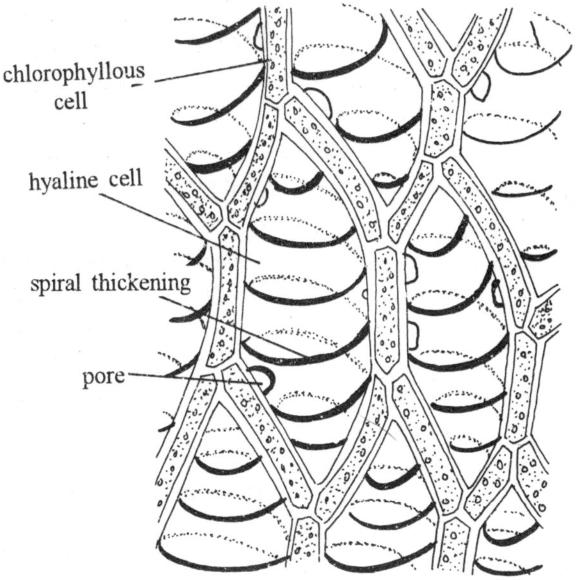

A portion of leaf under high power of microscope

Anatomy or the stem

1. In some submerged species there is absolutely no tissue differentiation.
2. In floating and terrestial species, some tissue differentiation is noticed.
3. A transverse section shows these regions, viz., cortex, hadrome and medulla.
4. There is a single layered epidermis as the outermost layer.

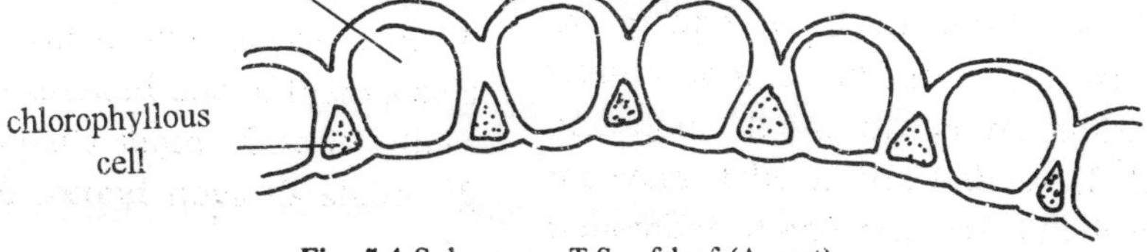

Fig. 5.4 Sphagnum. T.S. of leaf (A part)

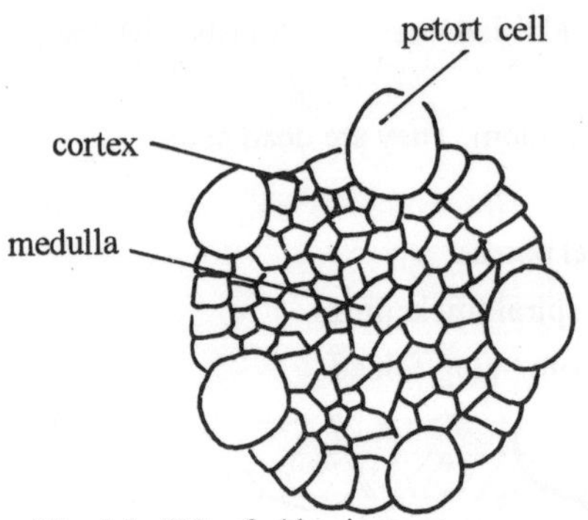

Fig. 5.5a T.S. of old axis

5. Next to the epidermis is the cortex (Hyalodermis). The cortical cells absorb and store water.

6. In some species, the epidermis of leaf insertions has flask shaped cells called retort cells.

7. The retort cells help in absorption and are usually inhabited by micro organisms.

8. The region next to the cortex is the hardroom. The cells are thick walled and give mechanical support to the stem.

9. The central region consists of thin walled medulla.

Vegetative propagation

1. Takes place by fragmentation or by deliberate separation of lateral branches which grow into new individuals.
2. Sex organs are antehridia (male) and archegonia (female) developed on separate branches.
3. In monoecious species, antheridia develop earlier than archegonia.

Antheridia

1. The antheridial branches are small and short.
2. They are covered by leaves which are red, purple, brown or yellow in colour.

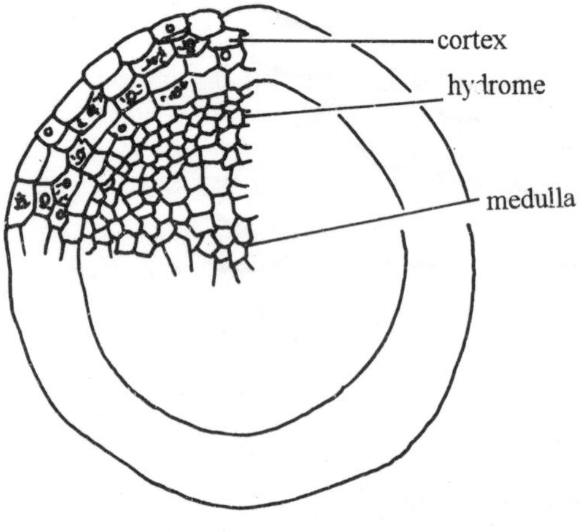

Fig. 5.6 Sphagnum. T.S. of old axis

3. A single antheridium arises from the axil or each leaf.
4. A mature antheridium has long multicellular stalk and an oval or globose antheridium proper.
5. The wall of the antheridium is made up of a single jayered jacket.
6. the antheridium has number of androcytes which metamorphose to form antherozoids the autherozoids are coiled and biflagellate.
7. The antherozoids will e suspended in a mucilagenous mass.

Archegonia

1. The archegonia are formed on separate branches called the archegonial branches.
2. the archegonial branches are very small having few compactly arranged leaves.
3. The leaves are large and green in colour.
4. These leaves protect the archegonia and also the ceveloping sporophyte.
5. the archegonia occur at the apex of the archegonial branch and range in number from 2 to 5. Normally they are 3 in number.
6. Central archegonium is called the primary archegonium as it would have directly developed from the apical cell.
7. The others are called secondary archegonia as they are formed from the segments cut off by the apical cell.
8. A mature archegonium has a basal slightly bugled venter and a narrow twisted neck.
9. The venter and lower portion of the neck have a 2-3 cell thick wall.

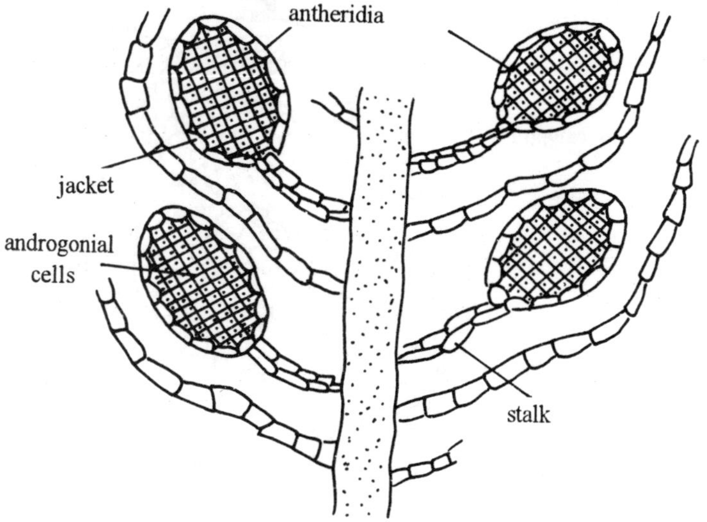

Fig 5.7L.S. of a portion of male branch passing through antheridia

10. Venter encloses a venter canal cell and an egg cell.
11. The neck encloses 8-9 neck canal cells and is covered by 8 or more cover cells.
12. In an archegonium ready for fertilization, all the cells except the egg dissolve.
13. The antherozoids enter the archegonium with one of them fertilizing the egg to form the zygote which develops into the sporophyte.

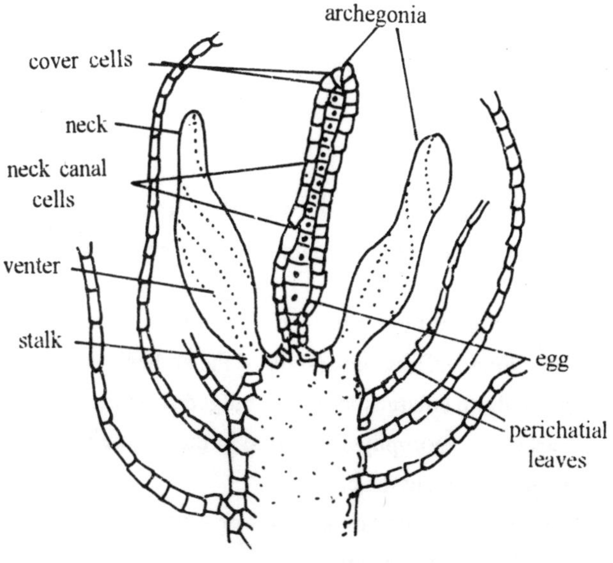

Fig. 5.8 Sphagnum. L.S. of female branch passing through archegonia

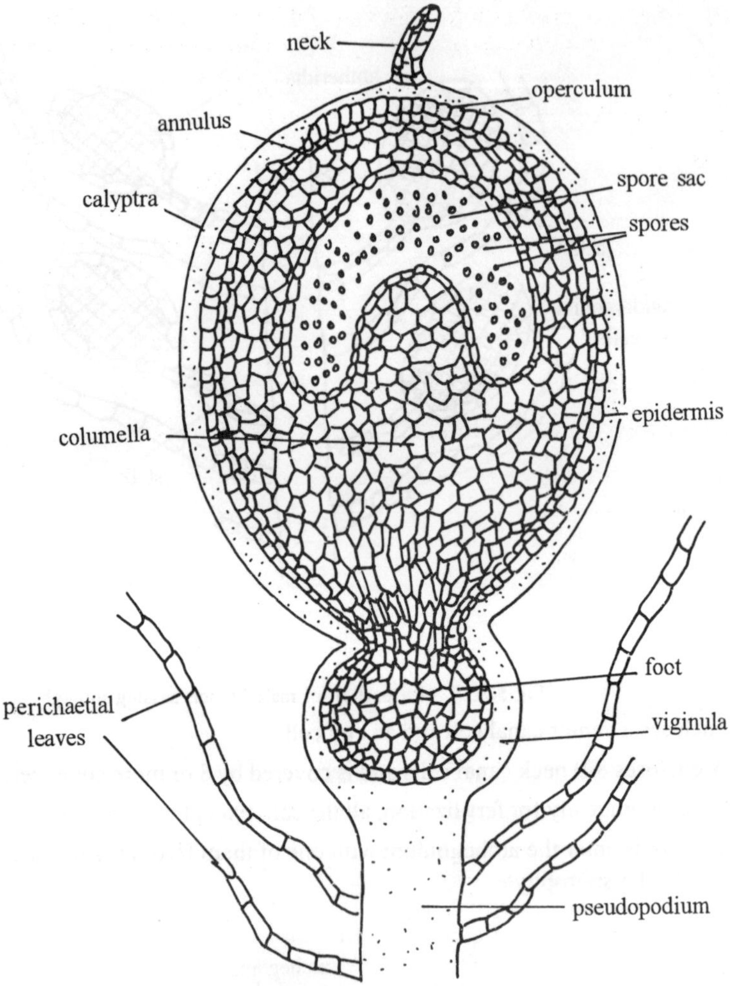

Fig. 5.9 Sphagnum. L.S. of sporogonium

Structure of the sporophyte

1. The sporophyte is covered by two protective sheaths, viz., calyptra & perichaetium.
2. Calyptra is formed by the venter wall of the archegonium, while perichaetium is formed by the perichaetial leaves.
3. The sporophyte is divided into ins usual three parts fix., foot, seta & causule.
4. The sporophyte is situated on a long leafless stalk (stalk of the archegonium) called pseudopoodium.
5. The upper end of the pseudopodium enlarges to form a bulged structure called vaginula.
6. The foot is the basal most part of the sporophyte is buried in the vaginula.
7. The foot is bulbous is made up of parenchyma cells.
8. The seta is highly reduced and is represented by a sinous or groove.
9. The terminal part of the sporophyte is the capsule.

10. The capsule is globose has a central hump of dome shaped tissue called columella.
11. Overarching the columella is the horse shoe shaped spore sac.
12. The spore sac at first has spore mother cells. All of them are fertile and undergo reduction division to produce tetrads of haploid spores.
13. A mature spore sac has only the spores and there are no elaters.
14. the wall of the capsule is made up of 4-5 layers of cells.
15. The outermost layer containing columnar cells is the epidermis.
16. At the apex of the capsule the wall has narrow, circular groove called the annulus.
17. The cells at the region of the annulus are thin walled.
18. The tip of the capsule above the annulus is called operculum.
19. In a mature sporophyte the annular cells break and the operculum comes out in the form of a hemispherical lid expressing the spores.
20. The spores are discharged because of the compressed air inside and the type of dispersal is called air gun mechanism.
21. The spores are spherical in shape and are surrounded by two wall layers namely, exine and intine.
22. The spores germinate on the availability of a suitably substratum not directly into gametophyte, but into a filamentous structure called Protonema.
23. The protonema are thalloid and lobed.
24. Rhizoidal branches arising from the ventral surface fix the protonema to the substratum.
25. The thalloid protonema may form marginal filaments called secondary protonema.
26. In some cases (s. girgensohni), secondary protonema might produce tertiary protonema.
27. Ultimately a bud in formed at the base of the protonema which develops into an erect gametophore.

Diagrams to be drawn

1. Habit sketch
2. A 'leaf' enlarged.
3. Transverse section of the stem.
4. L.S. of antheridial branch
5. L.S. of archegonial branch
6. L.S. of sporophyte
7. Protonema.

Systematic position and Identification

A. 1. Plant body is a parenchymotous gametophyte.

2. It is leafy or thalloid.

3. Vasculature and roots are absent.

4. Life cycle exhibits heteromorphic alternation of generations with a predominant gametophyte … Bryophyta.

B. 1. Gametophytes are leafy.

2. Rhizoids are multicellular with oblique septa

3. Columella present in the capsule of the spoorophyte… Bryopsida.

C. 1. Gametophytes are acquatic

2. 'Leaves' do not have a midrib but have two types of cells.
3. Branches arise in clusters ... Sphagnales

D. 1. Characters of family are the same as the those of the order. This is a single family .. Sphagnacae
2. Branches are of two types
3. Sprophyte has a pseudopodium
4. Foot in enclosed in the vaginula
5. Seta in extremely reduced.
6. Leaves of the gametophyte have two types of cells...Sphorgnum.

FUNARIA

Common species : *F. hygrometrica*

Occurrence: Funaria commonly called the moss. It grows as bright green velvety patches on damp soil, wet walls etc. The germs includes 117 species of which 15 have been reported from India. The most widely distributed species is F. hygrometrica.

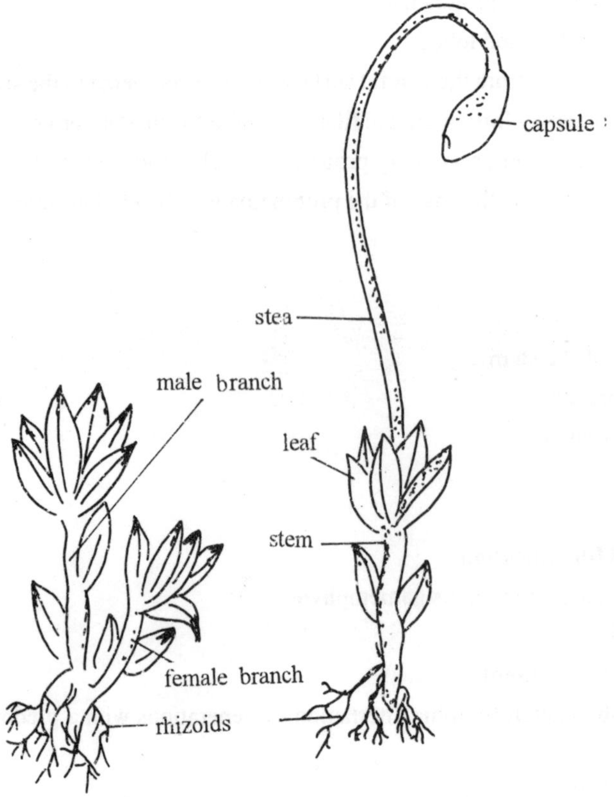

Fig. 5.10 Funaria. A plant showing branched axis A plant bearing sporophyte

Study: 1. External morphology
2. Anatomy of stem
3. Sexual reproduction
4. Rotonema
5. Systematic position and Identification

External Morphology:

1. The adult gametophytes arise from the prostrate thalloid protonema.
2. The gametophytes (gametophores) are erect, slender and reach a height of 2-4 cms.
3. The gametophyte has a central axis called stem or cauloid. The axis may or may not be branched.
4. The stem is anchored to the soil with the help of rhizoids. The rhizoids are anticellular with oblique cross walls.
5. The stem is surrounded by a number of spirally arranged 'leaves' or phylloids.
6. The leaves are sessile, simple, entire with distinct midrib.
7. The leaves are smaller at the lower region, larger and closely packed towards the upper portion.
8. The leaves at the apex are called perichaetial leaves.
9. The perichaetial leaves are very large and radially arranged in the form or a rosetle called the 'moss flower' within which are present sex organs.

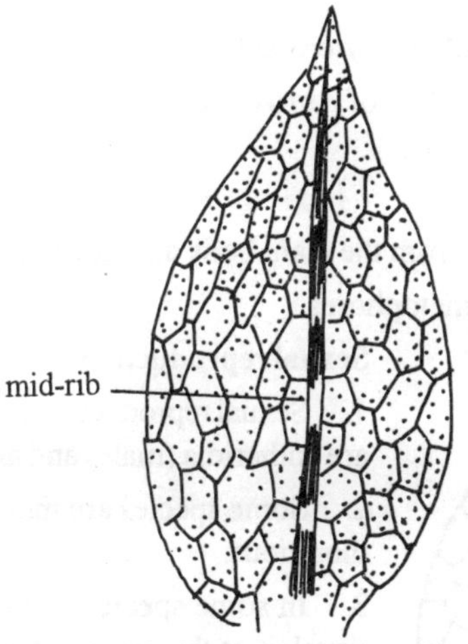

Fig. 5.11 Funaria. Single leaf

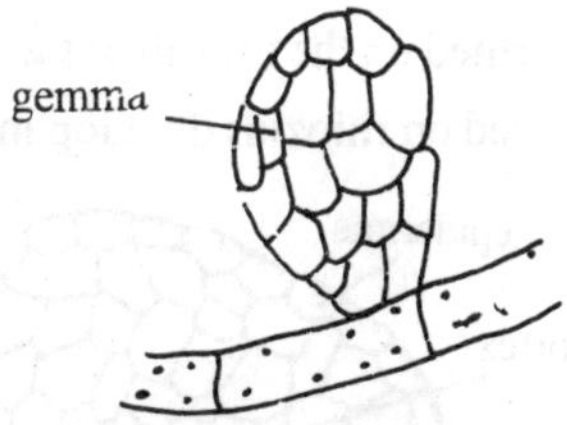

Single gemma

Anatomy of stem

1. A cross section of the stem shows three distinct region epidermis, cortex and central cylinder.
2. Epidermis is made up of a single layer of compactly arrange cells.
3. There are no stomata in the epidermis.
4. Epidermal cells at the base of the stem produce multicellular rhizoids.

5. Epidermal cells have many choloplasts .
6. Cortex is made up of many layers of polygonal cells.
7. In a young stem, the outer cortical cells are also chlorophyllous.
8. In older stems the outer cortical cells become thick walled and give mechanical support.
9. Central cylinder is composed of elongated compactly arranged thin walled dead cells. These cell also offer mechanical support to the gametophore.

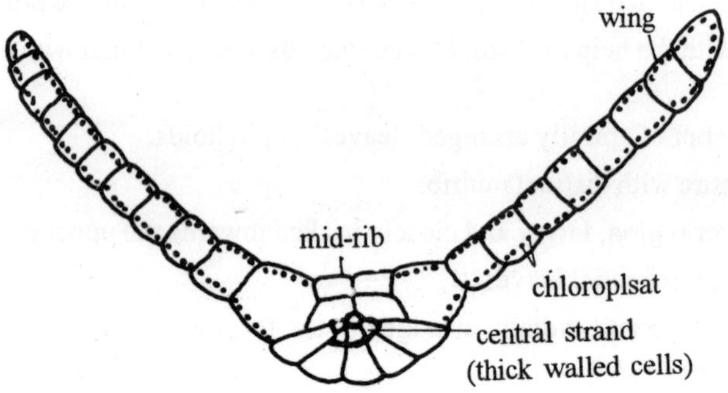

Fig. 5.12 Funaria. T.S. of leaf

Anatomy of leaf

1. Anatomically, the leaf shows a single layer of chlorophytons cells.
2. The midrib is made up of a central cylinder composed of narrow elongated thin walled dead cells. They are surrounded by a few large chlorophyllous cells.

Vegetative propagation

1. Gemmed formed on the midrib of the leaf detact from the plant body and develop into a new individual.
2. Bulbils formed on rhiozids develop into new gametophores.

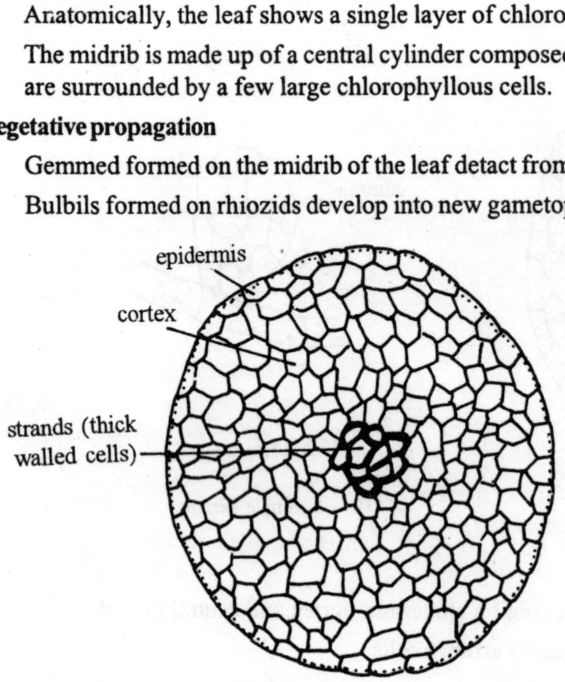

Fig. 5.13 Funaria. T.S. of axis

Sexual reproduction

1. Sexual reproduction is oogamous. Sex organs are antheridia (male) and archegonia (female).

2. Some species are monoecious and some are dioricious.

3. In some species, antheridia and archegonia develop at the apices of different branches of a cluster of the same gametophore. This condition is called Autoicous.

4. If antheridia and archegonia are produced on different branch clusters of the same gametophore the condition is said to be paraoicous.

5. If the sex organs are irregularly scattered at the tip of a gametophore, the conditions said to be *synaoicous*.

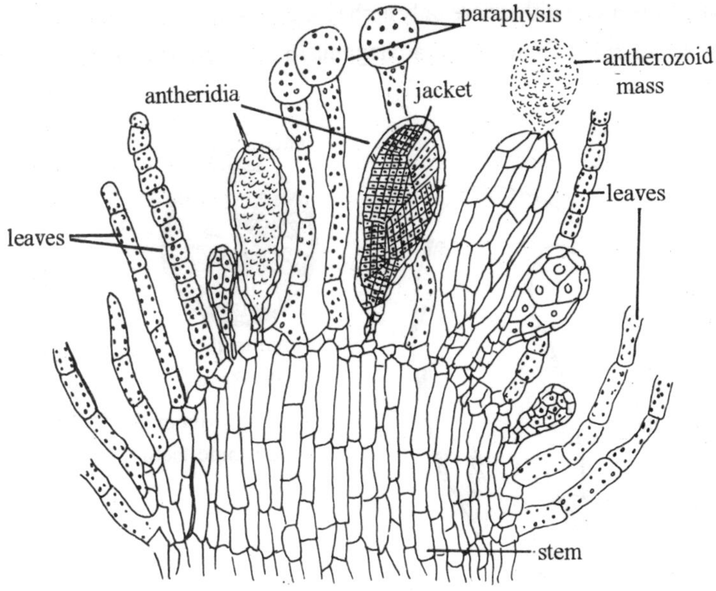

Fig. 5.14 Funaria. L.S. through the apex of male head

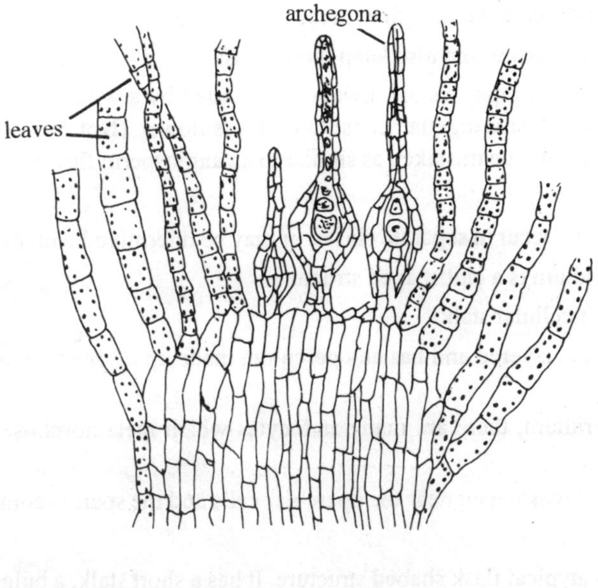

Fig 5.15 Funaria L.S through the apex of female head

Structure of a mass flower

1. A mass flower is nothing but a cluster of leaves surrounding the sex organs. If there are only antheridia, it is called male flower and if there are only archegonia it is said to be female flower.
2. The mass flower has either antheridia or archegonia intermixed with long or short filaments called paraphyses.

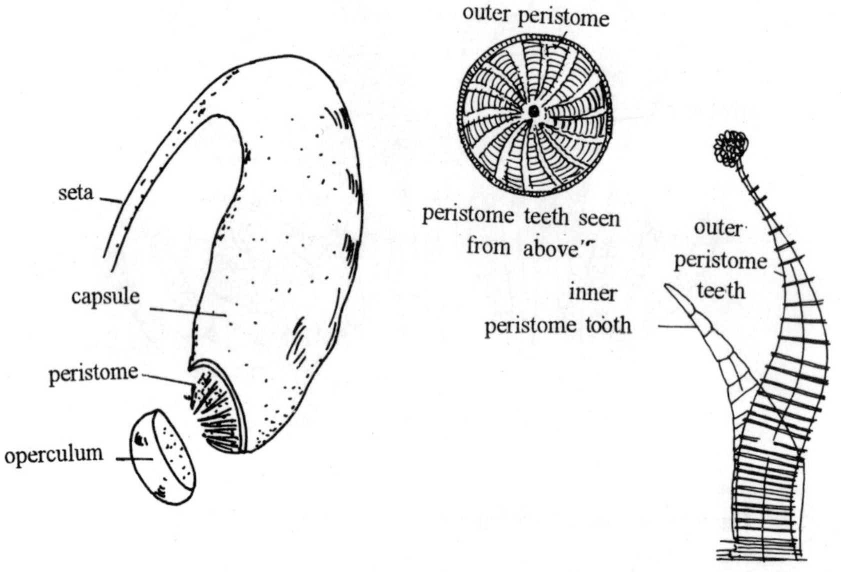

Fig. 5.16 Funaria. A mature capsule

Two peristome teeth

3. Each paralysis is 4-5 celled in length with the upper cellbecoming globose and filled with chloroplast.
4. The protect the sex organs and also keep them moist.
5. The sex organs and paraphyses are always surrounded by a whorl of large leaves called pinchaetial leaves. All these structures together constitute a moss flower. (Note: the term flower is used in a figurative sense, it should not be mistaken as similar to an angiosperm flower).

Antheridia

1. The antheridia may occur in separate cluster or may be intermixed with the archegonia.
2. A mature antheridium is a clubshaped structure.
3. It has a short multicellular stalk.
4. The jacket is single layered and has an operculum made up of one to two cells made up of one to two cells.
5. Inside the antheridium, there are many andocytes which metamorphose into spirally coiled biciliate sperms.
6. The antheridium breaks open near the opercular cells and the sperms come out in a mass.

Archegonia

1. Archegonium is atypical flask shaped structure. It has a short stalk, a bulged venter and a narrow neck.
2. The venter wall is two layered and it encloses a venter canal cell and an egg cell.
3. The neck is narrow and is made up of six rows of spirally twisted cells enclosing six or more neck canal cell.

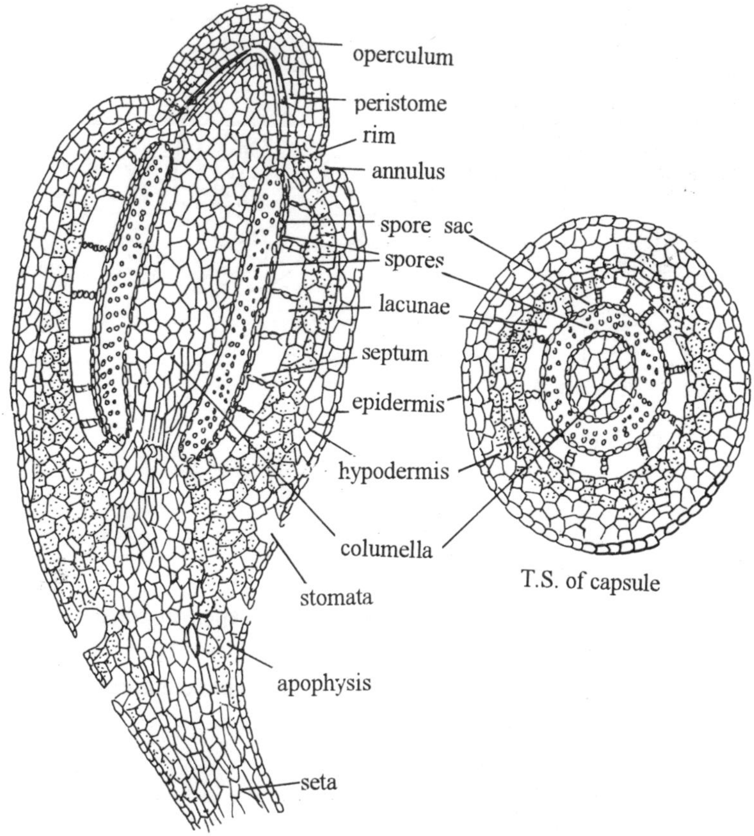

Fig. 5.17 Funaria. L.S. of capsule

4. the apex of the neck is covered by four large cover or lid cells.
5. The antherozoids enter the archegonium. The fertilized egg, i.e., zygote develops into the sporophyte.

Structure of sporophyte

1. A mature sporophyte is differentiated into 3 regions, viz., the basal foot, the middle seta and the terminal capsule.
2. The foot is very small and is embedded at the apex of the gametophore helping in absorption and anchorage.
3. Seta is very long and cylindrical and consists of central narrow elongated cells surrounded by a few layered cortex and single layered epidermis.
4. Seta helps in the conduction of water and mineral elements.
5. Seta also serves to keep the capsule in an elevated position.
6. The capsule is roughly beakshaped and bent towards one side.
7. In the beginning it is green in colour, later it turns to yellow and orange.
8. the capsule is differentiated into 3 regions, viz., the lower apophysis, the middle theca and the upper opercular region surrounded by persistent colyptra.

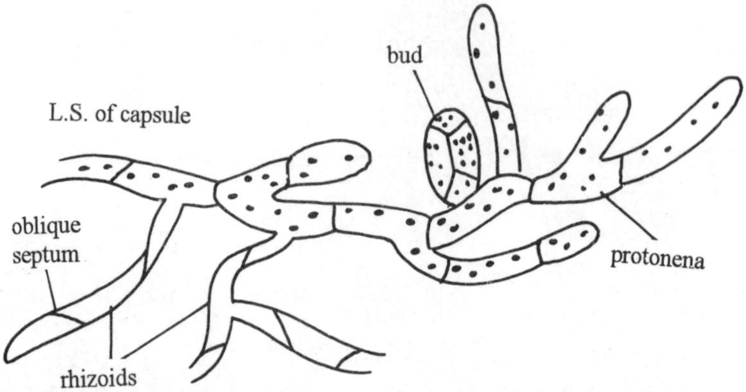

Fig. 5.18 Funaria. Protonema

Apophysis

1. the apophysis on the outside is covered by an epidermal layer.
2. the epidermis is provided with stomata Tach stoma has an aperture surrounded by 2 guard cells., which lead internally to an air space called stomatal cavity.
3. Immediately below the epidermis is a broad spongy zone. The cells are chlorophyllous, loosely arranged with intercellular spaces between them.
4. the spongy zone helps in photosythesis.
5. The spongy photosynthetic zone surrounds the central conducting strand made up of thin walled narrow, elongated cells without chloroplasts.
6. The central strand is continuous. It is connected with the central strand of seta below and liked with the columella above by means of a bundle of filaments.

Theca

1. Theca consists of a central sterile called columella.
2. Columella is amde up of thin walled parenchyma cells and projects above in the form of a cone in the poercular region.
3. Columella is surrounded by barrel shaped spore sac.
4. The inner wall of the spore sac is single layered while the outer wall is three layered.
5. the spore sac encloses numerous small haploid spores.
6. the outer wall of the spore sac gives rise to radiating filaments called trabeculae, each of which is made up of three to five chlorophyllous cells.
7. The trabeculae are separated from one another by large air chambers called lacunae.
8. The lacunae are surrounded by one to two layers of cholrenchyma cells.
9. The clorenchyma region is surrounded by hypodermis composed of two or more layers of thin walled colourless parenchyma cells.
10. Encircling the hypodermis is the single layered epidermis. The outer walls of epidermal cells are thickened.

Opercular region

1. The opercular region is made up of outer operculum and the inner peristome surrounding the central parenchymatous tissue.
2. Peristome consists of two sets of peristomial teeth. The outer set called exostome and the inner set endostome.
3. the peristomial teeth are narrow, long and tapering. They are compactly arranged and convergent in the form of a cone.
4. the peristomial teeth are hygroscopic and help in the dispersal of the spores.
5. The operculam is made up of three layers of cells.
6. The eperculum is attached to the theca region with the help of thin walled cells called annulus.
7. At maturity the annular cells tear open, with the operculum falling off. The peristomal teeth help in the dispersal of the spores.

Protonema

1. The spores released from the capsuleon germination do nto develop into the gametophore directly but develop first into filamentous structure called protonema.
2. The spore on germination develops into primary protonema.
3. the primary protonema has two types of branches, viz., chloronema(chlorophyllous branches) and Rhizonema (rhizoidal branches).
4. Rhizonema penetrates the substratum while chloronema are aerial.
5. Chloronemal penetrates the substratum while chloronema are aerial.
6. Many lateral buds are formed on the protonema. These develop into adult gametophores.

Diagrams to be drawn

1. Habit sketch
2. T.S. of 'stem'
3. T. S. of 'leaf'
4. L.S. of female 'flower'
5. L. S. of male 'flower'
6. L. S. of sporophyte
7. Peristomial teeth (enlarged)
8. Protonema showing buds.

Systematic position and Identification

A. 1. Plant body isa parechymatous gametophyte
   2. It is leafy or thalloid; vasculature and roots are absent.
   3. Sex organs surrounded by a sterile jacket layer.
   4. Life cycle exhibits heteromorphic alternation of generations with a predominant gametophyte ... Bryophyta.

B. 1. Gametophytes are leafy
   2. Rhizoids are multicellular with oblique septa
   3. Columella present in the capsule of the sporophyte... Bryopsida

C. 1. Leaves have a prominent midrib.

2. Seta is slender and elongated.
3. Air cavities present in the capsuel ... Bryidae

D. 1. Plants terrestrial in habitat.
2. Leaves clustered at the apices of the gametophore
3. Two concentric rings of peristomial teeth.
4. Colyptra is usually distended... Funariales

E. 1. Leaves many celled thick at the midrib, but one celled thick at sides.
2. Height of gametophytes rarely exceeds of few cms.
3. Colyptra has a long beak ... Funariaceae

F. 1. Leaves arranged spirally on the axis.
2. Anatomically the axis is highly differentiated; it has epidermis cortex and a central conducting strand.
3. Leaves at the apex cluster to form capitate structures. ...Funaria

POLYTRICHUM

Common species: *P. densifolumin, P. commune, P. xanthopilu, P. juniperinum,*

Occurrence: The genus with about 100 species is distributed universally, but majority of species show preference to cool temperate and tropical regions. Cool shady soil, marshes, peat bog, soil along pond banks etc. are the usual habitats where polytrichum grows.

Study: 1. External morphology
2. Anatomy of stem
3. Anatomy of leaf.
4. Sexual reproduction
5. Structure of the sporophyte.
6. Protonema.
7. Systematic position and Identification.

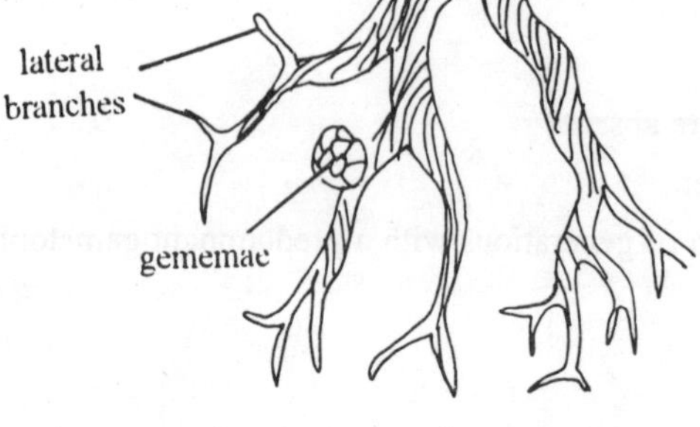

Fig. 5.19 Polytrichum. Rhizoidal strand with gemmae

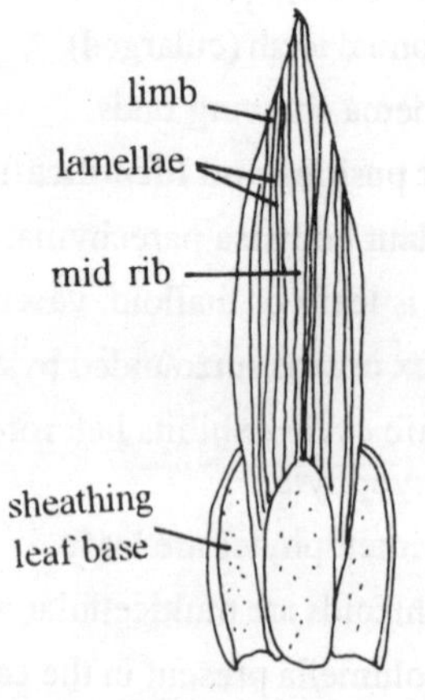

Single leaf

External Morphology:

1. The plant body if ploytrichum is a gametophyte.
2. It has an under ground rhizome from which spring up leafy shoots.
3. If fair environmental conditions, the leafy may each a height of 20 cms.

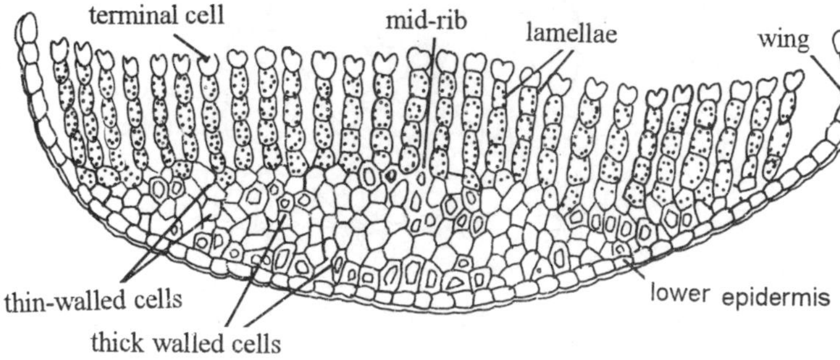

Fig. 5.20 Polytrichur T.S. of leaf

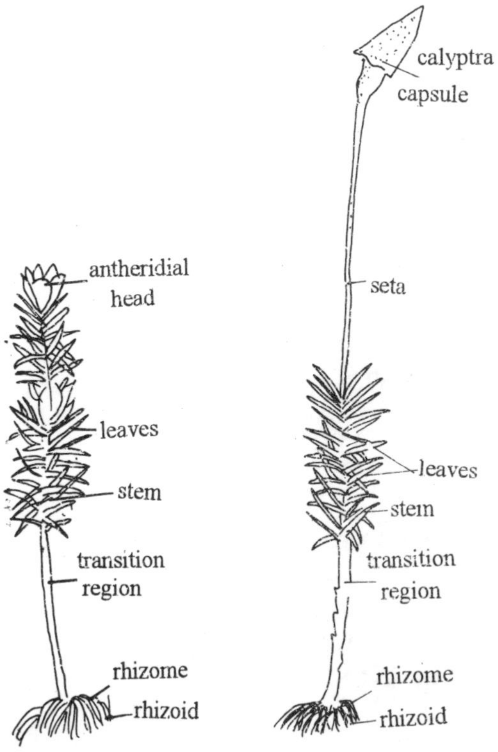

Fig. 5.21 Polytrichum. A male plant

A female plant with sporophyte

4. The stem is erect and cylindrical. It is branched.

5. In P. commune, a traditional region is found between the rhizone and the arial shoots.

6. The leaves are spirally arranged in three vertical rows.

7. The leaves on the transitional zone and rhizone and brown or colourless, those on the aerial shoots are large and rigid.

8. The leaves are provided with s sheathing membranes leaf base which narrows up into a limb.

9. The leaf has a broad midrib, which bears on it adaxial face many cell plates (lamellae). These are one celled thick and photosyhthetic.

10. The rhizome is anchored to the soil with the help of rhizoids.

11. The rhizoids are multicellular with oblique corss walls.

12. The rhizoids entangle themselves into loosely knit manes, thus helping in the external capillary movement of water.

13. The water can rise through the wick like rhizodial manes.

Anatomy of Rhizome.

1. A cross section of the rhizone is roughly circular in outline.

2. the superficial layer consists of thick or thin walled cells.
3. Internal to this is the cortex made up of 3-4 layers of cells.

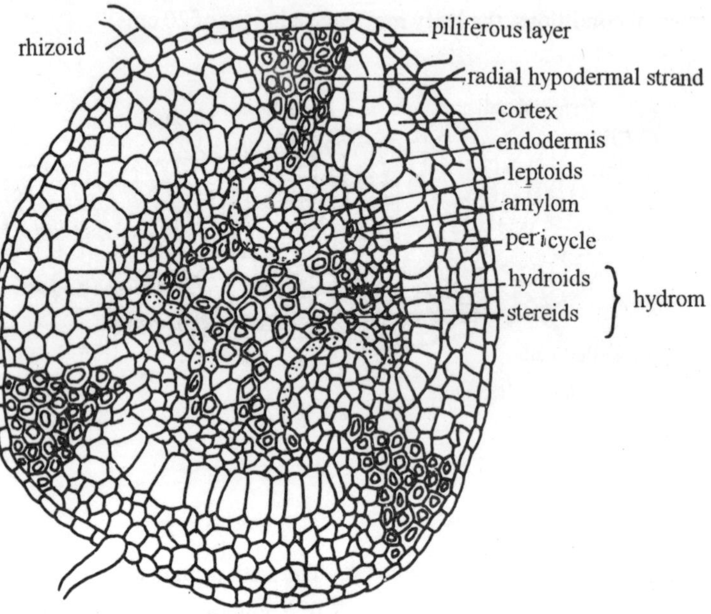

Fig. 5.22 Polytrichum. T.S. of rhizome

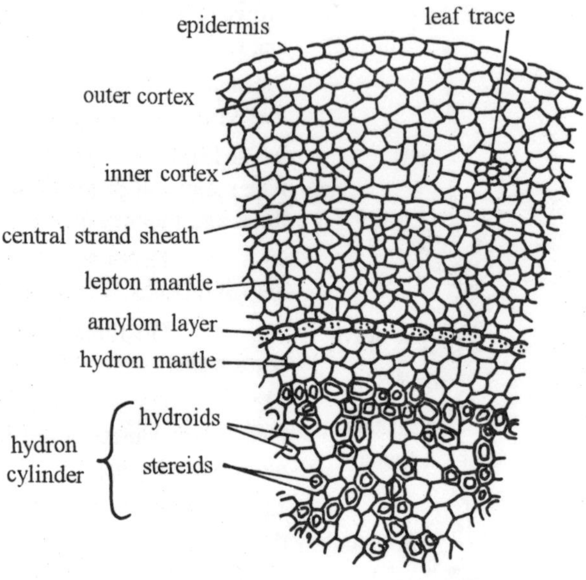

Fig. 5.23 Polytrichum. T.S. of aerial stem (a portion cellular)

4. The cortex is divided into 3 lateral chambers by the formation of hypodermal strands.
5. The cells of the hypodermal strands are thick walled living and sometimes filled with starch.
6. The cells of the hypodermal strands become broader with thinner walls near the central region and are called radial starnds.
7. The innermost layer of the cortex has superised raidal and tangential walls and is comparable to endodermis.
8. The layer however is broken because of the presence of radial strands.
9. Next to the endodermis is found a 2-3 layered penicycle. This layer also is discontinuous.
10. The internal to the pericycle is the central cylinder.
11. This is mostly composed of thickwalled but living cells often filled with starch.
12. These cells are individually called steroids, collectively constituting the Stereom.
13. Scattered among the cells of the stereom are present the hydroids, collectively known as hydrom.
14. the hydroids are deroid of cell contents and are believed to help in conduction of water.
15. The central mass of cells in the cylinder is triradiate.
16. Depressions found between the rays of the central mass of cells are filled up by a group of 6-8 polygonal cells containing protanaceous grannules.
17. These cells are comparable to the sieve elements and are called leptoids, together, known as leptom.
18. The leptom is surrounded by starchy cells forming the amytom.

Internal Structure of the aerial shoot.

1. The contour of transverse section looks irregular due to the attachment of leaves.
2. The epidermis is not clearly distinguishable.

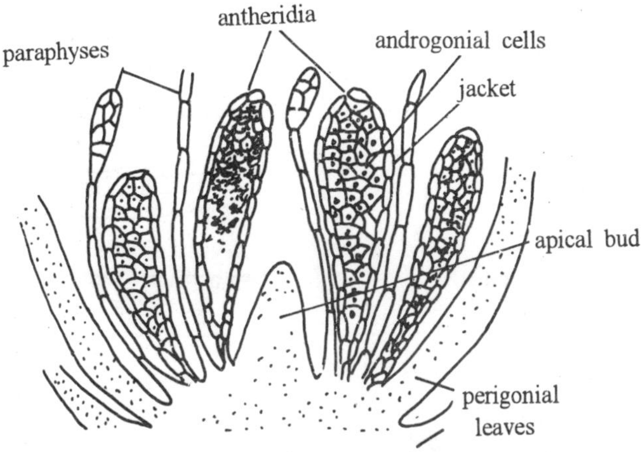

Fig. 5.24 Polytrichum. L.S. through the apex of male head

3. Internal to this is the cortex, divided into two zones, viz., outer prosenchymatous zone and inner parenchymatous zone.
4. The cortical cells are starch filled and many leaf traces are found in the cortex.
5. Internal to the cortex is an illdefined ruidmentary penicycle. It is discontinuous.

6. Next to the pericycle in the leptom mantle consisting of sieve tube like cells interrupted by starch filled cells.
7. The leptom mantle is lined internally by the hadrom sheath, where cells are subensed.
8.The hadrom sheath encloses the hydrom mantle conisiting of this walled cells.
9. Centrally located is the hydrome cylinder made up of thick walled cells.
10. this tissue helps in water conduction and is comparable to the xylem of vascular plants.

Internal structure of the leaf:

1. A vertical transverse section of the leaf shows a bifacial arrangement of cells.
2. the midrib is many celled thick and marges with the lamina at the margins.
3. The lower face of the leaf has a well marked single layered epidermis consisting of large thickwalled cells.
4. Closely attached to the lower epidermis is found a layer of scloritic cells.
5. The tissue next to this is made up of parenchyma cells.
6. On the upper face, the leaf has many parallely arranged plates of chlorophyllous cells arising from the lower layer of parenchyma cells.
7. In a sectorial view the chlorophyllous cell plates appear like lamellae.
8. Each lamella is made up of 5-8 cells in which the *distal* cell is papillate.
9. these cell plates besides helping in photosynthesis also help in the retention of water between them.

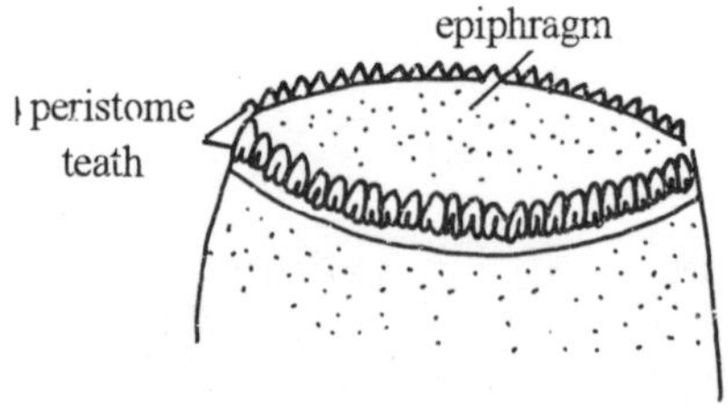

Fig. 5.25 Polytrichum. Dehisced capsule showing peristome

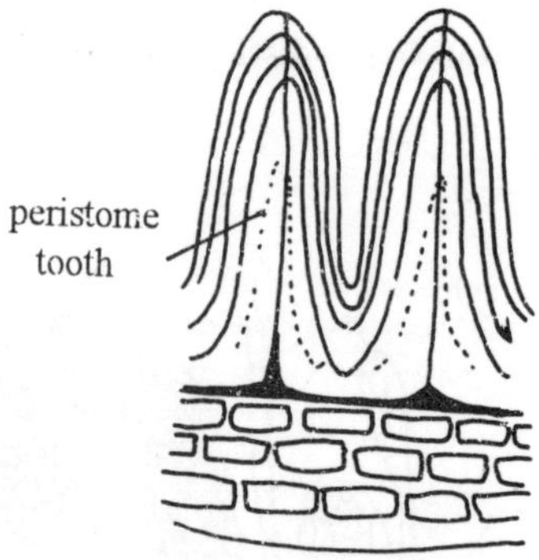

Two peristome teeth

Sexual reproduction

1. Sexual reproduction is oogamous. Sexoorgans are antheridia (male) and archogonia (female).
2. Polytrichum is generally dioecious.

Antheridia

1. These are generally borne is terminal clusters forming open flower like cups.
2. The cup is surrounded by perigonial leaves that are very much different from the vegetative leaves.
3. In some cases, the vegetative shoot apex may grow through the antheridial cluster producing one more group of antheridia.

Fig. 5.26 Polytrichum. T.S. of capsule passing through spore sac

4. Within the cup are found a number of antheridia intermingled with paraphyses.
5. Each antheridium is a clubshaped structure having a short stalk.
6. the jacket of the antheridium is made up of a single layer of cells.
7. the antheridium has a number of androcytes which metamorphose to give rise to coiled biciliate antherozoids.

Archegonia

1. Archegonia are also borne on terminal clusters.
2. There are surrounded by pink or are coloured perichaetial leaves.
3. the leaves are arranged in a rosette looking like a flower.
4. Usually three archegonia are found in each cluster as in P. commune.
5. With the development of the archegonia, the growth of the shoot stops.
6. Scattered among the archegonia are found a number of hair like structures called paraphyses.
7. Each archegonium has a stalk, a bulbous venter and a spirally twisted neck.
8. Rest of the structural details of archegonia are same as seen in Funaria.
9. Fertilization is as seen Funaria.

Sporophyte

1. The sporophyte has three parts, viz., basal foot, middle seta and the terminal capsule.
2. the foot is parenchymatous and is embedded in the archegonial cluster.
3. the seta is long and cylindrical. Anatomically it shows the following features.
4. The outermost layer of cells is thick walled.
5. Internal of this is band of sclerotic cells which gradually merge into the inner thin walled loosely arrangement chlorophyllous cells.

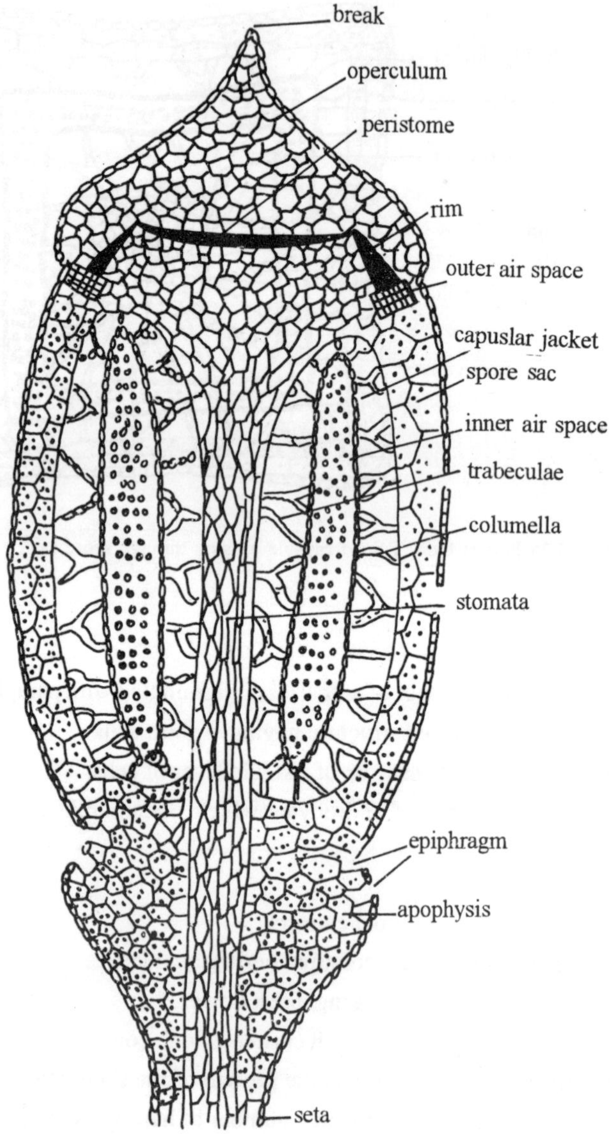

Fig. 5.27 Polytrichum. L.S. oi capsule

II

6. In the centre of the seta runs a conducting strand.
7. the seta terminates at the base of the capsule into a swollen structure called the apophysis.
8. the apophysis is separated from the rest of the capsule by constriction.
9. Automatically the apophysis has an epidermis interrupted with stomata.
10. The epidermis encloses a mass of green chorenchymatous cells.
11. In long section, the capsule shows an outer wall composed of several layers of cells.
12. The outermost layer of the wall is the epidermis.

13. Internal to the wall is an airspace (outer lacuna) separated into air chambers by means of parenchymatous partitions.
14. The cells of these partitions are chlorophyllous and are connected internally to the outer wall of the spore sac.
15. Next to the outer air chamber is found the spore sac which is covered on either side by means of a 2 layered wall.
16. the wall encloses the archesporium of 4-6 layers of cells which ungergo reduction division to form hapoid spores.
17. There are no elaters.
18. Internal to the spore sac is the inner air space (inner lacuna) which is similar in structure to the outer air space.
19. The partition walls of the inner air space are attached to the central columella.
20. The columella consists of vertically elongated parenchyma cells.
21. The capsule is covered on top by a lid like operculum.
22. The poerculum has beak or rostrum.
23. The annulus is not distinct, unlike in Funaria.
24. The poerculum at its base has the epiphtragm. This is made up of a transverse band of thin membranes cells without intercellular spaces.
25. this stretches like a drum head (tympanum) covering the opening of the capsule.
26. Outside the peiphragum is the peristome.
27. the peristome has 36-64 teeth which are connected to the capsule below and to the epiphragm above.
28. The spores are minute in size, haploid and uninucleate.
29. On the availability of the suitable substratum, the spores germinate to develop into protonema.

Protonema

1. The spore germinates to form a branched, septate filamentous protonema.
2. Some of the branches grow vertically above (chloronema) while some others grow down and help in anchorage.
3. Subsequently buds appear near the base of chloronema.
4. These buds develop into leafy gamelophytes of polytrichum.

Diagrams to be drawn

1. Habit sketch
2. Anatomy of Rhizome
3. Anatomy of aerial shoot.
4. anatomy of leaf.
5. L.S. of male flower.
6. L. S. of female flower.
7. L. S. Of sporophyte.
8. Protonema.

Systematic position and Identification:

A. 1. Plant body is a parenchymatous gametophyte.

2. It is leafy ar thalloid.
3. Vasculature and roots are absent
4. Sex organs surrounded by sterile jacket layer.
5. Life cycle exhibits a heteromarphic alternation of generations with a predominant gametophyte ... Bryophyta.

B. 1. Gametophytes are leafy.
2. Rhizoids are multicellular with oblique septa.
3. Columella present in the capsule of the sporophyte... *Bryopsida.*

C. 1. Leaves with distinct midrib.
2. Anatomically leaves show a high degree of tissue differentiation....*Polytrichidae.*

D. 1. Gametophytes perennial and quite tall.
2. Leaves have chlorophyllous cell plates on the dorsal surface.
3. Peristomial teeth 36-64 in number ....*Polytrichales.*

E. 1. Single family. Some characters as those of the order... *Polytrichaceae.*

F. 1. Capsule is angular in shape.
2. Rhizoids form a tangled mote.
3. Two rows of air spaces found in the capsule.
4. Operculum has a beak ... *Polytrichum.*

PART TWO
PTERIDOPHYTA

CHAPTER 6

CLASSIFICATION OF THE PLANT KINGDOM

The living world consists of myriads of plants and animals of various kinds. Even within the plant world many lakhs of different types of plants have been recognised and perhaps many more are yet to be discovered. If such huge number of individuals are not properly arranged their study will not only be difficult but also meningless because proper comprehension is impossible. Hence in order to bring an arrangement and discipline plants have been grouped into smaller or larger groups based on resemblences and differences (mostly morphological). This system of dividing plants into groups is known as "Plant classification or Systematics".

In general there are mainly two systems of classification viz.,

(1) Artificial system and

(2) Natural system.

In the artificial system of classification as proposed by earlier botanists importance is given to minor, superficial and frequently variable characters. For example division of plants based on their habit into herbs, shrubs and trees and using this as a criterion in the classification. The artificial system does not indicate the relationship existing between different plants.

The natural system of classification is based on the natural phylogenetic affinites listing between various types of plants. Natural system lays stress on the similarity in reproductive structures including anatomical characters, cell structure and physiology. Such characters are more stable than the ordinary vegetative characters and are not prone to frequent alterations by the environmental influence.

According to the natural system all the plants have been included in the plant kingdom. The plant kingdom is divided into *divisions*. Each division into several *sub-divisions*, each sub-division into *classes*, each class into *orders*, each order into *Families*, each family into *Genera* and each genera into *Species* and in some cases each species into *several varieties*.

Various natural systems of classifications have been proposed time and again. It is not the idea of the present attempt to give a detailed survey of all those systems. Only a few systems have been briefly discussed below so as to give an understanding to the students the relative positions of different groups of plants in the evolutionary sequence of the plant kingdom so that they will be in a better position to appreciate and understand the different groups of non-vascular and vascular plants.

In general two main categories can be identified in the *natural* systems. They are:-

(1) Old or Traditional system

(2) Revised or Modern system.

Traditional system.

Of all the old system perhaps the one that is followed even today is that of Eichles (1866).

He proposed that the plant kingdom be divided into -

A) Cryptogamae and B) Phanerogamae

(A) Cryptogamae

Cryptogams have their reproductive structures embeded in the plant body and do not bear flowers of seeds.

They include three divisions namely -

(1) Thallophyta

(2) Bryophyta and

(3) Pteridophyta

Thallophyta:- The division thallophyta is characterised by plants -

a) With an undifferentiated body called thallus (flat)

b) Absence of vascular tissue.

c) Unicellular reproductive structures.

d) Absence of embryonal stage in the life cycle.

e) Members generally aquatic (not a rule).

Thallophyta is further divided into two classes namely -

1) Algae and 2) Fungi.

Algae:- The class Algae includes majority of the members which are -

a) Aquatic in habitat.

b) Members are antotrophic because of chlorophyll green photosynthetic pigment, and other pigment.

c) Plant body may be unicellular or multicellular.

Fungi:- This includes all achlorophyllous thallophytes. As the members are pigmentless they are not antotrophic their mode of nurtition is heterotrophic. Fungi follow two methods of heterotrophic nutrition viz -

1) Parasitic and 2) Saprophytic.

In the parasitic method, fungi obtain their food nutrition from living organisms (plats and animals) while in saprophytic method fungi devise their nutrition from dead and decaying organisms. (For details see introductory chapter on Fungi).

Bryophyta:- Plants are amhibious in their habitat. Plant body is mostly an undifferentiated thallus except in some. They differ from thallophytes in the following respects -

1) Possession of multicellular sex organs.

2) Presence of an Embryonal stage in the life cycle.

3) Occurence of heteromorphic alternation of generations in the life cycle regularly.

4) Sporophyte being dependent on the gemetophyte.

Pteridophyta:- Pteridophytes are called vascular cryptogams. Because of all the cryptogams only they have the vascular tissue or conducting tissue which helps in the transportation of food materials and also offers mechanical support to the plant body. They are seedless. Unlike in Bryophytes, saprophyte is the predominant plant body but never the less the gametophyte is independent (with very few exceptions).

(B) Phanerogams:-

These are the plants in which the reproductive structures are not embeded. They produce seeds. Phanerogams are divided into two divisions namely -

Division (1) *Gymnospermae* - seeds are naked, not covered by any envelope, no fruit formation.

Divison (2) *Angiospermae* - Seeds enclosed by fruit wall, fruit being formed by the mature every which is a special envelope for the ovule (seed) Angiosperms are divided into two classes namely -

1) Dicotyledonae - seeds with two cotyledoms.
2) Monocotyledonae - seeds with one cotyledon.

Some Botanists divide the entire plant kingdom into four major divisions viz -

1) Thallophyta
2) Bryophyta
3) Pteridophyta
4) Spermatophyta (includes Gymnosperms and Angiospersm).

Modern system

According to According to one of the modern system there are only three major divisions in the plant kingdom. They are -

(1) Thallophyta
(2) Bryophyta
(3) Tracheophyta

The last division includes all vascular plants, which are classified into the following sub divisions -

1) Psilopsida
2) Lycopsida
3) Sphenopsida
4) Pteropsida

Pteropsida includes three classes viz -

a) Filicineae

b) Gymnospermae

e) Angiospermae

Angiospermae has sub class 1) Dicotyledonae and sub class 2) Monocotyledonae.

In another system only Thallophyta and Embryophyta have been recognised as two major divisions. In Embryophyta there is the formation of an embryonal stage in the life cycle while it is absent in thallophytes. Embryophyta includes bryophytes pteridophytes, Gymnosperms and angiosperms. (Details not discussed).

Many modern botanists consider that the plant kingdom consists of 18 major division. They are -

1. Chlorophyta
2. Charophyta
3. Englenophyta
4. Pyrrophyta
5. Chrysophyta
6. Phaeophyta
7. Cyanophyta
8. Rhodophyta

9. Schizomycophyta (Schizomycota)
10. Myomycophyta (Myxomycota)
11. Eurnycophyta (Eumoycota)
12. Bryophyta
13. Psilophyta
14. Lycophyta (Lepidophyta)
15. Arthrophyta (Calamophyta)
16. Filicophyta (Pterophyta)
17. Gymnophyta
18. Angiophyta (Anthophyta)

CHAPTER 7

GENERAL CHARACTERS AND CLASSIFICATION

Pteridophyta constitute a significant and important group in the plant kingdom. As the first true land plants they offer a very favourable material for the study of various adaptations that have made the colonization of the land possible for the plants.

Pteridophytes have a long geological history on our planet. They are known from as far back as 380 million years. Fossils of pteridophytes have been obtained from rock strata belonging to Silurian and Devonian periods of the Palaeozoic era.

The life cycle of a typical member comprises a regular heteromorphic alternation of generations in which both the gametophyte and the sporophyte exist as independent individuals. Nevertheless the sporophyte is the predominant generation. In terms of duration in the life cycle gametophyte is insignificant. In this respect pteridophytes are totally different from bryophytes. The reasons for the predominance of the sporophyte in the life cycle are not far to seek. Being a more robust body a sporophyte is suitable to effectively meet the challenges of a terrestrial environment.

As a group the position of the pteridophytes is between bryophytes and spermatophytes. They share the characters of both these groups in addition to their own unique features.

Sporophytic Generation

As has already been pointed out it is the sporophytic generation which constitutes the main plant body in pteridophytes. In any description whenever we refer to the plant body of a pteridophyte we mean the sporophyte.

There is a great variety in the nature and organiszation of the sporophytic plant body in pteridophytes. The simplest sporophyte in an extinct member is to be found in *Psilotum*. The plant body here is a nake branched axis with no evidence of roots. The axis is distinguishable into an underground prostrate system and an erect aerial system. Rhizoids help in the anchorage of the plant to the substratum. There are no leaves. The stem itself takes up the function of photosynthesis.

From such a type of simple plantbody further evolution (mainly necessiated by the environment) resulted in the complex and diverse types of sporophytes that we see today.

From a naked leafless plantbody the first step is the differentiation of photosynthetic laterals. (leaves) and roots.

The leaves of pteridophytes are basically of two types viz; microphylls and megaphylls (macrophylls).

Microphylls are those which are provided with a single unbranched mid vein. Further when a leaf trace departs from the main vascular cylinder to provide the leaf, no gap is left.

In megaphylls there is a branched venation. The branching may be dichotomous or reticulate. Further there invariably will be a leaf gap in the main vascular cylinder above the leaf trace.

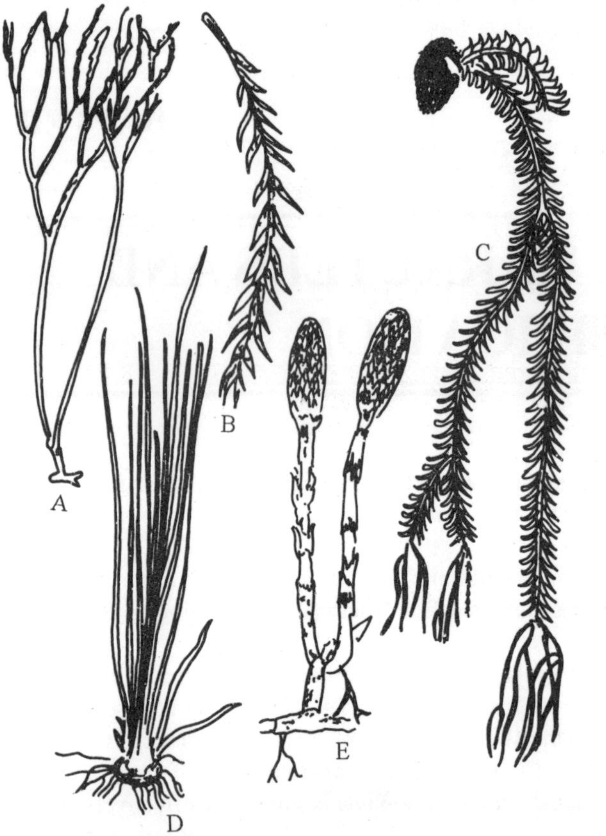

Fig. 7.1. Examples of some living Pteridophytes
A. Psilotum, B. Tmesepteris, C. Lycopodium, D. Isoetes, E. Equisetum

Microphylls are seen in members like *Lycopodium, selaginella* etc. Megaphylls are found in members like *Adiantum*, *Pteris* etc.

Vasculature in Pteridophytes.

All the Pteridophytes possess a vasculature in the center of the axis traversing from one end to the other end branching with all the branches.

The two main vascular elements are xylem and phloem. The xylem consists of mostly tracheids and rarely vessels (*Selaginella, Pteris* etc). The tracheids have different types of thickenings like scalariform, pitted, annular etc. The tracheids or vessels are of two types viz, protoxylem and metaxylem. The former matures early and has a narrow lumen while the latter matures later and has a wide lumen.

The relative distribution of protoxylem and metaxylem elements forms the basis of the classification of xylem group into three types viz exarch (centripetal protoxylem pointing towards the periphery), endarch (centrifugal- protoxylem pointing towards the center), and mesarch (protoxylem having metaxylem on either side).

The pholem is composed of sieve tubes and phloem parenchyma. A sieve tube consists of a series of long, living cells with sieve plates. Sometimes a distinction could be made between proto and metaphloem. But there is no pronounced morphological difference between the two.

Types of vasculature (Stele): A transverse section of the stem or root of the plantbody shows an outer epidermis middle cortex and central stele. The cortex is either parenchymatous or may have sclerenchyma also.

The central region consists of the concentration of the vasculature known as the stele. A stele may be defined as a vasculature sorrounded by pericycle and endodermis.

Originally plant anatomists thought that only the vascular bundles (xylem and phloem) formed the fundamental unit of the vasculature. Stele as the fundamental unit of the vasulature was first proposed by Van Tieghem and Douliot (1886). According to them the cortex and the stele are seperated by the endodermis. (The endodermis consists of barrel shaped cells and possesses thickenings known as Casparian thickenings on the radial and tangential walls).

There are mainly three types of steles in pteridophytes. These are 1) Protostele 2) Siphonostele and 3) Dictyostele. A protostele has the vasculature (xylem) occupying the central region of the stele. There is no pith in a siphonostele the vasculature is like a tube or siphon. Here the central region consists of non vasculated cells (parenchyma) forming the pith. In dictyostele the siphon breaks up into a number of individual bits called meristeles.

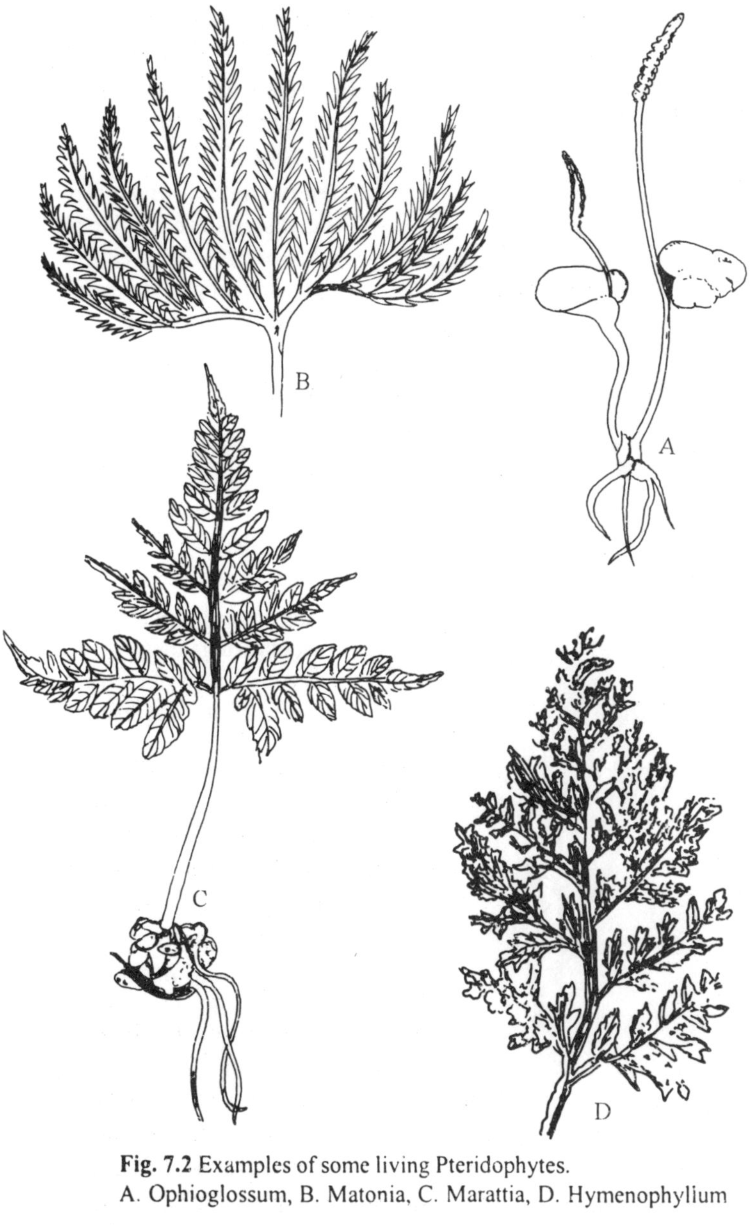

Fig. 7.2 Examples of some living Pteridophytes.
A. Ophioglossum, B. Matonia, C. Marattia, D. Hymenophylium

Protostele:- Basically all protostele are alike in not having a pith. They may be classified into the following types based mainly on the configuration of the xylem.

a) *Haplostele:-* This is the simplest type of stele that could be visualized for a vascular plant. The central region of the stele consists of a smooth core of xylem sorrounded by phloem. This type of stele is seen in *Rhynia*.

b) *Actinostele:-* The xylem is star shaped and the phloem completely surrounds the xylem. The contour of the stele is smooth. Eg. *Psilotum, Lycopodium phlegmaria* etc.

c) *Plectostele:-* The xylem breaks up into a number of bands or plates arranged parallely. The phloem not only surrounds the xylem but is distributed between the sylem plates. Eg. *Lycopodium wightianum*.

d) *Mixed Protostele:-* In this type the xylem breaks up into a number of small masses. The phloem is intermixed with the xylem. Eg. *Lycopodium cernuum*.

e) *Polystele:-* Generally the stem is traversed by a single stele. But in some cases as in some species of *Selaginella* several independent steles run parallely in the stem. Each stele has its own endodermis. The individual steles are haplostelic.

Siphonosteles:- The steles having a pith in the centre are reffered to as siphonosteles. These are further classified into two types viz, simple siphonostele and solenostele.

a) *Simple siphonostele:-* Here the central region of the stele consists of a parenchymatous or sclerotic pith. Leaf traces that depart from the main vascular cylinder do not leave any leaf gaps (a parencymatous cavity found immediately above the leaf trace in the vascular cylinder). This type of siphonostele is also called a medullated protostele or a cladosiphonic stele. Eg. *Psilotum* sp.

b) *Solenostele:*- In Solenonosteles leaf gaps are successive so that there is only one break in the vascular cylinder at any one given point. Solenosteles are also known as phyllosiphonic steles. Depending on the pattern of distribution of phloem solenosteles may be classified into two viz, *Amphipholic solenostele* in which the phloem lines the xylem both on its inner and outer face (Eg. *Adiantum, Marsilea* etc) and *Ectopholic solenostele* in which phloem lines the xylem only on its outer face (Eg. *Osmumda*).

There is also a third category of solenostele called polycyclic solenostele in which there will be more than one concentric ring of vasculature.

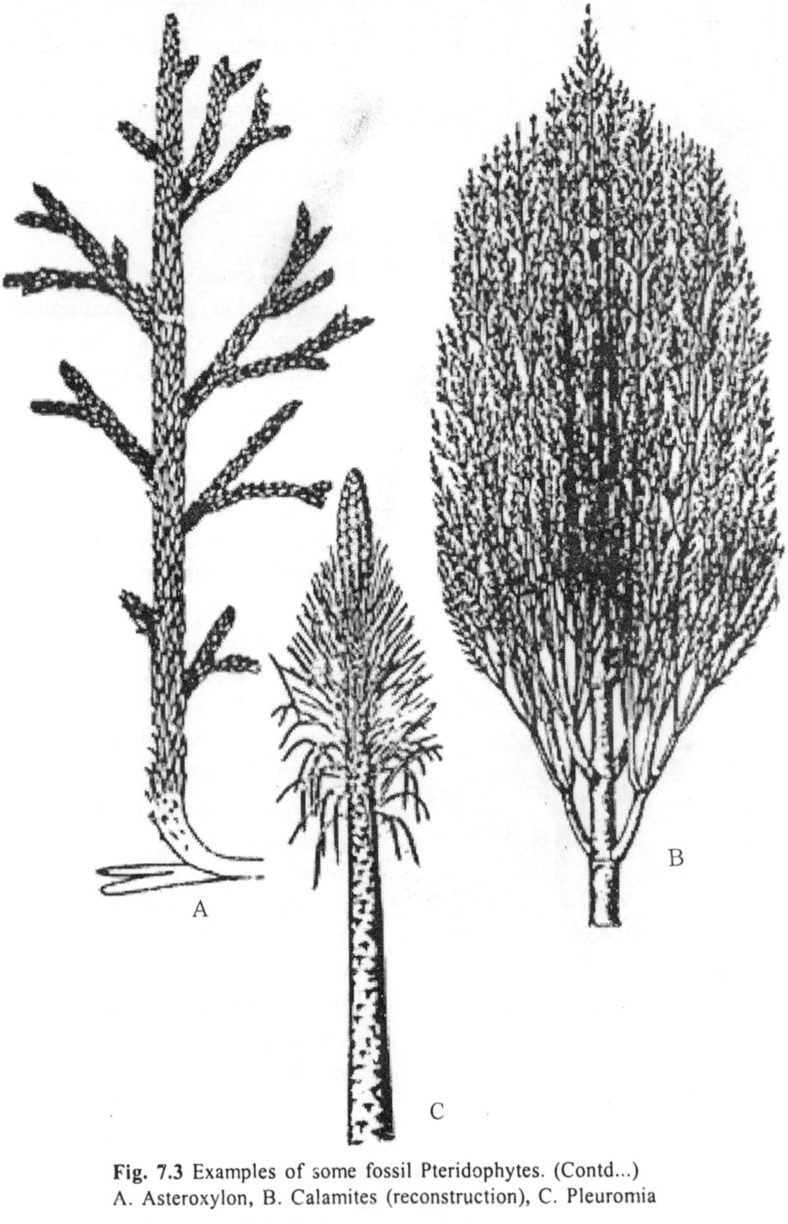

Fig. 7.3 Examples of some fossil Pteridophytes. (Contd...)
A. Asteroxylon, B. Calamites (reconstruction), C. Pleuromia

Dictyosteles:- Basically dictyosteles are similar to solenosteles in having a pith and leaf gap. In dictyosteles however the leaf gaps overlap (many occur at a point), as a result the vasculature ring breaks up into many arcs. Each arc is known as meristele. All the meristeles however are sorrounded by a common endodermis.

Based on the number of vascular rings dictyosteles may be classified into simple diotyosteles. in which there is only one ring and polycyclic dictyosteles in which there will be atleast two concentric rings of vasculature.

Origin of the vasculature:- The origin of the vascular tissue is one of the unsolved problems of plant morphology. It is closely linked with the problem of origin of land plants, which also is eluding a solution. If the algal ancestry for a land plant is proposed then there are no steps indicating the origin of vasculature. If one belives in the bryophytic ancestry of land plants, the columella in the sporophyte of *Anthoceros* seems to be a good starting point for the origin of vasculature. Indeed in its ideal location and to some extent in its function (though not in structure) the columella of the sporophyte of *Anthoceros* tempts any one to consider it as a precursor of the vasculature.

REPRODUCTION:- The sporophyte of pteridophytes generally reproduces by two methods viz, Vegetative propagation and Spore production.

Vegetative propagation takes place by a variety of methods such as fragementation, bulbil formation, resting bud formation, persistant apices etc. (The deatails of thes are given with the life histories of individual forms).

Spore production:- The characteristic method of reproduction is by the formation of spores. Spores are haploid and are produced in sporangia. The sporangia are generally borne on the adaxial or abaxial surface of a leafy appendage called the sporophyll.

Pteridophytes are homosporous (Isosporous) or heterosporous. Homosporous members produce only one type of a spore in the sporangium which develops into a gametophyte bearing both antheridia and archegonia. Heterosporous members produce two types of spores i.e., microspores and megaspores (macrospores). The microspores are smaller in size and are produced in microsporangia develop into the male gametophyte (microgametophyte). Megaspores are bigger in size and are produced in megasporangia borne on megasporophylls. On germination megaspores develop into the female gametophyte (mega cr macrogametophyte).

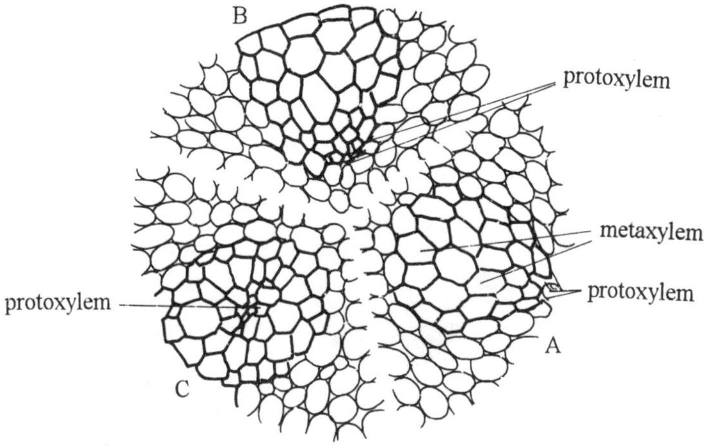

Fig. 7.4 . Direction of maturation of Xylem Elements
A. Exarch. B. Endarch, C. Mesarch.

Structure and development of a sporangium.

A sporangium usually has a stalk which bears the globular capsule region in which are produced the spores. The wall of the sporangium is one to many layered. With in the wall are found the spore mother cells which undergo reduction division to give rise to haploid spores.

There are mainly two types of sporangial development i.e., Ensporangiate type and Leprosporangiate type.

In Eusporangiate type the sporangium is derived from a group of cells. Further after the initial periclinal division the sporogenous tissue is derived from the inner daughter cell. In Leptosporangiate type, the sporangium is derived from a single initial cell. Further the sporangenous tissue is derived from the outer daughter cell of the sporangial initail.

A mature sporangium dehisces either vertically or transversely. Special types of cells may or may not be present in the wall to help in the dehisence. Usually in ferns the sporangia have a thick walled anmulus and a thin walled stomium. The annulus may be vertical, shield shaped or cap shaped. By the differential hygroscopic response of these cells the sporangium breaks open liberating the spores. The spores are usually wind disseminated.

Gametophytic generation.

The haploid spore is the starting point in the development of the gametophyte. In heterosporous forms two types of spores are formed (mega and microspores) hence there will be two types of gametophytes (male and female). In homosporous forms there will be only one type of gametophyte.

The spores rarely exceed a few mm in size. Their shape varies from triradiate to spherical. They have two layered wall. The outer sculptured layer is known as the exine (exospore) and the thin inner layer is known as

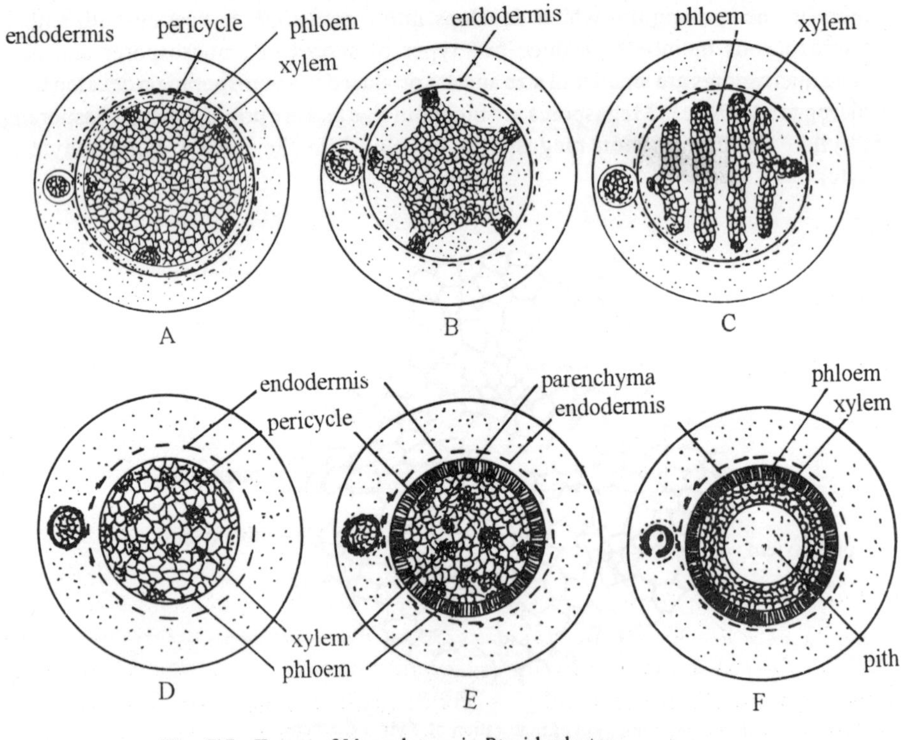

Fig. 7.5 Types of Vasculature in Pteridophytes.
A. Haplostele, B. Actinostele, C. Plectostele, D. Mixed Protostele,
E. Mixed Protostele (with pith), F. Ectopholic Siphonostele.

the intine (endospore). Sometimes there will be an outer most epispore and a middle mesospore in addition to exine and intine.

The germination and development the gametophyte is different between homosporous and heterosporous forms.

In heterosporous forms the gametophytes are extremely reduced and their size is limited to the confines of the spore. The development of the gametophyte here is called *endosporic*.

In homosporous forms the gametophytes are comparitively bigger and they break open the spore to grow independently on the soil. This type of development is said to be *exosporic*. Homosporous gametophytes are filamentous, cordate or tuberous in shape. Some of them exhibit dorsiventral symmetry. The gametophytes are composed mainly of parenchyma. The nutrition is either autotrophic (When chlorophyll is present) or saprophytic (when mycorrhiza is present).

Reproduction:- Gametophytes also reproduce by two methods viz, vegetative propagation and sexual reproduction. Vegetative propagation is uncommon and it takes place by the formation of gemmae or brood bodies (see *Lycopodium*).

Sexual reproduction takes place by the formation of antheridia (Male) and archegonia (female). Both the sex organs are extremely reduced in comparison with their counter part in Bryophytes.

An antheridium is a globose structure partially or completely embeded in the gametophytic tissue. It has a single layered Jacket enclosing the androgonial cells. The antherozoids are spirally coiled and may be biciliate or multiciliate.

An archegonium is a flask shaped structure having a basal venter and a short neck. Within the archegonium are found one or two neck canal cells one venter canal cell and an egg cell.

Fertilization:- Even in pteridophytes fertilization is dependent upon external moisture. The antherozoids that come out of the antheridium are attracted chemotactically by the nucilagenous mass produced by the disintegrating cells (neck canal and venter canal cells) of the archegonium. Many antherozoids enter the archegonium. Ultimately one succeeds in fusing with the egg resulting in the formation of a zygote.

Embryogeny:- The zygote represents the first cell of the sporophytic generation. The first division of the

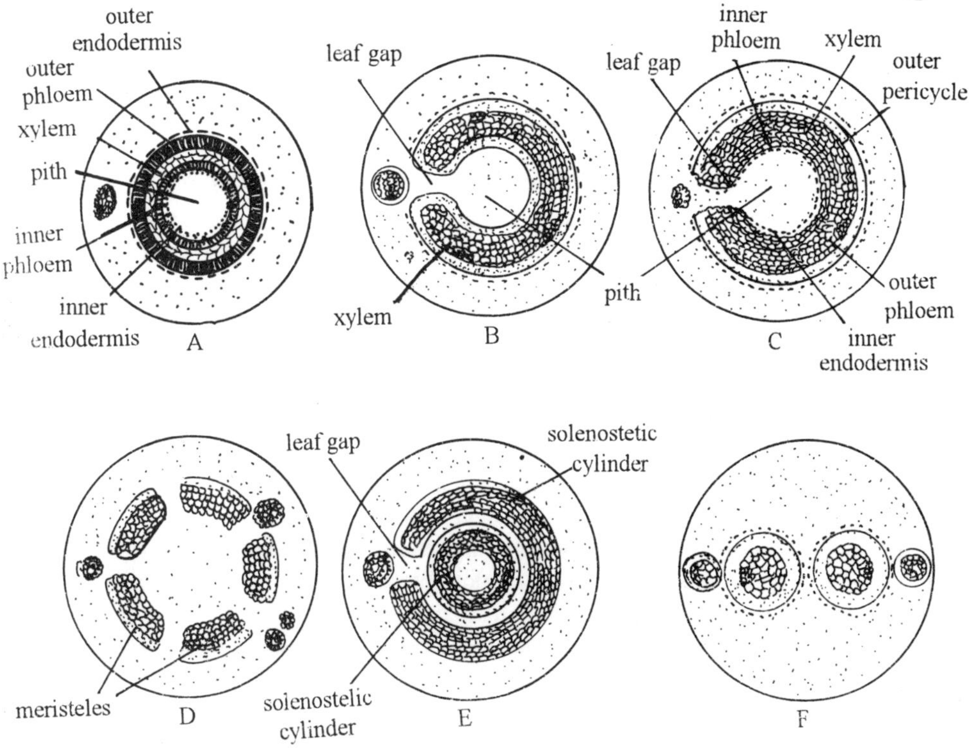

Fig. 7.6 Types of Vasculature in Pteridophytes (Contd....)
A. Amphipholic Siphonostele, B. Ectophloic Solenostele, C. Amphipholic Solenostele (with leaf gap), D. Dictyostele, E. Polycyclic Solenostele F. Polystele

zygote is either transverse or vertical. When the division is transverse two superposed cells are formed, of which the one nearer to the archegonial neck is the epibasal cell and the one away from the archegonial neck is the hypobasal cell. The embryo proper may be derived from either the epibasal cell (exoscopic) or the hypobasal cell (endoscopic). In the latter case the epibasal cell forms a tubular structure known as the suspensor. The function of the suspensor is to push the developing embryo deep into the gametophytic tissue so that it can easily absorb nutrition.

Further divisions in the embryonal cell are variable. After the first few divisions the characteristic parts of the embryo like foot, root, cotyledons and stem apex are differentiated. Untill the root establishes itself on the soil the foot behaving like a hanstorium absorbs food from the gametophyte. After the root grows into the soil the sporeling seperates itself from the gametophyte and grows independently.

Aberrvations in the life cycle.

In the normal type of life cycle showing alternation of generations the sporophyte gives rise to a gametophyte and vice-versa. Syngamy and reduction division occur at specific points in the life cycle so that the gametophyte is always haploid and sporophyte is always diploid.

In some cases there are deviations or aberrations in the life cycle disturbing the chromosome constitution of the sporophyte and the gametophyte. The aberrations are of three types viz, apogamy, apospory and parthenogenesis.

Apogamy:- This was first discovered by Farlow (1874) in *Pteris cretica*. Winkler (1908) defines apogamy as follows-it is the formation of sporophyte directly from the vegetative cells of the gametophyte without the act of syngamy or gametic fusion.

In nature apogamy has been reported in some twenty genera and fifty species of pteridophytes. Apogamy is of frequent occurrence if ferns. It is seen in *Pteris, Adiantum, Osmunda* etc.

Apogamy may also be experimentally induced as in *Lycopodium* (Bell 1959, Freeberg 1957) *Equisetum* (1959) etc.

Apogamous sporophytes may originate from one or more cells of the gametophyte.

Apospory:- The phenomenon of production of a gametophyte directly from the vegetative cells of the sporophyte without reduction division (spore formation) is called apospory. Apospory was first discovered by Drucry (1894) in a fern *Atheyrium fillex femina* var. *clarissima*. He observed the development of gametophytes from the sporangia. Subsequently apospory has been reported in several genera including *Osmunda*. Apospory may also be experimentally induced like apogamy. There are reports of induced apospory in *Pteridium* (Tak- ahashi, 1962) Cyclosorus (Hebra and Palta 1969(etc.

In apospory a filamentous or chordate gametophyte may be formed from one or more vegetative cells of the sporophyte. The aposporous structure may be an antheridium, a rhizoid or even an antherozoid.

Parthenogenesis:- The development of an unfertilized egg into an embryo is called parthenogenesis. This is different from apogamy here only the egg is capable of developing into an embryo. Parthenogenetic embryos are haploid. But in some cases when the spores are unreduced the resultant gametophytes are diploid. Consequently the eggs are also diploid and they develop into an embryo without fertilization. In these instances the parthenogenetic embryos are diploid. This type of parthenogenesis coupled with failure of reduction division during megaspore formation has been reported in *Marsilea drummondii* (Strassburger 1907).

Classification of Pteridophytes

Since a long time vascular plants are customarily divided into pteridophyta and spermtophyta. This classification was based on the assumption that the former lack the seeds while the latter produced them. But the discovery of pteridosperms (seed bearing ferns) broke down this artificial classification. Simmott (1935) has introduced the term Tracheophyta to include all vascular plants. Tracheophyta are further divided into four main groups viz, Psilopsida, Lycopsida, Sphenopsida and Pteropsida. But it is not certain whether these are divisions or classes. Hampt (1953) considers them as classes in the division Tracheophyta. While Limmermann (1930) Arnold (1974) Wardlaw (1952) etc., regard them as divisions.

Reimers (1954) considers pteridophyta as a division and divides it into five classe viz, Psilophytopsida, Lycopsida, Psilopsida, Artienlatae and Filices.

Smith (1955) Bold (1957) and others divide vascular cryptogams into five divisions viz Psilophyta, Lycophyta, Sphenophyta, Noeggerathiophyta and Pterophyta. Some people have changed the name Lycophyta to Lepidophyta (Smith 1955), Sphenophyta to Arthrophyta or Calamophyta (Smith 1955) and Pterophyta to Filicophyta (Parihar 1965).

In this book the classification is mainly based on the one proposed be Reimers in the 1954 edition of Engler's Syllubus der Pflanzen familien. This classification simple and easy to follow. Below is given the outline classification and the members discussed in this book.

Pteridophytes	
1. Psilophytopsida	
Psilophytales.	Eg: *Rhynia**, *Asteroxylon**.
2. Psilotopsida	
Psilotales	Eg: *Psilotum*
3. Lycopsida	
a) Protolepidodendrales	
b) Lycopodiales	Eg: *Lyocpodium, Phylloglossum.*
c) Lepidodendrales	Eg: *Lepidodendron**, *Lepidocarpon**.
d) Selaginellales	Eg: *Selaginella.*
e) Isoetales	Eg: *Isoetes.*
4. Sphenopsida	
a) Hygeniales	
b) Sphenophyllales	Eg: *Sphenophyllum*
c) Calamitales	Eg: *Calmites**, *Calamostachys*
d) Equisetales	Eg: *Equisetum*
5. Pteropsida	
A) Primofilices	Eg: *Zygopteris, Stauropteris, Botryopteris*
a) Cladoxylales	
b) Coenopteridales	
B) Eusporangiatae	
a) Ophioglossales	Eg: *Ophioglossum*
b) Marathiales	Eg: *Angiopteris*
C) Osmundidae	
Osmundales	Eg: *Osmumda*
D) Leptosporangiatae	
a) Filicales	Eg: *Hymenophyllum, Adiantum, Pteris, Pteridium*
b) Marsileales	Eg: *Marsilea, Azolla, Salvinia*
c) Salviniales	

(Asterisk mark indicates the fossil members).

CHAPTER 8

FOSSILS - A BRIEF ACCOUNT

A study of the Geological history of the Earth reveals that a sequenced succession of plants occurred on the face of this planet. During such a process diverse types of plant groups appeared and vanished in course of time. Some of them attained predominant position in the past flora, thereafter met an extinction while new plant groups emerged.

The study of succession of plants in the geological history forms an exciting study of botany called **Paleobotany**

Paleobotany in its preview includes the study of the Flora of the past. Such a study is made by the observations of plant remains called **fossils.**

What are fossils? It is difficult to define the word *fossil* precisely. The word fossil is derived from the Latin verb *foedre* meaning "to dig". Hence basically a fossil is meant to include anything that is excavated from the Earth. In the present usage however the term fossil is restricted to organic remains obtained from earth and excludes inorganic objects or objects fashioned by human beings.

A fossil includes the remains of a whole organism or some part of the organism or the direct evidence of the prior existence of some organism in the sediments of earth. In addition to this, any thing resulting from or indicating the prior existence of organisms such as the imprints tracks, trails, borings and in some cases even chemical precipitations may be regarded as fossils. Chemical fossils include things such as sulphur and iron deposits indicating the former bacterial acitivity and lime stones representing the deposition of calcium carbonate by algae.

TYPES OF FOSSILS:

Several types of fossils have been recognised based on the structure, process of fossilization etc. In general the following types of fossils have been recognised.

Compressions: Of all the types of fossils, these are the most common ones found in the rocks. As the very name suggests, the plantbody gets flattened due to the pressure of the sediments. Due to this a thin carbonaceous film is retained outlining the surface features of the plant parts. Generally little or no internal structure is retained. Rarely however the cell pattern of cutinised epidermis is retained. Sediments of various types are involved as matrices for compressing the plant parts. These are shale, sand stone, volcanic ash, diatomaceous earth etc. Compressions help us to understand the external features of the plant. They provide reliable clues in establishing the affinities.

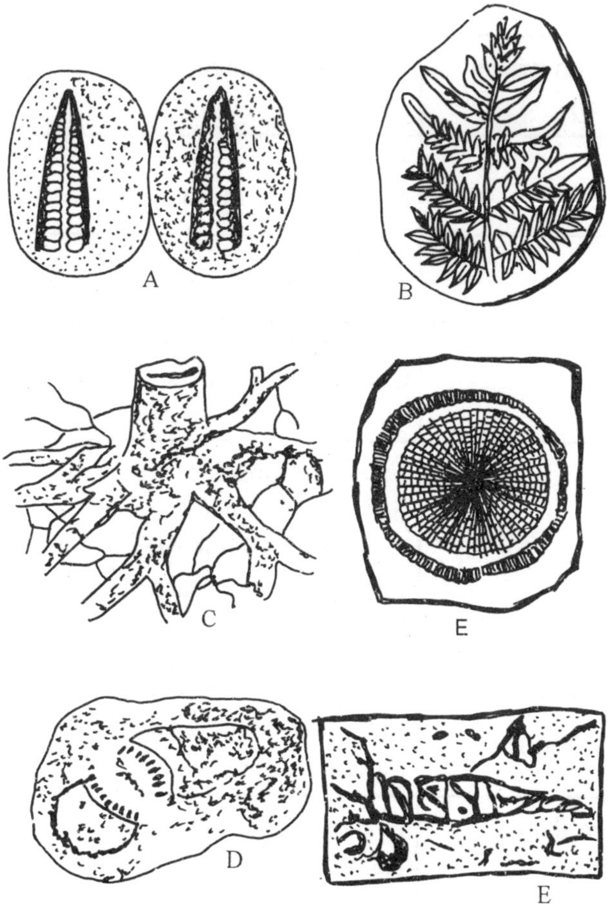

Fig. 8.1 Types of Fossils
A. Compression (cone of Lepidodendron), B. Impression (Leaf of Neuropteris), C. Cast (Stump of Stigmaria), D. Section of a coal ball showing transverse section of a petrified stem (Sphenophyllum). E. Sections of stems within coal balls. F. Compression (spike of Zosterophyllum)

Impressions: An impression may be defined as a negative of a compression. The rock strata that enclose the plant parts become hardened and when split open, they show the negative imprint of the plant part. There is no organic matter in an impression; therefore cellular details are not available. But the outlines of plantsparts are very well preserved. Of all the plant parts leaves yield good impressions.

Moulds, incrustations and casts: After the plant part is entombed in the sediment, if there is no compression, the surrounding sedimentary material hardens to form a three dimensional structure called *mould*.

Incrustations may be defined as external moulds of a plant, usually in some incompressible material like sand stone, iron stone or terfa.

After the formation of mould, sometimes the plant part may decay. This space is later filled by crystalline substances such as iron pyrites, sphalarites, chalerite, opal, agate etc. Such hardened structures are called casts. Casts exhibit the same external configurations of the original plant fragment. No organic material is seen in a cast.

Petrifactions: Of all the types of fossils, petrifactions are the best. Among the fossils they are also rare to obtain. In these fossils there is preservation of external form and internal structure and sometimes the substances of original plants are also preserved. Cellular details are preserved due to infilteration of some minerals into the plant tissues.

During petrifaction, the entombed plant is impregnated with about twenty minerals such as carbonates, sulphates, silicates, phosphates, iron pyrites etc. Solidification of plant tissues takes place immediately.

Petrifactions are of great importance, because they can be cut into sections for anatomical studies.

Geological time table: This is a table which gives in a chronological order, the different ages in the history of the Earth together with the fossils found in that period.

Era	Period	Age (in million years)	Type of vegetation
Cenozoic	Quaternary Tertiary	1 10	Morden Culmination of land flora
Mesozois	Cretaceous Jurassic Triassic	100 140 180	Flowering Cycads and conifers Cycads and conifers
Paleozoic	Carboniferous Devonian Silurain Ordovician Cambrain	250 300 250 425 500	Extensive forests of tree ferns horsetails and club mosses Land plant loke Rhynia Land plants like Baragwanthia Marine algae Marine algae
Proterozoic		550-2000	Fungi and bacteria are reported to be present
Archeozoid		more than 200	Evidence of the existence of life; but no fossils

CHAPTER 9

PSILOPHYTOPSIDA

The division psilophytopsida constitutes a very primitive group among pteridophytes. Members of the division are all fossils and no living members of the group are available.

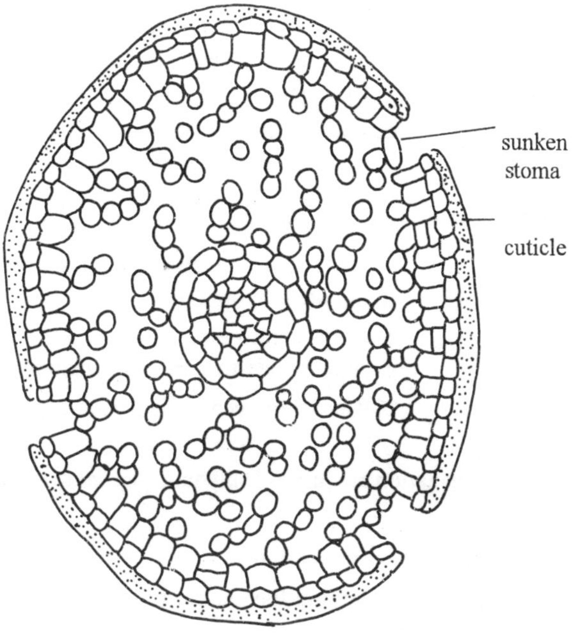

Fig.9.1:. T.S. of Aerial Shoot of *rhynia major*

Fossils of the group were first discovered in 1859 by Dawson a Canadian geologist. But the discoveries of Kidston and Lang in the year 1917 of well preserved plant specimens firmly established the affinities of Psilophytopsida with other pteridophytes. Kidston and Lang described certain petrified fossils from old red sand stone beds near Rhynie, in Aberdeenshire, Scotland.

The general characters of psilophytopsida may be listed as follows

1. Simple plantbody with no differntiation into root, stem and leaf.

2. The naked green stems are spiny having cuticles and stomata. Occasionally leaf like organs may be present.

3. Vasculature is of the simplest type.

4. Sporangia are borne at the tips of ultimate branches.

Classification: The class consists of a single order Psilophytales and five families viz, Rhyniaceae, Psilophytaceae, Zosterophyllaceae, Asteroxylaceae and

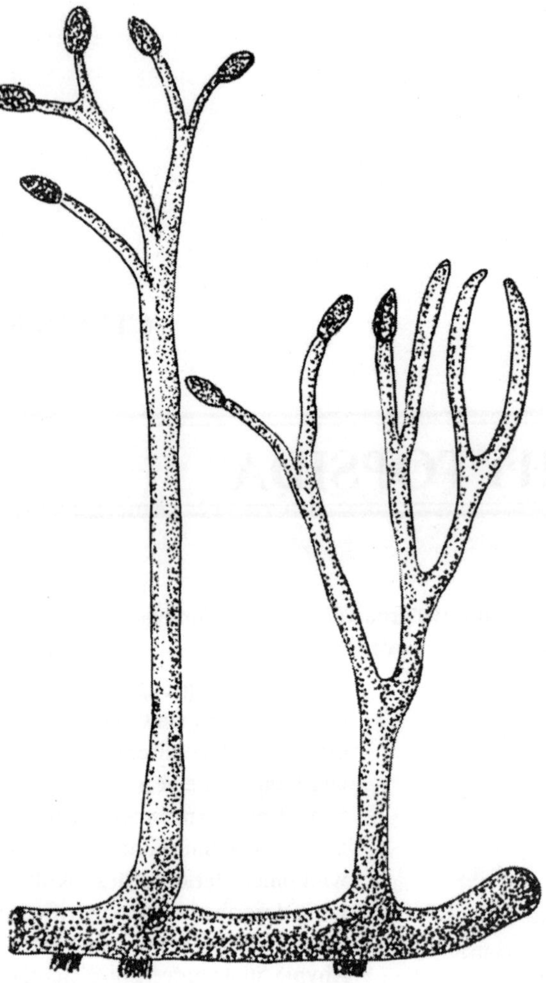

Fig. 9.2. Reconstruction of *Rhynia major* (After Kidston & Lang)

Pseudosporochnaceae. Of these families Rhyniaceae and Asteroxylaceae are discussed here.

RHYNIA

Occurrence and Distribution

Rhynia is a fossil member. The petrified fossils were first discovered in Rhynie chert in Aberdeen shire, Scotland from old Red sand stone beds.

Practical work

1. Study the external morphology from the model of the plant body.
2. Study the internal sturcture of the stem.
3. Study the reproductive structures (Sporangia) from a long section of the sporangium.

Morphology of the plantbody

1. The genus *Rhynia* consists of two species viz, *R.gwynnevaughnii* and *R.major*.
2. *R.major* is the larger and less advanced, while *R.gwynnevaughnii* is the smaller and more advanced.
3. The plant body in both the species is essentially similar and consists of rootless, subterranean creeping rhizome from which arise cylindirical erect leafless bifurcating shoots.
4. Some of the branches of the aerial shoot terminate in sporangia.
5. Roots are absent. Hair like structures called rhizoids arise from the rhizome.
6. *R.major* reached a height of 40cms while *R.gwynnevaughnii* measured 17-26cms in height.
7. Hemispherical bulges, probably helping in vegetative propagation are present only in *R.gwynnevaughnii*.

Anatomy

1. Internal structure of the aerial shoot may be studied from a transverse section.
2. A transection of the stem shows a smooth vascular cylinder occupying the central position in a wide cortex.
3. The stele is a protostelic haplostele with xylem forming a smooth central core.
4. There is no pith.
5. Phloem surrounds the xylem; pericycle and endodermis are absent.
6. The cortex is divided into a broad inner cortex and a narrow outer cortex.

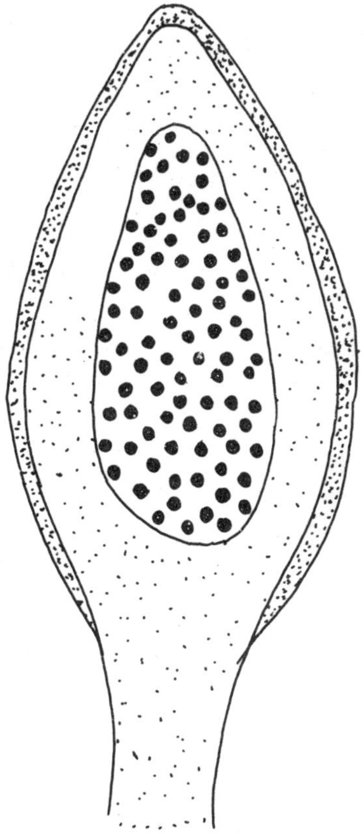

Fig. 9.3. A Sporangium of *Rhynia major*

7. The inner cortex consists of radially arranged rounded cells with intercellular spaces.

8. The inner cortical cells are chlorophyllous constituting the chief photosynthetic tissue of the plant.

9. The outer cortex is one to four layered thick and composed of large clear cells.

10. Outermost layer is the epidermis.

11. Stomata are present in the epidermal layer.

12. There is sub stomatal cavity which brings the assimilatory zone in direct communication with the stoma.

Reproduction

1. All the fossil specimens of *Rhynia* obtained so far are only sporophytes.

2. Gametophytic generation, has not been preserved as a fossil.

3. The sporophyte reproduces by means of sporangia. The sporangia are borne distally on certain branches of the stem.

4. The sporangium is cylindrical but tapers towards the distal end. It has a stout stalk at the base.

5. The sporangial wall is several layered thick.

6. Cuticle is deposited on the outermost layer (epidermis).

7. The innermost layer of the sporangial wall may be called tapetum.

8. A number of spores (often in tetrads) are found lying in the sporangial cavity.

9. The spores are homosporous.

10. Spores have cutinised cell walls.

Diagrams to be drawn

1. Habit sketch (Reconstruction)

2. Transverse section of Aerial shoot.

3. Long section of sporangium.

Systematic position

PSILOPHYTOPSIDA

PSILOPHYTALES

RHYNIACEAE

RHYNIA

ASTEROXYLON

(Astero = star)

Occurrence and distribution

Asteroxylon is a fossil member. The petrified fossils (*A.mackie*) have been obtained from *Rhynie* chert as well as from middle Devonian rock strata from Germany (*A.elberfeldens*)

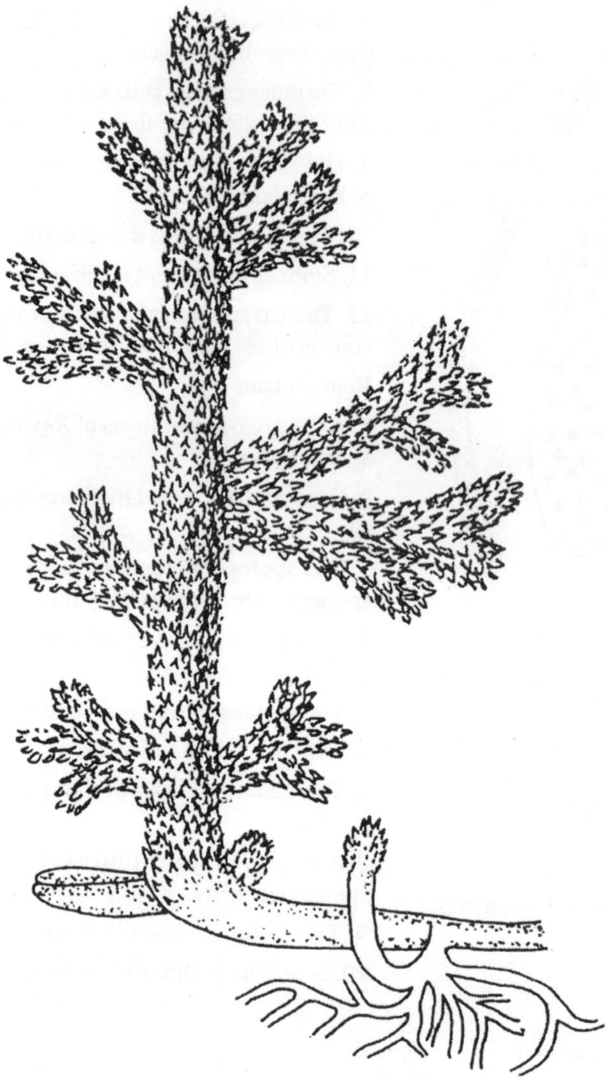

Fig. 9.4. Reconstruction of *Asteroxylon mackie*

Practical work

1. Study the external features of the plantbody from the reconstruction.
2. Roots as well as rhizoids are absent.
3. Some of the rhizomatous branches dichotomised freely and produced slender leaf like organs which penetrated the substatum.
4. The aerial branches have spirally arranged non casculated leaves.
5. At the base of the aerial shoot a transition zone is present studded with scales.
6. The branching of the aerial shoot is lateral and exogenous.

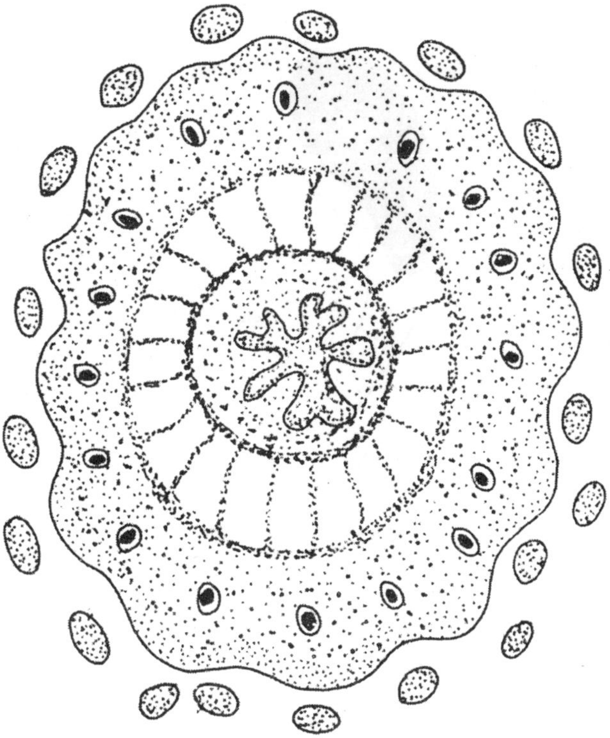

Fig. 9.5. T.S. of Aerial Shoot of *Asteroxylon mackie*

Anatomy

1. Internal structure of the aerial shoot can be studied in a cross section.
2. A cross section of the stem shows three regions viz., Epidermis, Cortex and Stele.
3. The outline of the section would be wavy due to the presence of leaf bases.
4. Epidermis is single layered. Stomata are present disturbing the continuity of the epidermis. Stomata are sunken unlike in *Rhynia*.
5. The cortex has three zones.
6. The inner cortex has compactly arranged parenchyma cells.
7. Middle cortex has elongated trabecular cells consisting of radiating plates of cells separating the lacunae.
8. Outer cortex has tangentially elongated cells.
9. Cortical cells contain aseptate fungal hyphae and resting spores belonging to the type genus *Paleomyces*.
10. The central region of the stem is occupied by the vascular cylinder.
11. Pericycle and endodermis are absent.
12. The stele is an actinostelic protostele.
13. The xylem strand is star like with four or more sometimes forked arms.
14. Protoxylem elements terminated the tips of the xylem arms.
15. The phloem surrounded the xylem completely with the result the outline of the stele is smooth.

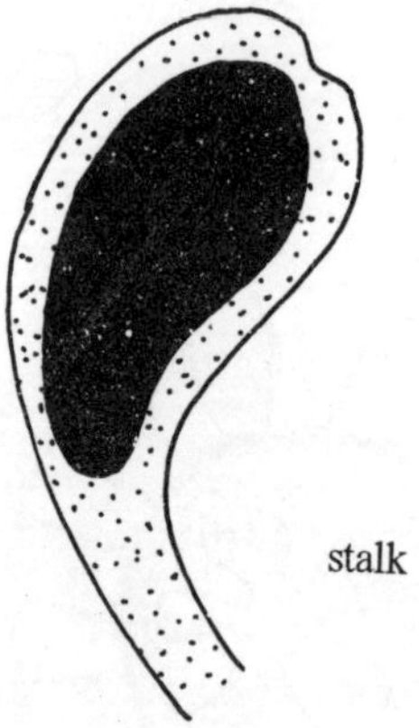

Fig. 9.6. A Sporangium of *Asteroxylon mackie*

16. Small leaf traces will be given off from the rays of the xylem.
17. The leaf traces however travel only up to the leaf base.
18. Pith is absent in *A.mackie*, while it is present in *A.elberfeldens*.

Reproduction

1. The fertile shoots were not in organic contact in *A.mackie*, while they were connected in *A.elberfeldens*.
2. The Sporangia are borne on slender naked leafless branches.
3. The sporangia are ovoid or ellipsoid in shape.
4. An annulus is present at the wider free end of the sporangium marking the line of dehiscence.
5. Spores are homosporous.

Diagrams to be drawn

1. Habit sketch (Reconstruction).
2. Transverse section of the stem.
3. Reproductive branches and sporangia.

Systematic position

PSILOPHYTOPSIDA

PSILOPHYTALES

ASTEROXYLACEAE

ASTEROXYLON

CHAPTER10

PSILOTOPSIDA

This class consists of most primitive living members among pteridophytes. The members show close resemblance to psilophytales.

In general, the class is characterised by the following features.

1. The sporophyte is a rootless, leafless structure. Rhizoids help in the absorption of water.
2. Branching of the sporophyte is always dichotomous.
3. The plantbody is usually differentiated into an underground prostrate system and an aerial erect shoot system.
4. The spore bearing organs are bi or trilobed and are borne in the axils of small leaf like appendages.
5. Gametophytes are cylindrical, branched subterranean and achlorophyllous. Nutrition is saprophytic.

The class Psilotopsida has a single order psilotales and a single family psilotaceae. The family includes two genera viz., *Psilotum* and *Tmesepteris*. Of the two genera *Psilotum* is discussed here.

PSILOTUM

Occurence and distribution:

The genus *Psilotum* has two species viz., *P.nudum* and *P.flaccidum* and is widely distributed in the tropics and subtropics. The genus is included in the list of endangered species in India.

Practical work

1. Observe the morphological structure of the plant.
2. Study the anatomy of rhizome and aerial shoot.
3. Study the structure of the reproductive organs.

Morphology of the plant

1. The plantbody is differentiated into two parts viz., a horizontal undergroundrhizome and an erect aerial shoot.
2. The rhizome is brownish in color and dichotomously branched.
3. Roots are absent. Rhizoids develop from the rhizome to help in absorption.
4. Some of the branches of the rhizome grow up and form erect or pendulous aerial shoots.
5. The aerial shoots are 20-70cms long usually ribbed and multiangular.

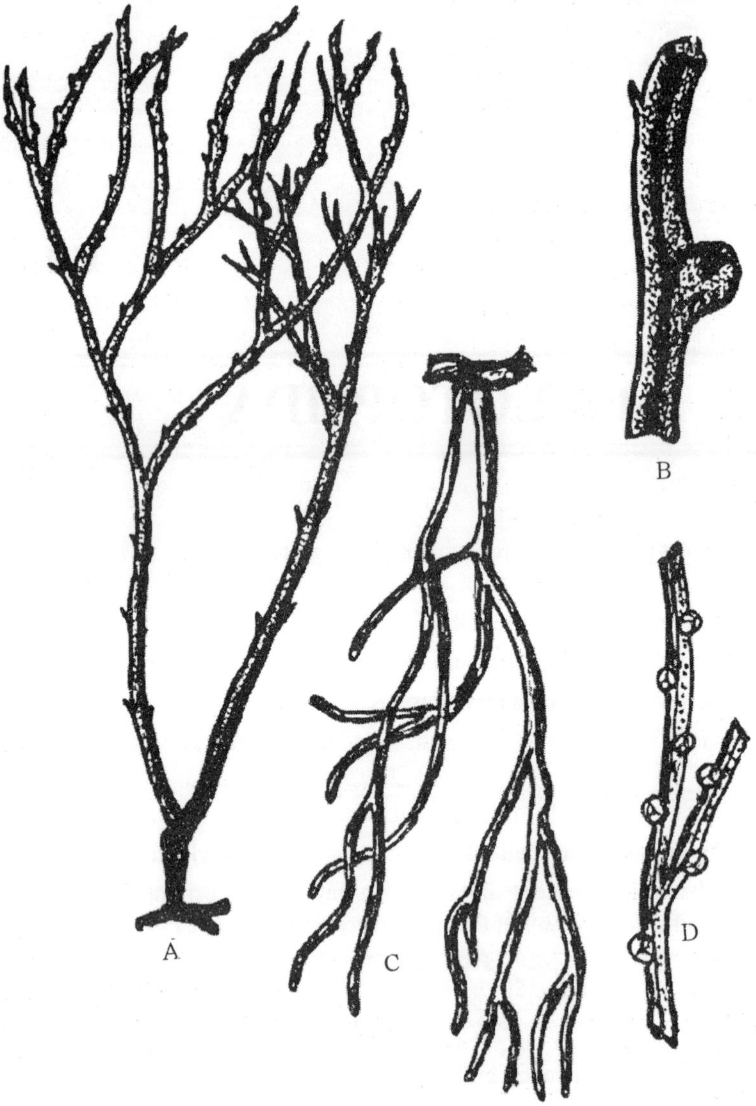

Fig. 10.1 *Psilotum* : Habit of *Psilotum nudum* (A) and a Part of the Shoot Bearing Synangia (B); Habit of *Psilotum flaccidum* (C) and a part of the Shoot bearing Synangia (D)

6. Here and there on the aerial shoot are found a number of scaly leaves.

7. The scaly leaves are of two types fertile and sterile.

8. The sterile ones are found all along the length of the aerial shoot while the fertile ones are restricted only to the upper portions.

9. The fertile ones bear in their axils, trilobed spore bearing structures called Synangia.

Anatomy of rhizome

1. A transverse section of the rhizome shows an epidermis, a cortex and stele.

2. The epidermis is single layered.

3. The cortex is divided into three zones. outer cortex is parenchymatous and the cells have the mycorrhizal association.

4. The middle cortex is also parenchymatous with cells rich in starch grains.

5. The inner cortical cells are brown in color due to the deposition of a compound called phlobaphene.

6. The stele is protostelic surrounded by pericycle and endodermis.

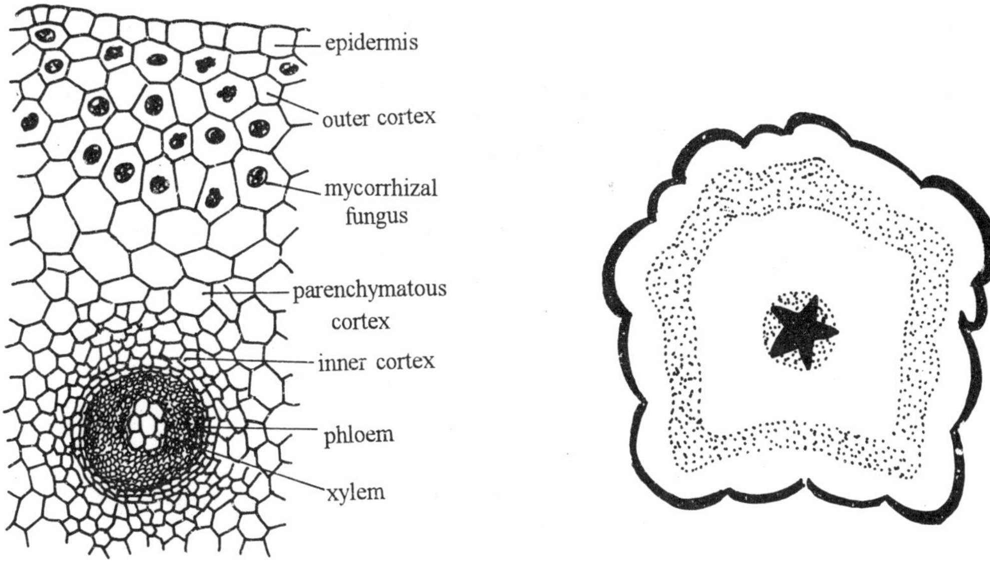

Fig.10.2 ***Psilotum*** **: T.S. of Rhizome.**
A. Sector Enlarged, B. Ground Plan.

7. The xylem is exarch and star shaped.

Anatomy of aerial shoot

1. A transverse section of the aerial shoot shows the usual three regions viz., epidermis, cortex and stele.

2. Epidermis is single layered and has closely packed cells.

3. Stomata are present in the epidermis and are restricted to the grooves. They are sunken in nature.

4. A thick cuticle is deposited above the epidermis.

5. Next to the epidermis is the cortex. It is divided into three zones.

6. The outer cortex is chlorenchymatous. It is 2-5 layered thick with loosely arranged cells. These cells constitue the chief photosynthetic tissue of the plant (leaves are reduced).

7. The middle cortex consistes of 4-5 layeres of sclerenchyma. This offers mechanical support to the plant.

8. The inner cortex has a few layers of parenchyma cells.

9. Occupying the central region of the stem is the stele. Outermost layer of the stele is endodermis.

10. Next to the endodermis is an ill defined pericycle.

11. The stele is usually an actinostelic protostele. But in the basal portions the stele may be a siphonostele with a sclerenchymatous pith.

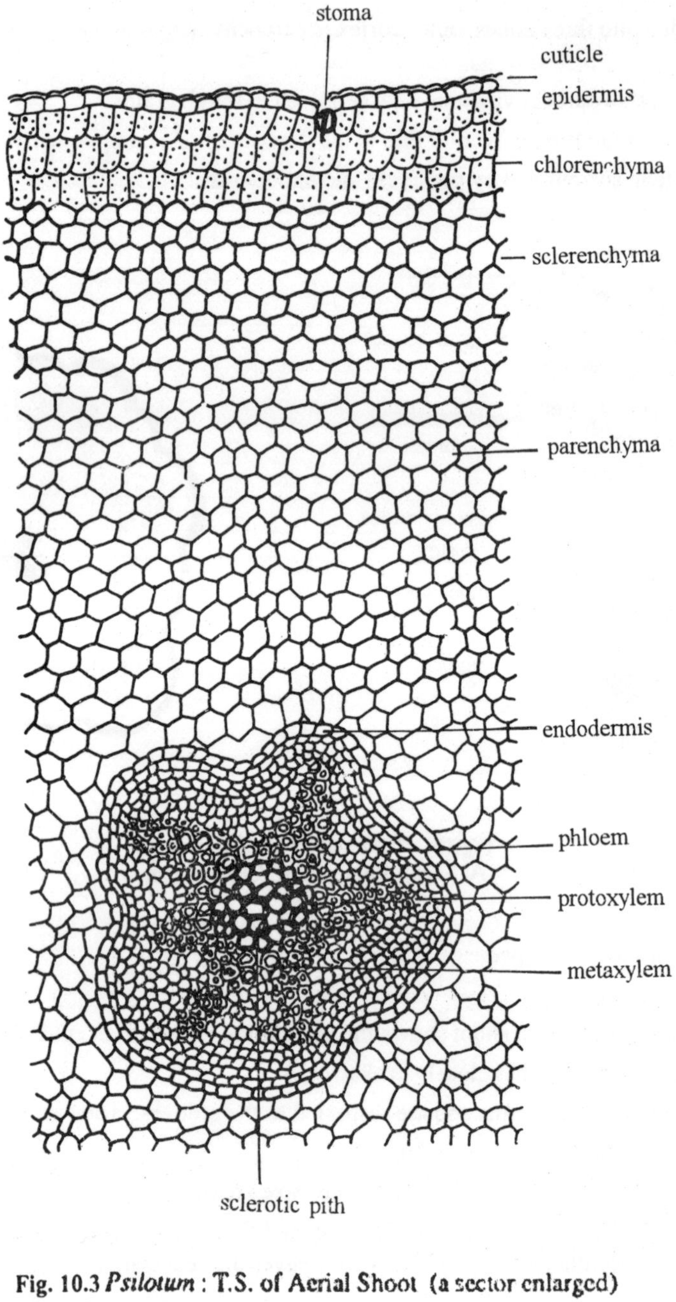

Fig. 10.3 *Psilotum* : T.S. of Aerial Shoot (a sector enlarged)

Reproduction

1. The sporophyte reproduces by means of spores. The spores are produced in trilobed structures called synangia.

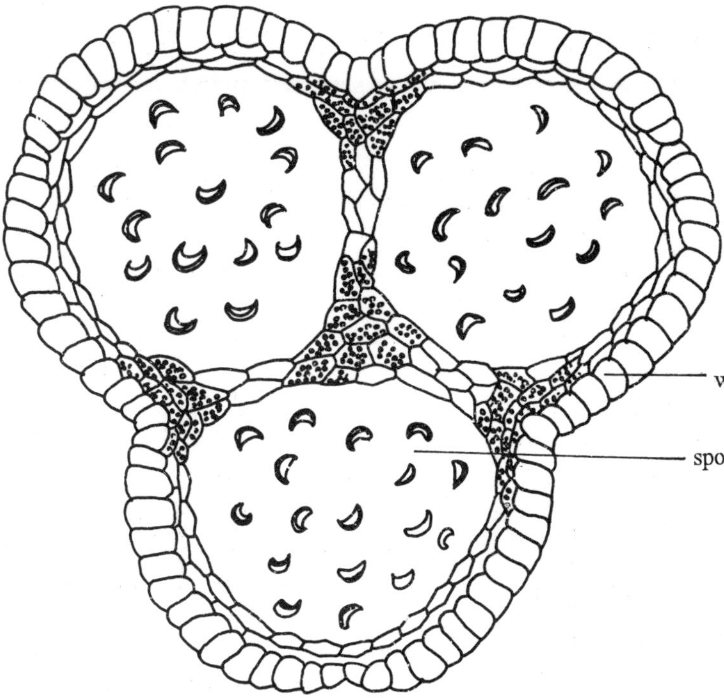

Fig.10.4 *Psilotum*, T.S. of Mature Synangium with Spores

2. The synangia are usually present in the upper part of the aerial shoots, in the axils of fertile scale leaves.

3. The wall of the trilobed synangium is made up of 4-5 layers of cells. The outermost layer of the wall is prismatic.

4. Within the synangium, ther are three chambers or spore cavities containing spores. The spores are homosporus.

5. The spores on germination develop into the gametophyte.

Structure of the gametophyte.

1. The gametophyte of Psilotum is a cylindrical structure with dichotomous branches.
2. The gametophyte is partly or totally subterranean.
3. The size of the gametophyte ranges from 0.5 to 2mm.
4. The gametophyte is usually dark brown in color. This is due to the mycorrhizal association.
5. Parenchyma is the only tissue found in the gametophyte. The cell walls of the outer layer are strongly cutinised.
6. A number of rhizoids develop from the outer layer of cells.
7. The cells are achlorophyllous and the nutrition of the gametophyte is saprophytic.
8. The gametophyte reproduces asexually as well as sexually.
9. Asexual reproduction of the gametophyte is brought about by means of gemmae.
10. Sexual reproduction is brought about by means of antheridia (male) and archegonia (female).

11. A mature antehridium is somewhat a spherical structure. It has a jacket (wall) surrounding mar antherozoids.

12. Archegonium is a small flask shaped structure. it has a short neck and a basal venter region.

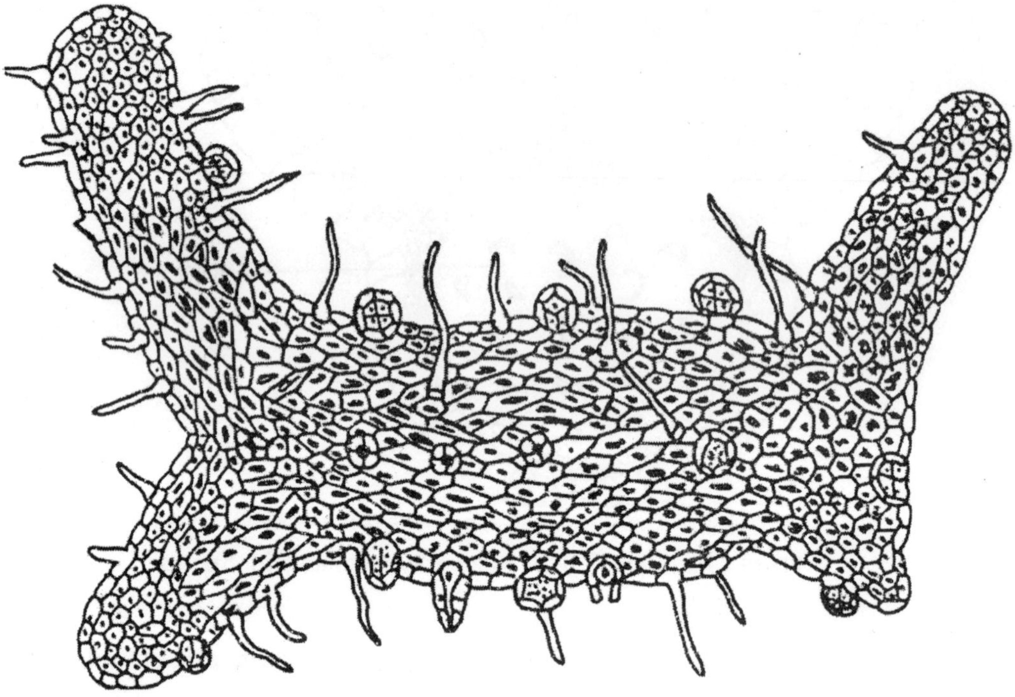

Fig.10.5 ***Psilotum***, Gametophyte of ***P.nudum***

13. Within the archegonium are found one or two neck canal cells, a venter canal cell and an egg cell.

14. Fertilization results in the formation of the zygote, which develops into the sporophyte.

Diagrams to be drawn

1. Habit of the plant.
2. A part of the aerial shoot with synangia.
3. Transverse section of the rhizome.
4. Transverse section of the aerial shoot.
5. Cross section of the syangium.
6. Habit of the gametophyte.

Systematic position

PSILOTOPSIDA

PSILOTALES

PSILOTACEAE

PSILOTUM

CHAPTER 11

LYCOPSIDA

The division Lycopsida includes both living as well as extinct plants. The Geological history of the group dates back to the Devonian period. In early periods, members of Lycopsids were tree like in habits. There were herbaceous members also.

The living members of the group are commonly called club mosses. There are only five living genera represented by over nine hundered species. They occupy a wide variety of habitat and enjoy universal distribution.

Lycopsids are characterised by the following features

1. Sporophytes are distinguished into roots, stems and leaves.
2. The branching of the stem as well as of the root is always dichotomous.
3. Leaves are spirally arranged and microphyllous (They have an unbranched midvein).
4. Leaves may have a small appendage (ligulate) occasionally. The leaves may be eligulate (no appendages) also.
5. Stele is protostelic with a number of variations in the configuration of the xylem.
6. Sporangia may or may not aggregate into strobili. They are always borne on the adaxial surface of the sporophyll.
7. Members are both homosporous as well as heterosporous.

Classification:

Lycopsida is classified into two sub classes Eligulopsida and Ligulopsida. In the former, the leaves are eligulate (a ligule is a small leafy outgrowth present on the adaxial surface of the leaves) and in the latter the leaves are ligulate.

Eligulopsida comprises two orders viz., Protolepidodendrales and Lycopodiales.

Ligulopsida has three orders viz., Lepidodendrales, Selaginellales and Isoetales.

In this book the following genera are discussed -

Lycopodium, Phylloglossum, Lepidodendron, Lepidocarpon, Selaginella and Isoetes.

LYCOPODIUM

Occurence and distribution

The genus *Lycopodium*, popularly called club moss, is world wide in distribution. The genus has 180 species. The species are so diverse in their nature, that many botanists divide the genus into two sub Genera viz., *Urostachya* and *Rhopalostachya*. The major differences between the two are as follows.

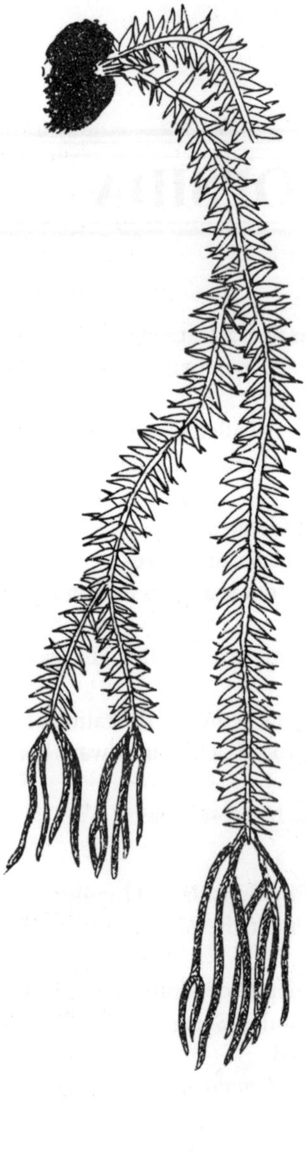

Fig. 11.1 *Lycopodium* : Habit of *L. phlegmaria* (*L. macrostachyum*)

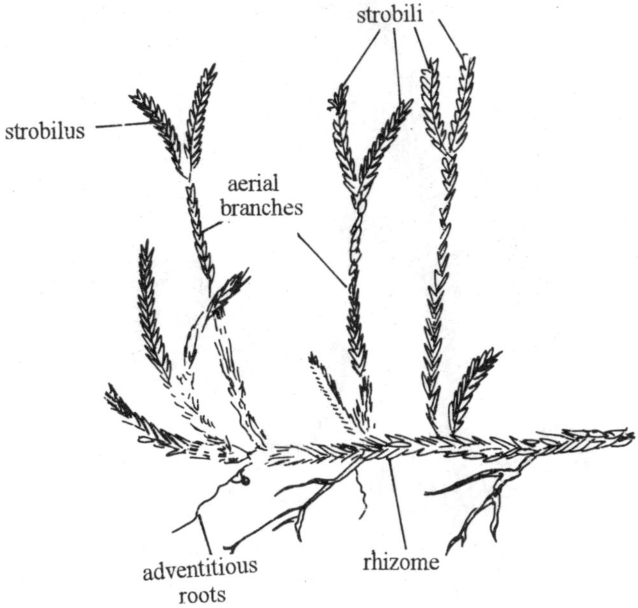

Fig. 11.2 *Lycopodium* : Habit of *L. wightianum*

UROSTASHYA		RHOPALOSTACHYA
Stem:	Erect or pendent	Prostrate or semi erect
Branching:	Dichotomous	Dichotomous are apparently monopodial.
Roots:	Arise at the base of the axis	Arise all along the length of the stem.
Strobilus:	May or may not be organised	Well organised.
Examples:	*L.selago, L.phlegmaria*, L. lucidulum etc.	L.wightianum, L.cernum, L.complanatum, L.inundatum etc.

Practical work

1. Study the morphological characters of various species.
2. Cut transverse sections of root, stem and leaf.
3. Mount a single sporangium and observe.
4. Study the details of sporangium from a long section of the strobilus.
5. Study the structure of the gametophyte.

Fig 11.3Lycopodium : Habit of L.cernuum

Morphological features of Lycopodium phlegmaria

1. The plantbody is differentiated into root stem and leaves.
2. The plant is an epiphyte and generally grows in humus packets in the crevices of the branches.
3. The stem is pendant hanging down from the branches of tress.
4. The stem is branched and the branching is dichotomous.
5. Roots arise only at the base of the stem. The primary root being short lived, the adult sporophyte has only adventitious roots.

Fig. 11.4. *Lycopodium* : Habit of *L. lucidulum*

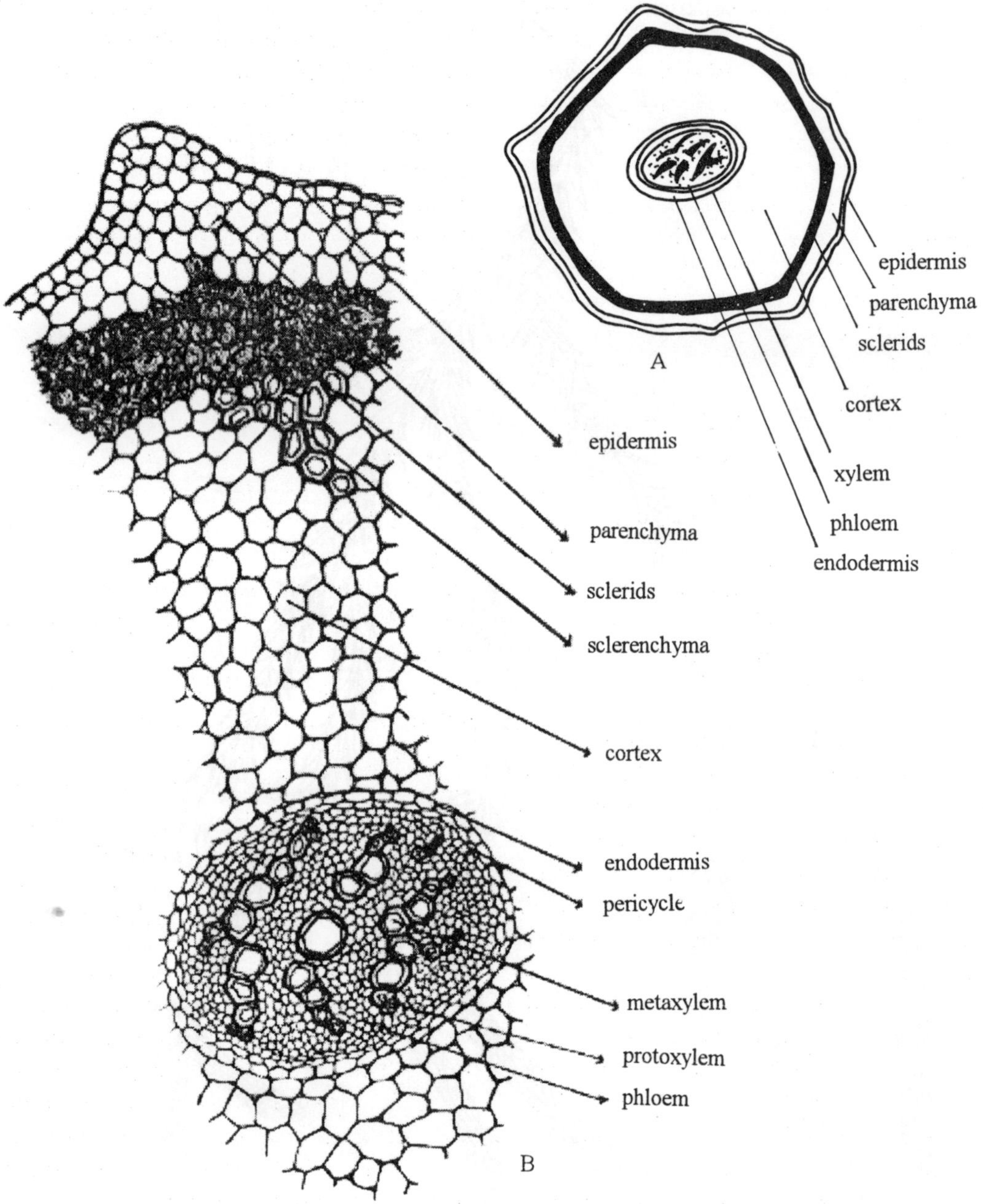

Fig. 11.5 *Lycopodium* : Internal Structure of Stem of *L. cernuum*
A. Ground Plan, B. A Sector Enlarged

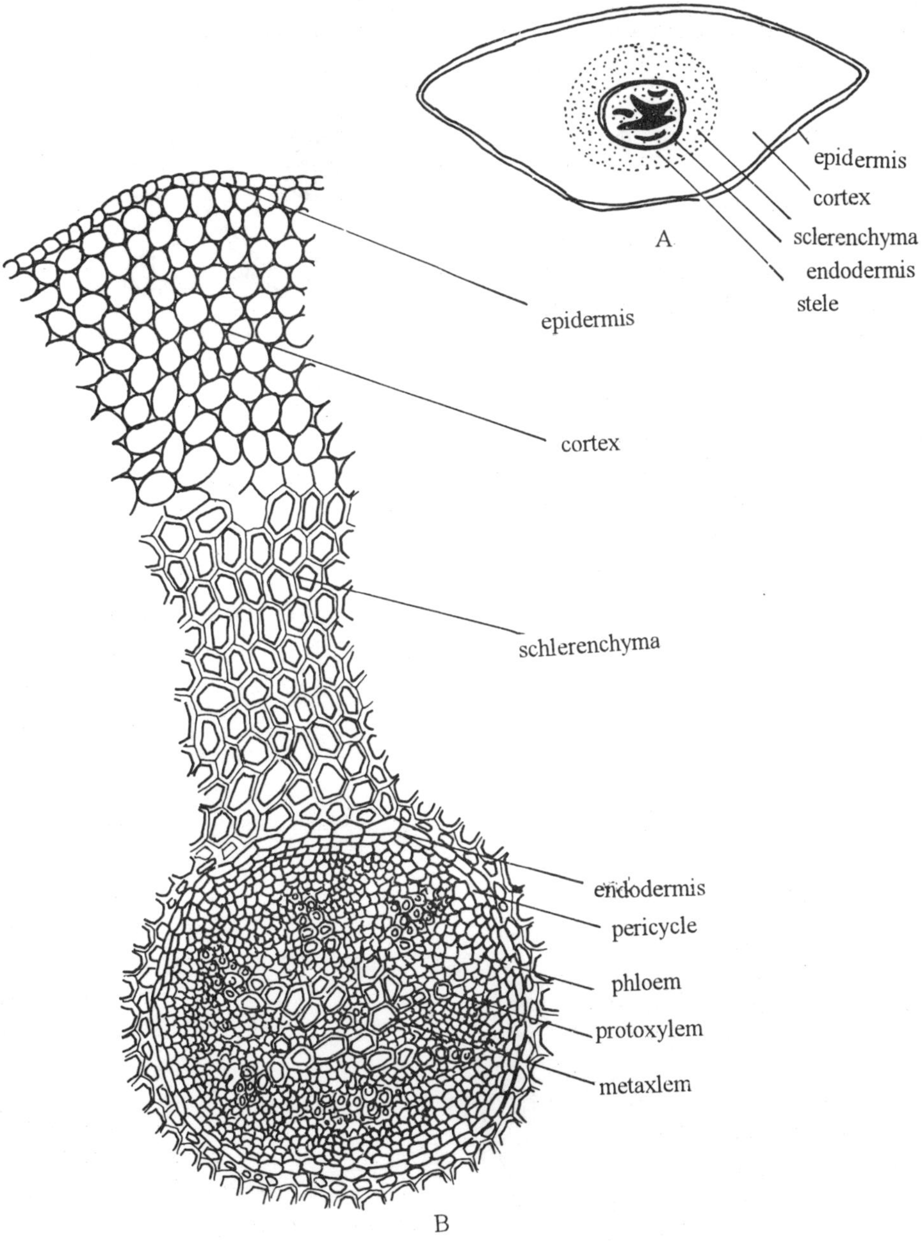

Fig.11.6 ***Lycopodium*** **: Internal Structure of Stem of** ***L. wightianum***
A. Ground Plan, B. A Sector Enlarged

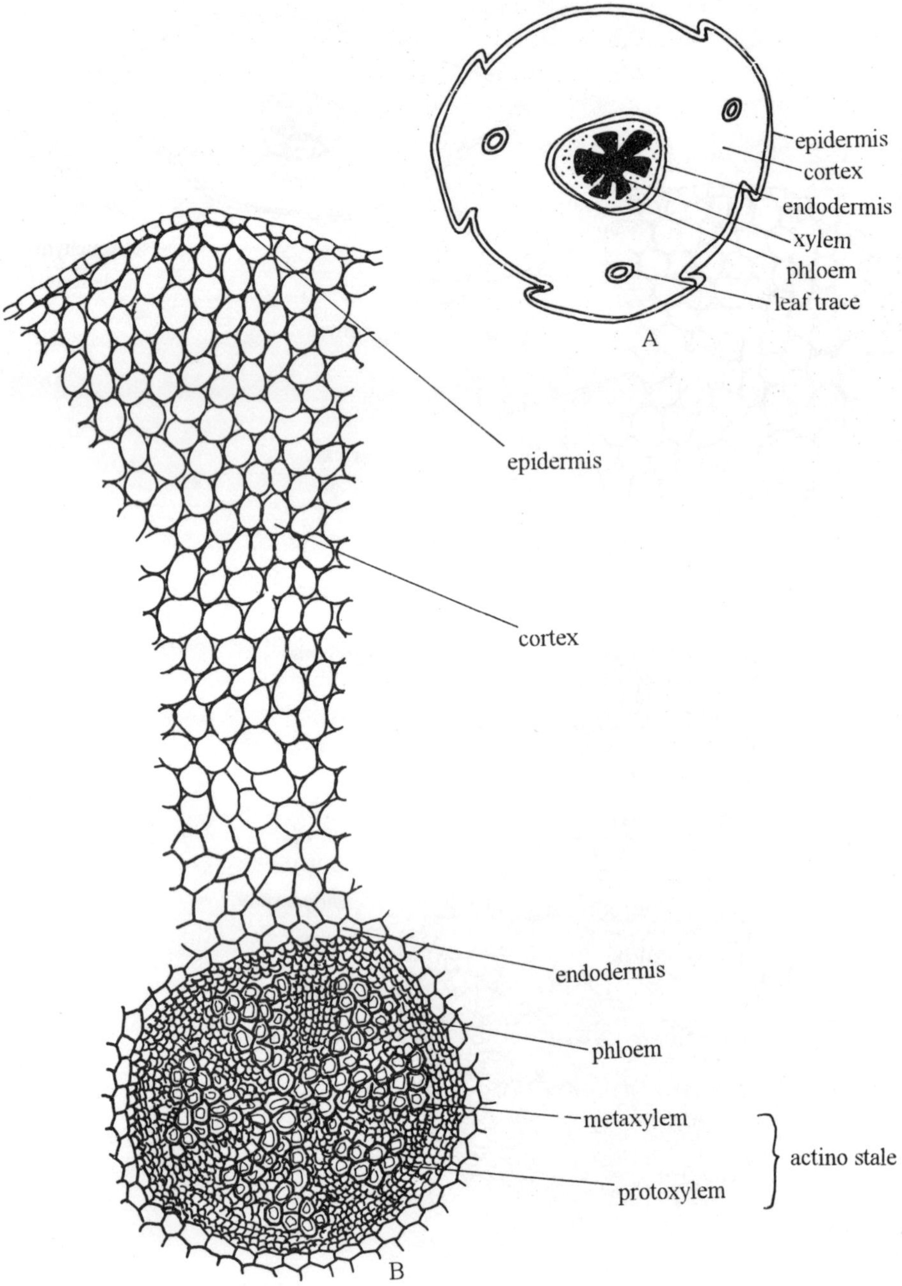

Fig. 11.7 *Lycopodium* : Internal Structure of Stem of *L. serratum*
A. Ground Plan, B. A Sector Enlarged

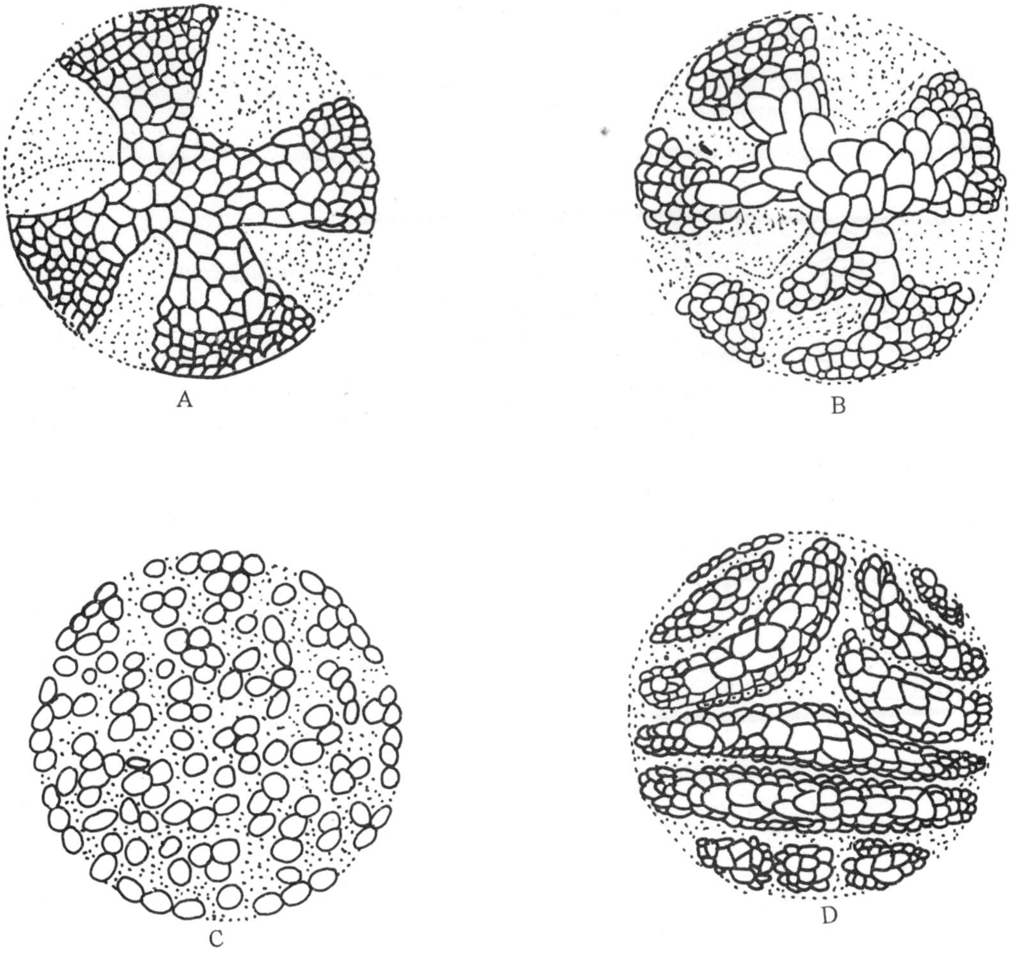

Fig. 11.8 *Lycopodium* : Various Types of Stelar Organisation in Stem
A. *L. serratum*, B. *L. lucidulum*, C. *L. cernuum* and D. *L. volubile*

6. Leaves arise all along the stem and are arranged in a spiral manner. They are microphyllous.

7. There is a clear distinction between foliage leaves and sporophylls.

8. The foliage leaves have a central unbranched vein. They are small, eligulate and sessile with a broad base.

9. The leaves in *L.phlegmaria* may sometimes be 2-3cms long.

10. The reproductive structures (sporangia) accumulate at the tips of branches in the form of strobili. The strobili are also branched.

Morphological features of Lycopodium cernuum

1. The plant belongs to the sub genus Rhopalostachya.

2. The plant body (sporophyte) is differentiated into root, stem and leaves.

3. The plant body is semi erect to erect in habit.

4. The plant grows in the soil and roots arise all along the length of the stem. They are adventitions.

5. The stem is branched and branching is dichotomous.
6. Leaves arise all along the stem in a spiral fashion. They are eligulate.
7. The leaves are extremely small and range in size from 2-10mm.
8. The strobili are well organized and occupy the tips of branches. They rarely exeed 1-1.5cms in length.

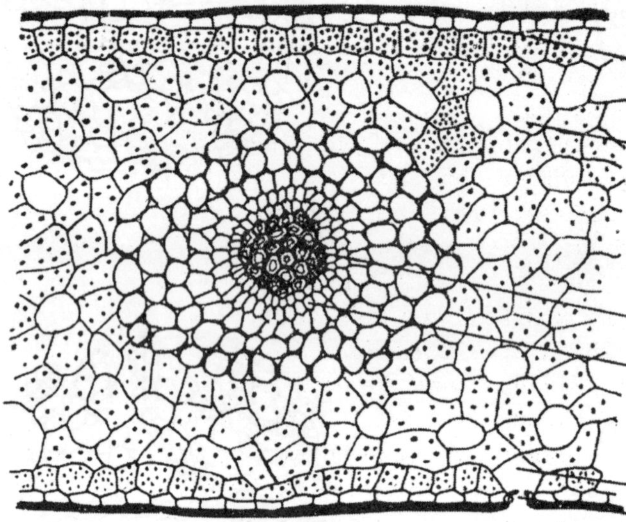

Fig. 11.9. *Lycopodium* : Internal Structure of Leaf (*L. phlegmaria*)

Morphological featrues of L.wightianum

1. The sporophytic plant body is distinguished into root, stem and leaf.
2. The stem (rhizome) is horizontal and it produces upright branches.
3. The adventitious roots arise all along the stem.
4. The plant belongs to the sub genus Rhopalostachya.
5. Leaves are small, eligulate, microphyllous (larger than those of *L.cernuum*) and spirally arranged.
6. Strobilus is clearly demarcated from the rest of the plant body.
7. Sporophylls are much smaller than those of foliage leaves and are compactly arranged.
8. The strobili are also branched.

Morphological features of L.hamiltonii

1. The plant belongs to the sub genus Urostachya.
2. The stem is erect and roots are produced at the basal region only.
3. The branching of the stem is dichotomous.
4. The leaves are small (but larger than those of L.cernuum), eligulate and microphyllous.
5. There is no distinction between the sporophylls and foliage leaves. No strobilus is found.
6. Some of the leaves bear the sporangia.

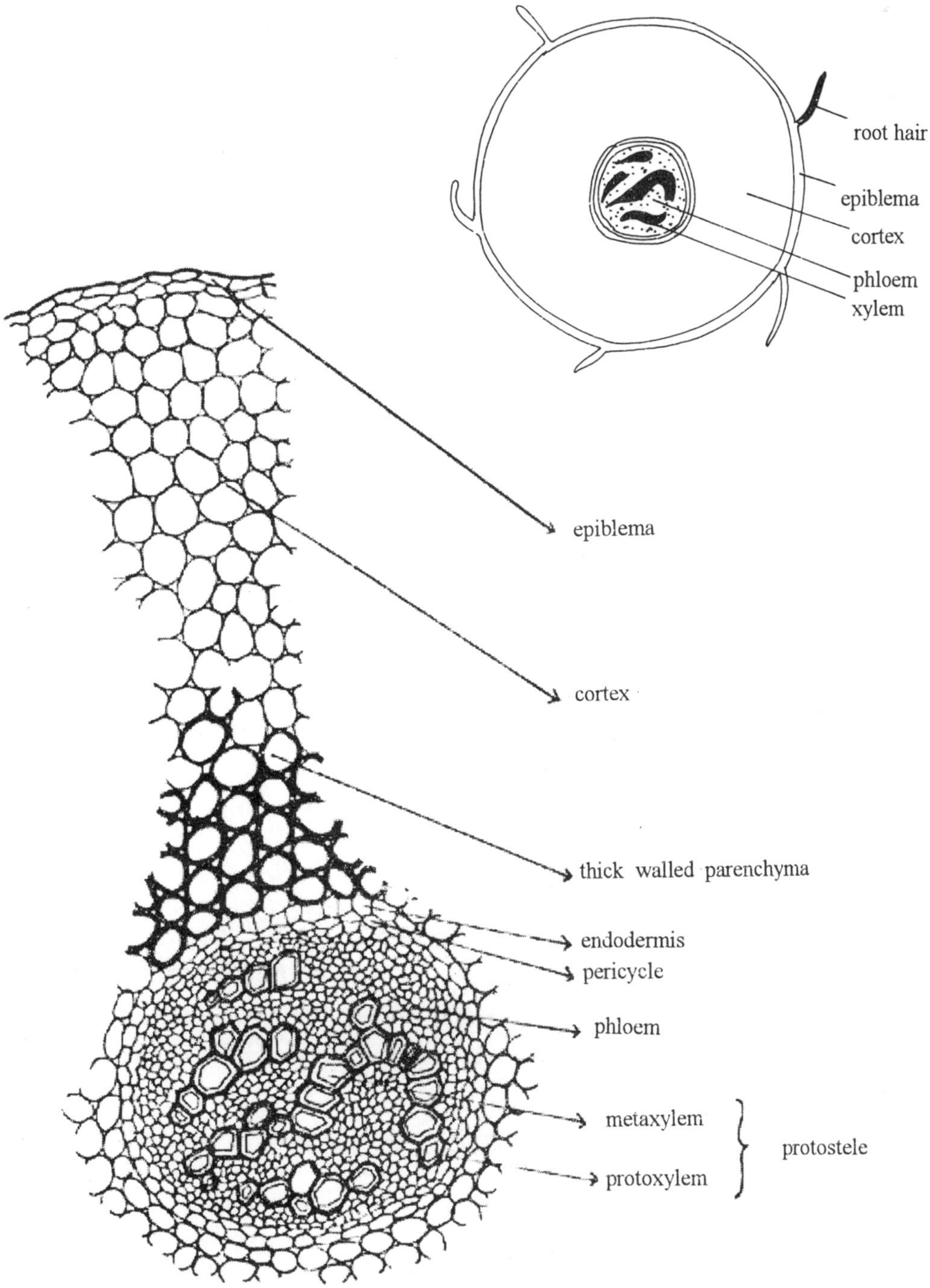

Fig.11.10. *Lycopodium* : Internal Structure of Root (T.S.)

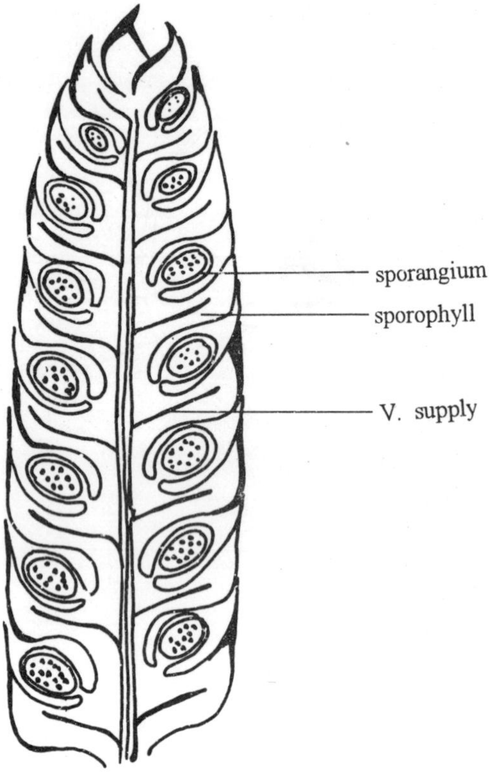

A

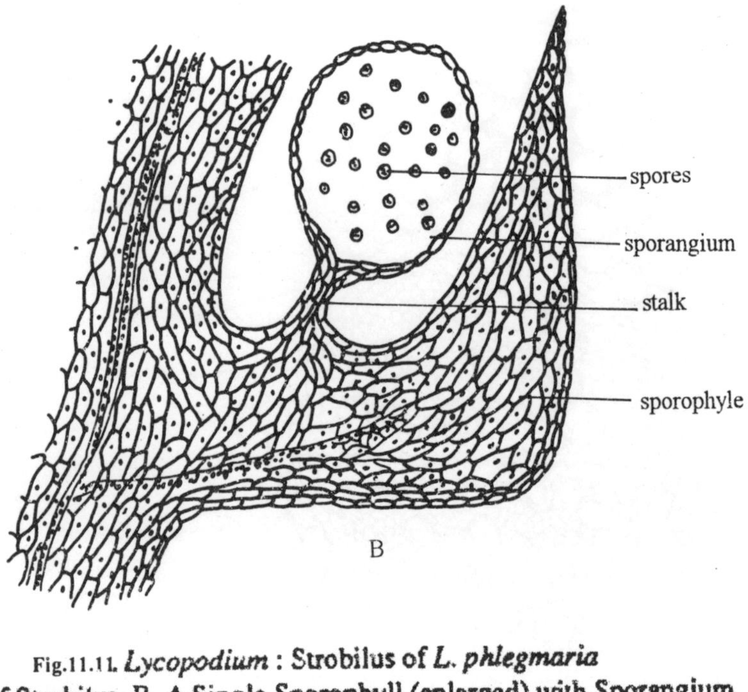

B

Fig.11.11. *Lycopodium* : Strobilus of *L. phlegmaria*
A. L.S. of Strobilus, B. A Single Sporophyll (enlarged) with Sporangium

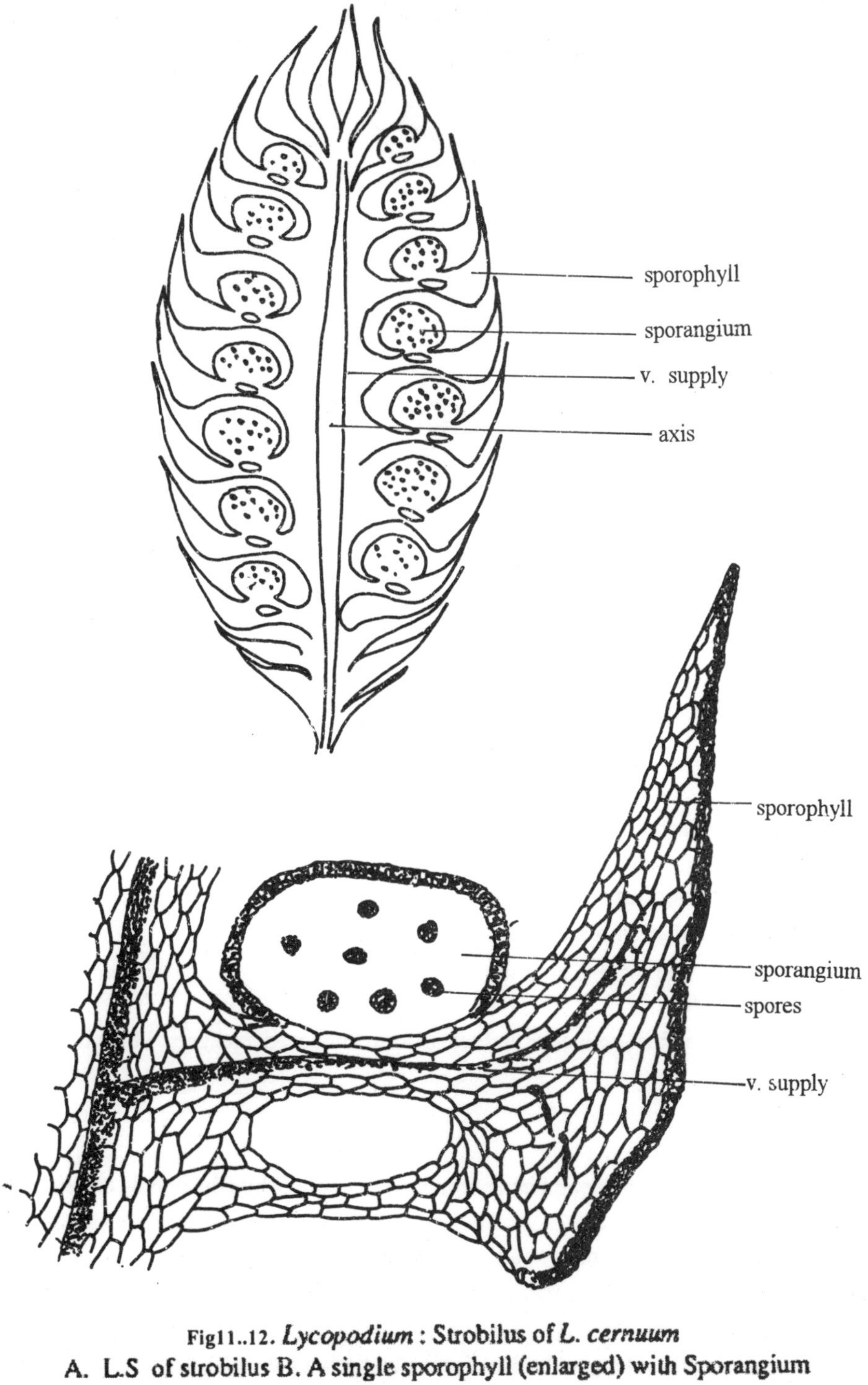

Fig11..12. *Lycopodium* : Strobilus of *L. cernuum*
A. L.S of strobilus B. A single sporophyll (enlarged) with Sporangium

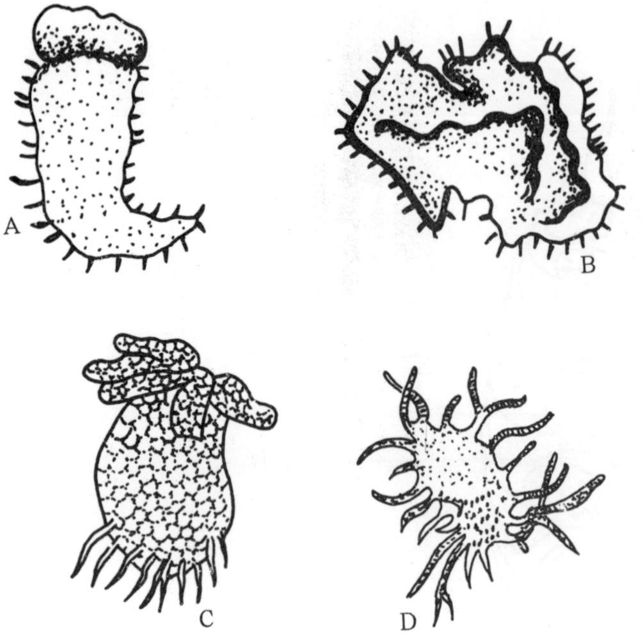

Fig.11.13. *Lycopodium* : Gametophytes
A. *L. complanatum*, B. *L. clavatum*, C. *L. cernuum*, D. *L. phlegmaria*

Mophological features of L.selago

1 to 6 of L.hamiltonii.

7. The leaves at the basal region are sterile. From the basal region upwards sterile and fertile regions alternate.

8. This type of sporangial distribution is called :Selago condition".

STEM ANATOMY

Lycopodium phlegmaria

1. A T.S of stem shows three regions. Epidermis, Cortex and Stele.

2. Epidermis is made up of a single layer of cutinised cells. Occasionally stomata are present in the epidermis.

3. Cortex is made up of parenchyma cells. The cells are polygonal and thin walled.

4. Stele is surrounded by a layer of endodermis followed by pericycle.

5. Stele is protostele. The xylem is stellate having a number of rays. The protoxylem is situtated at the tips of the rays (exarch). Phloem completely surrounds the xylem. This type of a protostele is called actinostele.

6. There is no pith.

Lycopodium cernuum

1 and 2 same as in *L.phlegmaria.*

3. Cortex is heterogenous. It is divided into three regions. Outer cortex is parenchymatous, middle cortex is sclerenchymatous, while inner cortex is again parenchymatous.

4. Stele is surrounded by endodermis and pericycle.

5. The stele is a protostele and there is no pith. The xylem and phloem are intermixed. Hence the stele is known as a mixed protostele.

6. Xylem is exarch and has annual, reticulate and scalariform thickened tracheids. Phloem has seive cells and phloem parenchyma.

Lycopodium wightianum

1 and 2 same as in *L.phlegmaria.*

3. Cortex has three regions. The outer and inner cortex are sclerotic while the middle cortex is parenchymatous.

4. Stele is surrounded by a layer of endodermis and pericycle.

5. Stele is a protostelic plectostele. Xylem is divided into a number of parallely arranged plate like structures. It is exarch.

6. Phloem not only surrounds xylem but is also distributed between the plates.

7. There is no pith.

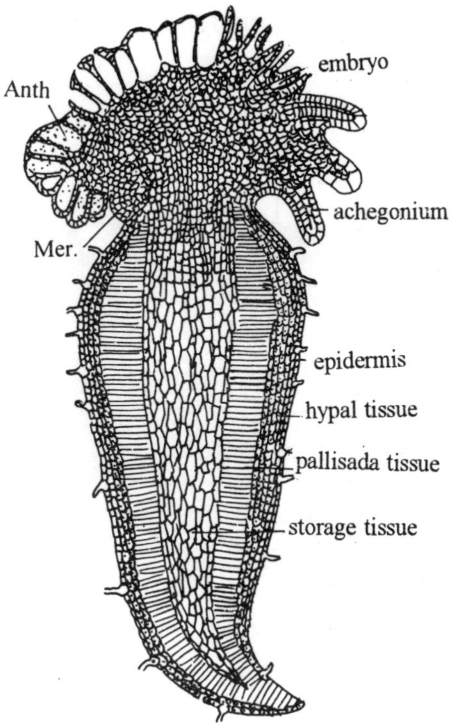

Fig.11.14. *Lycopodium* : Long Section of the Gametophyte of *L. complanatum*

Anatomy of leaf

1. T.S. of leaf shows epidermis, mesophyll and a concentric vein.

2. Upper and lower epidermis made up of a single layer of cells each.

3. Leaves may have stomata only on the lower side(hypostomatic) or have stomata on both surfaces (amphistomatic).

4. Mesophyll is undifferentiated. Cells may be angular or rounded.

5. The vein is concentric. Xylem is surrounded by phloem.

6. Vascular bundle may or may not be surrounded by endodermal sheath.

Anatomy of root.

1. A T.S of root shows epidermis cortex
and stele.

2. Epidermis is single layered and is composed of thin walled cells.

3. Unicellular root hairs arise from the epidermal cells.
4. Cortex is made up of many layers of parenchyma cells. In mature roots outer cortex ma· ·ave thick walled cells.
5. Stele is surrounded by a layer of endodermis and pericycle.
6. Stele is a protostele. Xylem may be monarch to polyarch.
7. When the xylem is diarch it is crescent shaped.

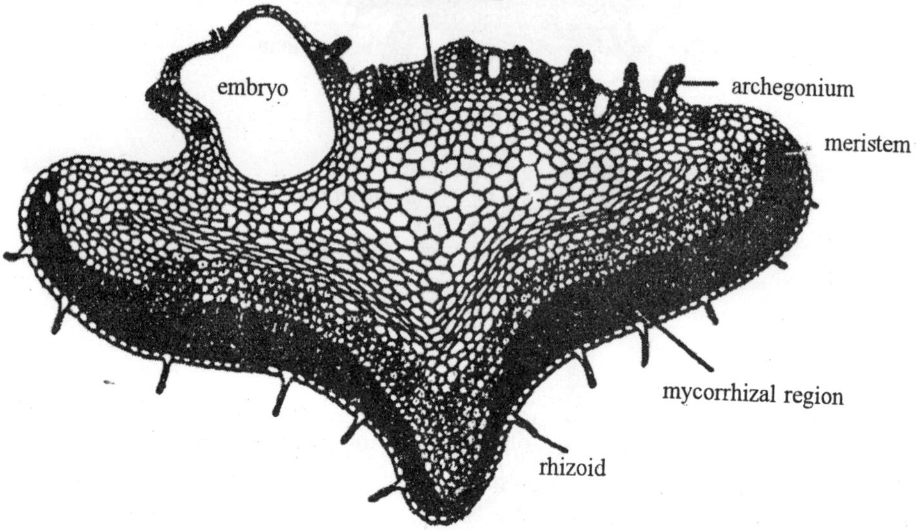

Fig.11.15. *Lycopodium* : L.S. of Gametophyte of *l. clavatum*

Stucture of the strobilus.

1. The strobilus or cone terminates the branches.
2. Each strobilus has a strobilar axis on which are arranged sporophylls in a spiral fashion.
3. Each sporophyll bears a single sporangium on its adaxial surface. The sporophyll are eligulate.
4. The sporangium has a short multicelluar stalk.

Long section of the strobilus.

1. A long section of the strobilus at the median region shows a central axis and the sporophylls on either side. Vasculature can be seen in the cone axis.
2. Each sporophyll is cut length wise and the top is folded up.
3. The sporophylls are eligulate.
4. Each sporophyll bears a single sporangium on its adaxial surface.
5. Each sporangium has a multicellular stalk.
6. The wall of the sporangium also is multicellular.
7. A number of spores (spore mother cells, or spore tetrads are found when the sporangium is young).
8. All the spores are of the same type (homosporous).
9. In a mature sporangium certain thin walled cells called stomial cells are present marking the future line of dehiscence.

Structure of the gametophyte.

Usually three types of gametophytes are found in *Lycopodium* based on life span, growth pattern and nutrition.

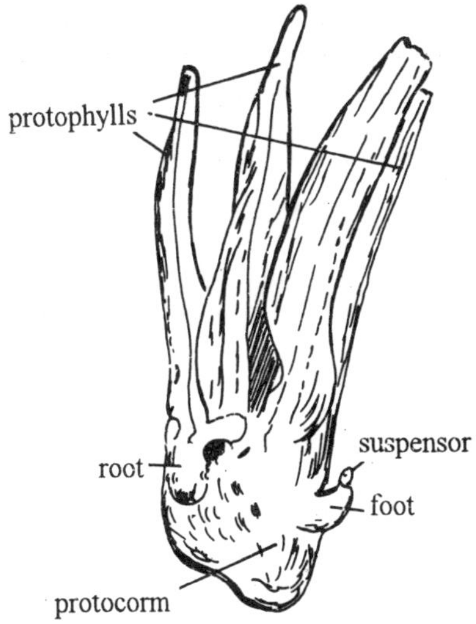

Fig.11.16. Protocorm of Lycopodium

First type (Eg: *Lycopodium cernuum, L.innundatum* etc)

1. The gametophyte grows up to only one season.
2. The size of the gametophyte is limited. It ranges from 2-3 mm.
3. The gametophyte usually grows above ground but the basal portion is buried in soil.
4. The upper part of the gametophyte has a number of small finger like projections forming the crown.
5. The finger like projections help in photosynthesis.
6. Basal portion of the gametophyte has a mycorrhizal association. Rhizoids also arise from this region.
7. Sex organs (archegonium and antheridium) are found in between the photosynthetic lobes.
8. Nutrition of the prothallus is both photosynthetic as well as saprophytic.

Second type (Eg: *L.clavatum, L.complanatum* etc).

1. Gametophytes are long lived. They grow for many seasons.
2. In size the gametophytes range from 1-2cms. They may be top shaped, disc shaped or carrot shaped.
3. The gametophytes are brownish to yellowish in colour and completely subterranean.
4. A number of rhizoids are produced from the lower surface.
5. Anatomically the gametophyte has only parenchyma cells but they vary in size, shape and structure.
6. Tissue diversificaion is seen in the lower cylindrical portion only, while the upper portion has got only polygonal parenchyma cells.
7. Central region has vertically elongated cells constituting the storage tissue.
8. Surrounding the storage tissue is the palisade tissue. The cells at this region are closely packed.
9. External to the palisade tissue is the hyphal tissue. The cells have fungal hyphae within them.
10. Outside the hyphal tissue is the epidermis. Rhizoids are produced from some of the epidermal cells.
11. Sex organs are borne on the upper surface of the gametophyte.

Third type (Eg: *L.phlegmaria*)

1. Gametophytes are small. They grow for only one growing season.
2. Nutrition is saprophytic. Parenchyma cells have the association of fungal hyphae in them.
3. Gametophytes are usually found on rotting barks close to the soil.
4. The gametophyte is an irregularly shaped tuber.
5. From this central tuber arise several cylindrical branches. The length of the branches varies from 1-6mm.
6. Sex organs (antheridia and archegonia) are borne on the upper surface of these branches.

Systematic position

LYCOPODIALES
LYCOPSIDA
ELIGULOPSIDA
LYCOPODIALES
LYCOPODIACEAE
LYCOPODIUM

Fig.11.17 *Lepidodendron* : Reconstruction of *L. obovatum*

LEPIDODENDRON

1. Like other ancient lycopods, *Lepidodendron* was also tree like in habit.
2. In general appearance it was not unlike that of present day *Lycopodium*.
3. But in size the genus enormously exceede the herbaceous *Lycopodium*.
4. The petrified trunks were sometimes as long as 100 feet.
5. Judging from this it may be safely assumed that the plant reached a height of over 120 feet.
6. The stem was erect and did not branch upto some distance from the ground.
7. The branching of the stem was typically dichotomous.
8. The ultimate dichotomies produced the leaves.
9. The branches and the foliage formed a sort of crown at the apex of the stem.
10. The leaves which clad the young stems and branches were acicular or linear in shape having a length of 5-9 inches.
11. The arrangement of the leaves was spiral or very rarely they showed a whorled arrangement.
12. The leaves were ligulate.
13. Each leaf had a single vein with the stomata situated in two bands on the ventral surface.
14. The leaves were deciduous.
15. Upon abscission a flat rhomboidal scar persisted on the stem resembling a small cushion.
16. The base of the stem had a stigmarian type of root system.

Internal Structure

Stem:

1. In majority of thespecies, secondary growth is characteristic.

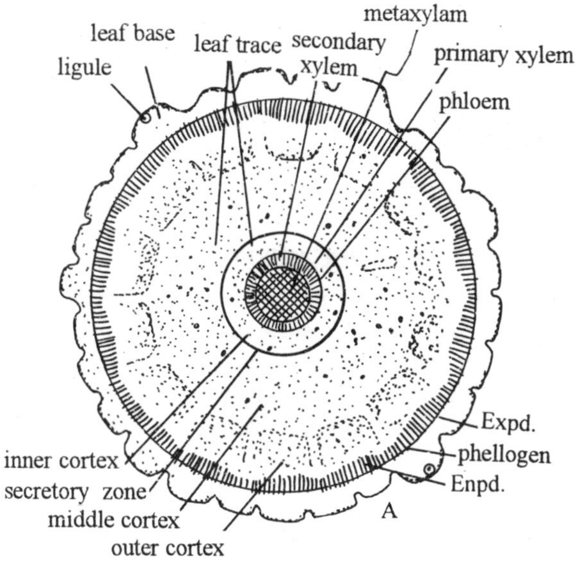

Fig. 11.18. ***Lepidodendron*** **: T.S. of Stem of *L. vasculare***

2. But some species seem to lack a cambium.
3. A transverse section of the trunk of L.vasculare shows three regions, stele, cortex and a periderm.
4. In the primary structure there was an epidermis but soon i.e., even before the initiation of secondary growth in the vasculature, it was replaced by the periderm.
5. The periderm was produced by a phellogen which produced phelloderm towards the interior and phellem towards the exterior.
6. The outline of the bark was wavy due to the presence of leaf bases.
7. Internal to this was the cortex.
8. It consisted of four regions viz. (1) outer cortex consisting of alternating bands of sclerotic and parenchymatous cells, (2) a middle cortex having a homogenous mass of parenchyma cells. Interspersed with the parenchyma cells were the leaf traces, (3) secretory zone consisting of glandular cells which were filled with a dark coloured substance. They pprobably secreted the waxy material which covered the surface of the stem, (4) An inner cortex having parenchyma cells.
9. The central region of the stem was occupied by the stele which was either protostelic or siphonostelic.
10. The protoxylem was exarch and polyarck.
11. In many species (*L. vasculature*) there was a secondary growth initiated by the cambium.
12. This produced secondary xylem to the interior and secondary phloem to the exterior.
13. The cambial activity was not uniform, as a result there was a tendency for the formation of an eccentric vascular ring.

14. The secondary xylem had radial rows of tracheids separated by *xylem rays* which were uniseriate.

Leaf:

1. Anatomically the leaves showed a single vascular bundle flanked on either side by *parichnos cavities*.
2. These are believed to the aerating organs.

Systematic position

LYCOPSIDA
LIGULOPSIDA
LEPIDODENDRALES
LEPIDODENDRACEAE
LEPIDODENDRON

LEPIDOCARPON

1. Lepidocarpaceae had many striking features not found in other families of the order.
2. The members attained a high degree of specialization in their reproductive features representing perhaps the climax of adaptation of the palaeozoic lycopods.
3. There are three genera referable to this family, all of them being known only from their fructifications.
4. Of the three genera *Lepidocarpon* is the most interesting.
5. Though *Lepidocarpon* is known only from its strobilus it has been widely held that it could have been borne on a *Lepidophlois* type of stem.
6. The fossils of *Lepidocarpon* have been obtained from the coal strata in England and U.S.A.
7. The strobili were mososporangiate though the plant was heterosporous.
8. The microsporophylls were like those of *Lepidostrobus*.
9. Each microsporophyll was peltate and had an elongate microsporangium.
10. The megasporophylls bore elongate microsporangia on their adaxial surface.
11. The jacket of the megasporangium extended at the apex to form a projection.
12. The sporangium was enveloped by the flaps of a tissue which is termed as integument.
13. But it is doubtful how far this terminology is appropriate.
14. Because in reality this inegument is nothing but an upturning of the megasporophyll (Amold 1947; Hosins and Cross 1941) though many people (Scot 1901, Schopf 1941) believe it to be a special outgrowth artising from the megasporophyll.
15. There were four spores in a sporangium of which only one developed into the female gametophyte.
16. The size of the gametophyte and the fact as to whether it completely or incompletely filled the sporangium are specific characters.
17. Andrews and Pannel (1941) have reported several fertile megagametophytes with archegonia at the apex.
18. The megasporangial cavity also showed many spores of which at least some of them were the microspres of *Lepidocarpon*.
19. This suggests that *Lepidocarpon* had something like an incipient pollination which is also seen in the extant genus *Selaginella*.
20. The one unique feature of *Lepidocarpon* which is not seen in any lycopods extinct or extant was the permanent retention of the megagametophyte within the sporangium thus marking a significant step towards the evolution of seed habit.
21. The megasporophylls were, ultimately liberated and on falling upon a suitable substratum the embryo developed into the sporophyte.
22. It will be very interesting to analyse whether *Lepidocarpon* could be regarded as having attained a seed habit.
23. Most of the attributes of *Lepidocarpon* - an integument, a permanent retention of the female game-

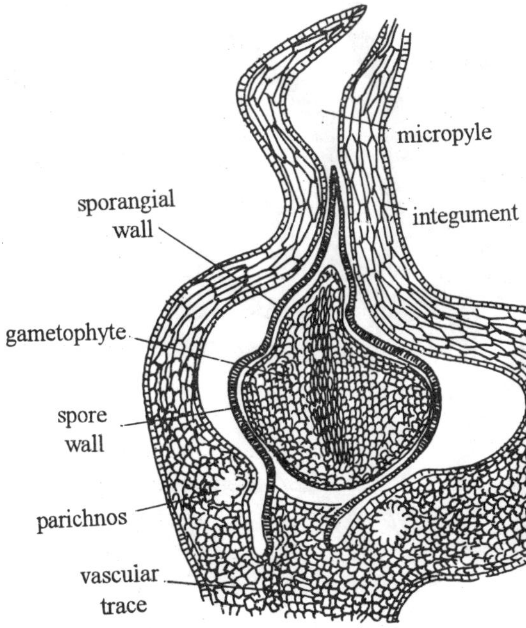

Fig.11.19. *Lepidocarpon* : L.S. of Seed (diagrammatic)

tophyte, development of only one spore in the megasporangium, incipient pollination, all point out towards the seed like habit.

24. But a closer scrutiny reveals that all the seed like features are only superficial, for, the so called integument is a part of the megasporophyll and not of the megasporangium.
25. Further there is no formation of pollen tube, no seed coat and in all probability the embryo did not have dormancy characteristic of true spermatophytes.
26. All these clearly show that though Lepidocarpaceae apparently possessed the attributes of a seed-habit (particularly in the permanent retention of the female gametophyte but the additional feature to support such a retention were not sufficient) it did not reach the level of a seed habit and hence it could at best be regarded as a pseudospermatophyte.

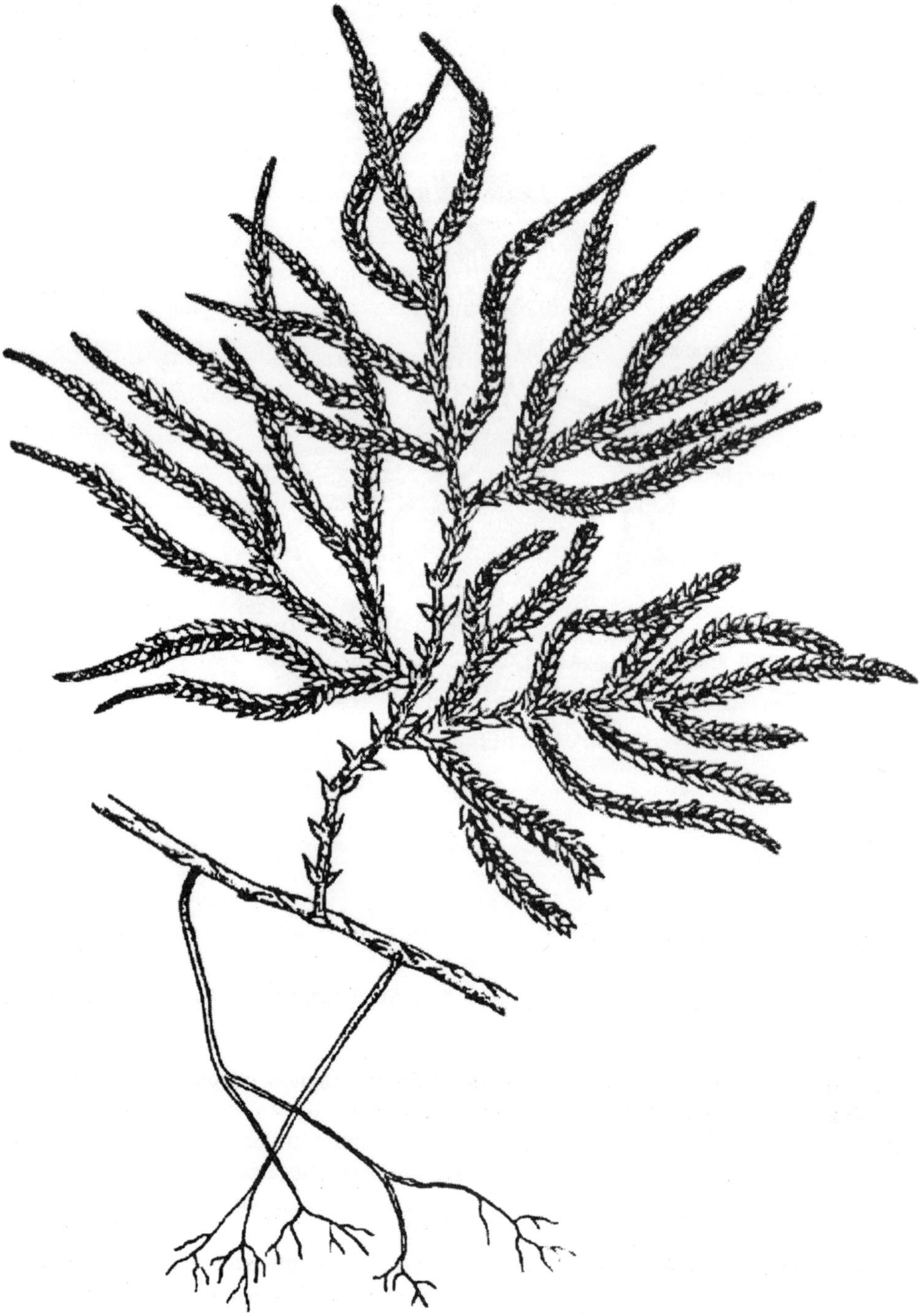

Fig. 11.20 Habit of Selaginella martant

Systematic position

LYCOPSIDA

LIGULOPSIDA

LEPIDODENDRALES

LEPIDOCARPACEAE

LEPIDOCARPON

SELAGINELLA

Selaginella is commonly called the spike moss. There are about 700 species in this genus with world wide distribution.

Practical work to be done

1. Study the habit of the plant.
2. Cut transverse sections of stem, root and leaf and observe under the microscope.
3. Observe the structure of the strobilus from a long section of the slide.
4. Observe the structure of the gametophyte from the slide.

Structure of the plantbody

1. The plantbody is herbaceous with a soft branched stem.
2. The branching of the stem may be dichotomous or lateral.
3. Leaves are found on the stem. These are small measuring a few mm in size.

Fig. 11.21. *Selaginella* : Habit of *S. kraussiana*

4. Leaves are spirally arranged.

5. In some species leaves are *dimorphic* i.e of two sizes. They are arranged in four rows; of these two rows are smaller and two rows are larger.

6. Leaves are simple and lanceolate to oval in shape.

7. In all the species leaves are ligulate. Ligule is a membranous outgrowth arising from the adaxial surface of the leaf.

8. Ligule has a basal region called glossopodium placed in a depression called ligular pit.

9. Ligule varies in shape. It may be tongue shaped, fan shaped or lobed.

10. The function of the ligule is not known. It may be secretory secreting water or mucilage.

11. Certain curious branches arise from the ramifications of the stem. These are called rhizophores. One to many rhizophores arise from a single point.

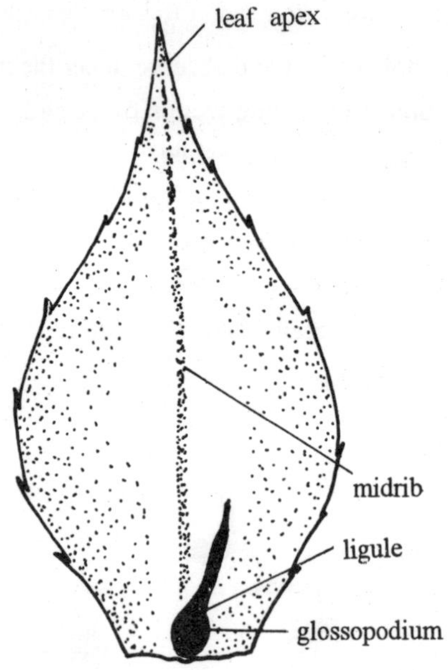

Fig. 11.22. *Selaginella* : Surface View of Leaf showing Ligule

12. The morphological nature of the rhizophore is controversial. It is regarded as a root, stem, organ of a special kind (organ sui generis) etc.

13. Rhizophores arise from special meristems called angle meristems.

14. Root system is adventitous. Roots may arise directly on the stem or on rhizophores. Branching of the root is dichotomous.

15. Reproductive structures generally aggregate at the apices of the branches to form the strobilus.

Classification of the genus

The genus is classified into two - *Homeophyllum* and *Heterophyllum*. The following table lists the differences between the two sections.

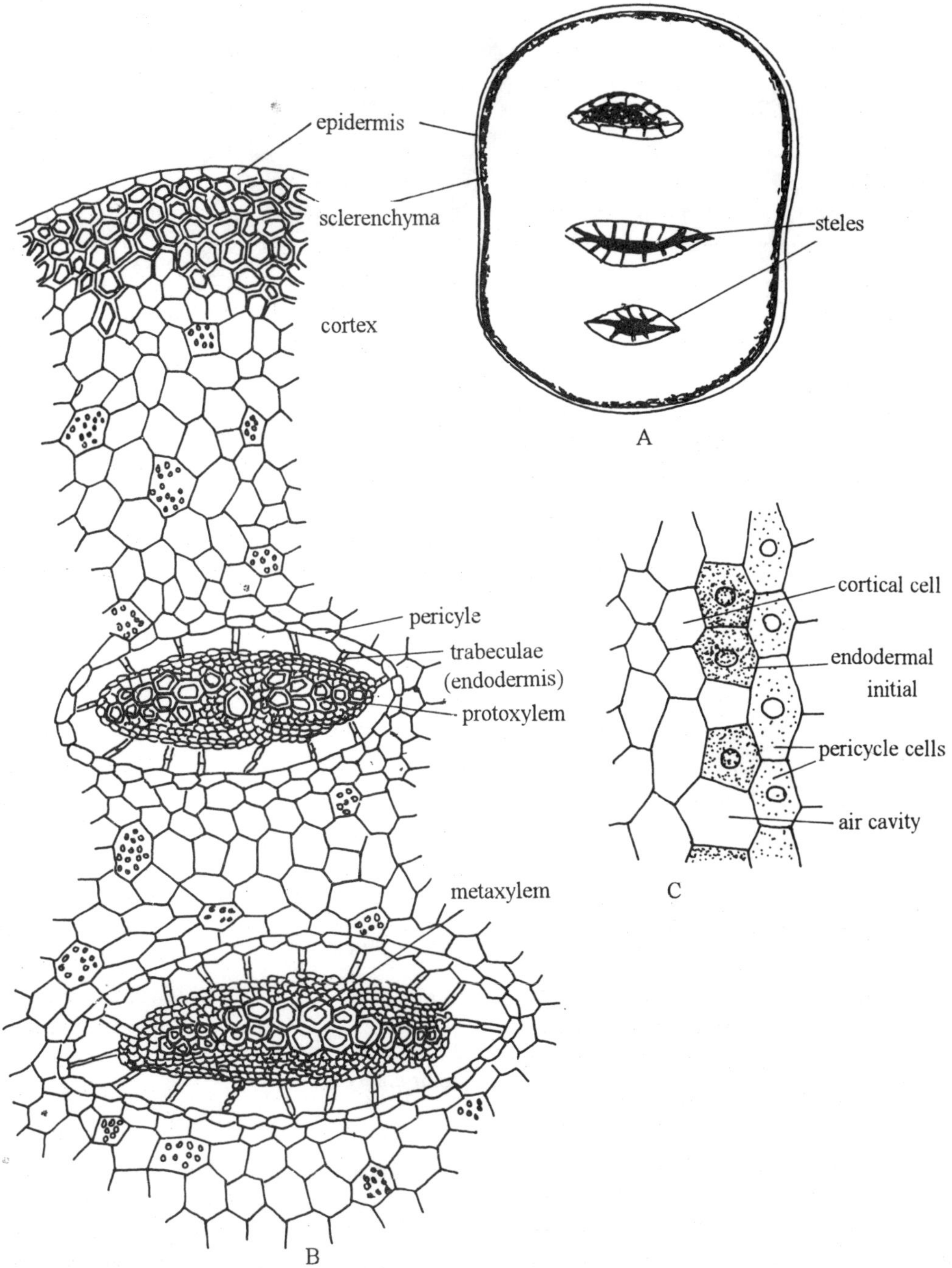

Fig. 11.23. Selaginella: Internal Structure of Stem
A. Ground Pland, B. A Sector enlarged, C. Development of Trabeculae (early stages);
Note the normal endodermal cells

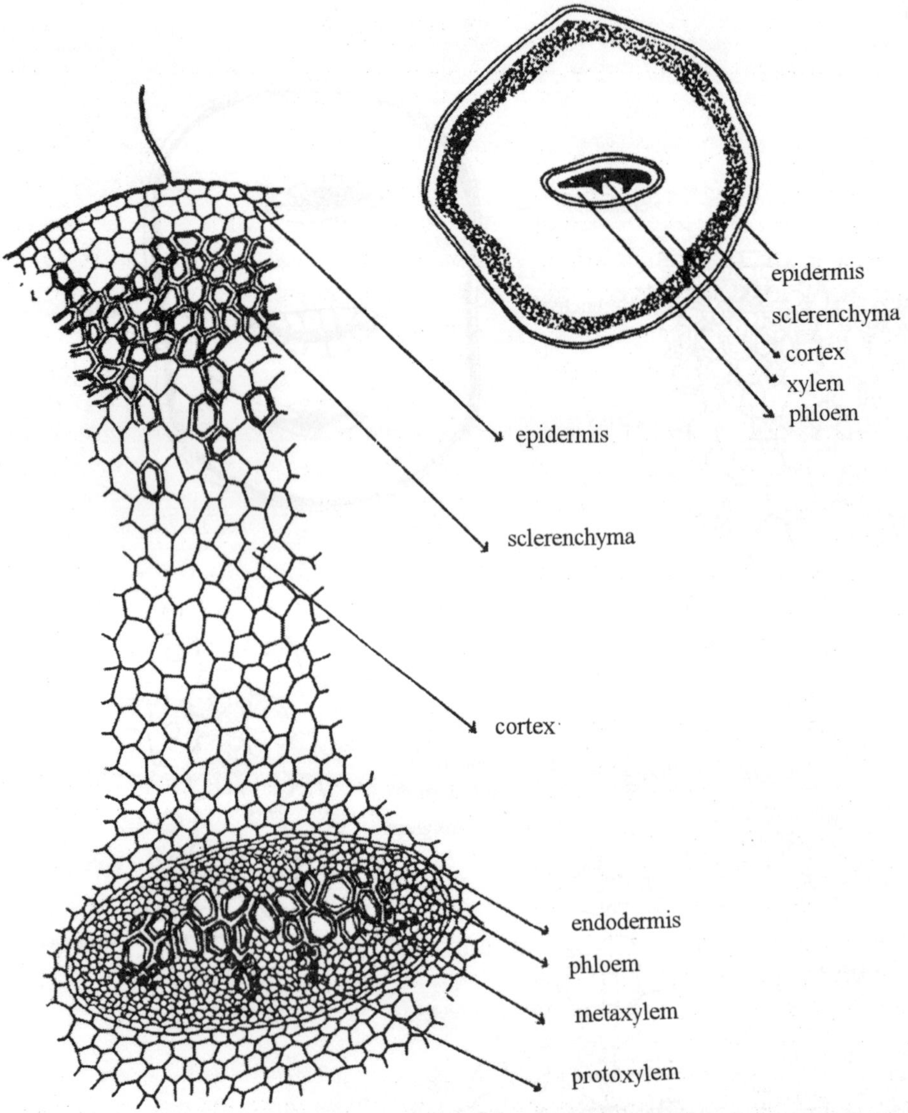

Fig. 11.24. *Selaginella* : Internal Structure of Root
A. Ground Plan, B. A Sector enlarged

	Homeophyllum	Hetrophyllum
1. **Examples**	S.rupestris, S.rugosa etc	S.kraussiana, S.lepidophylla etc.
2. **Leaves**	One type (isophyllous)	Dimorphic leaves.
3. **Phyllotaxy**	Spiral	Four rows.

Internal structure of stem

1. A transverse section shows three main regions namely epidermis, cortex and stele.

2. Epidermis is single layered. The cells are closely packed. Stomata are not present.

3. Cortex is parenchymatous completely (*S.flagellata*) or partially (*S.kraussiana*), when there will be a hypodermal sclerenchyma region.

4. Stele is protostelic. It has an outermost endodermis followed by pericycle and vasculature.

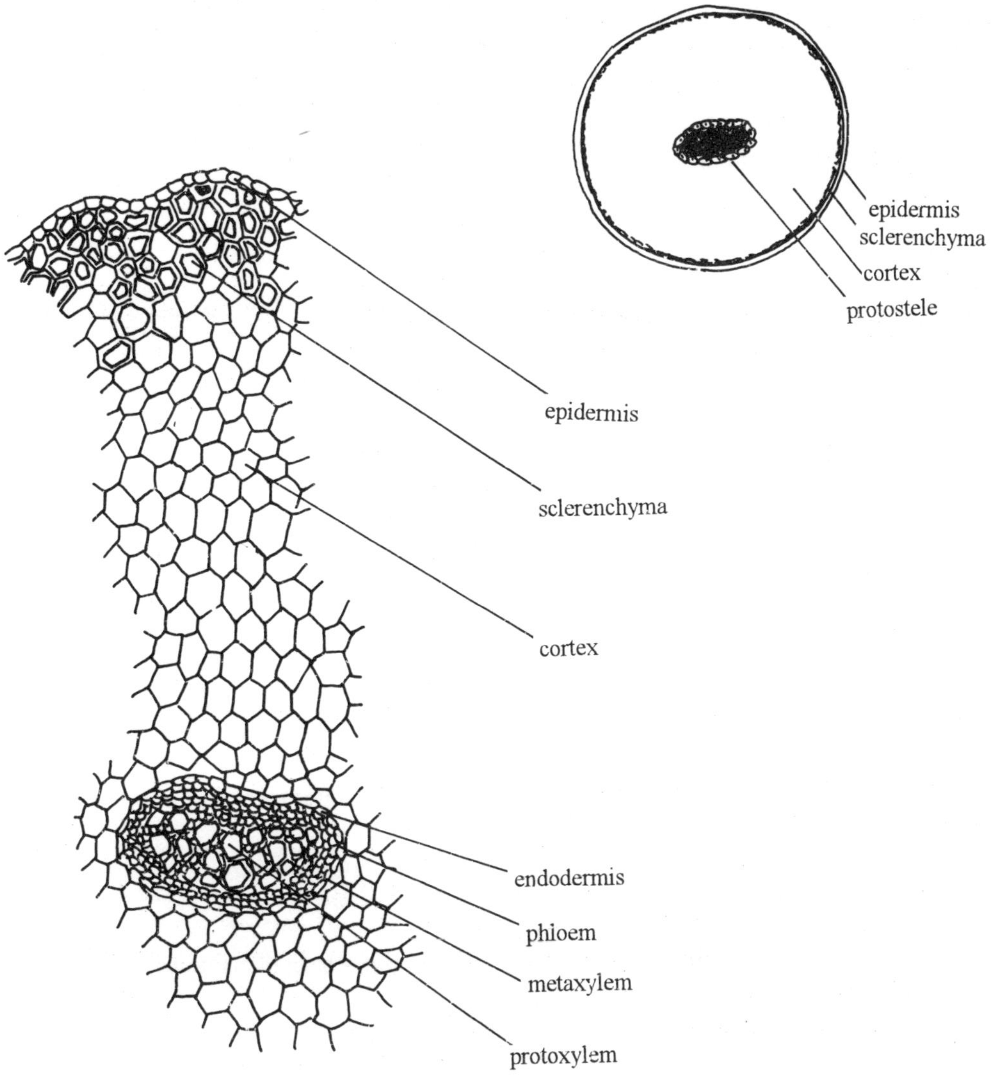

Fig. 11.25. ***Selaginella*** **: Internal Structure of Rhizophore A. Ground Plan, B. A Sector enlarged**

5. Endodermis is called trabeculae. There is a lacuna between the cortex and the stele.

6. The trabecular cells (endodermis) are radially elongated and are transversely divided.

7. The stele is normally polystelic. Usually 2-3 steles are present.

8. The xylem is exarch and diarch or rarely polyarch.

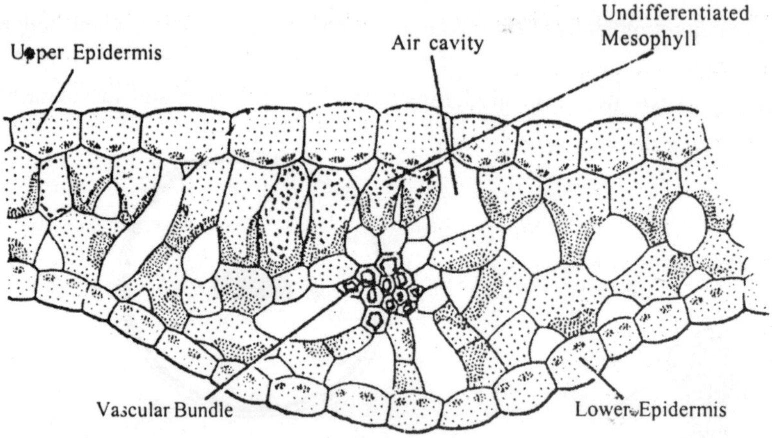

Fig.11 .26 Selaginella: T.S. of Leaf

9. Xylem has tracheids with annualar or scalariform thickenings.

10. Rarely vessels are also found surrounding the phloem parenchyma. Pericycle surrounds the phloem.

Internal structure of root.

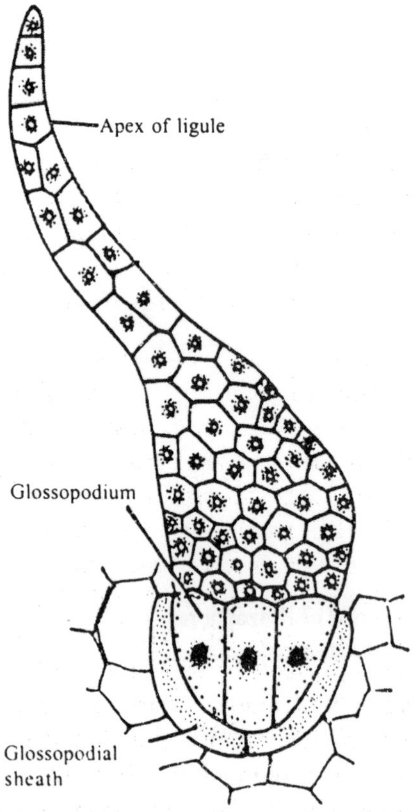

Fig.11.27. Selaginella: A Ligule enlarged

1. A transverse section of the root shows an epidermis cortex and stele.

2. The epidermis is made up of one layer of cells. Some of the epidermal cells give rise to root hairs.

3. Cortex may be wholly parenchymatous, wholly sclerenchymatous or partly parenchymatous and partly sclerenchymatous.

4. The stele is protostelic with monarch and exarch xylem. Surrounding the xylem is the phloem.

5. Pericycle is one to three layered. Endodermis may or may not be well defined.

Internal structure of rhizophore

1. A T.S. of rhizophore shows an epidermis cortex and stele.

2. Epidermis is single layered made up of compactly arranged cells.

3. Cortex is parenchymatous made up of several layers of cells.

4. The stele is surrounded by non trabecular endodermis followed by a layer of pericycle.

5. Stele is protostelic. There will be a single stele in the rhizophore even though the stem may be polystelic.

6. Xylem is exarch and monarch.

Internal structure of the leaf.

1. Anatomically the leaf shows epidermis, mesophyll and a vascular bundle.

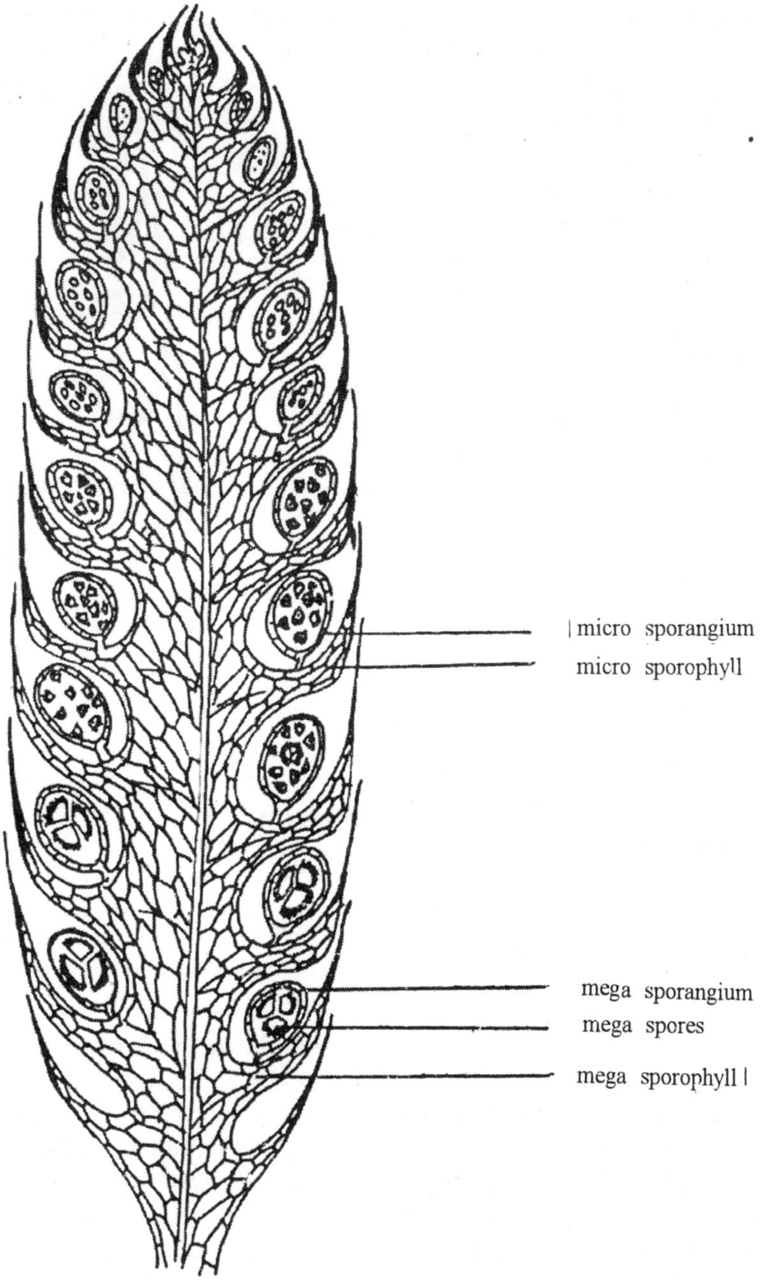

Fig. 11.28. Selaginella: L.S Strobilus (megasporangia at the base and microsporangia at the top

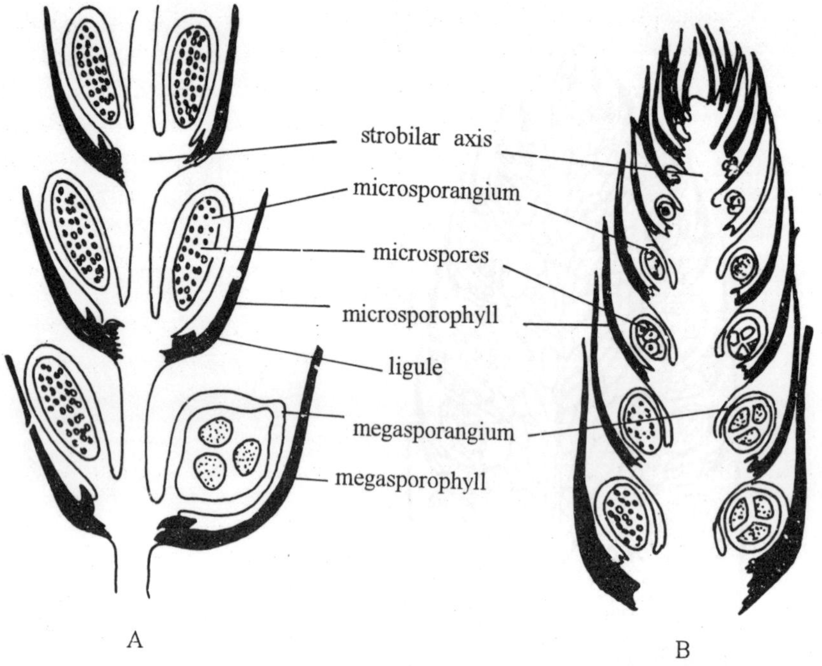

Fig.11 .29. Selaginella: L.S. of Strobili showing variety in Sporangial Arrangement
A.S. kraussiana, B. S. inaequalifolia

2. The upper and lower epidermis are made up of a single layer of cells.

3. Leaves may be hypostomatic (stomata only on the lower epidermis) or amphistomatic (stomata present on both the epidermal layers).

4. Mesophyll is undifferentiated. It is composed of isodiametric cells with intercellular spaces.

5. Mesophyll cells contain variable number of chloroplasts. In some cases there is a single cup shaped chloroplast in each cell.

6. Vascular bundle is concentric having a central xylem surrounded by the phloem.

Anatomy of ligule.

1. A transverse section reveals many thin walled cells with highly vacuolated protoplasts in the glossopodium.
2. The glossopodium is surrounded by a sheath called the glossopodial sheath.
3. Above the glossopodium the ligule narrows down into a long process. The cells at this region are polygonal and filled with dense cytoplasm.

Structure of the strobilus (Long section)

1. A long section of the strobilus shows a central axis on which are spirally arranged the sporophylls.

2. Each sporophyll bears a single stalked sporangium arising in its axil or a little above.

3. The sporangia are of two types megasporangia and microsporangia. Megasporangia have less number of larger spores, while microsporangia have large number of smaller spores.

Structure of a mature micro sporangium

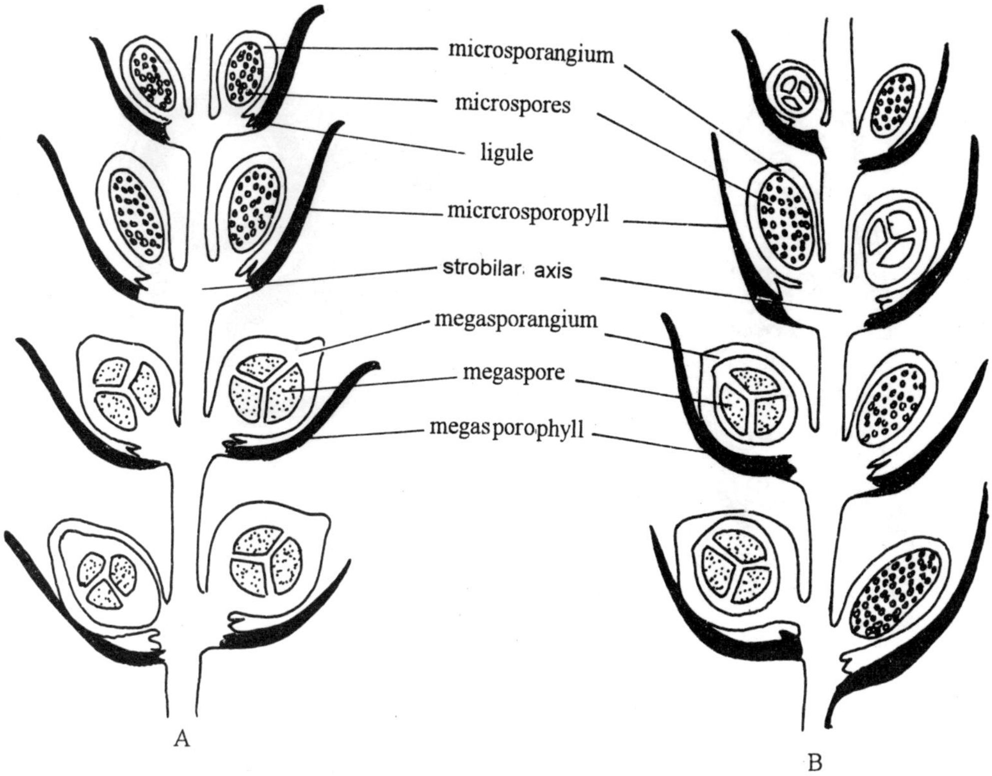

Fig.11 .30. Selaginella: L.S. of Strobili showing variety in Sporangial Arrangement (contd...)
A.S. rupestris, B. S. martensii

1. The sporangium is red, yellow or brown in colour.
2. The stalk of the sporangium is short.
3. The wall of the sporangia is two layered. The outer wall has columnar chlorophyllous cells. The inner layer has thin walled cells.
4. Sometimes a thin tapetal layer is persisting in the mature sporangium.
5. There are a large number of microspores in the sporangium. Their size ranges from 0.015 to 0.05mm.
6. The spore wall is two layered. The outer wall (exine) is very thick and sculptured. The inner wall is thin and delicate.
7. Reserve food material in the form of oil globules are found in the spore.

Structure of the mega sporangium

1. The mega sporangium is yellow orange or chalky white at maturity.
2. The stalk of the sporangium is short and multicellular.
3. The sporangial wall is two layered. The outer layer has columnar chlorophyllous cells. The inner layer has thin walled cells.
4. Sometimes a tapetal layer will be persisting in the mature sporangium.
5. The megaspores are usually 3-4 in number per mega sporangium.

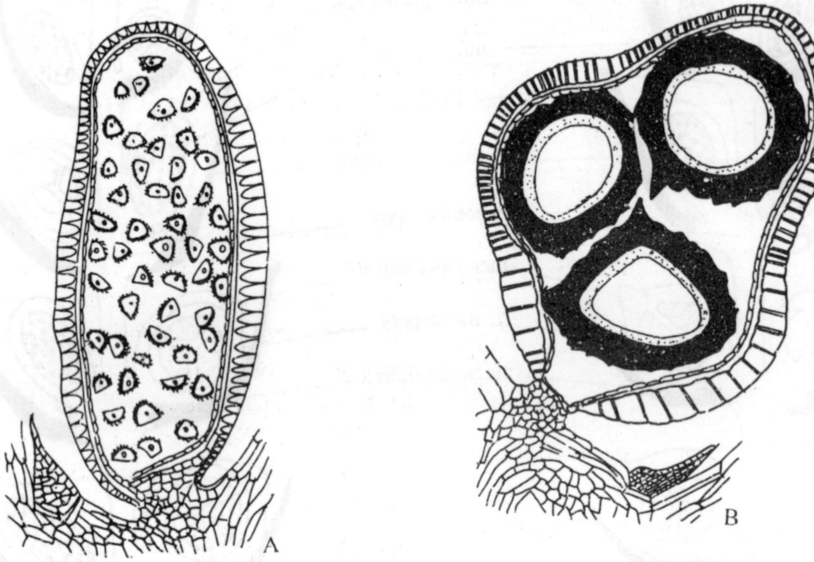

Fig.11 .31. Selaginella: Structure of a Mature Microsporangium (A) and Megasporangium (B)

6. Spores are bigger in size having a diameter of 1.5 - 5mm. Shape of the spore is triradiate or subspherical.

7. The wall of the sporangium is very thick and consists of a sculptured exine, a middle mesospore and a thin intine.

8. Cytoplasm of the spore consists of reserve food in the form of oil globules and nitrogenous matter.

Structure of a mature male gametophyte.

1. The female gametophyte is an extremely reduced structure. Its growth is confined to the spore interior.

2. A mature gametophyte has 13 cells - one prothallial cell, four primary androgonial cells and eight jacket cells.

3. The androgonial cells eventually give rise to sperms.

Structure of the female gametophyte.

1. A mature mega gametophyte has a basal free nuclear portion and an apical cushion.

2. The apical cushion is three layered thick towards the centre but one layered thick towards the margin.

3. The lower most cells of the apical cushion have a much thickened lower wall constituting the diaphragm.

4. The diaphragm separates the apical cushion from the lower free nuclear portion.

5. At later stages even the lower portion becomes cellular.

6. The cells in the apical cushion are uninucleate and polygonal.

7. The cells at the basal region are irregular in shape and sometimes multinucleate also these cells are rich in reseve food material.

8. Many archegonia are found in the apical cushion.

Systematic position

LYCOPSIDA
LIGULOPSIDA
SELAGINELLALES
SELAGINELLALES
SELAGINELLACEAE
SELAGINELLA

ISOETES

Pratical work to be done

1. Study the morphology of the plant.
2. Study the internal structure of corm, root and leaf with the help of slides.
3. Mount the sporangia and observe the spores.
4. Study the morphology of the male and female gametophytes from the slides.

Morphology of the plantbody

1. The plantbody consists of a condensed, lobed structure called the axis or corm.
2. The axis or corm has a complex morphological structure.
3. The corm on its upper surface bears a number of long quill like ligulate leaves.
4. The leaves are two to several cm long and are crowded in a closed spiral forming a sort of a fascilcle.
5. In some instances the leaves may be 0.5 meter (*Isoetes coromandalianum*), long.
6. The leaves have broad spoon shaped bases and tapering tips.
7. The outermost leaves are sterile; successively within them are found megasporophylls, microsporophylls and sporophylls with imature sporangia.
8. From the lower surface of the corm are produced a number of roots.
9. The roots usually take their origin from the grooves of the lobed corm. They are arranged in vertical rows at right angles to the groove.
10. In a tow lobed corm there are four series of roots, two on each side of the corm.

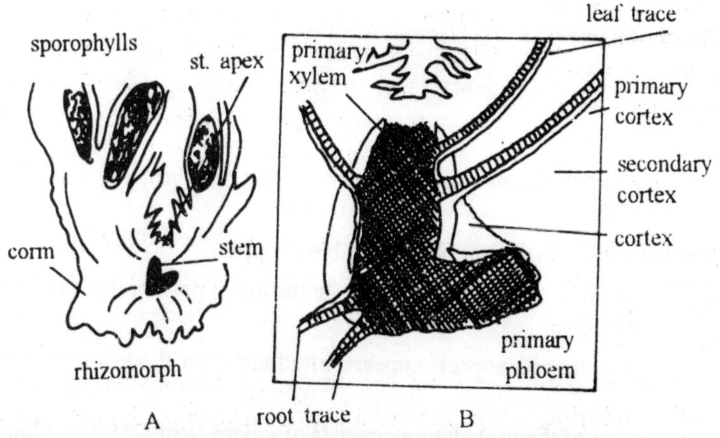

Fig.11.32. Isoetes: Structure of Rhizome
A. Vertical section of corm. B. A part of Vasculature enlarged

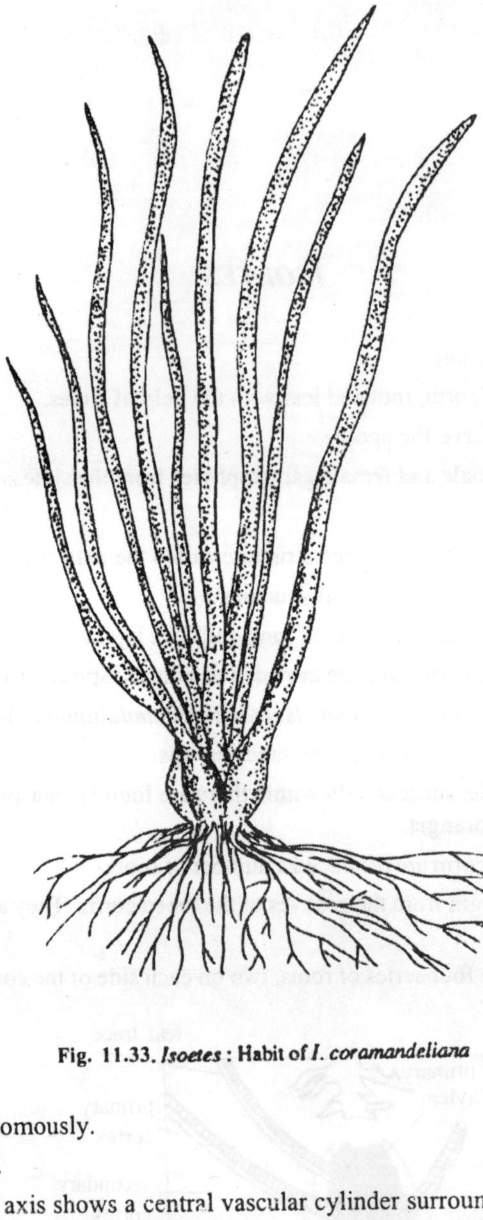

Fig. 11.33. *Isoetes* : Habit of *I. coramandeliana*

11. The roots branch dichotomously.

Internal structure of corm.

1. A vertical section of the axis shows a central vascular cylinder surrounded by a broad parenchymatous cortex.

2. The vascular cylinder in its shape resembles an anchor or a vegetable chopper.

3. In the upper part the stele is cylindrical connected with leaf traces. In the lower portion it is spade shaped flattened in the plane of the grooves of the corm.

4. In a transverse section the lower spade shaped portion appears triradiate even though there are four radiating arms.

5. The stele is protostelic; the central region of the vasculature consists of xylem ensheathed by phloem.

6. The xylem consists of parenchyma intermixed with tracheids.

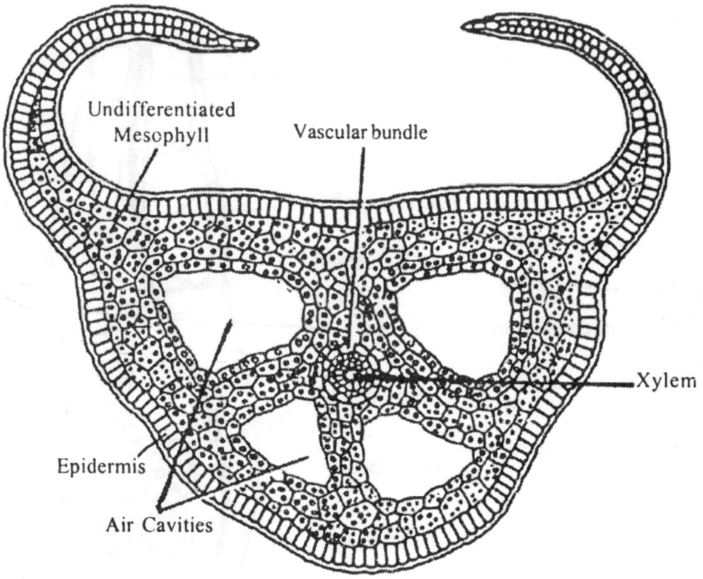

Fig.11.34. Isoetes: T.S. of Root

7. Leaf traces depart from the stele but there are no leaf gaps.

8. Cortex is parenchymatous and has starch filled cells.

9. Secondary growth seen in *Isoetes* is of a peculiar kind.

10. A cambium is seen external to the phloem. The outer derivatives of the cambium are parenchymatous constituting the secondary cortex.

11. The inner derivatives of the cambium are called prismatic layer. Prismatic layer is a mixture of tracheids and

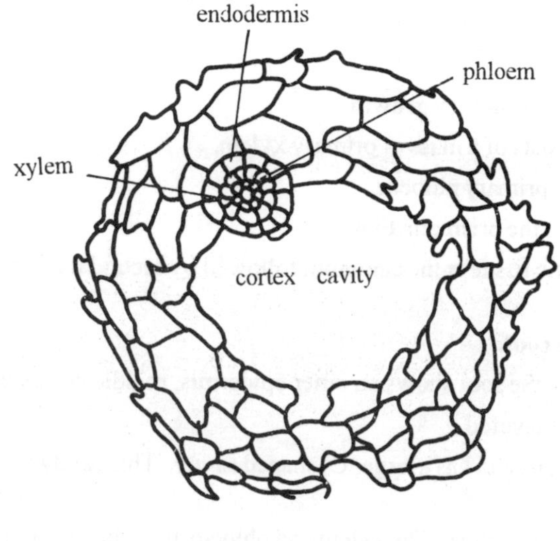

Fig. 11.35 Isoetes : T.S of Leaf

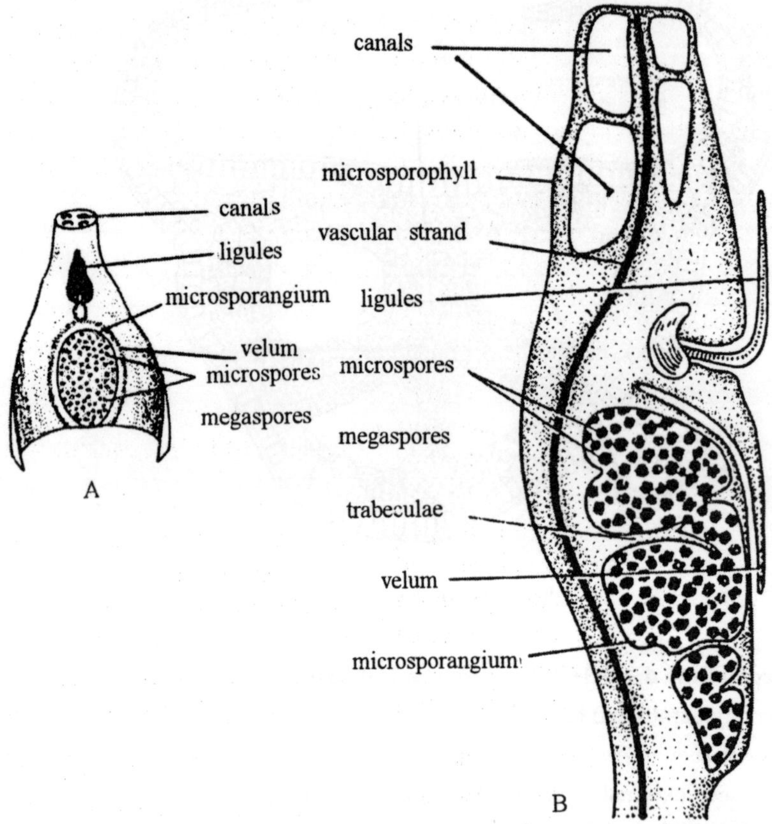

Fig. 11.36. *Isoetes* : Diagrammatic Representation of Microsporophyll (A) and Long Sectional View of the same (B)

seive elements.

Structure of a corm after secondary growth

1. The central region consists of a mass of primary xylem.

2. Next to the xylem is the primary phloem.

3. Outlining the phloem is the prismatic tissue.

4. External to the prismatic tissue is the cambium followed by secondary cortex and then the primary cortex, which is very broad.

Internal structure of the root.

1. A transverse section of the root shows an outer epidermis, middle cortex and central stele.

2. The epidermis is single layered.

3. Cortex is parenchymatous and has a large 'C' shaped cavity. This cavity is due to the breakdown of cortical cells.

4. The stele is a monarch protostele. The xylem and phloem are collateral and are arranged in such a way that the phloem is on the side away from the rhizome.

5. Surrounding the vasculature is a well defined endodermis.

Internal structure of the leaf

1. The leaf shows a single vascular bundle surrounded by an undifferentiated mesophyll.
2. The outermost layer is the epidermis in which stomata are found.
3. Mesophyll consists of parenchyma sometimes sclerenchyma may also be found.
4. There are four cylindrical air chambers in the mesophyll separated by parenchymatous partitions.
5. vascular bundle is collateral but may become concentric towards the tip of the leaf.

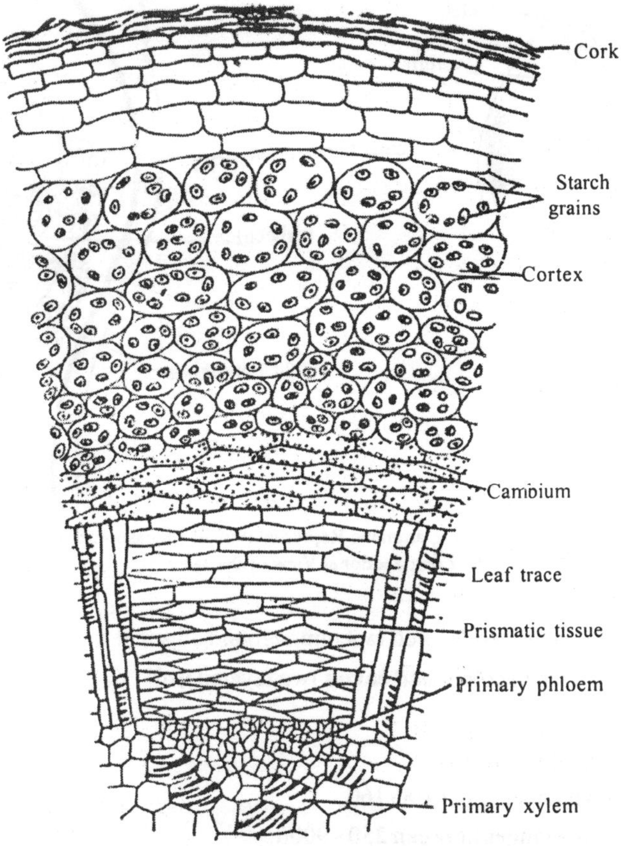

Fig.11.37. Isoetes: Structure of Rhizome; T.S. of Corm showing Cellular details

Structure of a microsporangium

1. The sporangium has a multilayered wall.
2. Above the sporangium is situated *velum,* a projection of cells between the ligule and the sporangium.
3. Within the sporangium are found a large number of microspores.
4. The sporangium is incompletely divided by partitions called trabeculae.

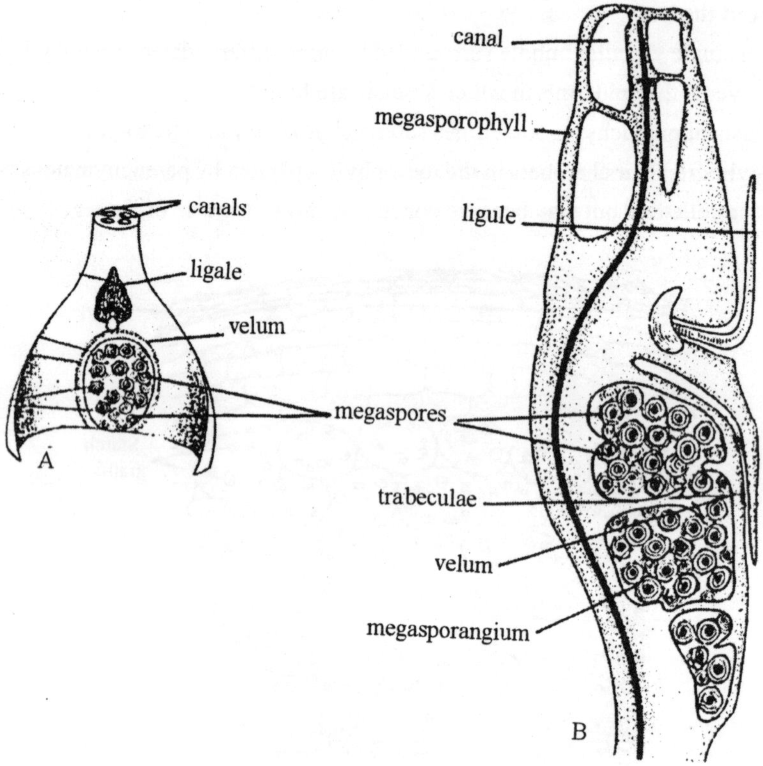

Fig. 11.38. *Isoetes* : Diagrammatic Representation of Megasporophyll (A) and Long Sectional View of the same (B)

5. When the spores are young a two layered tapetum is found adjacent to the trabeculae.
6. Spores are minute brownish in colour and have a two layered wall.

Structure of a megasporangium

Same as in microsporangium except.

1. Spores are large, less in number about 100-160.
2. The size of the megaspore ranges between 250 - 900μ.
3. Spores are white, grey or black in colour.

Systematic position

LYCOPSIDA

LIGULOPSIDA

ISOETALES

ISOETACEAE

ISOETES

CHAPTER 12

SPHENOPSIDA

Sphenopsida are commonly called horesetails because of the crowded secondary branches arising on stem. The group includes living as well as extinct members. Geologically the history of Sphenopsida extends to the Devonian period.

The Sphenopsida is divided into four orders - Hyeniales, Calmitales, Sphenophyllales and Equisetales. Calamitales includes *Calamites, Calamostachys* etc. *Sphenophyllales* includes Sphenophyllum, while Equisetales includes *Equisetum*.

In this book calamites, calamostachys, Sphenophyllum and Equisetum are discussed.

Sphenophyallales represent a small and compact group of sphenopsida. They have left no surviving representatives. The geological history of Sphenophyllales dates back to a time earlier than carboniferous period. Some of the genera like *Sphenophyllum* appeared during Devonian reaching maximum development during carboniferous period. A few of these Sphenophyllales survived upto Triassic.

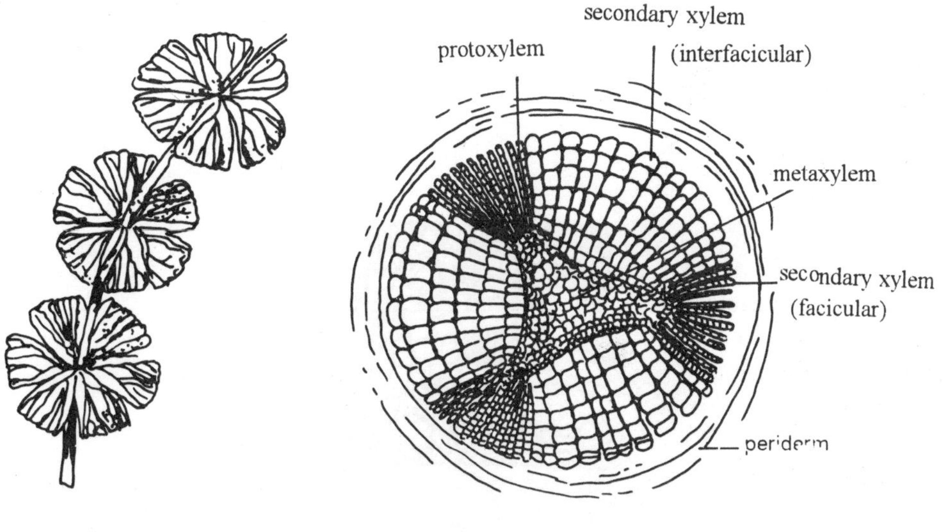

Fig. 12.1 . *Sphenophyllum* : Habit of Plant (A) and Cross Section of an Old Stem of *S. dawsonii* (B)

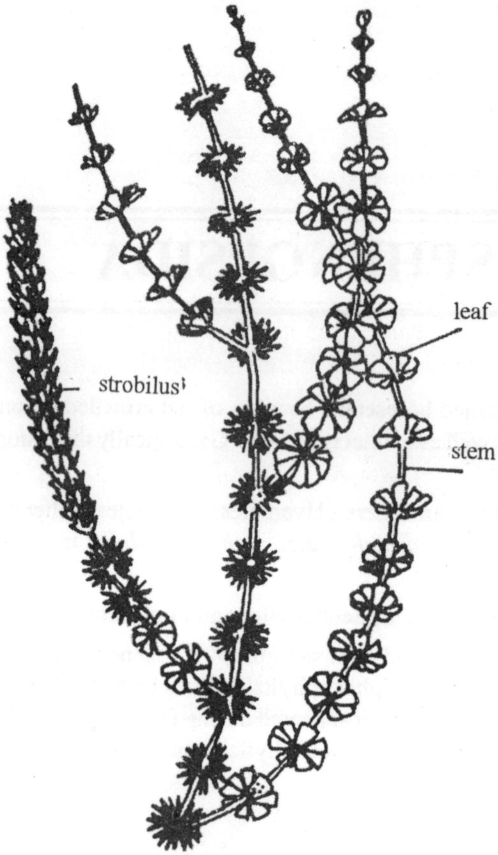

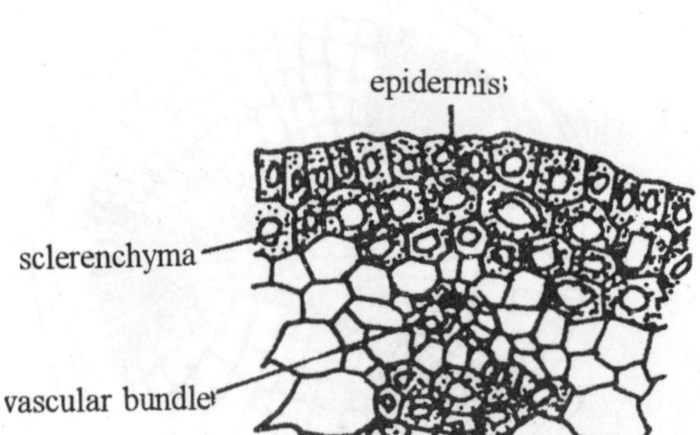

Fig. 12.2. *Sphenophyllum* : T.S. of Leaf

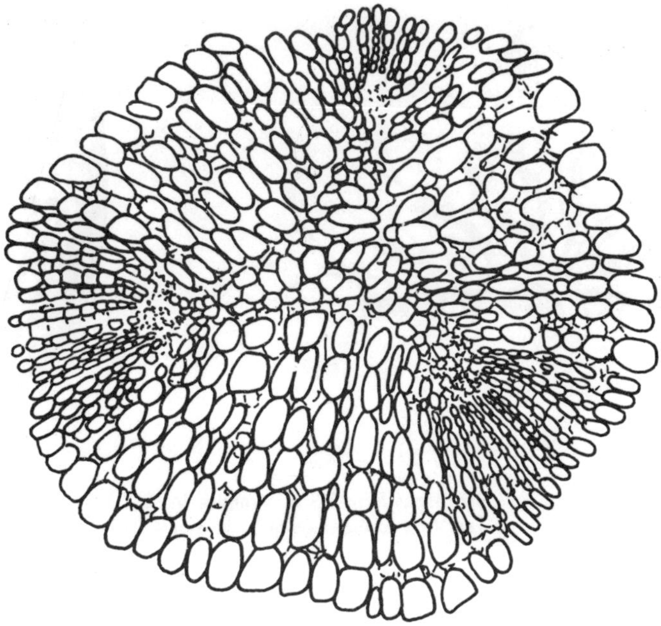

Fig. 12.3. *Sphenophyllum* : T.S. of Xylem Cylinder of *S. plurifoliatum*

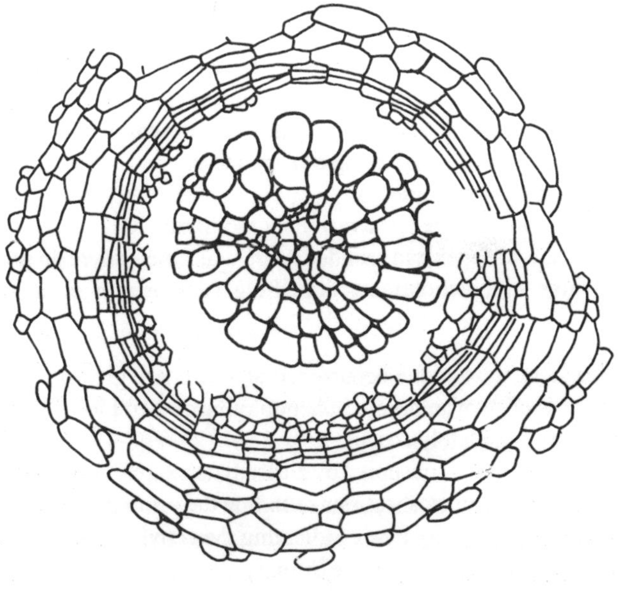

Fig. 12.4. *Sphenophyllum* : T.S. of Root

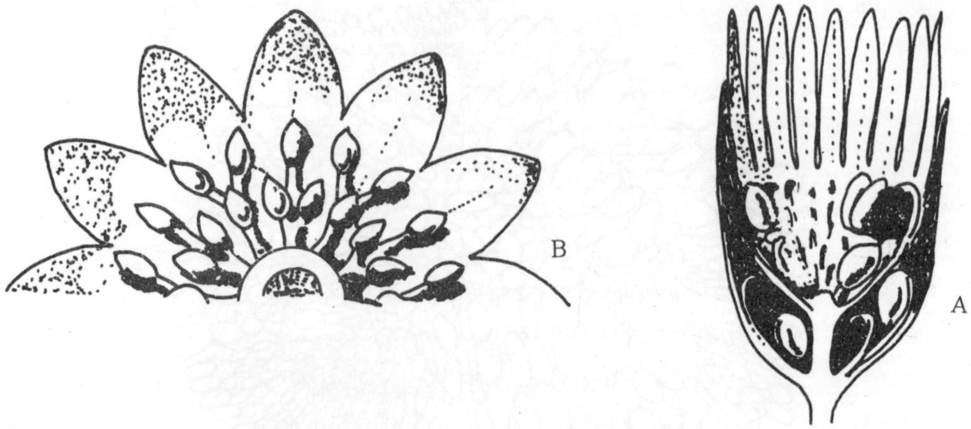

Fig. 12.5 Sphenophyllum: L.S of Strobilus (A) and Top View of the same of S. dawsonii(B)

1. The principal genus of the order is Sphenophyllum.
2. The compressed fossils of the plant body have been found extensively scattered in coal bearing formations belonging to the carboniferous period.
3. The plant body of Sphenophyllum was small and herbaceous.
4. There is a high probability that the plant had a climbing habit.
5. While, some of the morphological features indicate an aquatic habitat, anatomical features however, point out to a terrestrial habitat.
6. The plant body consisted of a main stem, which was slender and jointed, rarely exceeding 5 mm in diameter.
7. In its externals the stem recalls what is seen in Equisetum.
8. The stem was ribbed which did not alternate at nodes.
9. Leaves which are so typical of Sphenophyllum weie present in clusters at the nodal regions.
10. At each whorl the number of leaves ranged from 6 to 9.
11. Occasionally the number went upto 18.
12. The leaves were cuneate and the apex variously modified.
13. In S. emarginatum, the leaf apex had rounded teeth, while they awere pointed in S. cunifolium and S. majus.
14. The leaves had vains with dichotomous branching, each branch terminating in an apical tooth.

ANATOMY

1. Anatomically the stem showed the occurrence of secondary growth.
2. In some of the details of stem anatomy, Sphenophyllum resembled Calamites.
3. The Central pith was completely absent.
4. Instead, the central region was occupied by a triradiate exarch xylem mass.
5. Secondary xylem formed a thick sheath surrounding the primary xylem.
6. Cortical region exhibited a corky tissue indicating the activity of phellogen.
7. The phellogen probable arose deep in the primary cortex.

Roots:

1. The plant body was anchored to the soil with the help of adventitious roots that arose at the nodal

regions.

2. Anatomically the roots resembled the stems, expect that the former had a two angled xylem.
3. Secondary growth was also seen in root.

Reproduction

1. Sphenophyilum reproduced by means of sporangia borne on sporophylls.
2. The sporophylls aggregated to form long and slender cones terminating the stem apex.
3. The cones which are often given the name Bowmanites had a central axis on which were arranged many whorls of sporophylls.
4. The sporophylls of each whorl fused at their base to form a saucer like structure.
5. The sporophylls had two parts namely an adaxial fertile part and an abaxial sterile part.
6. The sporangia were borne on the fertile part either singly or in pairs.
7. In some cases the fertile lobe had two or three branches each having one to two sporangia.
8. The number and arrangement of sporangia on the lobe are taxonomic features used in the categorisation of the species.
9. The sporangia had a stalk which was highly reduced.
10. Most of the species of Sphenophyllum were homosporous.
11. Some species could have been heterosporous.

Systematic Position

SPHENOPSIDA
SPHENOPHYLLALES
SPHENOPHYLLACEAE
SPHENOPHYLLUM

CALAMITES

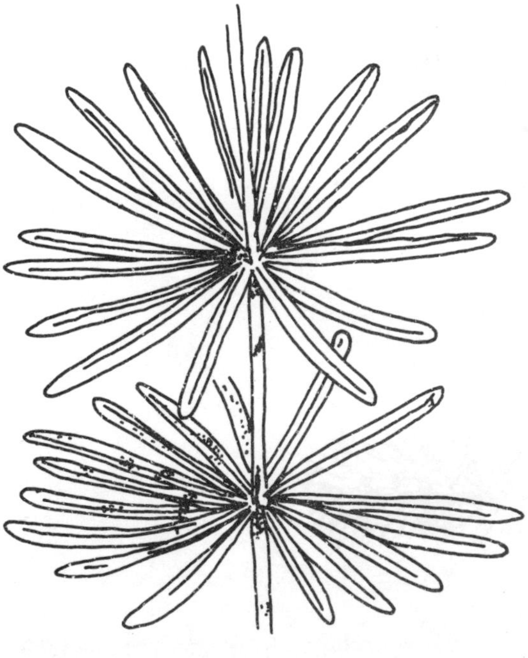

Fig. 12.6. *Calamites* : Leaves of *Annularia*

Practical work to be done

1. Study the external morphology from the reconstruction of the plantbody.
2. Study the internal structure of the stem from a slide.

Morphology of the plantbody

1. The plantbody is a small tree growing to a height of 20-30 meters.
2. The plant has an underground rhizome differentiated into nodes and internodes.
3. A whorl of adventitious roots arise at each node.
4. From the upper surface of the rhizome arise a number of aerial shoots.
5. Some of the aerial shoots produce roots at the lower node which indicates that the rhizome grew some distance below the soil.
6. Erect shoots are constricted at the point of their junction to the rhizome.
7. The aerial shoots have whorls of branches at the nodal region.
8. The surface of the stem has longitudinal ribs alternating with the grooves.

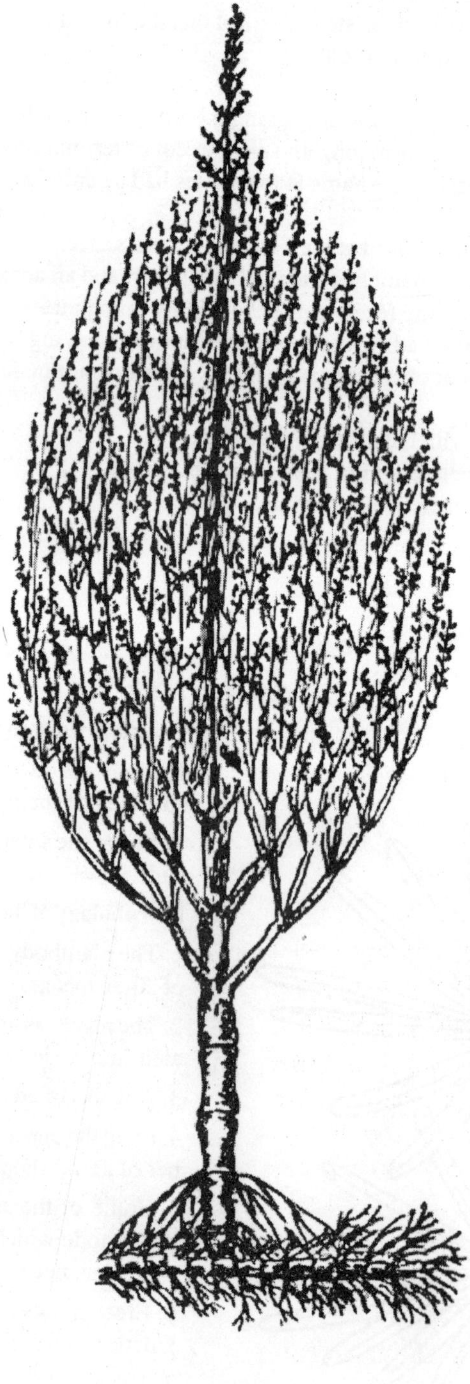

Fig. 12.7. *Calamites* : Reconstruction of Plant Body

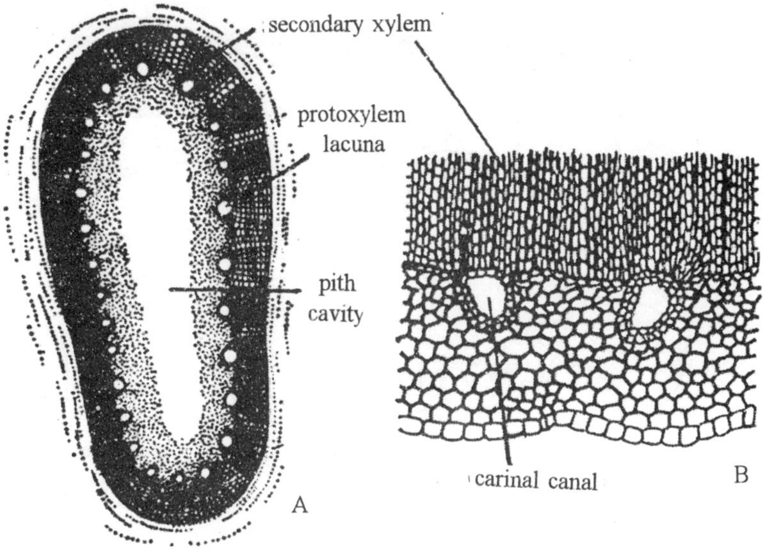

Fig. 12.8. Calamities: Anatomy of Stem (after secondary growth)
A.Ground Plan of Cross Section .B.A Sector enlaged

9. In some instances the ribs of successive internodes are directly above instead of alternating with the other.

Internal structure of the stem

1. A transverse section shows an epidermis, cortex and stele.
2. Epidermis is single layered.
3. The cortex has an outer sclerotic zone and an inner thin walled parenchymatous zone.
4. Stele is siphonostelic.
5. There is a pith in the nodal region in the centre but replaced by a central canal.
6. Vascular bundles have collaterally arranged xylem and phloem. The xylem is endarch.

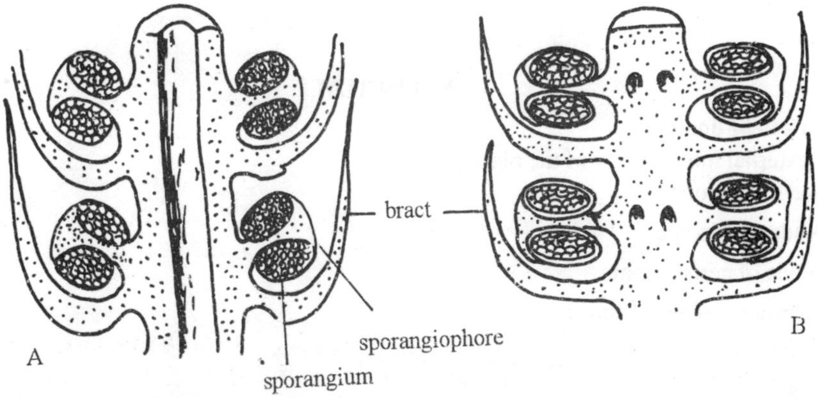

Fig. 12.9. *Calamites* : Reproductive Structure. Comparison of Strobili of *Palaeostachys* (A) and *Calamostachys* (B)

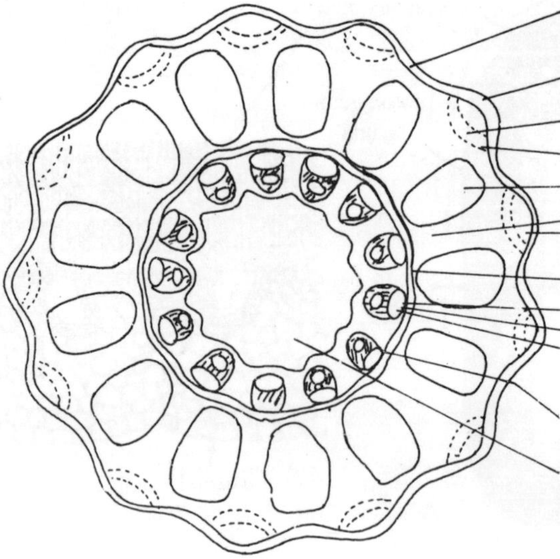

Fig. 12.10 *Equisetum* : T.S. of Internode of Aerial Shoot (ground plan)

7. The tracheids had scalariform thickenings in metaxylem but annular and spiral thickenings in the protoxylem.

8. In each vascullar bundle a carinal canal is found.

9. Outside the primary vascular bundles secondary xylem is seen.

10. There is no differentiation of annual rings in the secondary wood.

Systematic position

SPHENOPSIDA

CALAMITALES

CALAMITACEAE

CALAMITES

EQUISETUM

Practical work to be done

1. Study the external morphology of the plant.
2. Cut transverse sections of
    - a) Node of rhizome.
    - b) Node of aerial shoot.
    - c) Internode of aerial shoot.
    - d) Root.
    - e) Leaf.
    - f) T.S. and L.S. of cone.
3. Mount a few spores in water and observe.

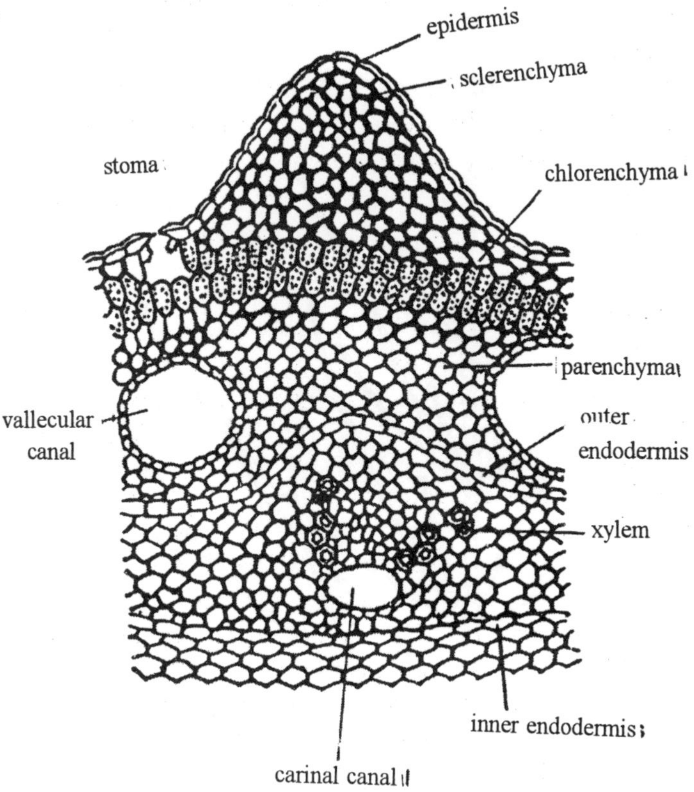

Fig. 12.11 *Equisetum* : T.S. of Aerial Shoot at Internode - A sector enlarged

4. Study the structure of the gametophyte.

External features

1. The plantbody consists of an underground prostrate rhizome from which spring up a number of aerial shoots.

2. Aerial shoots may be few cms long (*E.scirpoides*) or several meters long (*E.giganteum*).

3. The rhizome grows 1-2 feet below the soil and branches profusely. It is differentiated into nodes and internodes.

4. At each node are found a whorl of minute scale leaves. The leaves are connate at the base and form a sheath surrounding the rhizome.

5. Alternating with the leaves are found the buds and branch primordia. These buds develop either into a branch of the rhizome or the aerial shoot.

6. The aerial shoots arise from the nodal regions of the rhizome.

7. All the aerial shoots may be of the same type being chlorophyllous and much branched (*E.ramosissimum*) or may be of two type as in *E.arvense*. Here some aerial shoots are chlorophyllous and some achlorophyllous.

8. Achlorophyllous are less branched in comparison with the chlorophyllous branches. They bear reproductive structures - cones at their tips.

9. The aerial shoots produce secondary branches in whorls at the nodal regions. Usually they are 5-6 internodes in length.

10. Secondary branches arise from the branch primordia. Usually there are many of them resulting in the development of a whorl.

11. The aerial shoots are also differentiated into nodes and internodes with internodal regions having longitudinal ribs. The ribs alternate with the grooves at successive internodes.

12. Silica deposition is seen on the aerial shoot making the plantbody rough.

13. Leaves are found at the nodal regions. They are microphyllous and scaly. Phyllotaxy is whorled. Leaves are connate at the base.

14. The number of leaves at every node ranges between 3-40. Leaves have very little to do with photosynthesis. Much of the photosynthetic activity is carried out by aerial shoots.

15. Reproductive structures - cones are borne at the tip of aerial shoots.

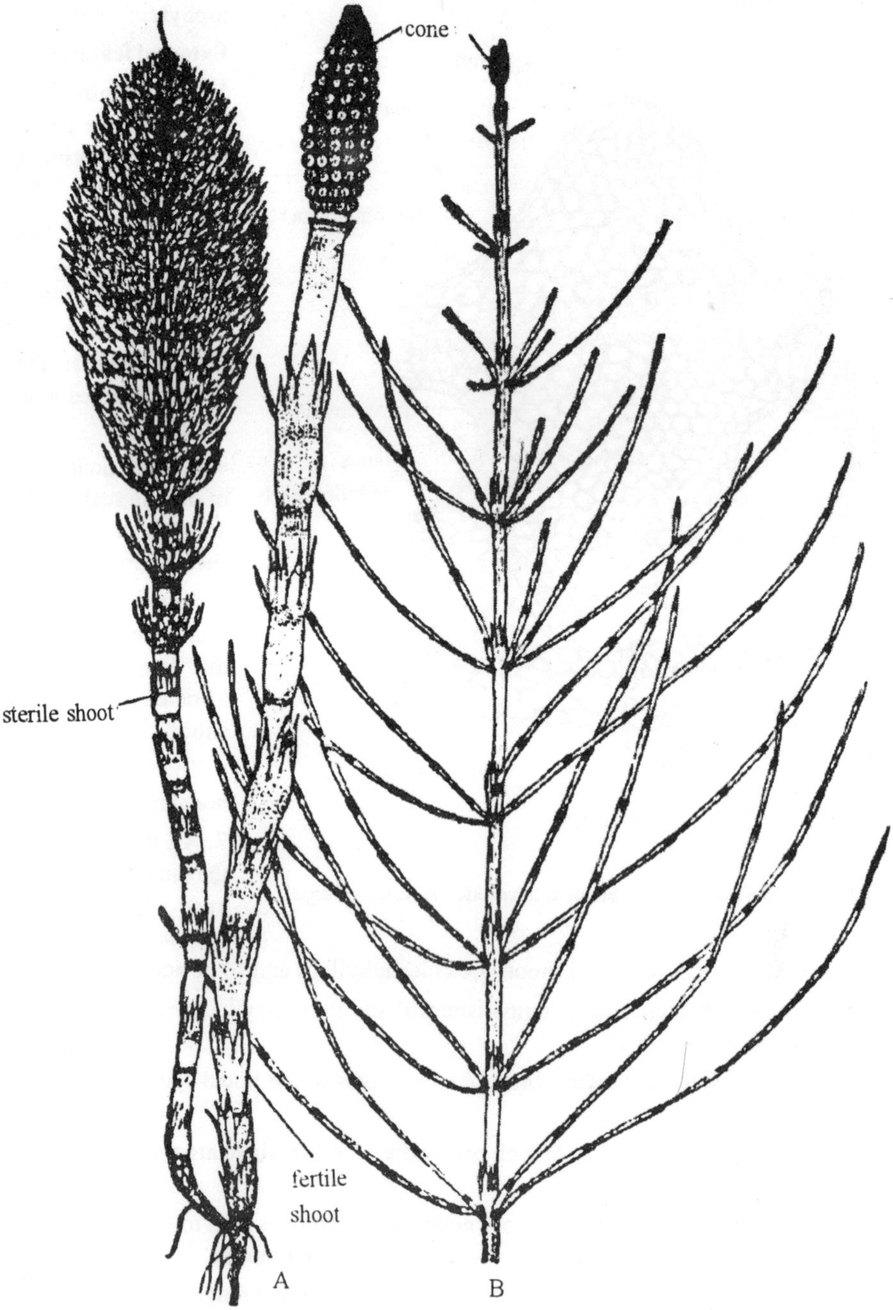

Fig. 12.12. *Equisetum* : Habit of *E. telemateia* and *E. palustrae*

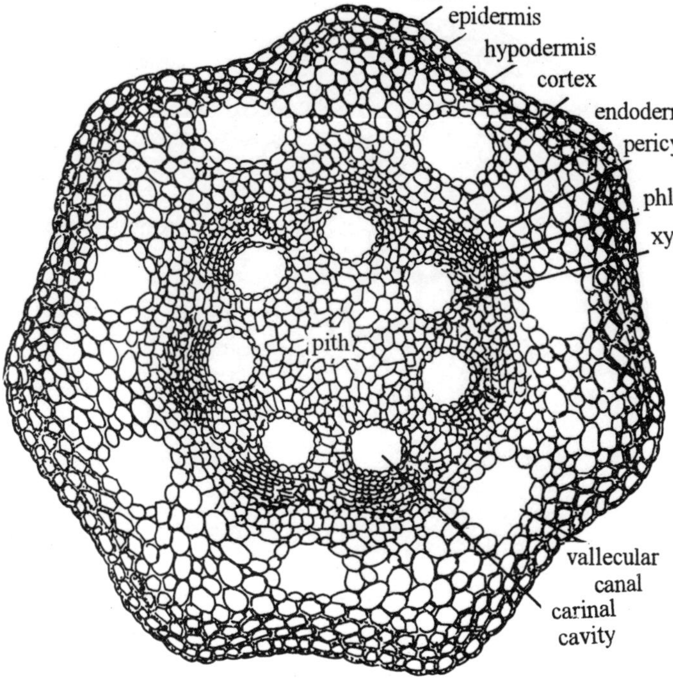

Fig.12.13 . *Equisetum* : T.S. of Rhizome

Internal structure of rhizome

1. A transverse section shows epidermis, cortex and stele. The outline of the section is wavy.

2. Epidermis is made up of a single layer of cells. There are no stomata in the epidermis.

3. Next to the epidermis is found hypodermis made up of 2-3 layers of sclerenchyma.

4. Internal to the hypodermis is the cortex. Cortex is made up of parenchyma cells. Vallecular canals are found in the cortex.

5. An endodermal layer surrounds the stele.

6. Next to the endodermis is a layer of pericycle.

7. Internal to the pericycle are seen a number of vascular bundles. These are arranged alternating with the vallecular canals.

8. Vascular bundles are collateral and endarch with poorly developed xylem. Some of the protoxylem elements degenerate to form a cavity called the carinal canal.

9. In the central region there is a parenchymatous pith.

Systematic position

SPHENOPSIDA

EQUISETALES

EQUISETACEAE

EQUISETUM

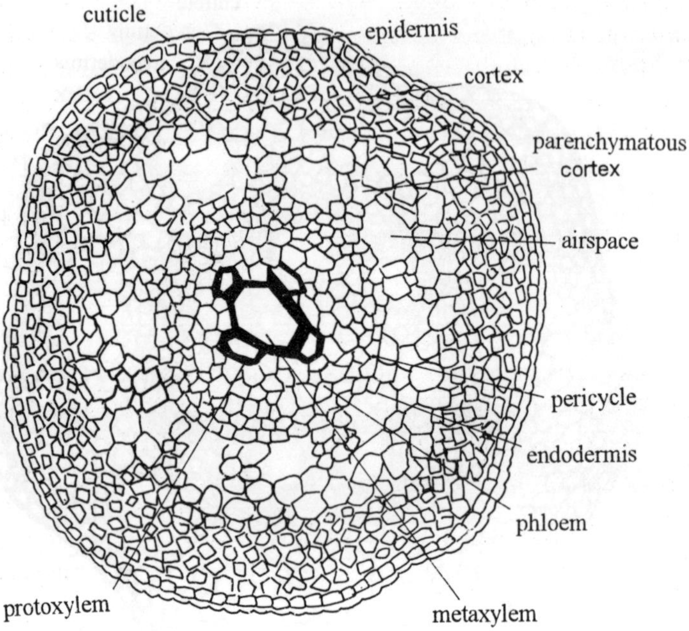

Fig.12.14 *Equisetum* : T.S. of Root

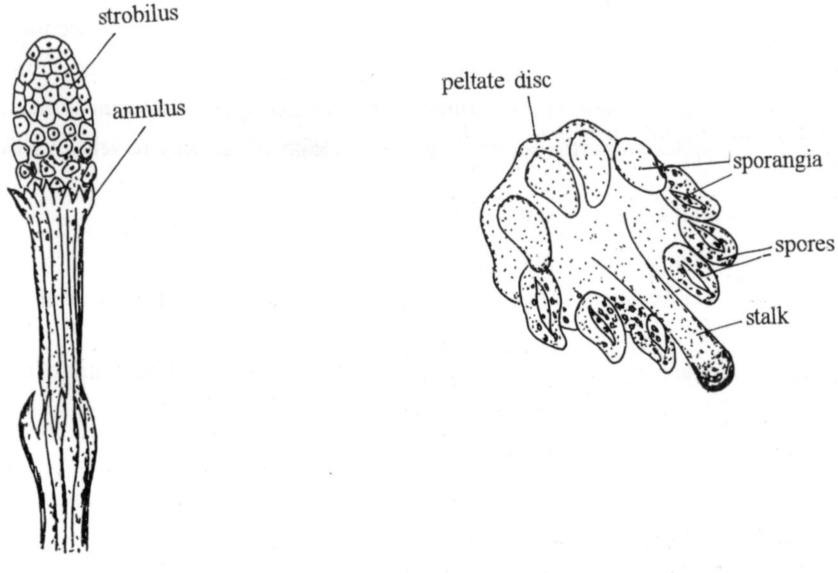

Fig. 12.15. Equisetum: Spore producing Organs
A. Cone, B. Ventral view of a Sporangiophore

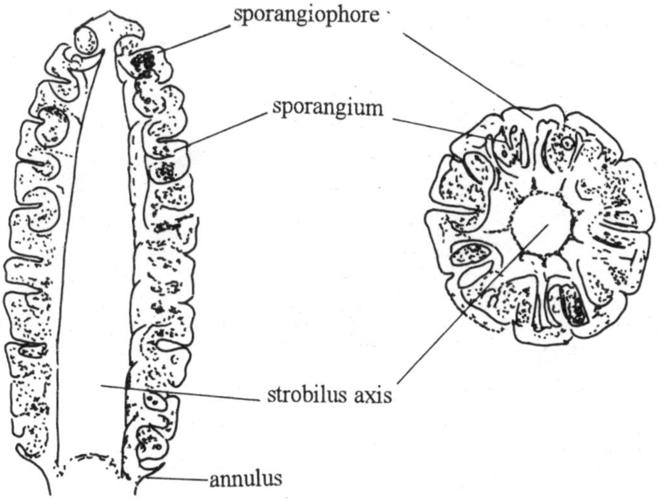

Fig.12.16 *Equisetum* : Spore Producing Organs
C. L.S. of Cone, D. T.S. of Cone

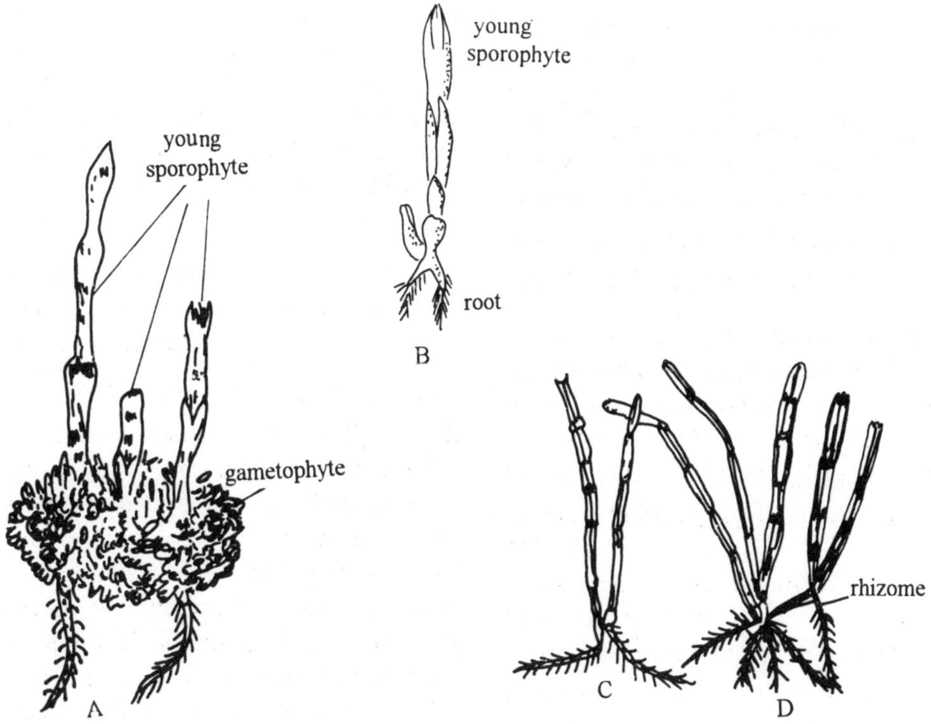

Fig.12.17. Equisetum: Young Sporophytes (A) on the Gametophyte and Establishment of Young Sporophytes on Soil (B-D)

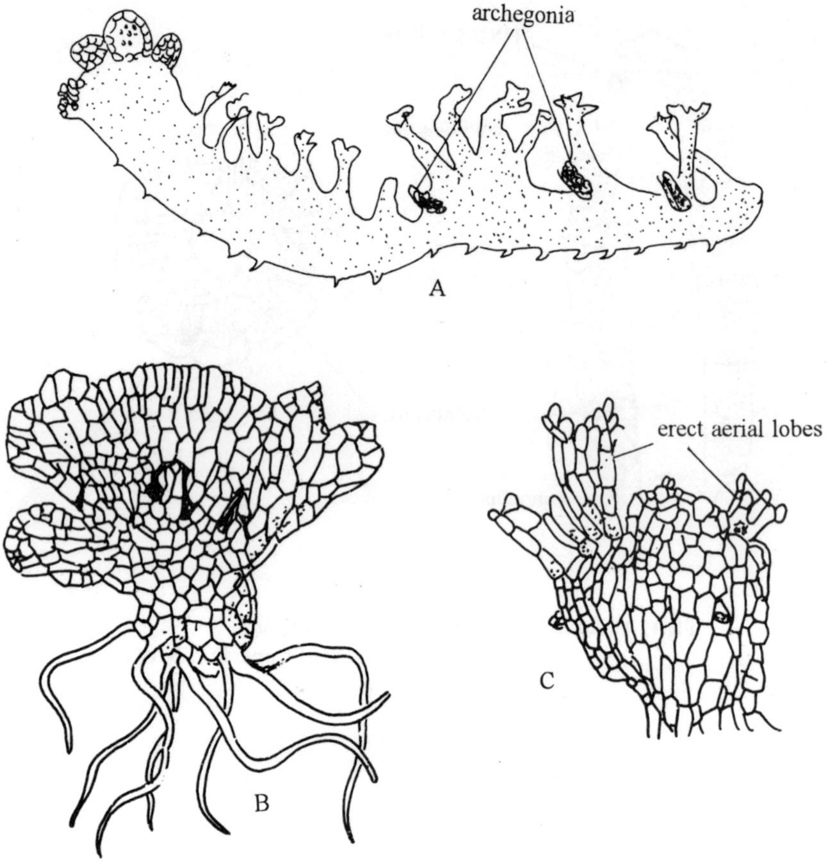

Fig. 12.18. *Equisetum* : Gametophytes
A. *E. arvense*, B. *E. telemateia*, C. *E. kansanum*

CHAPTER 13

PRIMOFILICOPSIDA

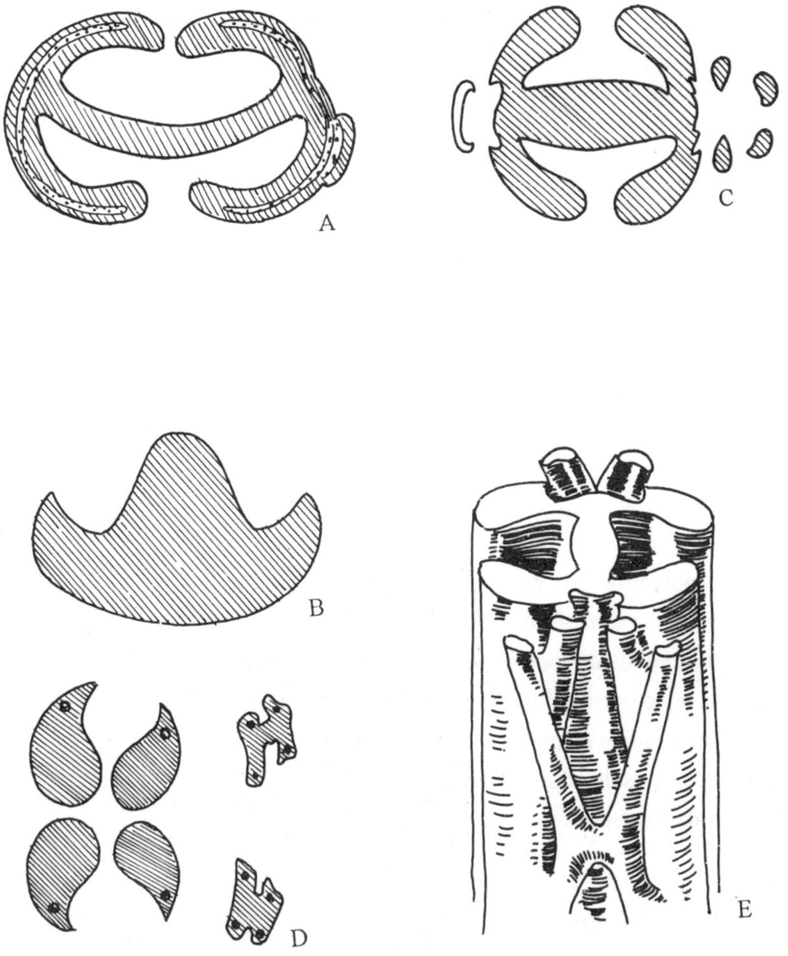

Fig.13.1 *Coenopteridales* : Stereodiagram of Steles
A. *Ankyropteris*, B. *Botryopteris*, C. *Stauropteris*, D-E. *Etapteris*

BOTRYOPTERIS

1. The genus *Botryopteris* is one of the best known among Coenopteridales.

2. It is the type genus of the family and has 5 species ranging from lower Carboniferous to the Permian.

3. The name *Botryopteris* is given to the fossil specimens of stem.

4. The stems are slender, cylindrical and few millimeters in diameter.

5. They are branched and bear spirally arranged fronds.

6. Anatomically, the main stem has a small mesarch protostele, which is surrounded by a broad cortex.

7. The cortex has a prominent band of sclerotic cells.

8. In the leaf stalk, the xylem strand is solid and has three prototoxylem points.

9. As in *B.forensis* the strand is deeply indented looking like a trident.

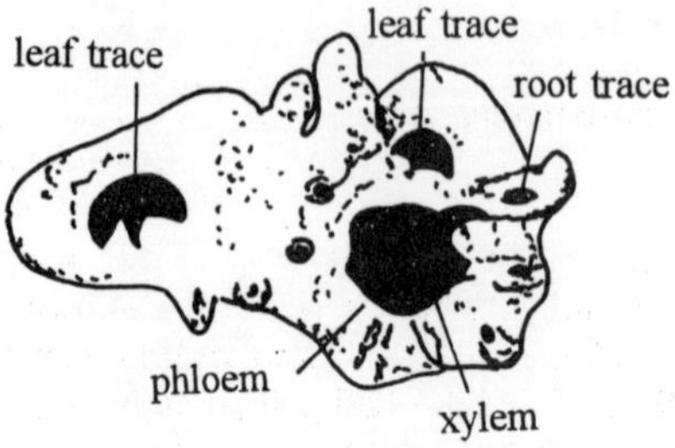

Fig.13.2 Coenopteridales:Diagramatic Transverse Section of Botryopteris ramasa

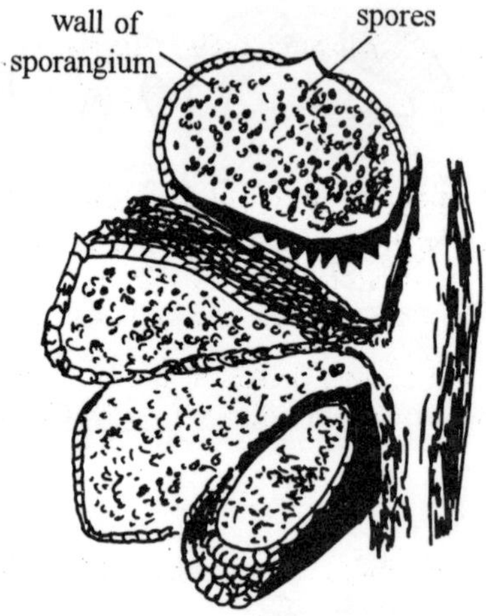

Fig 13.3 Coenopteridales:Sporangia of Botrypteris (sectional view)

10. In very few species of *Botryopteris*, sporangia have been found attached to the fronds.
11. The sporangia are found in clusters which is somewhat rare to ferns.
12. In *B.globosa* sporangial cluster has thousands of sporangia.
13. The sporangia themselves are small, oval to pyriform in shape.
14. They measure 2 mm in length and about 1 mm in diameter.
15. Each sporangium has a short stalk and a capsule which is somewhat oval in outline.
16. The wall of the capsule has a broad annulus represented by thick walled cells.
17. The spores are of the same type.

ZYGOPTERIS

1. The ferns belonging to the familuy Zygopteridaceae are more complex and older then Botryopteridaceae.
2. The fossil specimens belong to middle Devonian and possibly have connection with Psilophytales.
3. *Zygopteris*, the type genus of the family is the best known.

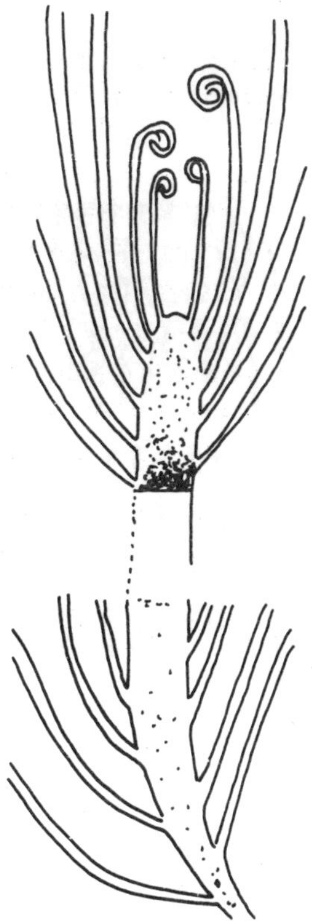

Fig.13.4 *Coenopteridales* : Reconstruction of Upper and Lower Portions of Shoots of *Zygopteris primaria* (after Sahni 1933)

4. It has several species of which *Z.primaria* has been studied extensively.
5. The plant body of *Zygopteris* is tree like with a trunk having a diameter of 20 cm.
6. The stem as such, is only about 1.5 cm in diameter being made up of an armour of leaf stalks and adventitious roots.
7. The stem of *Zygopteris* bears an elaborately branched frond having a number of leaf stalks.
8. These are usually cylindiracal and upto 2 cm in diameter.
9. The leaf stalks have a number of pinnae.
10. Occassionally the leaf stalks are given the name *Etapteris*.
11. Anatomically the stem of *Zygopteris* has a xylem cylinder consisting of scalariform tracheids.
12. An unusual feature here is the presence of a layer of secondary wood surrounding the primary xylem.
13. This is perhaps one of the rare instances of secondary vascular tissue in a fern extinct or extant.
14. The leaf stalk internally shows a vascular strand which is H-shaped with a straight median band with some what fixed lateral arms.
15. Two small protoxylem points lie in the shallow depressions at the end of the median band.
16. Various names have been given to the fructifications of *Zygopteris*.
17. *Corynopteris* is one such fructification genus.

18. The sporangia in the fructification are large and ovate.
19. They are usually *sessile* and are grouped into spherical sori.
20. The wall of the sporangium has a broad band like annulus.

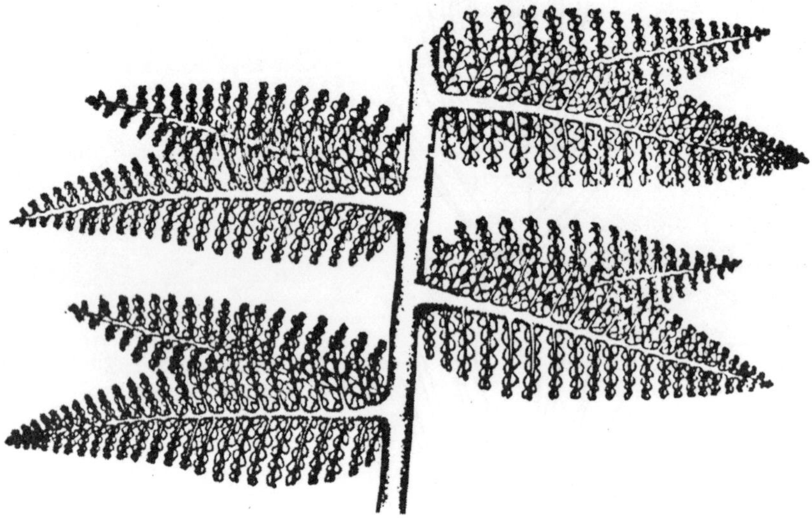

Fig.13.5. Coenopteridials: Reconstruction of leaves of Etapteris (leaves of Zygopteris)

CHAPTER 14

FILICOPSIDA

PTEROPSIDA

(Filicophyta or Pterophyta)

1. Pteropsida represent the most highly evolved group among the lower vascular plants.

2. In this group are included some of the most beautiful, and most familiar plants called *Ferns* which are the joy and pride of a gardener.

3. The delicate, varied and highly attractive foliage of ferns have made them a must in any garden.

4. Of all the pteridophytes, ferns are the most widely distributed.

5. The living ferns are represented by 305 genera and nearly 10,000 species (Holtum, 1949)

6. Geologically ferns are known from as far back as the Carboniferous period.

7. It is believed however that the group must have had its ancestors during the *Devonian* period itself.

8. A unique feature of ferns is their persistent basic characters yet sufficiently plastic to be receptive to the environmental fluctuations.

9. Pteropsida are distinct from lycopsida and sphenopsida in several characters.

10. Among the vegetative characters megaphyllous leaves with the attendant leaf gaps are most notable.

11. Among the reproductive features, (though some primitive members show some sort of a semblance to the strobilar organisation seen in the previous groups), the aggregation of sporangia on the abaxial or adaxial surface of the leaf into sori is the most significant.

12. Some of the characteristic features of ferns are given below :

PLANT BODY

1. Stem :

a) The sporophyte has an underground rhizomatous stem which may be elongated or tuberous.

b) The branching of the rhizome may not be dichotomous.

c) In some cases the rhizome is covered by hairs called "ramenta".

2. Leaves :

a) The leaves are compound and once or twice pinnate.

b) They are megaphyllous, having a dichotomous or reticulate type of venation.

c) The size of the leaves varies from few centimeters to several meters *(Angiopteris)*.

d) Usually only the leaves are aerial while the rest of the plant body is subterranean.

e) Some ferns show circinate vernation in the leaves i.e., the young growing parts are coiled inwards and uncoil as they grow.

3. Roots :

Like in other pteridophytes, roots are always adventitious.

EUSPORANGIATAE

1. This sub class includes all eusporangiate ferns.
2. The sprorangial wall is more than one layered.
3. Spore output is very high.
4. There are two orders in this sub class *viz.,* Ophioglossales and Marattiales.

OPHIOGLOSSALES

1. The order include herbaceous, fleshy sporophytes with a short rhizome.
2. Sporangia are borne on a separate outgrowth *called fertile spike.*
3. This arises at the junction of the leaf blade and lamina.
4. Sporangia have a multilayered wall with a high spore output.
5. There is no special dehiscence mechanism.
6. All the members of the order are homosporous.
7. Gametophytes are tuberous and saprophytic.
8. The order has a single family ophioglossaceae, with three genera - *Ophioglossum,Botrychium* and *Helminthostachys.*
9. The genus *Ophioglossum* is discussed here.

OPHIOGLOSSUM

Occurrence and Distribution :

1. Of all the genera of the family, *Ophioglossum* is the most widely distributed being found in almost all parts of the globe.
2. The genus is divided into three sub-genera.
   (a) *Euophioglossum* E.g., O. *moluccanum,* O. *vulgatum* etc.,
   (b) *Ophioderma* E.g., *o. pendulum* etc., and
   (c) *Cheiroglossa* E.g., *O.palmatum.*

Sporophyte

Morphology of the plant :

1. In most of the species, the plant body possesses a short, erect, tuberous rhizome exhibiting radial symmetry.
2. In *O. pendulum* and O. *intermedium* the rhizome is dorsiventral. Branching of the rhizome which is infrequent, may be dichotomous or lateral.
3. The upper surface of the rhizome produces leaves in an irregular spiral.
4. The number of leaves produced per season varies.
5. In temperate species only one leaf is formed per growing season.
6. In tropical species however, the number may be 4-5 per year.
7. The growth of the leaf is extremely slow.

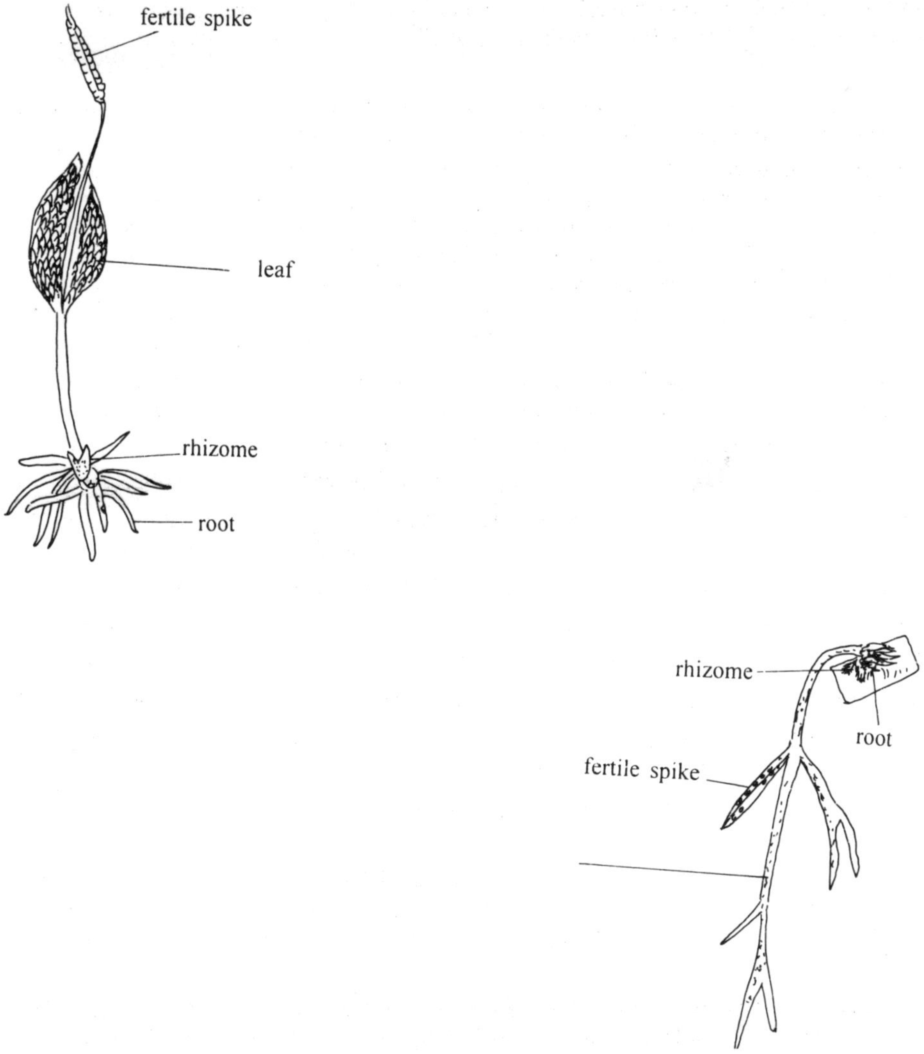

Fig. 14.1. *Ophioglossum* : Habit of *O. vulgatum* (left) and *O. palmatum* (right)

8. A leaf primordium is underground for the first three years and only in the fourth year it comes above the ground.

9. The leaves may be erect or pendant *(O. pendulum)*.

10. A great amount of diversity is exhibited in the nature of leaves.

11. Generally a leaf has two parts *viz.* petiole and lamina.

12. The petiole is erect and long.

13. It bears an expanded, simple and entire lamina.

14. At the junction of the lamina and the petiole arises the *fertile spike* - the reproductive structure of the plant.

15. In *O. pendulum* the distinction between the lamina and the petiole, is lost.

16. Here the leaf is long (1-1,5 meters) and dichotomously branched.

17. In *O.palmatumtbe* petiole bears at its tip a number of palmate lobes of the .lamina.

18. In this,below the lamina there are two rows of fertile spikes.

19. The lamina has reticulate venation but it is not closed.

20. The veinlets end blindly in the mesophyll.

21. There is no distinct midrib.

22. At the base of the petiole are found two stipules (forming a sheath) enclosing the young leaf of the next season.

23. The leaves of *Ophioglossum* do not exhibit circinate vernation that is so typical of ferns.

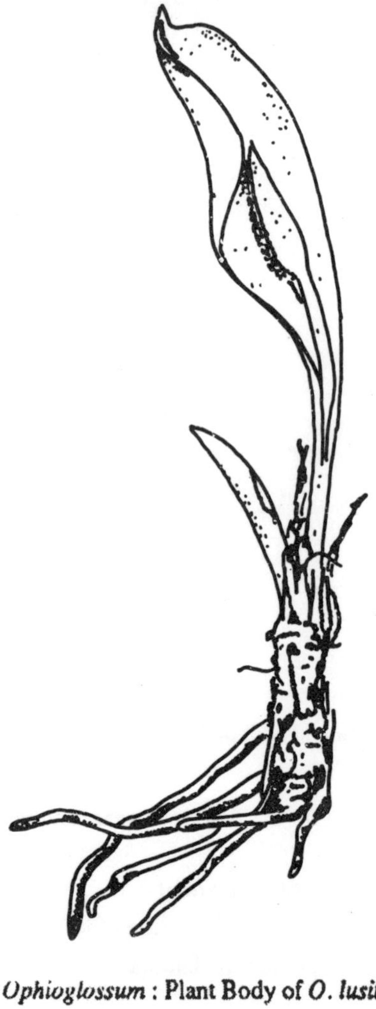

Fig. 14.2. *Ophioglossum* : Plant Body of *O. lusitanicum*

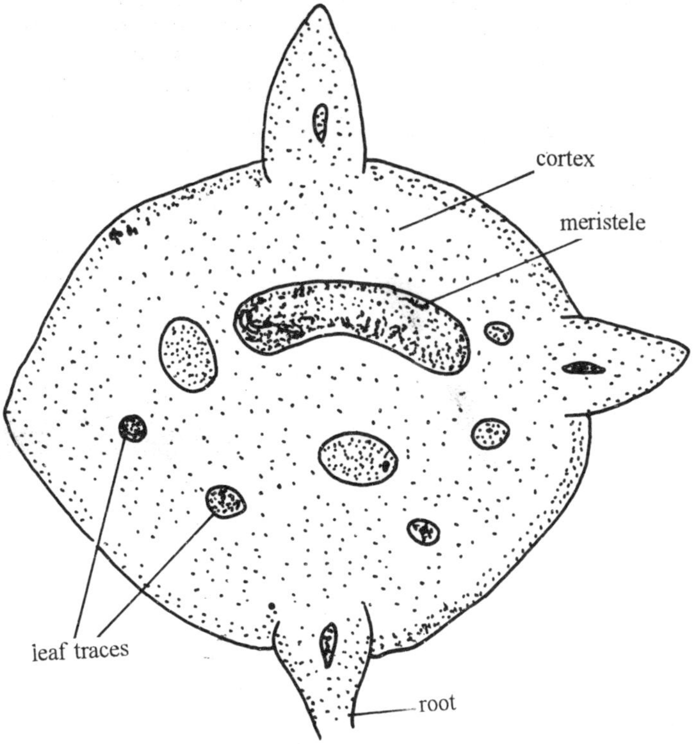

Fig. 14.3 Ohioglossum: T.S of Rhizome(ground plan)

24. The rhizome is anchored to the soil with the help of numerous adventitious roots that are produced from its (rhizome) under surface.

25. The roots lack root hairs.

26. The roots may be unbranched *(Euophioglossum),* infrequently branched *(Cheiroglassa)* or frequently branched *(Ophioderma).*

27. Usually only one root is formed at the base of the leaf of the current season.

28. Unlike leaves, the roots are perennial.

29. The roots have a mycorrhizal association which compensates the absence of root hairs.

INTERNAL STRUCTURE

Rhizome

1. A transverse section of the rhizome shows an epidermis, Cortex and the central stele.

2. The outline of the section will be irregular and discontinuous due to the presence of roots and leaf bases.

3. The epidermis is single layered. Cortex is wholly parenchymatons-The cells are loosely arranged with large number of intercellular spaces.

4. In older rhizomes epidermis is replaced by a periderm layer. But there is no cork cambium.

5. The presence of endodermis is doubtful in older rhizomes.

6. In younger rhizomes an endodermis has been identified (*O. pendulum).*

7. Pericycle may be present (Canrobell 1939).

8. The stelar organisation is variable.

9. In *0. reticulatum* the basal region of the rhizome is protostelic which becomes solenstelic as it travels up.

10. In majority of the species the stele is an ectophloic solenstele.

11. Occasionally leaf gaps may overlap resulting in a dictyostelic condition.

12. Some times in the same rhizome at different regions, the stele ranges from protostele to a dictyostele.

13. Each meristele is semi-lunar in shape with the concave side directcd towards the centre. Xylem is endarch.

14. In O. *palmatum,* the protoxylem elements are not well defined.

15. Phloem is collaterally arranged with reference to xylem.

16. Generally phoem consists of four to five layers of cells. But in *0. pendulum* phloem is single layered and is separated from xylem by a layer of Parenchyma cells.

17. Boodle C1899) has recorded a small amount of secondary thickening ill the rhizome of 0. *vulgatum.*

18. The pith is composed of thin walled parenchyma.

19. Mixed pith.has been reported in *O. pendulum* and *O. aitchisoni* where tracheids are intermixed with parenchyma cells.

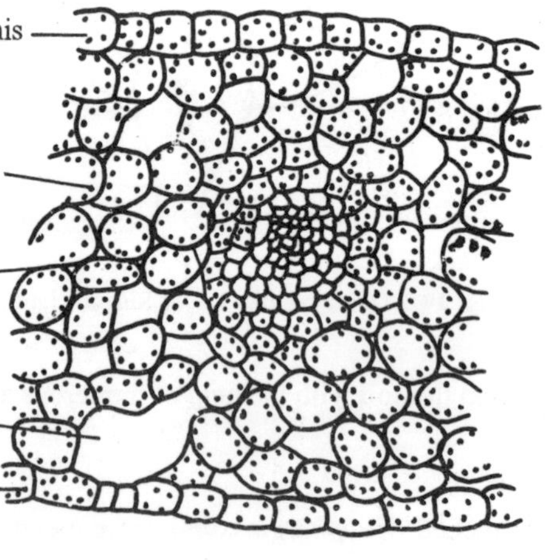

Fig. 14.4. *Ophioglossum* : Venation and Anatomy of Leaf
A. Lamina showing Venation, B. T.S. of Leaf

Leaf

1. From the rhizome a single leaf trace is given off but at the base of the petiole it divides into two strands and enters the leaf.

2. Bat in *O. palmatum* and several other species the leaf gap is associated with two leaf traces.

3. The leaf traces, further up in the petiole divide into a number of vascular bundles and are arranged in a 'C' shaped outline with the opening directed towards the adaxial surface.

4. The bundles are collateral with endarch xylem.

5. Phloem is on the outer surface of the xylem and is separated from the latter by a layer of parenchyma.

6. There is no bundle sheath.

7. Surrounding the vascular bundles is a ground tissue composed of parenchyma, a few outer layers of which have chloroplasts.
8. The ground tissue is protected by a single layered thick walled epidermis.
9. The lamina on either side is bounded by a single layered epidermis.
10. Leaves are amphistomatic, but the number of stomata in the upper and lower epidermis varies.
11. In *O. vulgatum* they are equal in number on both the surfaces; in others *(O. reticulatum)* stomata are more in number towards the lower epidermis.
12. The mesophyll is undifferentiated and is composed of thin walled isodiametric chlorophyllous cells.
13. The mesophyll consists of a number of veins (bundles) whose structure is similar to those of the petiole.

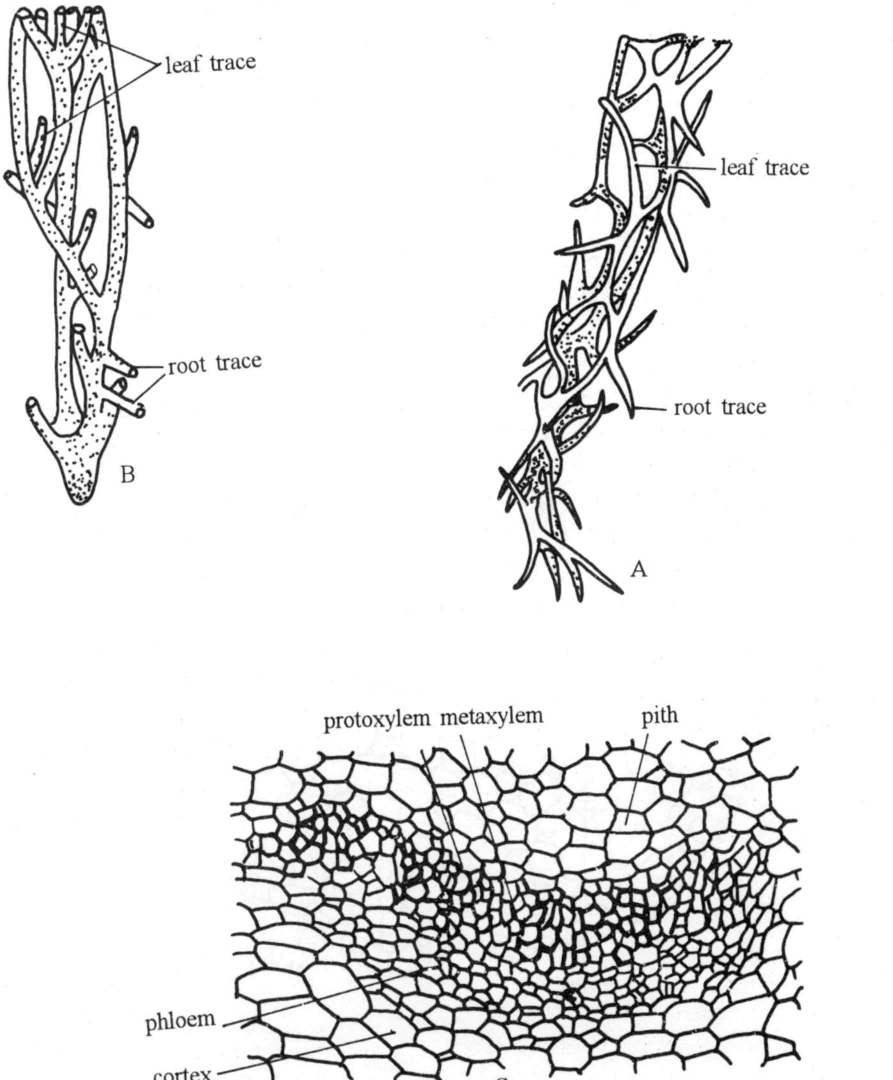

Fig 14.5 . Ohioglossum: Details of Vascular of Rhizome
A. Vascular of O. lusitanicum , B. Vasacular skeleton of O. vulgatum
C. Transverse section of a Meristcle

Root

1. A root in a transection shows a single layered epidermis, a massive parenchymatous cortex and central stele
2. The epidermis is short lived and is replaced by exodermis *(0. fibrosum),* or it may persist with suberization.
3. The cortex though wholly parenchymatous is divisible into two zones. In the outer zone, the cells are angular and closely packed without any intercellular spaces.
4. It is in this zone that the fungal hyphae are present.
5. The hyphae are aseptate.
6. The inner zone of cortex has loosely arranged parenchyma cells with no mycorrhizal fungus.
7. The stele of the root may be mono-, di-, tri-, tetra- or even pentarch. But the first two conditions are more frequent.

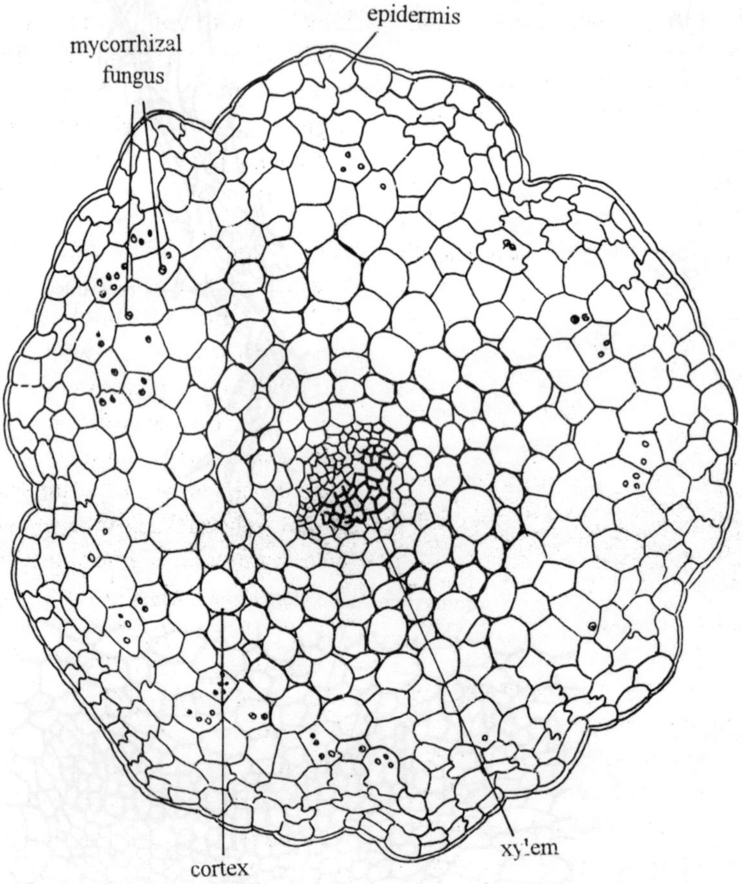

Fig. 14.6 Ophioglossum: T.S of Rhizome of O. lusitanicum

8. In diarch roots the xylem is plate like with a protoxylem *point* at each end of the plate.

9. In triarch stele, the xylem groups are separated by parenchyma.

REPRODUCTION :

1. As usual there are two methods of reproduction.

2. Vegetative propagation is brought about by the formation of adventitious buds (root tubers). E.g., *O. pendulum, O. vulgatum* etc.

Spore producing organs :

1. The spore producing organ is generally called a fertile spike.

2. The fertile spike is very much different from the strobili of *Lycopodium* or *Selaginella* in that the sporangia are completely embedded and there is no trace of any sporophyll.

3. There is a central axis like structure and on either side of this are found two rows of sporangia, arranged parallel to *the* long axis of the fertile spike.

4. The number of sporangia per fertile spike varies from 6-20.

5. Many vascular strands run into fertile spike and they frequently anastamose.

6. The lateral strands that arise from these main strands travel in between the sporangia take a 90^0 torn and reach the base of the sporangia.

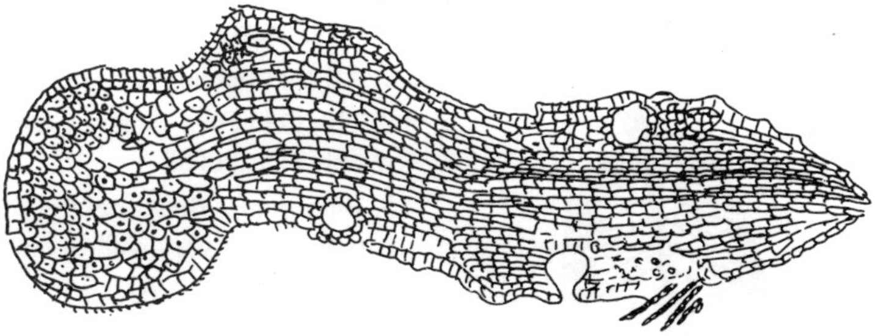

Fig. 14.7. *Ophioglossum* : Gametophyte of *O. vulgatum*

Structure of the mature prothallus :

1. The gametophyte of *Ophioglossum* like that of Psilotaceae and most of the species of *Lycopodium*, is a subterranean structure without any chlorophyllous tissue.

2. The shape of the prothallus varies in different species.

3. In *O. moluccanum* it is cylindrical and less than one cm., in length.

4. There is a slightly tuberous region at the base.

5. In *O. pendulum* the gametophyte is circular and has radiating branches.

6. The prothallus has only parenchyma.

7. The tuberous basal region consists of the mycorrhizal fungus. Prothalli may be, annual (*O. moluccanum)* or perennial *(O. pendulum)*. Rhizoids may or may not be present.

Reproduction :

1. There does not seem to be any vegetative propagation in the gametophyte.

2. Sexual reproduction is by the formation of antheridia and archegonia.

3. The prothalli are monoecious.

4. Sex organs are scattered all over the prothallus.

Systematic position

PTERROPSIDA
EUSPORANGIATAE
OPHIOGLOSSALES
OPHIOGLOSSACEAE
OPHIOGLOSSUM

MARATTIALES

1. The Marattiales include a primitive and compact group of ferns in which the sporangia are borne on the abaxial (away from the axis) surface of the leaf.

2. Sporangia are grouped into sorus.

3. Sometimes sporangia are fused laterally to form a synangium.

4. Leaves have a pair of stipules at the base of the petiole, which in not seen in the other groups of ferns.

5. All the members are homosporous.

ANGIOPTERIS

Occurrence :

1. The family Angiopteridaceae has seven genera and nearly 145 species.

2. All are tropical plants occurring in humid forests.

2. Of the seven genera *Marattia andAngiopteris* are most widely distributed.

3. *Angiopteris* flourishes very well in the tropical forests and is represented very well in the eastern hemisphere.

4. The number of species in *Angiopteris* is variable.

5. While some people recognise only one variable species *(A. evecta),* others recognise over a hundred species.

6. *A. evecta* is a common species found in India.

SPOROPHYTE

Morphology of the Plant

1. The sporophytic plant body consists of an upright, tuberous, conical, fleshy, rhizomatous stem.

2. The stem is quite thick and in habit the plant resembles a tree fern.

3. The stem is often called a **caudex** or trunk which may be a foot or two in height and almost the same girth.

4. The upper surface of the rhizome bears a crown of graceful, stately leaves.

5. The leaves are 5-6 meters long in a luxuriously growing plant, while a petiole is as thick as a man's arm.

6. The leaves are typically bipinnately compound.

7. The venation is of the open dichotomous type.

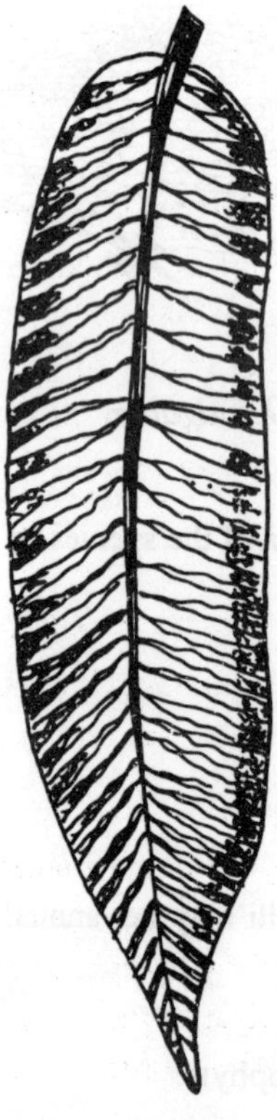

Fig. 14.8. Angiopteris : A Pinna enlarged

Fig. 14.9 Angiopteris : Habit of A. evecta

8. Pinnae are glabrous (smooth).

9. It is very typical of *Angiopteris* and other marattiaceous ferns to have a pair of thick fleshy stipules at the base of the leaf.

10. When the leaves fall off, these persist with the leaf bases and form a protective armour around the stem.

11. The base of the petiole and the stipules together appear like a horse's hoof. In some species *(A. evecta)*, at the base of the pinnae and the petiole there are swollen articulations (joints) which are apparently the regions of abscission.

12. The pinnae are long (more than 4-5" long) dorsiventrally flattened and have a long drawn tip.

13. The margin is serrate.

14. Roots are produced from the undersurface of the rhizome at the base of each leaf.

15. While leaves are deciduous, roots are perennial.

16. They are thick and have a mycorrhizal association.

17. The branching of the root may or may not be profuse.

18. Roots are endogenous in origin and pierce the stipules during the course of growth.

19. Root hairs are peculiar in being multicellular.

20. At the margins of the pinnae on the abaxial surface are borne, the sori.

21. There is no distinction into foliage leaves and sporophylls.

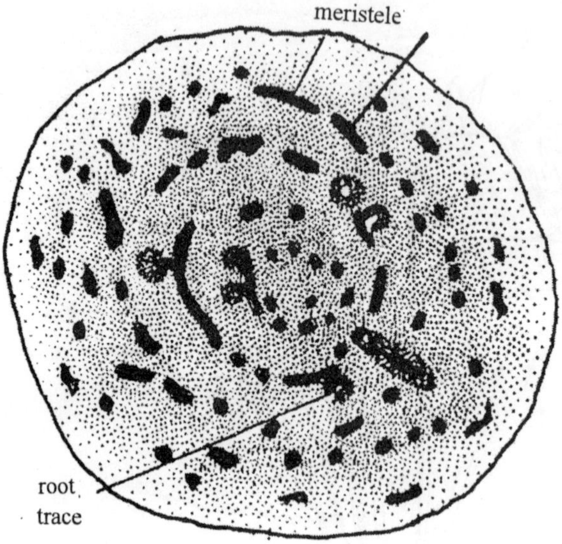

Fig. 14.10 Angiopteris: T.S of Rhizome (ground plan)

22. The sori occupy a near terminal position on the dichotomously branched veins.

INTERNAL STRUCTURE

Stem

1. A cross section of the stem shows an outer epidermis, a broad cortex and a central region of vasculature.
2. The cortex is wholly parenchymatous.
3. Here and there mucilage ducts are seen in the cortex.
4. In young stems an endodermis may be seen at the inner face of the cortex.
5. The vascular cylinder of the first formed part of a stem is protostelic.
6. At the level of the third or fourth leaf the stele changes into a siphonostele and then to a dictyostele.
7. There are many bundles arranged in several concentric rings.
8. These concentric circles represent the funnel shaped zones of anastamosing bundles.
9. Segments of the outer zone pass into the leaves as leaf traces and the gaps thus left are filled by segments of the next inner zone.
10. According to Campbell the concentric vasculature of the stem is built up by the union of leaf traces.
11. A very weak secondary development may occur in some cases. Tracheids have scalariform pitting.

Leaf

1. The petiole shows much the same structure as the stem.
2. In the hypodermal region there is a sclerenchymatous band.
3. The leaf trace may be 'C' shaped.
4. Lamina shows a prominent midrib.
5. There is a palisade parenchyma below the upper epidermis and a loose spongy parenchyma above the lower epidermis.

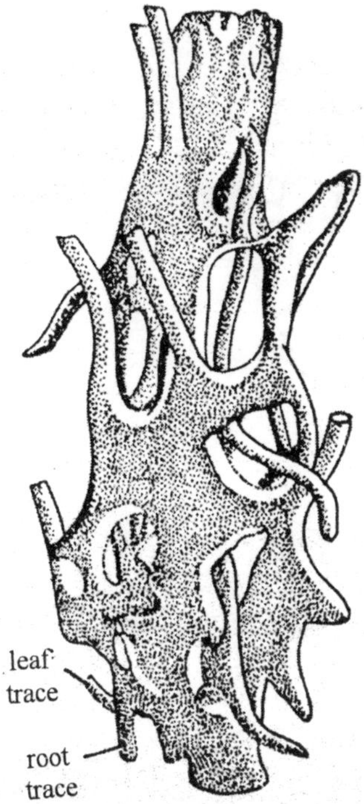

Fig.14.11. Angiopteris: Diagrammatic Representation of Stele in the Rhizome of A. evecta

6. Leaves are generally hypostomatic.

Root

1. A transverse section shows an epidermis, ccrtex and stele.
2. The cortex has parenchyma as well as sclerenchyma.
3. There is a conspicuous endodermis encircling the stele.
4. Stele is an exarch, polyarch, actinostele.
5. In the young roots a mycorrhizal fungus is present.

Reproduction

1. Sporophyte reproduces vegetatively as well as by spore production.
2. Dormant buds are formed on the rhizome where the stipules join the petiole.
3. When the stipules are shed ultimately from the plant, the buds develop into new individuals.

Spore producing organs

1. Spores are produced in sporangia which aggregate to form the sori.
2. The sporangia as has already been said, lie at the back of the veins.
3. The sporangia in the sori are arranged in two linear rows.
4. The sporangia may exhibit a certain degree of lateral fusion.
5. But this is not sufficient to call it a *Synangium*.
6. The synarigial nature in *Angiopteris* is not as evident as in *Marattia*.
7. In fact many people regard the sporangia *of Angiopteris* completely independent.

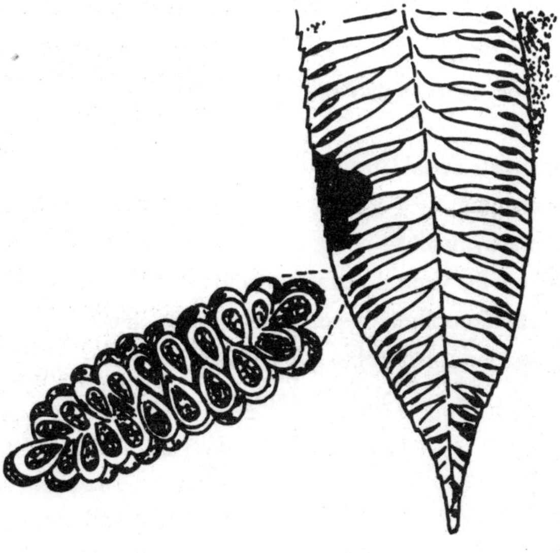

Fig14. .12. *Angiopteris* : A Part of Pinna showing Soral Arrangement

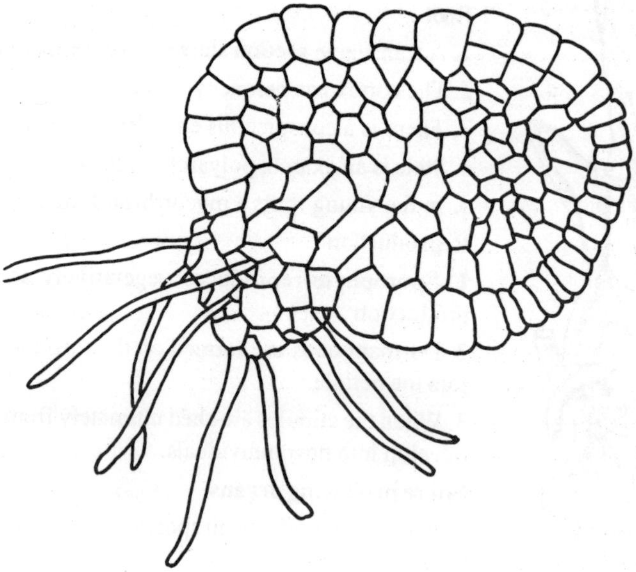

Fig. 14.13. Angiopteris: Gametophyte of A.evecta

8. There is no typical inducium.
9. Rarely inducial hairs may be present.
10. A nearly mature sporangium has a multilayered jacket enclosing a mass of spore mother cells, which undergo reduction division and produce haploid spores.
11. Spore output varies from 1450-7500, per sporaginium.
12. All spores are of the same type.
13. All the sporangia in a sorus mature together (simple sorus).

The mature prothallus :

1. The prothallus is a deep green, aerial (surface living) lobed or cushioned, thalloid structure.
2. There is a definite growing point.
3. The thallus has dorsiventral symmetry.
4. Rhizoids are produced from the ventral surface to help in anchorage and absorption.
5. In spite of the presence of chlorophyll, mycorrhizal association is always present.
6. To what extent this association is helpful to the thallus is doubtful.
7. All the cells in the prothallus are parenchymatous.

Systematic position

PTERROPSIDA

EUSPORANGIATAE

MARATTIALES

ANGIOPTERIDACEAE (Marattiaceae)

ANGIOPTERIS

OSMUNDIDAE

(Protoleptosporangiopsida)

1. Osmundidae are intermediate between eusporangiatae and leptosporangiatae in their morphological features.

2. The members are characterised by the hard, erect rhizomatous stem studded with persistent leaf bases.

3. Leaves exhibit circinate vernation typical to ferns.

4. They have winged base (as the stipular expansion) and covered with glandular hairs.

5. Stem has a complex anatomy and its cortical region is filled with 'C' shaped leaf traces.

6. Sporangia are borne on some of the pinnae of leaf which are very much crumpled and look like only aggregates of sporangia.

7. Soral organisation is not very conspicuous.

8. All the sporangia mature simultaneously.

9. Gametophytes are large and fleshy.

OSMUNDA

Occurrence and Distribution

1. *Osmunda* - the *Royal fern* is a common member of the green house.

2. It is commonly cultivated in gardens for its beautiful, attractive, lush green foliage.

3. Some of the common Indian species are *O. regalis, O. claytoniana, O. cinnamomea* etc. *Osmunda* is most widely distributed being found in the temperate and tropical regions of the southern and northern hemispheres.

4. The various species of *Osmunda* usually grow in cool, moist, shaded places such as the banks of streams, lakes, ponds etc., in the forests.

SPOROPHYTE

Morphology of the plant

1. All the species *of Osmunda* are moderately sized; a few like *O. cinnamomea* may reach a height of 2-3 metres.

2. The plant body consists of an upright rhizome which is mostly subterranean.

3. The rhizome is studded with a number of persistent, sclerotic leaf bases which are arranged in a rough spiral.

4. The rhizome is generally unbranched but occasionally a dichotomy may result.

5. From the lower surface of the rhizome are produced a number of brownish, adventitious roots.

6. Usually two roots arise at the base of each leaf.

7. The rhizome bears a crown of leaves.

8. The leaves are arranged spirally at the apex of the rhizome.

9. They are unipinnately *(O. claytoniana,* or bipinnately compound (*O. regalis)* and vary in length from 0. 30 to 3 metres.

10. Each leaf has a stout rachis with a wing like expansion (stipule) at the base covered with glandular hairs.

11. The leaves are deciduous and every year a fresh crown of leaves is produced.

12. But leaf bases are persistent and form a thick armour around the rhizome offering mechanical support.

13. The lamina has a leathery texture and is covered with brown hairs.

14. The venation is of the open dichotomous type.

15. The leaves exhibit circinate vernation characteristic of true ferns.

Fig.14.14 . *Osmunda* : Habit of *O. claytoniana*

16. The leaves may be **monomorphic** *(O. regalis* and *O. claytoniana)* i.e., there is no distinction into sterile and fertile leaves, or **dimorphic** *(O. cinnamomea* and *O. japanica).* i.e., fertile and sterile leaves are different.

17. In dimorphic species fertile leaves appear before the sterile ones, when a new crown of leaves is to be unfolded.

18. The chlorophyllous lamina is absent in the fertile leaves and they bear clusters of sporangia.

19. In monomorphic leaves at least the pinnae are dimorphic.

20. The fertile and sterile pinnae are variously distributed.

21. They may be at the base *(O. vachellii),*. tip *(O. regalis)* or in the middle (*O. claytoniana* and

INTERNAL STRUCTURE

Rhizome

1. A transverse section shows the usual three regions-epidermis, cortex and stele.
2. The outline of the section will-be irregular and discontinuous due to the presence of leaf bases and roots.
3. In a section taken in the middle region of the rhizome, the cortical region shows leaf traces or roots.
4. The cortex has two well marked zones.
5. The outer zone is mainly sclerotic and extensive, while the inner zone which is narrow, is composed of thin walled parenchyma. Numerous 'C' shaped leaf traces are found in both the cortical zones.

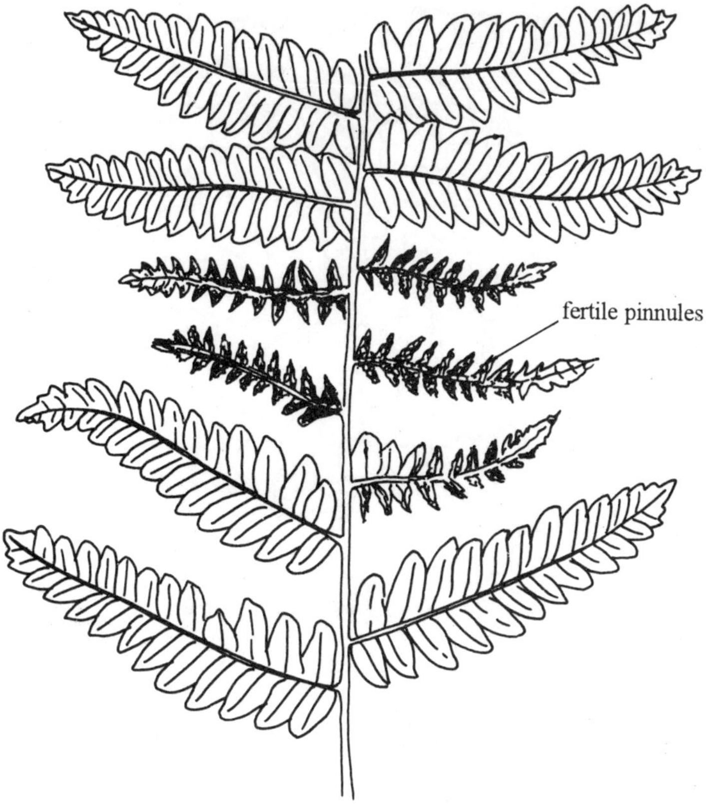

Fig.14.15 Osmunda : Part of the Leaf of O.dastoniana enlarged to show Fertile Pinnules

6. These arise obliquely from the main vascular cylinder.
7. Each leaf trace has a well defined endodermis which is lined internally by one or two layered pericycle.
8. The xylem mass is 'horse shoe shaped with the concave side pointing towards the main stele.
9. The xylem has scalariform tracheids constituting metaxylem; the protoxylem being situated in the depressions.
10. Internal to the cortex is the endodermis which is very prominent and continuous.
11. Only when a ramular gap (branch gap) is present, the continuity of the endodermis is disturbed.
12. In the young stem (sporeling) the stele is protostelic. Subsequently i.e after the development of few leaves, it becomes medullated.
13. The stele in the adult plant is an *ectopholic solenostele.*
14. The pith occupies a large portion in the centre.
15. It may be wholly parenchymatous *(O. claytoniana)* or mixed. In *O. claytoniana* the mixed pith has sclerenchyma intermingling with parenchyma, whereas in *O. regalis* pith may have tracheids and parenchyma.
16. Outlining the pith are present the meshes of xylem forming a reticulate cylinder.
17. In a transverse section xylem appears like a circle of horse shoe shaped or oval structures separated by parenchymatous rays.
18. These rays are nothing but leaf gaps.

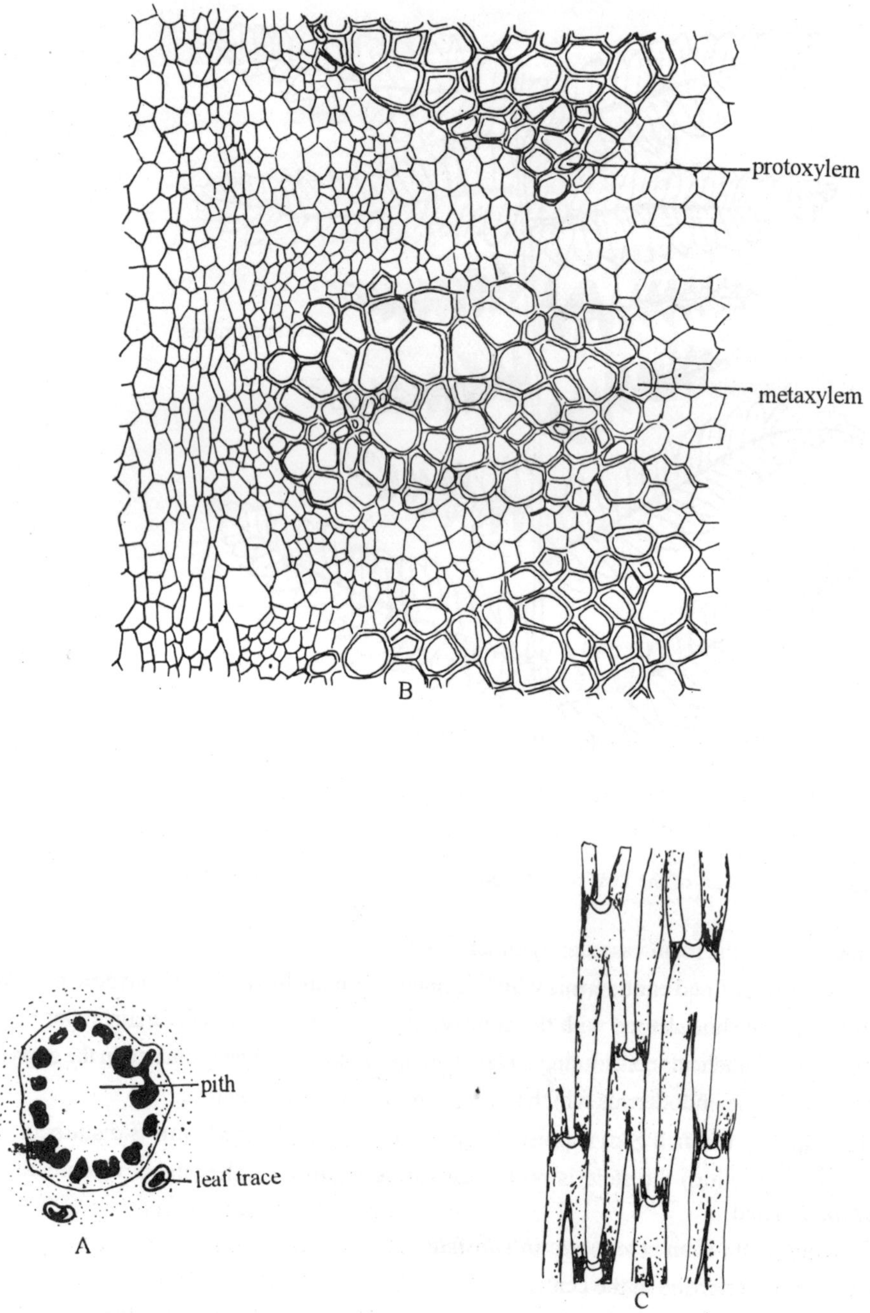

Fig. 14.16. *Osmunda* : Anatomy of Rhizome
A. Ground Plan of T.S., B. A Sector enlarged to show Steles,
C. Stereo Diagram of a part of the Stele

19. The number of xylem strands varies from 15 (*O. regalis)* to 40 *(O. claytoniana)*.

20. The protoxylem lies in the sinus fonned by the horse shoe shaped xylem mass.

21. Surrounding the xylem is a sheath formed of parenchyma.

22. External to this is the phloem. In *O. cinnamomea* an internal phloem has been reported at the region of branching.

23. Phloem consists of both proto and metaphloem.

24. External to the phloem is the pericycle, 2-4 layers in thickness.

Leaf

1. The petiole has a large concentric vasculature which is crescent shaped.
2. The two arms of the crescent are incurved.
3. The opening is towards the adaxial side.
4. Xylem consists of a number of protoxylem groups embedded in a continuous mass of metaxylem.
5. The protoxylem groups are arranged on the concave face alternating in position with groups of mucilage cells.
6. Phloem completely surrounds the xylem.
7. The endodermis is not very conspicuous.
8. The vasculature is embedded in a mass of parenchymatous tissue.

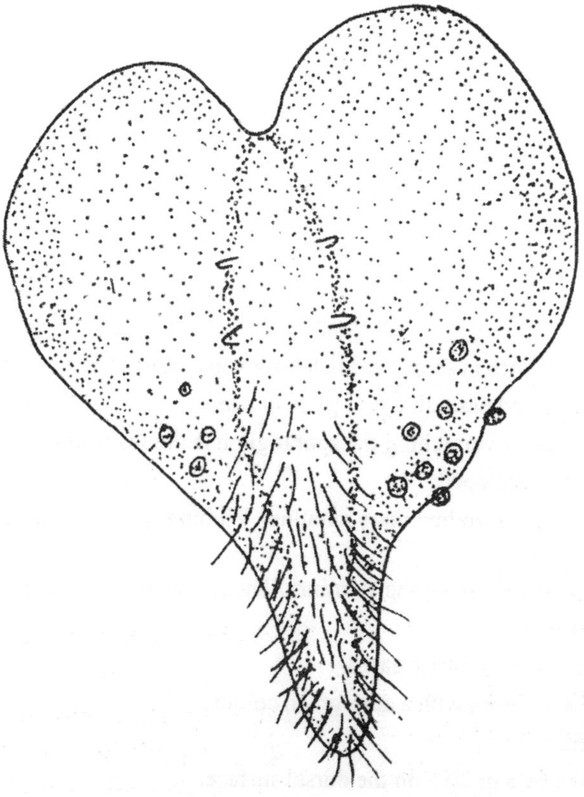

Fig. 14.17. Osmunda: Gametophyte

9. Some of the outer layers are thick walled.
10. The pinna anatomically shows bifacial arrangement.
11. The mesophyll is differentiated, but palisade parenchyma is not distinct.
12. Bounding the mesophyll on either side is the epidermis.
13. The lamina is hypostomatic.

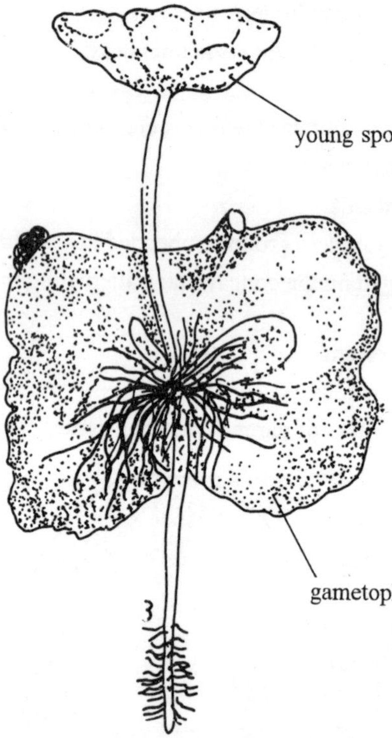

Fig. 14.18. Osmunda: Young Sporophyte emerging out of Gametophyte in O. claytoniana

14. The mature sporangium is pyriform in shape, reddish brown in colour and has a stout stalk.
15. The jacket is one layered thick and has a shield or plate like *annulus*.
16. Perpendicular to the annulus and overarching the sporangium is the *stomium.*
17. The annulus is not very well developed.
18. Hence it is often referred to as a 'rudimentary annulus'. Within the sporangium are the spores ranging in number from 256-512.
19. The sporangium dehisces vertically along the stomium. The spores are wind dispersed.

Structure of a mature prothallus

1. The protballus is cordate but somewhat elongated.
2. It is dorsiventrally flattened and fleshy with a dark green colour.
3. It grows on the surface of the soil.
4. There is a mid rib represented by a groove on the dorsal surface.
5. From the ventral surface of the prothallus are produced a number of rhizoids.
6. The prothalli are capable of indefinite growth and under favourable conditions may reach a size of 2-3 cms.

7. The prothallus is wholly parenchymatous.
8. The margin of the prothallus is thin; but towards the centre it may be 16-18 cells in thickness.
9. Nutrition is autotrophic.
10. The gametophytes are monoecious, but underdeveloped ones tend to produce only male sex organs.
11. Archegonia occur near the mid rib on the ventral surface and antheridia occur towards the fringes of the prothallus.

Systematic position

PTERROPSIDA

OSMUNDIDAE

OSMUNDALES

OSMUNDACEAE

OSMUNDA

LEPTOSPORANGIATAE

1. Among the ferns, leptosporangiatae represent a highly successful evolutionary line.
2. As the name indicates they have their sporangia traceable to a single initial cell.
3. Some of the other typical characters of leptosporangiatae are the single layered wall of the sporangium, definite number of spores per sporangium, a definite dehiscence mecnanism for the sporangium, a regular soral organisation, simple antheridia with relatively less number of antherozoids, early differentiation (at the quadrant stage) of the embryonal parts etc.
4. Leptosporangiatae are classified into three orders *viz.*, Filicales, Marsileales and Salviniales.

FILICALES

1. This is the largest order of leptosporangiatae and comprises all the homosporous members.
2. Bower has classified Filicales into three groups *viz., simplices* (simple *Sorus), gradate* (gradate sorus) and *mixtae* (mixed sorus).
3. Simplices includes a varying number of families like Schizaeaceae. Gleicheniaceae etc. Gradatae includes two families *viz.*, Hymenophyllaceae and Cyatheaceae.
4. In this book *Hymenophyllum, Adiantum* and *Pteris* are discussed.

HYMENOPHYLLUM

Members of Hymenophyllaceae are commonly called "Filmy ferns". Genera in the family is variable. Usually there are three well defined genera *viz., Hymenophyllum, Trichomanes* and *Cardiomanes*.

Occurrence and distribution

1. *Hymenophyllum* is distributed in humid, tropical and warm temperate regions of the world.
2. It is characteristically found in tropical rain forests.
3. In the west land forests of New Zealand it is especially abundant.
4. It usually grows on the forest floor or on the wet branches of trees as epiphytes.
5. In Karnataka State *Hymenophyllum* is available in Shiradi ghats near Sakleshpur (Hassan Dist.) where they grow on wet trunks of trees.
6. The number of species in the genus is variable.
7. It has about 200- 230 species.
8. Some of the common species are *H. australe, H. multifidum* and *H. recurvum* etc.

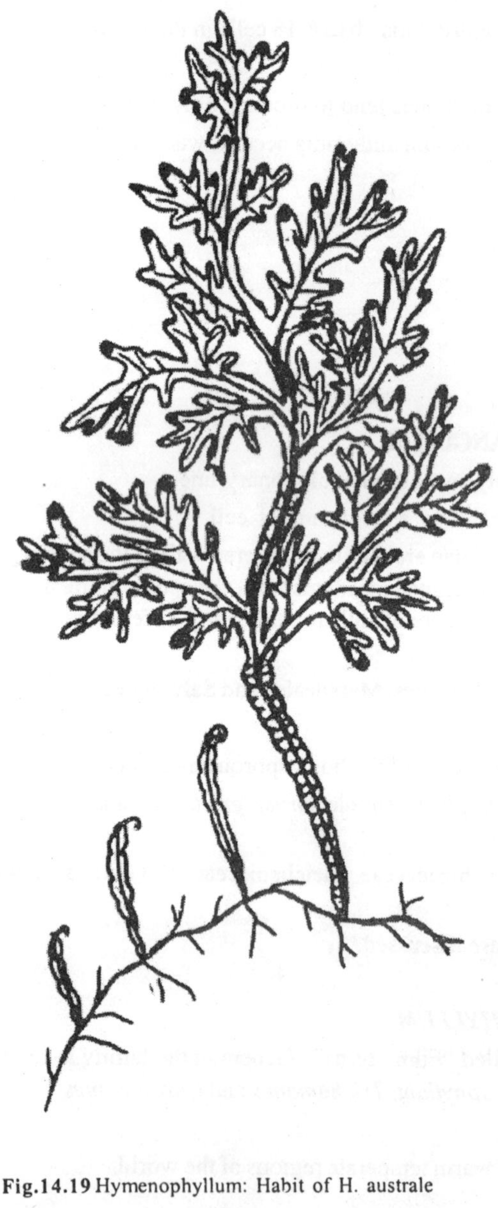

Fig.14.19 Hymenophyllum: Habit of H. australe

SPOROPHYTE

Morphology of the plant

1. The plant body rarely exceeds 3-4" in size But in *H. pulcherimum* the pendant fronds may be 2-3 feet in length.

2. It has a rhizome which is very slender and extensive branching of the rhizome is axillary.

3. Roots are adventitious (lacking in some smaller forms) and are produced in pairs at the base of each leaf.

4. But some times roots may be seen between the leaves.

5. This is due to the failure of a leaf primordium to develop above the region where the roots are produced.

6. The leaves of *Hymenophyllum* are very characteristic-true to the name filmy ferns and are objects of beauty.

7. The translucent leaves when growing in close mats may easily be mistaken for bryophytic thalli.

8. Leaves are produced in acropetalous succession and show circinate vernation.

9. But it is not uncommon to find some times the unfolding of a young leaf between two mature leaves.

10. This shows that many leaf primordia may remain dormant and may resume growth at a later stage.

11. Usually in *H. australe* one out of six primordia grows further.

12. The leaf primordia are formed in double series on the rhizome.

13. The leaves have a flattened petiole.

14. The lamina may be simple *(H.cruenta)* or dissected into unequal dichotomies as in the majority of species.

15. It is mostly one celled thick and has an open dichotomous venation.

16. In each shank of dichotomy there is a single vein.

17. In *H. cruenta* there are a series of veins reaching the margin of the simple lamina.

18. Sori are borne along the margin or at the apex of each dichotomy.

19. The leaves are pale green in colour and together with the roots they also seem to take part in absorption.

INTERNAL STRUCTURE

Rhizome

1. A transver.se section shows a relatively narrow cortex surrounding a central stele and being surrounded by an epidermis.

2. The cortex may be wholly sclerotic or partly thin walled (outer cortex) and partly thick walled (inner cortex).

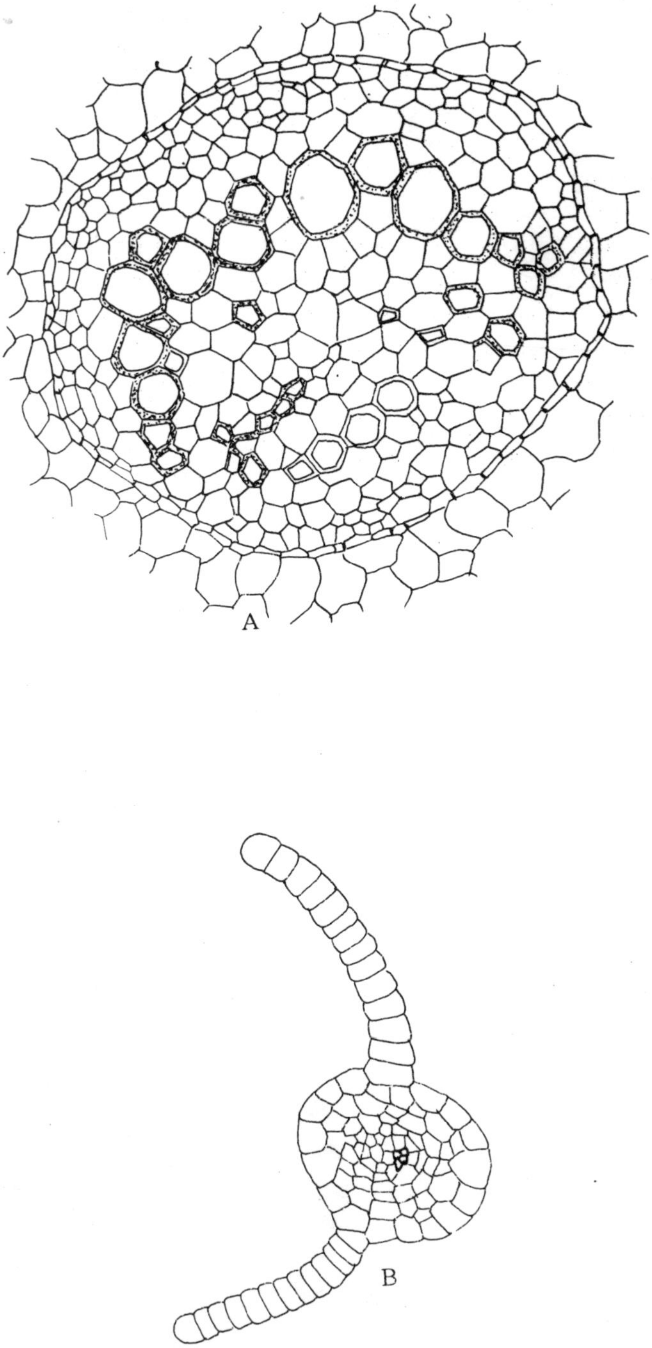

Fig. 14.20. Hymenophyllum: Anatomy of Stem and Leaf of H. australe
A. T.S. of Stele of Stem, B. T.S. of a Leaf

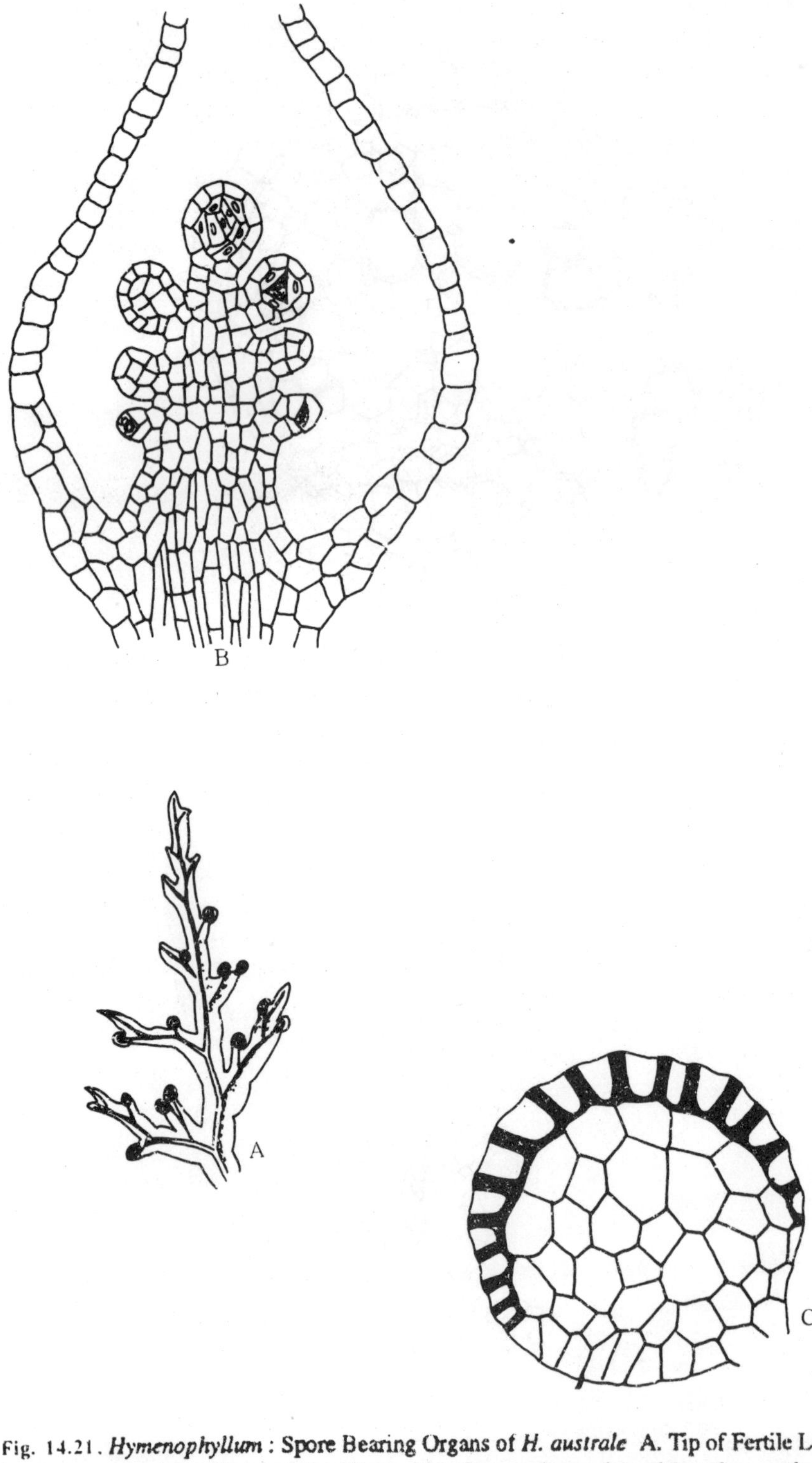

Fig. 14.21. *Hymenophyllum* : Spore Bearing Organs of *H. australe* A. Tip of Fertile Leaf. B. Long Section of Receptacle, C. A mature Sporangium enlarged (see the annulus)

3. Vascular cylinder is protostelic even in the mature plant, a feature unusual to ferns, but not unusual in the background of its habitat.
4. There is an outer endodermis surrounding the vasculature.
5. Internal to the endodermis is the pericycle one to many cells thick.
6. Phloem surrounds the xylem but the two are separated by conjunctive parenchyma.
7. The architecture of the xylem is variable in different species; in *H. demissum* there is a ring of metaxylem surrounding a parenchymatous region in which is embedded in the protoxylem, in *H. scabrum* the metaxylem ring is broken at two points to form two arcs, in *H. cruenta* one of the arcs (ventral) may be absent.
8. Leaf traces that depart from the vascular cylinder do not leave any gap.
9. Branch traces are connected to the leaf traces or more often arise from it.

Root

1. There is no root cap (Sadeback, 1898), the vascular cylinder is protostelic as in other leptosporangiatae.
2. Xylem may be monarch or diarch.

Leaf

1. Leaf blade is one cell in thickness towards the margin but many celled thick towards the central vein.
2. In *H. dilatatum* leaf blade also is many celled thick. The vein consists of poorly developed xylem and phloem.

REPRODUCTION

Vegetative propagation may be brought about by the fragmentation of the rhizome when each portion grows into a new individual.

Spore producing organs

1. The characteristic method of reproduction of the sporophyte is by the spore formation.
2. Sori are developed singly at the tip of each dichotomy.
3. In a dissected leaf blade sori are terminal and in an entire blade they are marginal.
4. The sorus has a fertile tissue called the receptacle from which sporangia are produced.
5. Receptacle development on the lamina is associated with the development of two flaps of tissue from the adaxial and abaxial surface of the leaf.
6. This is the two lipped inducium.
7. It is a true inducium as it does not represent the incurving margin of the lamina.
8. The inducium overarches the-sorus, offering protection to it.
9. The sporangia develop in a basipetalous succession true to the gradate nature of the sorus.

GAMETOPHYTE

1. A mature gametophyte of *Hymenophyllum* is filamentous, ribbon like and dichotomously branched or lobed.
2. Rhizoids are produced from the ventral surface.
3. Some of the branches of the gametophyte may grow erect.
4. Sex organs are mostly formed on the ventral side.
5. Antheridia are borne near the margins of the branches.
6. They *are* scattered all over the thallus.

Systematic position

PTERROPSIDA

LEPTOSPORANGIATAE

FILICALES - GRADATAE
HYMENOPHYLLACEAE
HYMENOPHYLLUM

ADIANTUM

Occurrence and distribution

1. *Adiantum* is popularly called *Maiden hair fern* because of the shiny black rachis of the leaves.
2. It is one of the most widely distributed genera of the family growing luxuriantly in both tropical and sub tropical regions of the world.
3. It grows ubiquitously wherever nature offers moist, shaded locality.

Fig. 14.22 Habit of A. capillus -veneris

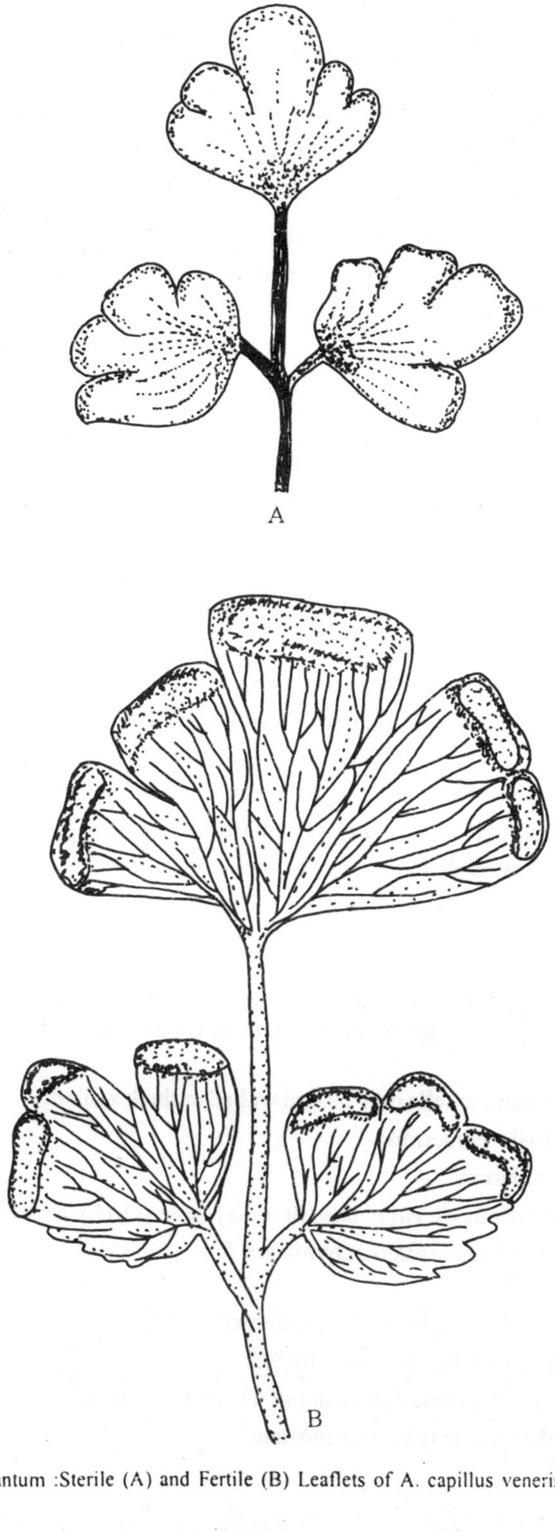

Fig. 14.23 Adiantum :Sterile (A) and Fertile (B) Leaflets of A. capillus veneris

SPOROPHYTE

Morphology of the plant

1. The sporophytic plant body consists of an underground rhizome from which are produced leaves and roots.
2. The rhizome is covered with chaffy scales *(Paleae).*
3. It may be erect *(A.caudatum),* semierect *(A.paedatum),* or creeping *(A. capillus - veneris).*
4. The rhizomes may be hard or soft and brown in colour.
5. The chaffy scales that cover the rhizome are of various shapes and sizes.
6. From the undersurface of the rhizome arise a number of adventitious roots.
7. The roots are stiff ánd black in colour.
8. Occasionally they may be branched.
9. The leaves. are produced in acropetalous succession on the creeping rhizome.
10. They show circinate vernation typical of ferns.
11. The rachis of the leaf is hard, wiry, shiny and black or dark brown in colour thus giving the name maiden hair fern.
12. The rachis has a median dorsal groove and is covered with paleae at the basal region.
13. In addition to this, glandular hairs may also be present.
14. The leaves may be unipinnate *(A. caudatum)* or bi or tripinnate as in *A. capillus-veneris.*
15. The pinnae are stalked and have a dichotomous venation.
16. The rachis may terminate in a pinna or may bear a bud.
17. In *A.capillus-veneris* the rachis divides pinnately and the ultimate branches bear pinnae in an alternate fashion.
18. There is no distinction between fertile and sterile leaves in *Adiantum.*
19. The whole leaf may be sporangiferous or only certain pinnae may bear sporangia.
20. The soral organisation is very evident.
21. Sori are borne on the ventral surface of the pinnae.

INTERNAL STRUCTURE

Rhizome

1. A transection reveals the usual three zones epidermis, cortex and stele. The outlines of the section would be wavy.
2. Epidermis is single layered and the cells may be thin walled or thick walled.
3. There is a cuticle external to the epidermis.
4. Cortex lies internal to the epidermis.
5. It may be wholly parenchymatous *(A. rubellum),* or it may have sclerenchyma and parenchyma.
6. In *A.pectinatum,* scattered masses ofsclerenchyma are found embedded in the parenchymatous ground tissue.
7. In *A. caudatum* sclerenchyma constitutes the hypodermal region.
8. The central vascular cylinder exhibits greatest variety.
9. In *A.capillus veneris,* it is a dictyostele consisting of a ring of meristeles.
10. In the young condition the stele may be asolenostele.

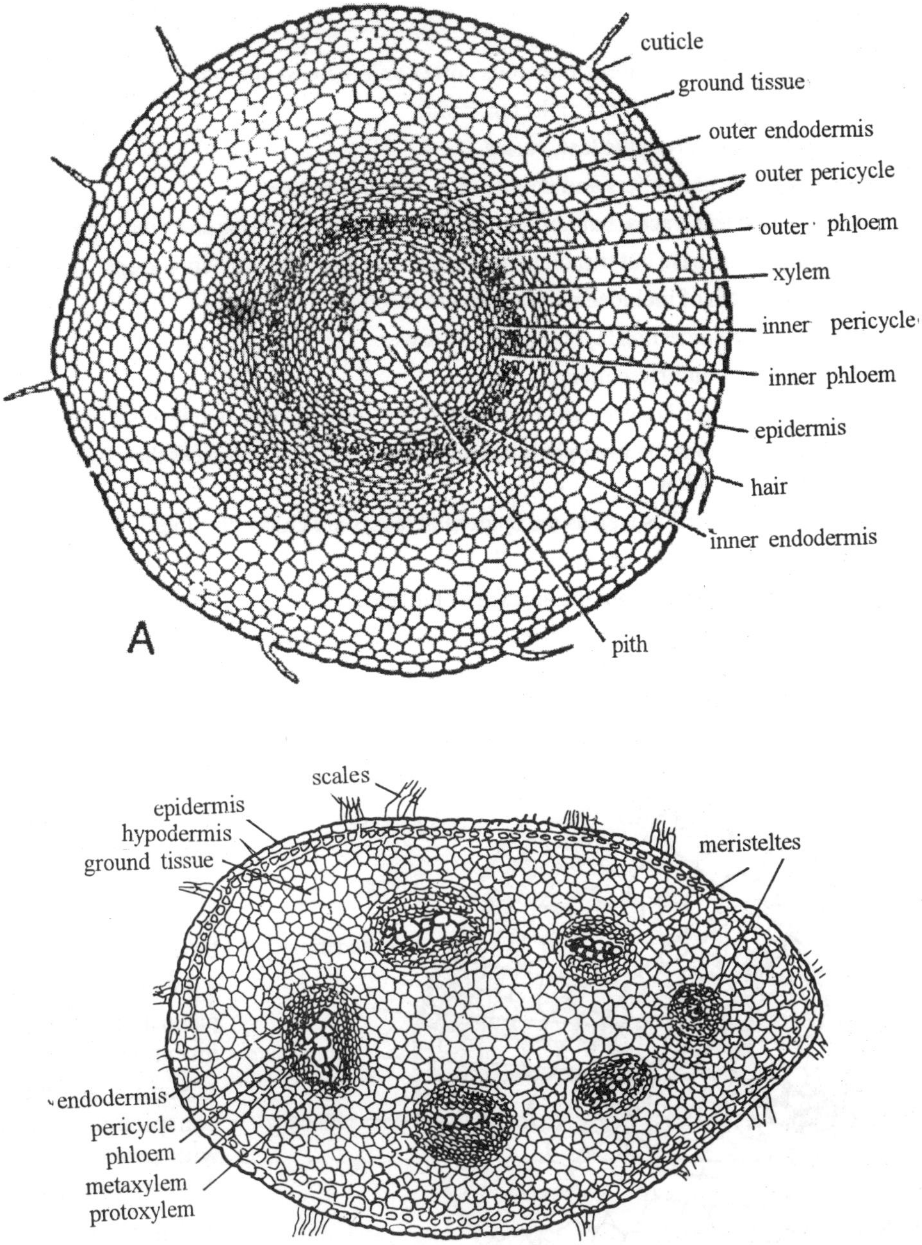

Fig. 14.24. *Adiantum* : Anatomy of Rhizome
A. Amphiphloic Siphonostele in *A. rubellum*, B. Dictyostele in *A. capillus veneris*

11. *In A. rubellum* the stele is a typical amphipbolic solenostele, with the characteristic features such as outer endodermis, outer pericycle, outer phloem, xylem, inner phloem, inner pericycle and inner endodemns lining the parenchymatous pith.

Leaf

1. The petiole shows an epidermis, parenchymatous cortex and the vascular trace.
2. There is a thick walled hypodermis next to the epidermis.
3. The number of leaf traces entering the leaf, varies.
4. It is single in *A. caudatum* and others and double in *A. capillus-veneris.*
5. Even when there are two leaf traces, both of them unite further up resulting in a single bundle.
6. The xylem is concave at the base but triradiate higher up with three protoxylem groups. Xylem is exarch.
7. In *A. bausei,* there is a patch of included parenchyma in the xylem.
8. The lamina shows the two epidermal layers upper and lower.
9. The mesophyll is generally undifferentiated.
10. It is highly reduced in *A.capillus-veneris, A. pedatum,* etc., having only two layers of cells.
11. In *A.pedatum,* in some regions the mesophyll is totally absent and at such places the two epidermal layers are closely appressed to each other.
12. The mesophyll (when present) as well as the epidermal layers are chlorophyllous.
13. The epidermal cells over the veins are thick walled.
14. The stomata are scattered throughout the surface of the leaf.
15. Palaeae or ramenta may be borne even on the epidermis of the lamina.
16. The vein may or may not have a bundle sheath. The vascular tissues show the characteristic X/P arrangement.

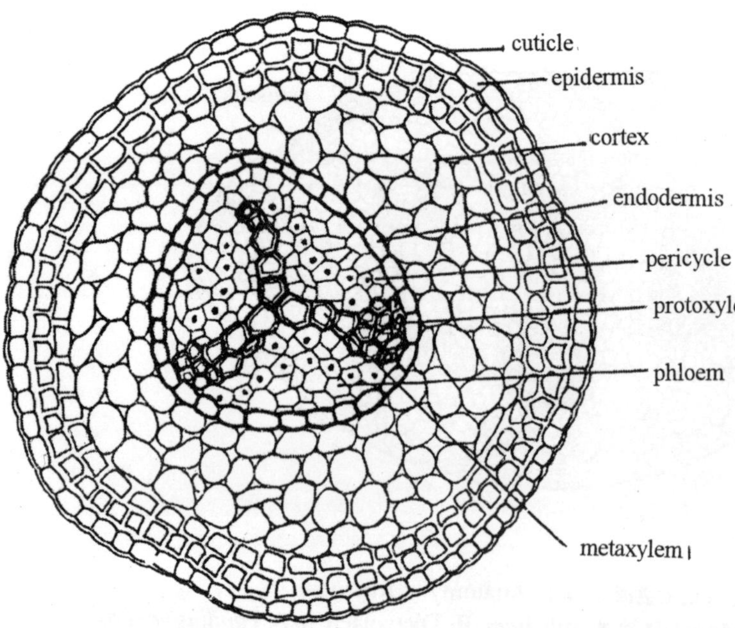

Fig. 14.25 . *Adiantum* : T.S. of Petiole

Root

1. A transection shows a very prominent piliferous layer, a two zoned cortex and the central protostele.
2. The piliferous layer has brown colored cell walls.
3. Cortex has an outer parenchymatous zone and an inner sclerotic zone.
4. Surrounding the stele is a conspicuous endodermis with prominent Casparian thickenings.
5. The xylem is exarch and diarch.
6. Phloem completely surrounds the xylem.
7. External to phloem is a single layered pericycle.

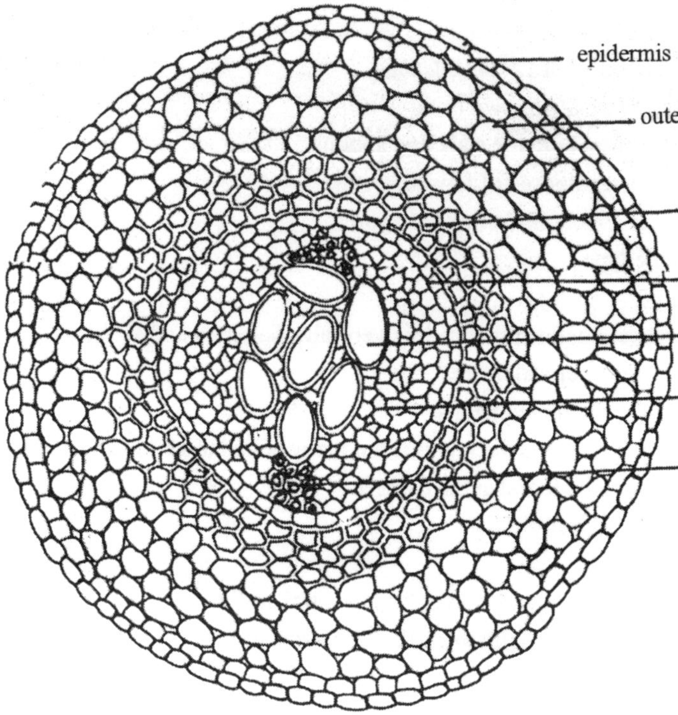

Fig. 14.26 Adiantum : T.S of Root

REPRODUCTION

1. Vegetative propagation is brought about by buds produced at the leaf tips.
2. The buds enter the ground when the leaf bends and touches the soil.
3. There, they develop into a new individual.
4. This, in turn may repeat the process leading to the "Walking habit".
5. Walking habit is seen in *A. caudatum.*

Spore producing organs

1. As has already been said there is no distinction into fertile and sterile leaves.
2. The sori are borne on the distal end of the pinnae.
3. But the sori are not exactly marginal.
4. They are borne a little behind the tip of the veins.
5. The sorus bearing margin of the leaf incurls and forms the false inducium.
6. In some cases sporangia may develop at the distal ends of the veins *(A. philippense).*
7. In the sori paraphyses may be present in between the sporangia as in *A. mbellum, A. tenerum,* etc.
8. The sorus is of the mixed type.

Development and structure of the sporangium

1. The development is the similar to what is seen in *Pteris.*
2. A mature sporangium has a stalk made up of three rows of cells.

3. The stalk terminates in a globose or biconvex capsule.
4. The wall is single layered.
5. There is an obliquely vertical annulus of 12-24 cells long.
6. The annulus is separated from the stalk by two or three cells.
7. The stomium also is separated from both the stalk and annulus.
8. The rest of the sporangial wall is composed of few large cells.
9. The sporangium dehisces transversely liberating the spores.
10. All the spores are of the same type.

GAMETOPHYTE

1. The mature prothallus is cordate, photosynthetic, dorsiventrally flattened and aerial.
2. The growing point is situated in the apical notch.
3. All the cells in the prothallus the parenchymatous.
4. The prothallus is one celled thick towards the margins but many celled thick towards the centre.
5. In some species collenchyma may be found at the corners.
6. Rhizoids are produced from the ventral surface.
7. The prothalli are monoecious.
8. Antheridia are found in between the rhizoids towards the ventral surface.
9. Archegonia are found near the growing point towards the ventral surface.

Systematic position

PTERROPSIDA

LEPTOSPORANGIATAE

FILICALES - MIXTAE

ADIANTACEAE

ADIANTUM

PTERIS

Occurrence and distribution

1. This is a cosmopolitan fern being distributed in almost all geographical regions.
2. *Pteris* however prefers tropical and sub tropical climates.
3. Plants usually grow in well drained places or in the crevices of rocks.
4. They are very common along the slopes of hills and can be seen even at 1200 metres above sea level.
5. There are about 250-280 species reported for the genus.
6. Some of the common Indian species are *P. quadriaurita, P. critica, P. vittata, P. pellucida, P. wallichiana, P. stenophylla, P. biaurita,* etc.

SPOROPHYTE

Morphology of the plant

1. The plant body has a rhizomatous stem that produces roots and leaves.
2. The rhizome may be creeping *(P.grandiflora)* or compact and erect as in *P. critica* and *P. longifolia.*
3. The rhizome may or may not show branching and is covered with scales.
4. Roots arise at the base of the leaf or all over the rhizome.
5. The growing point of the rhizome is covered with ramenta.

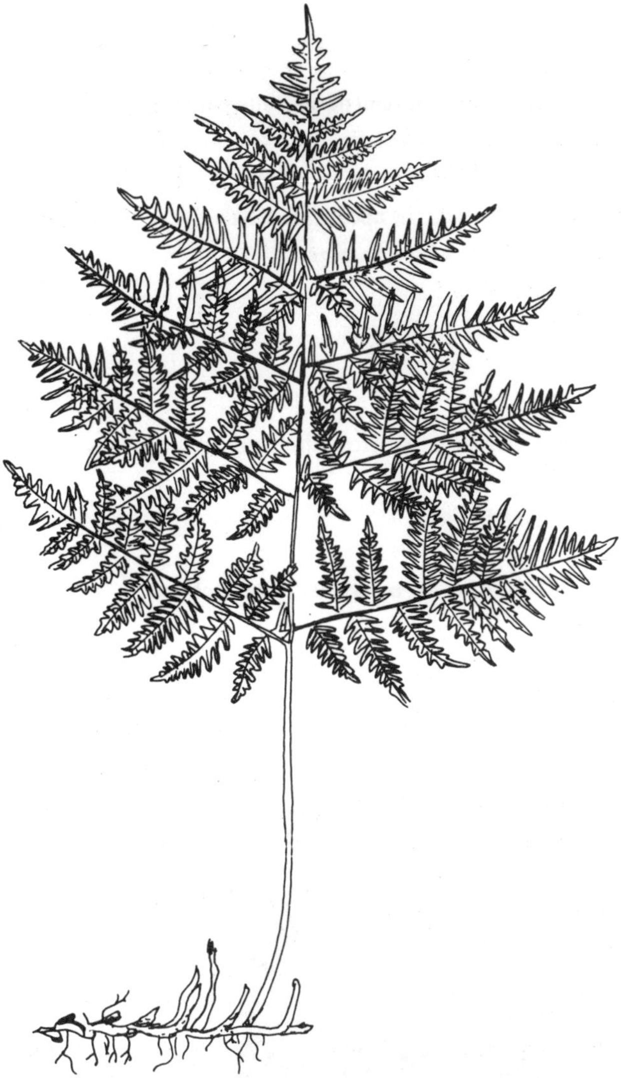

Fig. 14.27.Pteris: Habit

6. The leaves arise from the upper surface of the rhizome and have a long rachis.

7. They are unipinnately compound, decompound or multipinnately compound.

8. The dissections of the pinnae are not as deep as in *Pteridium.* Venation if of open dichotomous type.

9. The pinnae are small near the base, large towards the middle and once again small towards the apex as in *P.vittata.*

10. The leaves show circinate vernation that is typical to true ferns.

11. Very often pinnae are coriaceous (leathery).

12. The leaves bear the reproductive structures - sori along the ventral margin of the pinnae.

13. The sorus.is continuous along the margin but avoiding the apices of the segments and usually in the sinuses between them.

INTERNAL STRUCTURE

Rhyzome

1. The rhizome in a transection exhibits quite a good amount of diversity particularly in the stelar organisation.
2. It is solenostelic *(P.grandiflora, P. vittata),* or Dictyostelic.
3. In *P. biaurita* the rhizome shows a mixed protostele in the lower region.
4. It becomes siphonostelic a little higher up and near the apex it shows a polycyclic dictyostelic organisation.
5. The epidermis is single layered and it encloses a broad cortex.
6. There is a hypodermal sclerenchyma.
7. The bulk of the cortex however is made up of parenchyma.
8. The stele consists of a number of meristeles.
9. Usually there are two rings of meristeles.
10. The inner ring consist of 2 or 3 large meristeles whereas the outer ring consists of a number of small meristeles.
11. Each meristele has a band or plate like xylem mass.
12. Sometimes it is angular.
13. One or two protoxylem groups are embedded in the metaxylem (mesarch).
14. Surrounding the xylem is the phloem.
15. Each meristele may have its own endodermis.

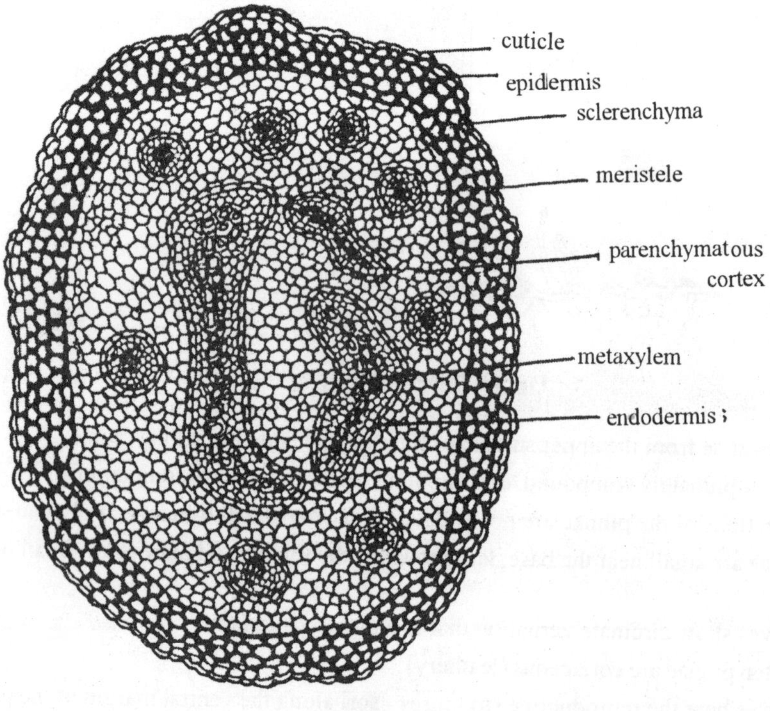

Fig. 14.28 Pteris : Anatomy of Rhizome

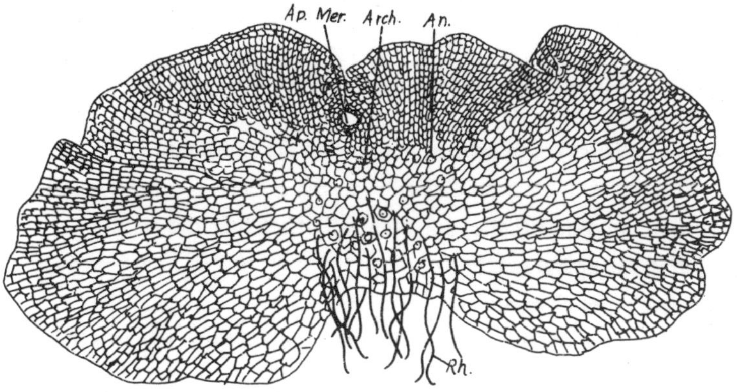

Fig. 14.29 .Pteris : Gametophyte

Leaf

1. The rachis is. traversed by a single leaf trace which is variable in shape. 'C' shaped leaf traces are found in *P. vittata.*
2. In *P. biaurita* the leaf trace is 'U"shaped while entering the leaf base, but further up it becomes 'V shaped.
3. The xylem strand appears hooked.
4. As usual xylem is surrounded by phloem, pericycle and endodermis.
5. The cortical region has an inner parenchymatous zone and an outer sclerotic zone.
6. Epidermis is single layered with a deposition of cuticle over it.
7. Ramenta arise from some of the epidermal cells.
8. The lamina appears bifacial.
9. It may or may not have a differentiated mesophyll.
10. Leaves are generally hypostomatic.
11. The midrib has a concentric vascular bundle *(P. vittata)* with a distinct endodermis (bundle sheath).

Root

1. Anatomically the root shows an outer piliferous layer and a cortex divided into an outer parenchymatous zone and an inner sclerotic zone.
2. The stele is protostelic and has an exarch, diarch xylem.

REPRODUCTION

This is brought about by spores.

Spore producing organs

1. The sorus is called a *Coenosorus* as it is a continuous one.
2. The coenosorus is borne on marginal connecting veins and is protected by the reflexed margin (false inducium) of the pinnae.
3. A true inducium is totally absent.
4. A section of the pinna passing through the sorus clearly shows the false inducium surrounding the sporangia in different developmental conditions.
5. Sorus is of the mixed type; sterile hairs being present in between the sporangia.

GAMETOPHYTE

1. The mature prothallus is cordate, photosynthetic, dorsiventrally flattened and aerial.
2. The growing point is situated in the apical notch.
3. All the cells in the prothallus are parenchymatous.
4. The prothallus is one celled thick towards the margins but many celled thick towards the centre.
5. In some species collenchyma may be found at the corners.
6. Rhizoids are produced from the ventral surface.
7. The prothalli are monoecious.
8. Antheridia are found in between the rhizoids towards the ventral surface.
9. Archegonia are found near the growing point towards the ventral surface.

Systematic position

PTERROPSIDA
LEPTOSPORANGIATAE
FILICALES - MIXTAE
DENNSTAEDTIACEAE
PTERIDOIDEAE
PTERIS

MARSILLEALES

1. These are heterosporous, leptosprangiate, aquatic or semiaquatic ferns.
2. Spore bearing organs are encased in a special bean shaped structure called the sporocarp.
3. The order comprises a single family Marseliaceae with three genera *viz., Marsilea, Pilularia* and *Regnellidint. Marsilea* is discussed here.

MARSILEA

Occurrence and distribution

1. It is one of the most widely distributed among ferns.
2. But the genus has a preference to the warmer parts of the world.
3. The plants grow in aquatic *(M. quadrifolia)* or semiterrestrial *(M. condensata)* habitats.
4. A few species are known to grow in dry soil.
5. The genus consists of about 53 species (Reed, 1954; Gupta 1962) of which about nine are recorded from India, some of the common species are *M. quadrifolia, M. hirsuta, M minuta, M. vestita, M. aegyptiaca* etc.

SPOROPHYTE

Morphology of tbe plant

1. The plant body consists of a creeping cylindrical rhizome which is partly or completely subterranean.

2. The rhizome has in definite length and under luxuriant growth conditions may reach a length of 25 metres.
3. Branching in the rhizome is quite frequent and it may be axillary or lateral.
4. The rhizome is distinguished into nodes and internodes.
5. The nodal regions are slightly swollen.
6. The internodes are longer in aquatic species *(M. quadrifolia)* and shorter in terrestrial species.
7. The leaves and roots are generally borne at the nodes.
8. The leaves have a long and slender petiole.
9. In aquatic species only the lamina floats on the surface of water.
10. The leaves are arranged alternately in two rows on the upper surface of the rhizome.
11. The leaves are compound and have a varying number of pinnae.
12. The number of pinnae per leaf may be 3, 5, 6, 7 or 8. Frequently it is four as see in n *M. quadrifolia*.
13. The leaves show circinate vernation.

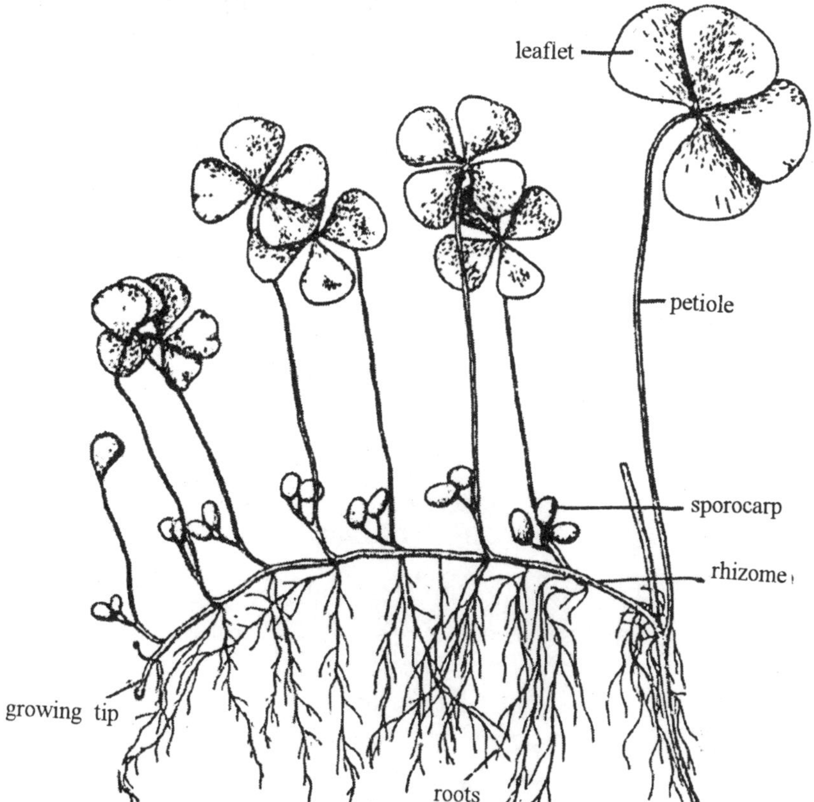

Fig. 14.30. Marsilea : Habit of M .quadrifolia

14. The pinnae vary in their shape.
15. They may be obovate, obcuneate or elliptical.
16. They may or may not be lobed.
17. The margin of the pinnae may be smooth or dentate (*M. minuta*).
18. The venation is of the closed reticulate type.
19. The pinnae exhibit sleeping movements.
20. Roots are borne on the undersurface of the rhizome at the nodal region.
21. But it is not uncommon to find roots emerging from the internodal region.
22. Roots develop in acropetalous succession (youngest near the growing tip of the rhizome).
23. The plant bears reproductive structures which are very special to Marsileales.
24. These are called the *sporocarps*.
25. The sporocarps have a stalk called the peduncle or the pedicel.
26. The peduncle may be attached to the stem or the petiole.
27. The sporocarps are bean shaped and bisporangiate.

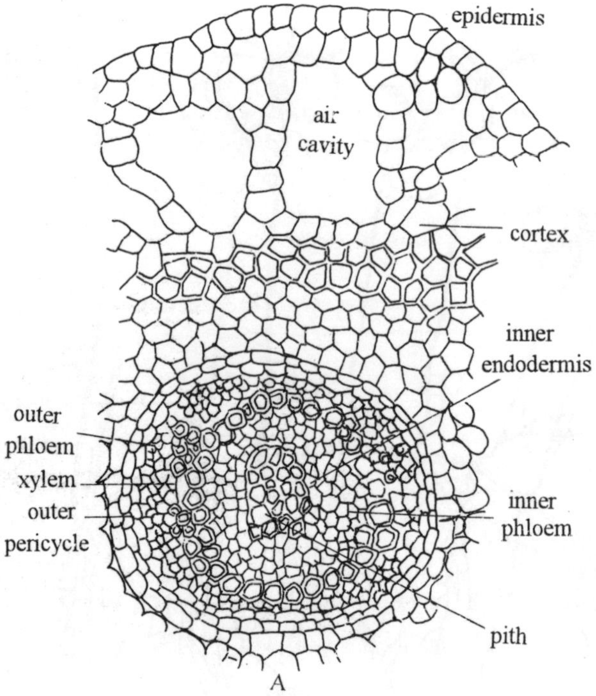

Fig. 14.31. *Marsilea* : T.S. of Rhizome

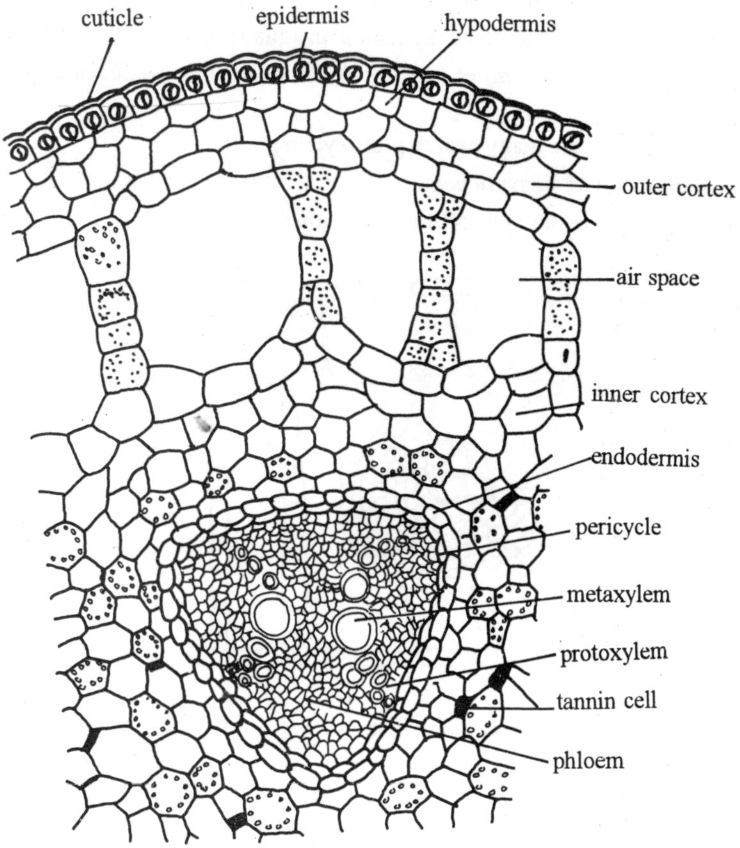

Fig. 14.32 Marsilea : T.S of Petiole

INTERNAL STRUCTURE

Rhizome

1. A transection shows the usual three regions *viz.,* epidermis, cortex and central stele.
2. Epidermis is single layered, continuous and lacks stomata.
3. The cortex has many regions.
4. The outermost cortex consists of compactly arranged parenchyma cells.
5. Next to this is the aerenchymatous cortex.
6. This consists of broad airspaces separated by parenchymatous portions.
7. Internal to the arenchyma is the sclerotic zone having one or two layers of thick walled cells.
8. The innermost region of the cortex is again parenchymatous.
9. In this region may be found a few tannin cells.
10. The central vasculature is an *amphiphloic solenostele.*
11. As is typical to such steles there is an outer endodermis, outer pericycle, outer phloem, xylem, inner phloem, inner pericycle and inner endodermis.

12. The composition of the pith varies with the habitat.
13. It is scelrotic in terrestrial species but parenchymatous in aquatic species.
14. In the xylem ring, there may *(M. vestita)* or may not be (*M. quadrifolia)* a distinction between proto and metaxylem.
15. When a distinction exists there are many exarch protoxylem points.
16. The xylem may be mesarch in *M. aegypticaca*.
17. Leaf traces depart from the vascular cylinder to provide the leaf.

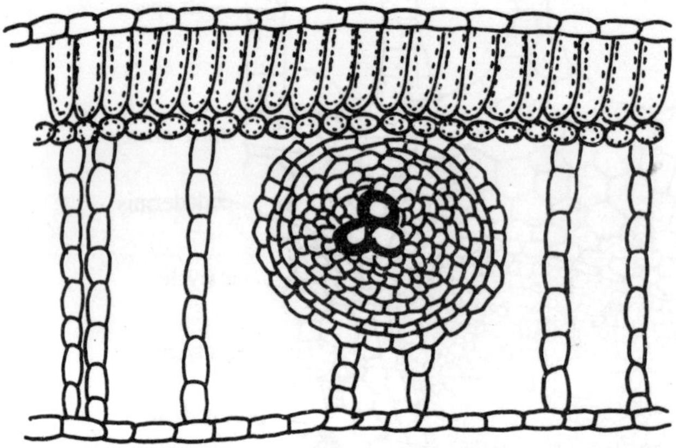

Fig. 14.33. Marisilea : T.S of Leaf

Root

1. Anatomically the root shows a single layered epidermis, a three zoned cortex and a central protostele.
2. The cortex has an outer aerenchymatous, a middle parenchymatous and in inner sclerotic zone.
3. Stele is bounded by an endodermis and pericycle.
4. Xylem consists of two large metaxylem tracheids and two protoxylem. tracheids.
5. It is exarch and diarch.
6. Vessels have been reported in *M. quadrifolia, M. drummnndii* and *M. hirstlta.*

Leaf

1. The petiole has an outermost epidermis.
2. The cortex has two zones *viz.,* an outer aerenchymatous and an inner parenchymatous zone.
3. The inner zone is rich in starch and tannin.
4. The vasculature consists of a 'V' shaped xylem with the two arms slightly bent outwards.
5. Some times the two arms may be separated at the base also.
6. The xylem is surrounded by phloem, pericycle and endodermis.
7. The lamina is bifacial.
8. It has an upper and lower epidermis bounding the mesophyll.
9. Leaves may be epistomatic (aquatic) or amphistomatic (terrestrial).
10. The mesophyll shows clearly defined palisade and spongy parenchyma.
11. Vascular bundles are concentric with the centrally situated xylem surrounded by phloem.

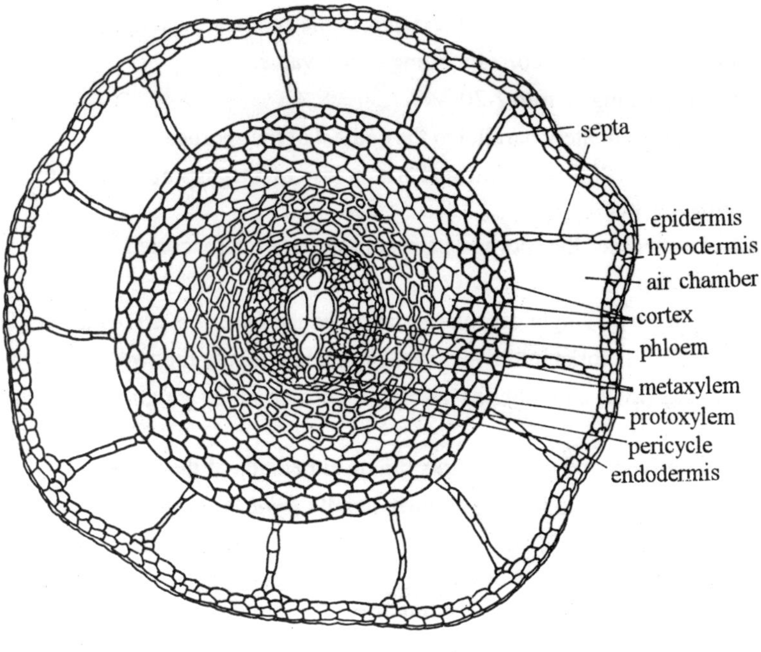

Fig. 14.34. Marsilea: T.S. of Root

REPRODUCTION

1. The sporophyte reproduces by vegetative propagation as well as by spore production.
2. Vegetative propagation is brought about *by tubers* (*M. hirsuta)*.
3. These are small, condensed side branches having leaf primordia.
4. The tubers persist even after a plant dies during unfavourable conditions and on the availability of favourable conditions, develop into a new plant.

Spore producing organs

1. As has already been pointed out, the spore bearing structures, are called sporocarps.
2. Each sporocarp has a short stalk - the pedicel
3. The mode of attachment of the pedicel to the plant is varied.
4. In *M.polycarpa* and *M. subangulata* the pedicels are directly attached to the petiole in a linear sequence.
5. In *M. quadrifolia* the pedicels of many sporocarps join together and have a common stalk joined to the petiole.
6. In *M. vestita* a single pedicel is attached to the base of the petiole.
7. The sporocarp is bisporangiate and is usually light green in colour.
8. At the point of attachment of the pedicel to the body of the sporocarp there are one to two teeth like projections called tubercles.

Structure of the mature sporocarp

1. The sporocarp is made up of two valves which can be split open along the dorsiventral plane so as to reveal the internal structure.
2. The wall of the sporocarp is very thick.

3. Adhering to the wall is a *gelatinous* ring that extends in the sporocarp cavity.

4. There are two rows of sori, one row corresponding to one valve.

5. The number of sori per row ranges from 2-20.

6. The sori are arranged approximately parallel to the pedicel of the sporocarp.

7. Each sorus is encased in a thin indicium and has a row of megasporangia on top and two rows of microsporangia towards the sides.

8. In *M. rajasthanensis* sometimes the sporocarps may have only microsporangia.

Vasculature in the sporocarp

1. The sporocarp is supplied by a single vascular trace called the *Dorsal median bundle* (DMB).

2. At the region of tubercle this forks into two and the branches supply the two valves.

3. From each branch of the DMB a number of laterals or commissurals are given out one corresponding to each sorus.

4. The laterals also fork halfway and the two forks run on either side of the sorus.

5. From the point of forking of the lateral, another bundle *(placental bundle)* starts and enters into the receptacle where it also forks.

6. The ends of the forks of the laterals are not free.

7. They fuse with those of preceding and succeeding forks forming a closed network.

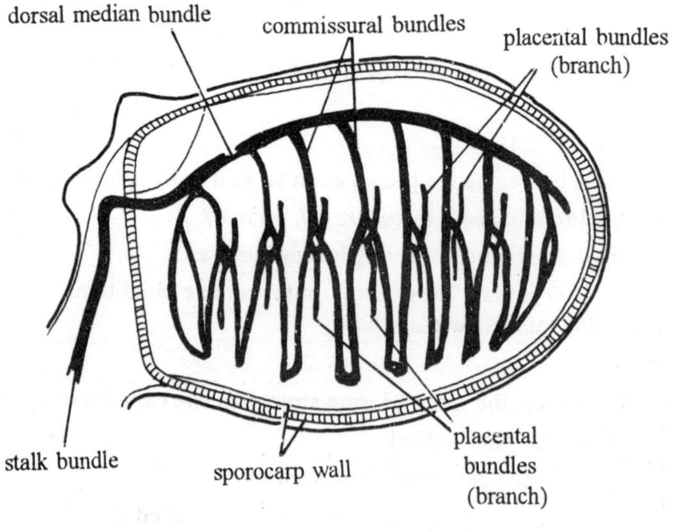

Fig. 14.35 .Marsilea: Vascular Supply of the Sporocarp

Detailed internal structure of the sporocarp

1. Sections cut at different planes clearly reveal the internal structure.

2. The following are the three planes of section:

   a) Vertical longitudinal section or section cut parallel to the flat surface of the sporocarp.

   b) Horizontal section or section perpendicular to the pedicel.

   c) Vertical transverse section or section perpendicular to the flat surface of the sporocarp.

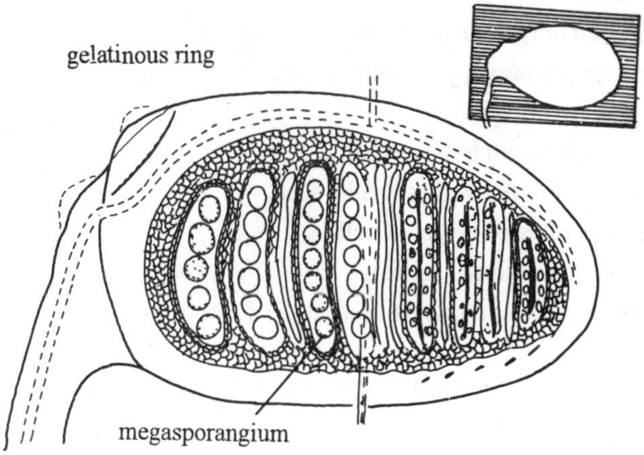

Fig. 14.36. Marsilea: Internal Structure of the Sporocarp (vertical long section parallel to the surface

Vertical longitudinal section

1. The section shows a wall made up of three layers of cells.
2. The outermost is epidermis which has stomata and two hypodermal layers.
3. The hypodermal cells are columnar.
4. Next to the wall is the continuous gelatinous ring.

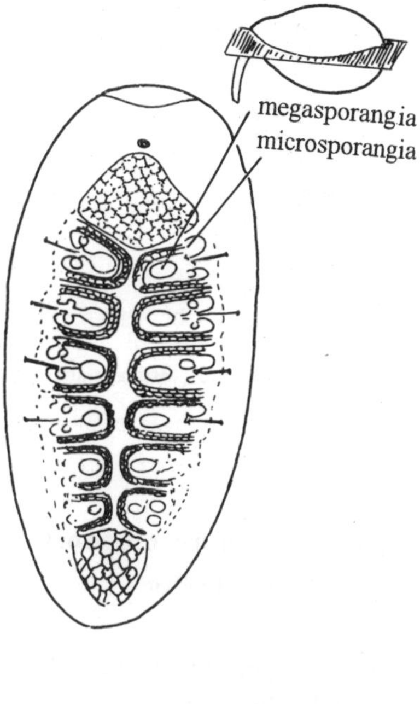

Fig. 14.37. Marsilea: Internal Structure of the Sporocarp (Horizontal Section)

5. Sori are cut lengthwise.

6. Sori of only one valve can be seen.

7. If the section is slightly superficial only microsporangia (two rows) are seen.

8. If the section is deeper only megasporangia (one row) are seen. An inducium envelopes all the sori.

Horizontal section

1. In horizontal section the pedicel will be cut transversely.

2. It is in this section that the sori of both the valves can be seen.

3. The wall of the sporocarp consists of three layers of cells.

4. Its structure is same as in vertical longitudinal section.

5. The gelatinous ring is cut transversely and can be seen in the form of two patches at either end of the section.

6. All the sori are cut transversely, so also the inducial covering and the lateral bundles which can be seen at the base of the sorus.

7. In each sorus can be seen a terminal megasporangium and two lateral microsporangia.

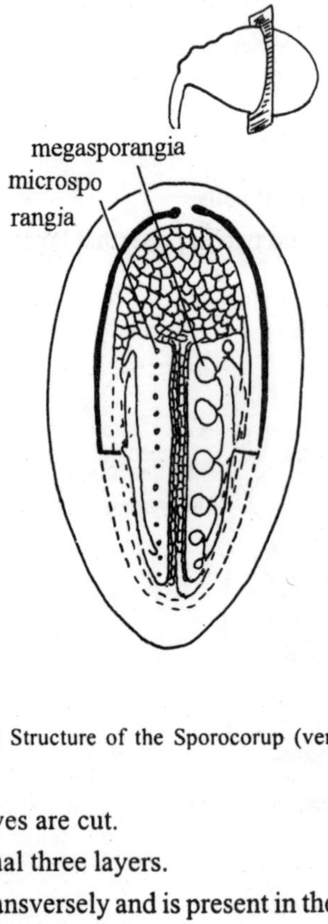

Fig. 14.38. Marsilea: Internal Structure of the Sporocorup (vertical transverse section)

Vertical transverse section

1. In this section both the valves are cut.

2. The wall consists of the usual three layers.

3. The gelatinous ring is cut transversely and is present in the form of two patches at either end of the section.

4. Sori are cut lengthwise and each sorus represents a valve.

5. Inducia are attached to this.

6. A slightly superficial section shows only microsporangia and a deeper section shows a row of megasporangia with one or two microsporangia at either end of the sours.

Systematic position

PTERROPSIDA

LEPTOSPORANGIATAE

MARSILEALES

MARSILEACEAE

MARSILEA

SALVINIALES

1. The order Salviniales belongs to sub class Leptosporangiate.
2. The order is closely related to Marsileales.
3. The members of the order are commonly referred to as 'water ferns'.
4. The orderis similar to Marsileales in having the reproductive structures encased within a bean shaped sporocarp.
5. The sporocarp in Salviniales is fundamentally different from that of Marsileales in that, in the latter the sporocarp represents a single sorus.
6. The order includes two families namely Salviniaceae and Azollaceae.
7. The family Salviniaceae has a single genus *Salvinia* with about 12 species and family Azollaceae includes one genus *Azolla* with 6 species.

SALVINIA

Occurrence And Distribution

1. The genus *Salvinia* has 12 species most of which are found to be distributed in the African continent.
2. The genus is represented in India by two species namely *S.natans* and *S.auriculata*. *S.natans* is found in Kashmir.

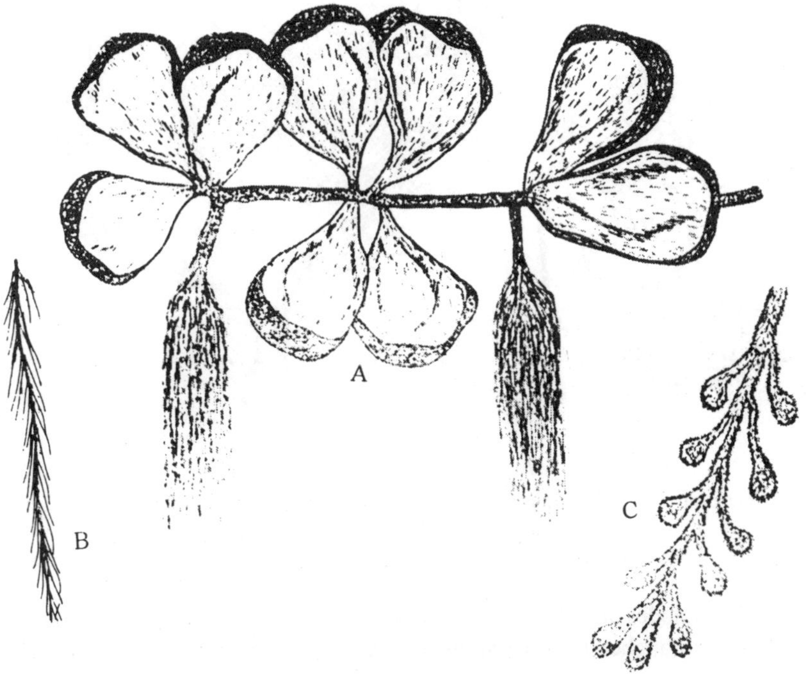

Fig. 14.39. Salvinia: Habit of S. auriculata
A. Plant Body (a portion), B. Submerged Leaf, c. Submerged Leaf with Sporocarp

3. Of the several specied of *Salvinia*, *S.natans* is the oniy annual.

SPOROPHYTE

Morphology of the Plant

1. The sporophytic plantbody consists of a long slender floating rhizome which grows horizonatally.
2. The rhizome is completely covered by the whorls of leaves.

Leaves

1. Leaves are oblong or sometimes hemispherical.
2. They usually arise in groups of three.
3. Of the three leaves, two lateral ones float on the surface of water while the third one is submerged.
4. There is a distinct morphological differences between the submerged and floating leaves.
5. The submerged leaves are highly dissected, long and filiform.
6. They very much resemble the roots.
7. But is it not certain whether they also function like roots.
8. It is very often suggested that the function of filliform leaves is
   (i) to proctect the sporocarp and
   (ii) to act as a ballast or a stabilizer in preventing undue drift of the plant.
9. The floating leaves are green and ovate or oblong.
10. They are covered withstiff hairs that prevent wetting.
11. Besides the hairs, papillate projections are also found.
12. The members of the leaf whorl alternate at successive nodes, so that there are actually six rows of leaves though the arrangement apparently suggests three rows.

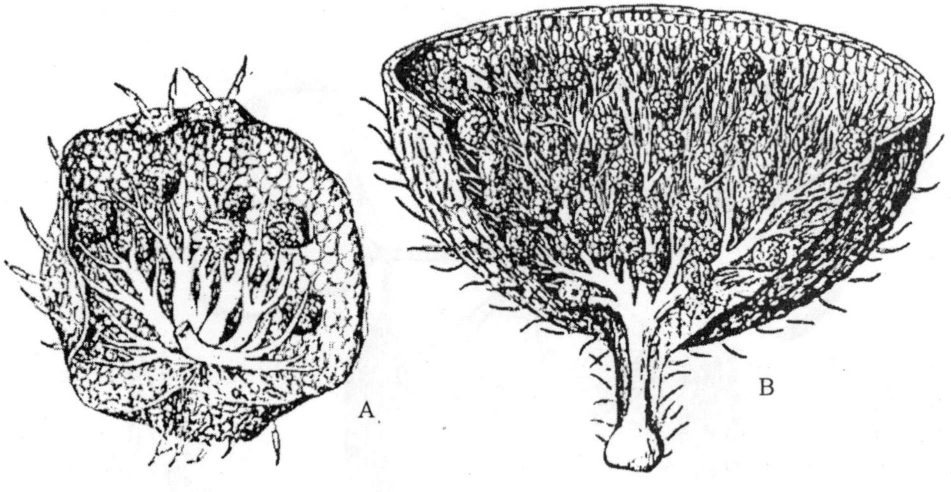

Fig. 14.40. *Salvinia* : Reproductive Structures
A. Sporocarp with Megasporangia. B. Sporocarp with Microsporangia

Stem

1. The stem is divided into nodes and internodes and a branch initial is always produced at every node.
2. But the stem is sparsely branched because of the non functioning of many of the branch initials.

Roots

1. There are no roots in the sporophyte.
2. Absorption is mainly carried out by the general surface.

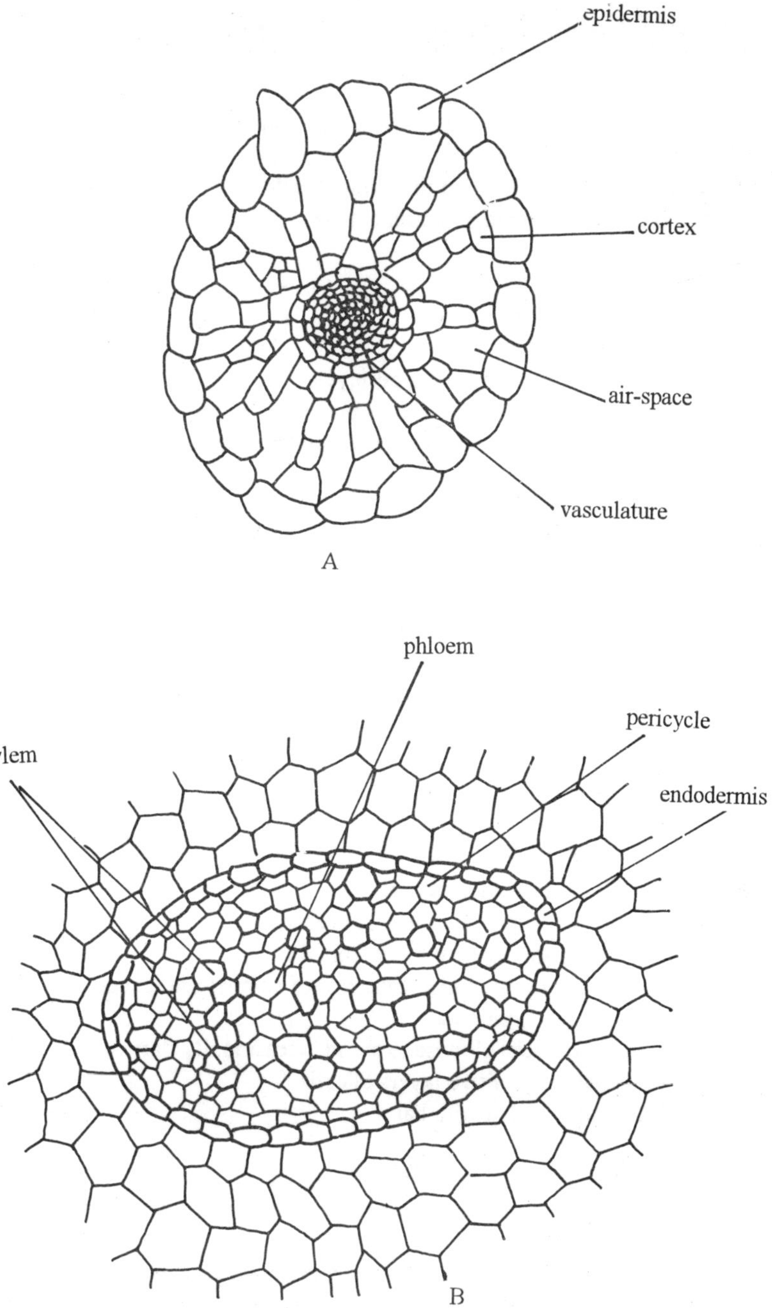

Fig. 14.41. *Salvinia* : Anatomy of Stems
A. T.S. of Stem (entire), B. T.S. of Stem (Stelar portion enlarged)

3. It is suggested that the hair like structures found on the submerged filiform leaves may help in the absorption of water and nutrients.
4. These hairs however are different from root hairs in being multicellular.
5. The plant body generally exhibits radial symmetry.

Internal Structure

1. A transverse section of the rhizome shows the following features.
2. Outermost layer is epidermis.
3. It is distinct, single layered and is made up of thin walled cells.
4. Epidermis on its outer surface is covered by a thin cuticle.
5. Stomata are absent.
6. Next to the epidermis is the cortex.
7. It consists of a number of lacunae or air cavities.
8. In the central region is found the stele.
9. The stele is surrounded by an endodermis and a pericycle.
10. True to the hydrophytic nature, the rhizome shows poorly developed vascular elements.
11. Central region of the stele consists of parenchyma within which are found many tracheids.
12. Xylem is surrounded by phloem.
13. The vascular arrangement suggests an ectophloic siphonostele.
14. In *S.auriculate*, the stele may be an amphiphloic siphonostele.
15. Anatomically, the floating leaves show a bifacial arrangement.
16. There are two epidermal layers surrounding a mesophyll with plenty of air cavities.
17. The mesophyll is undifferentiated.
18. The epidermal layers are studded with multicellular hairs.
19. The submerged leaves, anatomically exhibit a siphonostele near their base.
20. A little higher up, the single vascular strand first breaks up into two and ultimately into a number of small ectophloic steles.
21. Surrounding the vasculature, pericycle and endodermis are distinguishable.

Reproduction

1. The sporophyte reproduces by means of sporangia.
2. As in *Marselia*, the sporangia are aggregated into sporocarps.
3. The sporcraps appear in clusters ranging in number from 4 to 20.
4. The sporocarps are usually borne on the inner segments of submerged leaves.
5. In arrangement, the sporocarps are sympodial.
6. The shape of the sporocarp varies from globose to ovoid, occasionally flattened also with ridged surfaces.
7. The wall of the sporocarp is made up of two layers of cells.
8. The outermost layer when young is clad with hairs.
9. Externally all sporocarps are alike.
10. But unlike in *Marselia*, the sporocarps are monosporangiate.
11. The first one or two sporocarps in each cluster are megasporangiate whereas all the later formed ones are microsporangiate.
12. The sporocarp at its base has a stout columnar receptacle which carries the vascular supply into the sporocarp.
13. The receptacle is unbranched in the megasporangiate sporocarps of *S.natans*.

Structure of a Mature Sporocarp

1. A sectional view of mature sporocarp shows a well developed wall enclosing the, sporangia.
2. In a megasporangiate sporocarp, only about 25 megasporangia are seen whereas in a microsporangiate sporocarp, the number is much larger because of the branching of the receptacle.

Structure of Megasporangium

1. The megasporangium consists of a short stalk and a generally ovoid capsule.
2. The capsule is made up of a single layer of cells.
3. Internal to the wall of capsule is a layer of tapetum.
4. Enclosed by the tapetum are found 8 megasporocytes which divide meiotically to produce 32 megaspores.
5. Of these spores, all except one degenerate.
6. The degenerating spores and the tapetum harden and surround the functional megaspores in the form of a thick layer.
7. This is often called the 'perispore' or 'epispore'.
8. The perispore towards the apex of the sporocarp forms a traingular chamber, which very much simulates the pollen chambers of gymnosperms.

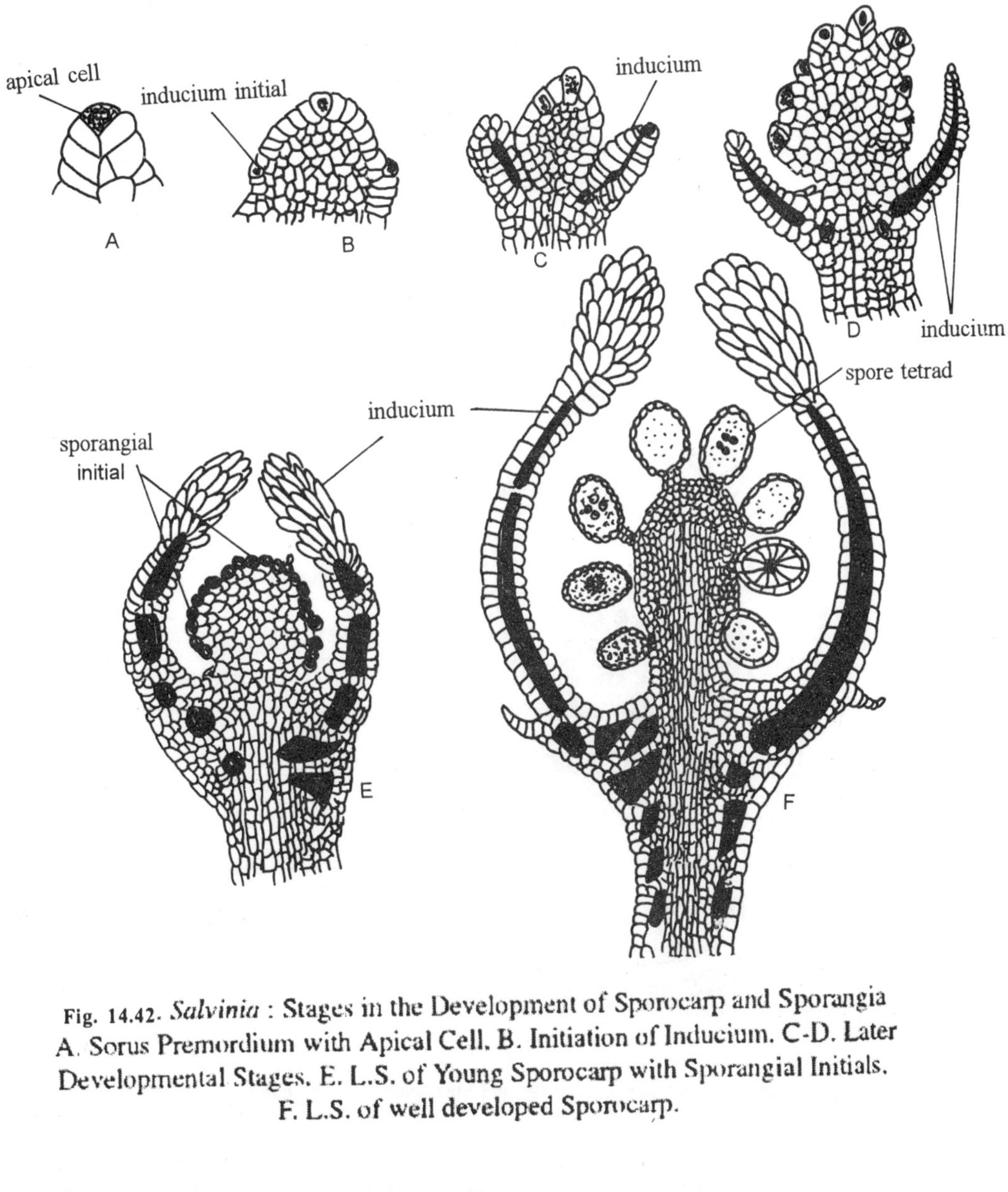

Fig. 14.42. *Salvinia* : Stages in the Development of Sporocarp and Sporangia A. Sorus Premordium with Apical Cell. B. Initiation of Inducium. C-D. Later Developmental Stages. E. L.S. of Young Sporocarp with Sporangial Initials. F. L.S. of well developed Sporocarp.

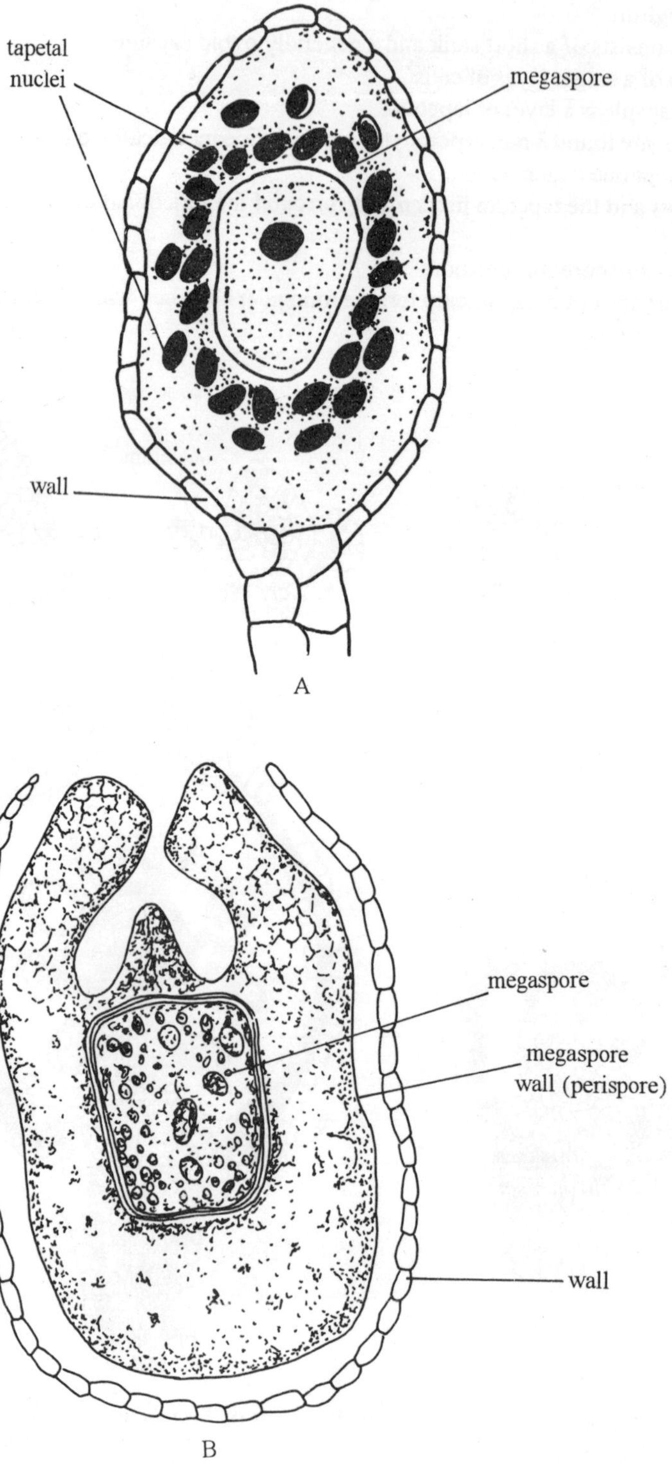

Fig. 14.43. Salvinia : Megaspoangium (A) and Megaspore (B) of S. natus

9. Arising from the floor of the pollen chamber is a central mound of tissue with three flaps.
10. These three flaps get separated during the germination of megaspore.

Structure of Microsporangium

1. The microsporangia are basically similar to megasporangia in their construction.
2. Enclosed in the sporangium are 16 microspore mother cells and consequently 64 microspores are formed.
3. All the spores attain maturity and survive.
4. The microspores are embedded in the cytoplasmic fluid derived from the tapetum.
5. As the spores mature, the tapetal cytoplasm encompassing the spores gets hardened and forms a mass called 'massula'.
6. The microspores are extermely small, triradiate in appearance and have two wall layers.

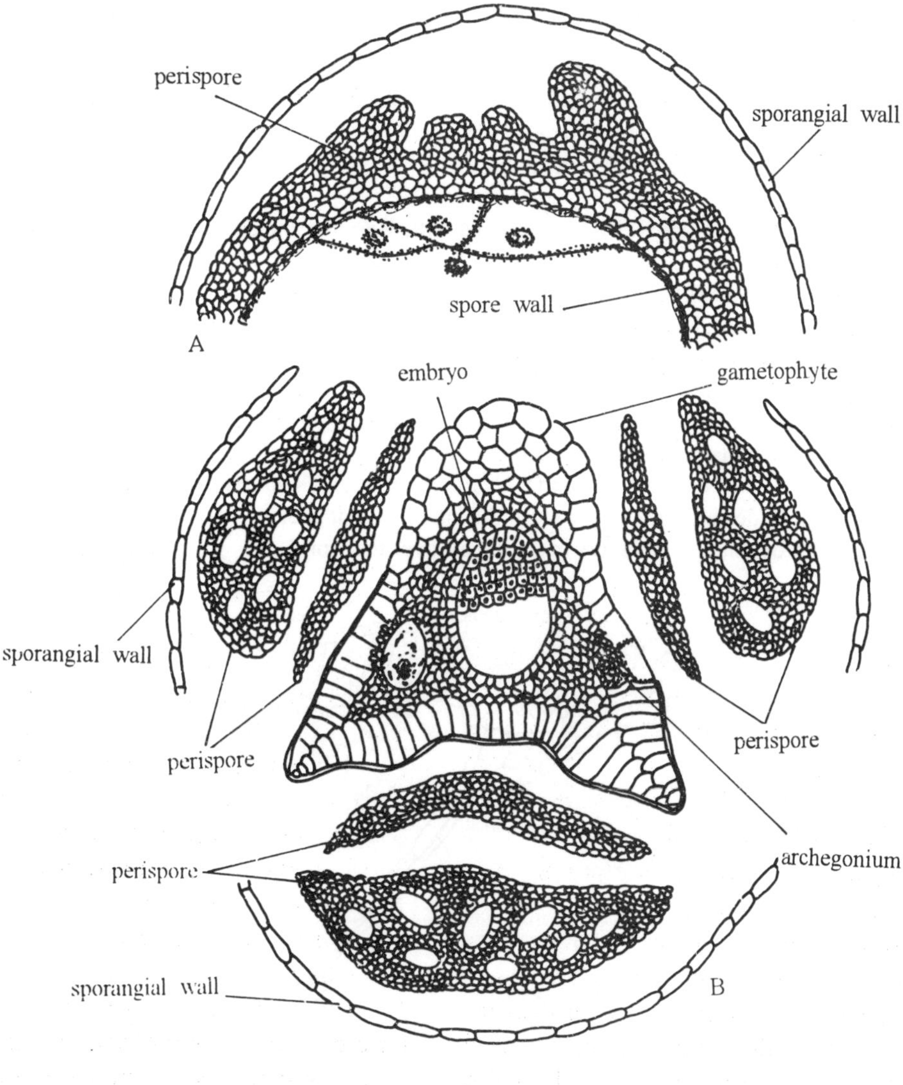

Fig. 14.44. Salvinia : Sectional views of Megaspore of S, natus
A. Median L. S through apex of Megaspore
B.T.S of Apical Region of germinated Megaspore

7. There is a single nucleus in each microspore.
8. Occasionally the number of microspores per microsporangia might vary.
9. In *S.auriculata*, there are only eight microspore mother cells and consequently only 32 microspores are formed.

Systematic position

PTERROPSIDA

LEPTOSPORANGIATAE

SALVINIALES

SALVINIACEAE

SALVINIA

AZOLLA

The genus *Azolla* has six species of which *A.filiculoides* is found in India. It is extensively distributed and is found floating on water in ponds, pools etc. The sporophyte grows profusely in undistrubed waters. Under favourable conditions the entire surface of water will be covered by the plant body. Young plants are bluish green, while the mature leaves are red in colour.

SPOROPHYTE

1. The sporophyte is extremely small when compared with *Marsilea* and *Salvinia*.
2. It is distinguishable into stem, leaves and roots.
3. The stem is often called the rhizome.
4. It is profusely branched and on its upper surface is covered with dense leaves.
5. The leaves are alternate and are arranged in two rows.
6. Each leaf has two lobes, the upper lobe being aerial and green in colour.
7. The lower lobe is thin and colourless, and is completely submerged in water.
8. The dorsal lobe encloses large mugilage filled cavities.
9. Inhabiting these mucilage cavities is found a Cyanophycean alga - *Anabaena azollae*.
10. While the alga provides nitrogen to the plant the latter gives it shelter.
11. The same species of *Anabaena* occurs in *Azolla* all over the world.

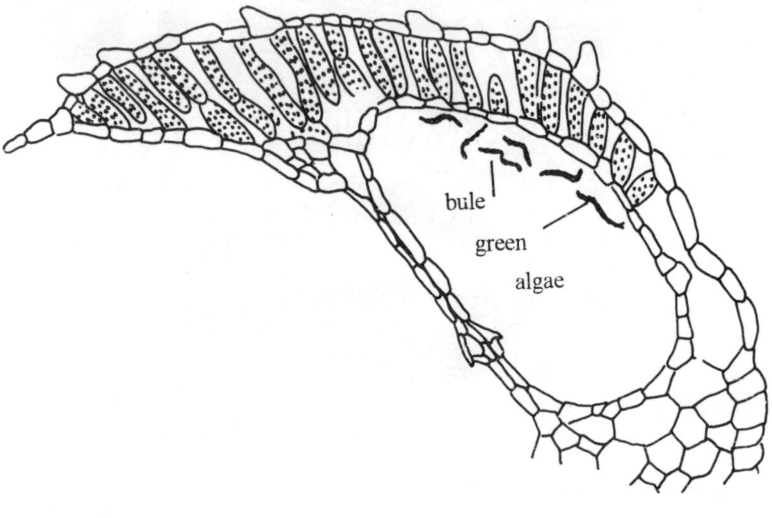

Fig. 14.45. Azolla: Anatomy of Leaf of A. filiculoides showing endophytic Blue Green Algae

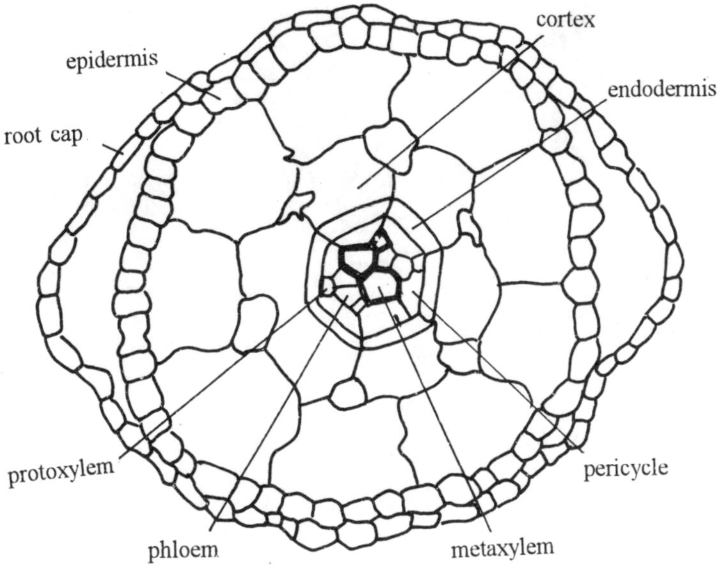

Fig. 14.46. *Azolla* : Anatomy of Root of *A. filiculoides*

12. The rhizome on its lower surface produces simple roots either singly or in clusters.
13. These roots help in stabilising the plants in water.

Anatomy

1. The rhizome, anatomically, resembles other ferns.
2. A cross section shows an epidermis, a middle cortex and a central stele.
3. Epidermis is single layered and it encloses a cortex which is 4-8 cells broad.
4. There are no air cavities in the cortex as in *Salvinia*.
5. The stele is surrounded by an indistinct endodermis.
6. Internal to the endodermis is the pericycle.
7. The vascular elements are greatly reduced.
8. There are a few tracheids surrounded by phloem elemnts.
9. It is difficult to determine the stelar type.
10. While Smith (1955) considers it protostelic, Eames (1936) says it is apparently siphonostelic.
11. Anatomically, the dorsal lobe of leaf shows two epidermal layers enclosing a thin mesophyll.
12. Stomata are found on both the epidermal layers.
13. The upper epidermis has a number of one to two celled layers.
14. The mesophyll is made up of loosely arranged cells.
15. Within the mesophyll is found a mucilage filled cavity containing *Anabaena*.
16. It has been suggested that the hormogones of *Anabaena* enter into the cavity through a small pore and get themselves established.

Reproduction

1. The main method of reproduction in *Azolla* seems to be vegetative fragmentation.
2. The lateral branches get separated and develop into new individuals.

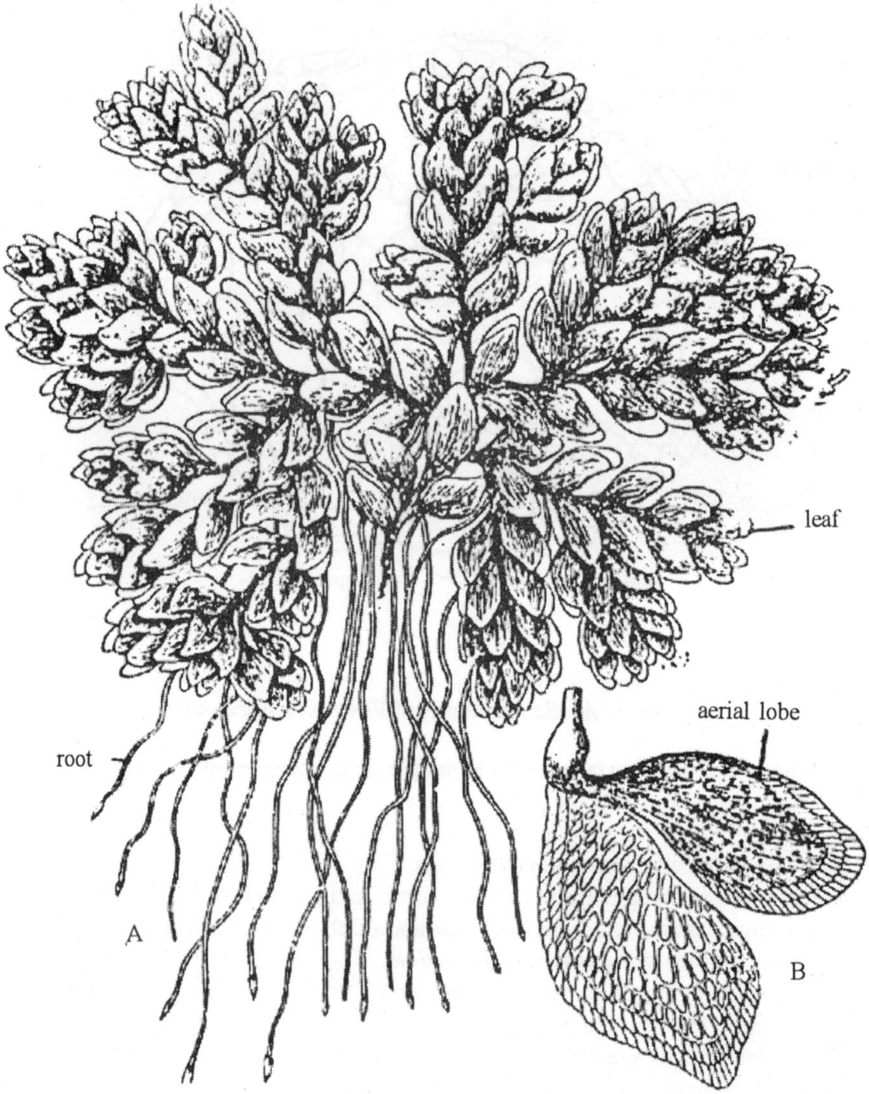

Fig. 14.47 Azolla: Habit of A. microphylla (A) and a Leaf Enlarged (B)

Spore Production

1. The spores in *Azolla* are produced in sporangia which in turn are enclosed in sporocarps as in *Marsilea* and *Salvinia*.
2. The sporocarps are usually borne on the first leaf of the lateral branch.
3. In fertile leaves, the submerged lobe is usually divided only twice, and on each a sporocarp is produced terminally.
4. The upper lobe of the leaf form a marginal flap covering the sporocarp.
5. The sporocarps are monosporangiate.
6. They have either microsporangia or megasporangia.
7. There is a size difference also between megasporocarps and microsporocarps.

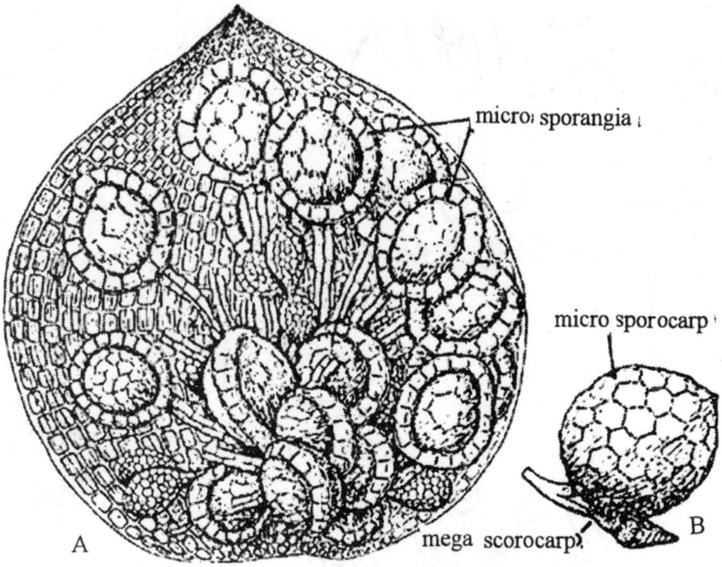

Fig. 14.48 Azolla: Sporocarps of A. microphylla
A. Sporocarp Split open to show Microsporangia
B. Leaf Segment with Macro and Microsporocarps

8. The former are small and have only one megasporangium, while the latter are large and have a number of microsporangia.
9. The wall of sporocarp is two layered.
10. In a megasporocarp, the megasporangium arises on a small receptacle at the base.
11. The sporangium is covered by a two layered inducium.
12. In a microsporocarp there is a central cushion like receptacle which gives rise to a number of microsporangia.

Systematic position

PTERROPSIDA

LEPTOSPORANGIATAE

SALVINIALES

AZOLLACEAE (SALVINIACEAE)

AZOLLA

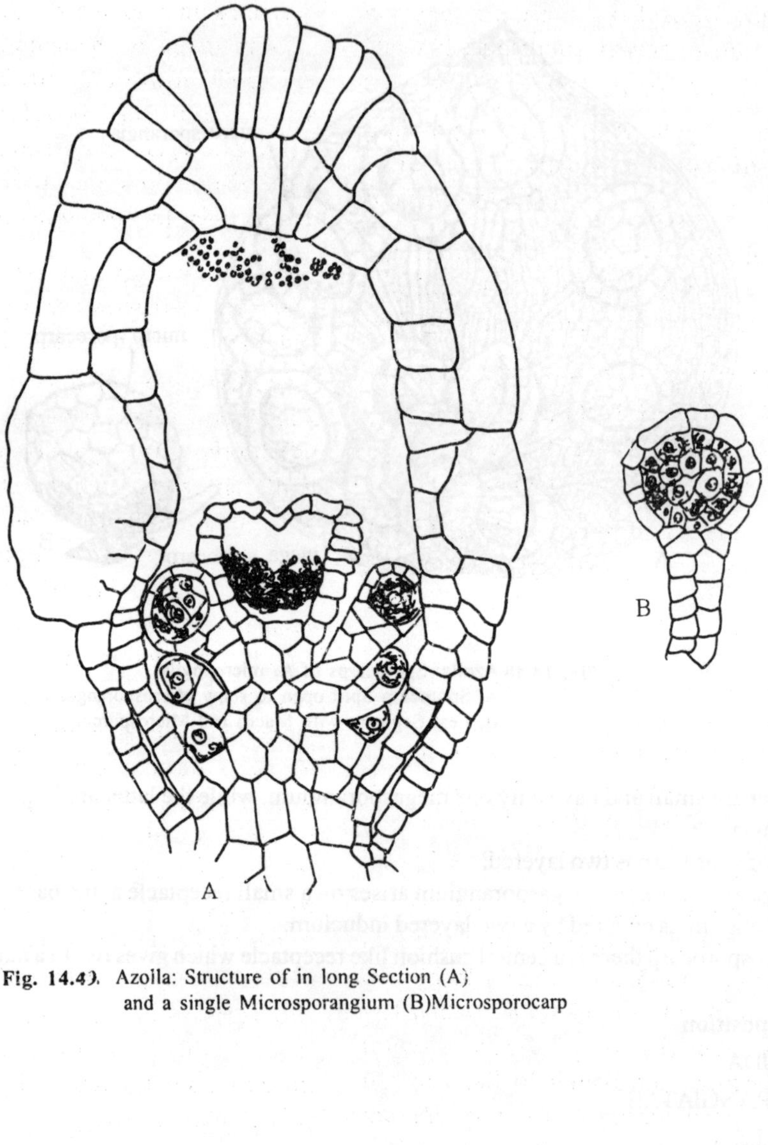

Fig. 14.43. Azolla: Structure of in long Section (A) and a single Microsporangium (B)Microsporocarp

PART THREE
GYMNOSPERMS

Chapter 15

GENERAL ACCOUNT AND CLASSIFICATION

The spermatophytes (seed bearing plants) are classified into two main groups *viz..,* *Gymnosperms* and *Angiospeims.* In Gymnosperms (Gr.gymnos = naked, *spermos* = seed) the seeds are exposed and do not have any covering. In angiospernis (Gr. *angio* = sac; *Spermos* = seed) the seeds are enclosed by the ovary wall, which (ovary) after fertilization becomes the fruit.

1. The first use of the word gymnosperms was by Theophrastus in 300 B.C.
2. In his book "Enquiry into plants" Theoprastus included all the plants with unprotected seeds under gymnosperms.
3. Geologically gymnosperms represent the most ancient among the seed plants.
4. Fossils of gymnosperms have been obtained from rock strata belonging to the late paleozoic era.
5. During mesozoic era there is an abundance of gymnospermous fossils indicating that they reached the climax of development.
6. In the present day flora however, gymnosperms - form a little part of the vegetation except perhaps for the conifers.
7. The life cycle of a typical gymnospermous plant comprises of a heteromorphic alternation of generations.
8. In the establishment of the sporophyte and in the reduction of the gametophyte, gymnosperms represent a further advancement over pteridophytes.
9. In gymnosperms, not only the sporophyte is the predominant generation, but the gametophytes have become microscopic and parasitic on the sporophyte due to extreme reduction.

Occurrence and distribution

1. Gymnosperms are found, both in tropical and temperate regions.
2. The same plants however are not found in both the regions.
3. Some are found even in arctic regions.
4. The living gymnosperms comprise approximately 70 genera and 725 species.

Morphology of the plant

1. The sporophytic plant body of gymnosperms is bushy or mostly a tall tree.

2. Some of the tallest and oldest trees are gymnospermous. *Sequoiadendron giganteum,* growing in the red wood park of California is 4000 years old and reaches a height of nearly 125 metres, with a girth of 30 metres.

3. The stem is anchored to the soil with the help of a tap root system.

4. Some of the secondary roots may have symbiotic association either with fungi (mycorrhizal.roots of *Pinus)* or blue green algae (coralloid roots of *Cycas).*

5. The stems are aerial, erect or climbing *(Gnetum),* branched or unbranched.

6. The stem is often surrounded by an armour of persistent leaf bases.

7. The branches of the stem are usually of two types *viz..,* branches of limited growth and branches of unlimited growth.

8. The former are short and usually bear the foliage leaves as in *Pinus;* the latter resemble the main shoot and in turn bear the branches of limited growth.

9. The leaves of gymnosperms are generally of two types - the green foliage leaves and the scaly leaves.

10. The green foliage leaves are evergreen and may be flattened or needle like.

11. The leaves may be simple or pinnately compound. Both microphyllous and megaphyllous leaves are seen in gymnosperms.

INTERNAL STRUCTURE

Stem

1. Anatomically the stem shows an epidermis, cortex and stele.

2. The epidermis is single layered and very soon replaced by a periderm. Cortex is parenchymatous.

3. Based on the relative quantity of the soft tissues and mechanical tissues gymnosperms are classified into *manoxylic* and *pycnoxylic* types.

4. In manoxylic forms (Cycads) the wood is not compact and has parenchyma.

5. In pycnoxylic forms wood is compact and relatively less amount of soft tissues is present.

6. The stele is surrounded by a typical endodermis followed by a layer of pericycle.

7. The vasculature consists of a number of vascular bundles arranged in the form of a ring surrounding a central pith.

8. The vascular bundles are collateral, endarch and open.

9. A stele of this type is given the name *Eustele.*

10. The xylem consists of mostly tracheids.

11. Vessels are found in the secondary wood as in *Gnetum.* Phloem consists of only sieve tubes and phloem parenchyma.

12. There are no companion cells.

13. Secondary growth is seen in all the members.

14. It is usually normal but anomalous secondary growth may also be found as in *Cycas* and *Gnetum.*

15. Normal secondary growth takes place due to the activity of a fascicular cambium, which produces the typical annual rings.

16. Occupying the central region of the stem is found the parenchymatous pith.

17. In some cases mucilage ducts or resin canals are found both in the pith and cortex.

Leaf

1. Anatomically the leaf exhibits xerophytic characters.

2. The epidermis is always covered by a cuticle.

3. The stomata are sunken and are often present in pits.

4. The mesophyll may be differentiated with palisade and spongy parenchyma *(Cycas)* or undifferentiated with arm parenchyma *(Pinus)*.

5. In some members *(Cycas)* with a single veined leaf, a *transfusion tissue* is present to help in lateral conduction.

6. Transfusion tissue may be regarded as a *paravascular tissue,* since it performs the function of a vascular tissue without having all the structural attributes of the latter.

7. The vascular bundles are collateral with a mesarch xylelm.

8. Vasculature shows the characteristic X/P arrangement:

Root

1. Most of the gymnospermous roots show a typical dicotyledonous root structure.

2. The xylem is either diarch or triarch.

3. Cortex is mostly parenchymatous.

4. As in the stem, secondary growth is seen both in the cortex and the stele.

REPRODUCTION

1. All the gymnosperms are heterosporcns.

2. Their typical mode of reproduction is by spore formation.

3. Vegetative propagation is very infrequent and is seen in some Cycads where it is brought about by the formation of *bulbils*.

4. The reproductive organs are generally grouped into strobili, except perhaps in the female reproductive structures of *Cycas*. Plants maybe monoecious or dioecious with reference to strobili.

5. Very rarely bisporangiate strobili may be found.

6. The male strobilus consists of a central axis on which are arranged spirally the microsporophylls.

7. These on their abaxial surface bear the microsporangia or the pollen sacs.

8. The number of pollen sacs per microsporophyll ranges from two *(Pinus)* to many *(Cycas)*.

9. The pollen sacs produce the pollen grains.

10. Within the pollen grains are found the male gametophytes.

11. The male gametophytes are incompletely developed when they come out of the pollen sacs.

12. Further development generally takes place within the ovule, after pollination.

13. The female strobilus consists of a number of megasporophylls arranged spirally on an axis.

14. In *Cycas* however the megasporophyus are not compact.

15. Each megasporophyll bears a varying number of ovules (megasporangia).

16. The ovules have a thick envelope (integument) which is generally three layered *viz.* .,outer fleshy, middle stony, and inner fleshy layers.

17. Internal to this integument, lies the *nucellus* which is the seat of development of the female gametophyte.

18. The female gametophyte arises by the development of one of the hypodermal cells of the nucellus.

19. Its development is either monosporic or tetrasporic.

20. The development of female gametophyte bears a resemblance to that of heterosporous pteridophytes.

21. Archegonia are highly reduced or they may be totally eliminated as in *Gnetum*.

22. Pollination is mostly anemophilous.

23. Only a part of the zygote develops into embryo. Such a development is said to be meroblastic.
24. Polyembrgony is of frequent occurrence.
25. The embryo usually has a suspensor.
26. The number of cotyledons varies from two to many.
27. Seed has a thick seed coat.
28. Germination is epigeal or hypogeal.
29. Growth of the seedling is slow.

GYMNOSPERMS

A. CYCADOPSIDA

1. PTERDIDOSPERMALES*
2. BENNETTITALES*
3. PENTOXYLALES*
4. CYCADALES

* Indicates fossil groups.

B. CONIFEROPSIDA

1. CORDIATALES*
2. CONIFERALES*
3. TAXALES
4. GINKGOALES

C. GNETOPSIDA

In this book the system of classification proposed by Sporne is followed.

AFFINITIES OF GYMNOSPERMS

Gymnosperms exhibit resemblances with pteridophytes as well as with angiosperms.

Resemblances with Pteridophytes

1. Life cycle shows heteromorphic alternation of generations with a reduced gametophyte.
2. Some of the gymnosperms (*Cycas*) exhibit circinate vernation.
3. There is no companion cell in phloem : occasionally vessels are present.
4. Development of the female gametophyte resembles that of heterosporous pteridophytes.
5. Motile sperms present in cycads recall what is seen in pteridophytes.
6. Gametophytes are endosporic as in heterosporous pteridophytes.

Differences with Pteridophytes

1. The sporophyte is much more complex.
2. Root system is always a tap root (adventitious in pteridophytes)
3. Stele is eustelic.
4. Seeds are produced.
5. Regular occurrence of secondary growth.
6. Permanent retention of female gametophyte in the megasporangium. Development of integument in the megasporangium.

Resemblances with Angiosperms

1. Tap root system
2. Possession of eustelic vascular cylinder.
3. Development of seed.
4. Permanent retention of the female gametophyte the ovule.
5. Siphonogamous fertilization.
6. Highly reduced male gametophytes.

7. Non motile male gametes (except in *Cycas)*.

Difliefences witb angiospenns :

1. Gymnosperms are mostly bushy or arborescent while it is not so in Angiospenns.
2. Phloem lacks the companion cells.
3. Flowers are bisexual or unisexual in Angiospenns, while strobili are mostly unisiexual in Gymnosperms.
4. Pollination is always anemophilous in Gymnosperms.
5. The ovules do not have a carpellary covering as is seen in angiosperms.
6. Archegonia are present in the female gymetophyte except in *Gnetum*.
7. There is no postfertilization development of endosperm. The female gametophyte itself is tbe endosperm. There is also no double fertilization.

CHAPTER 16

PTERIDOSPERMALES

The members of the order are commonly called seed bearing ferns or *Cycadofilicales*. Tbe order comprises many heterogeneous plants which appeared in the upper Devonian and extended through the Carboniferous and Permian to the Mesozoic.

The group was first discovered in 1877, by Grand ury. In India the group occurs in lower, middle and upper Gondwana systems belonging to Paleozoic and Mesozoic eras.

The group is subdivided into seven families *viz., Lyginopteridaceae, Medullosaceae, Calamopityaceae, Glossopteridaceae, Peltaspermaceae, Corystospermaceae* and *Caytoniaceae*. Of these, the family Lyginopteridaceae *(Lygnodendron, Lagenostoma* and *Heterangium)* is discussed here :

Fig. 16.1. **Lyginopteris.,**
Reconstruction of the plantbody of L. oldhamia.

LYGINOPTERIS

Lyginodendron is the stem genus of the plant *Lyginopteris oldhamia (Calymmatotheca hoeninghausi).* The stem genus is based upon the cortical impressions. This was first described in 1866 by Binney under the name *Dadoxylon oldhamia.* The various parts of the plant *Lyginopteris* have been described under the following names:

Stem *Lyginodendron*

Rachis *Rachiopteris aspera*

Root *Kaloxylon hookeri*

Frond *Sphenopteris hoeninghausi*

Pollen beanag organs *Crossotheca hoeninghausi*

Seed *Lagenostoma lomaxi*

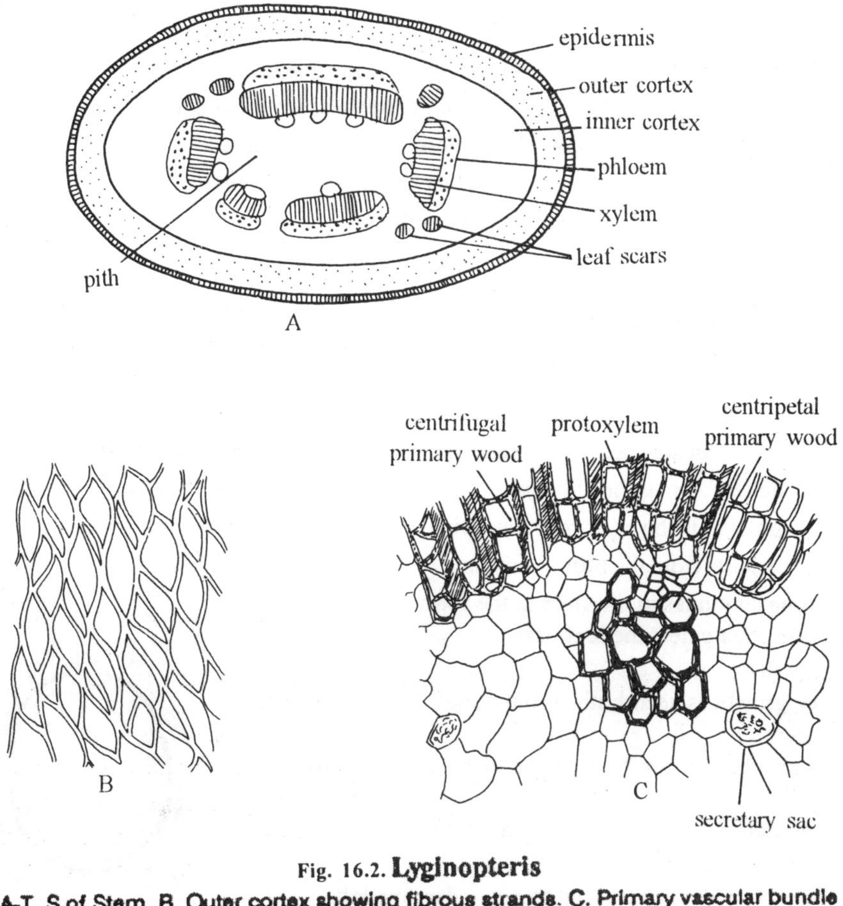

Fig. 16.2. **Lyginopteris**

A-T. S of Stem, B. Outer cortex showing fibrous strands. C. Primary vascular bundle (magnified)

Structure of the Plant

1. The plantbody of *Lyginopteris* had a long and slender stem.
2. The diameter of the stem varied from 2 mm. 4 cms.

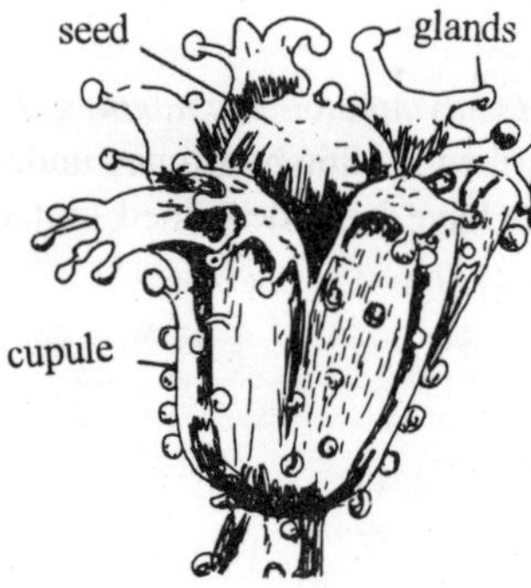

Fig. 16.3. **Lyginopteris**
Seed of lagenostoma lomaxii.

3. The stem was frequently branched and had adventitious roots which acted like prop roots.
4. The leaves were borne spirally and measured nearly 50 cms.
5. The rachis had glands similar to those on the seed.
6. The leaves were bipinnate to tripinnately compound and had pinnae at right angles to the rachis.
7. Pinnae were opposite to each other. Each pinnae in turn bore pinnules or leaflets which were connate and arranged alternately.
8. The stem was radially symmetrical showing that it grew erect, though it had to recline and seek the support of the neighbouring plants due to the heavy weight of the foliage.

Internal structure of the stem

1. A transverse section of the stem shows a central stele, middle cortex and an outermost epidermis.
2. In the primary structure, the epidermis is made up of a single layer of cells.
3. The cortex consists of two zones *viz.*, the outer cortex and the inner cortex.
4. The outer cortex consists of a network of sclerenchymatous fibres.
5. Within the meshes of the sclerenchyma, parenchyma cells are found.
6. The inner cortex is composed of parenchyma cells.
7. The vascular cylinder consists of a central parenchymatous pith.
8. Here and there in the pith are found masses of stone cells forming 'sclerotic nests".
9. Surroundiag the pith *are* found the primary vascular bundles.
10. These are separated by parenchymatous areas.
11. The vascular bundles are collateral with measarch xylem.
12. The protoxylem is located nearer the edge . Next to the xylem is the cambium which is rarely preserved.
13. External to the cambium is the phloem.
14. Outside the phloem is the pericycle consisting of short sclerotic cells with dark contents.
15. The leaf trace starts from the protoxylem and divides into two in the coctex.
16. Of these, one continues in the stem in the form of the Separator strand, while the other enters the leaf base.

Secondary growth

1. Secondary growth is seen both in the cortex as well as in the vascular cylinder.
2. In the cortex a periderm is formed due to the activity of the cork cambium.
3. In the vascular cylinder the fascicular cambium produces the secondary vascular tissues.
4. The secondary xylem is traversed by secondary medullary rays.
5. In addition to the medullary rays, leaf traces also disturb the vascular ring.
6. External to the secondary xylem is the secondary phloem.
7. The phloem cells are both large and small which are arranged alternately.

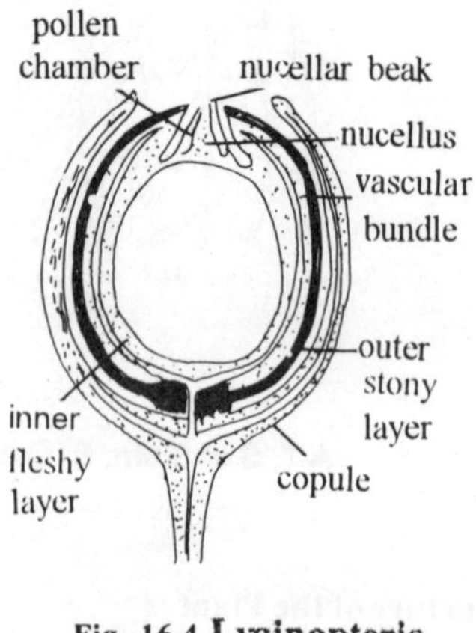

Fig. 16.4. **Lyginopteris**
Long section of the seed of lagenostoma lomaxii.

REPRODUCTION

1. The male and female reproductive structures probably arose on different plants.
2. The microsporangiate organ possessed fertile pinnules with each pinnule having six or more microsporangia.

SEED

1. The seed of *Lyginopteris* has been given the name *Lagenostoma lomaxi.*
2. In 1903, Oliver and Scott first showed the relationship between the stem and the seed.
3. The seed is terminal on the ultimate and naked branches of the frond.
4. Each seed is a small barrel shaped structure measuring approximately 4.25 mm. across and 53 mm. in length.
5. The ovule is orthotropous and 4.25 mm. across and *53* mm. in length and symmetrical.
6. Surrounding the ovule and covering it partially is a lobed husk or cupule.
7. The cupule has gland on the exterior resembling those present on the rachis.
8. A long section of the seed shows the following features.
9. The outermost layer is the integument which is traversed by nine vascular strands.
10. There are two regions in the integumer *viz.,* outer *sclerotesta* and inner *endotesta.*
11. The integument is fused with the nucellus except at the micropylar end where it forms a canopy that surrounds and overhangs the pollen chamber.
12. The canopy is divided into nine locules with a vascular strand extended into each.
13. It is believed that the loculed structure helps in the supply of water to the apex of the seed.
14. The pollen chamber is complex and in the apical cavity between the free part of the integument and the nucellus flask shaped prolongation arises from the nucellus and extends into the micropyle.
15. Within this prolongation arises a column of nucellar tissue for reception of the pollen.
16. This flask shaped structure called *Lagenostome* is responsible for the generic name.
17. Internal to the nucellus is the female gametophyte. A few archegonia have been reported by Lang (1944) in *L.ovoides.*

Systematic postion

CYCADOPSIDA

PTERIDOSPERMALES

LYGINOPTTERIDACEAE

LYGINOPTERIS

HETERANGIUM

Heterangium was first discovered from the coal balls of Bohemia, of middle carboniferous age (Corda 1845). The genus consists of about a dozen species of which the best known is *H. grievii.*

The stem of *H. grievii* was of considerable height and had a diameter of 1.5-2 cms. Most of them were straight, slender and unbranched. Very rarely branching was observed. The stem bore foliage known as *Sphenopteris elegans.*

Internal strucutre of the stem

1. A transverse section of the stem showed a large protostele.
2. The vasculature instead of forming a solid core at the centre had many gaps which were filled by parenchyma.
3. The tracheids had multiseriate bordered pits.

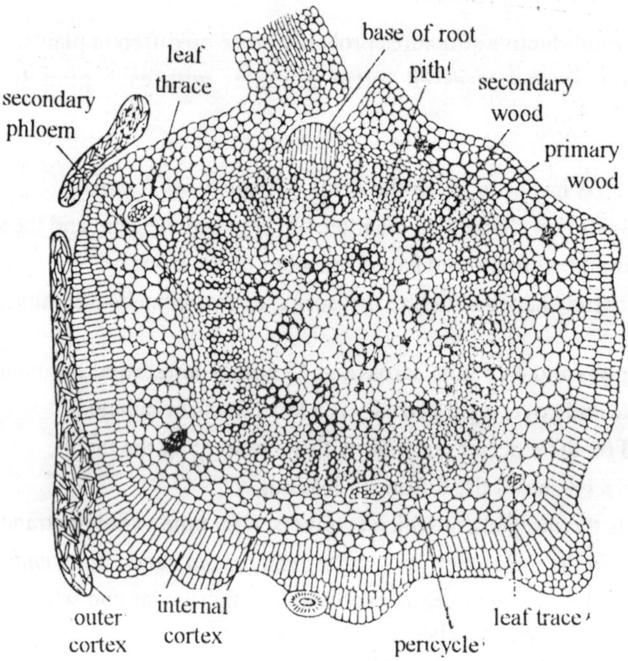

Fig. 16.5. Heteragium
T.S. of stem of H. grievil

4. The xylem groups (about twenty) were mesarch in arrangement.
5. The primary xylem was surrounded by a zone of secondary wood which was scarcely more than a few mm. thick.
6. The tracheids of the secondary wood also had multiscriate bordered pits.
7. The leaf traces started as a single strand having a single protoxylem point which immediately split into two.
8. The vasculature of the leaf trace was collateral at the base but concentric a little above.
9. Surounding the vasculature was a cortex whose outer zone had multiple sclerenchyma strands lying parallel to each other. Internal to this was a zone containing transverse plates of stone cells.
10. Next to this zone was the parenchymatous region of the cortex. The outermostt layer was the *epidermis.*

Systematic position

CYCADOPSIDA

PTERIDOSPERMALES

LYGINOPTERIDACEAE

HETERANGIUM

CHAPTER 17

CYCADOPSIDA

CYCADEOIDEA(=BENNETTITALES)

Of all the fossil gymnosperms no other genus perhaps has aroused so much of interest and attention as *Cycadeoidea*. This is because *Cycadeoidea* has reproductive organs that are very much similar to the flowers of angiosperms. This resemblance is the basis of a theory for the origin of angiosperms. (Arber and Parkin, 1907).

The genus *Cycadeoidea* consists of about 30 species being found in various places of the world in the form of petrified trunks. Geologically *Cycadeoidea* is traceable to the period between upper triassic and middle cretaceous. Some of the common species are *C. marylandica, C. dacotensis, C. wielandii etc.*

MORPHOLOGY

1. The plant body had a columnar or subspherical trunk with a diameter ranging from 5-60 cms. (Fig. 2.5).
2. The plant body attained a height of 3-3.5 meters.

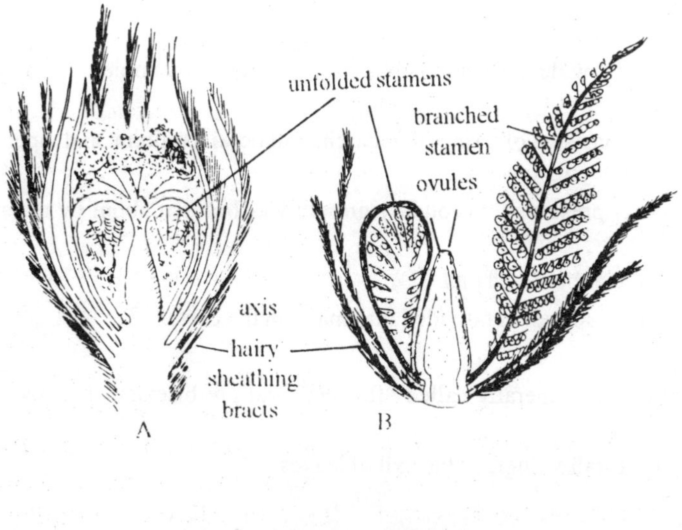

Fig. 17.1. CYCADEOIDEA
A-L.S. of strobilius, B-same with unfolded stamen (after Wieland)

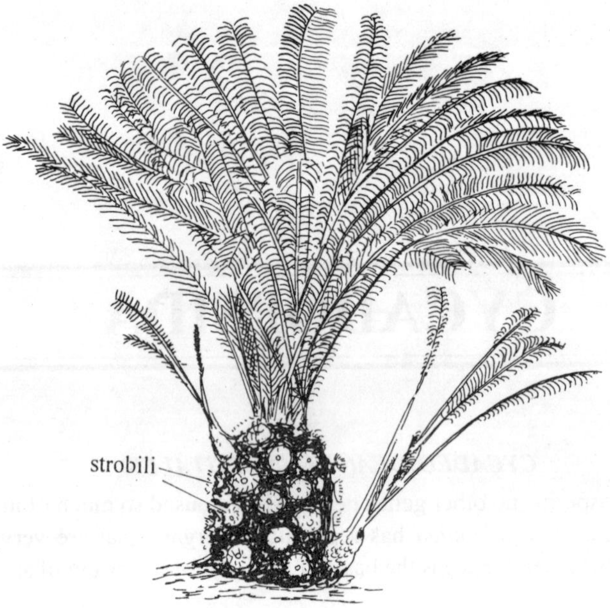

Fig. 17.2. **CYCADEOIDEA**
Reconstruction of the plantbody of C. dacotensis

3. The trunk was studded with persistent leaf bases as in living Cycads. Branching was very infrequent.
4. The stem had a crown of leaves at the apex which were pinnately compound.
5. The leaves were 2-3 meters long.

INTERNAL STRUCTURE

1. The stem internally showed a wide parenchymatous cortex traversed by numerous leaf traces and mucilage canals.
2. The primary vasculature consisted of a ring of endarch, collateral, conjoint and open vascular bundles surrounding a central parenchymatous pith.
3. The epidermis was poorly preserved. Secondary growth was brought about by the activity of vascular cambium.
4. Secondary xylem had circular bordered pits.
5. Leaf traces arose as a single strand, but divided into many in the cortex.

REPRODUCTION

1. The reproductive organs are generally called "flowers" and are bisexual in a majority of the species (unisexual in *C. wielandii).*
2. The flowers arise on short stalks singly at the axil of leaves.

Structure of a "flower"

1. The bisexual "flower" is also called a bisporangiate strobilus.
2. It has a short stalk extending laterally from the main stem.

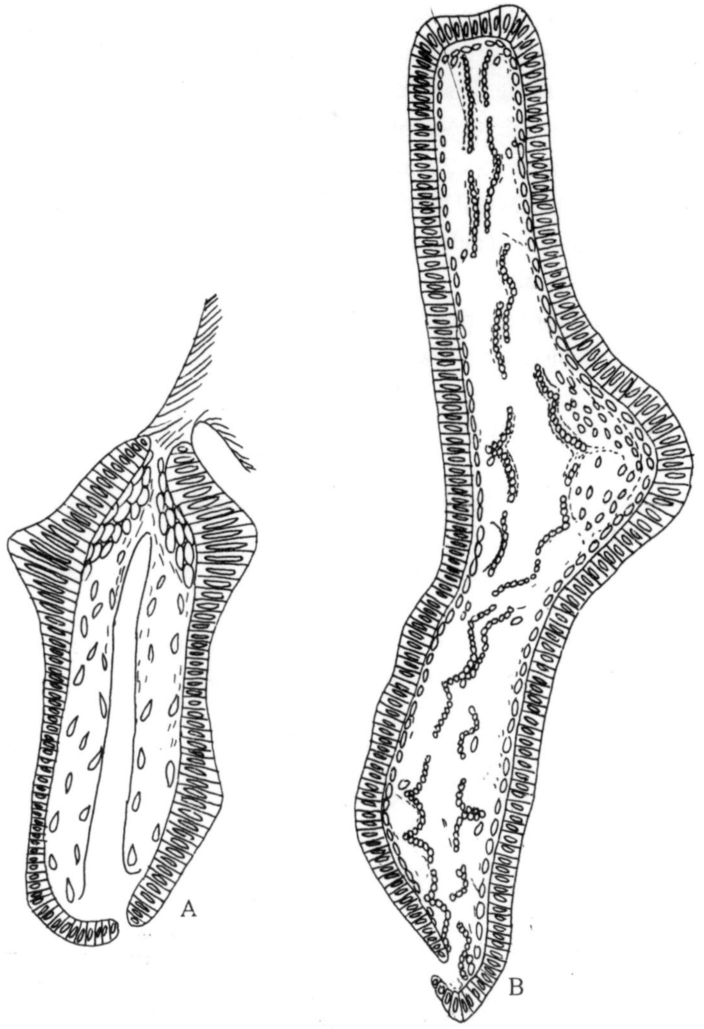

Fig 17.3.CYCADEOIDEA
Synangial structure of C. dacotenasis .A.L.S of synangium .B.T.S of synangium

3. A little below its tip are borne about 100- 150 hairy bracts.
4. These bracts protect the microsporangiate and megasporangiate organs.
5. At maturity the bracts expand to form cup shaped structures resembling the perianth of angiospernuc flowers.

Microsporophyll :

1. According to the interpretation of Wieland, each microsporophyll bears numerous alternating pinnae.
2. The pinnae bear two rows of kidney shaped, stalked pollen capsules or *synangia.*
3. According to Delevoryas (1963) the microsporophyll never expanded as interpreted by Wieland.
4. As Delevoryas points out, the tip of the microsporophyll was always fused to its base.

5. The tip of each microsporophyll is parenchymatous.
6. Within the fertile region there are trabeculae that connect the outer and the inner walls of microsporophyll.
7.Pollen capsules are borne along these trabeculae.
8. It is possible that the entire microsporophyll might have been shed as a single unit.
9. Each pollen capsule measures 0.5 x 2.5 mm and contains a varying number of microsporangia (20-30).
10. The synangium has a wall several layers in thickness.
11. The outer layer is composed of palisade like cells.
12. The inner layers are thin walled.
13. This is followed by a ring of microsporangia.
14. The microsporangia dehisced lengthwise and released the pollen into the synangium which opened by. an apical slit.
15. The microspores contain many cells indicating the incomplete development of the male gametophyte.

Gynoecium

1. The central region of the "flower" consists of a conical receptacle that bears many stalked ovules interspersed with many interseminal scales.
2. Each ovule is about one mm long and consists of a single integument which is fused with the nucellus except at its apex (Harris 1932).
3. Within the ovule is a pollen chamber above which there is a nucellar beak.
4. The ovules are orthotropous and have long micropylar beaks formed by the extended integument.
5. The tips of the interseminal scales fuse forming a protective covering surrounding the seeds.
6. Within the ovule is the nucellus.
7. The seed has two cotyledons and is almost oval in shape.
8. The stalked ovules have been interpreted as megasporophylls.
9. The interseminal scales are believed to be sterile ovules.

Systematic postion

CYCADOPSIDA
CYCADEOIDALES (=BENNETITITALES)
CYCADEOIDACEAE
CYCADEOIDEA (=BENNETTITES)

CYCAS

Occurrence and Distribirtiton

1. *Cycas* has about twenty species, of which about six are found to grow in our country.
2. These are C. *circinalis, C.heddornei, C. pectinata, C. rumphi, C. siamensis, and C. revoluta.*
3. The last two species are not found in wild condition in our country.
4. In Karnataka state *C. circinalis* grows wild near Nagamangala of Tumkur District.
5. *Cycas* is a very slow growing plant and grows under xerophytic conditions, like the exposed slopes of hills and other sunny places where there is scarcity of water.

Morphology of the plant

1. The plant body consists of an aerial, erect stem which is anchored to the soil with the help of a tap root as stout as the main stem.
2. The plant attains a height of 3-6 metres resembling a small palm tree.

3. Some of the species may reach a height of 20 metres.
4. The stem at its tip bears a crown of graceful leaves.
5. The stem has a diameter of 0.5-1 metre at the base.
6. It has a thick armour of persistent leaf bases.
7. The stem is generally unbranched but branched specimens are also not very rare.
8. Branching is dichotomous when it occurs.
9. Very rarely three branches may arise at a point.
10. The leaves are borne at the apex of the stem in a spiral manner.
11. They are of two types *viz., scale leaves* and *foliage leaves.*
12. Scales leaves are thick and short and brownish in colour.
13. In the axils of scale leaves, the foliage leaves develop.
14. The foliage leaves are more than a metre long when fully grown and exhibit *cininate vernation* which is unusual to gymnosperms and recalls a fern character.
15. The leaves are pinnately compound and have a stout rachis.

Fig. 17.4. CYCAS
Habit of C. rumphii (After Brown)

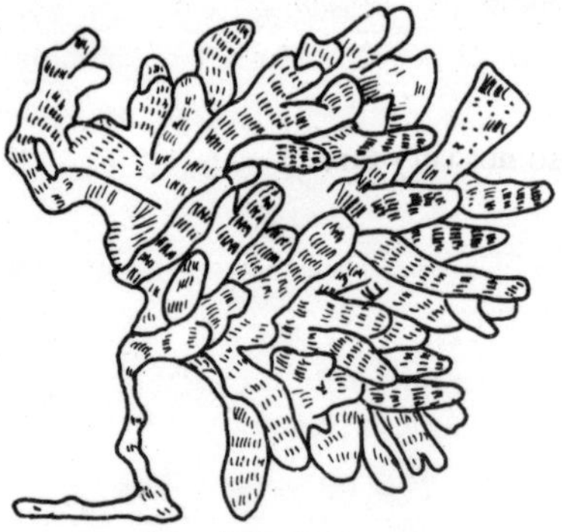

Fig. 17.5. **Cycas**
Coralloid roots

16. Some of the basal pinnae are modified into spines.
17. The pinnae are linear and have a central unbranched vien.
18. The stem is anchored to the soil with the help of a taproot which is as stout as the stem.
19. The tap root bears a number of secondary roots.
20. Some of these secondary roots are modified into *Coralloid roots*.
21. These are irregularly branched clusters of roots that grow apogeotropically and are found a few- inches beneath the soil.
22. The coralloid roots are greyish green in colour and owe their name to their irregular coral like branching.
23. On the surface of the root, small lentice like openings are found probably helping in respiration.
24. Internally the coralloid roots have an endophytic blue-green *(Anabena cycadacearum)* in the cortical region.
25. The relationship between the alga and the roots of *Cycas* is believed to be symbiotic.
26. While alga provides nitrogenous material (it has nitrogen fixing capacity) to the plant it gets protection and shelter in turn.
27. The manner in which a secondary root develops into a coralloid root is as follows.

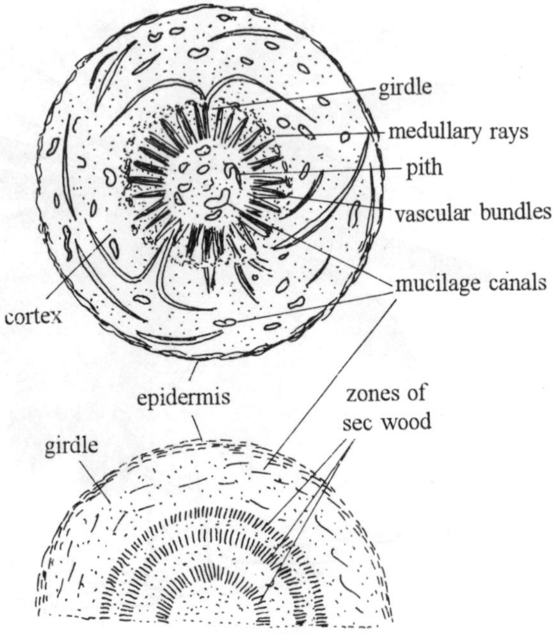

Fig. 17.6. **Cycas**
A- T.S. of young stem B- T.S. of old stem
(Note the rings of secondary Xylem)

28. Early in its development a secondary root gets the infection of the soil inhabiting bacterium.
29. Due to the infection of bacteria which come to lie in the cortical region, irregular branching sets in, producing the familiar coralloid roots.
30. The bacteria present in the cortex die very soon and the empty space left by them is filled by the blue green algae.

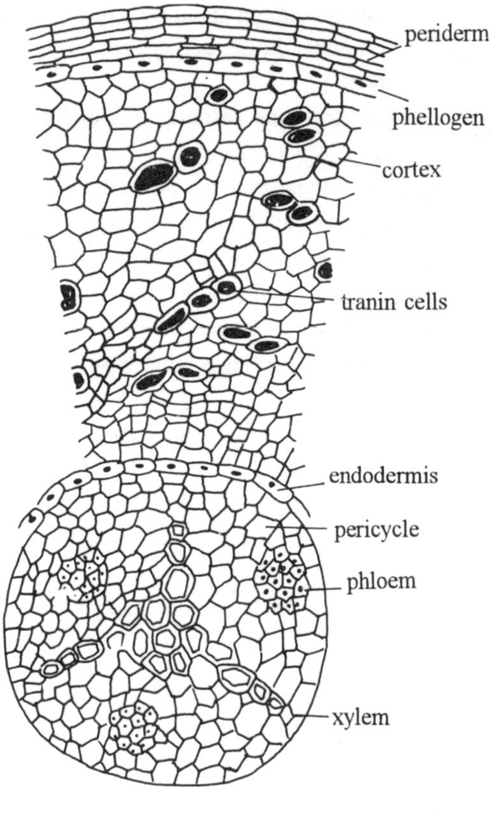

FIg 17.7 Cycas
T.S. of normal root
(A portion enlarged)

INTERNAL STRUCURE

Stem

1. A transverse section of a young stem shows the following features viz., epidermis, cortex and stele.
2. Epidermis is single layered and is discountinous due to the presence of leaf bases.
3. It is short-lived and very soon will be replaced by the periderm.
4. Cortex is very wide, surrounding a narrow vascular cylinder (manoxylic).
5. The entire cortical region is filled by thin walled parenchyma cells rich in starch.
6. The cortical region also consists of a number of mucilage ducts and leaf traces.
7. The leaf traces are of two types *viz., direct* or *radial traces* and *girdle traces.*
8. Direct traces arise from the primary vascular bundles, in duplicate and enter the leaf base.
9. The girdling traces which also arise in duplicate pass obliquely into the cortex, grow in opposite direction meet and then enter the leaf.
10. This forms a sort of girdle.
11. The girdle traces are linked by cross connections with each other, with other leaf traces and also with the radial traces forming a complicated network. Before entering into the leaf base the traces divide and form a large number of bundles.
12. The vasculature in the primary structure consists of a number of collateral, open vascular bundles with endarch xylem. 13. The vascular bundles surround a central parenchymatous pith and are separated by medullary rays.
14. Mucilage ducts are also found in the pith.

Secondary growth

1. This takes place both in the cortex as well as in the vasculature.
2. In the outer layers of the cortex, arise a *cork cambium* (phellogen) which produces the periderm.
3. Towards its inner face the cork cambium produces *phellem* or the cork tissue.
4. The three together constitute the periderm.
5. Secondary growth taking place in the vasculature produces many rings of vascular tissue.
6. But these are not annual rings as their development is irregular.
7. At the initiation of the secondary growth, the strips of fascicular cambia present in the vascular bundles

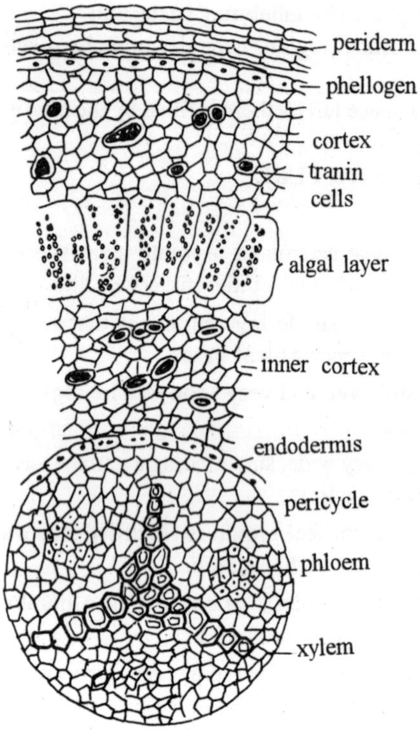

Fig. 17.8. **Cycas**
T.S. of coralloid root
(A portion enlarged)

radially extend and form a completely closed ring.

8. This produces secondary phloem to the exterior and secondary xylem to the interior.

9. Soon after the formation of the first vascular ring, the activity of the cambium ceases.

10. Further growth is brought about by a second cambial ring which arises either in the pericycle or in the inner layers of the cortex.

11. This also produces one vascular ring and ceases to function.

12. A third cambial ring is then produced in the cortex.

13. The process is repeated and sometimes as many as fourteen vascular rings may be formed as in *C.pectinata*.

14. Secondary growth in *Cycas* may be said to be both normal and anomalous.

15. The inner most vascular ring produced by the fasoeular cambium is nornal.

16. All the other rings are anomalous as they are produced by extrafascicular cambia.

17. Another important feature in the secondary vascular bundles produced towards the periphery is their concentric nature instead of the normal collateral.

18. The concentric bundles have mesarch xylem recalling the fern features.

Root

1. Anatomically the root resembles dicotyledons.

2. A transverse section shows an outermost epidermis (which is soon replaced by the periderm), cortex and stele.

3. The epidermis or epiblema is single layered.

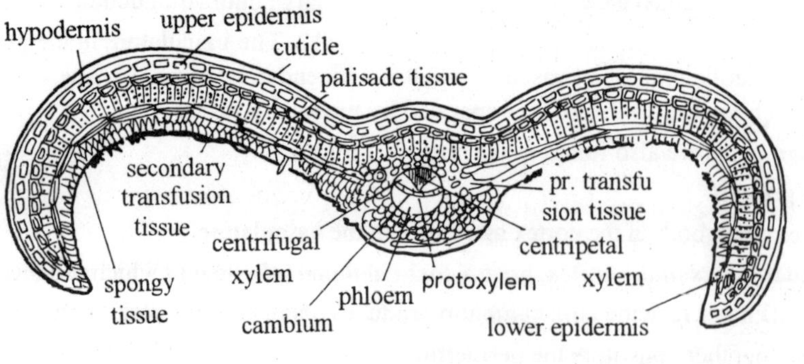

Fig. 17.9. **Cycas**
T.S. of leaflet of C. revoluta (entire)

4. The cortex consists of many layers of parenchyma cells interspersed with nucilage cells.
5. Tannin cells also found in the cortex.
6. The stele is surrounded by single layered endodermis followed by a layer of pericycle.
7. The vascular bundles are radial with diarch or triarch and exarch xylem.
8. There is no pith.
9. Conjunctive parechyma is present between phloem and xylem.
10. The xylem consists of tracheids only, with spiral, scalariform and pitted thickenings.
11. Phloem consists of sieve tubes and phloem parenchymma.
12. In older roots phloem may have bast fibres.
13. Secondary growth also takes place in the root much the same way as in a dicot root.

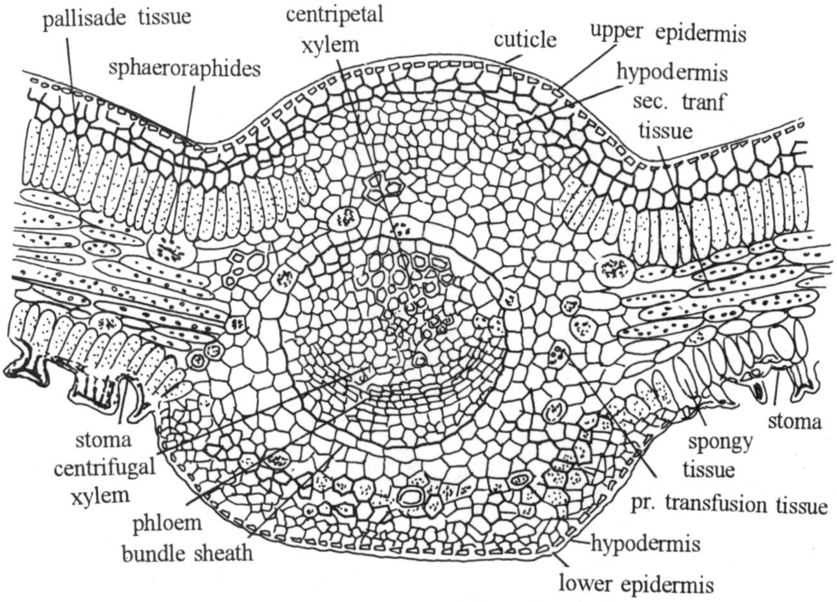

Fig. 17.10. **Cycas**
T.S. of leaflet of C. revoluta (a portion magnified)

14. In some cases, two peridsrm layers are found separated by the cortex.
15. This is due to the formation of two separate phellogens.
16. Lenticels are found in the periderm.

Coralloid root

1. It is similar to the normal roots in all respects except for the presence of an algal zone in the cortex.
2. The algal zone is known to consist of several blue-green algae *(Anabena cycadacearum, Nostoc punctiforme, Oscillatoria sp.,* etc.) and bacteria like *Azatobacter, Pseudomonas* etc.

LEAF

1. The pinnae arise on a stout primary rachis.

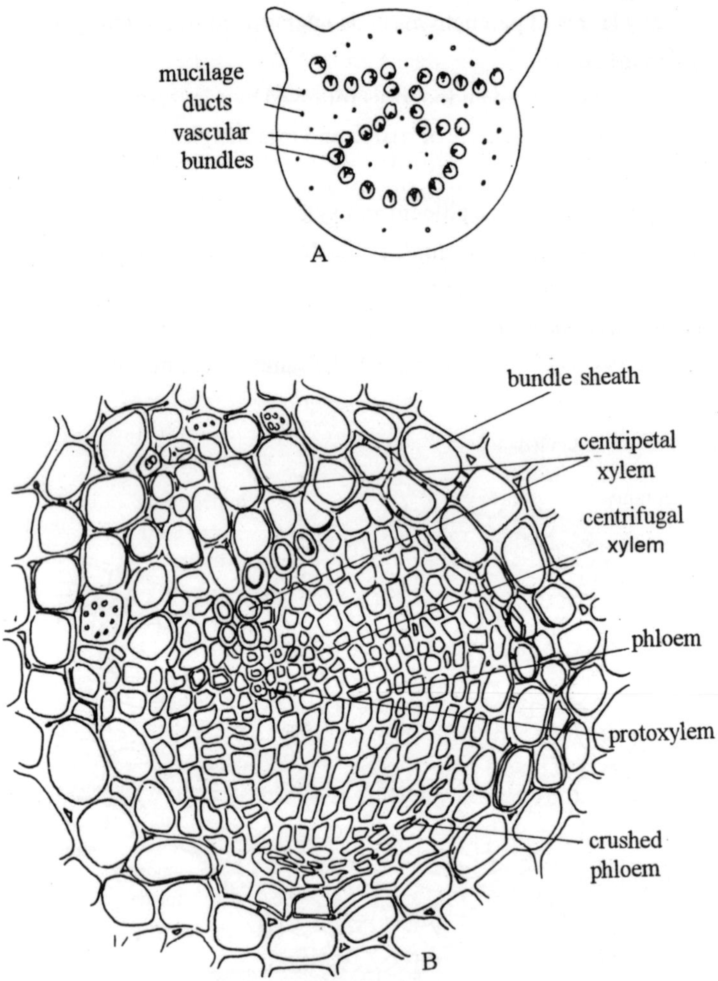

Fig. 17.11. Cycas
A- T.S. of rachis (ground plant) B- A vascular bundle (magnified)

2. The pinnae anatomically show bifacial arrangement.
3. It has mainly the following regions.

Cuticle: It is deposited above the upper and lower epidermal layers.

Upper- epidermis

It is single layered and consists of thick walled parenchyma cells.

Hypodermis

Internal to the epidermis is the one or two layered sclerenchymatous hypodermis.

Mesophyll

1. It is differentiated into *palisade* and *spongy parenchyma.*
2. The cells of the palisade parenchyma are columnar and are closer packed being present below the upper epidermis.

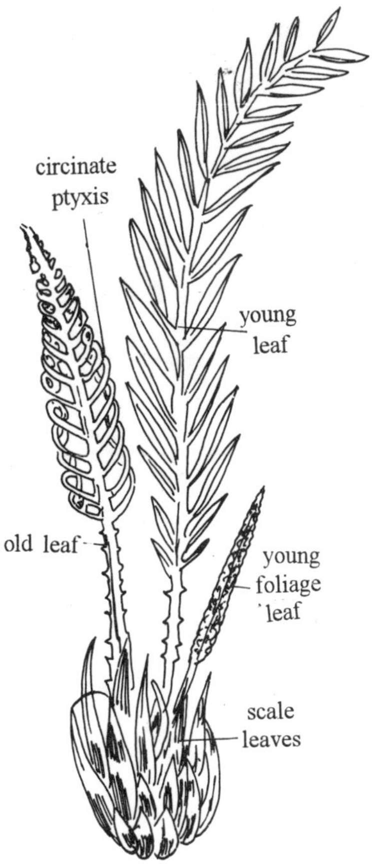

Fig. 17.12. CYCAS
A germinating bulbil

3. The cells of the spongy parenchyma are loosely arranged and are present above the lower epidermis.

4. The palisade parenchyma has plenty of chloroplasts and is the chief photosynthetic tissue of the plant.

5. A few chloroplasts are present in the cells of the spongy parenchyma also.

Vein

1. A vein cut in the transverse section shows an outermost bundle sheath surrounding the vasculature which shows the *X/P* arrangement.

2. The xylem consists of both centripetal (exarch) and centrifugal (endarch) elements and is mesarch.

3. The pholem is present below the xylem and consists of sieve tubes and phloem parenchyma.

Transfusion tissue

1. There are two types of transfusion tissues- primary and secondary.

2. The primary transfusion tissue consists of masses of tracheids having bordered pits on their walls.

3. It is present on the sides of the vascular bundle and embedded in the parenchyma that surrounds the vein.

4. According to Woresdell (1887), it is a lateral extention of the xylem.

5. According to Carter (1911) it is derived from leaf parenchyma.

6. In between the spongy and palisade parenchyma radially elongated cells are present constituting the secondary transfusion tissue.

7. These cells also have pits on their radial walls and are supposed to help in lateral conduction in the absence of lateral veins.

8. Some of the parenchyma cells surrounding the vasculatore have *spheroraphides*.

Lower epidermis

1. It is also single layered, but is discontinuous due to the preisence of small pits in which are found the the sunken stomata of *the amphicyclic* type.

2. The guard cells are surrounded by a ring of subsidiary cells.

Xerophytic feature seen in the leaflet

1. The leaflet shows many xerophytic adaptations.

2. These are : (a) thick cuticle, (b) thick walled epidermal cells, (c) presence of well developed mechanical tissue, (d) sunken stomata restricted to the ventral surface and (e) shiny and leathery nature of the leaflets.

Rachis

1. A transverse seclion show's an epidermis covered Itide.

2. Next to the epidermis is a band of chlorenchyma. Next of this, is a few layered sclerenchyma.

3. The entire inner region of the rachis is fillled by a mass of parenchymatous ground tissue in which are

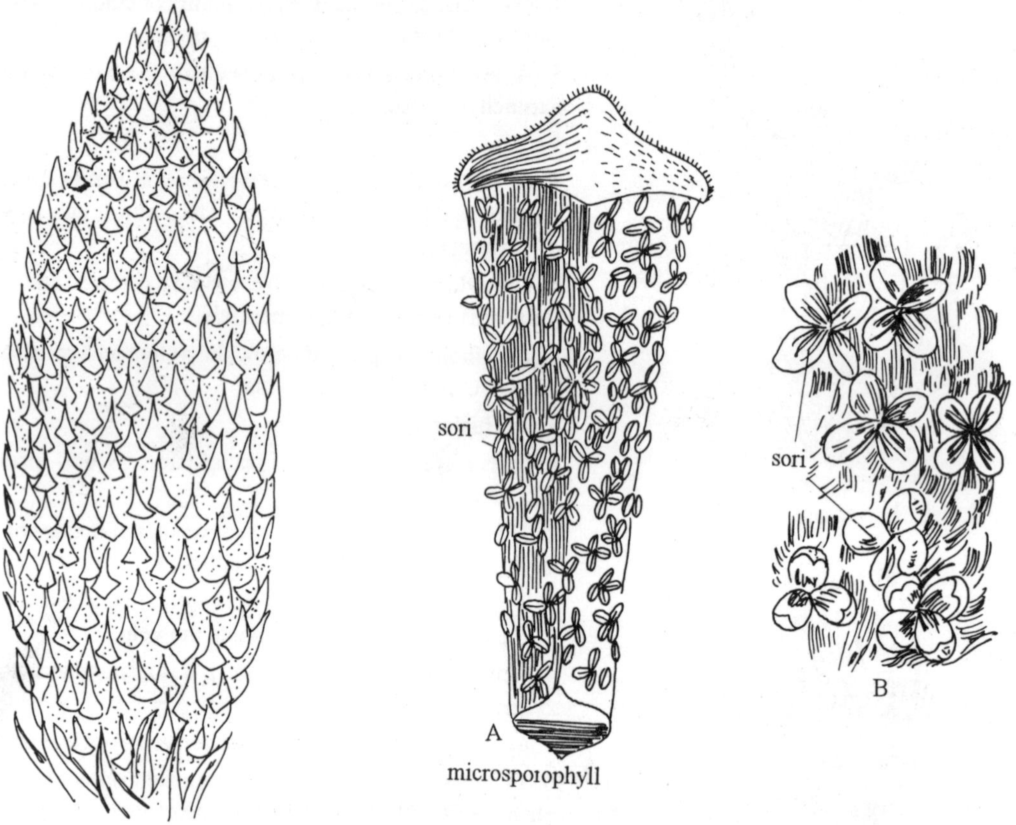

Fig. 17.13. Cycas
Male reproductive structure
Left - Entire male cone
Middle - microsporophyll
Right - Sori enlarged

embedded a number of vascular bundles arranged in an inverted omega fashion.

4. The vascular bundles are open and collateral.

5. The xylem lies on the inner side towards the centre of the petiole and consists of both centripetal and centrifugal elements (Diploxylic).

6. Below the xylem, phloem is found.

REPRODUCTION

1. The sporophytic plantbody reproduces by vegetative propagarion as well as by spore production.

2. Vegetative propogation is brought about by the formation of bulbills which are formed mostly in the basal part of the stem.

3. The bulbil consists of a condensed axis surrounded bv a few fleshy, scaly leaves.

4. On falling upon a suitable substratum a bulbil develops into an independent plant.

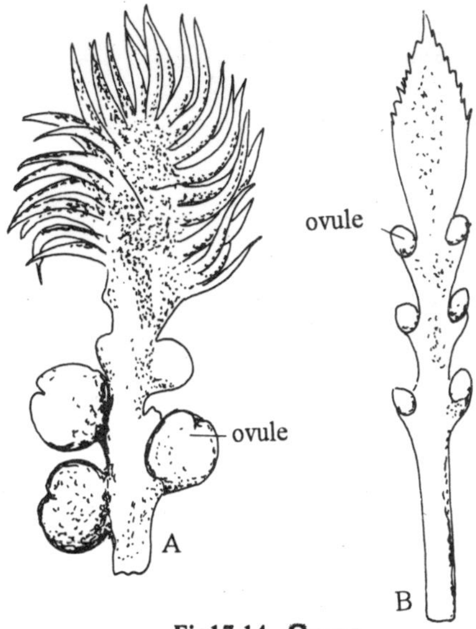

Fig17.14. **Cycas**
Megasporophylls
A- C. revoluta, B- C. circinalis

SPORE PRODUCTION

1. *Cycas* is strictly dioecious.
2. Micro and megasporophylls are borne on separate plants.
3. The growth pattern in the male and female plants differs.
4. The male cone is produced by the apical meristem and as a result growth comes to a stop.
5. Further growth is brought about by an accessory meristem (sympodial).
6. In the female plants, only the derivatives of the apical meristem develop into megasporophylls; hence the growth is uninterrupted (monopodial).

Male reproductive structures

1. The male cone is a compact structure and measures in length from 40-50 cms.
2. It has a central cone axis on which are spirally arranged the microsporophylls.
3. Each microsporophyll, is a flattened, wedge shaped structure consisting of a narrow lower portion which broadens above.
4. The upper portion is drawn into a projection called the *apophysis*.
5. The microsporophylls on their abaxial surface bear a number *of. pollen sacs* or *microsporangia.*
6. The microsporangia are present in two large goups separated by a median sterile line.
7. In each group there are a number of aggregations of microsporangia called *sori.*

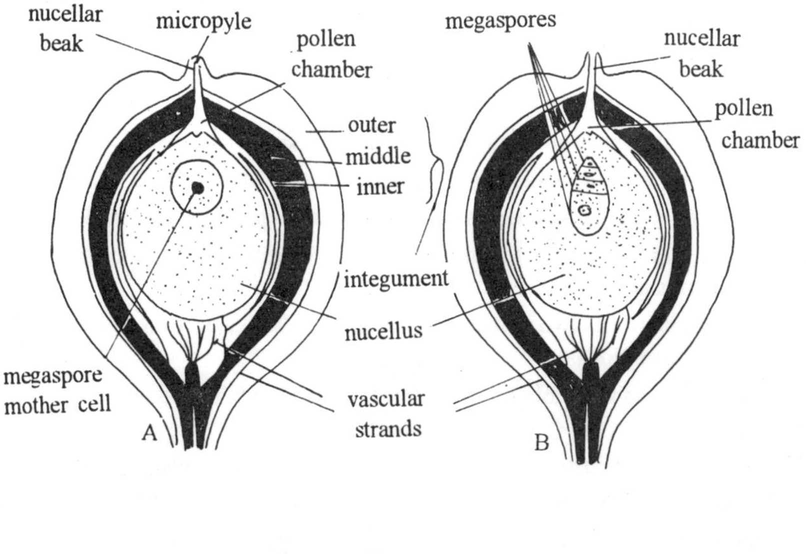

Fig17.15. **Cycas**
Long section of ovules. A- Ovule with megaspore mother cell. B-Ovule with megaspore

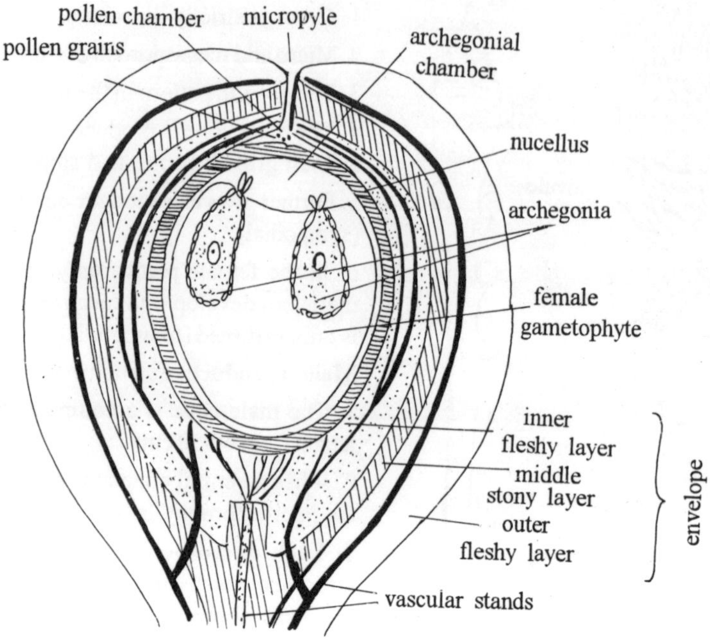

Fig17.16. CYCAS
L.S. of ovule of C. revoluta (before fertilisation)

8. Each sorus has an elevated central portion around which are found *3-6* micropsorangia.
9. Between the microsporangia, inducial hairs are present.
10. In each microsporangnim large number of dusty pollen are produced.
11. The soral organisation and the presence of inducial hairs indicate the fern ancestry of *Cycas*.
12. The microsporophylls present at the base and the apex of the cone are sterile.

Female reproductive structures

1. In the female plant there is no compact cone or strobilus.
2. At the apex of the stem the *megasporophylls* are spirally arranged.
3. They develop in acropetalous succession between foliage leaves and scale leaves.
4. Since only the derivatives of the apical meristem develop into the megasporophylla, the monopodial growth of the stem is unaffected.
5. Each megasporophyll has two portions *viz.*, a basal stalk like portion and an upper broad leaf like portion with a number of dissections.
6. The megasporophylls resemble the foliage leaves and are believed to be the modifications of the foliage leaves.
7. On the lower stalk like portion of the sporophyll two rows of ovules are borne.
8. The number of ovules per megasporophyll ranges from 2-10.
9. The ovules are believed to occupy the position of the pinnae.
10. Both the ovules and the sporophylls are covered by thick, yellowish, wooly outgrowth.
11. The ovule of *Cycas* has the size of a hen's egg and is green or orange red in colour.

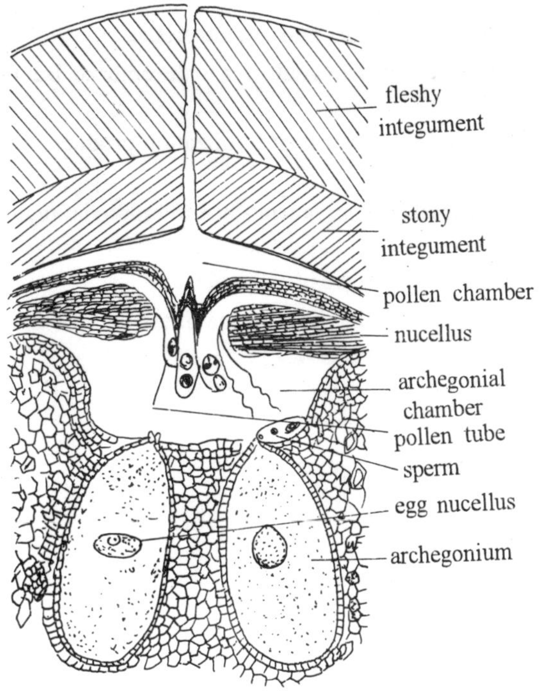

Fig17.17. Cycas Diagramatic L.S. of micropylar portion of ovule at the time of fertilisation (after Chamberlain)

12. It is the *biggest ovule* in the plant kingdom.

13. The different species of *Cycas* show variations in the nature of the megasporophyll and in the number of ovules per sporophyll.

14. The megasporophyll has a broad upper portion in *C. revoluta* whereas in *C.circinalU* the upper portion is also narrow.

Structure of a mature ovule

1. In a long sectional view the ovule shows the following features.

2. The integument is very thick and is made up of three layers.

3. The integument leaves a small opening called the micropyle.

4. Since the integument is very thick the micropyle forms a narrow tube called the micropylar canal.

5. The ovule is supplied by three bundles *viz.*, a central bundle, an outer bundle and an inner bundle.

6. The vasculature enters into the ovule and out of the three bundles two of them reach the tip of the integument.

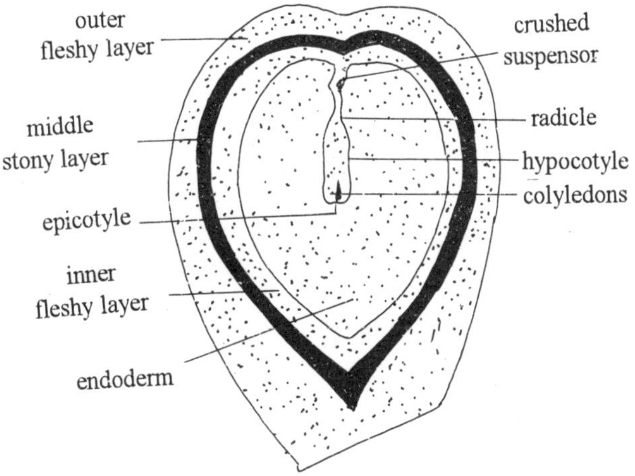

Fig17.18. Cycas
L. S. of seed

7. Internal to the integument is found the nucellus in the form of a few layers of cells, the bulk of which having been consumed during the development of the female gametophyte.

8. Internal to the nucellus is the large cellular female gametophyte occupying most of the interior of the ovule.

9. The cells of the female gametophyte are rich in starch content and other nutrient materials.

10. The female gametophyte is very often called *endosperm* since it provides nutrition to the embryo. But this term does not seem to be very appropriate.

11. It is totally, different from the endosperm of angiosperms which is a post fertilization product and is generally triploid.

12. The female gametophyte will be adhering to the nucellus at all regions except at the microphylar region, where a small space is left called the *archegonial chamber*.

13. In the micropylar region of the famale gametophyte two archegonia are usually seen.

14. The nucellus protrudes into the micrppylar canal to form the nucellar beak, some of the cells of which disintegrate to form the pollen chamber in which are lodged the pollen grains.

Structure of the seed :

1. The integument hardens to form the seed coat.
2. The outer fleshy layer gets dried up and adheres to the middle stoney layer.
3. The inner fleshy layer soon shrivels up.
4. Within the seed coat is found the female gametophyte.
5. Embedded in this is the embryo.
6. The radicile portion of the embryo is directed towards the micropyle.
7. The plumule is directed towards the chalzal region of the seed.

Systematic position

CYCADOPSIDA

CYCADALES

CYCADACEAE

CYCAS

CHAPTER 18

CONIFEROPSIDA

Of all the gymnosperms perhaps the most dominant and the most beautiful are the coniferous plants. The popular name *"Conifers"* is due to the cone like appearance of the plantbody. Tle branches are long towards the base, succeedingly they become shorter giving a cone like appearance to the plant body. The coniferous forests are quite extensive, particularly in the temperate regions of the world. Even in tropical countries the vegetation at higher altitudes always comprises coniferous plants. It has been estimated that conifers occupy about 8% of the land on Earth.

Geologically, conifers date back to the Carboniferous period. But some, like Cordaitales were found as early as the Devonian period.

Coniferopsida differ from Cycadopsida in having.(a) simple leaves (b) narrow pith and cortex, (c) active fascicular cambium producing annual rings, (d) pycnoxylic wood and (e) bilaterally symmetrical ovules and seeds.

Sporne, (1965) has classified coniferopsida into four orders *viz., Cordaitales, Ginkgoales, Coniferales,* and *Taxales.* Of these, Cordaitales and Confierales are discussed here.

CORDAITALES

1. This fossil order constitutes the most ancient and primitive order of the class.
2. Cordaitales are believed to have taken their origin during the Devonian period.
3. As revealed by the abundance of fossils, the order reached the peak of development during the Carboniferous and Permain periods, forming a dominant arborescent vegetation.
4. The plants were arborescent and had a monopodial branching.
5. Secondary wood was abundant and pycnoxylic.
6. Strobili were monosporangiate.

CORDAITES

The genus *Cordaites* is known from the assemblage of dismembered parts which are known under various names.

Stem and wood - *Cordaites, Dadoxylon, Mesoxylon, Parapitys* etc.

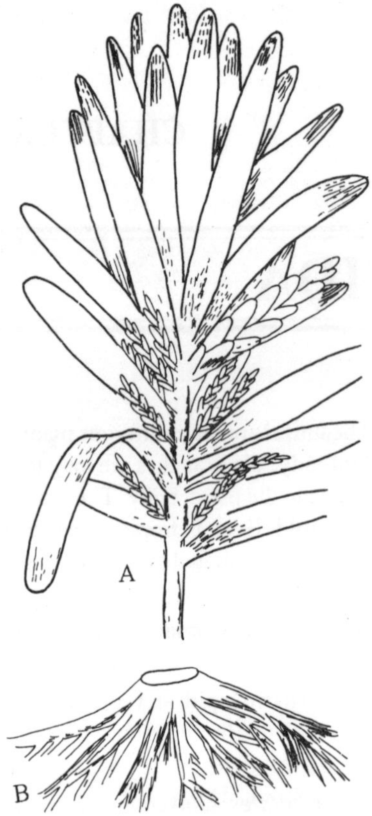

Fig. 18.1. Cordaites
Reconstruction of the plant body.
A. A brach bearing foliage and inflorescence,
B. Root system

Leaves - *Cordaites* - Defoliated branches - *Cordaicladus.*

Roots -*Amyelon.*

Strobilus - *Cordainthus.*

Seeds - *Cordiaicarpus.* (= *Cordiocarpus*).

The plant body of *Cordaites* was a tall tree reaching a height of 30 metres and grew in dense forests. The stems were slender and attained a diameter of 65-70 cms. The branching was lateral and arose only at the top part of the plant.

Leaves (compressions) were found on the branches and were spiral in arrangement. They were petiolate, subulate or spatulate or grass like and had an entire margin. Venation in the leaf was parallel. Based on the nature of the leaves the genus is divided into three sub-genera *viz., Eucordaites* (Leaves broad with a blunt apex *Dorycordaites* (leaves with acute apex) and *Poacordaites* (Leaves were grass like and small). .

The plant was anchored to the soil with the help of repeatedly branche root system.

INTERNAL STRUCTURE

Stem

1. As has alteady been said, Cordaitalean stems are known under various names, In *Cordaites* the stem has an extensive cortex which had parenchyma with interspersed scelernchyma.

2. Secretory sacs were also present in the pith. Pith had large air chambers separated by diaphragms.

3. Pith casts have given the name *Artisia.*

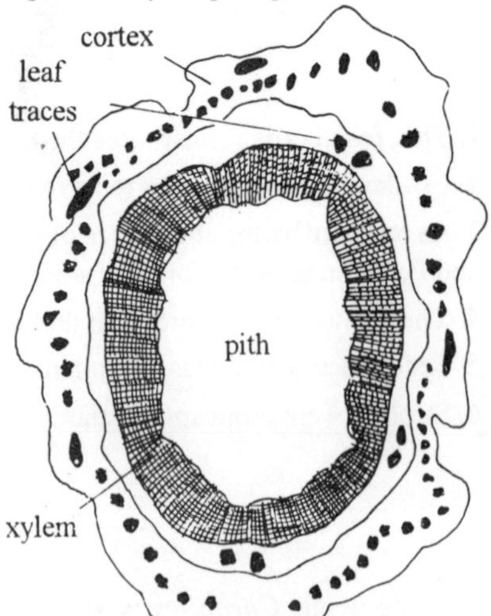

Fig. 18.2. Cordaites
T.S of stem

4. The vasculature in its primary structure consisted of a large number of endarch or mesarch protoxylem elements.

5. Secondary growth in the stem resulted in the formation of compact (pyconoxylic) wood.

6. The tracheids in the secondary wood were long and narrow and had uniseriate bordered pits.

7. Interspersed with the wood elements, were the parenchymatous rays which were simple and one cell wide.

8. Some of the cells of the ray had resinous material. However there were two resin canals as in *Pinus.*

9. The leaf traces resembled those of cycads in the *pseudomesarch* and *diploxylic* nature.

10. Each leaf was provided with a pair of traces.

Leaf

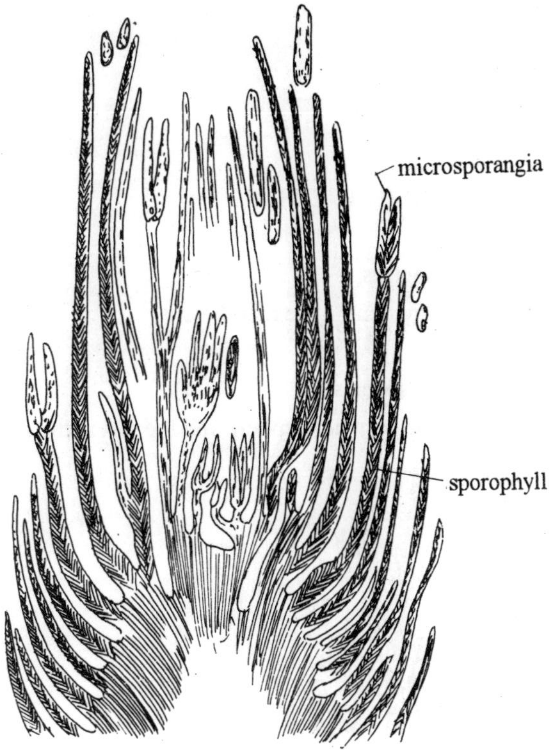

Fig. 18.3. Cordaites
L.S. of male reproductive structures
(Cordaianthus penjonii) (after Renault)

1. Anatomically the leaves exhibited differentiated mesophyll in most cases.
2. The epidermis had stomata distributed in longitudinal bands. Guard cells were surrounded by 4-6 subsidiary cells.

Root

1. Internally the root showed a diarch to tetrach xylem.
2. Stele was protostelic.

REPRODUCTION

1. Cordaitean plants were *dioecious*.
2. The strobili, called *Cordaianthus* were unisexual and 1 cm long.
3. The male strobili had a central axis which had a few microsporophylls and numerous sterile bracts. Each microsprophyll bore 1-6 pollen sacs.
4. The female strobili like the male ones had a stout central axis on which were borne spirally arranged bracts.
5. The length of the strobili varied from 25-30 cms.
6. In the axis of the bracts arose short, fertile, secondary shoots which had spirally arranged fertile and sterile *bracteoles* (megasporophylls).
7. Each megasporophyll had 2 or more ovules at its apex.

8. In the ovules, the nucellus remained free from the intergument for most of its length.

Systematic position

CONIFEROPSIDA

CORDIATALES

CORDIATACEAE

CORDAITES

CONIFERALES

1. The order *is* the largest among living gymnosperms.
2. The cone like appearance of the plant is very evident in the members.
3. The plants may be dioecious or monoecious.
4. The order is represented by 54 genera and 482 species.
5. Among these, about thirty genera are in the northern hemisphere and the rest in the southern hemisphere.

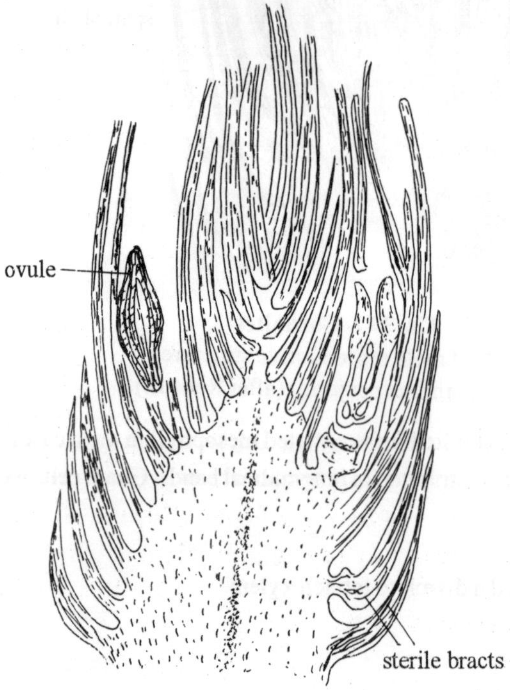

Fig. 18.4. Cordaites

L.S. of female reproductive structures, (Cordaianthus williamsonii) (after Renault)

6. The plants flourish very well in a temperate climate. In habit, the plants range from shrubs to gigantic trees.
7. Some of the tallest trees of the plant kingdom *(Sequoia, Sequoiadendron)* belong to this order.
8. Some of these are of great antiquity being more than 4000 years old.
9. Some of the pinp trees which are living even today were already old when the Roman empire was young.

PINUS

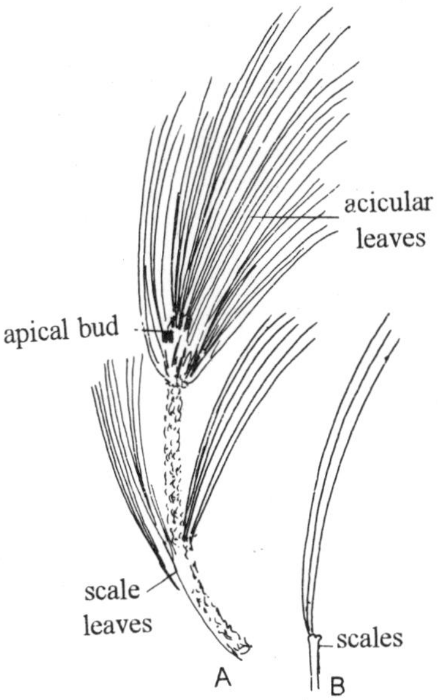

Fig. 18.5. **Pinus**
A. Shoot of unlimited growth.
B. Drawf shoot

Occurrence and distribution

1. The genus consists of 90-95 species spread over the northern hemisphere in temperate regions.

2. About 1/3 of the species are endemic to the old world and the rest in the new world.

3. The plants grow on the slopes of hills and form extensive evergreen forests.

4. Some of the species of *Pinus* are of great economic value.

5. In India, the genus is represented by six species *viz., P. wallichiana, P. armandi, P. gerardiana, P. insularis, P. roxburghii* and *P. merkusii.*

6. The classification of the Indian species is based on the number and form of leaves (Raizada and Sahni, 1960).

7. The Indian species of *Pinus* grow in the northeast and northwest Himalayas.

8. In the hills of south India *Pinus* is cultivated.

Morphology

1. The plant body *of Pinus* is a tall tree varying in height from 70-200 *feet.*

2. The girth of the stem maybe as much as 12 *feet.*

3. Some of the species of *Pinus (P. aristata)* growing in California are as old as 4000 years.

4. The stem is erect, woody and covered with rough scaly bark, which occasionally peels off.

5. The tip of the stem has a large apical bud which brings about the growth in length

6. The branching of the stem is monopodial.

7. It is very regular and symmetrical and arises laterally from the axis of scale leaves.

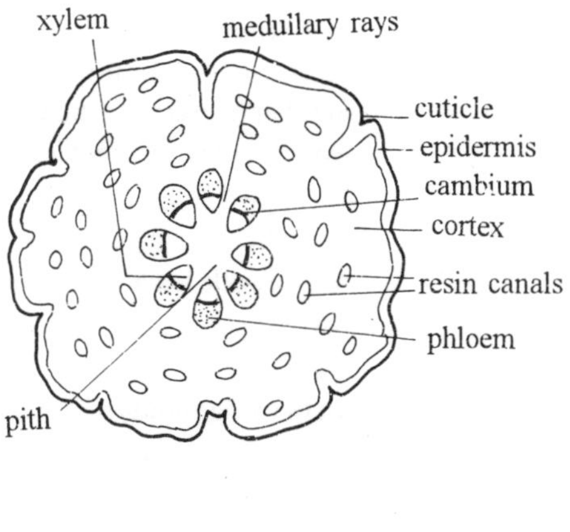

Fig. 18.6. **Pinus**
T.S. of young stem

8. The branches are longer on the lower part of the stem and become shorter towards the apex giving the familiar cone like appearance.

9. In some old stems the symmetry may be lost due to the death of some branches and irregular growth of the others.

10. The branches are dimorphic *viz.*, branches of *unlimited growth* or *long shoots* and branches of *limited growth* or *dwarf shoots.*

11. The long shoots arise from lateral buds on the main stem in the axils of scale leaves.

12. As the name indicates, they grow indefinitely by means of an apical bud (Fig. 2.34 A).

13. It is these branches which impart the characteristic appearance to the plant.

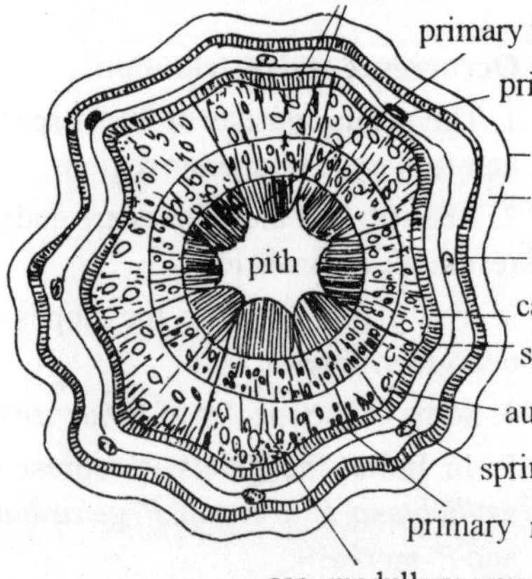

Fig. 18.7. **Pinus**
T.S. of old stem showing secondary growth

14. In some cases when the apical meristem of the main stem is injured, one of the long shoots at the apex becomes dominant and grows like a main stem.

15. The long shoots are covered with brown scale leaves in the axils of which arise the dwarf shoots.

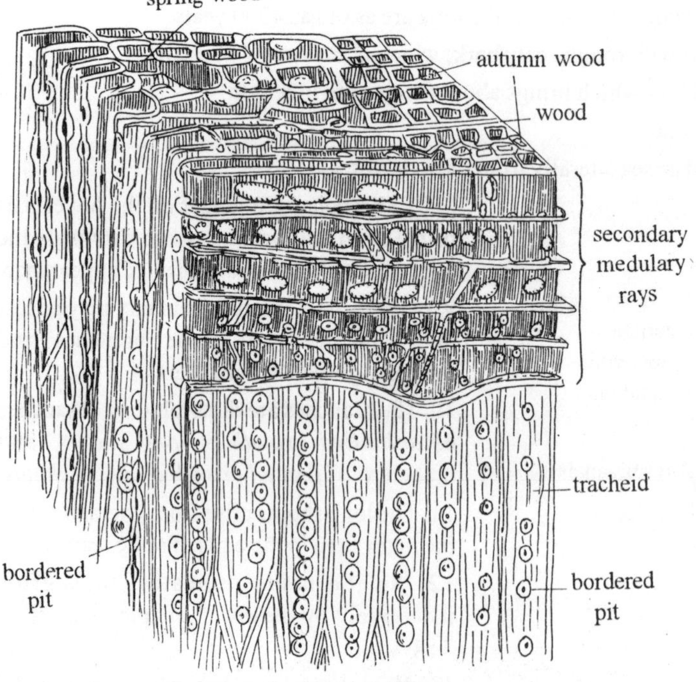

Fig. 18.8. Pinus
Three dimensional model of a block of wood

16. The drawf shoots never arise on the main stem.

17. They are numerous in number and as the name indicates, are of limited growth, because the apical bud is ephimeral.

18. Each dwarf shoot is about half an inch in length and bears a number of thin innmbricated scale leaves *(Cataphylls)* below and a fixed number of green, acicular cylindrical leaves called the *needles* above.

19. The leaves of *Pinus,* like the branches are also dimorphic *viz.,* brown scale leaves and green foliage leaves.

20. The scale leaves are borne both on the long shoots and dwarf shoots.

21. They usually fall off at the maturity of branches and are supposed to help in the conservation of water around the young branches.

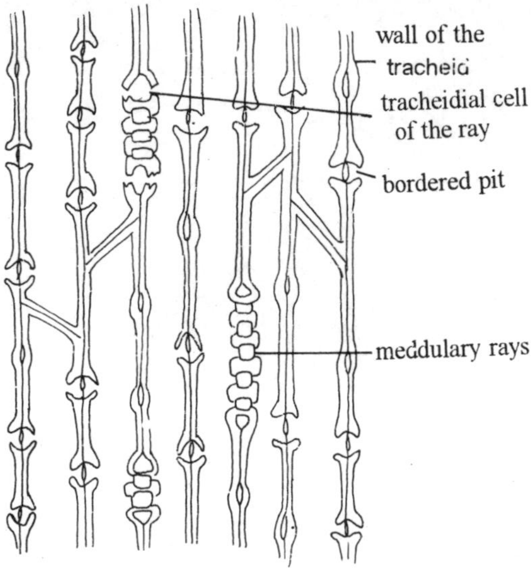

Fig. 18.9. Pinus
Longitudinal tangential section of wood.

22. The foliage leaves *i.e* needles are produced on the dwarf shoots only, persist for a number of years and fall off only when a *foliar spur* (dwarf shoot with needles) is shed.

23. The number of needles varies in different species of *Pinus.*

24. It may be three *(P. longifolia, P.gerardiana* etc.), four (*P. quardrifotia)* or five *(P. excelsa).* Even two leaved *(P.penaster)* and single leaved (*P. monophylla)* foliar spurs are also found.

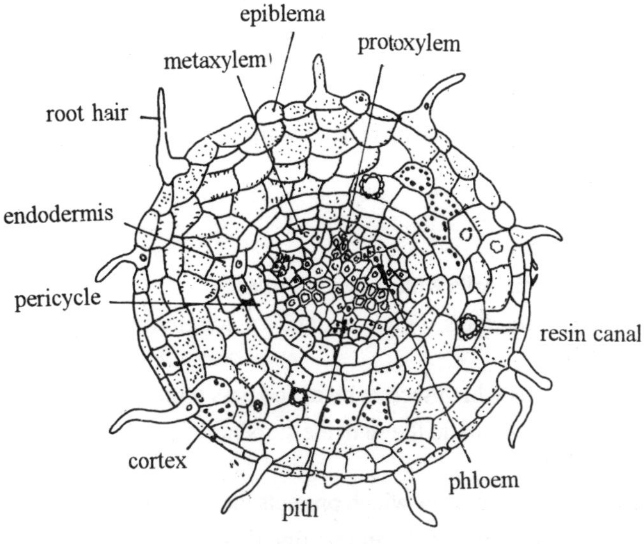

Fig.18.10 Pinus
T.S. of young root

25. The phyllotaxy depends on the number of needles. It may be opposite or whorled.

26. The plant is anchored to the soil with the help of a promiment tap root.

27. The tap root produces a number of well developed lateral roots.

28. The root system is extensive, but does not grow very deep in the soil.

29. The root hairs are not well developed and sometimes even absent on the ultimate branches.

30. The roots have the mycorrhizal association.

31. The fungus is ectotrophic and

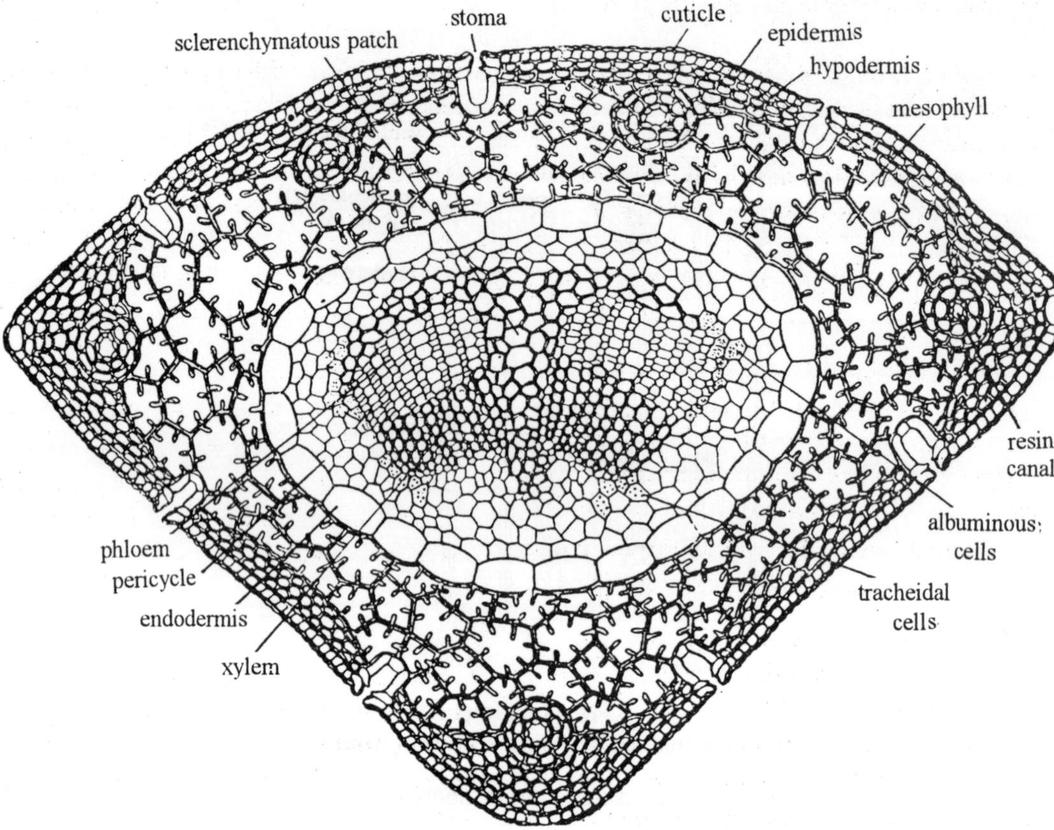

Fig. 18.11. Pinus. T.S. of needle of P. roxburghii

provides the root with an effective absorptive surface.

32. The relationship between the fungus and the root is symbiotic.

33. It is believed that the fungi provide some growth hormones to the plant.

INTERNAL STRUCTURE

Stem

1. Epidermis is irregular and made up of a single layer of cells whose outer walls are cutinised.

2. Occasionally scattered stomata are found on the epidermis.

3. Cortex has two regions.

4. The outer region *i.e* hypodermis has lignified cells.

5. The inner cortex consists of thin walled parenchyma cells with some chlorplasts.

6. Distributed in the cortex are found a number *of resin ducts,* which extend as longitudinal canals in the stem.

7. Each resin duct is lined by a layer of epithelial cells which secrete resin and pass it on to the duct.

8. The resin has *turpentine* and whenever there is an injury to the stem the turpentine comes out and forms a protective covering for the wound.

9. It is known that the turpentine has antibacterial properties which protects the stem from infection.

10. Economically, distillation of pine wood and leaves yields the natural turpentine.

11. The *stele is* surrounded by endodermis and pericycle, both of which are not clearly distinguishable.

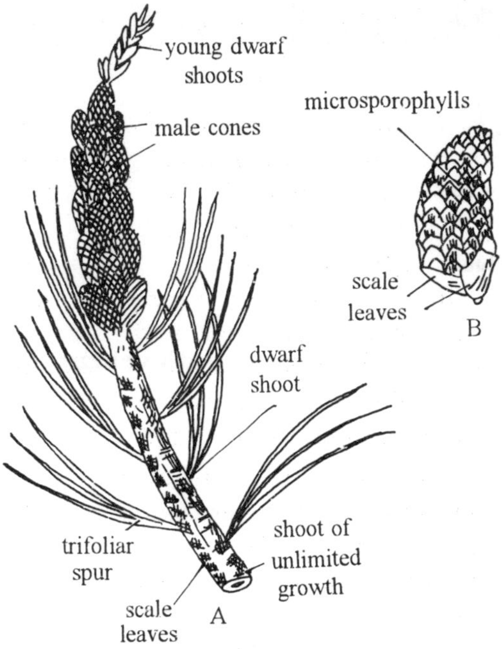

Fig. 18.12.Pinus
A. A shoot bearing male cones.
B. A male cone enlarged

12. The vascular tissue in its primary structure consists of 6 -8 collateral, conjoint, open, vascular bundles surrounding a central pith.

13 The xylem is endarch.

14. There are no vessles or wood fibres in the xylem.

15. The protoxylem consists of spiral and annular tracheids with a fewbordered pits.

16. The metaxylem trachieds have numerous bordered pits on-their radial walls.

17. A few wood rays are found in the xylem.

18. Phloem consists of sieve tubes, phloem parenchyma and abuminous cells.

19. The last mentioned cells are in lieu of companion of cells.

20. Between the vascular bundles are present very narrow primary medullary rays.

21. In its essentials, the anatomy of the stem resembles that of dicotyledons.

22. In a transverse section of the stem the primary structure reveals an epidermis, cortex and stele.

Secondary Growth

1. This is brought about by the activity of the fascicular cambium.

2. The cambium which is present in the form of strips in the primary vascular bundles extends radially and forms a closed ring of cambium, the cambial cells divide repeatedly in a tangential direction.

3. At each division two cells are formed. One remains cambial, while the other develops into either xylem or phloem.

4. Secondary xylem is formed towards the interior of the cambial cell while secondary phloem is formed towards the exterior.

5. As a result of secondary growth a completely closed ring of vasculature is formed.

6. Unlike in *Cycas* the cambium here remains as a permanent meristem.

7. Due to the activity of the cambium annual rings are produced.

8. Each annual ring has tracheids of two different sizes *viz.*, spring wood and autumn wood.

9. The former has abroad lumen while the latter has a narrow lumen in their tracheids.

10. Unlike in *Cycas,* the annual rings are regular and they give an approximate idea of the *age* of the plant.

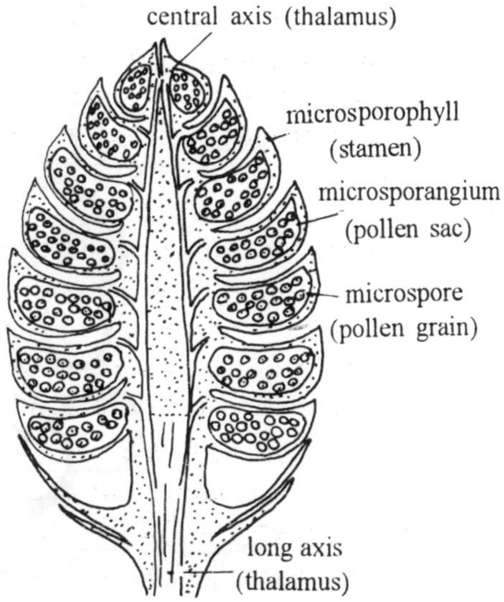

Fig. 18.13.Pinus
Long section of a male cone

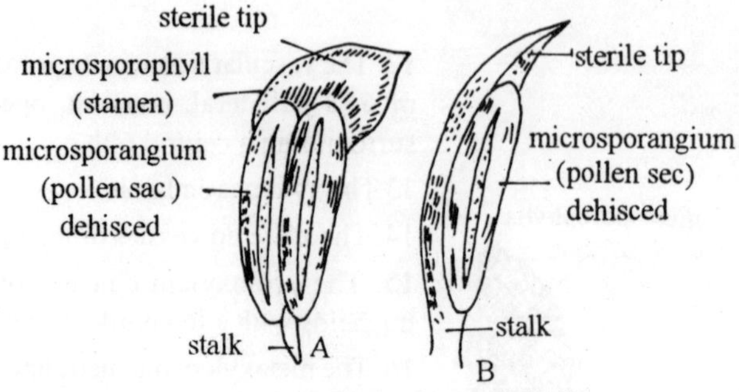

Fig. 18.14. Pinus

11. Risin ducts are present both in the primary and secondary wood.
12. The tracheids of autumn wood have small bordered pits while those of spring wood have larger pits.
13. The pits have a distinct torus.
14. Special thickenings of primary lamella surrounding the bordered pits are found.
15. These are called Bars or rims of sanio.
16. When the pits are too close to each other the pairs of Bars of sanio are seen round the rim. The phloem consists of radial rows of long sieve cells which taper on both the ends.
17. The sieve plates occur mostly on the radial walls.
18. The sieve pores may become covered with callus.
19. The primary medullary rays are replaced by secondary medullary rays.
20. Like the secondary vascular tissues they are also formed by cambium.
21. They travel through the secondary xylem and phloem at certain places.
22. The portion of the ray that passes through the secondary xylem is called the xylem ray and the one that passes through the secondary phloem is called the phloem ray.
23. The rays are one celled broad and 2 to 12 cells in height. Occasionally the rays may become multi seriate 'when they come in contact with a resin canal.
24. New medullary rays are formed every year.
25. The xylem rays are non living and have a lignified wall which may protrude into the lumen of the cells.

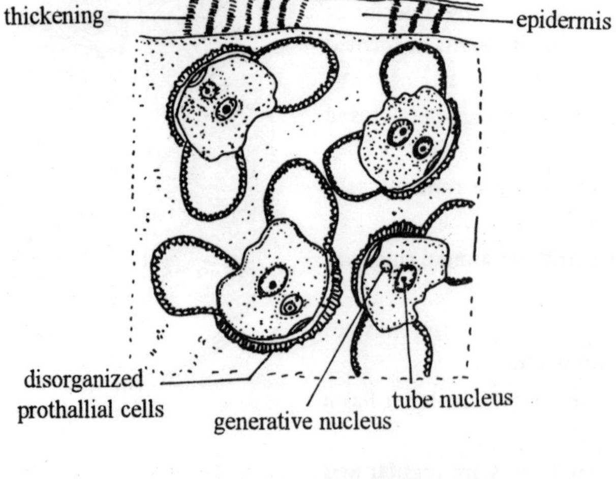

Fig. 18.15. Pinus pollen grains

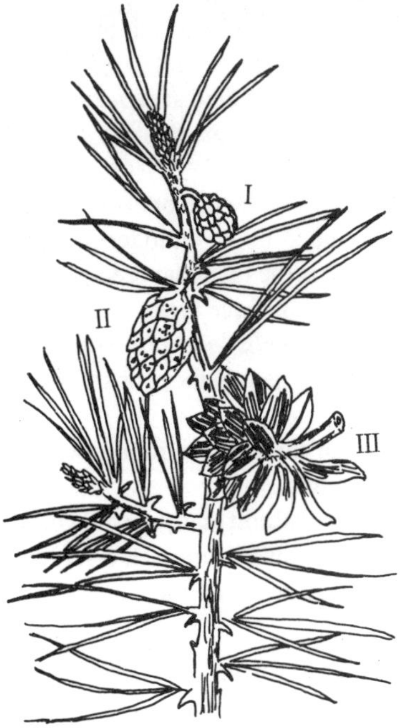

Fig. 18.16. **Pinus**
A shoot bearing female cones

26. The walls have bordered pits. In between the cells of the ray are situated a few living ray parenchyma cells.
27. The phloem rays have albuminous cells in place of ray tracheids.
28. They have a prominent nucleus and dense cytoplasm.
29. The albuminous cells function like the campanion cells of Angiosperms.
30. Between albuminous cells, occur cells of ray parenchyma.
31. The structure of the wood (pycnoxylic) and the secondary medullary rays maybe studied by taking sections of the stem in different planes.
32. These are, (1) Cross section, (2) Radial longitudinal section (RLS) and (3) Tangential longitudinal section (TLS).
33. In a *cross section*, both spring and autumn wood are seen cut transversely. Tracheids of autumn wood are squarish, while those of spring wood are polygonal in outline.
34. Wood rays are cut lengthwise and the breadth of the rays can be seen in this section.
35. In a **R.L.S.**, tracheids are cut lengthwise. Bordered pits are seen showing the surface view.
36. Secondary medullaiy ray is cut at its side, lengthwise.
37. In this, the height (thickness) of the rays can be seen.
38. Pits are seen in the medullary rays also. Both xylem rays and phloem rays may be seen.
39. In a **R.L.S.** the tracheids are cut lengthwise at their side.
40. Sectional views of the bordered pits, with torus is clearly noticeable.
41. Secondary growth in the cortex takes place by the activity of a cork cambium.
42. This produces the periderm tissue.

Root

1. A transverse section of the root shows that it is identical with the dicot root.
2. Epiblema is single layered, from which arise the root hairs.
3. This layer surrounds the parenchymatous cortex.
4. Here and there resin canals are seen. Bounding the stele is the single layered endodermis.
5. It is usually impregnated with tannin.
6. Internal to the endodermis is the many layered pericycle.
7. Vascular bundles are radial with exarch and diarch or triarc xylem.
8. The pith is very narrow.
9. In the mycorrhizal roots, the hyphae may be seen in the intercellular spaces of the cortex.
10. Secondary growth in the root takes place by the activity of interfascicular cambium in the stele, and by cork cambium in the cortex.

Leaf

1. The outline of the section depends of the number of needles in the foliar spur.
2. In single leaved spurs *(P. monophylla)* it is circular, in two leaved spurs-semicircular *(P. sylvestris)* and in three leaved spurs it is triangular *(P. longifolia)*.
3. The outermost layer is the epidermis whose cells have thick and heavily cutinised walls.
4. The continuity of the epidermis is interrupted by sunken stomata which are present on all sides of the leaves.

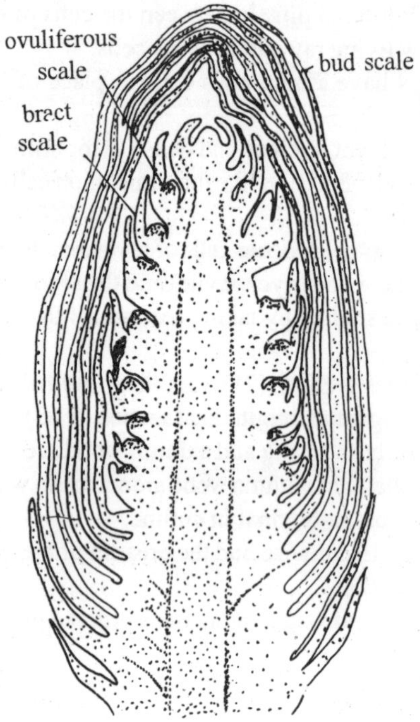

Fig. 18.17. Pinus
L.S of a young female cone of P. wallichiana

5. Internal to the epidermis is the hypodermis consisting of two to three layers of sclerenchyma.
6. The mesophyll is- undifferentiated.
7. It is made up of thin walled cells which project into the cavity in the form of infoldings.
8. Such a type of mesophyll is called arm *parenchyma.*
9. The cells of the arm parenchyma contain starch grains and chloroplasts.
10. Resin ducts are present in the mesophyll.
11. The vein is surrounded by an endodermis or bundle sheath, internal to which is the many layered pericycle.
12. The pericycle consists of three types of cells *viz., parenchyma cells, tracheidal cells, and albuminous cells.*
13. The last two types of cells together constitute transfusion tissue.
14. Lying in the centre of the vien are found two vascular bundles. They are collateral.
15. The leaves of *Pinus* exhibit many *xerophytic* characters such as thick cuticle, sunken stomata, sclerenchymatous hypodermis, transfusion tissue etc.

REFRODUCTION

There is no special method for vegetative propagation in *Pinus.*

Spore Production

1. The plants *otPinus* are monoecious, but the cones are monosporangiate.
2. The two cones are produced on different branches.

Male cone

1. The male cones are produced on the plant earlier than the females ones.
2. They are usually produced in the month of March or April in hills and January in plains.
3. The male cones are produced in clusters in the axils of scale leaves on the branches of the unlimited growth.
4. On the long shoots, the cones are found at the tip of the youngest shoots.
5. Since the male cones occupy the place of dwarf shoots they are supposed to be the modifications of the latter.
6. A young male cone is ovate in outline measuring about 3-4 cms in length and 0.65 cms in diameter.
7. The cone consists of a central axis on which are spirally arranged microsporophylls which are 60-135 in number.
8. Each microsporophylls is somewhat tri- angular in outline and has a short stalk.
9. The terminal portion of the sporophyll is sterile, pointed

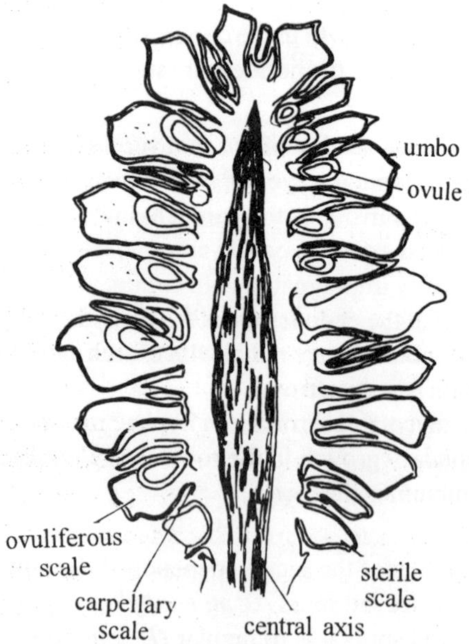

Fig. 18.18. Pinus
.L.S. of mature female cone of P. roxburghii

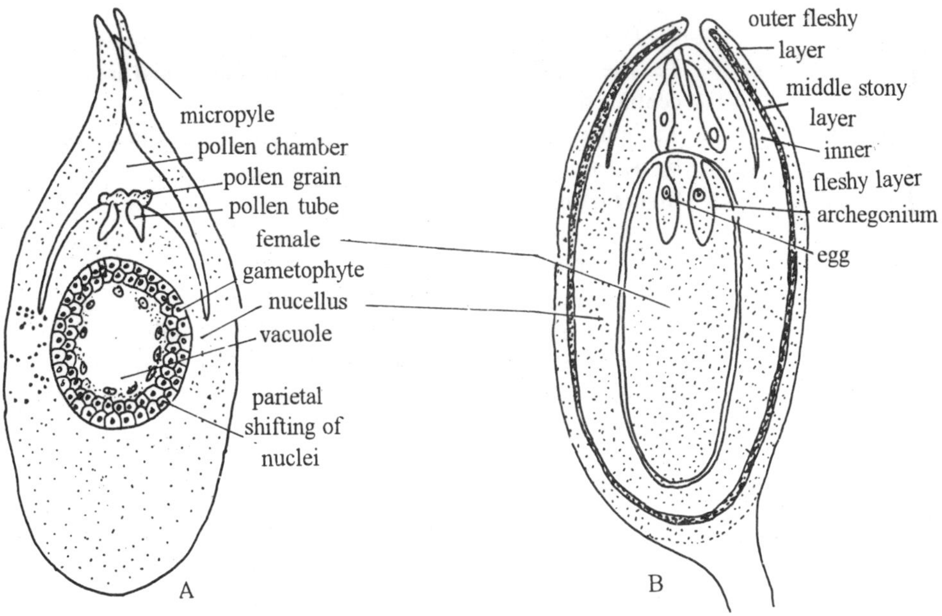

Fig 18.19 Pinus

L.S of ovules, A ovule with young female gametophyte, B ovule with female gametophyte ready for fertilization

and slightly bent upwards.

10. The basal portion of the sporophyll on its abaxial surface bears two microsporangia.

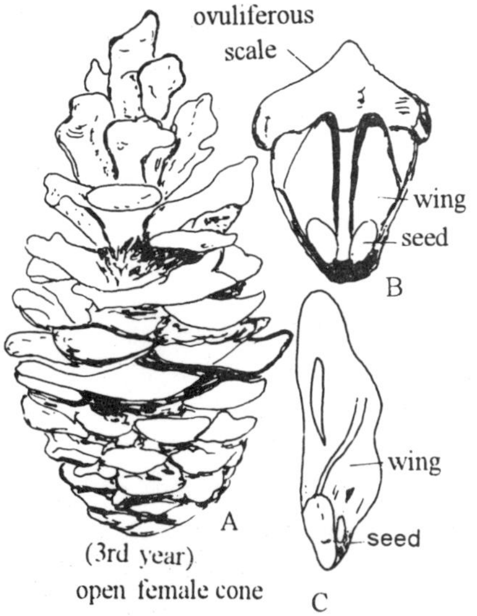

Fig. 18.20. Pinus
A Mature female cone with seeds ready for dispersal. B Ovuliferous scale bearing seeds C A winged seed

Female cone

1. This is also known as the- *ovulate cone.*
2. The female cones are produced on the uppoer part of the stem in the axils of scale leaves.
3. They take the position of a shoot of unlimited growth and are apparently the modifications of the latter.
4. On the newly formed long shoots which do not bear male cones, the female cones occur below the terminal bud, each representing a newly formed long shoot.
5. The young female cone is a small reddish green structure consisting of a central axis on which are arranged spirally the megasporophylls.
6. Each megasporophyll has two appendages *viz., the bract scale* (carpellary scale) and the *ovuliferous scale* (seed scale).
7. The bract scale is leathery and is directly attached to the cone axis.
8. The ovuliferous scale is larger than the bract scale and stouter and harder.
9. It develops from the upper surface of the bract scale and bears two ovules as swellings.
10. The ovuliferous scale is woody, wedge shaped and has a broad sterile end directed outwards, which is called the

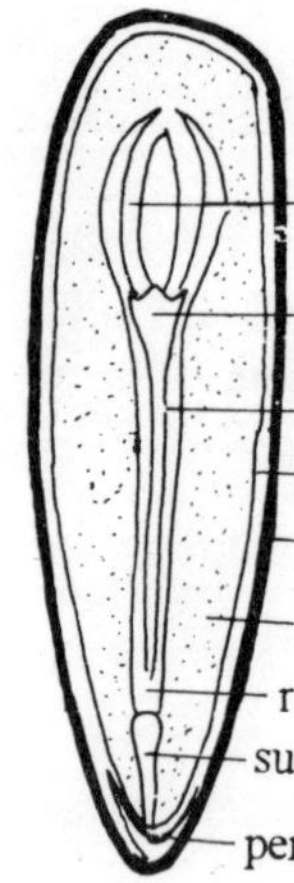

Fig. 18.21. **Pinus**
L.S. of seed of *P. gerardiana*.

apophysis.

11. Due to this, the bract scales are hidden from the view.
12. The surface of the cone is marked by rhomboidal areas, each with a central concial point called the *umbo*.
13. In the young condition the basal region of the female cone covered by bud scales, which project the growing cone.

Structure of the ovule

1. The ovule arises as a small hump of parenchyma cells on the upper surface of the ovuliferous scale.
2. This hump of parenchyma cells constitutes the nucellus.
3. From the base of the nucellus arises the integument.
4. The integument leaves an opening called the micropyle at one end.
5. The nucellus and the integument are fused below but free above.
6. The ovule does not develop a nucellar beak or a prominent pollen chamber.
7. Their absence is possibly related to the non production of ciliated sperms,
8. The integument is divisible into outer fleshy, middle stony and inner fleshy layers.
9. A mature ovule before fertilisation consists of a cellular female gametophyte surrounded by a thin layer of nucellus. 2-6 archegonia are found near the micropylar region of the female gametophyte.
10. A small space is left between the nucellus and the female gametophyte to form the archegonial chamber.
11. The pollen chamber is not properly defined.
12. Lodged in it are found a few pollen grains.
13. There is no cellular beak as in *Cycas*.
14. The integument is three layered.

Systematic position

CONIFEROPSIDA

CONIFERALES

PINACEAE

PINUS

CHAPTER 19

GNETOPSIDA

1. The class **Gnetopsida** comprises some of the most advanced members of gymnosperms.
2. The class consists of an order Gnetales.
3. But some botanists identify three orders namely *Ephedrales, Welwitschiales* and *Gnetales.*

GNETALES

1. The members of the order are woody plants: they may be trees, shrubs or lianes.
2. Many of the plants resemble dicotyledons in their general morphology.
3. Plants may be monoecious or dioecious and reproduce by means of "flowers".
4. Members have vessels in the secondary wood.
5. The order consists of three families *viz.*, *Gnetaceae, Welwitschiaceae* and *Ephedraceae.*

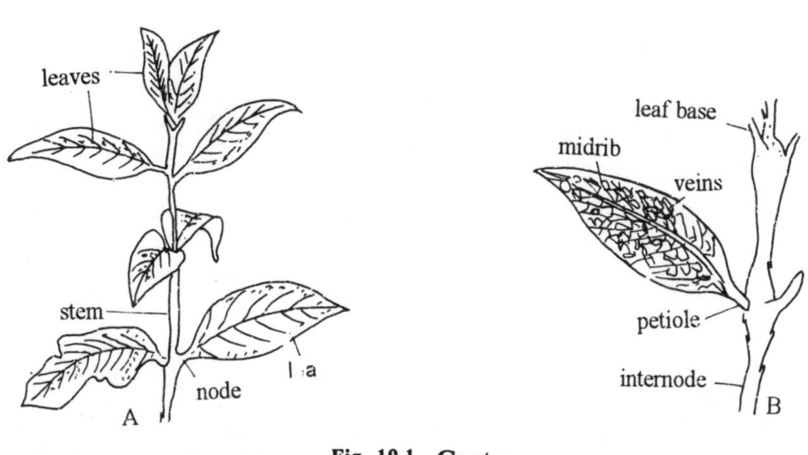

Fig. 19.1.. Gnetum

Habit of G. gnemon. A. A twig showing arrangement of leaves. B. A leaf showing venation.

GNETUM

The family *Gnetaceae* consists of a single genus *Gnetum,* with about *40* Species, which are distributed all over the tropics, mostly in Asia, America and Africa. In our country *Gnetum* is represented by six species (*G. montana, G. gnemon, G. latifolium, G. contractum* and *G.oblongum*). In India the distribution of *Gnetum* is mostly in the eastern Himalayas and southern peninsula between altitudes of 1000-2000 meters. In Karnataka, *G. ula* is common in western ghats.

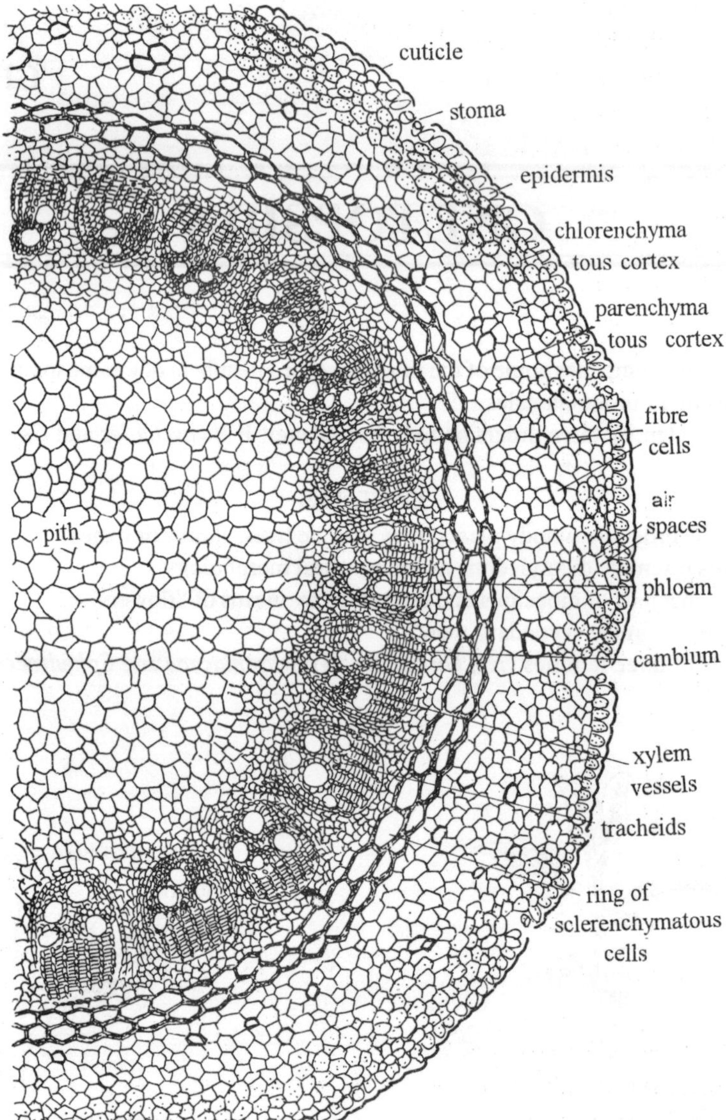

Fig. 19.2. Gnetum

T.S. of young stem of G. ula (a portion enlarged)

Morphology

1. In its general habit, *Gnetum* remarkably resembles a dicotyledolums plant.
2. The nature and habit of the plant is varied.
3. A great majority of the spedes are climbers with twining stems (G. *neglectum, G. latifolium* etc.), while others may be shrubs or trees.
4. Tree habit is found in G. *gnemon,* while G. *ula* is a liane (woody climber).
5. In the climbing species the lower part of the stem is usually devoid of leaves.
6. The stem bears two types of branches, *namely* branches of *unlimited growth* and branches of *limited growth.*
7. The difference between the two, is not as pronounced as in *Pinus.*
8. The stem is differentiated into nodes and internodes. Internodes are present even on the branches of limited growth.
9. The stem in several species is provided with articulations. It is at these joints that the branches are shed
10. The leaves of *Gnetum* are exactly similar to those of dicotyledons.
11. In climbing species, the foliage leaves are borne only on the branches of limited growth.
12. At each nodal region two opposite leaves are present.
13. About 9-10 leaves are arranged decussately on each branch.
14. Leaves are petiolate and have a large, oval and entire lamina with reticulate venation.
15. The leaves are leathery in texture.
16. The stem is anchored to the soil with the help of the tap root system.

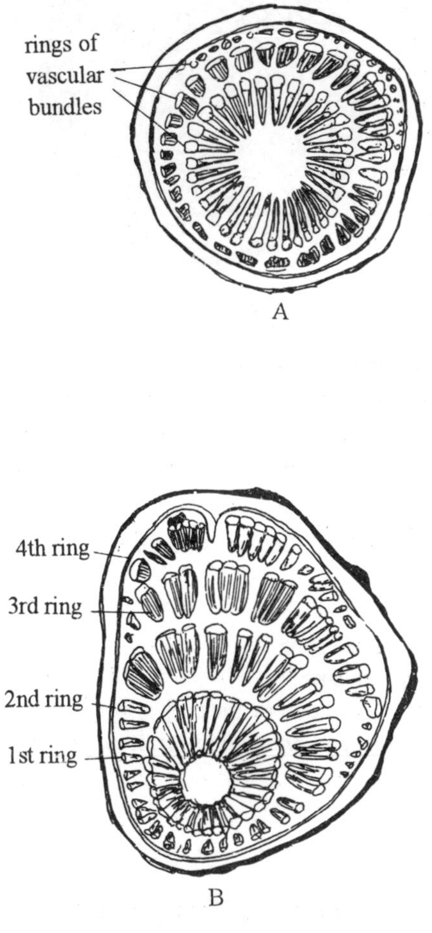

Fig. 19.3. Gnetum
Secondary growth. A. T.S. of stem showing concentric rings of vasculature. B. T.S. of stem showing eccentric rings of vasculature.

INTERNAL STRUCTURE

Stem

1. Transverse section of a young stem in its primary structure reveals the following features.
2. The outermost layer is the epidermis.
3. It is made up of rectangular cells which bear papillate outgrowths.
4. There is a deposition of the cuticle above the epidermis.
5. The continuity of the epidermis is interrupted by the presence of stomata.
6. Internal to the epidennis is the cortex, consisting of three distinct regions.
7. The outer cortex is chlorenchymatous and made up of 4-6 layers of oval chlorenchyma cells.
8. The middle cortex is parenchymatous, having compactly arranged, thin walled, polygonal cells.
9. The innermost cortex is made up of two to four layers of sclerenchymatous cells.
10. Next to the cortex, internally are present inconspicuous endodermis and pericycle.
11. The stele consists of a number of open, collateral vascular bundles arranged in the form of a ring surrounding a central pith. They xylem is endarch and consists of mostly tracheids and a few vessels.

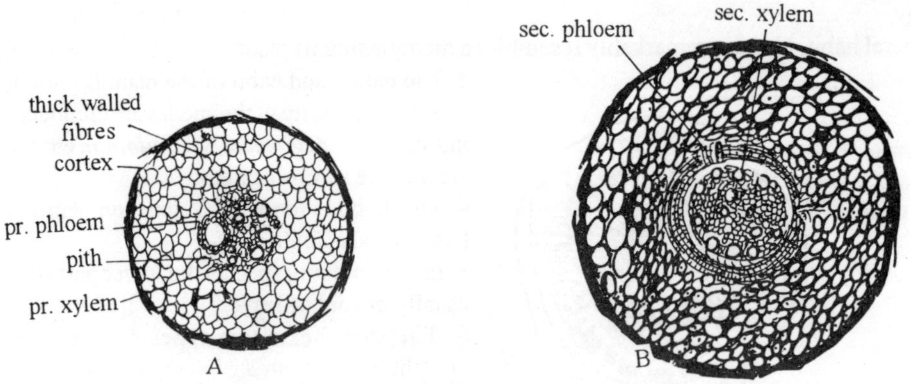

Fig. 19.4. **Gnetum**
Anatomy of root. A. T.S. of young root B. T.S. of old root.

12. Xylem fibres are absent and xylem parenchyma is scanty.
13. The protoxylem vessels have circular bordered pits. Metaxylem vessels have very few pores.
14. Primary phloem consists of sieve cells and phloem parenchyma.
15. Between xylem and phloem is present a strip of cambium.
16. The vascular bundles are separated from one another by broad medullary rays which are of a considerable height.

Secondary growth

1. Secondary growth takes place both in the cortex as well as in the stele.
2. In the cortex periderm formation occurs by the activity of a cork ambium.
3. Since the activity of the cork combium is not uniform all around the stem, the cork is not uniformly thick.
4. Secondary growth in the stele is both *normal* (trees or shrubs) and *anomalous* (climbing species).

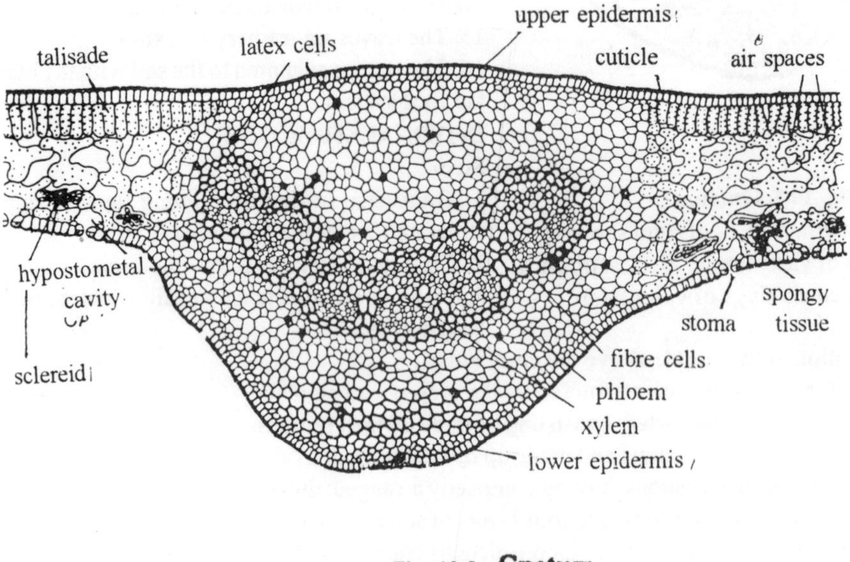

Fig. 19.5. **Gnetum**
T.S. of leaf

5. Normal secondary growth is brought about by the activity of a fascicular cambium.
6. Secondary medullary rays are present.
7. Secondary xylem consists of three types of elements *viz.*, *tracheids, xylem parenchyma,* and *vessels.*
8. Tracheids have circular bordered pits which are uniseriate in arrangement.
9. Xylem parenchyma cells are living thin walled cells.
10. Vessels are elongated cells with tapering ends and possess bordered pits.
11. The end walls are perforated.
12. The number of perforations varies with species.
13. In *G. ula* there is a single large perforation at the end walls.
14. The vessles develop from tracheids whose end walls have a number of bordered pits.
15. By the elimination of the end walls a vessel with a single large perforation is produced.
16. Secondary phloem also consists of three types of elements *viz.*, sieve cells, phloem parenchyma and albuminous cells.
17. Anomalous secondary growth is seen in climbing species like *G. ula.* Here the fascicular cambium ceases to function soon after the first vascular ring is formed.
18. Successive cambial rings arise at different positions in the cortex and produce the vascular rings.
19. Sometimes the cambia do not form a closed ring but remain as patches, due to the presence of secondary medullary rays with the result long vascular bundles with secondary xylem and phloem are formed.
20. These bundles in a transverse section resemble the canine teeth.
21. In some cases the cambium may form a closed ring but its activity is not uniform all round.
22. As a result, *eccentric* rings of vasculature is formed.

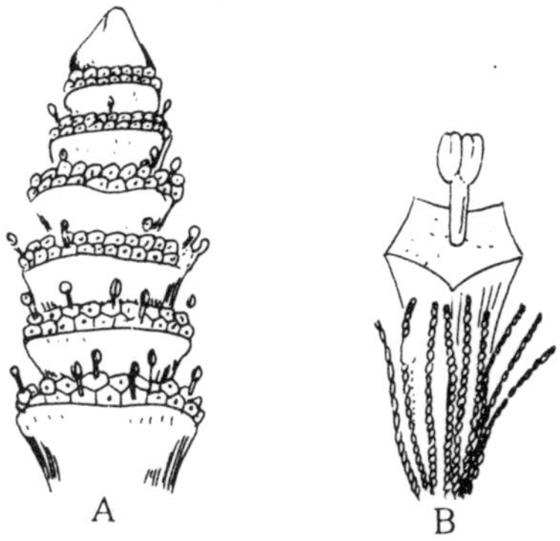

Fig. 19.6. **Gnetum**
Male reproductive structures, A. A portion of male cone, B. A single male flower.

Root

1. A transection of a young root shows a single layered epidermis surrounding a many layered parenchymatous cortex.
2. The cells are filled with starch.
3. The stele is surrounded by a pericycle and endodermis.
4. Vascular bundles are radial, two in number with exarch and diarch xylem.
5. Primary xylem consists of both tracheids and vessels.
6. There is a parenchymatous pith n the centre.
7. Secondary growth takes place by the activity of a fascicular cambium.
8. It is normal and forms a completely closed ring of vasculature.
9. The secondary xylem is similar to that of stem in its essentials.
10. Secondary medullary rays are multiseriate.

Leaf

1. A transverse section of the leaf shows a bifacial (dorsiventral) arrangement.
2. The upper epidermis is single layered and has cells with undulating walls; the cells of the lower epidermis are similar in structure.
3. There is a deposition of cuticle above the epidermal layers but it is comparatively thin towards the lower epidermis.
4. Leaves are hypostomatic.
5. The type of stomata may be *syndetocheilic* (a single mother cell produces both the guard cells and the two

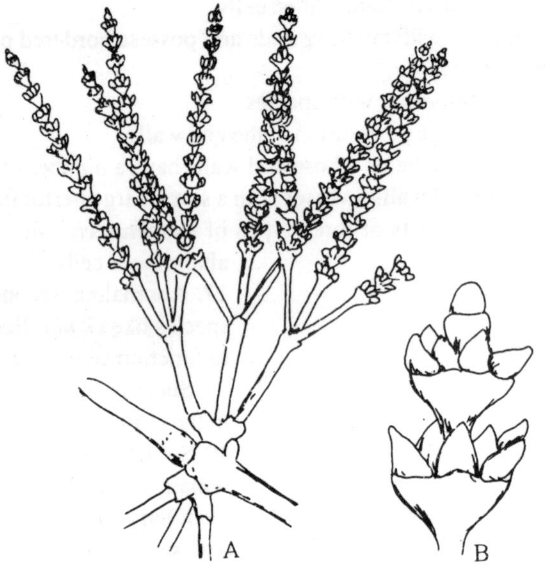

Fig. 19.7. Gnetum
Female reproductive structures of G. latifolium, A. Female cones.
B. A part ofthe female cone showing ovules.

subsidiary cells) as in *G.gnemon* or *haplocheilic* (stomatal initial produces only guard cells) as in *G. ula.*
6. The mesophyll consists of distinct palisade and spongy parenchyma.
7. However surrounding the midrib the cells are compact and the mesophyll is undifferentiated.
8. The vascular region in the midrib consists of an arc of vascular bundles.
9. In young leaves vascular bundles are not distinct.
10. The vascular tissues show x/p arrangement.
11. The bundles are conjoint, collateral and endarch.
12. In *G. ula,* each vascular bundle has a patch of stone cells below the phloem.
13. Around the midrib a few crystal cells occur.

REPRODUCTION

1. All the species of *Gnetum* are dioecious.
2. The reproductive organs are grouped into strobili or inflorescences.
3. The strobilus (cone) consists of a stout axis which bears at its base two opposite, connate bracts, in the axils of these bracts arise other inflorescences.

Male cone

1. The cone consists of an elongated axis and is distinguishable into nodes internodes.
2. At the nodal region are borne connate bracts forming a cup like structure called the *cupule* or the *collar.*
3. Above each collar, are found the male flowers arranged in three to six rings.
4. In each ring there will be a number of male flowers.
5. In some cases just above the male flowers there is a ring of imperfect female flowers.

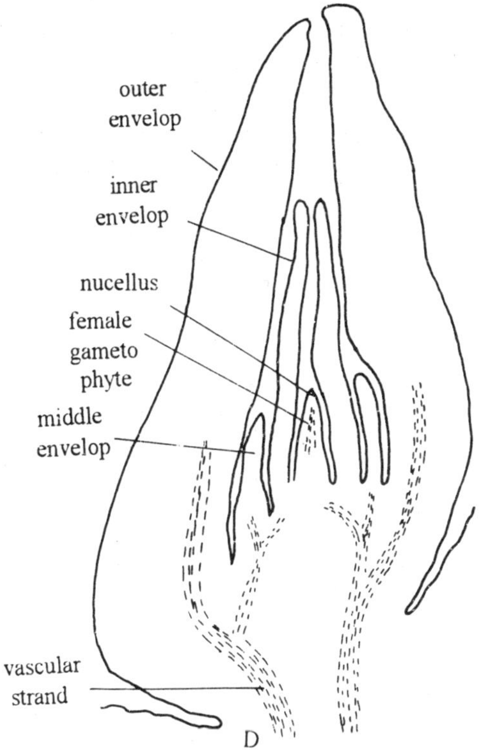

Fig. 19.8. **Gnetum**
Stages in the development of ovule in G. ula.

6. The male flower which is enclosed in a sheath like perianth consists of a stalk bearing two unilocular anthers or microsporangia.
7. Surrounding the perianth are found a number of uniseriate, multicellular hairs.
8. Perianth consists of sclerids and laticiferous tubes.
9. A young male cone is completely enclosed within the two basal bracts.
10. Later, the cone axis emerges piercing through the bracts.

Female cone

1. The female cone is also called the ovulate cone.
2. It is constructed on the same general plan as the male cone.
3. It consists of an axis in which are arranged a series of cupules formed by the connate bracts.
4. In the young condition it is difficult to distinguish between the male and female cones.

Within the cupule are found 4-10 female flowers (ovule) arranged in the form of a ring. Each ovule consists of a central mass of nucellus being surrounded by three envelops.

Structure of the ovule at the time of fertilisation

1. The ovule is surrounded by three envelopes. The morphological nature of these envelopes is uncertain.
2. Of the three envelopes, the innermost elongates to form a micropylar tube or style.
3. There is no poollen chamber or archegonial chamber.
4. The envelope encloses the nucellus embedded in which is found the embryo sac.
5. Two or three eggs are found near the micropylar region.
6. The lower portion is free nuclear. But in some cases the chalazal portion may become partly cellular.
7. Below the embryo sac, the nucellus forms a fan shaped tissue called the *pavement tissue*.

TEXTBOOK OF
PRACTICAL BOTANY

TEXTBOOK
OF
PRACTICAL BOTANY

Volume - III

(Taxonomy, Economic Botany)

By

Dr. S. Sundara Rajan

Principal

Reva Institute for Science and Technology Studies

R.T. Nagar

Bangalore – 560 024

ANMOL PUBLICATIONS PVT. LTD.

NEW DELHI - 110 002 (INDIA)

ANMOL PUBLICATIONS PVT. LTD.
4374/4B, Ansari Road, Daryaganj
New Delhi - 110 002
Ph.: 3261597, 3278000
Visit us: www.anmolbooks.com

Textbook of Practical Botany

First Edition, 2003

ISBN 81-261-1359-6 (Set)

PRINTED IN INDIA

Published by J.L. Kumar for Anmol Publications Pvt. Ltd., New Delhi and Printed at Mehra Offset Press, Delhi.

CONTENTS

PART ONE
TAXONOMY

srgemonemexicana (96-97); **Capparidaceae** - *Capparis decidua* (98-99); **Capparidaceae** - *Capparis spinosa* (100-101); **Capparidaceae** *Cleome viscosa* (102-103); **Capparidaceae** - *Clemoe gynandra* (104-105); **Cruciferae** - *Coronopus didymus* (106-107); **Cruciferae** - *Iberis amara* (108-109); **Cruciferae** - *Brassica campestris* (110-111); **Cruciferae** - *Brassica nigra* (112-113); **Cruciferae** - *Raphanus sativus* (114-115); **Caryophyllaceae** -*Dianthus caryophyllus* (116-117); **Caryophyllaceae** - *Stellaria media* (118-119) **Caryophyllaceae** - *Spergula arvensis* (120-121); **Portulacaceae** - *Portulaca oleracea* (122-123); **Malvaceae** - *Hibiscus rosa sinensis* (124-125); **Malvaceae** - *Althae rosea* (126-127); **Malvaceae** - *Malva sylestris* (128-129); **Malvaceae** - *Sida cordifolia* (130-131); **Malvaceae** - *Gossypium* (132-133); **Malvaceae** - *Abutilon indicum* (134-135); **Tiliaceae** - *Corchorus acutangula* (136-137); **Tiliaceae** - *Triumfetta rhomboidea* (138-139); **Sterculiaceae** - *Dombeya spectabilis* (140-141) **Sterculiaceae** - *Melochia corchorifolia* (142-143); **Sterculiaceae** - *Gauzuma tomentosa* (144 -145); **Sterculiaceae** - *Waltheria indica* (146 - 147); **Rutaceae** - *Ruta graveolens* (148-149); **Rutaceae** - *Murraya paniculata* (150-151); **Rutaceae** - *Citrus aurantium* (152-153); **Rutaceae** - *Toddalia asiatica* (154-155); **Meliaceae** - *Melia azadirachta* (156-157); **Anacardiaceae** - *Mangifera indica* (158 - 159); **Rosaceae** - *Rosa indica* (160-161); **Rosaceae** - *Pyrus malus* (162-163); **Rosaceae** - *Pyrus communis* (164 - 165); **Rosaceae** - *Prunus persica* (166-167); **Rosaceae** - *Potentilla supina* (168 - 169); **Rosaceae** - *Eriobotrya japonica* (170 - 171); **Papilionaceae** - *Trigonella foenumgraceum* (172-173); **Papilionaceae** - *Crotalaria juncea* (174-175) **Papilionaceae** - *Pisum sativum* (176-177); **Papilionaceae** - *Lathyrus odoratus* (178 - 179); **Papilionaceae** - *Dalbergia sissoo* (180-181); **Papilionaceae** - *Arachis hypogea* (182 - 183); **Papilionaceae** - *Melilotus indica* (184 - 185); **Papilionaceae** - *Indigofera linnaeis* (186-187); **Caesalpinaceae** - *Classia occipentalis* (188 - 189); **Caesalpinaceae** - *Cassia tora* (190 - 191); **Caesalpinaceae** - *Cassia fistula* (192 - 193); **Caesalpinaceae** - *Tramarindus indica* (194 - 195);**Caesalpinaceae** - *Caesalpinia pulcherrima* (196 - 197); **Caesalpinaceae** - *Bauhinia variegata* (198 - 199); **Mimoceae** - *Mimosa pudica* (200 - 201); **Mimoceae** - *Acacia nilotica* (202 - 203); **Mimoceae** - *Albizzia lebbek* (204 - 205); **Mimoceae** - *Prosopis juliflora* (206 - 207); **Myrtaceae** - *Eucalyptus lanceolatus* (208 - 209); **Myrtaceae** - *Callistemon lanceolatus* (210 - 211); **Myrtaceae** - *Psidium guajava* (212 - 213); **Myrtaceae** - *Eugenia jambolana* (214 - 215); **Combretaceae** - *Terminalia bellerica* (216 - 217); **Combretaceae** - *Quisqualis indica* (218 - 219); **Cucurbitaceae** - *Cucurbita maxima* (220 - 221); **Cucurbitaceae** - *Luffa cylindrica* (222 - 223); **Cucurbitaceae** - *Coccinea indica* (224 - 225); **Cucurbitaceae** - *Momordica charantia* (226 - 227); **Cucurbitaceae** - *Cucumis sativus* (228 - 229); **Umbelliferae** - *Coriandrum sativum* (230 -221); **Umbelliferae** - *Centella asiatica* (232 - 233); **Umbelliferae** - *Foeniculum vulgare* (234 - 235); **Umbelliferae** - *Carum copticum* (236 - 237); **Rubiaceae** - *Hamelia patens* (238 - 239); *Ixora parviflora* (240 - 241); **Rubiaceae** - *Morinda angustifolia* (242 - 243); **Rubiaceae** - *Musseanda frondosa* (244 - 245); **Rubiaceae** - *Oldenlandia corymbose* (246 - 247); **Rubiaceae** - *Richardia scabra* (248 - 249); **Compositae** - *Ageratum conyzoides* (250 251); **Compositae** - *Eclipta alba* (252 - 253); **Compositae** - *Helianthus annua* (254 - 255); **Compositae** - *Launea asplenifolia* (256 - 257); **Compositae** - *Lagascea mollis* (258 - 259); **Compositae** - *Sonchus oleraceous* (260 -261); **Compositae** - *Tagetes patula* (262 - 263); **Compositae** - *Tridax procumbens* (264 - 265); **Compositae** - *Xanthium xtrumarium* (266 - 267); **Sapotaceae** - *Achiras sapota* (268 -269); **Sapotaceae** - *Madhuca indica* (270 - 271); **Apocynaceae** - *Thevetia peruviana* (272 - 273); **Apocynaceae** - *Vinca rosea* (274 - 275); **Apocynaceae** - *Tabernemontana divaricata* (276 - 277); **Apocynaceae** - *Nerium odorum* (278 - 279); **Apocynaceae** - *Vallaris solanacea* (280 - 281); **Asclepiadaceae** - *Asclepias curassavica* (282 - 283); *Calotropis gigantea* (284 - 285); **Asclepiadaceae** - *Cryptostegia grandiflora* (286 - 287); **Convolvulaceae** - *Ipomoea palmata* (288 - 289); **Convolvulaceae** - *Convolvulus arvensis* (290 - 291); **Convolvulaceae** - *Cuscuta reflexa* (292 - 293); **Convolvulaceae** - *Quamoclit pinnata* (294 - 295); **Solanaceae** - *Solanum nigrum* (296 - 297); **Solanaceae** - *Solanum xanthocarpum* (298 - 299); **Solanaceae** - *Cestrum nocturnum* (300 - 301); **Solanaceae** - *Petunia hybrida* (302 - 303); **Solanaceae** - *Withania somnifera* (304 - 305); **Solanaceae** - *Datura stramonium* (306 - 307); **Scrophulariaceae** - *Lindenbergia indica* (308 - 309); **Scrophulariaceae**

PART IV: ECONOMIC BOTANY

Preface

Laboratory study is an integral part of the knowledge that a student acquires in the college. This is true not only for botany but for all science subjects. Theoretical study of a subject has to be supplemented with laboratory experiments to provide a comprehensive understanding.

One of the main advantages of laboratory or practical study is the first hand knowledge that a student acquires about the material of study.

A systematic, structured and disciplined method of work is necessary in the laboratory to obtain maximum benefit. In order to help the students realize their goal practical manuals indeed play an important role.

Textbook of Practical Botany aims to provide the students, a useful hand book to guide them in the proper study of the subject material. Theoretical details are kept at the minimum and more emphasis is laid on understanding the structural details to enable the student a proper identification. A large number of properly labelled original diagrams are incorporated in the book as they are integral to the understanding of the subject.

This work is organised in the following four volumes:

Volume One deals with Algae, Fungi, Bacteria, Lichens, Viruses and Microbiology.

Volume Two deals with Bryophytes, Pteridophytes and Gymnosperms.

Volume Three deals with Taxonomy and Economic Botany.

Volume Four deals with Anatomy and Embryology of Angiosperms, Ecology, Plant Physiology and Cytology.

I have taken utmost care in the presentation of matter, style of writing and choice of illustrations so that the volumes are user friendly. The examples chosen for study are such that the books should cater to the needs of all Indian Universities.

Constructive criticism and suggestions are most welcome.

I am thankful to Mr. J.L. Kumar, Managing Director of Anmol Publications Pvt. Ltd., New Delhi, for undertaking the publication of the book.

Dr. S. Sundara Rajan
Principal

Reva Institute for Science and Technology Studies
R.T. Nagar
Bangalore - 560 024

PART ONE
TAXONOMY

CHAPTER 1

INTRODUCTION

The study of science subjects is carried out at two levels,viz., *Theoretical* and *Practical*. Theoretical study is carried out in lecture classes when the students are imparted all the details regarding a particular topic. This has to be supplemented by practical or Laboratory studies, when the students come into active contact with the objects about which they would have heard in lecture classes. Therefore it may be safely said that laboratory studies form an integral part of a comprehensive study of any topic in science subjects.

In the study of plants, laboratory assumes a special significance as the students acquire a meaningful and firsthand knowledge about the plants which they study in lecture classes. In fact as far as botany is considered any study of a plant without being supplemented by laboratory observation is futile. It is only in the laboratory study that the student can dissect, observe and analyze the various features of the plant and it is only a study of this type that makes the student acquire a meaningful knowledge about the pant. Besides observation and study in the laboratory plants are also to be studied in field to obtain an account of their habitat, habit, association and such other details.

In this chapter an attempt is made to explain some of the essentials that a student should understand before he begins a practical work.

Cleanliness: This is the most important point that a student should keep in mind while in the laboratory. The working table and the surroundings should be kept dry and spotlessly clean.. All the requirements needed by the students such as petri dishes, slides, reagent bottles, stains etc. must be properly labeled and arranged neatly. The microscope and particularly the optical parts must be cleaned with a white lint free cloth. After the completion of the assigned work, the student should take care to clean the place, remove any debris of the plant material and put it into the dust bin.

The students should work in silence and independently unless the assigned experiment is to be carried out by a group. When a particular work is assigned to a student, he must carefully plan and divide the available time for experimentation, observation and recording of the results. Whenever a doubt arises, it is always better to consult the teacher in charge, than a classmate. It is only by a proper and systematic work, that a student can derive maximum benefit from the practical classes.

Laboratory requirements

The department provides certain amenities, instruments etc to each student for the purpose of practical work. But a student also should carry some of the requirements in order to be effective in the practical class.

One cardinal principle that every student should remember in a practical class is to have one's own requirements and not to depend on anyone. Never lend or borrow any requirements. Borrowing or lending a pencil, needle or blade will disturb the work of both.

In general every student should individually possess the following requirements

1. *Two sharp pointed needles* mounted on stainless steel or plastic handle. This is essential for the purpose of separating and teasing the microscopic material.

2. *A pair of forceps:* This is essential to remove the material either from the bottle or from petri dish.

3. A sharp properly honed razor to cut sections of the plant material.

4. A new safety razor blade. This will help either to cut section or to trim the material.

5. Two fine hair paint brushes. This will be of use in lifting the sections of plant (with out damage) from the petri dish on to the slide.

6. Two sharpened pencils (H, Hb or 2b).

7. A pencil eraser.

8. A pencil sharpener.

9. Glass droppers.

10. A piece of thin, white, lint free cloth to clean the optical parts of the microscope.

11. An observation book to note down immediately the points of interest.

12. A practical record.

13. Glass slides and cover slips(if they are not provided by the department).

Scheme of Practical Work

Practical work in the laboratory will be most useful and effective if the student keeps in mind the following:

1. Maintain silence during working. All the requirements for the practical work must be absolutely clean.

2. Every student should work independently, unless the experiment involves the participation of others.

3. Before beginning the practical work, listen to the instructions of the teacher attentively and work accordingly.

4. Follow a systematic procedure in a study.

5. If the work involved is the morphological study of an organism, follow the procedure given below:

(a) Study of external characters - for example whether a plant is microscopic or macroscopic, unicellular or multicellular, filamentous or colonial etc.

(b) Study of internal structure like details of cell, tissues etc. If the material requires sectioning, sections are to be cut and internal details should be observed.

(c) Study the reproductive details like asexual and sexual reproductive organs.

6. Mount the material after properly teasing and separating the filaments, mycelia etc. Stain the material wherever necessary and cover the preparation with a cover glass.

7. Observe the slide carefully under the microscope, first under low power and then under high power.

8. Make sure that you have observed all the details as instructed by the teacher.

9. After all the observations are over, record your findings, structural details etc. in the practical records in the form of diagrams. Draw proportionate diagrams in the proper place on the drawing sheet. Diagrams should never be drawn to the very edge of the drawing sheet. Write suitable explanatory notes pertaining to the diagrams on the interleaf.

Laboratory Equipments

In addition to the individual requirements carried by the student, the department generally provides the

following equipments/instruments etc.to the students. Sometimes these equipments may be given to each student .

These are :

(1) Students compound microscope .
(2) Dissection microscope.
(3) Binocular Research microscope.
(4) Watch glasses.
(5) Cavity blocks.
(6) Slides and cover glasses.
(7) Petri dishes.
(8) Enamel tray.
(9) Polythene/glass wash bottle.
(10) Spirit lamps.
(11) Glass marking pencil.
(12) Microtome.
(13) Staining /bottles.
(14) Stone/ for honing the razor.
(15) Chemicals and stains.
(16) Culture media(to be prepared by students).

In the list of requirements mentioned above, we will discuss some in detail.

Microscope

Of all the instruments used by students in the botany laboratory, the one that is most indispensable to the students of biology is the 'Microscope'. It is not an exaggeration to say that no biological study in the laboratory is possible without the microscope.

The function of the microscope is to enlarge the small objects, so that they become visible to the human eye. In reality the magnification power of the microscope lies in its resolving capacity between one particle and another. A microscope distinguishes one particle from another which are situated very closely. A human eye cannot resolve particles that are lying closer than 0.025 mm. In its function a microscope can be compared to a human eye, but much more effective than that.

There are different types of microscopes, but all of them are based on the same principle with variations in their construction. Some of the types of microscopes used in the laboratory are:

(1) Compound microscope.
(2) Binocular Research microscope.
(3) Dissection microscope.
(4) Stereoscopic dissection microscope.
(5) Phase contrast microscope.
(6) Electron microscope etc.

Of the six types of microscopes mentioned above the compound microscope and the dissection microscope are the ones that are most frequently used by students. The Electron microscope is very costly and a very highly sophisticated instrument.

In this book the compound microscope and the dissection microscope are explained in detail as the students should be familiar with them.

Compound Microscope

The microscope has a strong 'U' shaped or 'V' shaped base. It is also called the horse shoe shaped base. Attached to the base is a bent limb. This limb can be tilted in any direction as it is attached to the base with an inclination joint.

Attached to the limb is a stage at its lower end and a body tube at its upper end. The body tube is attached to the limb with the help of a rack and pinion so that it can move up and down. To help in the movement of the tube during focusing, there are 2 knobs viz.,*Coarse adjustment* knob and *Fine adjustment knob*.

The body tube at its top end consists of the ocular or eye piece lens. The eye pieces usually have three magnifications 5x,10x,and 15x, and can be used to get a low or high magnification. Attached to the lower end of the tube is a revolving nose piece with three objective lenses low (10x) medium (40x) and high (60x). Some times there will be a fourth objective of the magnification of 100x. This is also called the oil immersion lens (while viewing through an oil immersion lens a drop of oil-liquid paraffin has to be put on the viewing slide).

Below the objectives and attached to the limb is the *stage* of the microscope where the slides are kept for viewing. The stage may be circular, square or rectangular in shape. In order to keep the slides properly placed on the stage, there are two stage clips. In some of the costly microscopes, the stage has movable sliding system operated by knobs to move the slides. This is called the mechanical stage. The mechanical stage permits the movement of the slide backwards or forwards, left or right with the turning of the knobs provided for the purpose. A hole is provided in the stage to receive the light from below and illuminate the object. Attached to the limb, below the stage is a condenser (fixed or movable up and down) lens. Below the condenser lens is an Iris diaphragm to regulate the light passing through. Iris diaphragm is a metallic disc with an aperture. The aperture can be narrowed or widened by turning the lever. The condenser which is a system of lenses condenses the light rays passing through and converges them on the object.

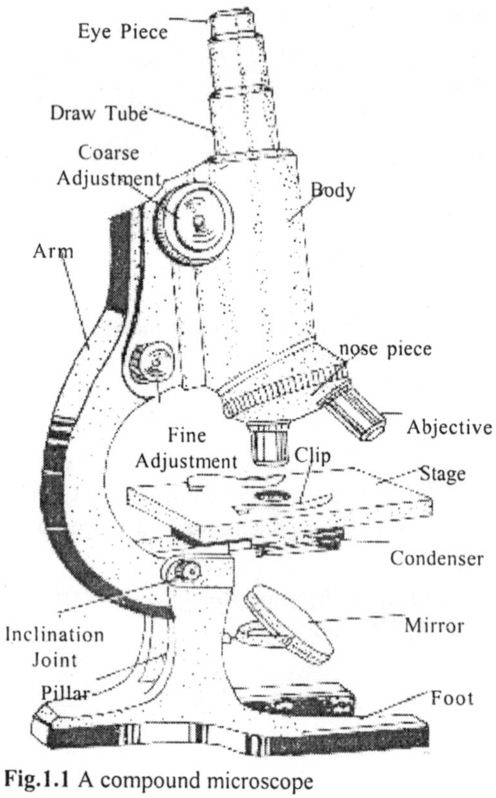

Fig.1.1 A compound microscope

The source of illumination in a student's microscope is a mirror which reflects the natural light or electric light. The mirror has two faces - plain and concave and is attached to the limb below the condenser. The mirror is adjustable and can be turned to the direction of light. When day light is the source of illumination, the concave side of the mirror is usually used to converge the rays of light. In some of the costly microscopes, instead of the mirror there is a built in illumination ,This is a tube with a lamp (9v or 12v) attached to a transformer, which can be plugged to any AC source.

Operation of the Microscope

Microscope should be placed in a well illuminated area but avoiding the direct light as it is harmful to the eyes. Generally northern light is preferable. When ' built in illumination' is employed, it is always better to use a blue or green filter to subdue the light.

The light has to be first focused on the condenser by adjusting the mirror towards the source of light. The amount of light entering the condenser may be regulated by closing or opening the iris diaphragm. The slide to be viewed is kept on the stage and fixed securely with the help of the stage clips. The object is first observed with a low power objective and then turned on to medium or high power, the specimen or section has to be covered with a cover slip. The slide should be observed with both the eyes open even while observing with monocular microscope.

Care of the Microscope

1. Clean all the parts of the microscope before and after use. The lenses and the mirror are to be cleaned with a lint free cloth or tissue paper.

2. Do not pull the microscope to the very edge of the table.

3. While lifting the microscope, use both hands. With one hand hold the base and with the other hold the arm.

4. The slide to be observed must be clean and dry (specially the under surface of the slide).

5. Do not remove the eyes piece or objectives from the microscope.

6. The microscope should be covered when not in use.

Binocular Research Microscope

This is like the ordinary compound microscope except that, it has binocular head for providing a natural stereoscopic view.

Dissection Microscope

This is a very simple microscope with only one set of lenses. Microscopes of this type will not give a very high magnification, but are very useful in dissecting flowers, plant parts etc.

The microscope has a 'V' shaped base and a limb. The stage is made up of a transparent glass plate. Below the stage is a mirror to focus the light. Attached to the limb is a mechanical arm which can be moved up and down with the help of a rack and pinion movement. The arm can also be moved horizontally. Attached to the free end of the arm is a holder to accommodate the lens (10x,15x etc.).

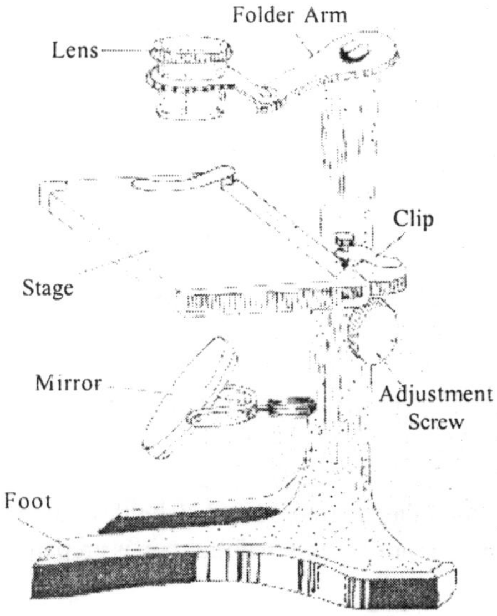

Fig.1.2 A dissecting microscope

Operation: The specimen to be observed / dissected is placed on the stage and located with the help of lens. Floral parts or others may be then dissected with the help of two needles, while still observing the specimen.

Stereoscopic Binocular Dissection Microscope

This is a combination of compound microscope and dissection microscope. While it has the higher magnification power of a compound microscope with two sets of lenses, it has retained the versatility and the manoeuvourability of a dissection microscope.

Phase contrast and Electron microscopes are not normally used by students.

Other Laboratory Equipments

Stains: These are colouring fluids. The specimens to be observed (algal or fungal material or sections of plant parts, would be generally decolourised due to preservation. These are to be coloured in order to observe them clearly. The following are the stains that are generally used in the laboratory.

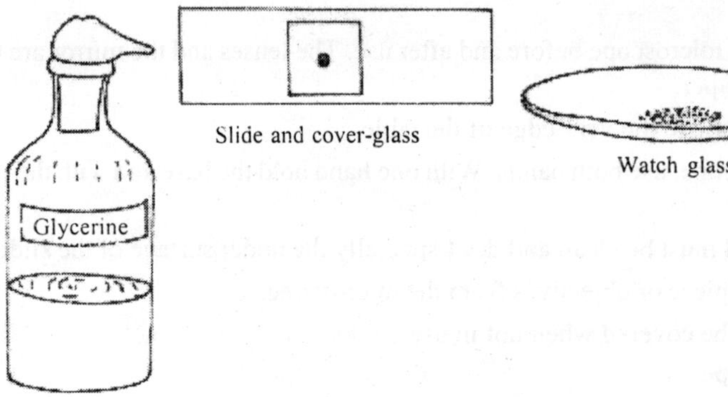

Fig.1.3 Dropping bottle

(a) Saffranin: This is the most commonly used stain. (1) It stains nucleus and lignified cell walls.

Preparation —

Saffranin —1gm.

95% (1) Ethyl alcohol –50ml.

Distilled water – 50ml.

An aqueous stain may also be prepared.

Saffranin – 18ml

Distilled water – 100ml.

(b) Fast Green

Fast green – 1gm.

90% Ethyl Alcohol – 100ml.

This stain is used to colour parenchyma cells and cytoplasm. Fast green may also be prepared in clove oil (75ml clove oil+25ml 100% alcohol).

(c) Cotton blue.

Cotton blue –1gm.

Lactophenol – 100ml.

(d) Haematoxylin

The dye is obtained from the heartwood of the tree *Hametoxylon campechianum,* There are two methods of preparation of the stain. These are:

1. *Heidenhain's Hametoxylin*

Dissolve 0.5 or 1.0 gm of the powder in 100 ml of distilled water. Filter and store in a dark bottle away from light. The stain is ready for use after 4-5 days of ripening. Ripening is not necessary if the stain is prepared in 70% ethyl alcohol.

2. *Delafelds Haematoxylin*

Sol.1 Aqueous solution of ferric ammonium sulphate 100ml).

Sol.2 One gm dye dissolved in 6ml of absolute alcohol.

mix solutions 1 and 2. To this mixture add 25 ml of glycerine and 25 ml absolute alcohol. The solution should be allowed to stand until it becomes dark red.

(e) Anilin blue

Also called cotton blue or China blue. It is prepared as follows:

Aniline blue –1gm.

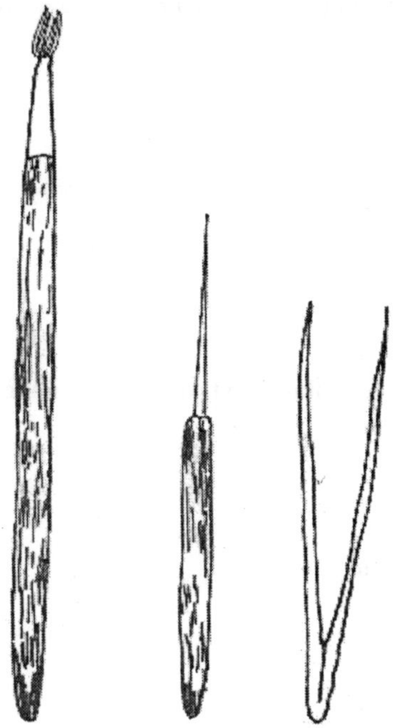

Fig.1.4 Brush Needle Forceps

Ethyl alcohol (90%)
or — 100ml.
Distilled water
Hydrochloric acid – 2drops.

(f) Gentian Violet (Crystal Violet)
Gentian Violet – 1gm.
90% Ethyl Alcohol 100ml.

(g) Erythrosine
Erythrosine –1gm.
90% Ethyl alcohol – 100ml.

Erythrosine may also be prepared in clove oil (Absolute alcohol 0.5 ml +95ml Clove oil).

(h) Light green
Light green –1gm.
Clove oil –95ml.
Absolute alcohol –5ml.

Light green may also be prepared without clove oil (in 100ml Absolute alcohol).

(i) Acetocarmine: The stain is prepared as follows:

Add 1 gm of stain to 100 ml of boiling acetic acid (45%)or propionic acid. Cool and decant, A few drops of saturated ferric acetate solution is added and the mixture is cooled over ice. The solution is allowed to stand for about 12 hours. Filter and store in a dark coloured bottle under refrigeration.

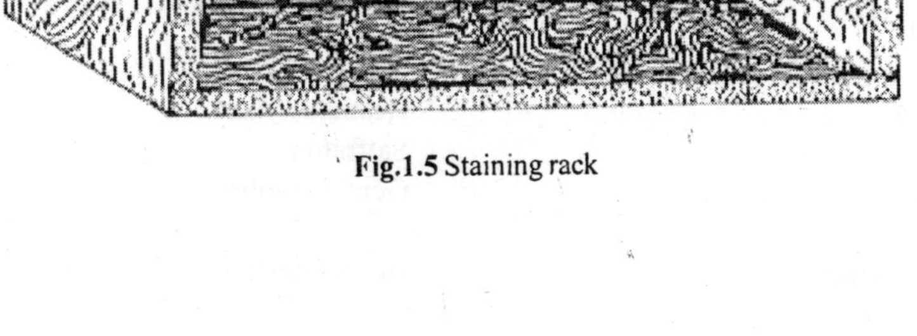

Fig.1.5 Staining rack

Specificity of Stains

Usually specific stains are employed in colouring different plant parts. The stains are used singly or in combination. The most popular stain combination are:

Saffranin - fast green.
Haematoxylin - Erythrosin.
Haematoxylin - light green.

While stain combinations are employed to study pteridophytes, gymnosperms and Angiosperms, single stains are used for Avascular cryptogams.

Green or Saffranin for Algae and Bryophytes.

Cotton blue for fungi.

The following table gives the specificity of stains to different plant parts or cell organelles.

Structure	Stain
Cellulose	Anilin blue Haematoxylin Fast green Light green
Cutinized cell wall	Gentian violet Erythrosine Saffranin
Lignified cell wall	Gentian violet Saffranin
Suberised cell wall	Saffranin
Cytopalsm	Fast green Light green Aniline blue
Chitin	Saffranin
Callose	Aniline blue
Achromatic figure	Aniline blue Erythrosine
Proteins	Saffranin
Nucleus	Heamatoxylin Saffranin Gentian violet
Chromonemata	Heamatoxylin

Chromosomes	Saffranin
Plastids	Gentian violet Haematoxylin
Mitochondria	Gentian violet
Flagella	Cotton blue

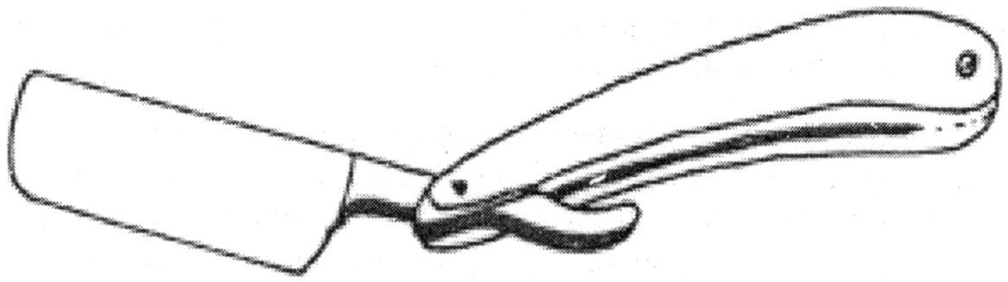

Fig.1.6 Razor

Fixing / Preserving Fluids

The plants or plant parts collected in nature are to be killed and preserved in the laboratory for the purpose of storage and study.

In killing the specimen, the various life processes are stopped at that instant, so that after preparation for microscopic study various stages may be observed.

When ever a material is collected, it is better to kill it in the field, so that there will be minimum or no distortion. Alternatively the material collected in the plastic bags or bottles should be brought to the laboratory and killed / preserved. In preservation, the material must be completely immersed in the fluid. The following are some of the killing/preservative fluids generally used:

1. *Formaline acetic alcohol (F.A.A)*

This is the most commonly used killing /preserving agent. It is popularly called standard preservative.

Ethyl alcohol (70%)	- 90ml.
Glacial acetic acid	- 5ml.
Formalin	- 5ml

2. *Formalin - Propiono - Alcohol*

Formalin	- 5ml
Propionic acid	- 5ml.
Ethyl alcohol 70%	- 90ml.

3. *Carnoy's Fluid - I*

Ethyl alcohol	- 30ml
Glacial acetic acid	- 10ml

4. *Carony's Fluid - II*

Ethyl alcohol	- 30ml.
Glacial acetic acid	- 30ml.
Chloroform	- 15ml.

Carony's fluid is widely used in killing root tips, flower buds etc. for the purpose of cytological study.

5. *Randolph's Modified Navashin Fluid*

This consists of two solutions A and B.

Solution A

Chromic acid	- 5.0 gms.
Glacial acetic acid	- 50ml
Distilled water	- 320ml.

Solution B

Commercial formalin	- 200ml.
Distilled water	- 175 ml.
Saponin	- 3gms.

Mix A and B in equal volumes before use.

Mounting media

These are chemicals used for the purpose of mounting (Temporary/Permanent) the specimens on the slide for microscopic observation. The following are some of the mounting media.

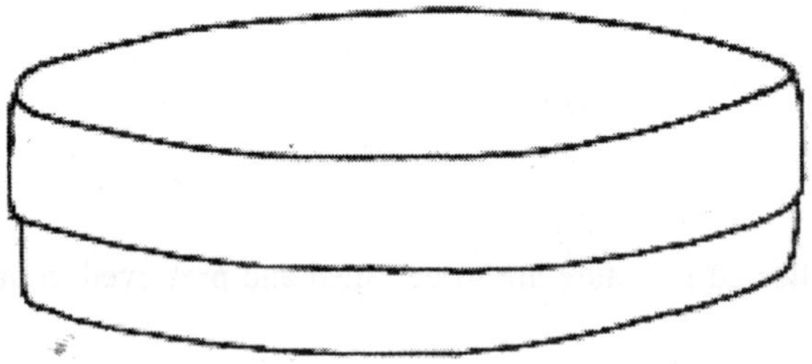

Fig.1.7 Petri dish

1. ***Acid water***

HCl – 10-15 drops

water – 100 ml.

2. ***Lactophenol***

Phenol	-100gms
Lactic acid	-100ml.
Glycerine	-100ml.
Distilled water	-100ml.

3. ***Glycerine***

Glycerine	- 15-20 drops
Formaline	- 1-2 drops
Distilled water	- 100ml.

4. ***Canada Balsam:*** This is a gum obtained from Gymnosperms. This fluid is readily available in the market.

5. ***Euparol:*** This is available in the ready to use form in the market.

Canada balsam and Euparol are used for permanent mounting, while the rest are used for temporary preparations.

Staining rack: This is a wooden rack used for holding the staining bottles.

Dropping bottles: This is a glass bottle having a narrow mouth fitted with a stopper. The stopper is grooved and has a beak. When the bottle is upturned the stain/fluid comes out in drops.

Slides: These are readily available in the market. The usual size 25mm x 75mm. The thickness is about 1mm. Special types of slides called cavity slides are available for specific study.

Cover slips (Cover glasses)

These are meant to cover the materials when the preparation is ready for observation. The coverslips may be round, square or rectangular.

Methods of Study of Plant Material

The following is the procedure for a scientific study in the laboratory.

Morphological study

The specimen given to study external detail should be observed either with or without a magnifying glass. Whatever details can be seen are to be recorded. If the specimen requires microscopic study, it should be suitably prepared for the purpose of microscopic observation. There are two types of microscopic preparation.

1. Whole Mount Technique

This is suitable for the purpose of external study of algal and fungal material. Take a little of the material, stain it with a suitable stain (staining procedure is explained later.) Wash the excess stain in water / alcohol. Mount in a suitable medium (temporary or permanent) and observe under the microscope.

2.Section cutting

For the purpose of anatomical study, sections are to be cut of plant parts. In algae, globule of *Chara*, stolons of *Caulerpa*., receptacles of *Sargassum* etc., can be cut. Similarly fruiting bodies of fungi, thalli of bryophytes also many be cut in a suitable plane. In higher plants cylindrical plant parts like root, stem, petiole etc. may be cut for the study of internal structure.

Planes of Cutting: In order to study the internal structure from various angles, sections are cut in various planes.

Transverse :The section is cut at right angles to the longitudinal axis.

Longitudinal : The section is cut parallel to the long axis (right angles to the transverse axis. The longitudinal section may be cut along a radius (Radial longitudinal section-RLS) or along a tangent (Tangential longitudinal section-TLS).

Cylindrical organs like roots and stems are cut transversely or longitudinally, while dorsiventral organs like thalli of bryophytes, leaves etc. are usually cut in vertical transverse plane.

Types of Section cutting

There are usually 2 types of section cutting methods employed for plant parts. These are (a)Free hand sectioning and (b) Microtome sectioning .

(a) Free hand Sectioning

For the purpose of laboratory study where slides are not made permanent, usually free hand sectioning method is employed. Free hand sections are cut mostly with the help of a good sharp razor (sometimes safety razor blades also may be used).

Honing the razor : The razor used for sectioning must have a sharp, even edge. There should be no nicks. Before cutting sections, the razor edge has to be honed to maximum sharpness either on a stone or on a leather strap. Oblique strokes, with the edge forward are to be given on the sone or strap. Usually oil drops are used on the honing medium to give maximum sharpness.

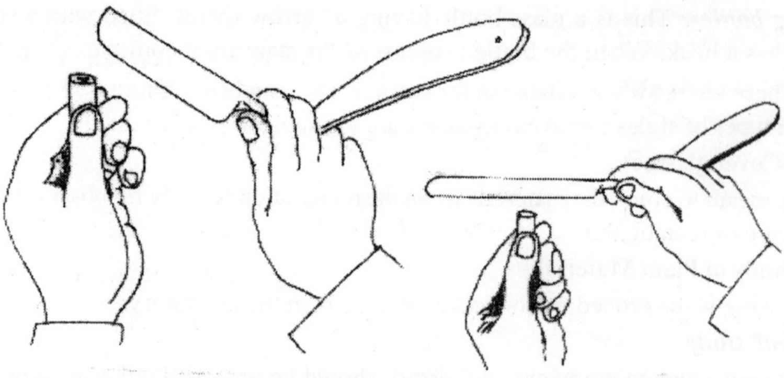

Fig.1.8 Method of holding pith | Methoding of holding razor | Method of section cutting

Method of cutting : Open out the razor and bend the handle backwards. Hold the razor between the index finger and thumb. The handle should be free. The remaining fingers of the hand will be touching the back edge of the razor blade.

In order to cut a transverse section of a cylindrical part like stem or root, follow the procedure given below.

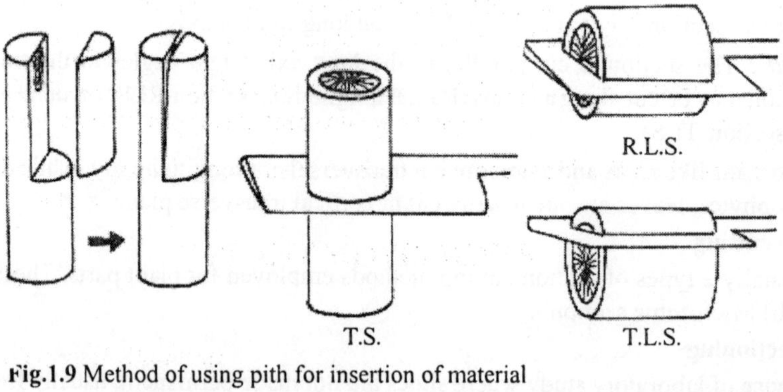

Fig.1.9 Method of using pith for insertion of material

The stem or root should be cut into a 2-3" long piece. It should be held in the hand in an erect position between the thumb and index finger. The material held in the hand should be exactly at right angles to the razor's edge. The thumb holding the specimen should be slightly at a lower level. Flood the edge of the razor with water. Now using the index finger as the platform move the razor quickly several strokes across the material. During this process sections are cut and they will be collected on the blade of the razor. The sections now may be carefully transferred with the help of a fine brush in to a petridish containing water.

For cutting sections of delicate and fragile materials like thalli or leaves, a suitable supporting medium has to be used. For this purpose a pith, carrot or potato may be used. These are to be cut into rectangular cubes and then split at one end. The material for sectioning should be inserted in the slit and sections are cut

After section cutting the razor should be wiped dry. It should be greased and encased.

Microtome Sectioning: This is generally employed very frequently in the study of bryophytes, pteridophytes and other higher plants. The microtome sectioning is a long and complicated process. It takes anywhere between 15-20 days to process a material and to obtain the sections.

Microtome sectioning involves the following steps:

1. Killing and preserving the material.
2. Dehydration and infilteration.
3. Paraffin embedding .
4. Section Cutting.

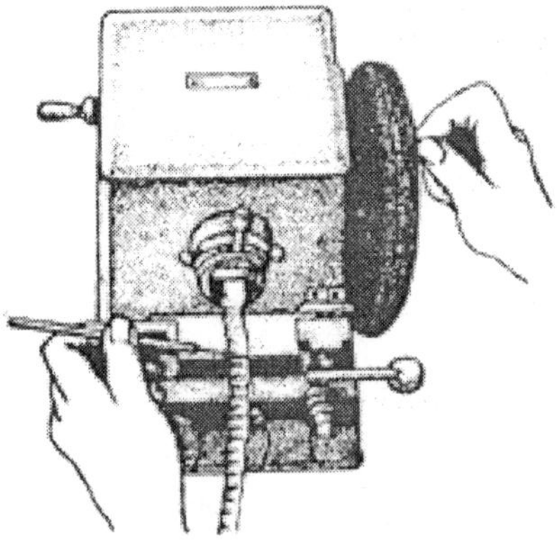

Fig.1.10 Rotary microtome

The material like stem, leaf, flower etc. is to be killed and preserved in a suitable fluid.

(a) Cut the material into bits of suitable size.

(b) Place the material in small vials containing a fixative or preservative.

(c) Label the vial giving the details of place of collection , name of the material, preservative used and date of collection.

The material should be in the killing fluid at least for six hours (preferably a day).

Dehydration: In order to dehydrate the material, remove it from the preservative and wash thoroughly in water and transfer it to a suitable tube or vial.

Dehydrate the material in alcohol series (Ethyl alcohol) as follows

Stage 1	50% Alcohol	First day
Stage 2	70% Alcohol	Second day
Stage 3	90% Alcohol	Third day
Stage 4	100% Alcohol	Fourth day
Stage 5	Absolute alcohol and xylol (50/50)	Fifth day
Stage 6	Absolute alcohol and xylol (25/75)	Sixth day
Stage 7	Xylol	Seventh day

When the material is transferred to pure xylol, there should be no fogging (i.e the solution should not turn milky). If there is fogging repeat the procedure starting from 4th stage.

The material in xylol will be completely dehydrated and looks transparent. It is now ready for infilteration by paraffin wax. The purpose of infilteration is to see that the wax enter into the material completely so that it can easily be cut.

Infilteration: Add a few pieces of paraffin wax everyday for about 4-5 days and then transfer the vial into a hot air oven, while temperature is adjusted to the melting point of wax. Everyday add a few pieces of wax. Adding of wax should be continued until all the xylol in the vial is replaced by paraffin wax. When the tube is smelt, there should be no smell of xylol. The material is now ready for embedding.

Paraffin embedding: In order to cut the material, it has to be embedded into a paraffin block. Prepare a paper boat or take a porcelain rectangular dish and smear glycerine to the inner surfaces. Pour molten paraffin into the porcelain dish and empty the contents of the vial into the dish. Before the paraffin solidifies, arrange

the material in proper rows and in such a manner as desired. Leave it for cooling for 6-8 hours. For quick cooling the dish may be transferred to a bowl containing ice water. Carefully remove the solidified paraffin block from the dish and cut it into suitable pieces. The material is now ready for sectioning.

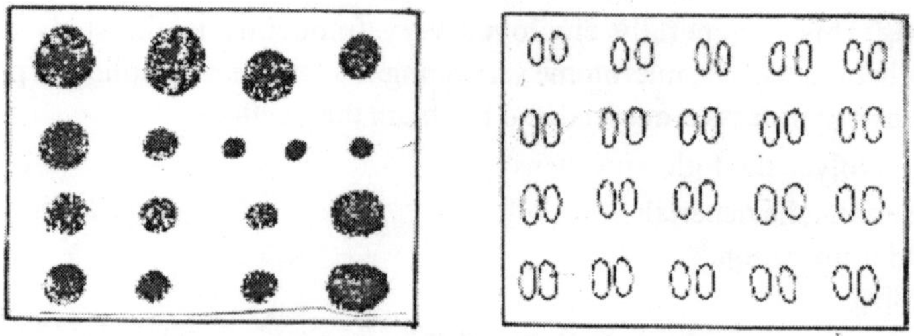

Fig.1.12 Arrangement of materials in paraffin blocks - A) Flower buds in groups B) flower buds in pairs

Sectioning: Sectioning is carried out with the help of an instrument called microtome. There are various types of microtomes like Rotary, sliding, Freezing etc. The one that is normally used is a rotary microtome. The following procedure can be adopted in the case of the rotary microtome.

1. The paraffin block is to be cut into small pieces each having the embedded material.

2. The pieces of paraffin containing the material is to be fixed to wooden 'riders'. For this purpose, heat the top of the rider and the base of the paraffin piece with a hot knife and fix. Allow it to cool. The paraffin piece is now firmly fixed to the rider. In order to make the edges of the paraffin even, cut the edges and make them parallel.

3. Fix the wooden rider into the microtome so that the paraffin piece with the material is exactly parallel to the edge of the microtome knife.

4. Adjust the knob of the microtome to give the required thickness to the sections.

5. Rotate the wheel of the microtome in a clockwise direction. The cylinder containing the rider moves up and down on the knife edge. Sections are cut in a sequence and get attached to each other forming a paraffin ribbon.

6. Keep a needle in your left hand to hold up the ribbon as it lengthens. When the ribbon is sufficiently long, cut it and transfer it on to a black cardboard. Cut the ribbons into suitable length and transfer to the slide.

7. Smear the slides with an adhesive like egg albumen or gelatin and rub it well.

8. Pour some water onto the slide with the help of dropping bottle.

9. Transfer the paraffin ribbons on to the slide and drain excess water.Retain a little water on the slide.

10. Place the slide on a slide warmer. The ribbon expands, and so also the sections of the material. Now draw out the water completely and allow the slides to cool.

11. The slides should be allowed to stand for atleast 24 hours whereupon the sections will stick to the slide permanently.

The slides are now ready for staining.

Staining procedure

The plant material (whole mount or sections) are to be stained and mounted properly before they are ready for microscopic observations.

Algal filaments, fungal mycelia, pollen grains and spores, prothalli of pteridophytes are usually stained with a single stain - saffranin or green. In some case Aniline blue is used to stain the filaments of Algae and mycelia of Fungi. The staining procedure is as follows.

Place the material either in a watch glass or cavity slide and add a few drops of stain. The period of staining varies with different materials. But usually a 5-10 minute period is sufficient. Drain out the stain and observe the material under the microscope. If the material has taken less stain, it may be kept in stain for some more period, or if it has taken excess stain it can be removed by washing either with water or 50% ethyl alcohol.

After the material is suitably stained pass it through a series of alcohol grades to dehydrate it. Usually different grades of alcohol (70%,90%,100%) taken in a filler are put on the material. After passing through absolute alcohol the material is kept in xylol and it is now ready for mounting.

Fungal mycelia can be stained with aniline blue. Usually the material taken on a clean slide is stained with a drop or two of the stain prepared in lactophenol. The material can bemounted straightway as lactophenol is also a mounting medium. If the stain is not prepared in lactophenol the material may be transferred to fresh lactophenol and mounted.

Double Staining Procedure

In the case of most of the algae and fungi there is little or no tissue differentiation and hence single staining is sufficient. In some cases of plants with relatively more tissue differentiation, a single stain is not sufficient to mark out the various tissues clearly. In such cases 2-3 stain combinations are used to stain the tissues. For example lignified and non-lignified tissues are stained with a combination of acidic and basic dyes.This type of staining is also called differential staining where different tissues take up different stains.

The following are some of the popular stain combinations.

Main stain	*Counter stain*
Haematoxylin	Saffranin
Haematoxylin	Erythrosine
Haematoxylin	Eosin
Haematoxylin	Orange G.
Saffranin	Fast green
Saffranin	Light green
Saffranin	Aniline blue
Saffranin	Crystal violet
Crystal violet	Erythrosine

Staining Procedure for free hand sections

(a) Procedure for temporary preparations

Free hand sections of root, stem etc., should be washed thoroughly in distilled water in a watch glass. After repeated washing the sections should be transferred to a watch glass containg the main stain. The sections should be kept in the stain for a period ranging from 30-40 minutes. After this period, the sections should be transfereed to a watch glass containing distilled water. The sections should be washed for 2-3 minutes whereupon excess of stain is removed. They should be observed under the microscope. If the main stain is saffranin only lignifed tissues should retain the stain, others should be destained. If the main stain is specific to non lignified tissues, only they should retain the stain. Then the sections should be washed thoroughly in water and repeatedly observed under the microscope. The sections can be mildly stroked with a fine hair brush, so that excess of stain comes out with distilled water. The sections are now ready for counter-staining. They should be transferred to a watch glass containing the counter stain (fast green etc). Counterstain acts quickly on the tissues. Hence one to two minutes of staining should be sufficient. The sections are now observed under the microscope. If there is excess of counter stain, it can be removed by

washing sections with 15-20%glycerine. After the final stianing the tissues whould be demarcated clearly with no mixing of the stains. The sections are now ready for mounting.

(b) Procedure for permanent preparaions

If the sections of plant materials are to be mounted and made into permanent slides, the procedure should be slightly altered so as to include dehydration and clearing with suitable agents. The procedure is as fallows:

1.Stain with the principle stain (Saffranin, Haemotoxylin etc).

2.Allow the sections to remain in stain for about 15-20 minutes.

3.Transfer the sections to water and wash thoroughly. The sections should be stroked mildly until no more stain comes out, and the water should remain clear. If the main stain like saffranin is produced in 50% alcohol then destaining (washing) should be done in 50% alcohol.

4.The sections should be now dehydrated in alcohol grades of 50%, 70% and 100%.

5.In each grade of alcohol keep the section for about 5 minutes.

6.Transfer the sections from one grade of alcohol to the other with the help of a fine brush. Never use needles as they injure the sections.

7.Transfer the sections to the counter stain (fast green etc). The counter stains are usually prepared in 90% alcohol.

8.Keep the section in counter stain for about a minute. Wash the sections in 90% alcohol and transfer them to 100% alcohol.

8a. In some case counter stains are prepared in clove oil. In such cases counter staining should be done after the sections are completely dehydrated with absolute alcohol.

1. Sections are transferred completely dehydrated to a solution of 50:50 Absolute alcohol and xylol followed by 25:75 Absolute alcohol and xylol and ultimately the sections are transfered to pure xylol. Now xylol should not turn milky or turbid. If it turns milky, dehydrate the sections again through combination of absolute alcohol : xylol series.

The sections are now ready for mounting.

Staining procedure for microtome sections

The slides with the paraffin ribbons adhering to them have sections of plant material.The processing and staining is as follows:

1. Immerse the slides in a coupling Jar containing xylol. Leave them for 10 minutes . The paraffin dissolves in xylol and only the sections remain on the slides.

2. The slides are now kept in 50:50 absolute alcohol and xylol.

3. Pass the slides through a down grade series of alcohol. 100%, 90% and 70%

4. Transfer the slides to pure water and wash them thoroughly to remove traces of alcohol.

5. Keep the slides in a mordant like 4% iron alum. This is necessary only when Haematoxylin is used. After about 10 minutes wash the slides in water.

6. Transfer the slides to the main stain Haematoxylin. Saffranin etc. and keep them for 15-30 minutes.

7. Remove the slides from stain and wash in water.

8. Observe the slides under microscope. Usually the sections are overstained.

9. If the main stain used in Hametoxylin, destaining can be done with 2% Iron alum or with water containing few drops of concentrated HCI. Repetaedly observe the slides under the microscope. Stop destaining when only the nuclei have retained the stain.

Caution: When destaining is done with HCI, not more than 5-6 drops should be used for 75ml of water.

Excess acid will eat away the adhesive and the sections come out of the slide.

10.Wash the slides in water overnight if the destaining agent is Iron alum, otherwise washing for 30 minutes is enough.

11.Now pass the slides through an upgrade series of alcohol (5 minutes in each grade) 70% 90% 100%

12. When the sections are completely dehydrated stain them with the counter stain (Fast green etc) for a minute or two. Wash the excess stain in clove oil and transfer the slides to pure xylol. The slides are now ready for mounting.

Mounting

(a) *Mounting Media*

Mounting media are of two categories: Temporary mounting media and permanent mounting media.

A temporary mounting medium may be just water or water mixed with glycerine. If the temporary sections are to be retained for 2-3 hours, dilute glycerine (15-20%) is recommended.

Permanent mounting media are used when the slides are to be retained indefinitely. The following are some of the permanent media.

(1) **Canada Balsam:** This is the most widely used mounting medium in botanical preparations. It is available in ready to use bottles. It is a gummy substance. If the solution is very thick add little xylol.

Canada balsam is a resinous product of a Gymnosperm, Abies balsamea. The gum has very good refractive index and as such it is highly reccommended.

(2) **Lactophenol:** Lactophenol can be prepared in the laboratorty. Mix equal quantites of phenol cystals, lactic acid glycerine and distilled water.

(3) **Glycerine gelly:B** This is prepared as follows:

Gelatine 1 part.
Water 6 parts.
Glycerine 7 parts.

Glycerine jelly is suitable even for mounting fresh specimens (even without dehyderation).

Add gelatine to water, stir thoroughly and warm. Add glycerine and cool the solution. After sometime the solution settles into a jelly.

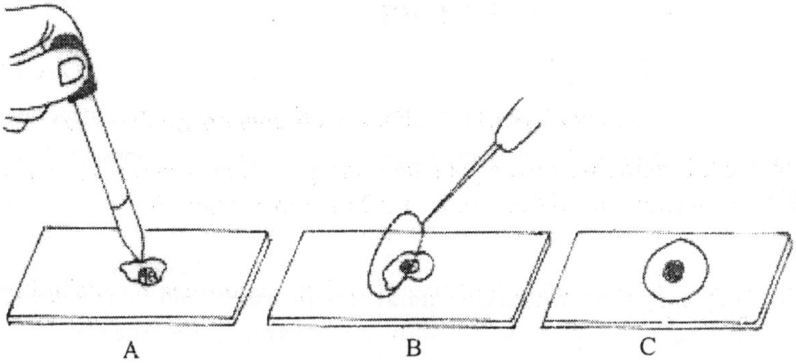

Fig.1.13 Method of mounting the material

(b) Mounting Method

The sections (free hand) are to be mounted on to clean slide in a specific mounting medium. If the sections are microtome preparations they would be already adhering to the slide. In any case the sections are to be covered with a thin coverslip or cover glass which are readily available in the market. They may be round, square or rectangular. Select the one that is suitable for your preparation.

The sections or objects should be mounted on the centre of the slide. In order to ensure placing of the object on the centre of the slide, the following procedure may be adopted. On a white paper draw a rectangle of the same size as the slide. Join the opposite corners by drawing diagonal lines. The centre of the slide lies where the two lines cross. Place the slide on the rectangle and material or objects may be kept on the central point.

Place a drop or two of the mounting medium. Hold the coverslip at the edges (see the figure) and allow it to touch the edge of the mounting medium at an angle of approximately 45°. Slowly lower the cover glass on the mounting medium with the help of a needle. Take care to see that no air bubbles enter the mounting medium. If there are air bubbles, lift the cover glass and mount again.

Excess mounting medium may be carefully wiped with the help of a clean tissue. Allow the slides to dry. If necessary the slides may be transferred to a hot air oven for drying. The slides should be labelled at one end as shown in the figure. They are now ready for microscopic observation.

Ringing of Slides

Temporary preparations may be kept for few days if the borders are sealed. This procedure is known as ringing . Ringing is done by applying wax, canada balsam or even nail polish to the edges. Ringing may be done manually or by using a ringing table having a rotary metallic disc.

Record of Practical Work

Practical work carried out in the laboratory should be recorded. The botany practicval record book has two types of sheets viz., Interleaves and drawing sheets. Interleaves are meant to write the notes pertaining to the diagrams. The drawing sheets which are thicker than the interleaves are used for diagrams. The following points may be useful in maintaining a good record.

1. Use good pencils (HHB) with sharp points and a good quality eraser.
2. Leave a minimum of one inch margin on all the sides.
3. Always write diagrams of only one plant in a page. Do not mix up diagrams of other plants.
4. On the left hand corner of the drawing sheet write the date of the experiment.
5. Give the complete systematic position of the member on the top of the drawing sheet.
6. For microscopic organisms draw first the habit sketch followed by low power and high power view/s. High power view should show the cellular details.
7. In the case of anatomical drawings, draw the ground plant first, followed by a sector enlargement to show the cellular details.
8. All parts of the diagrams must be properly labelled.
9. Stages of reproduction, if any, must be drawn after the details of vegetative structure.
10. The description corresponding to the diagrams must be written on the opposite interleaf.

Field Collection of plants

Specimens of algae, fungi and bryophytes can be collected in the field for the purpose of laboratory study. Field trip is an integral part of botanical study and all students must participate in it. It is in the field study that the students will come to understand the natural growth, habitat and environmental association of plants. During a field trip to collect specimens, the following items are to be carried:

1. Pocket lens.
2. Fixative F.A.A or Carnoy's fluid.
3. Collection bottles.
4. Blotting paper.
5. Polythene bags
6. Field note book
7. Labels.

When the specimens are collected they can be fixed and preserved for the purpose of laboratory study. Use the required preservative and immerse the specimens in it. Label the bottle showing details of

Date of collection

Location

Preservative fluid (name) used.

Name of the plant

As it is necessary for the purpose of identification, details of habitat, associations with other plants if any, colour of the fresh specimen etc. should be recorded in the field note book.

CHAPTER 2

CLASSIFICATION OF ANGIOSPERMS

MODERN systems of classification begin with the natural system proposed by George Bentham and Joseph Dalton Hooker which is in usage even now in many countries and has culminated in many types of classifications. The classification of Bentham and Hooker, Engler and Prantl, Hutchinson are discussed in detail in this chapter.

Bentham (1800-1884) and Hooker (1817-1911)

The two British botanists gave one of the most important classifications published in 3 volumes of their work *Genera plantarum'*. The three volumes together list about 97205 specimens. Hooker worked as Director, Kew botanic gardens. He has also written Flora of British India.

The system of classification proposed by Bentham and Hooker is Preevolutionary in concept and largely based on the Candollean systems. But it is a vast improvement over de Candolle's system. Groupings above family level greatly differ from the classification of deCandolle. Their system of classification is as follow

Spermatophyta (seed plants)

Dicotyledons	**Gymnospermae**	**Monocotyledons**
Polypetalae (Poly petalous Corolla, biseriate Perianth)	Gamopetalae (Sympetalous Corolla, biseriate perianth)	Monochlamydae (uniseriate perianth)

DICOTYLEDONS

A. POLYPETALAE: is divided into three series viz.,

(i) Thalamiflorae – Flowers hypogynous with a well developed receptacle, sepals free from the ovary.

(ii) Disciflorae – Sepals free or synsepalous; sometimes adnate to the ovary. A disc always surrounds the ovary. Ovary is superior.

(iii) Calyciflorae – Flowers perigynous or epigynous, sepals usually united and adnate to the ovary.

Series I THALAMIFLORAE

It has the following six cohorts (orders) :

1. Renales : It is characterised by indefinite stamens and apocarpous pistil. It has eight families Magnoliaceae, Anonaceae etc.

2. Parietales : It is characterised by unilocular ovary with a parietal placentation. It includes nine families such as Cruciferae, Capparidaceae etc.

3. Polygalinae : Includes four families such as Pittosporeae etc.

4. Caryophyllinae : Flowers have free central placentation. It has four families such as Caryophyllaceae etc.

5. Guttiferales : Includes six families such as Dipterocarpeae, Guttiferae etc.

6. Malvales : The Cohort is characterised by monadelphous stamens and axile placentation. It includes three families viz, Malvaceae, Tiliaceae and Sterculiaceae.

Series II DISCIFLORAE

It has four cohorts :

1. Geraniales : Flowers are usually irregular. An annular disc is present attached to the stamens. Placentation is axile. The order has eleven families like Geraniaceae, Rutaceae etc.

2. Olacales : It has the families Olacineae, Illicinae and Cyrilleae.

3. Celastrales : Has three families such as Rhamneae, Celastrinae etc.

4. Sapindales : Has three families Sapindaceae, Anacardiaceae and Sabiaceae.

5. Anomalous : Has two families - Coriarieae and Moringeae

Series III CALYCIFLORAE

It has five orders :

1. Rosales : Flowers are generally regular. Stamens and carpels many and free. There are nine families such as Connaraceae, Rosaceae, Leguminosae etc.

2. Myrtales : Flowers are characterised by epigynous condition. Placentation axile. There are six families such as Combretaceae, Myrtaceae etc.

3. Passiflorales : Has seven families such as Cucurbitaceae, Bignoniaceae etc.

4. Ficoidales : Has two families Cacteae, Ficoideae.

5. Umbellales : Flowers with inferior ovary axile placentation and one ovule in each locule. There are three familie: Umbelliferae, Araliaeae and Cornaceae.

B. GAMOPETALAE : There are three series in this, on the basis of number and position of the carpels

Series I INFERAE

Plants have inferior ovary. It has three orders.

1. Rubiales : Flowers regular, bisexual with epipetalous stamens. The order has two families viz., Caprifoliaceae and Rubiaceae.

2. Asterales : Flowers with bicarpellary, unilocular ovary with basal placentation. There are four families such as compositae, Dipsacaceae etc.

3. Companales : Has three families like Campanulaceae etc.

Series II HETEROMERAE

Flowers hypogynous with multi-carpellary ovary. There are three orders.

1. Ericales : There are six families like Ericaceae, Vaccineae etc

2. Primulales : There are three families Plumbaginaceae, Primulaceae, Myrsinaceae etc.

3. Ebenales : Has three families, Sapotaceae, Ebenaceae and Styracaceae.

Series III BICARPELLATAE

Flowers with superior ovary and two carpels. There are four orders.

1. Gentianales : Flowers actinomorphic, hypogynous with epipetalous stamens. It includes six families such as Gentianaceae, Apocynaceae, Asclepiadaceae etc.

2. Polemoniales : Flowers actinomorphic, hypogynous with multilocular ovary. It has five families like Solanaceae, Myrtaceae etc.

3. Personales : Flowers irregular with bilipped corolla. Stamens 2-4. There are eight families in this order. Ex:Scrophulariaceae, Acanthaceae, Bignoniaceae etc.

4. Lamiales : Corolla two lipped, ovary four locular. There are four families Ex: Verbenaceae, Labiatae etc.

Anomalous family : Plantagineae.

C. MONOCHLAMYDEAE (INCOMPLETAE): Perianth here is uniseriate. There are eight series.

Series I *Curvembryae* :

Embryo curved, flowers bisexual. There are seven families such as Amaranthaceae, Polygonaeae etc.

Series II *Multiovulatae aquaticae* :

Only one family Podostemaceae.

Series III *Multiovulatae terrestris* :

There are three families like Aristolochieae etc.

Series IV *Microembryae* :

Has four failies like piperaceae, Myristicaceae etc.

Series V *Daphnales* :

Five families like Proteaceae etc.

Series VI *Achlamydosporae* :

Includes three families such as Santalaceae, Loranthaceae etc.

Series VII *Unisexuales* :

All plants have unisexual flowers. There are nine families such as Euphorbiaceae, Urticaceae etc.

Series VIII *Anomalous families* :

There are four families such as Empetraceae, Salicaceae etc.

MONOCOTYLEDONS

The monocotyledons have been classified into seven series;

Series I Microspermae : Inner whorl of perianth petaloid, ovary inferior, Number of families three. Eg. Hydrocharideae Orchideae and Burmanniaceae.

Series II Epigynae : Includes seven families suas Scitamineae, Amaryllidaceae etc.

Series III Coronariae : Flowers Hypogynous. There are eight families such as, Liliaceae etc.

Series IV Calycineae : Perianth sepaloid. There are three families, Palmae etc.

Series V Nudiflorae : Members have achlamydeous flowes; ovary superior with united carpels. There are five families Aroideae etc.

Series VI Apocarpeae : Pistils apocarpous. There are three families such as Alismaceae etc.

Series VII Glumaceae : Flowers in spikelets with glames; ovary unilocular. The series has five families such as Cyperaceae, Gramineae etc.

The following table gives the summary of families, genera and species included in Genera plantarum.

Division (families)	Order	Genera	Species
Polypetalae	82	2,610	31,874
Gamopetalae	45	2,619	34,556
Monochlamydeae	36	801	11,784
Gymnosperms	3	44	415
Monocotyledons	34	1,495	18,576
Total	**200**	**7,569**	**97,205**

Salient features of Bentham and Hooker's Classification:

1. Plants are grouped into Dicotyledons, Gymnnosperms and Monocotyledons
2. Ranales are placed at the beginning and grasses at the end.
3. Gymnosperms are placed between Dicots and Monocots.
4. Dicotyledons are divided into polypetalae, Gamopetalae, and Monochlamydeae.
5. Monocotyledons are divided into seven series.

Merits :

1. The classification represents a careful and detailed analysis of flowering plants.
2. Monocotyledons are placed after dicotyledons.
3. Ranales, a primitive group is placed in the beginning
4. It is highly suitable for the field identification of plant groups.

Demerits :

1. Gymnosperms are placed between Dicots and Monocots.
2. The coherts (orders) of Monochlamydeae are given equal rank with the series of Polypetalae.
3. Cucurbitaceae, which has united petals is placed in Polypetalae.
4. The treatment of Apocarpae in Monocots is incorrect. They should have been placed in the beginning of Monocots.

Adolf Engler (1843 - 1930) and B. Prantl (1849 – 1893)

These two German Botanists published a 20 volume monumental work *'De Naturalischen Pflanzenfamilien'* wherein they have classified the entire plant kingdom from algae up to Angiosperms. They introduced certain changes in this classification in a later book *'Syllabus der Pflanzenfamilien'*

Engler and Prantl's system is an improvement over Bentham and Hooker's and Eichler's. They made a sincere effort to provide a phylogenetic system of classification. They distinguished between primitive and advanced plants. Their classification is widely adopted throughout the world. The outlines of their system of classification is as follows.

Embryophyta Siphonogama
(seed plants with fertilisation carried out by pollen tube)

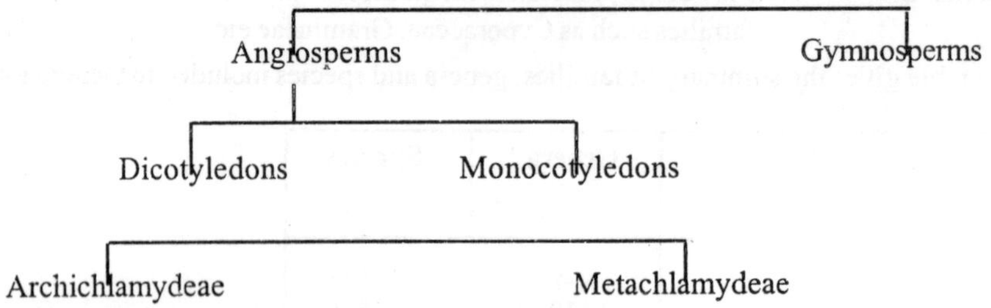

1. Monocotyledonae

Orders :
1. Pandanales (3)
2. Helobiae (7)
3. Triuridales (1)
4. Glumiflorae (2)
5. Principes (1)
6. Synanthae (1)
7. Spathiflorae (2)
8. Farinosae (13)
9. Liliflorae (9)
10. Scitamineae (4)
11. Microspermae (2)

II. Dicotyledoneae

Archichlamydeae

Orders :
1. Verticellatae (1)
2. Piperales (4)
3. Salicales (1)
4. Garryales (1)
5. Myricales (1)
6. Balanopsidales (1)
7. Litneriales (1)
8. Juglandales (1)
9. Batidales (1)
10. Julianales (1)

11. Fagales (2)
12. Urticales (3)
13. Proteales (1)
14. Santalales (8)
15. Aristolochiales (3)
16. Polygonales (1)
17. Centrospermae (9)
18. Ranales (18)
19. Rhoedales (6)
20. Sarraceniales (3)
21. Rosales (18)
22. Pandales (1)
23. Geraniales (20)
24. Sapindales (21)
25. Rhamnales (2)
26. Malvales (8)
27. Parietales (29)
28. Opuntiales 910
29. Myrtiflorae (19)
30. Umbelliflorae (3)

Metachlamydeae

1. Ericales (6)
2. Primulales (3)
3. Plumbaginales (1)
4. Ebenales (4)
5. Contortae (5)
6. Tubiflorae (20)
7. Plantaginales (1)
8. Rubiales (5)
9. Cucurbitales (1)
10. Campanulatae (6)

(The numbers in brackets indicate the number of families)

Salient Features of Engler and Prantl's Classification :

1. It is a phylogenetic system.
2. Monocots are regarded more primitive than dicots
3. Orchids are more evolved than grasses
4. Amentiferae group is regarded as primitive
5. Polypetalae and Apetalae are placed in one group, Archichlamydeae

Merits :

i. It is an excellent treatment of the seed plants.
2. Polypetalae and Apetalae are amalgamated.
3. Gymnosperms are placed as a distinct group
4. Classification of Monocotyledons is better than Bentham and Hooker's
5. In Archichlamydeae and Metachlamydeae plants with epigynous flowers are considered towards the end.

Demerits :

1. Monocots are placed earlier to dicots.
2. Amentiferae which is a simple group because of reduction is considered primitive.
3. Dichlamydeous flowers are regarded as primitive
4. Wind pollinated plants are regarded as primitive
5. Apetalous flowers are more primitive than petaloid flowers.

John Hutchinson

In his book 'Families of Flowering plants', Hutchinson has proposed a phylogenetic classification of Angiosperms. Hutchinson regarded that Angiosperms are monophyletic originating from hypothetical protoangiosperms. He has proposed a major division of Angiosperms into Lignosae (Trees) and Herbaceae (herbs).

Hutchinson's system is closer to Bessey's and is based on 24 principles or dicta. These are as follows :

General dicta :

1. Evolution is both ways i.e., upwards and downwards. While the former tends towards preservation, the latter brings degeneration.
2. Evolution will not uniformly affect all organs of an individual, while some organs are advancing, others may remain static or even undergo retrogression.
3. Evolution has normally been consistent, and when a particular progression or retrogression sets in, it continues in the same fashion till the end of the phylum.

Principles relating to general habit :

4. Trees and shrubs may generally be regarded more primitive than herbs in any particular group.
5. Trees and Shrubs are more ancient than climbers in any family or genus.
6. Annuals and biennials are more recent than perennials.
7. Epiphytic, Saprophytic or parasitic angiosperms are derived from normal autotrophic flowering plants. The same is true of aquatic flowering plants, which are derived from terrestrial forms.

Principles relating to the flowering plants :

8. Plants with collateral vascular bundles arranged in a ring form (Dicots) are more primitive than those (Monocots) with a scattered arrangement of vascular bundles. However, the former need not give rise to the latter directly.
9. Opposite and whorled arrangement of leaves is more advanced than spiral.
10. Simple leaves are more primitive than compound leaves.

Principles relating to flowers and fruits :

11. Dioecious plants are more recent than monoecious, unisexual flowers are derived from bisexual flowers.
12. Grouping of flowers is more advanced than solitary. Umbel and Capitulum are the more advanced types.
13. Whorled and valvate aestivation are more advanced than imbricate.
14. Oligomerous, floral structure follows the polymerous structure, accompanied by a progressive sterilisation of reproductive organs.
15. Apetalous flowers are derived from petaloid ones.
16. Polypetaly is more primitive than sympetaly.
17. Zygomorphy is more advanced than actinomorphy.
18. Epigyny and Perigyny are derivded from Hypogyny.
19. Apocarpy is primitive and Syncarpy is a derived condition
20. Polycarpy is more primitive than oligocarpy.
21. Albuminous seeds with a small embryo are more primitive than exalbuminous seeds with a large embryo.
22. Primitive flowers have more stamens and advanced flowers have few stamens.

23. Free stamens are more primitive than united ones.
24. Aggregate fruits are advanced. Simple fruits are primitive. Capsule preceeds the drupe or berry.

The system of classification

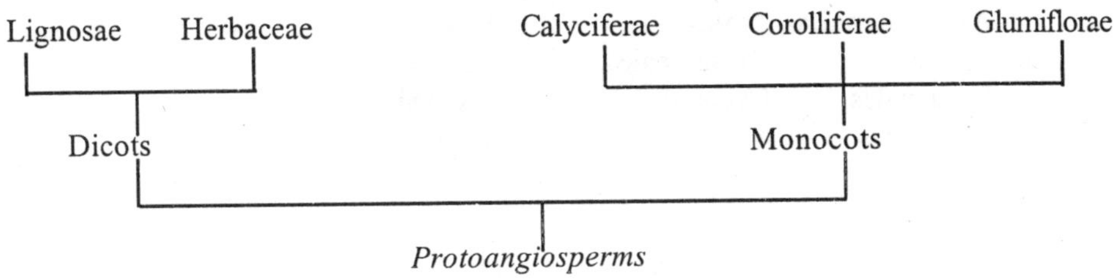

DICOTYLEDONS

Division I Lingosae :

Orders :

Mangnoliales,	Annonales,	Laurales,	Dilleniales,
Coriariales,	Rosales,	Leguminales,	Cunoniales,
Styracales,	Araliales,	Hamamelidales,	Salicales,
Lietneriales,	Myrcinales,	Balanopsidales,	Fagales,
Juglandales,	Casurinales,	Urticales	Bixales,
Thymeliales	Proteales,	Pittosporales,	Capparidales,
Tamaricales,	Violales,	Polygalales,	Rosales,
Passiflorales,	Cucurbitales,	Cactales,	Tiliales,
Malavales,	Malphigiales,	Euphorbiales,	Theales,
Ochnales,	Ericales,	Guttiferales,	Myrtales,
Celastarales,	Olacales,	Sanatalales,	Rhamnales,
Myricales,	Ebenales,	Rutales,	Meliales,
Sapindales,	Loganiales	Apocynales,	Rubiales,
Bignoniales,	Verbenales		(54 orders).

Division II Herbaceae :

Ranales,	Berberidales,	Aristolochiales,	Piperales,
Rhoedales,	Cruciferales,	Resedales,	Caryophyllales,
Polygonales,	Chenopodiales,	Plantaginales,	Piperales,
Primulales,	Saxifragales,	Lythrales,	Gentianales,
Podostemales,	Umbellales,	Valerinales,	Sarraceniales,
Goodeniales,	Asterales,	Solanagales,	Personalles,
Geraniales,	Polemoniales,	Boraginales,	Lamiales
			(38 Orders)

MONOCOTYLEDONS

Division I Calyciferae

Orders:

Butomales,	Alismatales,	Triuridales,	Juncaginales,
Aponogetanales,	Potamogetanales,	Najadales,	Commenlinales,

Xyridales,	Eriocaulales,	Bromeliales,	Zingiberales, *(12 orders).*

Division II Corolliferae :

Orders :

Liliales,	Alstroemeriales,	Arales,	Typhales,
Amarylidales,	Tridales,	Dioscoreales,	Agavales,
Palmales,	Pandanales,	Cyclanthales,	Haemodorales,
Burmanniales,	Orchidales		**(14 orders).**

Division III Glumiflora :

Orders :

Juncales,	Cyperales,	Graminales,	(3 orders)

The following table shows the total numberr of families, orders etc. in Hutchinson's classification of flowering plants.

	Division	Order	Family
	Dicotyledons		
	Lignosae	54	246
	Herbaceae	38	96
Monocotyledons			
	Calyiferae	12	29
	Corolliferae	14	34
	Glumiflorae	3	6
	Total	111	411

Sailent features of the Classification:

1. Dicotyledons are divided into Lignosae and Herbaceae.
2. Among Lignosae, Magnoliceae is the most primitive and Verbenaceae, the most advanced.
3. Among Herbaceae, Ranunculaceae is the most primitive and Labiatae the most advanced.
4. Butomaceae and Alismataceae are the most primitive among monocots and they are derived from Ranunculaceae.
5. Grasses are regarded as the most advanced among monocots.

Demerits :

1. Division of Dicots and Monocots based on habit is unnatural.
2. Monophyletic origin for angiosperms.
3. It is not a purely phylogenetic system.
4. Two lines of evolution suggested for dicots is not acceptable to many.

CHAPTER 3

HERBARIUM TECHNIQUES

Identification of the plant in the field when the specimen is fresh and live is the best method. But always such field identifications are not possible. Further when a floristic survey of an area is conducted a record has to be maintained for any future study. Hence it is imperative to preserve the plant specimens in the dry condition. The method of preservation of the plants is called herbarium technique. A herbarium sheet consists of a dried specimen mounted on it. The places where the herbarium sheets are stored are called herbaria. While in botanical garden live specimens are available for the study, in herbaria preserved specimens are available. Herbarium technique described in the following pages concerns mainly with the flowering plants, though it can be easily employed for pteridophytes and gymnosperms also.

Equipments and methods of field collection

Field collection of the plants in a particular locality is the most significant aspect of a taxonomic study. The plants collected for the purpose of study may not be in flowering or fruiting at a particular visit. Hence several visits to a particular "Ate may be necessary in order to collect the plant having ail essential parts and stages of growth. If the plant specimen is a herb, the entire plant has to be collected, if the specimen in question is a shrub or a tree, twigs of suitable length are to be collected. Leaves, branches (if any), flowers, fruits, etc., must be present in the plant specimen meant for preservation.

The following is a list of field equipment, a student of taxonomy must possess before proceeding for plant collection. I. Secature, 2. Digger, 3. Pen lmife, 4. Forceps, 5. Scisors, 6. Sickle with long handle, 7.Vasculum and polythene bags, 8. Plant press, 9. Labels with tags, 10. Field lens, 11. Preservative fluid, 12. Blotting paper and news paper, 13. Field note book. 14. Field binoculars and camera.

1. **Secature :** This is necessary to cut twigs or branches of trees and obtain specimens.

2. **Digger :** Many a time, it is necessary to dig the soil in order to obtain the plant together with its root system, rhizomes, bulbs, etc. This is particularly needed if the underground parts are of diagnostic value

3. **Pen Knife :** For cutting of branches or scraping, pruning etc, cutting knife or scalpel is necessary.

4. **Forceps :** Dissection forceps, needles etc, are necessary to hold or dissect plant parts.

5. **Scissors :** Tender plant parts may be cut with scissors.

6. **Sickle with long handle:** An acutely bent sickle attached to a long handle or pole, like the one

shepherds generally use is necessary to pluck twigs, flowers or fruits from a tree. Long handles may be prepared by attaching cylindrical rods of diminishing diametre, so that they can be telescoped (like the antenna of a transitor radio). This system would facilitate easy carrying of the sickles with a long pole.

7. Vasculum and Polythene bags: The plant specimens collected should ideally be pressed in the field itself. However it is not possible in all the situations. Hence vasculum comes in handy. Plants sprinkled with a little water and kept in a vasculum retain their freshness until they are pressed. The vasculum is a long oval ended aluminium container which has a hinged lid. It is 40-60 cms long, 12 cms in width and about 14-16 cms in depth. The hinged lid is almost air tight and prevents wilting of the plants by not allowing the escape of the moisture. At either end the vasculum is provided with buckles to be attached to a shoulder strap.

Polythene bags of suitable size are also useful in the collection and temporary preservation of specimens eventhough they are not as useful as the vasculum. One advantage with polythene bags is a large number of them can be carried easily.

8. Plant press: This helps in the quick drying and two dimensional flattening of the plant specimen. It is ideal to carry the plant press to the field and press the plant. If it is not possible, the specimen should be pressed as early as possible. Aplant press consists of two flat boards made of metal or wood with a mechanism to hold them together. The boards are rectangular and 18-24" in length and 8-10" in width. Usually the two boards can be tightly secured to each other by means of long (8-10") bolts and nuts. Between these two plates, plants are kept in folders or driers which are nothing but moisture absorbants like plotting sheets or newspapers. Corrugated sheets are kept in between driers, in order to provide easy ventialationfor the moisture released from the drying specimens.

9. Labels with tags: These are field labels which are small strips of paper with a tag or thread. As soon as the specimens are collected, they are given a number which is written on the label and tied to the plant. Multiple specimens of the same plant are given the same number when collected in the same locality. Usually it is necessary to have six tagged labels carrying the same number for each plant. This means ideally six specimens of a plant are to be collected.

10. Field lens: Pocket lens or preferably a dissection microscope with 10 X or 15 X lens is very helpful for field study.

11. Preservative fluid : In some instances it will be necessary to keep the flowers preserved in the hydrated condition for later study. Flowers or other parts are ideally preserved in Formalin acetic acid alcohol (FAA) or 70% ethyl alcohol.

12. Blotting paper or newspaper: Sufficient quantity of blotting sheets and newspapers are necessary to dry the plants. Two different specimens should not be kept in the same folder of the blotting sheet.

13. Field note book : This is the most important item in the field study. It is very essential to record, valuable field data at the time of collection. Note books of pocket size or slightly larger ones which are securely bound must be used as field note books. It is very essential to write the collection number of the plant (the number written on the label and tagged to the plant) in the field note book also, before recording the data concerning the plant.

What is the necessity of writing field notes? While it is true that plant specimens are preserved in the herbarium for a detailed study, many a time, there are lot of differences between a fresh plant and a dried specimen. There are many features which disappear on drying. The following data are to be recorded in the field note book.

a) Collection number
b) Date of collection

c) Place of locality of collection
d) Habitat
e) Special association with other plants if any. For e.g. if the specimen is a parasite or epiphyte, the identirty of the host.
f) Particulars of the soil nature, if they are of a special kind.
g) Special characters of the plant like colour of the flower, texture, presence or absence of latex, smell, etc.
h) Abudance of the plant in the locality.
i) The stage of the plant growth whether in flowering or fruiting stage etc. This facilitates in recording the data regarding seasonal growth.
j) Any association with other organisms like ants, insects, etc.
k) Local name/names ofthe plant (if known)
l) Botanical name ofthe plant together with family name to which it belongs (if known)
m) Any economic or other uses ofthe plant

It is preferable to maintain a duplicate ofthe field note book, for the field data are very valuable.

14. Field binoculars and camera : Binoculars are necessary for viewing the generally inaccessible plants from a distance. Camera is ideal for photographic record ofthe fresh plants.

Collection and pressing ofthe specimen :

A plant specimen should ideally have both flowers and fruits or atleast one. Plants consisting of only the vegetative stage should not be collected "as their identity cannot be established. If the specimen in question is a herb, entire plant together with its root system must be collected, but the roots should becleaned before the specimen is pressed. If the specimen is from a tree or shrub, a twig of about 25 cm length should be ideal. In case the leaves are multistichous and overlapping in arrangement, unwanted leaves must be removed so that, the specimen can be flattened easily during pressing.

The specimen should be carefully spread on the blotting sheet or newspaper and trimmed if necessary. Tom leaves, decayed parts etc., should be removed. If the specimen has any infection it should be discarded, unless the subject of interest is plant pathology. Extreme care should be taken at the initial stage of pressing as the beauty ofthe herbarium sheet is solely dependent on it. If the specimen is larger it should be bent in the form of 'V' or 'W' or 'N' shape. Sometimes it may be necessary to place cotton packing to preserve the shape ofthe corolla. After the plant is spread properly the folder should be closed. The folders and plants inside them should be kept in the press in such a way that it is of uniform thickness in the middle as well as the sides. In this way the specimens will be subjected to uniform pressure.

There are two methods of processing the plants further after they are kept in the folders (blotting sheets/ newspaper). These are - the **Wet method** and the **Dry method.**

WET METHOD: This is a convenient method to follow, but many a time the plant specimens turn black or decolorized. In this method which is easy to adopt particularly during long field trips, the specimens are kept in the folds ofthe newspapers and tied tightly into a flat bundle with the help of twine. The whole bundle is then dipped in 4% formaline solution. The bundle is then kept in a large polythene bag and sealed properly. The bundle can be opened after returning to the headquarters and dried. There is no need to change the newspapers soon after collection, in order to dry the plant. The purpose of dipping the plants in formalin solution is to kill them soom after collection, othterwise, the specimens within a day or two will lose all the leaves due to the development of abscission layer. A fresh specimen kept in the folds of the newspaper without changing, will have only the axis and no leaves.

DRY METHOD: This is a more cumbersome method and involves patience and care. Here the specimens are kept in the folds of blotting paper and tied in a press. Initially, the press should be opened and plants changed to fresh and dry blotting sheets at least two to three times a day. After about 3-4 days, it will be enough if plants are given one change daily. Any laziness to change the blotting sheets frequently at the initial stages, will render the specimen dark as well as loss of leaves due to the formation of abscission layer. The idea of frequent change to fresh blotting sheets is to see that, the moisture coming out of the plant, should not be reabsorbed by it.

In this method, specimens are preserved with their natural colours in tact. But there will also be the problem of loss of leaves due to abscission layer formation as the plants are not killed soon after pressing besides it is difficult to follow the dry method during a long field trip as the number of specimens will be large.

Drying the specimens: Plant press together with its herbarium sheets can be kept in sunlight for drying. In case wet method has been followed, the plants should be changed to fresh newspaper folds, twice or thrice daily for about 3-4 days and thereafter once daily until the plants are completely dry.

Plant specimens may also be dried artificially, instead of depending on sunlight. This method is employed particularly when the climate is damp or when the plants have to be dried within a short period. Presses with the plants are kept in large ventilated boxes with a provision for blowing of hot air. Alternatively presses may also be kept in a hot air oven. Asimplest method is to keep the plant presses in ventilated boxes fitted with one or two incandescent bulbs.

Small and delicate specimens may be dried easily by 'ironing' them with an electric iron box after they are kept in the folds of thick cloth.

Mounting of specimen :The dried specimens are now ready for mounting on herbarium sheets. Fixing the specimen either with glue or with stiches is called mounting. The first step in mounting is the selection of good quality herbarium sheet. The sheets must be heavy white papers or thin card board (KG card board) cut into standard size 28.76 cm by 41.25 cm (16" by 11"). As has already been mentioned the specimens may be glued with a suitable gum or fixed with the help of cellophane tape. The best metliod however is to stich the specimen to the herbarium sheet at several points with thread. Some of the parts of the plant which do not remain in tact with the plant, like fruits, seed etc., may be placed in a small envelope and the envelope itself is fixed to the herbarium sheet at the comers. Extreme care is necessary whiel mounting, as the dry specimens are likely to break or crumple.

Another precaution which may be taken to prevent fungal or other infections is to dip the specimens in a solution ofmercuric chloride in ethyl alcohol. As mercuric chloride is highly poisionous, maximum care is required in handling. Lauryl pentachlorophenate can also be used instead ofmercuric chloride.

Labelling ; Labelling the herbarium sheet is very necessary to establish the identity of the specimens. Printed labels (7-12 cms long) with the following information should be fixed securely on the right hand bottom comer of the herbarium sheet.

(1) Name of the Institution/Department etc. This should be at the top of the label.
(2) Name of the family
(3) Name of the plant (botanical) complete with genus and species followed by the authority (e.g. *Blainvillea latifolia*, DC)
(4) Place of collection
(5) Date of collection
(6) Brief remarks about the plant like habitat, habit, type of soil, abundance, distribution, any salient feature etc.
(7) Field number of the specimen
(8) Name of the collector
(9) Local names and uses of the plant, if any

Filing of specimens: Filing or storing of the specimens is another important task. Special care is required to store the herbarium sheets. These are stored in wooden or steel cabinets with a number of pigeon holes. The herbarium sheets are kept in the folders of thick sheets. Each pigeon hole may be labelled either for a family or for a genus depending on the necessity. The arrangement of the herbarium sheet may be alphabetical wise or better according to any particular system of classification. Sometimes folders of various colours are used indicating species, genus or a geographic area.

A proper indexing of all the specimens available at the herbarium (Institution) is also necessary.

Care of the specimen : Proper maintenance of the herbarium sheet is essential as they should lost long, and be available for all taxonomists for the purpose of comparison and study. Herbarium sheets should not be exposed to moisture, insects, fungal infection etc. Keeping insect repilenants, in the pegeon holes, fumigation of the herbarium sheets, coating the specimens with chemicals are some of the measures adopted to prevent damage to the plants. In addition to this, special care should be taken while handling the herbarium sheets. The sheets should not be bent or folded or any undue mechanical pressure applied on them. If any mechanical break appears in the specimen, the broken parts must be kept in an envelope and it should be glued to the herbarium sheet. Finally, the cabinets where herbarium sheets are stored should be fire proof. Rare or valuable and old collections, type specimens etc., must be treated as if they are treasures, for indeed they represent the plant wealth.

Botanical survey of India (BSI): This was established to coordinate the activities regarding the systematic study of plants. The BSI which started its work in 1890 has its central offices in Calcutta. There is a large botanical garden in Sibpur Calcutta, which also houses the Central National herbarium. BSI has also established several regional circles or herbaria at Dehradun, Coimbatore, Allahabad, Pune and Shillong. The regional herbaria have local collections of plants.

Role of herbaria in taxonomic research

Herbaria play a pivotal role in taxonomic research by providing ready reference material both in the form of herbarium specimens and available taxonomic literature. Taxonomists use the facility of herbaria to compare an unknown specimen with ones available in the herbaria. This helps in establishing the identity of the plant.

List of some of the Important National and International Botanical Gardens.

1. Botanical Garden, Calcutta.
2. National Botanical Garden, Lucknow.
3. Botanical Garden, Ooty.
4. Orto Botanico of Padua, Italy.
5. Pisa Botanical Garden, Italy.
6. Villa Taranto Botanical Garden, Italy.
7. Leyden University Botanical Garden, The Netherlands.
8. Royal Botanical Garden, Edinburgh, England.
9. The National Botanic Garden, Glasnevin, Ireland.
10. Botanical Garden, Munich, Germany.
11. Botanical Garden, Berlin, Germany.
12. Less Cedres Garden, France.
13. Oxford Botanic Gardens, England.
14. Royal Botanic Gardens, Kew, England
15. Amold Arboretum, USA.
16. Missouri Botanical Garden, St.Louis, USA
17. Botanical Garden, Moscow, USSR

CHAPTER 4

GLOSSARY OF TAXONOMIC TERMS

Taxonomic description of plants for the purpose of classification and identification requires a detailed account of the various parts of plants. Each part of the plant is carefully and precisely described so as to help in its identification. The branch of taxonomy dealing with the description of plant parts is called 'Phytography'. A through knowledge of Phytography is a preequisite for a taxonomic study. In this chapter, a detailed and illustrated glossary of all the technical terms involved in the description of plant parts is given.

Habitat and Habit

Habitat	:	The environment in which the plant lives.
Habit	:	External appearance of the plant body.
Herb	:	A small plant with soft and herbaceous stem.
Shrub	:	A bushy or woody plant taller than herb. Branching is profuse; a prominent trunk is absent.
Tree	:	A tall woody plant with an erect and hard stem.
Annual	:	Plants that complete their life cycle (Seed to Seed) in one year.
Biennial	:	Plants that complete their life cycle in two years.
Perennial	:	A plant that survives and grows for many years.
Ephermeral	:	A plant with a very short life cycle. There will be several generations within a year.
Parasite	:	A heterotrophic plant deriving nutrition from a living host.
Saprophyte	:	An organism obtaining nutrition from dead and decaying organisms.
Insectivorus	:	An autotrophic plant having an auxillary means of heterotrphic nutrition. Protein requirements of the plants are met with by the trapping of the insects.
Xerophyte	:	Plants living in the regions of water scarcity.
Hydrophyte	:	Aquatic plants.
Mesophyte	:	Plants that live in ordinary environment.

ROOTS

Root is the downward growing part of the primary axis of the plant and it generally develops from the radicle of the embryo.

Tap root system : The primary root gives rise to lateral roots and the tap root survives till the end of the life cycle of the plant. This root system is characteristic of dictyledons.

Adventitious root system : This is also called the fibrous root system. These are extra embryonal in origin. Here the radicle is short lived. This root system is characteristic of monocotyledons.

Root Modifications

Fusiform : Tap root modifications for food storage; root is spindle like due to accumulation of food.

Conical : Tap root modification for food storage, Root is top like.

Napiform : Tap root modification for food storage. Top portion of the root is globose; it suddenly becomes narrow.

Annualted : Adventitious root modification; made up of a number of rings or annuli.

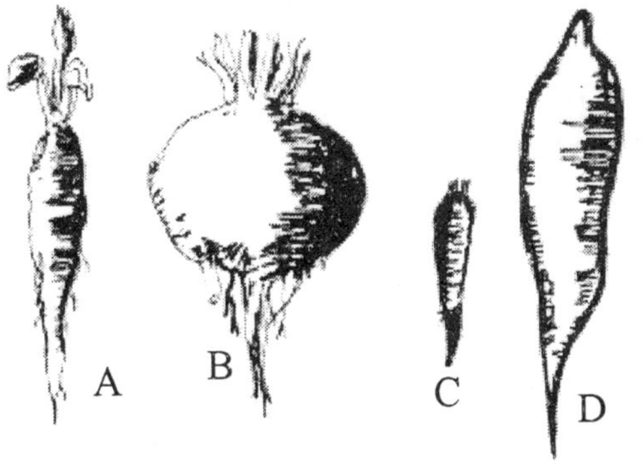

Fig. 4.1 Root - Tap root modification.
A. Fusiform (Raddish), B. Napiform (Beet root), C. (Carrot) D. Tuberous (Mirabilis)

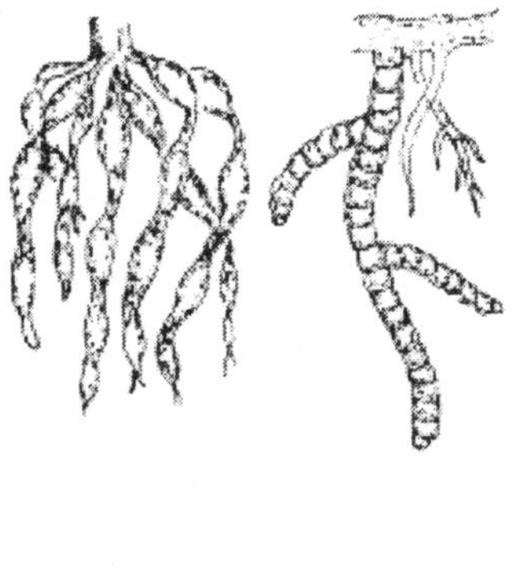

Fig. 4.2 Root - Adventitious root modifications
A. Moniliform (Bitter gourd), B. Annulated (pecac)

Nodulated : Adventitious root modification. Tips of roots have nodules.

Fasciculated : Cluster of adventitious roots storing food, present at the base of the stem.

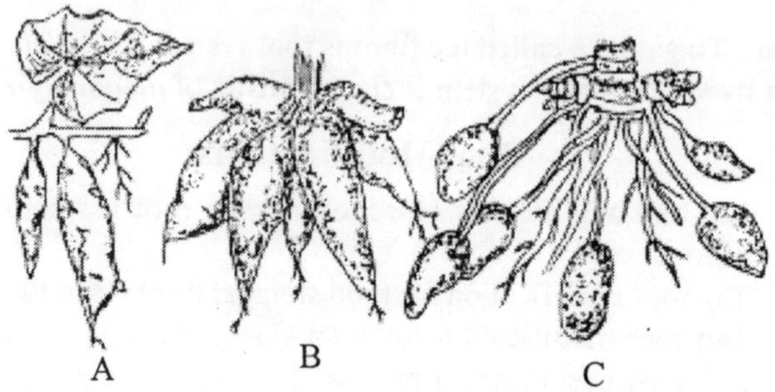

Fig 4.3 Root - Adventitious root modifications
A. Tuberous (Sweet potato) B. Fasciculated (Dhalia) C. Nodulose (Mango Ginger)

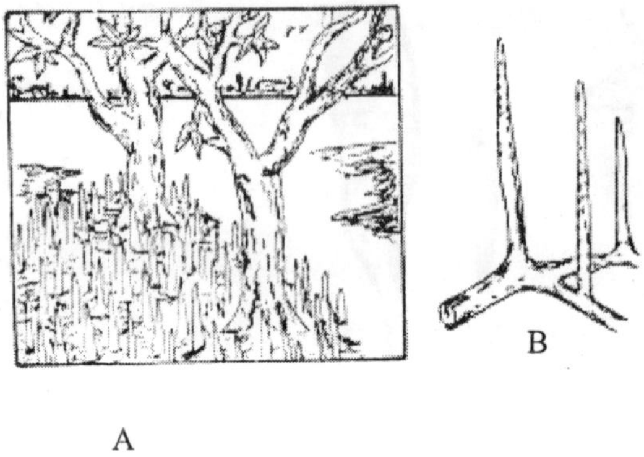

Fig 4.4 Root - Pneumatophores of *Avicennia*
A. Plant with pneumatophores, B. A root enlarged

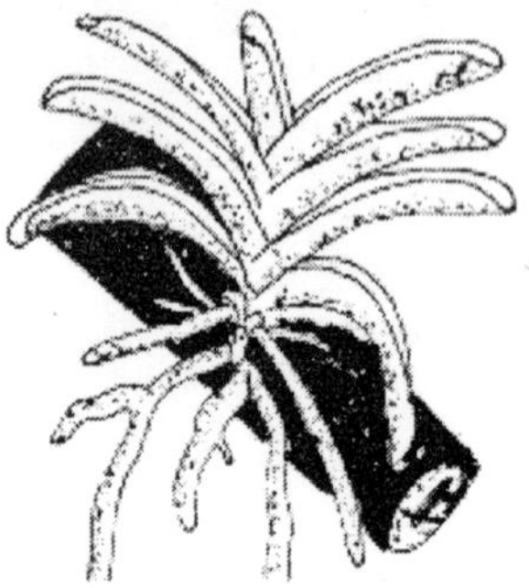

Fig 4.5 Root - Epiphytic roots of *Vanda*

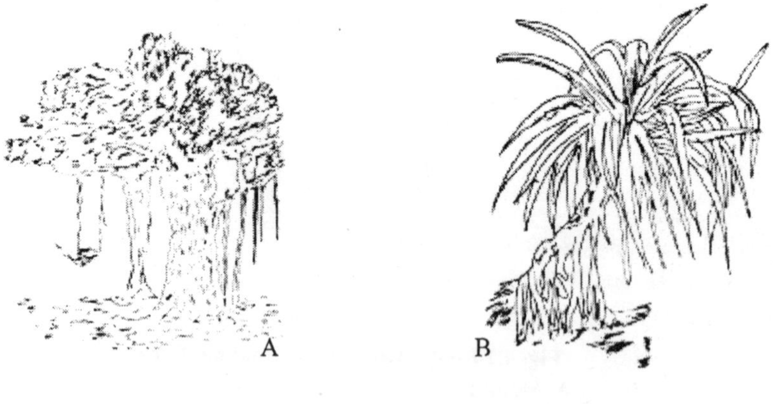

Fig 4.6 Roots - Modifications for mechanical functions
A. Prop roots of Banyan, B. Stilt roots of **Pandanus**

Moniliform	:	Adventitious roots appearing like a chain of beads.
Tuberous	:	Adventitious roots appearing singly. They store food materials.
Epiphytic	:	Roots that help in epiphytism. They are green in colour and help in photosynthesis. Epiphytic roots are aerial roots.
Parasitic	:	Roots that absorb nutrition from the host plant. They are also called haustorial roots.
Respiratory	:	Apogeotropic roots developed in mangroove plants. They are also called penumatophores and they help in respiration.
Prop	:	Roots arising from branches and grow down. They support the branches.
Stilt	:	In plants growing in loose soil, oblique roots grow from the stem and offer support.
Climbing	:	Roots developing from nodal regions helping in climbing.
Floating	:	Fleshy roots, storing air and helping the plant in buoyancy. These are found in some hydrophytes
Ballasting	:	Root clusters that give weight to the plants and maintain it in a particular place. This is found in some floating hydrophytes.

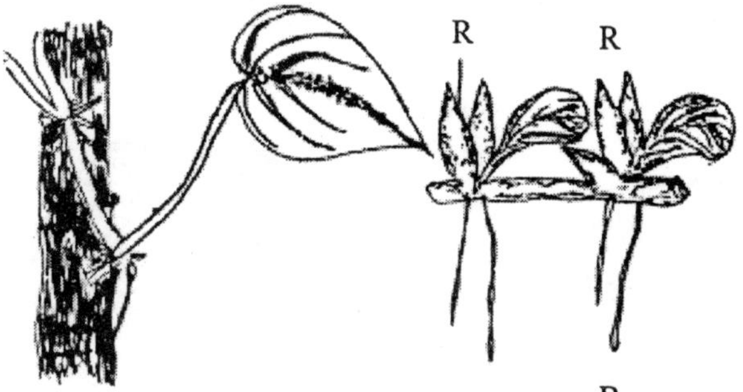

Fig 4.7 Roots. Modifications for mechanical functions
A. Climbing roots of Piper betle, B. Floating roots of Jussieua

STEM

Cylindrical	:	Cylinder like stem.
Triangular	:	Three angled stem.
Quadrangular	:	Four angled stem.
Solid	:	When it is completely filled up.
Fistular	:	Central portion is hollow.
Jointed	:	Distinct joints are seen at the nodes.
Herbaceous	:	Soft.
Woody	:	Hard
Branched	:	A stem with many branches.
Caudex	:	Woody base of an unbranched stem with a number of leaf scars.
Creeping	:	A weak stem growing parallel to the ground surface producing roots all along its length.

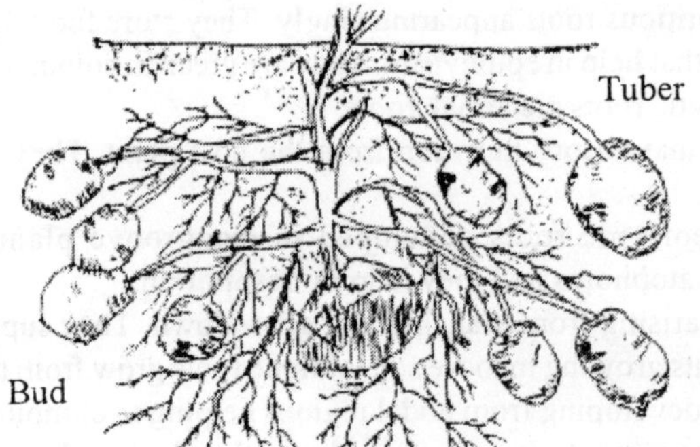

Fig 4.8 Stem - Tubers of potato

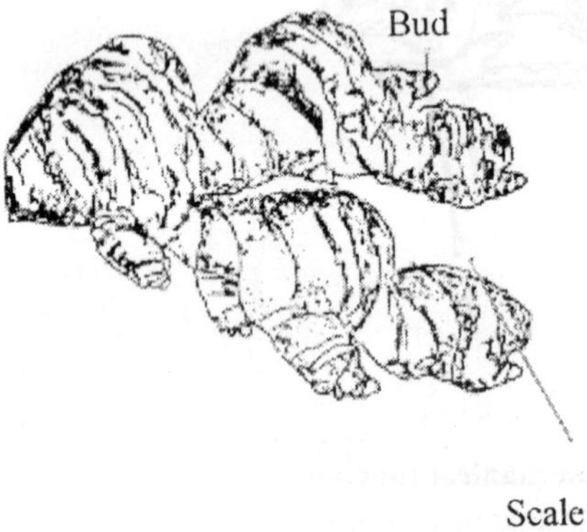

Fig 4.9 Stem - Rhizome of ginger

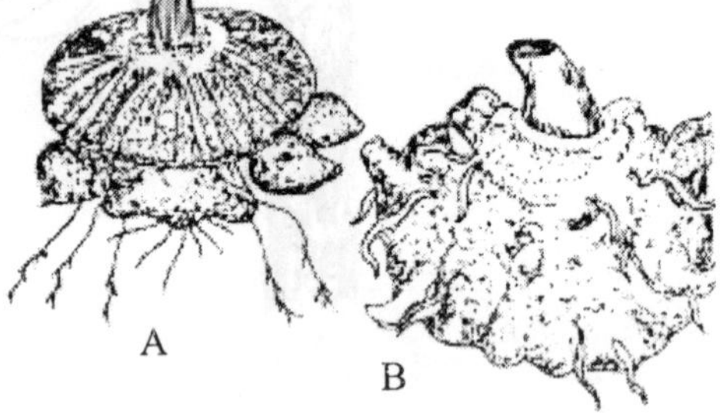

Fig 4.10 Stem - Corn of gladiolus
(A) and Amorphophallus (B)

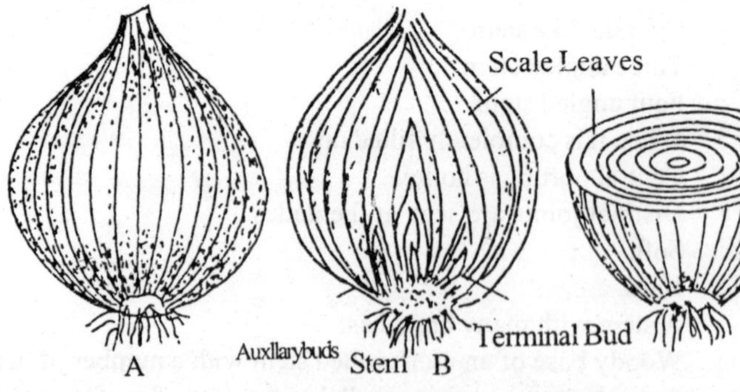

Fig. 4.11 Stem - Bulb of onion
A. Entire bulb, B. Long sections of the bulb, C. Cross section of the bulb

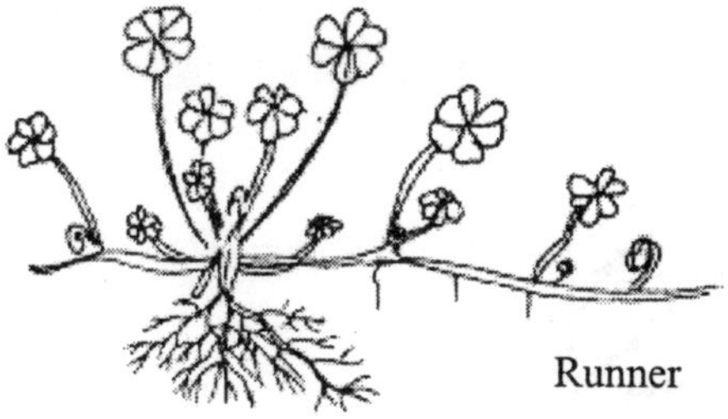

Fig. 4.12 Stem - Runner of Oxalis

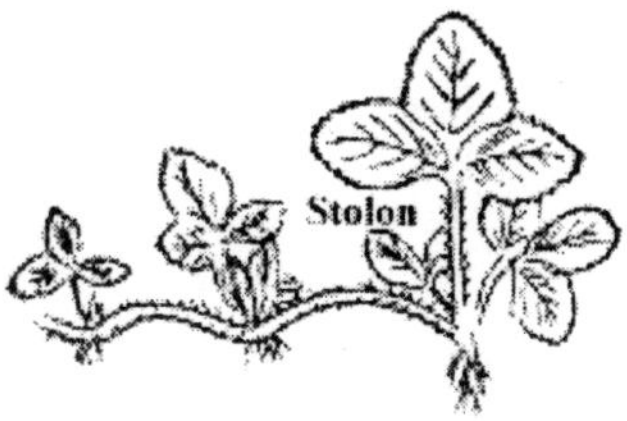

Fig. 4.13 Stem - Stolon of Fragaria

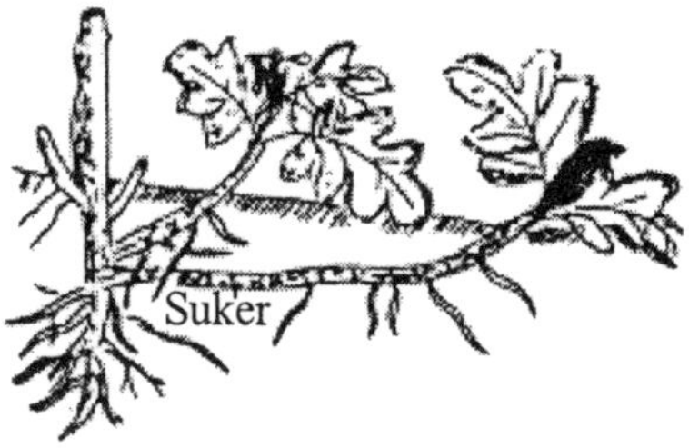

Fig. 4.14 Stem - Sucker of *Chrysanthemum*

Tendril

Fig. 4.15 Stem of *Pistia*

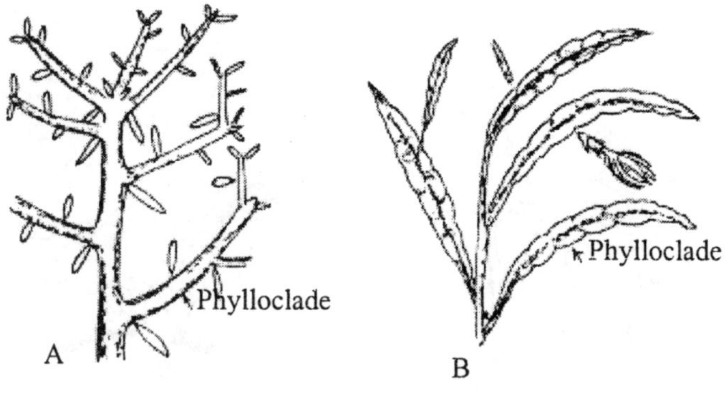

Fig. 4.16 Stem - **Phylloclades of** Ruscus (A) and Cladodes of **Asparagus** (B)

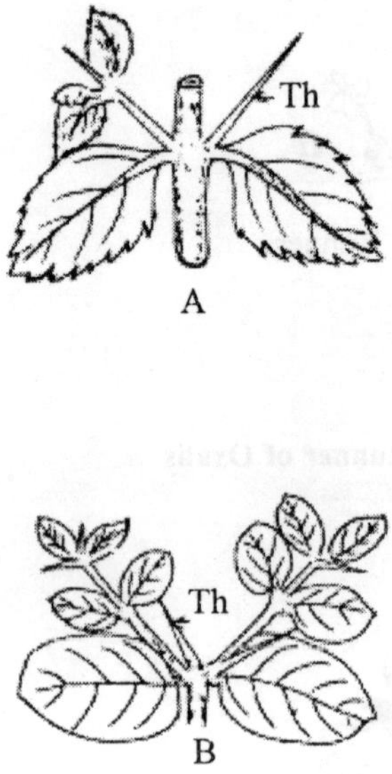

Fig. 4.17 Stem - Phylloclades of *Opuntia* (A) Muehlenbeckia (B) and Epithyllum (C)

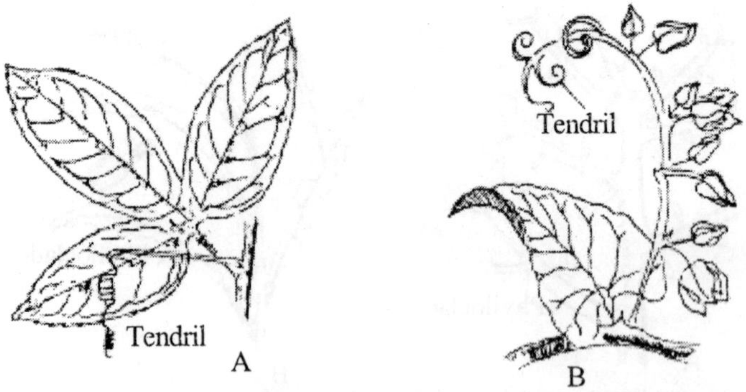

Fig. 4.18 Stem - Tendrils of *Passiflora* (A) and Antigonon (B)

Culm	:	A jointed stem with hollow internodes.
Prostrate	:	A stem parallel to ground. Roots at only one point i.e., at the base of the stem.
Decumbent	:	A prostrate stem with top becoming vertical.
Diffuse	:	Spreading branches found in stem.
Erect	:	Rigid vertically growing stem.
Glabrous	:	Smooth surfaced stem.
Pubescent	:	Stem having hairy outgrowths.
Internode	:	Part of the stem between two internodes.
Weak	:	A stem which cannot grow erect by itself.

Stem modifications

Underground

Stem tuber	:	Tuberous; part of the stem underground, scale leaves and buds are present. Ex: *Potato.*
Corm	:	Solid, globose part of the stem; underground, has buds; stores food. Ex. Yam.
Rhizome	:	Subterranean, food storing stem; branched leaves grow from this and become aerial. Ex. *Ginger.*
Bulb	:	Underground modification of stem. Stem reduced to a disc. Fleshy leaves store the food materials.Ex. Onion.

Sub aerial

Runner	:	Sub aerial modification, part of the stem grows horizontal to ground; tip produces a new plant Ex: *Oxalis.*
Stolon	:	Sub aerial modification. Exactly similar to runner, but sometimes shorter and partly underground Ex: *Colocasia.*
Offset	:	Sub aerial modification seen in water plants, and helps in vegetative propagation. It is like a runner. Ex: *Pistia.*
Sucker	:	An oblique branch arising from the underground part of the stem, grows up and gives rise to the aerial shoot Ex: *Chrysanthemum.*

Aerial

Phylloclade	:	Xerophytic stem, green in colour; has many internodes Ex: *Opuntia.*
Cladode	:	Xerophytic stem or branch; it may be regarded as a phylloclade of one internodal length.

LEAF

Leaf is a dorsiventrally flattened photosynthetic appendage of the plant bearing a bud in its axil. The normal function of a leaf is the manufacture of food.

Aphyllous	:	Leaves absent
Cauline	:	Leaves borne only on the main stem.
Ramal	:	Leaves borne on lateral branches.
Radical	:	Leaves produced at the ground level on a reduced stem.
Deciduous	:	Leaf falling off early.
Evergreen	:	Leaves persistent for several years.

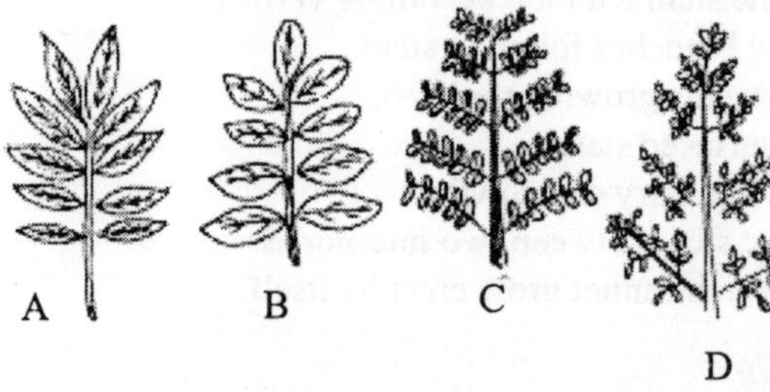

Fig. 4.19 Leaf - Pinnately Compound Leaves.
A. Unipinnate, (Paripinnate), B. Unipinnate (Imparipinnate), C. Bipinnate, D. Tripinnate

Phyllotaxy

Phyllotaxy is the various modes of arrangement of the leaves on the stem. The following types of phyllotaxy are generally seen in flowering plants:

Alternate : There will be a single leaf at each node. The leaves at successive nodes alternate their position.

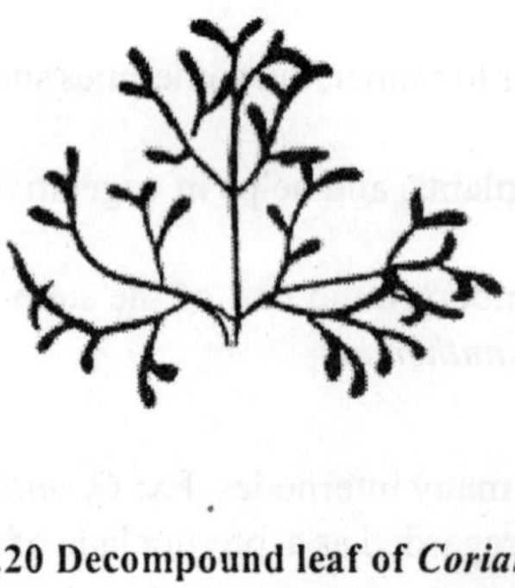

Fig 4.20 Decompound leaf of *Coriandrum*

Fig. 4.21 Bifoliately Compound leaf (Palmate)

Opposite : Two leaves are present at each node on opposite sides.

Decussate : The leaves are opposite and the leaves at successive nodes are at right angles.

Superposed : Pairs of leaves at successive nodes are on the same plane.

Whorled : Several leaves are present at each node.

Ternate : Three leaves at every node.

Distichous : Leaves are two ranked. Angle of divergence 180 degree.

Tristichous : Leaves are three ranked. Angle of divergence 120 degrees.

Pentastichous : Five ranked leaves. Angle of divergence 144 degrees. Every 5th leaf is on the same plane.

Octostichous : 8th ranked. Angle of divergence 135 degrees. Every eighth leaf is on the same plane.

Parastichous : Arrangement is irregular.

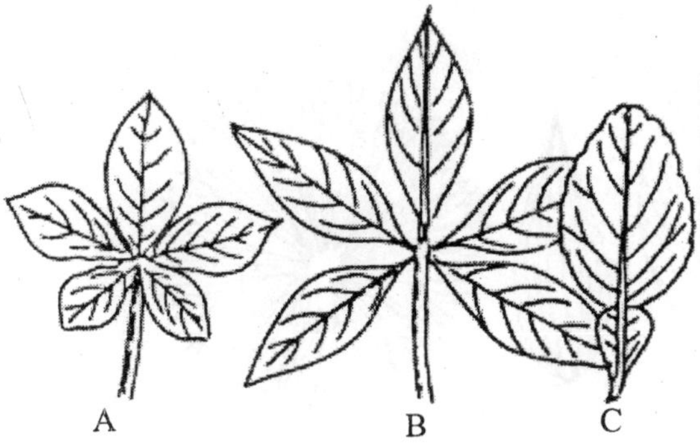

Fig. 4.21 Palmately Compound leaves
A - B - Digitately Compound leaves of **Gynandropsis** and **Bombax.** C. Unifoliate leaf of citrus

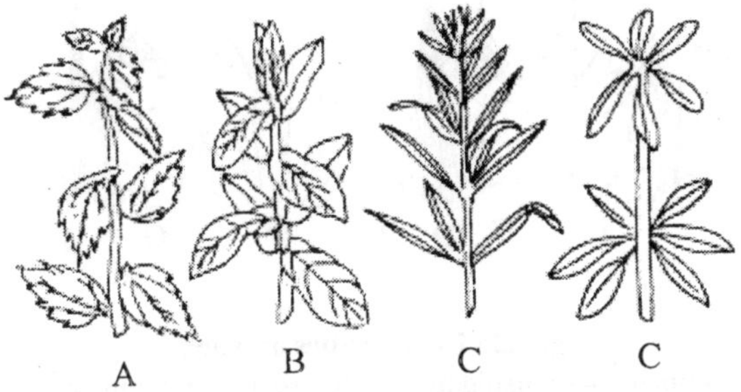

Fig. 4.22 Leaf - Phyllotaxy
A- Aiternate, B. Opposite decussate, CD - Whorled arrangement.

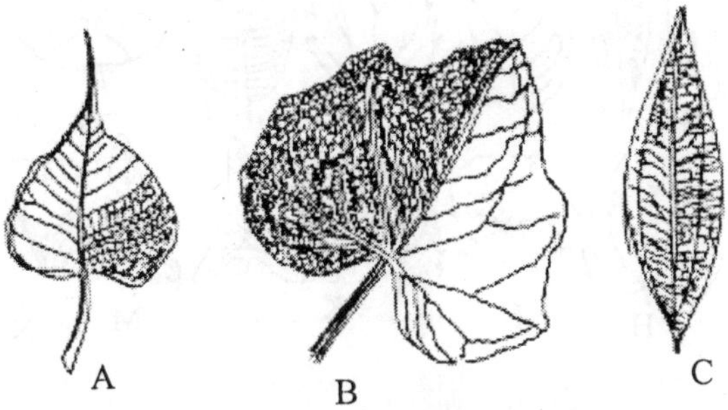

Fig. 4.23 Types of recticulate Venation

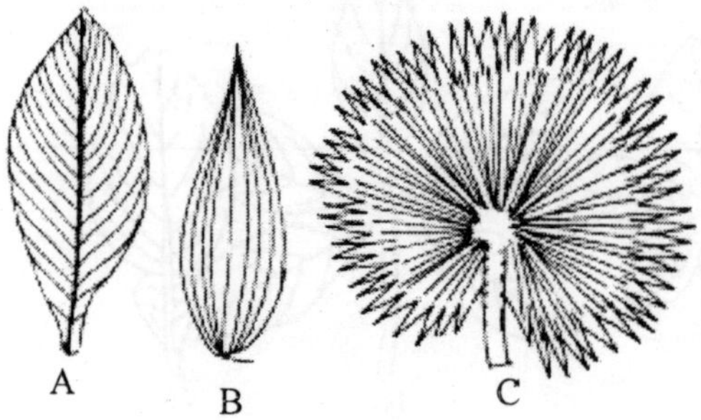

Fig 4.24 Leaf - Types of parallel venation
A- pinnate, B - Palmate Convergent, C - Palmate divergent.

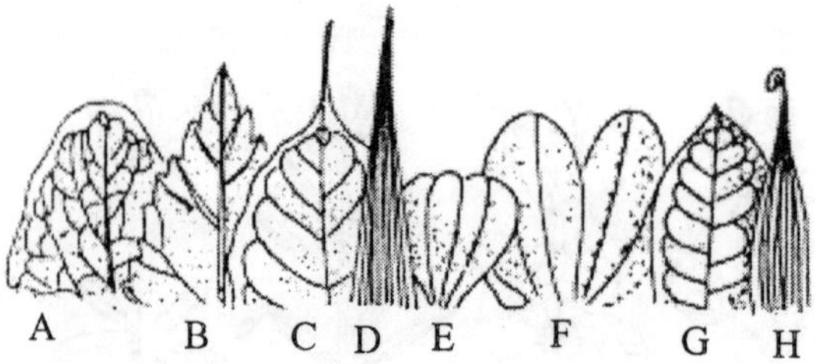

Fig. 4.25 Leaf - Types of Apex
1- Obtuse, 2 - Acute, 3 - Acuminate, 4 - Cuspidate, 5 - Retuse, 6 - Emarginate, 7. Mucronate, 8 - Cirrhose.

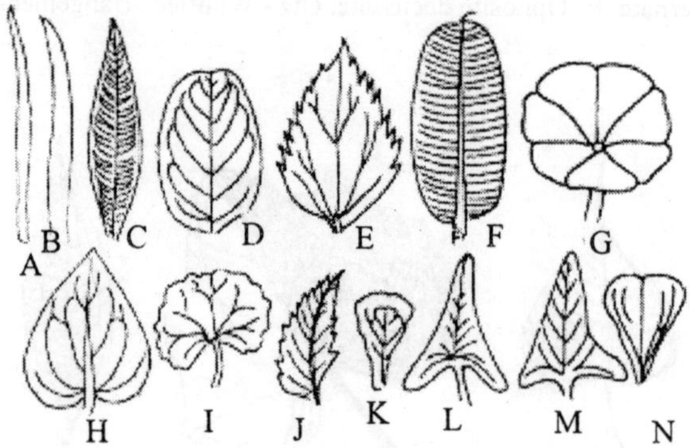

Fig. 4.26 Leaf - Various shapes
1. Acicular, 2. Linear, 3. Lanceolate, 4. Elliptical, 5. Ovate, 6. Oblong, 7. Rotund, 8. Cordate, 9. Reniform, 10. Oblique, 11. Spathulate, 12. Sagittate, 13. Hastate, 14. Cuneate.

Stipules

Stipule is a small leafy appendage present at the base of the leaf. It protects the leaf in the bud condition.

Stipulate : Leaf has a stipule
Exstipulate : Leaf has no stipule
Free lateral : Two independent stipules on either side of the leaf base.
Adnate : Two lateral stipules fused with petiole for some distance.
Interpetiolar : Stipule present in between the petiole of opposite leaves.
Intrapetiolar : Stipule pres`ent between the petiole and the stem axis.
Ochreate : Stipules form a hollow tube covering the stem for some distance from the node.
Foliaceous : Stipules very broad and leaf like.
Scaly : Papery thin stipules.
Tendrillar : Stipules modified into tendrils.
Spinous : Stipules modified into spines

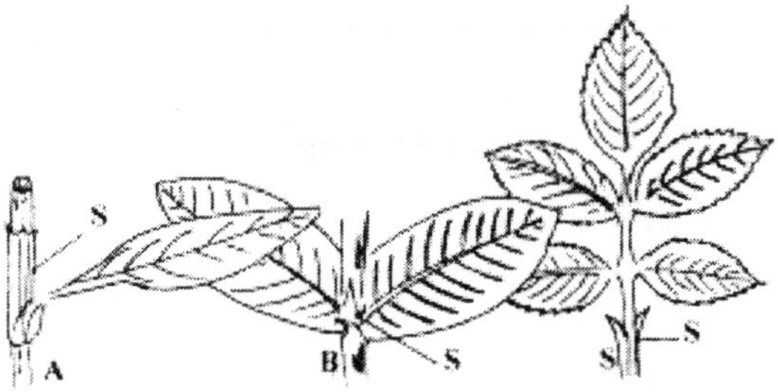

Fig. 4.27 Types of Stipules

Petiole

Petiole is the stalk of the leaf that joins the lamina with the axis.

Petiolate : Leaf has a petiole
Sessile : Leaf has no petiole
Striate : Petiole has longitudinal lines
Filiform : Elongated and slender petiole
Grooved : Petiole has a linear furrow.

Types of leaves

Leaves are generally classified into simple and compound leaves, based on the presence of one or more than one leaf blades.

Simple leaf : Leaf having single leaf blade.
Compound : Leaf has many leaf blades.
Rachis : Petiole of a compound leaf.
Leaflet : One of the leaf blases of a compound leaflet. It has no axillary bud.
Pinna : Same as a leaflet.
Pinnately : Compound leaf having pinnae throughout the rachis.

Compound		
Palmately	:	Pinnae arising from the top point of the rachis.
Compound		
Unipinnate	:	Rachis unbranched; leaflets arise directly on the rachis.
Bipinnate	:	Leaflets arise on the secondary rachis (first branch of the rachis)
Tripinnate	:	Leaflets arise on the secondary branch (Teritary rachis) of the rachis.
Decompound	:	Rachis branched many times. Leaflets on the ultimate branches.
Paripinnate	:	Compound leaf, ending in two pinnae.
Imparipinnate	:	Compound leaf ending with a single leaflet (Pinna)
Unifoliate	:	Palmately compound leaf having a single leaflet.
Bifoliate	:	Palmately compound leaf having two leaflets.
Trifoliate	:	Palmately compound leaf having three leaflets.
Quadrifoliate	:	Palmately compound leaf having four leaflets.
Multifoliate (Digitate)	:	Palmately compound leaf having five or more leaflets.
Pinnule	:	A secondary leaflet in a pinnately compound leaf.

Leaf shape

The leaves are of the following shapes:

Acicular	:	Needle like
Linear	:	Longer than broad.
Lanceolate	:	Lance shaped
Subulate	:	Awl shaped;tapering from base to apex.
Ovate	:	Egg shaped
Obovate	:	Inversely ovate; narrow towards the base.
Cordate	:	Heart shaped
Elliptical	:	Ellipse like; long circle shaped
Spathulate	:	Spoon shaped
Falcate	:	Sickel shaped
Cuneate	:	Wedge shaped; triangle like with pointed end towards base.
Reniform	:	Kidney shaped
Sagittate	:	Arrow head shaped
Cylindrical	:	Like a cylinder
Hastate	:	Arrow shaped with basal lobes turned outwards
Lyrate	:	Lyre shaped (Violin shaped)
Peltate	:	Shield shaped.

Leaf Margins

Entire	:	Uniform without notches
Undulate	:	Wavy margin.
Glandular	:	Sticky gland hairs present
Ciliate	:	Hairy margin
Serrate	:	With sharp teeth as is saw
Dentate	:	Tooth like projections present
Spiny	:	Pointed, spine like outgrowths present
Clefted	:	Divided into lobes separated by narrow sinuses.
Crispate	:	Curled

Fimbriate : Margin has long and slender projections
Inflexed : Margins turned inwards
Palmatifid : Cut about halfway as in a palmately compound leaf
Pinnatifid : Divided in a pinnate fashion.
Revolute : Margin rolled towards lower side (backwards).

Apex of Leaves

Acute : Ends into a sharp point with tapering sides
Acuminate : Tapers into a long narrow point
Obtuse : The tip is blunt and rounded
Mucronate : The tip abruptly terminates into a point
Retuse : It is like obtuse, but with a notch at the centre
Truncate : The tip ends as if it is cut transversely
Emarginate : A distinct notch is present at the tip
Cirrhose : The apex is tendrillar
Aristate : A stiff bristle present at the tip.
Caudate : The tip has a tail like appendage.

Leaf Bases

Acuminate : A prolonged and tapering base
Acute : Sharply pointed base
Auriculate : Leaf base has an ear like appendage
Cuneate : Leaf base is wedge shaped
Oblique : One side of the leaf base is larger than the other (unequal)
Obtuse : Blunt or rounded base

Vernation

The arrangement of leaves in the bud condition is of the following types:
Circinate : Growing tip coiled, with the youngest portion near the centre of the coil.
Complicate : The leaves in the bud are folded upon themselves
Conduplicate : Leaves folded together lengthwise
Convolute : Margins rolled; one inside and one outside.
Involute : Rolled in from the edges (both the margins)
Plicate : Folded on the ribs in the fashion of a hand fan (folded)
Revolute : Both margins rolled towards the mid rib on the lower surface.

Venation

The arrangement of the veins in the leaf is called venation. The following types of ventations are seen in the leaves:
Parallel : Veins unbranched. Arranged in parallel lines along the length of the leaf.
Parallel unicostate : There is a central mid rib from which arise veins all along the length of the mid rib.
Parallel multicostate : Many midribs. Parallel veins arise from the base of the multicostate leaf.

Reticulate : Venation in the form of network.

Reticulate unicostate : Single midrib. Veins arise all along the length of the midrib.

Reticulate multicostate : Many midribs arise from the base.

Reticulate multicostate Divergent : Many midribs diverging from the centre of the leaf

Reticulate multicostate Convergent : Many midribs converging towards the centre of the leaf.

Surface and Texture of leaf

Glabrous . Smooth
Pubsecent : Hairy outgrowths present.
Cereous : Waxy and shiny surface
Coriaceaous : Leathery
Corneous : Having a horny texture
Hyaline : Thin and transparent
Succulent : Juicy and felshy
Barbed : Hook (Retrose) like projections present
Ciliate : Small hairs present
Comose : Thin tuft of fine hairs present
Corrugate : Wrinkled surface
Echinate : Prickles present
Glochidiate : Surface with barbed hairs
Hirsute : Moderately sitff hairs
Hispid : Bristly haired
Lusturous : Glossy, Shiny surface
Punctate : Dotted with punctae (Depressions)
Stellate : Covered with branched hairs (branching is star like)
Scurvy : Small sclaes present
Tomentose : Densley woolly.
Velutinose : Velvety, having fine hairs.
Verrucose : Surface covered with wart like elevated portions
Lanate : Surface covered with matted hair.

INFLORESCENCE

An inflorescence may be defined as a collection or grouping of many flowers into a cluster.

Acropetalous : A type of growth; older flower towards the base,younger ones towards the tip.
Peduncle : The axis of the Inflorescence
Scape : A leafless peduncle arising from the ground
Rachilla : Secondary axis
Solitary : Single flower
Basipetalons : A type of growth older flower towards the tip younger ones towards the base.

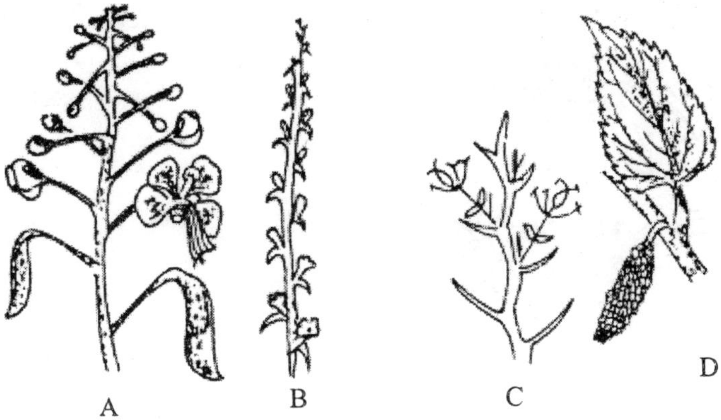

Fig. 4.28 Inflorescence, Racemose types
A. Simple raceme, B. Spike, C. Spikelet, D. Catkin.

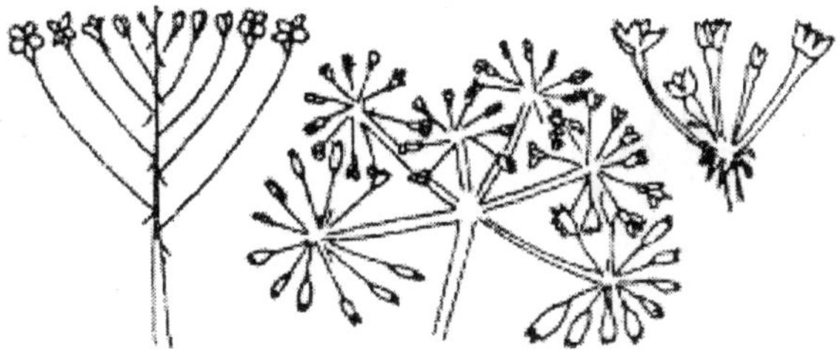

Fig. 4.29 Inflorescence types contd.
A. Corymb, B. Compound umbel, C. Simple umbel (All diagrammatic_

Terminal	:	Flower terminating the axis
Axillary	:	Borne in the axil of a leaf.
Racemose	:	An indeterminate type of inflorescence with flowers borne on acropetalous succession.
Simple raceme	:	Flowers borne directly on the peduncle in a racemose inflorescence.
Compound raceme (Panicle)	:	Branched raceme
Corymb	:	Racemose inflorescence in which older flowers have longer pedicels and younger flowers have shorter pedicels.
Spike	:	Like a raceme; flowers are sessile.
Spilelet	:	A secondary spike; a unit of compound inflorescence in grasses.
Catkin	:	A spike with a slender axis.

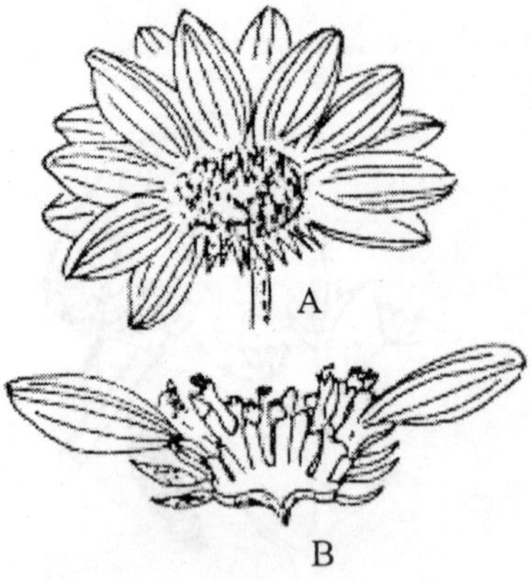

Fig. 4.30 Inflorescence - Capitulum
A entire, B. Long sectional view.

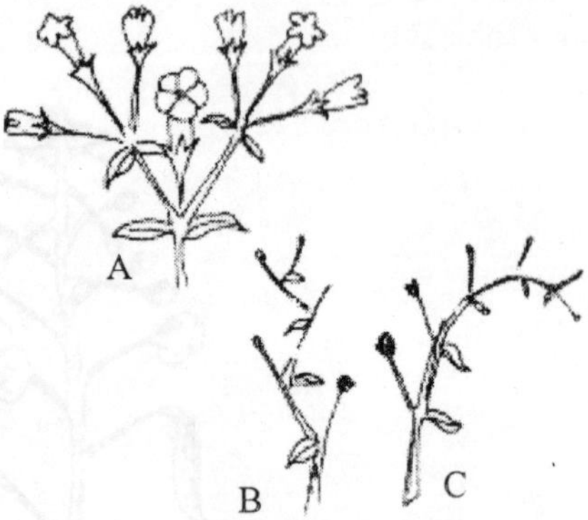

Fig. 4.31 Inflorescence - Cymose types
A. Biparous Cyme, B. Scorpioid Cyme, C. Helicoid Cyme.

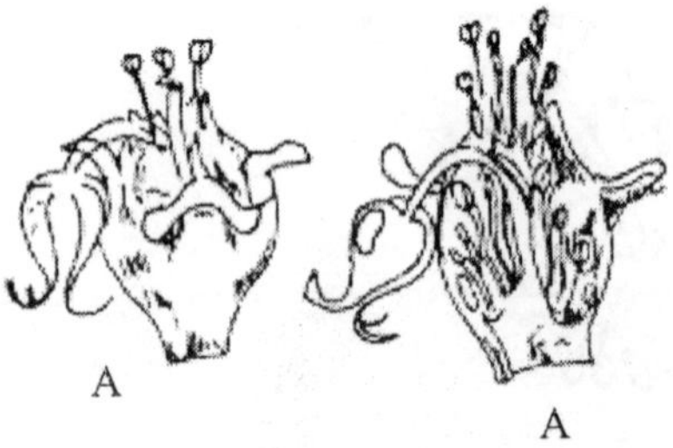

Fig. 4.32 Inflorescence - ***Cyathium***
A. entire, B. Long sectional view.

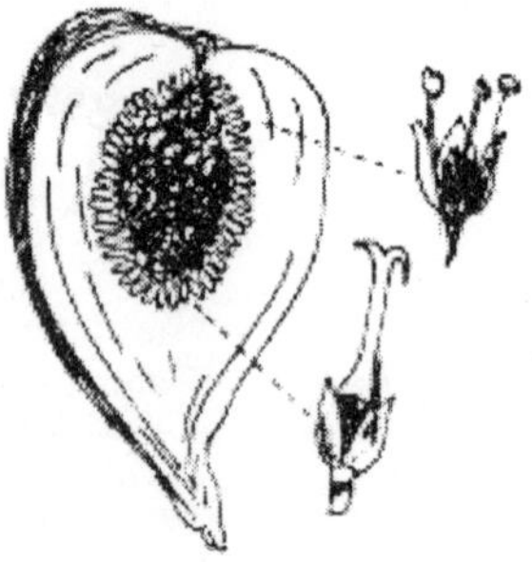

Fig. 4.33 Inflorescence - ***Hypanthodiium***
Male (above) and female (below) flowers are enlarged on the left side

Compound spike	:	A branched spike
Spadix	:	A fleshy spike subtended by a bract called the spathe
Umbel	:	A racemose inflorescence in which the flowers are borne in a cluster i.e., (simple) many flowers arise from the same point reaching the same height.
Compound umbel	:	A branched umbel
Head (Capitulum)	:	The tip of the peduncle is flattened into a receptacle on which the flowers are borne. The flowers are called florets.

Homogamous : Head inflorescence having only one type of florets ie. all the florets are bisexual

Heterogamous : Two types of florets are seen in the inflorescence viz., Ray florets (unisexual) and disc florets (bisexual).

Cymose : A determinate type of inflorescence in which flowers are borne in basipetalous succession.

Simple cyme : A cyme with only three flowers.

Monochacial Cyme : Flowers produced towards only one side of the axis.

Helicoid Cyme : A type of monochacial Cyme in which flowers are always towards one side.

Scorpioid Cyme : A type of monochacial Cyme in which flowers are produced alternately towards left and right.

Dichacial Cyme : Cymose inflorescence with branching on both sides.

Rhipidium : A fan shaped monochacial Cyme.

Cincinnus : A single branched monochacial Cyme.

Multichacial Cyme : A cymose infloresecene with a whorl of branches.

Verticellaster : A mixed cyme. First branching is dichacial: subsequent ones are monochacial.

Cyathium : A speical cyme having achlamydeous flowers. Seen in the genus *Euphorbia*; Flowers are unisexual.

Thyrsus : A mixture of raceme and Cyme. Main inflorescence is raceme. Branches are Cymose.

Hypanthodium : Inflorescence of Fig. Receptacle is cup like with an opening called ostiole.

FLOWER

A flower may be defined as a vegetative shoot modified for reproduction.

Pedicel : Stalk of the flower.

Thalamus (Receptacle) : Tip of the pedicel bearing floral parts.

Cotyloid Thalamus : Convex thalamus

Accessory organs : Non-essential organs; Calyx and corolla.

Essential organs : Androecium and Gynoecium.

Pedicellate : Flower having a stalk

Sessile : Stalkless flower

Bract : A leafy appendage protecting the floral bud

Bracteate : Flower with a bract

Bracteole : A small appendage below the Calyx

Bracteolate : Flower having a bracteole

Involucre : A whorl of bracts

Glume : Sterile bracts found in grasses

Complete : When a flower has all the organs (both accessory and essential)

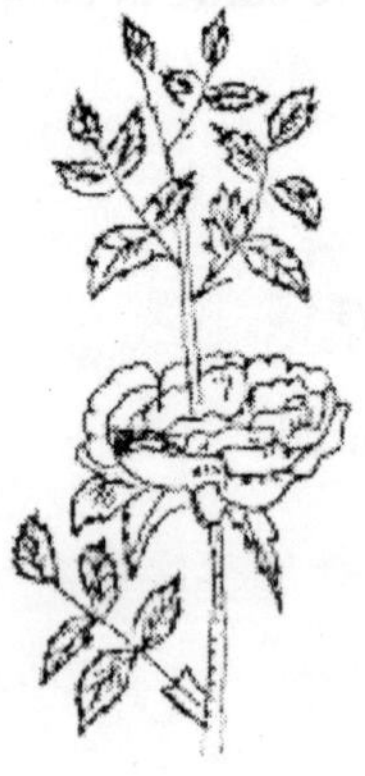

Fig. 4.34 Flower
Abnormal growth of vegetative shoot from the centre of the flower in Rose.

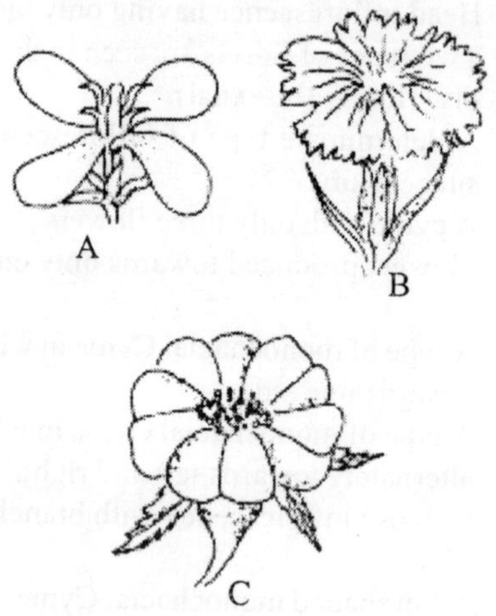

Fig. 4.35 Flower, Types of corolla
A. Cruciform, B. Caryophyllaceous, C. Rosaceous.

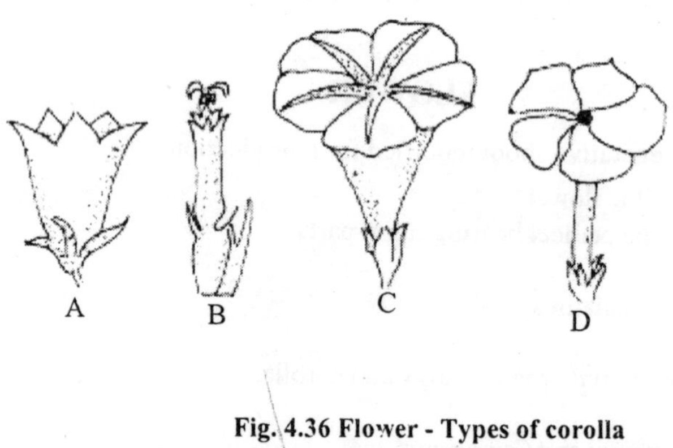

Fig. 4.36 Flower - Types of corolla
A. Campanulate, B. Tubular, C. Funnel shaped, D. Rotate

Incomplete	:	Any one of the floral organs is missing
Heterochalmydeous	:	Accessory organs present and distinguishable into calyx and corolla.
Homochlamydeous	:	Accessory organs present, but indistinguishable into Calyx and Corolla
Perianth (Tepal)	:	Accessory organs of a homochlamydeous flower.
Achlamydeous	:	A flower in which accessory organs are absent.
Acyclic	:	Floral organs arranged in a spiral
Cyclic (Whorled)	:	Floral organs arranged in a definite circle
Hemicyclic	:	Some organs are cyclic, some non cyclic
Dimerous	:	Two members in each whorl

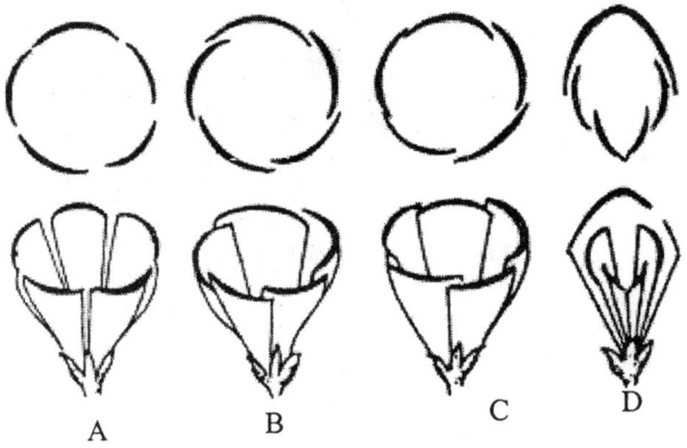

Fig. 4.37 Flower - Types of aestivation
A. Valvate, B. Imbricate, C. Ascendingly imbricate D. Vexillary

Trimerous	:	Floral organs in three or multiples of three
Tetramerous	:	Floral parts in four or multiples of four
Pentamerous	:	Floral parts in five or multiples of five
Bisexual (Hermaphrodite)	:	A flower having both male and female sexes

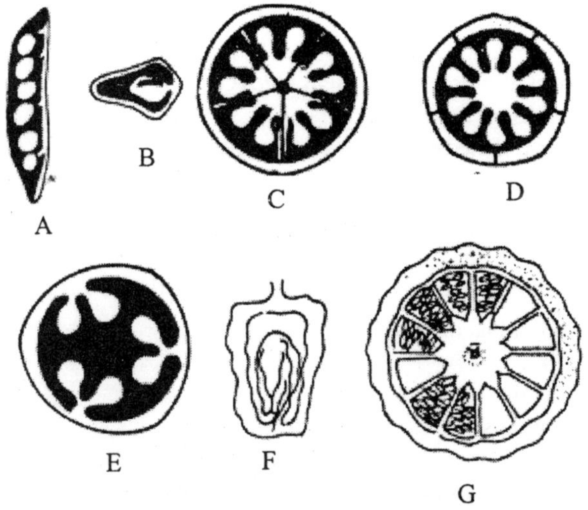

Fig . 4.38 Types of Placentation
A. Marginal (LS), B. Marginal (T.S), C. Axile (T.S), D. Free central (T.S.), E. Parietal (T.S.), F. Basal (L.S.), G. Superficial (T.S)

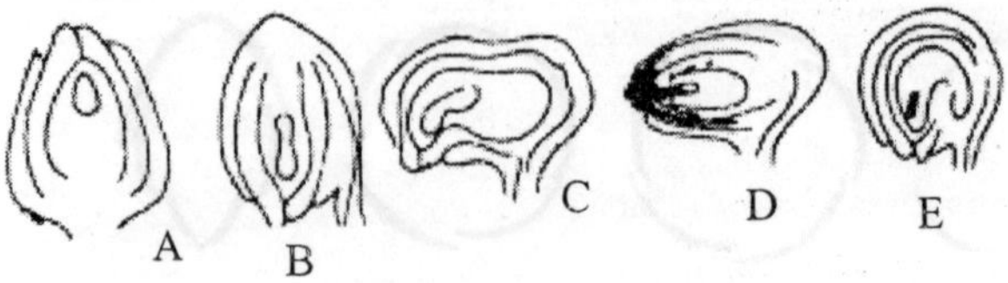

Fig. 4.39 Types of Ovules.

Unisexual	:	Only one sex is present in the flower
Pistillate	:	Female flower
Staminate	:	Male flower
Hypogynous	:	Flower with a superior ovary
Epigynous	:	Flower with an inferior ovary
Perigynous	:	Flower with a semi inferior ovary
Actinomorphic	:	Flower with radial symmetry (Regular)
Zygomorphic	:	Flower with bilateral symmetry (Irregular)
Monoecious	:	A plant having both male and female flowers
Dioecious	:	A plant having either male or female flowers
Polygamous	:	Plant having unisexual (male and female) as well as bisexual flowers

Calyx

Calyx is the lowermost whorl of floral leaves. It is generally green in colour

Sepal	:	The individual member of the Calyx
Polysepalous	:	Sepals free
Gamosepalous (Synseplaous)	:	Sepals fused
Epicalyx	:	A Whorl of bracteoles present below the Calyx
Caducous	:	Calyx falls off early
Persistant	:	Calyx is retained even after fertilisation

Corolla

This is the whorl of accessory organs next to Calyx. Its normal function is attraction of pollinating agents.

Petal	:	The individual member of the Corolla
Petaloid	:	Petal like.
Polypetalous	:	Free petals
Gamopetalous (Sympetalous)	:	Fused petals
Aestivation	:	Arrangement of petals (sepals also) in bud
Valvate	:	Arrangement without overlapping
Reduplicate	:	Valvate with margins rolled outwards
Imbricate	:	One in, one out, rest in and out
Quincouncial	:	2 in, 2 out, one in and out
Twisted	:	All are in and out

(Convolute)
Corrugate : Crumpled
Plicate : Folded lengthwise
Vexillary : 5 petals presnt, posterior outermost, anterior inner most, laterals in and out

Shape of Corolla

Cruciform : Arranged in the form of a cross, Each petal has a limb and claw.
Caryophyllaceous : Five petals with long claws and short limbs, which are placed at right angles to the claws
Rosaceous : Petals not distinguishable into limbs
Papilionaceous : Five petals, A large standard, two wings and two keel petals.
Campanulate : Bell shaped Corolla
Tubular : Corolla is tube like
Bilabiate : Corolla fused to form two lip like stuructures
Infudibuliform : Corolla is funnel like; petals five in number
Personate : Bilabiate corolla with pronounced plate
Rotate : Wheel shaped; circular and flat
Hypocrateriform : A slender tube and suddenly expanding top portion
Calcarate : Having spurs
Ligulate : Strap shaped
Urceolate : Urn shaped; like a pitcher.

Appendages of Corolla

Nectary : Nector secreting gland
Spur : A sac like projection; it may secrete nectar
Corona : An appendage or outgrowth in between petals and stamens
Gibbous : Swollen towards one side

Androecium

This is the whorl of male reproductive structures.

Stamen : The individual member of the whorl of androecium
Androphore : A stalk, bearing andreocium
Staminode : Sterile stamen
Pollinium : A mass of pollen grains
Filament : Stalk of the stamen
Anther : Terminal part of stamen bearing pollen grains
Connective : The part of the stamen connecting the anther and the filament
Polyandry : Indefinite number of stamens
Polyadelphous : Stamens grouped into many bundles
Diadelphous : Stamens grouped into two bundles
Monadelphatus : Stamens grouped into a single bundle
Syngenesious : Filaments free; all the anthers fused
Connivent : Stamens very close but not united
Synandry : All stamens completely fused (both filaments as well as anthers)

Tetra dynamous	:	Stamens six, four long and two short.
Didynamous	:	Stamens four, two longand two short
Gynandrous	:	Stamens fused with the carpels partly or wholly
Epipetalous	:	Stamens arising on the corolla
Epipetalous	:	Stamens arising on the perianth.
Diplostemnous	:	Stamens in two whorls, twice as many as petals. Outer whorl alternate with petals
Obdiplostemnous	:	Stamens in two whorls. Outer whorl opposite with the petals.
Inserted	:	Stamens as growing out of the corolla tube.
Exerted	:	Stamens projecting beyond corolla
Included	:	Stamens not projecting beyond corolla
Basifixed	:	Filament attached to the base of the anther
Dorsifixed	:	Filament attached to the back of the anther
Versatlie	:	Filament attached to the back of the anther at a central point so that the anther can freely oscillate in air
Adnate	:	Fillament attached to the anther throughout its length
Introrse	:	Anther dehiscing towards the centre of the flower
Extrose	:	Anther dehiscing towards the circumference of the flower
Dithecous anther	:	Anther 2 lobed (celled)
Monothecous anther	:	Anther single lobed

Gynoecium

This is the female reproductive organ of the flower

Gynophore	:	A stalk bearing gynoecium
Gynandrophore	:	A stalk bearing both androecium and gynoecium
Pistil	:	Gynoecium having ovary, style and stigma
Pistillode	:	A rudimentary pistil
Ovary	:	The basal part of the gynoecium
Style	:	The stalk that connects the stigma and the ovary. It is through this that the pollen tube grows
Stigma	:	Terminal part of the style; serves as a lodging place for the pollen grains
Stigma capitate	:	like a globose head
Stigma feathery	:	having hairy outgrowths
Stigma bifid	:	divided into two
Stigma sticky	:	producing a gummy substance
Style hollow	:	having a central cavity
Style solid	:	central region filled up
Style terminal	:	style at the tip and in vertical line with the ovary
Style lateral	:	style at right angles to the ovary
Style gynobasic	:	style arising from the base of the ovary or from the receptacle
Stylopodium	:	A disc surrounding the base of the style
Gynostegium	:	A sheath covering the gynoecium and anthers
Gynostermium	:	Column formed by the adnation of stamens, style and stigma
Carpel	:	The unit of ovary; megasporophyll
Monocarpellary	:	Ovary with one carpel
Bicarpellary	:	Ovary with two carpels

Tricarpellary	: Ovary with three carpels
Multicarpellary	: Ovary with many carpels
Syncarpous	: Carpels fused
Apocarpous	: Carpels free
Locule	: Chamber or partition within the ovary
Unilocular	: Only one locule in ovary
Bilocular	: Two locules in the ovary
Trilocular	: Three locules in the ovary
Multilocular	: Many locules in the ovary
Resupinate	: Ovary twisted 180 degrees (as in orchids)
Placenta	: A tissue attaching the ovule with the ovary
Placentation	: Mode of attachment of ovule to the placenta
Ovule	: A megasporangium, containing the embryosa covered by one or two integuments
Unitegmic	: One integument in the ovule
Bitegmic	: Two integuments in the ovule
Orthotropous	: Erect ovule with straight axis
Anatropous	: Ovule bent 180 degrees. Micropyle and funciculus side by side
Hemianotropous	: Ovule bent 90 degrees
Amphitropous	: An ovule with a curved funiculus so that the tip of the ovule and stalk are near each other.
Campylotropous	: An ovule, curved like a horse shoe very much similar to Amphitropous
Circinotropous	: An ovule taking a curvature of 360 degrees with micropyle pointing up as in orthotropous.

Types of placentation

Axile	: Ovules attached to a central axis
Parietal	: Ovules attached to the peripheral region of the ovary
Basal	: Single ovule arising from the base of the ovary
Freecentral	: Septa fusing to form a central hollow column with ovules attached to it
Apical	: One or two ovules arising from the top of the ovary
Lamellate	: A modified parietal placentation. Ovules are borne on lamellae (plates) within the ovary

FRUIT

A fruit is a fertilised ovary. After fertilisation the ovary ripens and develops into a fruit. The ovules inside become seeds

True fruit	: Fruit developed from the ovary
False fruit	: Any part of the flower, other than the ovary, becoming swollen and looking like a fruit
Simple fruit	: One flower developing into a single fruit
Multiple fruit	: A group of fruits developing from several Gynoecia of several flowers (may also be called compound fruit)

Aggregate : Clustering of many fruits, each one developing from *fruits* many gynoecia of a single flower.

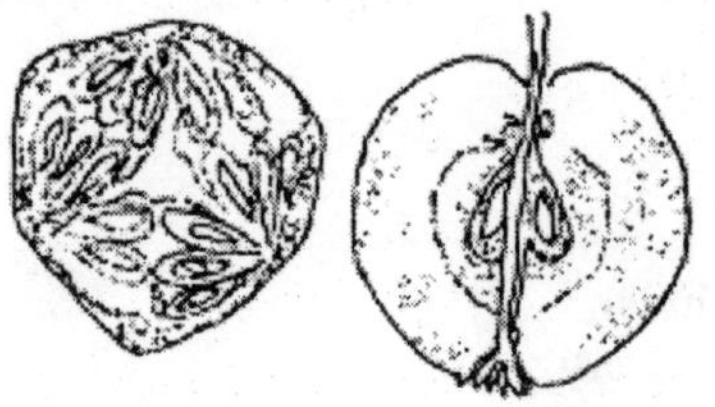

Fig. 4.40 Fleshy fruits
Pepo in cross section (left) Pome in long section (right)

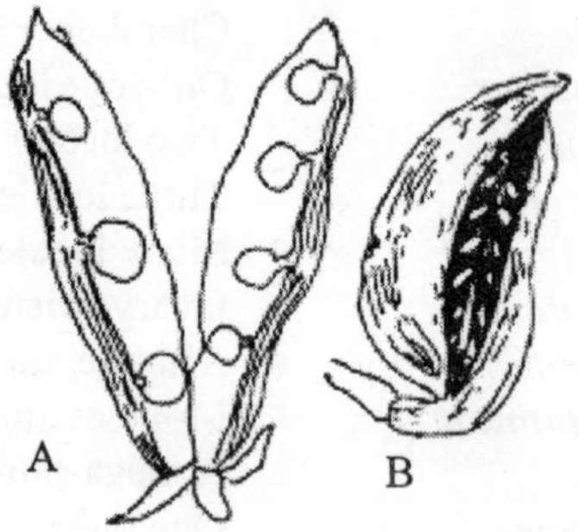

Fig. 4.41 Dry dehiscent fruits
A. Legume, B. Follicle

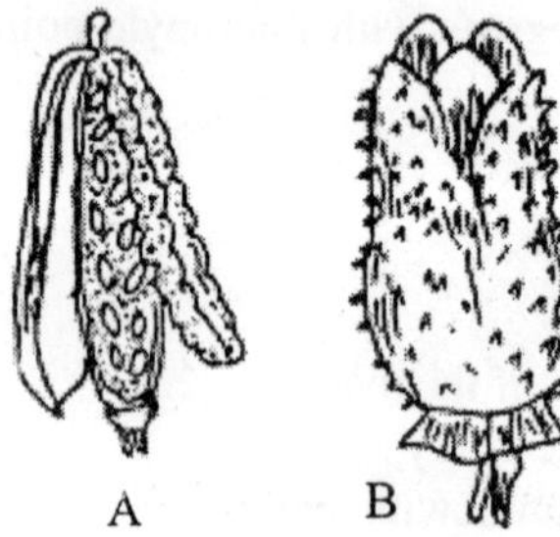

Fig. 4.42 Dry dehiscent fruits.
A. Siliqua, B. Capsule.

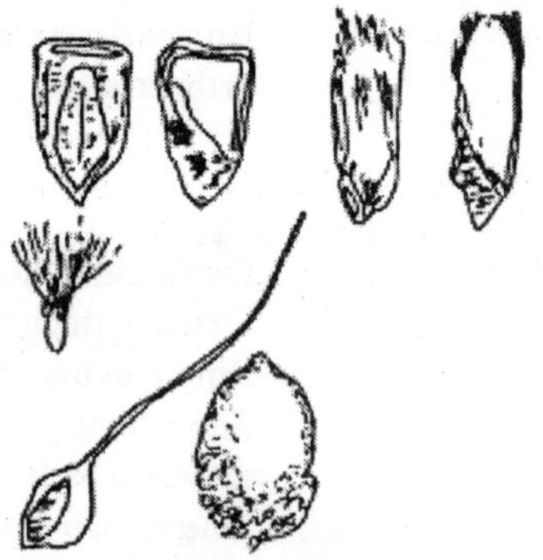

Fig. 4.43 Dry indehiscent fruits
Achene, caryposis, etc.

Simple Fruits

Fleshy fruits : Fruits with a fleshy juicy ovary
Berry (becca) : A fleshy, idehiscent fruit developing from many carpelled ovary.
Hesperidium : A berry with a leathery epicarp (rind)
Drupe : A fleshy fruit with one seed, enclosed in a hard endocarp
Drupelet : A drupe in an aggregate of drupes
Pyrene : The nutlet in the drupe (seed and the hard endocarp)
Pome : A fleshy fruit, with a swollen edible thalamus
Pepo : A pulpy fruit developing from an inferior many carpellary ovary

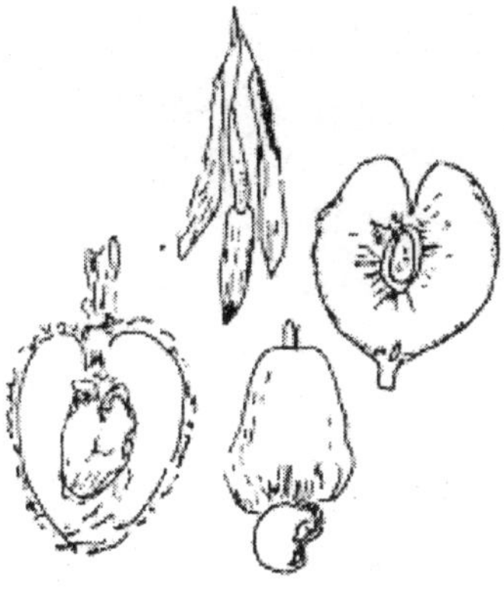

Fig. 4.44 Dry Indehiscent fruits
Winged fruits and Nut. (Anacardium)

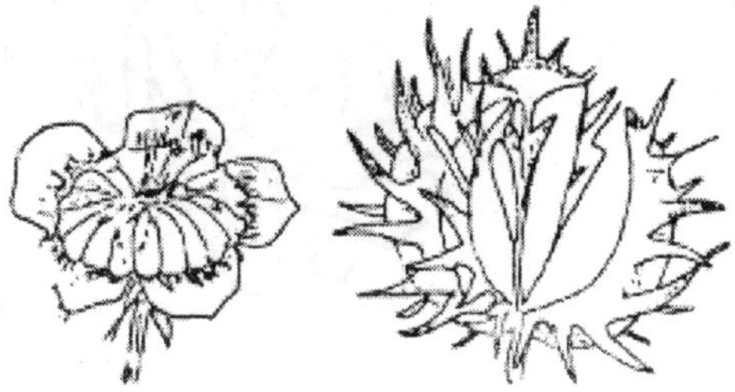

Fig. 4.45 Schizocarpic fruits
Left Carcerula. Right Regma.

Balusta : A fleshy fruit with a firm rind and with a corona of persistant calyx
Dry fruits : Fruits with a dry fruit wall
Dry dehiscent : Dry fruits which break open at maturity exposing the seeds
Legume : A dry dehiscent fruit developing from a monocarpellary ovary; opens on both valves.
Siliqua : A dry dehiscent fruit opening on both valves with seeds attached to a central replum

Fig. 4.46 Aggregate fruits
Etaerio of drupes, Left entire, right, long section.

Silicula : A short siliqua containing but few seeds
Follicle : A dry dehiscent fruit formed from a single carpel splitting along only one valve
Capsule : A dry dehiscent fruit formed from a multicarpellary superior ovary splitting in various ways
Pyxis : A type of Capsule with an upper lid and circumscissile dehiscence
Utricle : A single seeded fruit with a thin wall. A dehiscent fruit.

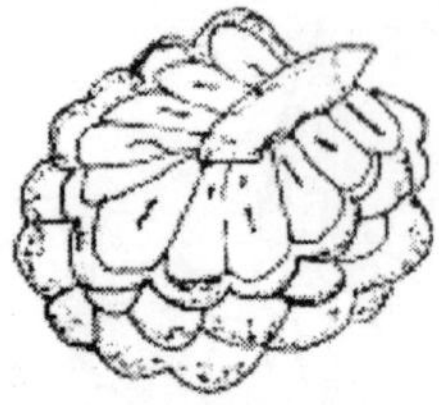

Fig. 4.47 Aggregate fruits
Etaerio of achenes (left) Etaerio of berries (right)

Dry Indehiscent	:	Fruit that does not break open even after maturity
Achene	:	A hard one seeded fruit, in which the ovary wall is free from the seed coat
Caryopsis	:	A hard dry indehiscent fruit in which the seed coat and fruit wall are fused
Cypsela	:	A dry indehiscent fruit devloping from an inferior ovary
Nut	:	A dry indehiscent fruit with a bony shell derived from a compound syconium but has only one seed

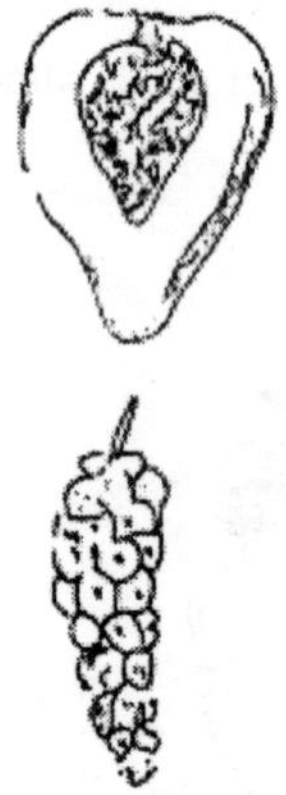

Fig. 4.48 Multiple fruits - syconus
(Top); Sorosis of Mulberry (below)

Fig. 4.49 Multiple fruits
Sorosis of Pineapple

Nutlet	:	A small nut
Schizocarpic fruit	:	A type of fruit in between dehiscent and indehiscent. The fruit breaks open into one seeded bits, but the seeds are intact within the locule wall.
Lomentum	:	A leguminous fruit dehiscing into one seeded bits

Cremocarp	:	A dry dehiscent fruit breaking into two one seeded mericarps
Samara	:	A winged, one seeded dry indehiscent fruit
Regma	:	A dry, splitting fruit with 3 or more elastic segments or cocci
Carcerula	:	It develops from a bi or multicarpellary ovary. The fruit opens into 4 or more bits
Cyamium	:	A type of follicle much like a legume
Cynarrhodion	:	A hollow fruit (like rose) enclosing many achenes
Schleranthium	:	An achene enclosed in a persistent calyx tube

Aggregate fruits

Etareio of follicles	:	Aggregate of follicles
Etareio of achenes	:	Aggregate of achenes
Etareio of Drupes	:	Aggregate of drupes
Etareio of berries	:	Aggregate of berries
Conocarpium	:	Aggregate of fruits on a conical receptacle

Multiple fruits

Sorosis	:	A flesy multiple fruit (Ex-mulberry)
Syconium	:	A fruit developing from a hypanthodium inflorescence as in fig.
Pericarp	:	Wall of the fruit (ovary)
Epicarp (Exocarp)	:	Outer layer of pericarp
Mesocarp	:	Middle layer of pericarp
Endocarp	:	Inner layer of pericarp
Replum	:	False septum found in a siliqua
Valve	:	A part of the fruit that gets separated from other parts
Suture	:	The line marking the dehiscence of the fruit

Dehiscence of fruits

Loculisidal	:	Dehiscence of a capsule in between the two septa.
Septisidal	:	Dehiscence of a capsule along the septa
Poricidal	:	Dehiscence through pores
Circumscissile	:	Opening by a transverse round split so as to re lease a lid like structure

SEED

Albuminous	:	Endospermous seed
Exalbuminous	:	Non-endospermous seed
Testa	:	Seed coat (outer)
Tegmen	:	Seed coat (inner)
Endosperm	:	Nutritive tissue present in albuminous seeds
Embryo	:	The epitome of the plant within a seed
Funciculus	:	Stalk of the seed
Raphe	:	The vestige of the funiculus on the seed
Hilum	:	Scar or point of attachment of funiculus to the seed

Micropyle	:	A small opening that helps in the exit of the radicle during germination
Chalaza	:	The part of the ovule (seed) which is exactly opposite to the micropyle
Aril	:	The third integument. An additional envelope on the seed
Caruncle	:	A protruding growth near the hilum
Cotyledon	:	Embryonic leaf (Seed leaf)
Strophiole	:	An appendage of the hilum seen in certain seeds

POLLINATION

It is the act of transference of pollen grains from the anther to the stigmatic surface.

Self Pollination	:	Pollen grains of a flower transfered on to the stigma of the same flower (Autogamy)
Cross pollination	:	Pollen grain of a flower transferred to the stigma of another flower. (Allogomy)
Xenogamy	:	Cross pollination in dioecious plants
Cleistogamy	:	Forced self-pollination due to the failure of the flower to open
Dichogamy	:	Sexes maturing at different intervals
Protandry	:	Androecium maturing early
Protogyny	:	Gynoecium maturing early
Dicliny	:	Separation of anther and stigma in space
Herkogamy	:	See dicliny
Heterostyly	:	The difference in style length and filament prevents self pollination
Self-sterility	:	Pollen of a flower failing to germinte on the stigma of the same flower

CHAPTER 5

DERIVATION OF FAMILIES

ANNONACEAE

1. Leaves reticulately veined; vascular bundles in the stem arranged in a ring; flowers tetra or pentamerous (rarely trimerous) .... *Dicotyledons*
2. Flowers generally heterochlamydeons; petals free .... *Polypetalae*
3. Flowers hypogynous; bisexual and actinomorphic; thalamus dome shaped .... *Thalimiflorae*
4. Stamens indefinite; carpels free .... Ranales
5. Trees or shrubs; flowers trimerous; leaves exstipulate, connective of anthers elongated .... *Annonaceae*

MAGNOLIACEAE

1-4 Same as in Annonaceae

5. Trees or shrubs; flowers trimerous; leaves stipulate but caducous leaving a scar on the petiole; connective of the anther not enlarged ....*Magnoliaceae*

RANUNCULACEAE

1-4 Same as in Annonaceae

5. Mostly herbs; flowers tetra or pentamerous; ovary superior; carpels many, apocarpous .... *Ranunculaceae*

PAPAVARACEAE

1-3 Same as in Announaceae

4. Ovary superior; carpels syncarpous; locule one; placentation parietal .... *Rhoedales*
5. Flowers zygomorphic, pentamerous; anterior petal spurred; fruit a loculicidal capsule ...*Papavaraceae*

CRUCIFERAE

1-3 As in Annonacene

4. As in Papavaraceae
5. Leaves alternate, exstipulate; petals clawed, cruciform; stamens usually tetradynamous. Ovary unilocular becoming bilocular due to septation; fruit a siliqua .... *Cruciferae*

CAPPARIDACEAE

1-3 as in Annonaceae

4. as in Papavaraceae
5. Leaves alternate, sepals 4, petals 4, gynandrophore or gynophore present, fruit berry or capsule .... *Capparidaceae*

VIOLACEAE

1-3 as in Annonaceae
4. as in Papavaraceae
5. Flowers zygomorphic, pentamerous; anterior petal spurred; fruit a loculicidal capsule .... *Violaceae*

CARYOPHYLLACEAE

1-3 as in Annonaceae
4. Inflorescence a cyme; ovary superior unilocular carpels syncarpous; free central placentation .... *Caryophyllineae*
5. Herbs; leaves opposite decussate; inflorescence a dichasial cyme; stamens obdiplostemnous; fruit a capsule .... Caryophyllaceae

PORTULACACEAE

1-3 in Annonaceae
4. as in Caryophyllaceae
5.

MALVACEAE

1-3 as in Annonaceae
4. Ovary multicarpellary, syncarpous, multilocular, placentation axile; stamens monadelphous and indefinite .... *Malvales*
5. Leaves stipulate; flowers pentamerous; epicalyx often present; anthers monothecous and reniform, fruit a capsule or schizocarp .... *Malvaceae*

STERCULIACEAE

1-3 as in Announaceae
4. as in Malvaceae
5. Anthers dithecous, stamens 5-10 more or less connate at the bose .... *Sterculiaceae*

TILIACEAE

1-3 as in Annonaceae
4. as in *Malvaceae*
5. Anthers dithecous, stamens many more or less;free .... *Tiliaceae*

ZYGOPHYLLACEAE

1-2 as in Annonaceae
3. Flowers hypogynous; a prominent disc is present below the ovary, ovary free or embedded in the receptacle disc .... *Disciflora*
4. Disc annular, attached to stamens; ovary superior; placentation axile, stamens twice as many as the petals .... *Geraniales*.

RUTACEAE

1-2 as in Annonaceae
3-4 as in Zygophyllaceae
5. Leaves exstipulate with aromatic oil glands; stamens free or polyadelphous; fruit a hesperidium .... *Rutaceae*

MELIACEAE

1-2 as in Annonaceae
3-4 as in Zygophyllacene
5. Trees; leaves alternate, compound and exstipulate; stamens 8-10, mostly form a staminal column; fruit a drupe .... *Meliaceae*

ANACARDIACEAE

1-2 as in Annonaceae
3 as in Zygophyllaceae
4. Disc well developed; Stamens inserted outside the disc; ovary many carpellary, syncarpous; placentation axile .... *Sapindales.*
5. Members trees; flowers polygamous; stamens very few; fruit a one seeded drupe .... *Anacardiaceae*

ROSACEAE

1-2 as in Annonaceae
3. Sepals fused, rarely free; flowers perigynous or epigynous; ovary occasionally superior also .... *Calyciflorae.*
4. Leaves alternate and stipulate; flowers peri or epigynous; carpels 1 to many .... *Rosales.*
5. Flowers actinomorphic; corolla rosaceous; Stamens numerous inflexed; carpels many, apocarpous .... *Rosaceae*

PAPILIONACEAE

1-2 asin Annonaceae
3-4 as in Rosaceae
5. Flowers zygomorphic or actinomorphic, fruit a legume, pod, or lomentum .... *Leguminosae*
6. Flowers zygomorphic, corolla papilionaceous; stamens 10, ... *Papilionaceae*

CAESALPINACEAE

1-2 as in Annonaceae
3-4 as in Rosaceae
5. as in Papilionaceae
6. Flowers zygomorphic, odd posterior petal smallest; stamens 10 free, but many of them sterile, aestivation ascendingly imbricate .... *Caesalpinaceae.*

MIMOCEAE

1-2 as in Annonaceae
3-4 as in Rosaceae
5. as in Leguminosae
6. Flowers actinomorphic, stamens 4 to many; fruit a lomentum .... *Mimoceae*

COMBRETACEAE

1-2 as in Annonceae
3 as in Rosaceae.
4. Leaves simple and entire; flowers 4 or 5 merous; ovary inferior; carpels fused; stamens free and numerous .... *Myrtales*
5. Ovary unilocular with 2-5 pendulous ovules .... *Combretaceae*

MYRTACEAE

1-2 as in Annonaceae
3 as in Rosaceae
4 as in Combretaceae
5. Flowers bisexual, leaves with pellucid aromatic glands and intramarginal venation .... *Mysrtaceae*

CUCURBITACEAE

1-2 as in Annonaceae
3. as in Rosaceae
4. Plants weak stemmed; flowers actinomorphic, unisexual, ovary tricarpellary and unilocular .... *Passiflorales*

5. Leaves simple alternate, tendrillar, flowers unisexual placentation parietal; fruit a pepo .... *Cucurbitaceae*

UMBELLIFERAE

1-2 as in Annonaceae
3. as in Rosaceae
4. Inflorescence umbel; ovary bicarpellary, syncarpous, bilocular, ovule pendulous, ovary inferior .... *Umbellales*.
5. Leaves alternate, exstipulate, decompound, stem fistular; styles two with stylopodium; fruit cremocarp .... Umbelliferae.

RUBIACEAE

1. as in Annonaceae
2. Flowers heterochlamydeous, petals united .... *Gamopetalae*
3. Ovary inferior, stamens as many as corolla lobes or flower than them .... *Inferae*
4. Leaves opposite; stamens free and epipetalous, ovary 2-8 called .... *Rubiales*.
5. Leaves opposite, with inter or intrapetiolar stipules, flowers actinomorphic, ovary bilocular .... *Rubiaceae.*

COMPOSITAE

1. as in Annonaceae
2-3 as in Rubiaceae
4. Inflorescence a capitulum, ovary unilocular, placentation basal fruit a cypsela .... *Asterales*.
5. Capitulum homogamous or hetrogamous, calyx reduced to pappus, stamens epipetalous with syngensious anthers ovary bicarpellary syncarpous .... *Compositae*

SAPOTACEAE

1. as in Annonaceae
2. as in Rubiaceae
3. Ovary superior, carpels more than two, stamens as many as corolla lobes, few .... *Heteromerae*
4. Corolla gamopetalous, stamens as many or twice the number of corolla lobes .... *Ebenales*
5. Plants laticiferous, stamens generally obdiplostemnous, carpels many, ovary superior .... *Sapotaceae*

APOCYNACEAE

1. as in Annonaceae
2. as in Rubiaceae
3. Ovary superior, carpels two, rarely less, stamens aiternipetalous and as many as petals .... *Bicarpellatae*
4. Flowers actinomorphic, stamens alternating with petals; leaves usually opposite .... *Gentianales*
5. Leaves normally opposite; latex present (watery or milky); Stamens 5, free and epipetalous; corolla has coronary outgrowths; carpels apocarpous, but united by styles .... *Apocynaceae*

ASCLEPIADACEAE

1. as in Annonaceae
2. as in Rubiaceae
3-4 as in Apocynaceae
5. Plants have milky latex; leaves opposite and decussate; inflorescence cymose; stamens 5 fused with the gynoecium to form gynostegium; pollen in bags called pollinia; staminat corona present; carpels two, apocarpous .... *Asclepiadaceae.*

CONVOLVULACEAE

1. as in Annonaceae
2. as in Rubiaceae
3. as in Apocynaceae
4. Flowers actinomorphic; leaves alternate, exstipulate; stamens alternipetalous and epipetalous; ovary

superior; carpels fused; placentation, axile .... *Polemoniales*

5. Flowers actinomorphic; corolla has starshaped thickening; stamens 5; filaments unequal; carpels two, superior; disc present at the base .... *Convolvulaceae.*

SOLANACEAE

1. as in Annonaceae
2. as in Rubiaceae
3. as in Apocynaceae
4. as in Convolvulaceae
5. Leaves simple, leaf base unequal, alternate phyllotaxy becomes opposite in the floral region; carpels two, superior, obliquely placed; placenta axile and swollen; fruit a berry .... *Solanaceae*

SCROPHULARIACEAE

1. as in Annonaceae
2. as in Rubiaceae
3. as in Apocynaceae
4. Flowers zygomorphic; corolla bilabiate; stamens 2-4, epipetalous; carpels uni to tetra locular .... *Personales*
5. Flowers in racemes or spikes; corolla personate, stamens 4, didynamous; ovary 2 celled; placenta axile .... *Scrophulariaceae.*

BIGNONIACEAE

1. as in Annonaceae
2. as in Rubiaceae
3. as in Apocynaceae
4. As in Scrophulariaceae
5. Flowers zygomorphic; trees, shrubs or climbers with compound leaves; seeds winged .... *Bignoniaceae*

ACANTHACEAE

1. as in Annonaceae
2. as in Rubiaceae
3. as in Apocynaceae
4. as in Scrophulariaceae
5. Herbs or shrubs; leaves siimple, seeds not winged capsule open elastically; seeds have hooks called retinaculum .... *Acanthaceae*

VERBENACEAE

1. as in Annonaceae
2. as in Rubiaceae
3. as in Apocynaceae
4. Leaves opposite or whorled; flowers zygomorphic and bilabiate; stamens 4, didynamous carpels 2-4 cell .... *Lamiales*
5. Leaves simple, style terminal; fruit a berry or drupe; corolla less strongly zygomorphic; calyx nearly regular not bilipped .... *Verbenaceae*

LABIATAE

1. as in Annonaceae
2. as in Rubiaceae
3. as in Apocynaceae
4. as in Verbenaceae
5. Style gynobasic; fruit usually of four one seeded nutlets, corolla bilabiate; often calyx also bilipped .... *Labiatae.*

AMARANTHACEAE

1. as in Annonaceae
2. Flowers homochlamydeous (i.e. sepals and petals not distinct), perianth sepaloid and in one whorl, sometimes absent .... *Monochlamydeae*
3. Ovary unilocular with a single ovule; embryo curved .... *Curvembryae*
4. Leaves simple, exstipulate and pubescent; flowers bisexual, rarely unisexual; bracts and bracteoles present; ovary superior; placentation basal .... *Amaranthaceae*

CHENOPODIACEAE

1. as in Annonaceae

2-3 as in Amaranthaceae

4. Stem and leaves fleshy and mealy; flowers monochlamydeous; stamens 5, antiepalous; fruit a nut .... *Chenopodiaceae*

POLYGONACEAE

1. as in Annonaceae

2-3 as in Amaranthaceae

4. Leaves with ochreate stipules; flowers bisexual; actinomorphic; fruit sided nut .... *Polygonaceae*

SANTALACEAE

1. as in Annonaceae
2. as in Amaranthaceae
3.
4. Santalaceae

EUPHORBIACEAE

1. as in Annonaceae
2. as in Amaranthaceae
3. Flowers unisexual; perianth reduced, sepaloid or even absent; ovary superior; 2 or 3 carpellary .... *Unisexuales*
4. Milky or watery latex present; ovary trilocular flowers unisexual; seeds carinculate .... *Euphorbiaceae*

MORACEAE

1 as in Announaceae

2 as in Amaranthaceae

3 as in Euphorbiaceae

4. Flowers unisexual; fruit compound .... *Moraceae*

URTICACEAE

1. as in Annonaceae
2. as in Amaranthaceae
3. as in Euphorbiaceae
4. Flowers unisexual; cystoliths present in the leaf; fruits simple .... *Urticaceae*

PALMAE

1. Leaves parallel veined; cotyledon one; flowers usually trimerous, homochlamydeous; vascular bundles in the stem scattered .... *Monocotyledons*
2. Ovary superior, endosperm copious; perianth inconspicuous or stiff .... *Calycineae*
3. Trees shrubs or spiny clibers; leaves pinnately or palmately compound; flowers in spadix; fruit a drupe .... *Palmae*

MUSACEAE

1. as in Palmae
2. Perianth slightly petaloid; flowers epigynous .... *Epigynae.*
3. Inflorescence spadix; zygomorphic; stamens 6 ovary tricarpellary syncarpous and trilocular .... *Musaceae*

CANNACEAE

1. as in Palmae
2. as Musaceae
3. Stamens petaloid; showy and connate at the base; only half stamen fertile; carpels 3, fused, 3 celled style petaloid; ovary wall warty .... *Cannaceae*

ZINGIBERACEAE

1. as in Palmae
2. as in Musaceae
3. placention axile; fertile stamen anlone others to petaloid .... *Zingiberaceae.*

LILIACEAE

1. as in Palmae
2. Flowers bisexual; inner perianth petaloid; ovary superior .... *Coronarieae*
3. Perianth 6 in 2 whorls of 3 each, ovary 3 carpellary syncarpous, trilocular; placentation axile .... *Liliaceae*

COMMELINACEAE

1. as in Palmae
2. as in Liliaceae
3. Perianth 6, leaf base sheathing; filaments of anthers bearded .... *Commelinaceae*

ARACEAE

1. as in Palmae
2. Ovary superior, perianth absent .... *Nudifloreae* Terrestrial oraantic herbs.
3. Perianth zero; anther 2-4 called, ovary sessile, 1 to 3 celled, fruit a berry .... *Araceae*

CYPERACEAE

1. as in Palmae
2. Flowers bracteate, bisexual, arranged in spikelets perianth absent or reduced .... *Glumaceae*
3. Leaf phyllotaxy ⅓, stem solid and triangular; leaf sheath fused with the stem; flowers in spikelets; spikelets not subtended by sterile bracts; perianth when present represented by bristles, hairs, scales; stamens 3; carpels 3 .... *Cyperaceae.*

GRAMINEAE

1. as in Palmae
2. as in Cyperaceae
3. Stem jointed and cylindrical; leaves alternate; phyllotaxy 2 ranked, ligule present; spikelets subtended bya pair of sterile bracts; perianth membranous; each flower is enclosed by a bract fruit a caryopsis .... Gramineae

CHAPTER 6

DESCRIPTION OF A FLOWERING PLANT

Generally a sequence or method is followed in the taxonomic description of a flowering plant. The following is the procedure :

Habitat : (a) The surrounding environment, Mesophyte, Xerophyte etc.

Habit : (a) Herb, Shrub, Tree.
(b) Annual, Biennial or perennial..

Root : (a) Type.
(b) Branched or not.
(c) Modifications if any.

Stem : (a) Posture, erect or weak etc.
(b) Modification if any.
(c) Branching and its type.
(d) Texture-soft or woody.
(e) Shape, Cylindrical etc.
(f) Surface - smooth, pubescent etc.

Leaf : (a) Simple or compound, types of compound.
(b) Whether radical, cauline etc. Whether stalked or sessile etc.
(c) Apex
(d) Base
(e) Margin
(f) Texture
(g) Stipule - present or absent
(h) Venation
(i) Phyllotaxy
(j) Shape of leaf

Inflorescence : Types of inflorescence

Flower : (a) Stalked or sessile
(b) Bracteate or not
(c) Bracteolate or not.

(d) Symmetry of the flower
(e) Arrangement of floral leaves, cyclic etc.
(f) Presence or absence of accessory organs.
(g) Position of Gynoecium
(h) Complete or incomplete.

Corolla : (a) Distinguishable into petals or not
(b) Whether perianth present
(c) Shape of corolla
(d) Aestivation
(e) Free or fused
(f) Number of sepals

Calyx : (a) Number of sepals
(b) Green or coloured
(c) Free or fused
(d) Aestivation
(e) Persistant or caducous

Androecium : (a) Number of stamens
(b) Free or fused
(c) Epipetalous or not
(d) Filament attachment to the anther
(e) Anthers mono or dithecous
(f) Anthers extrorse or introrse
(g) Mode of dehiscence

Gynoecium : (a) Position of ovary
(b) Number of carpels
(c) Syn or apocarpous
(d) Number of locules
(e) Placentation
(f) Number of ovules in each locule
(g) Nature of style and stigma
(h) Type of ovule

Fruit : Type and method of dehiscence
Seed : Structure and type etc.

At the end of the description Floral diagram has to be constructed and floral formula is to be written.

Classification :

The particular plant has to be assigned to its systematic position with reasons.

CHAPTER 7

SALIENT FEATURES OF FAMILIES

DICOTYLEDONS

ANNONACEAE

Trees or shrubs; some members may be climbing; wood aromatic; leaves alternate, simple, exstipulate. Flowers solitary, large, trimerous; perianth lobes spirally arranged. Stamens numerous, as also carpels which are apocarpous. Seeds have a ruminate endosperm.

MAGNOLIACEAE

Mostly trees or shrubs. No herbs, leaves alternate, stipulate (caducous) and simple. Flowers bisexual, actinomorphic and usually large in size. Perianth trimerous having 3 whorls. Receptacle cone like. Stamens indefinite spirally arranged. Carpels many, apocarpous and spirally arranged, fruit usually an etaerio of follicles or berries or could be a samara also.

RANUNCULACEAE

Generally erect herbs, rarely climbing shrubs (clematis), leaves alternate or opposite exstipulate flowers hypogynous, floral parts spiral on a convex thalamus. Sepals deciduous and petaloid often. Stamens many, indefinite. Corolla sometimes absent. Carpels many and apocarpous.

PAPAVERACEAE

Herbs with milky or coloured juice, leaves alternate, exstipulate,; Flowers large actinomorphic or zygomorphic. Sepals few, caducous, petals twice the number of sepals, stamens indefinite, free, ovary monolocular, placentation parietal. Fruit a poricidal or loculisidal capsule.

CRUCIFERAE

Mostly herbs, cultivated annuals or perennials. Inflorescence usually a raceme. Flowers heterochlamydeous, calyx free, corolla free, both 4 in number and cruciform. Stamens 6 in number and tetradynamous. Ovary bicarpellary and monococular but becomes bilocular due to the development of a false septum called replum. Fruit is either a siliqua or silicula.

CAPPARIDACEAE

Herbs, shrubs or rarely trees with a watery sap. Leaves alternate, simple or palmately compound; stipules spinescent. Flowers actinomorphic, bisexual and hypogynous. Androecium and gynoecium raised on stalks called androphore and gynophore respectively. Stamens 4-6 or many; ovary superior, bicarpellary, unilocular. Fruit, a bery or capsule.

CARYOPHYLLACEAE

Generally herbs, stems have swollen nodes. Leaves generally simple; phyllotaxy opposite decussate, occasionally the base of the opposite pair of leaves connate, usually exstipulate. Inflorescence cymose, flowers, penta or tetramerous, hypogynous, bisexual, rarely unisexual and heterochlamydeous. Androecium has 8-10 stamens in two whorls. Ovary superior 2-5 carpels uni or in completely bilocular; ovules many rarely less on free central or basal placenta. Fruit a capsule, embryo curved in seeds.

PORTULACACEAE

Herbs or undershrubs. Stem soft, semierect or prostrate, and succulent. Leaves, opposite or alternate, succulent and stipulate. Inflorescence cymose, racemose or solitary. Flowers actinomorphic bisexual and hypogynous. Sepals two free or connate; petals 4-5 free. Stamens 4 to many. Ovary superior, 3-5 carpels, unilocular ovules on basal placenta. Fruit a capsule. Seed endospermic.

MALVACEAE

Trees shrubs or herbs. Stem has mucilage and stellate hairs. Leaves alternate, simple, large, often variegated. Flowers large, showy with epicalyx. They are actinomorphic, bisexual and hypogynous. Calyx 5, corolla 5, imbricate or twisted. Stamens many; anthers monothecous. Ovary multicarpellary, syncarpous, multilocular with ovules on axile placenta. Fruit a capsule or schizocarp.

STERCULIACEAE

Herbs, shrubs or trees. Sttellate hairs commonly present on the plantbody. Mucilage sacs present in the stem;leaves alternate and stipulate. Flowers actinomorphic, hypogynous and bisexual. Calyx and corolla 5 each; aestivation valvate or imbricate. Stamens in two whorls, outer staminodal, inner connate. Carpels 5, syncarpous. Fruit schizocarpic

TILIACEAE

Trees or shrubs, herbs rare; leaves siimple, alternate and stipulate. Inflorescence cyme or a cymose panicle. Flowers, bisexual, rarely unisexual and hypogynous. Calyx and corolla 5 in number. Stamens 5-10 or many. Ovary bi or multicarpellary. Fruit berry, capsule or drupe.

PAPILIONACEAE

Herbs, shrubs or tall trees. Leaves compound with pulvinate leaf base. Flowers in racemes or panicles. Corolla papilionaceous with a large standard petal, two wing petals and a keel petal formed by the fusion of two petals. Stamens 10, ovary monocarpellary, placentation marginal, fruit a legume or pod.

CAESALPINACEAE

Herbs, shrubs or trees. Leaves pulvinate and pinnately compound. Inflorescence a corymb. Flowers zygomorphic, sepals and petals 5 in number odd posterior petal smallest; aestivation ascendingly imbricate, stamens 10, free, many are sterile; ovary monocarpellary; fruit a legume.

MIMOCEAE

Herbs, shrubs or trees. Stem often has epidermal prickles. Leaves pinnately compound with pulvinate leaf base. Inflorescence spike or globose head. Stamens 4 to indefinite. Ovary superior, monocarpellary with ovules on marginal placenta. Fruit a legume or lomentum.

ANACARDIACEAE

Mostly trees and shrubs. Leaves alternate, simple or pinnately compound. Stipules absent. Flowers polygamous found in panicles. Floral symmetry actinomorphic, stamens inserted at the base of an annular disc; twice as many as petals, occasionally less. Carpels 1-2, rarely may be 5 also. Ovary superior with a single pendulous ovule. Fruit is a drupe; endocarp is stony and hard.

RUTACEAE

Herb, shrubs or trees. Leaves gland dotted containing an essential oil. Flowers bisexual and hypogynous. Disc present below the ovary. Stamens obdiplostemnous. Fruit a berry, drupe or capsule.

MELIACEAE

Mostly trees or shrubs. Leaves aglandular, pinnately compound and exstipulate; phyllotaxy alternate. Flowers regular, bisexual or rarely unisexual. Stamens monadelphous, rarely free. A nector secreting disc present between petals and stamens;carpels 3-5, superior, 3-5 locular. Fruit berry, capsule or drupe.

CUCURBITACEAE

Prostrate or climbing weak stemed herbs, climb by means of tendrils; leaves cordately or palmately lobed. Leaf surface is studded with stiff hairs. Flowers large, monoecious or dioecious; calyx and corolla 5 each, stamens usually 3, occasionally 5, anthers free or fused. Ovary inferior, 3 locular with parietal placentation. Fruit a berry; cotyledons oily.

ROSACEAE

Herbs, shrubs or trees; climbers are also seen. Leaves stipulate with adnate stipules. Flowers regular, bisexual, perigynous, hypogynous or even epigynous. Stamens indefinite; petals variously coloured. Ovary mono or polycarpellary and apocarpous or syncarpous. Placentation basal, axile or parietal. Seeds exalbuminous.

COMBRETACEAE

Mostly trees, occasionally climbers. Leaves simple, flowers 4-5 merous; sepals and petals 4-5 in number sometimes petals absent. Stamens twice as many as sepals arranged in two whorls. Ovary inferior, monocarpellary, unilocular with 4-6 ovules suspended from the locule apex. Fruit one seeded, multiangled nut.

MYRTACEAE

Shrubs or trees; leaves opposite, gland dotted and with intramarginal venation. Stipules absent. Leaf margin entire. Flowers in cymes, actinomorphic, pentamerous and epigynous. Calyx, 5, united, occasionally comes out like a lid. Corolla 5, free or united to form a calyptra which also comes out like a lid. Stamens indefinite; ovary inferior, 2-5 carpellary. Fruit a berry, drupe, capsule or nut.

UMBELLIFERAE

Mostly herbs; annuals, biennials or perennials. Stem soft and fistular; oil glands present in all the aerial parts; leaves compound generally with a broad sheathing leaf base; no stipules. Flowers small, aggregated into simple or compound umbels. They are heterochlamydeous, actinomorphic or zygomorphic, pentamerous except for carpels. Ovary inferior. Fruit a schizocarp splitting into two mericarps.

RUBIACEAE

Herbs, shrubs or trees; leaves alternate or opposite, decussate with inter or intrapetiolar stipules. Inflorescence is a cymose cluster. Flowers regular, bisexual, epigynous and tetra or pentamerous. Calyx and corolla 5 each. Corolla gamopetalous and tubular. Stamens 4-5 free and epipetalous, ovary inferior, bicarpellary, syncarpous and bilocular with many ovules in each locule on axile placenta. Fruit a capsule or berry.

COMPOSITAE

Mostly herbs, some are climbers. Trees and shrubs very rare. Leaves simple or pinnately divided; stipules absent; phyllotaxy alternate or opposite. Inflorescence typically a head. Sometimes capitula may be arranged in corymbs. Head simple or compound; homogamous or heterogamous. Flowers epigynous, actinomorphic, or zygomorphic, unisexual or bisexual. In heterogamous heads two types of florets are present viz., disc florets and ray florets. Disc florets bisexual, while ray florets are unisexual and female. Stamens 5, epipetalous; anthers syngenesious. Ovary inferior, bicarpellary, syncarpous and bilocular. Ovule one on basal placenta. Fruit a cypsela.

SAPOTACEAE

Mostly trees and shrubs. Laticiferous vessels present in the stem. All the aerial parts are often tomentose. Inflorescence a cymose cluster. Flowers bisexual, pentamerous and hypogynous. Sepals 2-8, in one or two whorls; petals 4-8, gamosepalous. Stamens as many as petals in one or two whorls. Ovary superior and multicarpellary. Fruit a berry.

APOCYNACEAE

Herbs, shrubs or trees; twiners also present. Leaves simple and entire. Phyllotaxy opposite or whorled latex present. Flowres bisexual, actinomorphic, and hypogynous. Sepals and petals 5 and united;stamens also 5, epipetalous. Anthers sagittate. Ovary superior, bicarpellary and apocarpous. Styles two, free basally but united at the top by stigma which is green and dumb bell shaped. Fruit follicle or berry. Seeds have a coma of long hairs.

ASCLEPIADACEAE

Members with thick milky latex. Herbs, shrubs, climbers or small trees. Leaves usually simple, opposite decussate or even whorled. Inflorescence is usually a di or monochasial cyme or even an axillary umbel. Flowers bisexual, regular, pentamerous and hypogynous. Sepals 5, free, imbricate aestivation; petals 5, gamopetalous, twisted. Stamens 5, filaments short or none, fused with the gynoecium to form gynostegium. Pollen in 1-2 sacs called pollinia. Ovary superior, carpels 2, apocarpous. Fruit a pair of follicles; seeds winged.

CONVOLVULACEAE

Herbs, shrubs or climbers, watery or milky sap often present. Bicollateral vascular bundles present in the stem. Some are parasitic and leafless. Leaves simple, entire or bilobed or digitately compound. Inflorescence is a cymose cluster. Flowers regular, bisexual and hypogynous; calyx 5, free; corolla five fused and banded. Stamens 5, epipetalous, ovary superior, bicarpellary and placed on a disc. Fruit is a loculisidal capsule. Seeds with folded or bilobed cotyledons.

SOLANACEAE

Plants herbs or shrubs; occasonally climbers also. Anatomically stem, has bicollateral vascular bundles. Vascular bundles congenitally fused at the nodes. Leaves simple, alternate in the vegetative region but opposite in the floral region. Leaf base unequal. Inflorescence is a cyme. Flowers actinomorphic bisexual and hypogynous. Calyx and corolla five each and fused. Stamens 5, epipetalous. Carpels two, superior and obliquely placed. Placentation parietal. Fruit a berry or capsule.

BIGNONIACEAE

Members are lians, shrubs or trees. Leaves compound, flowes irregular; anthers connivent in pairs. Ovules many. Fruit a woody capsule. Mature seeds exalbuminous.

ACANTHACEAE

Herbs or climbers. Cystoliths present. Leaves simple, opposite, decussate and exstipulate. Inflorescence cymose. Flowers bisexual zygomorphic and hypogynous. Sepals and petals 4-5 corolla bilipped. Stamens 2-4, epipetalous. Ovary superior, bicarpellary, syncarpous and bilocular with ovules on axile placenta. Fruit a capsule; seeds provided with retinacula.

SCROPHULARIACEAE

Mostly herbs; shrubs and trees rare: A few are climbing. Leaves alternate or opposite and exstipulate. Inflorescence racemose or cymose. Flowers bisexual, zygomorphic or somewhat actinomorphic and hypogynous. Calyx 5, fused, corolla 5 fused often bilabiate. Stamens epipetalous, 4 in number and didynamous; occasionally stamens vary in number from 2-5. Carpels two, superior, syncarpous with many ovules on axile placenta. Fruit, a berry or capsule with persistent calyx.

VERBENACEAE

Herbs, shrubs or trees. Stem erect and woody in trees. Leaves simple, or pinnately or palmately compound. Stipules absent. Inflorescence cyme, raceme or spike. Flowers bisexual, zygomorphic and hypogynous. Sepals 4-5, persistent, petals 4-5 fused, occasionally bilipped. Stamens 4 epipetalous ovary superior, carpels two, syncarpous. Fruit drupaceous containing four or less single seeded nutlets. Seeds endospermous.

LABIATAE

Herbs or shrubs. Stem quadrangular; leaves simple, opposite and decussate. Inflorescence verticellaster or Thyrsus. Flowers zygomorphic, bisexual and hypogyonous. Corolla bilabiate. Stamens didynamous and epipetalous; ovary bicarpellary and tetralocular. Style gynobasic. Fruit nutlets, schizocarpic or carcerulus.

POLYGONACEAE

Herbs, occasionally climbers. Leaves have ocharaceous stipules; Stipules interpetiolar and sheathing around the swollen nodes. Inflorescence a spike or a panicle. Flowers small, clusters into bunches, homochlamydeous, actinomorphic and hypogynous. Perianth 2-3 merous. Stamens 6-9. Ovary superior, 2-3 carpellary with a single basal ovule. Fruit a nut. Seed has a shiny testa; convex or traingular.

AMARANTHACEAE

Mostly herbs or shrubs; leaves usually hairy. Inflorescence spike, head or cymose panicle. Flowers bisexual with membranous perianth, bracts and bracteoles. Perianth 5, antitepalous. Ovary superior, bicarpellary, syncarpous, monolocular with a single basal ovule. Fruit, a one seeded nut or utricle. Seeds have a mealy endosperm.

CHENOPODIACEAE

Herbs, (some halophytic), occasionally shrubs or trees. Plant parts pubescent. Leaves simple, alternate, sometimes succulent. Inflorescence cymose; flowers monochlamydeous, bi or unisexual and irregular. Perianth 2-5, succulent, stamens antitepalous as many or fewer than tepals. Ovary superior, 2-3 carpellary, unilocular with a single basal ovule. Fruit a nut or utricle with a persistent perianth. Seeds albuminous.

EUPHORBIACEAE

Herbs, shrubs and trees. Milky or watery latex present. Leaves alternate, simple or compound. Inflorescence highly variable; flowers always unisexual, heterochlamydeous,homochlamydeous or achlamydeous. Perianth 3-5, stamens one to many. Ovary superior, tricarpellary, syncarpous, trilocular; placentation axile. Fruit a capsule or schizocarp.

URTICACEAE

Herbs, rarely shrubs or soft stemmed trees. Stinging hairs present; inflorescence cyme; flowers unisexual and homochlamydeous; stamens antitepalous; ovary monocarpellary, unilocular with a single basal ovule. Fruit an achene.

MORACEAE

Trees or shrubs. Milky latex present. Flowers unisexual, monoecious or dioecious and arranged in catkins. Leaves stipulate, perianth 2-6, free or connate or sometimes absent. Stamens 4, antitepalous and usually inflexed in bud.

MONOCOTYLEDONS

PALMAE

Mostly shrubs and trees. Some are spiny lianous climbers. Leaf pinnately or palmately divided; leave a scar on the stem. A crown of leaves usually terminates the columnar stem. Stem generally unbranched. Inflorescence is a spadix with a woody bract enclising the flowers. Flowers unisexual with a fleshy perianth. Both male and female flowers are trimerous. Androecium has six free stamens. Ovary superior and tricarpellary. Fruit berry, drupe or nut. Endosperm abundant, seed smal.

MUSACEAE

Members perennial herbs with underground rhizome. Sheathing leaf bases produce an aerial pseudostem. Leaves simple and large. Inflorescence is a spadix. Flowers homochlamydeous, zygomorphic and bisexual. Perianth 5+1, fused or free. Stamens 5 with one staminode. Ovary inferior and tricarpellary. Fruit a berry.

ORCHIDACEAE

Herbaceous perennials, climbers, terrestrial herbs or even epiphytes. Some are saprophytic. Roots fibrous. Assimilatory aerial roots found in epiphytes. Stem either aerial or an underground tuber; often a pseudobulb. Leaf simple linear or broad. Inflorescence racemose type. Flowers conspicuous, attractively coloured, bisexual, zygomorphic and epigynous. Perianth biseriate with 6 members. Median lobe of inner whorl modified to form labellum. Stamens usually one, sometimes two. Staminodes also two. Ovary inferior, tricarpellary, syncarpous, trilocular; ovules on parietal placenta stigmatic surface three, only two are receptive, third sterile one forms rostellum. Stamens and stigma fuse to form gynostegium. Fruit a capsule. Seeds exalbuminous.

CANNACEAE

Perennial herbs with rhizomatous stem. Leaves large paralley veined; leaf base sheathing. Inflorescence raceme or panicle. Flowers large, bright coloured, bisexual, zygomorphic and epigynous. Perianth biseriate, outer sepaloid, inner petaoid. Stamens 6, but majority reduced to staminodes; only one half of the anther is fertile. Ovary inferior, tricarpellary, syncarpous, trilocular with ovules on axile placenta. Fruit a capsule.

ZINZIBERACEAE

Herbs; stem underground rhizome. Inflorescence spike or raceme. Flowers bisexual, zygomorphic and epigynous. Calyx and corolla three each. Stamen one fertile, rest (4-5) staminodal. Ovary inferior, tricarpelary, syncarpous, trilocular. Fruit a capsule.

COMMELINACEAE

Herbs with jointed stem. Leaves alternate and sheathing. Inflorescence a cincinnus. Flowers bisexual, actinomorphic and hypogynous. Perianth in two whorls of 3 each. Stamens also in two whorls of 3 each.: some of the stamens absent are reduced. Ovary superior, 3 carpels with axile placentation. Fruit a capsule. Seed endospermic.

LILIACEAE

Herbs and small shrubs. Some are climbers. Stem rhizomatous or aerial. Leaves variously modified or reduced. Inflorescence racemose. Flowers trimerous, homochlamydeous, regular and bisexuali perianth 6 and biserate. Stamens 6 free. Ovary superior, tricarpellary. Placentation axile;fruit a berry or capsule. Seeds albuminous.

ARACEAE

Terrestrial or acquatic herbs with tuberous or rhizomatous stem. Leaves simple or with divided margins; cauline or radical having a sheathing leaf base. Inflorescence a spadix. Flowers small, di or trimerous and bisexual. Perianth 4-6. Stamens 2-8 forming a synandrium. Ovary superior, monocarpellary. Fruit a berry.

GRAMINEAE

Mostly herbs. Shrubs and trees not frequent. Stem hollow, cylindrical; internodal region covered with sheathing leafbase. Leaves simple, ligulate with clefted sheath. Inflorescence is a compound spike. Flowers bisexual or occasionally unisexual; glumes subtend the flowers. Perianth minute, called lodicules which are 2-3 in number. Stamens 3 with versatile anthers. Carpels usually one, superior with a single basal ovule. Fruit a dry indehiscent caryopsis.

CYPERACEAE

Perennial herbs; look like grasses or rushes;stem solid, three angled. Leaves have ⅓ arrangement; ligule absent; leaf base sheathing and margins fused. Inflorescence spike like cymes. Flowers arise in the axil of a bract (glume); bracteoles and perianth normally absent. Perianth when present six and arranged in two whorls or reduced to bristles and hairs. Ovary superior 2-3 carpels, syncarpous, unilocular with a single basal ovule. Fruit an achene. Testa does not adhere to the pericarp. Seeds endospermous; embryo is surrounded at its base by the endosperm.

DESCRIPTION OF PALNTS IN SEMITECHNICAL TERMS

CHAPTER 8

DESCRIPTION OF PLANTS

MAGNOLIA GRANDIFLORA

(Magnoliaceae)

Habitat: Terrestrial, Mesophyte

Habit: A medium sized evergeen tree

Root: Tap root system, extensively branched and deep feeder type.

Stem: Aerial, erect, woody and branched;stem is solid with soft and light wood.

Leaf: Ramal, alternate phyllotaxy, stipulate with large stipules enclosing the bud, leaf margin entire apex acute; lanceolate, glabrous with reticulate unicostate venation

Inflorescence: Flowers large, solitaary and terminal

Flower: Bracteate, ebracteolate, pedicellate, complete regular, hamochlamydeous, actinomorphic, bisexual, hypognous, Spirocyclic with an elongated cone like thalamus.

Perianth: White, fragrant showy, attractive, free, 9 in number arranged in three whorls of three each, valvate in each whole but overlapping together and inferior.

Androecium: Stamens many, polyandrous spiral on the thalamus. Each stamen has a short filament and a long conical anther. Anthers 2 celled (dithecous) with immersed pollen sacs. Dehiscence longitudinal, introrse; pollen with a single aperture.

Gynoecium: Carpels many in number, apocarpous with superior ovary, carpels spirally arranged on the thalamus with a basal swollen ovary, a short curved style and a beak shaped or simple stigma. Carpel unilocular with two ovules on a marginala placenta.

Fruit: Etaerio of follicles.

Seed: Large, endospermic with copions endosperm embryo small and erect. An additionaal fleshy sarco testa envelops the seed.

Floral formula: ✝ ⚥ $P_{3+3+3}\ A_{\infty}\ \underline{G}_{\infty}$

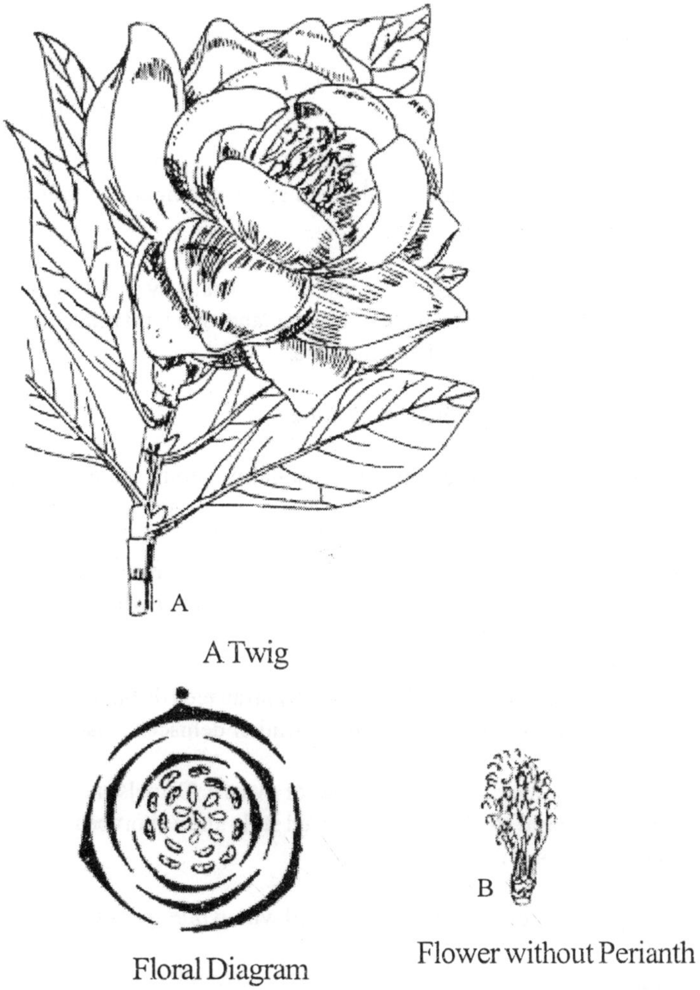

Fig 8.1 ***MAGNOLIA GRANDIFLORA*** **(Magnoliaceae)**

MICHELIAA CHAMPAKA

(Magnoliaceae)

Habitat : Evergreen mesophyte

Habit : A tall graceful tree

Root : Tap root system; profusely branched

Stem : Aerial, erect, branched, woody and solid; often green in colour; a number of lenticels are found on the stem.

Leaf : Simple, Ramal and alternate, large with some whaat leaathery structure (coriaceans); stipulate with convolute and caducaus stipules; petiolate ovate or oblong lanceolate; margin entire with acute apex. Colour of the lamina dark green; venation: Unicostate reticulate.

Inflorescence: Flowers solitary and axillary in position.

Flower : Bracteaate, ebracteolate, pedicellate, heterochalmydeous, complete, regular, actinomorphic spinocyclis, bisexuaal and hypogynons. Flowers are large, showy and sweet scented.

Perianth : Distinguished somewhat into caalyx and corolla. Calyx three in number free, greenish, aestivation valvate. Corolla – made up of two whorls of 3 petals each, free, outer whorle valvate, inner whorle twisted. Petals pale yellow in colour and sweet scented.

Androecium: Stamens numerous, polyandrous, spirally arranaged distinguishable into filament, connective and anther. Anthers dithecous basifixed and extrorse. Anther dehiscence is longitudinal.

Gynoecium : Carpels many, apocarpous with superior ovaary. Carpels are spirally arranged on a cone like receptacle. Each carpel has a basal swollen ovary, curved style and a short beaked stigma. Carpels one celled (unilocular) with many ovules on a marginal placenta.

Fruit : Aan etaerio of follicles somewhat dark coloured with white specks.

Seed : Large, albuminons with copious endosperm; embryo small.

Floral formula : † ♀ $K_3 A_{3+3} \underline{G}_{\infty}$

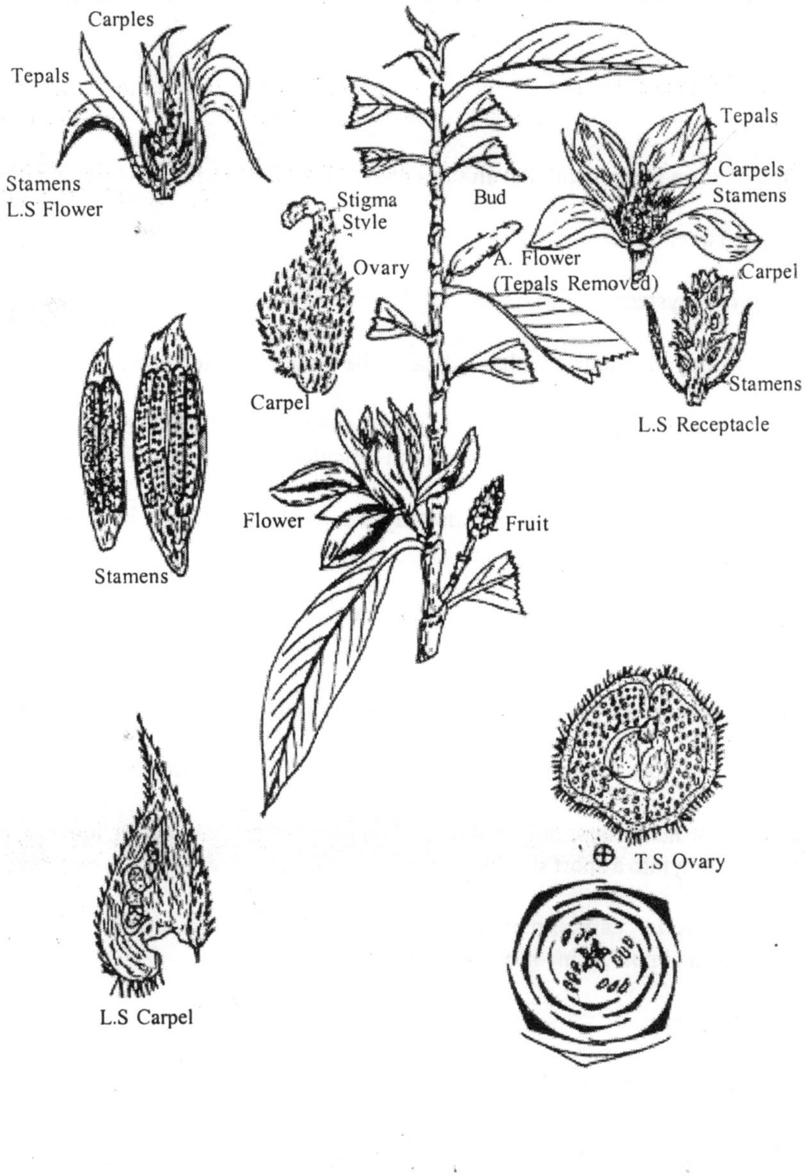

Fig 8.2 ***MICHELIAA CHAMPAKA*** **(Magnoliaceae)**

ANONA SQUAMOSA

(Annonaceae)

Habitat: Mesophyte growing generally at alitudes of about 1050m. and in well drained areas.

Habit: A medium sized tree.

Root: Branched tap root system.

Stem: Aerial, erect, woody, deciduous with oil glands. The trunk is about 6 m tall.

Leaf: Simple, ramal, alternate, petiolate and exstipulate. Lamina ovate lanceolate,; margin entire with acute apex. Leaf surfaace glabrous and gland dotted; venation reticulate unicostate.

Inflorescence: Flowers generally solitary; ocassionally groups of flowers arise in axillary position. Rarely cauliflorous also.

Flower: Bracteate, ebracteolate, pedicellate, helerochalamydeous, regular, actinomorphic, spirocyclic, bisexual and hypogynous.

Calyx: Sepals 3 in number, polysepalous fleshy; aestivation valvate.

Corella: Petals six in number, free arranged in two whorls of 3 each. Aestivation valvate.

Androecium: Staamens numerous, polyandrous filaments shorter than anther lobes with a thick and broad connective protruding into a short sterile structure beyond the anther. Anthers dithecous and extrorse.

Gynoecium: Ovary superior, carpels many, apocarpous; each carpel unilocular with ovules on marginal placenta. Style short with a simple or inconspicuous stigma.

Fruit: An Etaerio of achenes.

Seed: Dark coloured, large with a hard and shiny seed coat. Seeds are albuminous.

Floral formula: † ♀ $K_3\ C_{3+3}\ A_\infty\ \underline{G}_\infty$

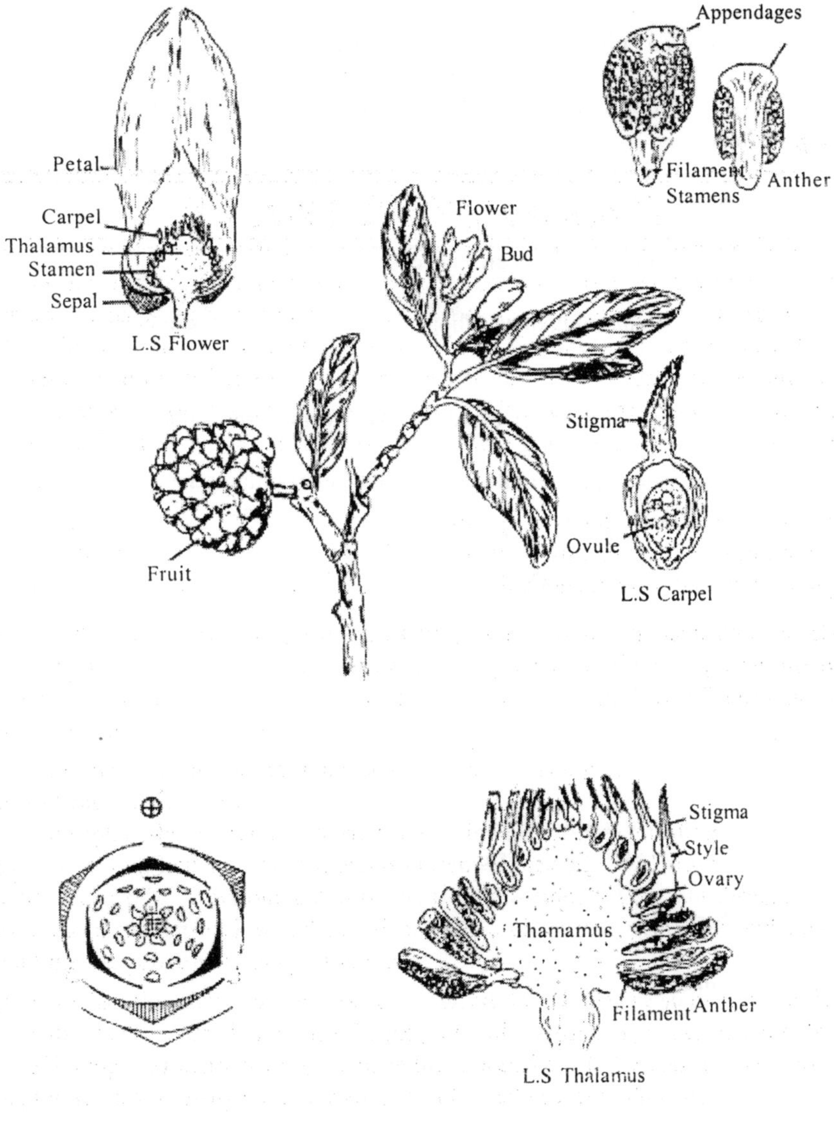

Fig 8.3 *ANONA SQUAMOSA* (Annonaceae)

ARTABOTRYS ODORATISSIMUS

(Annonaceae)

Habitat: Mesophyte; often cultivated in gardens as ornamental.

Habit: A medium sized woody climber, appears bushy. Climbing is facilitated by hooks developed as modification of peduncle

Root: Much branched tap root

Stem: Aerial, woody, climbing, branched and sweet scented.

Leaf: Simple, alternate, exstipulate petiolate, elliptical, entire with acute apex: Lamina gland dotted, venation reticulate unicostate.

Inflorescence: Flowers solitary arising on a peduncle with curves into a hook to help in climbing.

Flower: Pedicellate Ebracteate, ebracteolate, somewhat heterochlamydeous, regular, actinomorpyhic, spirocyclic, bisexual and hypogynous

Calyx: Sepals greenish, 3 in number, free, aestivaation valvate.

Corolla: Petals pale greenish turning into pale yellow later, polypetalous, six in number arranged in two whorls of 3 each. Aestivation valvate. Petals with glands of aromatic oil.

Androecium: Stamens indefinite polyandrous distinguishable into a short filament, connective and bilobed anther. Anthers dithecous and extrorse.

Gynoecium: Ovary superior, multicarpellary, apocarpous, unilocular with ovules on marginal placenta. Styles simple and separate for each carpel; stigma simple

Fruit: An etaerio of berries

Seed: Large and albuminous.

Floral formula: ✝ ♀ $K_3\ C_{3+3}\ A_\infty\ \underline{G}_\infty$

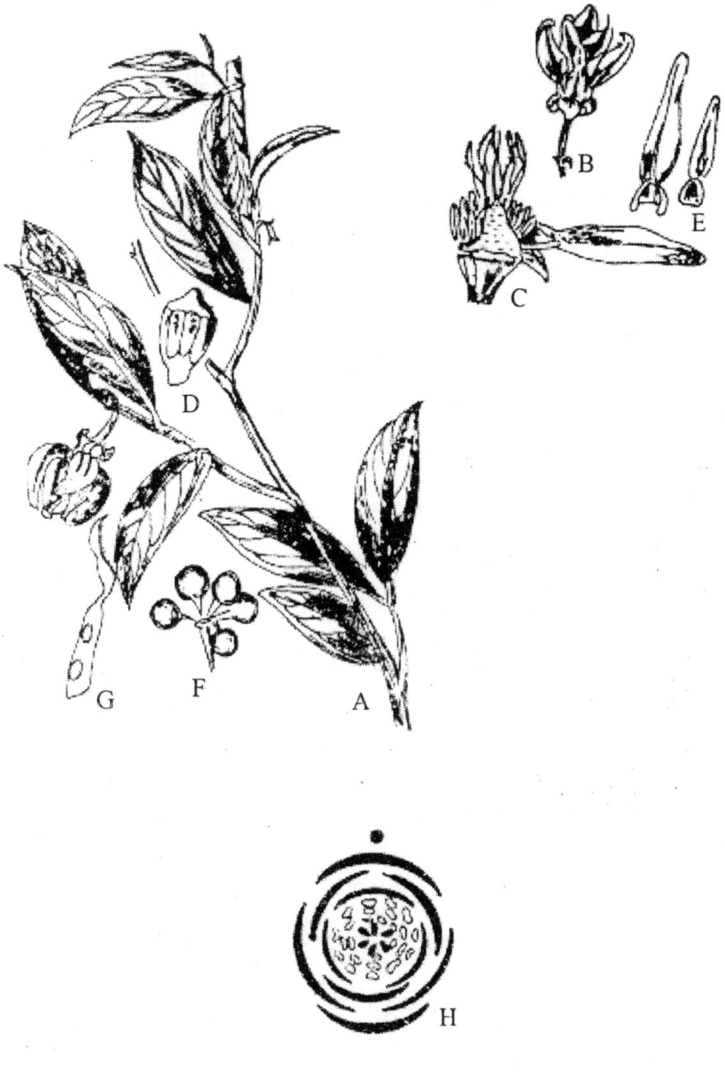

Fig 8.4 *ARTABOTRYS ODORATISSIMUS* (Annonaceae)

POLYTHIA LONGIFOLIA.

(Annonaceae)

Habitat : A mesophyte; cultivated as an avenue tree.

Habit : A medium sized evergreen tree sometimes reaching to a height of 25-30 feet or more.

Root : Branched tap root.

Stem : Aerial, erect, woody and branched which generally droop down.

Leaf : Simple, petiolate alternate long and lanceolate. Margin undulating, surface waxy.

Inflorescence : Yellowish green flowers are clustered in dense umbels.

Flower : Pedicellate, bracteate, ebracteolate, regular, spirocyclic, actinomorphic, bisexual and hypogynous.

Calyx : Sepals 3 in number, free and valvate, traingular in shape

Corolla : Petals six in number arranged in two whorls of 3 each. Aestivation valvate petals subequal , linear tapering to a point.

Androecium : Stamens many, cuneate connective dialated beyond anther. Anthers dithecous and extrorse.

Gynoecium : Carpels one to many, apocarpous with superior ovary. Carpels unilocular with 1-2 basal ovules. Style oblong

Fruit : An etaerio of berries.

Seed : Large albuminous smooth and shiny

Floral formula : † ♀ $K_3\ C_{3+3}\ A_{\infty}\ \underline{G}_{\infty}$

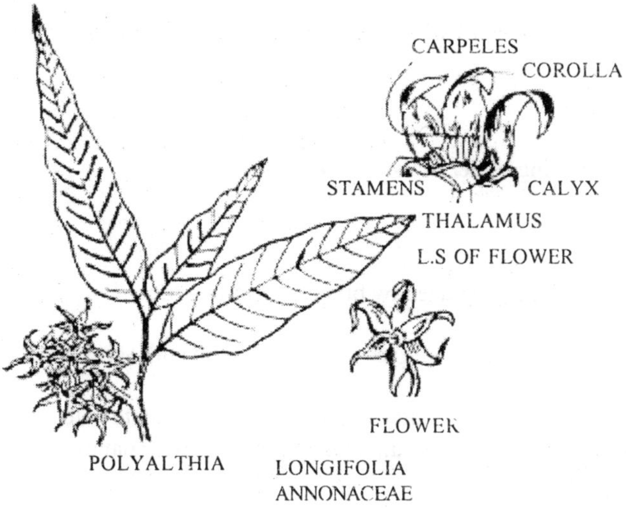

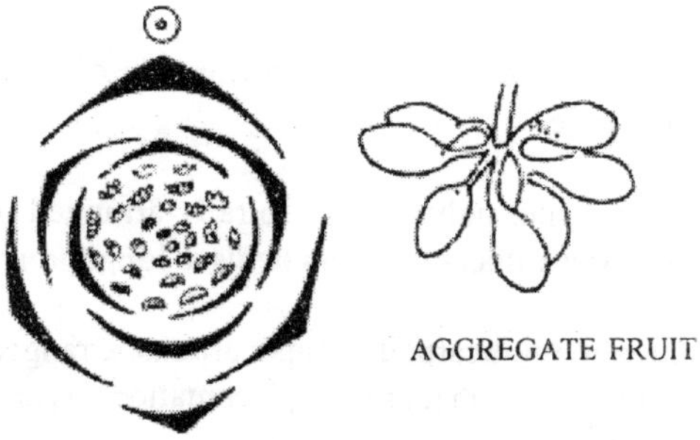

Br,Ebrl,$\oplus$⚥,P_{3+3+3},A_{00},G_{00}

Fig 8.5 *POLYTHIA LONGIFOLIA* (Annonaceae)

RANUNCULUS SCLERATUS

(Ranunculaceae)

Habitat: A mesophyte growing under moist conditions

Habit: An annual erect herb.

Root: Orinally a tap root gradually replaced by branched fibrous root

Stem: Herbaceous, cylindrical, aerial, erect green in colour; solid with acrid juice

Leaf: Radical, cauline and ramal, simple, alternate, petiolate, exstipulate with sheathing leaf base. Lamina palmately tripartite with each lobe further cut into cuneate segments; venation multicostate reticulate.

Inflorescence: Flowers in cymose clusters.

Flower: Bracteate or ebracteate, ebracteolate, pedicellate, regular, heterochalamydeous, complete, actinomorphic, bisexual and hypogynous.

Corolla: Petals 5, polypetalous, shining yellow towards inside. Aestivation imbricate. Each petal has a pocket shaped nectary at the base.

Androecium: Stamens indefinite, polyandrous spirally arranged on an elongated thalamus. Filaments long bearing dithecous basifixed anthers. Anthers dehisce longitudin:.lly and are extrorse.

Gynoecium: Ovary superior, multicarpellary, apocarpous, arranged spirally on a conical thalamus. Each carpel unilocular with only one ovule on marginal placentation; style very short; stigma beaked.

Fruit: An etaerio of one seeded achenes.

Seed: Albuminous, embryo small and straight

Floral formula: ✝ ♀ $K_5\ C_5\ A_\infty\ \underline{G}_\infty$

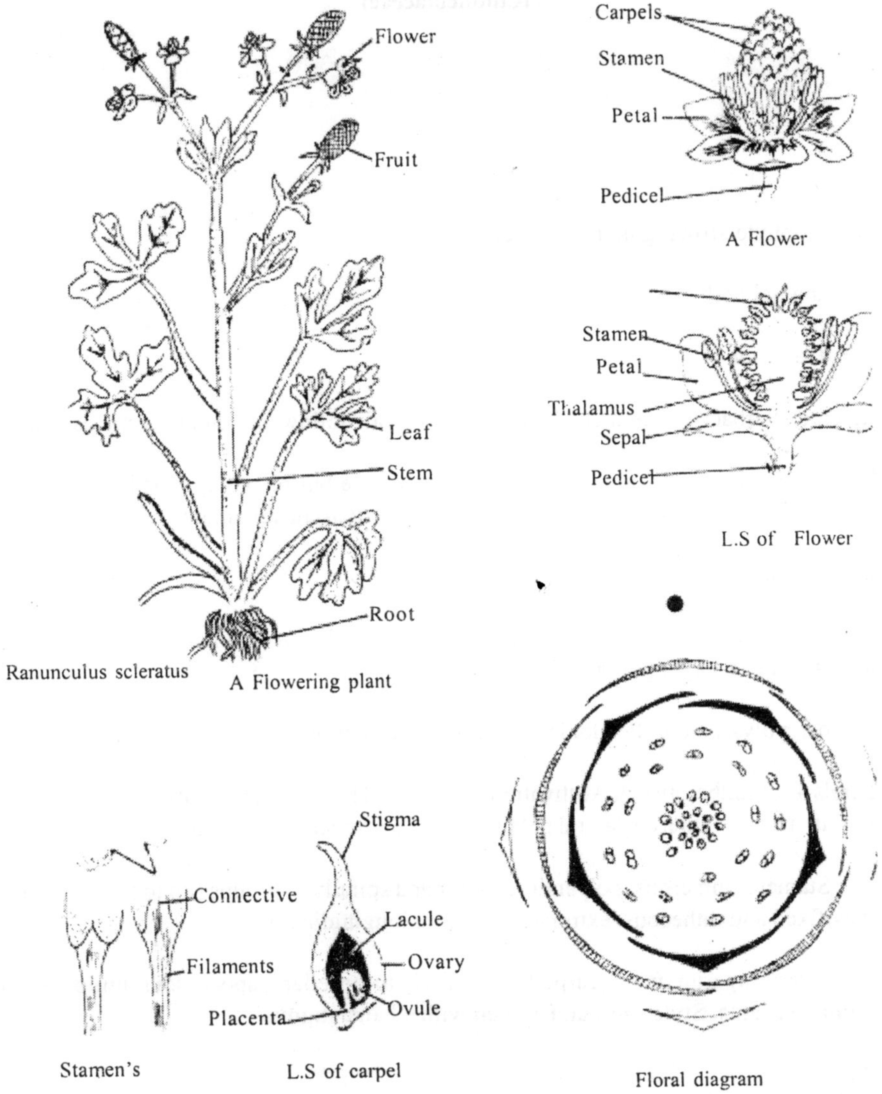

Fig 8.6 *RANUNCULUS SCLERATUS* (Ranunculaceae)

DELPHINIUM AJACIS

(Ranunculaceae)

Habitat : A mesophyte growing in moist places.

Habit : An erect annual herb

Root : Branched tap root.

Stem : Aerial, erect, herbaceous, branched, green cylindrical, hollow and minutely pubescent

Leaf : Cauline and ramal sessile, alternate exstipulate, simple but pinnately lobed with each linear lobe ending in an acute apex. Leaf surface glabrous. Venation musticostate reticulate

Inflorescence : A simple recame.

Flower : Bracteate (lobed and green bracts), bracteolate, complete, heterochlamydeous, regular, pentamerous, actinomorphic, bisexual and hypogynous.

Calyx : 5 in number, polysepalous, Petaloid with posterior sepal prolonged into a spur. Aestivation imbricate.

Corolla : Petals 4 in number, fused, Aestivation imbricate. The two posterior petals form a spur extending into the Calyx spur. The two anterior/lateral petals enclose androecium and gynoecium.

Androecium : Stamens numerous, polyandrous arranged spirally in groups (up to 5) of 3 each. Filaments long; anthers basifixed and dithecous, extrorse, dehiscing longitudinally.\

Gynoecium : Ovary superior, mono carpellary or rarely bicarpellary, apocarpous, unilocular with many ovules on marginal placenta. Style short and curved with a hairy stigma.

Fruit : A follicle.

Seed : Albuminons, embryo small.

Floral formula : ✝ ♀ $K_5\, C_{(4)}\, A_{\infty}\, \underline{G}_1$

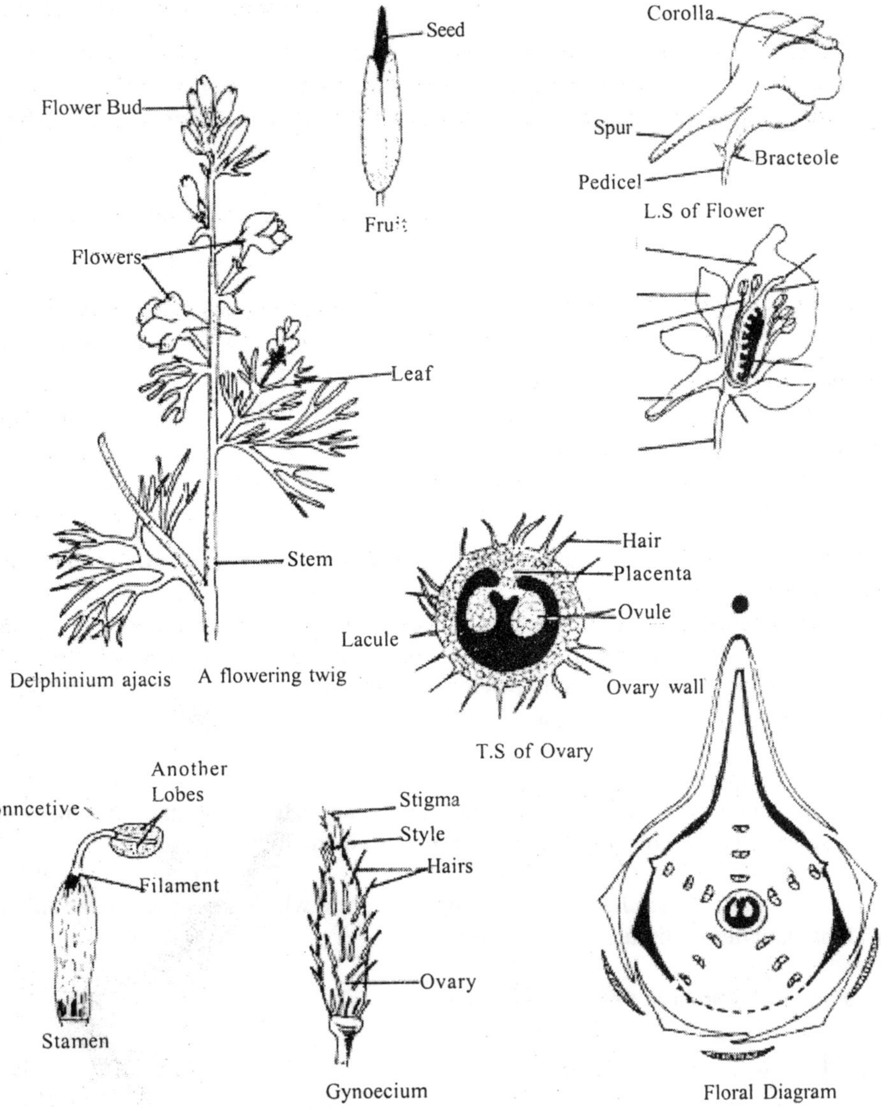

Fig 8.7 ***DELPHINIUM AJACIS*** **(Ranunculaceae)**

PAPAVER SOMNIFERUM

(Papaveraceae)

Habitat: Mesophyte growing is moist places

Habit: An annual herb

Root: Branched tap root

Stem: Aerial, erect, herbaceous, cylindrical, branched, glabrous or occasionally pubescent, contains milky latex.

Leaf: Cauline and ramal, simple, sessile, exstipulate, semiamplexicaul, phyllotaxy alternate venation unicostate reticulate.

Inflorescence: Flowers large, solitary terminal or axillary.

Flower: Pedicellate, ebracteate, ebracteolate, regular, complete, actinomorphic bisexual and hypogynous.

Calyx: Sepals two in number, free, hairy, caducous with overlapping arrangement.

Corolla: Petals four in number, free (polypetalous) arranged in several alternating whorls. Anthers basifixed, dithecous, extrorse with longitudinal dehiscence.

Gynoecium: Ovary superior, polycarpellary syncarpous unilocular with ovules arranged on parietal placenta. The placenta projects into the cavity of the ovary incompletely dividing it into many chambers, style not distinct, stigma discoid, lobed or rayed and persistant in the fruit.

Fruit: A poricidal capsule. The seeds come out through the pores which appear below the stigmatic rays. The capsule wall has laticiferous tissues yielding latex.

Seed: Small, albuminous, embryo small, seeds contain oil.

Floral formula: † ♀ $K_2\ C_{2+2}\ A_{\infty}\ \underline{G}_{(8)}$

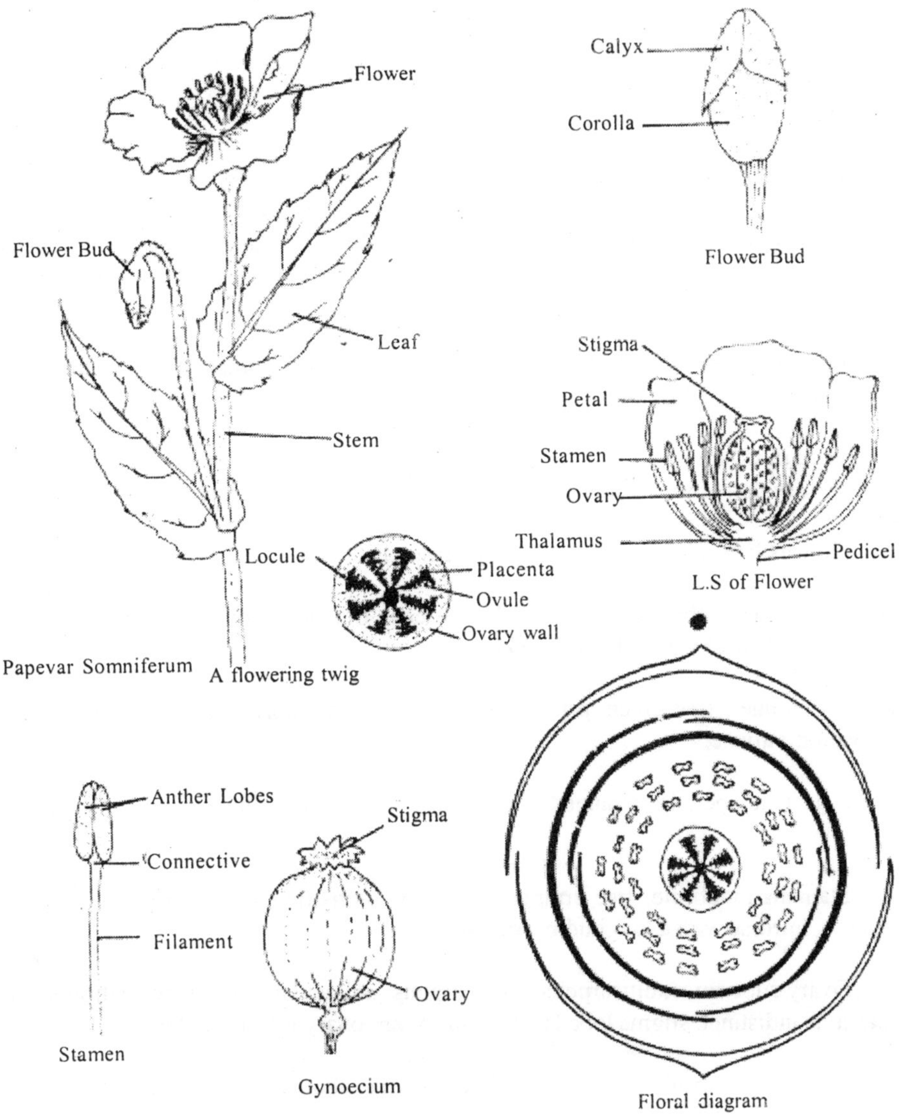

Fig 8.8 ***PAPAVER SOMNIFERUM*** **(Papaveraceae)**

ARGEMONE MEXICANA

(Papaveraceae)

Habitat: An annual weed found growing in dry places.

Habit: An erect prickly herb sometimes may reach a bushy habit. The plant contains yellowish latex.

Root: Much branched tap root

Stem: Aerial, erect, sparingly branched, herbaceous, occasionally becomes woody below, cylindrical, spinous, solid, glaucous with yellow latex.

Leaf: Ramal and cauline, simple, alternate, exxtipulate, sessile glaucous, margin deeply dissected with spiny teeth. Leaf surface dotted with white spots. Venation reticulate unicostate.

Inflorescence: Flowers large and attractive; solitary terminal

Flower: Pedicellate, ebracteate, ebracteolate, heterochlamydeous, trimerous, complete, actinomorphic, bisexual and hypogynous, pedicel short studded with prickles.

Calyx: Sepals 3 in number, free green, prickly and bear a distinct horny outgrowth at the apex sepals caducous. Aestivation valvate.

Corolla: Petals six in number, free arranged in two whorls of three each, shiny yellow in colour. Aestivation imbricate.

Androecium: Stamens indefinite, free arranged in many alternating whorls; filaments long, Anthers basifixed, dithecous, introrse with a longitudinal dehiscence

Gynoecium: Ovary superior, multicarpellary syncarpous unilocular with many anatropous ovules on parietal placenta style indistinct; stigma lobed and red in colour; ovary wall is prickly.

Fruit: A prickly septicidal capsule dehicing from apex down to about half the length.

Seed: Small, numerous and albuminous

Floral formula: ✝ ⚥ $K_3\ C_{3+3}\ A_\infty\ \underline{G}_{(3-7)}$

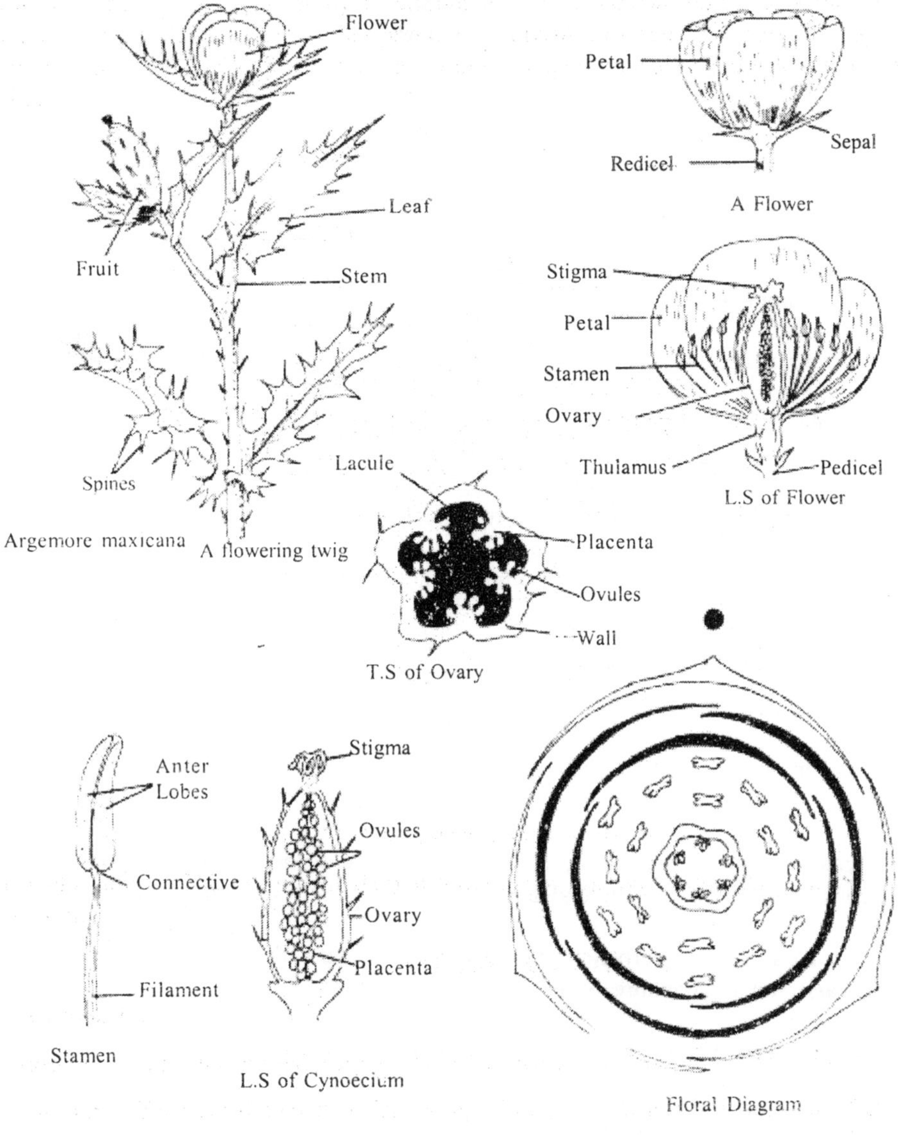

Fig 8.9 *ARGEMONE MEXICANA* (Papaveraceae)

CAPPARIS DECIDUA
(= C. aphylla)
(Capparidaceae)

Habitat: Wild, Xerophytic perennial.

Habit: Much branched shrub, occasionally reaching the height of a small tree.

Root: Deep feeding, branched tap root.

Stem: Aerial, erect branched somewhat woody, solid glabrous green and photosynthetic.

Leaf: Ramal, simple, alternate, petiolate stipulate with stipules modified into spines; leaves absent on older branches, caducous when present leaf has entire margin with acute apex. Venation reticulate unicostate.

Inflorescence: Corymbose raceme arising on short lateral branches.

Flower: Pedicellate, ebracteate, ebracteolate, complete, heterochlamydeous, zygomorphic bisexual and hypogynous. Gynophore present.

Calyx: Sepals greenish, 4 in number, free in two whorls of two each. Anterior and posterior two sepals form the outer whorl, while the lateral two form the inner whorl.

Corolla: Petals 4 in number, free, scarlet or reddish brown in colour. Aestivation twisted, petals oval in shape.

Androecium: Stamens many, polyandrous, inserted on a disc at the base of the gynophore. Anthers basifixed, dithecous introrse with longitudinal dehiscence.

Gynoecium: Ovary superior, bicarpellary syncarpous, unilocular with many ovules arranged on parietal plancenta. Ovary raised on a gynophore. Style absent, stigma bifid.

Fruit: Globose many seeded berry. Red in colour when ripe.

Seed: Large, exalbuminous

Floral formula: † ♀ $K_{2+2}\ C_4\ A_{\infty}\ \underline{G}_{(2)}$

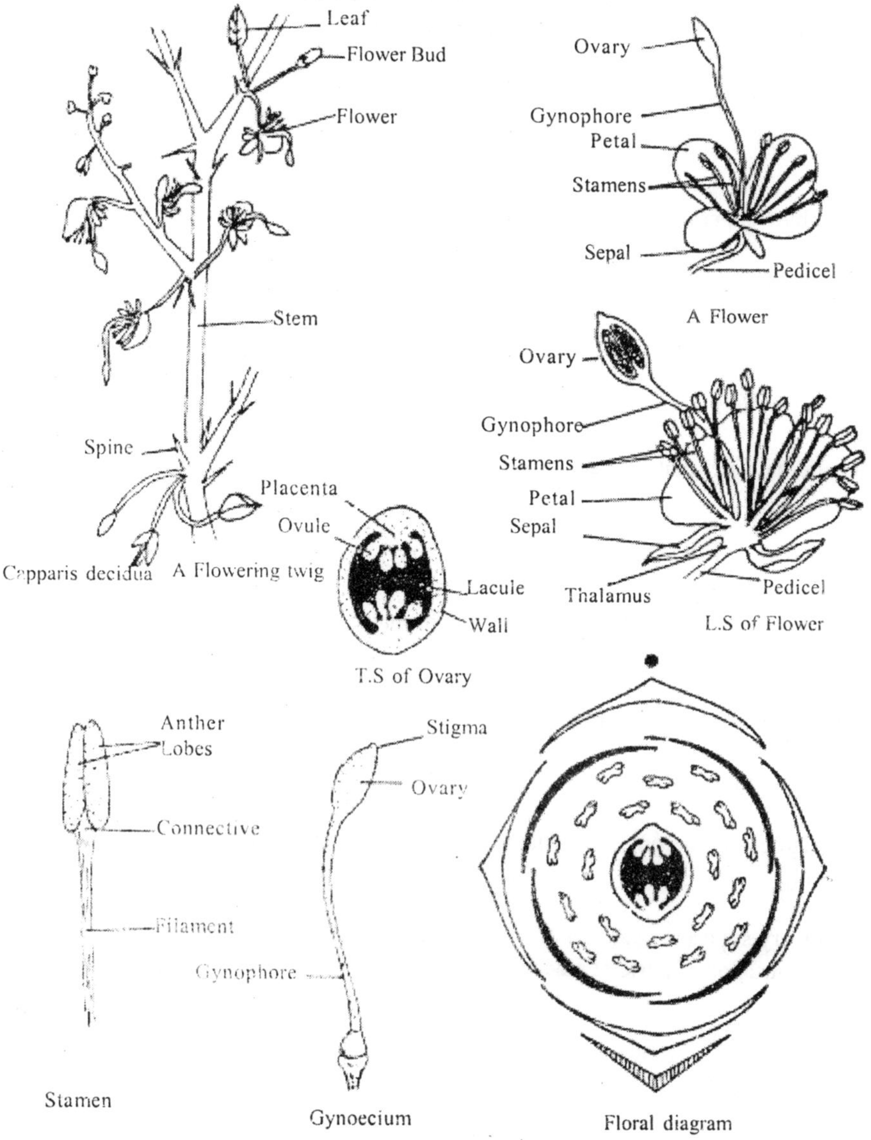

Fig 8.10 ***CAPPARIS DECIDUA*** (= ***C. alphylla***) **(Capparidaceae)**

CAPPARIS SPINOSA

(Capparidaceae)

Habitat : Mesophyte generally, grows in dry places.

Habit : A diffuse prostrate shrub.

Root : Branched tap root.

Stem : Aerial, erect, cylindrical somewhat woody with white pubescence specially on buds.

Leaf : Simple, alternate, petiolate, lovate, orbicular, stipulate with the two stipules being modified into two hooked spines

Inflorescence : Flowers solitary and axillary.

Flowers : Pedicellate, ebracteate, ebracteolate, zygomorphic, bisexual and hypogynous. Gynophore present. Flowers are large and showy.

Calyx : Sepals four in number, free, unequal, arranged in two whorls of two each.

Corolla : Petals four in number, free, brightly coloured, arranged in a single whorl. Aestivation twisted.

Androecium : Stamens numerous, free, filaments purple coloured and attractive; anthers dithecous with longitudinal splitting.

Gynoecium : Ovary superior made up of two to four carpels, syncarpous, unilocular sometimes many locular due to the protrusion of parietal placentae. Ovules many; ovary is borne on a gynophore.

Fruit : An ovoid, ribbed berry.

Seed : Reniform, small in size.

Floral formula : † ♀ $K_{2+2} C_4 A_{\infty} \underline{G}_{(2-4)}$

Fig 8.11 ***CAPPARIS SPINOSA*** **(Capparidaçeae)**

CLEOME VISCOSA

(Capparidaceae)

Habitat: Mesophyte usually found as weed in fields.

Habit: A small annual herb, appearing during rainy season.

Root: Branched tap root

Stem: Aerial, erect, soft often greenish in colour, grooved and pubescent.

Leaf: Petiolate, alternate, pinna 3-5 partite with elliptic oblong leaflets, apex of the leaflets acute.

Inflorescence: Simple raceme

Flower: Pedicellate, bracteate, ebracteolate, heterochlamydeous, actinomorphic bisexual and hypogynous.

Calyx: Sepals 4 in number free or connate at the base. They are lanceolate in shape and hairy outside sepals are arranged in two whorls of two each.

Corolla: Petals four in number, free, clawed, limb ovate. Aestivation valvate.

Androecium: Stamens range in number from 4-20, free. Anthers dithecous with longitudinal dehiscence.

Gynoecium: Ovary superior, bicarpellary, syncarpous, unilocular with ovules on parietal placenta. Gynophore absent, style short with simple stigma.

Fruit: A siliqua.

Seed: Small, black, sub globose in shape.

Floral formula: ✝ ⚥ $K_{2+2}\ C_4\ A_{4-20}\ \underline{G}_{(2)}$

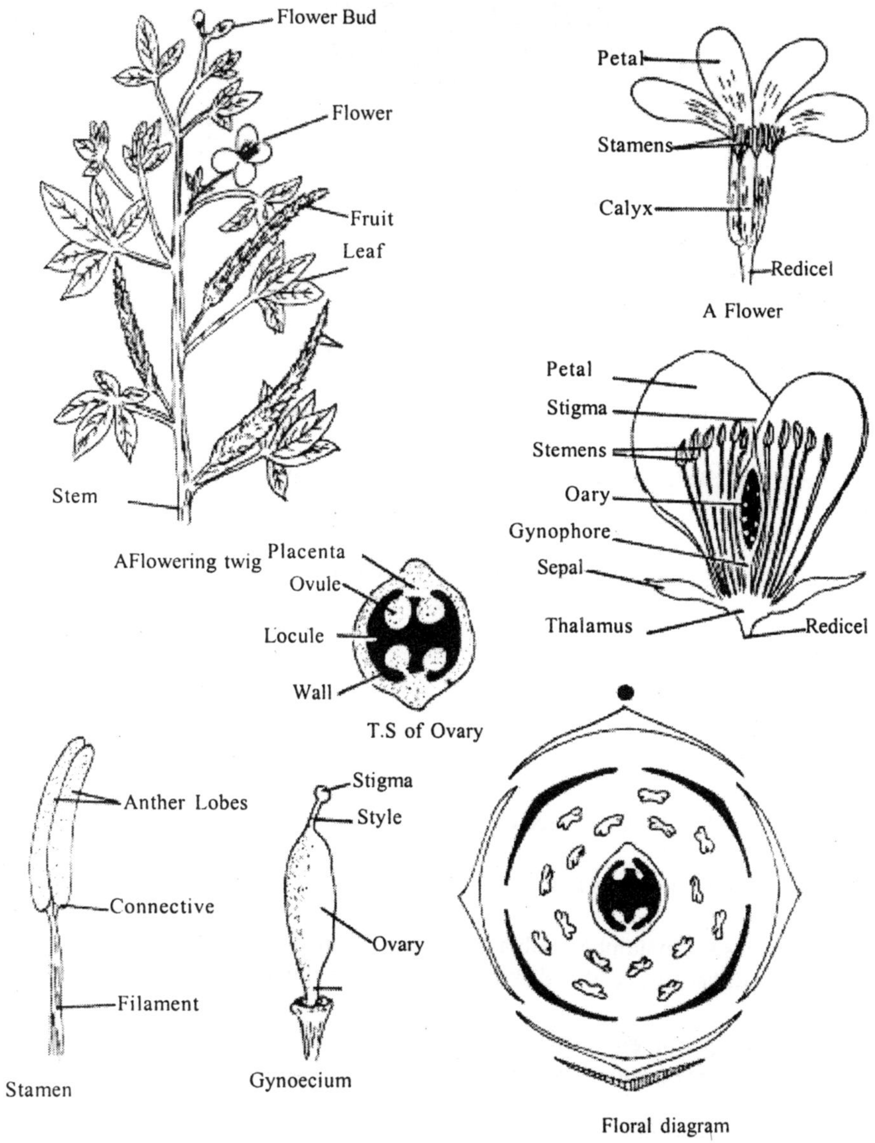

Fig 8.12 *CLEOME VISCOSA* (Capparidaceae)

CLEMOE GYNANDRA

(= Gynandropsis pentaphylla)

(Capparidaceae)

Habitat: Mesophyte, grows wild in open places.

Habit: Glandular, pubescent annual herb growing up to a height of 1-3 feet.

Root: Tap root.

Stem: Aerial, erect, branched, cylindrical, soft greenish above, somewhat woody at the base and glandular or viscoid.

Leaf: Ramal and cauline, petiolate, alternate, palmately compound with 3-5 leaflets. Leaflets sessile broadly ovate or elliptic oblong; venation unicostate reticulate.

Inflorescence: Corymbose raceme.

Flower: Pedicellate, bracteate (leaf bracts) ebracteolate, heterochlamydeous, actinomorphic bisexual and hypogynous. The thalamus is elongated into a long androgynophore, which is white or purplish in colour.

Calyx: sepals four in number, free, arranged in two whorls of two each. Aestivation imbricate.

Corolla: Petals four, free, arranged in valvate manner. Limb and claw can be distinguished in petals.

Androecium: Stamens six in number, polyandrous arising on an androphore which separates the petals from stamens. Anthers dithecous, basifixed extrorse with longitudinal dehiscence.

Gynoecium: Ovary superior, bicarpellary, syncarpous unilocular with many ovules arranged on parietal placentation. Pistil is located on a long gynophore; style short; stigma bifid. Ovules campylotropous.

Fruit: A capsule which looks like a siliqua.

Seed: Exalbuminous, reniform or orbicular with brown or black testa

Floral formula: ✝ ♀ $K_{2+2}\ C_4\ A_6\ \underline{G}_{(2)}$

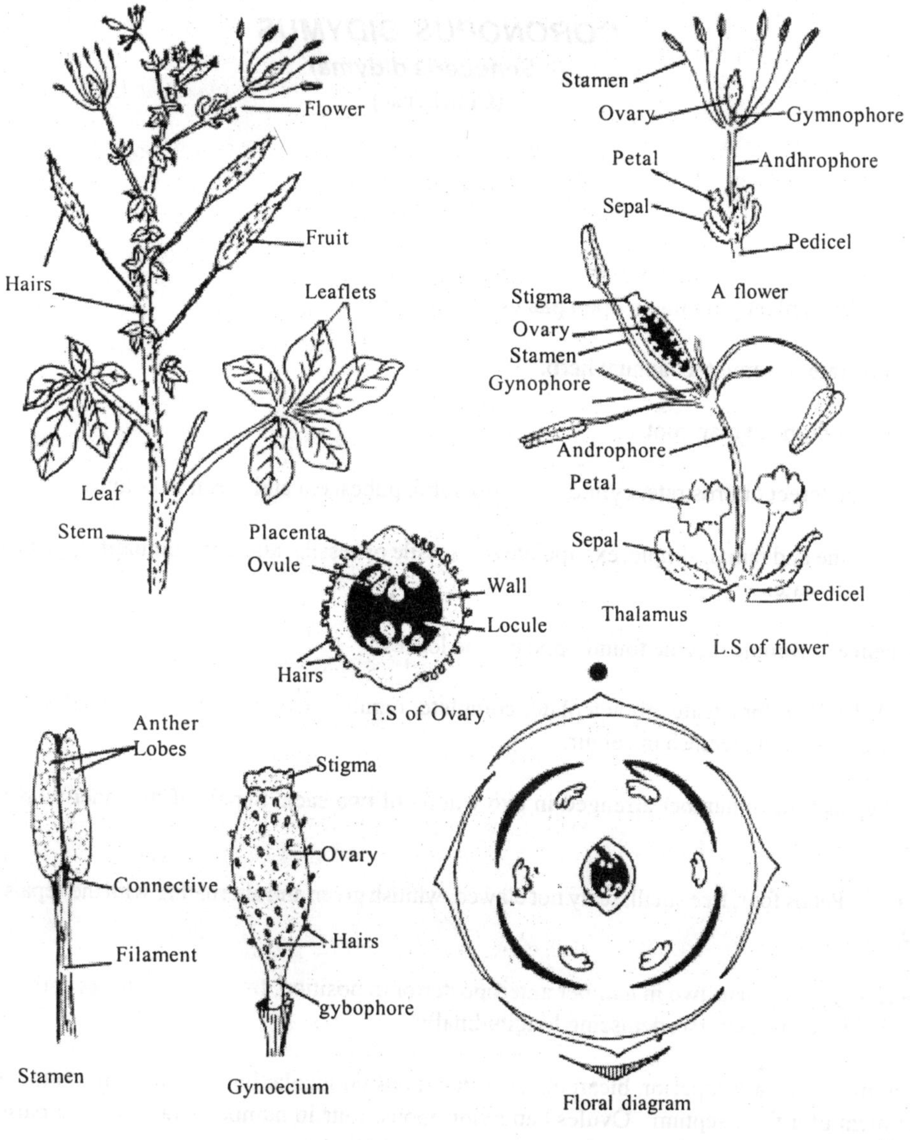

Fig 8.13 ***CLEMOE GYNANDRA (= Gynandropsis pentaphylla)*** **(Capparidaceae)**

CORONOPUS DIDYMUS

(= Seneberia didyma)

(Cruciferae)

Habitat : Wild plant growing in open places

Habit : prostrate or semierect annual herb.

Root : Much branched tap root.

Stem : Aerial, erect or prostrate, cylindrical, soft solid, pubescent and green in colour.

Leaf : Cauline and ramal, simple, exstipulate sub sessile or sessile, alternate, pinnatifid, glabrous, venation reticulate unicostate.

Inflorescence : Simple raceme found opposite the leaves.

Flower : Pedicellate, bracteate, ebracteolate, complete, regular, actinomorphic, bisexual and hypogynous. Flowers are small and greenish in colour.

Calyx : Sepals four in number arranged in two wholrs of two each. Sepals of the inner whorl (lateral) not saccate.

Corolla : Petals four, free small, scaly not clawed, whitish green and alternating with the sepals. Aestivation valvate.

Androecium : Stamens two in number anteroposterior in position, free with nectaries at the base. Anthers basifixed, dithecous, extrorse, dehiscing longitudinally.

Gynoecium : Ovary superior, bicarpellary, syncarpous unilocular but later becomes bilocular due to the development of a false septum. Ovules campylotropous, four in number arranged on a parietal placenta. Ovary short and laterally compressed

Fruit : A silicula (small siliqua)

Seed : Small exalbuminous

Floral formula : ✝ ⚥ $K_{2+2}\ C_4\ A_2\ \underline{G}_{(2)}$

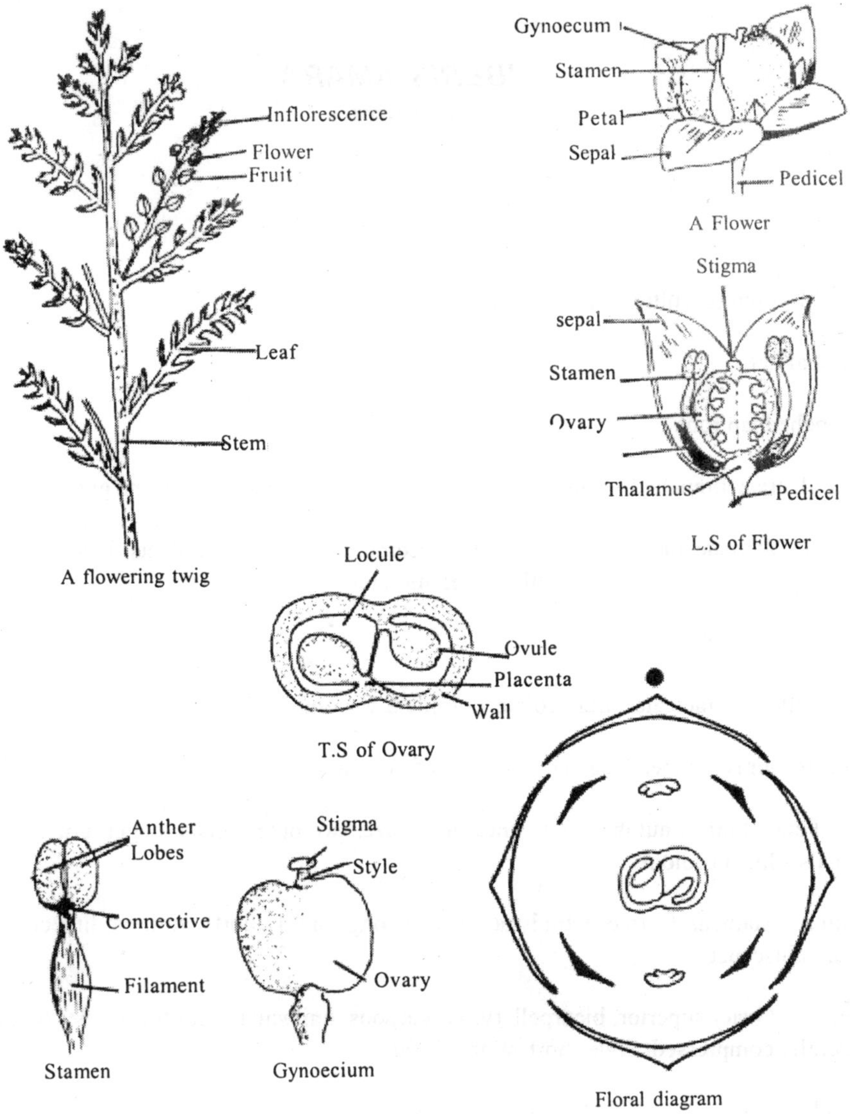

Fig 8.14 ***CORONOPUS DIDYMUS (= Seneberia didyma)*** **(Cruciferae)**

IBERIS AMARA

(Cruciferae)

Habitat : Mesophyte, cultivated as ornamental in gardens.

Habit : Annual herb.

Root : Branched tap root

Stem : Acrial, erect, branched, cylindrical, hollow soft above, woody below and green in colour.

Leaf : Simple, alternate, radical in young plants, becomes cauline in older plants. Leaves variously dissected lyrate, or ovate lanceolate, sessile, auriculate. Venation reticulate unicostate.

Inflorescence : Corymb.

Flower : Pedicellate, ebracteate, ebracteolate, complete, zygomorphic , bisexual and hypogunous.

Calyx : Sepals four is number, free in two whorls of two each. Aestivation imbricate.

Corolla : Petals four in number, free, unequal in size, two outer petals longer, cruciform. Aestivation valvate. Petals white in colour.

Androecium : Stamens 6, free, tetradynamous (2 long and 4 short), anthers dithecous, extrorse with longitudinal dehiscence.

Gynoecium : Ovary superior, bicarpellary, syncarpous, unilocular with few ovules on parietat placenta ovary is laterally compressed. Style short, stigma bifid.

Fruit : A Silicula

Seed : Exalbuminous, small.

Floral formula : † ♀ $K_{2+2}\ C_4\ A_{2+4}\ \underline{G}_{(2)}$

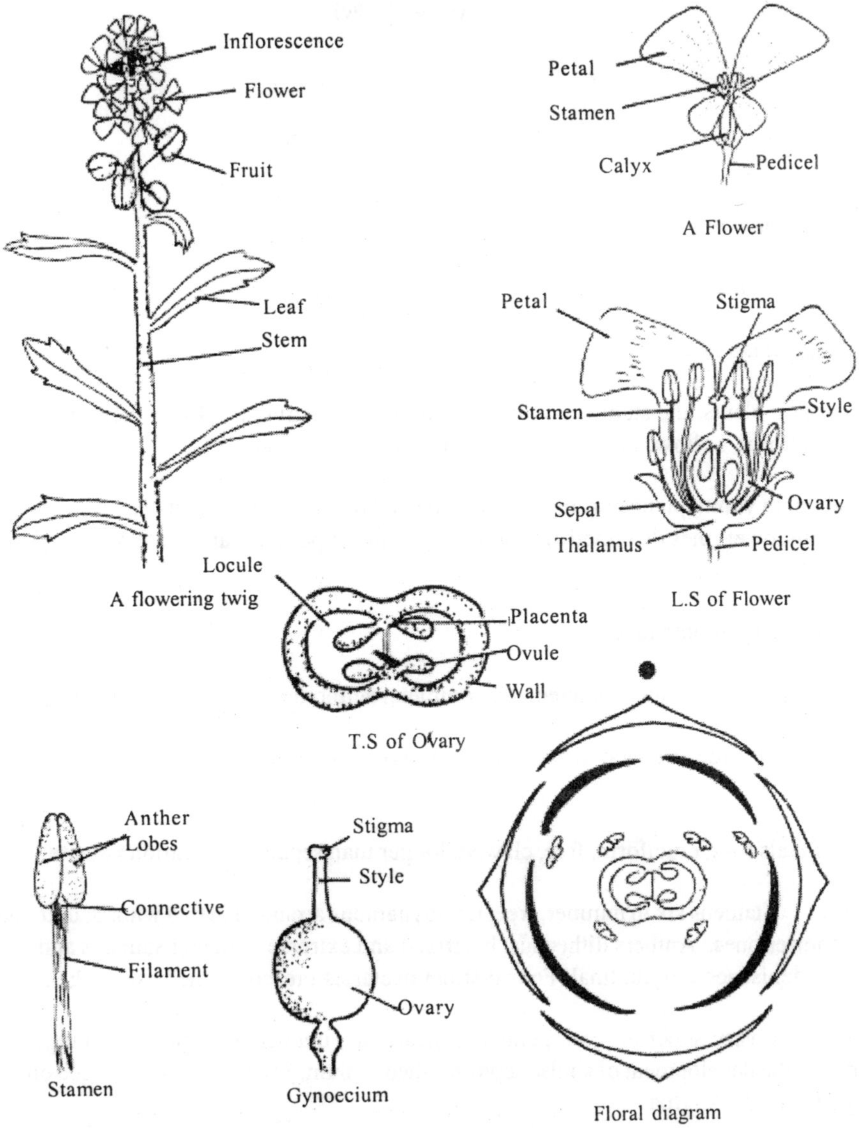

Fig 8.15 *IBERIS AMARA* (Cruciferae)

BRASSICA CAMPESTRIS
(Cruciferae)

Habitat: Mesophyte growing in open fields

Habit: An annual or biennial herb

Root: Branched tap root

Stem: In young plants, the stem is a reduced to a disc like structure. Later the stem becomes aerial and erect. It is cylindrical, hollow, soft above and somewhat woody below.

Leaf: Radical in young plants turning into cauline and ramal in older plants. Leaves are simple, alternate, exstipulate with lower ones dissected and lyrate while the upper ones are ovate or ovate lanceolate. Leaves are sessile also.

Inflorescence: Corymbose raceme

Flower: Pedicellate, ebracteate, ebracteolate, complete, actinomorphic, bisexual and hypogynous.

Calyx: Four sepals, free arranged in two whorls of two each. The anteroposterior sepals constitute the outer whorl.

Corolla: Petals four, cruciform, free, clawed, longer than sepals. Aestivation valvate

Androecium: Stamens six in number, free, tetradynamous arranged in two whorls; outer whorl consists of two lateral shorter ones. Anthers dithecous, basifixed and extrorse in longer stamens and introrse in shorter ones. Anther dehisence longitudinal. Four distinct nectaries present at the base of the stamens.

Gynoecium: Ovary superior, bicarpellary, syncarpous, unilocular in the beginning but later becomes bilocular due to the development of a false septum called replum. Ovules many, arranged on parietal placenta, style short with a bifid stigma.

Fruit: A siliqua

Seed: Exalbuminous, cotyledons conduplicate

Floral formula: ✝ ♀ $K_{2+2}\ C_4\ A_{2+4}\ \underline{G}_{(2)}$

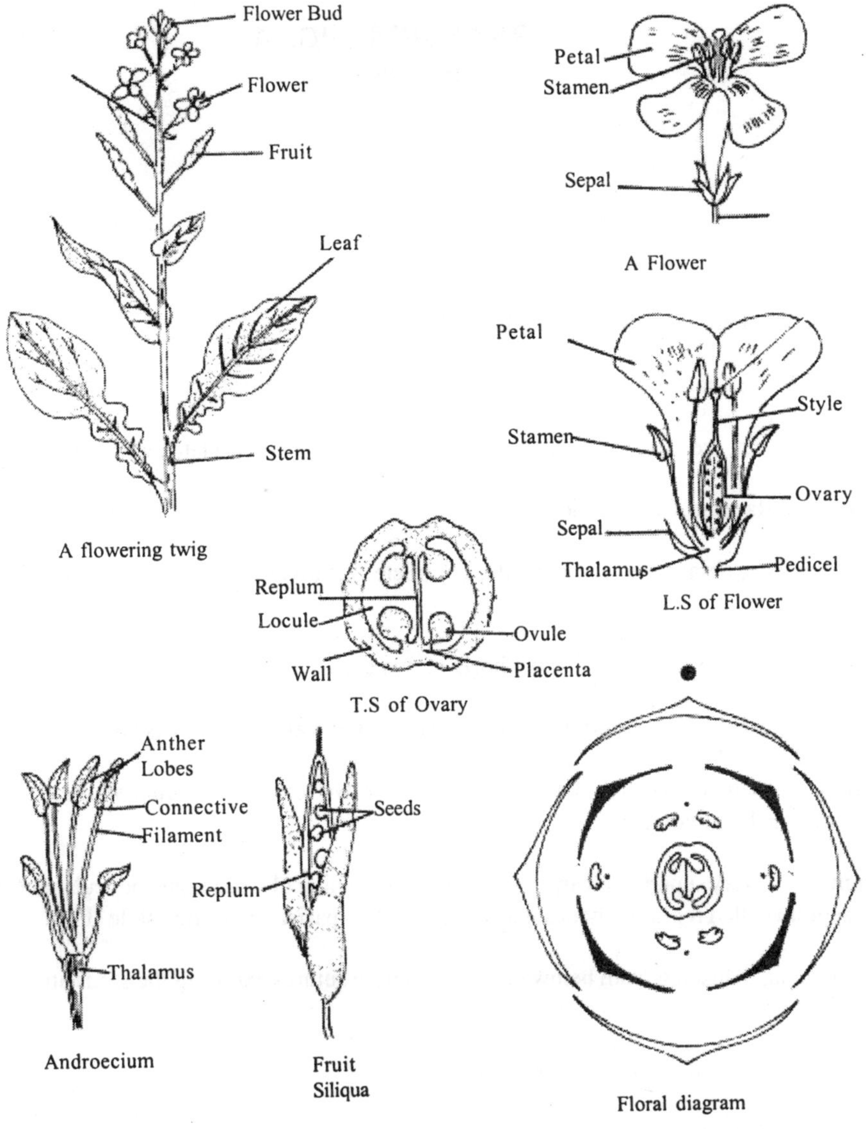

Fig 8.16 ***BRASSICA CAMPESTRIS*** **(Cruciferae)**

BRASSICA NIGRA
(Cruciferae)

Habitat: Mesophyte, grows in open fields

Habit: An annual herb growing to a height of about 2-3 feet.

Root: Branched tap root.

Stem: Aerial, erect, green, cylindrical, hollow shining, with yellowish latex with sulphurous odour.

Inflorescence: Raceme; Corymbose.

Flower: Pedicellate, ebracteate, ebracteolate, complete, actinomorphic, heterochlamydeous, bisexual and hypogynous.

Calyx: Sepals four free arranged in two whorls of two each.

Corolla: Petals four, free, cruciform, clawed. Aestivation valvate; yellowish in colour

Androecium: Stamens six, free, tetradynamous outer two longer, inner four shorter. Anthers dithecous, basifixed, splitting longitudinally.

Gynoecium: Ovary superior, bicarpellary, syncarpous, unilocular becoming bilocular due the development of a false septum called replum. Ovules many, arranged on parietal placenta, style short.

Fruit: A siliqua, dehiscing from below upwards on both sutures exposing the seeds attached to the central replum.

Seed: Non endospermic, cotyledons conduplicate

Floral formula: ✝ ♀ $K_{2+2}\ C_4\ A_{2+6}\ \underline{G}_{(2)}$

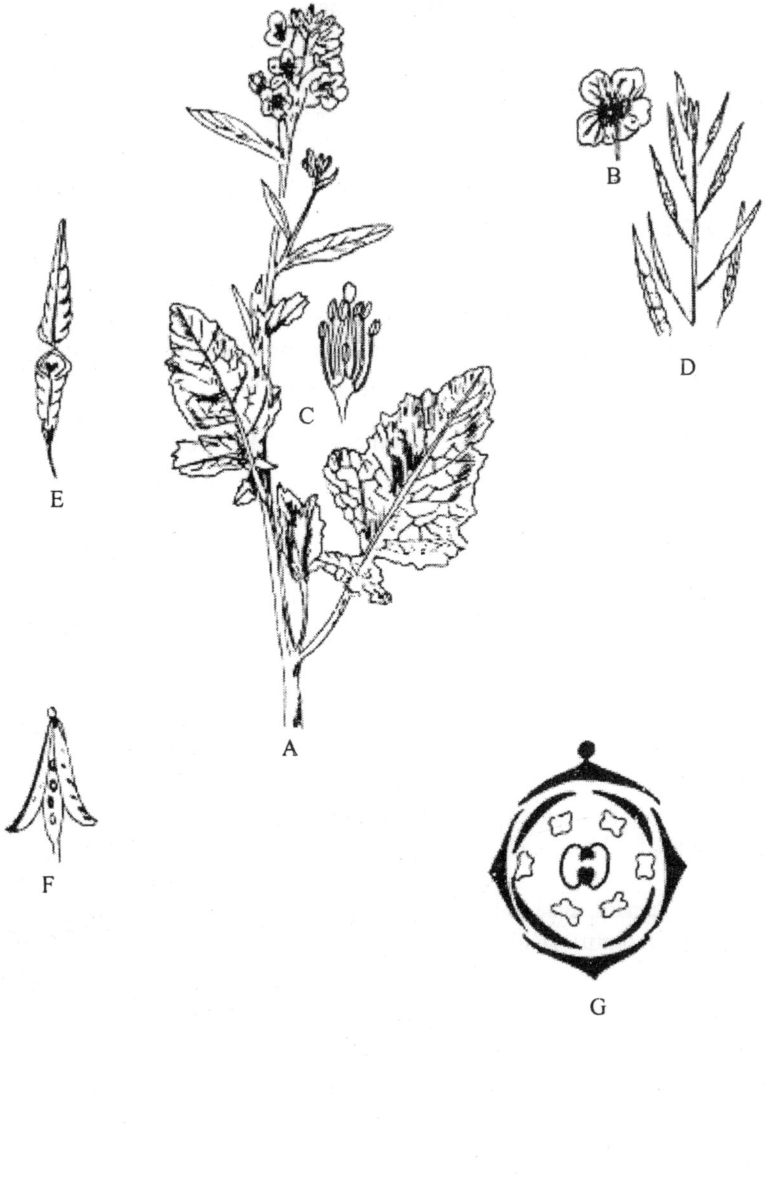

Fig 8.17 ***BRASSICA NIGRA*** **(Cruciferae)**

RAPHANUS SATIVUS

(Cruciferae)

Habitat: Mesophyte, growing in moist soils.

Habit: An annual herb.

Root: Tap root, branched; the tap root however becomes modified to store food. It is a swollen, white, spindle shaped structure called fusiform root (used as a vegetable)

Stem: Discoid; reduced in the young plant but becomes aerial later, green in colour with yellowish sulphur odoured latex.

Inflorescence: Raceme, corymbose.

Flower: Pedicellate, ebracteate, ebracteolate, actinomorphic, heterochlamydeous, bisexual and hypogynous.

Calyx: Four sepals, free arranged in two whorls of two each.

Corolla: Petals four free cruciform, white in colour. Each petal has a broad limb and a narrow claw. Aestivation valvate

Androecium: Stamens six, free, tetradynamous, outer two longer, inner four shorter. Anthers dithecous, basifixed with a longitudinal dehiscence.

Gynoecium: Ovary superior, bicarpellary, syncarpous, unilocular but becomes bilocular due to the formation of false septum (replum). Ovules many, arranged on parietal placenta. Style short.

Fruit: A siliqua, splitling on both valves from below upwards exposing the seeds attached to the central replum.

Seed: Non endosperomic, carpels conduplicate

Floral formula: $\dagger$ ♀ $K_{2+2}\ C_4\ A_{2+6}\ \underline{G}_{(2)}$

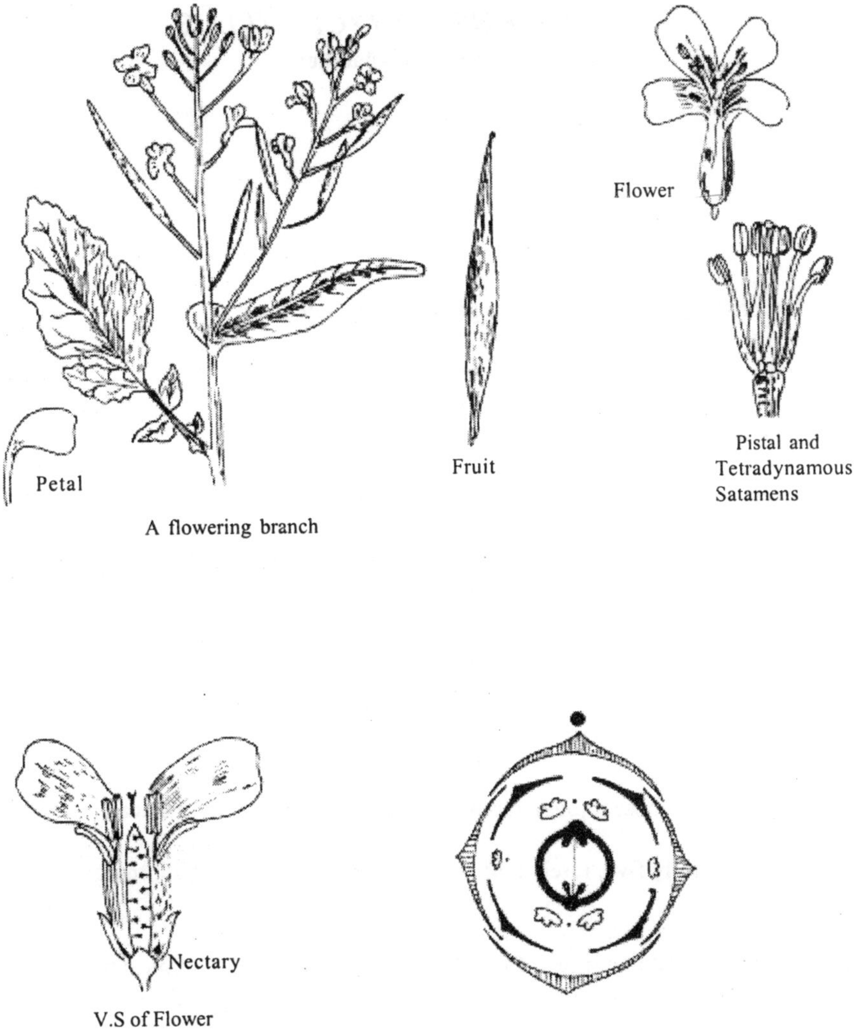

Fig 8.18 *RAPHANUS SATIVUS* (Cruciferae)

DIANTHUS CARYOPHYLLUS

(Caryophyllaceae)

Habitat : Mesophyte, grows in fields and gardens, cultivated.

Habit : An annual herb flowering during winter

Root : Branched tap root

Stem : Aeral, erect, branched, cylindrical, solid, green colour and glabrous. Silica present in the epiderms.

Leaf : Simple cauline, opposite decussate, sessile, exstipulate, entire margin, linear lanceolate and somewhat coriaceous. Venation multicostate and reticulate.

Inflorescence : Cymose cluster (Dichacial) or solitary terminal

Flower : Pedicellate, bracteate (many bracts), bracteolate, heterochlamydeous, cyclic, complete, actinomorphic bisexual and hypogynous.

Calyx : Sepals 5, fused (gamoseaplous), tubular, persistant. Aestivation imbricate.

Corolla : Petals generally five, rarely more, free spirally arranged in bud, margin lobed, pink in colour. Aestivation twisted.

Androecium : Stamens 10 in number, free in two whorls of 5 each, obdiplostemnous, filaments slender. Anthers dorsifixed, dithecous, adnate to the petals, introrse with a longitudinal dehiscence.

Gynoecium : Ovary superior, bicarpellary, syncarpous, incompletely bilocular with many ovules on a free central placentation; styles two, stigma simple.

Fruit : A capsule.

Seed : Albuminous, embryo straight.

Floral formula : † ♀ $K_{(5)}\ C_5\ A_{5+5}\ \underline{G}_{(2)}$

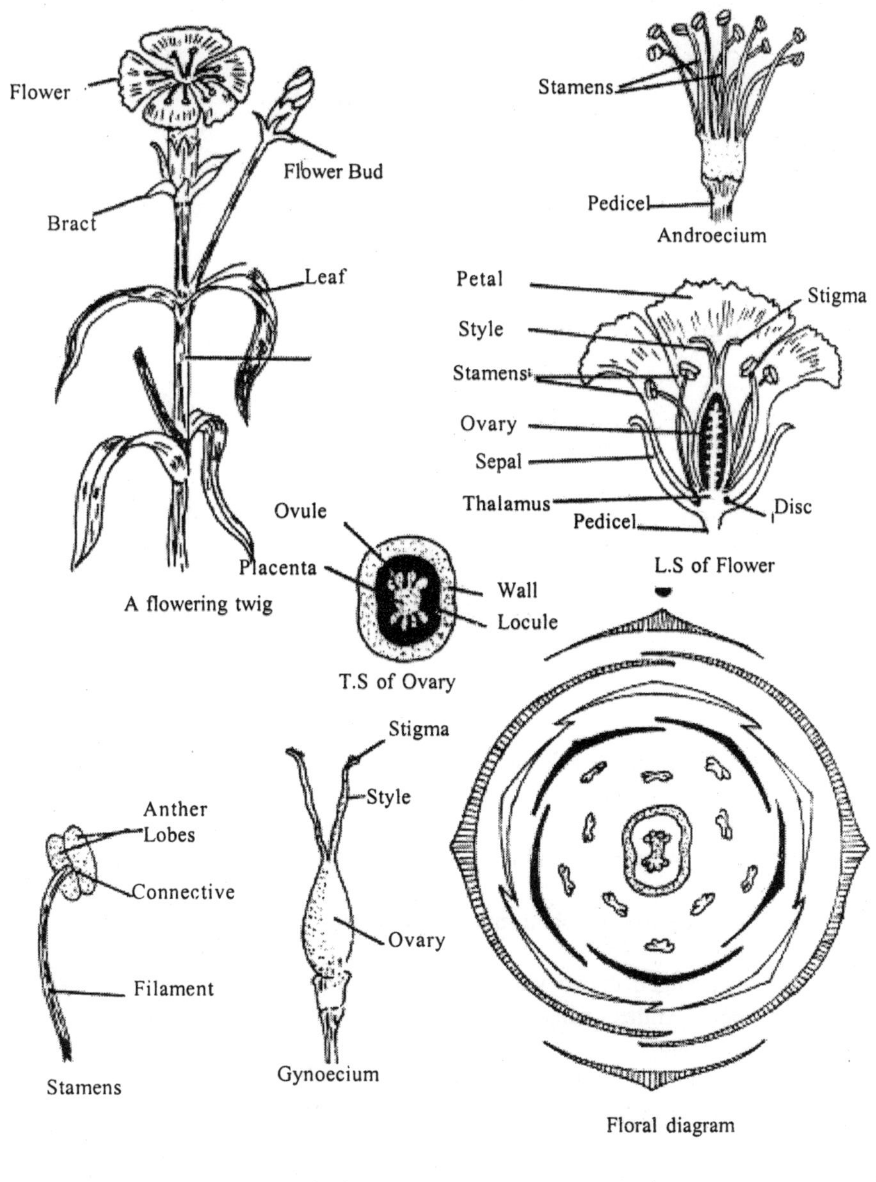

Fig 8.19 ***DIANTHUS CARYOPHYLLUS*** **(Caryophyllaceae)**

STELLARIA MEDIA

(Caryophyllaceae)

Habitat : A common weed growing is wasteland

Habit : An annual herb.

Root : Branched tap root.

Stem : Prostrate or erect, decumbent, much branched, green, soft cylindrical, upto two feet in height and slightly swollen at the nodes with occasional pubescence.

Leaf : Simple, ramal and cauline, opposite and decustate with ovate or cordate lamina in the lower leaves and lanceolate shape in upper leaves; venation reticulate unicostate.

Inflorescence : Dichacial Cyme

Flower : Pedicellate (glandular), bracteate, ebracteolate, complete, actinomorphic, pentamerous, bisexual, hypogynous.

Calyx : Sepals 5, free, glandular. Aestivation imbricate (quincuncial)

Corolla : Petals 5, rarely 4, free, deeply lobed shorter than sepals. Aestivation imbricate.

Androecium : Stamens 10 arranged in two whorls of 5 each, occasionally 8 in two whorls of 5+3 in outer and inner and obdiplostemnous. Anthers adnate, dithecous with longitudinal dehiscence.

Gynoecium : Ovary superior, tricarpellary, syncarpous, unilocular with many ovules on free central placentation. Remnants of septa incompletely dividing the ovary into 3 locules are visible. Styles 3, stigma minute.

Fruit : Ovoid capsule opening on 3 valves.

Seed : Albuminous, with curved embroyo

Floral formula : ✝ ♀ $K_5\ C_5\ A_{5+5 \text{ or } 5+3}\ \underline{G}_{(3)}$

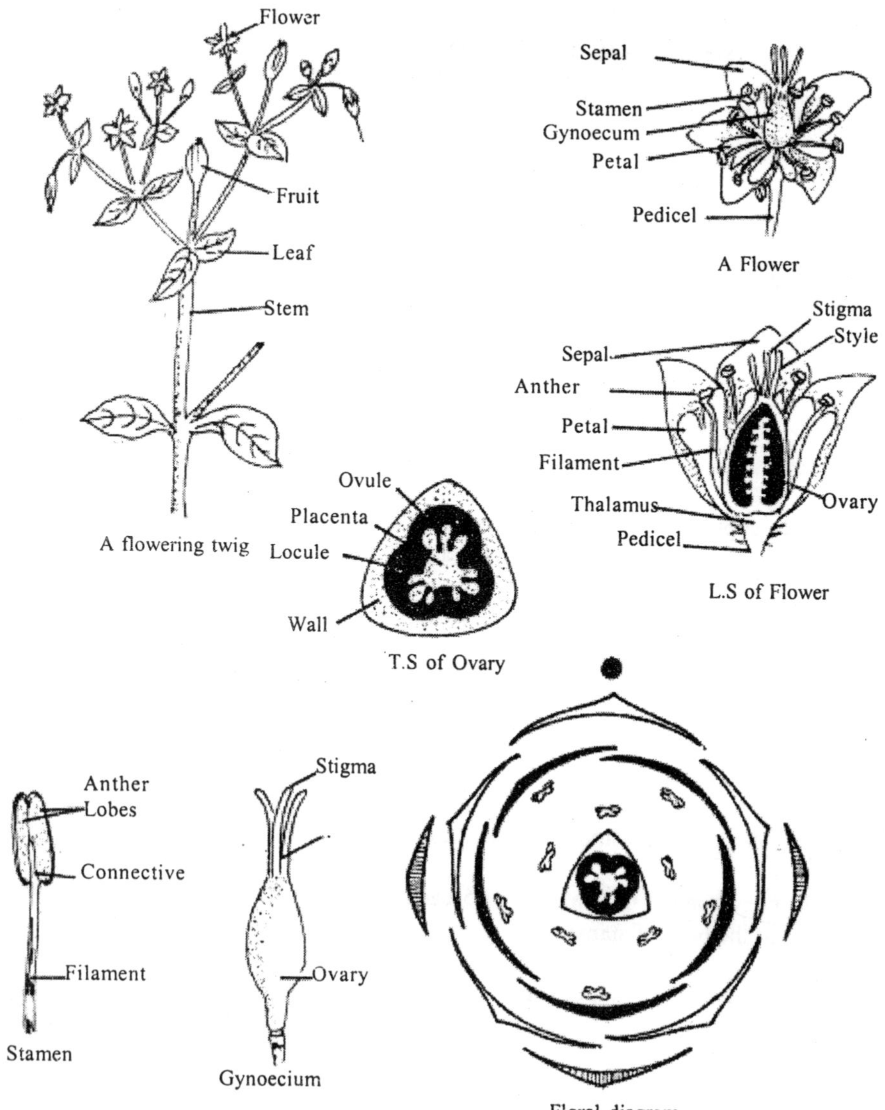

Fig 8.20 *STELLARIA MEDIA* (Caryophyllaceae)

SPERGULA ARVENSIS

(Caryophyllaceae)

Habitat: Mesophyte, growing in open fields

Habit: Annual herb.

Root: Branched tap root.

Stem: Aerial, erect, cylindrical, solid, soft, branched and green in colour.

Leaf: Simple, sessile, opposite, stipulate with minute and scarious stipules, leaves linear and acicular, and fleshy; venation reticulate unicostate; leafy buds present in the axils.

Inflorescence: Dichacial cyme; branches of the peduucle are suppressed on one side.

Flower: Pedicellate, bracteate, bracteolate (2), complete, pentamerous, actinomorphic, bisexual and hypogynous. Flowers greenish white in colour.

Calyx: Sepals 5, free. Aestivation quincuncial

Corolla: Petals 5, free, caryophyllaceous, white in colour. Aestivation imbricate.

Androecium: Stamens 10, free arranged in 2 whorls of 5 each, obdiplostemnous; anthers dorsifixed, dithecous, introrse with glands at the base of stamens.

Gynoecium: Ovary superior, pentacarpellary, syncarpous, unilocular with ovules on free central placentation.

Fruit: A Unilocular capsule

Seed: Small endospermic

Floral formula: ✝ ♀ $K_5\ C_5\ A_{5+5}\ \underline{G}_{(5)}$

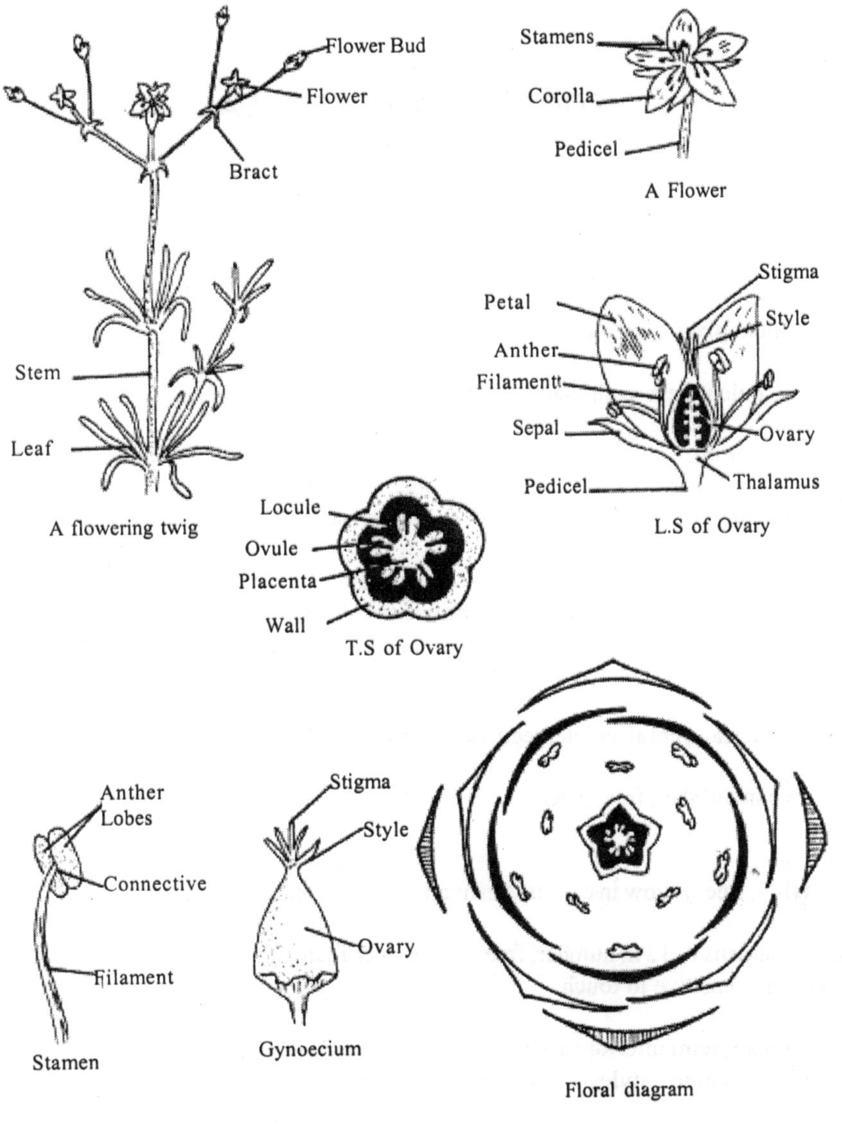

Fig 8.21 ***SPERGULA ARVENSIS*** **(Caryophyllaceae)**

PORTULACA OLERACEA
(Portulacaceae)

Habitat : Succulent plant growing as weed in waste places.

Habit : Annual succulent herb with many prostrate and ascending branches.

Root : Branched tap root.

Stem : Aerial, creeping, herbaceous, branched, succulent, solid and glabrous.

Leaf : Cauline, stipulate (hairy), opposite or sub opposite, rarely alternate, sessile or sub sessile, and spathulate with obtuse or truncated apex and margin entire. Venation reticulate unicostate.

Inflorescence : Flowers in terminal clusters; rarely solitary.

Flower : Sessile, ebracteate, ebracteolate, regular, complete, cyclic, bisexual and perigynous.

Calyx : Sepals two in number, free, placed in an anteroposterior position with anterior one overlapping the posterior.

Corolla : Petals 5, free, yellow in colour. Aestivation imbricate.

Androecium : Stamens 8+12 in number, free with slender filaments and dithecous, longitudinally splitting anthers. Stamens are sensitive to touch.

Gynoecium : Ovary semi inferior carpels 3-5, syncarpous unilocular with many campylotropous ovules on a free central placentation. Styles 3-5 with a minute stigma.

Fruit : Dyxidium (A many seeded capsule dehiscing by transverse slits)

Floral formula : ✝ ♀ $K_2 C_5 A_{8+12} \underline{G}_{(3-5)}$

Fig 8.22 ***PORTULACA OLERACEA*** **(Portulacaceae)**

HIBISCUS ROSA SINENSIS
(Malvaceae)

Habitat : Mesophyte, grows in open lands.

Habit : Perennial Shrub reaching a height of 8-10 feet.

Root : Branched tap root.

Stem : Aerial, erect, branched, solid, cylindrical and woody.

Leaf : Ramal and cauline simple, petiolate, alternate, stipulate, ovate, serrate margin, apex acute, venation reticulate unicostate.

Inflorescence : Solitary axillary flowers, with jointed peduncles.

Flower : Pedicellate, ebracteate, ebracteolate, complete, actinomorphic, cyclic, bisexual, protandrous and hypogynous.

Calyx : Outside the calyx there is a whorl of 5 epicalyx lobes; calyx 5, gamesepalous. Aestivation valvate.

Corolla : Petals 5, Polypetalous, obovate with sinuous upper margins. Petals are of various shades of red and other colours and mucilagenous. Aestivation twisted.

Androecium : Stamens many; monadelphous epipetalous. The staminal tube surrounds the style and red in colour. Anthers yellow, reniform, monothecous, extrorse dehiscing longitudinally.

Gynoecium : Ovary superior, pentacarpellary, syncarpous, pentalocular with many ovules on axile placenta. Style tubes united at base, free above with 5 stigmatic surfaces. The stigmas are valvety and red in colour.

Fruit : 5 valved, loculicidal capsule.

Seed : Albuminous, hairy.

Floral formula : ✝ ♀ $Epik_2\ K_{(5)}\ C_4\ A_{\infty}\ \underline{G}_{(5)}$

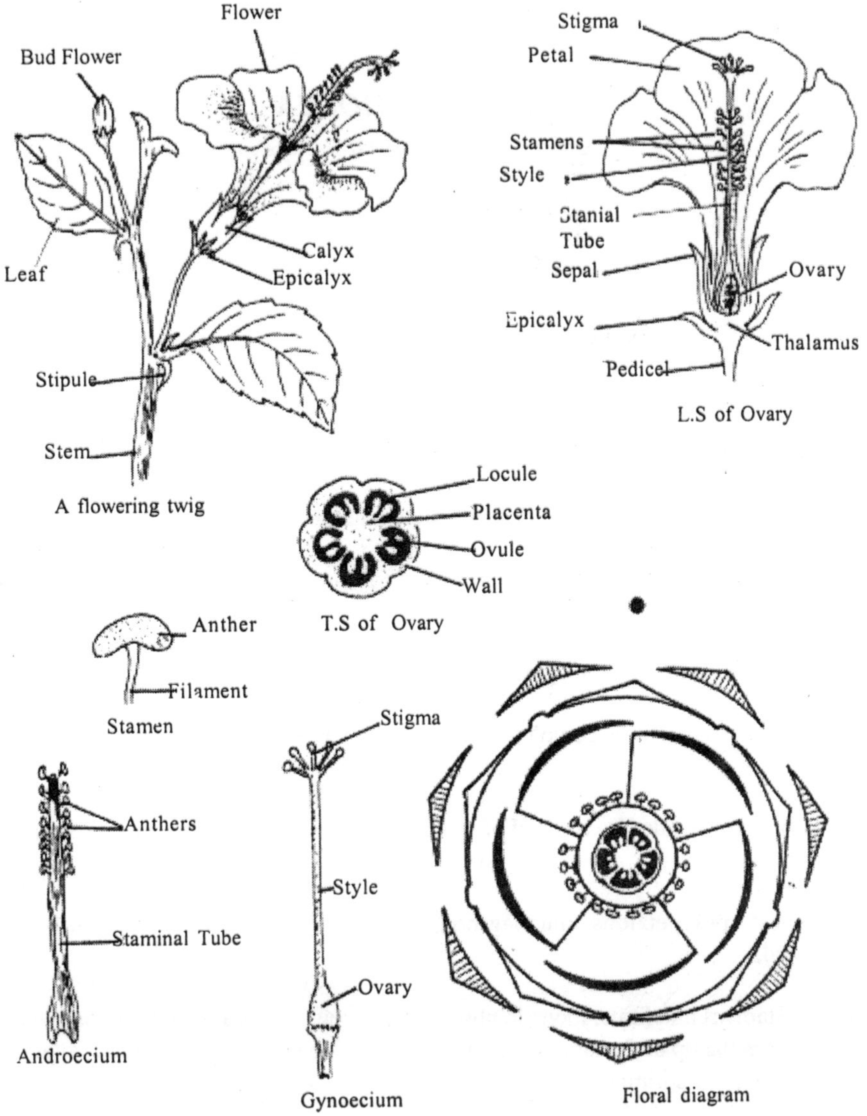

Fig 8.23 ***HIBISCUS ROSA SINENSIS*** (Malvaceae)

ALTHAEA ROSEA

(Malvaceae)

Habitat: Mesophyte, growing in open places.

Habit: An annual herb sometimes reaching a height of 6-7 feet.

Root: Tap root, the mucilagenous root extract is edible as a cooling drink.

Stem: Aerial, erect, cylindrical, branched, green, mucilagenous, studded with stellate hairs.

Leaf: Ramal and cauline, simple, alternate, stipulate (hairy), palmately lobed, hairy. Venation reticulate multicostate divergent.

Inflorescence: Solitary axillary or in cymose clusters.

Flowers: Pedicellate, ebracteate, complete, actinomorphic, heterochlamydeous, protandrous, bisexual and hypogynous.

Calyx: The calyx is surrounded by a whorl of 6-9 fused epicalyx lobes. Calyx itself is made up of 5 fused sepals. Aestivation valvate.

Corolla: Petals 5, polypetalous, mucilaginous, obovate with sinuate apex and brightly coloured. Aestivation twisted.

Androecium: Stamens indefinite, monadelphous, epipetalous. The staminal tube completely encircles the ovary and style. From the tip of the tube filaments arise carrying monothecous anthers attached transversely to the tip. Anthers extrorse splitting longitudinally.

Gynoecium: Ovary Superior, polycarpellary, syncarpous, multilocular with one ovule in each locule on axile placenta. Styles as many as carpels; free for most of their length except at the base.

Fruit: A carcerula.

Seed: Endosperm scanty, reniform in shape

Floral formula: ✝ ♀ $Epik_{(6-9)}\ K_{(5)}\ C_5\ A_{(\infty)}\ \underline{G}_{(\infty)}$

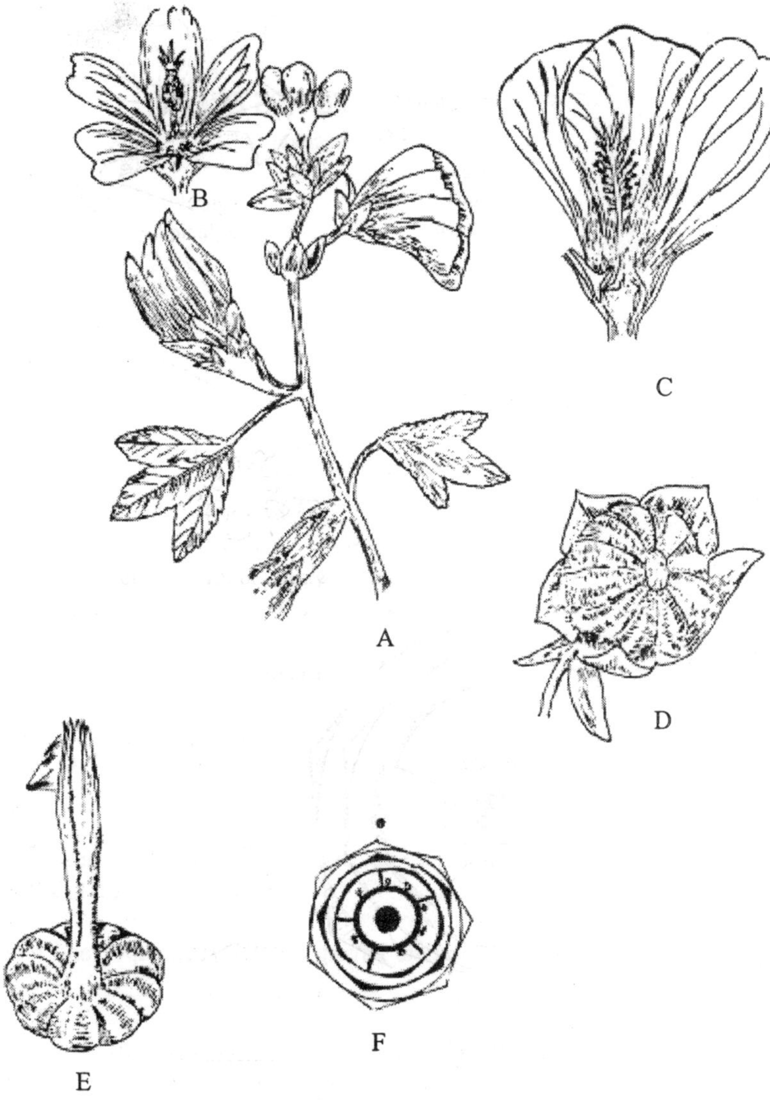

Fig 8.24 ***ALTHAEA ROSEA*** **(Malvaceae)**

MALVA SYLYESTRLS
(Malvaceae)

Habitat: Mesophyte, weed, grows in wasteland.

Habit: Annual herb, 3-5 feet in height.

Root: Branched tap root.

Stem: Aerial, erect, branched, cylindrical, solid covered with stellate hairs, green in colour and herbaceous.

Leaf: Simple, alternate, petiolate, petioles as long as the lamina, stipulate, cordate, rounded or reniform, palmately lobed and covered with hairs; venation reticulate multicostate.

Inflorescence: Axillary cymose cluster, each cluster has 4-5 flowers arising on a peduncle. Rarely flowers may be solitary axillary.

Flower: Pedicellate (hairy), bracteate, bracteolate (epicalyx), complete, cyclic, actinomorphic, heterochalmydeous, bisexual and hypogynous.

Calyx: The calyx is surrounded by three free epicalyx lobes. Sepals 5, gamosepalous, odd sepal anterior. Aestivation valvate.

Corolla: Petals 5, polypetalous, spathulate having a notch at the apex and reddish yellow in colour. Aestivation twisted.

Androecium: Stamens many, monadelphons forming a column around the gynoecium. Filaments arise from the tip of the column. Anthers monothecous, dorsifixed, extrorse with a longitudinal dehiscence.

Gynoecium: Ovary superior, multicarpellary, syncarpous, multilocular with one ovule in each locule on axile placenta. Styles as many as carpels

Fruit: A carcerula splitting into many one seeded mericarps.

Seed: Albuminous with scanty endosperm flattened with a 'v' shaped slit on one side.

Floral formula: ✝ ⚥ $Epik_3\ K_{(5)}\ C_5\ A_{(\infty)}\ \underline{G}_{(10)}$

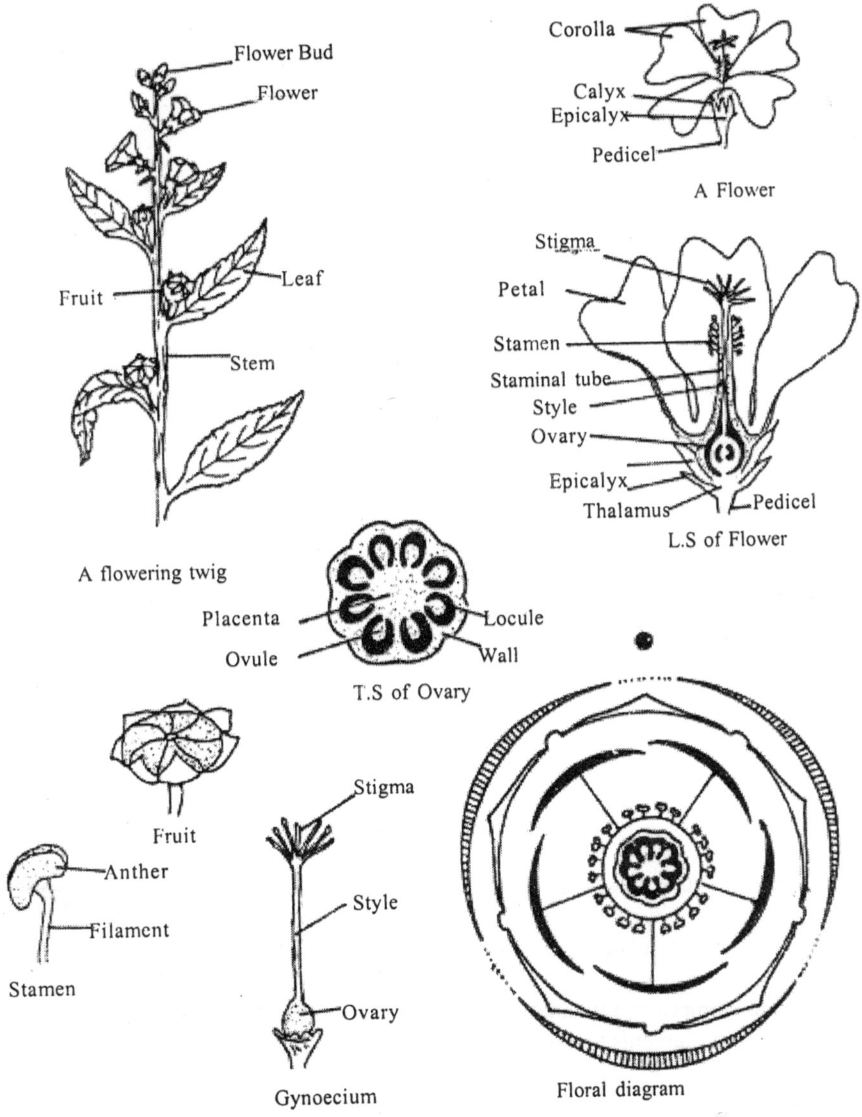

Fig 8.25 ***MALVA SYLYESTRLS*** **(Malvaceae)**

SIDA CORDIFOLIA
(Malvaceae)

Habitat: Mesophytic weed growing in dry open places.

Habit: A perennial undershrub

Root: Branched tap root.

Stem: Aerial, erect, branched solid, soft at the top, woody below.

Leaf: Simple, petiolate, alternate, margin entire at the base but serrate above, cordate, acute apex with both surfaces hairy.

Inflorescence: Flowers axillary solitary

Flower: Pedicellate, bracteate, complete, cyclic, actinomorphic, mucilagenous, heterochlamydeous, bisexual and hypogynous. There is no epicalyx.

Calyx: Sepals 5, fused persistant Aestivation valvate.

Corolla: Petals 5, free poly petalous, yellow occassionally connate at the base. Aestivation twisted.

Androecium: Stamens numerous, monadelphous fused to form a column surrounding the gynoecium. The column gives rise to free filaments bearing monothecous, reniform, vertically splitling anthers.

Gynoecium: Ovary superior, multicarpellary, syncarpous, multilocular with many ovules in each loucle on axile placenta. Style apically divided into as many branches as carpels each terminating in a capitate stigma.

Fruit: A Schizocarp, splitling into mericarps, mericarps are awned.

Seed: Small, black in colour with scanty endosperm.

Floral formula: ✝ ⚥ $K_{(5)}\ C_{(5)}\ A_{(\infty)}\ \underline{G}_{(\infty)}$

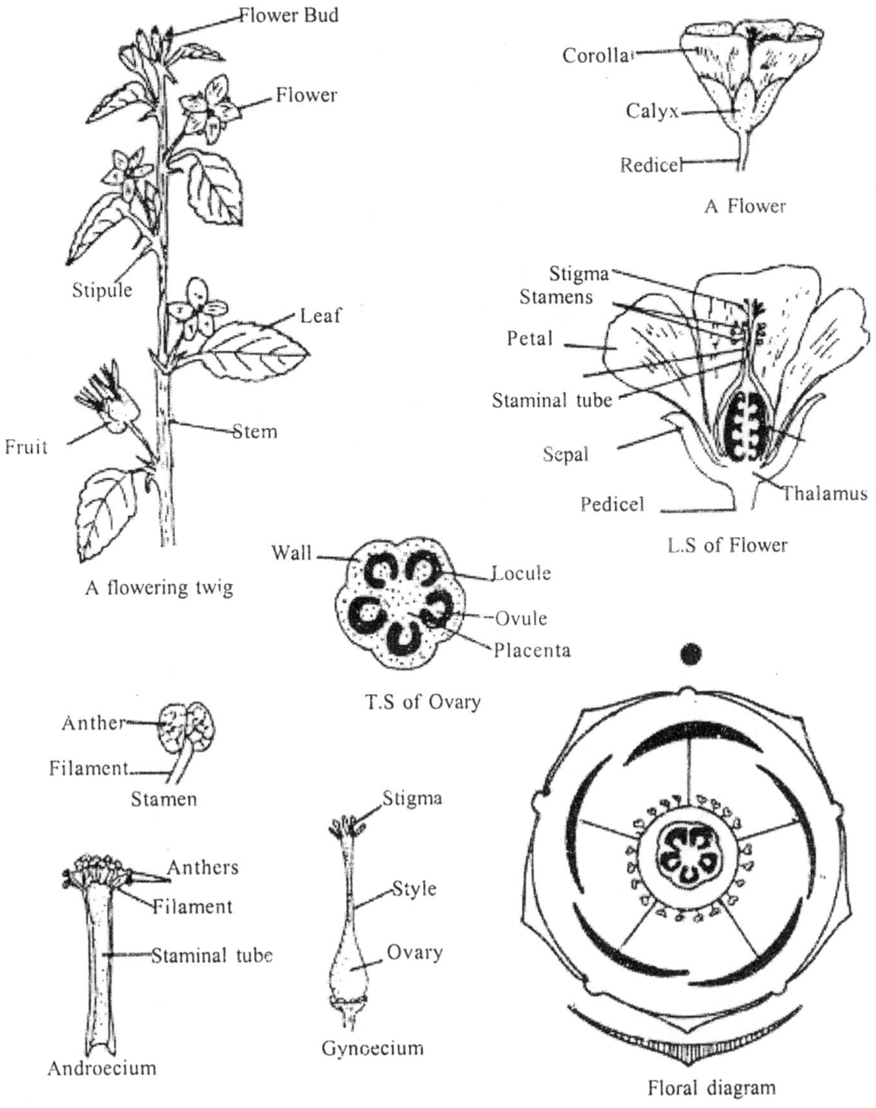

Fig 8.26 ***SIDA CORDIFOLIA*** **(Malvaceae)**

GOSSYPIUM
(Malvaceae)

Habitat : Mesophyte, grows in black dry soils

Habit : An annual shrub, may become perennial also reaching the size of a small tree.

Root : Branched tap root, deep feeder

Stem : Aerial, erect, branched, woody, hard, with stellate hairs.

Leaf : Ramal and cauline, simple, alternate, stipulate, petiolate (hairy), palmately lobed. Venation reticulate multicostate divergent.

Inflorescence : Flowers solitary, axillary, occasionally flowers may be grouped into cymose clusters.

Flower : Pedicellate, ebracteate, complete, actinomorphic, heterochlomydeous, cyclic, bisexual and hypogynous. Flowers are large yellow having a central dark eye.

Calyx : The calyx is surrounded by three fused lobes of epicalyx with deeply cut margins. The calyx is made up of 5 fused sepals with valvate aestivation. Odd sepal anterior

Corolla : Petals 5, free, obovate with sinuate upper margin, mucilaginous and yellow in colour. Aestivation twisted.

Androecium : Stamens numerous, monadelphous, epipetalous, anthers monothecous and transversely attached to the filaments, extrorse with longitudinal dehiscence. Pollen grains spinous.

Gynoecium : Ovary superior, tri or pentacarpellary, syncarpous 3-5 locules with many ovules in each locule on axile placenta. Style simple, stigma as many as carpels.

Fruit : A loculisidal capsule. The capsule is 3-5 valved, opens along the locule exposing 5-7 seeds in each locule.

Seed : Dark coloured with one end densley covered with epidermal hairs (constitutes the cotton fibres of commerce) which are cellutosic.

Floral formula : ✝ ⚥ $Epik_{(3)}\ K_{(5)}\ C_5\ A_{(\infty)}\ \underline{G}_{(\infty)}$

Fig 8.27 *GOSSYPIUM* (Malvaceae)

ABUTILON INDICUM
(Malvaceae)

Habitat : Mesophyte, A common weed growing in dry open places.

Root : An erect herb.

Stem : Much branched tap root.

Stem : Aerial, erect, cylindrical, hairy, solid, branched.

Leaf : Ramal and cauline, simple, alternate, petiolate, stipulate, ovate, serrate, and acute; venation multicostate reticulate.

Inflorescence : Flowers, solitary axillary.

Flower : Pedicellate, bracteate, ebracteolate, heterochlamydeous, pentamerous, bisexual and hypogynous.

Calyx : Sepals 5, fused, persistent; valvate aestivation.

Corolla : Petals 5, free, fused only at the base with the staminal column. Aestivation twisted.

Androecium : Stamens numerous, monadelphous forming a column surrounding the style. Anthers reniform, monothecous, dorsifixed, extrorse with longitudinal splitting.

Gynoecium : Ovary superior, multicarpellary, syncarpous, multilocular with one ovule in each locule on axile placenta. Style long, stigma as many as carpels.

Fruit : A schizocarp which splits into as many one seeded units as there are carpels. Each unit is indehiscent and is called a mericarp.

Seed : Small non endospermic.

Floral formula : † ♀ $K_{(5)}\ C_5\ A_{(\infty)}\ \underline{G}_{(\infty)}$

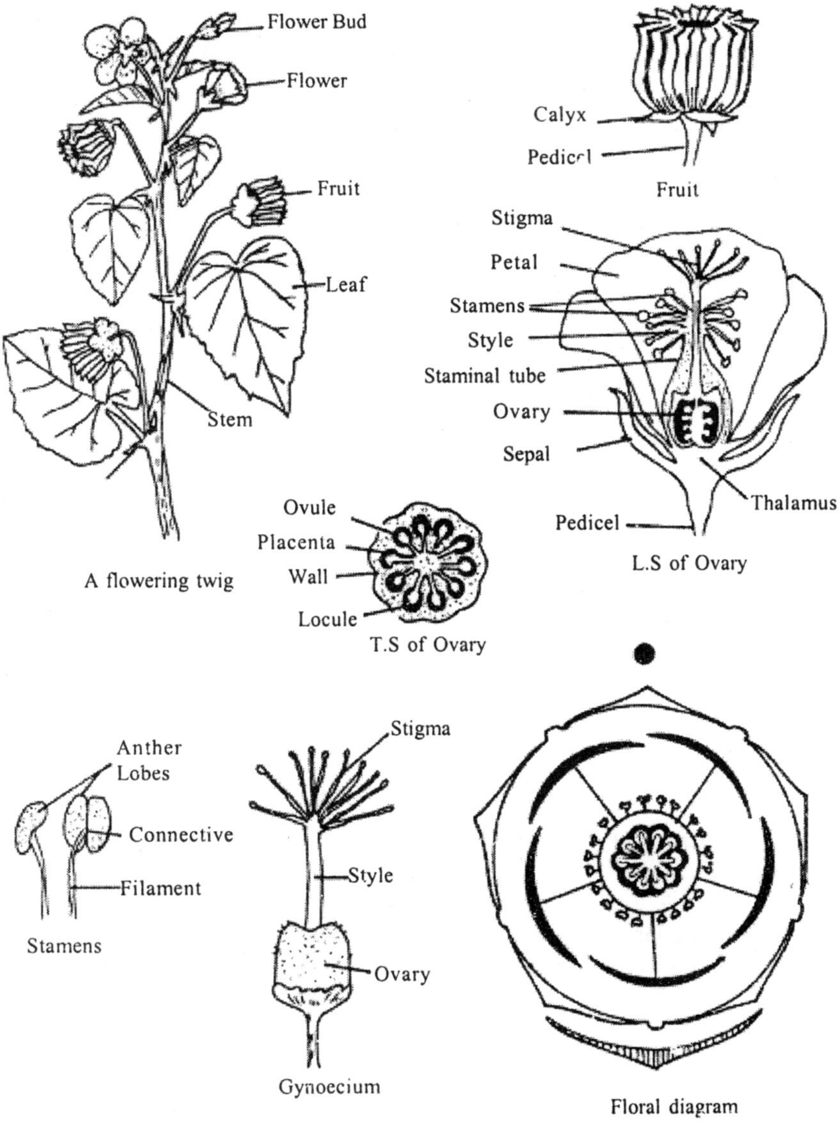

Fig 8.28 ***ABUTILON INDICUM*** (Malvaceae

CORCHORUS ACUTANGULA

(Tiliaceae)

Habitat : Mesophyte, grows in open places.

Habit : Erect herb.

Stem : Aerial, erect, branched, cylindrical, soft, solid and pubescent.

Leaf : Ramal and cauline, Alternate, stipulate. Simple, petiolate, ovate, margin serrate, acute, somewhat coriaceous. Venation reticulate unicostate.

Inflorescence : Axillary cymose cluster.

Flower : Pedicellate, bracteate, bracteolate, complete, heterochlamydeous, actinomorphic, bisexual and hypogynous.

Calyx : Sepals 5, thin, free. Aestivation valvate.

Corolla : Petals 5, free, yellow Aestivation valvate.

Androecium : Stamens many (15-20) arranged in whorls of 5 each, tree. Anthers dithecous, dorsifixed, introrse and splitting longitudinally.

Gynoecium : Ovary superior, tricarpellary, syncarpous, trilocular with two ovules in each locule on axile placentation; style short, stigma trifid with each lobe bifurcating again.

Fruit : A winged capsule

Seed : Non endospermic.

Floral formula : ✝ ♀ $K_{(5)}\ C_5\ A_{15\text{-}20}\ \underline{G}_{(3)}$

Fig 8.29 ***CORCHORUS ACUTANGULA*** **(Tilliaceae)**

TRIUMFETTA RHOMBOIDEA

(Tiliaceae)

Habitat : Mesophytic plant growing in open dry land.

Habit : A large annual herb reaching upto 1 metre height.

Root : Branched tap root.

Stem : Aerial, erect, branched, hairy with stellate hairs.

Leaf : Simple, alternate, stipulate, somewhat rounded, generally 3 lobed, upper lobe narrow and acute, tomentose, irregularly toothed, venation reticulate unicostate.

Inflorescence : Extra axillary raceme, could be axillary also.

Flower : Pedicellate, bracteate, ebracteolate, cyclic, heterochlamydeous, pentamerous, actinomorphic, bisexual and hypogynous.

Calyx : Sepals 5, free, linear, apically calyptrate, longer than corolla, villous on the outer surface; valvate aestivation.

Corolla : Petals 5, free, ablong, yellow, imbricate aestivation.

Androecium : Stamens usually 10, short androgynophore present; anthers dorsifixed, dithecous splitting longitudinally.

Gynoecium : Ovary superior usually bi or tetra carpellary syncarpous bilocular with two ovules per locule on axile placenta. Style single splits into two linear stigmas ovary wall studed with hooked warty projections.

Fruit : An oblong capsule covered with hooked projection.

Seed : Non endospermic

Floral formula : ✝ ♀ $K_{(5)}\ C_5\ A_{10}\ \underline{G}_{(2)}$

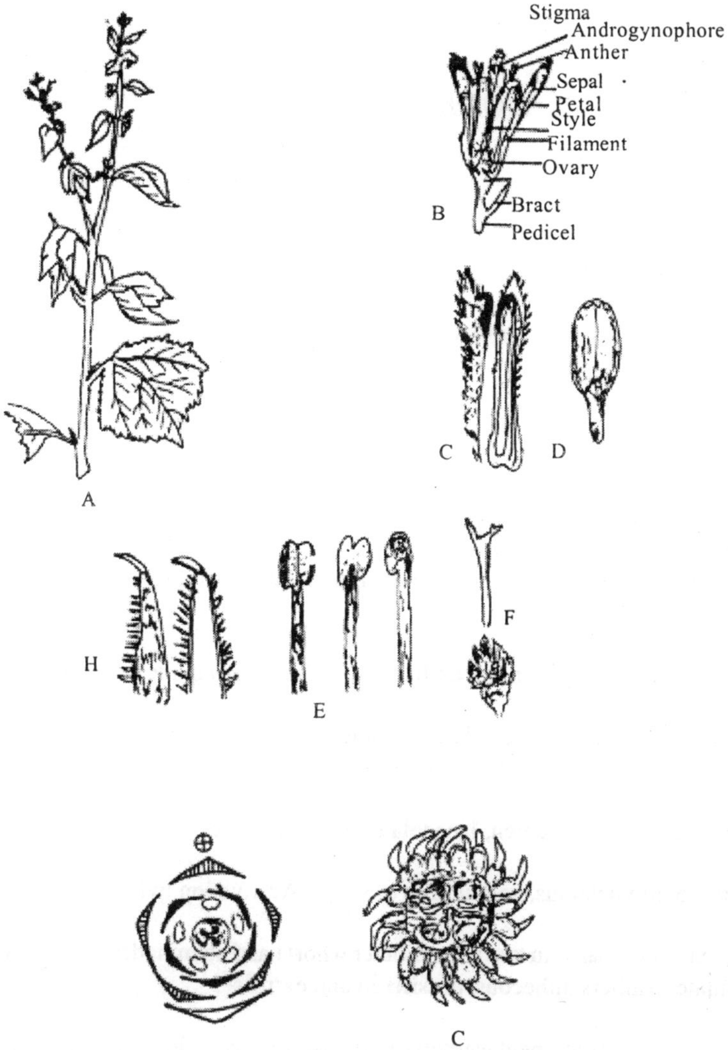

Fig 8.30 ***TRIUMFETTA RHOMBOIDEA*** **(Tilliaceae)**

DOMBEYA SPECTABILIS

(Sterculiaceae)

Habitat : A mesophyte; grows (cultivated) in open areas with moisture.

Habit : A shrub.

Root : Much branched tap root

Stem : Woody below and herbaceous above, erect, cylindrical, solid, branched and pubescent.

Leaf : Simple, alternate, stipulate (leafy stipules), petiolate, dentate (margin), acuminate (apex), cordate (base), venation reticulate unicostate.

Inflorescence : Flowers arranged in two axillary clusters appearing like an umbel.

Flower : Pedicellate, bracteate, ebracteolate, complete, heterochlamydeous, actinomorphic, bisexual and hypogynous; colour of the flowers white or pink.

Calyx : Sepals 5, polysepalous, green, lanceolate, pubescent. Aestivation valvate.

Corolla : Petals 5 polypetalous, colour pink or white. Aestivation twisted.

Androecium : Stamens many, in two whorls;outer whorl has 5 staminodes, inner whorl fertile consisting of 10 stamens; connate. Anthers dithecous, dorsifixed and extrorse.

Gynoecium : Ovary superior, pentacarpellary, syncarpous, pentalocular with one ovule in each locule on axile placenta; style one; stigma 5; ovary hairy.

Fruit : Schizocarpic.

Seed : Non, endospermic.

Floral formula : ✝ ♀ $K_{(5)}\ C_5\ A_{(5+10)}\ \underline{G}_{(5)}$

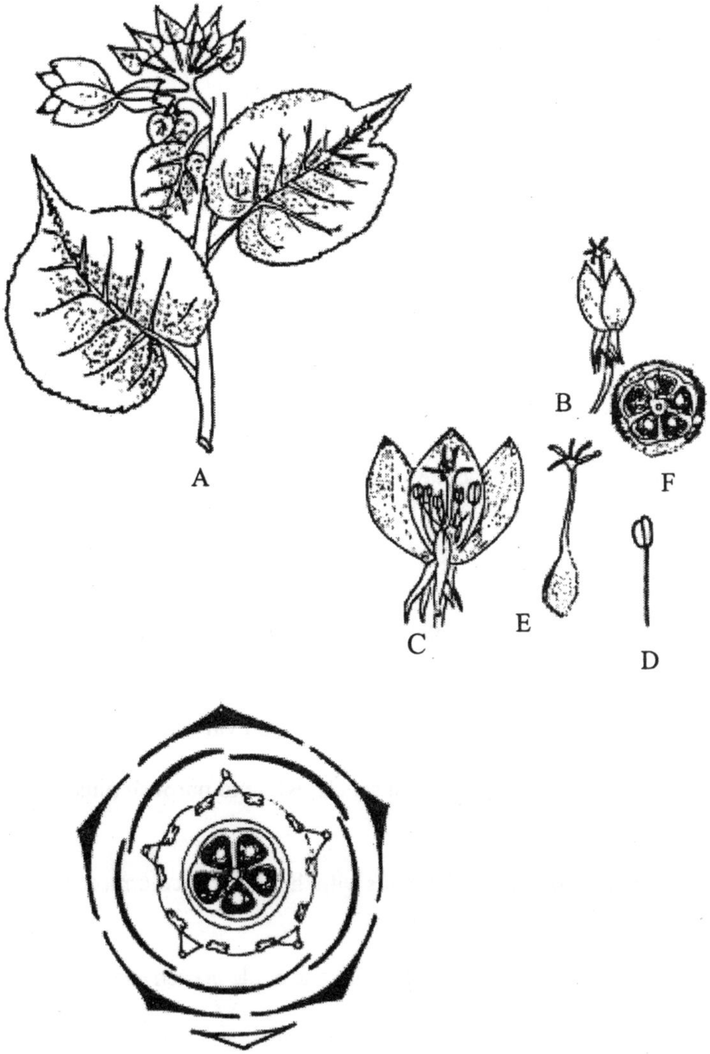

Fig 8.31 ***DOMBEYA SPECTABILIS*** (Sterculiaceae)

MELOCHIA CORCHORIFOLIA

(Sterculiaceae)

Habitat: Mesophyte, grows in open wasteland.

Habit: An erect herb or undershrub.

Root: Branched tap root.

Stem: Aerial, erect, branched cylindrical, hairy with stellate hairs.

Leaf: Simple, alternate, stipulate (hairy), petiolate, oblong, base truncate, undersurface studded with stellate hairs, serrate, tip acute venation retianlate unicostate.

Inflorescence: Usually a terminal cyme.

Flower: Bracteate, bracteolate, pedicellate, complete, cyclic, heterochlamydeous, actinomorphic, bisexual and hypogynous.

Calyx: Sepals 5, gamosepalous, cup shaped and hairy; aestivation valvate.

Corolla: Petals 5, free (Polypetalons), longer than sepals, white to purplish pink in colour. Aestivation twisted.

Androecium: Stamens 5, antipetalous, monadelphous with the staminal column enclosing the gynoecium. Anthers dorsifixed, dithecous and extrorse.

Gynoecium: Ovary superior, 5 carpellary 5, locular with one ovule in each locule on axile placenta. Styles 5, ovary wall hairy.

Fruit: A bispid 5 valved loculisidal capsule.

Seed: Small, non endospermic.

Floral formula: † ♀ $K_{(5)} C_5 A_{\infty} \underline{G}_{(5)}$

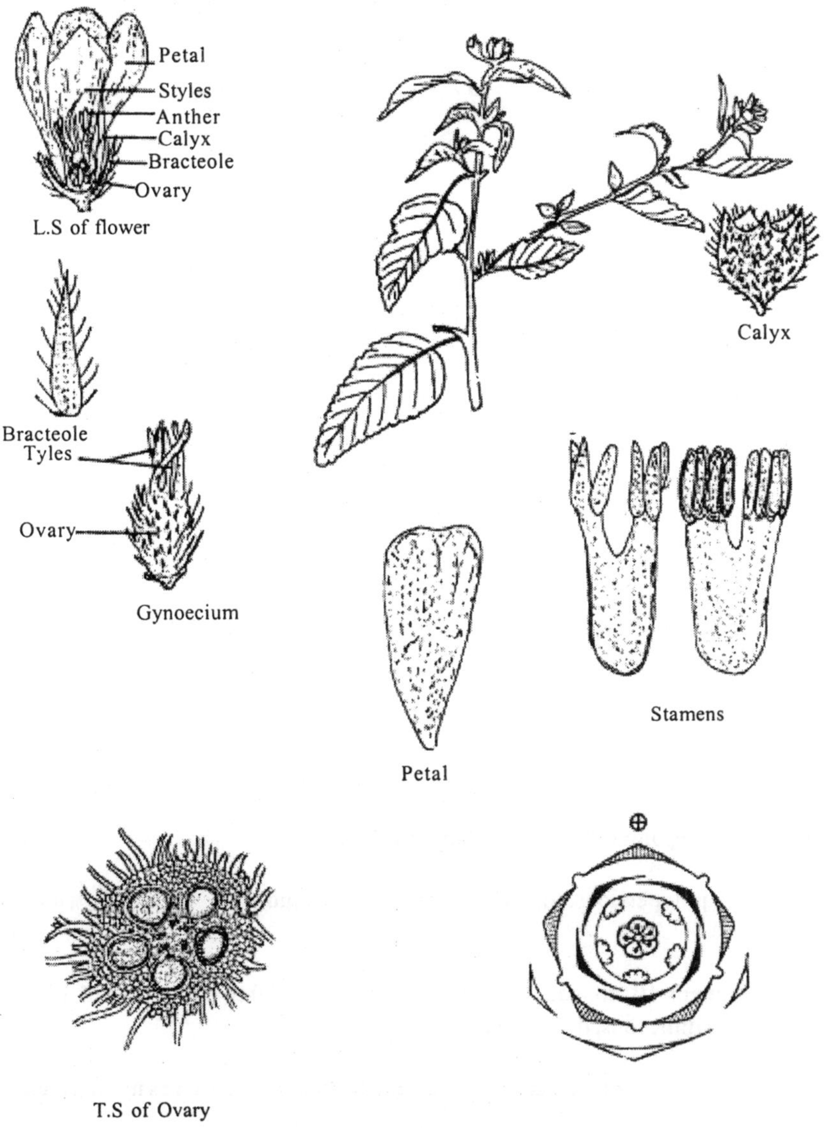

Fig 8.32 ***MELOCHIA CORCHORIFOLIA*** **(Sterculiaceae)**

GAUZUMA TOMENTOSA

(Sterculiaceae)

Habitat : Mesophyte; grows in open forests.

Habit : A medium sized tree

Root : Tap root system.

Stem : Aerial, erect, woody, cylindrical, hairy and solid.

Leaf : Simple, alternate, petiolate, stipulate (stipules caducous), ovate, serrate, apex acute, hairy on lower surface. Venation reticulate unicostate.

Inflorescence : A much branched cymose panicle.

Flower : Pedicellate, bracteate, bracteolate, cyclic, complete, actinomorphic, bisexual and hypogynous.

Calyx : Sepals 5, polysepalous, hairy. Aestivation valvate.

Corolla : Petals 5, polypetalous, top of the petals inflexed and bifid. Yellowish with red tips. Aestivation twisted.

Androecium : Stamens 10, connate at the base (form a tube) free above, out of 10 stamens 5 are sterile (stamimode), Anthers dithecous, dorsifixed and extrorse.

Gynoecium : Ovary superior, pentacarpellary, pentalocular, with 2 ovules in each locule on axile placenta.

Fruit : A capsule

Seed : Small non endospermic.

Floral formula : ✝ ⚥ $K_{(5)}$ C_4 $A_{(5+5)}$ $\underline{G}_{(5)}$

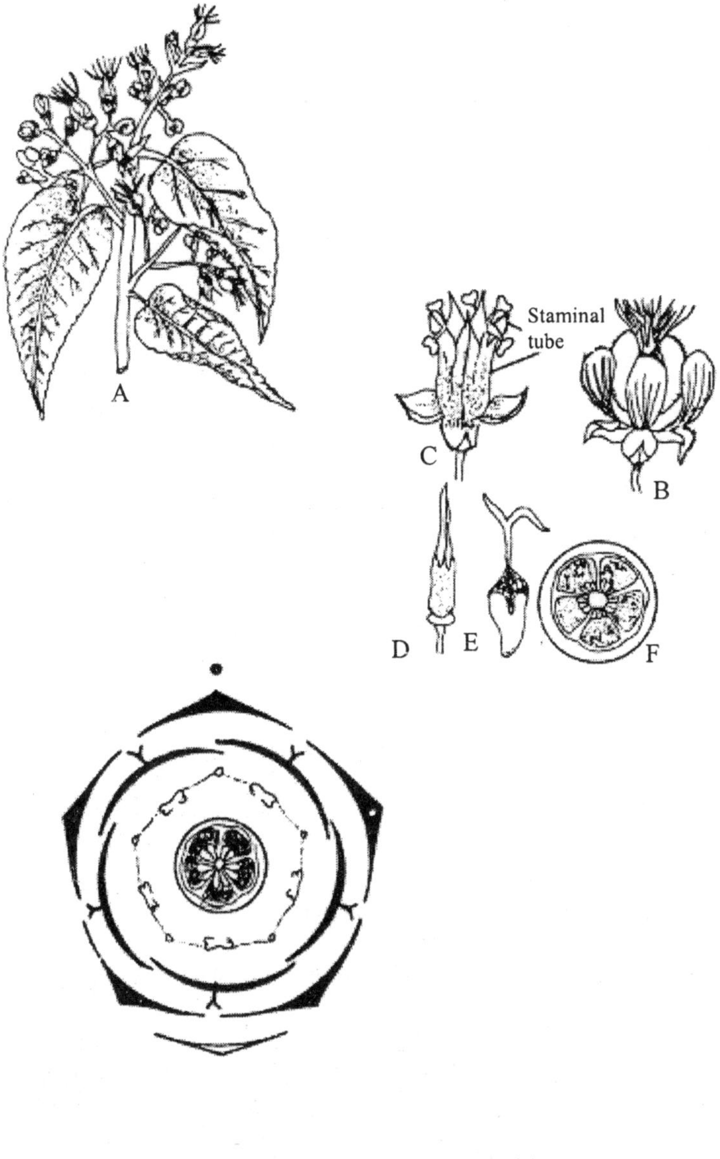

Fig 8.33 *GAUZUMA TOMENTOSA* (Sterculiaceae)

WALTHERIA INDICA

(= W. americana)

(Sterculiaceae)

Habitat: Mesophyte, weed, grows in open wasteland.

Habit: Herb or undershrub.

Root: Branched tap root.

Stem: Aerial, erect, cylindrical, branched, stellate hairs present.

Leaf: Simple, alternate, stipulate, petiolate, crenate, serrate, ovate, cordate at base, venation reticulate unicostate.

Inflorescence: Cymose cluster.

Flower: Pedicellate, bracteate, ebracteolate, complete, cyclic, heterochlamydeons, actinomorphic, bisexual and hypogynous.

Calyx: Sepals 5, gamosepalous, Aestivation valvate.

Corolla: Petals 5, gamopetalous, clawed, Aestivation twisted.

Androcium: Stamens 4-5, monadelphous, filaments form a column around the gynoecium, no staminodes anthers dithecous with longitudinal splitting.

Gynoecium: Ovary superior, monocarpellary, monolocular with 1-2 ovules on basal placenta.

Fruit: A berry.

Seed: Exalbuminous

Floral formula: † ♀ $K_{(5)}\ C_4\ A_{(4\text{-}5)}\ \underline{G}_{(1)}$

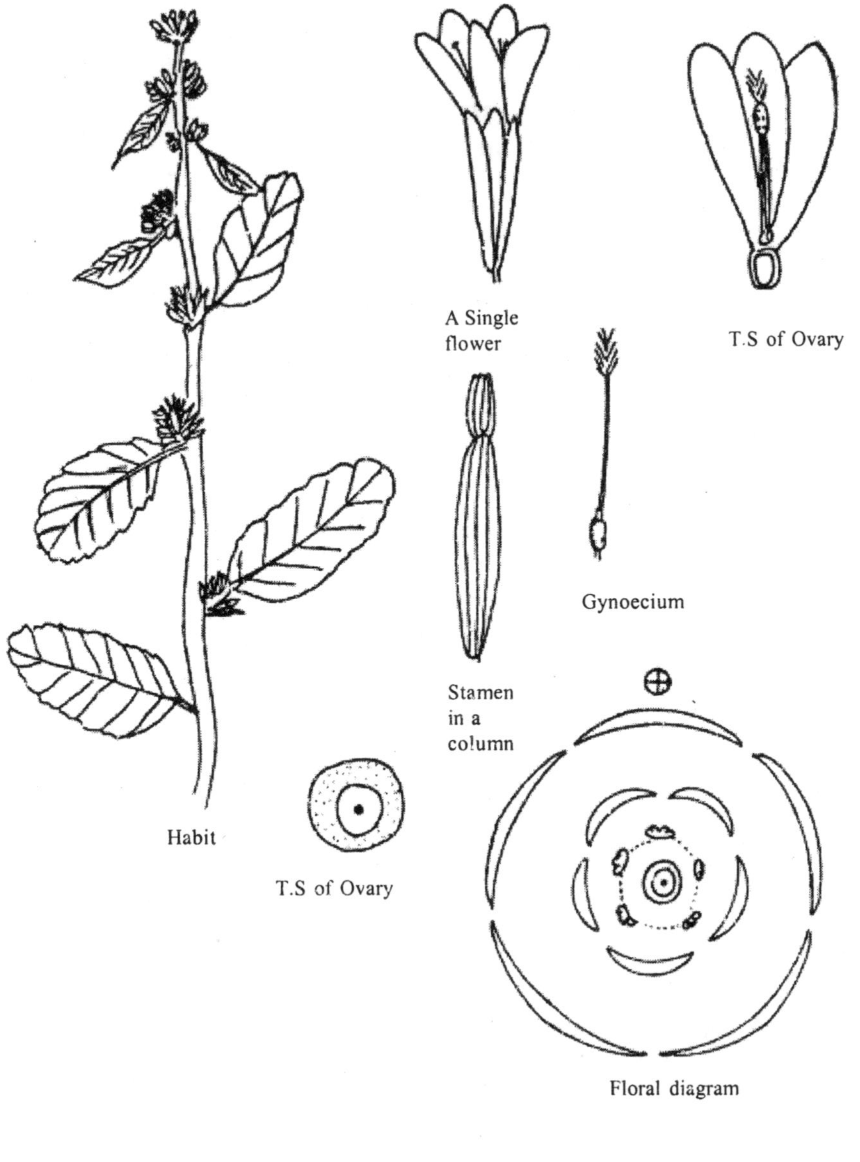

Fig 8.34 ***WALTHERIA INDICA (= W. americana)*** (Sterculiaceae)

RUTA GRAVEOLENS

(Rutaceae)

Habitat: Mesophytic herb, grows in moist places.

Habit: A fragrant herb.

Root: Branched tap root.

Stem: Aerial, erect, branched, cylindrical, soft, contains yellowish juice, gland dotted.

Leaf: Alternate, ramal and cauline, bipinnately compound, leaflets petiolate or subsessile, gland dotted.

Inflorescence: Flowers solitary.

Flower: Pedicellate, ebracteate, ebracteolate, cyclic, heterochlamydeons, actinomorphic, bisexual and hypogynous.

Calyx: Sepals 4, synsepalous, Aestivation valvate.

Corolla: Petals 4 , Aestivation valvate; polypetalons.

Androecium: Stamens twice as many as petals, arranged in two whorls, outer whorl antisepalous, inner whorl antipetalous. Anthers dithecous, introrse and dehisce longitudinally.

Gynoecium: Ovary superior, tetra carpellarly, syncarpous, tetralocular with one ovule in each locule on axile placenta. A disc is present below the ovary; style simple.

Fruit: A berry, carpels remain distinct in the fruit.

Seed: Non endospermic

Floral formula: ✝ ⚥ $K_{(4)}\ C_4\ A_{4+4}\ \underline{G}_{(4)}$

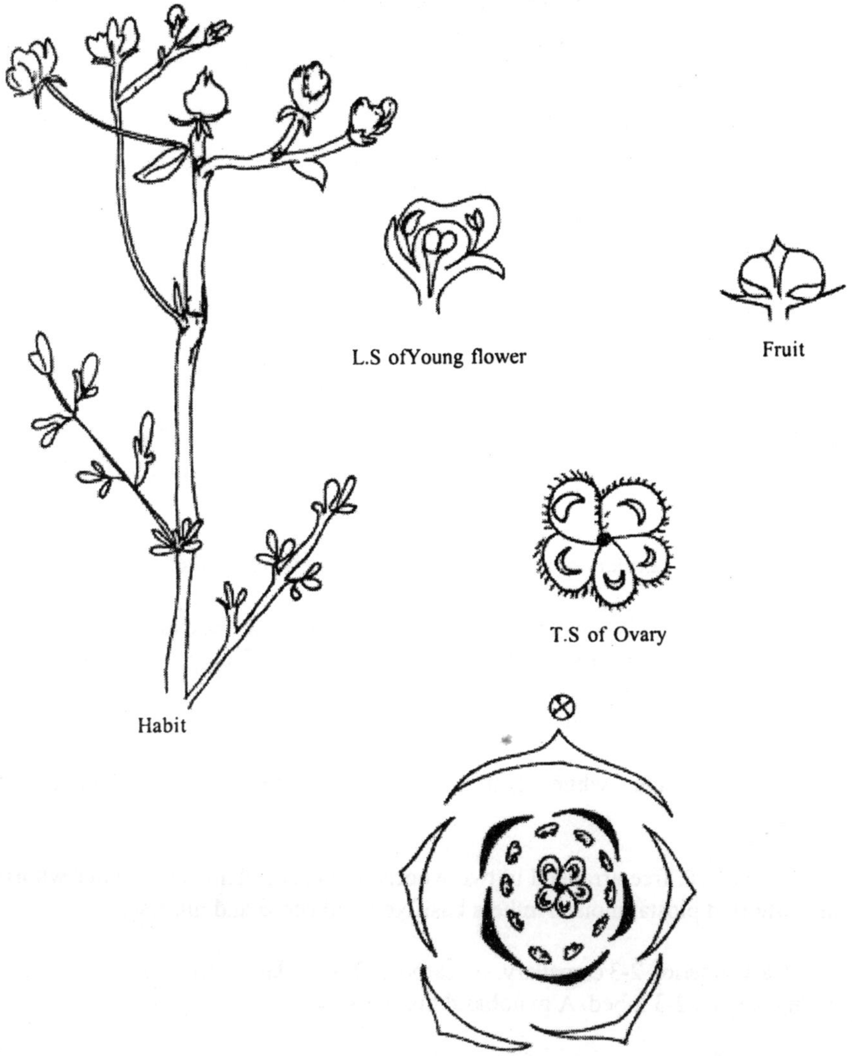

Fig 8.35 *RUTA GRAVEOLENS* (Rutaceae)

MURRAYA PANICULATA

(Rutaceae)

Habitat : Mesophyte grows in open somewhat moist soil.

Habit : An evergreen shrub or small tree

Root : Branched tap root.

Stem : Aerial, erect, cylindrical, branched woody, glabrous in older parts, pubescent in younger parts.

Leaf : Ramal and cauline, compound (imparipinnate), alternate, exstipulate, 4-5" long, leaflets 3-7, (alternate, short stalked, obovate or rhomboidal, entire, acute or abtuse apex, coriaceous and dark green, venation reticulate unicostate).

Inflorescence : Monochacial, axillary or terminal, few to many flowered cyme.

Flower : Pedicellate, ebracteate, ebracteolate, complete, actinomorphic, bisexual, heterochlamydeous and hypogynous. There is a distinct nectar secreting disc below the ovary. Flowers white in colour.

Calyx : Sepals 5, fused, Aestivation valvate

Corolla : Petals 5, polypetalous, white, gland dotted, oblong lanceolate, reflexed and expanding above. Aestivation imbricate.

Androecium : Stamens 10, free arranged in two whorls of 5 each, filaments of outer whorl shorter than those of the inner whorl, diplostamnous. Anthers basifixed, dithecous and introrse.

Gynoecium : Ovary superior, 2-3 carpellary, syncarpous, 2-3 locular, with one ovule in each locule on axile placenta. Style long, stigma 2-3 lobed. A gynobasal disc present.

Fruit : An ovoid red coloured berry, rugose and pointed.

Seed : Albuminous, testa hairy.

Floral formula : ✝ ⚥ $K_{(5)}\ C_5\ A_{5+5}\ \underline{G}_{(2-3)}$

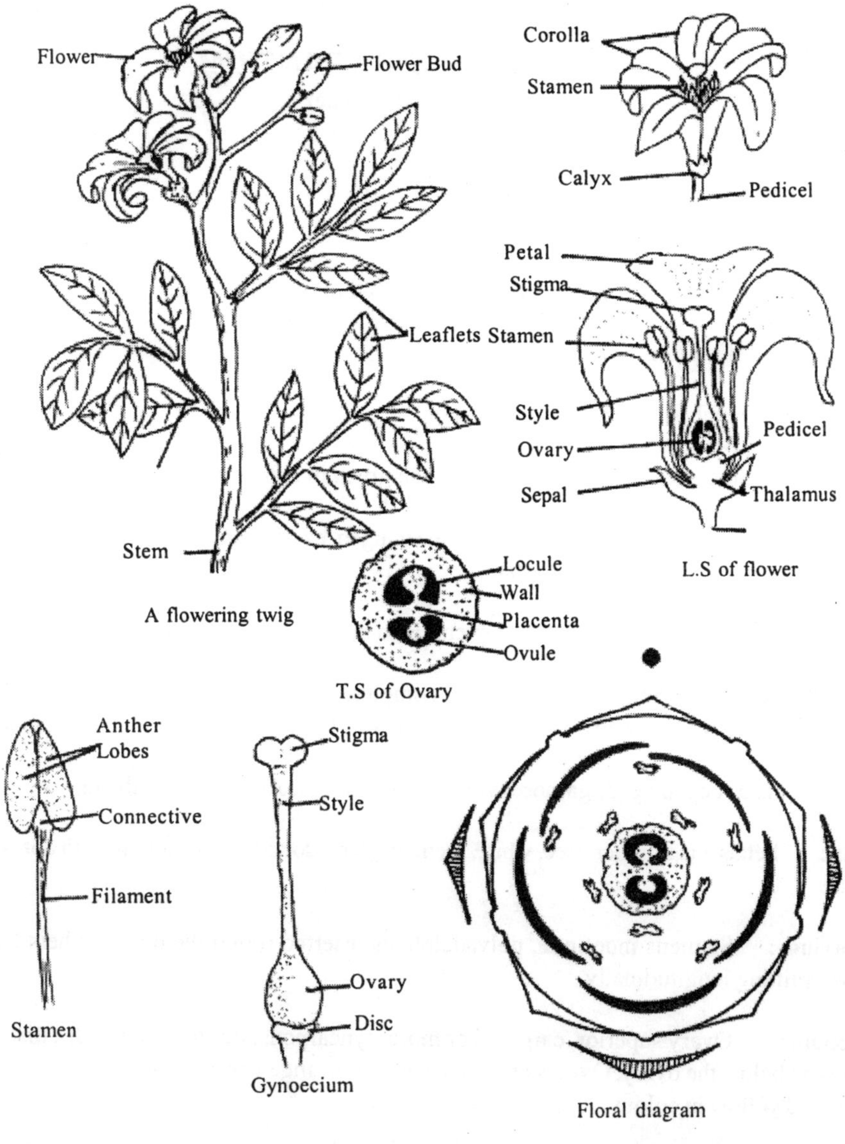

Fig 8.36 ***MURRAYA PANICULATA*** **(Rutaceae)**

CITRUS AURANTIUM
(Rutaceae)

Habitat : Cultivated, mesophyte, grows in open and moist well drained soil.

Habit : A small sized tree

Root : Much branched, deep feeding tap root

Stem : Aerial, erect, branched, woody, solid, ash green in colour and glabrous.

Leaf : Petiolate, unifoliately compound, petiole winged, exstipulate, terminal leaflet elliptic or ovate, serrate, acute apex, venation reticulate unicostate.

Inflorescence : Small axillary cymes, flowers could be solitary axillary also.

Flower : Pedicellate, ebracteate, ebracteolate, complete, actinomorphic, heterochlamydeous, pentamerous, bisexual and hypogynous. Flowers white in colour and fragrant. Gynobasal disc present.

Calyx : Sepals 5, cup shaped, gamosepalous, gland dotted, Aestivation valvate.

Corolla : Petals 5 in number, free, white, scented, gland dotted, ovoid oblong with obtuse apex. Aestivation imbricate.

Androecium : Stamens indefinite, polyadelphous inserted round the disc. Anthers basifixed, dithecous, introrse, splitting longitudinally.

Gynoecium : Ovary superior, carpels 5 or more, syncarpous, multilocular with a distinct nectar secreting disc present below the ovary. Ovules one in each locule arranged on axile placenta. Style single, green; stigma capitate and yellow in colour.

Fruit : Hesperidium (a kind of berry)

Seed : Exalbuminous, contains an oil.

Floral formula : ✝ ♀ $K_{(5)}\ C_5\ A_{(\infty)}\ \underline{G}_{(5-\infty)}$

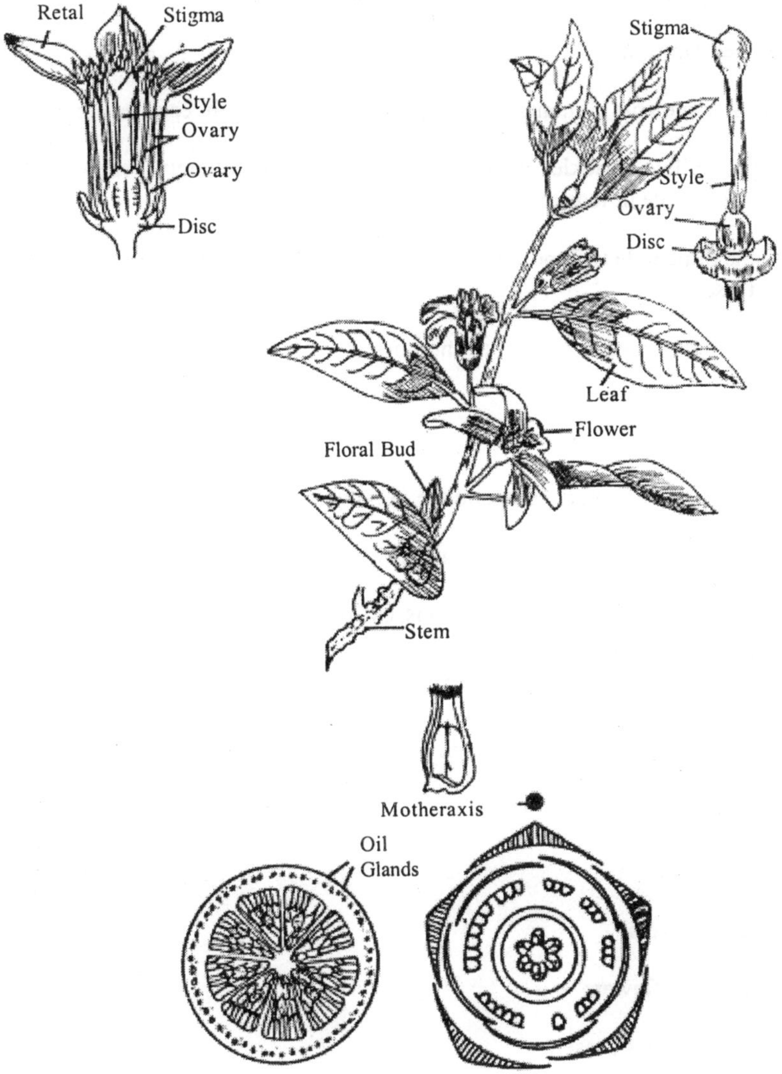

Fig 8.37 *CITRUS AURANTIUM* (Rutaceae)

TODDALIA ASIATICA

(Rutaceae)

Habitat: Autotrophic, mesophyte, grows wild in dry open land.

Habit: Erect or climbing shrub armed with epidermal prickles.

Root: Much branched tap root.

Stem: Aerial, erect or somewhat climbing, green, cylindrical armed with recurved epidermal prickles.

Leaf: Alternate, pedicellate, prickles present on the pedicel, trifoliately compound, leaflets sessile, coriaceous. Venation reticulate unicostate.

Inflorescence: Cyme or panicle with polygamous flowers i.e. flowers are of three types – unisexual male, unisexual female and bisexual.

Flower: Pedicellate, ebracteate, ebracteolate, actinomorphic, polygamous, hypogynous. Disc present in the female flower.

Male flower: (Staminate)

Calyx: sepals 5, gamosepalous, small, campanulate. Aestivation vlavate.

Corolla: Petals 5, polypetalous, white. Aestivation valvate or imbricate.

Androecium: Stamens 4-3, free, Anthers dithecous, extrorse, splitting longitudinally.

Gynoecium: Pistillode present , simple or 4 partite.

Fruit: Absent.

Floral formula: † ♂ $K_{(5)}\, C_5\, A_5\, \underline{G}_0$

Female flowers: (Pistillate)

Calyx: Same as in male flowers

Corolla: Same as in male flowers

Androecium: 4-5 staminodes present

Gynoecium: Ovary superior, pentacarpellary, syncarpous, pentalocular with one or two ovules in each locule on axile placenta. Style short or absent, stigma capitate.

Fruit: A berry, fleshy or coriaceous.

Seed: Exalbuminous, angular remiform.

Floral formula: † ♀ $K_{(5)}\, C_5\, A_0\, \underline{G}_{(5)}$

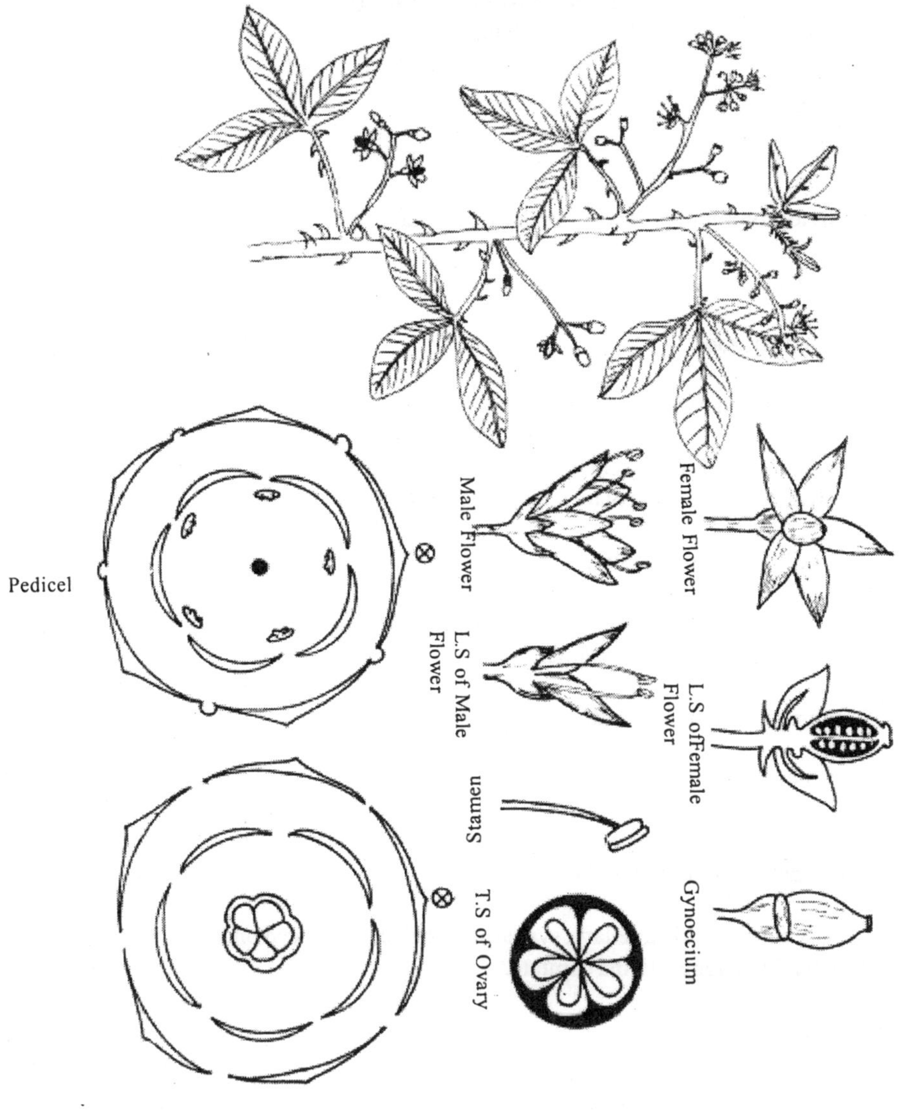

Fig 8.38 ***TODDALIA ASIATICA*** **(Rutaceae)**

MELIA AZADIRACHTA.
(= Azadirachta indica)
(Meliaceae)

Habitat: Mesophyte, commonly growing in dry open land often cultivated as avenue tree.

Habit: A tall, evergreen tree.

Root: Much branched tap root.

Stem: Aerial, erect, branched, woody, cylindrical, solid and glabrous.

Leaf: Alternate, exstipulate, pinnatelty compound (imparipinnate), leaflets with oblique base, serrate margin, pulvinate leaf base, ovate or ovate lanceolate acuminate apex. Venation reticulate unicostate.

Inflorescence: An axillary cymose panicle

Flower: Pedicellate, ebracteate, bracteolate, complete, cyclic, heterochlamydeous, bisexual and hypogynous. A disc is present between the petals and stamens.

Calyx: Sepals 5, polypetalous, white. Aestivation valvate.

Corolla: Petals 5, polypetalous, white. Aestivation imbricate.

Androecium: Stamens 10, monadelphous, staminal column covers the gynoecium. Anthers dithecous, basifixed, introrse and dehisce longitudinally.

Gynoecium: Ovary superior, tricarpellary, syncarpous, trilocular, placed on a disc, with 1-2 ovules in each locule on axile placenta. Style long, stigma capitate.

Fruit: One seeded drupe.

Seed: Exalbuminous.

Floral formula: † ♀ $K_{(5)}\ C_5\ A_{(10)}\ \underline{G}_{(3)}$

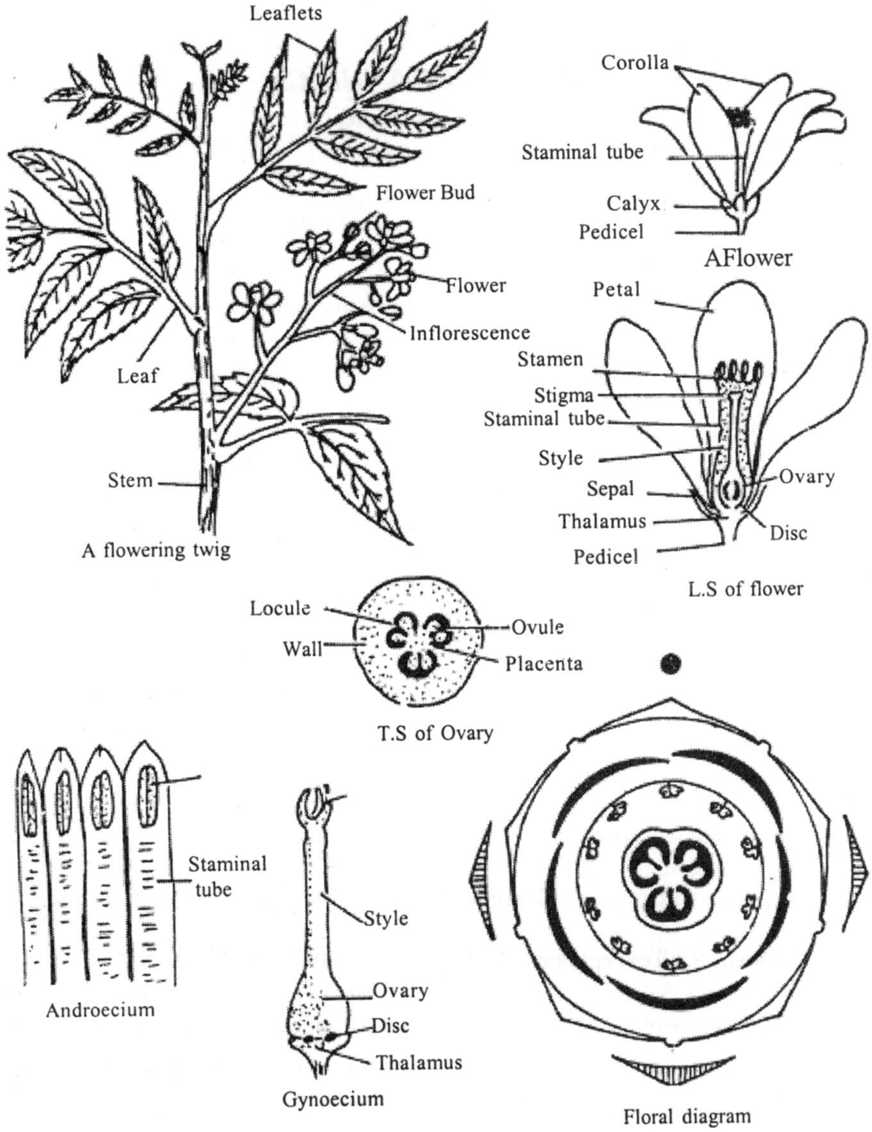

Fig 8.39 ***MELIA AZADIRACHTA (= Azadirachta indica)*** **(Meliaceae)**

MANGIFERA INDICA

(Anacardiaceae)

Habitat: Mesophyte, evergreen tree grows well in open dry lands.

Habit: A large tree.

Root: Much branched tap root.

Stem: Erect, aerial, cylindrical, hard and stout.

Leaf: Simple alternate, exstipulate, long (15-30cm), oblong or oblong lanceolate, coriaceous, glabrous, shining, entire, acute or acuminate, usually clustered at the tips of branches, petiolate, petiole swollen at the base. Venation reticulate unicostate.

Inflorescence: Panicle.

Flower: Pedicellate, bracteate, ebracteolate, cyclic, complete, heterochlamydeous, actinomorphic, bisexual and hypogynous . Fleshy Disc present between stamens and petals.

Calyx: Sepals 5, gamosepalous, deciduous Aestivation imbricate.

Corolla: Petals 5, free, with orange coloured ridges on the inner face. Aestivation imbricte.

Androecium: Stamens 5, all except one reduced or have imperfect anthers.

Gynoecium: Ovary superior, monocarpellary, unilocular oblique with a single pendulous ovule; style short, stigma lobed.

Fruit: A fleshy drupe, mesocarp edible.

Seed: Exalbuminous, fleshy cotyledons, with papery testa.

Floral formula: ✝ ♀ $K_{(5)} C_5 A_{1+4} \underline{G}_1$

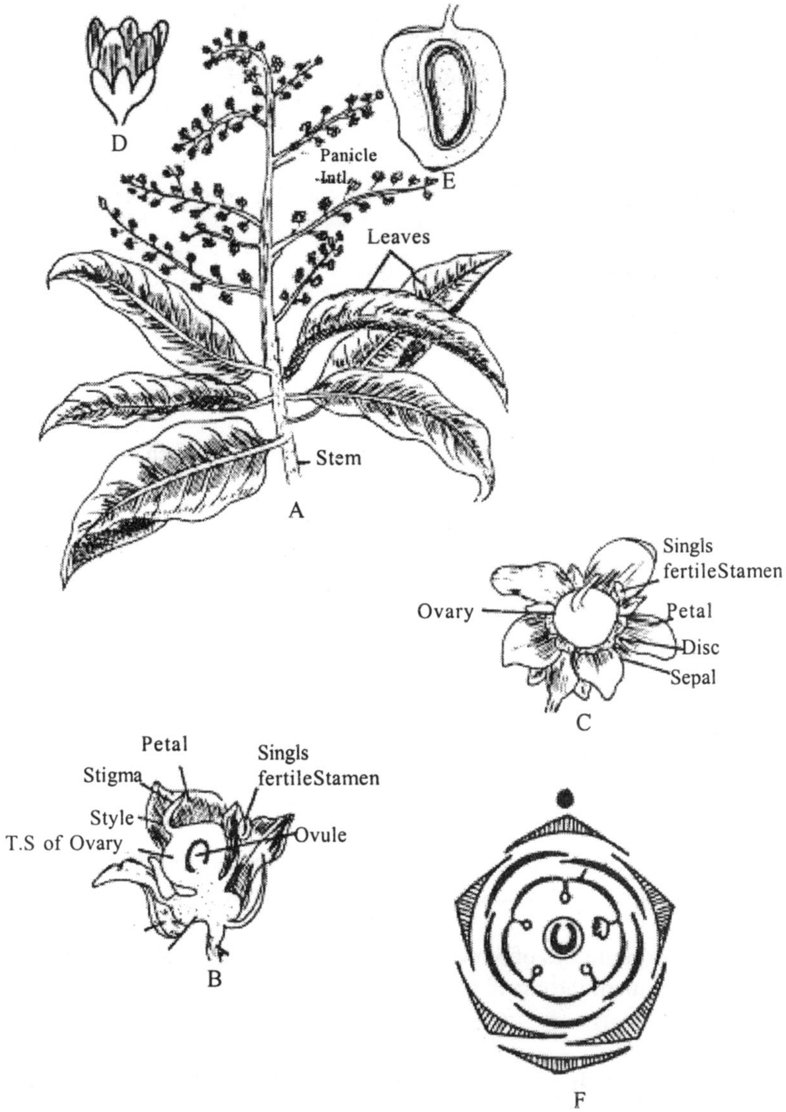

Fig 8.40 ***MANGIFERA INDICA*** **(Anacardiaceae)**

ROSA INDICA
(Rosaceae)

Habitat : Mesophyte, growing in well aerated, moist soils.

Habit : Perennial prickly shrub.

Root : Branched tap root system.

Stem : Aerial, erect, woody, cylindrical, solid, glabrous, green and armed with sharp epidermal emergences called prickles. These prickles do not have a vascular supply.

Leaf : Petiolate, stipulate (stipules adnate), alternate, imparipinnately compound, leaflets 3-7, ovate, serrate, acute with reticulate unicostate venation.

Inflorescence : Solitary axillary flowers, sometimes grouped in cymose clusters.

Flower : Pedicellate, bracteate or ebracteate, ebracteolate, complete, actinomorphic, heterochlamydeous, bisexual and perigynous. The thalamus is cup like from the floor of which arises the gynoecium. Other floral parts are situated above the gynoecium arising from the rim of the thalamus cup. Flowers of many colours and sweet scented.

Calyx : Sepals 5 rarely 4 in number, gamosepalous, lanceolate dentate sometimes enlarged and foliar, urceolate or tubular, hairy with drawn out apex, odd sepal and posterior. Aestivation imbricate (quincuncial)

Corolla : Petals 5, rarely 4 or multiples of 5, polypetalous, obovate, large, showy, fragrant. Aestivation imbricate. Petals inserted at the rim of the cup like thalamus. There is transition between petals and stamens and vice versa.

Androecium : Stamens indefinite, free, unequal inserted at the throat lining the calyx tube. Anthers introrse,basifixed, dithecous with longitudinal dehiscence.

Gynoecium : Ovary seminferior, polycarpellary, apocarpous arising from the floor of the thalamus cup; each carpel having a single basal ovule. Style short rarely rising above the cup, stigma thickened.

Fruit : An etaerio of achenes. Each carpel develops into an achene; several achenes are enclosed in the fleshy red coloured thalamus cup. The whole structure is called a hip.

Seed : Exalbuminous.

Floral formula : † ⚥ $K_{(5)}\ C_{5-\infty}\ A_{(\infty)}\ \underline{G}_{(\infty)}$

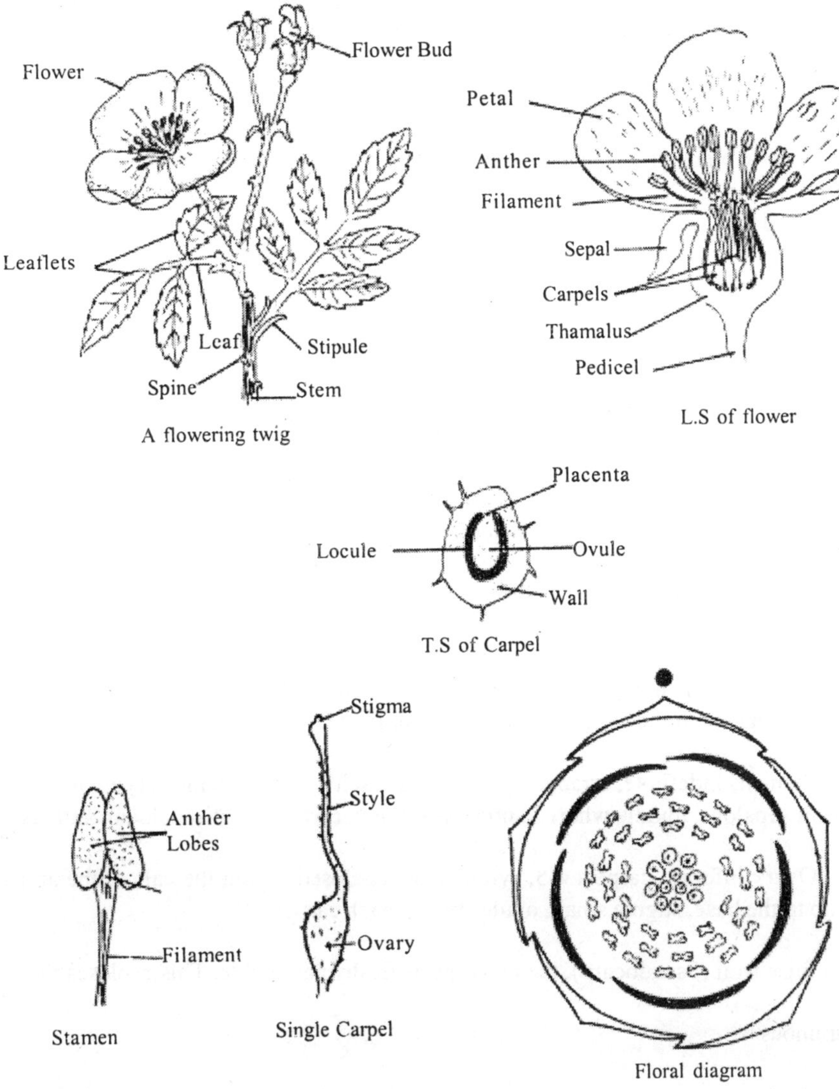

Fig 8.41 ***ROSA INDICA*** **(Rosaceae)**

PYRUS MALUS
(Rosaceae)

Habitat: Mesophytic plant, grows in temperate climate.

Habit: Medium sized tree.

Root: Branched tap root.

Stem: Aerial, erect, woody, branched.

Leaf : Petiolate, simple, stipulate with caducous stipules. Venation reticulate unicostate.

Inflorescence: Umbellate cyme.

Flower: Pedicellate, ebracteate, ebracteolate, heterochlamydeous, actinomorphic, bisexual and epigynous.

Calyx: Sepals 5, gamosepalous, reflexed, bell shaped, attached to the ovary, becomes enlarged and fleshy in fruit. Aestivation valvate.

Corolla: Petals 5, white or slightly pinkish in colour, orbicular, shortly clawed, arising on the rim of the epigynous cup; alternates with the sepals. Aestivation imbricate.

Androecium: Stamens indefinite, arranged in 3 whorls, each whorl contains varying number of stamens (outer whorl 10 antisepalous, middle whorl 5 antipetalous, inner whorl 5 antisepalous), anthers recurved.

Gynoecium: Ovary inferior, carpels 4-5, syncarpous, enclosed within the calyx tube and adnate to it, styles 4-5, free up to the base, stigma small, ovules two in each carpel.

Fruit: Pome, a false fruit (Pseudocarp), the thalamus is fleshy and edible. This encloses the seeds.

Seed: Exalbuminous

Floral formula: ✝ ♀ $K_{(5)}\ C_5\ A_{(\infty)}\ \underline{G}_{(4\text{-}5)}$

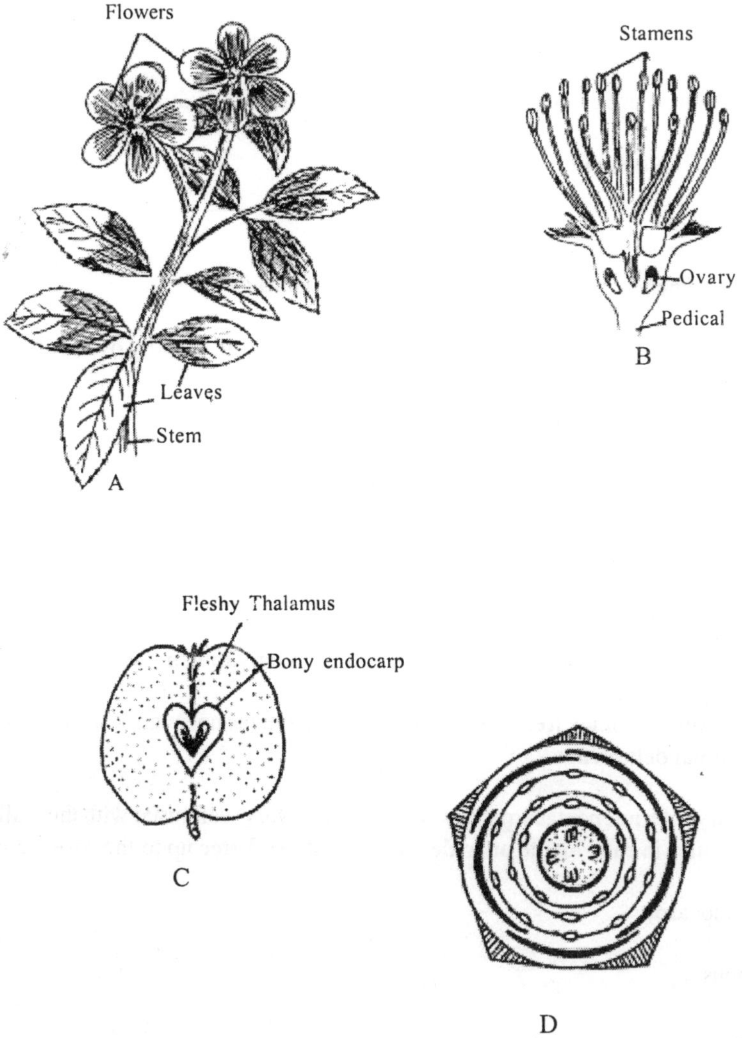

Fig 8.42 *PYRUS MALUS* (Rosaceae)

PYRUS COMMUNIS

(Rosaceae)

Habitat : A mesophytic plant growing in temperate climate at higher altitudes.

Habit : A perennial, glabrous small tree

Root : Branched tap root

Stem : Aerial, erect, branched, solid, cylindrical, hard and glabrous.

Leaf : Simple, alternate, petiolate (Petiole as long as lamia) stipulate, broadly ovate, abtusely serrate, apex acute, venation reticulate unicostate.

Inflorescence : Terminal umbels or cymes

Flower : Pedicellate (Pedicel hairy), ebracteate, ebracteolate, complete, actinomorphic, pentamerous, bisexual and epigynous.

Calyx : Sepals 5, gamosepalous, persistant. Aestivation vlavate.

Corolla : Petals 5, polypetalous, obovate, white in colour. Aestivation imbricate.

Androecium : Stamens indefinite, free arranged in whorls, filaments unequal. Anthers dorsifixed, dithecous, introrse with longitudinal dehiscence.

Gynoecium : Ovary inferior, pentacarpellary, syncarpous, ovary wall fused with the wall of the thalamus, 5 locular, with two ovules in each locute on axile placenta; styles 5 free up to the base, stigma not distinct.

Fruit : Pome (pseudocarp)

Seed : Exalbuminous

Floral formula : † ♀ $K_{(5)}\, C_5\, A_{(\infty)}\, \underline{G}_{(5)}$

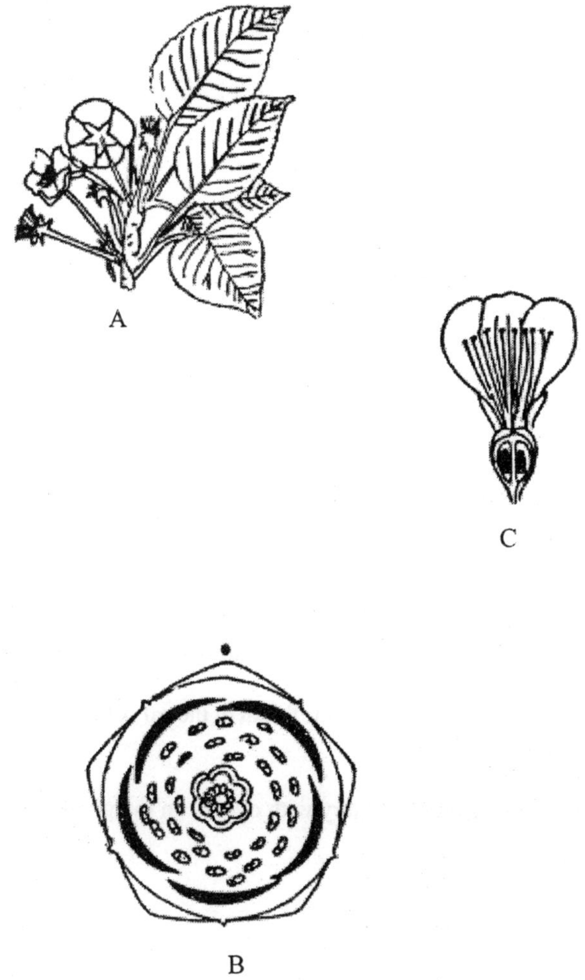

Fig 8.43 ***PYRUS COMMUNIS*** **(Rosaceae)**

PRUNUS PERSICA

(Rosaceae)

Habitat: Mesophyte; cultivated at higher altitudes in a cool climate

Habit: A medium sized evergreen tree.

Root: Branched tap root.

Stem: Aerial, erect, branched, woody, special dwarf, shoots arise from the branches bearing flowers.

Leaf: Short petiolate (sub sessile), simple, alternate, stipulate, oval with a pair of flangs present at the base of each leaf.

Inflorescence: Flowers may be solitary or borne on cymose clusters.

Flower: Pedicellate or sub sessile, ebracteate or ebracteolate, complete, actinomorphic, bisexual, perigynous with a cup shaped thalamus.

Calyx: Sepals 5, gamosepalous, companulate, wooly, caducous. Aestivation valvate.

Corolla: Petals 5, free, pink in colour. Aestivation imbricate.

Androecium: Stamens indefinite, free, inserted at the throat of the calyx tube. Anthers dorsifixed, dithecous, introrse with longitudinal splitting.

Gynoecium: Ovary seminferior, monocarpellary with ovules on marginal placenta style simple, stigma capitate.

Fruit: A one seeded drupe.

Seed: Exalbuminous.

Floral formula: † ♀ $K_{(5)} C_5 A_{(\infty)} \underline{G}_1$

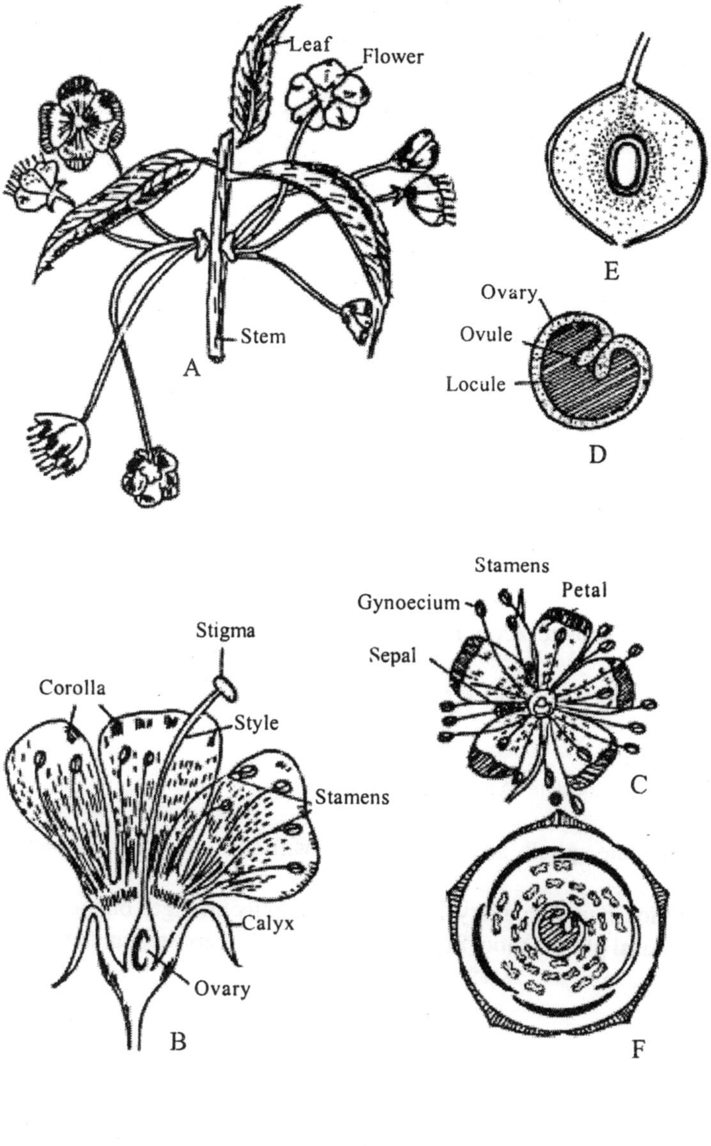

Fig 8.44 *PRUNUS PERSICA* (Rosaceae)

POTENTILLA SUPINA

(Rosaceae)

Habitat: Mesophyte, growing in higher altitudes.

Habit: An annual herb

Root: Branched tap root

Stem: Aerial, erect or diffusely spreading, branched, with branches being hairy.

Leaf: Petiolate, alternate, compound with 3-9 leaflets. Leaflets obovate, serrate, with adnate stipules.

Inflorescence: Flowers solitary axillary or terminal.

Flower: Pedicellate, ebracteate, complete, actinomorphic, heterochlamydeous, bisexual and hypogynous.

Calyx: There is a whorl of epicalyx with 5 lobes; sepals 5, connate, pubescent, persistant, acute. Aestivation valvate.

Corolla: Petals 5, free yellow, obovate. Aestivation valvate.

Androecium: Stamens indefinite, spirally arranged; anthers basifised, dithecous, with longitudinal splitting.

Gynoecium: Ovary superior, multicarpellary, apocarpous, spirally arranged on the elongated thalamus. Ovule one in each carpel on basal placentation. Style short and lateral, persistent in the fruit, stigma minute and pointed.

Fruit: An etaerio of achenes

Seed: Exalbuminous

Floral formula: ✝ ⚥ $Epik_5\,K_{(5)}\,C_5\,A_{(\infty)}\,\underline{G}_{(\infty)}$

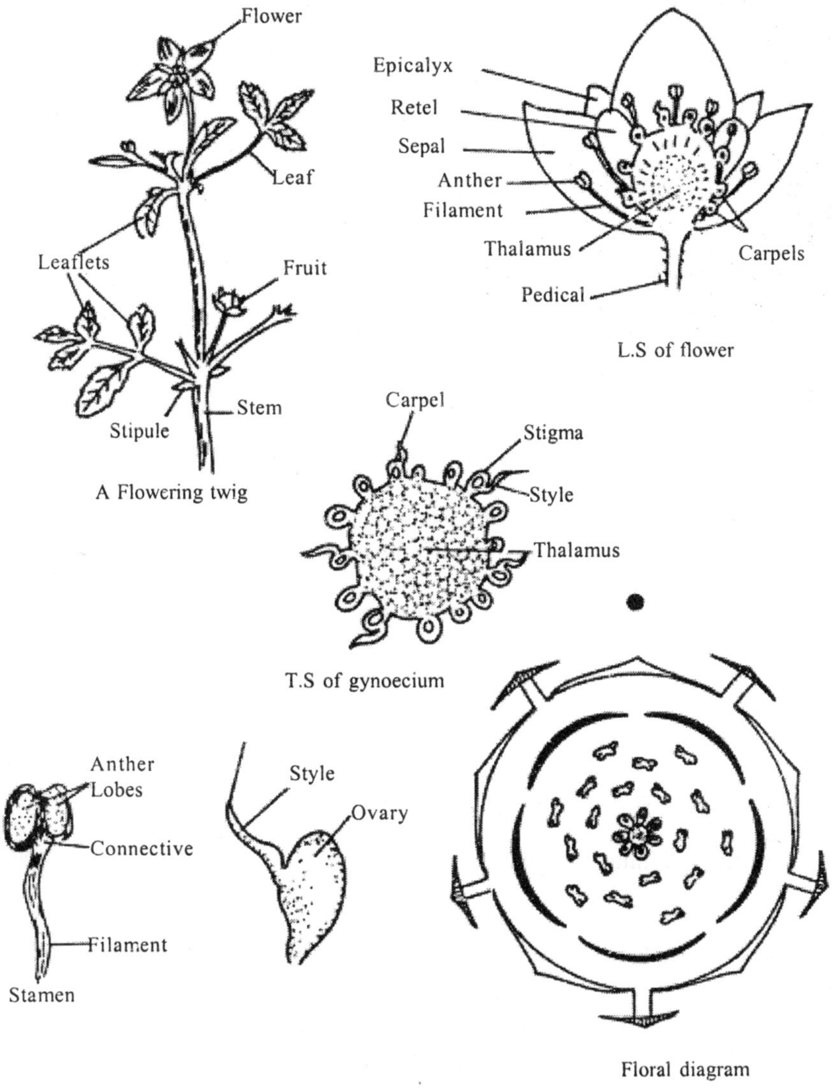

Fig 8.45 *POTENTILLA SUPINA* (Rosaceae)

ERIOBOTRYA JAPONICA

(Rosaceae)

Habitat : Mesophyte, grows (cultivated) in cool places.

Habit : A small tree

Root : Tap root, branched.

Stem : Aerial, erect, woody, branched and cylindrical

Leaf : Simple stipulate alternate, leathery (coriaceous), large 6-12" in length, subsessile, lanceolate or oblanceolate, acuminate, dentate or serrate margin, wooly below; venation reticulate unicostate.

Inflorescence : Flowers grouped in dense, wooly panicles.

Flower : Pedicellate, bracteate, bracteolate, actinomorphic bisexual and eipgynous.

Calyx : Sepals 5, gamosepalous, wooly, arranged on the rim of the thalamus.

Corolla : Petals 5, free, obovate or orbicular, notched at margins, wooly. Aestivation contorted or imbricate

Androecium : Stamens indefinite, free, curved inwards in the bud. Anthers dithecous, dorsifixed splitting longitudinally.

Gynoecium : Carpels 5, enclosed in the hypanthium pentalocular with two ovules in each locule on axile placenta.

Fruit : The fruit is a pyriform berry. It is succulent and 1-5 seeded.

Seed : Exalabuminous, large

Floral formula : ✝ ⚥ $K_{(5)} C_5 A_{(\infty)} \underline{G}_5$

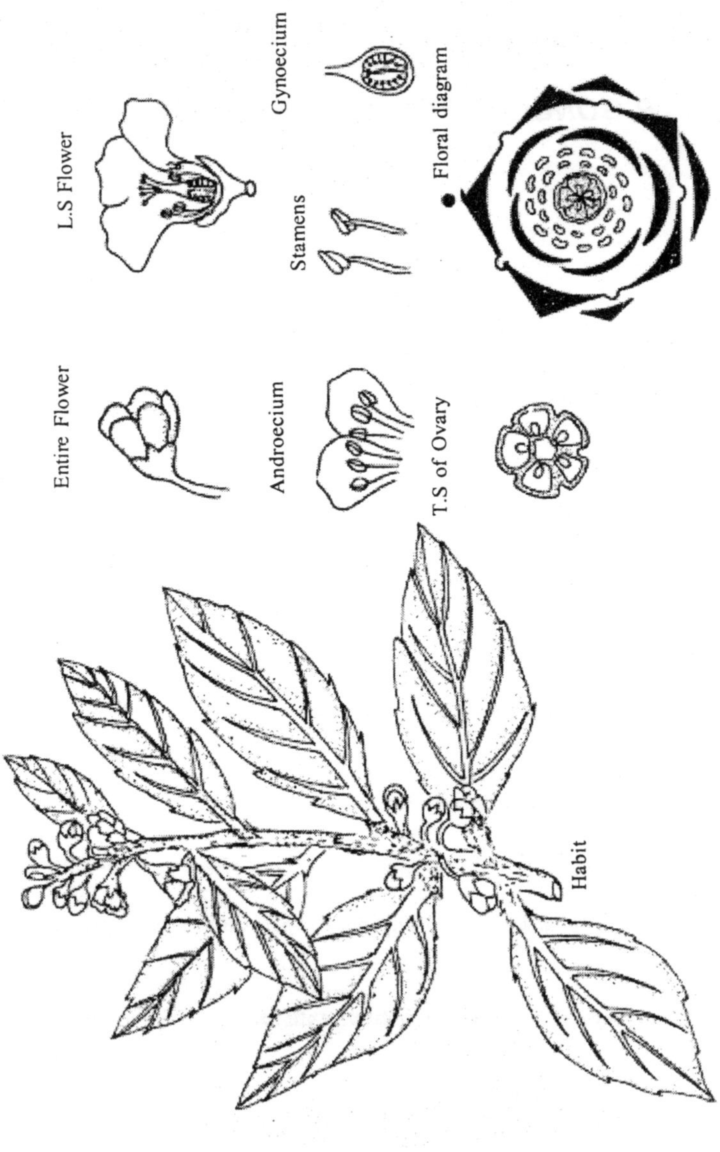

Fig 8.46 *ERIOBOTRYA JAPONICA* (Rosacceae)

TRIGONELLA FOENUMGRACEUM

(Papilionaceae)

Habitat : Autorophic, mesophyte, grows in moist soil.

Habit : An annual herb usually cultivated

Root : Tap root, bacterial root nodules present

Stem : Erect, branched, aerial, green, soft

Leaf : Petiolate stipulate alternate, trifoliately compound, fragrant.

Inflorescence : Raceme

Flower : Pedicellate, bracteate, bracteolate, heterochlamydeous, complete, zygomorphic, bisexual and hypogynous.

Calyx : Sepals 5, gamosepalous, odd sepal anterior, Aestivation valvate.

Corolla : Petals 5, papilionaceous (For details see Page No.174) Aestivation descendingly imbricate.

Androcium : Stamens 10, diadelphous with 9+1 grouping. Anthers dithecous. Staminal column surrounds the pistil.

Gynoecium : Ovary superior, monocarpellary, unilocular with many ovules on marginal placenta. Style simple with a terminal stigma.

Fruit : A legume, dehiscing on only one valve.

Seed : Non endospermous.

Floral formula : † ⚥ $K_{(5)} C_5 A_{9+1} \underline{G}_1$

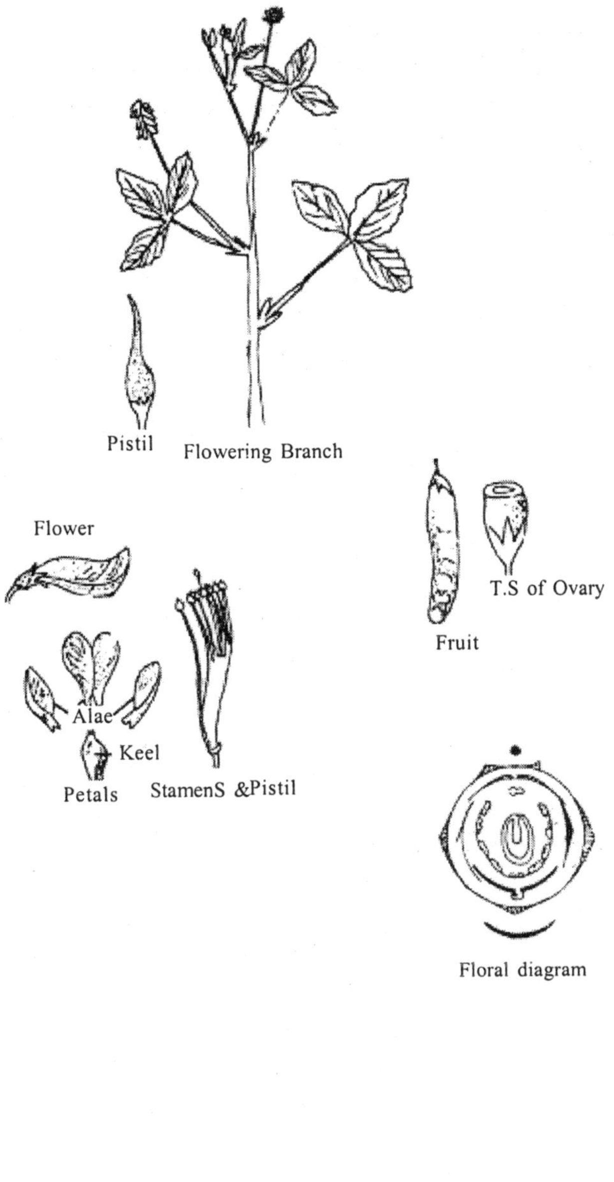

Fig 8.47 ***TRIGONELLA FOENUMGRACEUM*** **(Papilionaceae)**

CROTALARIA JUNCEA
(Papilionaceae)

Habitat: Mesophyte, grows in dry open lands

Habit: A herb, sometimes becomes shrubby

Root: Branched tap root, with branches containing bacterial root nodules helping in N_2 fixation

Stem : Aerial, erect, branched cylindrical, pubescent, green and sometimes grooved. Stem is generally soft may become hard at the base

Leaf: Pedicellate, alternate, stipulate, or trifoliately compound, rarely leaflets may be more in numbewr, leaflets linear oblong with silky pubescence on both surfaces. Leaf base pulvinate venation reticulate unicostate.

Inflorescence : Simple raceme, racemes may be terminal, lateral or leaf opposed.

Flower : Pedicellate, bracteate, bracteolate, heterochlamydeous, pentamerous, complete, zygomorphic, bisexual and hypogynous.

Calyx : Sepals 5, gamosepalous. Aestivation valvate; odd sepal anterior; sepals inconspicuously bilabiate upper lip bipartite, lowerlip tripartite.

Corolla : Petals 5, polypetalous, papilionaceous, consists of a large standard petal (*vexillum*) in anterior position, two lateral petals (*alae* or wings) and at the posterior position the two petals fuse to form a boat shaped keel petal or *carina*. Aestivation descendingly imbricate i.e. standard outermost, keel petal innermost and wings in and out.

Androecium : Stamens 10 in number monadelphous, connate, the staminar column encloses the ovary. Anthers dithecous splitting longitudinally.

Gynoecium : Ovary superior, monocarpellary unilocular with many ovules arranged on marginal placenta. Style long incurved or reflexed; stigma small, obliquely placed.

Fruit : A legume, the fruit breaks open on the ventral suture exposing the seeds.

Seed : Exalbuminous, small

Floral formula : † ♀ $K_{(5)} C_5 A_{(10)} \underline{G}_1$

Fig 8.48 ***CROTALARIA JUNCEA*** **(Papilionaceae)**

PISUM SATIVUM

(Papilionaceae)

Habitat: Mesophytic plant growing in moist soil

Habit: Annual herb

Root: Branched tap root. Bacterical root nodules present, They help in N_2 fixation

Stem: Aerial, soft, climbing with the help of leaf drils, cylindrical, branched and smooth.

Leaf: Alternate, stipulate, stipules two, modified into foliaceous structures, unipinnately compound, some of the terminal leaflets and the tip of the primary rachis modified into tendril to help in climbing; leaflets ovate, smooth, entire margin, mucronate apex. Leaf base pulvinate.

Inflorescence: A simple raceme

Flower: Pedicellate, bracteate, bracteolate, heterochlamydeous, complete, pentamerous, zygomorphic bisexual and hypogynous.

Calyx: Sepals 5 in number, gamosepalous, Aestivation valvate; odd sepal anterior

Corolla: Papilionaceous (For details see Page No. 174) petals 5, free white or red in colour. Aestivation descendingly imbricate.

Androecium: Stamens 10, diadelphous with 9+1 arrangement, anthers basifixed, dithecous and introrse splitting open longitudinally.

Gynoecium: Ovary superior, monocarpellary unilocular with several ovules on marginal placenta, style long and curved with a hairy stigma.

Fruit: A Legume

Seed: Non endospermous

Floral formula: ✝ ⚥ $K_{(5)}C_5A_{9+1}\underline{G}_1$

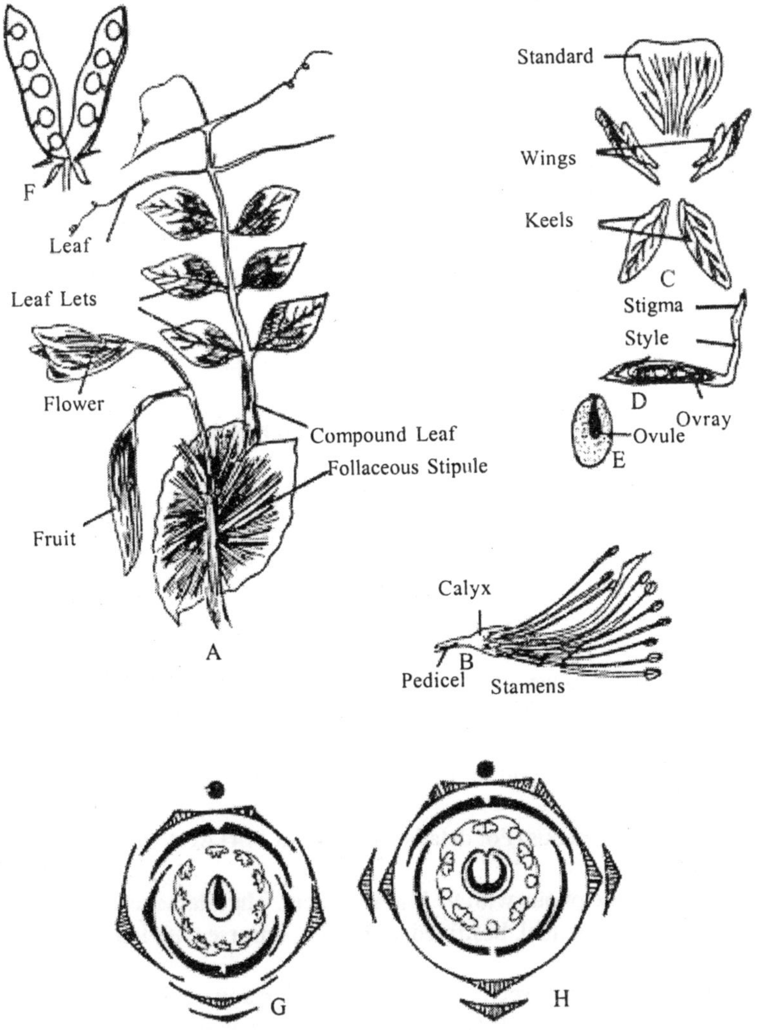

Fig 8.49 ***PISUM SATIVUM*** **(Papilionaceae)**

LATHYRUS ODORATUS

(Papilionaceae)

Habitat : Mesophytic plant, growing in irrigated open land.

Habit : An annual climbing herb

Root : Branched tap root, with bacterial root nodules helping in N_2 fixation

Stem : Aerial, soft, weak, green, glabrous, flattened and winged climbing with the help of leaf tendrils.

Leaf : Petiolate, alternate, stipulate (stipules foliaeous), imparipinnately compound 5-7 pairs of leaflets found, upper pairs as well as the top of the primary rachis modified into tendrils to help in climbing. Leaflets sessile, opposite, oval, margin entire, acute, glabrous, venation unicostate reticulate. Leaf base pulvinate

Inflorescence : Simple raceme flowers may be rarely solitary axillary.

Flower : Pedicellate, bracteate, bracteolate, heterochlamydeous, pentamerous, cyclic, zygomorphic, bisexual and hypogynous. Flower butterfly shaped and of various hues of colour.

Calyx : Sepals 5, green, gamosepalous, odd sepal anterior, Aestivation valvate.

Corolla : Petals 5, free, papilionaceous (For details see Page No.174) Aestivation descendingly imbricate

Androecium : Stamens 10, diadelphous with 9+1 arrangement. The filaments of stamens fuse to form a tube enclosing the ovary. Anthers unequal, basifixed, dithecous, introrse with longitudinal splitting.

Gynoecium : Ovary superior, monocarpellary, unilocular with many ovules on marginal placenta ovary almost transversely located with a short stalk. Style short and curved, hairy; stigma simple and hairy.

Fruit : A legume

Seed : Non endospermic

Floral formula : † ♀ $K_{(5)} C_5 A_{9+1} \underline{G}_1$

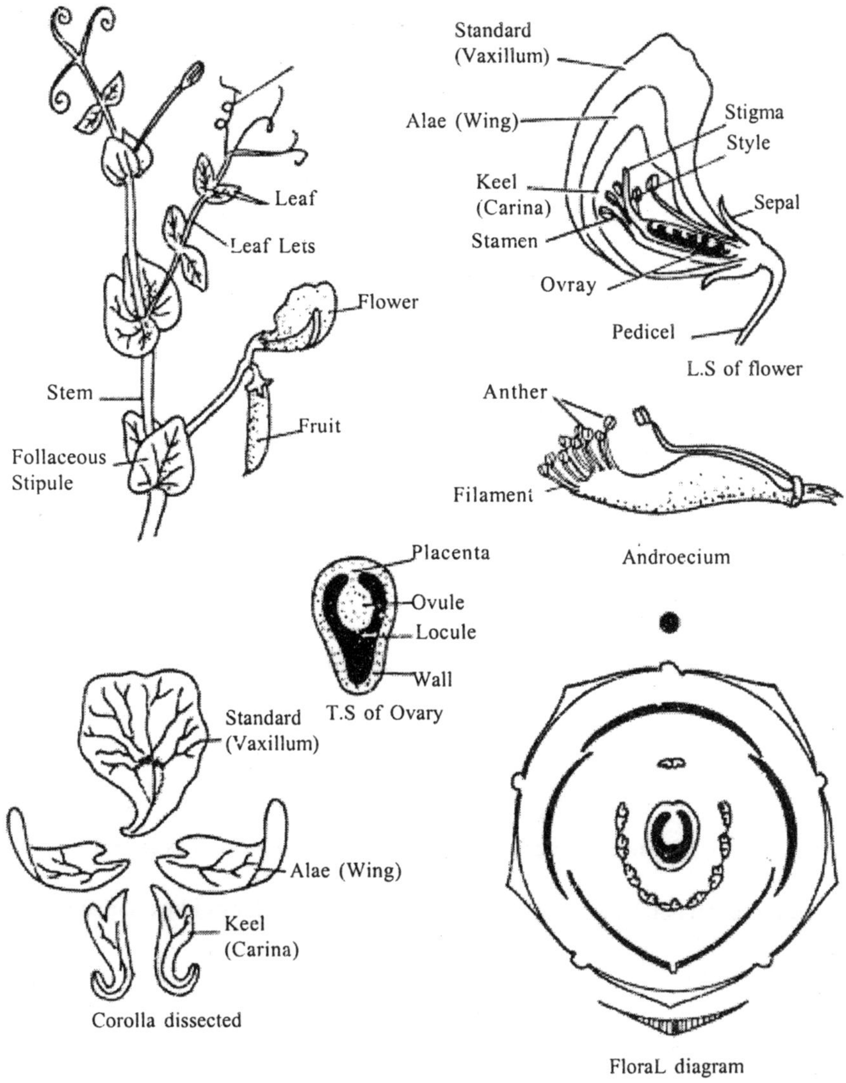

Fig 8.50 ***LATHYRUS ODORATUS*** **(Papilionaceae)**

DALBERGIA SISSOO

(Papilionaceae)

Habitat: Common avenue tree; mesophyte

Habit: A deciduons tree reaching a maximum height of 100 feet and a circumference of 8 feet.

Stem: Aerial, erect, woody, branched

Leaf: Petiolate, distichous, alternate, pulvinate leaf base, imparipinnately compound with usually 3-5 leaflets, rachis zigzag, stipulate, stipules lanceolate, leaflets orbicular, acuminate and entire.

Inflorescence: Axillary panicle

Flower: Pedicellate, bracteate, ebracteolate, heterochlamydeous, pentamerous, cyclic, zygomorphic, bisexual and hypogynous.

Calyx: Sepals 5, gamosepalous, odd sepal anterior Aestivation valvate.

Corolla: Petals 5, free, papilionaceous (For details see Page No.174), pale yellow in colour.

Androecium: Stamens 9, monadelphous forming a column surrounding the ovary.

Gynoecium: Ovary superior, monocarpellary, unilocular with 1-4 ovules on marginal placenta.

Fruit: A legume called pod (few seeds). The fruit is flattend and looks like a samara

Seed: Non endospermic

Floral formula: † ⚥ $K_{(5)} C_5 A_9 \underline{G}_1$

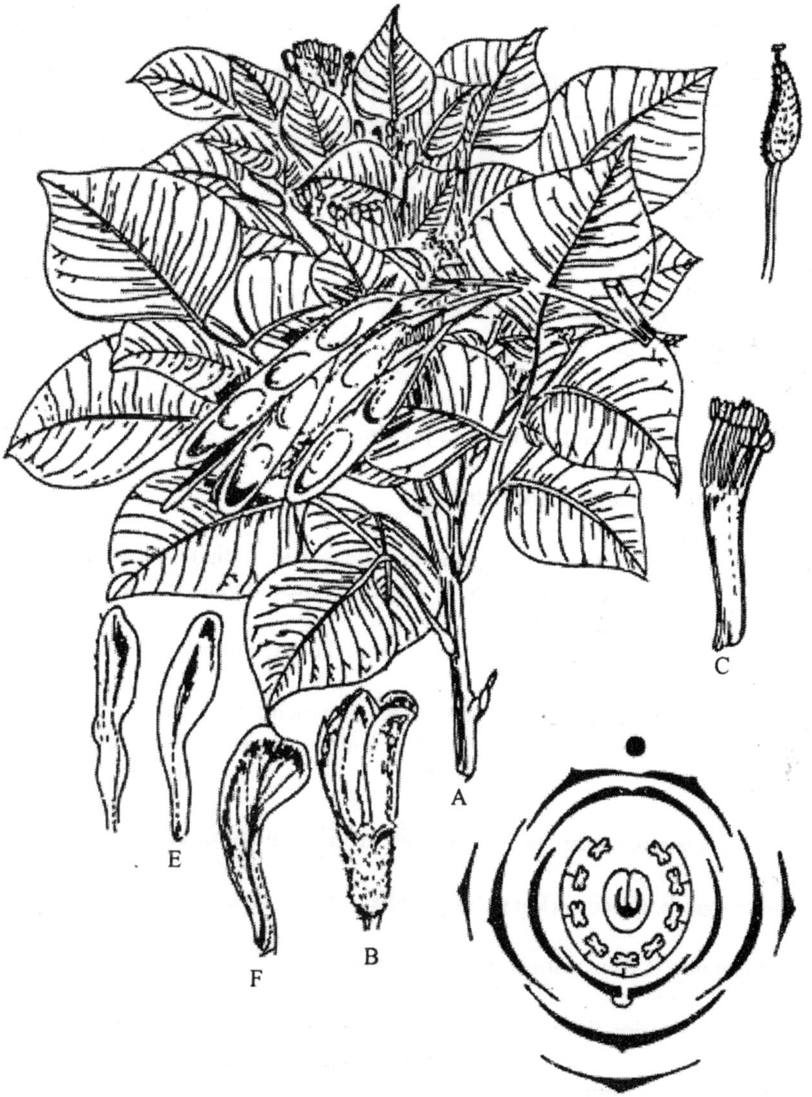

Fig 8.51 *DALBERGIA SISSOO* (Papilionaceae)

ARACHIS HYPOGEA

(Papilionaceae)

Habitat : Mesophyte, grows in loose moist soil.

Habit : Annual herb

Root : Branched tap root, bacterial root nodules present.

Stem : Aerial, erect or prostrate, cylindrical, hairy

Leaf : Alternate, pulvinate, pinnately compound, stipulate with long stipules, leaflets usually four in number in two opposite pairs, entire, obovate, pubescent; venation reticulate unicostate.

Inflorescence : Solitatary flowers or in groups of three in the axils of leaves.

Flower : Pedieliate, bracteate, ebracteolate, complete, heterochlamydeous, zygomorphic, bisexual and hypogynous flowers; develop at the base of the plant and sometimes partially buried in the soil. Flowers cleistogamous

Calyx : Sepals 5, gamosepalous forming a calyx tube. Aestivation imbricate.

Corolla : Petals 5, papilionaceous (For details see Page No.174) Aestivation descendingly imbricate.

Androecium : Stamens 10, (two stamens sterile.) monadelphous.

Gynoecium : Ovary superior, monocarpeliary, unilocular with 2-5 ovules on marginal placenta.

Fruit : A pod (few seeded legume) develops underground

Seed : Exalbuminons, cotyledons oily.

Floral formula : † ⚥ $K_{(5)} C_5 A_{10} \underline{G}_1$

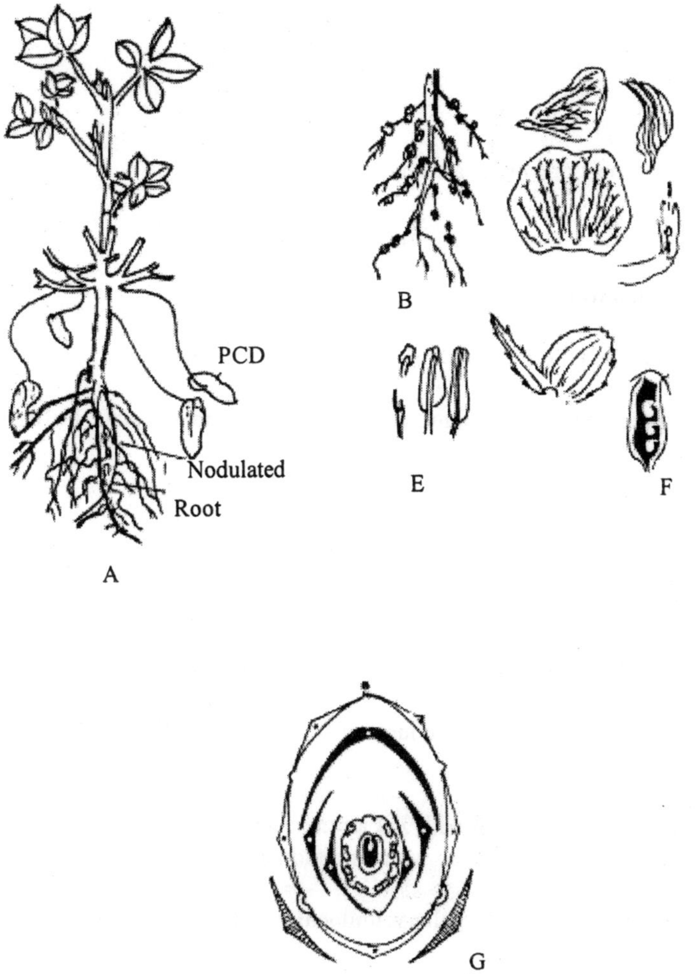

Fig 8.52 ***ARACHIS HYPOGEA*** **(Papilionaceae)**

MELILOTUS INDICA
(Papilionaceae)

Habitat : Mesophyte, grows in well aerated moist soil

Habit : An annual herb

Root : Much branched tap root

Stem : Aerial, erect, branched, herbaceous, cylindrical green and glabrous.

Leaf : Ramal and cauline, alternate, petiolate, trifoliately compound, stipulate, leaflets sub sessile, ovobate, serrate. Venation reticulate unicostate.

Inflorescence : Simple raceme

Flower : Pedicellate, bracteate, ebracteolate, heterochlamydeous, complete, zygomorphic, bisexual and hypogynous.

Calyx : Sepals 5, gamosepalous, bell shaped, green, odd sepal anterior. Aestivation valvate.

Corolla : Petals 5, polypetalous, eapilionaceous (For details see Page No.174) yellow in colour, Aestivation descendingly imbricate

Audroecium : Stamens 10, diadelphous with 9+1 arrangement 9 stamens form a staminal tube surrounding the gynoecium. Anthers diothecous, introrse with longitudinal splitting

Gynoecium : Ovary superior, monocarpellary, unilocular with many ovules an marginal placenta

Fruit : A legume

Seed : Non endospermous

Floral formula : † ♀ $K_{(5)}C_5A_{9+10}\underline{G}_1$

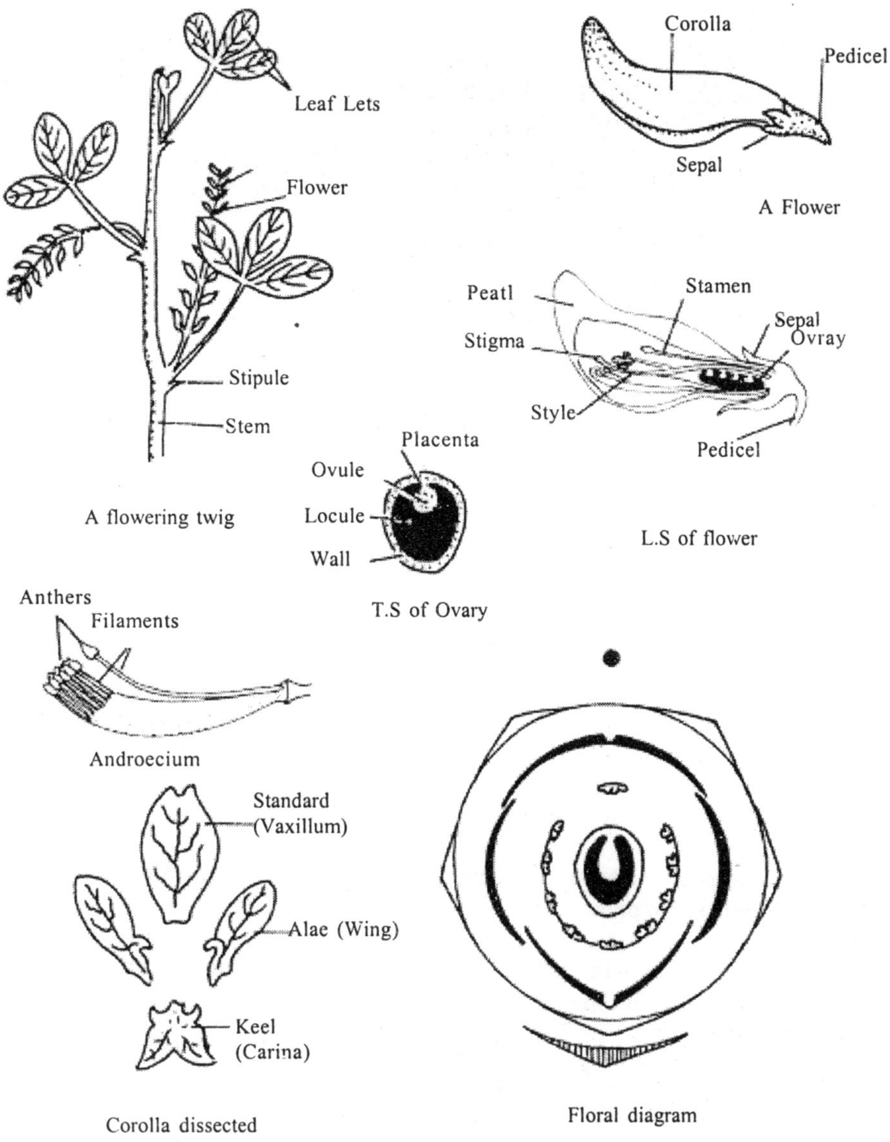

Fig 8.53 *MELILOTUS INDICA* (Papilionaceae)

INDIGOFERA LINNAEIS

(Papilionaceae)

Habitat : Mesophyte, grows in dry open lands amidst grasses. Found in lawns

Habit : A small semiprostrate herb.

Root : Much branched tap root.

Stem : Aerial, semierect, green, soft, branched, solid, hairy and cylindrical.

Leaf : Ramal and cauline, alternate, petiolate, imparipinnately compound, stipulate. Leaflets sessile, small, oblanceolate, entire margin, venation reticulate unicostate.

Inflorescence : Axillary raceme

Flower : Pedicellate, bracteate, ebracteolate, complete, heterochlamydeous, zygomorphic, pentamerous, bisexual and hypogynous.

Calyx : Sepals 5, gamosepalous, green, odd sepal anterior. Aestivation valvate

Corolla : Petals 5, polypetalous, papilionaceous (For details see Page No.174), Aestivation descendingly imbricate, colour of the petals bright red or pink.

Androecium : Stamens 10, diadelphons with 9+1 arrangement. 9 stamens form a column around the gynoecium. Anthers dithecous, basifixed, introrse and split longitudinally

Gynoecium : Same as in *Melilotus indica.*

Fruit : A legume

Floral formula : † ♀ $K_{(5)} C_5 A_{9+1} \underline{G}_1$

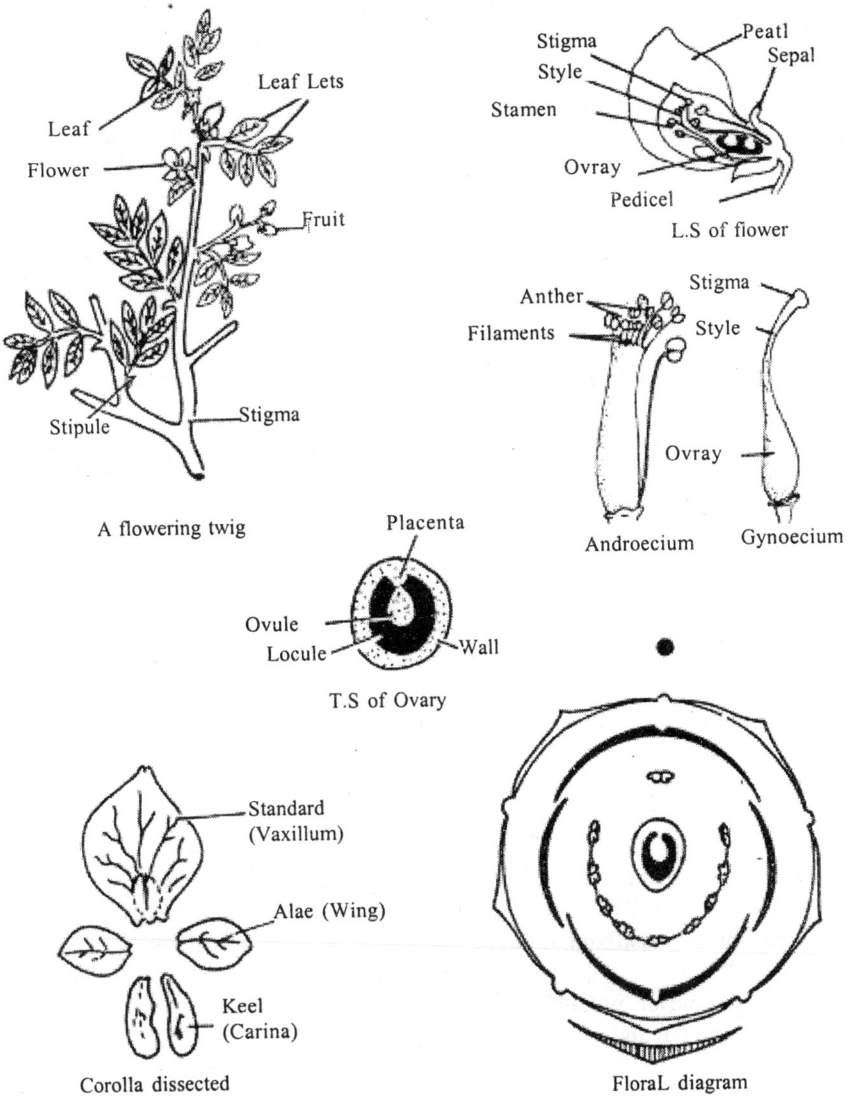

Fig 8.54 ***INDIGOFERA LINNAEIS*** **(Papillionaceae)**

CLASSIA OCCIPENTALIS

(Caesalpinaceae)

Habitat: Mesophytic weed, grows in open dry land

Habit: A well grown herb or undershrub

Root: Branched tap root

Stem: Aerial, erect; Herbaceous, cylindrical, branched, solid and reddish in colour

Leaf: Alternate, stipulate, unipennate, imparipinnately compound, pinnae 3-5 pairs; leaflets elliptic lanceolate, entire, acute. Venation reticulate unicostate. Leaf base pulvinate. A dark coloured gland present near the axil of the rachis

Inflorescence: Axillary corymbose raceme

Flower: Pedicellate, bracteate, ebracteolate, complete, heterochlamydeous, zygomorphic, bisexual and hypogynous

Calyx: Sepals 5, polysepalous, boat shaped, odd sepal anterior; aestivation imbricate or quincuncial.

Corolla: Petals 5, polysepalous, yellow in colour, petals unequal in size, odd posterior sepal smallest; aestivation ascendingly imbricate.

Androecium: Stamens 10, free arranged in two whorls of 5 each, in the outer whorl one anterior and two posterior stamens and in the inner whorl one posterior stamen reduced to staminodes. Stamens of differing heights, anthers long curved, basifixed, dithecous and dehiscence by apical pores.

Gynoecium: Ovary superior, monocarpellary, unilocular with many ovules on marginal placentation. Style short, curved; stigma flat and capitate. Ovary wall pubescent.

Fruit: A legume

Seed: Exalbuminous

Floral formula: † ♀ $K_{(5)} C_5 A_{5+5} \underline{G}_1$

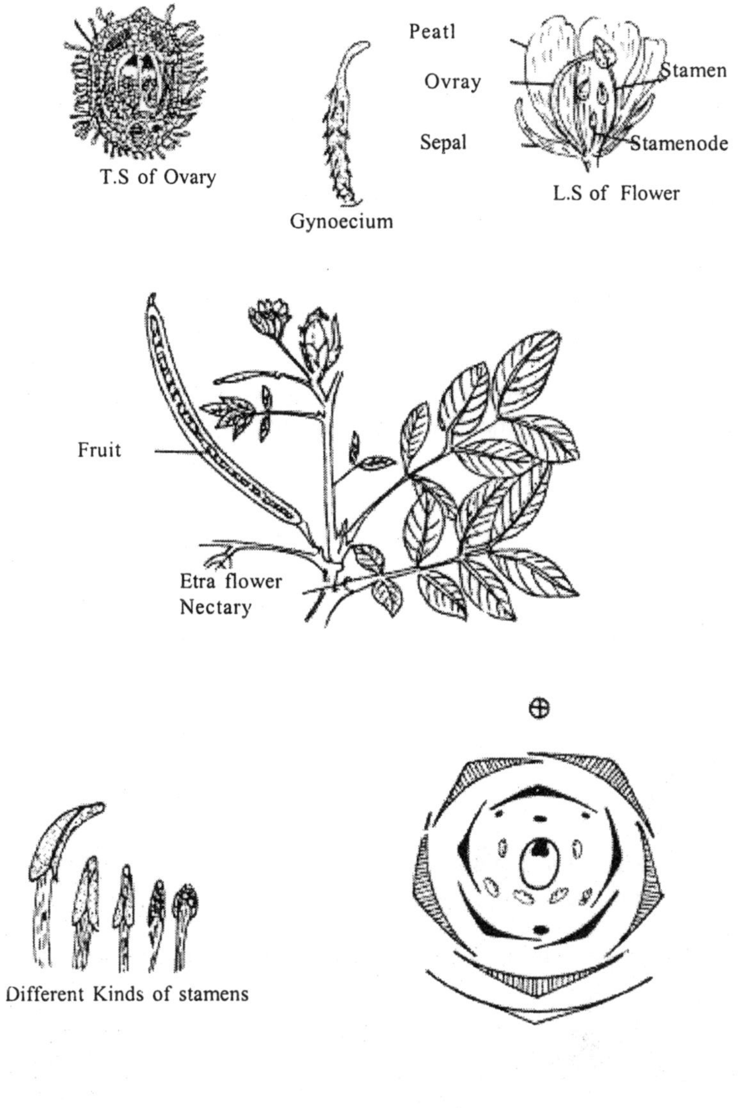

Fig 8.55 *CLASSIA OCCIPENTALIS* (Caesalpinaceae)

CASSIA TORA

(Caesalpinaceae)

Habitat : Mesophyte, grows in plains and also at elevated regions.

Habit : Annual shrubby herb.

Root : Branched tap root

Stem : Aerial, erect, somewhat woody, branched, cylindrical

Leaf : Alternate, petiolate, unipinnate, paripinnately compound, pulvinate, stipulate (caducous), leaflets in 3-5 pairs, ovate, entire, acute, glabrous venation reticulate, unicostate, subulate, glands present between the leaflets.

Inflorescence : racemose, axillary

Flower : Pedicellate, bracteate, bracts linear and caducous, complete, heterochlamydeous, zygomorphic, bisexual and hypogynous.

Calyx : Sepals 5, polysepalous, yellowish green in colour, odd sepal anterior, Aestivation quincuncial or imbricate.

Corolla : Petals 5, polypetalous, yellow, clawed odd petal posterior and smallest. Aestivation ascendingly imbricate

Androecium : Stamend 10, free, unequal in length, 3 posterior stamens reduced to stamens reduced to staminodes. Anthers basifixed and dithecous.

Gynoecium : Ovary superior, monocarpellary, unilocular with many ovules arranged on marginal placenta, style short, stigma hairy.

Fruit : A legume

Seed : Exalbuminous

Floral formula : † ⚥ $K_{(5)} C_5 A_{3+7} \underline{G}_1$

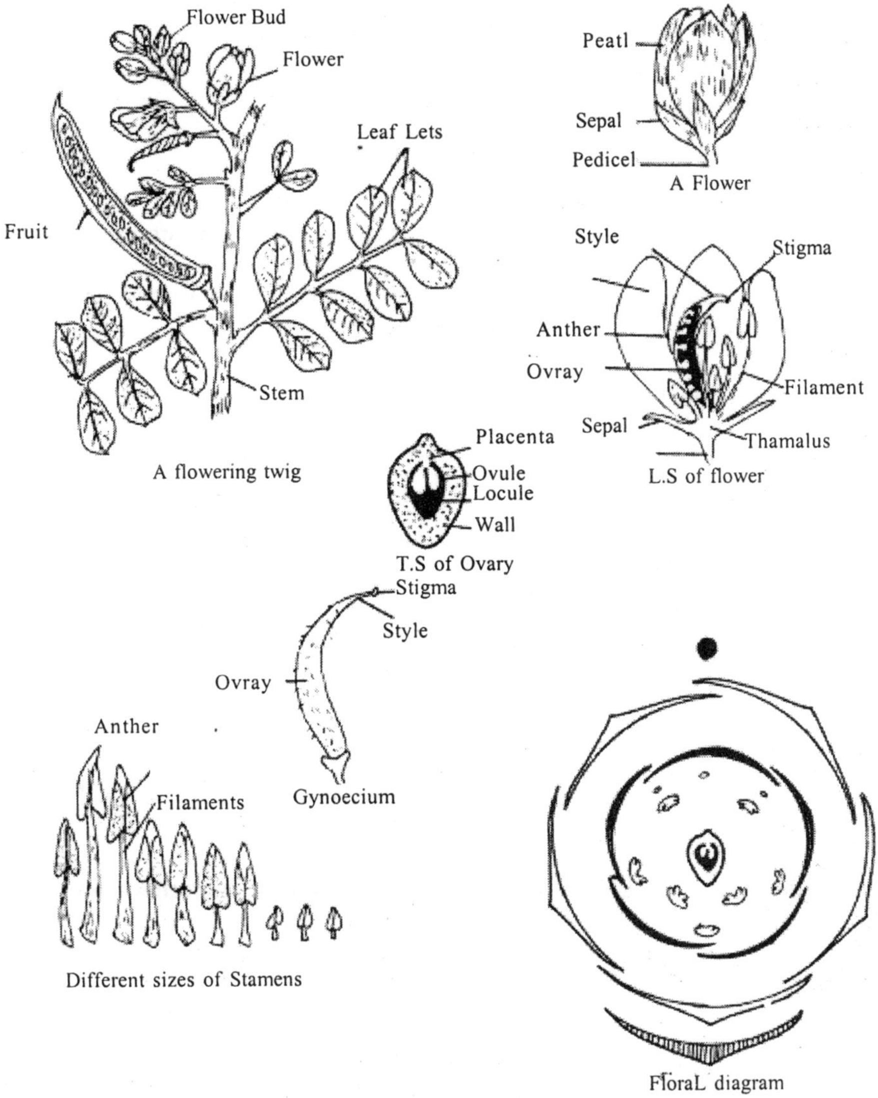

Fig 8.56 *CASSIA TORA* (Caesalpinaceae)

CASSIA FISTULA
(Caesalpinaceae)

Habitat : Mesophytic tree, grows abundantly in open deciduous forests.

Root : Branched tap root

Stem : Aerial, erect, woody, branched, cylindrical, solid and glabrous; younger parts green in colour.

Leaf : Ramal and cauline, alternate, petiolate, exstipulate unipinnate, paripinnately compound, leaflets ovate, entire or wavy, acute, glabrous coriaceous; venation reticulate unicostate

Inflorescence : Long, pendulous, axillary or extra axillary raceme.

Flower : Pedicellate, bracteate, ebracteolate, complete, heterochlamydeous, zygomorphic, bisexual and hypogynous.

Calyx : Sepals 5, polysepalous, yellowish green in colour, odd sepal anterior Aestivation quincuncial

Corolla : Petals 5, polypetalous, odd posterior petal smallest, bright yellow; aestivation ascendingly imbricate.

Androecium : Stamens 10, arranged in two whorls of 5 each, 3 upper (posterior) stamens reduced to staminodes. Fertile stamens are unequal in length, Anthers dorsifixed, dithecous and introrse.

Gynoecium : Ovary superior, monocarpellary, unilocular, with many ovules on marginal placenta. Style short, stigma capitate and hairy.

Fruit : A legume, cylindrical in shape and divided into segments. Fruits are green when young and dark when ripe.

Seed : Exalbuminous, large, circular in shape

Floral formula : † ♀ $K_{(5)} C_5 A_{10} \underline{G}_1$

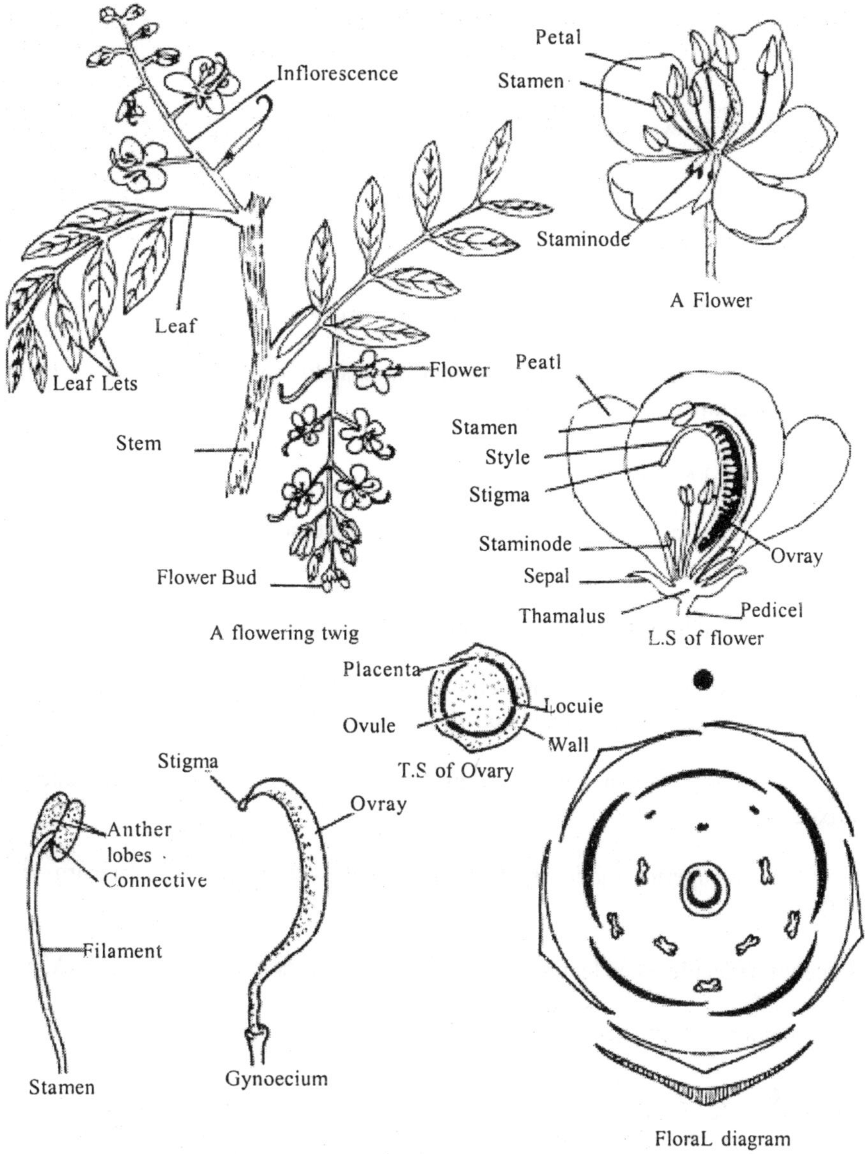

Fig 8.57 ***CASSIA FISTULA*** **(Caesalpinaceae)**

TRAMARINDUS INDICA

(Caesalpinaceae)

Habitat : Mesophyte, grows in dry land

Habit : A large perennial tree

Root : Branched tap root

Stem : Aerial, erect, woody, cylindrical, rough, brownish in colour.

Leaf : Ramal and cauline, alternate, unipinnate and paripinnately compound, exstipulate, petiolate, elliptical, margin entire, venation unicostate reticulate

Inflorescence : Axillary raceme

Flower : Pedicellate bracteate, bracteolate, cyclic, complete, heterochlamydeous, zygomorphic, bisexual and hypogynous.

Calyx : Sepals 5, polysepalous, odd posterior sepal larger, greenish yellow in colour; aestivation imbricate

Corolla : Petals 5, polypetalous, brightly coloured, two anterior petals reduced to scales. Aestivation ascendingly imbricate; odd posterior petal hooded, yellow incolour

Androecium : Stamens 7, monadelphons forming a column; of the 7 stamens 4 are reduced to staminodes; anthers basifixed, dithecous and extrorse

Gynoecium : Ovary superior, monocarpellary, unilocular, with many ovules on marginal placenta; style long, stigma flat.

Fruit : A lomentum (deeply lobed legume), epicarp thin and crustaceous, mesocarp thick, pulpy and acidic.

Seed : Dark in colour, flat, and hard, non endospermic

Floral formula : † ⚥ $K_{(5)} C_5 A_{3+4} \underline{G}_1$

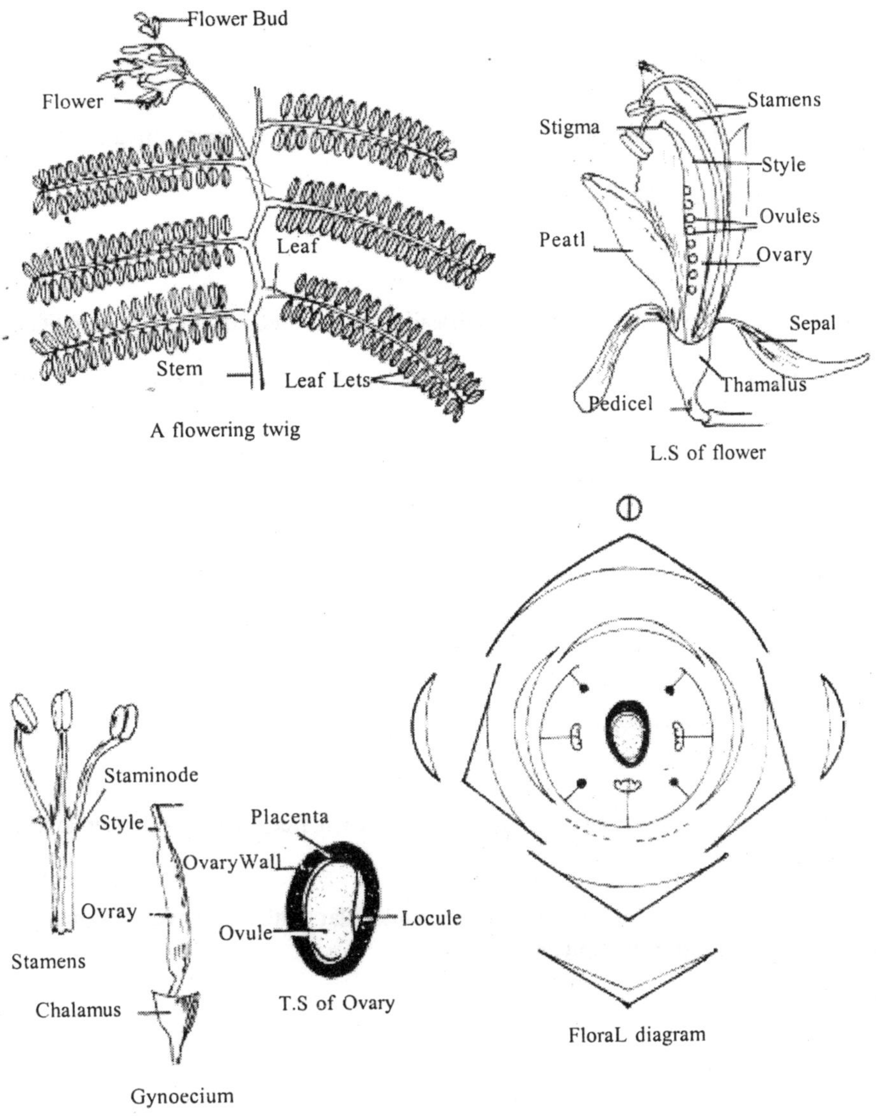

Fig 8.58 *TRAMARINDUS INDICA* (Caesalpinaceae)

CAESALPINIA PULCHERRIMA

(Caesalpinaceae)

Habitat: Mesophyte, commonly found in dry decidous forests, often cultivated.

Habit: A large shrub or small tree

Root: Branched tap root

Stem: Aerial, erect, somewhat woody, cylindrical, solid and green in colour in younger parts

Leaf: Ramal and cauline, alternate, petiolate, bipinnately (Paripinnate) compound, stipulate, pulvinate, leaflets elliptical, obtuse, venation reticulate unicostate.

Inflorescence: Axillary or extraaxillary raceme

Flower: Pedicellate, bracteate, ebracteolate, complete, cyclic, heterochlamydeous, pentamerous, zygomorphic, bisexual and hypogynous. Flowers large and showy, reddish yellow colour, thalamus has slight depression

Calyx: Sepals 5, polysepalous (occasionally fused at the base); odd posterior sepal large and hooded. Aestivation imbricate

Corolla: Petals 5, polypetalous, unequal in size, clawed, margin undulate, orange red or yellow in colour; aestivation ascendingly imbricate.

Androecium: Stamens 10, arranged in two whorls of 5 each, free, filaments slender, long, anthers dithecous, dorsifixed and introrse

Gynoecium: Ovary superior, monocarpellary, unilocular, with many ovules on marginal placenta; style long, somewhat curved; stigma capitate

Fruit: A legume

Seed: Non endo spermous

Floral formula: † ⚥ $K_{(5)} C_5 A_{5+5} \underline{G}_1$

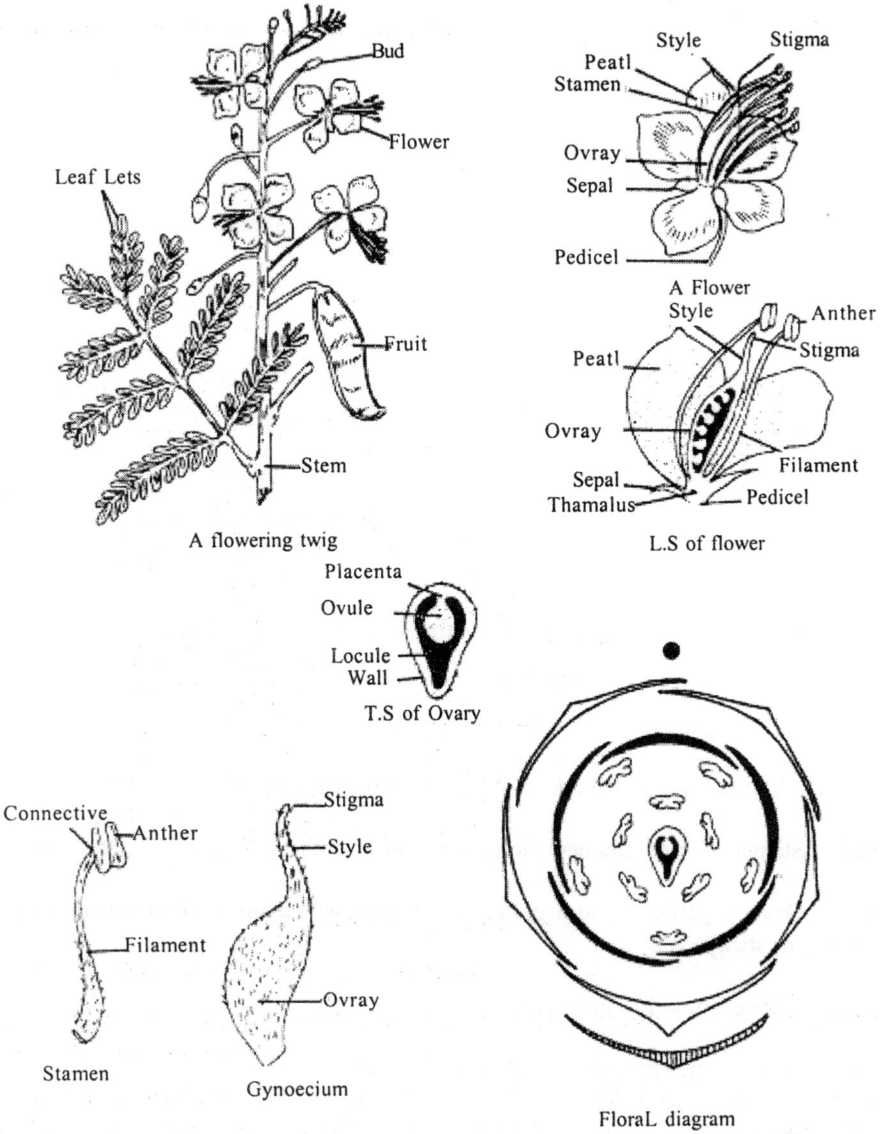

Fig 8.59 ***CAESALPINIA PULCHERRIMA*** **(Caesalpinaceae)**

BAUHINIA VARIEGATA

(Caesalpinaceae)

Habitat: Mesophyte, common avenue tree, grows in deciduous forests.

Habit: A decidnous tree; bushy

Root: Tap root

Stem: Aerial, erect, cylindrical, woody, solid, greenish in colour

Leaf: Alternate, ramal and cauline, simple, bilobed, petiolate, margin entire, leaf base pulvinate; venation reticulate multicostate.

Inflorescence: Axillary raceme.

Flower: Pentamerous, heterochlamydeous, zygomorphic, bisexual and hypogynous (could be slightly perigynous)

Calyx: Sepals 5, free (slightly fused at base), odd sepal anterior; aestivation imbricate

Corolla: Petals 5, free white or pink in colour; aestivation descendingly imbricate.

Androecium: Stamens 5, free, filaments long and slender; anthers dithecous, versatile and introrse

Gynoecium: Ovary superior, monocarpellary, unilocular, with many ovules on marginal placenta; style long, curved; stigma sticky.

Fruit: A legume, long; about 12" in length

Seed: Exalbuminous, somewhat remiform, brownish papery testa.

Floral formula: † ⚥ $K_{(5)} C_5 A_5 \underline{G}_1$

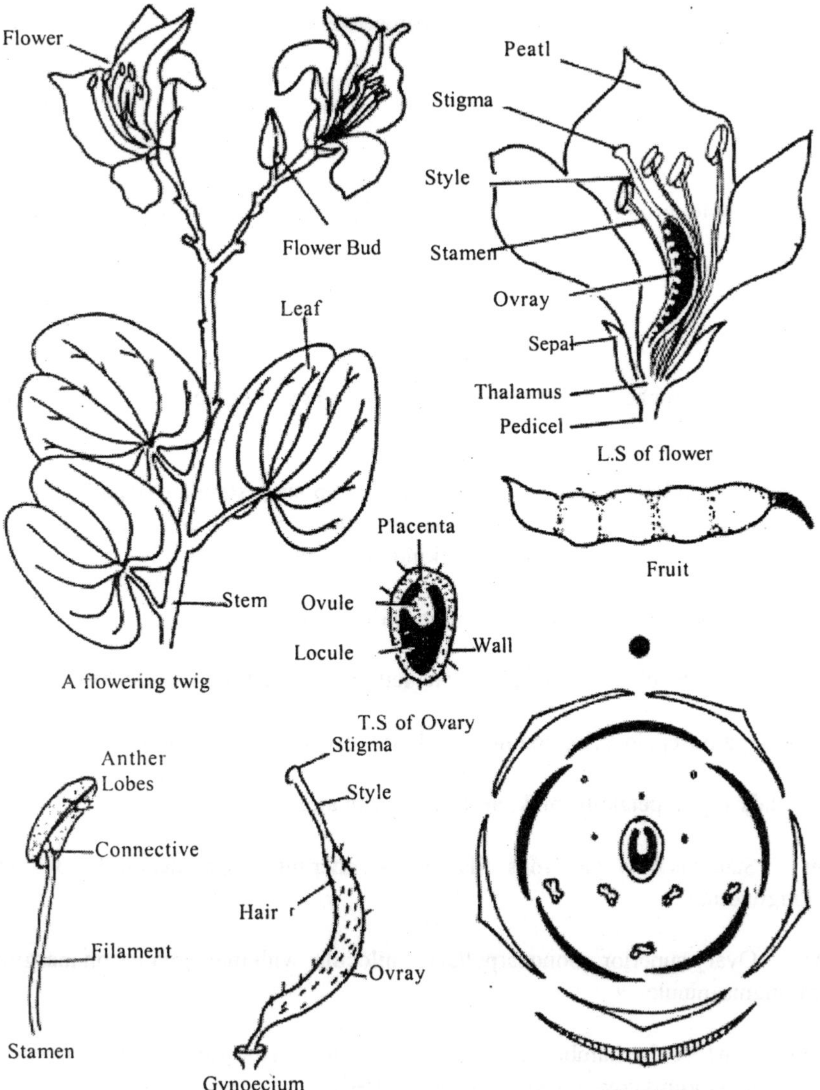

Fig 8.60 ***BAUHINIA VARIEGATA*** **(Caesalpinaceae)**

MIMOSA PUDICA
(Mimoceae)

Habitat: Mesophyte, grows in open dry waste lands

Habit: A prickly herb

Root: Tap root

Stem: Aerial, erect, branched, pubescent, greenish or brown in colour. Epidermal prickles found on the stem.

Leaf: Alternate, pedicellate, pulvinate, bipinnately compound, stipulate (stipules modified into spines), leaflets small, sessile, oblong, acute, hairy, venation reticulate unicostate. Leaflets are sensitive to touch (touch me not) and they drop down exhibiting a kind of sleeping movement.

Inflorescence: A globose head, usually found in axillary pairs

Flower: Sessile, bracteate, ebracteolate, complete, tetramerous, actinomorphic, bisexual and hypogynous.

Calyx: Sepals 4, gamosepalous, campamulate, shortly dentate, aestivation valvate

Corolla: Petals 4, polypetalous, pink; aestivation valvate.

Androecium: Stamens 4, polyandrous, filaments slender pinkish, anthers dithecous, versatile, introrse; dehiscence longitudinal.

Gynoecium: Ovary superior, monocarpellary, unilocular with many ovules on marginal placenta; style long, filiform; stigma minute.

Fruit: A lomentum (deeply chambered) produced in clusters on the globose infructuscence. Fruits, greenish in colour, flat with 3-7 joints covered with yellowish bristles.

Seed: Non endospermic

Floral formula: ✝ ♀ $K_{(4)} C_4 A_4 \underline{G}_1$

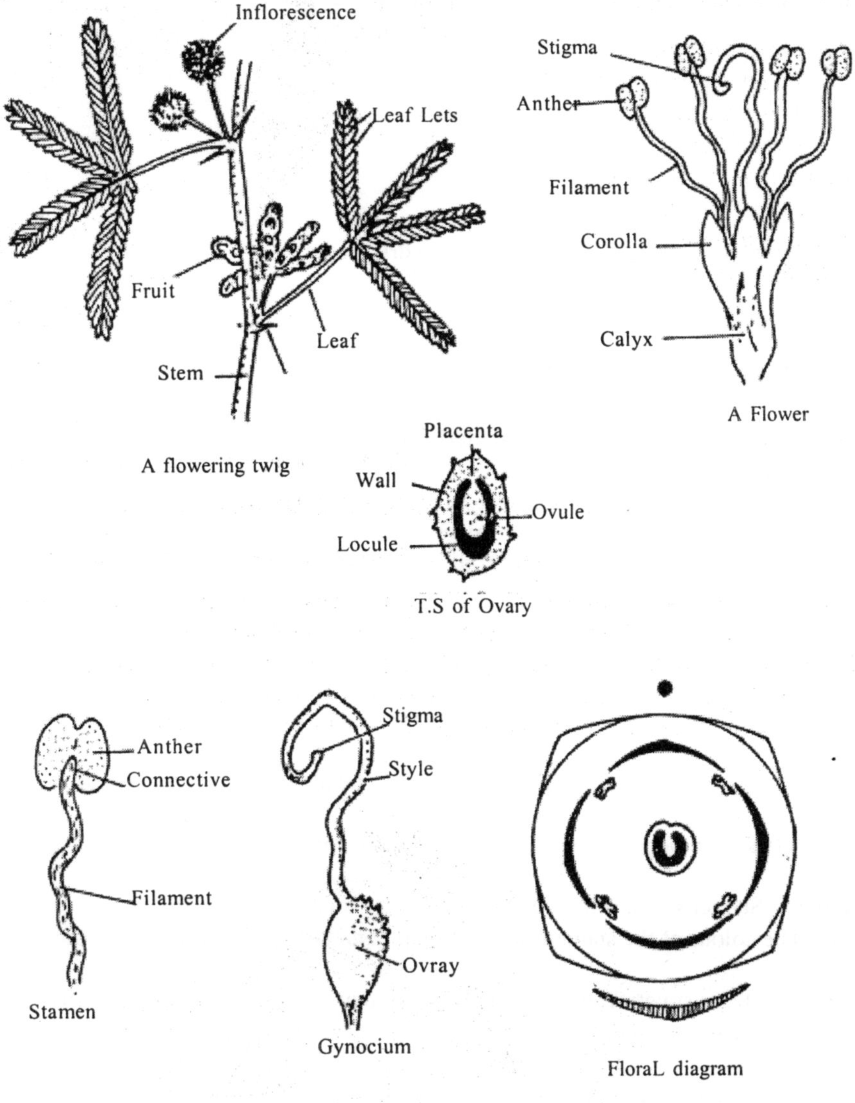

Fig 8.61 *MIMOSA PUDICA* (Mimoceae)

ACACIA NILOTICA
(= A. Arabica)
(Mimoceae)

Habitat : Xerophytic plant, wild or cultivated in dry places

Habit : A medium sized tree

Root : Much branched, deep feeder tap root

Stem : Aerial, erect, branched, cylindrical, dark brown in colour (bark), bark exudes a gummy substance (gum arabic)

Leaf : Ramal and cauline; alternate, petiolate, stipulate stipules modified into spines (1-15cm long), bipinnately compound, leaflets small in 10-20 pairs on the rachilla, subsessile, opposite, oblong, entire, venation reticulate unicostate.

Inflorescence : Axillary cymose head, borne on short peduncles; heads about 1-2 cm in diametre

Flower : Sessile, bracteate, ebracteolate, heterochlamydeous, pentamerous, bisexual and hypogynous.

Calyx : Sepals 5, gamosepalous, campanulate, slightly petaloid; aestivation valvate.

Corolla : Petals 5, gamopetalous, tubular; aestivation valvate

Androecium : Stamens indefinite, free, filaments long and slender, yellowish; anthers dithecous, versatile and introrse. The colour of the stamens is preeminent and attracts the attention of the pollinating agents

Gynoecium : Ovary superior, monocarpellary, unilocular with many ovules on marginal placenta, style long and filiform; stigma minute.

Fruit : A lomentum, 8-15 cm long; fleshy and somewhat grey in colour

Seed : Non endospermic,

Floral formula : ✝ ⚥ $K_{(5)} C_5 A_{\infty} \underline{G}_1$

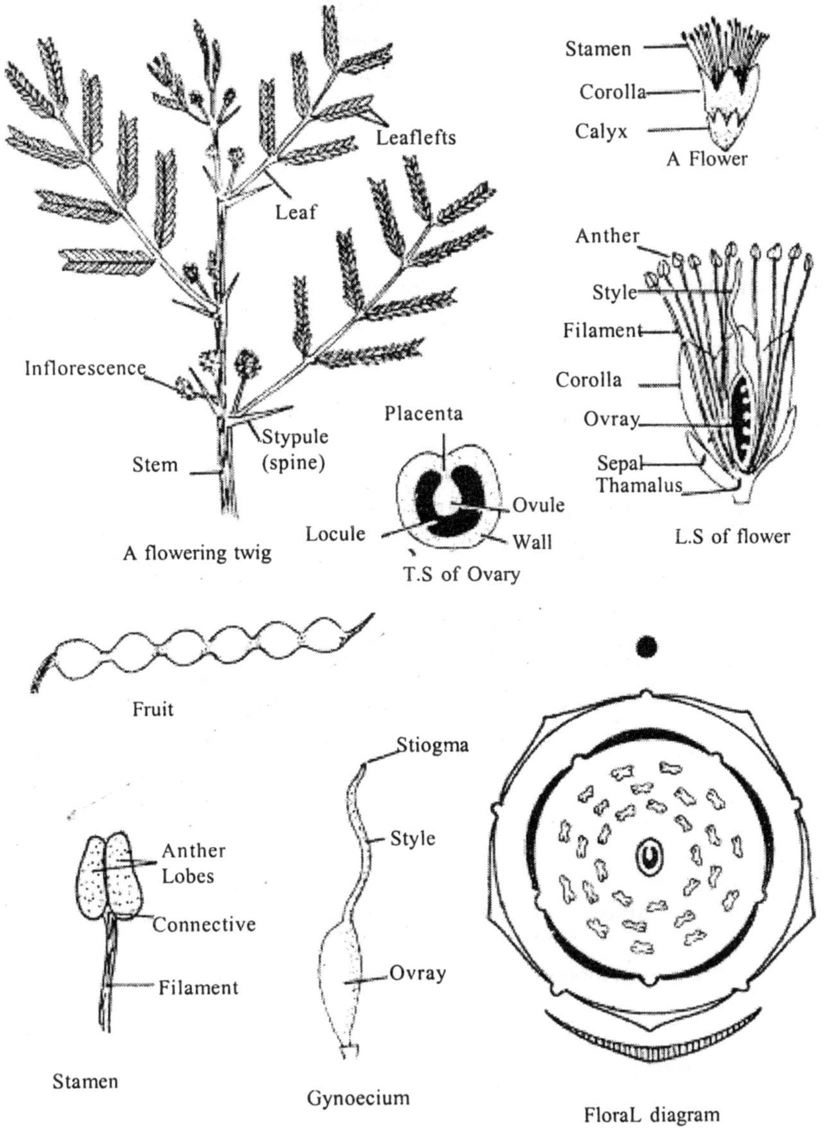

Fig 8.62 ***ACACIA NILOTICA (= A. arabica)*** **(Mimoceae)**

ALBIZZIA LEBBEK

(Mimoceae)

Habitat: A deciduous tree growing in dry places

Habit: A mesophytic tree shows tendency of xerophytism

Root: Branched tap root

Stem: Aerial, erect, cylindrical, hard, bark greyish. Gum exudes from bark

Leaf: Alternate, petiolate, pulvinate, bipinnately compound, stipulate (linear and caducous); a large gland present at the base of the petiole, leaflets in 5-9 pairs, opposite, obovate-oblong, entire, obtuse or slightly retuse; venation reticulate unicostate.

Inflorescence: Cymose heads; heads have a long peduncle; arise singly or in groups of 2-4 at the axils of leaves. Rarely clusters of heads may be corymbose.

Flower: Bracteate, ebracteolate, pedicellate, complete, actinomorphic, cyclic, bisexual and hypo or perigynous

Calyx: Sepals 5, gamosepalous, campanulate, greenish white in colour; aestivation valvate.

Corolla: Petals 5, poly or gamopetalous, tubular, longer than calyx; white in colour; aestivation valvate.

Androecium: Stamens indefinite monadelphous, filaments slender connate at the base; attractively coloured with a *large of rose*; anthers dithecous, versatile, introrse, dehiscence longitudinal

Gynoecium: Ovary superior, monocarpellary, unilocular with many ovules on marginal placenta; style slender; stigma capitate.

Fruit: A legume (Pod), oblong, 4-12" long, flat, pale yellow, smooth, blunt at both ends, shining,

Seeds: 4-12 per fruit, pale brown, exalbuminous.

Floral formula: ✝ ⚥ $K_{(5)} C_5 A_{\infty} \underline{G}_1$

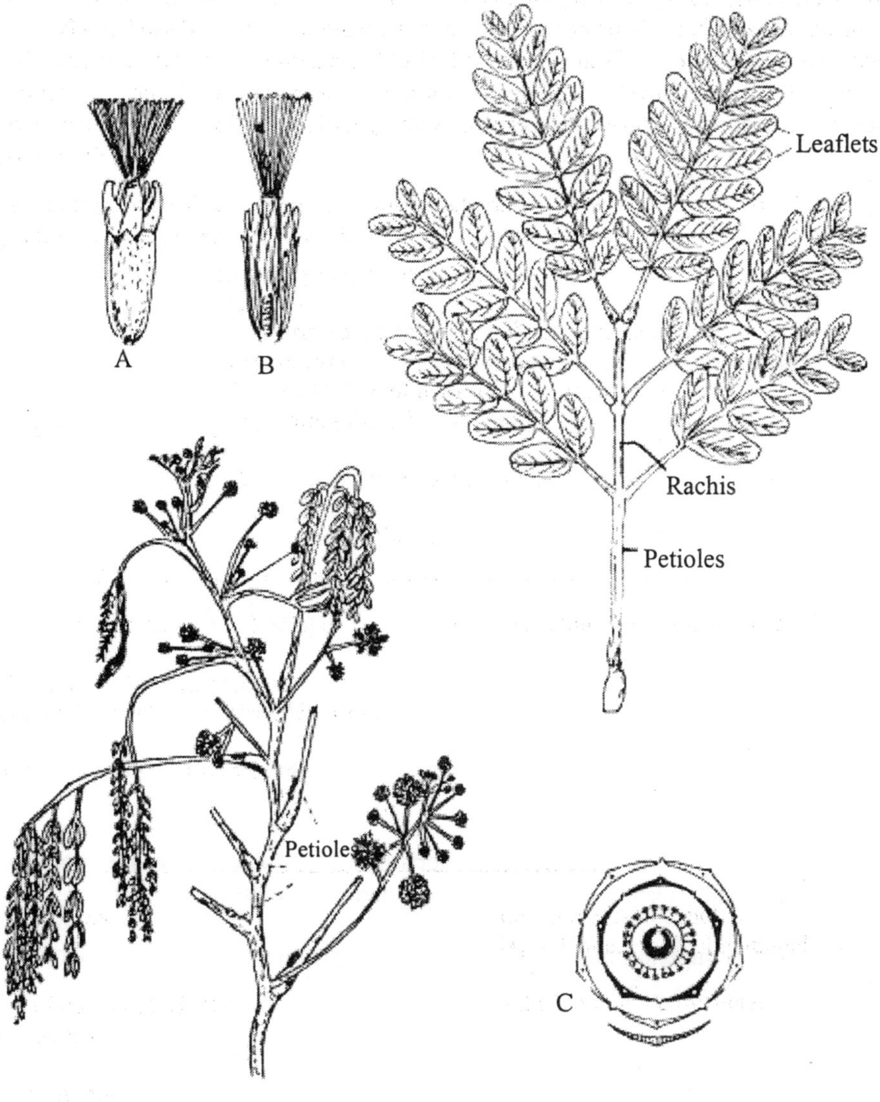

Fig 8.63 ALBIZZIA LEBBEK (Mimoceae)

Fig 8.63 ***ALBIZZIA LEBBEK*** **(Mimoceae)**

PROSOPIS JULIFLORA

(Mimoceae)

Habitat: Mesophyte, grows in dry open lands

Habit: A shrub

Root: Branched tap root

Stem: Aerial, erect, somewhat woody, hollow, cylindrical occasional lenticels present.

Leaf: Alternate, petiolate, stipulate; stipules modified into spines, pulvinate, bipinnately compound (Paripinnate, a gland is found at every branching of the primary rachis, leaflets linear, stalked, apex mucronate; venation reticulate unicostate

Inflorescence: A spike

Flower: Pedicellate, ebracteate, ebracteolate, regular, complete, cyclic, pentamerous, heterochlamydeous, bisexual and hypogynous

Calyx: Sepals 5, gamosepalous, inner surface pubescent, campanulate; aestivation valvate.

Corolla: Petals 5 in numbger, polypetalous (appears gamopetalous), greenish yellow, soft white pubescent on the inner face; aestivation valvate

Androecium: Stamens 10, arranged in two whorls of 5 each, outer whorl alternates with the corolla; filaments white; anthers dithecous, dorsifixed, introrse and gland crested.

Gynoecium: Ovary superior, monocarpellary, unilocular with many ovules on marginal placenta style thick and bent, stigma pointed, carpel wall white pubescent.

Fruit: A legume

Seed: Non endospermic and flat.

Floral formula: † ♀ $K_{(5)} C_5 A_{5+5} \underline{G}_1$

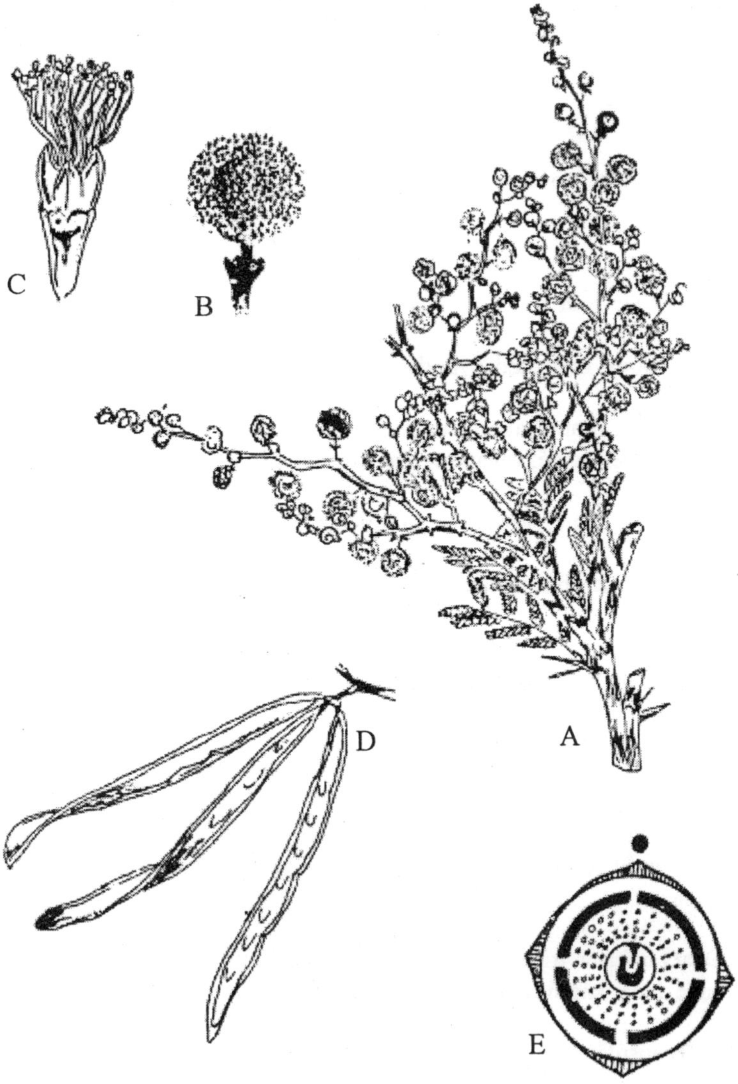

Fig 8.64 ***PROSOPIS JULIFLORA*** **(Mimoceae)**

EUCALYPTUS LANCEOLATUS

(Myrataceae)

Habitat : A mesophytic plant, grows very well in areas of water scarcity.

Habit : A tall tree

Root : Much branched tap root

Stem : Aerial, erect, woody, branched, cylindrical, solid, glabrous, bark smooth, white and shiny

Leaf : Ramal and cauline, simple, alternate, petiolate, exstipulate, lanceolate-ovate, smooth, gland dotted aromatic; margin entire, apex acute; venation reticulate unicostate; intramarginal venation present.

Inflorescence : Umbellate cyme; flowers are grouped in clusters of three.

Flower : Pedicellate, ebracteate, ebracteolate, pentamerous, cyclic, heterochlamydeous, actinomorphic, bisexual and hypogynous.

Calyx : Sepals 5, gamosepalous, reduced fused with the corolla forming a lid; aestivation valvate.

Corolla : Petals 5, gamopetalous, reduced, fused with the calyx and forms a lid called 'calyptra' which falls off during anthesis (opening of the flower) Aestivation valvate.

Androecium : Stamens indefinite, free, attached to the rim of the calyx onp, densley packed in bud condition; filaments greenish white in colour, slender, unequal in length; anthers dithecous, basifixed, introrse with longitudinal dehiscence.

Gynoecium : Ovary inferior, tricarpellary, syncarpous, trilocular with many ovules in each locule on axile placenta. Style simple, long; stigma minute, carpel wall gland dotted.

Fruit : A loculicidal capsule.

Seed : Non endospermic

Floral formula : ✝ ♀ $K_{(5)} C_5 \underline{G}_{(3)}$

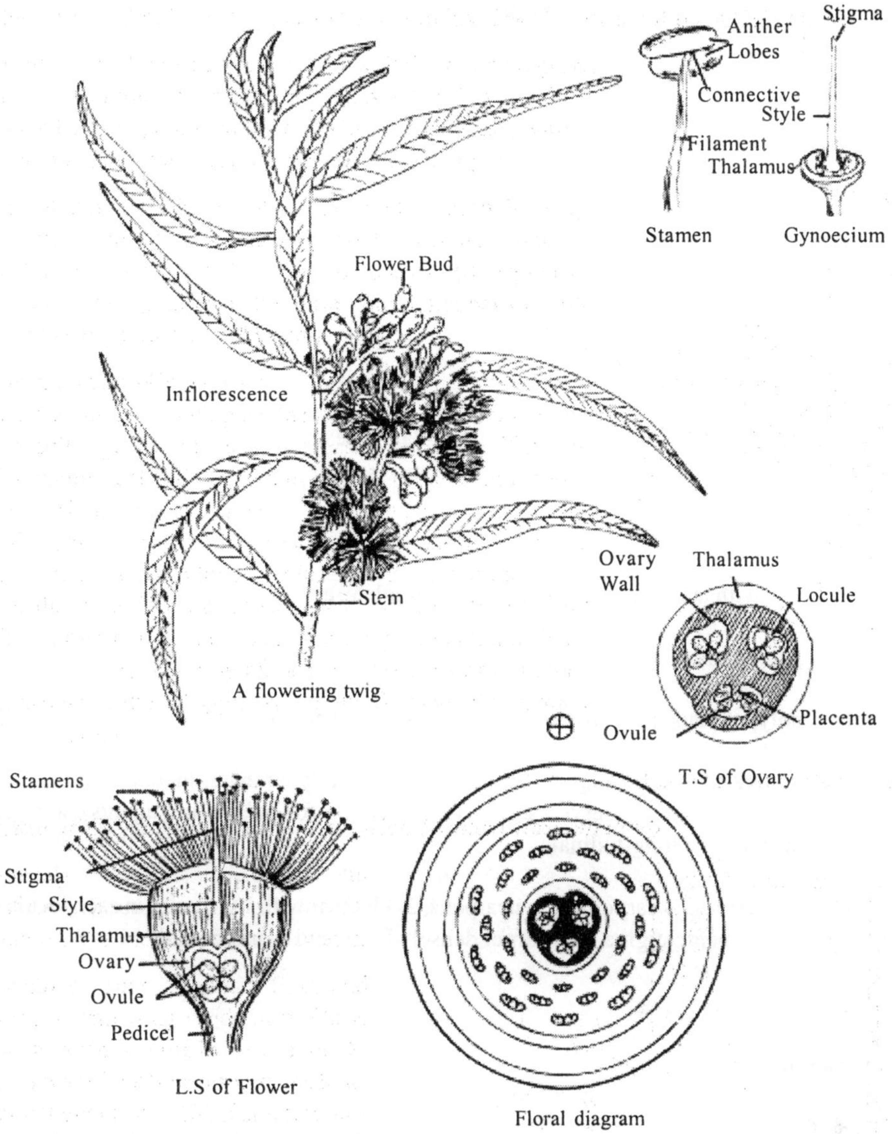

Fig 8.65 *EUCALYPTUS LANCEOLATUS* (Myrataceae)

CALLISTEMON LANCEOLATUS

(Myrtaceae)

Habitat: Mesophytic plant, cultivated in gardens

Habit: A medium sized tree

Root: Branched tap root

Stem: Aerial, erect, woody, solid, cylindrical hairy

Leaf: Simple, alternate, exstipulate, petiole short, lanceolate, apex acute gland dotted; venation reticulate unicostate; intramarginal vein present

Inflorescence: Spike, long, red coloured attractive

Flower: Sessile, bracteate with leafy bracts, ebracteolate, complete, actinomorphic, bisexual and epigynous.

Calyx: Sepals 5, free, green in colour, adnate to the ovary wall; aestivation quincuncial

Corolla: Petals 5, polypetalous, greenish white, aestivation quincuncial

Androecium: Stamens indefinite, arranged in many whorls, basally connate, incurved in bud, filaments long and slender, scarlet red in colour (serve as the chief attraction for pollinating agents), anthers small, dorsifixed, dithecous; dehiscence longitudinal

Gynoecium: Ovary inferior, 2-4 carpellary, syncarpous, 2-4 locular with 1-2 ovules in each locule on axile placenta, style long, slightly bent; stigma flat, carpel densely hairy and gland dotted.

Fruit: A capsule

Seed: Non endospermic

Floral formula: † ♀ $K_{(5)} C_5 A_{\infty} \underline{G}_{(2-4)}$

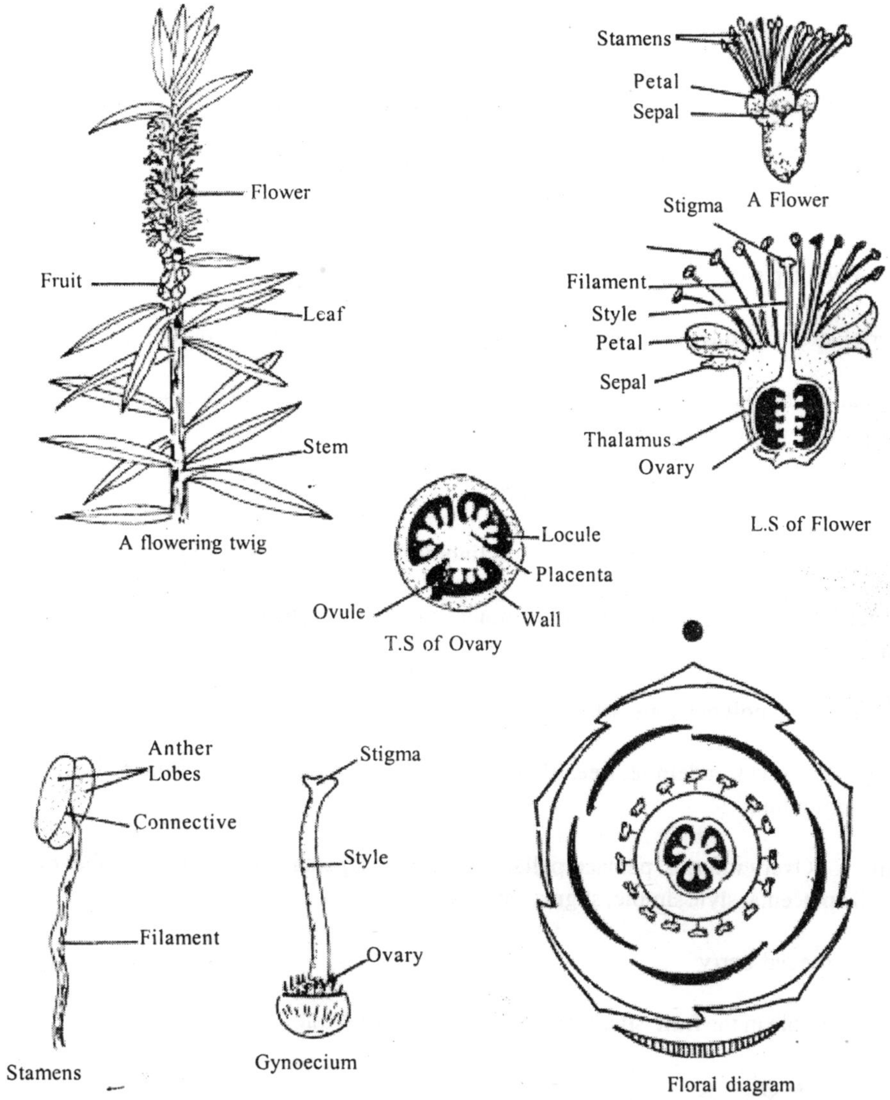

Fig 8.66 ***CALLISTEMON LANCEOLATUS*** **(Myrtaceae)**

PSIDIUM GUAJAVA

(Myrtaceae)

Habitat : Mesophyte, commonly cultivated in gardens

Habit : A medium sized tree, evergreen

Root : Branched tap root

Stem : Aerial, erect, woody, branched, cylindrical, whitish or brownish coloured bark which peels off.

Leaf : Ramal, exstipulate petiolate, phyllotaxy opposite decussate, simple, gland dotted, slightly aromatic, coriaceous, oblong-ovate, entire, acute with reticulate unicostate venation.

Inflorescence : Flowers are solitary and axillary.

Flower : Pedicellate bracteate, complete, cyclic actinomorphic, heterochlamydeous, bisexual and epigynous.

Calyx : Sepals 5 in number, gamosepalous, forming a cap like calyptra in bud, greenish in colour; aestivation quincuncial

Corolla : Petals 5, polypetalous, white, gland dotted; aestivation imbricate

Androecium : Stamens indefnite, free, filaments whitish in colour, slender; anthers dorsifixed, dithecous, dehiscence longitudinal

Gynoecium : Ovary inferior, pentacarpellary, syncarpous, pentalocular with a number of ovules in each locule on axile placenta, style simple; stigma capitate

Fruit : Many seeded berry

Seed : Innumerable in the fruit, small, testa hard non endospermic.

Floral formula : † ⚥ $K_{(5)} C_5 A_{\infty} \underline{G}_{(5)}$

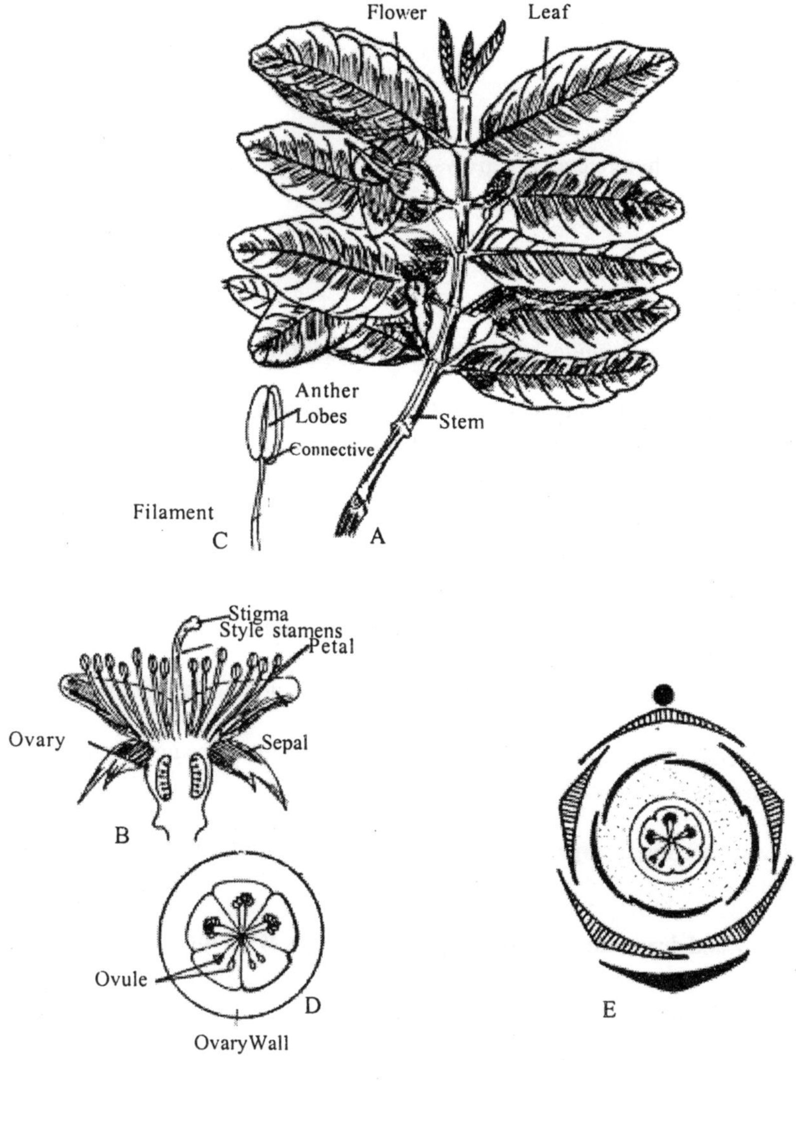

Fig 8.67 *PSIDIUM GUAJAVA* (Myrtaceae)

EUGENIA JAMBOLANA
(= Syzygium cumini)
(Myrtaceae)

Habitat: Evergreen mesophyte, generally grows in moist deciduous or rain forests, often cultivated

Habit: A large tree

Root: Tap root, branched.

Stem: Aerial, erect, woody, bark white, branched, smooth and solid.

Leaf: Ramal and cauline, simple, opposite, petiolate, gland dotted, leathery (coriaceous) oval or elliptic oblong, glaucous; venation reticulate unicostate. Intra-marginal vein present.

Inflorescence: Flowers aggregated in paniculate cymes.

Flower: Pedicellate or sub sessile, ebracteate, ebracteolate, complete, actinomorphic, bisexual and epigynous.

Calyx: Sepals 4-5 gamosepalous, adnate to the ovary forming a tube; aestivation valvate.

Corolla: Petals 4, forms a cap like structure called calyptra, deciduous; aestivation imbricate

Androecium: Stamens indefinite, inserted at the rim of the calyx tube; arranged in many whorls, filaments filiform; anthers dithecous, versatile and introrse.

Gynoecium: Ovary inferior, bicarpellary, syncarpous, bilocular with many ovules on axile placenta. Style filiform, stigma simple.

Fruit: A berry, juicy and deep purple in colour; persistant calyx at the top.

Seed: One in each furit, non endospermic, often green in colour.

Floral formula: † ♀ $K_{(5)} C_4 A_{\infty} \underline{G}_{(2)}$

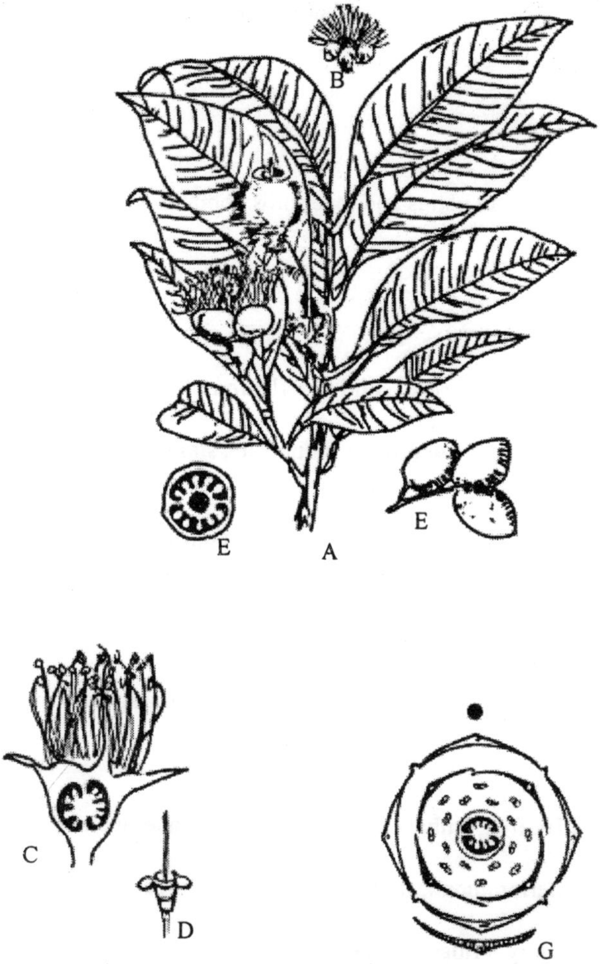

Fig 8.68 ***EUGENIA JAMBOLANA (= Syzygium cumini)*** (Myrtaceae)

TERMINALIA BELLERICA

(Combretaceae)

Habitat : Mesophyte, grows in plains and lover hills common in deciduous forests

Habit : A tall tree

Root : Branched tap root

Stem : Aerial, erect, woody, cylindrical, solid and glabrous; bark contains tannin

Leaf : Ramal and cauline alternate or rarely subopposite, simple broadly obovate, entire, somewhat acute; venation reticulate unicostate

Inflorescence : Spikes, clustered towards apex; often as long as 23cm.

Flower : Sessile, creamy white in colour, incomplete, actinomorphic, bisexual and epigynous occasionally flowers may be unisexual (staminate)

Calyx : Sepals 5, gamosepalous, limb campanulate with short traingular lobes. Aestivation valvate.

Corolla : Petals absent

Androecium : Stanies 10, free, anthers dithecous, versatise and open by longitudinal slits.

Gynoecium : Ovary inferior, monocarpellary, unilocular with 2-6 pendulous ovules.

Fruit : Drupe, ovoid, slightly 5 ridged, brown pubescent

Seed : Exalbuminous

Floral formula : † ♀ $K_{(5)} C_0 A_{10} \underline{G}_1$

Fig 8.69 ***TERMINALIA BELLERICA*** **(Combretaceae)**

QUISQUALIS INDICA

(Combretaceae)

Habitat : Mesophyte, cultivated in gardens; grows well in open mosit soil

Habit : A creeping, scandent shrub

Root : Tap root

Stem : Aerial, weak, green in colour, cylindrical, soft

Leaf : Simple, opposite, ovate, entire, exstipulate, venation reticulate unicostate

Inflorescence : Racemose clusters

Flower : Pedicellate or sub sessile pentamerous bracteate, ebracteolate, bisexual, regular, and epigynous

Calyx : Sepals 5, campanulate forming a tube which is adnate to the ovary; aestivation valvate

Corolla : Petals 5, free, white first later turns to pink; aestivation imbricate.

Androecium : Stamens 10, free in two whorls of 5 each, outer whorl antipetalous, inner antisepalous filaments long and slender; anthers dorsifixed, dithecous and introrse.

Gynoecium : Ovary inferior, monocarpellary, unilocular, with many pendulous ovules hanging from the top of the ovary; ovary 5 angled; style long, filiform, stigma minute and capitate.

Fruit : 5 angled achene

Seed : Only one in the fruit, small and exalbuminous.

Floral formula : † ♀ $K_{(5)} C_5 A_{5+5} \underline{G}_1$

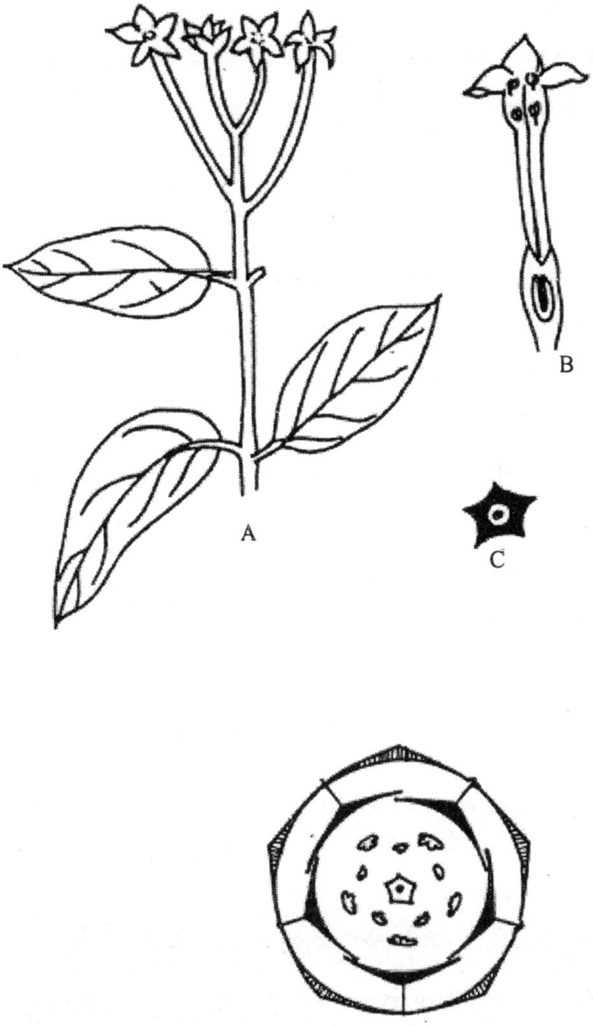

Floral diagram

Fig 8.70 ***QUISQUALIS INDICA*** **(Combretaceae)**

CUCURBITA MAXIMA

(Cucurbitaceae)

Habitat : Mesophyte, grows in moist cool soil

Habit : A climbing herb, tendrils present

Root : Branched tap root

Stem : Aerial, slender, pubescent or hirsute, 5 angled, soft and fistular. The stem has a watery juice

Leaf : Petiolate (long cylindrical petiole) alternate, simple, hairy, broad exstipulate and palmately veined

Inflorescence : Flowers large and solitary, yellow in colour and unisexual, but the plant monoecious

Male flower : Pedicellate, ebracteate, ebracteolate, pentamerous, heterochlamydeous, cyclic, incomplete, actinomorphic and unisexual (staminate)

Calyx : Sepals 5, gamosepalous, leafy or linear; hairy; aestivation valvate

Corolla : Petals 5, gamopetalous, campanulate, yellow; aestivation valvate

Androecium : Stamens 5, four united in two pairs and the remaining free; anthers connate and twisted spirally; one anther 2 celled and the other two to four celled. The condition is known as <u>synandry</u> (where both filaments and anthers of individual stamens are connate)

Gynoecium : Absent

Fruit : Absent

Female flowers : Pedicellate, ebracteate, ebracteolate, heterochlamydeous, pentamerous, cyclic, actinomorphic, unisexual, pistillate and epigynours

Calyx and corolla : Same as in male flowers

Androecium : Absent

Gynoecium : Ovary inferior, tricarpellary, syncarpous, trilocular, with many ovules in each locule on parietal placenta. Style single; stigma trifid

Fruit : A large fleshy pepo

Seed : Many in the fruit, compressed, non endospermic

Floral formula : Male flower : ✝ ♀ $K_{(5)}C_5A_{(5)}\underline{G}_0$

Female flower : ✝ ♀ $K_{(5)}C_5A_0\underline{G}_{(3)}$

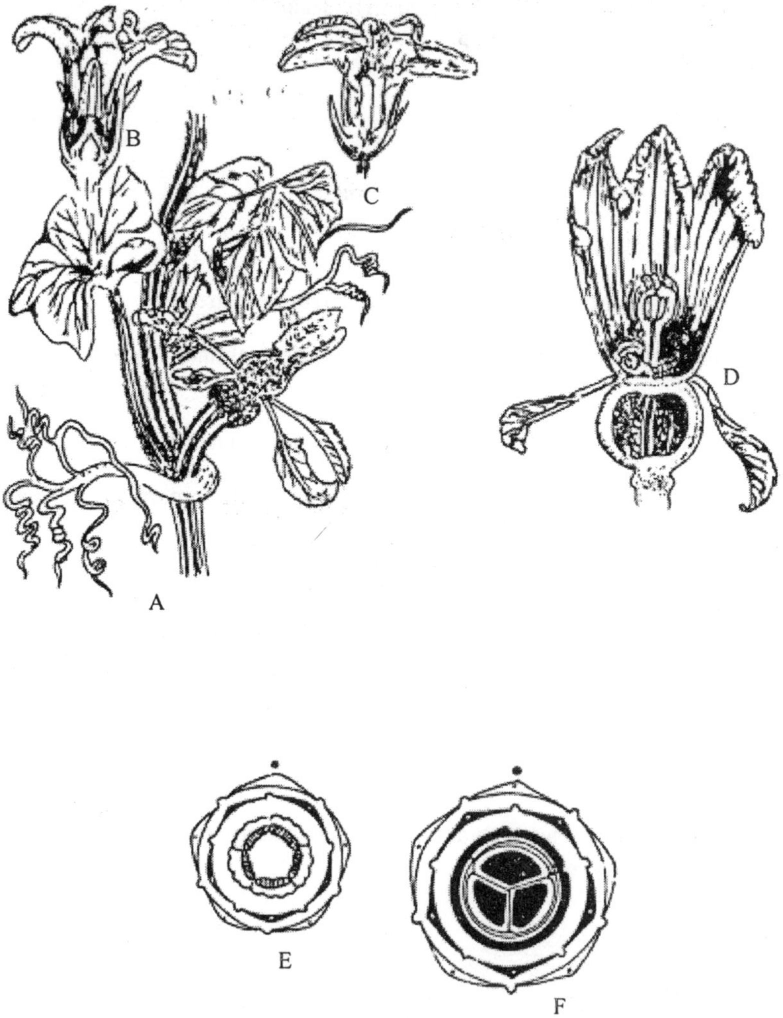

Fig 8.71 ***CUCURBITA MAXIMA*** (Cucurbitaceae)

LUFFA CYLINDRICA

(= L aegyptiaca)

(Cucurbitaceae)

Habitat: Mesophyte, cultivated, grows in most soil

Habit: A climbing tendrillar herb

Stem: Aerial, soft, weak, angular, branched, rough, hirsute and green

Leaf: Ramal and cauline, alternate, exstipulate, simple, palmately lobed, hairy, venation reticulate multicostate.

Inflorescence: Female flowers solitary, male flowers grouped in racemes

Male flower: Pedicellate, bracteate, ebracteolate, incomplete, actinomorphic, pentamerous, cyclic unisexual and staminate

Calyx: Sepals 5, polysepalous, or basally connate; aestivation valvate

Corolla: Petals 5, polypetalous, sometimes basally connate; aestivation imbricate or quincuncial

Androcium: Stamens 5, free, adnate to the corolla at the base; anthers monothecous, basifixed and extrorse.

Gynoecium: Absent

Female flower: Pedicellate, bracteate, ebracteolate, heterochlamydeous, pentamerous, cyclic, actinomorphic, unisexual, pistillate and epigynous.

Calyx and corolla:

Androecium: Absent

Gynoecium: Ovary inferior, tricarpellary, syncarpous, unilocular with many ovules on parietal placentae which project into the loculus; styles 3; stigmas 3, lobed.

Fruit: A pepo

Seed: Flat, many, non endospermic

Floral formula: Male flower: ✝ ♀ $K_{(5)} C_5 A_{(5)} \underline{G}_0$

Female flower: ✝ ♀ $K_{(5)} C_5 A_0 \underline{G}_{(3)}$

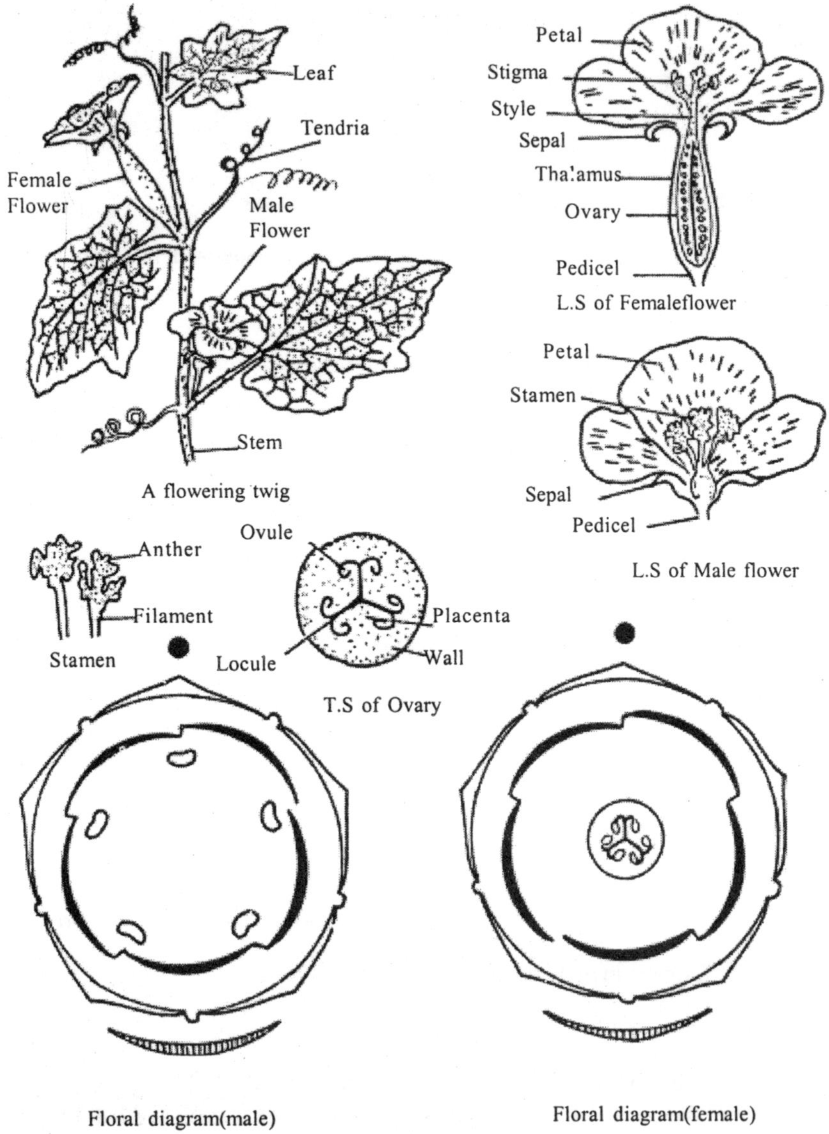

Fig 8.72 *LUFFA CYLINDRICA (= L. aegyptiaca)* (Cucurbitaceae)

COCCINEA INDICA

(Cucurbitaceae)

Habitat: Mesophyte, cultivated in gardens, grows in moist soil

Habit: A climbing herb; tendrils help in climbing, tendrils leaf opposed, not branched

Root: Tap root branched

Stem: Aerial, soft, weak, climbing, angular, branched solid and green.

Leaf: Ramal and cauline, alternate, exstipulate, simple, petiolate, palmately, divided, lobes denticulate, acute glabrous; venation reticulate multicostate diverging

Inflorescence: Flowers solitary, axillary, unisexual; plant monoecious.

Flower: Pedicellate, ebracteate, ebracteolate, cyclic, heterochlamydeous, incomplete, actinomorphic, pentamerous, unisexual and staminate

Calyx: Sepals 5, gamosepalous; aestivation valvate

Corolla: Petals 5, gamopetalous white in colour, campanulate; aestivation valvate

Androecium: Stamens 5, united in groups of 2+2+1, synandrous; anthers monothecous and exrorse.

Gynoecium: Absent

Fruit: Absent

Female flower: Same as in male flower except flowers are – unisexual, pistillate and epigynous

Calyx and Corolla: Same as in male flowers

Androecium: Absent

Gynoecium: Ovary inferior, tricarpellary, syncarpous, unilocular, with many ovules on parietal placenta. Placentae project into the cavity of the loculus. Styles short; stigma trifid and feathery.

Fruit: A pepo

Seed: Small, numerous, flat, non endospermic

Floral formula: Male flower: ✝ ♀ $K_{(5)} C_5 A_{2+2+1} \underline{G}_0$

Female flower: ✝ ♀ $K_{(5)} C_5 A_0 \underline{G}_{(3)}$

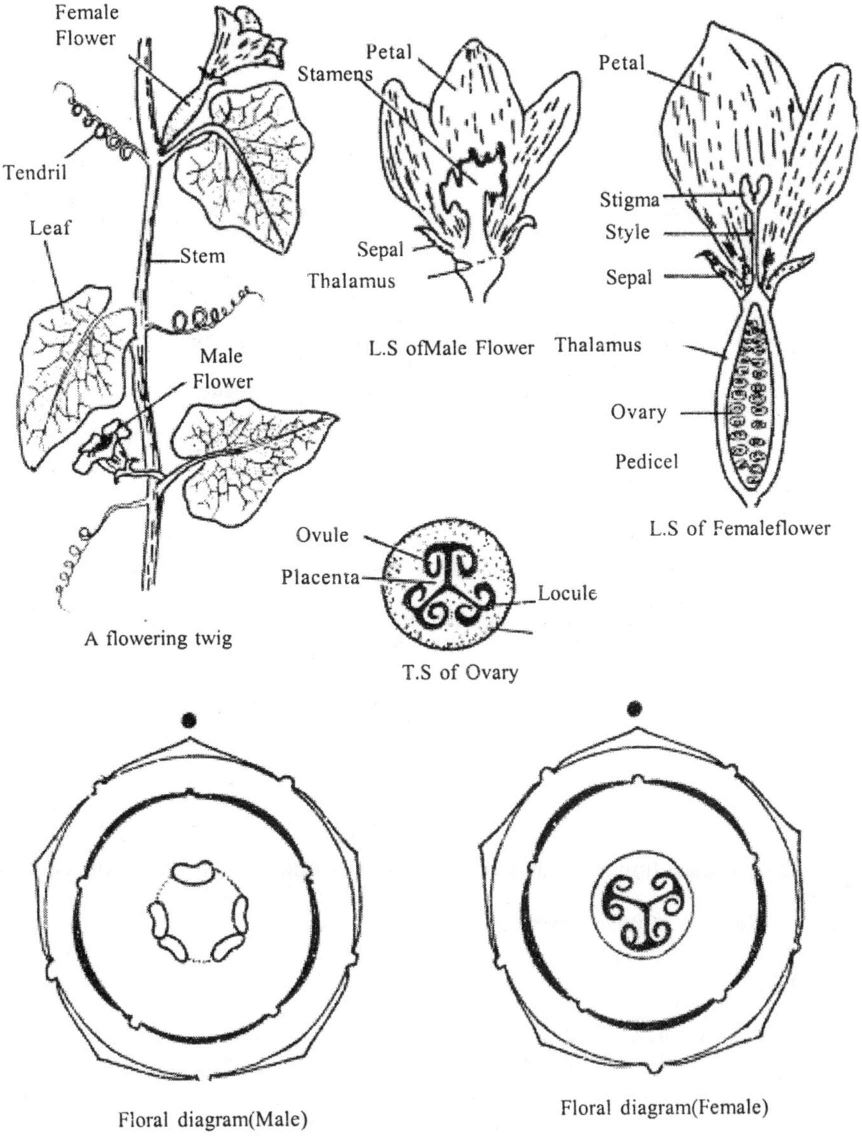

Fig 8.73 ***COCCINEA INDICA*** **(Cucurbitaceae)**

MOMORDICA CHARANTIA

(Cucurbitaceae)

Habitat : Mesophyte, cultivated in gardens, grows in loose, moist soil.

Habit : A herbaceous tendrillar annual climber

Root : Tap root branched

Stem : Aerial weak, cylindrical, solid, soft, green, branched, soft pubescent towards the tip; climbs with the help of simple, unbranched leaf opposed tendrils.

Leaf : Ramal and cauline, ,simple alternate, petiolate, exstipulate, deeply divided into 5-7 lobes; lobes sinuate dentate, smooth or slightly pubescent; venation reticulate multicostate.

Inflorescence : Flowers solitary, axillary; unisexual, but the plant monoecious

Male flower : Pedicellate, bracteate (bract large leafy and attached towards the base or middle of the peduncle and cordate in shape), ebracteolate, pentamerous, cyclic, actinomorphic, yellow in colour, unisexual and staminate.

Calyx : Sepals 5, gamosepalous, campanulate, ovate, top acute; aestivation valvate.

Corolla : Petals 5, polypetalous or slightly connate at the base, yellow in colour; aestivation quincuncial

Androecium : Stamens 5, in three groups of 2+2+1, synandrous; anthers extrorse, splitting by longitudinal division

Gynoecium : Absent

Fruit : Absent

Female flower : Same as male flower (no androecium) except – unisexual, pistillate and eipgynous.

Calyx and Corolla : Same as in male flower

Androecium : Absent

Gynoecium : Ovary inferior, tricarpellary, syncarpous, unilocular with many ovules arranged on parietal placentae which intrudes into the loculus incompletely dividing into chambers; style long stigma 3.

Fruit : A berry, dark green or somewhat pale yellow in colour, splitting by 3 valves. The fruit is externally ribbed with rows of tubercles.

Seed : Many large, flat, hard testa, non endospermic. When ripe, seeds are embedded in a red pulp.

Floral formula : Male flower : ✝ ♀ $K_{(5)} C_5 A_{2+2+1} \underline{G}_0$

Female flower : ✝ ♀ $K_{(5)} C_5 A_0 \underline{G}_{(3)}$

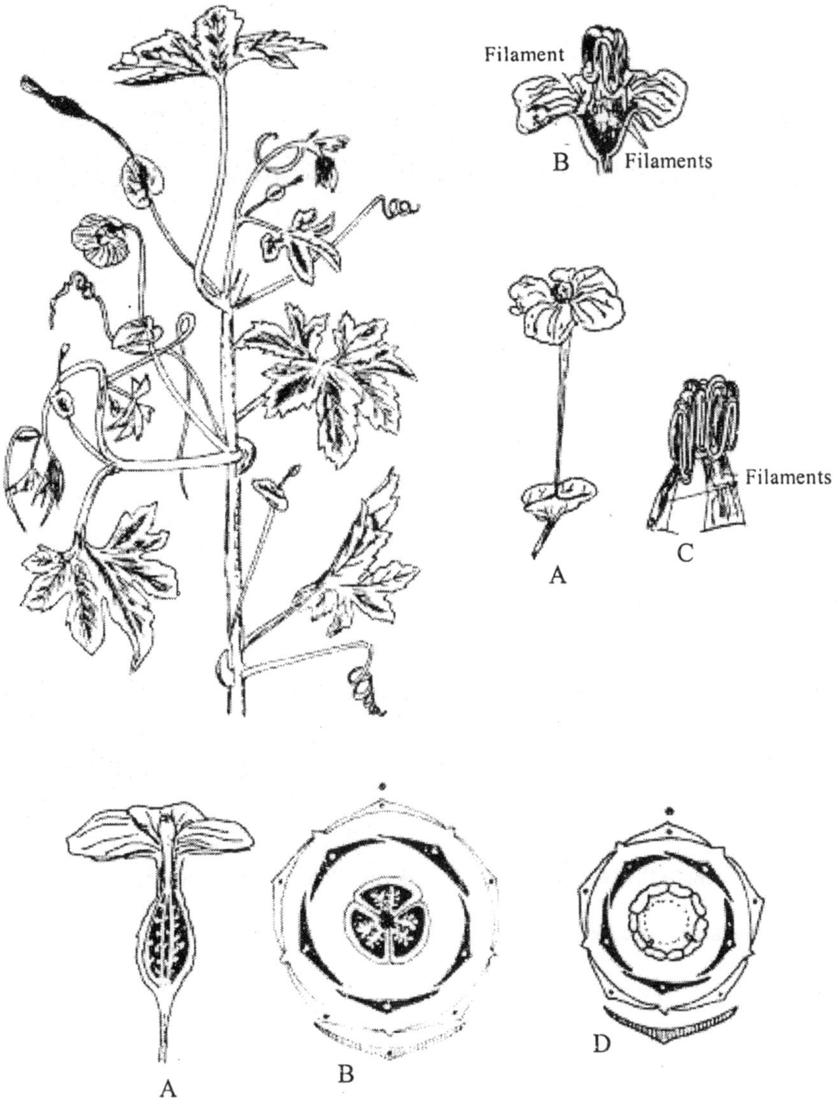

Fig 8.74 *MOMORDICA CHARANTIA* (Cucurbitaceae)

CUCUMIS SATIVUS
(Cucurbitaceae)

Habitat: A cultivated mesophyte, grows in loose, moist soil.

Habit: A straggling or climbing herb that generally spreads on the ground

Root: Tap root branched

Stem: Aerial, soft, weak, climbing by means of simple, unbranched, leaf opposed tendrils, angular, hollow, densely hirsute and green in colour

Leaf: Ramal and cauline, simple, petiolate, exstipulate alternate, palmately lobed with dentate lobes; leaf surface hairy, leathery, hirsute; venation reticulate unicostate.

Inflorescence: Flowers solitary axillary or rarely in small axillary fascicles, unixesual but the plant is monoecious.

Male flower: Pedicellate, bracteate, ebracteolate, cyclic, heterochlamydeous, pentamerous, unisexual and staminate.

Calyx: Sepals 5, gamosepalous, green, hairy, campanulate; aestivation valvate

Corolla: Petals 5, gamopetalous, yellow, distinctly nerved; aestivation twisted

Androecium: Stamens 5, arranged in 3 groups in 2+2+1 manner; inserted at the rim of the thalamus, synandrous; anthers hairy, sigmoid, extrorse, with a connective elongated into a sterile structure.

Gynoecium: A Pistillode is present

Female flower: Same as in male flowers except unisexual, pistillate and epigynous.

Calyx and Corolla: Same as in male flower

Androecium: Absent

Gynoecium: Ovary inferior, tricarpellary, syncarpous, unilocular, with many ovules an parietal placenta. Thalmas hairy and fused with carpels; style short; stigma papillate.

Fruit: An indehiscent berry

Seed: Numerous, white testa – flat and albuminous, endosperm has a long tail.

Floral formula: Male flower: ✝ ♀ $K_{(5)} C_5 A_{2+2+1} \underline{G}_0$

Female flower: ✝ ♀ $K_{(5)} C_5 A_0 \underline{G}_{(3)}$

Fig 8.75 ***CUCUMIS SATIVUS*** **(Cucurbitaceae)**

CORIANDRUM SATIVUM

(Umbelliferae)

Habitat: Mesophyte, annual, grows in moist soil

Habit: An aromatic annual herb

Root: Branched tap root.

Stem: Aerial, erect, herebaceous, cylindrical, fistular, glabrous, aromatic and green.

Leaf: Ramal and cauline, alternate, exstipulate, petiolate, decompound, (multipinnately compound); leaf base sheathing; upper pinnae ovate lanceolate, margin serrale; venation reticulate unicostate.

Inflorescence: A compound umbel; Involucre absent

Flower: Pedicellate, ebracteate, ebracteolate, cyclic, heterochlamydeous, pentamerous, peripheral flowers in the inflorescence zygomorphic, while central flower are actinomorphic, bisexual and epigynous.

Calyx: Sepals 5, gamosepalous, slightly petaloid; aestivation valvate.

Corolla: Petals 5, polypetalous, each petal bilobed and equal in central actinomorphic flowers, in the peripheral zygomorphic flowers – one anterior petal with equal large lobes, two lateral petals with unequal lobes and two posterior petals in the small equal lobes. Aestivation valvate.

Androecium: Stamens 5, free, alternate with petals, filaments same as or shorter than petals in length anthers dithecous, dorsifixed, introrse and split longitudinally.

Gynoecium: Ovary inferior, bicarpellary, syncarpous bilocular, with one ovule in each locule in a pendulous position, styles 2, gynobasic, arising from a disc called stylopodium; stigma capitate

Fruit: A schizocarp called cremocarp. It splits into two valves called mericarps. Ridges and grooves are present on the fruit

Seed: Albuminous and concave

Floral formula: Male flower: † ♀ $K_{(5)} C_5 A_{(5)} \underline{G}_{(2)}$

Female flower: † ♀ $K_{(5)} C_5 A_5 \underline{G}_{(2)}$

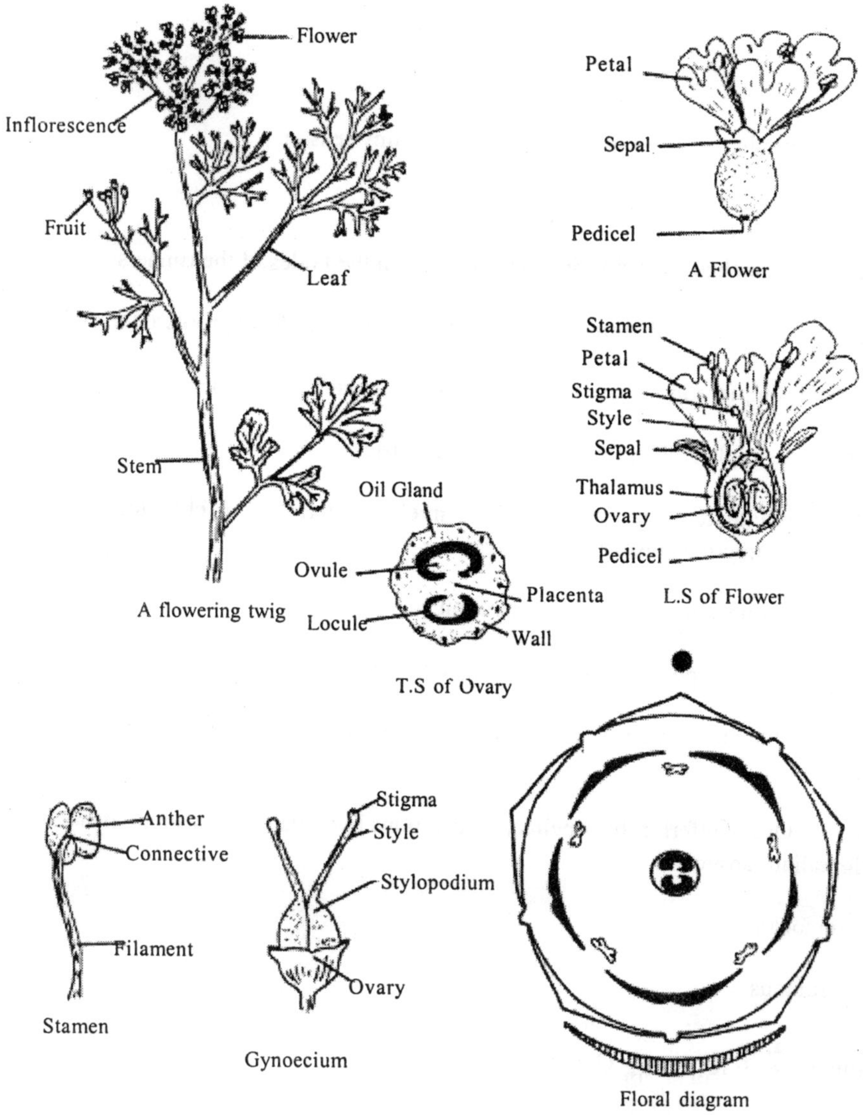

Fig 8.76 *CORIANDRUM SATIVUM* (Umbelliferae)

CENTELLA ASIATICA

(Umbelliferae)

Habitat: A mesophyte; grows in moist soil during rainy season

Habit: An annual prostrate herb

Root: Tap root, branched; adventitions roots arise from the nodes of the runners

Stem: Partly aerial, horizontal, runner, cylindrical, branched, producing roots at the nodes; green in colour.

Leaf: Petiolate, simple, stipulate, reniform, orbicular, crenate; venation reticulate multicostate.

Inflorescence: A simple umbel; flowers 3-6 in each cluster

Flower: Sessile, bracteate (small ovate bracts envelop the flowers), ebracteolate, complete, cyclic, actinomorphic, pentamerous, bisexual and epigynous

Calyx: Sepals 5, gamosepalous, green, aestivation valvate

Corolla: Petals 5, polypetalous, ovate, small, acute; aestivation valvate

Androecium: Stamens 5, free, alternate with corolla lobes; anthers dithecous, dorsifixed, dehiscence longitudinal.

Gynoecium: Ovary inferior, bicarpellary, syncarpons bilocular, with a single pendulous ovule in each locule; stylopodium absent.

Fruit: A cremocarp

Seed: Albuminous

Floral formula: † ♀ $K_{(5)} C_5 A_{(5)} \underline{G}_{(2)}$

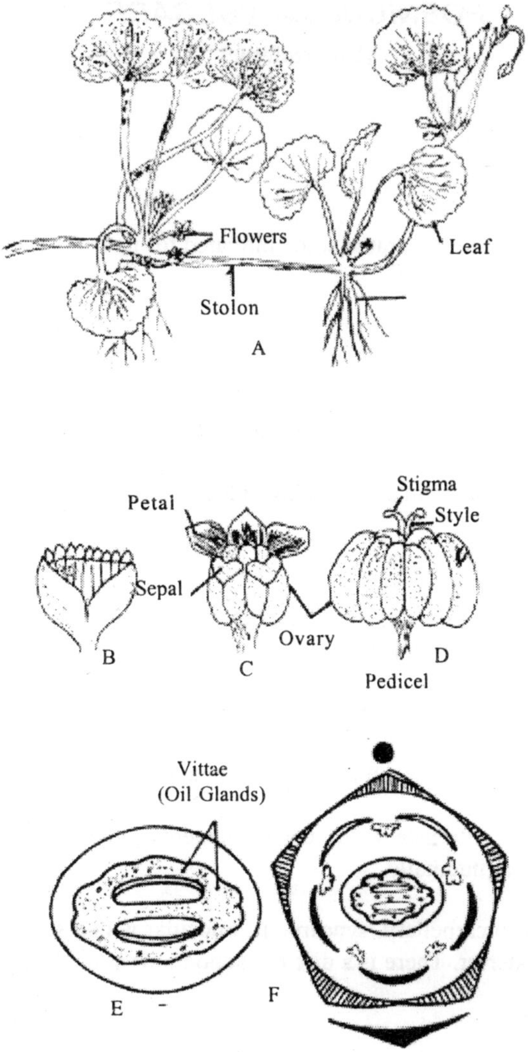

Fig 8.77 ***CENTELLA ASIATICA*** **(Umbelliferae)**

FOENICULUM VULGARE

(Umbelliferae)

Habitat : Mesophyte, grows in moist, aerated soil; cultivated for its fruits used as spice.

Habit : A tall annual herb

Root : Branched tap root

Stem : Aerial, erect, soft, green, ribbed, fistular, branched and glabrous

Leaf : Ramal and cauline, alternate, petiolate, pinnatisect leaf base sheathing, exstipulate and aromatic

Inflorescence : A compound umbel

Flower : Pedicellate, ebracteate, ebracteolate, complete, heterochlamydeous, pentamerous, actinomorphic, bisexual and epigynous. Flowers yellow and aromatic.

Calyx : Sepals 5, gamosepalous, forming a tube; tube adnate to the ovary wall; aestivation valvate

Corolla : Petals 5, polypetalous, yellow, notched at the tip; aestivation valvate

Androecium : Stamens 5, free, alternate with petals; filaments longer than petals; anthers dithecous basifixed, introrse and dehisce longitudinally.

Gynoecium : Ovary inferior, bicarpellary, syncarpous, bilocular with a single pendulous ovule in each locule. Ovary placed antero posterior. There is a disc (Stylopodium) at the base of the style which is very short; stigma capitate.

Fruit : A cremocarp

Seed : Albuminous

Floral formula : † ♀ $K_{(5)} C_5 A_{(5)} \underline{G}_{(2)}$

Fig 8.78 ***FOENICULUM VULGARE*** **(Umbelliferae)**

CARUM COPTICUM
(= Trachyspermum amni)
(Umbelliferae)

Habitat : Mesophyte; grows in moist well aerated soil.

Soil : Cultivated as a spice

Habit : An annual herb

Root : Branched tap root

Stem : Aerial, erect, branched, cylindrical, fistular, ribbed and green with swollen nodes.

Leaf : Ramal and cauline, alternate, petiolate, exstipulate; has a sheathing leaf base; multipinnately compound, segments (pinnae) wedge shaped at the base, cut at the apex.

Inflorescence : A compound umbel (racemose); there is an involucre of bracts surrounding the flowers.

Flower : Pedicellate, ebracteate, ebracteolate, pentamerous, cyclic, heterochlamydeous, actinomorphic, bisexual and epigynous.

Calyx : Sepals 5, gamosepalous, tubular, adnate to the carpels; aestivation valvate.

Corolla : Petals 5, polypetalous, yellow; aestivation valvate.

Androecium : Stamens 5, free, alternate with the petals, filaments long; anthers dorsifixed, dithecous and introrse with longitudinal splitting.

Gynoecium : Ovary inferior, bicarpellary, syncarpous, bilocular, with one pendulous ovule in each locule. Disc present; style extremely short; stigma bilobed.

Fruit : A cremocarp

Seed : Endospermic

Floral formula : ✝ ♀ $K_{(5)}\ C_{(5)}\ A_5\ \underline{G}_{(2)}$

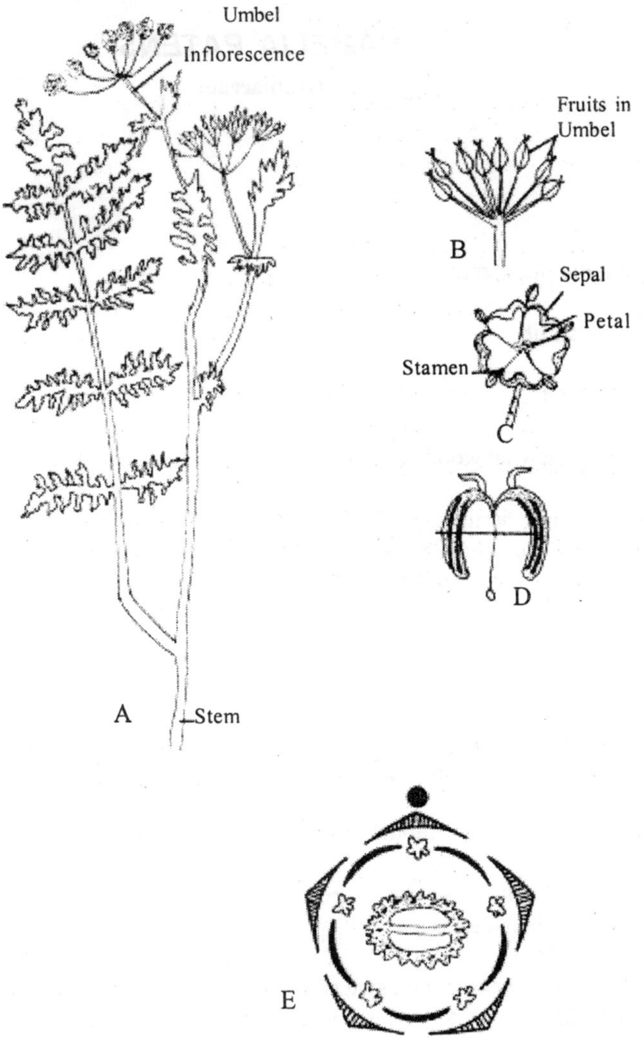

Fig 8.79 ***CARUM COPTICUM (= Trachyspermum amni)*** **(Umbelliferae)**

HAMELIA PATENS

(Rubiaceae)

Habitat: Mesophytic plant, cultivated in gardens.

Habit: An evergreen shrub

Root: Branched tap root

Stem: Aerial, erect, somewhat woody cylindrical, branched, solid and hairy.

Leaf: Ramal and cauline, simiple, whorled, 3 or more leaves found at each node, stipulate; stipules interpetiolate. Petiolate, ovate, entire, acute; venation reticulate unicostate

Inflorescence: Basically a dischasial or polychasial cyme, with each branch a monochasial helicoid cyme

Flower: Pedicellate, ebracteate, ebracteolate, cyclic, pentamerous, heterochlamydeous, complete, actinomorphic, bisexual and epigynous.

Calyx: Sepals 5, gamosepalous, petaloid, adnate to the carpels; aestivation valvate

Corolla: Petals 5, gamopetalous, tubular, red in colour; aestivation valvate or twisted

Androecium: Stamens 5, alternate with the petals, epitalous, filaments short; anthers dithecous, basifixed, introrse with longitudinal dehiscence

Gynoecium: Ovary inferior, bicarpellary, syncarpous, bilocular, with many ovules arranged on axile placenta style filiform; stigma bifid.

Fruit: A berry

Seed: Non endospermic

Floral formula: ✝ ♀ $K_{(5)}\ C_{(5)}\ A_5\ \underline{G}_{(2)}$

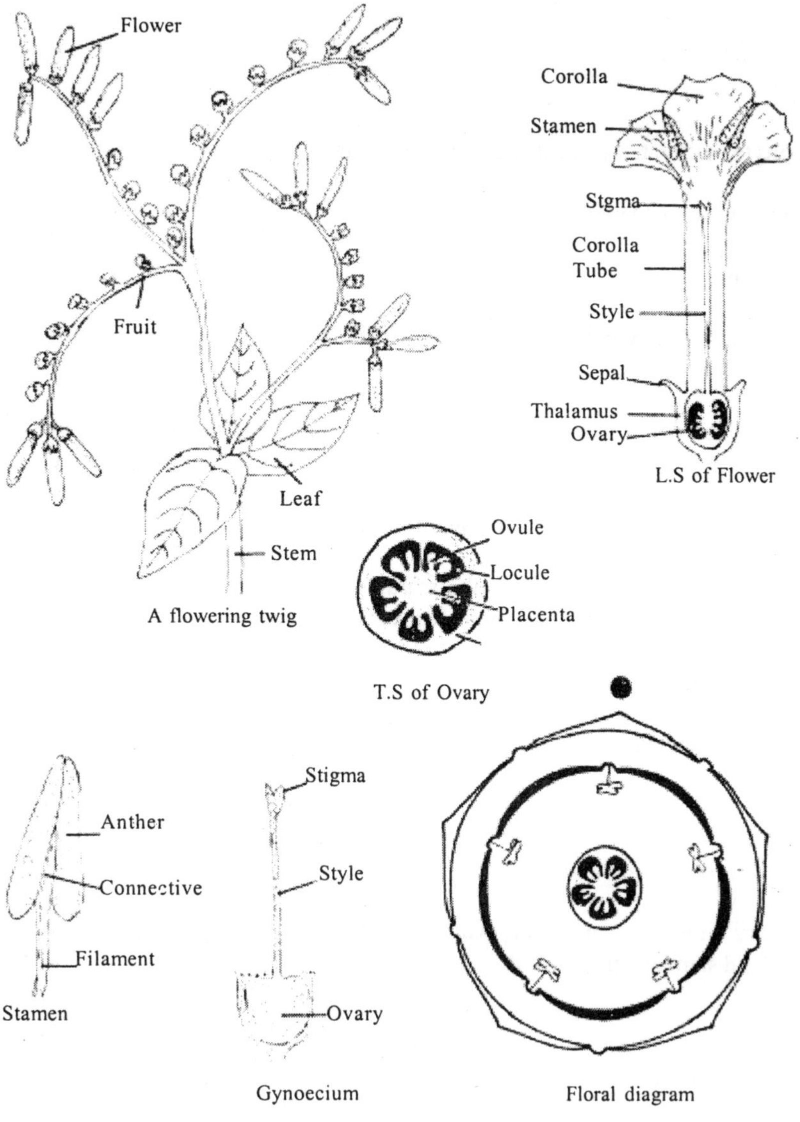

Fig 8.80 ***HAMELIA PATENS*** **(Rubiaceae)**

IXORA PARVIFLORA

(Rubiaceae)

Habitat: Mesophyte, cultivated in garden as an ornamental

Habit: An evergreen shrub or a small tree

Root: Branched tap root

Stem: Aerial, erect, somewhat woody below and soft in the young branches, cylindrical, solid, smooth and greenish in colour.

Leaf: Ramal and cauline, opposite decussate, sestile, simple, stipulate, stipules interpetiolar and cuspidate, elliptical, entire, acute, venation reticulate unicostate

Inflorescence: Trichotomously branched cymose inflorescence; terminal in position

Flower: Pedicellate, bracteate (bracts appear like stipules), bracteolate red or white in colour, cyclic, tetramermous, heterochlamydeous, actinomorphic, bisexual and epigynous;

Calyx: Sepals 4, gamosepalous, calyx tube ovoid, aestivation valvate

Corolla: Petals 4, gamopetalous, forming a long tube, lobes spreading above white or pink in colour; aestivation twisted.

Androecium: Stamens 4, free, alternate with the petals, inserted at the throat of the corolla tube, filaments short; anthers dithecous, dorsifixed and introrse

Gynoecium: Ovary inferior, bicarpellary, syncarpous bilocular, with one ovule in each locule on axile placenta Style hairy and slender, stigma bifid

Fruit: Berry

Seed: Small, non endospermic

Floral formula: ✝ ♀ $K_{(4)}\ C_4\ A_4\ \underline{G}_{(2)}$

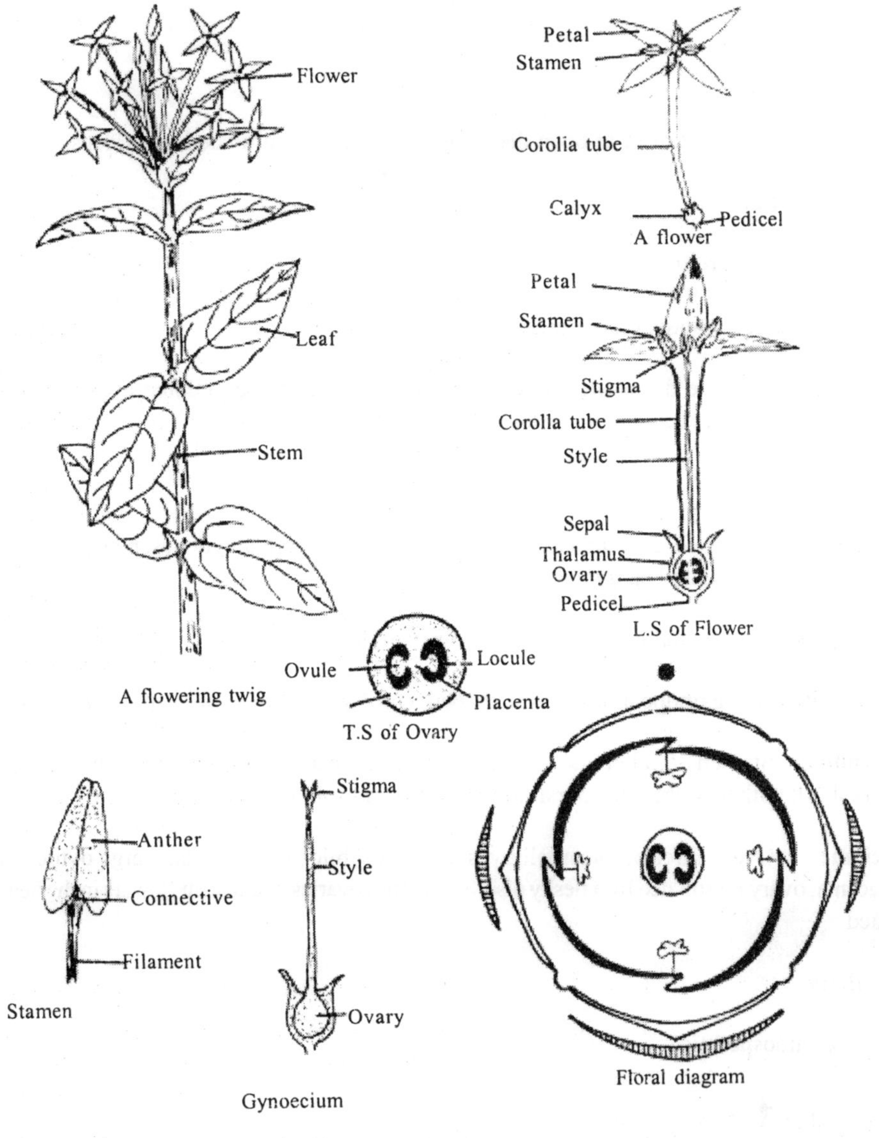

Fig 8.81 ***IXORA PARVIFLORA*** **(Rubiaceae)**

MORINDA ANGUSTIFOLIA

(Rubiaceae)

Habitat: Mesophyte, grows in open deciduous forests, also cultivated in gardens.

Habit: A small tree

Root: Branched tap root

Stem: Aerial, erect, woody, branched and cylindrical

Leaf: Simple, opposite decussate, petiolate, stipulate, stipules interpetiolar, entire, acute, intramarginal vein present; venation reticulate unicostate.

Inflorescence: Globose head

Flower: Sessile, ebracteate, ebracteolate, complete, cyclic, heterochlamydeous, pentamerous, salvar shaped, actinomorphic, bisexual and epigynous

Calyx: Sepals 5, gamopetalous, green valvate aestivation

Corolla: Petals 5, gamopetalous, salvar shaped, corolla tube long, lobes spreading, aestivation valvate

Androecium: Stamens 5, alternate with the petals, epipetalous, inserted near the throat of the corolla; filaments short; anthers dithecous, dorsifixed, introrse and split longitudinally

Gynoecium: Ovary inferior, bicarpellary, syncarpous, bilocular, with one large ovule in each locule on axile placenta; ovary is crowned by a fleshy disc. Sometimes ovaries of adjacent flowers in the head inflorescence get united

Fruit: Berry

Seed: Non endospermic

Floral formula: † ♀ $K_{(5)} C_{(5)} A_5 \underline{G}_{(2)}$

Fig 8.82 ***MORINDA ANGUSTIFOLIA*** **(Rubiaceae)**

MUSSEANDA FRONDOSA

(Rubiaceae)

Habitat: Mesophyte, cultivated in the gardens, also grows wild in moist deciduous forests.

Habit: A small perernial shrub

Root: Tap root, branched

Stem: Aerial, erect, cylindrical, somewhat woody, branched, hairy, solid and green.

Leaf: Ramal and cauline, simple, opposite decussate, sub sessile, stipulate (interpertiolar), entire, acute venation unicostate reticulate

Inflorescence: A dichasial cyme

Flower: Pedicellate, bracteate, bracteolate, complete, cyclic, pentamerous, actinomorphic, but older flowers may become zygomorphic due to the enlargement of sepals, bisexual and epigynous.

Calyx: Sepals 5, gamosepalous, in some flowers one of the sepals enlarges becomes foliar, yellow coloured and serves to attract the pollinating agents, aestivation valvate.

Corolla: Petals 5, gamopetalous, tubular, tube long, lobes spreading, a hairy corona present at the mouth of the tube; aestivation valvate

Androecium: Stamens 5, free, alternate with the petals, epipetalous, inserted at the throat of the corolla; filaments short; anthers dithecous, dorsifixed, introrse and dehisce longitudinally

Gynoecium: Ovary inferior, bicarpellary, syncarpous, bilocular with many ovules in each locule on axile placenta; style long; stigma bifid.

Fruit: A berry

Seed: Small, non endospermic

Floral formula: † ♀ $K_{(5)}\ C_{(5)}\ A_5\ \underline{G}_{(2)}$

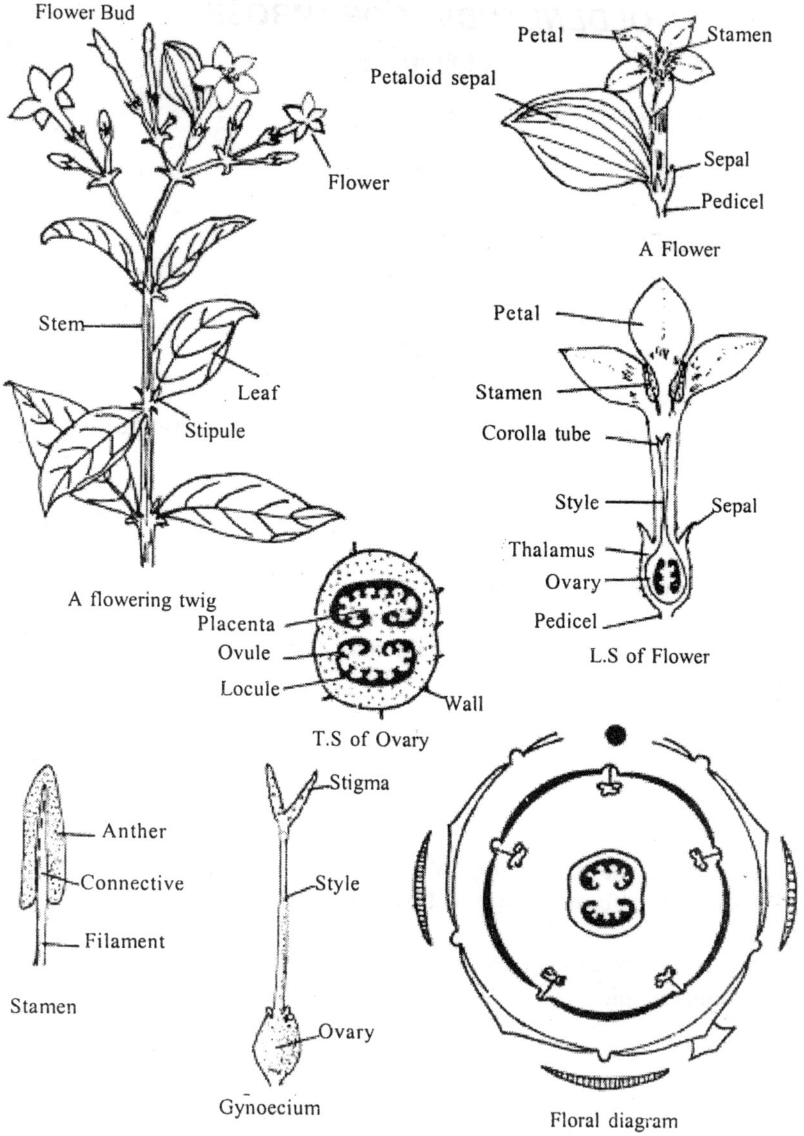

Fig 8.83 ***MUSSEANDA FRONDOSA*** **(Rubiaceae)**

OLDENLANDIA CORYMBOSE

(Rubiaceae)

Habitat : Mesophyte; grows in open dry land

Habit : A small prostrate or semierect herb

Root : Tap root, branched

Stem : Aerial, weak, prostrate and trailing, angular, branched, solid, soft hairy and green

Leaf : Ramal and cauline, simple opposite decussate, interpetiolar stipulate, sessile, elliptic lanceolate, entire acute, minutely hairy; venation reticulate unicostate

Inflorescence : Axillary dichasial cyme or umbel

Flower : Pedicellate, bracteate, bracteolate, cyclic, tetramerous, heterochlamydeous, complete, actinomorphic bisexual and epigunous

Calyx : Sepals 4, gamosepalous; aestivation valvate

Corolla : Petals 4, gamopetalous, funnel shaped, white or violet coloured; aestivation valvate

Androecium : Stamens 4, epipetalous, alternipetalous, inserted at the throat of the corolla; filaments short; anthers sagittate, dithecous, dorsifixed, introrse; dehiscence longitudinal

Gynoecium : Ovary inferior, bicarpellary, syncarpous, bilocular with many ovules in each locule on axile placenta. Style short; stigma simple

Fruit : A berry

Seed : Exalbuminous

Floral formula : ✝ ⚥ $K_{(4)}\ C_4\ A_4\ \underline{G}_{(2)}$

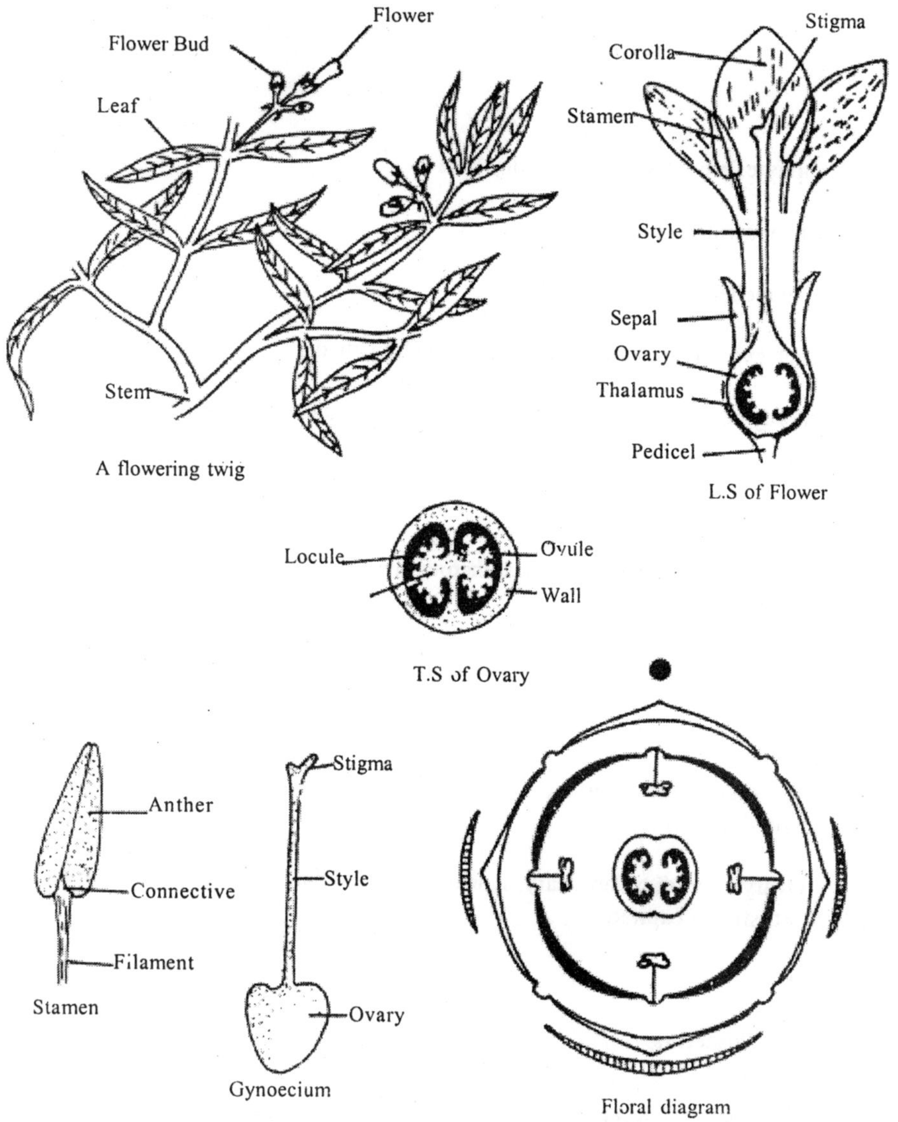

Fig 8.84 *OLDENLANDIA CORYMBOSE* (Rubiaceae)

RICHARDIA SCABRA

(Rubiaceae)

Habitat : Mesophyte, grows wild in open dry land

Habit : A prostrate or semierect herb

Root : Top root, branched

Stem : Aerial, erect, cylindrical, hairy, green, solid and branched

Leaf : Simple, shortly petiolate, hairy, opposite, stipules persistent and interpetiolar, obovate, entire, venation reticulate unicostate

Inflorescence : Cymose cluster

Flower : Pedicellate, ebracteate, ebracteolate, hexamerous, heterochlamydeous, actinomorphic, complete, cyclic, bisexual and epigynous

Calyx : Sepals 6, gamosepalous; aestivation valvate

Corolla : Petals 6, gamopetalous, white; aestivation valvate

Androecium : Stamens 6, free, alternate with the petals, epipetalous; filaments short; anthers dithecous, introrse with a longitudinal dehiscence

Gynoecium : Ovary inferior, tricarpellary, syncarpous, trilocular, with many ovules in each locule on axile placenta, style short; stigma capitate

Fruit : A capsule

Seed : Small, non endospermic

Floral formula : † ♀ $K_{(6)}\ C_{(6)}\ A_6\ \underline{G}_{(3)}$

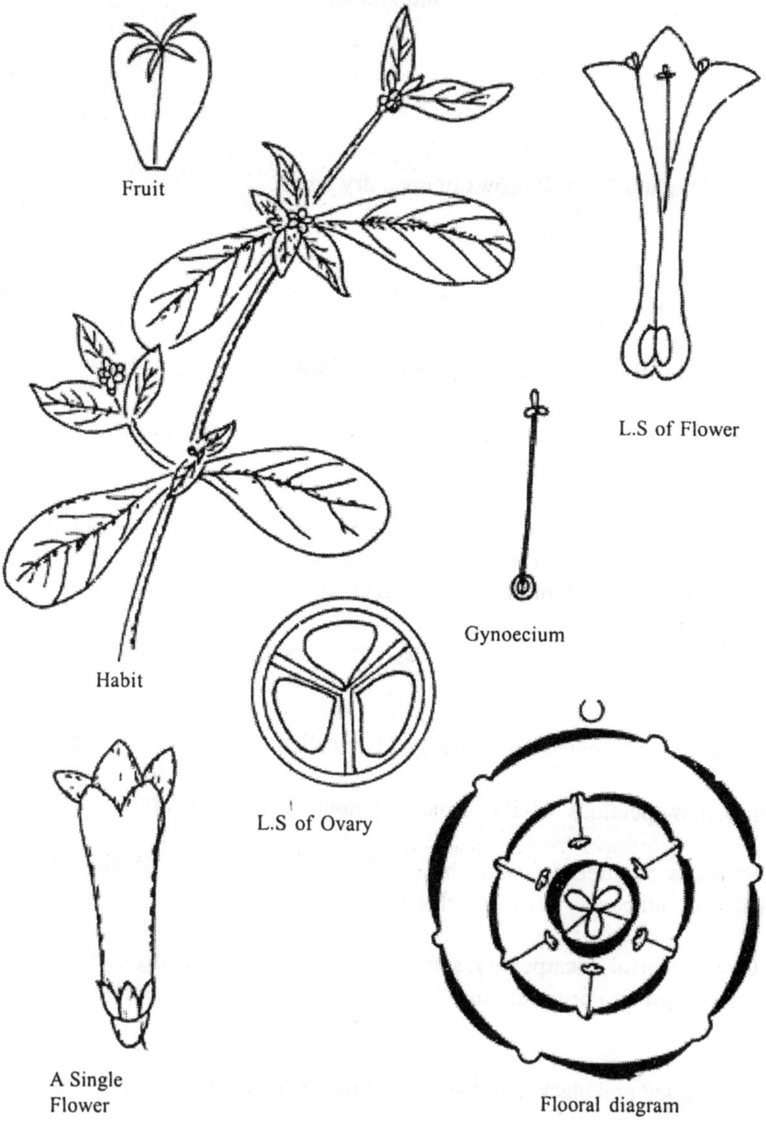

Fig 8.85 *RICHARDIA SCABRA* (Rubiaceae)

AGERATUM CONYZOIDES

(Compositae)

Habitat : Mesophytic; annual weed; grows in open dry lands

Habit : Annual herb

Root : Tap root, branched

Stem : Aerial, erect, cylindrical, branched, solid, hairy and soft.

Leaf : Ramal and cauline, simple, petiolate, exstipulate, opposite, ovate, serrate, acute and hairy, venation reticulate unicostate

Inflorescence : The primary aggregation of flowers is a head (capitulum). The capitula are clustered in a corymbose inflorescence. The corymbs are terminal. Heads are homogamous i.e. there is no distinction into ray florets and disc florets; involucre of bracts green in colour.

Flower : Sessile, bracteate, ebracteolate, complete, cyclic, pentamerous, actinomorphic, bisexual and epigynous; flowers violet in colour

Calyx : Sepals basically 5, but reduced to scaly structure called pappus; aestivation valvate

Corolla : Petals 5, gamopetalous, tabular, violet-in colour; aestivation valvate

Androecium : Stamens 5, alternate with the petals and epipetalous; filaments short; anthers long syngenesious, basifixed, bithccous, introrse, dehiscing longitudinally

Gynoecium : Ovary inferior, bicarpellary, syncarpous unilocular with only one ovule on basal placenta. Style long and divided into two arms; stigmas long flattened at the tip; glandular papillae present on the stigma

Fruit : A cypsela, 5 angled and black in colour. At the tip of the fruit is present 5 scales of the calyx

Seed : Exalbuminous

Floral formula : † ♀ $K_{(5)}$ $C_{(5)}$ $A_{(5)}$ $\underline{G}_{(2)}$

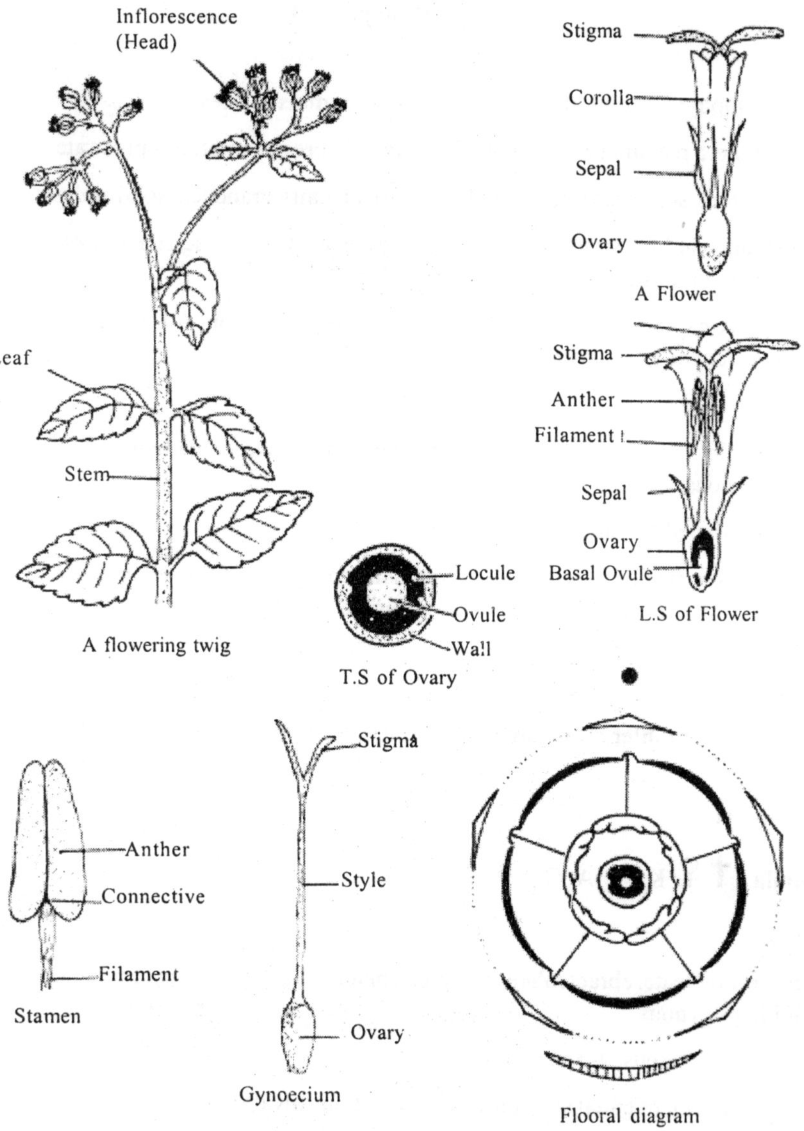

Fig 8.86 ***AGERATUM CONYZOIDES*** **(Compositae)**

ECLIPTA ALBA
(= E. Prostrata)
(Compositae)

Habitat: Mesophyte, grows along gutters and other moist places

Habit: A small herb reaching a height of 15 cm; mostly erect, sometimes prostrate

Stem: Aerial, erect, sometimes horizontal, cylindrical, hairy branched, solid and green

Leaf: Ramal and cauline, simple, opposite decussate, exstipulate, sessile, elliptic lanceolate acute, hairy; venation reticulate unicostate

Inflorescence: A capitulum capitula are heterogamous with about 6-8 ray florets and a number of disc florets. An involucre of green bracts surrounds the inflorescence

Ray florets:

Flower : Sessile, bracteate, ebracteolate, pentamerous, heterochlamydeous, zygomorphic, unisexual, female and epigynous.

Calyx: Sepals reduced to pappus

Corolla: Petals basically 5, two petals towards the centre of the flower are reduced, while the remaining three (sometimes two) are fused to form a large ligule or ray. (Sometimes petals may be four in number) Aestivation valvate

Androecium: Absent

Gynoecium: Ovary inferior, bicarpellary, syncarpous, unilocular with a single basal ovule. Style short stigma bifid

Fruit: A cypsela

Floral formula: ✝ ♀ $K_{(4)}\ C_4\ A_0\ \underline{G}_{(2)}$

Disc florets:

Flower: Sessile, bracteate, ebracteolate, complete, heterochlamydeous, cyclic, pentamerous, actinomorphic, bisexual and hypogynous

Calyx: Reduced to pappus, 4-5 in number

Corolla: Petals 4-5, of equal size, tubular, aestivation valvate

Androecium: Stamens 4-5, filaments free, alternate with mthe petals and epipetalous; anthers syngensious, dthecous, basifixed and introrse with longitudinal dehiscence

Gynoecium : Same as in ray florets

Fruit: Cypsela

Floral formula: ✝ ♀ $K_{(4\text{-}5)}\ C_{(4\text{-}5)}\ A_{4\text{-}5}\ \underline{G}_{(2)}$

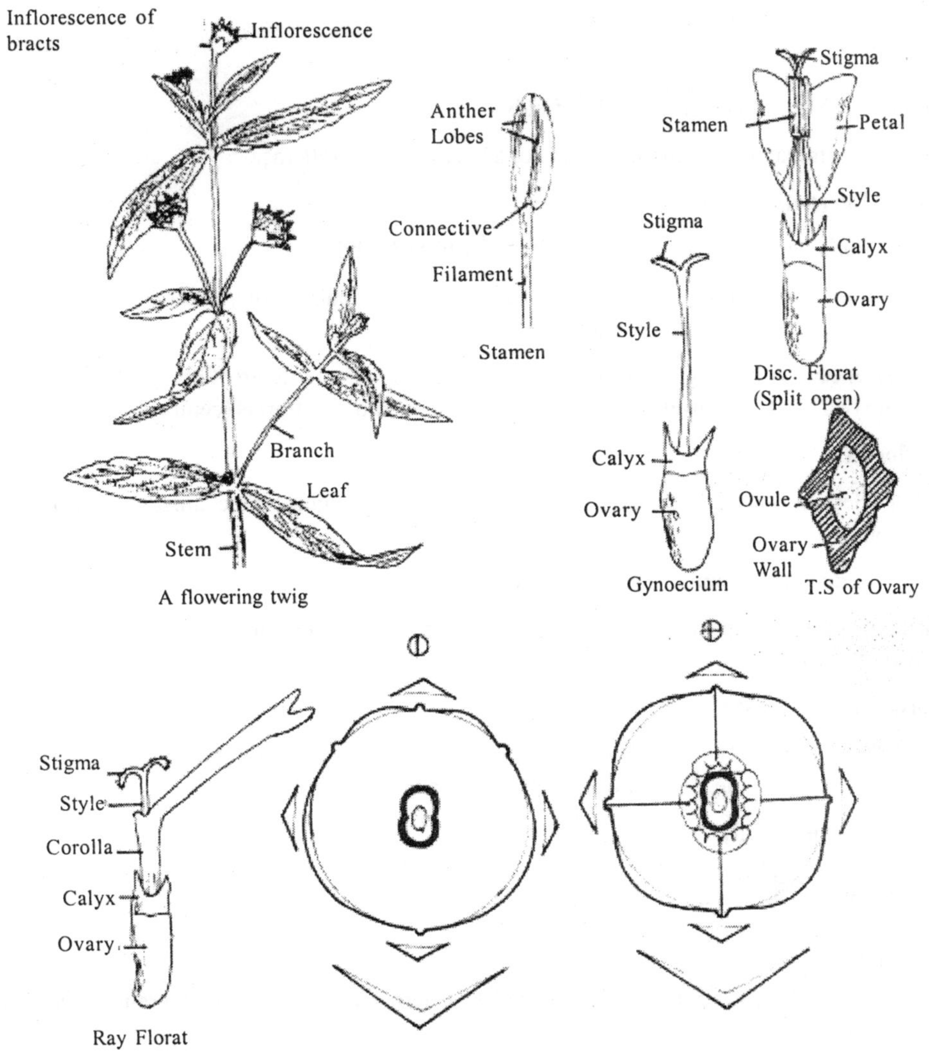

Fig 8.87 ***ECLIPTA ALBA (= E.prostrata)*** **(Compositae)**

HELIANTHUS ANNUA

(Compositae)

Habitat : Mesophyte, cultivated in fields

Habit : An annual herb, sometimes reaches a height of 150-160cms.

Root : Tap root, branched

Stem : Aerial, erect, green, cylindrical, hairy, soft, and solid.

Leaf : Ramal and cauline, simple, petiolate, hairy, exstipulate, alternate or sub opposite, ovate, serrate acute; venation reticulate unicostate.

Inflorescence : A heterogamous head; 2 or 3 rows of green involucre of bracts surround the inflorescence. There are two kinds of flowers – Ray florets (marginal) and disc florets (central).

Ray florets :

Flower : Sessile, bracteate, ebracteolate, incomplete cyclic, zygomorphic, unisexual or sometimes sterile and epigynous

Calyx : Reduced to pappus 2 or 3 in number

Corolla : Petals 5 in number, gamopetalous, 2 reduced, 3 join together to form a large yellow ligule; aestivation valvate

Androecium : Absent

Gynoecium : (Soometimes absent as in sterile flowers) ovary inferior bicarpellary, syncarpous, unilocular with a single basal ovule. Style simple, long, hairy; stigma bifid.

Fruit : A cypsela; oily

Floral formula : † ♀ $K_{2\text{-}3}\ C_5\ A_0\ \underline{G}_{(2)}$

Disc Florets :

Flower : Sessile, bracteate, ebracteolate, complete, heterochlamydeous, pentamerous, actinomorphic, bisexual and epigynous

Calyx : 2-3, scale like; hairy

Corolla : Petals 5, gamopetalous, equal in size yellow, tubular; aestivation valvate

Androecium : Stamens 5, alternate with the petals, epipetalous, filaments short; anthers syngenesious, bithecous, basifixed, with longitudinal dehiscence.

Gynoecium : Same as in ray florets

Fruit : Cypsela

Floral formula : † ♀ $K_{2\text{-}3}\ C_5\ A_5\ \underline{G}_{(2)}$

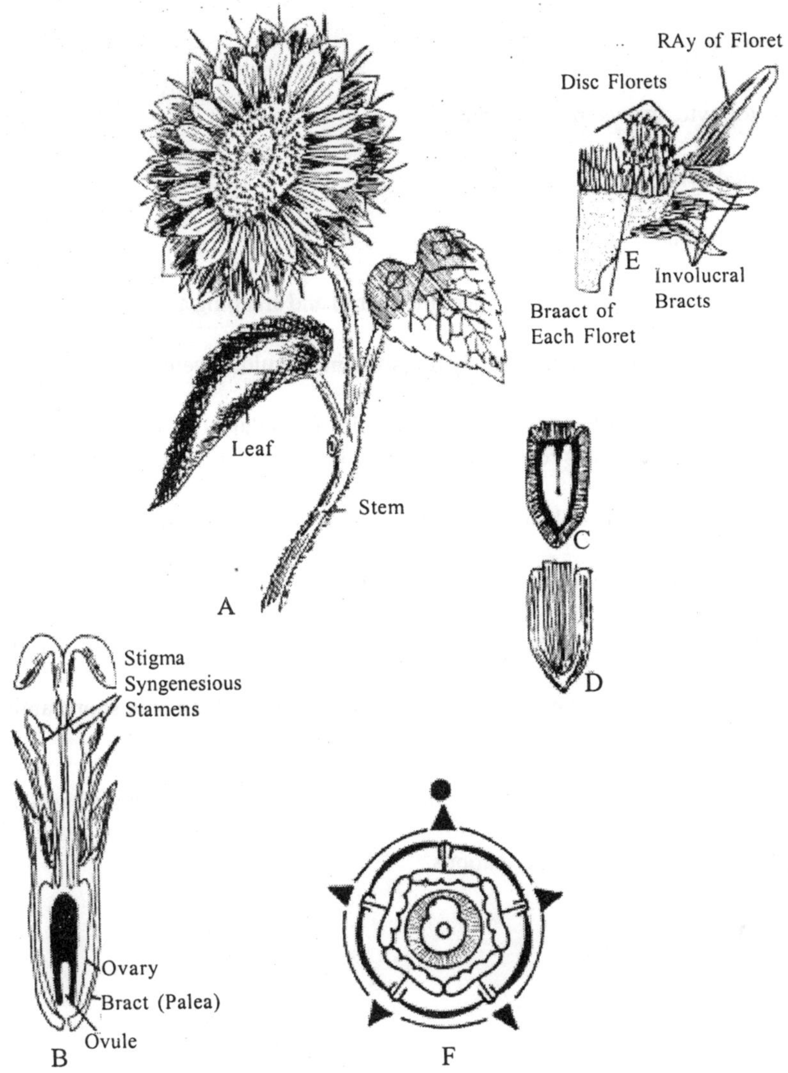

Fig 8.88 ***HELIANTHUS ANNUA*** **(Compositae)**

LAUNEA ASPLENIFOLIA

(Compositae)

Habitat: Mesophyte, grows in waste lands

Habit: A perennial herb

Root: Tap root, branched

Stem: Extremely reduced, underground; becomes aerial and creeping at the time of flowering.

Leaf : Radical sessile, simple, exstipulate, lyrate, margin spinulose; venation reticulate unicostate.

Inflorescence: Homogamous heads on a long peduncle. Secondarily heads form a panicle. All the flowers rayed (ligulate). Multiseriate bracts form a campanulate involucre surrounding the head.

Flower: Sessile, bracteate, ebracteolate, complete, pentamerous, zygomorphic, bisexual and epigynous.

Calyx: Reduced to many hair like structures called pappus.

Corolla: Petals 5, gamopetalous, ligulate with 5 petals forming a ray, towards the margin; yellow in colour

Androecium: Stamens 5, alternate with the petals, epipetalous, filaments short; anthers syngenesious, basifixed, bithecous, sagittate with longitudinal dehiscence.

Gynoecium: Ovary inferior, bicarpellary, syncarpous, unilocular with a single basal ovule.

Fruit: A cypsela; covered with a silky pappus.

Floral formula: † ♀ $K_0\ C_5\ A_5\ \underline{G}_{(2)}$

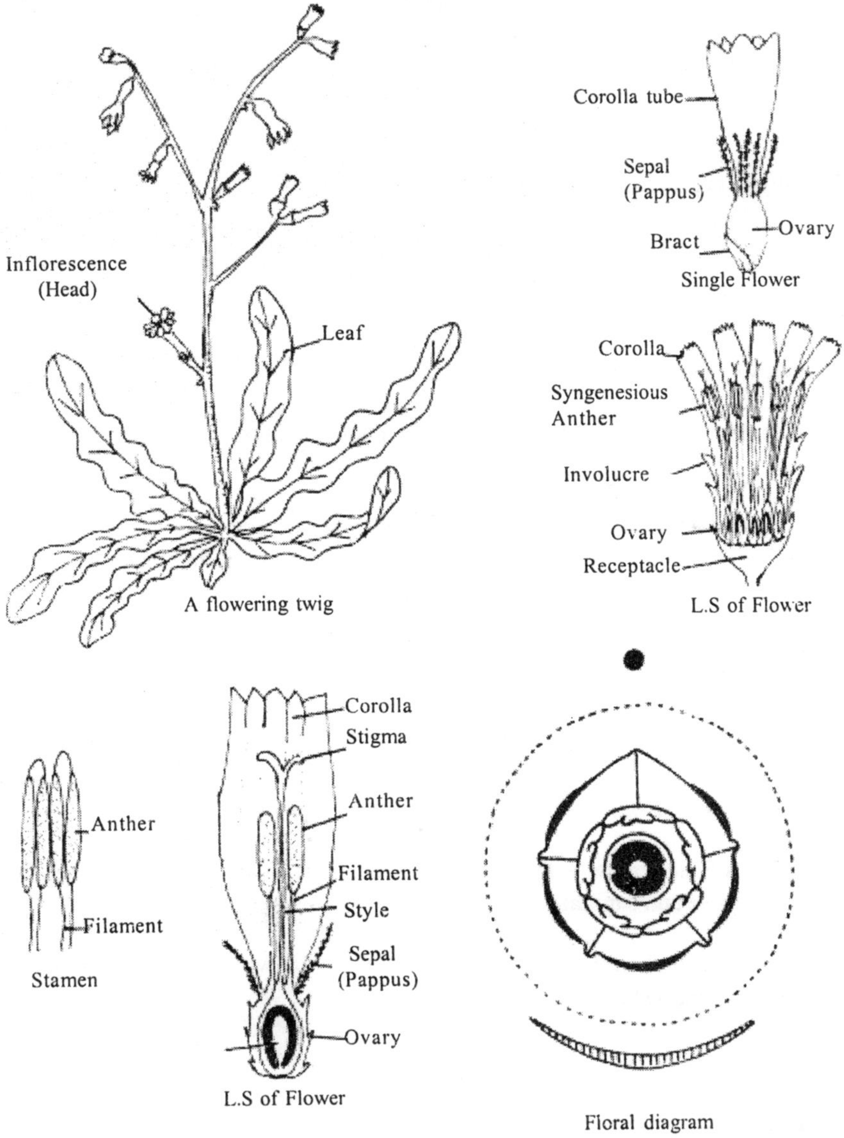

Fig 8.89 ***LAUNEA ASPLENIFOLIA*** **(Compositae)**

LAGASCEA MOLLIS

(Compositae)

Habitat: Mesophyte, grows in open lands.

Habit: A small herb

Root: Tap root, branched

Stem: Aerial, erect, soft, pubescent, cylindrical, solid, and green.

Leaf: Ramal and cauline, simple, alternate, exstipulate, petiolate, hairy, margin entire, obovate; venation reticulate unicostate.

Inflorescence: A compound head; each individual head is represented by a single flower as it has its own involucre of bracts. The compound head is again encircled by a common involucre of bracts. The head is homogamous. There is no distinction into ray and disc florets.

Flower: Sessile, bracteate, ebracteolate, pentamerous, cyclic, heterochlamydeous, complete, actinomorphic, bisexual and eipgynous.

Calyx: Reduced to pappus

Corolla: Petals 5, gamopetalous, tubular, white; aestivation valvate

Androecium: Stamens 5, altnernate with the petals, epipetalous; filaments short; anthers syngenesious, basifixed, bithecous, introrse and sagittate.

Gynoecium: Ovary inferior, bicarpellary, syncarpous, unilocular with a single anatropus ovule on basal placentation. Style simple, stigma bifid.

Fruit: A cypsela

Floral formula: ✝ ♀ $K_0 C_5 A_5 \underline{G}_{(2)}$

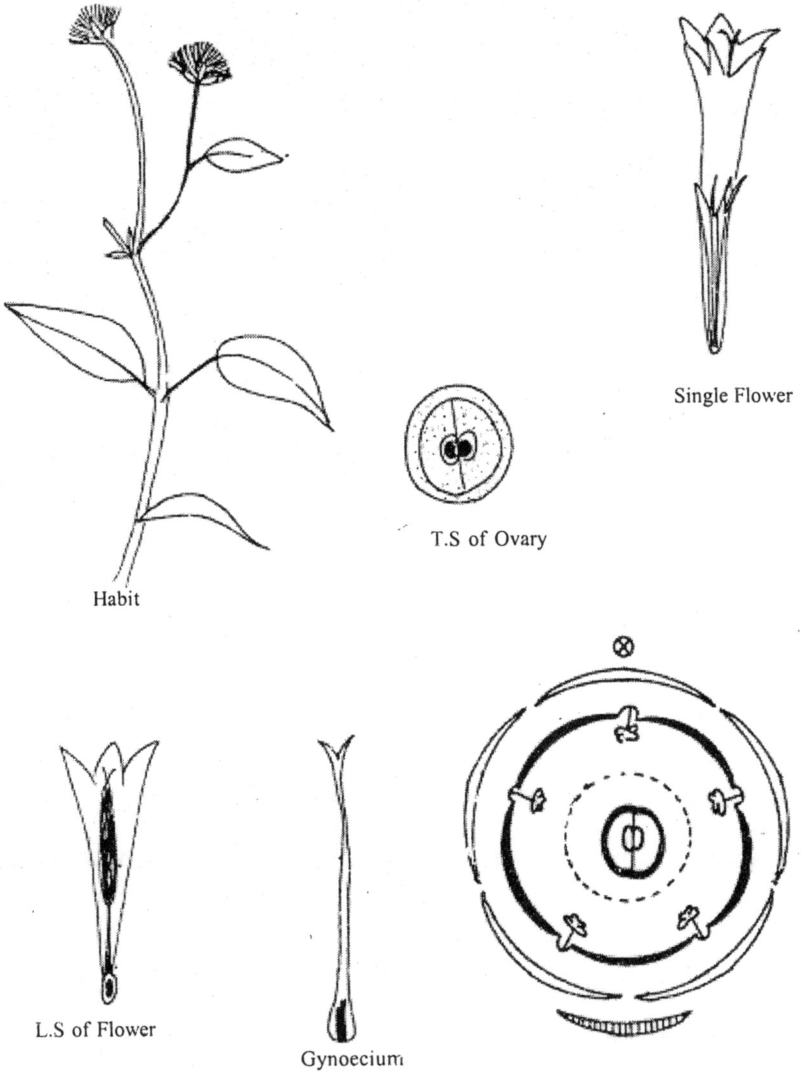

Fig 8.90 ***LAGASCEA MOLLIS*** **(Compositae)**

SONCHUS OLERACEOUS

(Compositae)

Habitat: Mesophyte; weed; grows in waste land

Habit: A wild annual herb with milky latex

Root: Tap root, branched

Stem: Aerial, erect, branched, cylindrical, hollow, green, hairy, glandular and consists of milky latex.

Leaf: Radical in young plants and cauline and ramal in older plants; sessile, alternate, simple, exstipulate, lanceolate, dentate, teeth long and glandular hairy, apex acute; venation reticulate unicostate.

Inflorescence: Capitulum (head), homogamous; the heads cluster to form a secondary umbellate inflorescence. There is a whorl of involural bracts which is green and somewhat leafy.

Flower: Sessile, bractreate, ebracteolate, cyclic, complete, heterochlamydeous, pentamerous, zygomorphic, bisexual and epigynous.

Calyx: Reduced to pappus, indefinite, white and slightly connate at the base

Corolla: Petals 5, gamopetalous, ligulate, strap shaped, yellow; aestivation valvate

Androecium: Stamens 5, alternate with the petals, epipetalous; filaments short; anthers syngenesious, dithecous, dorsifixed, sagittate, introrse with longitudinal dehiscence

Gynoecium: Ovary inferior, bicarpellary, syncarpous, unilocular with a single basal ovule; style long and simple; stigma bifid.

Fruit: A cypsela

Floral formula: † ♀ $K_0\ C_5\ A_5\ \underline{G}_{(2)}$

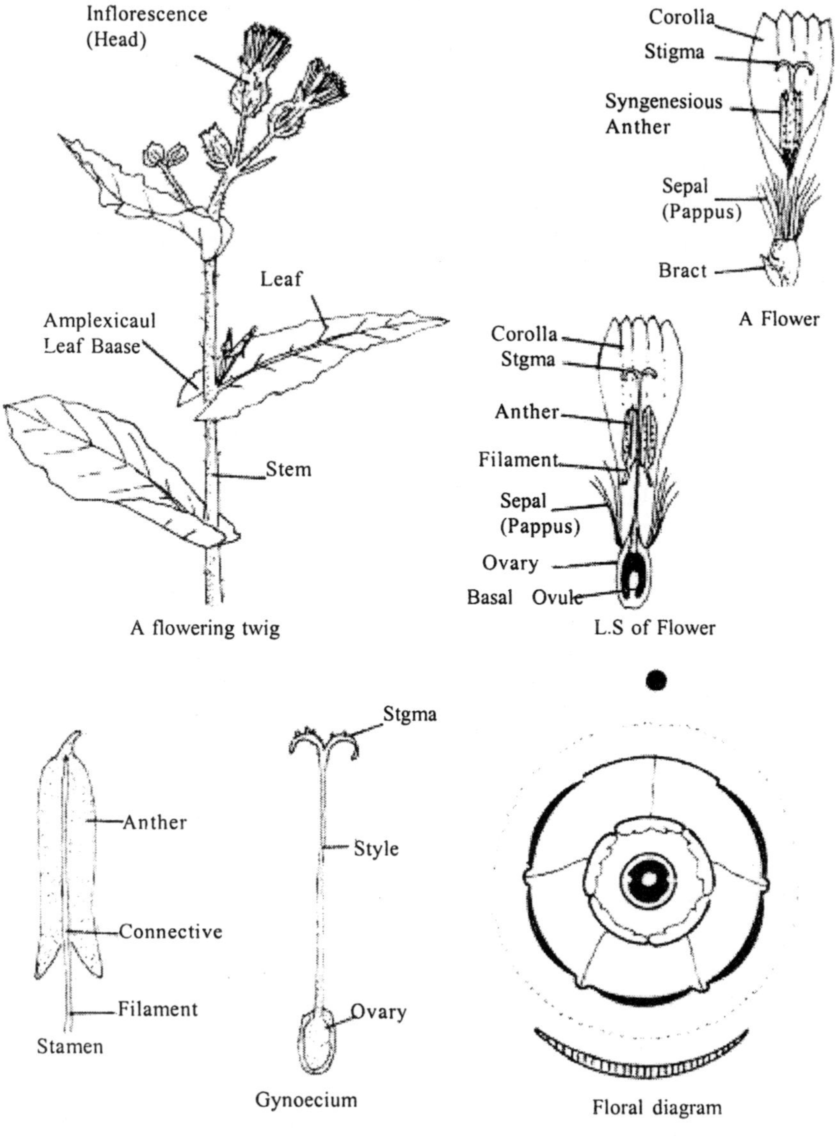

Fig 8.91 *SONCHUS OLERACEOUS* (Compositae)

TAGETES PATULA

(Compositae)

Habitat : Mesophyte, cultivated annual

Habit : An annual herb

Root : Branched tap root

Stem : Aerial, erect, cylindrical, herbaceous above, somewhat woody below; solid and green in colour.

Leaf : Petiolate, alternate or opposite, exstipulate, multipinnately imparipinnately compound; pinnae lanceolate, serrate, acute and glabrous; venation reticulate unicostate.

Inflorescence : A capitulum; heterogamous with marginal deep yellow ray florets and central disc florets. A single whorl of bracts surrounds the inflorescence.

Ray florets :

Flower : Sessile, bracteate, ebracteolate, cyclic, incomplete, zygomorphic, ligulate, unisexual, female and epigynous.

Calyx : Reduced to pappus

Corolla : Petals basically 5, 2 are highly reduced while the remaining 3 unite to form a yellow strap shaped ligule or ray; aestivation valvate.

Androecium : Absent

Gynoecium : Ovary inferior, bicarpellary, syncarpous, unilocular with a single basal ovule, style long; stigma two, hairy.

Fruit : A cypsela

Floral formula : † ♀ $K_0 C_5 A_0 \underline{G}_{(2)}$

Disc floret :

Flower : Sessile, bracteate, ebracteolate, complete cyclic, heterochlamydeous, pentamerous, actinomorphic, bisexual and epigynous.

Calyx : Same as in ray florets

Corolla : Petals 5, gamopetalous, equal, tubular yellow; aestivation valvate

Androecium : Stamens 5, alternate with the petals epipetalous; filaments short; anthers syngenesious bithecous, dorsifixed, introrse and dehiscence by apical pores.

Gynoecium : Same as in ray florets

Fruit : A cypsela

Floral formula : † ♀ $K_0 C_5 A_5 \underline{G}_{(2)}$

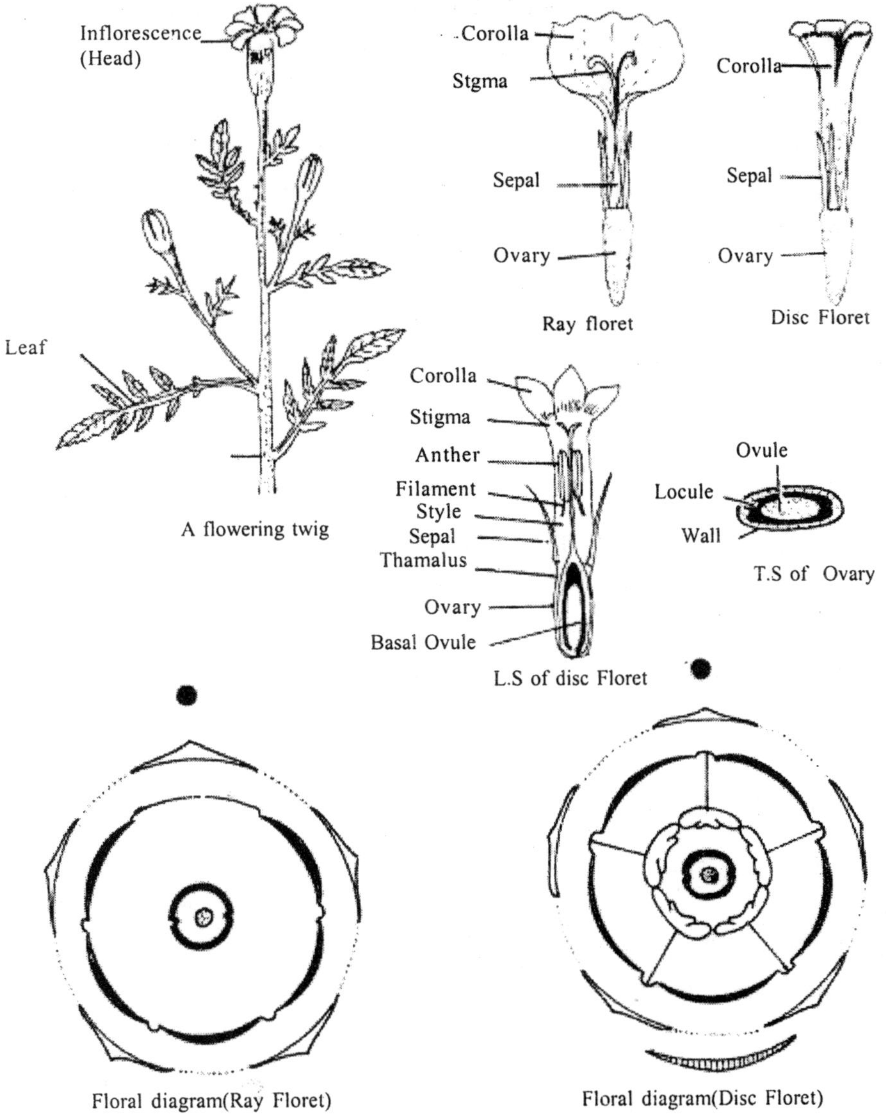

Fig 8.92 *TAGETES PATULA* (Compositae)

TRIDAX PROCUMBENS

(Compositae)

Habitat: Mesophyte; grows wild in wasteland and open places.

Habit: A perennial weed

Root: Branched tap root

Stem: Aerial, partially erect, decumbent, cylindrical, solid, hairy, green and herbaceous.

Leaf: Ramal and cauline, simple, opposite exstipulate, toothed or lobed, hairy; venation reticulate unicostate

Inflorescence: Capitulum (head); a single head terminates a long hairy peduncle. Heads heterogamous with ligulate ray florets and tubular disc florets; bracts biseriate, involucre campanulate; receptacle convex.

Ray florets:

Flower: Sessile, bracteate, bracteolate, incomplete, heterochlamydeous, zygomorphic, unisexual, female and epigynous.

Calyx: Reduced to pappus

Corolla: Petals basically 5, gamopetalous, two petals towards the centre reduced, while the remaining 3 towards the periphery join to form a ligule or ray; ligule pale yellow in colour; Aestivation valvate

Androecium: Absent

Gynoecium: Ovary inferior, bicarpellary, syncarpous, unilocular with a single anatropus ovule on basal placenta. Style long, stigma bifid and hairy.

Fruit: A cypsela

Floral formula: † ♀ $K_0\,C_5\,A_0\,\underline{G}_{(2)}$

Disc florets:

Flower: Sessile, bracteate, bracteolate, cyclic, complete, pentamerous, heterochlamydeous, actinomorphic, bisexual and epigynous.

Calyx: Same as in ray florets

Corolla: Petals 5, gamopetalous, tubular, equal in size, deep yellow in colour; aestivation valvate.

Androecium: Stamens 5, alternate with the petals, epipetalous, filaments short; anthers syngenesious bithecous, dorsifixed and introrse.

Gynoecium: Same as in ray florets

Fruit: A cypsela

Floral formula: † ♀ $K_0\,C_5\,A_5\,\underline{G}_{(2)}$

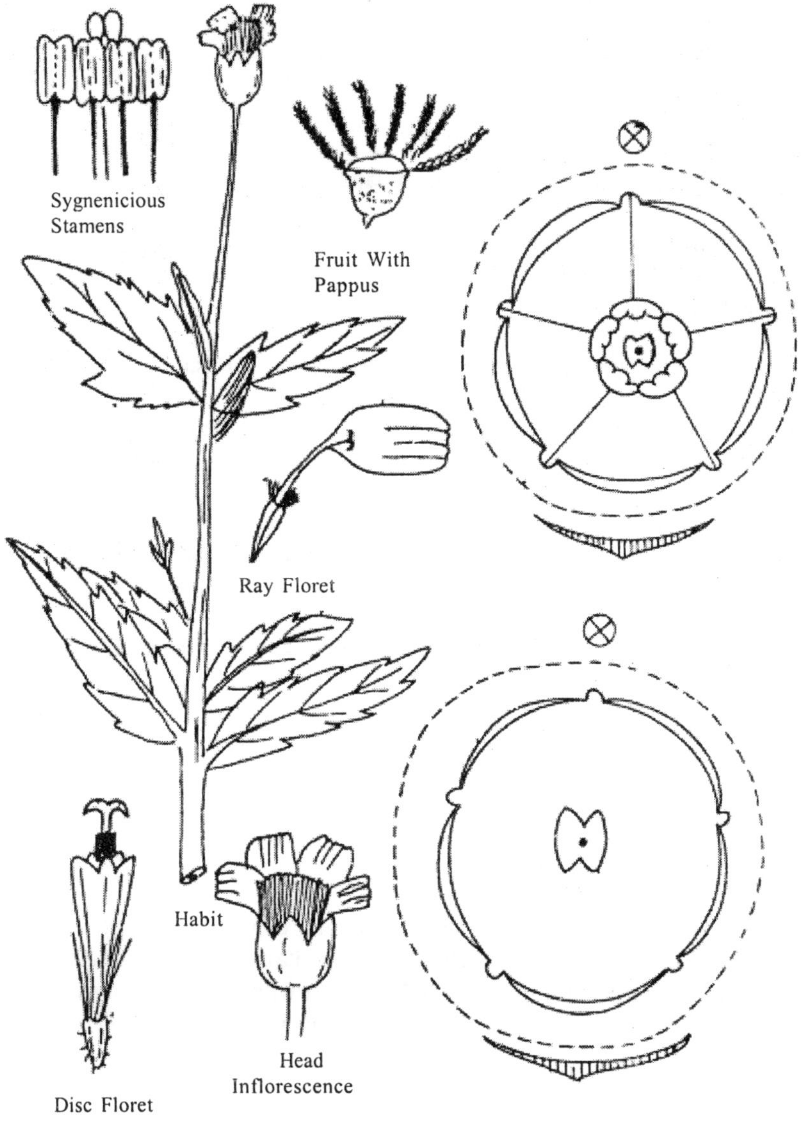

Fig 8.93 ***TRIDAX PROCUMBENS*** **(Compositae)**

XANTHIUM STRUMARIUM

(Compositae)

Habitat: Mesophyte, weed, grows in dry open land

Habit: An annual herbaceous weed

Root: Branched tap root

Stem: Aerial, erect, branched, cylindrical, hairy, herbaceous, solid and green.

Leaf : Simple, alternate, exstipulate, somewhat palmately lobed, margin serrate, acute tip, scabrous on both the surfaces; venation reticulate multicostate.

Inflorescence : Basically a capitulum, secondarily the capitula form terminal or axillary racemes. While the plants are monoecious, inflorescences are unisexual i.e have only male or only female flowers. Male heads above and female heads below in the plant.

Male flowers : Bracteate, ebracteolate, incomplete, actinomorphic, cyclic, unisexual, staminate, several male flowers borne on a columnar receptacle.

Calyx: Absent

Corolla : Petals 5, gamopetalous, tubular, spinous; aestivation; valvate, anthers, dithecous, exerted.

Gynoecium: Represented by a pistillode

Female flowers : Only 2 in a head, enclosed by an ovoid, cupule formed by involucre of bracts; bristles present outside. The cupule is 2 chambered, each chamber having a naked female flower. Calyx and corolla absent. Gynoecium is bicarpellary, syncarpous, unilocular with a single basal ovule. Style curved emerges out of the cupule through an apical opening and then splits into two linear stigmas.

Fruit: A utricle

Floral formula : Male flower : ✝ ♀ $K_0\ C_5\ A_{(5)}\ \underline{G}_0$

Female flower : ✝ ♀ $K_0\ C_0\ A_0\ \underline{G}_{(2)}$

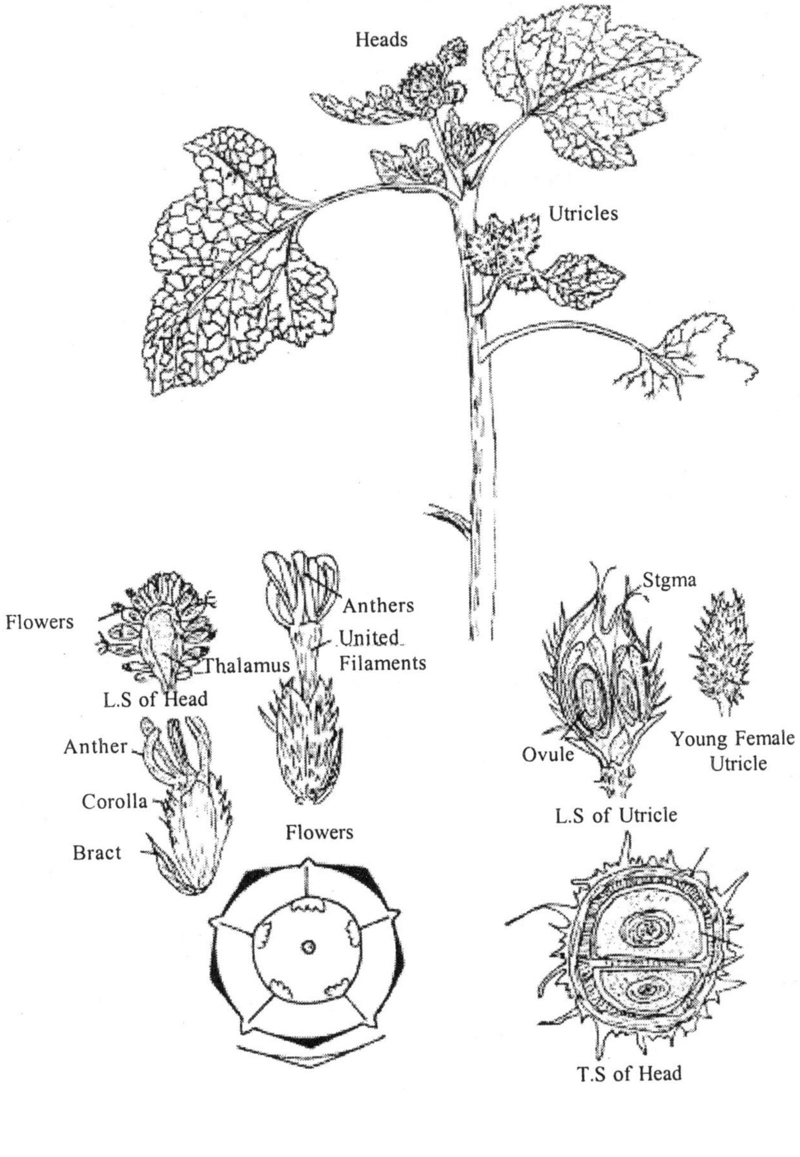

Fig 8.94 ***XANTHIUM STRUMARIUM*** **(Compositae)**

ACHRAS SAPOTA

(Sapotaceae)

Habitat: Mesophyte, cultivated tree

Habit: A medium sized tree, cultivated for its edible fruits

Root: Tap root branched

Stem: Aerial, erect, cylindrical, woody, branched, hairy

Leaf: Ramal and cauline, simple, petiolate, alternate, obovate, entire, acute, smooth, and coriaceous. Leaves generally crowd near the tip of branches

Inflorescence: Flowers – Solitary axillary

Flower: Pedicellate, ebracteate, ebracteolate, complete, cyclic, heterochlamydeous, actinomorphic, bisexual and hypogynous.

Calyx: Sepals 6, free, arranged in two whorls of 3 each; aestivation imbricate

Corolla: Petals 6, gamopetalous, in a single whorl, small white, tube small, aestivation imbricate

Androecium: Stamens 6, free, antitepalous and epitepalous; filaments short; anthers long, dithecous and introrse. Alternating with the stamens 6 scale like structures are found and these are regarded as staminodes.

Gynoecium: Ovary superior, multicarpellary (six) syncarpous, multilocular, with one ovule in each locule on axile placenta. Style and stigma simple; some locules abort in the fruit. Fruit is laticiferous, latex milky.

Fruit: A fleshy berry

Seed: Dark, large, with a pronounced raphe.

Floral formula: ✝ ♀ $K_{3+3}\ C_6\ A_{6+6}\ \underline{G}_{(6)}$

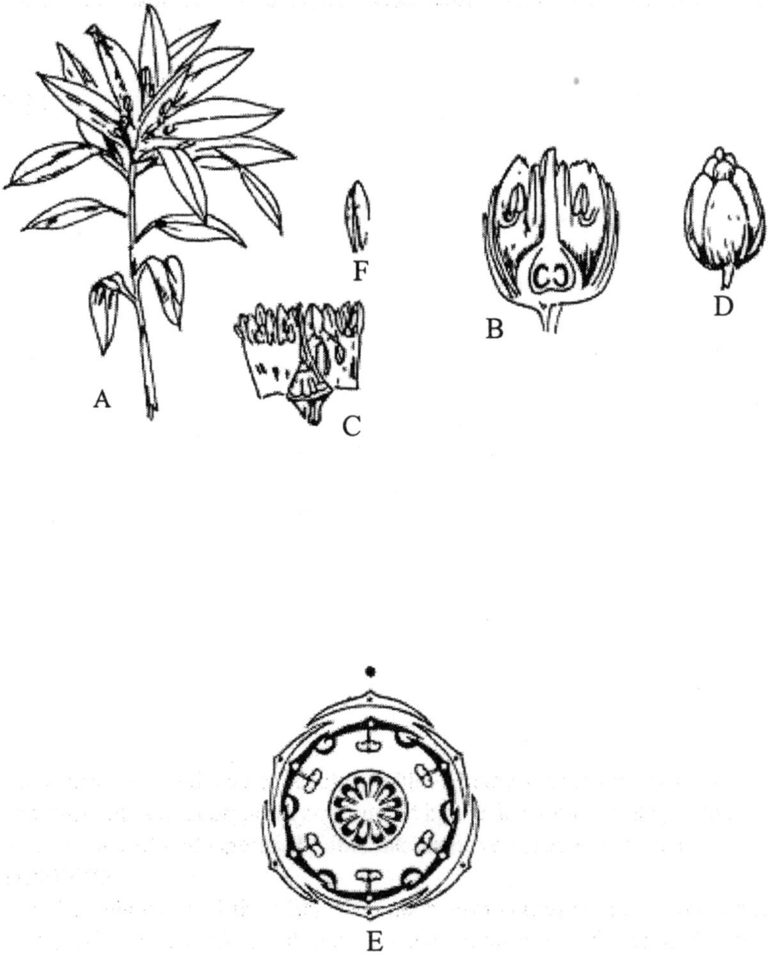

Fig 8.95 *ACHRAS SAPOTA* (Sapotaceae)

MADHUCA INDICA

(Sapotaceal)

Habitat: Mesophyte, found in deciduous forests, often cultivated

Habit: A small deciduous tree

Root: Tap root, branched

Stem: Aerial, erect, cylindrical, woody, branched, rough sparsley tomentose

Leaf: Ramal, simple, alternate, stipulate, petiolate, broadly elliptic, rounded at both ends or sub cordate at base, margin entire, silky or tomntose; venation reticulate unicostate.

Inflorescence: Terminal clusters (cymes)

Flower: Pedicellate, bracteate, ebracteolate, complete, cyclic, heterochlamydeous, actinomorphic, bisexual and hypogynous. Flowers may be cauliflorus.

Calyx: Sepals 4, in two whorls of 2 each (biseriate) outer valvate, thicker, inner thinner and aestivation imbricate. Sepals ovate, acuminate and pubescent.

Corolla: Petals generally 8, gamopetalous one seriate or 2 seriate (arranged in 1-2 whorls), white, lobes ovate acuminate

Androecium: Stamens indefinite (16-20), all fertile, arranged in 2-3 whorls; filaments short; anthers dithecous.

Gynoecium: Ovary superior, multicarpellary (eight), syncarpous, multilocular with one ovule in each locule on axile placenta. Style persistent; stigma minute and sticky.

Fruit: An ovoid berry, fleshy, brown tomentose

Seed: Non endospermic

Floral formula: ✝ ♀ $K_{2+2}\ C_8\ A_{\infty}\ \underline{G}_{(8)}$

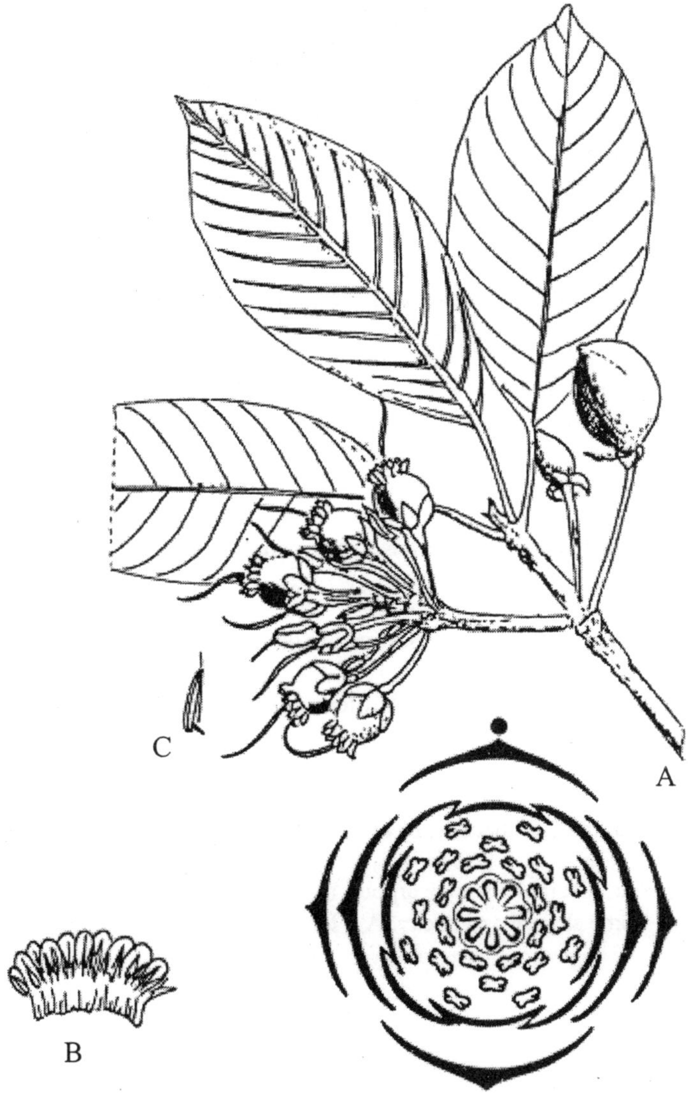

Fig 8.96 ***MADHUCA INDICA*** **(Sapotaceal)**

THEVETIA PERUVIANA

(Apocynaceae)

Habitat: Mesophyte, grows in dry soil, open lands and forests; often cultivated.

Habit: An ervergreen small tree

Root: Branched tap root.

Stem: Aerial, erect, woody below, soft in younger branches, cylindrical, smooth, greyish green, milky latex present.

Leaf: Ramal and cauline; simple alternate or whorled, subsessile, exstipulate, linear lanceolate, margin entire, apex acute, milky latex present; leaf base pulvinate; venation reticulate unicostate.

Inflorescence: Axillary dichasial cyme

Flower: Pedicellate, bracteate, bracteolate, pentamerous, cyclic, complete, heterochlamydeous, actinomorphic, bisexual and hypogynous.

Calyx: Sepals 5, polysepalous, green, persistent, aestivation quincuncial.

Corolla: Petals 5, gamopetalous, yellow, infundibuliform, hairy, coronary out growth present near the mouth of the corolla; aestivation twisted.

Androecium: Stamens 5, free, alternate with the petals, epipetalous, inserted at the throat of the corolla tube; filaments short; anthers sagittate, dithecous, basifixed and introrse with longitudinal dehiscence.

Gynoecium: Ovary superior, bicarpellary, syncarpous, bilocular, with two ovules in each locule on axile placenta. Style long; stigma dome shaped and sticky. A 5 lobed nectar secreting disc is situated below the ovary.

Fruit: A drupe

Seed: Non endospermic

Floral formula: ✝ ⚥ $K_5\ C_{(5)}\ A_5\ \underline{G}_{(2)}$

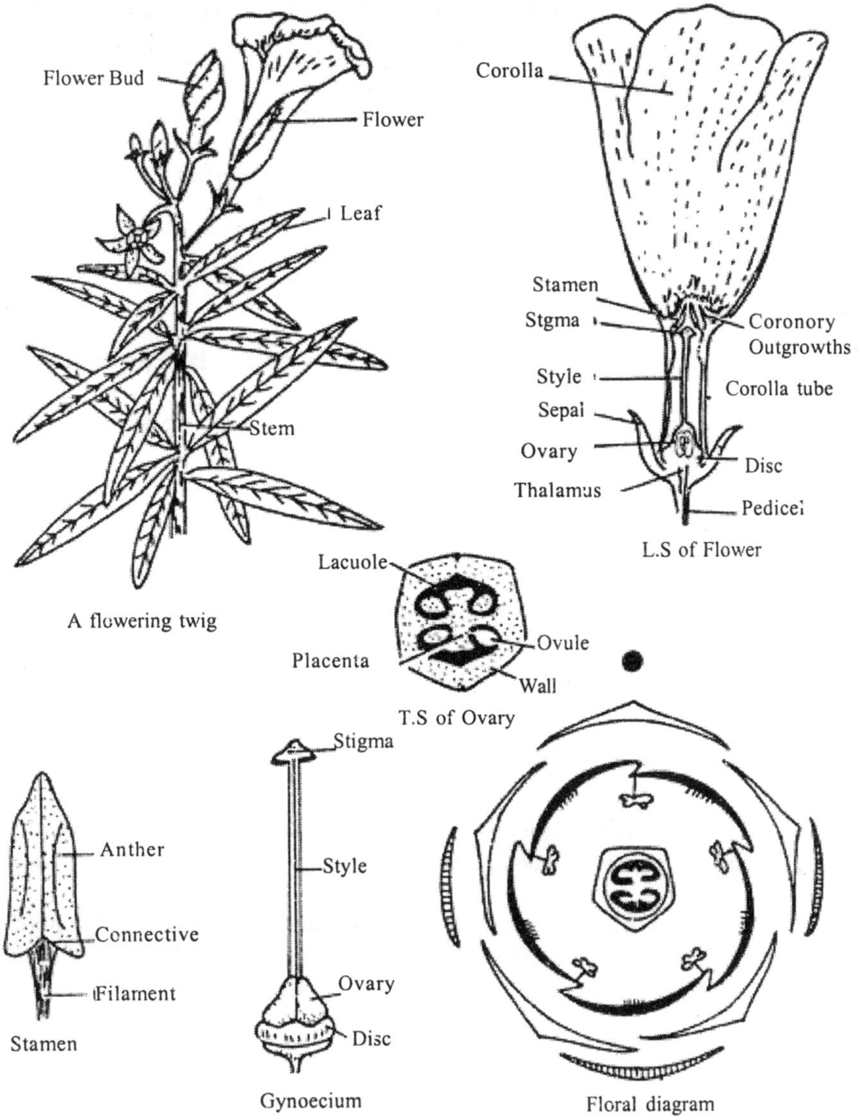

Fig 8.97 *THEVETIA PERUVIANA* (Apocynaceae)

VINCA ROSEA

(= Catharanthus roseus)

(Apocynaceae)

Habitat: Mesophyte, common weed, cultivated for its medicinal properties.

Habit: A perennial herb with pink flowers reaching a height of about 15 cms.

Root: Tap root, branched.

Stem: Aerial, erect, soft, branched, angular purplish or green solid; milky latex present.

Leaf: Ramal and cauline, simple, petiolate, stipulate, (interpetiolar), opposite decussate, elliptic obovate, entire, acute or mucronate, glaucous; venation reticulate unicostate.

Inflorescence: Flowers may be axillary solitary or in axillary dichasial cymes.

Flower: Pedicellate, ebracteate, ebracteolate, cyclic, complete, heterochlamydeous, pentamerous, actinomorphic, bisexual and hypogynous.

Calyx: Sepals 5, polysepalous, persistent, green; aestivation valvate.

Corolla: Petals 5, gamopetalous, hypocraterformi; hairy corona present at the throat of the corolla tube; pink or white in colour; aestivation twisted.

Androecium: Stamens 5, free, alternate with the petals, epipetalous, inserted at the mouth of the corolla tube; filaments short; anthers sagittate, dithecous, dorsifixed, introrse and dehiscence by longitudinal splitting.

Gynoecium: Ovary superior, bicarpellary, partially syncarpous (i.e. carpels free but joined by a single style) unilocular with many ovules on marginal placenta; style long; stigma dumb bell shaped and green in colour. Two honey secreting nectaries present on the anterior and posterior side of he ovary.

Fruit: An etaerio of two follicles

Seed: Albuminous

Floral formula: ✝ ♀ $K_5\ C_{(5)}\ A_5\ \underline{G}_{(2)}$

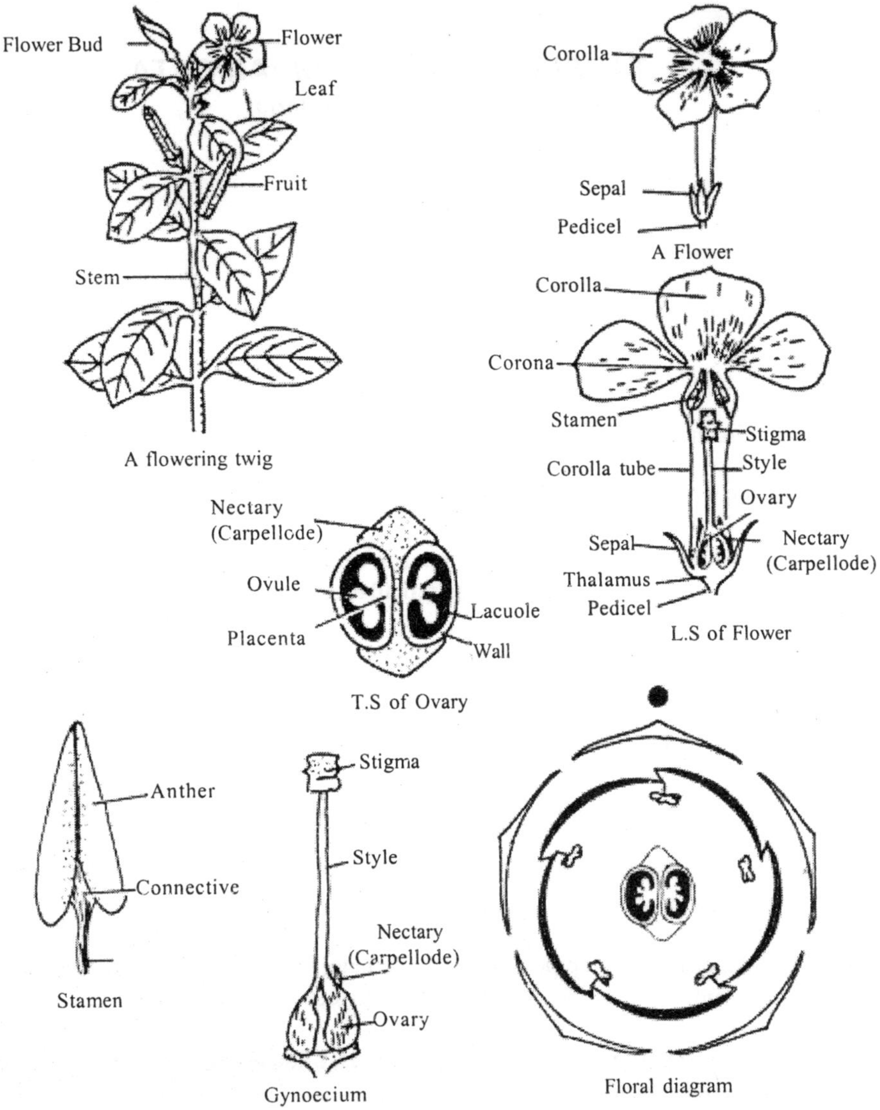

Fig 8.98 ***VINCA ROSEA (= Catharanthus roseus)*** **(Apocynaceae)**

TABERNEMONTANA DIVARICATA
(= Ervatamia Coronaria)
(Apocynaceal)

Habitat: Mesophyte, cultivated, perennial

Habit: A cultivated evergreen shrub

Root: Tap root branched

Stem: Aerial, erect, woody below, herbaceous and green above, cylindrical, branched with milky latex.

Leaf: Ramal and cauline, simple, petiolate, opposite decussate, stipulate (interpetiolar) ovate lanceolate entire, acute, glabrous with milky latex; venation reticulate unicostate.

Inflorescence: Terminal cyme.

Flower: Pedicellate, bracteate, ebracteolate, complete, cyclic, heterochlamydeous, pentamerous, actinomorphic, bisexual and hypogynous.

Calyx: Sepals 5, polysepalous or basally connate, green, persistent; aestivation valvate.

Corolla: Petals 5, gamopetalous white tubular, hypocrateriform, a corona present at the mouth of the corolla tube; aestivastion twisted.

Androecium: Stamens 5, free, epipetalous and alternate with the petals, filaments short, inserted at the throat of the corolla; antehrs dithecous, basifixed and introrse.

Gynoecium: Ovary superior, bicarpellary, syncarpous, bilocular with many ovules per locule on axile placenta. Style long; stigma bifid.

Fruit: A berry

Seed: Endospermic

Floral formula: ✝ ♀ $K_5\ C_{(5)}\ A_5\ \underline{G}_{(2)}$

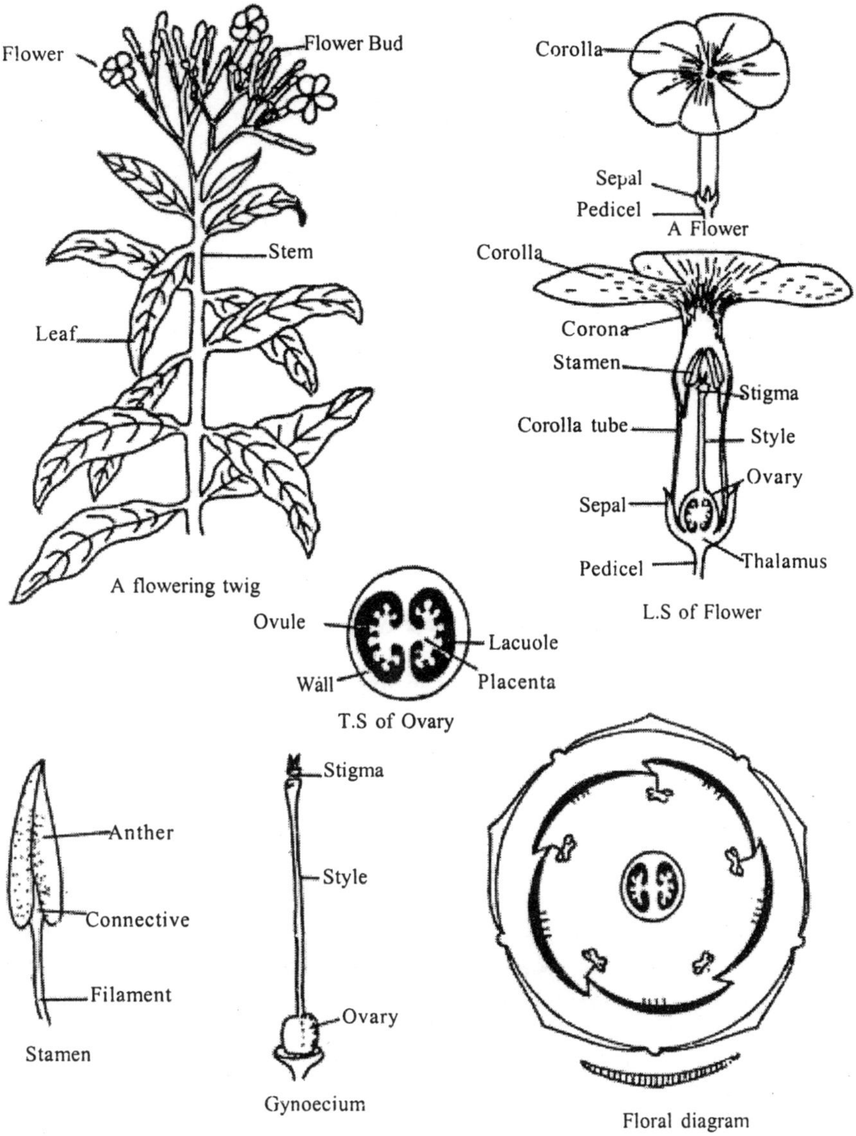

Fig 8.99 ***TABERNEMONTANA DIVARCATA (= Ervatamia Coronaria)*** **(Apocynaceal)**

NERIUM ODORUM

(= N. Indicum)

(Apocynaceal)

Habitat : Mesophyte, evergreen, grows in dry land.

Habit : An evergreen, cultivated shrub or a small tree.

Root : Branched tap root

Stem : Aerial, erect, branched, cylindrical, woody, green with milky latex.

Leaf : Ramal and cauline, simple, subsessile, whorled or opposite decussate, stipules present and interpetiolar, lanceolate, entire, acute, glabrous, coriaceous, leaf base pulvinate, milky latex present; venation reticulate unicostate.

Inflorescence : Terminal, dichasial or paniculate cyme

Flower : Pedicellate, bracteate, bracteolate, pentamerous, cyclic, complete, heterochlamydeous, actinomorphic, bisexual and hypogynous.

Calyx : Sepals 5, polysepalous, persistent; purple red in colour; aestivation imbricate.

Corolla : Petals 5, gamopetalous, white or pink in colour, rotate, corona present at the throat of the corolla tube; aestivation twisted.

Androecium : Stamens 5, free, alternate with the petals, epipetalous, located at the throat of the corolla around the stigma; filaments short; anthers sagittate; dithecous, basifixed, introrse; connective appendiculate and all the connectives twist around one another to form a coiled thread.

Gynoecium : Ovary superior, bicarpellary, syncarpous, bilocular with many ovules in each locule on axile placenta; style filigform, stigma dome shaped.

Fruit : A Follicle

Seed : Linear, ribbed and has a coma of villans hairs endospermic.

Floral formula : ✝ ⚥ $K_5\ C_{(5)}\ A_5\ \underline{G}_{(2)}$

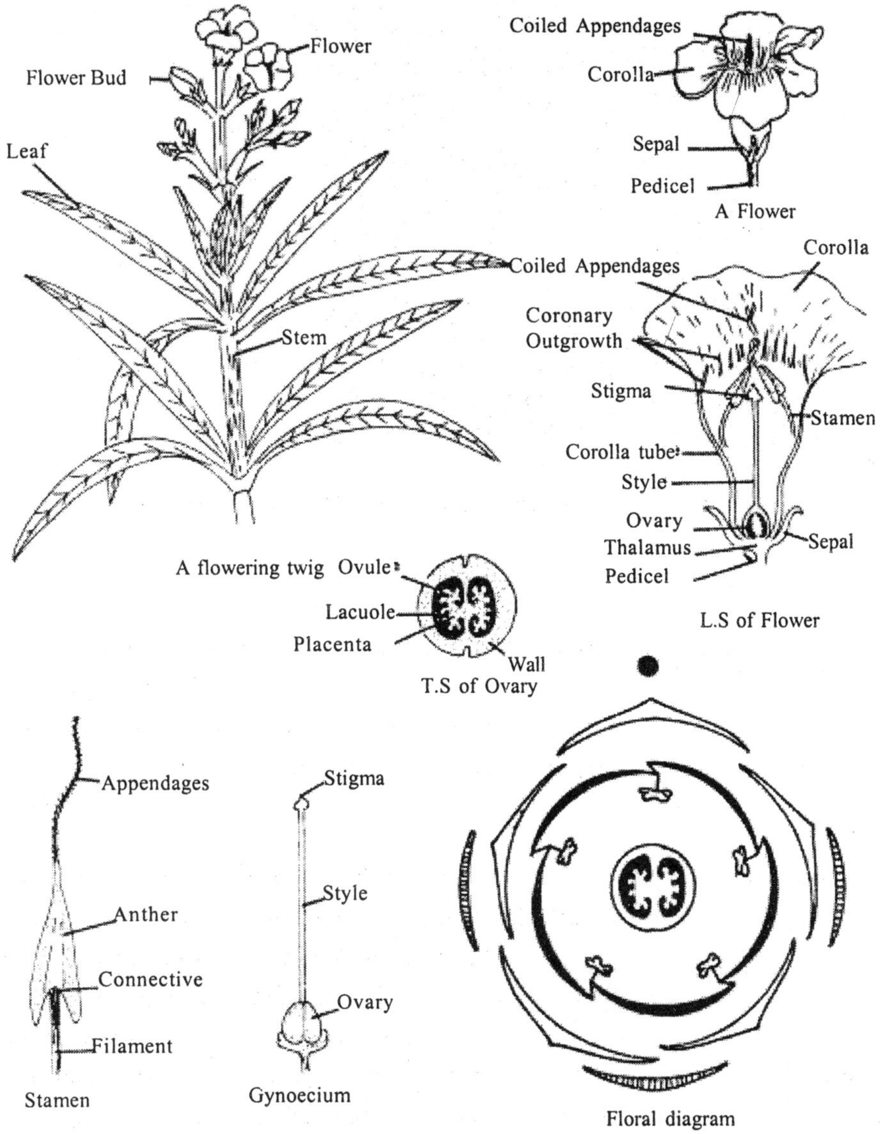

Fig 8.100 ***NERIUM ODORUM (= N. Indicum)*** **(Apocynaceal)**

VALLARIS SOLANACEA

(Apocynaceae)

Habitat: Mesophyte, cultivated plant

Habit: A cultivated climbing shrubby perennial

Root: Branched tap root

Stem: Aerial, weak, climbing, green, cylindrical and branched

Leaf : Ramal and cauline, simple, petiolate, opposite obovate, entire and acute; venation reticulate unicostate.

Inflorescence: Axillary cyme

Flower: Pedicellate, bracteate, ebracteolate, cyclic, pentamerous, heterochlamydeous, complete, actinomorphic, bisexual and hypogynous.

Calyx: Sepals 5, gamosepalous, green; aestivation valvate.

Corolla: Petals 5, gamopetalous, tubular, greenish white in colour; scaly corona present at the throat; aestivation twisted.

Androecium: Stamens 5, free, alternate with the petals, epipetalous, inserted at the throat of the corolla tube; filaments short; anthers sagittate, dithecous, basifixed, introrse and covnverge around the stigma.

Gynoecium: Ovary superior, bicarpellary syncarpous, bilocular with ovules many on axile placenta. Style long; stigma green coloured and dumb bell shaped.

Fruit: A drupe

Seed: Endospermic

Floral formula: ✝ ⚥ $K_5\ C_{(5)}\ A_5\ \underline{G}_{(2)}$

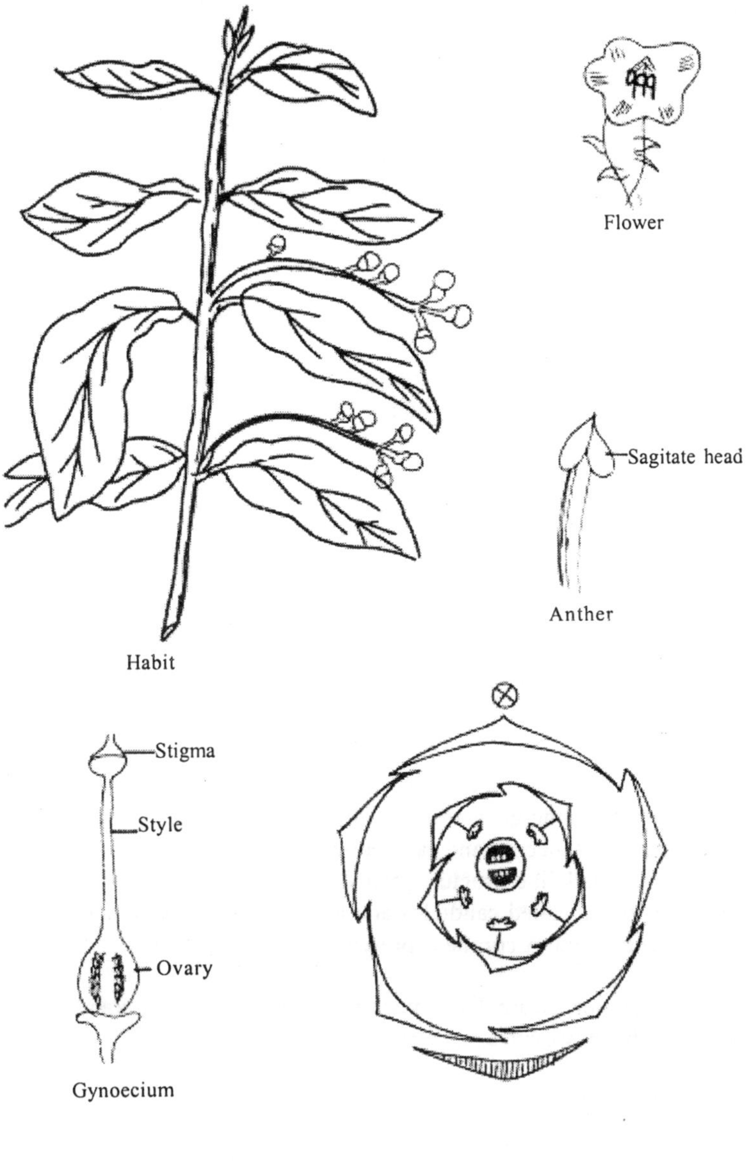

Fig 8.101 *VALLARIS SOLANACEA* (Apocynaceae)

ASCLEPIAS CURASSAVICA

(Asclepiadaceae)

Habitat: Mesophyte, perennial, grows in waste land; often cultivated

Habit: A small herb with yellowish red or white flowers.

Root: Branched tap root

Stem: Aerial, erect, herbaceous, branched, cylindrical, green, glabrous; milky latex present.

Leaf: Ramal and cauline, simple, exstipulate, opposite decussate, petiolate, linear lanceolate, entire, acuminate, glabrous, venation reticulate unicostate.

Inflorescence: Extra axillary umbellate cyme

Flower: Pedicellate, bracteate, ebracteolate, cyclic, complete, pentamerous, heterochlamydeous, actinomorphic, bisexual and hypogynous.

Calyx: Sepals 5, free; aestivation valvate or quincuncial

Corolla: Petals 5, gamopetalous, reflexed and dropping below orange red in colour, a corolline corona is present; aestivation twisted.

Androecium: Stamens 5, fused with the stigma to form a 5 angled disc like structure called gynostegium, epipetalous. The anthers have undergone some modifications and represented by two sac like structures called pollinia; the anthers are dithecous and each pollinium represents a lobe of the anther. The pollinia of adjacent anthers are fused to stalk like structures called translators. Each translator has a basal adhesive disc called corpusculum, two stalks called caudicles and a spoon shaped terminal part called shovel through which it is attached to the pollinia. A corona is present behind each stamen.

Gynoecium: Ovary superior, bicarpellary apocarpous below but syncarpous above (joined by the style), each carpel unilocular with many ovules on marginal placentation; style and stigma fused

Fruit: An etaerio of two follicles

Seed: Endospermic, each seed has a crown of hygroscopic hairs.

Floral formula: ✝ ⚥ $K_5 C_{(5)} A_{(5)} \underline{G}_{(2)}$

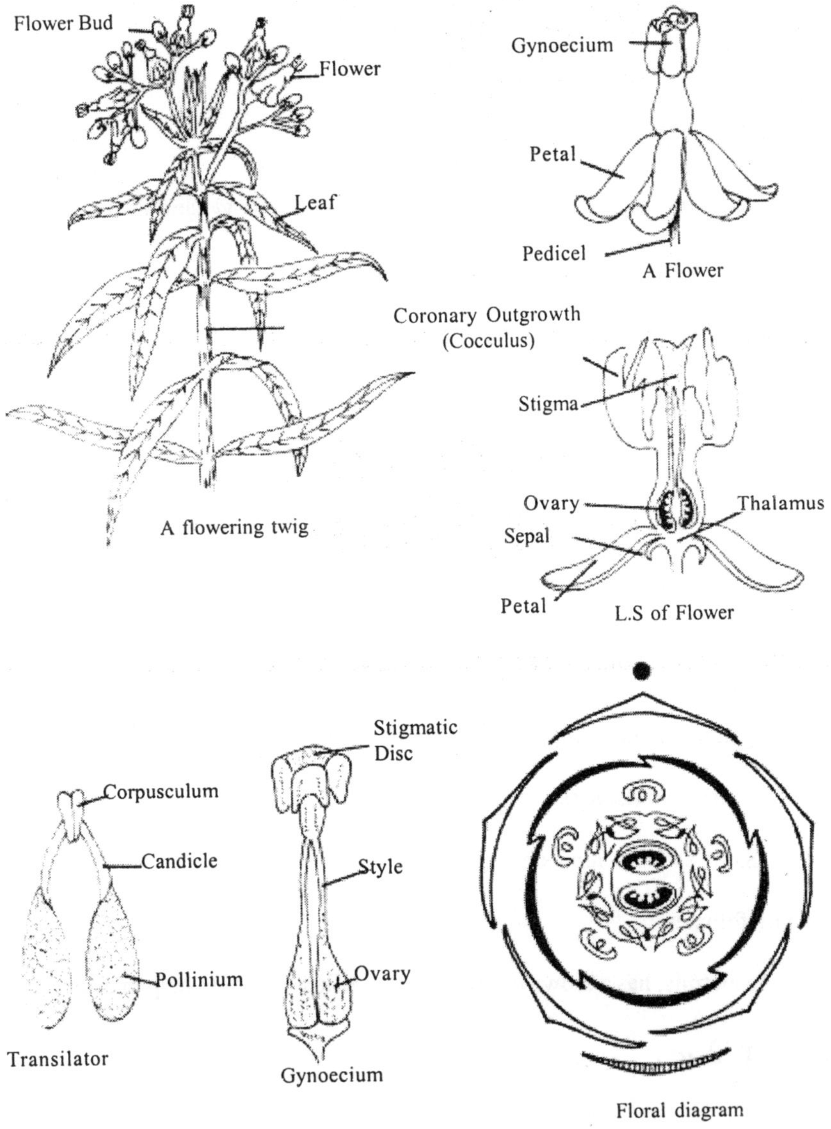

Fig 8.102 ***ASCLEPIAS CURASSAVICA*** **(Asclepiadaceae)**

CALOTROPIS GIGANTEA

Habitat: Mesophyte, but grows in dry soils

Habit: A perennial shrub; clothed with soft white pubescence all over and with milky latex in all parts.

Root: Tap root, branched; deep feeder

Stem: Aerial, erect, soft at top, woody below, branched, cylindrical younger parts densely clothed with soft white tomentum; pale green; contains milky latex.

Leaf: Ramal, cauline, simple, sessile, opposite decussate, leaf base semiamplexicaul, obovate oblong, entire, acute, thick, clothed with dense wooly hairs, latex present; venation reticulate unicostate

Inflorescence: Axillary umbellate cyme clothed with dense wooly tomentum

Flower: Pedicellate, bracteate, bracteolate, cyclic, complete, pentamerous, heterochlamydeous, actinomorphic, bisexual and hypogynous.

Calyx: Sepals 5, free, ovate linear, pubescent, glandular; greyish green; aestivation valvate.

Corolla: Petals 5, gamopetalous, campanulate, purplish red in colour; aestivation twisted.

Androecium: Same as in *Asclepias curassavica*

Gynoecium: Same as in *Asclepias curassavica.*

Fruit: A pair of follicles.

Seed: Flat, endospermic, has a crown of villous hairs called 'coma' to help in dispersal.

Floral formula: ✝ ⚥ $K_5\ C_{(5)}\ A_5\ \underline{G}_2$

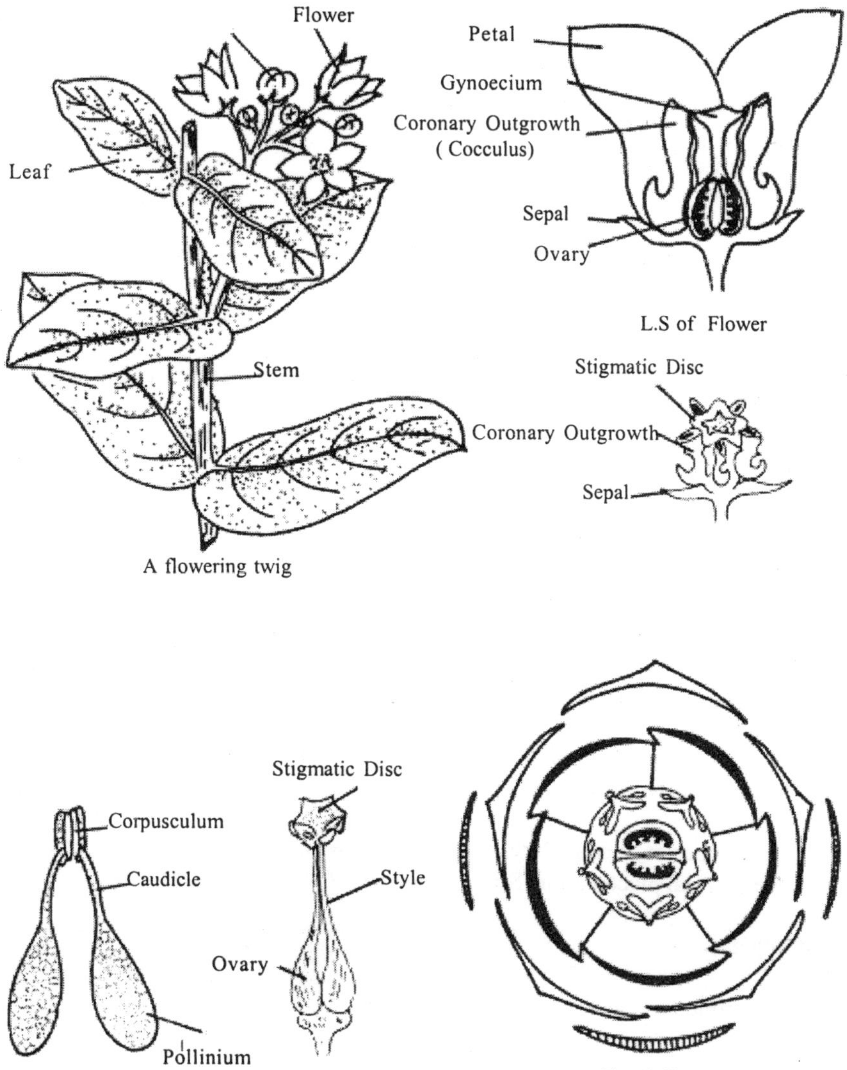

Fig 8.103 *CALOTROPIS GIGANTEA*

CRYPTOSTEGIA GRANDIFLORA

(Asclepiadaceae)

Habitat : Mesophyte, cultivated in gardens

Habit : An ornamental climbing shrub

Root : Branched Tap root

Stem : Aerial, weak, climbing by twining round the support, cylindrical, solid, lower part woody, upper part soft, greenish, milky latex present.

Leaf : Ramal and cauline, opposite decussate, simple, petiolate, exstipulate, pulvinate, elliptic ovate, entire, acute, glaucous; venation reiculate unicostate

Inflorescence : Dichasial cyme

Flower : Pedicellate, bracteate, bracteolate, complete, cyclic, pentamerous, heterochlamydeous, actinomorphic, bisexual and hypogynous.

Calyx : Sepals 5, polysepalous, margins membranous, aestivation quincuncial

Corolla : Petals 5, gamopetalous, five corolline coronae present (violet coloured); aestivation twisted

Androecium : Stamens 5, free, epipetalous and alternate with the petals. Pollen are shed into the translator which is attached by an adhesive disc in between the two stamens.

Gynoecium : Ovary superior, bicarpellary, syncarpous above as the free carpels are held by a single style, with many ovules on marginal placentation. Stigma knob like. Gynostegium present.

Fruit : A pair of follicles

Seed : Small, endospermic

Floral formula : ✝ ⚥ $K_5\ C_{(5)}\ A_5\ \underline{G}_2$

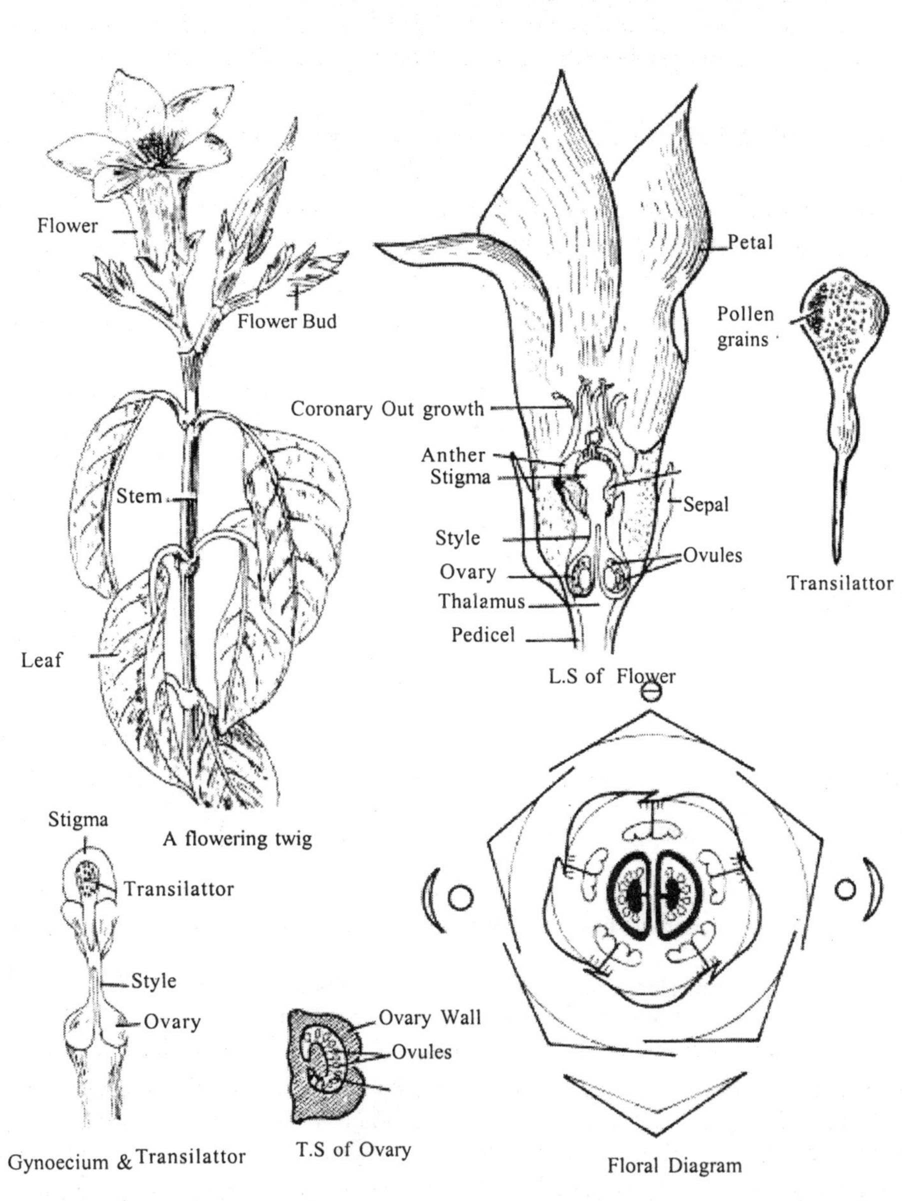

Fig 8.104 *CRYPTOSTEGIA GRANDIFLORA* (Asclepiadaceae)

IPOMOEA PALMATA

(Convolvulaceae)

Habitat : Mesophyte, grows in open land; commonly called railway creeper

Habit : An ornamental twining herb.

Root : Branched tap root

Stem : Aerial, weak, twining, soft, branched, cylindrical solid, glabrous and green.

Leaf : Ramal and cauline, simple, alternate, petiolate, palmately lobed (5-7 lobes), exstipulate glabrous, entire, apex of lobes acute; venation reticulate multicostate.

Inflorescence : Dichasial cyme; rarely solitary

Flower : Pedicellate, bracteate, bracteolate, cyclic, complete, pentamerous, bluish, heterochlamydeous, actinomorphic, bisexual and hypogynous.

Calyx : Sepals 5, polysepalous, broad, obtuse, persistent; aestivation imbricate or quincuncial

Corolla: Petals 5, gamopetalous, infundibuliform, with a long tube and spreading top, blue in colour, the petals have traingular thickenings at their centre which appears starshaped; aestivation twisted.

Androecium : Stamens5, free, alternate with the petals, epipetalous; 2 long and 3 short, inserted deep in the corolla tube; anthers sagittate, basifixed, dithecous and introrse with longitudinal splitting

Gynoecium : Ovary superior, bicarpellary, syncarpous, bilocular with two ovules in each locule on axile placenta; style long; stigma bilobed. A nectar secreting disc is found below the ovary.

Fruit : A capsule

Seed : Endospermic, clothed with silky hairs

Floral formula : ✝ ⚥ $K_5\ C_{(5)}\ A_5\ \underline{G}_{(2)}$

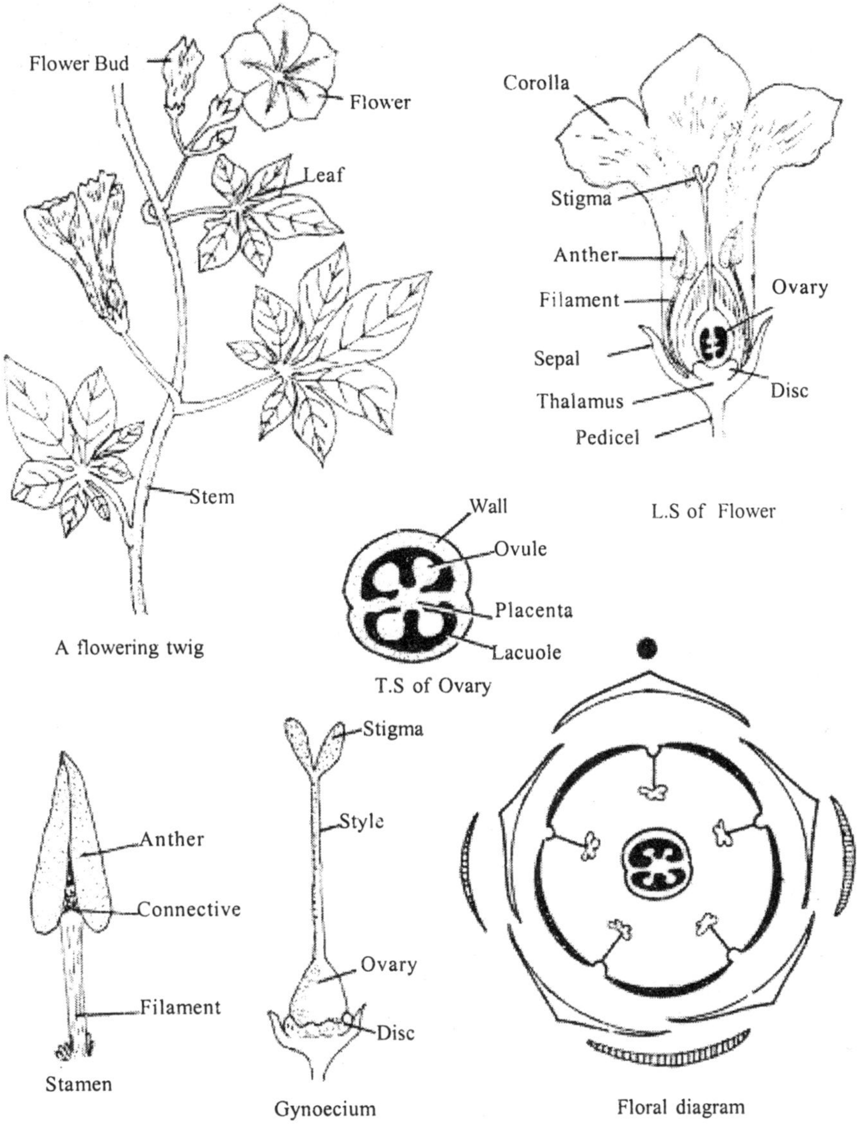

Fig 8.105 *IPOMOEA PALMATA* (Convolvulaceae)

CONVOLVULUS ARVENSIS

(Convolvulaceae)

Habitat: Mesophyte, wild, grows in dry open lands.

Habit: A twining herb

Root: Tap root, branched, adventitious roots also present

Stem: Aerial, weak, soft, climbing (twiner), rarely prostrate also

Leaf: Ramal and cauline, simple, alternate, exstipulate, ovate or hastate, entire, acute, glabrous, venation reticulate unicostate

Inflorescence: Dichasial cyme or occasionally solitary.

Flower: Pedicellate, bracteate, bracteolate, cyclic, complete, heterochlamydeous, pentamerous, actinomorphic, bisexual and hypogynous.

Calyx: Sepals 5, polysepalous, connate at the base slightly; aestivation imbricate

Corolla: Petals 5, gamopetalous, infundibuliform, tube not very long, limb spreading, white or pink in colour; aestivation valvate; star shaped thickenings present on the petals.

Androecium: Stamens 5, free, alternate with the petals, epipetalous; filaments unequal in length; anthers dithecous, basifixed, introrse with longitudinal dehiscence.

Gynoecium: Ovary superior; bicarpellary, syncarpous, bilocular with two ovules in each locule on axile placenta. Style long; stigma bifid nectar disc present underneath the ovary.

Fruit: A capsule

Seed: Endospermic

Floral formula: † ⚥ $K_5\ C_{(5)}\ A_5\ \underline{G}_{(2)}$

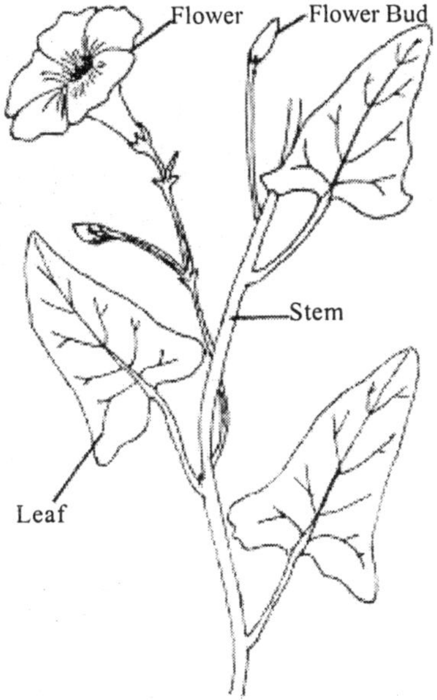

A flowering twig

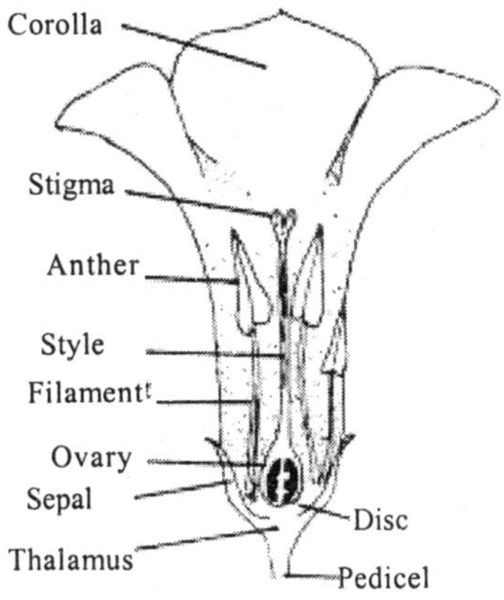

L.S of Flower

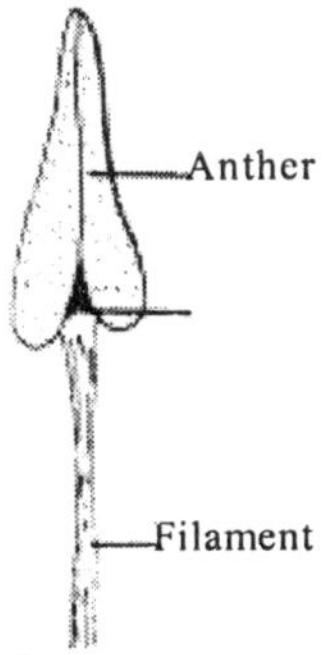

Stamen

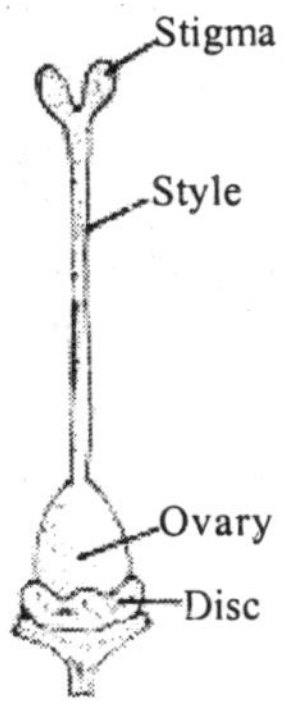

Gynoecium

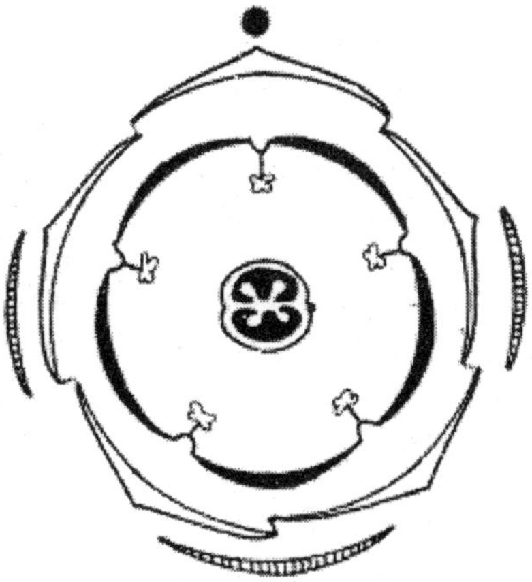

Floral diagram

Fig 8.106 *CONVOLVULUS ARVENSIS* (Convolvulaceae)

CUSCUTA REFLEXA

(Convolvulaceae)

Habitat: A parastic plant which grows as a stem twiner on the host plant

Habit: A perennial twiner

Root: Absent, but haustoria present

Stem: Aerial, weak, twining, parasitic, yellowish in colour, cylindrical, solid; glabrous. The stem produces haustorial roots which penetrate the host stem and suck the food materials.

Leaf: Absent

Infloresence: Flowers may be solitary or in cymose clusters.

Flower: Pedicellate, bracteate, ebracteolate, cyclic, complete, pentamerous, heterochlamydeous, actinomorphic, bisexual and hypogynous.

Calyx: Sepals 5, polysepalous, slightly connate at the base; aestivation quincuncial

Corolla: Petals 5, gamopetalous, campanulate, petals small and reflexed, corona like outgrowth is present at the base, white in colour, aestivation imbricate.

Androecium: Stamens 5, polyandrous, alternate with the petals and epipetalous; filaments short; anthers dithecous, basifixed and introrse with longitudinal dehiscence.

Gynoecium: Ovary superior, bicarpellary, syncarpous, bilocular with many ovules on axile placenta; style short or may be absent; stigma bifid and reflexed; a nectar secreting disc is present below the ovary.

Fruit: A capsule

Seed: Non endospermic

Floral formula: ✝ ⚥ $K_5\ C_{(5)}\ A_5\ \underline{G}_{(2)}$

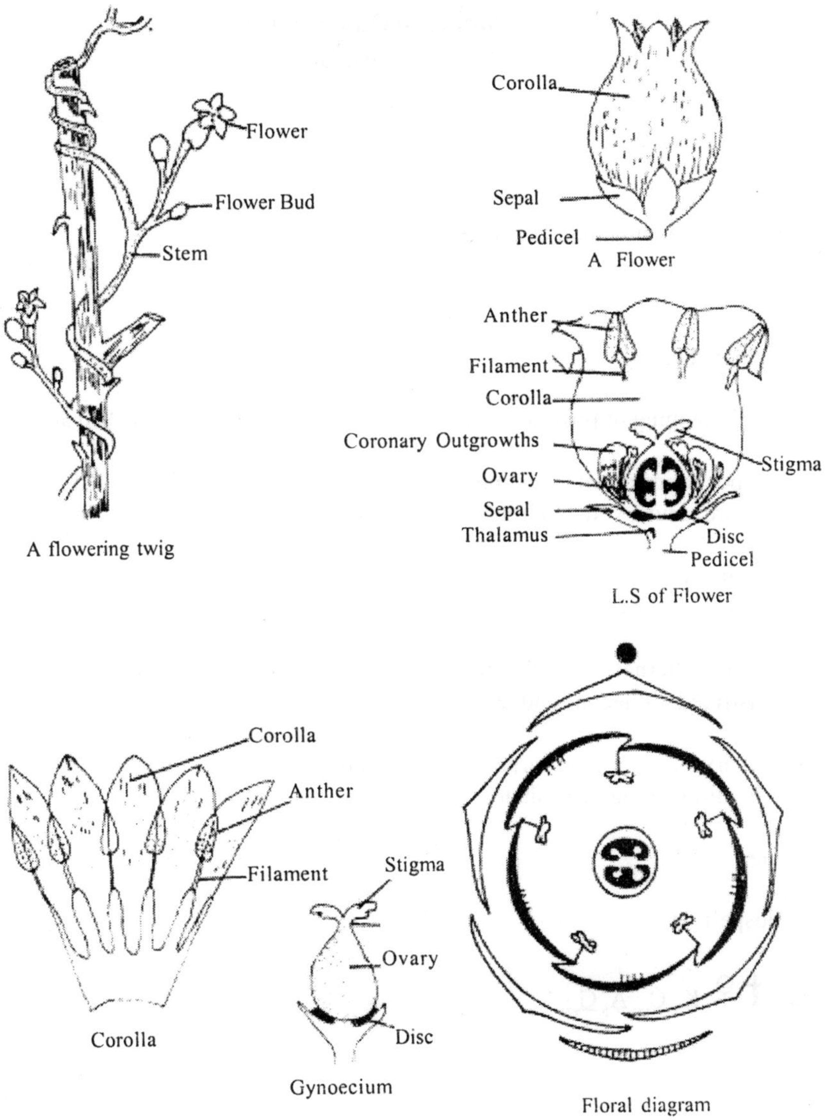

Fig 8.107 *CUSCUTA REFLEXA* (Convolvulaceae)

QUAMOCLIT PINNATA
(= Ipomoea Quamochit)
(Convolvulaceae)

Habitat : Mesophyte, ornamental, cultivated

Habit : An annual climber

Root : Tap root, branched

Stem : Aerial, weak, twining or prostrate cylindrical, herbaceous, green, branched and glabrous.

Leaf : Sessile, opposite, decussate, exstipulate, pinnately compound and divided into a number of thin segments, linear, entire, acute, venation reticulate unicostate.

Inflorescence : Pedicellate, bracteate, ebracteolate, complete, cyclic, heterochlamydeous, pentamerous, actinomorphic, bisexual and hypogynous.

Calyx : Sepals 5, free, alternate with the petals and epipetalous, included in the corolla tube; filaments slender; anthers basifixed, dithecous and introrse with longitudinal dehiscence.

Gynoecium : Ovary superior tetra carpellary, syncarpous, tetra locular with one ovule in each locule on axile placenta; style filiform; stigma capitate.

Fruit : A capsule

Seed : Non endospermic

Floral formula : ✝ ♀ $K_5\ C_{(5)}\ A_5\ \underline{G}_{(2)}$

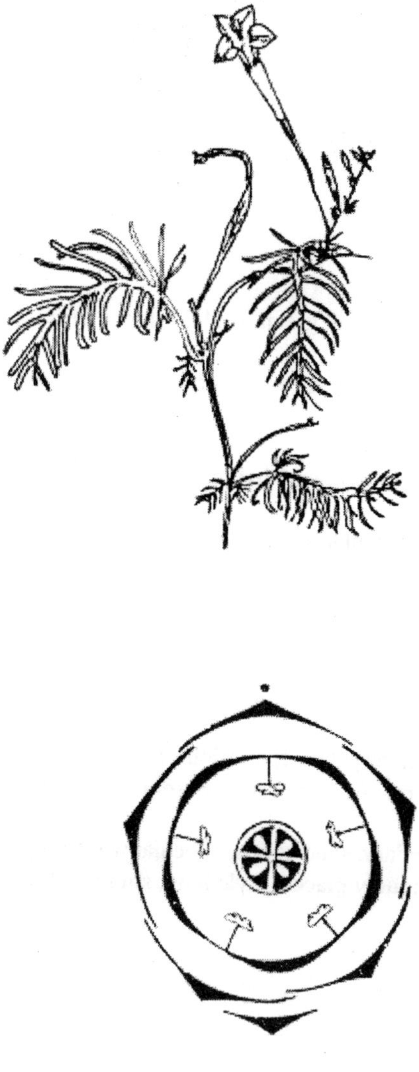

Fig 8.108 *QUAMOCLIT PINNATA (= Ipomoea Quamochit)* (Convolvulaceae)

SOLANUM NIGRUM

(Solanaceae)

Habitat: A wild plant growing in waste land and open places during most part of the year

Habit: An annual herb

Root: A branched tap root

Stem: Aerial, erect, branched, cylindrical solid, green, slightly ridged throughout, and herbaceous.

Leaf : Ramal and cauline, simple,alternate (opposite in the floral region), petiolate, exstipulate, hairy; venation reticulate unicostate.

Inflorescence : Extra axillary scorpioid cyme called Rhypidium

Flower : Pedicellate, bracteate, ebracteolate, pentamerous, cyclic, complete, heterochlamydeous, actinomorhic, bisexual and hypogynous.

Calyx: Sepals 5, gamosepalous, bell shaped, green, coriaceous, persistent; aestivation valvate

Corolla : Petals 5, gamopetalous, white, rotate, campanulate; aestivation valvate.

Androecium : Stamens 5, alternate with the petals, epipetalous, filaments short; anthers dithecous, basifixed, long, yellow and dehiscing by apical pores. Anthers connate around the style to form a cone like covering.

Gygnoecium : Ovary superior, bicarpellary, syncarpous, bilocular with many ovules in each locule on axile placenta,; placenta swollen; carpels obliquely placed; style long; stigma bifid green capitate.

Fruit : A berry

Seed : Albuminous

Floral formula : ✝ ♀ $K_5\ C_{(5)}\ A_5\ \underline{G}_{(2)}$

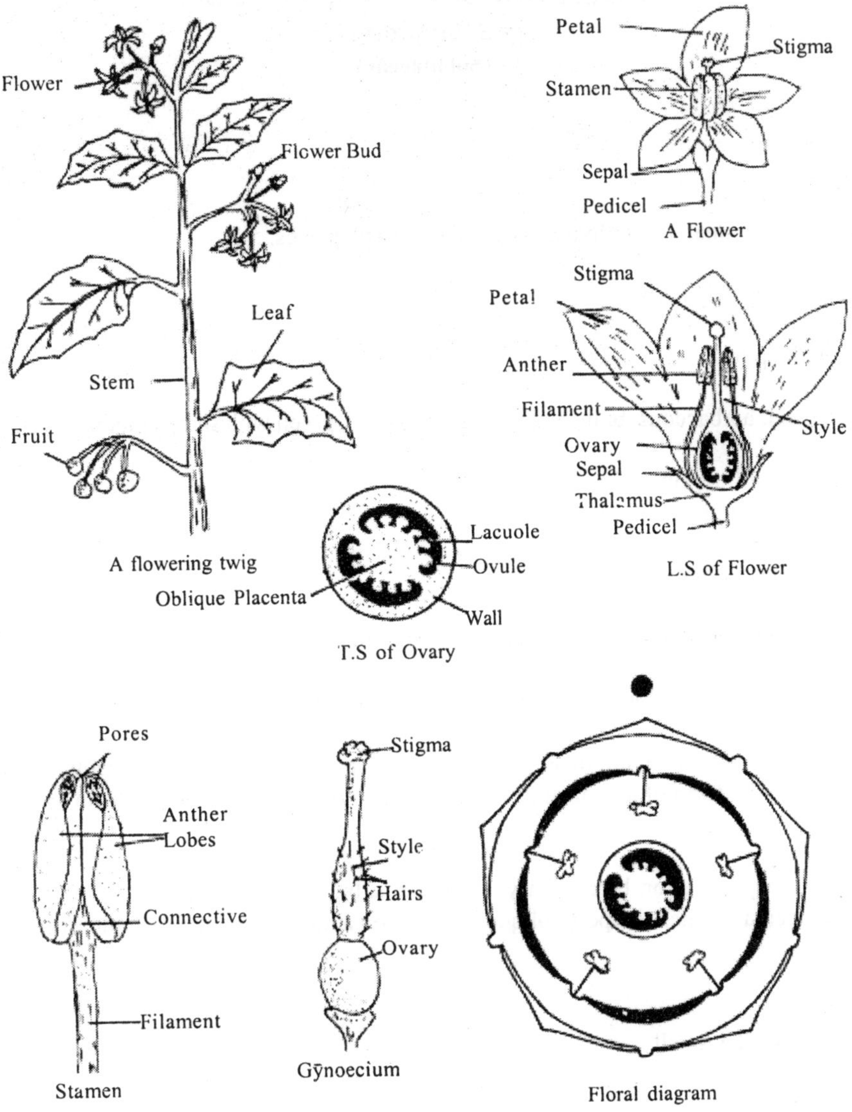

Fig 8.109 *SOLANUM NIGRUM* (Solanaceae)

SOLANUM XANTHOCARPUM
(= S. Surattense)
(Solanaceae)

Habitat : Wild, perennial, mesophytic, grows in dry waste places.

Habit : A perennial prickly herb

Root : Branched tap root.

Stem : Aerial, erect, herbaceous, branched, cylindrical, solid, green with sharp epidermal prickles all over.

Leaf : Ramal and cauline, simple, petiolate, alternate, exstipulate, deeply lobed, margin serrate with sharp points, acute, venation reticulate unicostate

Inflorescence : Extra axillary cymose cluster

Calyx : Sepals 5, gamosepalous, green, persistent; aestivation valvate.

Corolla : Petals 5, gamopetalous, purplish blue, rotate; aestivation twisted.

Androecium : Stamens 5, free, alternate with the petals, epipetalous, filaments unequal; anthers yellow, surround the style in the form of a conical envelope, dithecous, basifixed, introrse with apical dehiscence, through a pore.

Gynoecium : Ovary, superior, bicarpellary, syncarpous, bilocular with many ovules in each locule on axile placenta; placenta swollen; carpels obliquely placed on a honey disc, style simple; stigma bifid.

Fruit : A berry

Seed : Endospermic, flat, small

Floral formula : ✝ ♀ $K_{(5)}\ C_{(5)}\ A_5\ \underline{G}_{(2)}$

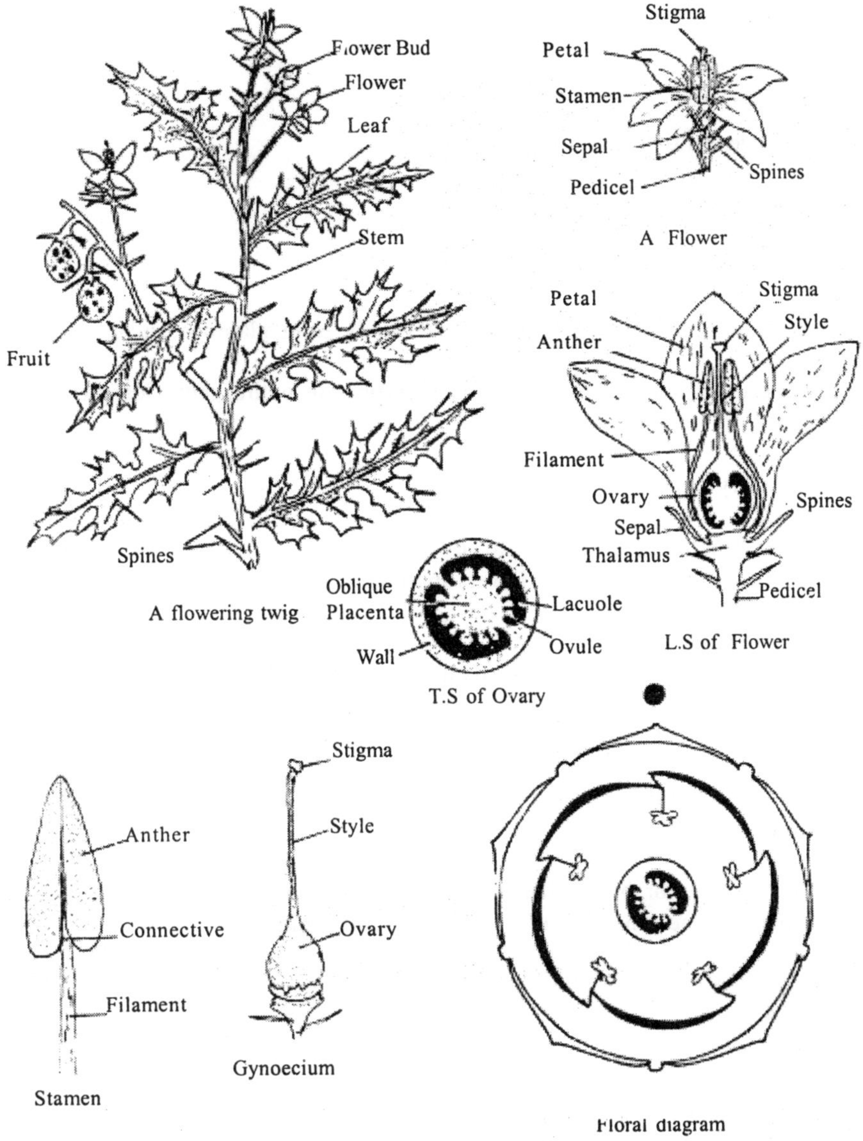

Fig 8.110 ***SOLANUM XANTHOCARPUM (= S. Surattense)*** **(Solanaceae)**

CESTRUM NOCTURNUM

(Solanaceae)

Habitat: Mesophyte, cultivated

Habit: A perennial shrub

Root: Tap root, branched cylindrical

Stem: Aerial, erect, branched, cylindrical somewhat woody, solid, glabrous, and green in colour.

Leaf: Ramal and cauline, simple, shortly petiolate, exstipulate, entire, acute, leaf base unequal; venation reticulate unicostate.

Inflorescence: Axillary cyne

Flower: Pedicellate, bracteate, ebracteolate, cyclic, complete, pentamerous, heterochlamydeous, actinomorphic, bisexual and hypogynous. Flowers are fragrant (mostly during night, the flowers bloom) and white in colour.

Calyx: Sepals 5, gamosepalous; aestivation valvate.

Corolla: Petals 5, gamopetalous, tubular, white fragrant, aestivation valvate.

Androecium: Stamens 5, free, alternate with the petals, epipetalous; filaments short; anthers dithecous basifixed and introrse.

Gynoecium: Ovary superior, bicarpellary syncarpous, bilocular with many ovules on axile placentation; carpels obliquely placed; placenta swollen, style long; stigma capitate.

Fruit: A berry

Seed: Endospermic

Floral formula: ✝ ⚥ $K_{(5)}\ C_{(5)}\ A_5\ \underline{G}_{(2)}$

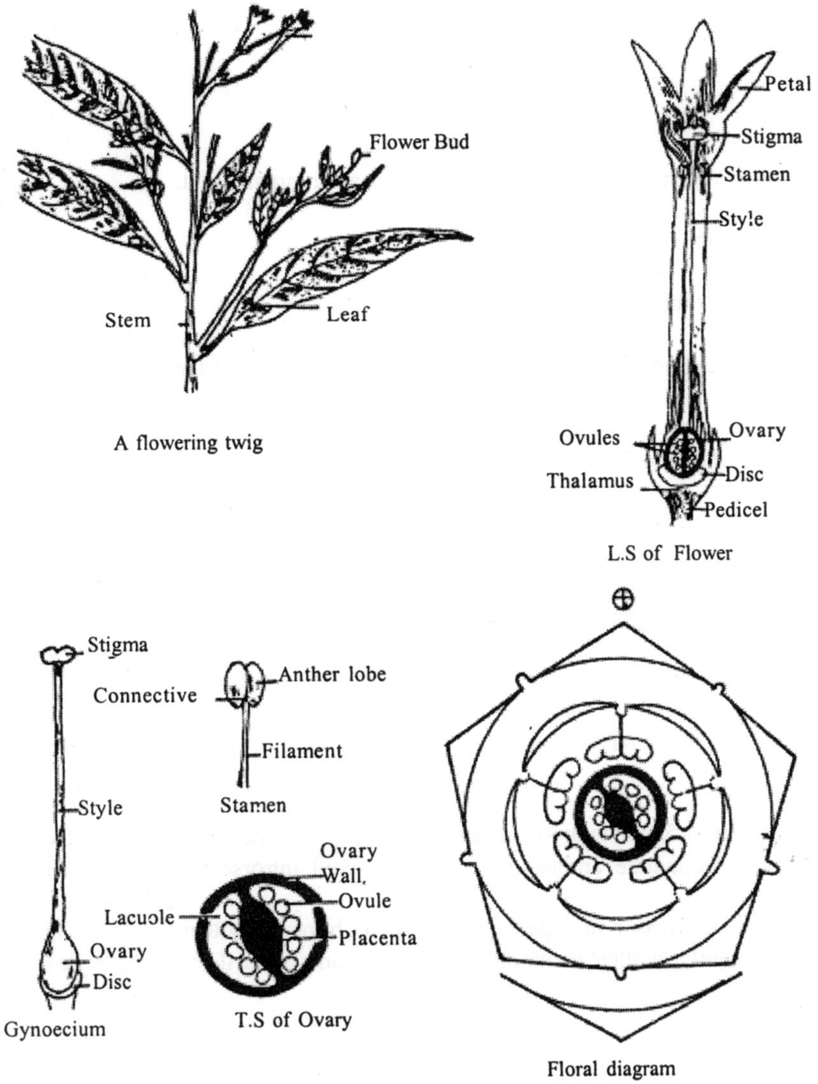

Fig 8.111 ***CESTRUM NOCTURNUM*** **(Solanaceae)**

PETUNIA HYBRIDA

(Solanaceae)

Habitat: A cultivated mesophyte; grown in gardens as an ornamental.

Habit: An annual herb

Root: Tap root, branched

Stem: Aerial, erect, cylindrical, branched, solid, green herbaceous and hairy.

Leaf : Ramal and cauline, simple, alternate in the lower region but opposite in the floral region, ovate, entire, hairy; venation reticulate unicostate.

Inflorescence: Solitary, axillary or terminal; rarely in small clusters.

Flower: Pedicellate, bracteate, ebracteolate, cyclic, complete, pentamerous, heterochlamydeous, actinomorphic, bisexual and hypogynous.

Calyx: Sepals 5, gamosepalous, campanulate, hairy, aestivation imbricate.

Corolla: Petals 5 gamopetalous, tubular, long and spreading into a broad limb at the top; aestivation twisted.

Androecium: Stamens 5, polyandrous, alternate with the petals and epipetalous; filaments unequal, inserted at the middle of the corolla tube; anthers dithecous, basifixed, introrse with a longitudinal dehiscence.

Gynoecium: Ovary superior, bicarpellary, syncarpous, bilocular with many ovules in each locule on axile placenta. Placenta slightly swollen, ovary obliquely placed on a nectariferous disc; style simple, hairy; stigma bilobed.

Fruit: A capsule

Seed: Eudospermous

Floral formula: ✝ ♀ $K_{(5)}\ C_{(5)}\ A_5\ \underline{G}_{(2)}$

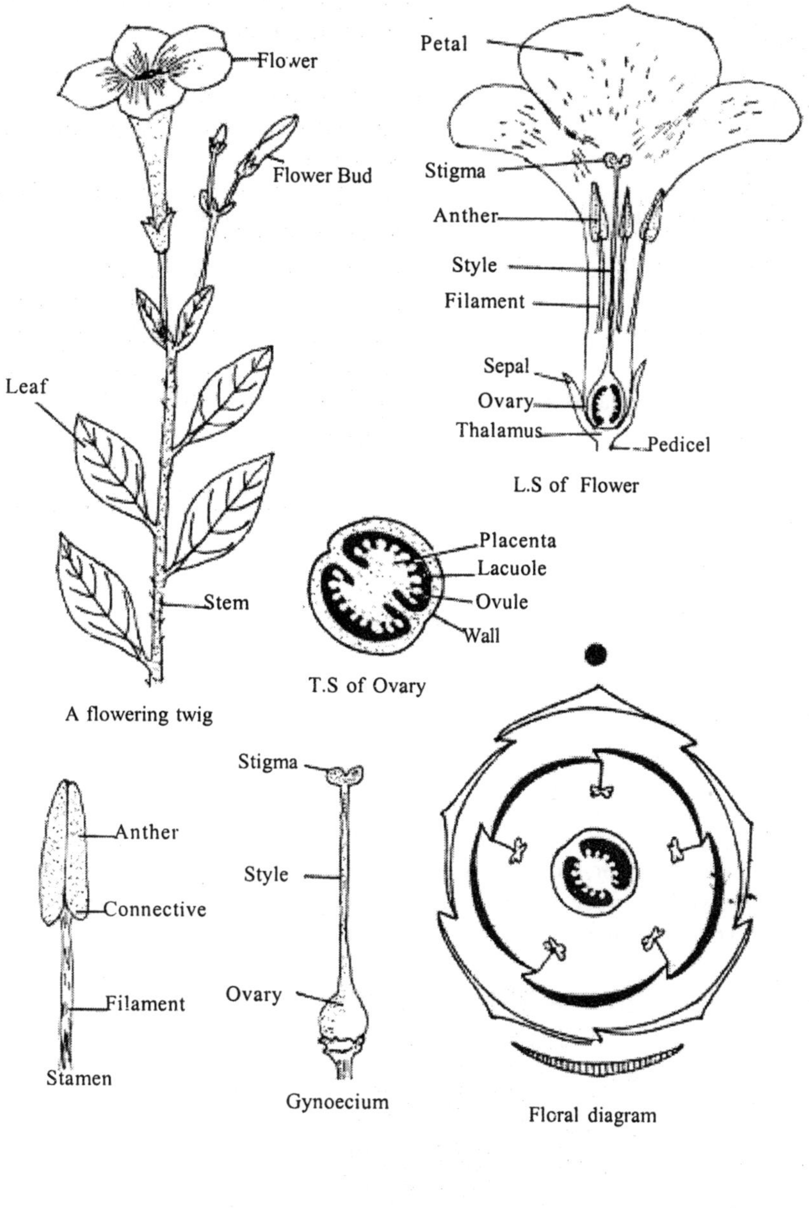

Fig 8.112 ***PETUNIA HYBRIDA*** **(Solanaceae)**

WITHANIA SOMNIFERA

(Solanaceal)

Habitat: Mesophyte, wild, often cultivated, grows in open dry land.

Habit: A perennial herb with red coloured berries

Root: Branched tap root

Stem: Aerial, erect, cylindrical, somewhat woody, branched, hairy and green in colour.

Leaf: Ramal and cauline, simple, petiolate, exstipulate, alternate or sub opposite, leaves at the nodal region are unequal in size, ovate, leafbase unequal, hairy, entire, acute, venation reticulate unicostate.

Inflorescence: Axillary cyme.

Flower: Pedicellate, ebracteate, ebracteolate, cyclic, complete, pentamerous, heterochlamydeous, actinomorphic, bisexual and hypogynous.

Calyx: Sepals 5, gamosepalous, persistent in the fruit, hairy, campanulate; aestivation valvate.

Corolla: Petals 5, free, alternate with the petals and epipetalous, inserted near the base of the corolla tube; anthers dithecous, basifixed, introrse with longitudinal dehiscence.

Gynoecium: Ovary superior, bicarpellary, syncarpous, bilocular with many ovules in each locule on axile placenta. Placenta swollen, ovary oblique, hairy and situated on a nectariferans disc; style simple, stigma bifid, papillate and green in colour.

Fruit: A berry, green when young, turns to yellow or red when mature; enveloped by presistent calyx.

Seed: Endospermous

Floral formula: ✝ ♀ $K_{(5)}\ C_{(5)}\ A_5\ \underline{G}_{(2)}$

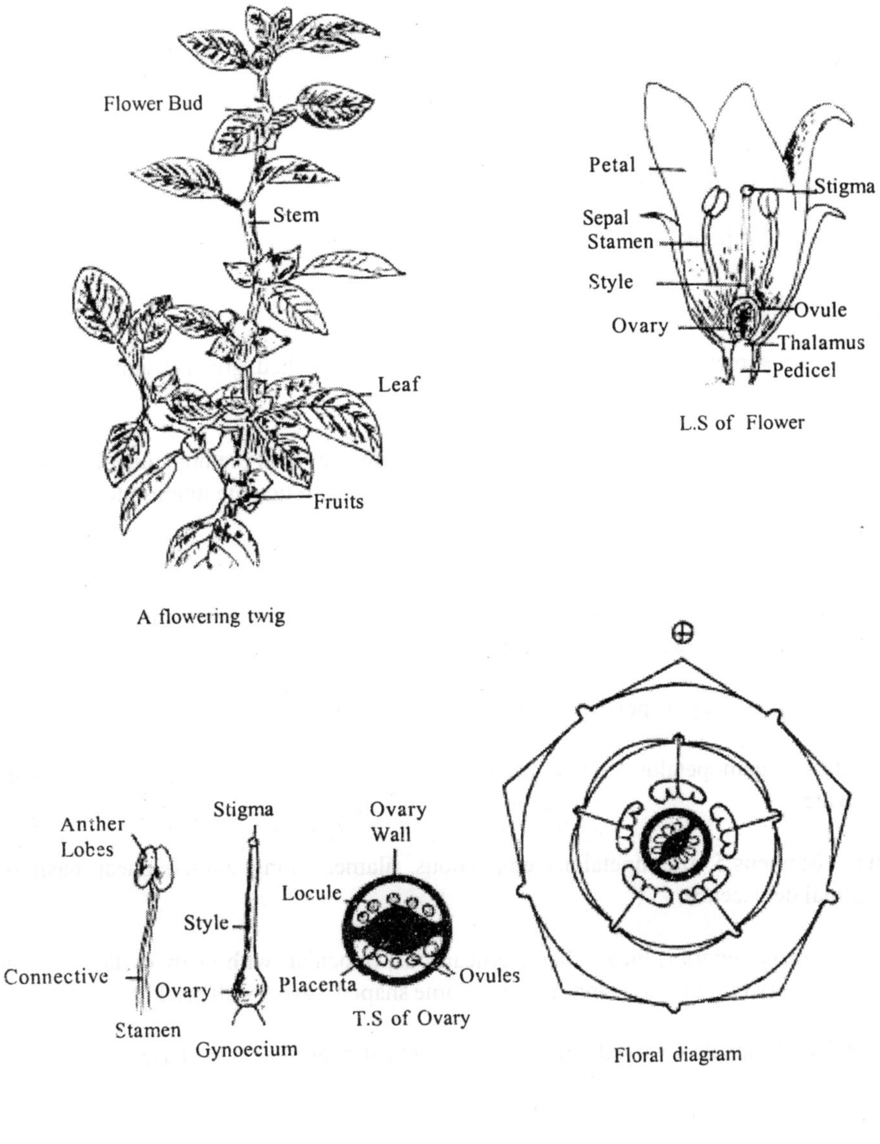

Fig 8.113 ***WITHANIA SOMNIFERA*** **(Solanaceae)**

DATURA STRAMONIUM

(Solanaceae)

Habitat : Mesophyte, weed, grows in open dry land.

Habit : An annual stout herb

Root : Branched tap root.

Stem : Aerial, erect, somewhat woody below and soft above, branched, green or violet coloured, hairy and solid.

Leaf : Ramal and cauline, simple, alternate or sub opposite; leaves at the node unequal in size, leaf base unequal, petioles adnate to the axis, entire, ovate, acute, venation reticulate unicostate.

Inflorescence : Solitary, terminal

Flower : Pedicellate, ebracteate, ebracteolate, complete cyclic, pentamerous, heterochlamydeous actinomorphic bisexual and hypogynous. Flowers large and white in colour.

Calyx : Sepals 5, gamosepalous, persistent, tubular, long 5 toothed, aestivation twisted.

Corolla : Petals 5, gamopetalous, white, corolla tube long and trumpet shaped, longitudinally ribbed; aestivation valvate.

Androecium : Stamens 5, alternipetalons, epipetalous, filaments long; anthers linear, basifixed, introrse with a longitudinal dehiscence.

Gynoecium : Ovary superior, bicarpellary, syncarpous, bilocular, with many ovules on axile placenta. Placenta swollen, ovary oblique; style long; stigma dome shaped, carpel wall prickly.

Fruit : A septifragal capsule 4 valved, prickly, persistent calyx present at the base.

Seed : Scantily endospermous.

Floral formula : ✝ ⚥ $K_{(5)}\ C_{(5)}\ A_5\ \underline{G}_{(2)}$

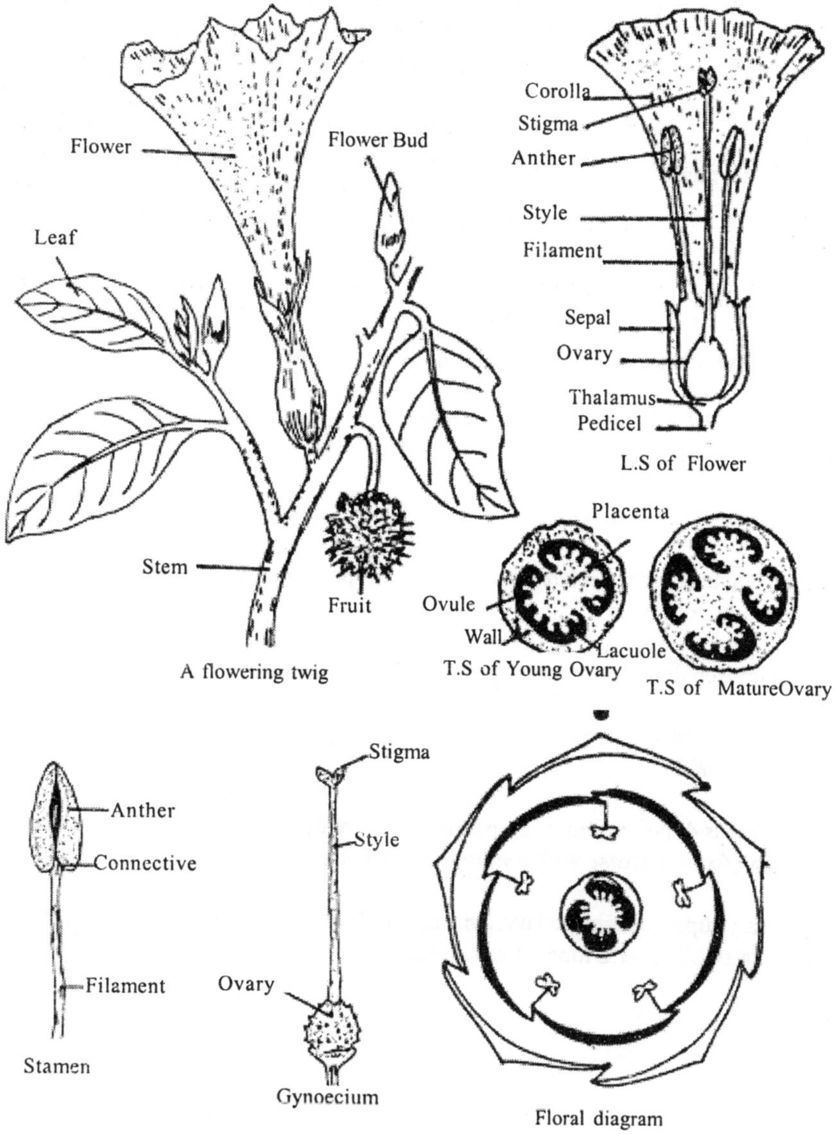

Fig 8.114 ***DATURA STRAMONIUM*** **(Solanaceae)**

LINDENBERGIA INDICA

(Scrophulariaceae)

Habitat: Mesophyte, annual, grows wild in waste land.

Habit: An annual herb

Root: Tap root, branched

Stem: Aerial, erect, cylindrical, branched, herbaceous, hairy green and solid.

Leaf: Ramal and cauline, simple, petiolate, exstipulate, opposite, decusste, ovate, serrate, acute, hairy; venation reticulate unicostate.

Inflorescence: Cymose or solitary axillary.

Flower: Pedicellate, bracteate, ebracteolate, complete, cyclic, pentamerous, heterochlamydeous, zygomorphic, bilipped, bisexual and hypogynous.

Calyx: Sepals 5, gamosepalous, hairy, persistent; aestivation imbricate.

Corolla: Petals 5, gamopetalous, bilabiate, personate with upper lip bilobed and lower lop 3 lobed, yellow in colour; aestivation imbricate

Androecium: Stamens 4, polyandrous, epipetalous, didynamous, filaments short; connective elongated; anthers dithecous, basifixed, introrse with longitudinal dehiscence.

Gynoecium: Ovary superior, bicarpellary, syncarpous, bilocular with many ovules in each locule on axile placentation; placenta swollen style long; stigma bifid.

Fruit: A capsule

Seed: Non endospermic

Floral formula: † ♀ $K_{(5)}\ C_{(5)}\ A_4\ \underline{G}_{(2)}$

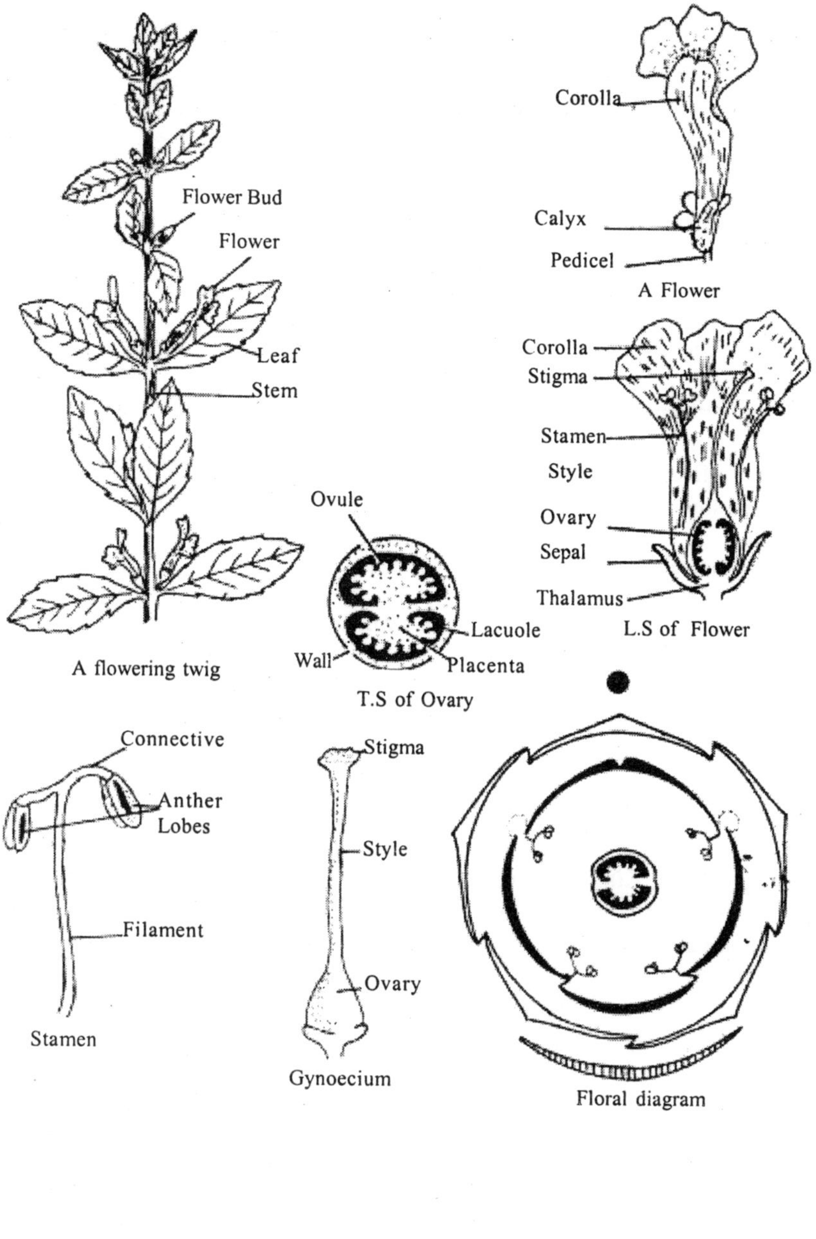

Fig 8.115 ***LINDENBERGIA INDICA*** **(Scrophulariaceae)**

ANTIRRHINUM MAJUS

(Scrophulariaceae)

Habitat : Mesophyte, cultivated, grows in moist places.

Habit : An annual herb

Root : Tap root, branched

Stem : Aerial, erect, branched, cylindrical, herbaceous, green, hairy and solid.

Leaf : Ramal and cauline, simple, sessile, exstipulate opposite in the lower part and alternate in the upper part of the plantbody, smooth or hairy, linear lanceolate, entire, acute, venation reticulate unicostate.

Inflorescence : A simple raceme

Flower : Shortly pedicellate, bracteate, ebracteolate, cyclic, complete, pentamerous, zygomorphic, bilipped, bisexual and hypogynous.

Calyx : Sepals 5, gamosepalous, pubescent, persistent; aestivation imbricate

Corolla : Petals 5, gamopetalous bilabiate with upper lip formed by 2 petals and lower lip by 3 petals, personate, variousluy coloured, anterior petal gibbous; aestivation imbricate.

Andriecium : Stamens 4, fifth one absent or represented by a staminode, didynamous, alternipetalous and epipetalous, inserted at the base of the corolla tube; filaments unequal; anthers dithecous, basifixed, introrse with longitudinal splitting.

Gynoecium : Ovary superior, bicarpellary, syncarpous, bilocular with many ovules in each locule on axile placenta. Placenta swollen, style simple; stigma bifid and capitate

Fruit : A capsule

Seed : Endospermic, scanty endosperm

Floral formula : ✝ ⚥ $K_{(5)}\ C_{(5)}\ A_4\ \underline{G}_{(2)}$

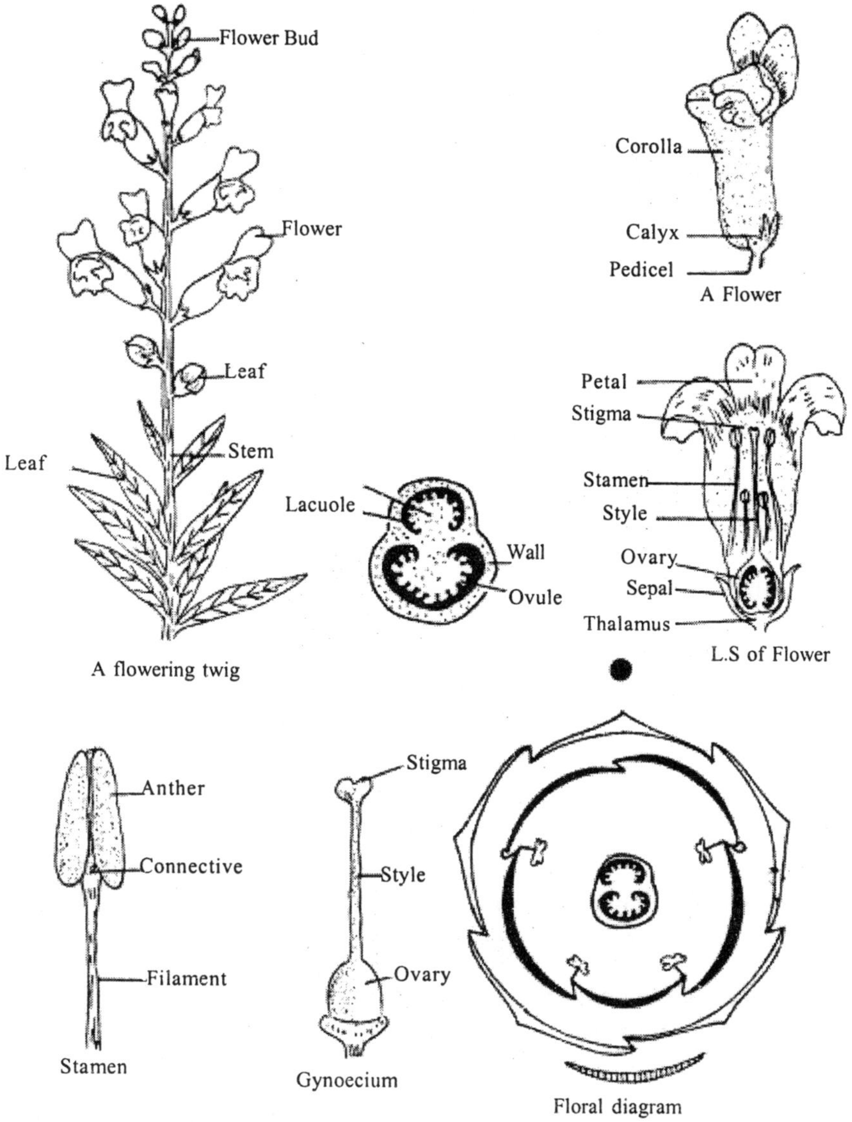

Fig 8.116 *ANTIRRHINUM MAJUS* (Scrophulariaceae)

RUSSELIA EQUISETIFORMIS

(Scrophulariaceae)

Habitat : Mesophyte, oftem cultivated also as an ornamental.

Habit : A perennial shrub

Root : Branched tap root

Stem : Aerial, erect, cylindrical, woody, branched, glabrous and green.

Leaf : Ramal and cauline, simple, exstipulate; opposite, ternate or whorled, ovate, serrate, acute; venation reticulate unicostate.

Inflorescence : Axillary cymose panicle

Flower : Pedicellate, bracteate, ebracteolate, cyclic, complete, pentamerous, heterochlamydeous, zygomorphic, slightly bilabiate, bisexual and hypogynous.

Calyx : Sepals 5, gamosepalous, lobes ovate; green; aestivation imbricate.

Corolla: Petals 5, gamopetalous, tubular, tube long, cylindrical, 2 lipped; upper lip 2 lobed, lower lip 3 lobed, red coloured; aestivation imbricate

Androecium : Stamens basically one staminode, fertile stamens alternipetaluns, epipetalous, didynamous; filaments unequal; anthers dithecous, basifixed introrse with longitudinal splitting.

Gynoecium : Ovary superior, bicarpellary, syncarpous, bilocular with many ovules in each locule on axile placenta. Placenta slightly swollen; style long; stigma linear.

Fruit : A capsule

Seed : Scantily endospermous

Floral formula : † ⚥ $K_{(5)}\ C_{(5)}\ A_4\ \underline{G}_{(2)}$

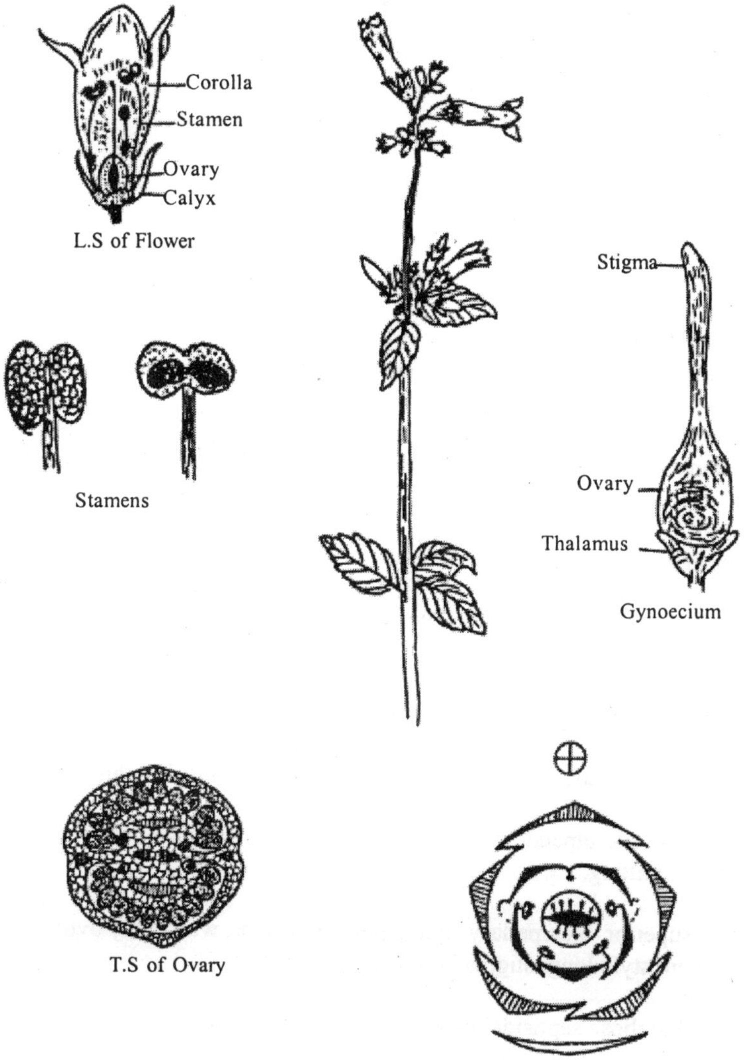

Fig 8.117 *RUSSELIA EQUISETIFORMIS* (Scrophulariaceae)

SCOPARIA DULCIS

(Scrophulariaceae)

Habitat: Mesophyte, weed, grows in open dry land.

Habit: A perennial stout herb or undershrub.

Root: Branched tap root

Stem: Aerial, erect, angled, branched, somewhat woody, glabrous or sparsley hairy and green.

Leaf: Ramal and cauline, simple, opposite and decustate at the base of the stem, but ternate or whorled apically, exstipulate, pinnately divided, elliptic lanceolate or elliptic obovate, serrate, acute; venation reticulate unicostate.

Inflorescence: Flowers solitary or in pairs, axillary

Flower: Pedicellate, bracteate, ebracteolate, complete, cyclic, tetramerous, heterochlamydeous, actinomorphic, bisexual and hypogynous.

Calyx: Sepals 4, gamosepalous, oblong, green, aestivation imbricate.

Corolla: Petals 4, gamopetalous, corolla tube short, densley wolly inside, white, aestivation imbricate.

Androecium: Stamens 4, alternipetalous, epipetalous, filaments unequal; anthers dithecous, dorsifixed, introrse with longitudinal splitting.

Gynoecium: Ovary superior, bicarpellary, syncarpous, bilocular, with many ovules per locule on axile placenta; placenta swollen; style long; stigma notched.

Fruit: A globose or subglobose septisidal capsule with persistent stigma.

Seed: Scantily endospermons, pitted.

Floral formula: ✝ ♀ $K_{(4)}\ C_{(4)}\ A_4\ \underline{G}_{(2)}$

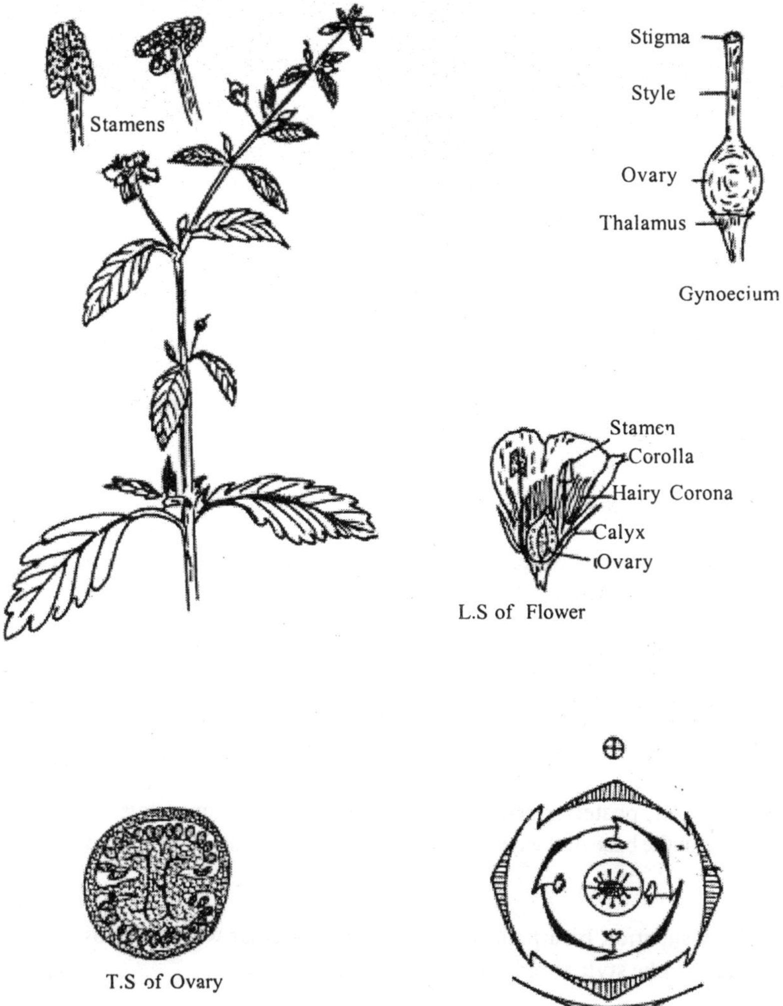

Fig 8.118 *SCOPARIA DULCIS* (Scrophulariaceae)

BACOPA MONNIERI

(Scrophulariaceae)

Habitat : Mesophyte, frequent in wet or marshy places.

Habit : A prostrate herb with ascending branches.

Root : Branched tap root.

Stem : Aerial, partially erect, cylindrical, soft smooth, solid, branched and green.

Leaf : Ramal and cauline, sessile, opposite, simple, exstipulate, entire, oblong or obovate, thick, venation reticulate unicostate; veins obscure.

Inflorescence : Flowers solitary, axillary

Flower : Pedicellate, bracteate, bracteolate, cyclic, complete, pentamerous, heterochlamydeous, zygomorphic, bilabiate, bisexual and hypogynous.

Calyx : Sepals 5, gamosepalous, tubular, green, aestivation imbricate.

Corolla : Petals 5, gamopetalous, bilipped, upper lip exterior in bud,made up of 2 petals, lower lip made up of 3 petals, white or lilac; aestivation imbricate

Androecium : Stamens 4, posterior stamen absent alternate with the petals, epipetalous, didynamous; filaments short, inserted at the base of the corolla tube; anthers dithecous, basifixed, introrse with a longitudinal dehiscence.

Gynoecium : Ovary superior, bicarpellary, syncarpous, bilocular with many ovules in each locule on a fleshy, swollen axile placenta; style simple; stigma capitate.

Fruit : A capsule, style persistant

Seed : Scantily endospermic, reticulate

Floral formula : † ♀ $K_{(5)}\ C_{(5)}\ A_4\ \underline{G}_{(2)}$

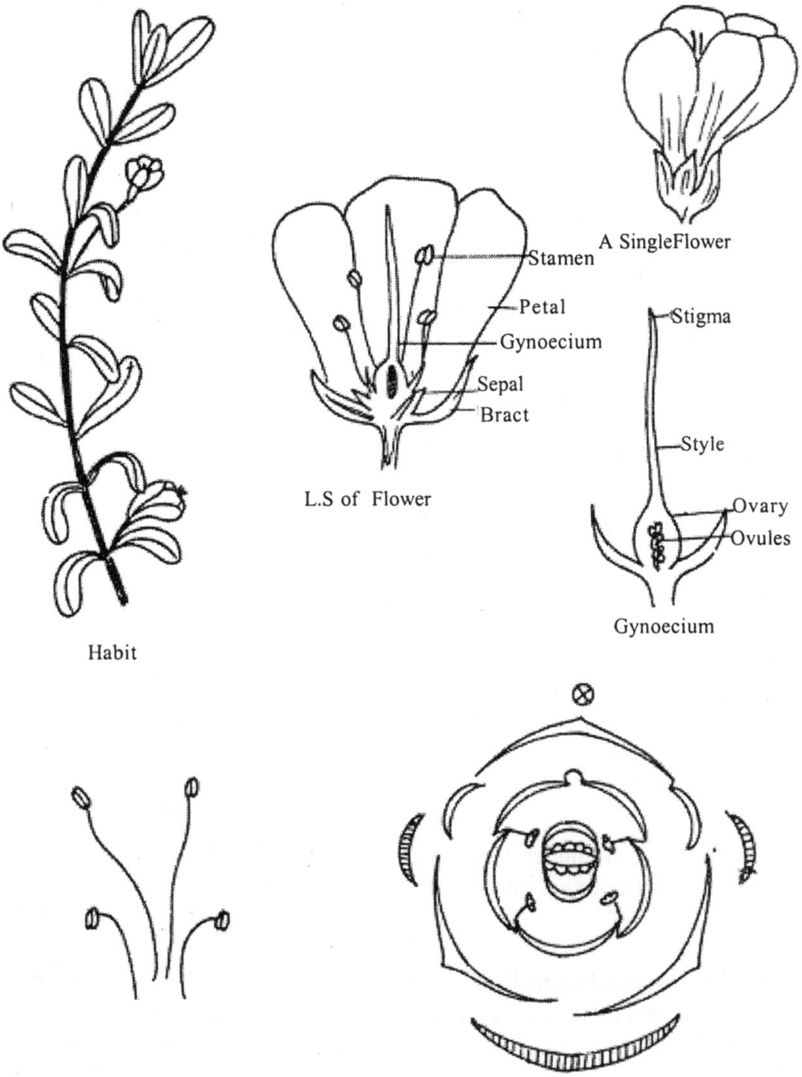

Fig 8.119 ***BACOPA MONNIERI*** **(Scrophulariaceae)**

ADHATODA VASICA

(Acanthaceae)

Habitat: Mesophyte, evergreen, grows in open dry land.

Habit: A perennial shrub

Root: Branched tap root

Stem: Aerial, erect, herbaceous above, woody below, green, branched, cylindrical, nodes swollen, solid and glabrous.

Leaf: Ramal and cauline, simple, opposite decussate, exstipulate, lanceolate, entire, acute, glabrous venation reticulate unicostate.

Inflorescence: A spike

Flower: Sessile, bracteate, bracteolate, (bracteoles leafy) cyclic, complete, pentamerous, heterochlamydeous, zygomorphic, bisexual and hypogynous.

Calyx: Sepals 5, polysepalous (slightly connate at the base) green; aestivation quincuncial

Corolla: Petals 5, gamopetalous, tubular, tube short, bilipped (upper lip 2, lower lip3), coronary outgrowths present at the throat of the corolla, yellowish white in colour, personate; aestivation valvate.

Androecium: Stamens 2, staminodes 3, polyandrous, alternate with petals, epipetalous; filaments thick and long; anthers dithecous, basifixed, introrse, spurred; anther lobes unequal in size; anther dehiscence by vertical splits.

Gynoecium: Ovary superior, bicarpellary, syncarpous, bilocular, with two ovules in each locule on axile placenta. A disc is present below the ovary; style long and curved; stigma capitate.

Fruit: A capsule

Seed: Endospermic, usually rugose.

Floral formula: † ⚥ $K_{(5)}\ C_{(5)}\ A_2\ \underline{G}_{(2)}$

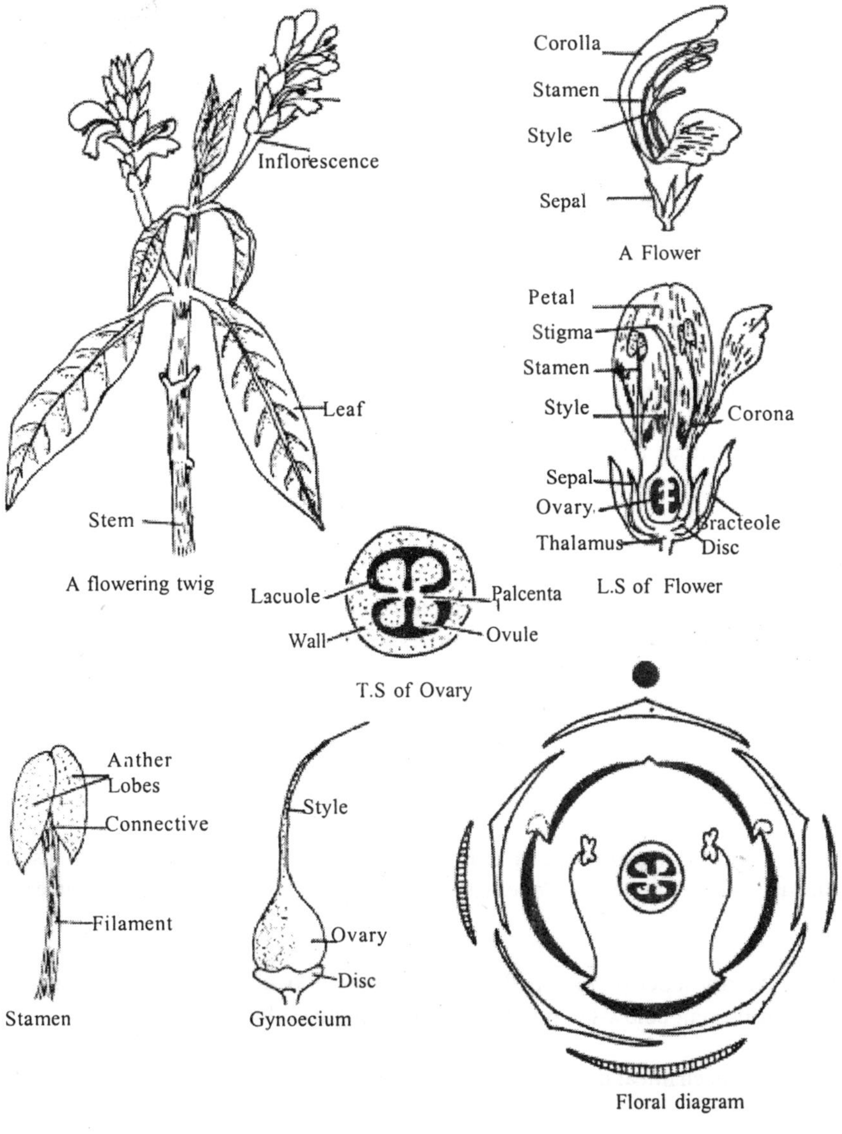

Fig 8.120 ***ADHATODA VASICA*** **(Acanthaceae)**

RUELLIA TUBEROSA

(Acanthaceae)

Habitat: Mesophyte, wild, grows in moist places

Habit: Annual herb

Stem: Branched tap root

Stem: Aerial, erect, soft, green, branched, swollen at the nodes, hairy, cylindrical and solid.

Leaf: Ramal and cauline, opposite decussate, simple, petiolate, exstipulate, ovate with sinuous margin, acute, glabrous; venation reticulate unicostate.

Inflorescence: A dichasial cyme

Flower: Pedicellate, bracteate, bracteolate, complete, cyclic, pentamerous, heterochlamydeous, zygomorphic funnel shaped, bluish pink in colour bisexual and hypogynous.

Calyx: Sepals 5, gamosepalous, persistent; aestivation valvate.

Corolla: Petals 5, gamopetalous, funnel shaped, bluish pink in colour, blipped (upper 2 lobes, lower 3) aestivation twisted.

Androecium: Stamens 4, polyandrous, alternipetalous, epipetalous, didynamous, inserted at the middle of the corolla tube; anthers dithecous, lobes situated at unequal levels, basifixed, introrse with a longitudinal splitting.

Gynoecium: Ovary superior, bicarpellary, syncarpous, bilocular with one ovule in each locule on axile placenta; a disc is present below the ovary; style long; stigma bilobed.

Fruit: A capsule; mechanical dehiscence, splits open with a noise that can be heard.

Seed: Scantily endospermic

Floral formula: † ♀ $K_{(5)}\ C_{(5)}\ A_4\ \underline{G}_{(2)}$

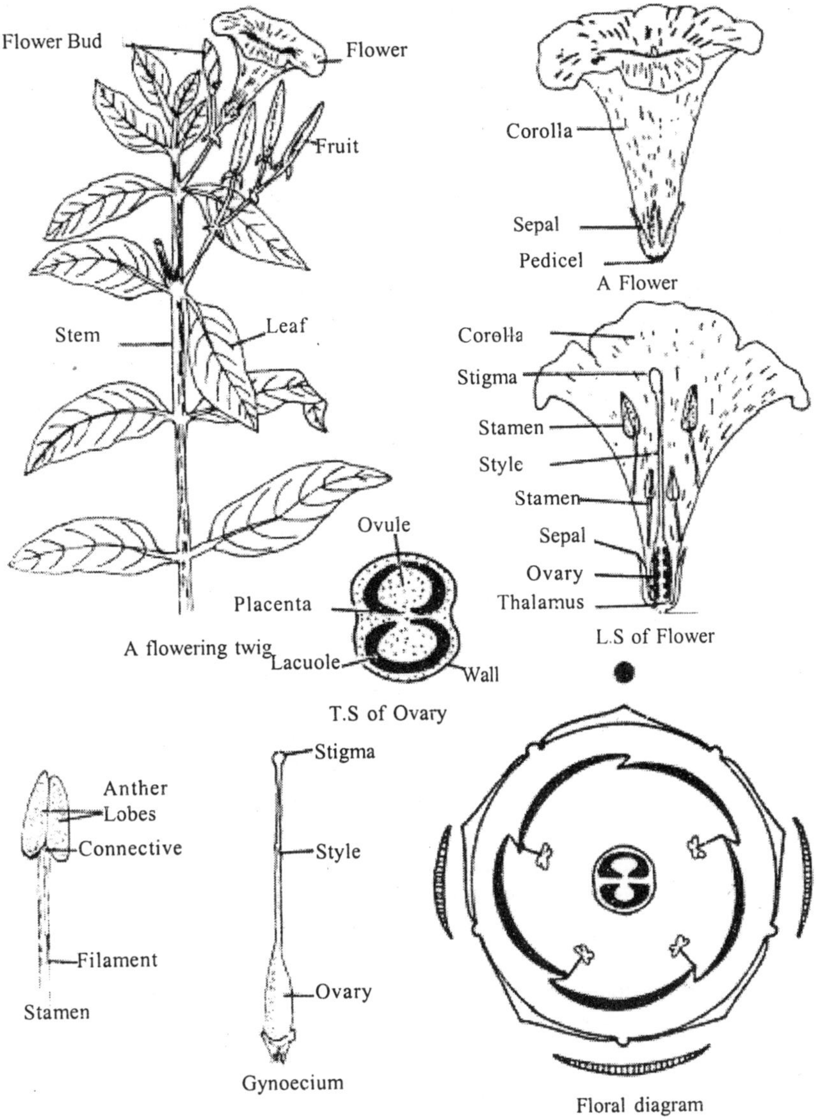

Fig 8.121 *RUELLIA TUBEROSA* (Acanthaceae)

JUSTICIA GENDURASSA

(Acanthaceae)

Habitat: Mesophyte, wild, often cultivated as hedge plant.

Habit: A perennial undershrub

Root: Branched tap root.

Stem: Aerial, erect, branched, herbaceous generally or lower parts may be woody, cylindrical nodes swollen and flat, greenish brown, solid and glabrous.

Leaf: Ramal and cauline, simple, petiolate, opposite decussate, stipulate, lanceolate, entire, acute, glabrous; venation reticulate unicostate.

Inflorescence: Basically a dichasial cyme, arranged in a racemose fashion.

Flower: Pedicellate, bracteate, bracteolate, cyclic, complete, pentamerous, heterochlamydeous, zygomorphic, bisexual and hypogynous. Flowers yellowish white in colour.

Calyx: Sepals 5, gamosepalous; aestivation valvate.

Corolla: Petals 5, gamopetalous, bilipped (upper lip 2 lobed and lower 3 lobed), yellowish white in colour aestivation imbricate.

Androecium: Stamens 2, free, anterior in position epipetalous, inserted halfway in the corolla tube, filaments long; anthers dithecous, basifixed, introrse, lobes situated at unequal heights, lower lobe appendicular (has an appendage); deliscence longitudinal.

Gynoecium: Ovary superior, bicarpellary, syncarpous, bilocular, with one ovule in each locule on axile placenta. Style long; stigma simple.

Fruit: Capsule

Seed: Endospermous

Floral formula: † ♀ $K_{(5)}\ C_{(5)}\ A_2\ \underline{G}_{(2)}$

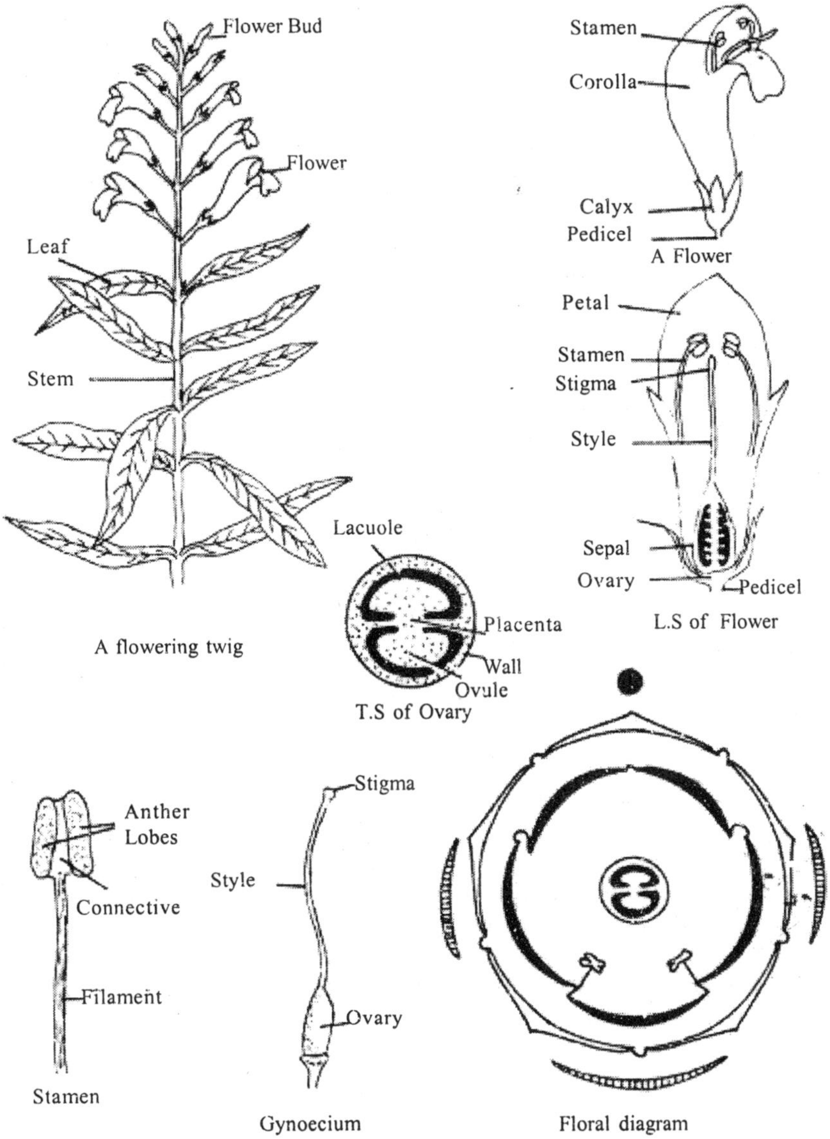

Fig 8.122 *JUSTICIA GENDURASSA* (Acanthaceae)

PERISTROPHE BICALYCULATA

(Acanthaceae)

Habitat : Mesophyte, wild, found in open places.

Habit : A perennial herb

Root : Branched tap root

Stem : Aerial, erect, angular, herbaceous, nodes swollen, solid, hairy and puberulous, green and branched.

Leaf : Ramal and cauline, simple, opposite decussate, petiolate, exstipulate, ovate, entire, acute, hairy; venation reticulate unicostate.

Inflorescence : A dichasial or polychasial cyme.

Flower : Pedicellate, bracteate, bracteolate, with a whorls of 5 bracteoles, complete, cyclic, pentamerous, heterochlamydeous, zygomorphic, bisexual and hypogynous.

Calyx : Sepals 5, polysepalous, slightly connate at the base, whitish, odd posterior petal larger than the rest; aestivation valvate.

Corolla : Petals 5, gamopetalous, bilipped (upper 2 lobes, lower 3 lobes), reddish violet in colour, aestivation valvate.

Androecium : Stamens 2, free, epipetalous, inserted halfway in the corolla tube; filaments long and hairy; authers dithecous, basifixed introrse; anther lobes situated at different levels.

Gynoecium : Ovary superior, bicarpellary, syncarpous, bilocular with two ovules in each locule on axile placenta. Style long and hairy; stigma bifid.

Fruit : Capsule

Seed : Endospermic

Floral formula : † ♀ $K_{(5)}\ C_{(5)}\ A_2\ \underline{G}_{(2)}$

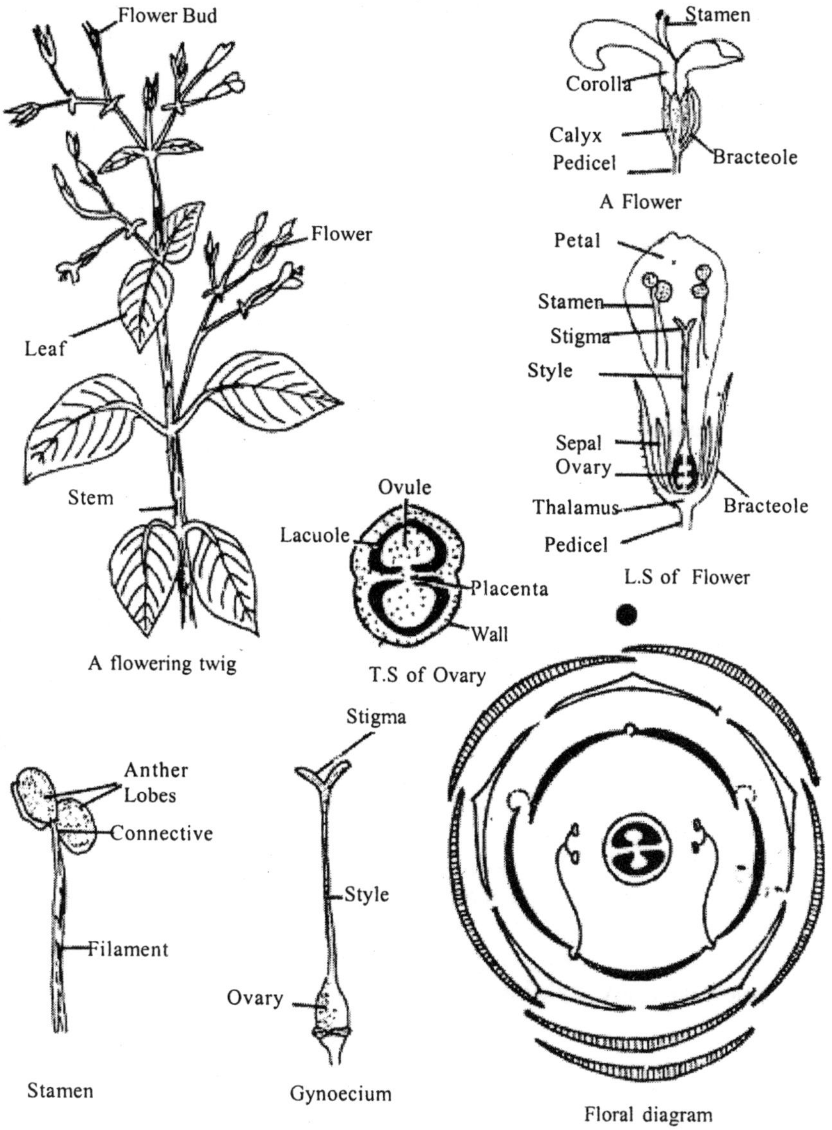

Fig 8.123 ***PERISTROPHE BICALYCULATA*** **(Acanthaceae)**

BARLERIA PRIONITIS

(Acanthaceae)

Habitat : Mesophyte, wild, grows in dry areas.

Habit : A perennial and thorny undershrub

Root : Branched tap root

Stem : Aerial, erect, woody, angular, solid, glabrous, green and branched.

Leaf : Ramal and cauline, simple, petiolate, opposite decussate, stipulate, stipules interpetiolar and modified into spines, lanceolate, entire, acuminate, venation reticulate unicostate.

Inflorescence : Terminal spike.

Flower : Sessile, beacteate, bracteolate, complete, cyclic, pentamerous, heterochlamydeous, zygomorphic, bisexual yellow and hypogynous.

Calyx : Sepals 4, polysepalous arranged in 2 whorls of 2 each; aestivation imbricate

Corolla : Petals 5, gamopetalous, bilipped (upper 4, lower 1 lobed), yellow; aestivation imbricate.

Androecium : Stamens 4, free, alternipetalous, epipetalons, didynamous, 2 posterior stamens sterile, filaments long; anthers dithecous, versatile and introrse.

Gynoecium : Ovary superior, bicarpellary, syncarpous, bilocular with a single ovule in each locule on axile placenta. Disc present below the ovary; style long; stigma bifid.

Fruit : A capsule

Seed : Endospermic

Floral formula : † ♀ $K_4\ C_{(5)}\ A_4\ \underline{G}_{(2)}$

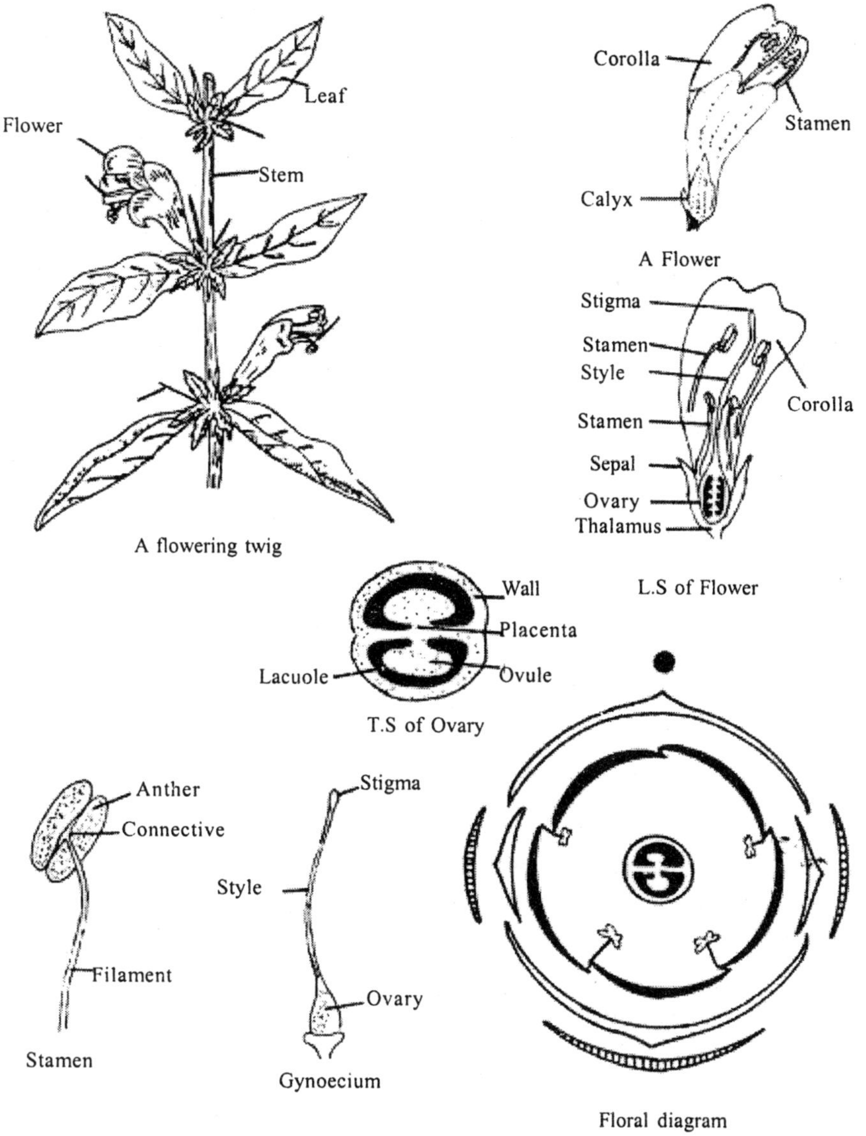

Fig 8.124 *BARLERIA PRIONITIS* (Acanthaceae)

THUNBERGIA ALATA

(Acanthaceae)

Habitat : Mesophyte, wild, grows in open forests.

Habit : A perennial twiner

Root : Branched tap root

Stem : Aerial, weak, herbaceous, green, twining and branched.

Leaf : Ramal and cauline, simple, opposite, exstipulate, ovate, entire, acute, venation reticulate unicostate, petiole is winged.

Inflorescence : Flowers solitary and axillary

Flower : Pedicellate (with along pedicel), breacteate, (leafy bracts), bracteolate (large bracteoles), cyclic, complete, pentamerous, heterochlamydous, zygomorphic, bisexual and hypogynous.

Calyx : Petals 5, gamopetalous, tubular, bright yellow with a dark centre; aestivation twisted.

Androecium : Stamens 4, free, epipetalous and alternipetalous, posterior stamen usually abortive; didynamous; anthers dithecous, lobes slightly at unequal levels, basifixed and introrse. Anther lobes spurred below and have glandular hairs.

Gynoecium : Ovary superior, bicarpellary, syncarpous, bilocular with two ovules in each locule on axile placenta. Style long; stigma funnel shaped.

Fruit : A pubescent capusule, splits into two valves.

Seed : Endospermic

Floral formula : † ♀ K_{10} $C_{(5)}$ A_4 $\underline{G}_{(2)}$

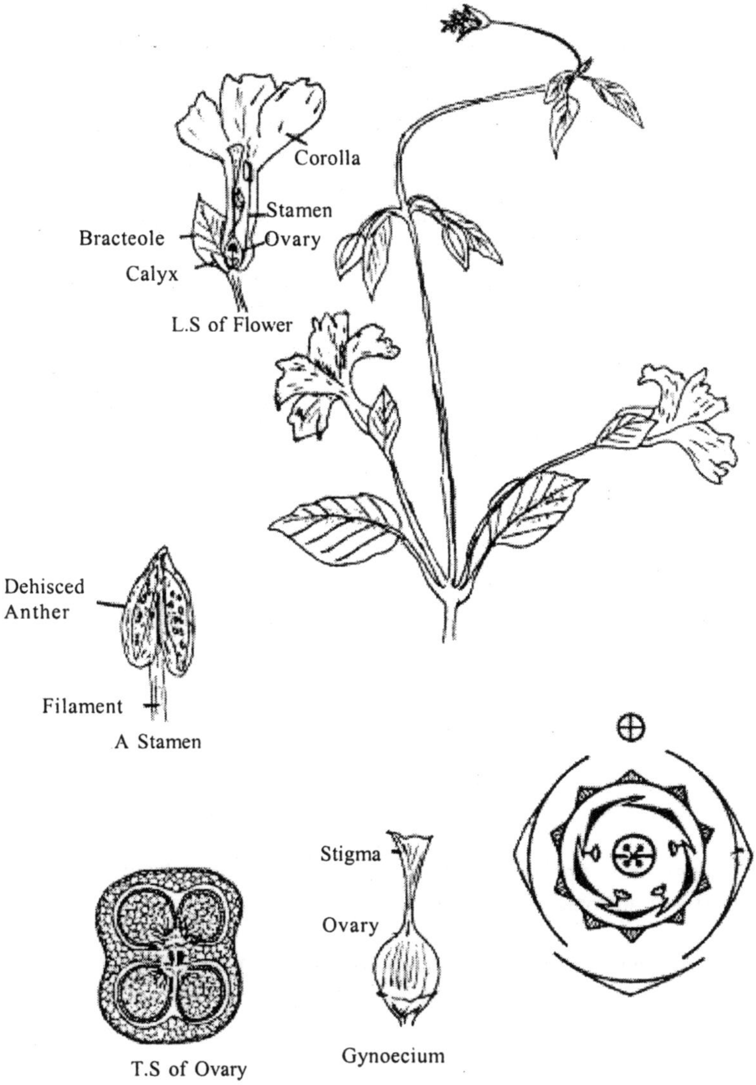

Fig 8.125 *THUNBERGIA ALATA* (Acanthaceae)

DURANTA REPENS

(Verbenaceae)

Habitat: Mesophyte, commonly cultivated as hedge plant.

Habit: Perennial shrub

Root: Branched tap root.

Stem: Aerial, erect, woody, branched, green, cylindrical and solid.

Leaf: Ramal and cauline, simple, subsessile, opposite decussate, exstipulate, ovate, denate, acute; venation reticulate unicostate.

Inflorescence: Racemose panicle, axillary.

Flower: Pedicellate, bracteate, bracteolate, cyclic, complete, pentamerous, heterochlamydeous, slightly zygomorphic, bisexual, pinkish blue and hypogynous.

Calyx: Sepals 5, gamosepalous; aestivation valvate

Corolla: Petals 5, gamopetalous, hypocrateriform, bilipped (upper 2 lobed, lower 3 lobed), anterior 3 petals (of the lower lip) large, blue, corolla tube hairy inside aestivation quincuncial

Androecium: Stamens 4, free, eipetalous, didynamous, inserted almost near the throat of the corolla tube; anthers dithecous, basifixed, introrse and sagittate.

Gynoecium: Ovary superior, bicarpellary, syncarpous, bilocular with two ovules in each locule on parietal placenta. The parietal placenta appears at later stages of the ovular growth as axile due to inward growth. Style short; stigma capitate.

Fruit: A drupe

Seed: Endospermic

Floral formula: † ⚥ $K_{(5)}\ C_{(5)}\ A_4\ \underline{G}_{(2)}$

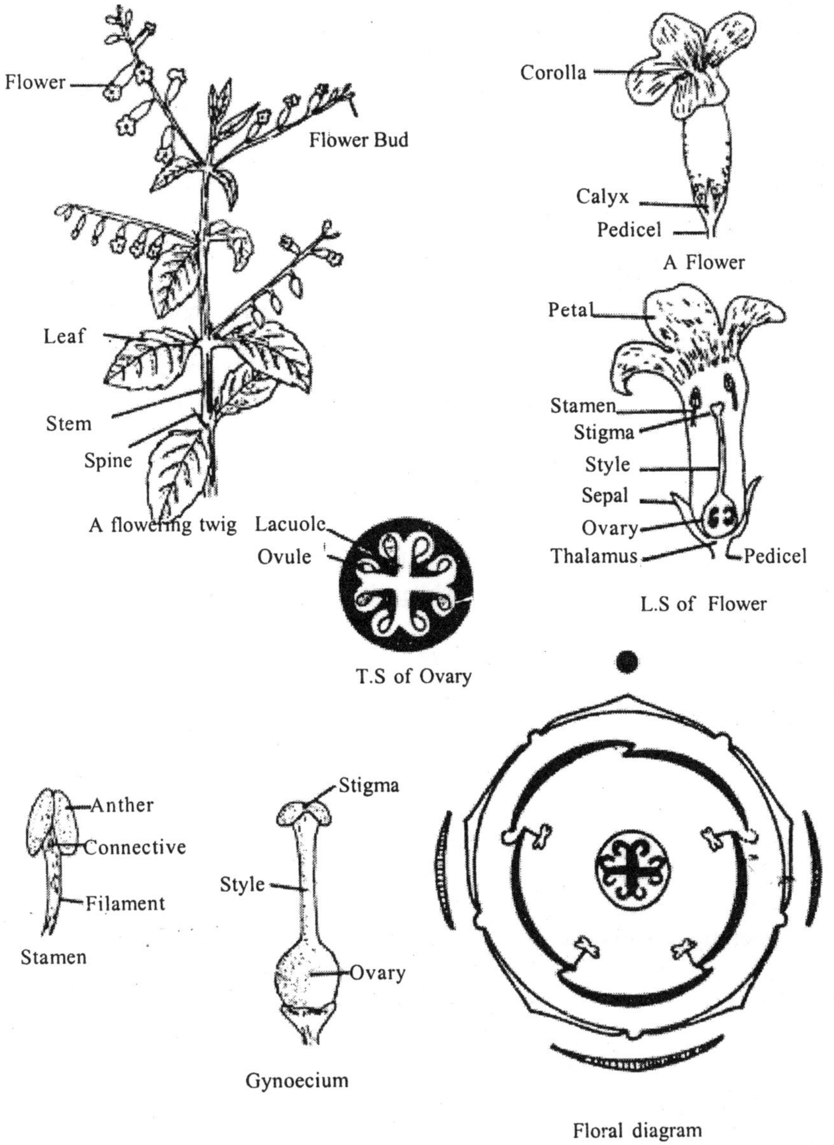

Fig 8.126 ***DURANTA REPENS*** **(Verbenaceae)**

LANTANA INDICA

(Verbenaceae)

Habitat : Wild, shrub, grows in dry land.

Habit : A perennial shrub

Root : Branched tap root

Stem : Aerial, erect, woody in lower parts, herbaceous in upper parts, branched, quandrangular, rough, greyish green, prickly with recurved prickles, hairy and solid.

Leaf : Ramal and cauline, simple, petiolate, opposite decussate, exstipulate, ovate, dentate, acute, coriaceous, darkgreen, rough surface venation reticulate unicostate.

Inflorescence : Head or spike, axillary

Flower : Sessile, bracteate, bracteolate, complete, cyclic, pentamerous, heterochlamydeous, zygomorphic, bisexual and hypogynous. Colour of the flowers, white, pink or yellow.

Calyx : Sepals 5, gamosepalous; aestivation valvate.

Corolla : Petals 5, gamopetalous, slightly bilabiate (upper lop 4, lower lip 1 lobed), anterior petal large, tubular, coronary outgrowth present at the throat of the corolla tube; aestivation quincuncial

Androecium : Stamens 4, free, alternipetalous, epipetalous, didynamous, inserted halfway in the corolla tube; anthers dithecous, adnate, basifixed and introrse

Gynoecium : Ovary superior, bicarpellary, syncarpous, bilocular, with one ovule in each locule on axile placenta. Style simple; stigma knob like.

Fruit : A drupe

Seed : One in the fruit, non endospermic

Floral formula : † ♀ $K_{(5)}\ C_{(5)}\ A_4\ \underline{G}_{(2)}$

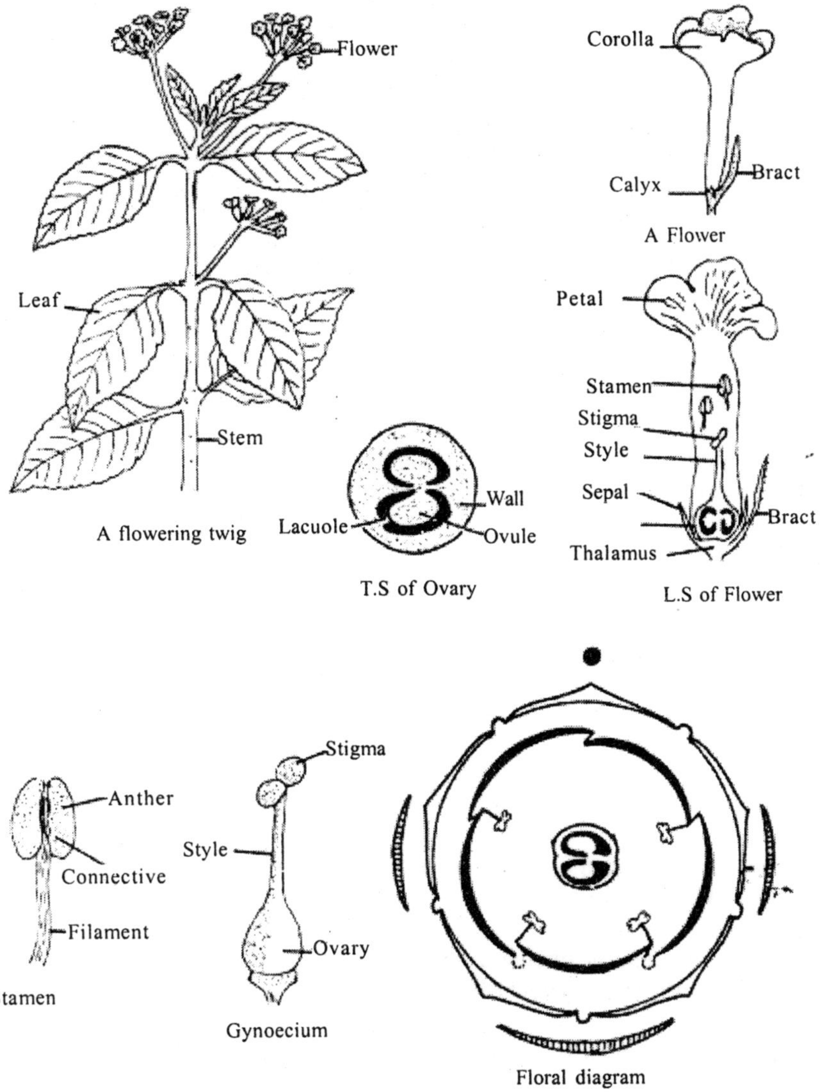

Fig 8.127 *LANTANA INDICA* (Verbenaceae)

STACHYTARPETA INDICA

(Verbenaceae)

Habitat: Mesophyte, wild, grows in dry open land

Habit: An annual shrub

Root: Branched tap root

Stem: Aerial, erect, woody, branched, quandrangular, solid, greyish green and pubescent.

Leaf: Ramal and cauline, simple, petiolate, stipulate, opposite decussate, ovate, crenate, acute, base obtuse; venation reticulate unicostate.

Inflorescence: Spike, terminal in position

Flower: Pedicellate, bracteate, ebracteolate, cyclic, complete, pentamerous, heterochlamydeous, zygomorphic bisexual and hypogynous. Flowers violet coloured

Calyx: Sepals 5, gamosepalous, membranous; aestivation valvate.

Corolla: Petals 5, gamopetalous, bilabiate (upper 2, lower 3 lobed), tubular, salvar shaped, deep violet in colour; aestivation valvate.

Androecium: Stamens 2, free, epipetalous inserted at the base of corolla tube staminodes 3, anthers bithecous, basifixed and introrse.

Gynoecium: Ovary superior, bicarpellary, syncarpous, bilocular with only one ovule in each locule on axile placenta. Style long and filiform; stigma thick.

Fruit: Drupe, included in the persistent calyx

Seed: Non endospermic

Floral formula: † ♀ $K_{(5)}\ C_{(5)}\ A_2\ \underline{G}_{(2)}$

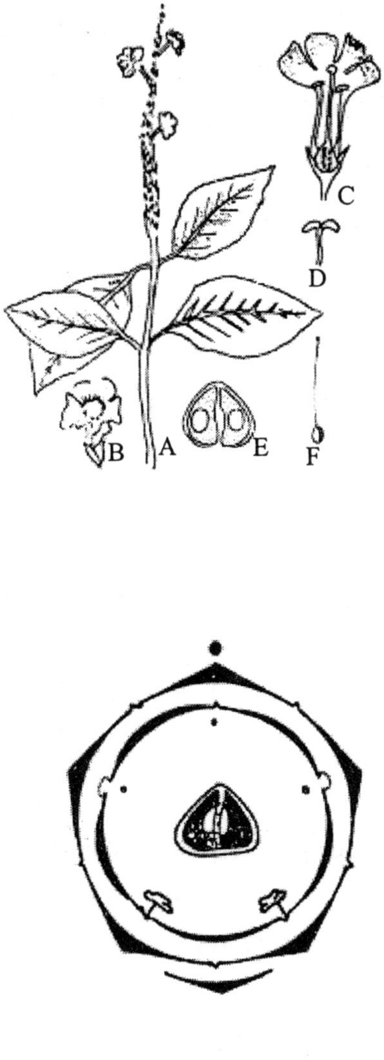

Fig 8.128 ***STACHYTARPETA INDICA*** **(Verbenaceae)**

TECTONA GRANDIS

(Verbenaceae)

Habitat: Mesophyte, deciduous tree grows in open dry deciduous forests; often cultivated.

Habit: A large tree growing to a height of 90 feet

Root: Branched tap root

Stem: Aerial, erect, woody, hard, hairy, cylindrical or squarish angled.

Leaf: Ramal and cauline, simple, very large, petiolate (petiole, up to 6 cm long and stout), opposite or ternate, elliptic obovate, entire, acute, truncate at base, rough and hairy; venation reticulate unicostate.

Inflorescence: Large terminal hairy cymose panicles.

Flower: Sessile, bracteate, bracteolate, complete, cyclic, pentamerous, actinomorphic, bisexual and hypogynous.

Calyx: Sepals 5-6 gamosepalous, bell shaped, hairy, accrescent, enlarged and looks like bladder in fruit and encloses it, ovoid; aestivation valvate.

Corolla: Petals 5-6, gamopetalous, white, unequal, aestivation imbricate.

Androecium: Stamens 5, free, alternipetalous, epipetalous, inserted at the base of the corolla tube; anthers dithecous and introrse

Gynoecium: Ovary superior, bicarpellary, syncarpous, tetralocular with one ovule in each locule on axile placenta. Style slendor, stigma bifid.

Fruit: A subglobose drupe enclosed in inflated calyx, fruit wall studded with stellate hairs

Seed: Non endospermic

Floral formula: ✝ ♀ $K_{(5-6)}\ C_{(5-6)}\ A_5\ \underline{G}_{(2)}$

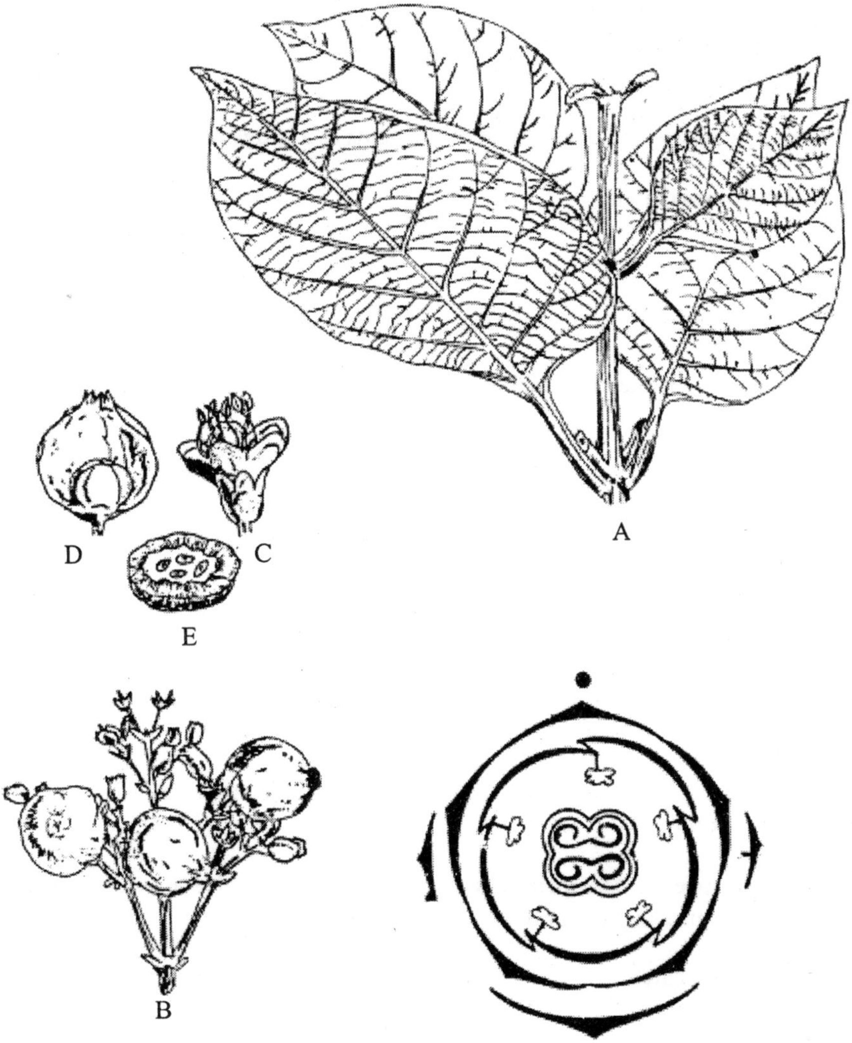

Fig 8.129 *TECTONA GRANDIS* (Verbenaceae)

VITEX NEGUNDO

(Verbenaceae)

Habitat : Mesophyte, grows wild, cultivated in dry fields as a hedge plant

Habit : A large shrub or small tree

Root : Branched tap root.

Stem : Aerial, erect, cylindrical, hard at base, soft above and covered with fine white pubescence.

Leaf : Ramal and cauline, palmately compound, (leaflets 3-5), exstipulate, obovate, entire, acute; venation reticulate unicostate.

Inflorescence : Cymose panicle

Flower : Pedicellate, bracteate, bracteolate, complete, cyclic, pentamerous, heterochlamydeous, zygomorphic, bisexual and hypogynous.

Calyx : Petals 5, gamopetalous, cupular, toothed, aestivation valvate

Corolla : Petals 5, gamopetalous, bilabiate, tube short, with a ring of hairs at the base of the stamens, upper lip 2 lobed, lower lip 3 lobed; aestivation imbricate.

Androecium : Stamens 4, free, didynamous, alternipetalous and epipetalous, filaments smooth or hairy; anthers dithecous and extrorse.

Gynoecium : Ovary superior, bicarpellary, syncarpous, 4 locular with one ovule in each locule on axile placenta. Style simple; stigma slightly bifid.

Fruit : Drupe, enclosed in calyx.

Seed : Non endospermic

Floral formula : ✝ ♀ $K_{(5)}\ C_{(5)}\ A_5\ \underline{G}_{(2)}$

Fig 8.130 ***VITEX NEGUNDO*** **(Verbenaceae)**

CLERODENDRON INERME

Habitat : Mesophyte, grows wild, often cultivated as a hedge plant.

Habit : A perennial shrub with aerid smell.

Stem : Aerial, erect, herbaceous above, woody below slightly angular, branched, solid, pubescent and green.

Leaf : Ramal and cauline, simple. opposite decussate, exstipulate, petiolate, elliptical, entire obtuse, pubescent; venation reticulate unicostate. Leaves gland dotted.

Inflorescence : Axillary dichasial cyme, occasionally terminal

Flower : Pedicellate, bracteate, bracteolate, cyclic, complete, pentamerous, heterochlamydeous, zygomorphic, bisexual and hypogynous.

Calyx : Sepals 5, gamosepalous, hairy; aestivation valvate.

Corolla : Petals 5, gamopetalous, hypocrateriform, tube long, limbs spread out; aestivation imbricate corolla slightly bilipped and white in colour.

Androecium : Stamens 4, free, alternipetalous, epipetalous, didynamous; filaments long, exserted; anthers dithecous, dorsifixed and introrse.

Gynoecium : Ovary superior, bicarpellary, syncarpous, bilocular (rarely tetralocular) with one ovule in each locule on parietal placenta. Style simple, long; stigma bifid.

Fruit : A drupe

Seed : Non endospermic

Floral formula : † ♀ $K_{(5)}\ C_{(5)}\ A_4\ \underline{G}_{(2)}$

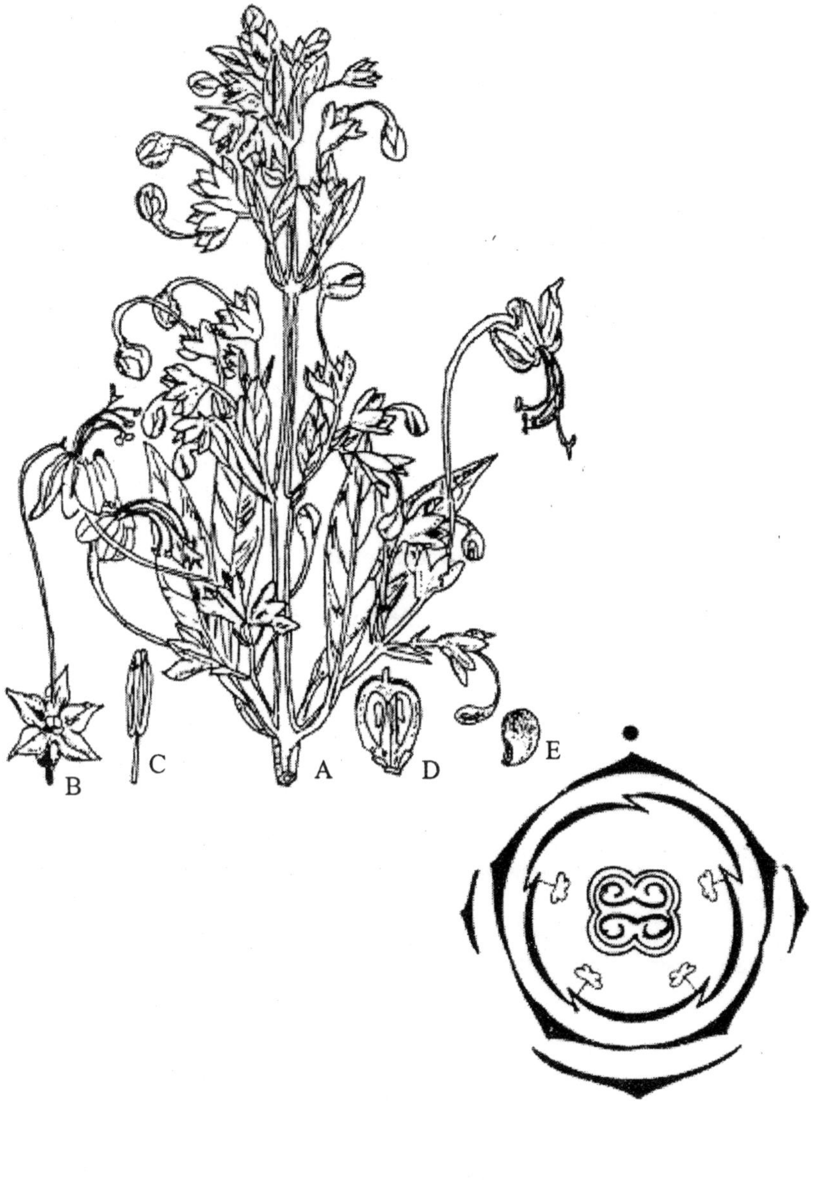

Fig 8.131 *CLERODENDRON INERME*

LIPPIA NODIFLORA

(= Phyla Nodiflora)

(Verbenaceae)

Habitat : Mesophyte, generally grows in moist soil.

Habit : Slender diffuse herbs often rooting at lower nodes.

Root : Tap root at the base of the stem. Adventitious roots arise from the lower nodes occasionally.

Stem : Aerial, procumbent, soft, cylindrical, solid, glabrous, branched and green.

Leaf : Ramal and cauline, simple, opposite decussate, petiolate or sub sessile, obovate, coarsely serrate towards the apex; venation reticulate unicostate.

Inflorescence : A condensed terminal spike, flowers minute.

Flower : Sessile, bracteate, ebracteolate, cyclic, complete, pentamerous, heterochlamydeous, zygomorphic, bisexual and hypogynous.

Calyx : Sepals 5, small, fused, membranous, deeply bipartite; aestivation valvate.

Corolla : Petals 5, gamopetalous, bilabiate, upper lip 2 lobed rounded or emarginate, lower lip 3 lobed with the middle lobe being large. Aestivation imbricate.

Androecium : Stamens 4, free, alternipetalous and epipetalous, didynamous, filaments short; anthers dithecous and introrse.

Gynoecium : Ovary superior, bicarpellary, syncarpous, bilocular with one ovule in each locule on axile placenta.

Fruit : Drupe

Floral formula : † ♀ $K_{(5)}\ C_{(5)}\ A_4\ \underline{G}_{(2)}$

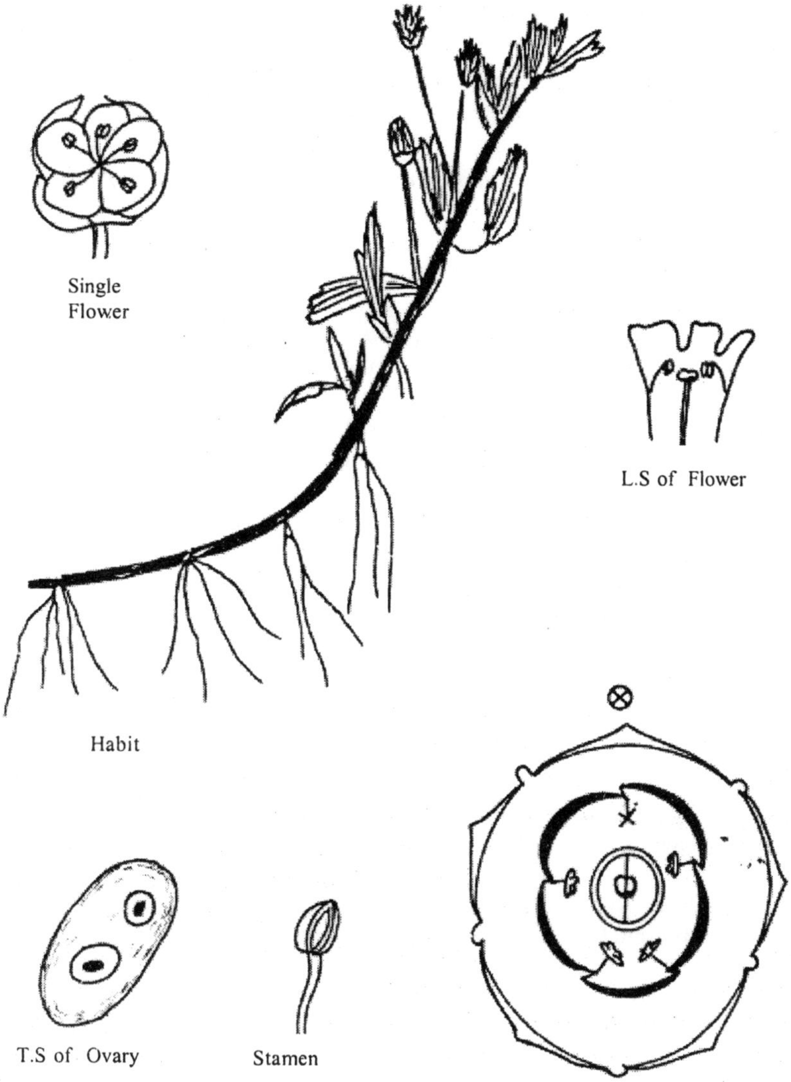

Fig 8.132 ***LIPPA NODIFLORA (= Phyla Nodiflora)*** **(Verbenaceae)**

OCIMUM SANCTUM

(Labiatae)

Habitat : Mesophytic plant, grows in open dry land; also cultivated for its medicinal properties.

Habit : A perennial herb, aromatic

Root : Tap root, branched.

Stem : Aerial, erect, soft, greenish, pubescent, solid, quadrangular, and branched.

Leaf : Ramal and cauline, simple, petiolate, exstipulate, opposite decussate, ovate, serrate, acute, pubescent, gland dotted, aromatic, venation reticulate unicostate.

Inflorescence : Pedicellate, bracteate, ebracteolate, cyclic, complete, heteroclamydeous, pentamerous, zygomorphic, bisexual and hypogynous.

Calyx : Sepals 5, gamosepalous, bilabiate (upper lip one lobe, lower lip four lobes), posterior lobe large, persistent, aestivation valvate or imbricate.

Corolla : Petals 5, gamopetalous, bilabiate (upper lip 4, lower lip one); corolla tube short, pinkish white; aestivation imbricate.

Androecium : Stamens 4, polyandrous, alternipetalous, epipetalous, didynamous; anthers dithecous, dorsifixed and introrse.

Gynoecium : Ovary superior, bicarpellary, syncarpous, tetralocular with one ovule in each locule on axile placenta. A disc is present below the ovary. Style gynobasic, stigma bifid.

Fruit : Carcerculus, consists of four nutlets.

Seed : Non endospermic.

Floral formula : † ♀ $K_{(5)}\ C_{(5)}\ A_4\ \underline{G}_{(2)}$

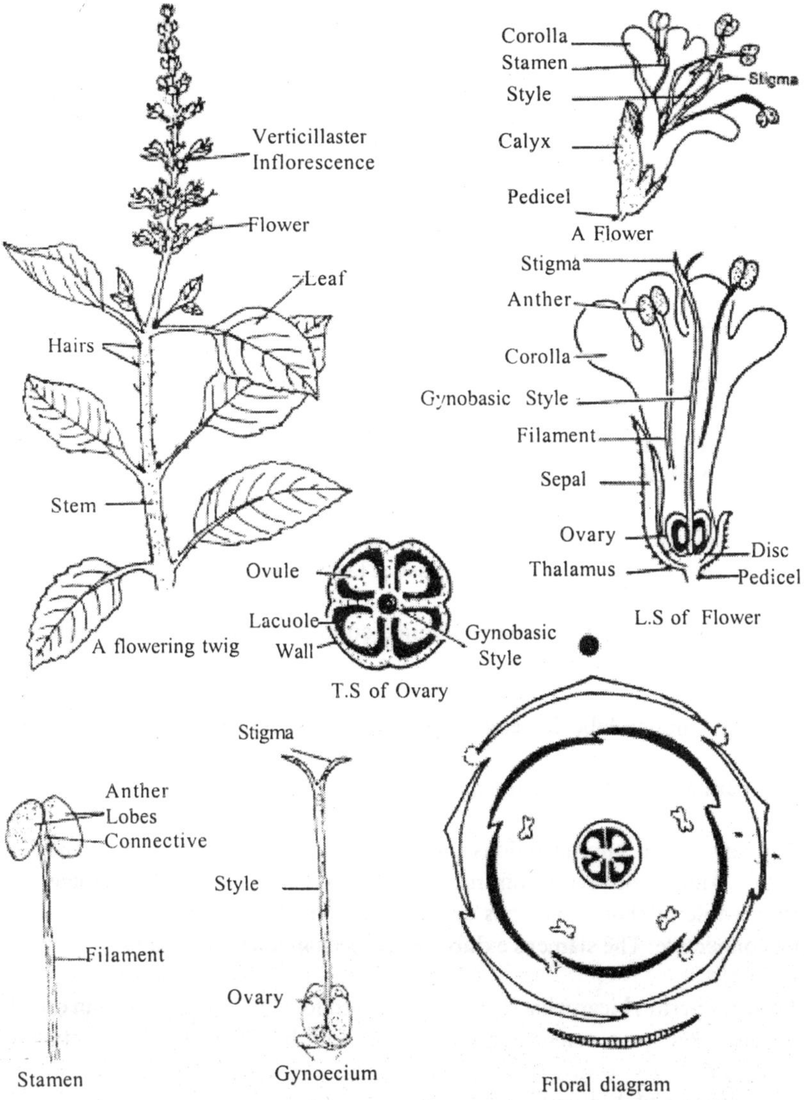

Fig 8.133 *OCIMUM SANCTUM* (Labiatae)

OCIMUM BASILICUM

(Labiatae)

Habitat : Mesophyte, wild, grows in wasteland, often cultivated.

Habit : A tall aromatic herb.

Root : Tap root, branched.

Stem : Aerial, erect, green, branched, pubescent, quadrangular, solid and herbaceous.

Leaf : Ramal and cauline, simple, petiolate, exstipulate, opposite and decussate, ovate, serrate ,acute, hairy, aromatic, gland dotted; venation reticulate unicostate.

Inflorescence : A verticellaster. The inflorescence is present at the nodal regions on either side of the node, there is a dichasial cyme whose branches are monochasial. The two clusters together are called a verticellaster.

Flower : Pedicellate, ebracteate, ebracteolate, cyclic, complete, pentamerous, heterochlamydeous, zygomorphic, bisexual and hypogynous.

Calyx : Sepals 5, bilipped (upper, lobe, lower 4 lobes) tubular, violet green coloured; aestivation valvate.

Corolla : Petals 5, gamopetalous, bilabiate (upper lip 4, lower lip one), pink, aestivation valvate.

Androecium : Stamens 4, free, alternipetalous and epipetalous, didynamous (2 long and 2 short), inserted a little above the ovary. Anthers dithecous introrse, basifixed or versatile. In taller stamens both the lobes of the anther are fertile, while in shorter stamens upper lobe sterile and the lower fertile, the two are being separated by a long connective. The stamens exhibit lever mechanism in pollination.

Gynoecium : Ovary superior, bicarpellary, syncarpous, tetralocular with one ovule in each locule on axile placenta. Style long and gynobasic, comes out between the four locules of the ovary; stigma bifid.

Fruit : Carcerculus, consists of 4 nutlets.

Seed : Non endospermic.

Floral formula : $\dagger$ ⚥ $K_{(5)}\ C_{(5)}\ A_4\ \underline{G}_{(2)}$

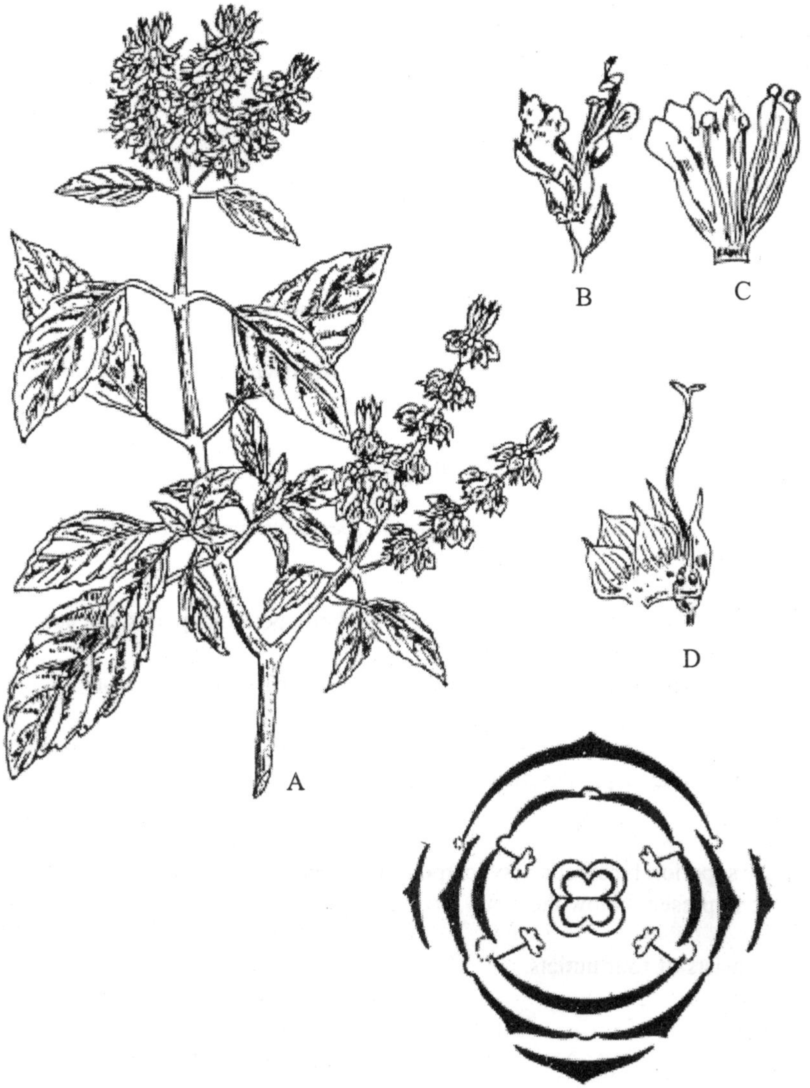

Fig 8.134 *OCIMUM BASILICUM* (Labiatae)

SALVIA OFFICINALIS

(Labiatae)

Habitat : Mesophyte, cultivated.

Habit : An annual herb.

Root : Tap root, branched

Stem : Aerial, erect, soft, branched, green, quadrangular, solid and hairy.

Leaf : Ramal and cauline, simple, petiolate, exstipulate, opposite decussate, ovate, serrate, acute, glabrous; venation reticulate unicostate.

Inflorescence : A verticellaster terminating the stem apex.

Flower : Pedicdellate, bracteate, ebracteolate, complete, cyclic, pentamerous zygomorphic heterochlamydeous, bisexual and hypogynous.

Calyx : Sepals 5, gamosepalous, bilipped (upper 3, lower 2 lobed), petaloid, persistent; aestivation valvate.

Corolla : Petals 5, gamopetalous, bilipped (upper 2, lower 3 lobed), scarlet red; aestivation imbricate.

Androecium : Stamens 2, free, epipetalous, inserted near the mouth of the corolla, anthers dithecous, with one lobe sterile, connective long, separates the two lobes forming a lever mechanism, Anthers versatile and split longitudinally.

Gynoecium : Ovary superior, bicarpellary syncarpous, tetralocular, with one ovule in each locule on axile placenta. A nectar disc in present below the ovary, style gynobasic; stigma bifid.

Fruit : Carcerulus, consists of four nutlets.

Seed : Non endospermic

Floral formula : † ♀ $K_{(5)}\ C_{(5)}\ A_2\ \underline{G}_{(2)}$

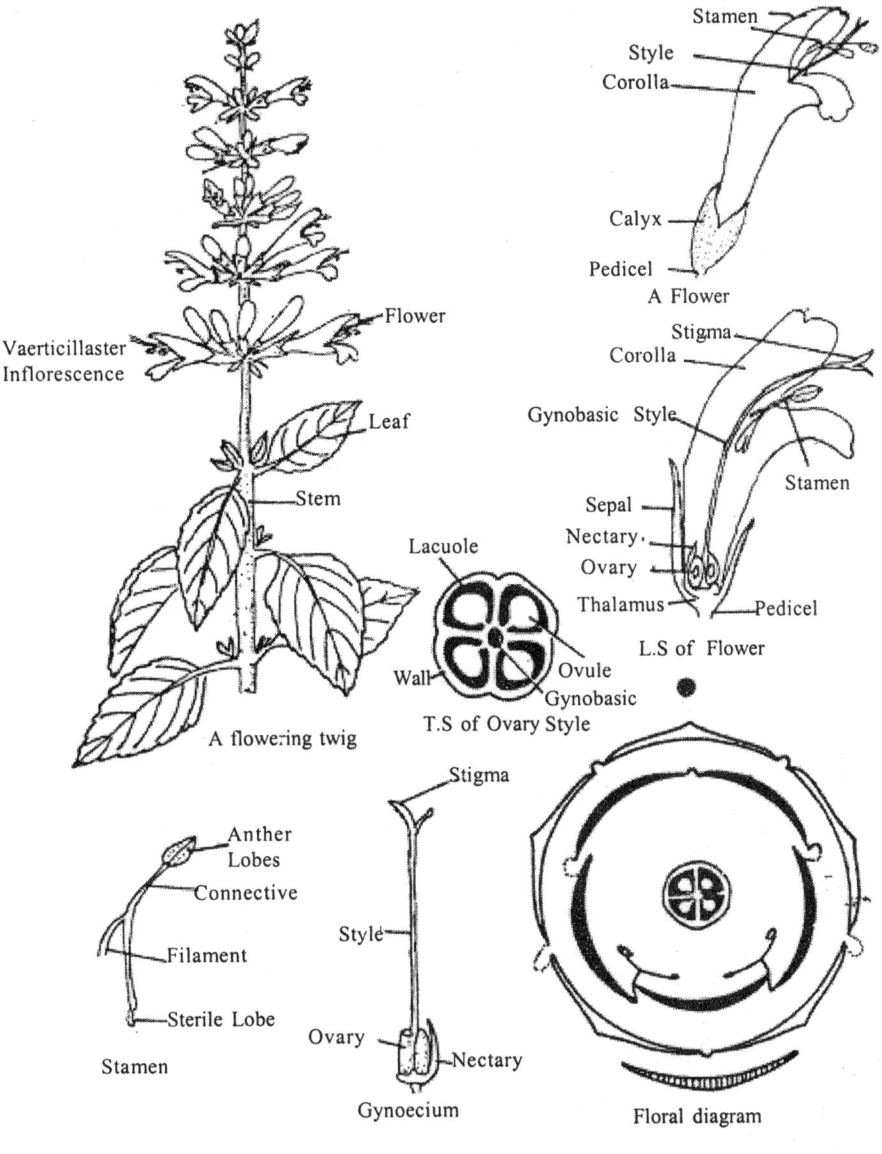

Fig 8.135 *SALVIA OFFICINALIS* (Labiatae)

LEUCAS ASPERA

(Labiatae)

Habitat: Mesophyte, grows in open dry land.

Habit: A small white flowered herb.

Root: Tap root, branched

Stem: Aerial, erect green, pubescent, rough, quadrangular, branched, solid with bicollateral vascular bundles and also four cortical vascular bundles.

Leaf: Ramal and cauline, simple, subsessile, exstipulate, opposite decussate, lanceolate, crenate, acute, pubescent; venation reticulate unicostate.

Inflorescence: A Verticellaster

Flower: Subsessile, bracteate, ebracteolate, complete, cyclic, heterochlamydeous, pentamerous, zygomorphic, bisexual and hypogynous.

Calyx: Sepals 10, gamosepalous, small; aestivation valvate.

Corolla: Petals 5, gamopetalous, bilabiate, (upper 2, lower 3 lobed), white small; aestivation imbricate.

Androecium: Stamens 4, free, epipetalous, didynamous, anthers dorsifixed and introrse

Gynoecium and fruit: Same as in <u>Salvia.</u>

Floral formula: † ♀ $K_{(10)}\ C_{(5)}\ A_4\ \underline{G}_{(2)}$

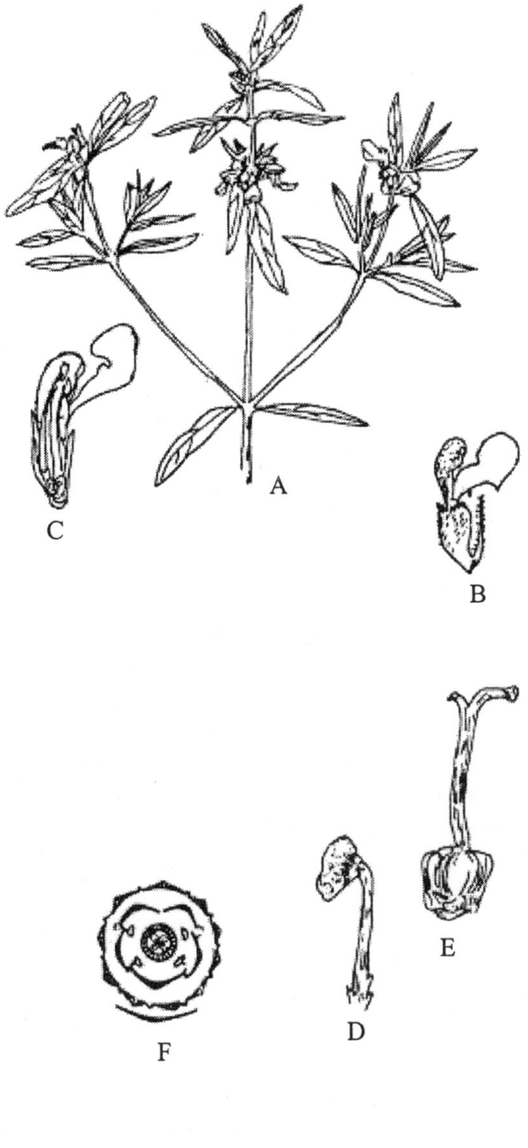

Fig 8.136 ***LEUCAS ASPERA*** **(Labiatae)**

POLYGONUM BARBATUM

(Polygonaceae)

Habitat: Mesophyte, usually grows in moist or mainly areas like drainages.

Habit: An annual herb.

Root: Branched tap root

Stem: Aerial, erect, or rarely prostrate, herbaceous, branched, cylindrical, swollen at nodes, glabrous, green, solid. A part of the stem is an underground shizome.

Leaf: Ramal and cauline, simple, sessile, alternate with achreate stipules, lanceolate, entire, acute; venation reticulate unicostate.

Inflorescence: Racemose panicle

Flower: Pedicellate (short and jointed), bracteate, ebracteolate, cyclic, incomplete, pentamerous, homochlamydeous, actinomorphic, bisexual and hypogynous.

Perianth: Tepals 5, polytepalous, pink, aestivation quincuncial

Androecium: Stamens 7-8, free arranged in 2 whorls, outer whorl of introrse inner whorl of 3 extrorse. Anthers dithecous.

Gynoecium: Ovary superior, tricarpellary, syncarpous, unilocular with a single ovule on basal placenta. A disc is present below the ovary; style 3; stigma capitate.

Fruit: A 3 angled nut

Seed: Non endospermic

Floral formula: † ♀ $P_{(5)}\ A_{7\text{-}8}\ \underline{G}_{(3)}$

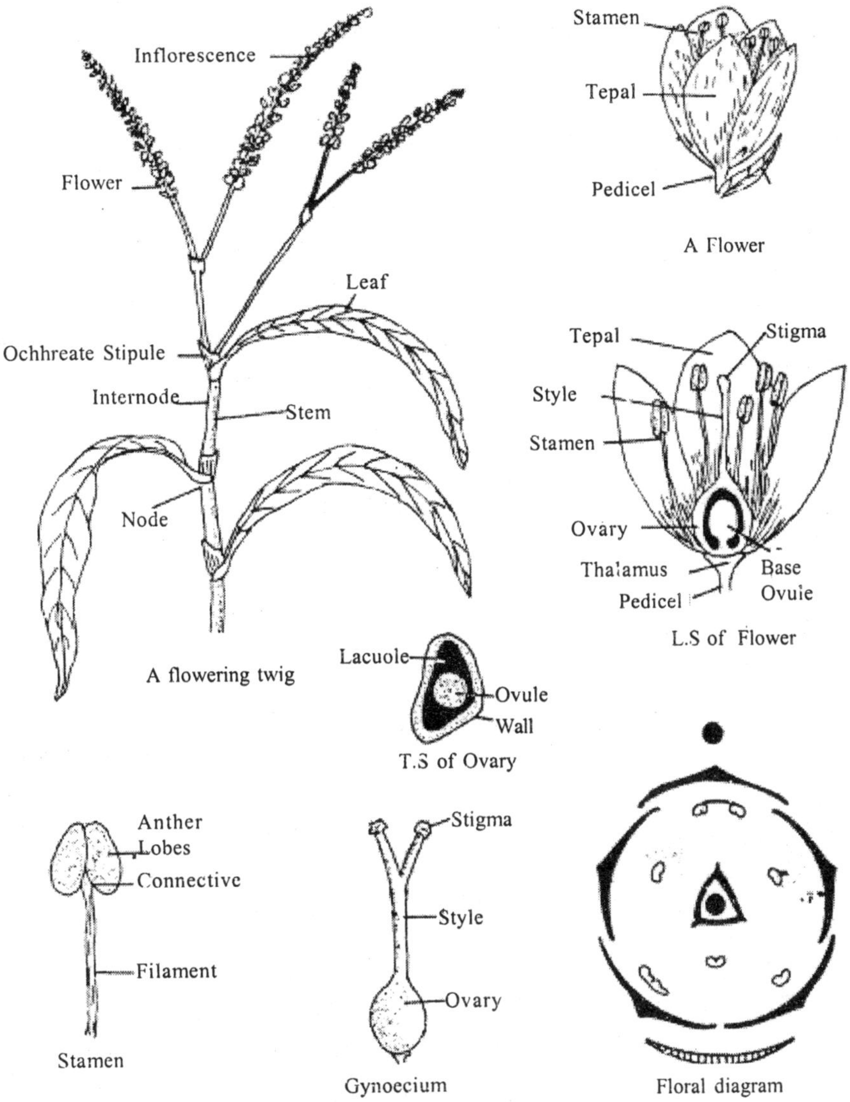

Fig 8.137 ***POLYGONUM BARBATUM*** **(Polygonaceae)**

RUMEX DENTATUS

(Polygonaceae)

Habitat : Mesophyte, weed, grows in moist places.

Stem : Aerial, erect, branched, angular, green, glabrous, and solid.

Leaf : Ramal and cauline, simple, petiolate, alternate, stipulate with ochreate stipules, elliptic lanceolate, undulate, acute glabrous; venation reticulate unicostate

Inflorescence : Racemose pancile

Flower : Pedicellate, ebracteate, ebracteolate, complete, cyclic, trimerous, homochlamydeous, actinomorphic, bisexual and hypogynous.

Perianth : Tepals 6, arranged in two whorls of 3 each, polytepalous; inner whorl large and persistent. Aestivation valvate.

Androecium : Stamens 6 free arranged in 2 whorls of 3 each outer whorl opposite to outer whorl of perianth and inner whorl opposite to inner whorl of perianth, filaments long; anthers dithecous, basifixed and introrse.

Gynoecium : Ovary superior, tricarpellary, syncarpous, unilocular with a single basal ovule; style short, stigma 3 drooping down.

Fruit : A single seeded nut.

Seed : Non endospermic, embryo curved.

Floral formula : ✝ ⚥ $P_{3+3}\ A_{3+3}\ \underline{G}_{(3)}$

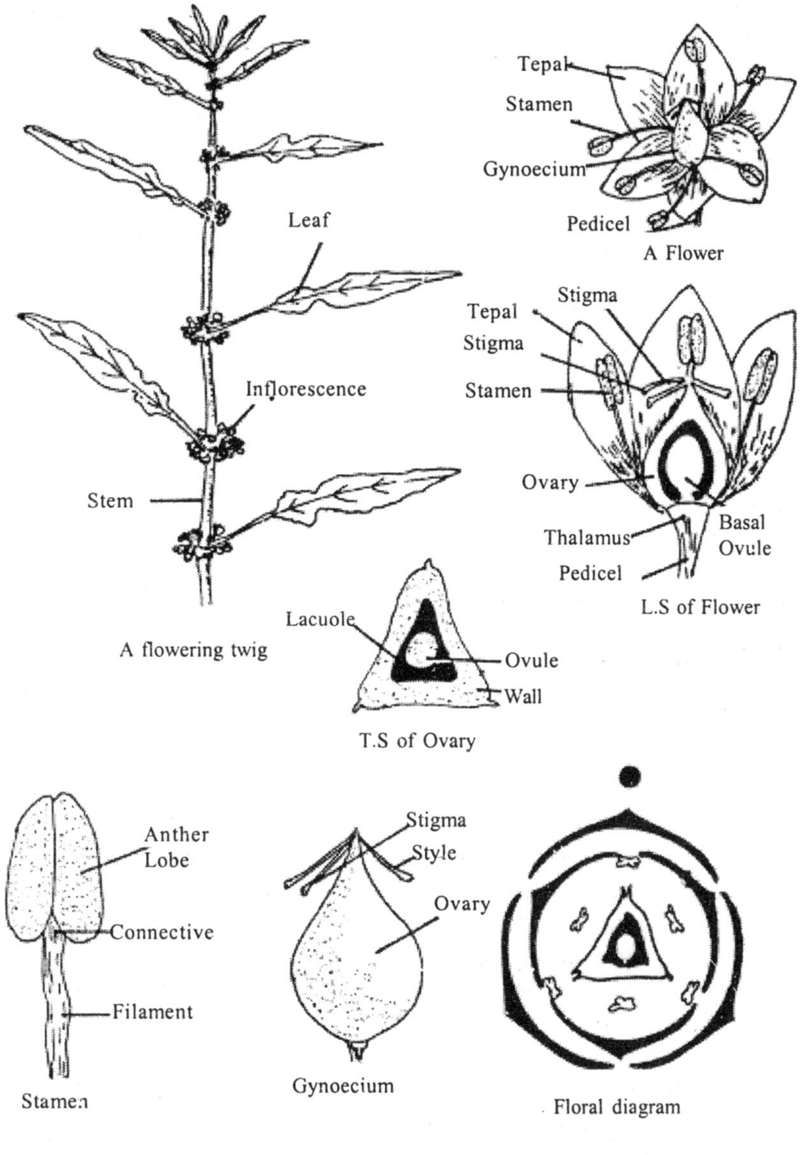

Fig 8.138 ***RUMEX DENTATUS*** **(Polygonaceae)**

ANTIGONON LEPTOPUS

(Polygonaceae)

Habitat: Mesophyte, cultivated as an ornamental climber

Habit: A perennial climbing shrub

Root: A branched tap root.

Stem: Aerial, weak, climbing by tendrils; tendrils formed by peduncle and some terminal flowers, cylindrical, ribbed weakly, branched, solid, rough and green coloured.

Leaf: Ramal and cauline, simple, petiolate, stipulate, stipules reduced to wedge like structures, alternate, cordate, ovate cordate, entire, acuminate, rough; venation reticulate unicostate. Ochreate stipular sheath characteristic of polygonaceae is not found here.

Inflorescence: An axillary raceme; tip of the peduncle and some terminal flowers are modified into tendrils to help in climbing.

Flower: Pedicellate, bracteate, ebracteolate, complete, cyclic, homochlamydeous, zygomorphic, bisexual and hypogynous.

Perianth: Tepals 5, polytepalous, attractive, pink coloured, 3 outer ones large and ovate cordate, 2 inner ones small and long; aestivation quincuncial.

Androecium: Stamens 7-8, monadelphous, arranged in two whorls (outer whorl 5; inner 2-3), filaments papillate; anthers dithecous, dorsifixed and introrse.

Gynoecium: Ovary superior, tricarpellary, syncarpous, unilocular with a single basal ovule. A nectar disc in present below the ovary, styles 3 recurved; stigma 3 and capitate.

Fruit: One seeded nut

Floral formula: † ⚥ $P_{(5)}\ A_{7\text{-}8}\ \underline{G}_{(3)}$

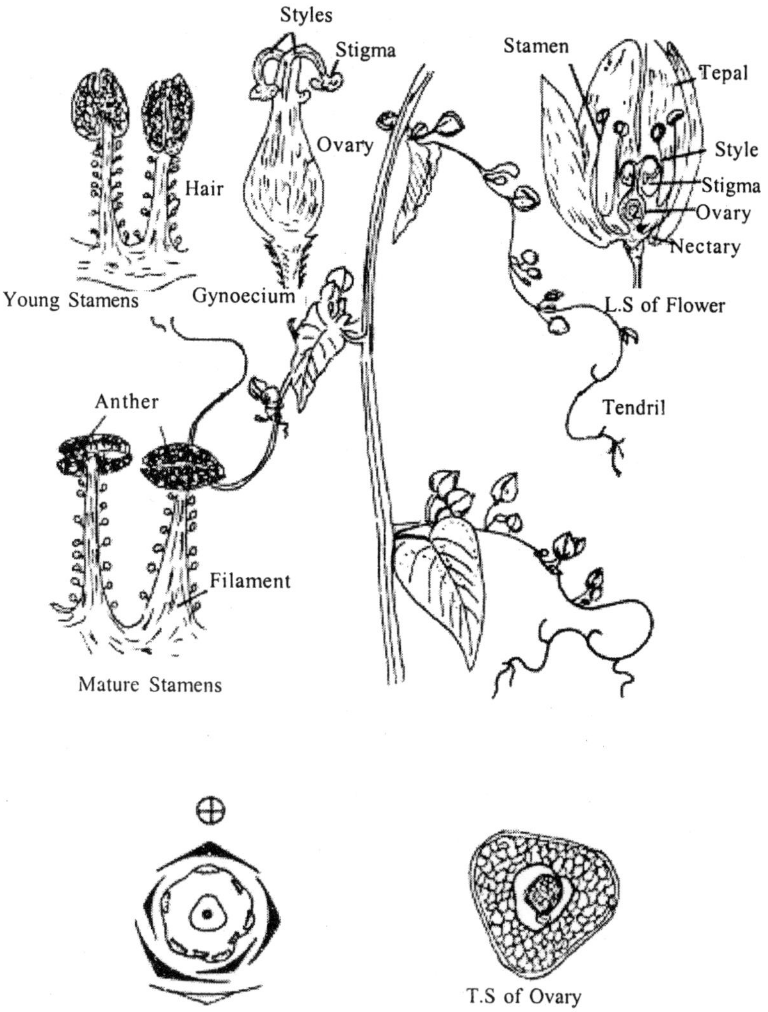

Fig 8.139 *ANTiGONON LEPTOPUS* (Polygonaceae)

ACHYRANTHES ASPERA

(Amaranthaceae)

Habitat : Mesophyte, commonly grows in dry land weed.

Habit : An annual or occassionally perennial herb.

Root : Tap root, branched

Stem : Aerial, erect, branched, cylindrical or upper branches quadrangular, solid, pubescent, herbaceous and green.

Leaf : Ramal and cauline, simple, subsessile, exstipulate, wavy or entire, elliptic ovate, acute, hairy, coriaceous; venation reticulate unicostate.

Inflorescence : Spike, flowers deflexed.

Flower : Sessile, bracteate, bracteolate, (bracts and bracteoles have spinous apex), complete, cyclic, homochlamydeous, actinomorphic, bisexual and hypogynous.

Perianth : Tepals 5, polytepalous, membranous, sepaloid, aestivation quincuncial or contorted.

Androecium : Stamens 10, out of which 5 are sterile, antitepalous, monadelphous forming a membranous tube, the staminodes alternating with the stamens are represented by fimbriate scales; filaments short; anthers bithecous and dorsifixed.

Gynoecium : Ovary superior, bicarpellary, syncarpous, unilocular with a single basal ovule. Style single; stigma capitate.

Fruit : An indehiscent utricle enclosed by the persistant perianth and spiny bracteoles.

Seed : Endospermous, brownish in colour.

Floral formula : ✝ ♀ $P_{(5)}\ A_5\ \underline{G}_{(2)}$

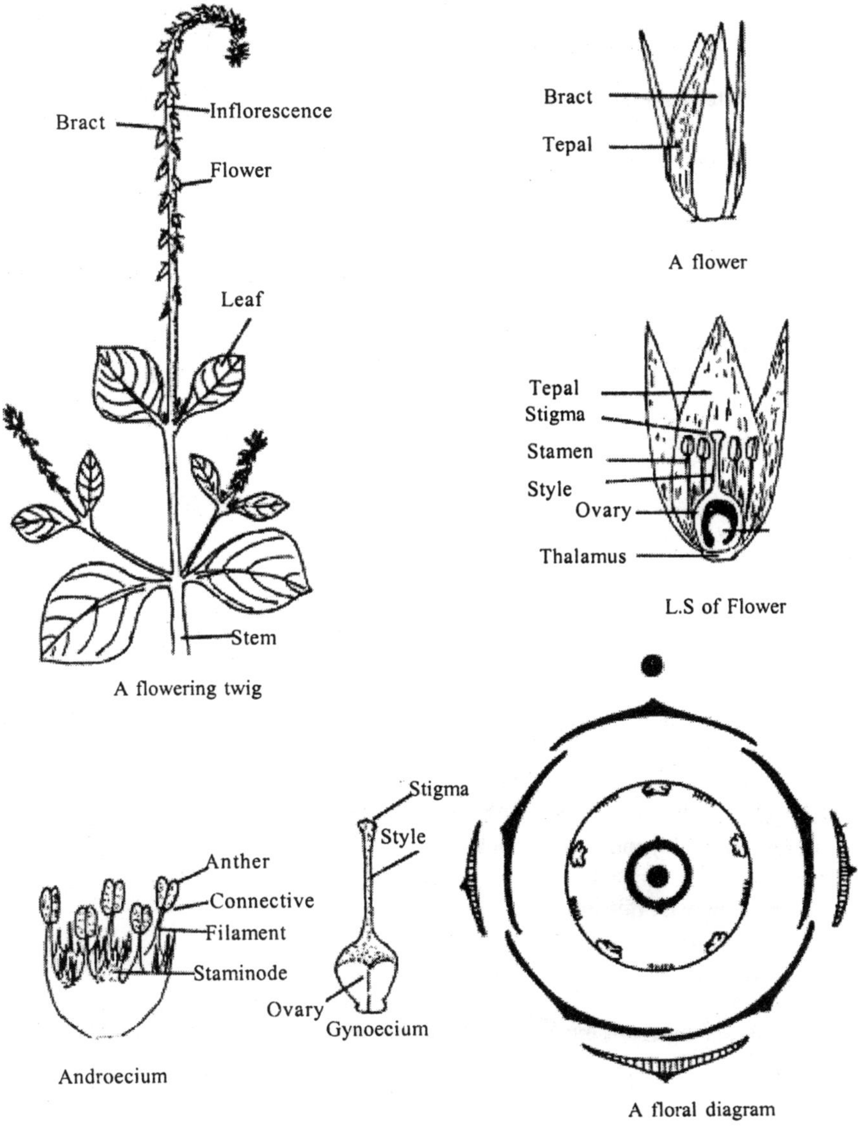

Fig 8.140 ***ACHYRANTHES ASPERA*** **(Amaranthaceae)**

AMARANTHUS SPINOSUS

(Amaranthaceae)

Habitat: Mesophyte, wild, grows in dry open places.

Habit: An annual herb.

Root: Branched tap root

Stem: Aerial, erect, soft, green, angular (terete), Solid, branched, smooth and spiny.

Leaf: Ramal and cauline, simple, petiolate, alternate, exstipulate, ovate, entire, acute; venation reticulate unicostate. There are 2 spines in the axil of each leaf which represent the modified axillary branch.

Inflorescence: A spike, terminal or axillary, flowers unisexual but the plant monoecious.

Male flower: Sessile, bracteate, bracteolate, incomplete, actinomorphic, cyclic, homochlamydeous pentamerous, unisexual and staminate.

Perianth: Tepals 5, polytepalous, membranous and chaffy; aestivation quincuncial

Androecium: Stamens 4-5, free, opposite to the petals, filaments slender and long; anthers dithecous versatile and introrse.

Gynoecium: Absent.

Female flower: Sessile, bracteate, bracteolate, incomplete, homochlamydeous, pentamerous, actinomorphic unisexual, pistillate (female) and hypogynous.

Perianth: Same as in male flowers.

Androecium: Absent

Gynoecium: Ovary superior, bicarpellary, syncarpous, unilocular with a single basal ovule. Style short stigma bifid.

Fruit: A utricle

Seed: Dark, shiny and flat, embryo curved, endosperm absent.

Floral formula: Male flower: † ♀ $P_{(5)} A_4 \underline{G}_0$

Female flower: † ♀ $P_{(5)} A_0 \underline{G}_{(2)}$

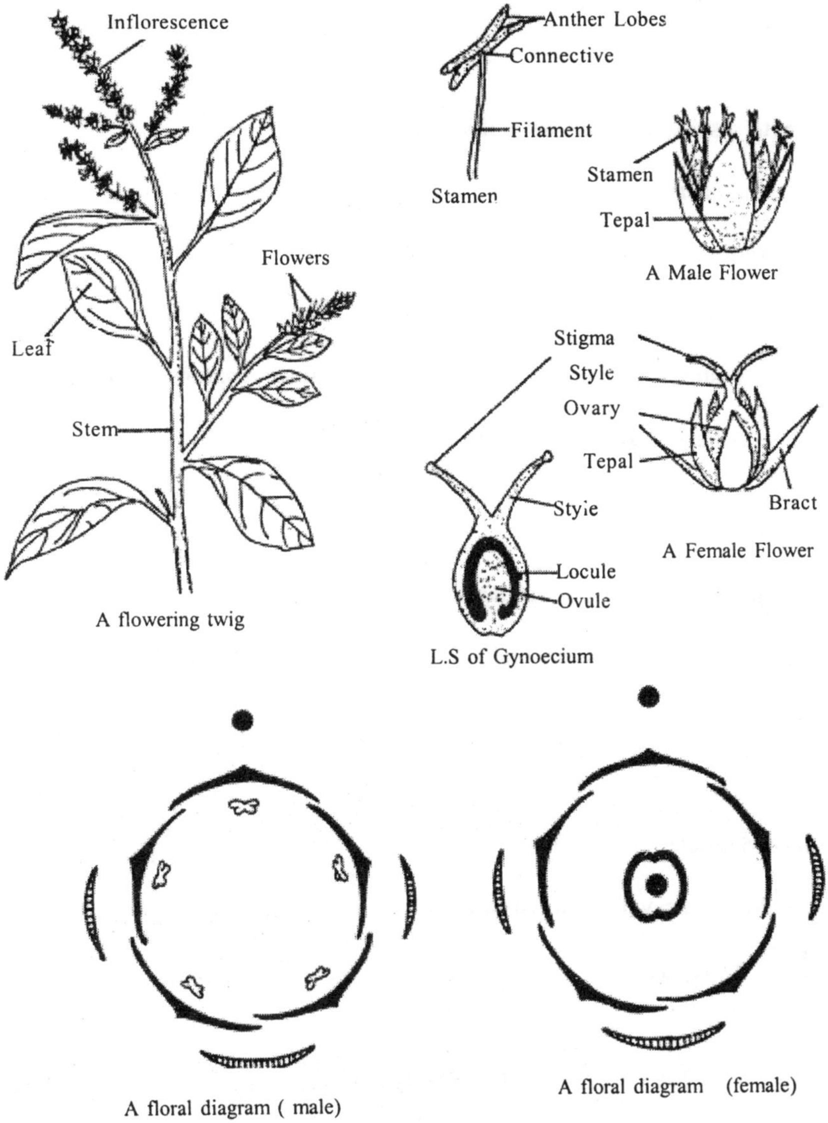

Fig 8.141 ***AMARANTHUS SPINOSUS*** **(Amaranthaceae)**

ALTERNATHERA SESSILIS
(Amaranthaceae)

Habitat: Mesophyte, a spiny weed grows in dry open land.

Habit: A prostrate annual herb.

Root: Branched tap root

Stem: Aerial, erect, herbaceous, green, cylindrical branched, solid and glabrous.

Leaf: Simple, sessile, exstipulate, opposite, ovate lanceolate, entire; venation reticulate unicostate.

Inflorescence: Dense axillary spikes with white flowers.

Flower: Sessile, bracteate, bracteolate, cyclic, complete, pentamerous, homochlamydeous, actinomorphic, bisexual and hypogynous.

Perieanth: Tepals 5, polytepalous, white, aestivation quincuncial.

Androecium: Stamens 3, 2 staminodes (totally 5), free opposite to the tepals; filaments short; anthers dithecous, dorsifixed and introrse.

Gynoecium: Ovary superior, bicarpellary, syncarpous, unilocular with a single basal ovule. Style short, stigma simple.

Fruit: A single seeded utricle

Seed: Dark, non endospermic, with a curved embryo

Floral formula: ✝ ♀ $P_{(5)} A_5 \underline{G}_{(2)}$

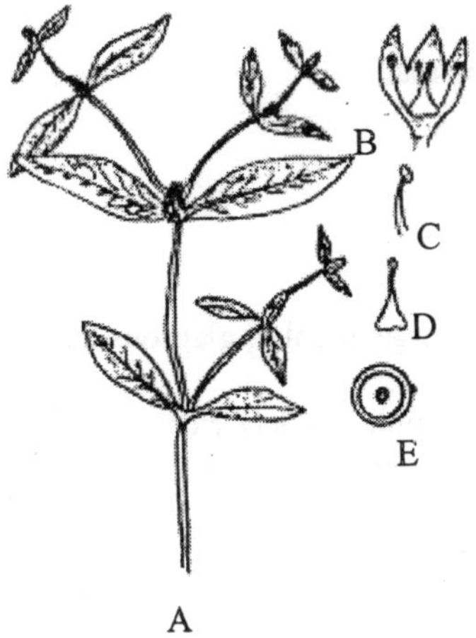

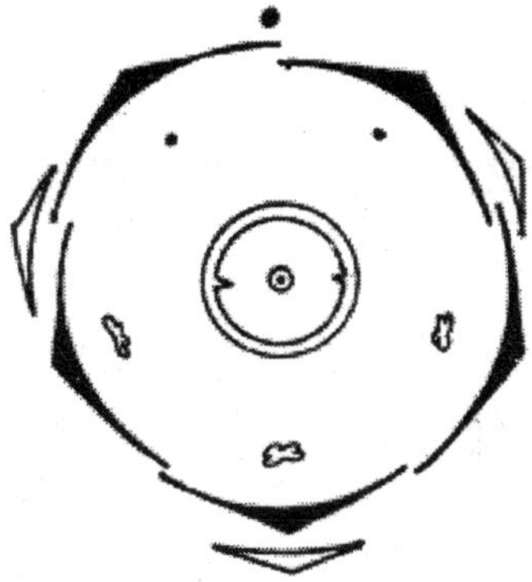

Fig 8.142 ***ALTERNATHERA SESSILIS*** **(Amaranthaceae)**

CHENOPODIUM ALBUM

(Chenopodiaceae)

Habitat: Mesophyte, weed, grows in open land.

Habit: An annual herb.

Root: Branched tap root.

Stem: Aerial, erect, branched, herbaceous, angular solid, glabrous and green.

Leaf: Ramal and cauline, simple, petiolate, petiole slender), exstipulate, alternate, elliptic lanceolate, sinuous margin, acute, fleshy and covered with white mealy powder; venation reticulate unicostate.

Inflorescence: A condensed cyme.

Flower: Sessile, bracteate, ebracteolate, complete, cyclic, pentamerous, homochlamydeous, actinomorphic, bisexual and hypogynous.

Perianth: Tepals 5, polytepalous, sepaloid and membranous, small; aestivation quincuncial

Androecium: Stamens 5, free, opposite to perianth lobes; filaments long; anthers dithecous, basifixed and introse.

Gynoecium: Ovary superior, bicarpellary, syncarpous unilocular with a single basal ovule; style short and stigma bifid.

Fruit: A utricle, single seeded.

Seed: Small, non endospermic.

Floral formula: ✝ ♀ $P_{(5)}\ A_5\ \underline{G}_{(2)}$

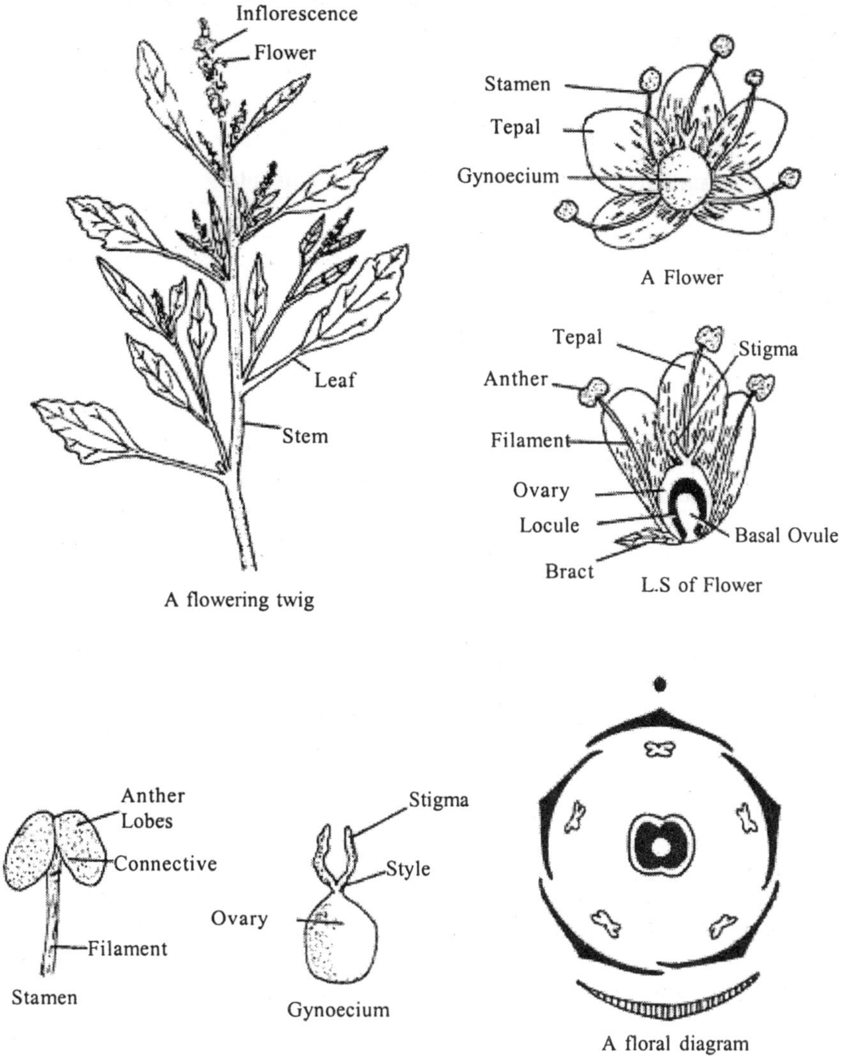

Fig 8.143 ***CHENOPODIUM ALBUM*** **(Chenopodiaceae)**

EUPHORBIA PULCHERRIMA
(= Poinsettia Pulcherrima)
(Euphorbiaceae)

Habitat : Mesophyte, commonly cultivated is gardens as an ornamental

Habit : A perennial shrub.

Root : Tap rot, branched.

Stem : Aerial, erect, herbaceous at the top and woody below, branched, angular, solid, rough, green with milky latex.

Leaf : Ramal and cauline, simple, petiolate, alternate, stipulate, (glandular stipules), ovate, entire, acute, smooth, soft, milky latex present; venation reticulate unicostate.

Inflorescence : A special type of condensed cyme called cyathium. The inflorescence is surrounded by two kinds of bracts – large foliar but deep red coloured bracts (these are the chief organs of attraction for pollinating agents) and small green bracts (five in number) which form a cup like structure surrounding the inflorescence. There is a yellow coloured extrafloral bilabiate nectary attached to the bracts.

The flowers in the inflorescence are unisexual but the same cyathium has both male and female flowers. There are five groups of male flowers arranged in monochasial scorpioid cymes opposite the bracts while there is a single female flower in the centre. In both male and female flowers accessory organs (calyx, corolla) are absent.

Male flower : Pedicellate, bracteate, ebracteolate, incomplete, achlamydeous, unisexual and male (staminate). Each male flower is represented by a single stamen. There is a short pedicel and a joint on which arises the filaments of the stamen. Anthers monothecous, basifixed and introrse.

Floral formula : ✝ ♀ $K_0 C_0 A_1 G_0$

Female flower : Pedicellate, bracteate, ebracteolate, incomplete, achlamydeous, unisexual, female (pistillate) and hypogynous. The female flower has no calyx or corolla and is represented by a pistil which is raised on a gynophore. Ovary superior, tricarpellary, syncarpous, trilocular with one ovule in each locule on axile placenta. Style short; stigma 3 and each bifid.

Fruit : A Capsule.

Seed : Non – endospermic.

Floral formula : ✝ ♀ $K_0 C_0 A_0 \underline{G}_{(3)}$

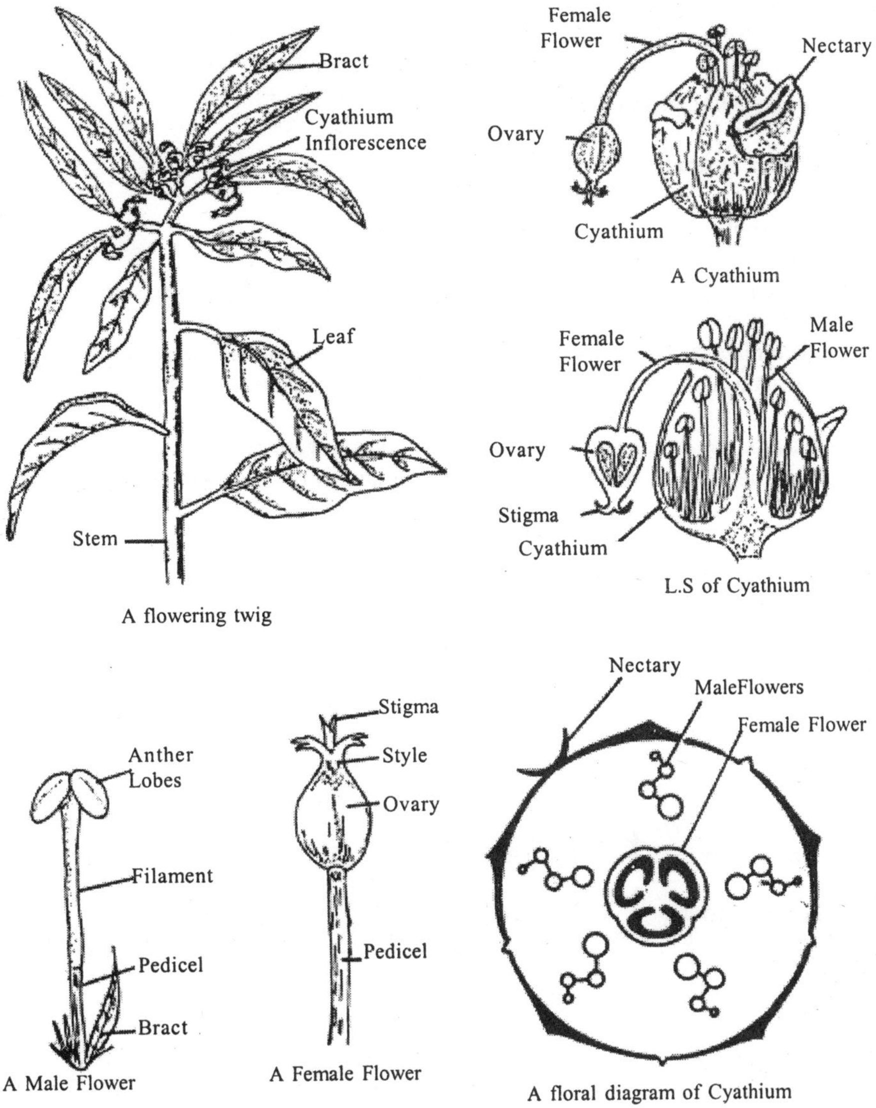

Fig 8.144 ***EUPHORBIA PULCHERRIMA (= Poinsettia Pulcherrima)*** **(Euphorbiaceae)**

EUPHORBIA HIRTA

(Euphorbiaceae)

Habitat : Mesophyte, wild, grows in dry land.

Habit : An annual erect or semierect herbaceous weed.

Root : Branched tap root.

Stem : Aerial, erect or semierect, branched or unbranched, cylindrical, hairy, green, solid, milky latex present

Leaf : Ramal and cauline, simple, subsessile, stipulate (stipules caducous), opposite, superposed, narrow base, lanceolate, serrulate, acute, milky latex present; venation reticulate unicostate.

Inflorescence : A cyathium, cyathia are smaller in size and aggregate into axillary cymes. 4-5 small honey glands are present at the apex of the bracts. (Rest of the details of cyathium are same as in *Euphorbia pulcherrima*).

Male flower : Short pedicellate, bracteate, ebracteolate, incomplete, unisexual, anchlamydeous and staminate. Each male flower is represented by a single stamen. Anther 2 lobed, basifixed and introrse.

Female flower : Pedicellate, bracteate, ebracteolate, incomplete, achlamydeous unisexual, pistillate and hypogynous. The female flower is represented by a pistil at the centre of the inflorescence and is raised on a gynophore. (Rest of the details of female flower is same as in *Euphorbia pulcherrima*)

Fruit : A capsule

Floral formula : Male flower : ✝ ♀ $K_0 C_0 A_1 \underline{G}_0$

Female flower : ✝ ♀ $K_0 C_0 A_0 \underline{G}_{(3)}$

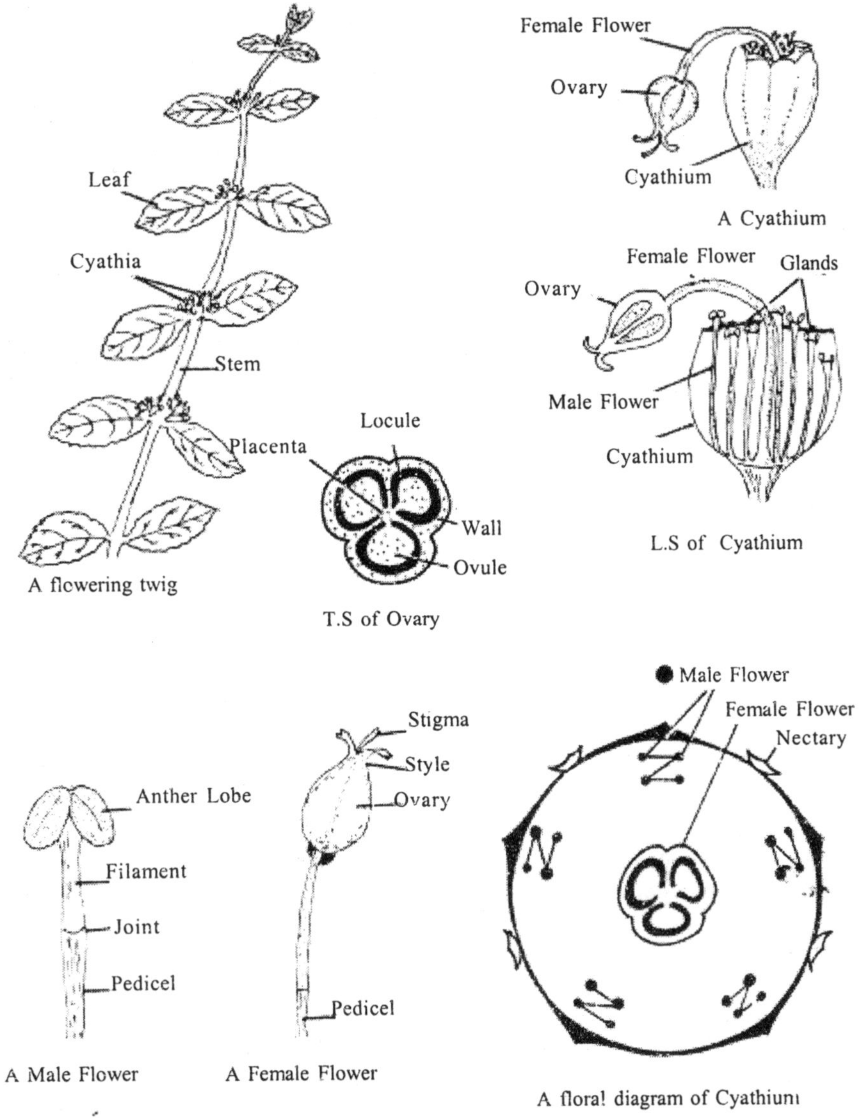

Fig 8.145 *EUPHORBIA HIRTA* (Euphorbiaceae)

RICINUS COMMUNIS

(Euphorbiaceae)

Habitat : Mesophyte, cultivated plant.

Habit : An annual or biennial shrub or a small sized tree.

Root : Branched tap root

Stem : Aerial, erect, cylindrical, herbaceous or partly woody, branched, younger portion solid while older portion hollow, greenish purple in colour, nodes somewhat swollen and watery latex present.

Leaf : Ramal and cauline, simple, alternate, stipulate (stipule convolute), sometimes stipules absent, petiolate, (petiole long and reddish) palmately lobed, lobes many, margin serrate, apex acute, glabrous, latex present; venation reticulate multicostate. A gland is present at the base of the petiole.

Inflorescence : A cymose pancile. Flowers unisexual, but the same inflorescence has both flowers; male flowers below and female above.

Male flower : Pedicellate, bracteate, ebracteolate, cyclic, incomplete, pentamerous, homochlamydeous, actinomorphic, unisexual and staminate (male)

Perianth : Tepals 5, polytepalous, slightly connate at the base; aestivation valvate.

Androecium : Stamens 5, free, antitepalous, each stamen is profusely branched appearing like a miniature tree ultimate braches of the filaments bear an anther each; anthers monothecous, basifixed and introrse.

Gynoecium : Absent

Female flower : Pedicellate, bracteate, ebracteolate, incomplete, cyclic, pentamerous, homochlamydeous, unisexual, pistillate and hypogynous.

Perianth : Tepals 3, polytepalous, caducous, sepaloid; aestivation valvate.

Androecium : Absent

Gynoecium: Ovary superior, tricarpellary, syncarpous, trilocular with one ovule in each locule on axile placenta – styles 3, stigma 3; each bifid, carpels have appendages on the outer wall.

Fruit : A schizocarpic capsule splitting into 3 cocci.

Seed : Large, testa ornamented and shing, albuminous, A caruncle is present at one end of the seed.

Floral formula : Male flower : ✝ ♀ $P_5 A_5 \underline{G}_0$

Female flower : ✝ ♀ $P_3 A_0 \underline{G}_{(3)}$

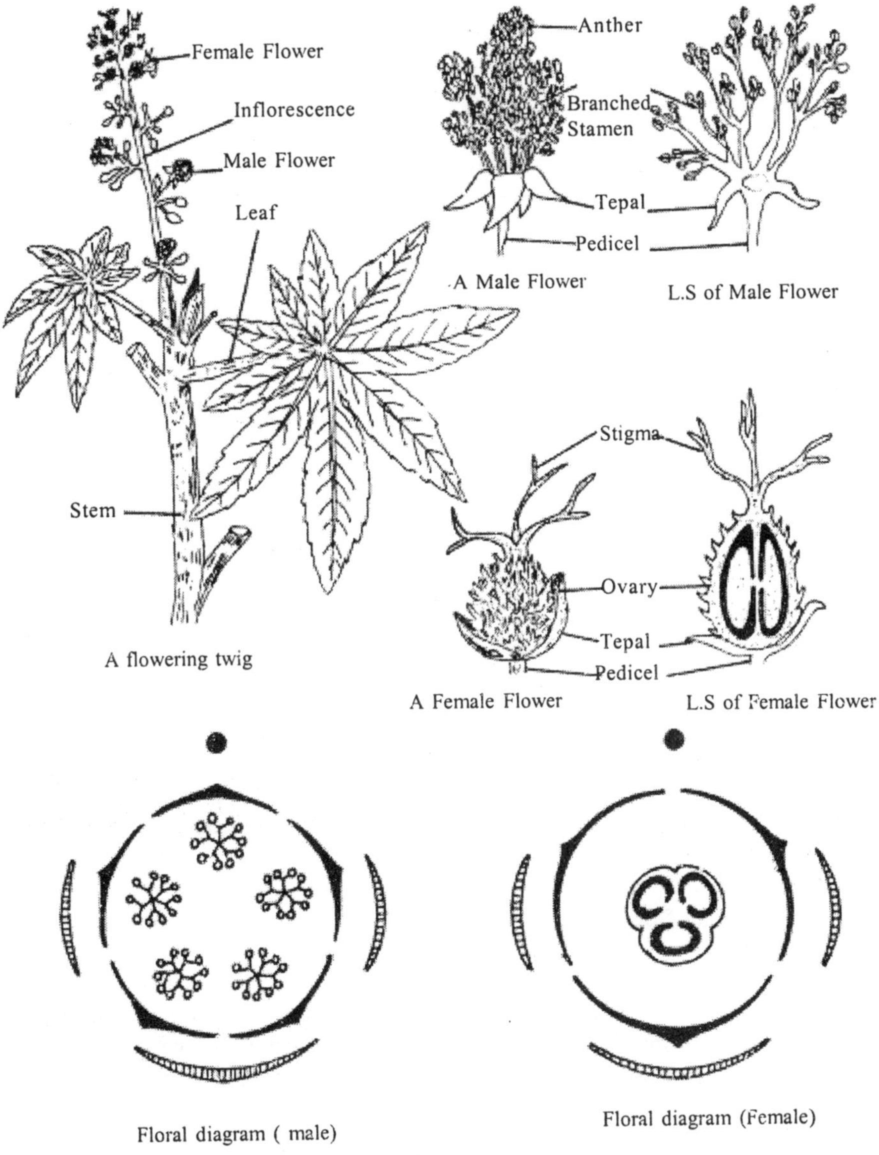

Fig 8.146 *RICINUS COMMUNIS* (Euphorbiaceae)

PHYLLANTHUS NIRURI

(Euphorbiaceae)

Habitat : Mesophyte, grows in open dry places.

Habit : A small annual wildly growing herb.

Root : Branched tap root, surface feeding

Stem : Aerial, erect, slender, herbaceous, cylindrical, smooth green and hollow.

Leaf : Ramal and cauline, siimple, alternate, small, stipulate (stipules free lateral), subsessile, elliptical oblong, entire, apex rounded, pale green on dorsal surface, glacous on the ventral surface, venation reticulate unicostate unisexual.

Inflorescence : Flowers unisexual small and solitary axillary; plant is monoecious. In the same branch female flowers are towards the apex while male flowers are towards the base. The flowers are present on the lower side of the branches.

Male flower : Small, pedicellate, bracteate, ebracteolate. incomplete, cyclic, trimerous, homochlamydeous, actinomorphic, unisexual and staminate.

Perianth : Tepals 6, arranged in two whorls of 3 each, polytepalous, sepaloid; aestivation valvate. Minute nectaries are present at the base of the tepals.

Androecium : Stamens 3, monadelphous, filaments short, fuse to form a column; anthers monothecous, basifixed and introrse with longitudinal splitting

Gynoecium : Absent

Fruit : Absent

Female flower : Pedicellate, bracteate, ebracteolate, incomplete, trimerous, actinomorphic, homochlamydeous, unisexual, pistillate and hypogynous.

Perionth : Same as in male flowers.

Androecium : Absent

Gynoecium : Ovary superior, tricarpellary, syncarpous, trilocular with two ovules in each locule on axile placenta; style short or absent; stigma 3 each bifid and sticky. Mature ovary becomes 6 locular.

Fruit : A capsule

Seed : Endospermous and traingular

Floral formula : Male flower : ✝ ♀ $P_{3+3} A_{(3)} \underline{G}_0$

Female flower : ✝ ♀ $P_{3+3} A_0 \underline{G}_{(3)}$

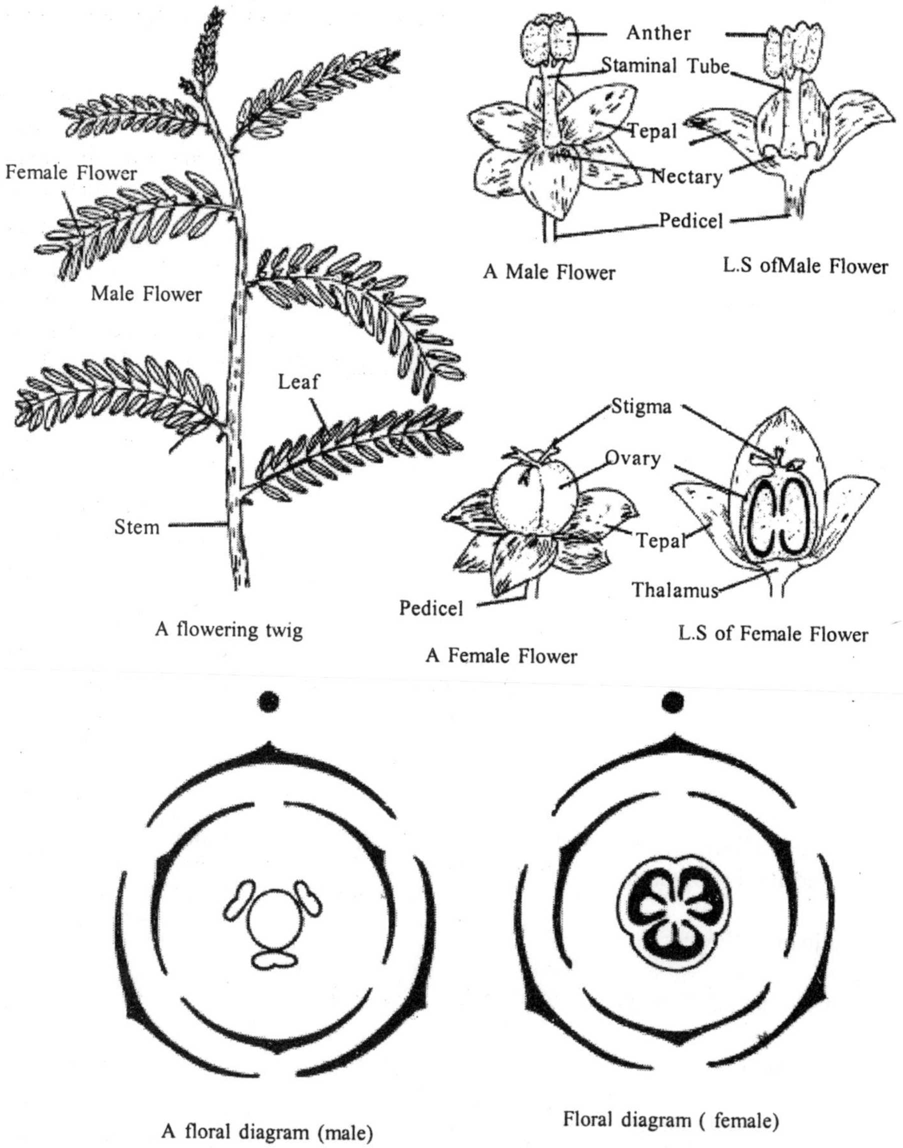

Fig 8.147 ***PHYLLANTHUS NIRURI*** **(Euphorbiaceae)**

JATROPHA GOSSYPIFOLIA

(Euphorbiaceae)

Habitat : Mesophyte, wild, growing in wasteland.

Habit : A perennial shrub

Root : Branched tap root

Stem : Aerial, erect, woody with young branches being herbaceous, cylindrical, green in lower parts, dark red in upper portions, studded with glandular hairs.

Leaf : Ramal and cauline, simple, alternate, exstipulate, petiolate (slender petiole) deeply palmately lobed, covered with glandular hairs, lobes elliptic ovate, margin finely serrate, acute, glaucous, shiny, coriaceous; venation reticulate multicostate.

Inflorescence : A cymose panicle, terminal. Flowers unisexual, but plant monoecious, male and female flowers present in different cymes.

Male flower : Pedicellate, bracteate, ebracteolate, cyclic, incomplete, heterochlamydeous, pentamerous unisexual and staminate. Five nectar secreting discs are present alternating with the petals

Calyx : Sepals 5, green, glandular, polysepalous, persistent; aestivation quincuncial.

Corolla : Petals 5, polypetalous, dark red; aestivation twisted.

Androecium : Stamens 9 (rarely 10), monadelphons, in two whorls of 4-5 each, filaments short; anthers dithecous, dorsifixed and introrse.

Gynoecium : Absent

Female flower : Pedicellate, bracteate, ebracteolate, cyclic, incomplete, pentamerous, actinomorphic, unisexual, pistillate and hypogynous.

Calyx : Same as in male flowers

Corolla : Same as in male flowers

Androecium : Absent

Gynoecium : Ovary superior, tricarpellary, syncarpous, trilocular with one ovule in each locule on axile placenta. Styles 3: stigmas 3, each bifid.

Fruit : A regma (capsule) splitting into 3 cocci

Seed : Endospermous

Floral formula : Male flower : † ♀ $K_5 C_5 A_{9-10} \underline{G}_0$

Female flower : † ♀ $K_5 C_5 A_0 \underline{G}_{(3)}$

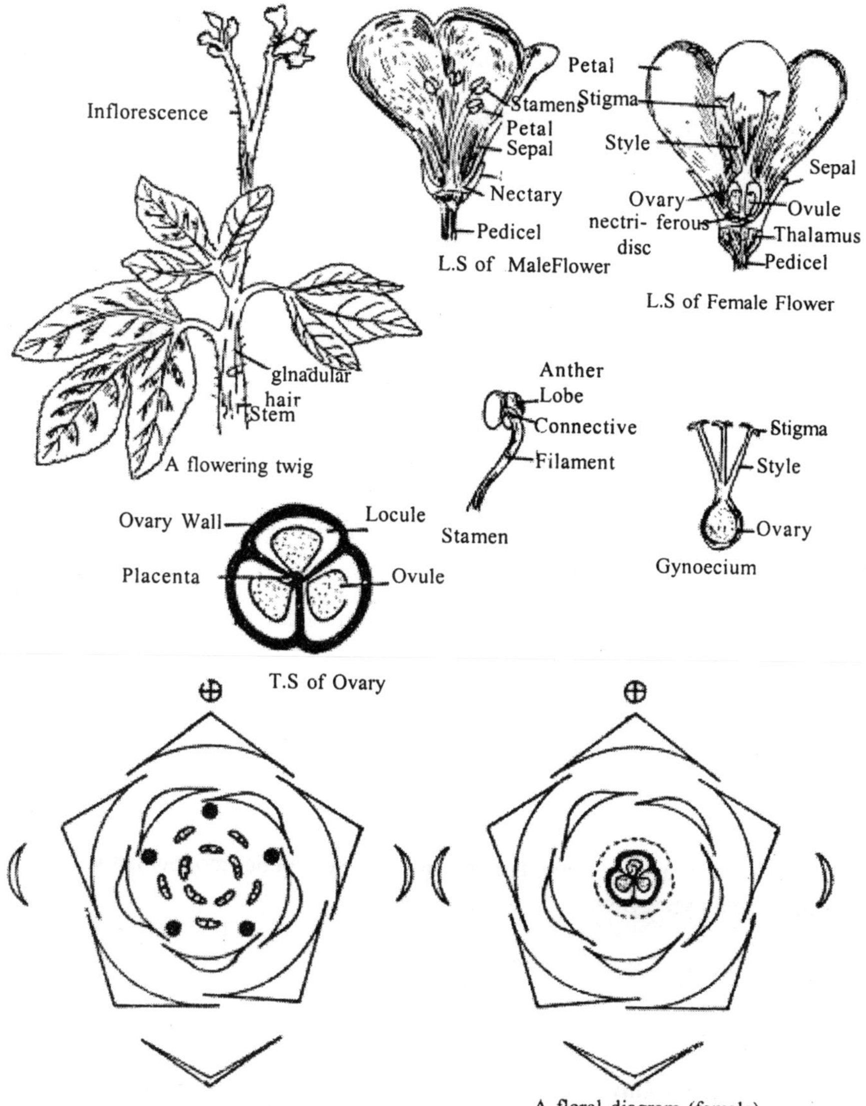

Fig 8.148 ***JATROPHA GOSSYPIFOLIA*** **(Euphorbiaceae)**

CROTON SPARCIFLORUS

(Euphorbiaceae)

Habitat : Mesophyte, wild, grows in wasteland

Habit : An annual herb reaching a height of 2 feet.

Root : Tap root, branched

Stem : Aerial,erect, green, herbaceous, basal portion somewhat woody, cylindrical, branched, solid and hairy.

Leaf : Ramal and cauline, simple, petiolate, stipulate, whorled above and alternate below, oblong lanceolate, serrate, acuminate; venation reticulate unicostate.

Inflorescence : A simple raceme, flowers unisexual, plant monoecious, the same inflorescence has both male (above) and female (at the base) flowers.

Male flowers : Pedicellate, bracteate, ebracteolate, incomplete, cyclic, pentamerous, heterochlamydeous, actinomorphic, unisexual and staminate.

Calyx : Sepals small, 5 in number, green, polysepalous; aestivation imbricate.

Corolla: Petals 5, polypetalous, white, aestivation imbricate.

Androecium : Stamens indefinite, free, adnate, filaments short; anthers monothecous.

Gynoecium : Absent

Female flower : Sessile, bracteate, ebracteolate, incomplete, cyclic, monochlamydeous, actinomorphic, unisexual, pistillate and hypogynous.

Calyx : Sepals 5, polysepalous; aestivation valvate

Corolla: Absent

Androecium : Same as in *Jatropha gossypifolia*

Fruit : A capsule, breaks open into 3 cocci

Floral formula : Male flower : † ♀ $K_5 C_5 A_\infty \underline{G}_0$

Female flower : † ♀ $K_5 C_0 A_0 \underline{G}_{(3)}$

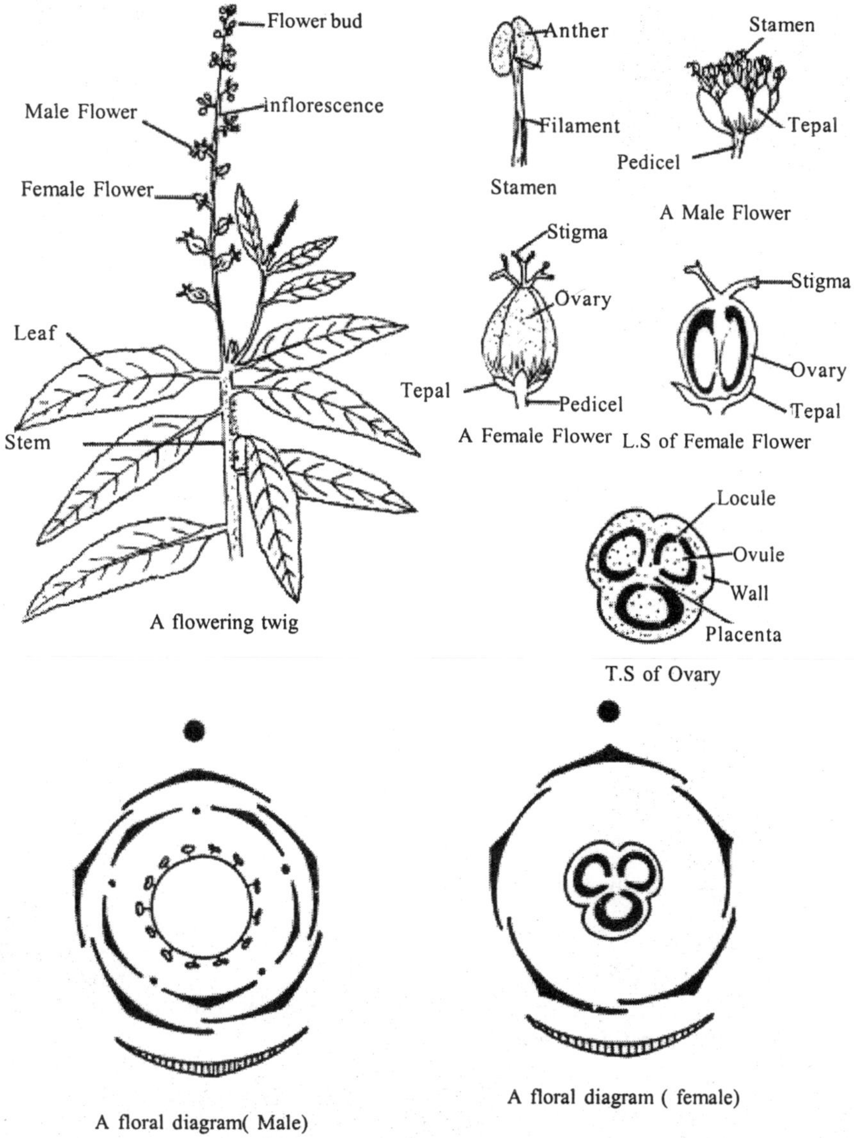

Fig 8.149 ***CROTON SPARC!FLORUS*** **(Euphorbiaceae)**

ACALYPHA INDICA

(Euphorbiaceae)

Habitat: Mesophyte, cultivated as an ornamental

Habit: A perennial shrub

Root: Tap root, branched

Stem: Aerial, erect, basal region woody, upper part herbaceous, cylindrical, hairy, branched and solid.

Leaf: Ramal and cauline, simple, alternate, petiolate, stipulate, rhomboid, ovate, margin serrate, obtuse, venation reticulate unicostate.

Inflorescence: An axillary catkin. Flowers unisexual; male and female flowers in separate inflorescences, but the plant is monoecious.

Male flower: Sessile, ebracteate, ebracteolate, incomplete, actinomorphic, homochlamydeous, unisexual and staminate.

Perianth: Tepals 4, gamotepalous, aestivation valvate.

Androecium: Stamens 8-9, filaments short, anthers twisted.

Gynoecium: Absent or short pedicellate

Female flower: Sessile bracteate, ebracteolate, incomplete, actinomorphic, homochlamydeous unisexual, pistillate and hypogynous.

Perianth: 3-4 tepals, free, aestivation imbricate.

Androecium: Absent

Gynoecium: Ovary superior, tricarpellary, syncarpous, trilocular with one ovule in each locule on axile placenta. Style short, stigma trifid and feathery.

Fruit: A capsule

Seed: Scantily endospermous

Floral formula: Male flower: ✝ ♀ $P_4 A_{8\text{-}9} \underline{G}_0$

Female flower: ✝ ♀ $P_{3\text{-}4} A_0 \underline{G}_{(3)}$

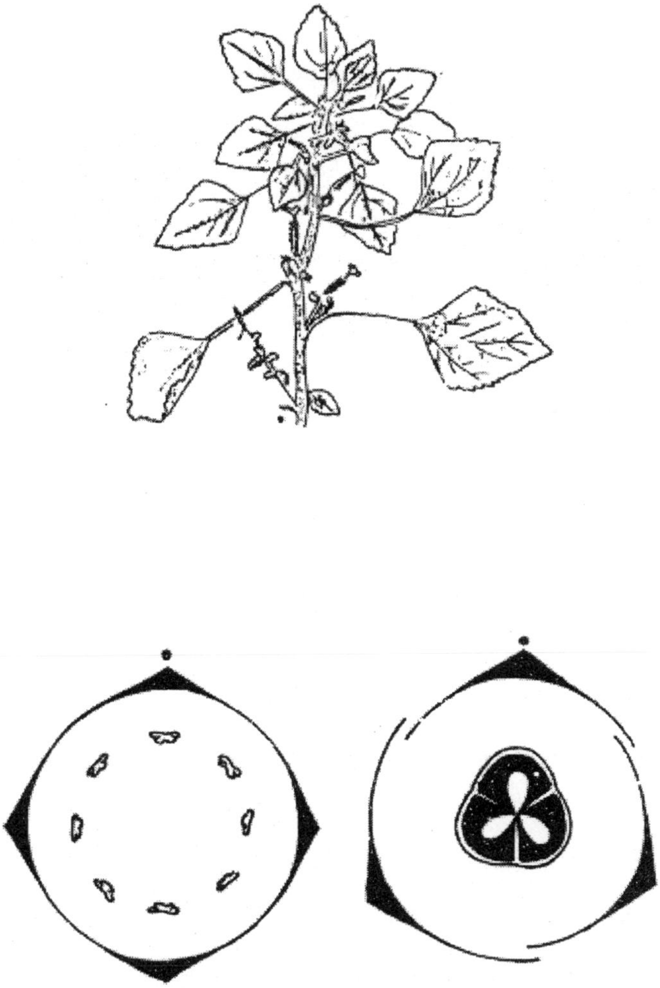

Fig 8.150 ***ACALYPHA INDICA*** (Euphorbiaceac)

POUZOLZIA INDICA

(Urticaceae)

Habitat: Mesophyte, grows on the banks of streams and other moist areas.

Habit: An annual herb

Root: Tap root system

Stem: Aerial, erect or prostrate, cylindrical, branched, solid, rough and green in colour.

Leaf : Ramal and cauline, simple, petiolate, stipulate, alternate, ovate, lanceolate; venation reticulate unicostate, cystoliths are present in the leaf.

Inflorescence : Axillary cyme, flowers unixexual

Male flower : Sessile, bracteate, ebracteolate, tetramerous, incomplete, cyclic, actinomorphic, monochlamydeous, unisexual and staminate.

Perianth : Tepals 2+2, polytepalous, membranous; aestivation imbricarte.

Androecium : Stamens 4, free, antitepalous, filaments long; anthers dorsifixed and introrse.

Gynoecium : Absent

Female flower : Sessile, bracteate, eracteolate, cyclic, incomplete, actinomorphic, tetramerous, unisexual, pistillate and hypogynous.

Perianth : Same as in male flowers.

Androecium : Absent

Gynoecium : Ovary superior, monocarpellary, unilocular, with a single basal ovule. Style short, stigma simple.

Fruit : A nut

Floral formula : Male flower : ✝ ♀ $P_{2+2} A_4 \underline{G}_0$

Female flower : ✝ ♀ $P_{2+2} A_0 \underline{G}_{(1)}$

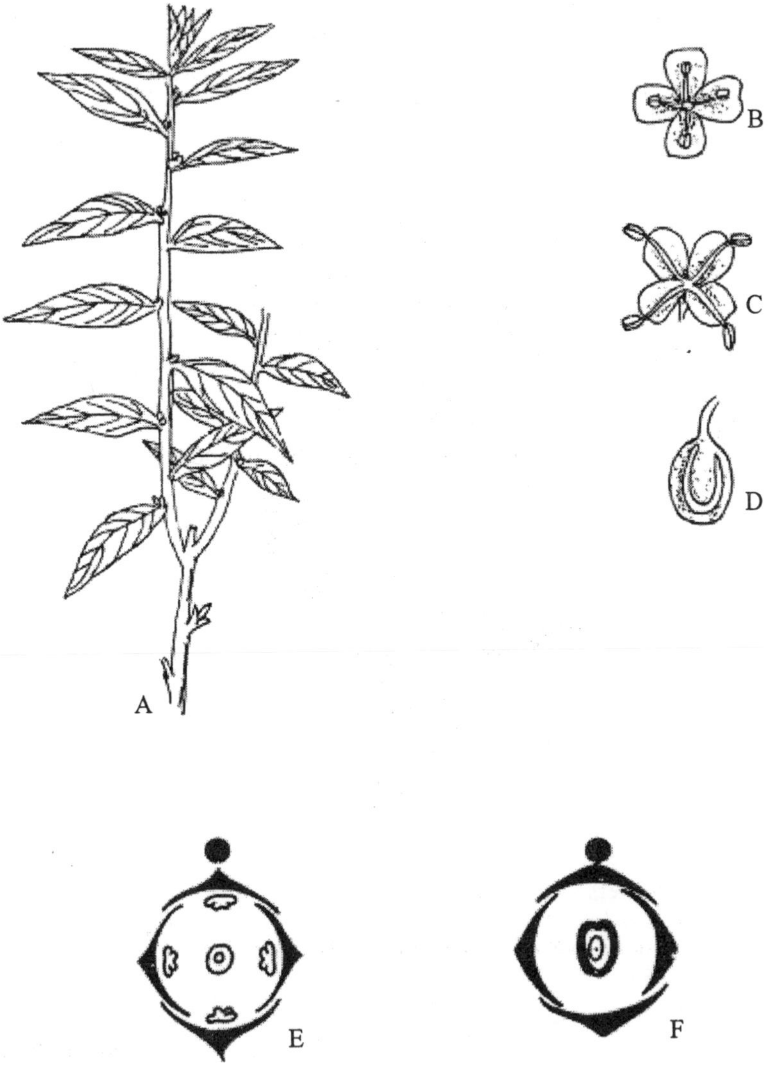

Fig 8.151 ***POUZOLZIA INDICA*** **(Urticaceae)**

URTICA DIOICA

(Urticaceae)

Habitat: Mesophyte, common in open forests, weed.

Habit: A herbaceous plant with stinging hairs.

Root: Tap root, branched

Stem: Aerial, erect, soft, hairy with stinging hairs, green, rough, somewhat cylindrical.

Leaf: Ramal and cauline, simple, alternate, large, coriaceous, rough with stinging hairs, stipulate, ovate, serrate; venation reticulate unicostate.

Inflorescence: Axillary cymes. Flowers unisexual, plants are also dioecious.

Male flower: Sessile, ebracteate, ebracteolate, incomplete, cyclic, actinomorphic, monochlamydeous, unisexual and staminate.

Perianth: Tepals 2+2, membranous, free, aestivation valvate

Androecium: Stamens 4, free, antitepalous, filament elastic and jointed.

Gynoecium: Absent, pistillode present

Female flower: Sessile, bracteate, ebracteolate, incomplete, cyclic, actinomorphic, monochlamydeous, unisexual pistillate and hypogynous.

Perianth: Same as in male flowers

Androecium: Absent.

Gynoecium: Ovary superior, monocarpellary, unilocular with a single basal ovule. Style simple; stigma large and feathery.

Fruit: An achene, enclosed by the persistent perianth.

Seed: Albuminous

Floral formula: Male flower: ✝ ♀ $P_{2+2} A_4 \underline{G}_0$

Female flower: ✝ ♀ $P_{2+2} A_0 \underline{G}_1$

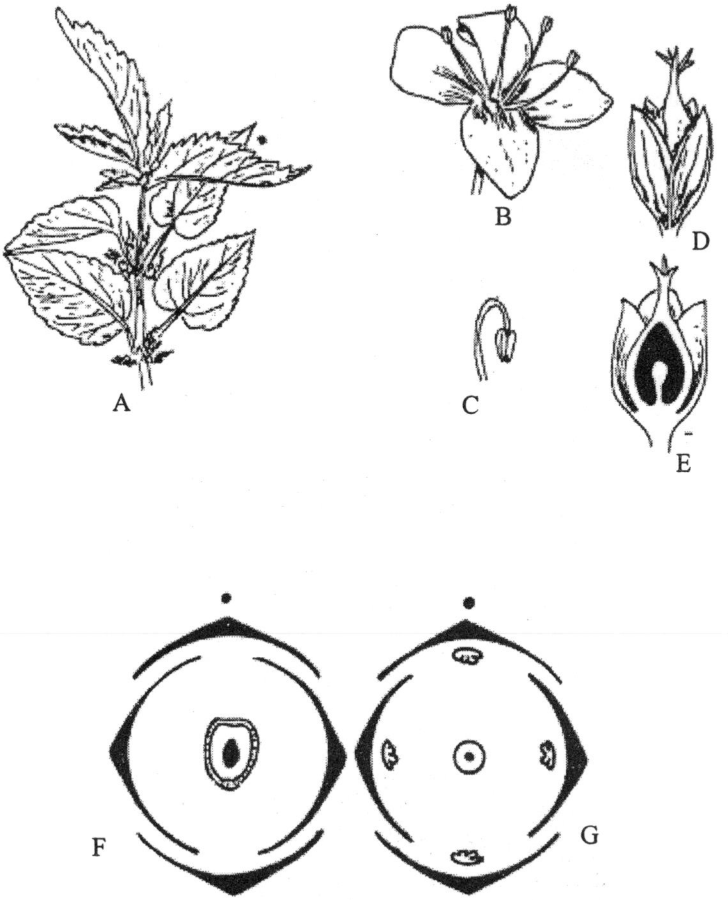

Fig 8.153 ***URTICA DIOICA*** (Urticaceae)

MORUS ALBA

(Moraceal)

Habitat : Mesophyte, cultivated, in gardens, fields etc.

Habit : A small sized tree

Root : Tap root, branched

Stem : Aerial, erect, woody, branched, solid, cylindrical smooth and greyish green.

Leaf : Ramal and cauline, simple, petiolate, stipulate, alternate, ovate, serrate, acute, watery latex present; venation reticulate unicostate.

Inflorescence : A cat-kin; flowers unisexual distributed in separate inflorescence; plant monoecious.

Male flower : Sessile, ebracteate, ebracteolate, incomplete, cyclic, monochlamydeous, actinomorphic, tetramerous, unisexual and staminate.

Perianth : Tepals 4, polytepalous, arranged in 2 whorls of 2 each; aestivation valvate.

Androecium : Stamens 4, free, antitepalous, filaments long, exserted; anthers dithecous, basifixed and introrse

Gynoecium : Absent

Female flower : Sessile, ebracteate, ebracteolate, cyclic, incomplete, monochlamydeous, actinomorphic, unisexual, pistillate and hypogynous.

Perianth : Same as in male flower

Androecium : Ovary superior, bicarpellary, syncarpous, unilocular with a single basal ovule. Style siimple; stigma forked.

Fruit : A compound fruit, sorosis

Seed : Scantily endospermic

Floral formula : Male flower : † ♀ $P_{2+2} A_4 \underline{G}_0$

Female flower : † ♀ $P_{2+2} A_0 \underline{G}_{(2)}$

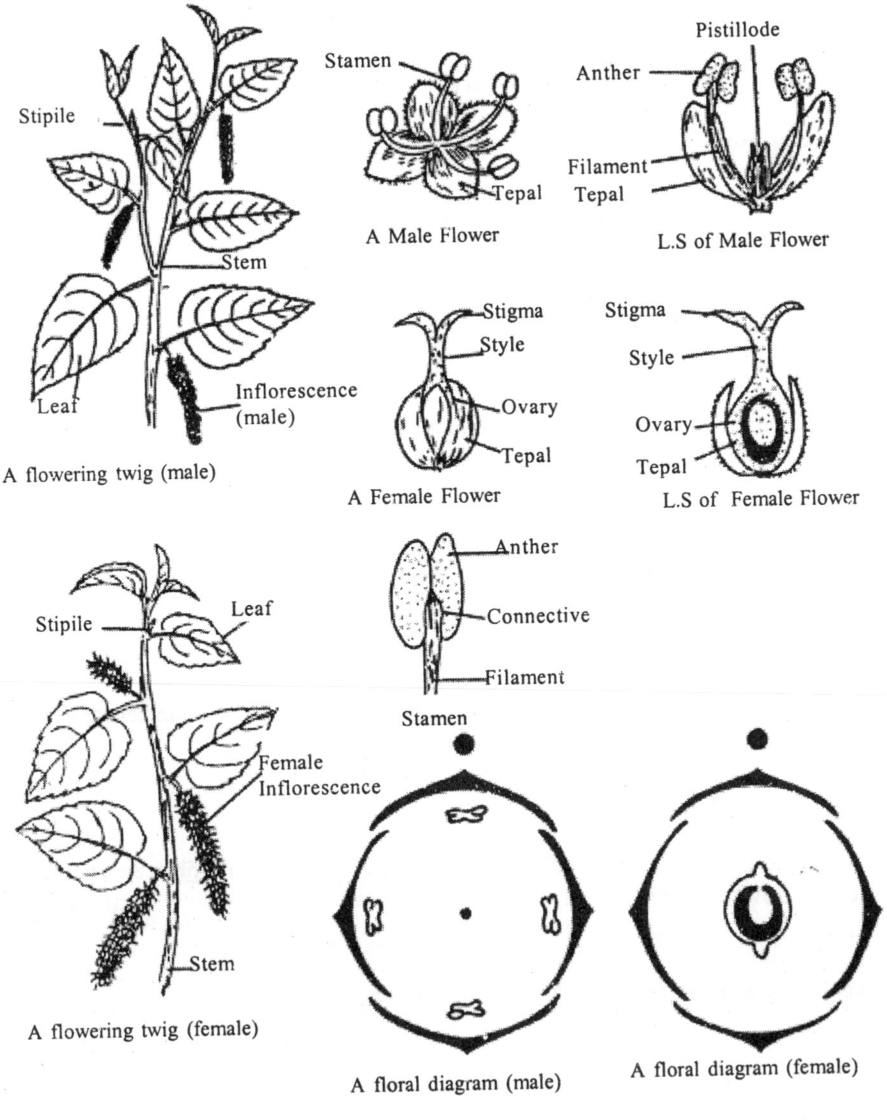

Fig 8.153 *MORUS ALBA* (Moraceal)

FICUS GLOMERATA

(Moraceae)

Habitat : Mesophyte, commonly growing tree in dry deciduous forests.

Habit : A tall tree with grey coloured bark.

Root : Branched tap root

Stem : Aerial, erect, woody, branched, cylindrical, solid, greyish green, milky latex present.

Leaf : Ramal and cauline, simple, petiolate, stipulate (stipules enclose young buds and caducous), alternate, ovate oblong, entire, acute, coriaceous, venation reticulate unicostate.

Inflorescence : A hypanthodium borne in bunches on the main stem as well as on large branches. Flowers unisexual. The same hypanthodium has both male and female flowers. Some flowers are devoid of both sex organs.

Male flower : Sessile, ebracteate, ebracteolate, incomplete, cyclic, actinomorphic, unisexual and staminate. The flowers are borne near the mouth of the hypanthodium

Perianth : Tepals 4, polytepalous, membranous, pinkish; aestivation imbricate.

Androecium : Stamens 2, free, antitepalous and anteroposterior, filaments bent ; anthers dithecous, basifixed and introrse

Gynoecium : Absent

Female flower : Sessile, ebracteate, ebracteolate, incomplete, cyclic, actinomorphic, unisexual, pistillate and hypogynous. Female flowers are present all around the inner wall of the receptacle, particularly in the lower region.

Perianth : Tepals 4-5, gamopetalous, membranous, pink; aestivation valvate.

Androecium : Absent

Gynoecium : Ovary superior, bicarpellary, syncarpous, unilocular with a single pendulous ovule. Style lateral; stigma clavate.

Fruit : Sorosis, individual fruits one seeded achenes.

Floral formula : Male flower : ✝ ♀ $P_4 A_2 \underline{G}_0$

Female flower : ✝ ♀ $P_4 A_0 \underline{G}_{(2)}$

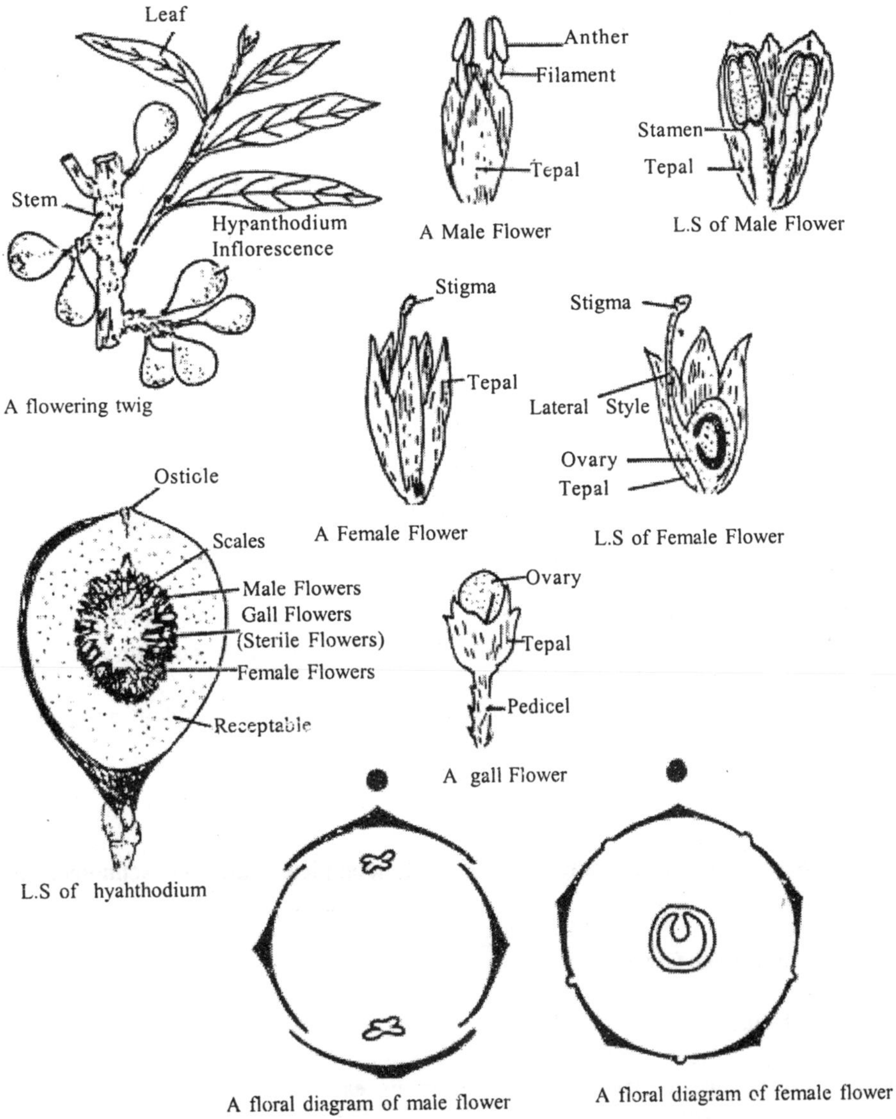

Fig 8.154 *FICUS GLOMERATA* (Moraceae)

PHOENIX SYLVESTRIS

(Palmae)

Habitat: Mesophyte, a wild growing tree in dry open areas.

Habit: A medium sized tree with a crown of leaves. Male and female plants are different.

Root: Fibrous root system.

Stem: Aerial, erect, woody, cylindrical, columnar with leaf scars, rough, and brownish grey in colour. The stem is called a caudex.

Leaf: Cauline, forms a dense crown of leaves at the apex of the stem. Individual leaf is compound, unipinnate, exstipulate, large; pinnae 6-18" long, entire, acute, subsessile, thick, rough con duplicate (folded up), smooth with unicostate parallel venation.

Inflorescence: A spadix. Flowers unisexual and the plant is also dioecious. The male and female inflorescences are enclosed in a woody spathe.

Male flower: Tepals 6, arranged in two whorls of 3 each, outer whorl of tepals connate at the base, inner free, outer tepals small valvate inner whorl of tepals larger and twisted. Tepals of both whorls ivory coloured.

Androecium: Stamens 6, free in two whorls of 3 each, antitepalous, filaments short; anthers erect, dithecous, basifixed and introrse.

Gynoecium: Absent.

Female flower: Sessile, bracteate, ebracteolate, cyclic, incomplete, trimerous, actinomorphic, unisexual, pistillate and hypogynous.

Perianth: Same as in male flower.

Androecium: Absent

Gynoecium: Ovary superior, tricarpellary, apocarpous, each carpel unilocular with one ovule in basal placenta. Style very short; stigma flat.

Fruit: A berry developed out of only one carpel; the remaining two carpels abortive.

Floral formula: Male flower: † ♀ $P_{3+3}\, A_{3+3}\, \underline{G}_0$

Female flower: † ♀ $P_{3+3}\, A_0\, \underline{G}_{(3)}$

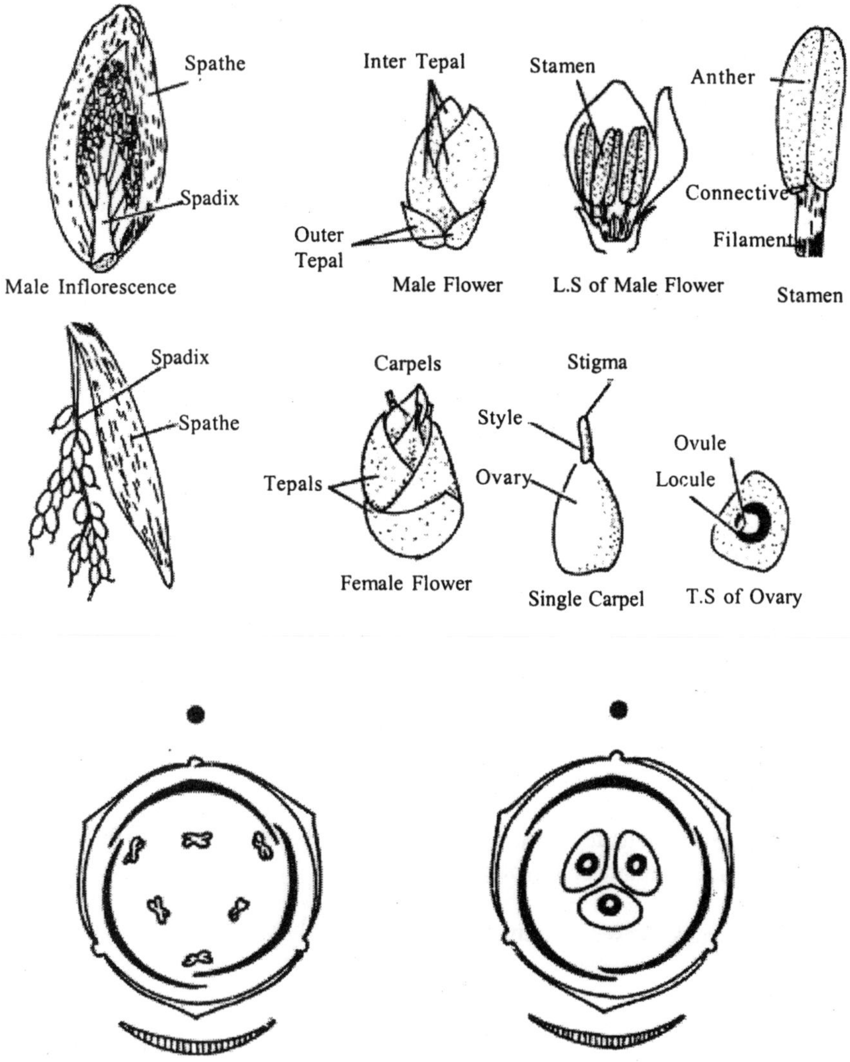

Fig 8.155 ***PHOENIX SYLVESTRIS*** **(Palmae)**

COCOS NUCIFERA

(Palmae)

Habitat : Mesophyte, essentially native of coastal region, but cultivated all over.

Habit : A tall graceful tree, growing up to a height of 80-90 feet bearing a crown of leaves at the top.

Root : Fibrous root system, spreading from the base of the stem.

Stem : Aerial, erect, cylindrical with leaf scars, hard, unbranched, glabrous, solid, rough.

Leaf : Failry large 2-3m long, petiolate, thick flat and solid, exstipulate, pinnately compound, leaf base sheathing and fibrous, leaf base persistent leaving a prominent scar on the stem.

Inflorescence : A compound spadix. The flowers are unisexual, but both male and female flowers are found in the same inflorescence. Each branch of the spadix has 1-3 female flowers at the base and an indefinite number of male flowers throughout the length of the peduncle.

Male flower : Sessile, ebracteate, ebracteolate, cyclic, incomplete, trimerous, homochlamydeous, actinomorphic, unisexual and staminate.

Perianth : Tepals 6, free, arranged in two whorls of 3 each, tough, leathery or fleshy; outer whorl smaller than the inner; Aestivation valvate.

Androecium : Stamens 6, free, arranged in two whorls of 3 each, antitepalous; filaments short; anthers dithecous, introrse dehiscing by longitudinal split.

Gynoecium : Absent

Female flower : Sessile, ebracteate, ebracrteolate, cyclic, incomplete, trimerous, homochlamydeous, actinomorphic, unisexual, pistillate and hypogynous.

Perianth : Same as in male flowers. Aestivation sometimes imbricate.

Androecium : Absent

Gynoecium : Ovary superior, tricarpellary, syncarpous, trilocular with one ovule in each locule on axile placenta. But during development two carpels abort and the mature ovary consists of only one carpel. The presence of 3 carpels can be seen by three ridges and 3 pores in the endocarp of the fruit.

Fruit : A drupe

Floral formula : Male flower : † ♀ $P_{3+3} A_{3+3} \underline{G}_0$

Female flower : † ♀ $P_{3+3} A_0 \underline{G}_{(3)}$

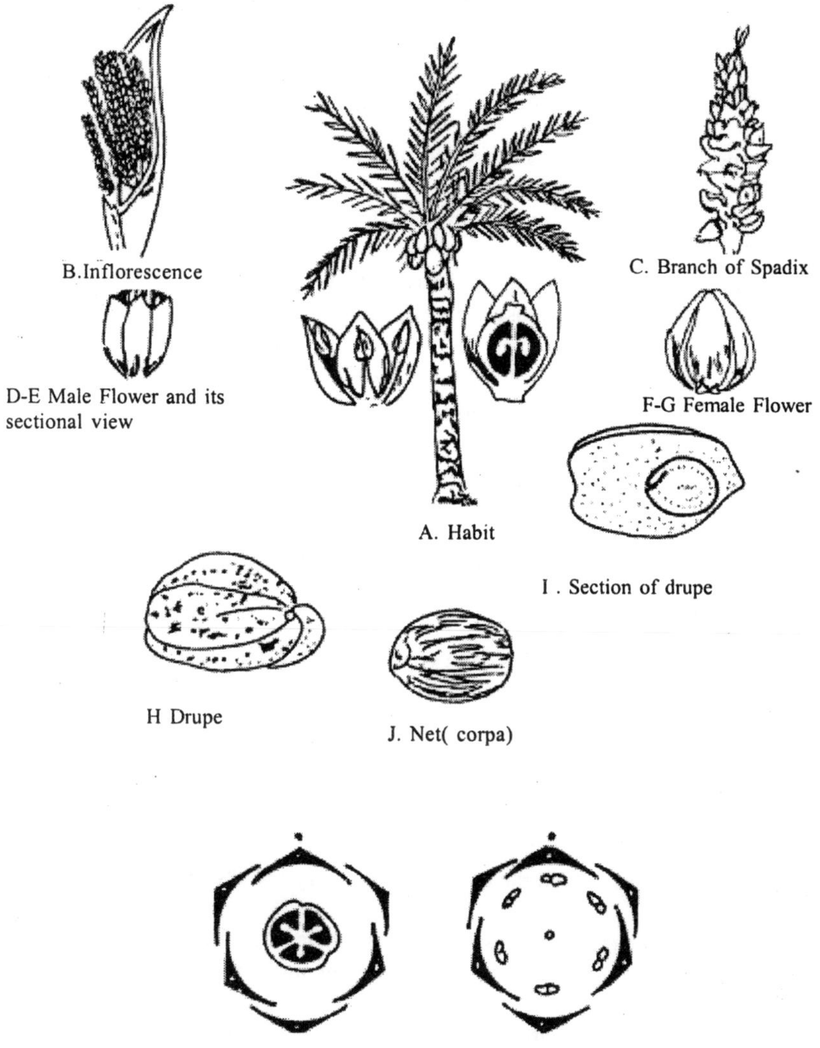

Fig 8.156 *COCOS NUCIFERA* (Palmae)

MUSA PARADISIACA

(Musaceae)

Habitat: Mesophyte, cultivated, grows in moist soil

Habit: A herb reaching the height of 15-18 feet and appears like a tree.

Root: Adventitious root system, arising all along the stem (underground)

Stem: The actual stem of Musa is an underground rhizome. The aerial, erect columnar structure (pseudostem) is actually the erect axis surrounded by closely appressed sheathing leaf bases.

Leaf: Cauline, simple, alternate, exstipulate, petiolate, leaf base sheathing, large, broadly elliptical in shape, entire, obtuse, glabrous; venation unicostate parallel.

Inflorescence: A spadix with a thick pink coloured sapthe. Peduncle long. There is only a single inflorescence for the aerial axis.

Flower: Sessile, bracteate, ebracteolate, complete, cyclic, trimerous, homochlamydeous, zygomorphic, bisexual and epigynous.

Perianth: Tepals 6, gamotepalous, arranged in two whorls of 3 each, inner posterior tepal is free, large and boat shaped; the remaining 5 united to form a tubular structure, petaloid; aestivation imbricate.

Androecium: Stamens 6, free, arranged in 2 whorls of 3 each, antitepalous, the posterior stamen of the inner whorl abortive; filaments long; anthers dithecous, dorsifixed and introrse.

Gynoecium: Ovary inferior, tricarpellary, syncarpous, trilocular with many ovules in each locule on axile placenta; style long; stigma capitate. All the ovules in the ovary are abortive.

Fruit: Berry, it is a parthenocarpic fruit covered by the thalamus (skin of banana); epicarp, fibrous and attached to the inside of thalamus, meso and endocarp fleshy and edible.

Seed: Abortive, as there is no fertilization

Floral formula: † ♀ $P_{3+3} A_{3+3} \underline{G}_{0}$

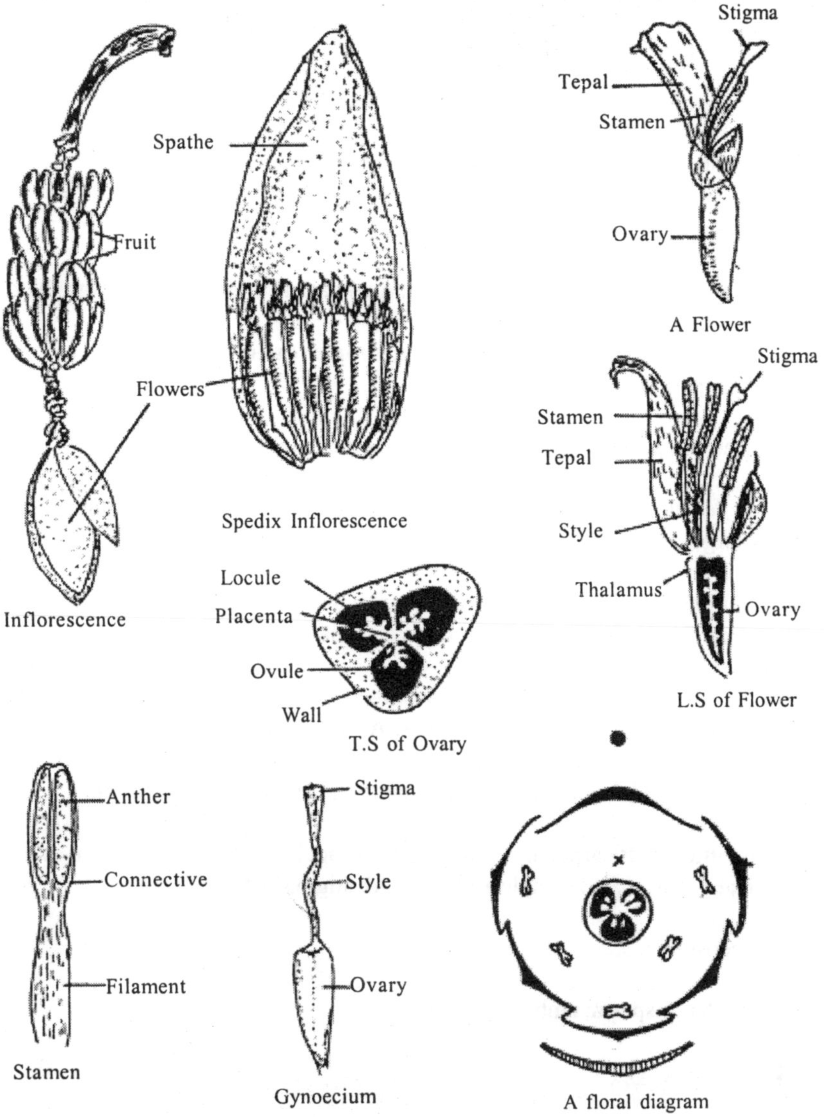

Fig 8.157 ***MUSA PARADISIACA*** **(Musaceae)**

CANNA INDICA

(Cannaceae)

Habitat: Mesophytic, culivated in moist soil.

Habit: Herb, reaching a height of 3-4 feet.

Root: Fibrous root system

Stem: An underground rhizome; scale leaves present; many aerial branches arise from the rhizome.

Leaf: Cauline, sessile, alternate, leaf base sheathing, simple, exstipulate, ligulate, lamina large, broad, entire; venation pinnately parallel.

Inflorescence: Raceme or a panicle. Rarely a 2 flowered cincinnus.

Flower: Pedicellate or sessile, bracteate, bracteolate, complete, trimerous, zygomorphic, bisexual and hypogynous. Flowers large, bright coloured, yellow or red.

Periath: Tepals 6, arranged in 2 whorls of 3 each, outer whorl sepaloid, green, free or connate at the base; aestivation imbricate inner whorl is petaloid, large, connate at the base, bright coloured, aestivation imbricate.

Androecium: Highly modified, 6 or rarely 4-5, arranged in 2 whorls of 3 each; all except the medium posterior one of the inner whorl reduced to staminodes. All the staminodes and the fertile stamen petaloid, outer 3 staminodes free and imbricate; in the inner whorl 1-2 staminodes roll on and form the labellum. The fertile stamen is also petaloid, it has a liner anther adnate to the outer margin at the apex.

Gynoecium: Ovary inferior, tricarpellary, syncarpous, trilocular with many ovules in each locule on axile placenta. Ovary green and warty. Style single and petaloid; stigma flattened.

Fruit: A loculisidal capsule

Seed: Albuminous with perisperm; embryo straight

Floral formula: ✝ ♀ $P_{3+3}\,A_{1/2}\,\underline{G}_{(3)}$

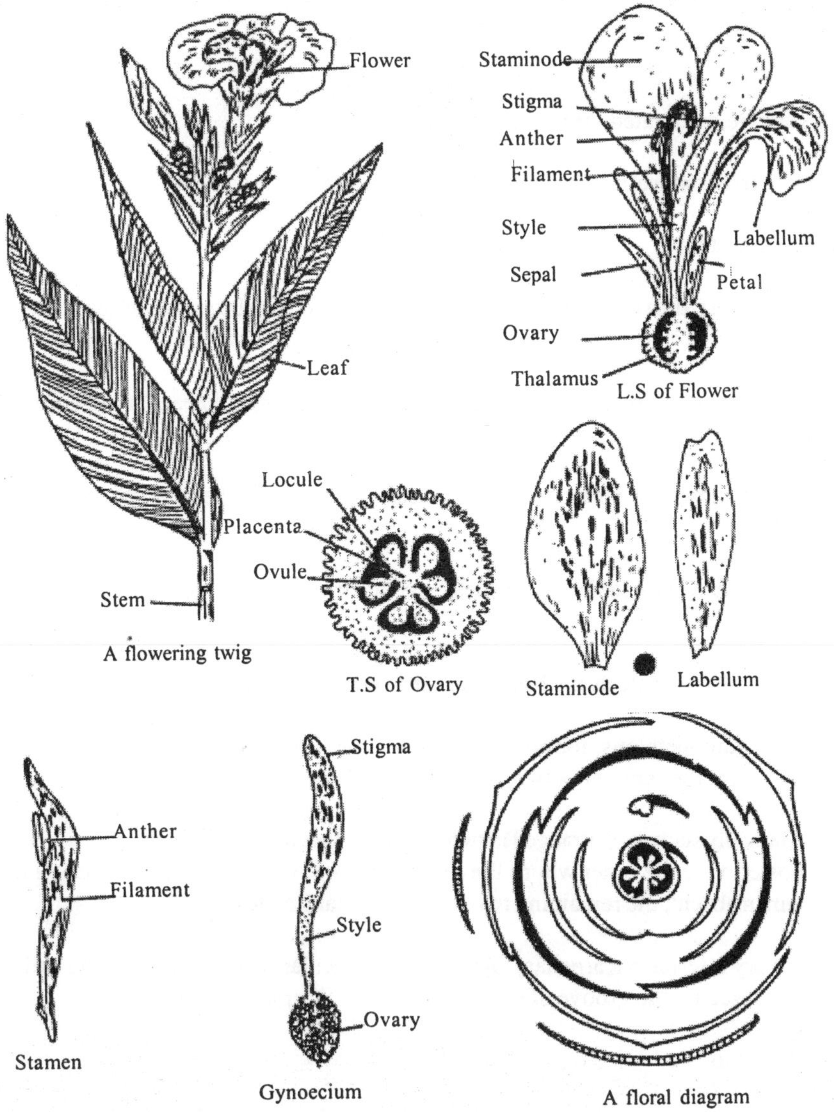

Fig 8.158 ***CANNA INDICA*** **(Cannaceae)**

ZINGIBER OFFICINALE

(Zingiberaceae)

Habitat: Mesophyte, cultivated, grows in moist places.

Habit: Aromatic herb perenneating through underground rhizome

Root: Fibrous root system

Stem: Underground horizontal rhizome, jointed, nodes covered with brown scale leaves with roots arising on the lower side.

Leaf: Radical and cauline, simple, subsessile, leaf base sheathing, entire, linear, lanceolate, ligulate, venation unicostate parallel.

Inflorescence: Bracteate spikes borne on a leafless scape arising from the rhizome. Bracts stiff and closely overlap on each other.

Flower: Sessile, bracteate, ebracteolate, complete, cyclic, trimerous, zygomorphic, heterochlamydeous, bisexual and epigynous.

Calyx: Sepals 3, slightly connate at the base, odd sepal anterior; aestivation valvate.

Corolla: Petals 3, gamopetalous, forming a corolline tube, and separating into 3 limbs above; posterior petal largest and it covers the remaining two; aestivation valvate.

Androecium: Basically stamens 6 arranged in two whorls of 3 each. All except one are staminodes. Functional stamen median, posterior of the inner whorl, the remaining two of inner whorl form labellum. In the outer whorl, anterior stamen absent, the remaining two form leafy staminodes.

Gynoecium: Ovary inferior, tricarpellary, syncarpous, trilocular with many ovules on axile placenta; style long, simple and enclosed by the groove of the fertile stamen. Stigma simple capitate.

Fruit: A loculisidal capsule

Seed: Endosperm mealy, embryo straight.

Floral formula: † ⚥ $K_3 C_3 A_1 \underline{G}_{(3)}$

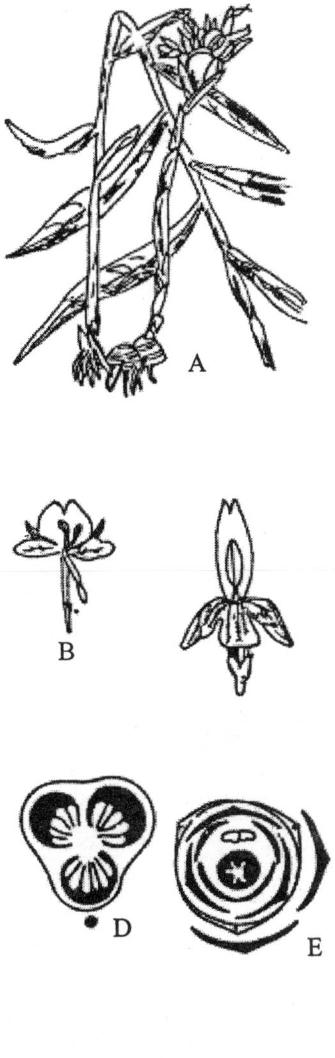

Fig 8.159 *ZINGIBER OFFICINALE* (Zingiberaceae)

ZEUXINE SULCATA

(Orchidaceae)

Habitat : Grows commonly in moist places and canal banks.

Habit : A small terrestrial herb, 2-10" high

Root : Adventitious root system.

Stem : Small underground rhizome

Leaf : Cauline, 1-2" long, linear, sessile, leafbase sheathing, alternate, exstipulate, simple, somewhat thick, glabrous, venation parallel unicostate.

Inflorescence : A spike, densely crowded with flowers.

Flower : Sessile, bracteate, ebracteolate, complete, cyclic, zygomorphic, homochlamydeous bisexual and epigynous, resupinate when mature, white or pink in colour.

Perianth : Tepals 6, in 2 whorls of 3 each, polytepalous, anterior tepal of the outer whorl large and concave while posterior tepal of the inner whorl modified into labellum Aestivation outer whorl valvate; inner whorl imbricate.

Androecium : Stamen one, staminodes two, stamen united with the style to give rise to a columnar gynostegium opposite the labellum; Anther represented by a pair of pollinia connected by caudicle to the corpusculum; stamen is connected to the ovary by rostellum.

Gynoecium : Ovary inferior, tricarpellary, syncarpous, trilocular with many ovules in each locule on parietal placentation; stigma 2.

Fruit : A capsule.

Floral formula : † ♀ $P_{3+3}\ A_1\ \underline{G}_{(3)}$

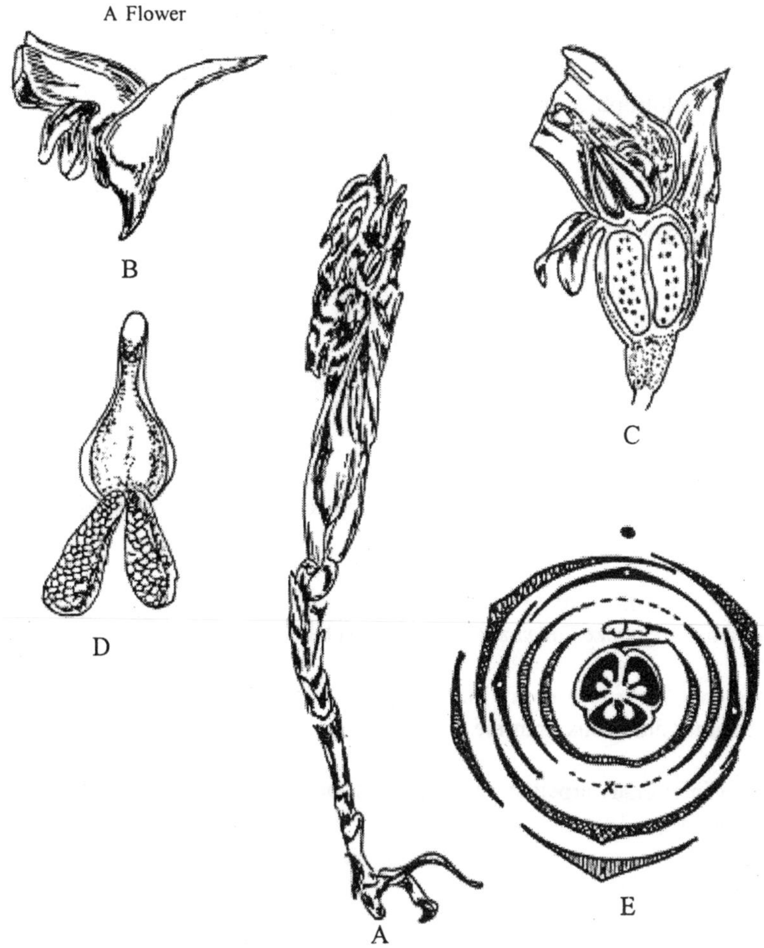

Fig 8.160 ***ZEUXINE SULCATA*** **(Orchidaceae)**

SPATHOGLOTTIS PLICATA

(Orchidaceae)

Habitat: Mesophyte, autotrophic, grows in shady places.

Habit: A perennial herb

Root: Adventitious root system

Stem: An underground horizontal rhizome. The rhizome gives rise to aerial shoot one per year. The food is stored in the pseudobulb (swollen region of the stem)

Leaf: Cauline, simple, shortly petiolate, exstipulate, linear, entire, acute; venation parallel unicostate.

Inflorescence: A raceme

Flower: Pedicellate, bracteate, ebracteolate, cyclic, complete, trimerous, homochlamydeous, zygomorphic, bisexual and epigymous.

Perianth: Tepals 6, in two whorls of 3 each, perianth lobes of outer whorl equal; while in the inner whorl one of the lobes is of a different size and shape and is called labellum.

Androecium: Stamens 3, one fertile, two staminodes. A gynostegium is formed by the fusion of style stigma and stamen ; anther dithecous, introrse

Gynoecium: Ovary inferior, tricarpellary, syncarpous, trilocular with many ovules in each locule on parietal placenta. Stigmatic lobes 3.

Fruit: A capsule

Floral formula: † ♀ $P_{3+3} A_1 \underline{G}_{(3)}$

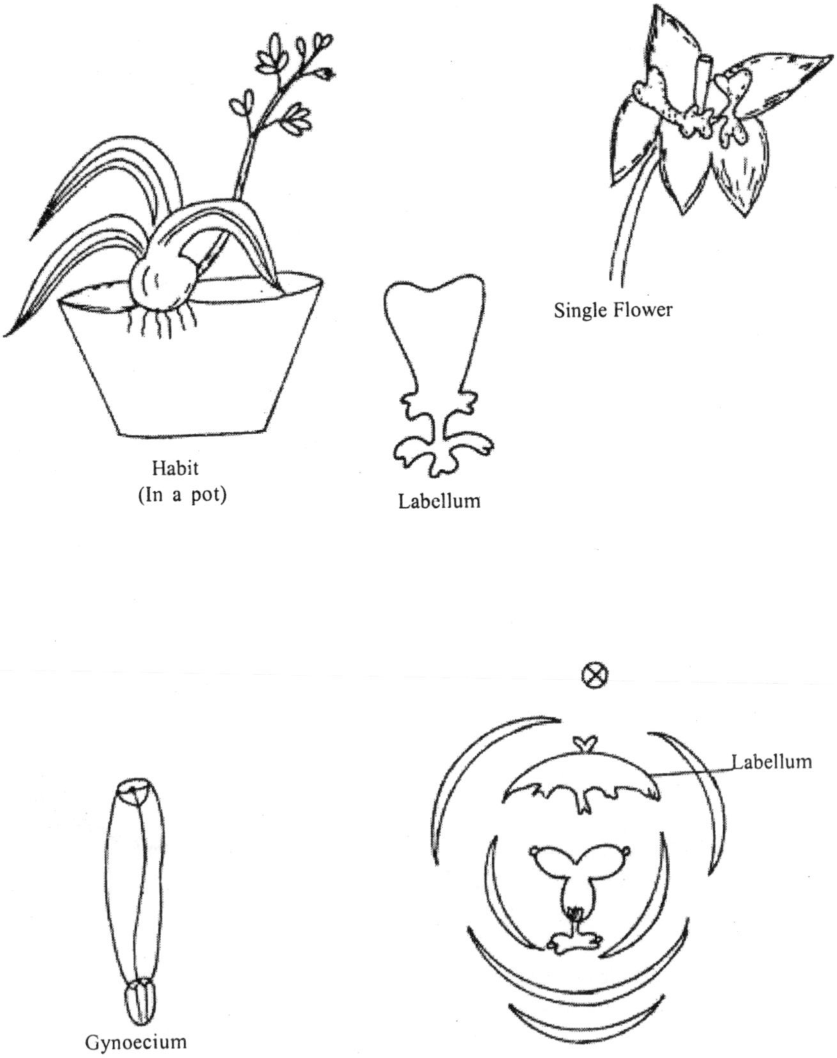

Fig 8.161 *SPATHOGLOTTIS PLICATA* (Orchidaceae)

VANDA ROXBURGHI

(Orchidaceae)

Habitat: An epiphyte, usually grows on the branches of trees.

Habit: An epiphytic herb

Root : Adventitious root system. There are two kinds of roots clinging roots, that hold on to the supporting branch of the host tree and hanging aerial roots – that are green in colour. The hanging roots have a special tissue called velamen to absorb moisture from the atmosphere

Stem : Foliar or scapose, branching monopodial, produces leaves continuously with axillary flower clusters.

Leaf : Cauline, simple, alternate and distichons, linear and fleshy, distichous, leaf base sheathing, as the axis is short, it appears as if the leaf directly arises from the substratum. Leaf margin entire, apex acute; venation parallel.

Inflorescence : A raceme

Flower : Pedicellate, bracteate, ebracteolate, complete, cyclic, trimerous, homochlamydeous, zygomorphic, bisexual and epigynous.

Perianth : Tepals 6, arranged in 2 whorls of 3 each, outer whorl sepaloid, inner petaloid, polypetalous, aestivation imbricate, tepals unequal in size different in shape; one of the lobes of the inner whorl gets enlarged and becomes a labellum. The labellum is spurred.

Androecium : Stamens basically 6, in two whorls of 3 each; all but one are absent; the single stamen unites with the extension of the gynoecium forming a column; anthers diehecous, introrse splitting open by longitudinal slit pollen grains collect together to form two pollinia.

Gynoecium : Ovary inferior, tricarpellary, syncarpous, unilocular with many ovules in parietal placenta. Stigma 3 lobed, sterile and 2 fertile, sterile stigmatic lobe forms a rostellum.

Fruit : A capsule

Seed : Numerous, very small.

Floral formula : † ♀ $P_{3+3}\, A_1\, \underline{G}_{(3)}$

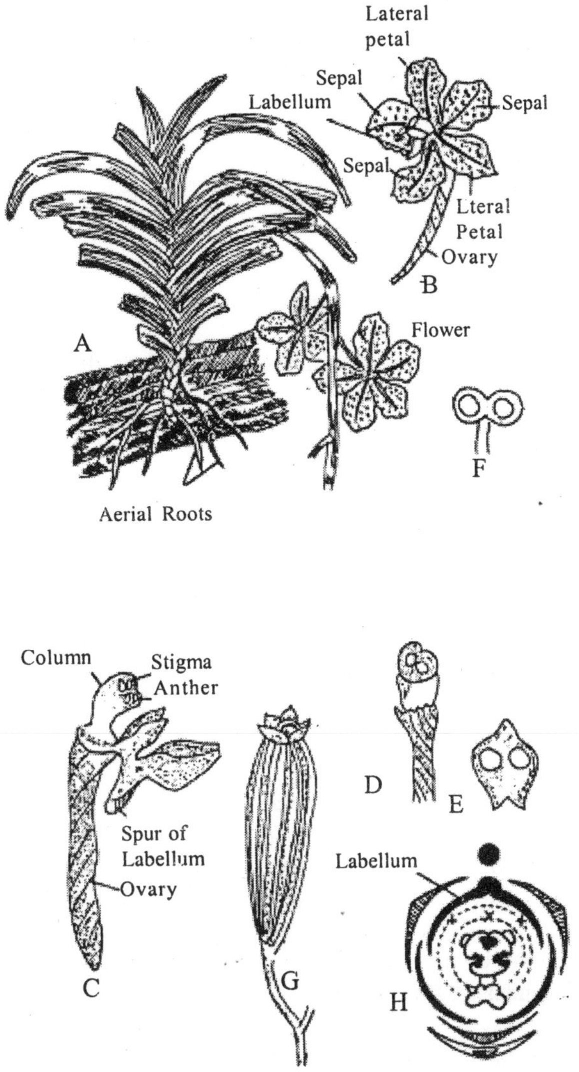

Fig 8.162 ***VANDA ROXBURGHI*** **(Orchidaceae)**

COMMELINA BENGALENSIS

(Commelinaceae)

Habitat : Mesophyte, grows in moist soil.

Habit : Annual herb

Root : Adventitious, arises at the nodal regions as well as at the base of the stem.

Stem : Aerial, weak, prostrate, creeping, rooting at the nodal regions, dichotomously branched.

Leaf : Sessile or shortly petiolate, broadly ovate or oblong obtuse, hairy on both surfaces; leaf base oblique and sheathing

Inflorescence : Biparous cyme, arising from the conduplicate spathe (bract) one at a time. In the two branches of the cyme, the flowers of upper branch small and deciduous, of the lower branch fertile. Bracts 1-3 in number, auricled on one side and hirsute.

Flower : Pedicellate, bracteate, ebracteolate, cyclic, complete, heterochlamydeous, trimerous, zygomorphic, bisexual and hypogynous. Flowers blue in colour. Some flowers cleistogamous.

Calyx : Sepals 3, free, small, oblong, hairy, aestivation valvate.

Corolla: Petals 3, free, blue, longer; one largest and clawed; aestivation valvate.

Androecium : Stamens 6, of which 3 are sterile, free; filaments long; anthers bithecous, introrse and have a longitudinal dehiscence.

Gynoecium : Ovary superior, tricarpellary, syncarpous trilocular with 1-2 ovules in each locule on axile placenta. Style simple and slender; stigma capitate.

Fruit : A capsule; hidden in spathes and 3 valved.

Seed : Endospermic

Floral formula : † ⚥ $K_3\ C_3\ A_{3+3}\ \underline{G}_{(3)}$

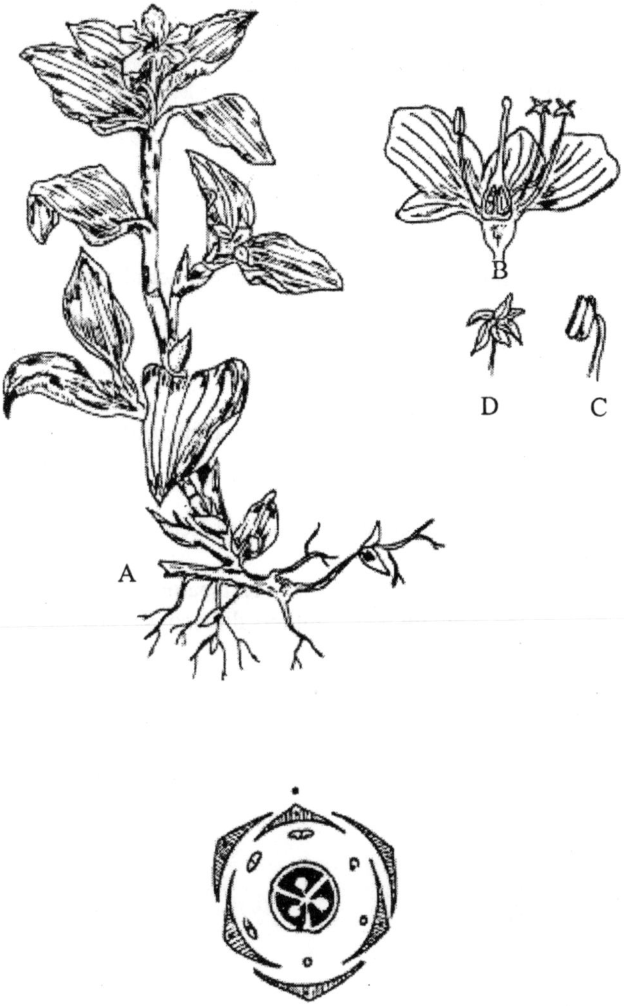

Fig 8.163 ***COMMELINA BENGALENSIS*** **(Commelinaceae)**

ASPHODELUS TENUIFOLIUS

(Liliaceae)

Habitat: Mesophytic, grows in moist soil.

Habit: Annual herb

Root: Adventitions

Stem: Underground, reduced, it produces leaves as well as the aerial scape bearing flowers. The scape cylindrical, green and branched.

Inflorescence: A raceme

Flower: Pedicellate, bracteate, ebracteolate, complete, cyclic, trimerous, homochlamydeous, bisexual and hypogynous; flowers white in colour.

Perianth: Tepals 6 in number, free, arranged in two whorls of 3 each; aestivation imbricate.

Androecium: Stamens 6, arranged in 2 whorls of 3 each, epitepalous; filaments of inner whorl shorter than those of the outer whorl; base of the filament hairy; anthers dithecous, versatile and introrse.

Gynoecium: Ovary superior, tricarpellary, syncarpous, trilocular with two ovules in each locule on axile placenta. Style slender; stigma bilobed.

Fruit: A capsule

Seed: Small, rugose.

Floral formula: ✝ ⚥ $P_{3+3}\ A_{3+3}\ \underline{G}_{(3)}$

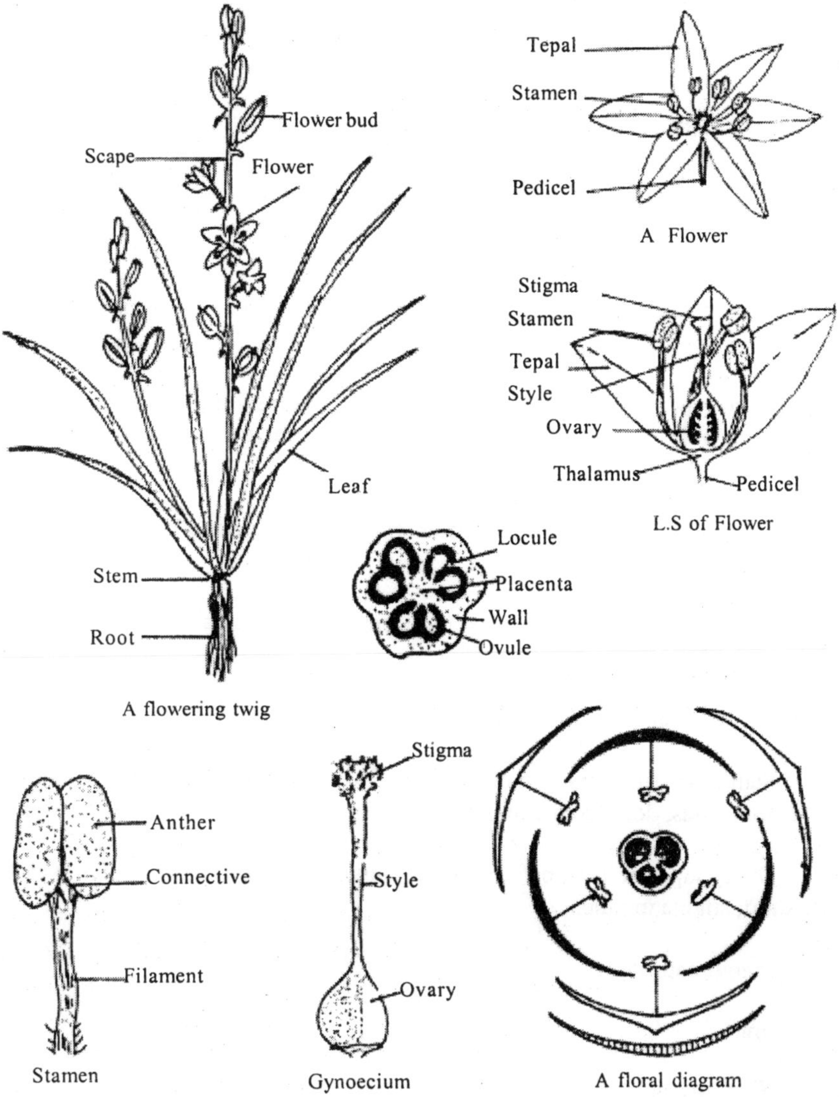

Fig 8.164 ***ASPHODELUS TENUIFOLIUS*** **(Liliaceae)**

ALLIUM CEPA

(Liliaceae)

Habitat: Mesophyte, grows in well aerated, moist soil; cultivated.

Habit: An annual herb with pungent aroma.

Root: Adventitious root system

Stem: Underground, concentric bulb; the stem is highly reduced and is surrounded by fleshy storage leaves.

Leaf: There are two types of leaves – storage leaves and photosynthetic leaves. Storage leaves reddish white and surround the bulb. Photosynthetic leaves – green, radical, cylindrical, leafbase sheathing, fistular, venation unicostate parallel.

Inflorescence: Cymose umbel several monochasial cymes aggregate in an umbellate fashion and enclosed by 2-3 membranous bracts.

Flower: Pedicellate, bracteate, ebracteolate, cyclic, complete, homochlamydeous, trimerous, bisexual and hypogynous.

Perianth: Tepals 6, free, arranged in two whorls of 3 each, connate at the base, white, aestivation valvate.

Androecium: Stamens 6, free, arranged in 2 whorls of 3 each antitepalous; filaments narrow and dilated at the base; anthers dithecous, dorsifixed and introrse.

Gynoecium: Ovary superior, tricarpellary, syncarpous, trilocular with two ovules in each locule on axile placenta. Style short; stigma minute.

Fruit: A membranous capsule

Seed: Endospermic

Floral formula: ✝ ♀ $P_{3+3}\ A_{3+3}\ \underline{G}_{(3)}$

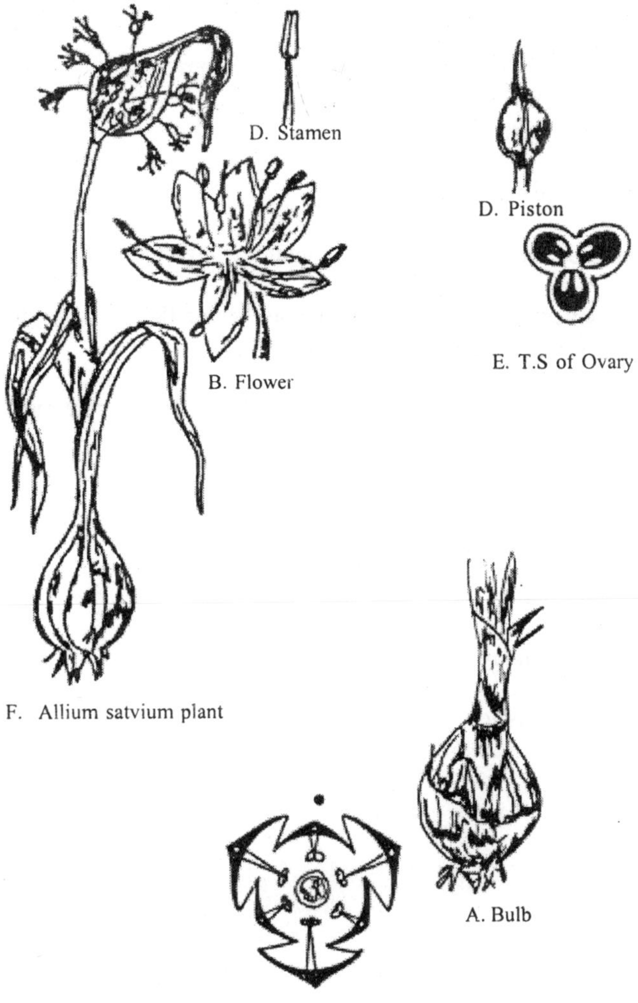

Fig 8.165 *ALLIUM CEPA* (Liliaceae)

ASPARAGUS OFFICINALIS

(Liliaceae)

Habit : Xerophytic, grows in dry soil.

Habit : A prickly ornamental large herb or under shrub

Root : Adventitious, some roots store food, white in colour are aggregated into 'fasciculated roots'.

Stem : Aerial, branched, grooved. At the nodes the branches becom flattened, green, grow up to one inter nodal length and form cladodes which look like leaves.

Leaf : Reduced, small scales found at the base of the cladodes, the stipules are modified into spines.

Inflorescences : Racemose panicle

Flower : Same as in *Asphodelus*

Perianth : Same as in *Allium cepa*

Androecium : Same as in *Asphodelus*

Gynoecium : Same as in *Asphodelus*

Fruit : A berry

Seed : Endospermic

Floral formula : ✝ ⚥ $P_{3+3}\,A_{3+3}\,\underline{G}_{(3)}$

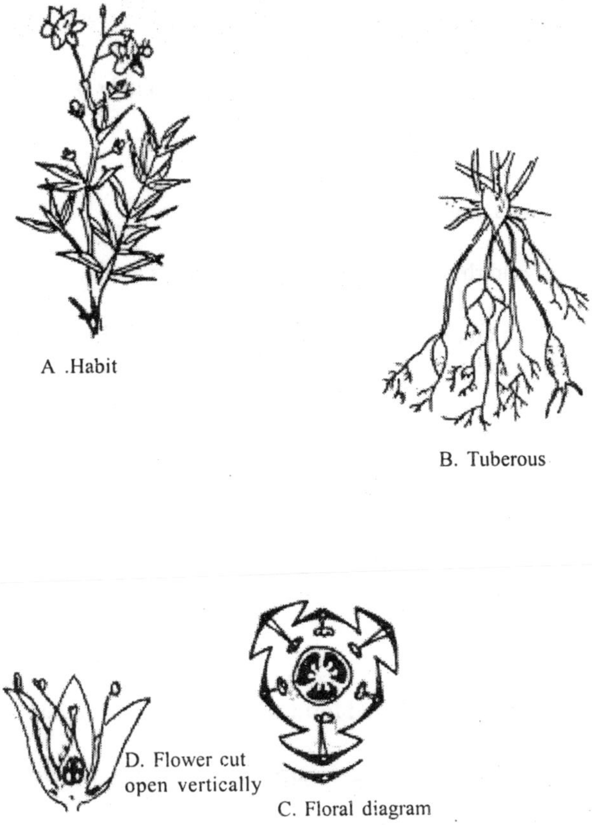

Fig 8.166 *ASPARAGUS OFFICINALIS* (Liliaceae)

ALOCASIA INDICA

(Araceae)

Habitat : Mesophytic, but grows in moist, cool and shady areas; cultivated for foliage and edible rhizome.

Habit : Annual herb,

Root : Adventitious; some roots also arise at the nodes.

Stem : Underground rhizome, but a part of it becomes aerial.

Leaf : Large, petiolate, petiole very long swollen at the base; lamina ovate, cordate or sagittate, usually peltately atached to the petiole; venation parallel, Raphides present in the cells of the leaf.

Inflorescence : A spadix; covered by a leafy spathe. Spadix shorter than the spathe; Flowers unisexual, but both male and female flowers found on the same spadix; female above and male below, neutral flowers found in between male and female flowers.

Flower : Sessile, bracteate, ebracteolate, incomplete, achlamydeous, actinomorphic, unisexual (pistillate or staminate) and hypogynous.

Perianth : Absent

Androecium : Stamens 3-5 fused forming a synandrium, anthers dithecous, dehiscing by pores.

Gynoecium : Ovary superior, 2-4 carpels, syncarpous, unilocular with few basal ovules. Style short; stigma lobed (2-4).

Fruit : A berry, red coloured, few seeded

Seed : Black coloured, albuminous, testa smooth.

Floral formula : Male flower : ✝ ♀ $P_0 A_{(3-5)} \underline{G}_0$

Female flower : ✝ ♀ $P_0 A_0 \underline{G}_{(2-4)}$

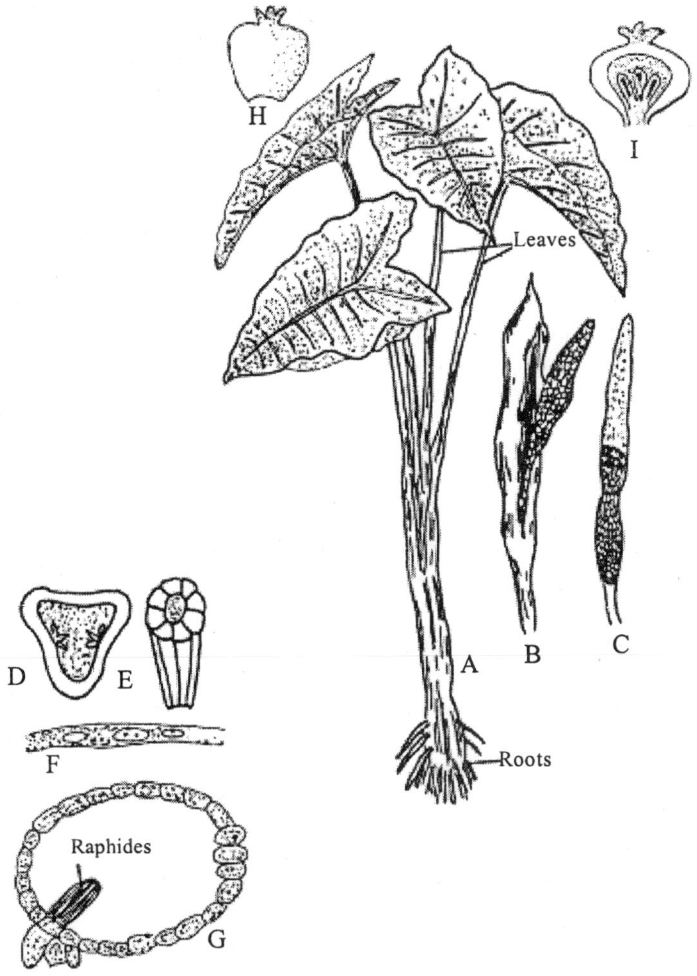

Fig 8.167 ***ALOCASIA INDICA*** (Araceae)

PISTIA STRATIOTES

(Araceae)

Habitat: Hydrophyte, found floating in tanks, ponds etc.

Habit: Aquatic, floating herb

Root: Adventitious, no root caps, root packets present.

Stem: Modified, stolon, horizongtal on the surface of water, cylindrical, soft, contains a watery sap, sap is pungent, spongy and green in colour.

Leaf: Radical simple, sessile, clustered, large, coriaceous, spongy; venation parallel; heavily pubescent.

Inflorescence: A spadix, enclosed by a spongy spathe.

Flower: Sessile, small, bracteate, ebracteolate, actinomorphic, trimerous, cyclic, incomplete, unisexual, (staminate or pistillate) and hypogynous (in female flower). In the spadix the female flowers are towards the base and male flowers above, in between, the neutral flowers are present.

Perianth: Tepals 4-6, cupulate, small, connate at the base.

Androecium (in male flowers): Anthers two, filament zero, fused to form a synandrium

Gynoecium: Ovary superior, monocarpellary, unilocular with many ovules in each locule on basal placenta. Style short; stigma disc shaped or lobed; pistils are embedded in the fleshy spathe.

Fruit: A berry, many seeded

Seed: Endospermic

Floral formula: Male flower: † ♀ $P_{4-6} A_2 \underline{G}_0$

Female flower: † ♀ $P_{4-6} A_0 \underline{G}_1$

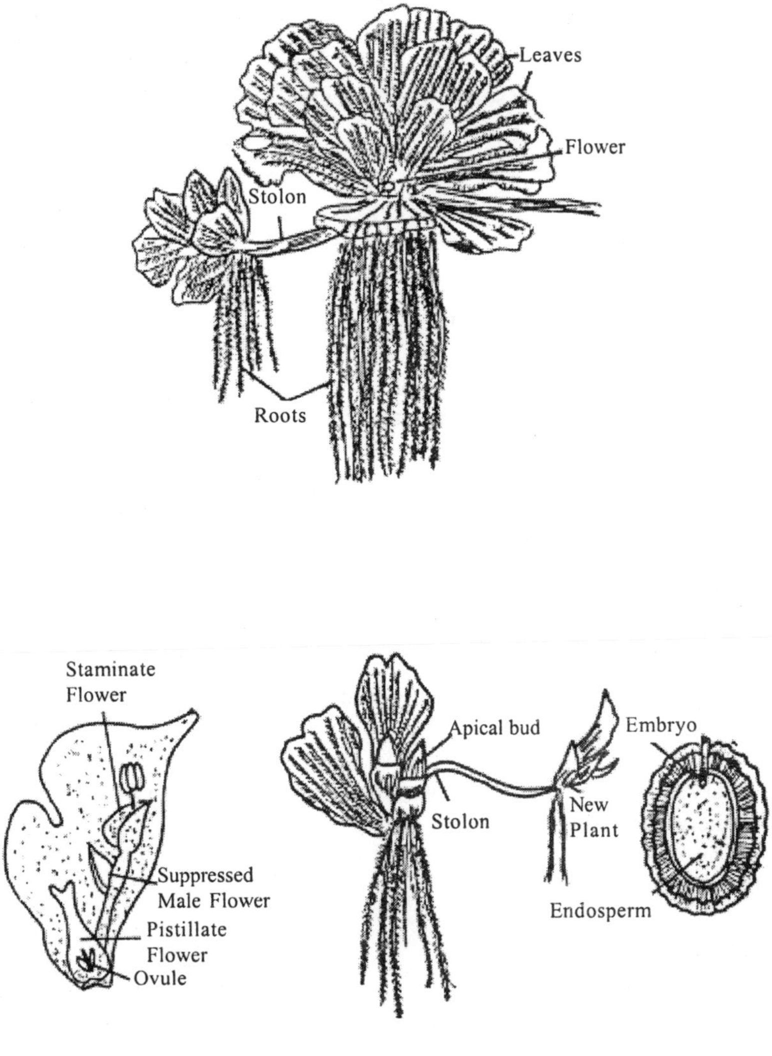

Fig 8.168 *PISTIA STRATIOTES* (Araceae)

CYPERUS ALTERNIFOLIUS

(Cyperaceae)

Habitat : Plants grow in water logged soil or marshy region

Habit : A perennial herb

Root : Adventitious root system

Stem : Underground, rhizomatous, partly aerial and erect also, herbaceous, unbranched, triangular, solid and glabrous. Aerial part of the stem is called a culm.

Leaf : Cauline, simple, alternate, 3 ranked ⅓ sessile, exstipulate, sheathing leaf base, complete and fused, margin entire, linear, acute or acuminate; venation parallel multicostate.

Inflorescence : Racemose; spike with a number of spikelets surrounded by glumes

Flower : Sessile, bracteate, (glume) bracts dry, greenish brown, boat shaped) incomplete, trimerous, cyclic, zygomorphic, bisexual and hypogynous

Perianth : Absent

Androecium : Stamens 3, free; filaments long, flattened, exserted; anthers linear, basifixed, diethecous and have a pointed apex.

Gynoecium : Ovary superior, tricarpellary, syncarpous, unilocular with a single basal ovule; style single; stigma long and feathery, projecting out.

Fruit : A nut, compressed, 3 angled

Seed : Endospermous

Floral formula : † ♀ $P_0 A_3 \underline{G}_{(3)}$

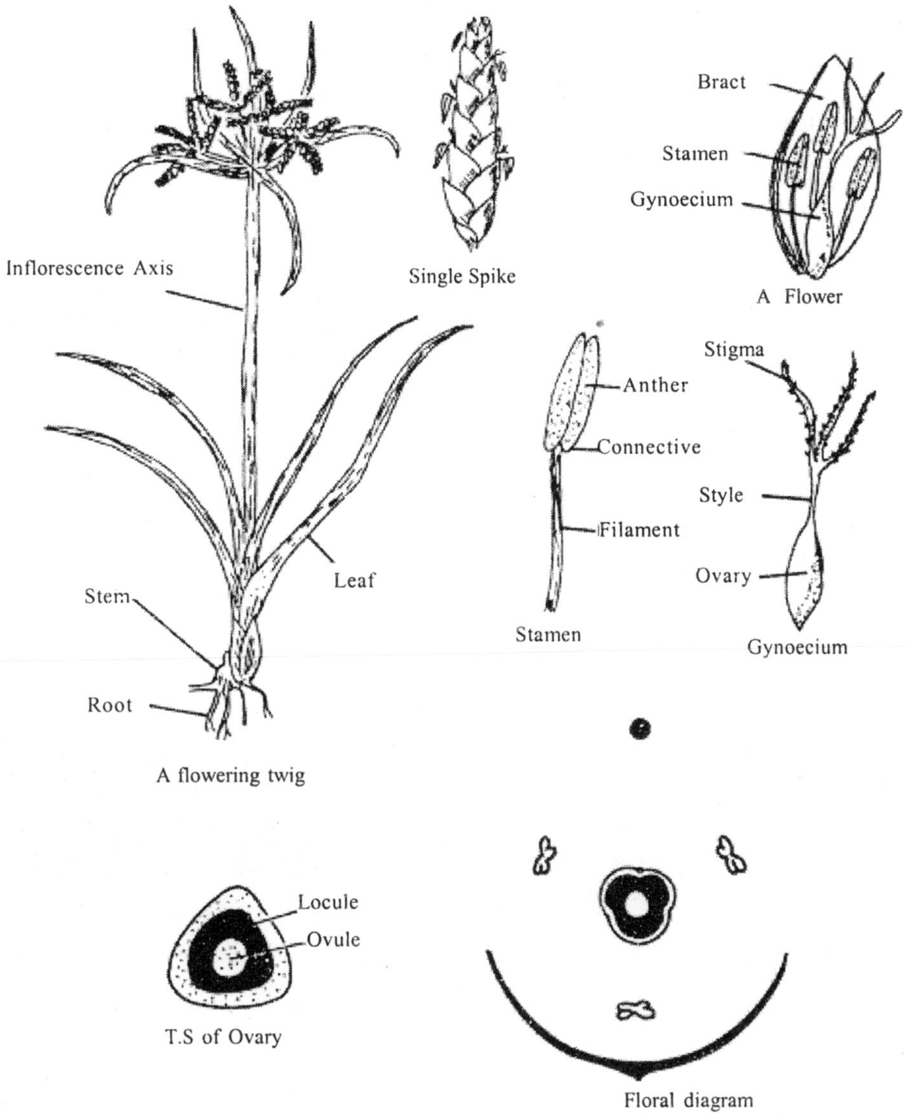

Fig 8.169 ***CYPERUS ALTERNIFOLIUS*** **(Cyperaceae)**

SCIRPUS ROYELEI

(Cyperaceae)

Habitat : Grows in moist soil, marshy area etc.

Habit : A perennial herb

Root : Adventitious arising from the rhizome

Stem : Underground, rhizomatous, terete, nodes and inter nodes not very prominent, a cluster of spikelets are borne on the culm (the aerial part of the stem which is hollow)

Leaf : Sessile, simple, leafbase sheathing, linear, rough, entire, apex pointed, venation parallel.

Inflorescence : Spike, consists of a number of spikelets borne on the green traingular culm.

Flower : Sessile, bracteate (bract is represented by the glume), small, incomplete, trimerous, cyclic zygomorphic, bisexual and hypogynous.

Perianth : Reduced to bristles

Androecium : Stamens 3, free, anthers linear, large and basifixed

Gynoecium : Ovary superior tricarpellary, syncarpous, unilocular with a single basal ovule. Style single, stigmas 3, long and feathery.

Fruit : A nut, trigonous.

Seed : Endospermic

Floral formula : † ♀ $P_0 A_3 \underline{G}_{( )}$

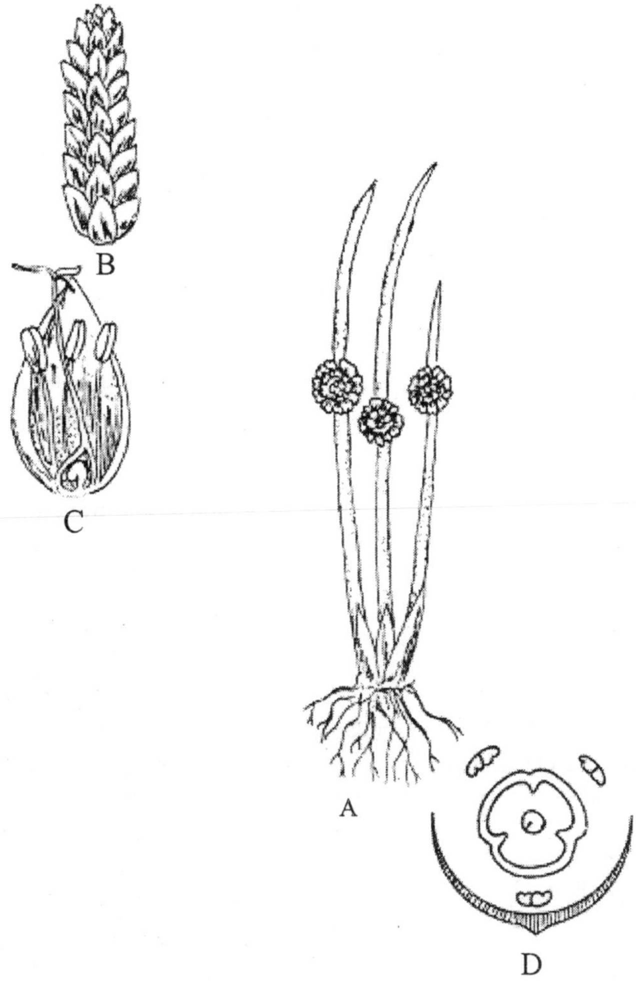

Fig 8.170 ***SCIRPUS ROYELEI*** **(Cyperaceae)**

TRITICUM ANGARE

(Gramineae)

Habitat : Grows in cultivated fields

Habit : Annual herb

Root : Adventitious root system

Stem : Aerial, erect, herbaceous, cylindrical, hollow, green, branched, branching by tillers, nodes solid and swollen, glabrous.

Leaf : Cauline, simple, alternate, sessile, two ranked, ligulate, exstipulate, leaf base sheathing, margins of the sheath do not encircle the stem, lamina-linear lanceolate, entire, acute; venation parallel unicostate. Leaf surface rough due to the presence of silica.

Inflorescence : A spike; branched, consists of many spikelets, each spikelet has 3-4 flowers and has at its base two sterile glumes, one lower lemma (palea), and one upper lemma (palea). Glumes are regarded as bracts and lemma as bracteoles.

Flower : Sessile, bracteate, bracteolate, cyclic, complete, zygomorphic, trimerous, bisexual and hypogynous.

Prianth : Tepals 2, reduced, membranous, free and are called lodicules.

Androecium : Stamens 3, free, (one anterior, two lateral) filaments slender; anthers dithecous, versatile and introrse

Gynoecium : Ovary superior, monocarpellary, unilocular with a single ovule on marginal placenta style absent, stigma two, long and feathery.

Fruit : Caryopsis.

Floral formula : † ♀ $P_0 A_3 \underline{G}_{(3)}$

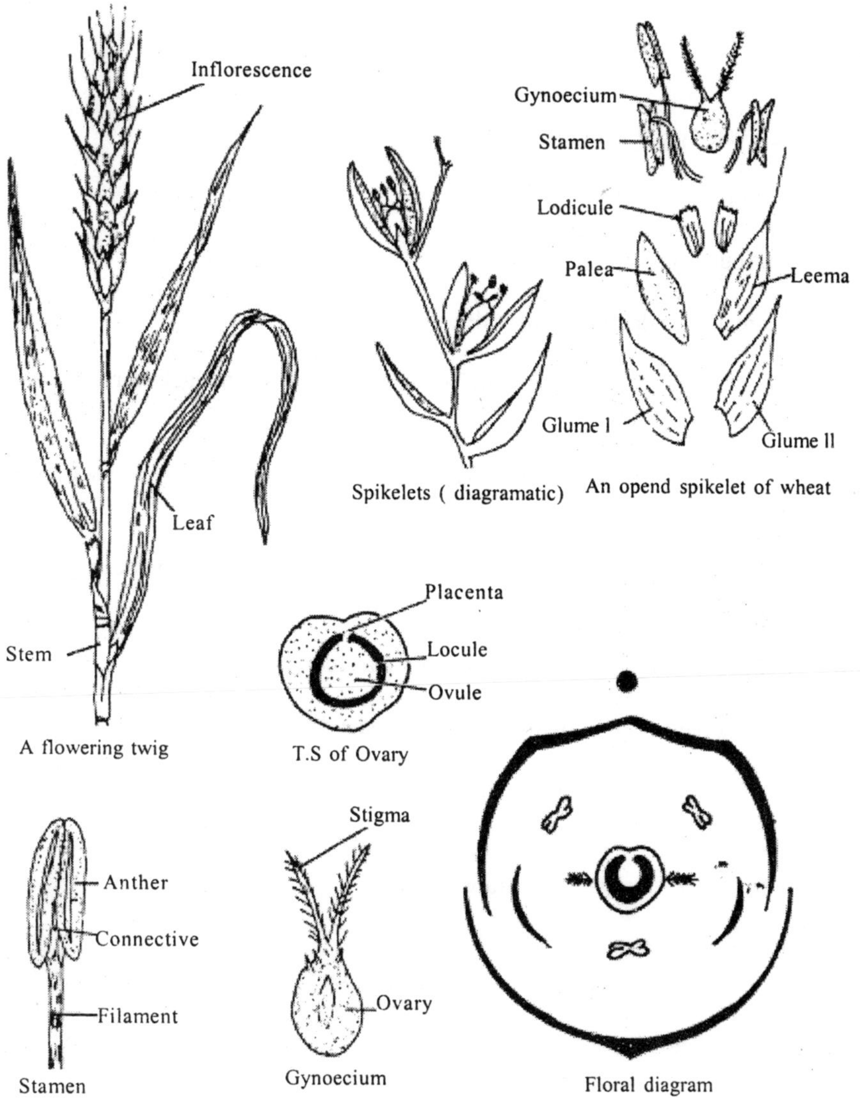

Fig 8.171 *TRITICUM ANGARE* (Gramineal)

AVENA SATIVA

(Gramineae)

Habitat: Cultivated plant in fields

Habit: Annual herb

Root: Adventitious root system

Stem: Aerial, erect, cylindrical, green, nodes swollen, fistular.

Leaf: Cauline, alternate, exstipulate, sessile, ligulate, leaf base sheathing, linear, acute; venation multicostate parallel.

Inflorescence: Spike, branched in a paniculate manner consisting of spikelets; each spikelet has 2-3 flowers enclosed between 2 glumes.

Flower: Sessile, bracteate, (there are two palea equivalent to bracteoles), cyclic, zygomorphic, bisexual and hypogynous.

Perianth: Reduced, represented by two membranous lodicules.

Androecium: Stamens 3, free; filaments long and slender anthers dithecous, versatile and introrse.

Gynoecium: Ovary superior, monocarpellary, unilocular, with a single basal ovule; style short, single, stigmas two and feathery.

Fruit: A single seeded carypsis, Albuminous.

Floral formula: † ♀ $P_2 A_3 \underline{G}_1$

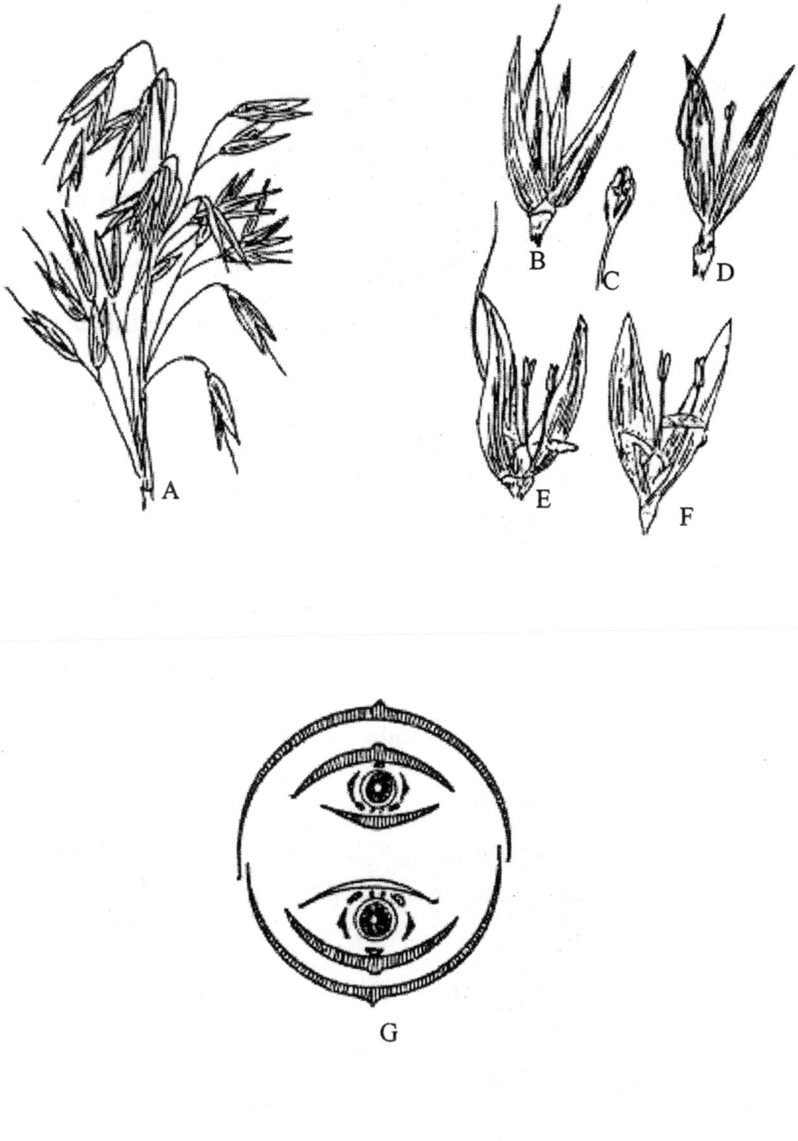

Fig 8.172 *AVENA SATIVA* (Gramineae)

ZEA MAYS

(Gramineae)

Habitat: Cultivated in open fields, during rainy season

Habit: Annual herbaceous plant, reaching 5-6 feet.

Root: Adventitious; in addition stilt roots develop from the lower nodes of the stem.

Stem: Aerial, erect, solid, with prominent nodes which are swollen.

Leaf: Cauline simple, alternate ligulate, long, flat, sheathing base, amplexicaul, linear, margin entire; venation parallel multicostate.

Inflorescence: Flowers unisexual, but the plant is monoecious; male (staminate) inflorescenceis a terminal panicle. The panicle has a main axis with several lateral branches, many spikelets are borne on the lateral branches. Each spikelet is 2 flowered. Pistillate (female) inflorescence is a spadix commonly called cob, enclosed by a number of large foliaceous bracts; female inflorescences are axillary.

Male flower: Male spikelets are subtended by two glumes. Each maleflower has a pair of palea, (superior and inferior), flowers sessile, incomplete, unisexual and staminate.

Perianth: Reduced, represented by a pair of lodicules.

Androecium: Stamens 3, free; filaments long; anthers linear, verstaile and introrse.

Gynoecium: Represented by a pistillode.

Female flower: Arranged compactly in several long rows on the fleshy axis. Each spikelet is 2 flowered, and subtended by 2 glumes. Each female flower is enclosed by a pair of palea. Flowers sessile, incomplete unisexual, pistillate and hypogyneous.

Perianth: As in the male flowers.

Gynoecium: Ovary superior, monocarpellary, unilocular with a single ovule. Style filiform; stigma long and feathery, comes out of the bracts from the top and droops down like a tuft.

Fruit: A caryopsis. Seed endospermous.

Floral formula: Male flower: † ♀ $P_2 A_3 \underline{G}_0$

Female flower: † ♀ $P_2 A_0 \underline{G}_1$

Fig 8.173 ***ZEA MAYS*** **(Gramineae)**

PART TWO
ECONOMIC BOTANY

CHAPTER – 9

ECONOMIC BOTANY

FIBRE YIELDING PLANTS

Fibre yielding plants provide fibres for various uses – clothing, rope and mat manufacturing etc. The following is a brief description of various plant products yeilding fibres.

Cotton

(a) *Gossypium arboreum, G. herbaceum*
(b) Malvaceae
(c) Epidermal hairs arising on the seed coat provide the cotton fibres of commerce; these hairs are cellulosic.
(d) Cotton fibre find extensive use in textile industry; It is the chief material used by human beings for clothing.

Kapok

(a) *Ceiba pentandra*
(b) Malvaceae
(c) Woody outgrowth of the fruit (ovary wall), single celled lustrous hairs
(d) Commonly called silk cotton, it is extensively used as filing fibres for pillows, cushion beds etc.

Coir

(a) *Cocos nucifera*
(b) Palmae (Arecaceae)
(c) Mesocarp of the fruit contains sclerenchyma fibres.
(d) The mosocarp consists of course fibres which constitute the coir of commerce. Fibres are exensively used in making ropes, cricket rugs, door mats, mattresses etc.

Red silk cotton

(a) Salmalia malabarica
(b) Bombacaceae
(c) Seed contains fibres
(d) Commonly called Indian Kapok, the soft fibres are used for stuffing sofas, mattresses etc.

White Silk Cotton

(a) Cochlospermum religiosum
(b) Cochlospermaceae
(c) ....
(d) The silky fibres are used for stuffing in sofa, pillows, cushions etc.

Fig 9.1 Flowering shoot of sea island cotton (Gossypium barbadense)

Manila Hamp

(a) *Musa textilis*
(b) Musaceae
(c) Fibres are obtained from leaf stalks (sheathing leaf bases)
(d) The long and shiny fibres are used for cordage (rough weaving); fibres finding an extensive demand on marine cables, binder twines etc.

Agave fibres

(a) *Agave sisalana, A. cantala, A. funkiana* etc.
(b) Agavaceae
(c) Fibres are obtained from the fleshy leaves
(d) Used in making of ropes, bags etc.

Bowstring Hemp

(a) *Sansevieria zeylanica, S.longifolia* etc.
(b) Liliceae
(c) Fibres obtained from the fleshy leaves
(d) Used for cordage and matting

Pine Apple Fibre

(a) *Ananas Cosmosus*
(b) Bromeliaceae
(c) Leaves yeild a good quality fibre
(d) The fibres are shiny white, durable and water resistant. In philippines a variety of cloth is manufactured from this fibre.

Wicker work fibre

(a) *Calamus rotang*
(b) Palmae
(c) Twining stem is fibrous
(d) Stem is stripped and the fibres so obtained are used in making baskets, chairs, cots, divans etc. Older stems are used as ropes, walking sticks etc.

Brush Fibres

(a) *Caryota urens*
(b) Palmae
(c) ....
(d) Used for making brushes

Jute

(a) *Corchorus capsularis*
(b) Tiliaceae
(c) Bast or phloem fibres obtained from the stem, it consists of many schlerenchyma cells.
(d) Used in the manufacture of sacks, bags etc. Popularly called gunny bags; these are used for packing onion, sugar etc.

Flax

(a) *Linum usitatissimum*
(b) Linaceae
(c) Phloem fibres from the stem are 1-3 feet long threads. Threads are also used for fishing lines nets etc.

Hemp

(a) *Cannabis sativa*
(b) Cannabinaceae
(c) Fibres are obtained from stem; cells of pericycle become lignified and form long fibres
(d) Hemp fibres are used in the manufacture soft ropes, twines, sacks, tarpaulins etc. Fibres are also used for rough weaving.

Sunn Hemp

(a) *Crotalaria juncea*
(b) Fabaceae (Leguminosae) – Papilionaceae
(c) Fibres are present in the stem
(d) Used for making ropes, mats etc. Also used for making canvas, rope soles of shoes and sandals, etc.

Ramie Fibre

(a) *Boehmeria nivea*
(b) Urticaceae
(c) Fibres obtained from 10 month old stems. The fibre is the longest, toughest and most silky of all the plant fibres.
(d) Used in Europe for sacks, sail cloth, belting, table cloth, sheeting, nets, cordage, paper etc. Gas mantles of good quality is also manufactured from the ramie fibre

Roselle Hemp

(a) *Hibiseus sabdariffa*
(b) Malvaceae
(c) Stem fibres; obtained after soaking and retting the stem.
(d) Used for sacks, cordage, ropes, fishing nets. Bags are also manufactured for packing sugar.

Congo Jute

(a) *Urena lobata*
(b) Malvaceae
(c) Stem fibres; fibres become free after soaking in water and retting
(d) Grown as a fibre crop in Brazil, it is also called aramina fibre. Used for making ropes, carpets lessam etc.

OIL YIELDING PLANTS

FATTY OILS

Coconut Oil

(a) *Cocos nucifera*
(b) Palmae
(c) The endosperm (meat of coconut) on maturity contains the oil. On crushing the oil is released.
(d) Used in cooking in coastal areas of our country. Also used in hair oils, shampoos, soaps, shaving creams etc. The oil is a good lubricant. Oil cake is a good cattle feed.

Palm Oil

(a) *Elaeis guinensis*
(b) Palmae
(c) Fibrous pulp of the fruit contains oil; on boiling the oil is released.
(d) Used in the manufacture of soaps and margarines. It is also a good lubricant

Mahua Oil

(a) *Madhuca indica*
(b) Sapotaceae
(c) Seeds contain the oil; the oil solidifies during cold weather
(d) Used in the manufacture of washing soaps. The oil is also a good remedy for rheumatism.

Nutmeg Butter

(a) *Myristica fragrans*
(b) Myristicaceae
(c) Seeds contain up to 40% of a yellow fat called nutmeg butter
(d) The flat is used in soaps, perfumery and candles. It is also good for rheumatism.

Pongam Oil

(a) Pongamia pinnata
(b) Leguminosae – Papilionaceae
(c) Seeds (Cotyledons) contain oil
(d) Used as a lubricant and illuminant, used in the treatment of rheumatism and skin ailments; also used in soaps.

Linseed Oil

(a) *Linum Usiatissinum*
(b) Linaceae
(c) Seeds are the source of oil. The oil content is about 33-47%.
(d) Used in the manufacture of soaps, paints, varnishes, linoleum, printers ink etc. The oil is also used for seasoning of sports goods. Oil cake is a good cattle feed.

Safflower Oil

(a) *Carthamus tinctorius*
(b) Compositae (Asteracea)
(c) Achenes contain oil up to 24-36%
(d) The oil has very little saturated fatty acids. It is a good cooking oil as it has very little cholesterol. It is also used in paints and varnishes. Oil cake is a good cattle feed.

Niger seed oil

(a) *Guizotia abyssinica*
(b) Compositae (Asteraceae)
(c) Achenes contain the oil
(d) Used in cooking. The achenes themselves are occasionally fried and eaten. Oil is also used in pharmaceutical industry. Oil cake is a good cattle feed as well as manure.

Soya bean oil

(a) Glycine max
(b) Leguminosae – Papilionaceae
(c) Colyledons in the seed have on oil content of 22%
(d) The oil is used in paints soaps and varnishes. Soya oil cake is used in plastics, adhesives etc. The cake is a good livestock feed. Soyabean flour is obtained from dried cotyledons and is used as a fertilizer.

Walnut Oil

(a) Juglans Regia
(b) Juglandaceae
(c) Kernels in the fruit are the oil source.
(d) The oil is edible and is also used in paints soaps and printing ink.

Tung oil

(a) *Alcurites moluccana*
(b) Euphorbiaceae
(c) Seeds constitute the source of oil
(d) This is used extensively in waterproofing.

Cotton seed oil

(a) Gossypiium spp
(b) Malvaceae
(c) Seeds contain up to 35% oil.
(d) The oil has very little quantity of satureted fatty acids and is perhaps the best cooking medium. The oil is also used in soaps and margarines. The oil cake is a good cattle feed.

Sesame Oil

(a) *Sesamum indicum*
(b) Pedaliaiceae
(c) Seeds contain up to 52% oil.
(d) Used in cooking soap manufacture, paints, hydrogenation, perfumery etc. Seeds are also used in sweet meats. The oil cake is rich in proteins, calcium and phosphorus; used as cattle feed.

Rape Oil

(a) *Brassica campestris*
(b) Cruciferae
(c) Seeds contain 3-48% oil.

(d) Oil as well as the seed is used in seasoning for flavouring curries, pickles and vegetables. Rape oil is also used as a lubricant, in the manufacture greases, quenching steel plates and also in the manufacture of soft soaps. The oil cake is a good cattle feed and manure.

Mustard Oil

(a) *Brassica juncea, B. nigra*
(b) Crucifieiirae
(c) Seeds have the oil up to 48%
(d) Used as a condiment in the preparation of pickles. It is also used for seasoning.

Sunflower oil

(a) *Helianthus annus*
(b) Compositae (Asteracene)
(c) The Achenes contain up to 50% oil.
(d) Used as a cooking medium; Also used in the manufacture of vanaspati.

Corn Oil

(c) *Zea mays*
(b) Gramineae (Poaceae)
(c) Embryo contains about – 75% oil.
(d) Used in cooking, soap manufacture etc. It is also a good lubricant

Groundnut or Peanut-oil

(a) *Arachis hypogea*
(b) Leguminosae – Papilionaceae
(c) The underground pods have 2-4 seeds which are rich in oil. The seeds are edible. The oil content is about 40-50%
(d) Used as a cooking medium; also used in the manufacture of vanaspati after hydrogenetion. Poorer grades of oil used in soap manufacture. Oil cake is a good fertilizer as well as manure.

Castor Oil

(a) *Ricinus communis*
(b) Euphorbiaceae
(c) Seeds have 35-58% oil which is thick and non drying.
(d) In a pure condition castor oil is used as pergative; also used as lubricant for aircraft engines. Soaps, paints, plastic, varnishing etc are other products where castor oil is used.

Olive Oil

(a) *Olea europoea*
(b) Oleaceae
(c) Oil is contained in the mesocarp of the fruit (75%) as well as in the kernel (12-28%)
(d) Green fruits are pickled and canned. After removing the bitter glucoside 'oleuropein' the fruits are pickled in brine solution. The oil is used in cooking in the mediterranean and European contries. It is also used in fish cunning. In textile industry olive oil is helpful in wool combing. Cosmetics, toiletries and creams are other products where olive oil is used extensively. Oil cake is a good manure.

Taramira Oil

(a) *Eruca Sativa*
(b) Cruciferae
(c) Seeds are the source of oil.
(d) Used as a burning oil for lamps in villages. Oil cake is a good cattle feed.

Cacao butter

(a) *The obroma cacoa*
(b) Sterculiaceae
(c) Cocoa butter is obtained from the seeds. The butter is pale yellow and brittle. It is composed of glycerides of palmitic, stearic and oleic acids
(d) Cocoa butter is used in the manufacture of chocolate, confectimery. Cosmetic and some medicinal preparations.

Shea butter

(a) *Butryospermum paradoxum*
(b) Sapotaceae
(c) Seeds contain 45-60% oil; it is solid at temperature below 32°C.
(d) Used for cosmetics and illumination. It is also edible. Often used as a substitute for cocoa butter. Oil cake is used as cattle feed.

ESSENTIAL OILS

Rose Oil

(a) *Rosa damascena, R.alba R.centifolia*
(b) Rosaceae
(c) Petals contain oil; oil is extracted by distillation or solvent extraction.
(d) Used in perfumery and cosmetics.

Geranium Oil

(a) Pelargonium spp
(b) Geraniaceae
(c) Oil is distilled from the leaves; yield of 1gm/$_{1000}$ gms of leaf.
(d) Used in the manufacture of perfumes and soaps.

Cananga Oil

(a) *Cananga odorata*
(b) Annonaceae
(c) Fragrant flowers are the source of essential oil.
(d) Used in perfumery and toilet soaps.

Citrus Oil

(a) *Citrus aurantifolia*
(b) Rind of the fruit consists of oil
(c) Used in perfumery

Lavender Oil

(a) *Lavendula officinalis*
(b) Labiatae
(c) Flowers are the source of oil
(d) Used in perfumery, cosmetics and toilet soaps.

Lemongrass Oil

(a) *Cymbopogon flexuosus, C. citratus*
(b) Gramineae (poaceal)
(c) The leaves contain the essential oil which is extracted by water distillation

(d) The lemongrass oil is reddish yellow in colour with a strong odour and taste of lemon. It is used in perfumery and toilet soaps and also is a source of 'innoe' from which synthetic violet is made. It is also widely used in medicine as a flavouring agent. In recent years vitamin 'A' is also extracted from the oil.

Oil of citronella

(a) *Cymbopogan nardus*
(b) Gramineae (Poaceae)
(c) Leaves of the grass contain the oil
(d) This pale yellow oil is used extensively in inexpensive soaps and perfumes and also as an insect repellant. It is also used as a substitute for otto of roses as it contains 8—90% geranoil.

Mecassar Oil

(a) *Schliechera trijuga*
(b) Sapindaceae
(c) Seeds contain the oil which is released by solvent extraction method.
(d) The oil is used for culinary and lighting purpose in the malabar region of Kerala state. The oil has stimulating and cleaning properties to the scalp and as such it promotes hair growth. It is also used to cure itch and ache. Santal tribals use the bark to releive pain in the back. Other medicinal uses of the oil include treatment of skin ailments, rheumatism and headache.

Jasmine Oil

(a) *Jasminum officinale* var. grandiflorum
(b) Rubiaceae
(c) Flowers constitute the source of oil
(d) Used in perfumery, cosmetics and soaps

Sandal wood oil

(a) *Santalum album*
(b) Santalaceae
(c) Sandal wood oil is obtained by distillation of heartwood. The oil consists of α and β santanol
(d) The oil is used in perfumery, cosmetics and soap manufacture. It is an excellent fixature and used extensively in blends. The sweet scented wood is used in making boxes chests and also in carving statuettes.

Orange blossom oil (oil of Neroli)

(a) *Citrus aurantium*
(b) Rutaceae
(c) Orange blosom oil is obtained from flowers, and is different from citrus oil obtained from the rind of the fruits. Oil is obtained chiefly by steam distillation and occasionally by solvent extraction. Chemically the oil has linalol and various terpenes.
(d) The oil is chiefly used as an extender and as an ingredient in many synthetic perfumes as well as in eau de cologne.

Saffron Oil

(a) *Crocus sativus*
(b) Iridaceae
(c) Dried stamens and stigmas of the flower constitute the chief source of the oil.
(d) Used in perfumery, soaps and also for flavouring of sweatmeats.

Camphor Oil

(a) *Cinnamomum camphora*
(b) Lauraceae

(c) Twigs, barks and leaves are the source of oil which is obtained by steam distillation
(d) Used chiefly in medicine

Clove Oil

(a) *Eugenia caryophyllata*
(b) Myrtaceae
(c) Fragrant flower buds contain the oil.
(d) The oil is used in medicine as an effective remedy agair ooth ache. It is also used in microscopic studies as it has a high refractive index.

Cedar wood oil

(a) *Juniperus verginiana*
(b) Cupressaceae (Gymnosperms)
(c) Heartwood is the source of oil. It is obtained by steam distillation.
(d) The oil has a high refractive index and is used in microscopic studies

Champak Oil

(a) *Michelia champaka*
(b) Magnoliaceae (Labiatae)
(c) Perianth lobes of the flower contain this fragrant oil which is obtained by solvent extraction.
(d) Used in perfumery, cosmetics etc.

Patchouly Oil

(a) *Pogostemon Perilloides*
(b) Lamiaceae
(c) The essential oil is distilled from the fleshy leaves and young buds of the plant. The oil is dark brown in colour and is similar to sandal wood.
(d) Used in perfumes, soaps, hair oils and tobacco preparations.

Oil Rosemar

(a) *Rosemarinus officinalis*
(b) Labiatae
(c) The oil is distilled from the fresh flowering tops of the plant.
(d) Used in medicine, perfumery, deodourants and soaps.

Pippermint Oil

(a) *Mentha piperita*
(b) Labiatae
(c) Oil is distilled from the shoots and leaves.
(d) Used in confectionery, pharmaceutical and perfumery industry. Also used in the preparation of liquors such as Crème de mentha. The oil is a sources of menthol, very useful in the treatment of colds. Menthol is also used in tooth pastes.

Carway Oil

(a) *Anethum graveolens*
(b) Umbelliferae
(c) Leaf, Stem and seed on distillation yield the caraway oil. Caraway oil is also known as 'Dil Oil'
(d) The oil is used in flavouring pickles. Caraway is also a spice.

Encalyptus Oil.

(a) *Encalyptus globulus*
(b) Myrtaceae

(c) Leaves on distillation yield the oil.
(d) The oil is used in medicine, primarily for the treatment of cold related oilments.

TIMBER YIELDING PLANTS

Teak

Botanical Name : *Tectona grandis*
Family : Verbenaceae

(1) Teak is a large deciduos tree that grows on a variety of soils but requires perfect drainage.
(2) The tree is indigenous to India, Java and other islands of Indian archipelago. In our country teak grows naturally in Western Ghats and Madhya Pradesh.
(3) Leaves are elliptic, underside clothed with dense hairs; branchlets are quadrangular; flowers small and white; fruit a nut.
(4) Wood is hard and strong scented due to the presence of an oil which resists attack by fungi and white ants
(5) Sapwood is white and little in extent; heart wood dark golden yellow turning dark brown to black with age.
(6) Annual rings marked by one or more lines of regularly arranged pores, Medullary rays slightly broad giving a nice silver grain of elongated plates.
(7) The wood is highly durable; it does not shrink, crack or alter its shape. The wood takes a fine polish and does not corrode in contact with metals.
(8) Teak finds extensive use in building constuction, door and window frames, ship and boat building, interior construction of rail compartments, railway sleepers. A variety of house hold furniture, like chairs, tables sofas etc are made out of teak.

Pine

Botanical Name : *Pinus roxburghii; P.wallichiana, P. excelsa* etc
Family : Pinaceae (Gymnosperms)

(1) Natural pine forests are found in temperate Himalayas
(2) Pine is a tall tree attaining a height of 45 metres or more; leaves are acicular and in clusters arising on the dwarf shoot.
(3) Bark is greyish brown; wood moderately hard; sapwood white; heartwood light red; annual rings marked by dense autumn wood. Medullary rays nemerous. Resin ducts scattered and numerous.
(4) Wood is scented; used for doors, windows, caskets, cabinets, boxes and matches. Resin extracted from the wood yields terpentine and is used in medicine.

Deodar

Botanical Name : *Cedrus deodara*
Family : Pinaceae (Gymnosperms)

(1) Deodar is a tall majestic tree reaching a height of up to 75 metres. It is found distributed mainly in western Himalayas
(2) Bark is greyish brown in colour and has a number of shallow vertical fissures. Wood is moderately hard, strong scented and oily. Sapwood is white while hard wood is yellowish brown. Medullary rays numerous resulting in a silver grain of small plates. Resin ducts absent.
(3) Timber is used for construction and railyway sleepers. Deodar oil is fragrant and used in perfumery. All medicine (skin ailments, theumatism, piles etc)

Shisham

Botanical Name : *Dalbergia sissoo*

Family : Leguminosae, Papilionaceae

(a) A large deciduous tree with grey bark found growing in Punjab, Bengal, Assam etc.
(b) The wood is very hard and close grained; heart wood is brown; annual rings are not very distinct.
(c) Timber is in great demand for furniture and cabinet making. It is also used for interior decoration, boat building, carts, agricultural implements etc.

Rose wood

Botanical Name : *Dalbergia latifolia*
Family : Leguminosae, Papilionaceae

(1) A large deciduous tree growing in the whole of peninsular India reaching a height of 25 meters.
(2) Wood is extremely hard and close grained – sapwood is yellow and little in extent, while heartwood is dark purple with black longitudinal streaks. Annual rings not distinct. Medullary rays fine and numerous.
(3) The wood is highly priced and used for cabinet, costly furniture, pians, scientific instruments etc

Red Silk Cotton tree

Botanical Name : *Salmalia malabarica*
Family : Bombacaceae

(1) The plant native to India and Burma is a very large tree attaining a height of 45 metre, with a girth of 12 meters.
(2) Wood is greyish white, grained and porous. It is soft white when fresh but turns dark later. There is neither heart wood nor annual rings.
(3) Apart from the major use of silk cotton (see fibre yielding plants), the soft wood is used for packing cases, toys, match boxes, brush handles, fishing floats, rafts etc.

Bamboo

Botanical Name : *Bambusa aurundinacea*
Family : Gramineae.

(1) Many species (about 141) of bamboo are native to India (69 and grow well in open deciduous forests.
(2) Bamboos are tall (6-15 meters) arborescent grasses with woody jointed stems called culms which arise from woody rhizomes. Bamboos grow very fast, about 7.5 cms per day. Each clump of a plant has about 30-100 culms. The culms are cylindrical and smooth with hollow internodal regions. The bamboos flower only once in their lifetime.
(3) Young succulent shoots are edible. Bamboos are also used for construction, scaffolding, Articles of common use such as tool handles, yokes, cordage, tentpoles, brushes, tap etc are made out of bamboo. Bamboos also provide the most important raw material in paper manufacture.

Mahogany

Botanical Name : *Switentia mahogany*
Family : Meliaceae

(1) In India Mahogany plantations are situated in Kerala state
(2) The wood is diffused porous, considerasly harder with moderately broad to fine medullary rays.
(3) Mahogany is one of the finest timbres; It gives a good finish taking a high polish. As such it is in great demand for cabinet work, ship building, furniture, musical instruments jewellery boxes etc.

Neem

Botanical Name : *Azadirachta indica*
Family : Meliaceae

(1) In India neem grows wild in the forests of Andhra and Tamilnadu. It is also planted as an avenue tree.
(2) Neem is a moderately tall tree with a yellow sap wood and reddish brown heart wood. The wood is diffused porous with distinct growth rings.
(3) The wood is used in carts, carving, and manufacture of agricultural implements. It is not easily attacked

by insects and undergoes good seasoning. Hence it is used for door and window frames in building construction.

Sale

Botanical Name : *Shorea robusta*

Family : Dipterocarpaceae

(1) It is a tall decideous tree reaching a height of 37-46 metres, with a girth of 3.7 metres

(2) The sap wood is white with brownish tinge while heart wood is reddish brown.

(3) The wood is moderately hard and heavy. It is diffuse porous to ring porous in structure. Annual rings are not distinct.

(4) The wood is in great demand as a structural timber in building construction beings used for door and window frames, beams; planks etc.

Mango

Botanical Name : *Mangifera indica*

Family : Anacardiaceae

(1) This cultivated medium sized tree also yields a timber of moderately good quality.

(2) The wood is used in building construction to provide for doors, windows etc.

RUBBER YIELDING PLANTS

Para Rubber

(a) *Hevea brasiliensis*

(b) Euphorbiaceae

(c) Incisions made on the stem to collect the latex oozing out. The latex is seieved to remove the impurities and then poured into a pan for coagulation. After adding acetic acid and formic acid, it is pressed between the rollers to produce sheet rubber. This is crude rubber. It is further refined to produce the rubber of commerce

(d) Rubber finds an extensive array of uses. Bicycle, automobile and aircraft tyres and tubes are made of rubber. Sports goods, footwear, water proofing clothes, toys, handgloves and other insulation material are other products where rubber is used.

Guayule Rubber

(a) *Parthenium argentatum*

(b) Compositae (Asteraceae)

(c) The plants are pulled out; both the root as well as the shoot are crushed in a pebbel mill, where upon the particles of rubber float. They are collected compressed and further processed.

(d) Same as for para rubber

Assam Rubber

(a) *Ficus elastica, F. Krishna*

(b) Moraceae

(c) Latex is tapped from the stem

(d) Same as for para rubber

MEDICINAL PLANTS

Fig 9.2 A twig of para rubber tree
(Heva Brasiliensis)

Aconite

(a) *Aconitum napellus*
(b) Ranunculaceae
(c) Roots contain several alkaloids, notable among them being aconitine which is poisonous.
(d) Externally aconite is used against neuralgia and rheumatism. It is also an analgesic and antipyretic.

Rowolfia (sarpagandhi)

(a) *Rawolfia serpentina*
(b) Apolynaceae
(c) Roots of the plants contain several alkaloids of which the most important is reserpine whose pharmacological property has been studied in detail.
(d) Reserpine has a depressant action on the nervous system and generally administered in hypertension and high blood pressure. It is also a sedative. A kind of insanity is also treated with reserpine

Quinine (peruvian bark)

(a) *Cinchona succirubra*
(b) Rubiaceae
(c) Bark is the source of medicine. The alkaloids are extracted from the bark by solvent extraction. There are four alkaloids – quinine, quinidine, cinchonine and cinchonidine. Of these quinine is the most import. It is the most important antimalaria drug.

Ephedrine

(a) *Ephedra equisetina and E.sinica*
(b) Ephedraceae (Gnetales, Gymnosperms)
(c) The plant is a leafless xerophyte. The alkaloid ephedrine is present in the entire plant.
(d) The drug is useful in the treatment of bronchial asthama, allergic cough, fevers etc.

Coccaine

(a) *Erythroxylon coca*
(b) Erythroxylaceae
(c) The drug is extracted from the leaves by the process of solvent extraction. Trees aged about 2-3years are suitable for picking the leaves.

Fig 9.3 A branch of coca plant (Erythroxylon coca)

(d) The drug coccaine is a local anaesthetic. Taken as an injection the drug is supposed to give a 'ride' producing hallucinations. Leaves are chewed as stimulants. A mild form of coccaine is used in some soft drinks.

Belladona

(a) *Atropa belladona*
(b) Solanaceae
(c) Leaves are the chief source of the drug though the other parts also contain the same. The alkaloids atropine, hyoscymine, scopalamine, belladonin, atropine, hyoscine etc are released by solvent extraction of the leaves.
(d) Atropine, hyoscymine and beiladonine are used as stimulants. Atropine is used in the treatment of eye ailments. It dilates the pupil of the eye. The alkaloids also counteract muscle spasms, stimulate circulation and relieve joint pain.

Digitaline

(a) *Digitalis purpurea*
(b) Scrophylariaceae
(c) Leaves constitute the source of the drug which is obtained by solvent extraction. The glycosides present in the leaves are digitoxin, digitalin and degiton.
(d) Digitalin is a potent stimulant of the heart muscle. By increasing the entry of calcium ions into the heart muscles, the drug regularises their contraction. It is used in the treatment of heart ailments.

Fig 9.4 A branch of belladona plant (Atropa belladona)

Sweet Flag

(a) *Acorus calamus*
(b) Araceae
(c) Rhizome is the medicinal source.
(d) The rhizomes are emetic, expecorant, carminative and antispasmodic. They are also mildly laxative

Colchicine

(a) *Colchicum autumnale*
(b) Liliaceae
(c) Bulbs contain the alkaloid colchicine
(d) The drug is used in the treatment of gout and rheumatism. It is also used to induce polyploidy in plants.

Asafoetida

(a) *Ferula asafoetida*
(b) Umbelliferae

(c) Tuberous roots of this perennial plant yield the latex which is dried and processed to yield the asafoetida of commerce. It contains dimethyl sulfoxide (DMSO)

(d) Asafoetida is useful against indigestion, dyspepsia, flatulence etc. It is also carminative.

Licorice

(a) *Glycyrrhiza glabra*

(b) Leguoninosae – Papilonaceae

(c) Roots of the plant contain several alkaloids and they are sweet in taste. They (roots) are dried and cut into small pieces.

(d) Licorice is used to relieve symptoms of thirst in fever, pain, cough and breathlessness. The drug promotes urination.

Aswagandh

(a) *Wythania somnifera*

(b) Solanaceae

(c) Roots and leaves are the source of several alkaloids including scopalamine

(d) The total decoction of the plant is used in Ayrvedic medicine. The roots relieve fever, rheumatic pain and inflammation of joints.

Rhubrab

(a) *Rheum officinale*

(b) Polygonaceae

(c) Roots and rhizomes

(d) Useful in idigestion and also as a tonic and laxative.

Ivory tree

(a) *Holarrhena antidysenterica*

(b) Apocynaceae

(c) Leaves, seed and bark are the source of medicine

(d) Holarhena is antidysenteric, antihelmentic and carminative. It releives stomach pain.

Opium

(a) Papaver somniferum

(b) Papaveraceae

(c) Latex coming out of the sufficiently ripe capsules when cut is coagulated and constitutes the crude opium. Opium consists of several alkaloids like morphine, codeine, narcotine, papaverine etc.

(d) As a drug morphine is sedative. It is also narcotic and antispasmodic. High dosages of opium produces hallucinations. Highly purified form of opium is 'heroin' the most notorious drug.

Nux Vomica

(a) *Strychnos nux vomica*

(b) Loganiaceae

(c) Seeds of this tree contain the alkaloids strychnine and Brucine.

(d) The alkaloids are used in the treatment of nervous disorders and paralysis. They are also used as stimulants. Higher doses of alkaloids are deadly poisonous.

Senna

(a) *Cassia angustifolia*

(b) Leguminosae – Caesalpinaceae

(c) Leaves are the source of drug – senna

(d) The leaf extract is used in the treatment of stomach pain and colic. It is also a mild pergative.

Fig 9.5 A twig of Nux vomica (Strychnos Nux vomica)

Eucalyptus Oil

(a) *Eucalyptus globulus*
(b) Myrtaceae
(c) Leaves contain an essential oil which is cleared on steam distillation.
(d) The eucalyptus oil is used in the treatment of cold, nasal congestion, fever etc.

Aloe

(a) *Aloe barbadense*
(b) Liliaceae
(c) The juice of this xerophytic plant has medicinal properties. The active chemicals found in the juice are aloin, isobarbation, emodin etc.
(d) The juice of the plant is used in the treatment of stomach ache, piles and rectal fissures. It is also a purgative and autihelmentic.

Aristolochin

(a) *Aristolochia indica*
(b) Aristolochiaceae
(c) Dried stems and roots contain the drug
(d) The drug promotes digestion and regulates menstruation in small doses. Larger doses however are harmful as they cause irritations of the digestive system and kidneys. It may even bring about abortion.

Mamira

(a) *Copis teeta*
(b) Ranunculaceae
(c) The rhizome contains berberine and coptine
(d) The drug is antidiabetic, carminative, expectorant, febrifuge, opthalmic and stomachic. Used in nervous disorders; atonic dyspepsia, asthama and cough. It is also a rejuvinator acting against debility.

Isbogol (Aswagol)

(a) *Plantago ovata*
(b) Plantaginaceae
(c) Fruit husk is fibrous and constitutes the drug.
(d) It is aphrodisiac, cooling, demulcent, diuretic and expectorant. Useful in chronic diarrhoea, constipation gastrointestinal and genitourinal troubles, rheumatism and gout.

Peepal

(a) *Ficus religiosa*
(b) Moraceae
(c) All parts are medicinal (bark, root, fruit, bitter, acid and cooling.
(d) Useful in diseases of blood, vagina, uterus problems such as leucorrhea. Ripe fruit is cooling, good for burning sensation, biliousness. The root is good for gout, seeds are useful in the treatment of urinary discharge. Old leaves soaked in water control vomiting. The wood purgative.

Wormwood

(a) *Artemesia vulgaris*
(b) Compositae
(c) Extract of the whole plant is medicinal
(d) Juice of the herb is applied to the head to prevent convulsions; also helpful in ear pain.

Ergot

(a) *Claviceps purpurea*
(b) Clavicipitaceae (Ascomycetis, Fungi)
(c) Purple or dark coloured sclerotia contain the alkaloids ergotomine, ergotoxin, ergometrine, lysergic acid etc. Ergot alkaloids affect smooth muscles and also the nervous system.
(d) Medicinally ergot is used to check uterine haemorrhage during child birth, block pain and eliminate nervous inflammations. In recent years it is also used as a psychotropic drug.

Mistletoe

(a) *Viscum album*
(b) Loranthaceae
(c) The whole plant is used in medicine
(d) It is a laxative, The extract of the plant clears the system of bile and phlegm and is a ramedy for lumbago and piles. Inhomoepathy, mistletoe is used in the tretment of epilepsy, sterility, vertigo, depression, hypertension and menstrual disorders.

Vasaka

(a) *Adathoda vasaca*
(b) Acanthaceae
(c) The leaves contain the alkaloid vascine which is a powerful expectorant and antispasmodic
(d) Decoction of the leaves is used in chest diseases althama and chronic bronchitis.

Datura

(a) *Datura stramonium*
(b) Solanaceae
(c) Whole plant, leaves, seeds; roots and fruits are used in medicine.
(d) Seeds are useful in bites of mad dogs, purulent discharges from the ear, elephanti asis; Roots boiled in milk and administered with clarified butter and used to treat insanity. Leaves and seeds are useful in the treatment of fevers with cerebral complications, diabrhoea, skin ailments etc. Extracts of leaves are useful in reducing pain in rheumatic swellings, boils and tumours.

Black Myrobalan

(a) *Terminalia chebula*
(b) Combretaceae
(c) Fruits are rich in tannin and extensively used in Ayurvedic medicine as one of the famous triphalas.
(d) The extract of fruits can be applied externally on chronic ulcers, wounds. Inflammation of the mucuous membrane is also treated with the extract. Extract of myrobalans are used to treat blood pressure and cardiac ailments.

Myrobalan

(a) *Terminatia bellerica*
(b) Combretaceae
(c) Fruit is rich in tannin and also a source of medicine. It is one of the triphalas of Ayurvedic medicine
(d) The dried fruits constitute the drug batiera, useful in stomach disorders such as indigestion, diarrhoea, piles, leprosy, dropsy and fever. It is also a brain tonic. It is also used in eye lotion.

BEVERAGE YIELDING PLANTS

Coffee

(a) *Coffee a arabica* (Arabian coffee), *C.robusta* (robust coffee), *C.liberica* (liberian coffee)
(b) Rubiaceae
(c) The dried, roasted and powdered seeds are used as coffee powder. Coffee seeds contain the alkaloid caffeine.
(d) The decoction of the powder may be drunk straight as black coffee. Usualy sugar and cream or milk is added to the decoction before drinking. The following are the effects of coffee on human system.
    (a) Increase of adrenaline output
    (b) Slight increase of blood pressure
    (c) Increase of breathing rate
    (d) Acting as a stimulant

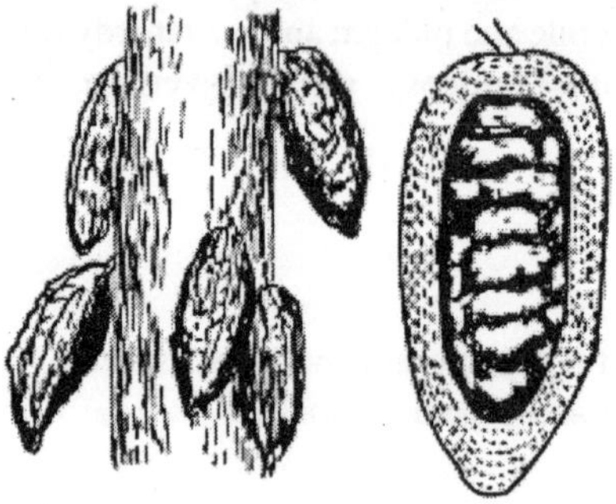

Fig 9.6 Stem of cocoa plant bearing fruits (left) and a fruit cut open (right)

Tea

(a) *Camellia cinensis*
(b) Theaceae
(c) Tannin content which is the source of the drink is contained in the leaves; usually the bud and two leaves associated with it have the highest tanin content. Commercially four types of tea are recogonised. These are black tea, oblong tea, brick tea and leppet tea. Chemically tea contains – Theine 2-5%, Tannin 13-18% traces of caffeine and volatile oils.

Fig 9.7 A fruiting branch of Coffee plant (Coffee Arabica)

(d) It is the most popular non alcoholic beverage the world over. It releives fatigue and acts as a stimulant and refresher.

Chocolate

(a) *Theobroma cacao*
(b) Sterculiaceae
(c) Seeds contain chocolate
(d) Used for manufacture of toffees, drinks, sweets cakes etc.

SPICES AND CONDIMENTS

Asafoetida

(a) *Ferula asafoetida*
(b) Umbelliferae
(c) Roots yield a gum resin containing sulphorous compounds
(d) Dried resin is used for flavouring food

Greater Galangal

(a) *Alpinia galanga*
(b) Zingiberaceae
(c) Reddish brown rhizome is the source
(d) Used as a condiment as well as a source of essential oil

Ginger

(a) *Zingiber officinalis*
(b) Zingiberaceae
(c) Rhizome is the source of condiment
(d) Used to flavour food, beverages etc

Turmeric

(a) *Curcuma longa*
(b) Zingiberacene
(c) Rhizomatous stem is the turmeric of commerce.
(d) Dried powdered rhizome is used as a condiment

Horse Radish

(a) *Avmoracia lapathifolia*
(b) Cruciferae
(c) Large fleshy roots contain a pungent glucoside
(d) Used as a condiment

Sarasaparilla

(a) *Similax aristolochiaefolia, S. macrophylla*
(b) Liliaceae
(c) Roots of this plant are bitler and pungent
(d) Used as a condiment as well as a pickel;

Angelica

(a) *Angelica archangelica*
(b) Umbelliferae
(c) Root is the source of the condiment
(d) Used to flavour candies, cakes, perfumery etc.

Fig 9.8 A flowering twig of Tea plant (Camellia Sinesis)

Fig 9.9 Rhizome and aerial shoots of Giner plant (Zingiber Officinalis)

Cinnamon

(a) *Cinnamomum zeylanicum*
(b) Lauraceae
(c) Leaves and bark of the tree are aromatic. Strips of bark are peeled of from the plant. They contain 0.5% of oil caryophyllene an ester of isobutyric acid.
(d) Cinnamon bark is the most favoured flavouring agent in Indian cooking. Leaf also is used as a flavouring agent. Oil extracted from the leaf is used in perfumery also.

Sassafras

(a) *Sassafras albidum*
(b) Lauraceae
(c) Stem and other parts contain an essential oil commonly called oil of sassafras.
(d) Used for flavouring medicines, gums, soaps etc.

Cloves

(a) *Eugenia coryophyllata*
(b) Myrtaceae
(c) Dried unopened flower buds constitute the cloves of commerce. They contain an essential oil – eugenol.

Fig 9.10 A branch of Eugenia caryobhyllata (Cloves)

Fig 9.11 A branch of Eugenia Caryophyllata (Cloves)

(d) Used as a flavouring agent in the cooking of spicy floods and sweets.

Capers

(a) *Capparis spinosa*
(b) Capparidaceae
(c) Flower buds pickled in salt and vinegar is the commercial source of capers
(d) Used as a flavouring agent in cooking savouries.

Saffron

(a) *Crocus sativus*
(b) Iridaceae
(c) Dried stamens and stigmas constitute the saffron of commerce.
(d) It is used for flavouring sweet dishes

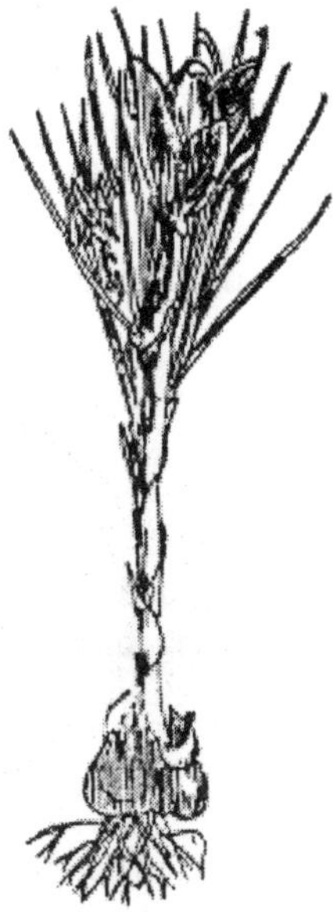

Fig 9.12 Saffron plant with flowers

Black Pepper

(a) *Piper nigrum*
(b) Piperaceae
(c) Fruits contain 5-8% of an alkaloid – piperine and pepper oils like pinene, limonene and sabinene
(d) Mainly used as a condiment. It is also useful as a remedy against cold related ailments.

Long Pepper

(a) *Piper longum*
(b) Piperaceae
(c) And (d) same as in pepper

Fig 9.13 A fruiting branch of pepper plant

Chillies

(a) *Capsicum annum*
(b) Solanaceae
(c) The capsules are green when young, turning to red as they ripe. The inner wall of the fruit and seeds have the pungent principle
(d) Dried and powdered fruits used for flavouring food.

Star anise

(a) Illicium verum
(b) Magnoliaceae
(c) Stellate fruits are reddish brown in colour and aromatic
(d) They are used as culinary spice. Star anise oil is carminative and used in medicine, perfumary

Vanilla

(a) *Vanilla planifolia*
(b) Orchidaceae
(c) The unripe fruits contain the flavouring principle vanilline
(d) Used for flavouring toffees, ice creams, cakes etc.

Coriander

(a) *Coriandrum sativum*
(b) Umbelliferae
(c) The fruit (called cremocarp botanically) is the source of condiment of commerce
(d) Used as a condiment in cooking; carminative as well as digestive.

Fennel

(a) *Foeniculum vulgare,*
b, c and d same as in coriander

Cumins

(a) Cuminum cyminum
b, c and d same as in coriander

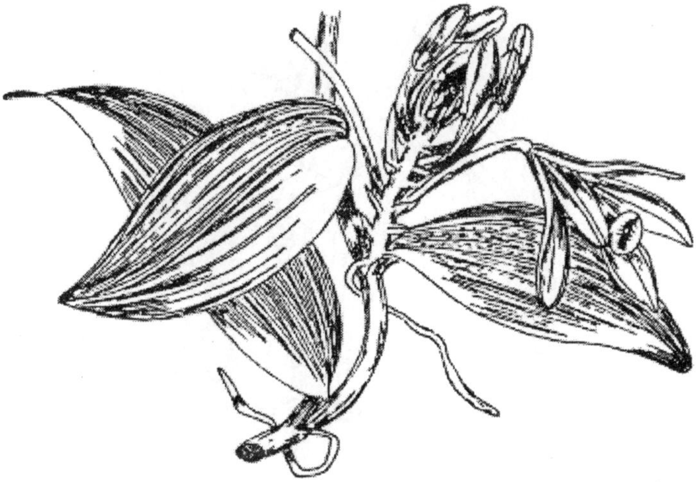

Fig 9.14 Flowering branch of Vanilla planifolia with aerial roots

Black caraway

(a) *Bunium persicum*
b, c and d same as in coriander

Anise

(a) *Pimpinella anisum*
b, c and d same as in coriander

Celery

(a) *Apium graveolens*
b, c and d same as in coriander

Dill

(a) *Anethum graveolens*
b, c and d same as in coriander

Cardamom

(a) Elettaria cardomomum
(b) Zingiberaceae
(c) Seeds contain an essential oil (5-60%) which gives the characteristic aroma
(d) Used for flavouring beverages, sweet dishes etc.

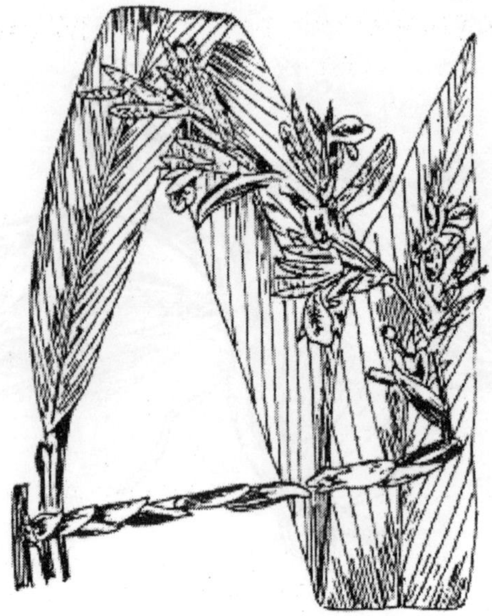

Fig 9.15 A leaf and flowering branch of Cardamomum (Elettaria Cardamomum)

Fenugreek

(a) *Trigonella foenumgraceum*
(b) Leguminosae – papilionaceae
(c) The pods of the plant have a number of yellowish seeds which constitute the fenugreek seeds of commerce
(d) Used as a condiment; also used as tonic, astringent and carminative.

Mustard

(a) *Brassica nigra*
(b) Cruciferae
(c) Seeds contain an edible oil
(d) Used as a conditment mainly for seasoning.

Nutmeg and Mace

(a) *Myristica fragrans*
(b) Myristicaceae
(c) The fruits have brownish seeds covered with a golden brown aril. Both the seed kernel (nutmeg) and aril (mace) are aromatic
(d) Used as flavouring agent; useful medicinally also.

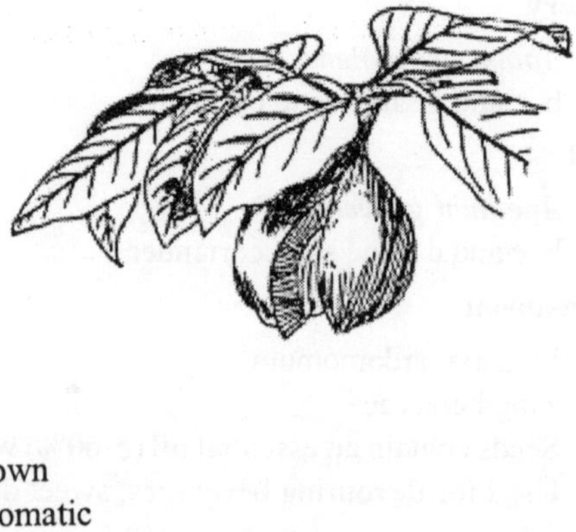

Fig 9.16 A fruiting branch of nutmeg (Myristica fragrans)

Bengal Cardamum

(a) *Amonum aromaticum*
(b) Zingiberaceae
(c) Seeds are the source of condiment
(d) Seeds and sometimes the fruit are used as flavouring agents

Basil

(a) *Ocimum basilicium*
(b) Labiatae
(c) Leaves contain an aromatic oil
(d) Used as flavouring agent

Indian Cassia

(a) *Cinnamomum tamala*
(b) Lauraceae
(c) Leaves are the source of condiment
(d) Used as flavouring agent is cooking spicy foods

Balm

(a) Melissa officinalis
(b) Labiatae
(c) Leaves taste like lemon
(d) Used for flavouring food

Sweet Maforam

(a) *Majorana hortensis*
(b) Labiatae
(c) Leaves flowers etc are used as a source of conditment
(d) Used for flavouring food

Sage

(a) *Salvia officinalis*
(b) Labiatae
(c) The plant is cultivated for its aromatic leaves used as spice
(d) Used for flavouring meat preparations and sausages.

CEREALS AND PULSES

Rice

(a) *Oryza sativa*
(b) Gramineae
(c) The fruit of the plant is a dry indehiscent fruit called caryopsis. This is the edible part. The food is stored in the endosperm. It has 90% carbohydrates. 11% fat and about 1% minerals.
(d) Rice is the staple food for majority of the Asian countries Cooked rice and rice gruel are the major preparations. Rice bran oil is used as a salad oil.

Wheat

(a) *Triticum vulgare*
(b) Gramineae

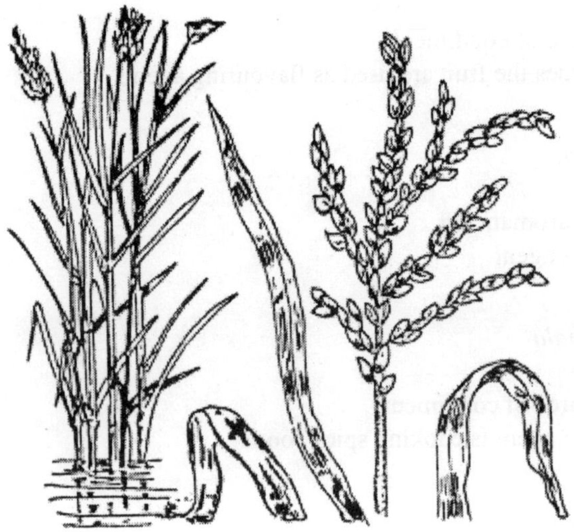

Fig 9.17 Rice plant (Oryza sativa), Habit (left), A part of the inflorescence (right)

Fig 9.18 Wheat plant - Habit

(c) The edible part is the caryopsis, the endosperm of which stores starch. Surrounding the endosperm is the proteinaceous aleurone layer.

(d) Wheat grains are powdered into a fine flour called atta and used in the making of bread, cake biscuits, macaroni, noodles, chapathi etc. Soji a form of wheat in coarse particles is another wheat product used in cooking.

Maize

(a) *Zea mays*

(b) Gramineae

(c) Fruit of the plant (yellowish in colour), caryopsis is the edible part. The grains are produced in linear rows on the cob. Maize grains are rich in easily digestible carbohydrates, proteins and fats.

(d) Maize has several uses. The flour is used for making bread, chapati etc. Fried grain is called 'popcorn' or popular snack. Corn flakes is an excellent breakfast food.

Oats

(a) *Avena sativa*

(b) Gramineae

(c) The caryopsis (fruit) is a highly nutritious cereal containing 16% protein, 8% fat and many minerals.

(d) Oats is used for making porridges, biscuits cakes etc. Oat flakes is an excellent break fast food.

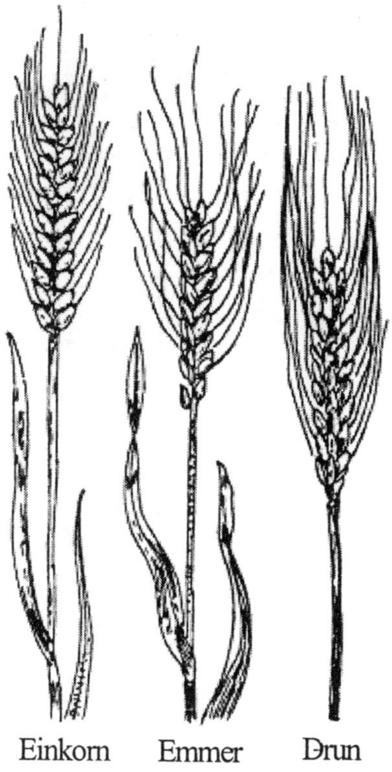

Fig 9.19 Important types of wheat
Einkorn (triticum monococeum) Emmer (T. Dicocum)
Eurum (T. Turgidum var. durum)

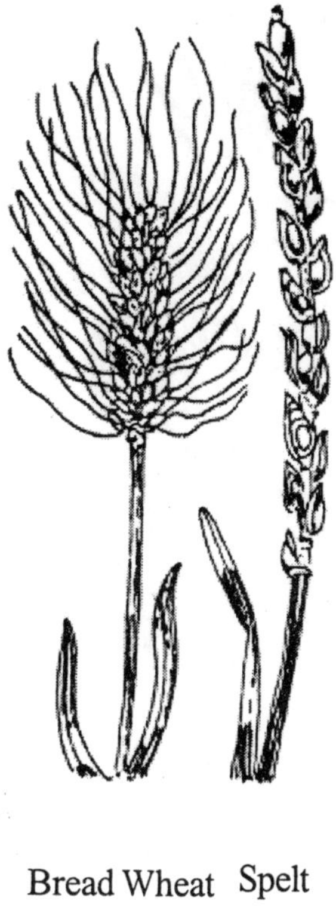

Fig 9.20 Important types of wheat (contd..)
Bread wheat (T. Aestivum)
Spelt wheat (T. Speltoides)

Barley

(a) *Hordeum vulgare*
(b) Gramineae
(c) The grains (fruit) are the edible part. There are two kinds, two rowed and six rowed.
(d) Used for malts, porridges etc.

Sorghum (Indian corn, jowar)

(a) *Sorghum vulgare*
(b) Gramineae

(c) Caryopsis is the edible part
(d) Corn flour is used in making a kind of bread in many parts of south India. Straw of the plant is a good cattle feed.

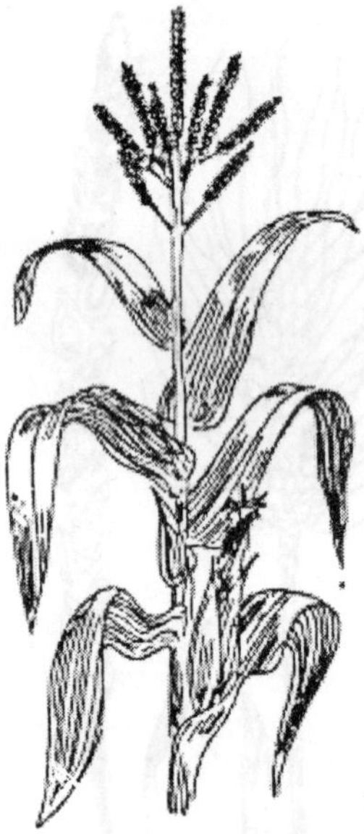

Fig 9.21 Maize plant - Habit

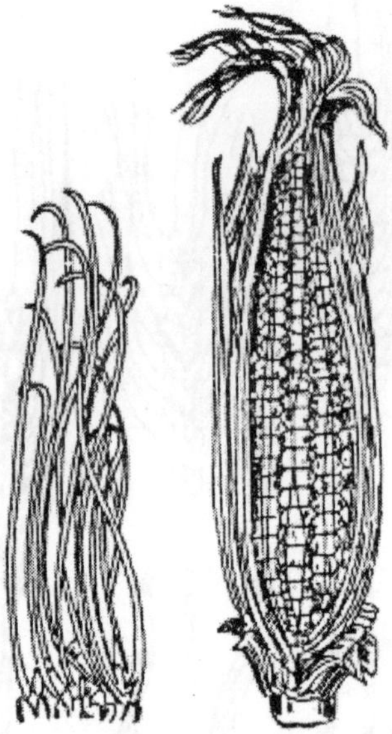

Fig 9.22 Tip of the female inflorescence of Maize (left Maize cobe with grains (right

Rye

(a) Secale cereale
(b) Gramineae
(c) Rye grains (caryopsis) are edible. The grains are milled and made into rye flour.
(d) The chief use is in the making of the rye bread. Straw is used as fodder, making hats and manufacturing paper.

Foxtail or Italian millet

(a) *Setaria italica*
(b) Gramineae
(c) The grains are white, yellow, brown or red in colour.
(d) The grains are cooked like rice and made into porrdge.

Proso or bread millet

(a) *Panicum miliaceum*
(b) Gramineae
(c) Cultivated for grains as well as hay
(d) Dehusked grain is cooked and eaten or used for making bread

Fig 9.23 Barley plant Habit

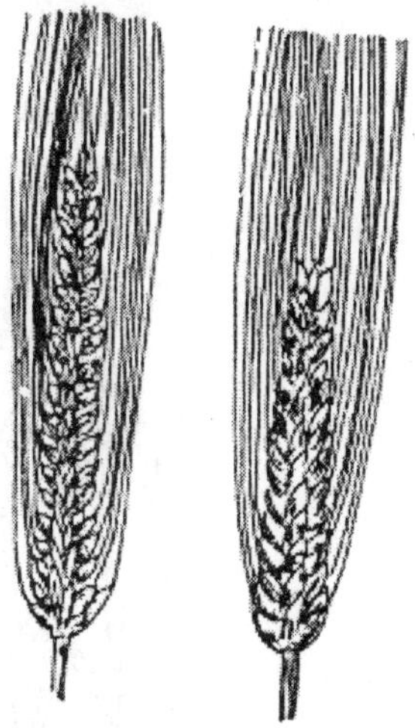

Fig 9.24 Two rowed barley (left Six rowed barley (right

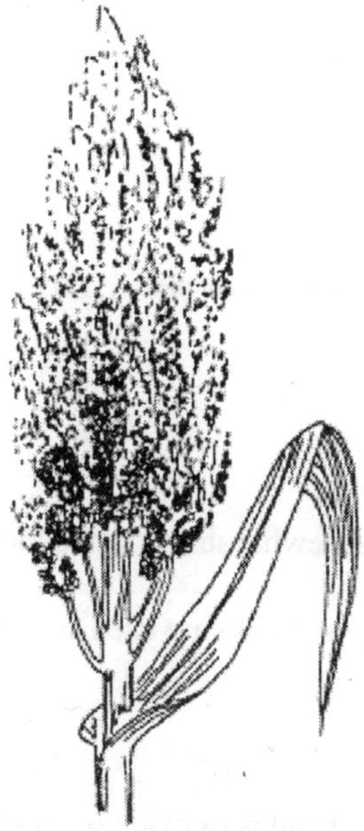

Fig 9.25 Indian corn (Sorghum vulgare)

Fig 9.26 Indian corn - a portion of the head magnified

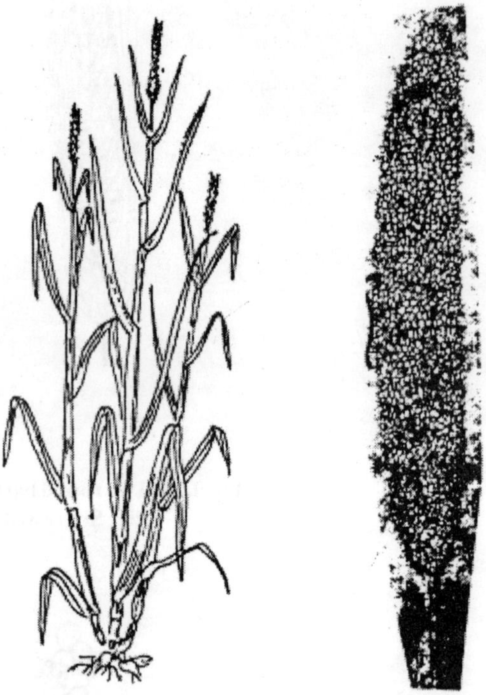

Fig 9.27 Foxtail millet (Setaria italica) Habit of plant (left) Inflorescence (right)

Bulrush millet (pearl millet)

(a) *Pennisetum typhoideum*

(b) Gramineae

(c) The small grains (caryopsis) are arranged in spikelets. They are somewhat shiny, hence the name pearl millet.

(d) Used for making bread; cooked and eaten in the same manner as rice. Stalks are used as fodder.

Ragi (Finger millet)

(a) *Eleusine coracana*

(b) Gramineae

(c) The ragi grain is the fruit. It is not a typical caryopsis as the seed coat and pericrp are separate. The seeds are small, rounded and reddish black in colour. The food is stored in the endosperm.

(d) Ragi is a very useful crop as it can be stored for 4-5years without the loss of nutritive value. It is a staple food of rural South India. Ragi malt obtained from the germinated seeds is a refreshing and nutritious beverage.

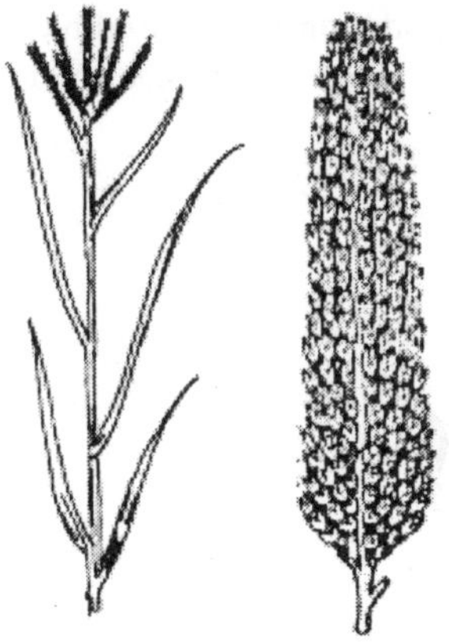

Fig 9.28 Ragi (Elensine coracana)
A part of the plant (left)
Inflorescence enlarged (right

Kodo millet

(a) *Paspalum scorbiculatum*
(b) Gramineae
(c) The grain is the caryopsis and is the source of kodo millet.
(d) Used as food.

Barnyard millet (Japanese millet)

(a) Echinochloa colona
(b) Gramineae
(c) Grains (fruit) is the millet
(d) Grains are used as food, usually mixed with rice in porridges.

Bengal Gram

(a) *Cicer arietinum*
(b) Leguminosae – Papilionaceae
(c) The seeds are the source. Food is stored in the two cotyledons.
(d) Gram seeds are etaen raw as well as cooked. The gram flour is used for various types of snacks both sweet as well as hot. The chaff is a good cattle feed.

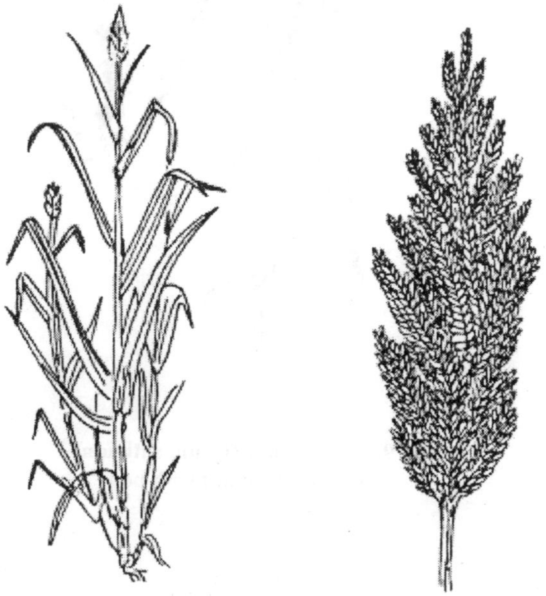

Fig 9.29 Barnyard millet (Echinocloa colona) Habit of plant (left)
Inflorescence enlarged (right)

Sweet Pea

(a) *Pisum sativum*
(b) Leguminosae – Papilionaceae
(c) The round or wrinkled and yellow or green coloured seeds found in pods constitute the edible sweet pea
(d) Fresh seeds are used as vegetable. Dried seeds are fried and used as a snack. Fresh peas are dehydreted and canned.

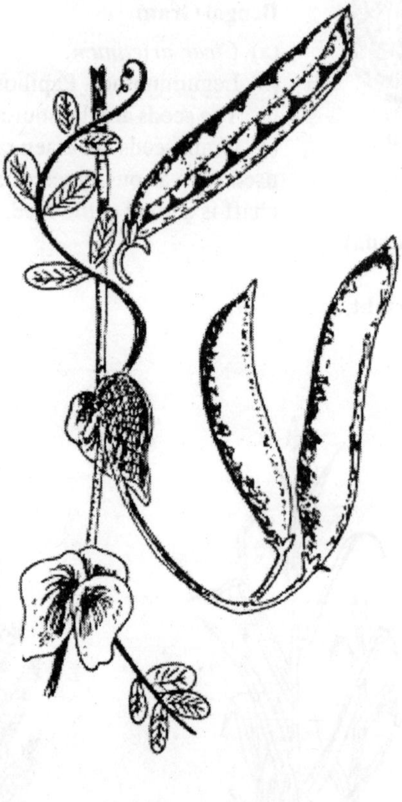

Fig 9.30 Sweet pea (Pisum sativum)
Part of the plant with pods

Pigeon Pea (Tur dhal)

(a) Cajanus cajana
(b) Leguminosae – Papilionaceae
(c) The seeds are dried and the cotyledons separated. This constitutes the tur dhal of commerce.
(d) Pigeon pea is cooked and eaten as a vegetable in the fresh state. The dried cotyledons are used as dhal in curries, soups etc. The plant is used as a forage for cattle.

Lentils

(a) *Lens esculenta*
(b) Leguminosae – Papilionaceae
(c) The seeds are the source, unripe pods are also a source of vegetable.
(d) Lentils are cooked in soup. Seeds are rich in protein and rank equal to pigeon pea.

Fig 9.31 A part of lentil plant (Lens Esculenta)

Mung bean (green gram)

(a) *Phaseolus aureus*
(b) Leguminosae – Papilionaceae
(c) The pods are slender and 3-4" in length. The seed has a green seed coat while the cotylledous are yellow.
(d) Young pods are used as vegetable. Seeds are used as pulse. Germinating seeds constitute a good source of nutrition.

Black gram (urd dhal)

(a) *Phaseolus mungo*
(b) Leguminosae - Papilionaceae

(c) Fruits are long, black coloured and somewhat curved. Seeds are black in colour while the cotyledons are yellowish white.
(d) The seeds (cotyledons) are used as pulse. Many fried snacks are prepared using black gram flour.

Moth bean

(a) *Phaseolus aconitifolius*
(b) Leguminosae – Papilionaceae
(c) The fruits are 1-2" in length with cylindrical seeds
(d) Seeds are used as pulse while the unripe fruits are used as vegetable

Soyabean

(a) *Glycine max*
(b) Leguminosae – Papilionaceae
(c) Legumes are 1-4" long and have a hairy exterior; seeds contain oil (20-33%) as well as protein (39-40%)
(d) Soya seeds are a good source of oil and proetin. Seeds are also consumed fresh, dry or fried. Soya flour is a good nutritive food while soya oil is a good cooking medium.

Fig 9.32 Soyabean (Glycine max)
A. Habit B. Part of the plant with fruits C. Seeds

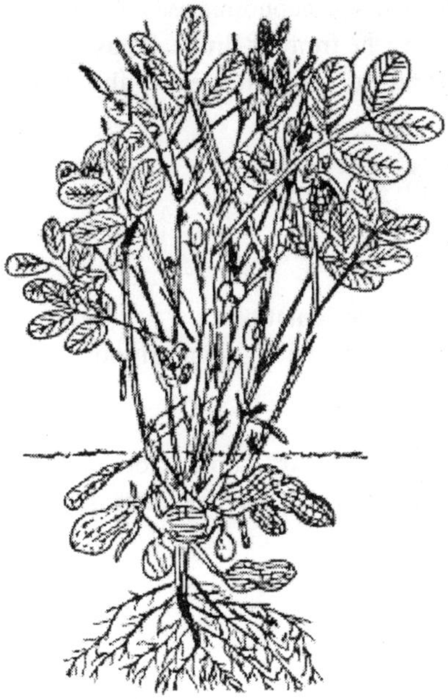

Fig 9.33 Ground nut (Arachis hypogea) plant.

Lab Lab

(a) *Dolichos lab lab*
(b) Leguminosae – Papilionaceae
(c) Pods are flat – and long; they have several white, black, red or mottled seeds
(d) Young pods are cooked as vegetable. Dried seeds are used as pulse.

Ground Nut

See oil yielding plants
(a) *Cyamopsis tetragonoloba*
(b) Leguminosae – Papilionaceae
(c) Pods are produced in clusters
(d) Young pods are used as vegetable, mature seeds are used as a substitute

Cowpea

(a) Vigna sinensis
(b) Leguminosae – Papilionaceae
(c) Seeds and young pods.
(d) Young pods are used as vegetable while dried seeds constitute the pulse.

Broad bean

(a) *Vicia faba*
(b) Leguminosae – Papilionaceae
(c) Seeds and young pods are the useful parts
(d) Young pods and seeds are used as vegetable.

NUTS

Botanically a nut is a one celled, one seeded dry fruit. Some of the popular nuts however are not nuts from the point of view of botanical definition. The following are some of the popular nuts.

Cashew Nut

(a) *Anacardium occidentale*
(b) Anacardiaceae
(c) The 'fruit' is the useful part. The fruit is both a false fruit and a true fruit. The fleshy portion that looks like fruit (false fruit) is the swollen pedicel. The cashew nut of commerce is the reniform seed like structure. Botanically it is the true fruit. The pericarp is dark in colour and has a single seed with brownish seed coat. It has two white cotyledons.
(d) Cashew nuts are most nutritious. They have 47% fat, 21% protein and 22% carbohydrates. The nuts are also rich in calcium, phosphorous, iron and vitamin B, and riboflavin.
Cashew nuts are eaten raw, roasted and salted. They are used in the preparation of several sweet dishes and sauvouries. Cashew oil is very nutritious. The false fruit on fermentation yeilds 'feni' or cashew wine, a popular alcoholic beverage in Goa.

Coconut

(a) *Cocos nucifera*
(b) Palmae
(c) The coconut palm is the most verstaile tree from the point of view of economic uses. All parts of the plant – trunk, leaves, inflorescence and fruit are useful. Botanically the fruit is a drupe. It has a thin epicarps, a fibrous mesocarp and a hard endocarp. The fleshy portion i.e. coconut meat is the endosperm; the coconut water is also the endosperm.
(d) **Uses of coconut**
    (i) The tree trunk is a good timber
    (ii) The leaves are used for thatching
    (iii) The Petioles of the leaf are used as broom
    (iv) The inflorescence is often tapped to yield a beverage on fermentation
    (v) Tender coconut i.e. coconut water is a good refreshing drink and highly nutritious. It is sterile and bacteria free. Hence used in tissue culture experiments.
    (vi) The coconut meat is used in cooking
    (vii) The dried endosperm often called copra contains as much as 70% of oil. Coconut oil has a high quantity of saturated fats. It is used as cooking medium. It is also used in soaps, shampoos etc.

Almond

(a) *Prunus amygdalus*
(b) Rosaceae
(c) The fruit is a drupe. It has a fibrous rind surrounding the hard endocarp. The seeds are edible.
(d) The seeds are green when young and can be eaten raw. Usually the seeds are dried, roasted, salted or made into paste to be used in making cake. Almond oil extracted from the seeds has many medicinal uses.

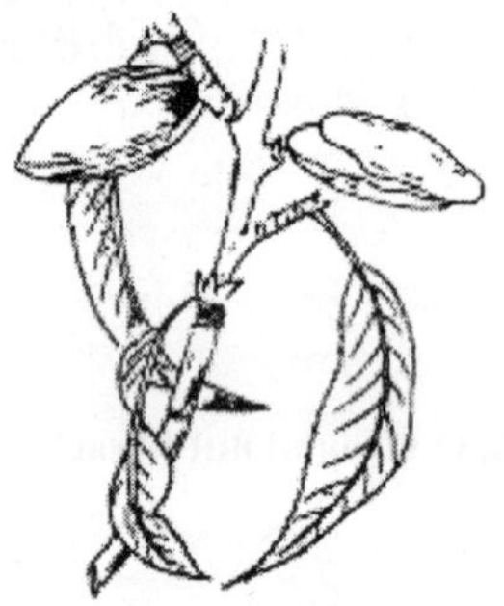
Fig 9.34 Almond (Prunus Amygdalus) a part ofthe plant

Walnut

(a) *Juglans regia, J. cinera* and *J.Nigra*
(b) Juglandaceae
(c) The fruits are green in colour and the outer husk is removed to expose the 'walnut'. There are three kinds of walnuts viz., Black walnut (*J.nigra*), butternut (*J. cinerea*) and English walnut (*J. regia*) The kernels (cotyledons) are rich in oil.
(d) Black walnut and butternut are used in candy and icecream industries. The kernels of english walnut are characteristically furrowed and are eaten raw, fried, salted; they are also used for garnishing. Sweet dishes.

Pistachio Nut

(a) *Pistacia vera*
(b) Anacardiaceae
(c) The fruit is a drupe consisting of a single seed (nut) which constitutes the nut of commerce. The seeds contain two large cotyledons with reddish covering.
(d) The nuts are salted in brine before consumption. They are eaten raw, roasted, salted or used in cooking sweet dishes and in ice creams.

Fig 9.35 A shoot of walnut (Juglans regia)

Pine nut

(a) *Pinus edulis*, *P.Monophylla*, *P.Gerardiana*
(b) Pinaceae (Gymnosperms)
(c) The plant produces female cones within which are found the ovules which on maturity develop into seeds (nuts). The nuts are of the size of a bean seed and two in number per seed scale. The female gametophyte which is often erroneously termed endosperm is the edible part. It has 60% fat.
(d) The kernels are eaten raw or roasted

Chestnut

(a) *Castanea dentata*, *C.sativa C.crenata*
(b) Fagaceae
(c) Chestnuts are the fruits (nut) of the plant
(d) The nuts are eaten raw or roasted or boiled

FUMITORIES AND MASTICATORIES

Tobacco

(a) *Nicotiana tabacum*
(b) Solanaceae

(c) The leaves constitute the tobacco of commerce. The fresh leaves are harvested and the cured in barons when they change their colours from green to yellow. Curing is necessary to impart the required texture, aroma and colour. Chemically the major constituent is the alkaloid nicotine

(d) Tobacco leaves are mainly used in smoking in the form of cigars, cigarettes, beedies, other uses are chewing, snuff, hookah etc. Extract of tobacco leaves is a good insecticide.

Betel Leaves

(a) *Piper betel*

(b) Piperaceae

(c) The ovate leaves constitute the betel-leaves of commerce. Leaves are consumed raw or bleached when the taste and quality is said to improve

(d) Betel leaves are chewed along with several nuts and spices like nuts, cloves, cinnamon, fennel etc. It is supposed to be carminative and promotes digestion.

Arecanut

(a) *Areca catechu*

(b) Palmae (Arcecaceae)

(c) The fruit is a small drupe. It has a hard and fibrous mesocarp. The seed constitutes the arecanut. The food is stored in the endosperm which is very hard. Several alkaloids are present in the arecanut. These are – arecoline, arecaidine, guvacoline etc. Arecanut is used either raw or after curing. The de shelled nuts are boiled in water and subsequently air dried for about 2-3 weeks. This will remove excess of tannins.

(d) Arecanut is chewed along with betel leaves, continuous chewing may be harmful as it may produce carcinoma of mouth. The juice of tender nuts acts as a laxative.

Hemp

(a) *Cannabis sativa*

(b) Cannabinaceae

(c) Dried flowering tops yield a narcotic drug in the form of a resin. The resin is popularly called charas or hashish. Bhang is obtained from the leaves. Ganja is the dried female inflorescence. The active chemical ingredient is cannabinol.

(d) The different forms of hemp generally have hallucinatory effects acting as a psychotropic drug. They bring about an ecstatic feeling by excessively stimulating the nervous system. Long range use of this however has degenerative effect as the individuals become addicts.

Bhang chewing is a common practise among the Indians particularly the 'Sadhus' inhabiting the Himalayan region.

Jimson weed

(a) *Datura stramonium*

(b) Solanaceae

(c) The leaves and fruits contain an alkaloid scopalamine

(d) Leaves and tops are mixed with another herb *Lobelia inflata* and used for smoking.

Coca

(a) *Erythroxylon coca*

(b) Erythroxylaceae

(c) Leaves constitute the source of coccaine, an alkaloid which is known to be a stimulant

(d) Coca leaves are chewed either in the raw condition or in powder form along with lime and alkaline ash of plants like *Chenopodium quinoa*. Use of coca leaves is known to resist mental fatigue. Coccaine acts on the centrol nervous system directly resulting is an ecstaticfeeling.

FRUITS

Mango

(a) *Mangifera indica*
(b) Anacardiaceae
(c) The fruit is a drupe. It has a fleshy as well as fibrous mesocarp which is edible.
(d) Unripe mango fruits are pickled either wholly or made into pieces. Ripe mango fruits are highly delicious. Mango pulp has 20% sugars, 1% proteins and 0.2 – 0.5% acids. Medicinally mango juice is diuretic and laxative.

Banana

(a) *Musa paradisiaca*
(b) Musaceae
(c) Banana is a parthenocarpic fruit. The outer skin is the thalamus. Fibrous structure attached to the skin inside is the epicarp. Mesocarp and endocarp form the pulp of banana and are edible.
(d) Unripe banana is used as a vegetable. It is also used in the making of 'banana wafers'(chips). Banana pulp is dried and made into flour.

Fig 9.36 Top portion of banana plant (Musa Paradisiaca)

Guava

(a) *Psidium guajava*
(b) Myrtaceae

(c) Guava fruits are of different shapes – long, rounded, oval etc.
(d) The entire fruit is edible. It is a rich source of vitamin C.

Papaya

(a) *Carica papaya*
(b) Caricaceae
(c) Fruit is highly nutritive. It has 0.5% proteins, and an equal amount of phosphorous, Iron and calcium, Vit A and C. Milky latex is the source of papain a protein hydrolyzing enzyme.
(d) Unripe fruit is cooked as vegetable. Ripe fruit is very delicious. The enzyme obtained from the latex is used in brewing industries.

Grapes

(a) *Vitis vinifera*
(b) Vitaceae
(c) Fruits are highly delicious. The edible parts are pericarp and placenta
(d) Fruits are consumed and are also used in wine making. Fruit juice is a good beverage.

Apple

(a) *Pyrus malus*
(b) Rosaceae
(c) Apple is a false fruit (pome) the juicy edible portion is the thalamus; the seed like structures are actually achenes i.e. botanical fruits.
(d) Nutritionally apple has 84% water, 0.3.% proteins, 0.4% fat, 11% sugar and 0.5% acid. Fruits are eaten raw or made into jams, jellies and juice.

Pear

(a) *Pyrus communis*
(b) Roasceae
(c) The fruit is pyriform and the edible portion is the thalamus. It is juicy and consists of many stone cells. The false fruit is called pome.
(d) It is eaten raw as table fruit. It is also canned.

Apricot

(a) *Prunus armeniaca*
(b) Rosaceae
(c) Apricot fruit is a none seeded drupe with a fleshy mesocarp, outer thin epicarp and a hard stony endocarp. The seed also is edible. The kernels are rich in oil.
(d) Fruits are eaten raw as well as dried.

Cherry

(a) *Prunus ovium*
(b) Rosaceae
(c) The fruit is a drupe
(d) Cherry is a favourite table fruit, the fleshy mesocarp being edible. Fruits are also canned. Cherry juice and cherry brandy (an alcoholic drink) are also popular.

Peach

(a) *Prunus persica*
(b) Rosaceae
(c) Fruits are spherical and velvety; the hard endocarp is compressed; mesocarp is the edible part
(d) Peach is a table fruit, it is eaten either raw or dried. Seeds have a volatile oil.

Plum

(a) *Prunus domestica*
(b) Rosaceae
(c) Mesocarp and endocarp are fleshy and edible.
(d) Fresh fruits are eaten. Fruits are also used in the preparation of jams.

Fig 9.37 A fruiting branch of plum (Prunus Domestics)

Muskmelon

(a) *Cucumis melo*
(b) Cucurbitaceae
(c) Fruit is a pepo; fleshy mesocarp and endocarp are edible
(d) The fruit is eaten raw. It is good for constipation.

Watermelon

(a) *Citrullus vulgaris*
(b) Cucurbitaceae
(c) Fruit is a pepo; meso and endocarp are edible
(d) Fruits are reddish inside and very sweet. They are eaten raw.

Mulberry

(a) *Morus alba*
(b) Moraceae
(c) Fruit is a compound one called sorosis. The actual fruits are small nutlets surrounded by fleshy sepals around a fleshy axis
(d) The fruits are eaten raw

Strawberry

(a) Fragaria chiloensis, *F.vesca*
(b) Rosaceae
(c) There are two types of straw berries – garden strawberry (*F.chiloensis*) and alpine strawberry (*F.vesca*). Fruit is an etaerio of dry achenes embedded on the surface of a fleshy receptacle.
(d) Straw berries are eaten raw or used in jams, jellies, syrups, wines, ice creams etc. Straw berry flavour is used as a flavouring agent.

Raspberry

(a) *Rubus biflorus*, *R.fruticosa* etc.
(b) Rosaceae
(c) The edible part is the fruit. It is an aggregate fruit called etaerio of drupes
(d) Raspberry fruits are eaten fresh or used in jams and other preserves.

Date Palm

(a) *Phoenix dactylifera*
(b) Palmae
(c) The fruit is a drupe. The fleshy pulp of the fruit is the pericarp and it is edible. Fruits have 54% sugar 70% proteins besides pectins and gum. It is also rich in Iron.

(d) Fruits are eaten either raw ordried. They are also used in jams

Sweet Orange

(a) *Citrus sinensis*
(b) Rutaceae
(c) The fruit is a hesperidium. It is a type of berry. The juicy placental hairs constitute the edible portion.
(d) Fruits are eaten raw and also in the form of juice. Oil from the rind is used in perfumery.

Sour Orange

(a) *Citrus Aurantium*
b, c and d same as in sweet orange.

Lemon

(a) Citrus limon
(b) Rutaceae
(c) Same as in sweet orange
(d) Lemon is used for its juice, lemonade and for pickling.

Lime

(a) *Citrus aurantifolia.*
b and c same as in sweet orange
(d) The fruit is not suitable for eating. The juice is very rich in citric acid.

Citron

(a) *Citrus media*
b and c same as in sweet orange
(d) Rind is very thick and is used for candies, marmalades etc.

Shaddock

(a) *Citrus decumana*
b and c same as in sweet orange
(d) Fruits are very big in size (size of a pumpkin) and are eaten raw.

Custard apple

(a) *Anona squamosa*
(b) Annonaceae
(c) A compound fruit (etaerio of berries) developed from apocarpous pistils. Fleshy covering on the seeds are edible.
(d) Fruits are eaten raw as they are very sweet.

Loquat

(a) *Eriobotrya japonica*
(b) Rosaceae
(c) Botanically the fruit is a pome and the edible part is the elongated stalk of the fruit.
(d) Fruit are consumed fresh are made into jams, jellies, pies and sauces.

Mangosten

(a) *Garcinia mangostama*
(b) Guttiferae (clusiaceae
(c) Fruits borne on this evergreen tree are dark purple in colour with persistent calyx. The pulp consisting the mesocarp and endocarp is edible.
(d) Fruits are eaten raw. The pulp has a good taste and flavour.

Pomegranate

(a) *Punica granatum*
(b) Punicaceae
(c) The fruit is a balusta with a tough pericarp and two rows of carpels with irregularly arranged seeds. Calyx is persistent. Testa on the seed is juicy and constitutes the edible portion.
(d) Fruits are eaten raw or made into Juice. Dried seeds are used for flavouring food.

Fig

(a) *Ficus carica*
(b) Moraceae
(c) The fruit is a compound structure called syconus wherethe entire inflorescence develops into a fruit. The fleshy receptacle covering the fruit is the edible portion.
(d) Fig is eaten fresh. It is also one of the popular dried fruits. Fruits are also used in flavouring ice creams jams etc.

Pine apple

(a) *Ananas cosmosus*
(b) Bromeliaceae
(c) Fruit is a sorosis. It is a compound fruit developing from the entire inflorescence. The edible portion is the fleshy tissue formed by succulent bracts, sepals, petals and inflorescence axis.
(d) Fruits are eaten raw. They are also canned. Pine apple juice is a good beverage.

Litchi

(a) *Litchi chinensis*
(b) Sapindaceae
(c) The fruit is a one seeded nut. The fleshy aril of the seeds constitutes the edible part.
(d) Fruits are eaten fresh as well as canned.

Sapota

(a) Achras sapota
(b) Sapotaceae
(c) Fruit is a berry. The edible portion is the pulp formed by mesocarp and endocarp.
(d) Fruits are eaten fresh.

Jack Fruit

(a) Atrocarpus integrifolia
(b) Moraceae
(c) Jackfruit is a multiple fruit called sorosis. The thalamus is spiny. Periath of the individual flowers is fleshy, juicy and edible.
(d) Young unripe fruits are used as vegetable. Ripe fruits are eaten. Dried perianth lobes are used as wafers. Seeds are also cooked.

PLANT DYES

Cutch

(a) *Acacia catechu*
(b) Mimoceae
(c) The heartwood of the tree yields a brown coloured dye. Wood pieces are boiled in water and the dye is extracted by evaporation.
(d) The dye is used to colour fabrics like khaki, canvas etc. Crude form of the dye is used for chewing along with betel leaves.

Logwood (Hametoxylin)

(a) *Haematoxylon campechianum*

(b) Caesalpinaceae

(c) Heartwood of the tree yields the dye haematoxylin on steam distillation.

(d) The dye is used as a nuclear stain in biological preparations

Indigo

(a) *Indigofera tinctoria*

(b) Papilionaceae

(c) Leaves of this plant contain a soluble glucoside. Indican which oxidizes in water to form the dye indigo.

(d) The dye is extracted by soaking and crushing the leaves in water.

(e) Indigo is used for colouring fabrics. Recently a synthetic dye has replaced natural indigo.

Fig 9.38 A flowering branch of logwood tree (Hametoxylon campechianum)

Fig 9.39 Habit of Indigo plant (Indigofera tionctoria)

Henna

(a) *Lawsonia inermis*
(b) Lythraceae
(c) Leaves contain the orange dye which is extracted after crushing the leaves.
(d) Henna dye is used for colouring hair, eyebrows, fingers, palms etc. Fabrics are also dyed with henna.

Litmus and Orchill

(a) *Rocella tinctoria* and *Lacanora* spp
(b) Lichens
(c) The thalli contain the dyes litmus and orchill.
(d) Orchill is used for dyeing woolen and silk fabrics. Orcein a purified form of orchill is used as a microscopic stain-litmus – a chemical indicator is used to detect acids and alkalies.

TEXTBOOK OF
PRACTICAL BOTANY

TEXTBOOK
OF
PRACTICAL BOTANY

Volume - IV

(Anatomy, Embryology, Ecology,
Physiology, Cytology)

By

Dr. S. Sundara Rajan

Principal
Reva Institute for Science and Technology Studies
R.T. Nagar
Bangalore – 560 024

ANMOL PUBLICATIONS PVT. LTD.

NEW DELHI - 110 002 (INDIA)

ANMOL PUBLICATIONS PVT. LTD.
4374/4B, Ansari Road, Daryaganj
New Delhi - 110 002
Ph.: 3261597, 3278000
Visit us: www.anmolbooks.com

Textbook of Practical Botany

First Edition, 2003

ISBN 81-261-1359-6 (Set)

PRINTED IN INDIA

Published by J.L. Kumar for Anmol Publications Pvt. Ltd., New Delhi and Printed at Mehra Offset Press, Delhi.

CONTENTS

PART FIVE

PART FIVE

Preface

Laboratory study is an integral part of the knowledge that a student acquires in the college. This is true not only for botany but for all science subjects. Theoretical study of a subject has to be supplemented with laboratory experiments to provide a comprehensive understanding.

One of the main advantages of laboratory or practical study is the first hand knowledge that a student acquires about the material of study.

A systematic, structured and disciplined method of work is necessary in the laboratory to obtain maximum benefit. In order to help the students realize their goal practical manuals indeed play an important role.

Textbook of Practical Botany aims to provide the students, a useful hand book to guide them in the proper study of the subject material. Theoretical details are kept at the minimum and more emphasis is laid on understanding the structural details to enable the student a proper identification. A large number of properly labelled original diagrams are incorporated in the book as they are integral to the understanding of the subject.

This work is organised in the following four volumes:

Volume One deals with Algae, Fungi, Bacteria, Lichens, Viruses and Microbiology.

Volume Two deals with Bryophytes, Pteridophytes and Gymnosperms.

Volume Three deals with Taxonomy and Economic Botany.

Volume Four deals with Anatomy and Embryology of Angiosperms, Ecology, Plant Physiology and Cytology.

I have taken utmost care in the presentation of matter, style of writing and choice of illustrations so that the volumes are user friendly. The examples chosen for study are such that the books should cater to the needs of all Indian Universities.

Constructive criticism and suggestions are most welcome.

I am thankful to Mr. J.L. Kumar, Managing Director of Anmol Publications Pvt. Ltd., New Delhi, for undertaking the publication of the book.

Dr. S. Sundara Rajan
Principal

Reva Institute for Science and Technology Studies
R.T. Nagar
Bangalore - 560 024

CHAPTER 1

INTRODUCTION

The study of science subjects is carried out at two levels,viz., *Theoretical* and *Practical*. Theoretical study is carried out in lecture classes when the students are imparted all the details regarding a particular topic. This has to be supplemented by practical or Laboratory studies, when the students come into active contact with the objects about which they would have heard in lecture classes. Therefore it may be safely said that laboratory studies form an integral part of a comprehensive study of any topic in science subjects.

In the study of plants, laboratory assumes a special significance as the students acquire a meaningful and firsthand knowledge about the plants which they study in lecture classes. In fact as far as botany is considered any study of a plant without being supplemented by laboratory observation is futile. It is only in the laboratory study that the student can dissect, observe and analyze the various features of the plant and it is only a study of this type that makes the student acquire a meaningful knowledge about the pant. Besides observation and study in the laboratory plants are also to be studied in field to obtain an account of their habitat, habit, association and such other details.

In this chapter an attempt is made to explain some of the essentials that a student should understand before he begins a practical work.

Cleanliness: This is the most important point that a student should keep in mind while in the laboratory. The working table and the surroundings should be kept dry and spotlessly clean.. All the requirements needed by the students such as petri dishes, slides, reagent bottles, stains etc. must be properly labeled and arranged neatly. The microscope and particularly the optical parts must be cleaned with a white lint free cloth. After the completion of the assigned work, the student should take care to clean the place, remove any debris of the plant material and put it into the dust bin.

The students should work in silence and independently unless the assigned experiment is to be carried out by a group. When a particular work is assigned to a student, he must carefully plan and divide the available time for experimentation, observation and recording of the results. Whenever a doubt arises, it is always better to consult the teacher in charge, than a classmate. It is only by a proper and systematic work, that a student can derive maximum benefit from the practical classes.

Laboratory requirements

The department provides certain amenities, instruments etc to each student for the purpose of practical work. But a student also should carry some of the requirements in order to be effective in the practical class.

One cardinal principle that every student should remember in a practical class is to have one's own requirements and not to depend on anyone. Never lend or borrow any requirements. Borrowing or lending a pencil, needle or blade will disturb the work of both.

In general every student should individually possess the following requirements

1. *Two sharp pointed needles* mounted on stainless steel or plastic handle. This is essential for the purpose of separating and teasing the microscopic material.

2. *A pair of forceps:* This is essential to remove the material either from the bottle or from petri dish.

3. A sharp properly honed razor to cut sections of the plant material.

4. A new safety razor blade. This will help either to cut section or to trim the material.

5. Two fine hair paint brushes. This will be of use in lifting the sections of plant (with out damage) from the petri dish on to the slide.

6. Two sharpened pencils (H, Hb or 2b).

7. A pencil eraser.

8. A pencil sharpener.

9. Glass droppers.

10. A piece of thin, white, lint free cloth to clean the optical parts of the microscope.

11. An observation book to note down immediately the points of interest.

12. A practical record.

13. Glass slides and cover slips(if they are not provided by the department).

Scheme of Practical Work

Practical work in the laboratory will be most useful and effective if the student keeps in mind the following:

1. Maintain silence during working. All the requirements for the practical work must be absolutely clean.

2. Every student should work independently, unless the experiment involves the participation of others.

3. Before beginning the practical work, listen to the instructions of the teacher attentively and work accordingly.

4. Follow a systematic procedure in a study.

5. If the work involved is the morphological study of an organism, follow the procedure given below:

(a) Study of external characters - for example whether a plant is microscopic or macroscopic, unicellular or multicellular, filamentous or colonial etc.

(b) Study of internal structure like details of cell, tissues etc. If the material requires sectioning, sections are to be cut and internal details should be observed.

(c) Study the reproductive details like asexual and sexual reproductive organs.

6. Mcunt the material after properly teasing and separating the filaments, mycelia etc. Stain the material wherever necessary and cover the preparation with a cover glass.

7. Observe the slide carefully under the microscope, first under low power and then under high power.

8. Make sure that you have observed all the details as instructed by the teacher.

9. After all the observations are over, record your findings, structural details etc. in the practical records in the form of diagrams. Draw proportionate diagrams in the proper place on the drawing sheet. Diagrams should never be drawn to the very edge of the drawing sheet. Write suitable explanatory notes pertaining to the diagrams on the interleaf.

Laboratory Equipments

In addition to the individual requirements carried by the student, the department generally provides the

following equipments/instruments etc.to the students. Sometimes these equipments may be given to each student .

These are :

(1) Students compound microscope .
(2) Dissection microscope.
(3) Binocular Research microscope.
(4) Watch glasses.
(5) Cavity blocks.
(6) Slides and cover glasses.
(7) Petri dishes.
(8) Enamel tray.
(9) Polythene/glass wash bottle.
(10) Spirit lamps.
(11) Glass marking pencil.
(12) Microtome.
(13) Staining /bottles.
(14) Stone/ for honing the razor.
(15) Chemicals and stains.
(16) Culture media(to be prepared by students).

In the list of requirements mentioned above, we will discuss some in detail.

Microscope

Of all the instruments used by students in the botany laboratory, the one that is most indispensable to the students of biology is the 'Microscope'. It is not an exaggeration to say that no biological study in the laboratory is possible without the microscope.

The function of the microscope is to enlarge the small objects, so that they become visible to the human eye. In reality the magnification power of the microscope lies in its resolving capacity between one particle and another. A microscope distinguishes one particle from another which are situated very closely. A human eye cannot resolve particles that are lying closer than 0.025 mm. In its function a microscope can be compared to a human eye, but much more effective than that.

There are different types of microscopes, but all of them are based on the same principle with variations in their construction. Some of the types of microscopes used in the laboratory are:

(1) Compound microscope.
(2) Binocular Research microscope.
(3) Dissection microscope.
(4) Stereoscopic dissection microscope.
(5) Phase contrast microscope.
(6) Electron microscope etc.

Of the six types of microscopes mentioned above the compound microscope and the dissection microscope are the ones that are most frequently used by students. The Electron microscope is very costly and a very highly sophisticated instrument.

In this book the compound microscope and the dissection microscope are explained in detail as the students should be familiar with them.

Compound Microscope

The microscope has a strong 'U' shaped or 'V' shaped base. It is also called the horse shoe shaped base. Attached to the base is a bent limb. This limb can be tilted in any direction as it is attached to the base with an inclination joint.

Attached to the limb is a stage at its lower end and a body tube at its upper end. The body tube is attached to the limb with the help of a rack and pinion so that it can move up and down. To help in the movement of the tube during focusing, there are 2 knobs viz.,*Coarse adjustment* knob and *Fine adjustment knob*.

The body tube at its top end consists of the ocular or eye piece lens. The eye pieces usually have three magnifications 5x,10x,and 15x, and can be used to get a low or high magnification. Attached to the lower end of the tube is a revolving nose piece with three objective lenses low (10x) medium (40x) and high (60x). Some times there will be a fourth objective of the magnification of 100x. This is also called the oil immersion lens (while viewing through an oil immersion lens a drop of oil-liquid paraffin has to be put on the viewing slide).

Below the objectives and attached to the limb is the *stage* of the microscope where the slides are kept for viewing. The stage may be circular, square or rectangular in shape. In order to keep the slides properly placed on the stage, there are two stage clips. In some of the costly microscopes, the stage has movable sliding system operated by knobs to move the slides. This is called the mechanical stage. The mechanical stage permits the movement of the slide backwards or forwards, left or right with the turning of the knobs provided for the purpose. A hole is provided in the stage to receive the light from below and illuminate the object. Attached to the limb, below the stage is a condenser (fixed or movable up and down) lens. Below the condenser lens is an Iris diaphragm to regulate the light passing through. Iris diaphragm is a metallic disc with an aperture. The aperture can be narrowed or widened by turning the lever. The condenser which is a system of lenses condenses the light rays passing through and converges them on the object.

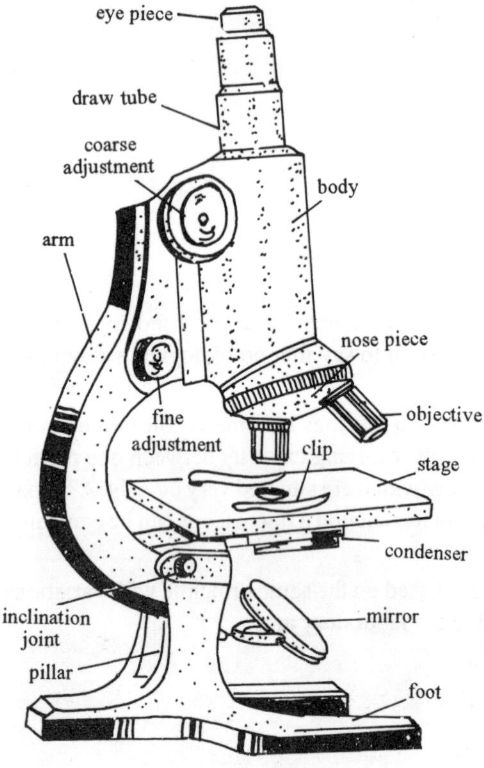

Fig.1.1 A compound microscope

The source of illumination in a student's microscope is a mirror which reflects the natural light or electric light. The mirror has two faces - plain and concave and is attached to the limb below the condenser. The mirror is adjustable and can be turned to the direction of light. When day light is the source of illumination, the concave side of the mirror is usually used to converge the rays of light. In some of the costly microscopes, instead of the mirror there is a built in illumination ,This is a tube with a lamp (9v or 12v) attached to a transformer, which can be plugged to any AC source.

Operation of the Microscope

Microscope should be placed in a well illuminated area but avoiding the direct light as it is harmful to the eyes. Generally northern light is preferable. When ' built in illumination' is employed, it is always better to use a blue or green filter to subdue the light.

The light has to be first focused on the condenser by adjusting the mirror towards the source of light. The amount of light entering the condenser may be regulated by closing or opening the iris diaphragm. The slide to be viewed is kept on the stage and fixed securely with the help of the stage clips. The object is first observed with a low power objective and then turned on to medium or high power, the specimen or section has to be covered with a cover slip. The slide should be observed with both the eyes open even while observing with monocular microscope.

Care of the Microscope

1. Clean all the parts of the microscope before and after use. The lenses and the mirror are to be cleaned with a lint free cloth or tissue paper.

2. Do not pull the microscope to the very edge of the table.

3. While lifting the microscope, use both hands. With one hand hold the base and with the other hold the arm.

4. The slide to be observed must be clean and dry (specially the under surface of the slide).

5. Do not remove the eyes piece or objectives from the microscope.

6. The microscope should be covered when not in use.

Binocular Research Microscope

This is like the ordinary compound microscope except that, it has binocular head for providing a natural stereoscopic view.

Dissection Microscope

This is a very simple microscope with only one set of lenses. Microscopes of this type will not give a very high magnification, but are very useful in dissecting flowers, plant parts etc.

The microscope has a 'V' shaped base and a limb. The stage is made up of a transparent glass plate. Below the stage is a mirror to focus the light. Attached to the limb is a mechanical arm which can be moved up and down with the help of a rack and pinion movement. The arm can also be moved horizontally. Attached to the free end of the arm is a holder to accommodate the lens (10x,15x etc.).

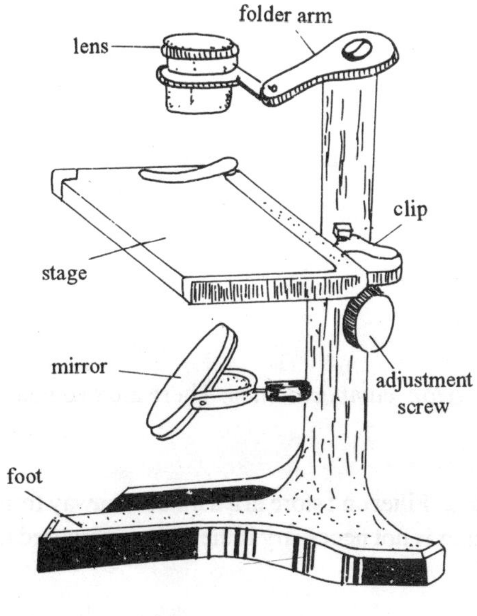

Fig.1.2 A dissecting microscope

Operation: The specimen to be observed / dissected is placed on the stage and located with the help of lens. Floral parts or others may be then dissected with the help of two needles, while still observing the specimen.

Stereoscopic Binocular Dissection Microscope

This is a combination of compound microscope and dissection microscope. While it has the higher magnification power of a compound microscope with two sets of lenses, it has retained the versatility and the manoeuvourability of a dissection microscope.

Phase contrast and Electron microscopes are not normally used by students.

Other Laboratory Equipments

Stains: These are colouring fluids. The specimens to be observed (algal or fungal material or sections of plant parts, would be generally decolourised due to preservation. These are to be coloured in order to observe them clearly. The following are the stains that are generally used in the laboratory.

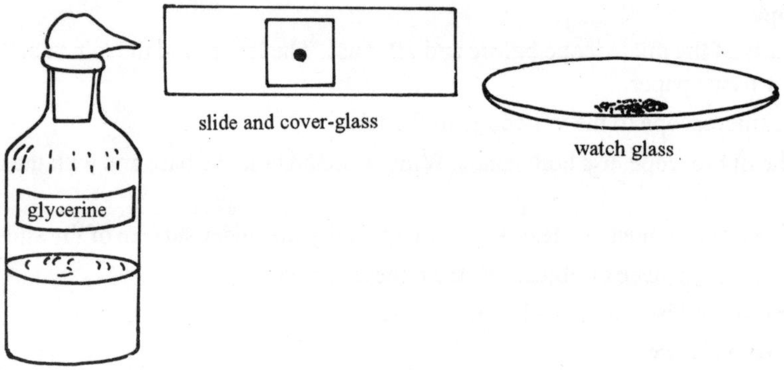

Fig.1.3 Dropping bottle

(a) Saffranin: This is the most commonly used stain. (1) It stains nucleus and lignified cell walls.

Preparation —

Saffranin —1gm.

95% (1) Ethyl alcohol –50ml.

Distilled water – 50ml.

An aqueous stain may also be prepared.

Saffranin – 18ml

Distilled water – 100ml.

(b) Fast Green

Fast green – 1gm.

90% Ethyl Alcohol – 100ml.

This stain is used to colour parenchyma cells and cytoplasm. Fast green may also be prepared in clove oil (75ml clove oil+25ml 100% alcohol).

(c) Cotton blue.

Cotton blue –1gm.

Lactophenol – 100ml.

(d) Haematoxylin

The dye is obtained from the heartwood of the tree *Hametoxylon campechianum,* There are two methods of preparation of the stain. These are:

1. *Heidenhain's Hametoxylin*

Dissolve 0.5 or 1.0 gm of the powder in 100 ml of distilled water. Filter and store in a dark bottle away from light. The stain is ready for use after 4-5 days of ripening. Ripening is not necessary if the stain is prepared in 70% ethyl alcohol.

2. *Delafelds Haematoxylin*

Sol.1 Aqueous solution of ferric ammonium sulphate 100ml).

Sol.2 One gm dye dissolved in 6ml of absolute alcohol.

mix solutions 1 and 2. To this mixture add 25 ml of glycerine and 25 ml absolute alcohol. The solution should be allowed to stand until it becomes dark red.

(e) Anilin blue

Also called cotton blue or China blue. It is prepared as follows:

Aniline blue –1gm.

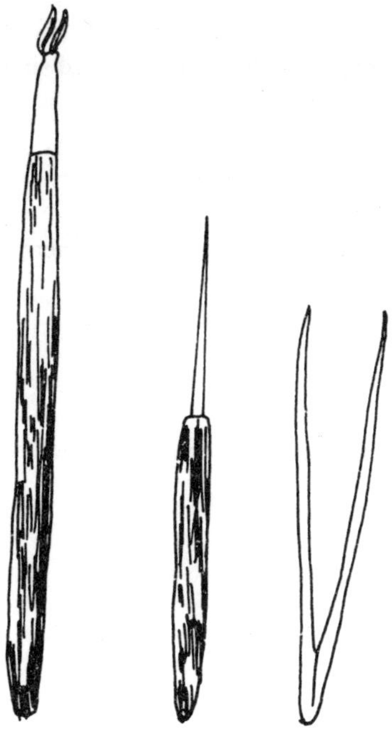

Fig.1.4 Brush Needle Forceps

Ethyl alcohol (90%)
or — 100ml.
Distilled water
Hydrochloric acid – 2drops.

(f) Gentian Violet (Crystal Violet)

Gentian Violet – 1gm.
90% Ethyl Alcohol 100ml.

(g) Erythrosine

Erythrosine –1gm.
90% Ethyl alcohol – 100ml.

Erythrosine may also be prepared in clove oil (Absolute alcohol 0.5 ml +95ml Clove oil).

(h) Light green

Light green –1gm.
Clove oil –95ml.
Absolute alcohol –5ml.

Light green may also be prepared without clove oil (in 100ml Absolute alcohol).

(i) Acetocarmine: The stain is prepared as follows:

Add 1 gm of stain to 100 ml of boiling acetic acid (45%)or propionic acid. Cool and decant, A few drops of saturated ferric acetate solution is added and the mixture is cooled over ice. The solution is allowed to stand for about 12 hours. Filter and store in a dark coloured bottle under refrigeration.

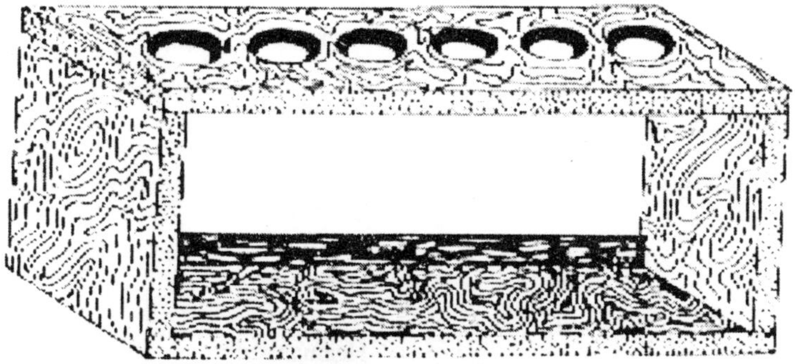

Fig.1.5 Staining rack

Specificity of Stains

Usually specific stains are employed in colouring different plant parts. The stains are used singly or in combination. The most popular stain combination are:

Saffranin - fast green.
Haematoxylin - Erythrosin.
Haematoxylin - light green.

While stain combinations are employed to study pteridophytes, gymnosperms and Angiosperms, single stains are used for Avascular cryptogams.

Green or Saffranin for Algae and Bryophytes.

Cotton blue for fungi.

The following table gives the specificity of stains to different plant parts or cell organelles.

Structure	Stain
Cellulose	Anilin blue Haematoxylin Fast green Light green
Cutinized cell wall	Gentian violet Erythrosine Saffranin
Lignified cell wall	Gentian violet Saffranin
Suberised cell wall	Saffranin
Cytopalsm	Fast green Light green Aniline blue
Chitin	Saffranin
Callose	Aniline blue
Achromatic figure	Aniline blue Erythrosine
Proteins	Saffranin
Nucleus	Heamatoxylin Saffranin Gentian violet
Chromonemata	Heamatoxylin

Chromosomes	Saffranin
Plastids	Gentian violet Haematoxylin
Mitochondria	Gentian violet
Flagella	Cotton blue

Fig.1.6 Razor

Fixing / Preserving Fluids

The plants or plant parts collected in nature are to be killed and preserved in the laboratory for the purpose of storage and study.

In killing the specimen, the various life processes are stopped at that instant, so that after preparation for microscopic study various stages may be observed.

When ever a material is collected, it is better to kill it in the field, so that there will be minimum or no distortion. Alternatively the material collected in the plastic bags or bottles should be brought to the laboratory and killed / preserved. In preservation, the material must be completely immersed in the fluid. The following are some of the killing/preservative fluids generally used:

1. *Formaline acetic alcohol (F.A.A)*

This is the most commonly used killing /preserving agent. It is popularly called standard preservative.

Ethyl alcohol (70%)	- 90ml.
Glacial acetic acid	- 5ml.
Formalin	- 5ml

2. *Formalin - Propiono - Alcohol*

Formalin	- 5ml
Propionic acid	- 5ml.
Ethyl alcohol 70%	- 90ml.

3. *Carnoy's Fluid - I*

Ethyl alcohol	- 30ml
Glacial acetic acid	- 10ml

4. *Carony's Fluid - II*

Ethyl alcohol	- 30ml.
Glacial acetic acid	- 30ml.
Chloroform	- 15ml.

Carony's fluid is widely used in killing root tips, flower buds etc. for the purpose of cytological study.

5. *Randolph's Modified Navashin Fluid*

This consists of two solutions A and B.

Solution A

Chromic acid	- 5.0 gms.
Glacial acetic acid	- 50ml
Distilled water	- 320ml.

Solution B

Commercial formalin	- 200ml.
Distilled water	- 175 ml.
Saponin	- 3gms.

Mix A and B in equal volumes before use.

Mounting media

These are chemicals used for the purpose of mounting (Temporary/Permanent) the specimens on the slide for microscopic observation. The following are some of the mounting media.

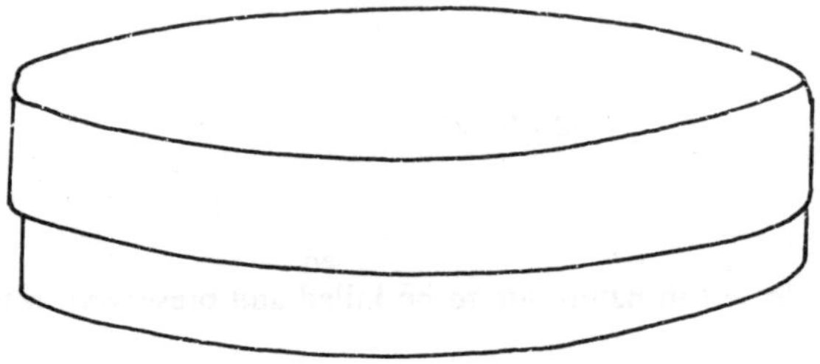

Fig.1.7 Petri dish

1. ***Acid water***

HCl – 10-15 drops
water – 100 ml.

2. ***Lactophenol***

Phenol	-100gms
Lactic acid	-100 ml.
Glycerine	-100 ml.
Distilled water	-100ml.

3. ***Glycerine***

Glycerine	- 15-20 drops
Formaline	- 1-2 drops
Distilled water	- 100ml.

4. ***Canada Balsam:*** This is a gum obtained from Gymnosperms. This fluid is readily available in the market.

5. ***Euparol:*** This is available in the ready to use form in the market.

Canada balsam and Euparol are used for permanent mounting, while the rest are used for temporary preparations.

Staining rack: This is a wooden rack used for holding the staining bottles.

Dropping bottles: This is a glass bottle having a narrow mouth fitted with a stopper. The stopper is grooved and has a beak. When the bottle is upturned the stain/fluid comes out in drops.

Slides: These are readily available in the market. The usual size 25mm x 75mm. The thickness is about 1mm. Special types of slides called cavity slides are available for specific study.

Cover slips (Cover glasses)

These are meant to cover the materials when the preparation is ready for observation. The coverslips may be round, square or rectangular.

Methods of Study of Plant Material

The following is the procedure for a scientific study in the laboratory.

Morphological study

The specimen given to study external detail should be observed either with or without a magnifying glass. Whatever details can be seen are to be recorded. If the specimen requires microscopic study, it should be suitably prepared for the purpose of microscopic observation. There are two types of microscopic preparation.

1. Whole Mount Technique

This is suitable for the purpose of external study of algal and fungal material. Take a little of the material, stain it with a suitable stain (staining procedure is explained later.) Wash the excess stain in water / alcohol. Mount in a suitable medium (temporary or permanent) and observe under the microscope.

2.Section cutting

For the purpose of anatomical study, sections are to be cut of plant parts. In algae, globule of *Chara*, stolons of *Caulerpa*., receptacles of *Sargassum* etc., can be cut. Similarly fruiting bodies of fungi, thalli of bryophytes also many be cut in a suitable plane. In higher plants cylindrical plant parts like root, stem, petiole etc. may be cut for the study of internal structure.

Planes of Cutting: In order to study the internal structure from various angles, sections are cut in various planes.

Transverse :The section is cut at right angles to the longitudinal axis.

Longitudinal : The section is cut parallel to the long axis (right angles to the transverse axis. The longitudinal section may be cut along a radius (Radial longitudinal section-RLS) or along a tangent (Tangential longitudinal section-TLS).

Cylindrical organs like roots and stems are cut transversely or longitudinally, while dorsiventral organs like thalli of bryophytes, leaves etc. are usually cut in vertical transverse plane.

Types of Section cutting

There are usually 2 types of section cutting methods employed for plant parts. These are (a)Free hand sectioning and (b) Microtome sectioning

(a) Free hand Sectioning

For the purpose of laboratory study where slides are not made permanent, usually free hand sectioning method is employed. Free hand sections are cut mostly with the help of a good sharp razor (sometimes safety razor blades also may be used).

Honing the razor : The razor used for sectioning must have a sharp, even edge. There should be no nicks. Before cutting sections, the razor edge has to be honed to maximum sharpness either on a stone or on a leather strap. Oblique strokes, with the edge forward are to be given on the sone or strap. Usually oil drops are used on the honing medium to give maximum sharpness.

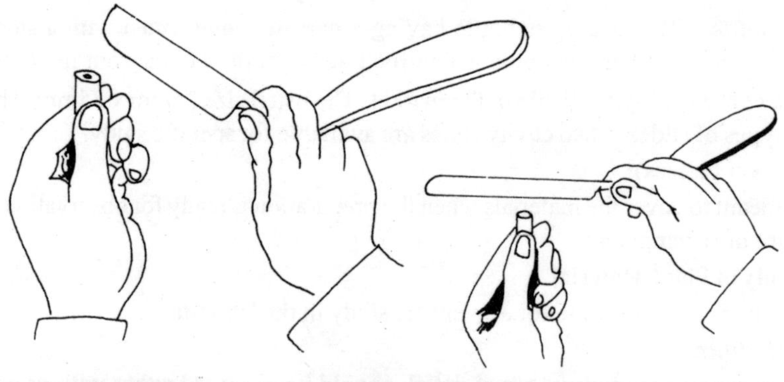

Fig.1.8 Method of holding pith — Methoding of holding razor — Method of section cutting

Method of cutting : Open out the razor and bend the handle backwards. Hold the razor between the index finger and thumb. The handle should be free. The remaining fingers of the hand will be touching the back edge of the razor blade.

In order to cut a transverse section of a cylindrical part like stem or root, follow the procedure given below.

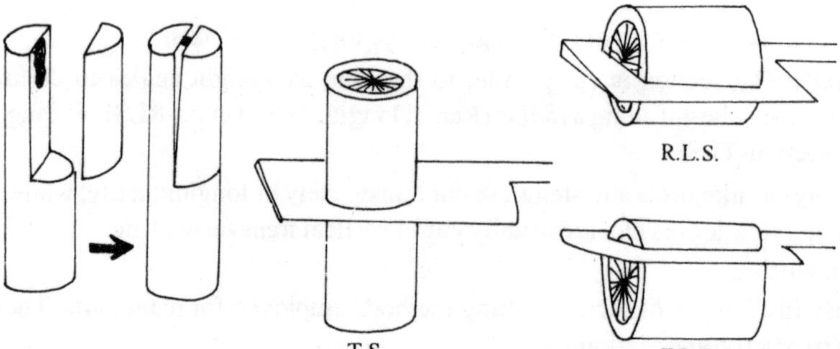

Fig.1.9 Method of using pith for insertion of material

The stem or root should be cut into a 2-3" long piece. It should be held in the hand in an erect position between the thumb and index finger. The material held in the hand should be exactly at right angles to the razor's edge. The thumb holding the specimen should be slightly at a lower level. Flood the edge of the razor with water. Now using the index finger as the platform move the razor quickly several strokes across the material. During this process sections are cut and they will be collected on the blade of the razor. The sections now may be carefully transferred with the help of a fine brush in to a petridish containing water.

For cutting sections of delicate and fragile materials like thalli or leaves, a suitable supporting medium has to be used. For this purpose a pith, carrot or potato may be used. These are to be cut into rectangular cubes and then split at one end. The material for sectioning should be inserted in the slit and sections are cut

After section cutting the razor should be wiped dry. It should be greased and encased.

Microtome Sectioning: This is generally employed very frequently in the study of bryophytes, pteridophytes and other higher plants. The microtome sectioning is a long and complicated process. It takes anywhere between 15-20 days to process a material and to obtain the sections.

Microtome sectioning involves the following steps:

1. Killing and preserving the material.
2. Dehydration and infilteration.
3. Paraffin embedding .
4. Section Cutting.

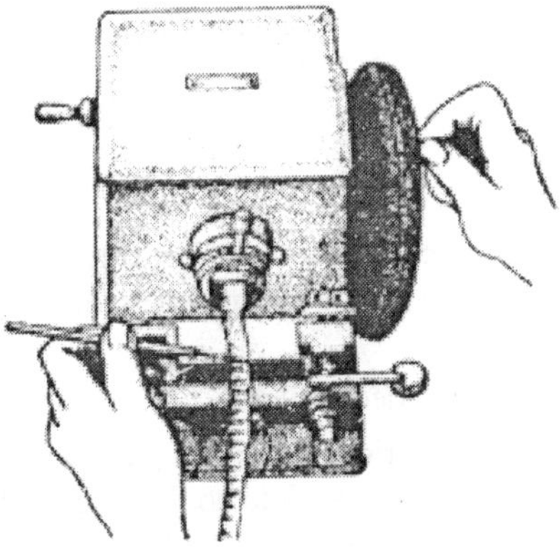

Fig.1.10 Rotary microtome

The material like stem, leaf, flower etc. is to be killed and preserved in a suitable fluid.

(a) Cut the material into bits of suitable size.

(b) Place the material in small vials containing a fixative or preservative.

(c) Label the vial giving the details of place of collection , name of the material, preservative used and date of collection.

The material should be in the killing fluid at least for six hours (preferably a day).

Dehydration: In order to dehydrate the material, remove it from the preservative and wash thoroughly in water and transfer it to a suitable tube or vial.

Dehydrate the material in alcohol series (Ethyl alcohol) as follows

Stage 1	50% Alcohol	First day
Stage 2	70% Alcohol	Second day
Stage 3	90% Alcohol	Third day
Stage 4	100% Alcohol	Fourth day
Stage 5	Absolute alcohol and xylol (50/50)	Fifth day
Stage 6	Absolute alcohol and xylol (25/75)	Sixth day
Stage 7	Xylol	Seventh day

When the material is transferred to pure xylol, there should be no fogging (i.e the solution should not turn milky). If there is fogging repeat the procedure starting from 4th stage.

The material in xylol will be completely dehydrated and looks transparent. It is now ready for infilteration by paraffin wax. The purpose of infilteration is to see that the wax enter into the material completely so that it can easily be cut.

Infilteration: Add a few pieces of paraffin wax everyday for about 4-5 days and then transfer the vial into a hot air oven, while temperature is adjusted to the melting point of wax. Everyday add a few pieces of wax. Adding of wax should be continued until all the xylol in the vial is replaced by paraffin wax. When the tube is smelt, there should be no smell of xylol. The material is now ready for embedding.

Paraffin embedding: In order to cut the material, it has to be embedded into a paraffin block. Prepare a paper boat or take a porcelain rectangular dish and smear glycerine to the inner surfaces. Pour molten paraffin into the porcelain dish and empty the contents of the vial into the dish. Before the paraffin solidifies, arrange

the material in proper rows and in such a manner as desired. Leave it for cooling for 6-8 hours. For quick cooling the dish may be transferred to a bowl containing ice water. Carefully remove the solidified paraffin block from the dish and cut it into suitable pieces. The material is now ready for sectioning.

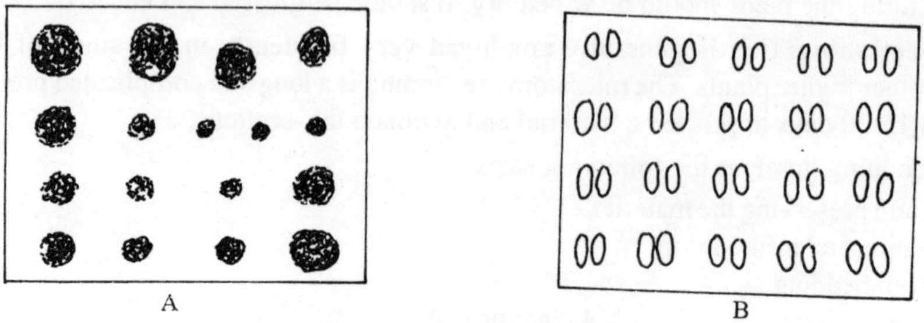

A B

Fig.1.12 Arrangement of materials in paraffin blocks - A) Flower buds in groups B) flower buds in pairs

Sectioning: Sectioning is carried out with the help of an instrument called microtome. There are various types of microtomes like Rotary, sliding, Freezing etc. The one that is normally used is a rotary microtome. The following procedure can be adopted in the case of the rotary microtome.

1. The paraffin block is to be cut into small pieces each having the embedded material.

2. The pieces of paraffin containing the material is to be fixed to wooden 'riders'. For this purpose, heat the top of the rider and the base of the paraffin piece with a hot knife and fix. Allow it to cool. The paraffin piece is now firmly fixed to the rider. In order to make the edges of the paraffin even, cut the edges and make them parallel.

3. Fix the wooden rider into the microtome so that the paraffin piece with the material is exactly parallel to the edge of the microtome knife.

4. Adjust the knob of the microtome to give the required thickness to the sections.

5. Rotate the wheel of the microtome in a clockwise direction. The cylinder containing the rider moves up and down on the knife edge. Sections are cut in a sequence and get attached to each other forming a paraffin ribbon.

6. Keep a needle in your left hand to hold up the ribbon as it lengthens. When the ribbon is sufficiently long, cut it and transfer it on to a black cardboard. Cut the ribbons into suitable length and transfer to the slide.

7. Smear the slides with an adhesive like egg albumen or gelatin and rub it well.

8. Pour some water onto the slide with the help of dropping bottle.

9. Transfer the paraffin ribbons on to the slide and drain excess water.Retain a little water on the slide.

10. Place the slide on a slide warmer. The ribbon expands, and so also the sections of the material. Now draw out the water completely and allow the slides to cool.

11. The slides should be allowed to stand for atleast 24 hours whereupon the sections will stick to the slide permanently.

The slides are now ready for staining.

Staining procedure

The plant material (whole mount or sections) are to be stained and mounted properly before they are ready for microscopic observations.

Algal filaments, fungal mycelia, pollen grains and spores, prothalli of pteridophytes are usually stained with a single stain - saffranin or green. In some case Aniline blue is used to stain the filaments of Algae and mycelia of Fungi. The staining procedure is as follows.

Place the material either in a watch glass or cavity slide and add a few drops of stain. The period of staining varies with different materials. But usually a 5-10 minute period is sufficient. Drain out the stain and observe the material under the microscope. If the material has taken less stain, it may be kept in stain for some more period, or if it has taken excess stain it can be removed by washing either with water or 50% ethyl alcohol.

After the material is suitably stained pass it through a series of alcohol grades to dehydrate it. Usually different grades of alcohol (70%,90%,100%) taken in a filler are put on the material. After passing through absolute alcohol the material is kept in xylol and it is now ready for mounting.

Fungal mycelia can be stained with aniline blue. Usually the material taken on a clean slide is stained with a drop or two of the stain prepared in lactophenol. The material can bemounted straightway as lactophenol is also a mounting medium. If the stain is not prepared in lactophenol the material may be transferred to fresh lactophenol and mounted.

Double Staining Procedure

In the case of most of the algae and fungi there is little or no tissue differentiation and hence single staining is sufficient. In some cases of plants with relatively more tissue differentiation, a single stain is not sufficient to mark out the various tissues clearly. In such cases 2-3 stain combinations are used to stain the tissues. For example lignified and non-lignified tissues are stained with a combination of acidic and basic dyes.This type of staining is also called differential staining where different tissues take up different stains.

The following are some of the popular stain combinations.

Main stain	*Counter stain*
Haematoxylin	Saffranin
Haematoxylin	Erythrosine
Haematoxylin	Eosin
Haematoxylin	Orange G.
Saffranin	Fast green
Saffranin	Light green
Saffranin	Aniline blue
Saffranin	Crystal violet
Crystal violet	Erythrosine

Staining Procedure for free hand sections

(a) Procedure for temporary preparations

Free hand sections of root, stem etc., should be washed thoroughly in distilled water in a watch glass. After repeated washing the sections should be transferred to a watch glass containg the main stain. The sections should be kept in the stain for a period ranging from 30-40 minutes. After this period, the sections should be transfereed to a watch glass containing distilled water. The sections should be washed for 2-3 minutes whereupon excess of stain is removed. They should be observed under the microscope. If the main stain is saffranin only lignifed tissues should retain the stain, others should be destained. If the main stain is specific to non lignified tissues, only they should retain the stain. Then the sections should be washed thoroughly in water and repeatedly observed under the microscope. The sections can be mildly stroked with a fine hair brush, so that excess of stain comes out with distilled water. The sections are now ready for counter-staining. They should be transferred to a watch glass containing the counter stain (fast green etc). Counterstain acts quickly on the tissues. Hence one to two minutes of staining should be sufficient. The sections are now observed under the microscope. If there is excess of counter stain, it can be removed by

washing sections with 15-20%glycerine. After the final stianing the tissues whould be demarcated clearly with no mixing of the stains. The sections are now ready for mounting.

(b) Procedure for permanent preparaions

If the sections of plant materials are to be mounted and made into permanent slides, the procedure should be slightly altered so as to include dehydration and clearing with suitable agents. The procedure is as fallows:

1.Stain with the principle stain (Saffranin, Haemotoxylin etc).

2.Allow the sections to remain in stain for about 15-20 minutes.

3.Transfer the sections to water and wash thoroughly. The sections should be stroked mildly until no more stain comes out, and the water should remain clear. If the main stain like saffranin is produced in 50% alcohol then destaining (washing) should be done in 50% alcohol.

4.The sections should be now dehydrated in alcohol grades of 50%, 70% and 100%.

5.In each grade of alcohol keep the section for about 5 minutes.

6.Transfer the sections from one grade of alcohol to the other with the help of a fine brush. Never use needles as they injure the sections.

7.Transfer the sections to the counter stain (fast green etc). The counter stains are usually prepared in 90% alcohol.

8.Keep the section in counter stain for about a minute. Wash the sections in 90% alcohol and transfer them to 100% alcohol.

8a. In some case counter stains are prepared in clove oil. In such cases counter staining should be done after the sections are completely dehydrated with absolute alcohol.

1. Sections are transferred completely dehydrated to a solution of 50:50 Absolute alcohol and xylol followed by 25:75 Absolute alcohol and xylol and ultimately the sections are transfered to pure xylol. Now xylol should not turn milky or turbid. If it turns milky, dehydrate the sections again through combination of absolute alcohol : xylol series.

The sections are now ready for mounting.

Staining procedure for microtome sections

The slides with the paraffin ribbons adhering to them have sections of plant material.The processing and staining is as follows:

1. Immerse the slides in a coupling Jar containing xylol. Leave them for 10 minutes . The paraffin dissolves in xylol and only the sections remain on the slides.

2. The slides are now kept in 50:50 absolute alcohol and xylol.

3. Pass the slides through a down grade series of alcohol. 100%, 90% and 70%

4. Transfer the slides to pure water and wash them thoroughly to remove traces of alcohol.

5. Keep the slides in a mordant like 4% iron alum. This is necessary only when Haematoxylin is used. After about 10 minutes wash the slides in water.

6. Transfer the slides to the main stain Haematoxylin. Saffranin etc. and keep them for 15-30 minutes.

7. Remove the slides from stain and wash in water.

8. Observe the slides under microscope. Usually the sections are overstained.

9. If the main stain used in Hametoxylin, destaining can be done with 2% Iron alum or with water containing few drops of concentrated HCI. Repetaedly observe the slides under the microscope. Stop destaining when only the nuclei have retained the stain.

Caution: When destaining is done with HCI, not more than 5-6 drops should be used for 75ml of water.

Excess acid will eat away the adhesive and the sections come out of the slide.

10.Wash the slides in water overnight if the destaining agent is Iron alum, otherwise washing for 30 minutes is enough.

11.Now pass the slides through an upgrade series of alcohol (5 minutes in each grade) 70% 90% 100%

12. When the sections are completely dehydrated stain them with the counter stain (Fast green etc) for a minute or two. Wash the excess stain in clove oil and transfer the slides to pure xylol. The slides are now ready for mounting.

Mounting

(a) *Mounting Media*

Mounting media are of two categories: Temporary mounting media and permanent mounting media.

A temporary mounting medium may be just water or water mixed with glycerine. If the temporary sections are to be retained for 2-3 hours, dilute glycerine (15-20%) is recommended.

Permanent mounting media are used when the slides are to be retained indefinitely. The following are some of the permanent media.

(1) **Canada Balsam:** This is the most widely used mounting medium in botanical preparations. It is available in ready to use bottles. It is a gummy substance. If the solution is very thick add little xylol.

Canada balsam is a resinous product of a Gymnosperm, Abies balsamea. The gum has very good refractive index and as such it is highly reccommended.

(2) **Lactophenol:** Lactophenol can be prepared in the laboratorty. Mix equal quantites of phenol cystals, lactic acid glycerine and distilled water.

(3) **Glycerine gelly:B** This is prepared as follows:

Gelatine 1 part.
Water 6 parts.
Glycerine 7 parts.

Glycerine jelly is suitable even for mounting fresh specimens (even without dehyderation).

Add gelatine to water, stir thoroughly and warm. Add glycerine and cool the solution. After sometime the solution settles into a jelly.

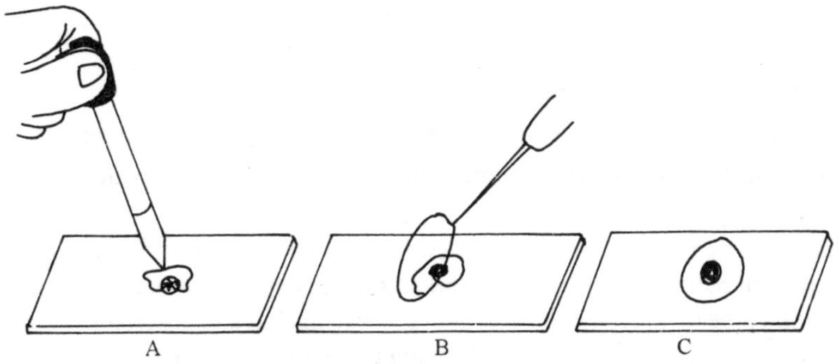

Fig.1.13 Method of mounting the material

(b) Mounting Method

The sections (free hand) are to be mounted on to clean slide in a specific mounting medium. If the sections are microtome preparations they would be already adhering to the slide. In any case the sections are to be covered with a thin coverslip or cover glass which are readily available in the market. They may be round, square or rectangular. Select the one that is suitable for your preparation.

The sections or objects should be mounted on the centre of the slide. In order to ensure placing of the object on the centre of the slide, the following procedure may be adopted. On a white paper draw a rectangle of the same size as the slide. Join the opposite corners by drawing diagonal lines. The centre of the slide lies where the two lines cross. Place the slide on the rectangle and material or objects may be kept on the central point.

Place a drop or two of the mounting medium. Hold the coverslip at the edges (see the figure) and allow it to touch the edge of the mounting medium at an angle of approximately 45°. Slowly lower the cover glass on the mounting medium with the help of a needle. Take care to see that no air bubbles enter the mounting medium. If there are air bubbles, lift the cover glass and mount again.

Excess mounting medium may be carefully wiped with the help of a clean tissue. Allow the slides to dry. If necessary the slides may be transferred to a hot air oven for drying. The slides should be labelled at one end as shown in the figure. They are now ready for microscopic observation.

Ringing of Slides

Temporary preparations may be kept for few days if the borders are sealed. This procedure is known as ringing . Ringing is done by applying wax, canada balsam or even nail polish to the edges. Ringing may be done manually or by using a ringing table having a rotary metallic disc.

Record of Practical Work

Practical work carried out in the laboratory should be recorded. The botany practicval record book has two types of sheets viz., Interleaves and drawing sheets. Interleaves are meant to write the notes pertaining to the diagrams. The drawing sheets which are thicker than the interleaves are used for diagrams. The following points may be useful in maintaining a good record.

1. Use good pencils (HHB) with sharp points and a good quality eraser.
2. Leave a minimum of one inch margin on all the sides.
3. Always write diagrams of only one plant in a page. Do not mix up diagrams of other plants.
4. On the left hand corner of the drawing sheet write the date of the experiment.
5. Give the complete systematic position of the member on the top of the drawing sheet.
6. For microscopic organisms draw first the habit sketch followed by low power and high power view/s. High power view should show the cellular details.
7. In the case of anatomical drawings, draw the ground plant first, followed by a sector enlargement to show the cellular details.
8. All parts of the diagrams must be properly labelled.
9. Stages of reproduction, if any, must be drawn after the details of vegetative structure.
10. The description corresponding to the diagrams must be written on the opposite interleaf.

Field Collection of plants

Specimens of algae, fungi and bryophytes can be collected in the field for the purpose of laboratory study. Field trip is an integral part of botanical study and all students must participate in it. It is in the field study that the students will come to understand the natural growth, habitat and environmental association of plants. During a field trip to collect specimens, the following items are to be carried:

1. Pocket lens.
2. Fixative F.A.A or Carnoy's fluid.
3. Collection bottles.
4. Blotting paper.
5. Polythene bags
6. Field note book
7. Labels.

When the specimens are collected they can be fixed and preserved for the purpose of laboratory study. Use the required preservative and immerse the specimens in it. Label the bottle showing details of

Date of collection

Location

Preservative fluid (name) used.

Name of the plant

As it is necessary for the purpose of identification, details of habitat, associations with other plants if any, colour of the fresh specimen etc. should be recorded in the field note book.

PART ONE
ANATOMY

CHAPTER 2

STUDY OF PLANT CELL

Purpose : To study the structure of a typical plant cell under microscope by using thin layer of onion peel.

Materials : Onion bulbs, slides, coverslip, microscope and stain (saffranin).

Procedure : Carefully peel off a thin layer from the bulb of the onion. Initially it may be difficult to get a thin layer but after the practice it will be possible to get a thin layer which is suitable for microscope examination. For this purpose carefully separate one fleshy leaf of the onion bulb. Bend it in the middle so that it incompletely breaks. When you separate the two pieces a thin layer will come out. Immediately after removing place the onion peel in a petridish containing water.

Cut a small piece of the onion peel suitable for observation and place it in a drop of water in the centre of a clean glass slide. Add a drop of stain, put a cover glass and observe under microscope.

Observation and conclusions. The study shows the following structural details of the plant cell.

1. A number of rectangular cells can be seen under the field of the microscope.
2. The cells are many times longer than broad and are joined end to end and side to side to form a layer.
3. The cell wall which is the outermost layer can be clearly demarkated.
4. The cell membrane is not very clearly visible.

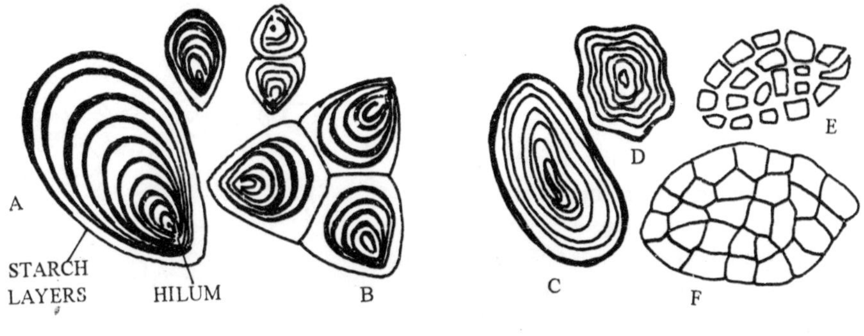

Fig. 2.1 Plant cell

Starch grains, A.-B. Potato, C. Pea, D. Maize, E. Paddy, F. Oats.

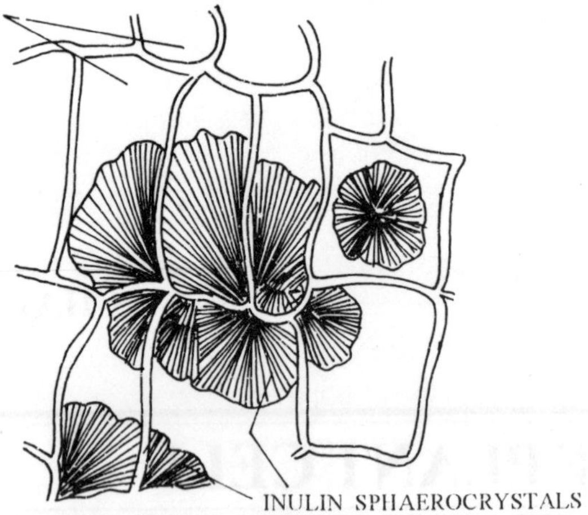

Fig. 2.2 **Plant cell**
Sphaerocrystals of inulin in the roots of Dahlia.

5.Occupying the centre of the cell is a large vacuole.
6.Cytoplasm is peripheral and occupies the space between the cell membrane and the vacuole.
7.Lying in the cytoplasm is a large vacuole which is red coloured (because it has taken saffranina stain).
8.Also distributed in the cytoplasm are seen a number of leucoplasts.

ERGASTIC SUBSTANCES (Non living inclusions):

Many kinds of organic and inorganic compounds occur as non living inclusions in cytoplasm. These are produced during the metabolic activities of the cell and are also known as ergastic substances. Principally, there are three types of ergastic substances, *viz., Reserve food products, Secretory products* and *Waste products.*

1. RESERVE FOOD PRODUCTS: These are synthesised in cytoplasm and are stored as reserve food material. These are classified into three categories *viz., Carbohydrates* (non nitrogenous products), *nitrogenous products* and *fats* and *fatty oils.*

A. Carbohydrates:

1. These are the principal types of food substances containing carbon, hydrogen and oxygen generally in the ration 1: 2: 1.
2. Carbohydrates may or may not be soluble in water.
3. Some of the important carbohydrates are: Starch, Inulin, Hemicellulose, Cellulose and Sugars.

Starch:

1. It is an insoluble carbohydrate and is found in the cells in the form of various types of grains.
2. Starch grains consist of a substances called *amylose* which turns blue on contact with iodine.
3. These are usually present in cereals and underground storage roots and stems.
4. In each starch grain there is a central spot called *hilum* on which are deposited starch layers in the form concentric rings.
5. Based on the structure, starch grains may be classified into *concentric, eccentric, simple* and *compound.*
6. In concentric grains (peas) the rings are concentric.
7. In eccentric grains the hilum is towards one side (Potato), in simple grains (wheat) there is a single hilum and in compound grains (rice) there is more than one hilum.

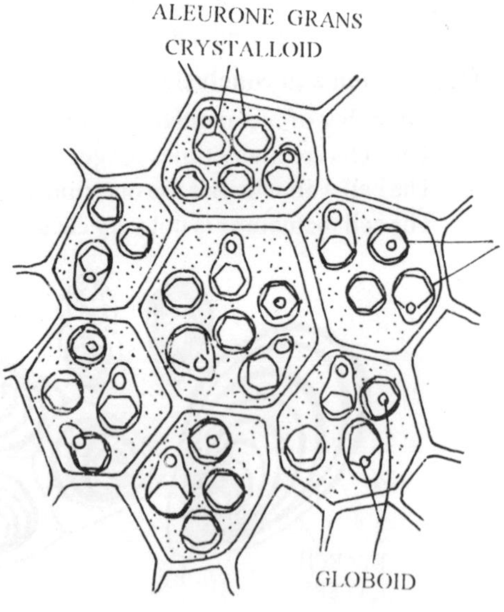

Fig. 2.3 **Plant cell**
Aleurone grains in the endosperm of Ricinus communis (Castor)

Inulin:

1. This is a carbohydrate, soluble in water but insoluble in alcohol.

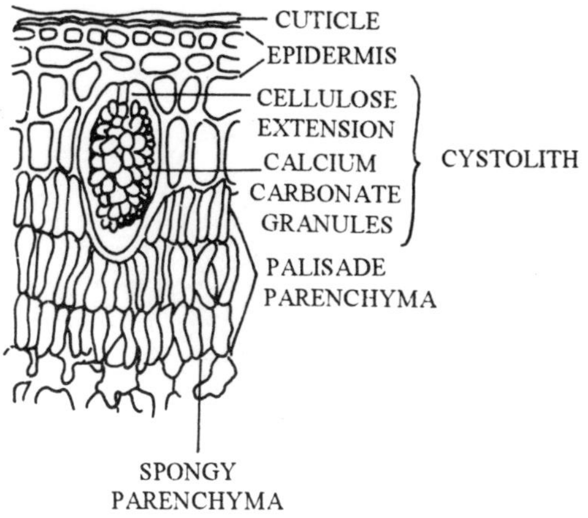

Fig. 2.4. Plant cell
Calcium carbonate cystolith in the leaf of Ficus elastica.

2. It is generally found in the tuberous roots of the members of the family *Asteraceae* such as *Dahlia, Helianthus, Cichorium* etc.
3. In these members inulin forms fan shaped cyrstals.
4. The inulin of *Dahlia* tubers can be precipitated by keeping them in alcohol.

B. Nitrogenous Products:

1. These are reserve food materials containing nitrogen.
2. There are two types of nitrogenous food materials viz., proteins and amino compounds.
3. *Proteins* are very complex macromolecules having carbon, hydrogen, oxygen and nitrogen.
4. Sometimes, in addition to these, sulphur and phosphorous are also present.
5. Proteins are stored in seeds or grains in the form of *aleurone crystals.*
6. These are present in the endosperm of castor, pericarp of maize grain etc.
7. Each aleurone grain (crystal) is a solid, oval or rounded body enclosing a small crystal called *crystalloid* and a small globular body called *globoid.*
8. The crystalloid is proteinaceous and occupies the major portion of the grain, while the globoid consists of phosphates of calcium and magnesium and occupies the rest of the grain.

Amino Compounds occur in the form of amino - acids and amines in the cell sap. These are more frequently present in the growing parts rather than in storage tissues.

C. Mineral Crystals:

1. Various types of mineral crystals are found in plant cells.
2. They may be present either in cell walls or in a cavity in the cytoplasm.
3. Majority of the crystals contain calcium carbonate or calcium oxalate.

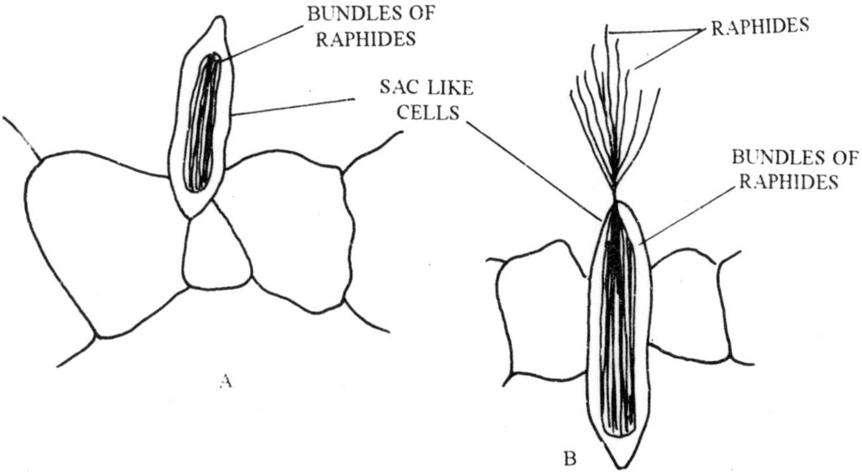

Fig. 2.5 **Plant cell**
Raphides - A raphides in colocasia, B. ejection of a raphide.

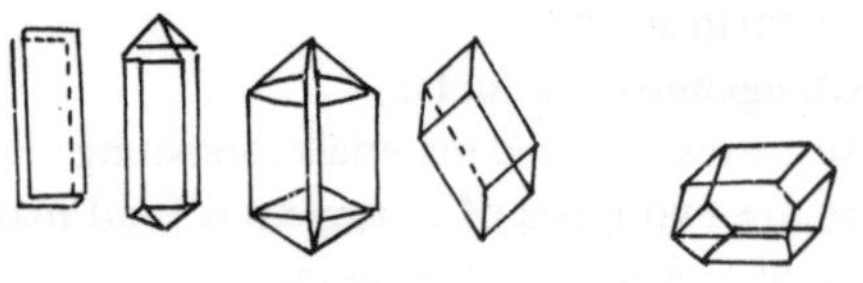

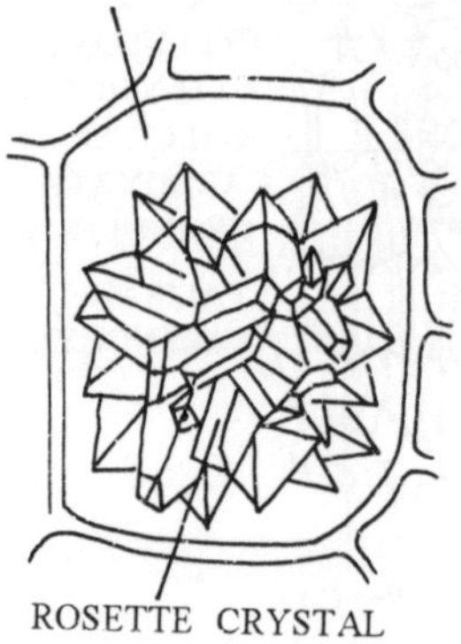

Fig. 2.6 **Plant cell**
Calcium oxalate crystals. Prismatic crystals (left) and Rosette crystal (right)

4. The following are some of the types of mineral crystals.

1. Cystoliths:

1. These are crystals of calcium carbonate, generally found in the leaves of various flowering plants.
2. The members of the family Moraceae, Urticaceae etc., have cystoliths.
3. Each cystolith has a peg like projection of the cell wall into the cell cavity.
4. Cystoliths consist of many small crystals grouped together like a bunch of grapes.

2. Calcium oxalate crystals:

1. These are found in the vacuoles and are of the following types.

(a) Raphides:

1. These are long needle like crystals arranged parallely in the form of stacks or bundles.
2. These are found in sacs and when the sacs are ruptured come out with a jerk.
3. Raphides are commonly seen in the leaves of *Colocasia*, *Pistia* etc.

(b) Druses, Rosette or cluster crystals:

1. These are compound crystals arranged in the form of a rosette.
2. These are found in the leaves of *Eucalyptus*, *Ixora* etc.
3. Crystals found in globose masses are called druses.
4. Shining crystals are called prismatic crystals.

(c) Idioblasts:

1. These are stellate crystals and are usually found in water plants (*Limnanthemum*, *Nymphea* etc.)

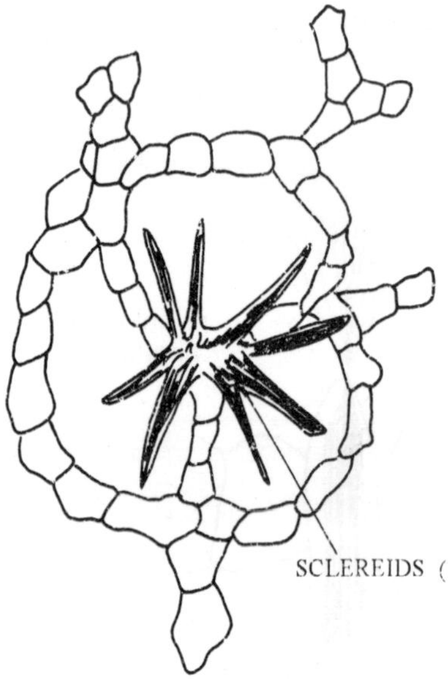

Fig. 2.7 **Plant cell**
Idioblast in a leaf cell

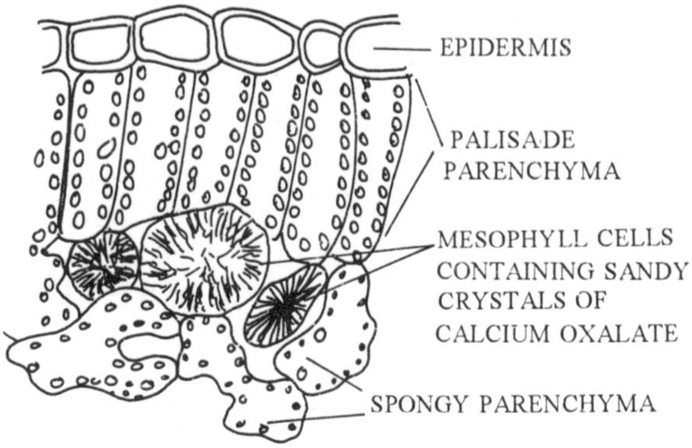

Fig. 2.8 **Plant cell**
Sand crystals of Atropa belladona.

offering mechanical support.

(d) Sand crystals:

1. These are found in masses packed into a cell.
2. Such crystals are seen in the roots and leaves of *Atropa belladona*.

(e) Latex:

1. In some plants, a milky or watery juice comes out of latex vessels or cells.
2. Latex contain starch grains proteins, oils, tannins, gums, resins, alkaloids, salts, enzymes and many poisonous substances.
3. Laticiferous cells are seen in *Carica, Euphorbia, Ficus, Hevea, Calatropis* etc.
4. In some cases the dried and processed latex is the source of natural rubber.

(f) Essential oils:

1. These are volatile oils and are secreted by oil glands.
2. Oil glands are found in almost all parts of the plants, specially in leaves and in the skin of fruits (*Citrus* fruits).
3. Essential oils are also found in tl

Systematic position
GNETOPSIDA
GNETALES
GNETACEAE
GNETUM

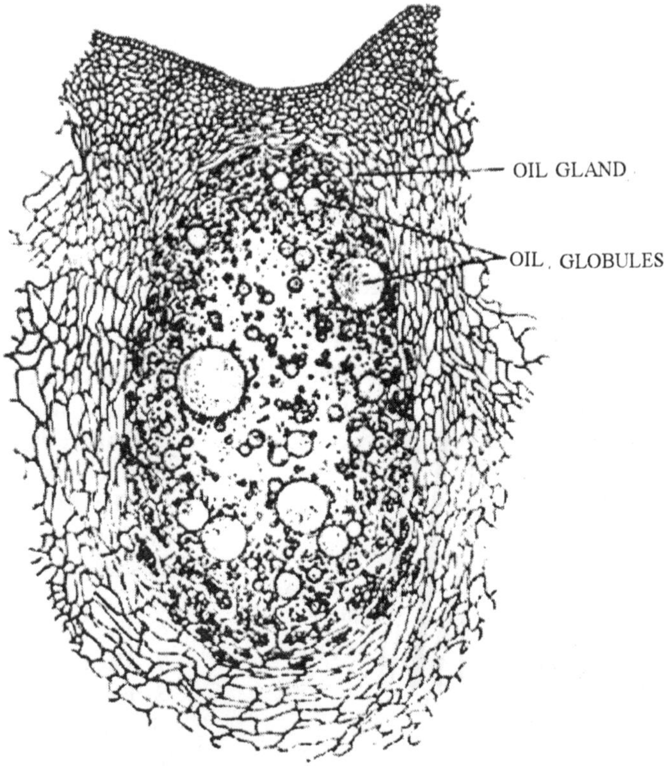

Fig. 2.9 **Plant cell**
Long section of an oil gland from Cirtus aurantiaca.

CHAPTER 3

TISSUES

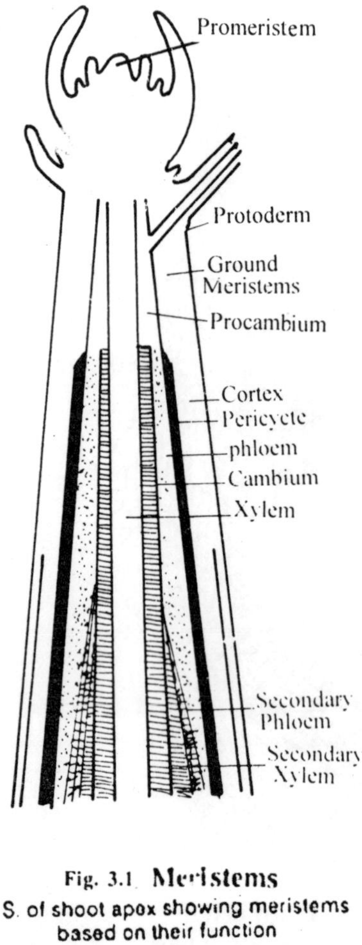

Fig. 3.1 **Meristems**
L.S. of shoot apex showing meristems based on their function

Cells are grouped together to form tissues. Generally cells which are alike in origin, form and functions constitute a tissue. In higher plants like, Gymnosperms and to a certain extent in Pteridophytes cells are of diverse types. They vary in their origin, shape, size, structure etc., Consequently the assemblage of these different types of cells form various types of tissues.

Basically tissues are classified into meristematic and permanent tissues. The basis for this classification is the stage of development of the cells. In a meristematic tissue the cells are actively dividing and have not attained maturity. The cells of a permanent tissue onthe other hand have lost the capacity of cell division and have attained their mature form.

MERISTEMATIC TISSUES

1. The cells of a meristematic tissue are immature and are in an active state of division.
2. The cells are usually closely packed with no intercellular spaces.
3. The shape of the meristematic cells varies.
4. They may be oval, polygonal rounded or rectangular in shape.
5. The cell inclusions will be in their precursor stage.
6. The nuclei are very prominent.
7. The plane of division in the cells varies.
8. It may be periclinal (parallel to the surface) or anticlinal (Perpendicular to the surface).

Classification of meristems:

1. According to the location, there are three types of meristems, viz., apical, intercalary and lateral.

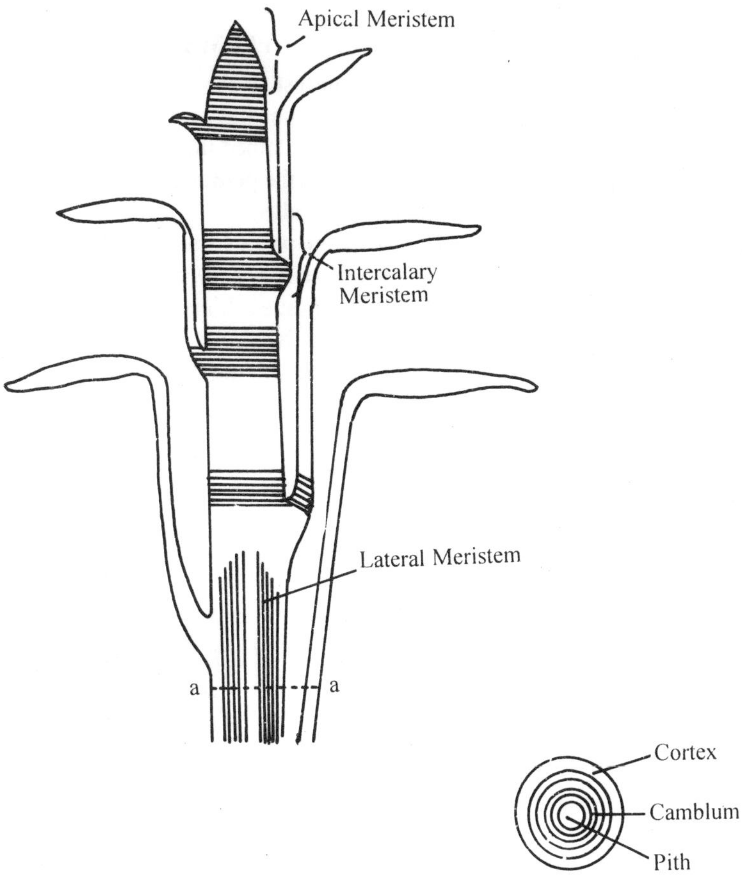

Fig. 3.2 Meristems

A. Diagrammatic L.S. of shoot apex showing the positions of meristems.

B.T.S. of shoot at A - A.

2. Apical meristems as the very name suggests are located at the apices of shoot, root and other organs.
3. They constitute the growing points of plants.
4. The growth in length of the plant and plant organs is due to apical meristems.
5. Growth is initiated by one or more cells of the apical meristems.
6. Apical meristem consists of two zones, promeristem and meristem.
7. Promeristem consists of actively dividing cells called apical initials.
8. The meristematic zone is classified into protoderm, procambium and ground meristems.
9. Intercalary meristems represent the detached portions of the apical meristems seperated due to the growth of an organ.
10. Intercalary meristems are always internodal.
11. In the early stages of development of the plant, whole or part of the internode is meristematic.
12. Subsequently a portion of it matures, as a result, patches of meristems are found embedded in permanent tissues.
13. Lateral meristems are present along the side of the stem.
14. The cells divide in only one plane and add to the girth of the tissues.
15. Typical example of a lateral meristem is the vascular cambium.
16. Vascular cambium produces secondary Xylem and Phloem thus bringing about secondary growth.

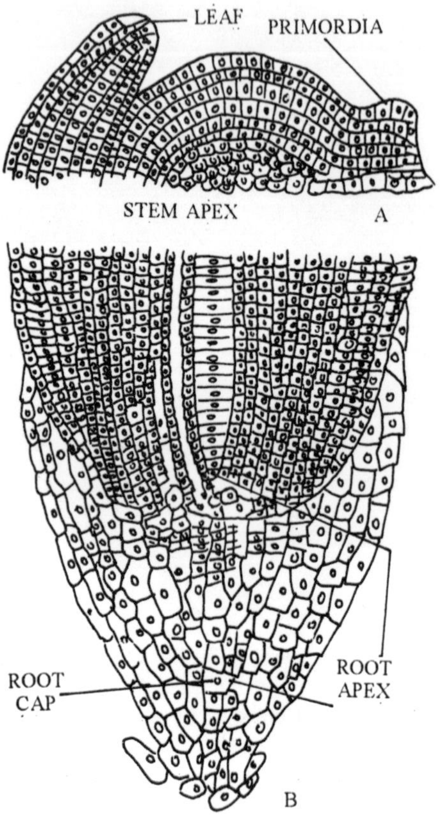

Fig. 3.3 **Meristems**
A. L.S. of shoot apex showing structure.
B. L.S. of root apex showing structure

7. But these cells coordinate and function like a single unit.
8. Generally one or more types of simple tissues combine together to form a complex tissues.

SIMPLE PERMANENT TISSUES:

There are further classified into Parenchyma, Collenchyma and Sclerenchyma.

PARENCHYMA:

1. Parenchyma refers to a tissue composed of living cells variable in their morphology and physiology.
2. The word parenchyma is derived from the Greek *para*, beside and *en-chein*, to pour, meaning a semi-liquid substance poured beside other tissues.

PERMANENT TISSUES

1. As a contrast to the meristematic tissues, the cells of permanent tissues have no capacity of division.
2. They have attained their mature form and shape.
3. The cells of the permanent tissues may be living or dead, thin or thick walled.
4. Permanent tissues may be classified into simple and complex, based on their composition.
5. A simple tissue is homogenous i.e., it is made up of only one type of cells.
6. In a complex tissue there will be at least two different types of cells.

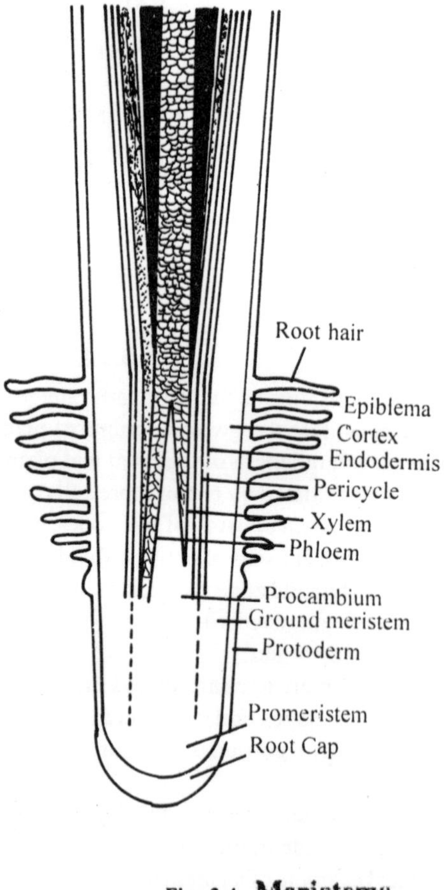

Fig. 3.4 **Meristems**
L.S. of root apex showing meristems

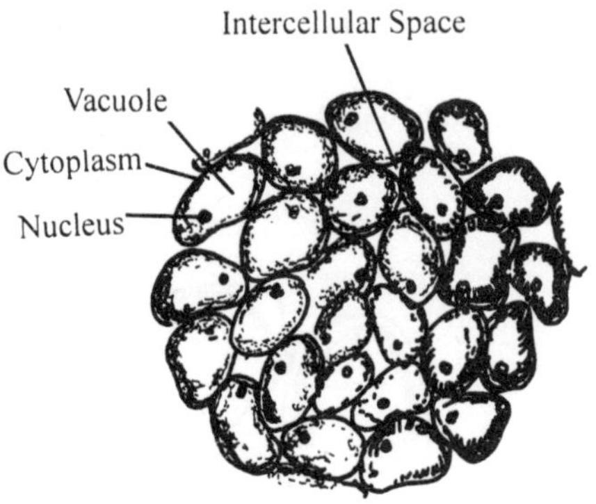

Fig. 3.5 **Permanent tissues.,**
Parenchyma

Distribution:

1. Parenchyma cells are present in almost all parts of the plant, like the pith and cortex of rocts and stems, mesophyll of leaves, the flesh of succulent roots, the endosperm of seeds etc.
2. As components of heterogeneous tissues parenchyma cells form vascular rays.

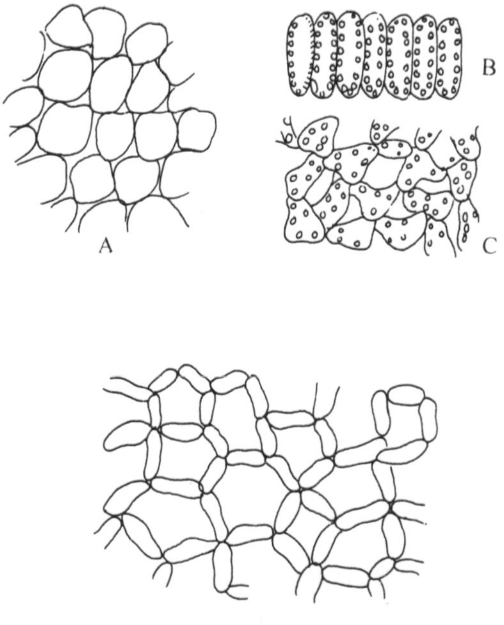

Fig. 3.6 **Permanent tissues.,**
Types of Parenchyma. A. Simple, B. Palisade, C. Spongy, D. Aerenchyma.

Cell shape:

1. Parenchyma cells are spherical or polygonal in shape.
2. Sometimes they may bee elongated and columnar as in the leaves.
3. Studies based on geometric models have indicated that closely packed parenchyma cells have the shape of a polyhedron with approximately fourteen faces.
4. Parenchyma cells are closely packed or arranged loosely with large number of intercellular spaces.

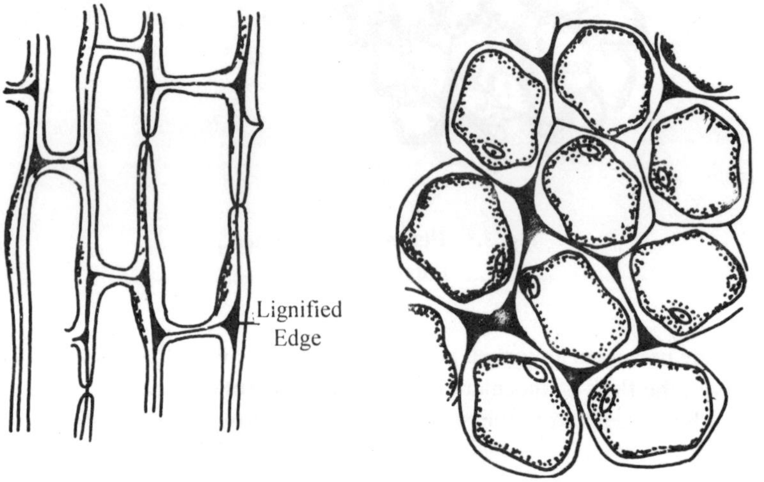

Fig. 3.7 **Permanent tissues.,**

Collenchyma. A-L.S.B. - Ts.

Cell structure:

1. The cells have thin walls and have dense cytoplasm with a prminent nucleus.
2. Parenchyma cells in storage organs have abundant reserve food materials like proteins, fats and carbohyderates.
3. Parenchyma cells found in leaves have number of choloroplasts.
4. These help in photosynthesis.

Cell types:

1. Based on their structure and function parenchyma cells are classified into the following types.
2. *Simple parenchyma* which forms the ground tissue, *Chlorenchyma* which is the chief photosynthetic tissue in the plants, and *Aerenchyma* which has parenchyma cells with large intercellular spaces storing air.
3. Aerenchyma is present in hydrophytes and helps in respiration and buoyancy.
4. Chlorenchyma cells found in dorsiventral leaves are of two types viz., *Palisade parenchyma* and *spongy parenchyma*.
5. Cells of palisade parenchyma are columnar and are filled with chloroplasts.
6. Spongy parenchyma cells also have chloroplasts, but they are rich in starch grains.
7. Parenchyma cells found in Xerophytes (*Aloe, Agave* etc.) act as water storage tissues.

COLLENCHYMA:

1. Collenchyma is a living tissue and is composed of more or less thick primary non lignified walls.
2. Just as in parenchyma, even in collenchyma the protoplast is living.

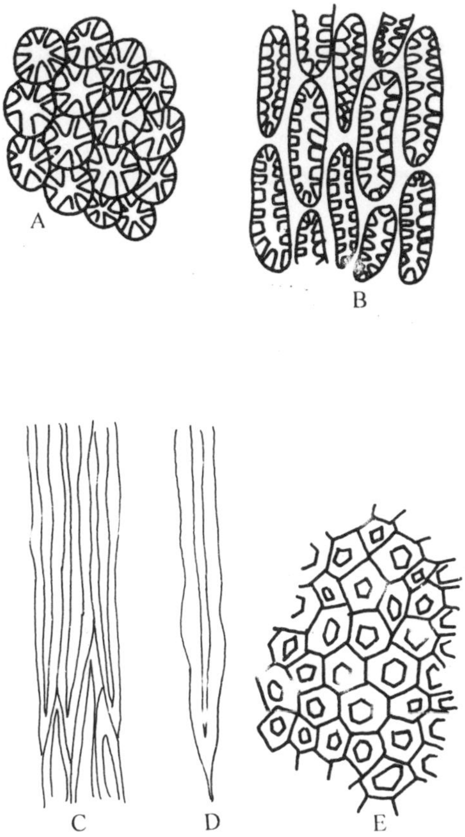

Fig. 3.8 **Permanent tissues.,**
Sclerenchyma. A., T.S. of sclereids. B., L.S. of the same. C. and D., L.S. of fibres. E. T.S. of Fibres

3. The word collenchyma is derived from the Greek word *colla*, meaning glue, which refers to the thick glistening wall

Distribution:

1. Collenchyma characterstically occurs in the pripheral position of stems, leaves, floral parts etc.
2. It may also occur in root cortex particuiarly if it is exposed to light.
3. In the stems and leaves of dicotyledons, collenchyma is hypodermal situated immediately below the epidermis or seperated from it by one or two layers of parenchyma.
4. Collenchyma cells are considerably elongated.
5. Sometimes, cells are 2 mm, long resembling fibres.
6. The shorter collenchyma cells are prismatic and appear polygonal in a transection.
7. In a long section the cells appear rectangular.

Cell structure and function:

1. The structure of the cell wall is the most characterstic feature of the collenchyma cells.
2. The thickenings are deposited unevenly, mainly deposited at the corners where several cells join together.
3. In long section, collenchyma cells show thin and thick wall portions dependng on the direction of the cut.
4. The walls containing mainly of cellulosic and pectic compounds and consist of much water.
5. Collenchyma cells have living protoplasts and may have chloroplastss.

6. Chloroplastss are less in number in long and narrow cells and more in cells, which approach parenchyma in form. Tannin may be present in collenchyma cells.
7. Collenchyma is primarily a mechanical tissue adapted for support of growing organs.
8. The thick walls and close packing, make collenchyma a strong tissue.
9. At the same time, the peculiarities of growth and the structure of its walls impart a considerable tensile strength with flexibility and plasticity.

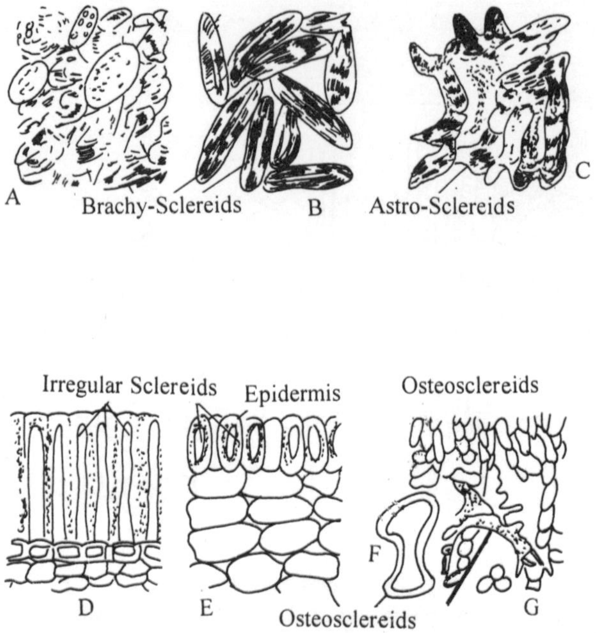

Fig. 3.9 **Permanent tissues.,**
Types of Sclereids. A and B. Brachysclereids, C. Irregular selereids, D and E. Macrosclereids, F. Osteosclereid, G. Astrosclereid.

SCLERENCHYMA:

1. This tissue consists of thick walled cells, (lignified), their main function being mechanical.
2. The word is derived from the Greek *Sclerous*, hard and *Enchyma* an infusion which emphasizes the hardness of the cell walls.
3. Sclerenchyma cells show variety in form, structure, origin and development , which the different types often integrating.
4. Generally, sclerenchyma cells are classified into *fibres* and *sclereids*.
5. Fibres are considered as long cells while sclereids, are relatively short cells.

Fibres:

1. Fibres are long narrow, thick, lignified cells with tapering ends.
2. The wall thickenings sometimes even block the lumen of the cells.
3. The pits of fibres are simple, small round and often oblique.
4. The middle lamella is conspicuous.

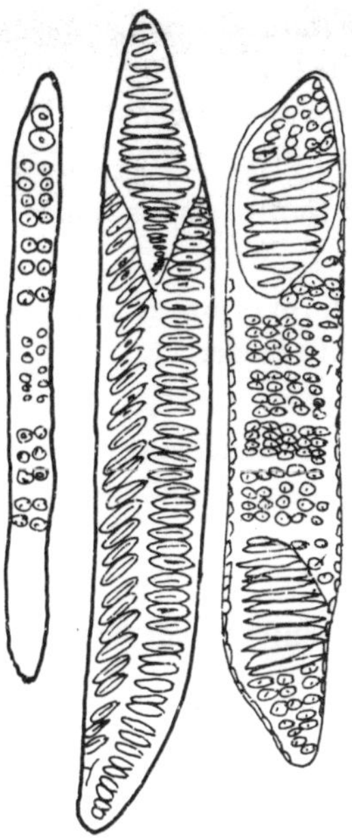

Fig. 3.10 **Permanent Tissues**
L.S. & T.S. of tracheids from Quercus alba

Sclereids:

1. These are short isodiametric cells but sometimes may be elongated also.
2. The cell walls are very thick and lignified but the cells do not have tapering ends.
3. The cells are dead and have a very narrow lumen.
4. The secondary walls show prominent pits which are simple.
5. Sometimes there is slight overarching of the secondary wall over the pit chamber.
6. The lignin deposition is uniform, but may be uneven as in the seed coats of some Leguminosae.
7. Sclereids are widely distributed in the plant body.
8. The cortex and the pith of gymnosperms and dicotyledons very often contain sclereids arranged singly or in groups.
9. Xylem and phloem may often contain sclereids besides fibres.
10. Sclereids may be present in leaves, seeds and fruits.
11. In fruits they are sometimes dispersed in the soft tissues.
12. Sclereids are classified into the following types viz., brachysclereids or stone cells (short isodiametric cells widely distributed in the flesh of fruits), macrosclereids long rod like cells found in the epidermis of leguminous seeds Osteosclereids (bone like cells) Astrosclereids (star like cells) Bizarre sclereids (Irregular sclereids), Filiform sclereids (long cells resembling fibres and trichosclereids branched, thin walled sclereids).

COMPLEX PERMANENT TISSUES

1. The vascular tissues are the most important among the complex tissues.
2. There are mainly two components in the vascular tissue viz., xylem and phloem.

XYLEM:

1. This is mainly a conducting tissue of water and mineral nutrients.
2. Structurally the xylem is most complex, because it consists of various types of living and non-living cells.
3. The xylem is mainly classified into primary and secondary xylem.
4. The former differentiates in the primary plant body and is derived from the procambium, while the latter originates from a fascicular cambium during secondary growth.

Elements of xylem:

1. The xylem is composed of different kinds of elements viz., tracheids, fibre tracheids, vessels or tracheae, wood fibres and wood parenchyma.

(a) Tracheids:

1. A tracheid is an elongated tube like cell with tapering rounded oval ends.
2. The tip of the tracheid of a secondary xylem may be chisel-like.
3. The walls are thickened due to the deposition of lignin, which are of various types.

Thickening of the cell wall:

1. The secondary wall acquires different types of thickenings.
2. This is specially seen in cells which are to differentiate into trachea**r**y elements i.e, vessels and tracheids.
3. In these cells, the thickening is due to the deposition of lignin.
4. The deposition of lignin is not uniform, hence we find different types of thickenings.
5. Chiefly, the following types of thickening are seen in xylem.

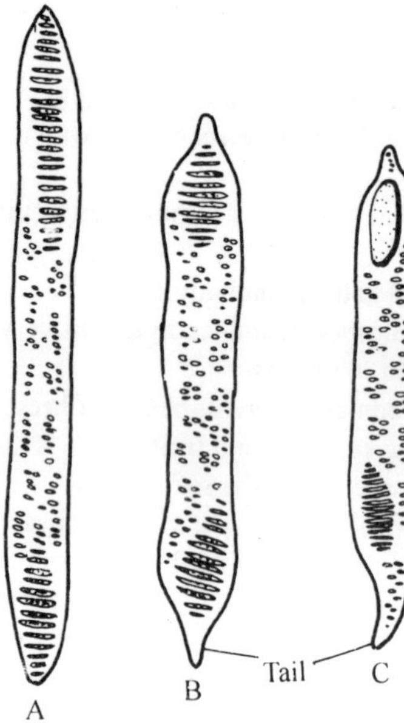

Fig. 3.1. **Permanent Tissues**

Type of vessels Simple C and Scalariform (A-B

Annular:

1. Lignin deposition is in the form of annuli or rings.
2. These rings are unconnected.
3. The space between the rings is unthickened.

Spiral:

1. Lignification occurs in the form of a spiral band.

Scalariform (ladder like):

1. Wall thickening is in the form of transverse rods, very much resembling the rungs of a ladder.

Reticulate:

1. Lignin is deposited in the form of network or reticulum.
2. The unthickened areas of the cell wall are irregular in shape.

Pitted:

1. Lignin deposition is almost uniform over the wall, except for small pores here and there.
2. These pores are called pits.

Structure of pit:

1. Pits are formed on either side of the wall.
2. In any area there will be two pits lying opposite to each other.
3. These are usually termed as "pit pairs".
4. There is a space inside each pit called pit cavity or pit chamber.
5. There is a thin membrane which seperates the pit cavities of the opposite pits.
6. It is called pit membrane.
7. The opening to a pit is called the pit aperture.
8. A pit membrane is common to both the pits.
9. There are two types of pits viz. *simple pits* and *bordered pits*.
10. A simple pit is just an opening in the wall.
11. Whereas in a bordered pit, the pit aperture is overarched by the projection of the secondary wall.
12. In a pit pair both may be simple, both may be bordered or one may be simple and the other one bordered.
13. The lost mentioned one is generally called a half bordered pit pair.
14. In a bordered pit, the pit membrane becomes thick at the centre.
15. The thickening is known as the torus.
16. Tracheids are dead cells and like sclerenchyma have neither the nucleus nor the protoplast.
17. The lumen of the cell is large.
18. The wall of the tracheids is provided with a number of bordered pits.
19. In a cross section tracheids appear polygonal or rectangular.

(b) Fibre tracheids :

1. During the development of fibres the wall thickening increases whereas the lumen decreases.
2. In many cells, the lumen becomes so narrow that it is altogether obliterated.
3. In such cells there is little or no conduction and they may be said to have reached the stage of true fibre.

(c) Vessels:

1. These are evolved from tracheids and are present mainly in Angiosperms.
2. Vessels arise through the dissolution of the end walls of tracheids.
3. Vessels are composed of rows of cells placed one above the other with the transverse walls being eliminated.
4. This facilitates the easy flow of water.

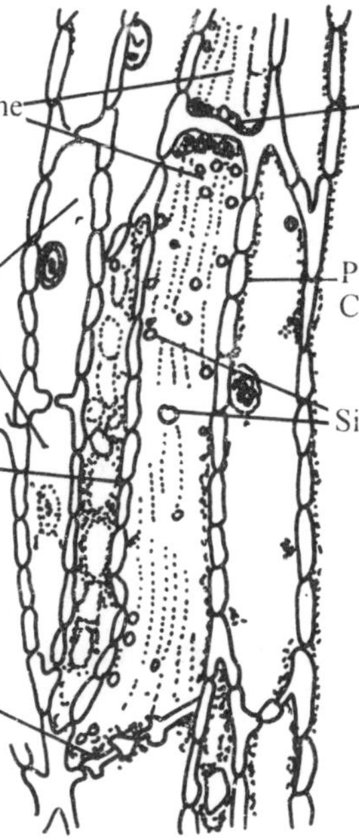

Fig.3.12 Permanent tissues.,
Phloem. Long sectional View.

5. Vessels (tracheids also) may be of two types protoxylem and metaxylem.
6. The former matures a little early in the ontogeny and has a narrow lumen while the latter matures later and has a broad lumen.

(d) Wood fibres:

1. They are also called xylaiy fibres.
2. Sclerenchyma fibres associated with xylem are termed wood fibres.
3. They possess very thick walls and few small pits.
4. They originate from the derivatives of the cambium and have varied shapes.
5. In wood fibres, known as libriform fibres, the wall is extremely thick and lumen is almost absent.
6. Like the tracheids , wood fibres are also dead and their main function is mechanical support.

(e) Wood Parenchyma:

1. Parenchyma cells assicoated with xylem are called wood parenchyma.
2. These are living cells and remain so, as long as the tissue in which they lie conducts water.
3. The cells of wood parenchyma may be thin or thick walled have plenty of storage food and may help in conduction.

4. In secondary xylem wood parenchyma forms the rays which may help in radial conduction.

PHLOEM:

1. The phloem is the principal food conducting tissue in vascular plants.
2. Unlike xylem all the elements of phloem (except for phloem fibres) are living.
3. Like the xylem phloem is also of two types of viz., primary phloem (derived from procambium) and secondary phloem (derived from fascicular cambium).
4. The basic components of phloem are sieve elements, companion cells, phloem parenchyma and phloem

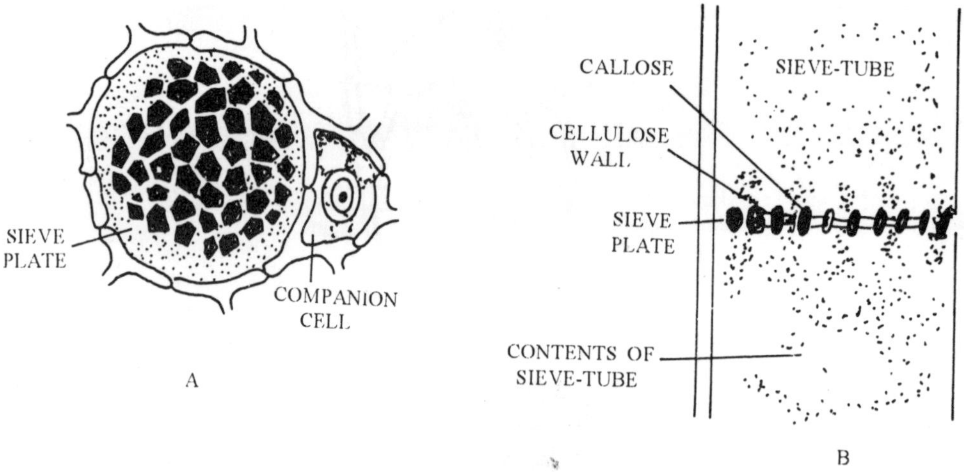

Fig. 3.13 **Permanent tissues.,**
Phloem. A. T.S. of sieve plate, B., L.S. of sieve tube.

fibres.

5. Occasionally sclereids, laticiferous cells and idioblasts are also found in phloem.

(a) Sieve elements:

1. What the tracheids are to xylem sieve elements are to phloem.
2. Sieve elements may be seggregated into relatively less specialized sieve cells (found in pteridophytes) and more specalized sieve tubes.
3. Seive tubes are living cells and have a thin cellulose wall.
4. In angiosperms they are enucleate, while they are nucleate in gymnosperms.
5. The protoplast of a sieve tube has a big central vacuole which is filled with viscous and albuminous proteins.
6. Leucoplasts are also reported in the vacuole.
7. The cross walls between the seieve tubes are porus and are called sieve plates as they resemble a sieve.
8. Sieve plates are usually of two types viz., simple and compound.
9. In simple sieve plates, the pores are large and the plate is horizontal or transverse in position.
10. The compound sieve plate is somewhat oblique.
11. The sieve pores are large and the plate is horizontal or transverse in position.
12. The compound sieve plate is somewhat oblique.
13. The sieve pores are very often plugged with a shining substance called *callus pad* which is usually deposited at the end of the spring season.
14. Chemically callus pads are made up of a carbohydrate called *callose*.

15. At the approach of the spring, the callus dissolves.
16. Through the sieve pores, thin cytoplasmic strands come out and enter into other cells.

(b) Companion cells:

1. These are specialised small cells usually associated with the sieve tubes.
2. These are living, nucleate cells and are supposed to compensate the absence of nuclei in sieve tubes.
3. Companion cells are absent in gymnosperms, where albuminous clls are present.
4. The cytoplasm of cmpanion cell is dense and granular with a prominent nucleus.
5. Ontogenetically both sieve tubes and companion cells are derived from the same mother cell.

(c) Phloem parenchyma :

1. Parenchyma cells in association with phloem are called phloem parenchyma.
2. They are no different from other ordinary parenchyma cells.
3. In shape, the cells may be elongated, pointed, broadly cylindrical or even subspherical.
4. The cells contain oil, starch mucilage, latex etc.
5. Phloem parenchyma is absent in most of the monocotyledons.

(d) Phloem fibres:

1. These are also called bast fibres.

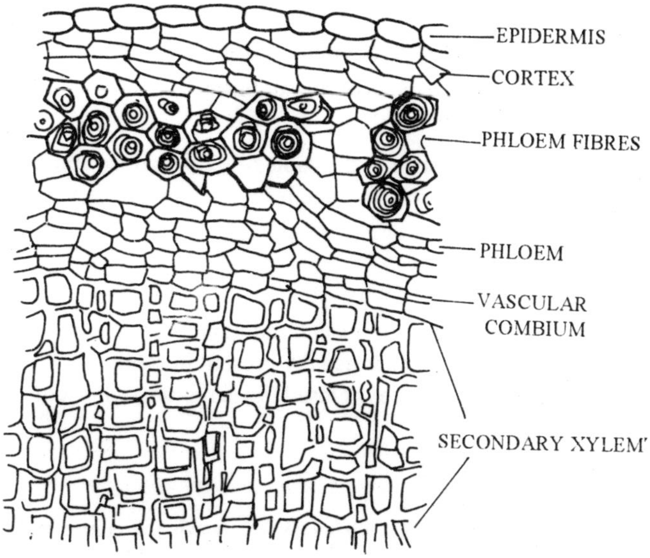

Fig. 3.14 **Permanent tissues.,**
Phloem fibres in the stem of Linum.

2. These are found both in primary and secondary phloem.
3. The walls have simple pits.
4. The bast fibres may be lignified or non lignified.

Types of Vascular bundles

1. A vascular bundle may be defined as grouping together of vascular elements namely xylem and phloem to form a functional unit.
2. Within a vascular bundle the two major components - xylem and phloem are arragned in various ways.
3. Depending on their relative position vascular bundles are categorized into the following types - cambium

which brings about secondary growth may or may not be present.

Open vascular bundles :

1. This is seen in dicotyledons.
2. Here between xylem and phloem, cambium (secondary meristem) is present.
3. As a result of this the vascular bundle is open to secondary growth.
4. There is no bundle sheath surrounding the vascular bundle.

Closed vascular bundle :

1. This is seen in monocotyledons.
2. Here there is no cambium in the vascular bundle.
3. Hence there is no secondary growth.
4. Further a bundle sheath surrounds the vascular bundle.
5. Based on the relative positions of xylem and phloem vascular bundles are classified into- *concentric*, *radial* and *conjoint* types.

Concentric vascular bundle:

1. In concentric vascular bundles xylem and phloem form concentric rings; either xylem surrounds the phloem or phloem surrounds the xylem.
2. Concentric vascular bundles are of two types-

   Amphivasal (Leptocentric): Here phloem is in the centre surrounded by xylem.

   Amphicribal (Hadrocentric) : Here xylem is in the centre surrounded by phloem.

   Concentric bundles are always closed ie., they have no cambium.

Radial bundles:

1. This kind of a vascular bundle is usually seen in the roots of dicotyledons and monocotyledons.
2. Here the xylem and phloem together will not form a single bundle, instead they form alternating bundles and are present on different radii.
3. This is regarded as the most primitive type.

Conjoint bundles :

1. In this type xylem and phloem together form compact bundles and they are present on the same radius.
2. Conjoint bundles may be open or closed and are usually present in stems.
3. Conjoint vascular bundles of monocots are always closed (no cambium) while those of dicots are always open (cambium present).
4. Conjoint vascular bundles are of two types.

Collateral :

1. Here xylem and phloem lie on the same radius with xylem towards the centre and phloem towards the periphery.
2. In otherwords there is one group of phloem and one group of xylem.
3. In dicotyledons cambium is present in between xylem and phloem while it is absent in monocotyledons.
4. Collateral vascular bundles are found in stems of Angiosperms.

Bicollateral :

1. In these bundles there is one group of xylem surrounded by two groups of phloem - outer and inner.
2. Similarly two strips of cambia are also present.
3. Here the arrangement of various elements are as follows- inner phloem, inner cambium, xylem, outer cambium and outer phloem.
4. Bicollateral vascular bundles are usually seen in the members of the family Cucurbitaceae.

CHAPTER 4

ANATOMY OF ROOTS

Root of Helianthus

A transverse section of a young sunflower root shows the following features.

Epidermis:

1. This is the outermost layer.
2. This is also known as epiblema or piliferous layer and is made up of closely packed cells with no intercellular spaces.

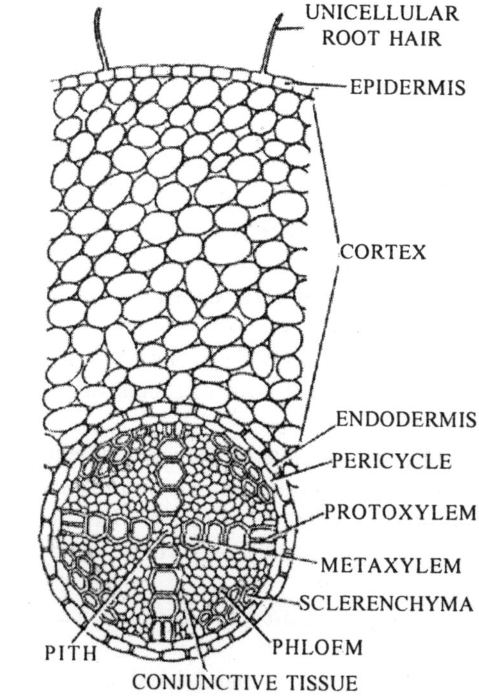

Fig. 4.1 Anatomy of root. T.S. of dicot root. Detail of a sector.

3. Some of the epidermal cells produce unicellular extensions called root hairs.
4. These help in the absorption of nutrients from the soil.

Cortex:

1. Internal to the epidermis is the cortex.
2. It is made up of several layers of thin walled spherical or oval parenchyma cells with large number of intercellular spaces.
3. Some of the outer layers of the cortex may become thick walled and function as exodermis.

Stele:

1. Occupying the central region of the root is the vascular cylinder.
2. The outermost layer of the stele is the endodermis.
3. It is made up of a single layer of barrel shaped cells., closely packed without any intercellular spaces.
4. The endodermal cells have thickenings on their radial walls called casparian thickenings.
5. In between endodermal cells thin walled cells are present.
6. They are called the passage cells.
7. Usually they are present opposite the protoxylem.
8. Internal to the endodermis is the pericycle.
9. It is made up of a single layer of rough, thin walled parenchyma cells.
10. Surrounded by the pericycle are the vascular bundles.

11. They are radial (Xylem and Phloem form separate alternating bundles) with *exarch* (protoxylem pointing towards the periphery) and *tetrarch* (four protoxylem points) xylem.
12. Xylem consists of annular and sprial vessels with metaxylem being made up of reticulate and pitted vessels.
13. Occupying the central region of the stele is a narrow parenchymatous pitch.
14. Sometimes the metaxylem groups join together so that pith is obliterated.
15. Between the radial vascular bundles is present parenchyma constituting the conjuctive tissue.

BEAN

A transverse section of a young bean root shows the following features.

Epidermis:

1. This is the outermost layer.
2. This is also known as *Epiblema* or *Piliferous layer* and is made of closely packed cells with no intercellular spaces.
3. Some of the epidermal cells produce unicellular extensions called root hairs.
4. These help in the absorptionof nutrients from the soil.

Cortex:

1. Internal to the epidermis is the cortex.

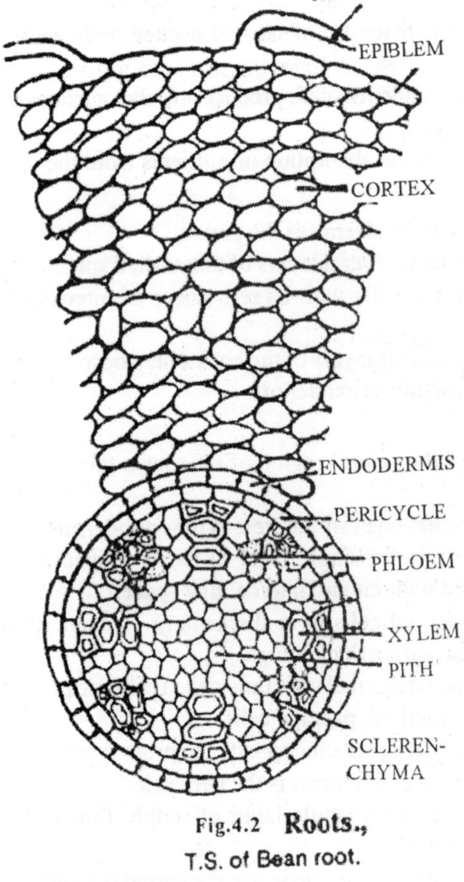

Fig.4.2 **Roots.,**
T.S. of Bean root.

2. It is made up of several layers of thin walled spherical or oval parenchyma cells with large number of intercellular spaces.
3. Some of the outer layers of the cortex may become thick walled and function as **exodermis**.

Stele:

1. Occupying the central region of the root is the vascular cylinder.
2. The outermost layer of the stele is the *endodermis.*
3. It is made up of a single layer of barrel shaped cells closely packed without any intercellular spaces.
4. The endodermal cells have thickenings on their radial walls called *Casparian thickenings*.
5. A few of the endodermal cells here and there are not thickened.
6. They are called the *passage cells*. Usually they are present opposite the protoxylem.
7. Internal to the endodermis is the *pericycle*.
8. It is made up of a single layer of rough, thin walled parenchyma cells.
9. Surrounded by the pericycle are the vascular bundles.
10. They are *radial* (xylem and phloem form separate alternating bundles), with *exarch* (protoxylem pointing towards the periphery) and *tetrarch* (Four protoxylem points) xylem.
11. Phloem consists of sieve tubes, companion cells and phloem parenchyma.
12. Protoxylem consists of annular and spiral vessels while Metaxylem is made up of reticulate and pitted vessels.
13. Occupying the central region of the stele is a narrow parenchymatous pith.

14. Sometimes the metaxylem groups join together so that pith is obliterated.
15. Between the radial vascular bundles is present parenchyma, constituting the conjunctive tissue.

RANUNCULUS

1. A T.S shows three regions - Epiblema, cortex and stele.
2. Epiblema is made up of a single layer of cells.
3. Some cells of epiblema produce unicellular root hairs.
4. Cortex is made up of many layers of rounded or polygonal parenchyma cells.
5. Some of the layers of cells immediately below the epiblema constitute exodermis.
6. The cells of the cortex is loosely arranged with intercellular spaces.
7. The cells of the cortex are rich in starch grains.
8. The stele is externally bounded by endodermis, the cells of which have casparian thickenings.
9. Next to the endodermis internally is the single layered peri cycle.
10. The vascular bundles are radial.
11. The xylem is tetrarch (four strands).
12. The xylem is exarch (protxylem points towards the perrphery).
13. Alternating with the xylem are found bundles of pholem.
14. Phloem consists of seive tubes, companion cells and phloem parenchyma.
15. Conjunctive parenchyma is present between the vascular bundles.
16. There is no pith.

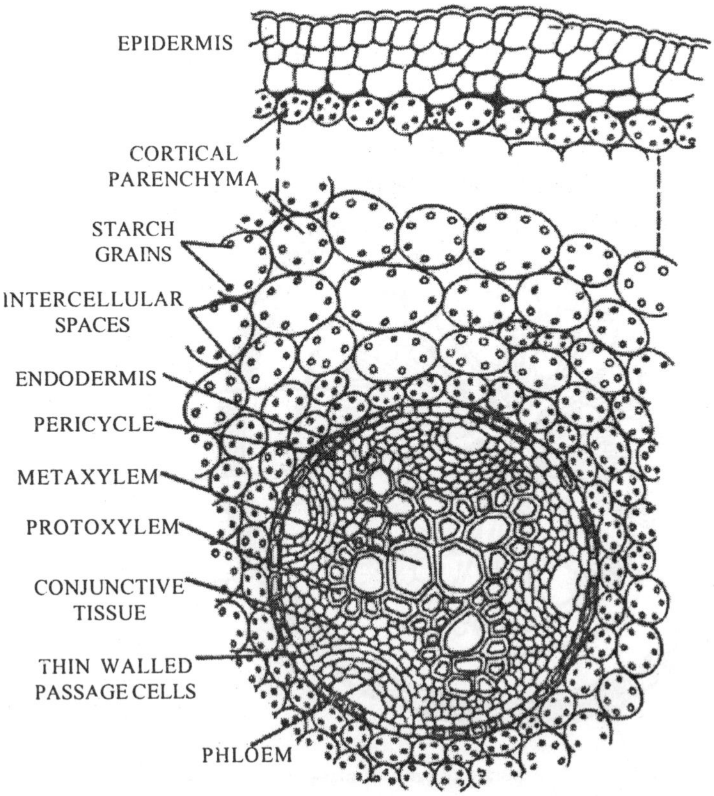

Fig. 4.3 T.S. of Root of Ranunculus - A part enlarged

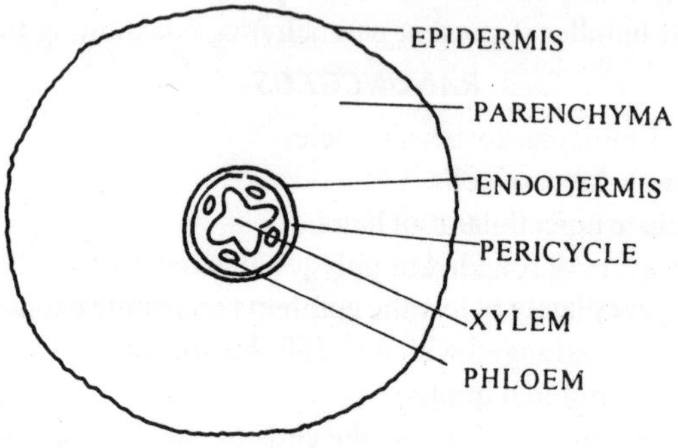

Fig. 4.4 T.S. of Root of Ranunculus - Ground plan

TINOSPORA

1. A transfer section of the root shows rhizodermis, cork, cork cambiam and vasculature.
2. The rhizodermis is the outermost layer and is made up of rectangular cells.
3. Immediately below the rhizodermis is found the cork which is made up of few layers of rectangular cells.
4. The cells of the cork are dead.
5. Internal to the cork is found the cork cambium or phellogen.
6. The cells of phellogen are actively dividing and produce cork to the exterior and secondary cortex towards the interior.
7. Endodermis and pericycle are not clearly distinguishable.

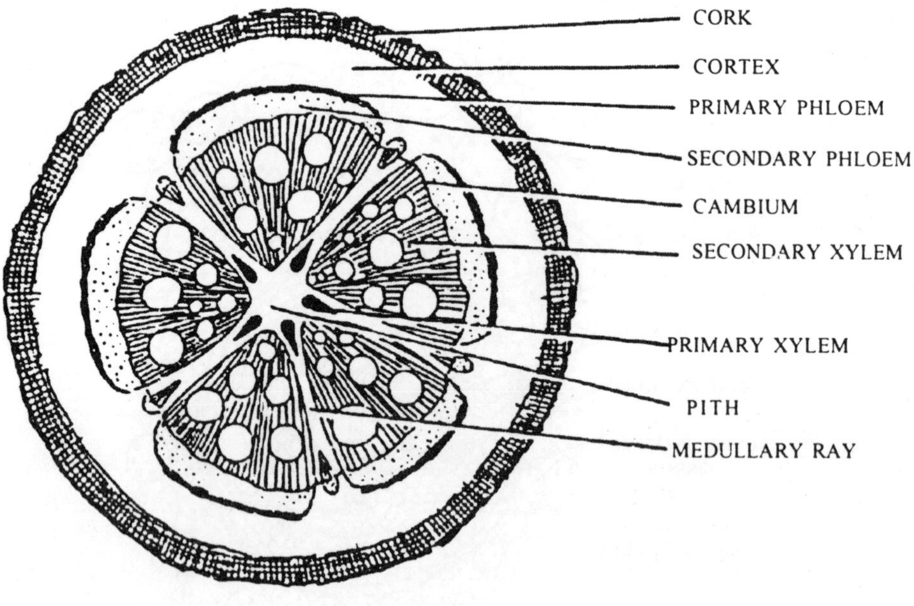

Fig. 4.5 T.S. of Tinospora root (ground plan)

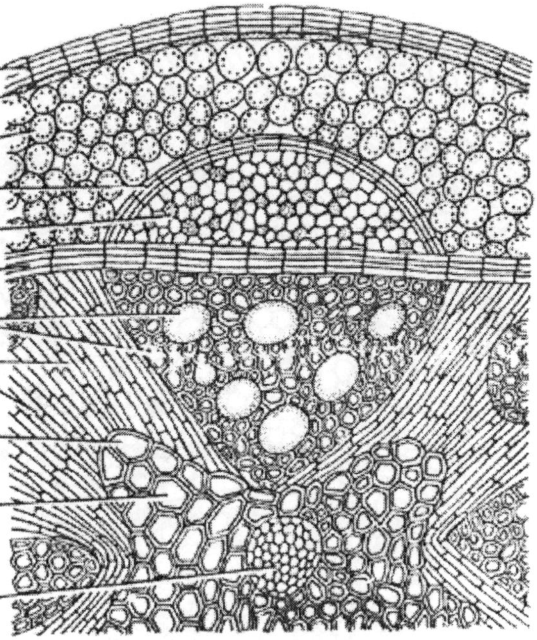

Fig. 4.6 T.S. of Tinospora root (A sector enlarged)

8. The vascular consists of primary phloem secondary phloem and secondary xylem.
9. The primary phloem is crushed and is found next to the secondary cortex.
10. Next to the primary phloem is secondary phloem consisting of seive tubes companion cells and phloem parenchyma.
11. Next to the secondary phloem is the fascicular cambium.
12. Internal to the cambium is the secondary xylem consisting of large vessels.
13. Very close to the centre are found primary xylem elements.
14. The primary xylem is exarch and tetrarch.
15. Pith is very narrow and found in centre.
16. Between the vascular bundles are found medullary rays.

***PROP ROOT OF BANYAN** (Ficus bengalensis)*

1. A transverse section of the aerial root of *Ficus* in the primary structure shows an outermost epliblema followed by cortex and stele.
2. The cortex consists of parenchymatous cells which may have chloroplasts.
3. The stele is surrounded by endodermis and pericycle.
4. The vascular bundles are *radial* and six to nine in number (*polyarch*).
5. In the central region there is a well developed parenchymatous pith.
6. After the initiation of secondary growth which takes place as in other dicotyledonous roots, the following features are seen

Cork or phellem:

1. This replaces the epibelma and consists of 4-6 layers of suberiseed cells.
2. Certain cells of the cork contain tannin. Lenticels are also present.

Cork cambium:

1. This is present below the cork. It is this which produces the cork tissue.
2. The cells of the cork cambium are rectangular and actively dividing.

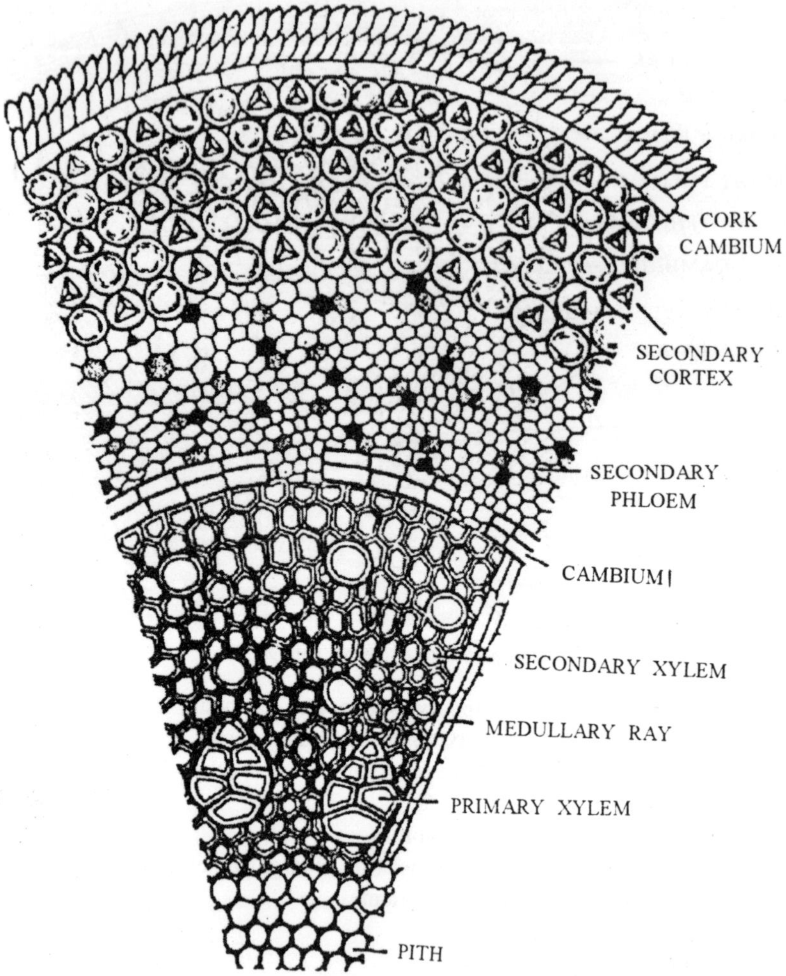

Fig. 4.7 T.S of Banyan root (A part enlarged)

Secondary cortex:

1. This is also produced by the cork cambium on its inner face.
2. The cells are thin walled parenchyma, containing chloroplasts.

Vascular tissues:

1. After the initiation of secondary growth also, both primary and secondary vasculature can be seen.
2. The primary vascular bundles are radial with exarch, polyarch xylem.
3. Secondary vasculature forms a complete ring and is produced due to the activity of cambium.
4. In a transverse section, cambium is seen between secondary xylem and secondary phloem.
5. Fibres may be seen in secondary phloem.
6. Secondar xylem consists of vessels, tracheids and wood parenchyma.
7. In the centre there is a parenchymatous pith.
8. Medullary rays are present traversing across the secondary vascular tissues.

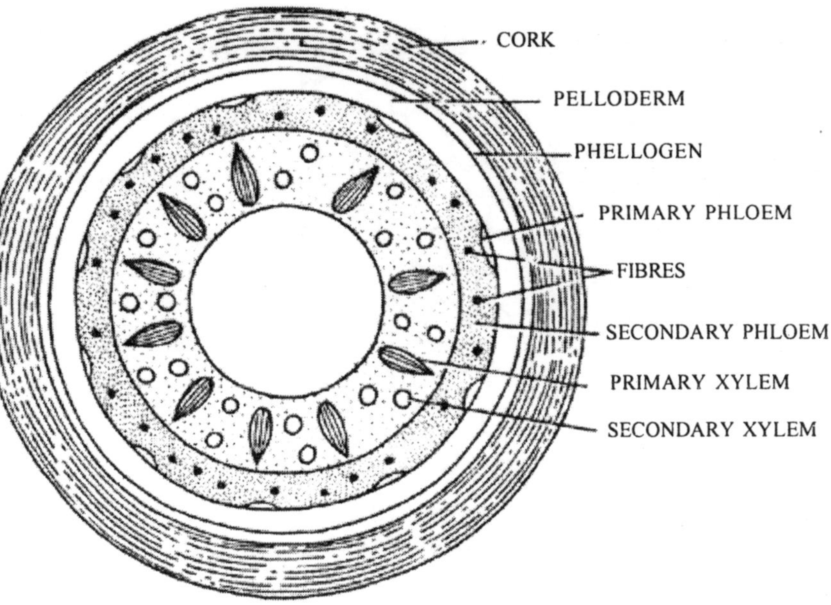

Fig. 4.8 T.S. of Prop root of Banyan (ground plan)

SMILAX

1. A transfer section shows epiblema, cortex, and stele.
2. Epibelma is single layered and it produces a number of unicellular root hairs.
3. Next to the epibelma is exodermis consisting of a single layer of thick walled cells.
4. The cortex is many layered consisting of loosely arranged thin walled parenchyma cells.
5. The cells of cortex are oval in shape and have a number of starch grains.
6. The stele is externally bounded by endodermis.
7. The cells of endodermis are barrel shaped and have casparian thickenings on their radial and tangential walls.
8. Next to the endodermis is single layered peri cycle.
9. Vascular bundles are radial and they are many in number.
10. Xylem and phloem form seperate alternative bundles.
11. The xylem is polyarch (many bundles) and exarch.
12. Phloem consists of seive tubes companion cells and phloem parenchyma.
13. Occupying the central region of the root is the pith consisting of loosely arranged parenchyma cells.

CANNA

1. A transverse section shows epibelma, cortex and stele.
2. Epibelma is single layered and it produces a number of unicellular root hairs.
3. Next to the epibelma is exodermis consisting of a single layer of thick walled cells.
4. The cortex is many layered consisting of loosely arranged thin walled parenchyma cells.
5. The cells of cortex are oval in shape and have a number of starch grains.
6. The stele is externally bounded by endodermis.
7. The cells of endodermis are barrel shaped and have casparian thickenings on their radial and tangential walls.
8. Next to the endodermis is single layered peri cycle.
9. Vascular bundles are radial and they are many in number.

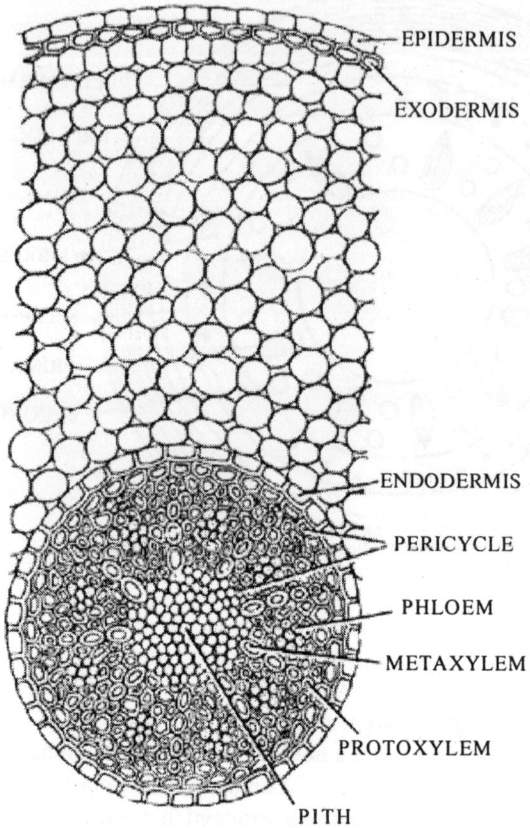

Fig. 4.9 Anatomy of Smilax root (A sector enlarged)

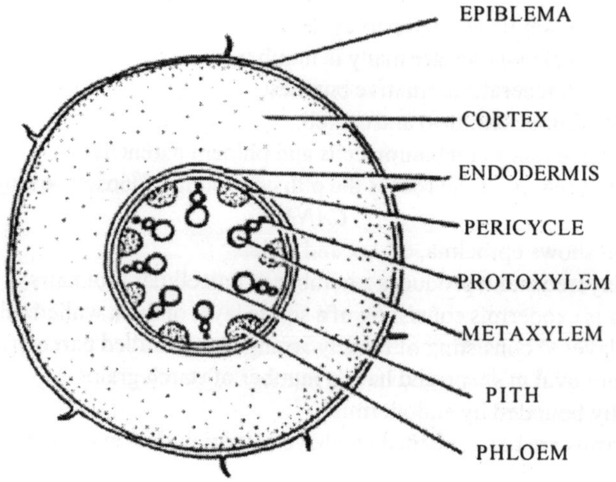

Fig. 4.10 T.S of Cana root (diagrammatic)

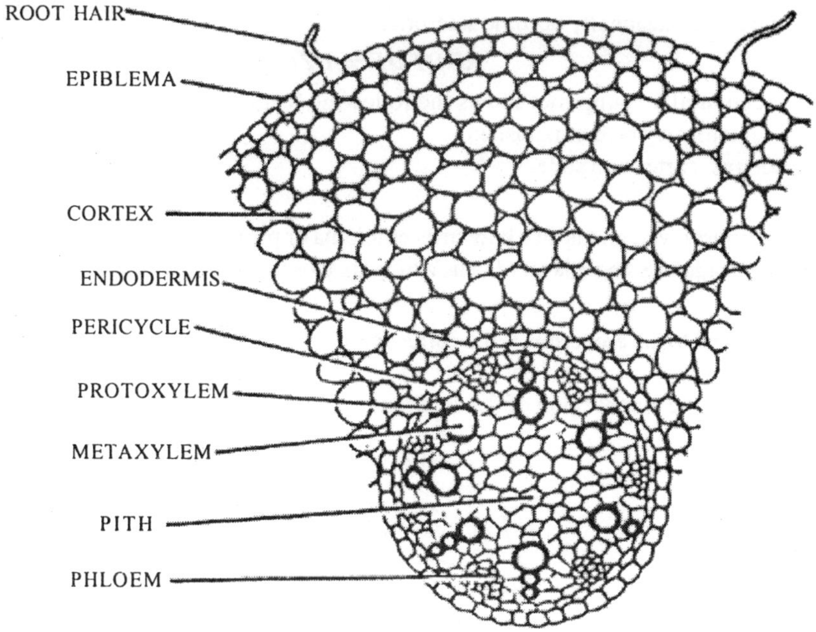

Fig. 4.11 T.S of Cana root (A part cellular)

10. Xylem and phloem form seperate alternative bundles.
11. The xylem is polyarch (many bundles) and exarch.
12. Phloem consists of seive tubes companion cells and phloem parenchyma.

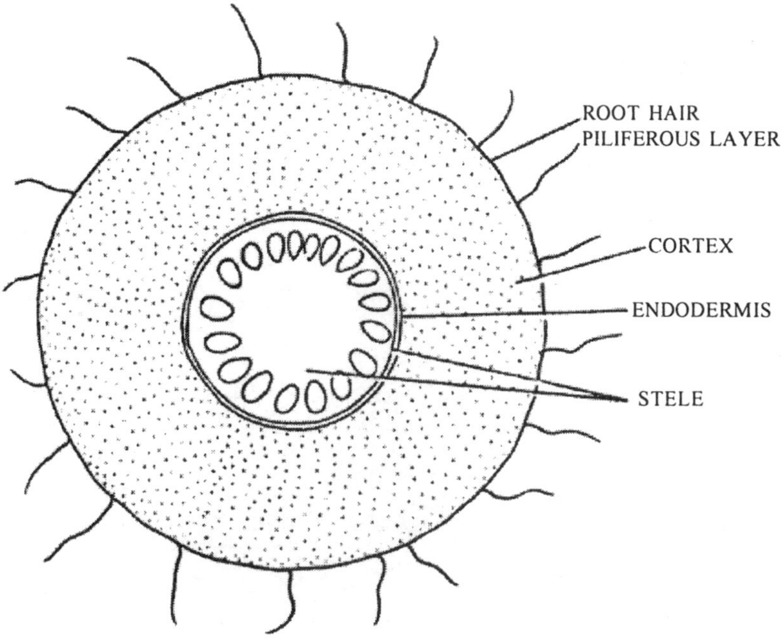

Fig. 4.12 T.S of Maize root (Diagrammatic)

13. Occupying the central region of the root is the pith consisting of loosely arranged parenchyma cells.
14. The cells of the pith occasionally become sclerotic.

ROOT OF MAIZE

1. A transverse section of a young root shows the following features.
2. Mainly three regions may be distinguished.
3. These are epiderms, cortex and stele.

Epidermis:

1. This is the outermost layer, and is also known as epiblema or piliferous layer.
2. The epidermis is made of a single layer of cells which are closely packed without any intercellular spaces.

Cortex:

1. Immediately internal to the epidermis is the cortex which is made up of many layers of loosely arranged parenchyma cells.
2. Some of the cells in the outer region of the cortex may be closely packed constituting the hypodermis.

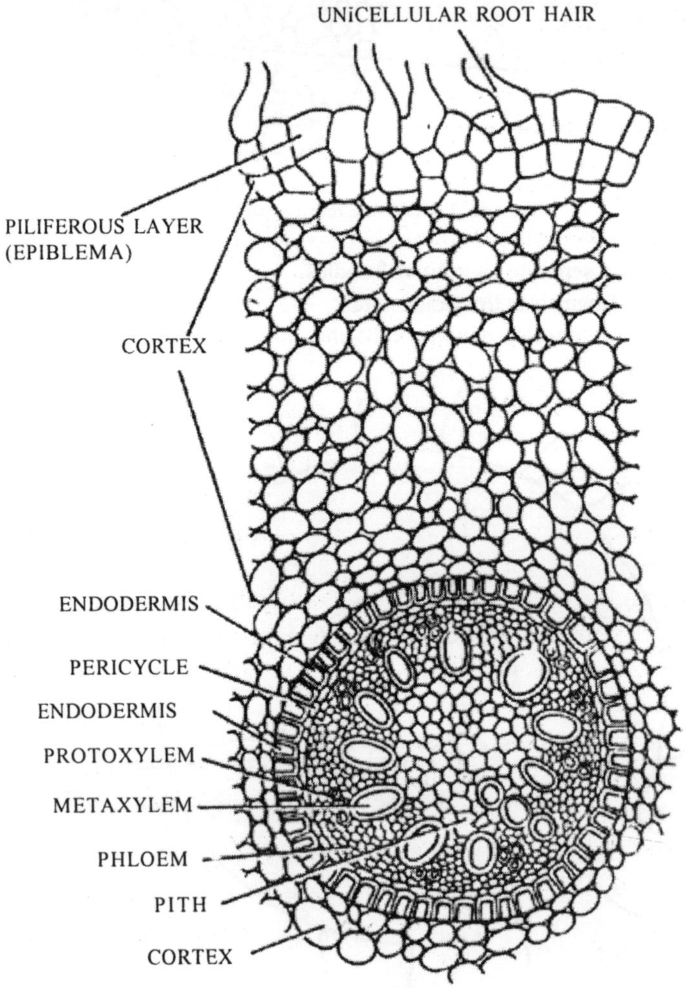

Fig. 4.13 T.S of Maize root (Detail Structure)

Stele:

1. The stele is bounded by an endodermis and pericycle.
2. Endodermal cells are barrel shaped and have casparian thickenings on their radial and tangential walls.
3. Pericycle is represented by a single layer of thin walled parenchyma cells.
4. The vascular bundles are many in number (polyarch) and radial with exarch xylem.
5. In the centre of the stele is a parenchymatous pith.
6. The main points of difference between the dicotyledonous root and the monocotyledonous root are the number of xylem groups and the presence or absence of pith.

MUSA ROOT

1. A transverse section of the root shows as usual epidermis, cortex and stele.
2. Epidermis or epiblema is single layered.
3. Internal to the epidermis is the cortex which is made up of many layers of loosely arranged parenchyma cells.
4. The stele is bounded by an endodermis and pericycle.
5. The vascular bundles are radial, with exarch and polyarch xylem.
6. There is a central pith.
7. In between the vascular bundles is the conjunctive tissue.
8. The conjunctive tissue as well as pith are lignified and have sclerotic cells.
9. Lying scattered in the pith are many isolated metaxylem vessels.
10. Additional phloem groups are also found scattered in the pith.

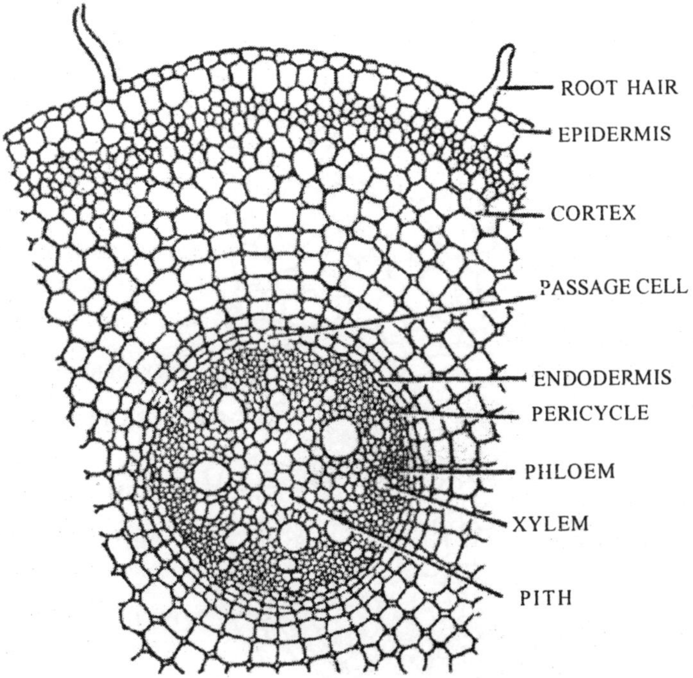

Fig. 4.14 T.S of Musa root (A portion enlarged)

ORCHID ROOT (*Vanda*)

1. The epiphytic roots of orchids have certain special features.
2. This is associated with their special function.
3. The epiphytic roots are also known as assimilatory roots isnce they help in photosynthesis.
4. A transverse section of the root of *Vanda* shows the following details:

Velamen:

1. This is the outermost region, and may also be calleα a multilayered epidermis.
2. The cells here are non living and elongated radially.
3. Arrangement of the cells is compact.
4. Velamen cells have the capacity to retain air and water.
5. Sometimes the cell walls develop fibrous thickenings.

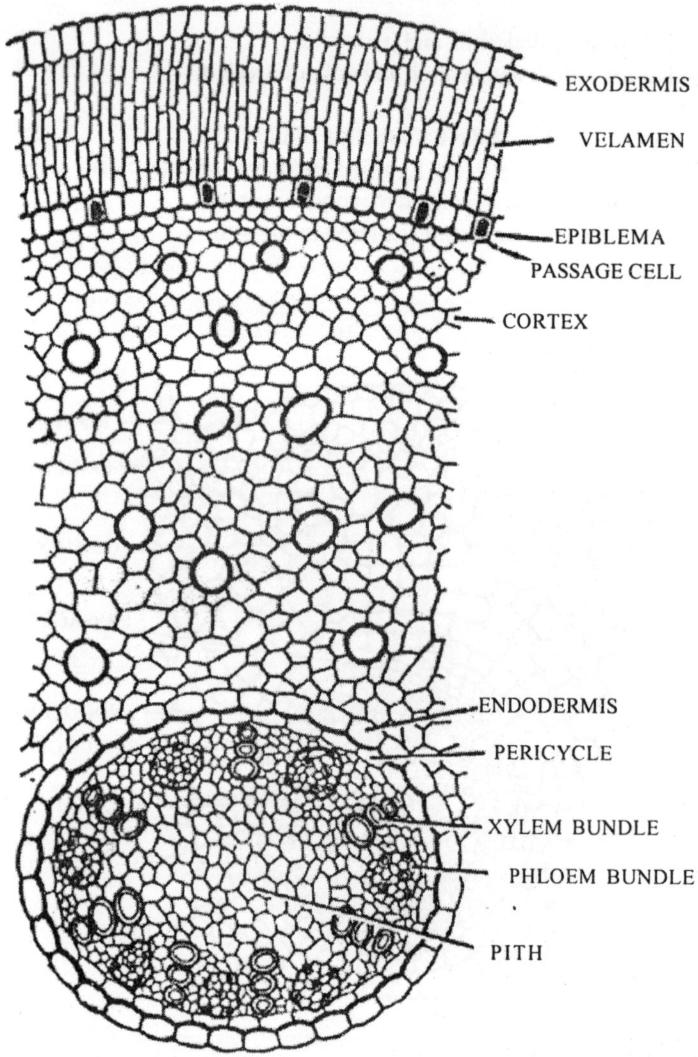

Fig4.15 T.S of Orchid root (A portion enlarged)

Anatomy of Roots

Exodermis:

1. The outermost layer of the cortex is called exodermis.
2. It is made up of single layer of cells with thickened outer and lateral walls.
3. Here and there some thin walled cells are present called passage cells.

Cortex:

1. Internal to the exodermis is the multilayered cortex.
2. It consists of many layers of thin walled loosely arranged parenchyma cells.

Stele:

1. Endodermis and pericycle are present.
2. There structure is as in other roots.
3. The vascular bundles are radial with *exarch* and *polyarch* xylem.
4. There is a well developed parenchymatous pith in the centre.
5. Conjuctive parenchyma is present between the vascular bundles.

CHAPTER 5

ANATOMY OF STEMS

HELIANTHUS (Sunflower):

A transverse section of young sunflower stem shows an outmost epidermis, a middle cortex and a central stele.

Epidermis:

1. This is single layered with closely packed rectangular cells.
2. Above the epidermis there is a well defined cuticle.
3. Some of the epidermal cells produce multicellular hairs.
4. Occasionally stomata may be seen in the epidermis.

Cortex:

1. There are two regions in the cortex.

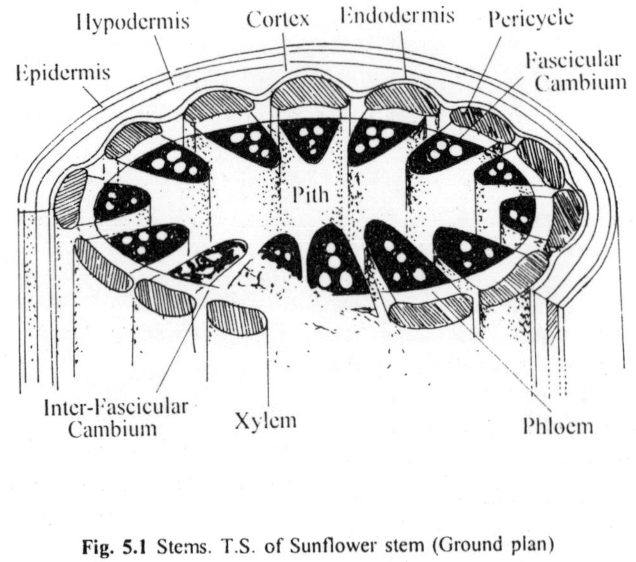

Fig. 5.1 Stems. T.S. of Sunflower stem (Ground plan)

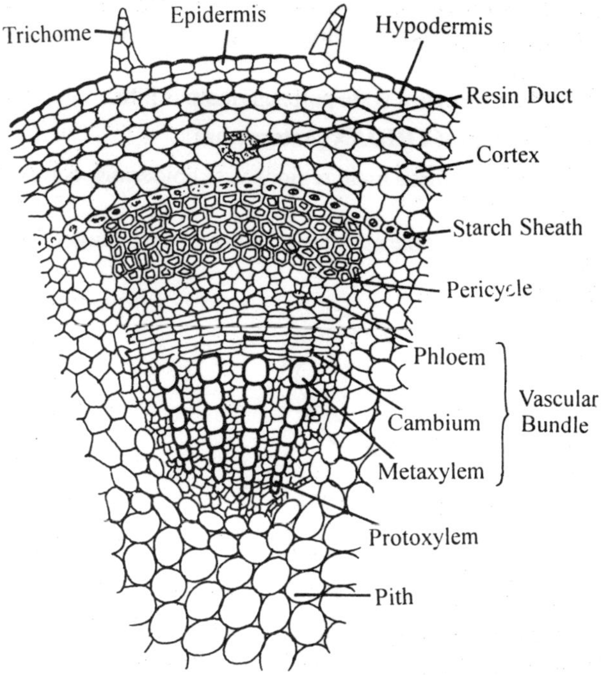

Fig. 5.2 Stems. T.S. of Sunflower stem (A portion enlarged)

2. The outer cortex (immediately below the epidermis) consists of a few layers of collenchyma whose corner walls are thickened by the deposition of pectin and cellulose.
3. Next to the collenchyma is parenchyma constituting the inner cortex.
4. This consists of several layers of parenchyma , some of the outer layers of which are chlorophyllous.

Stele:

1. The outermost layer is the endodermis whose cells are rich in starch grains; hence also known as the starch sheath.
2. It has the thickening on radial walls called casparian thickenings.
3. Next to the endodermis is the pericycle.
4. It is represented by patches (cap like) of sclerenchyma present immediately above each vascular bundle.
5. The vascular bundles are many in number, arranged in the form of a ring surrounding a central pith.
6. Each vascular bundle is conjoint (xylem and phloem combining to form a single bundle), collateral (xylem and pholem on the same radius with x/p arrangement), open (open to secondary growth i.e. cabium is present) with endarch (protoxylem pointing towards the pith) xylem.
7. Phloem consists of seive tubes companion cells and phloem parenchyma .
8. In between phloem and xylem is present cambium which is said to be intrafascicular (present within a vascular bundle).
9. The cambial cells are thin walled, rectangular and actively dividing.
10. Next to the cambium, internally is the xylem, with the protoxylem facing the pith.

11. Xylem consists of vessels, trachieds, wood fibres and xylem parenchyma.
12. Between the vessels are found the wood parenchyma.
13. Between the vascular bundles are present radially elongated parenchyma cells constituting the primary medullary rays.
14. Occupying the central region of the stem is a wide parenchymatous pith.

Stem of TRIDAX (Asteraceae)

A transverse section shows as usual the three main regions viz., epidermis, cortex and stele.

Epidermis:

1. It is same as in sunflower stem, except that the epidermal hairs are much longer and two cells wide at the base.

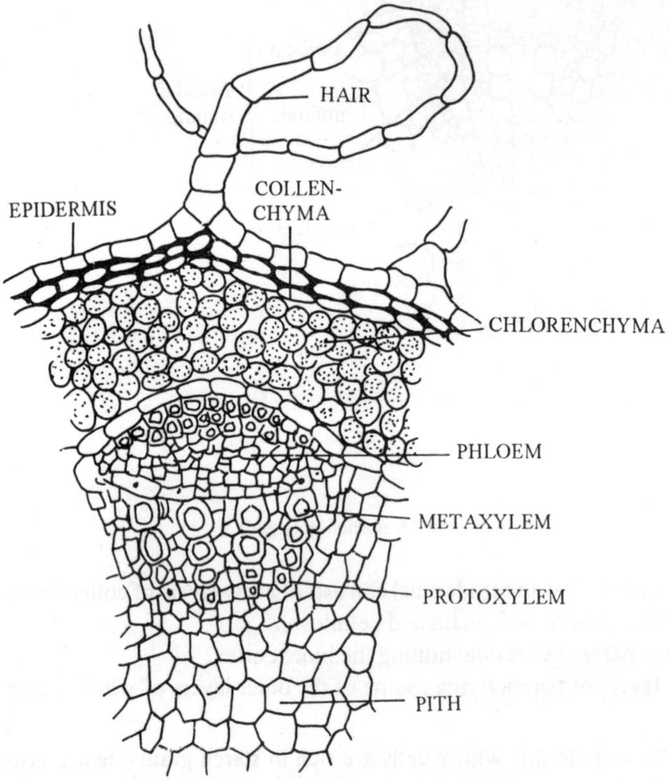

Fig. 5.3 Stems. T.S. of Tridax stem (A portion enlarged)

Cortex:

1. It is also same as in sunflower stem.
2. All the cells of parenchyma are chlorophyllous.
3. Endodermis, pericycle, vascular bundles, pith and medullary rays are same as in sunflower stem.

XANTHIUM

1. A transverse section of a young stem shows epidermis, cortex and stele.
2. Epidermis is made up of a single layer of closely packed cells.
3. Some of the cells of epidermis produce multicellular hairs.

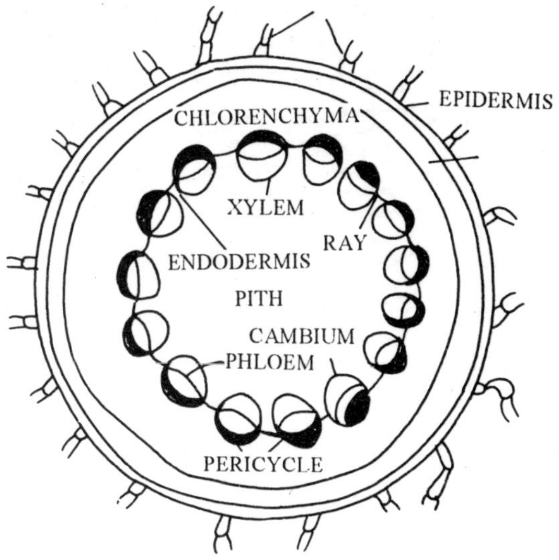

Fig. 5.4 Stems. T.S. of Tridax stem (ground plan)

4. Some times there is deposition of cuticle above the epidermis.
5. Cortex is very broad and is made up of two zones.
6. The outer zone of the cortex consists of two to three layers of collenchyma cells.
7. The collenchyma cells are closely packed and have thickened corner walls.
8. The inner zone of the cortex consists of many layers of parenchyma cells.
9. The parenchyma cells are loosely arranged and oval in shape.
10. Bounding the stele externally is the endodermis which is not as conspicuous as in the root.
11. Next to the endodermis is the pericycle which is discontinous and found in the form of patches of sclerenchyma immediately above the vascular bundles.
12. The vascular bundles are many in number arranged in the form of a ring surrounding a central pith.
13. The vascular bundles are collateral, conjoint, and open.
14. Between xylem and phloem is present the fascicular cambium.
15. The cambial cells are narrow rectangular, and actively dividing.
16. The cambium produces secondary xylem to the interior and secondary phloem to the exterior.
17. Medullary rays are found between the vascular bundles.
18. The pith is broad and made up of loosely arranged rounded or oval parenchyma cells.

RICINUS

1. A transverse section of a young stem shows epidermis, cortex and stele.
2. Epidermis is made up of a single layer of closely packed cells.
3. Some of the cells of epidermis produce multicellular hairs.
4. Some times there is deposition of cuticle above the epidermis.
5. Cortex is very broad and is made up of two zones.
6. The outer zone of the cortex consists of two to three layers of collenchyma cells.
7. The collenchyma cells are closely packed and have thickened corner walls.
8. The inner zone of the cortex consists of many layers of parenchyma cells.
9. The parenchyma cells are loosely arranged and oval in shape.

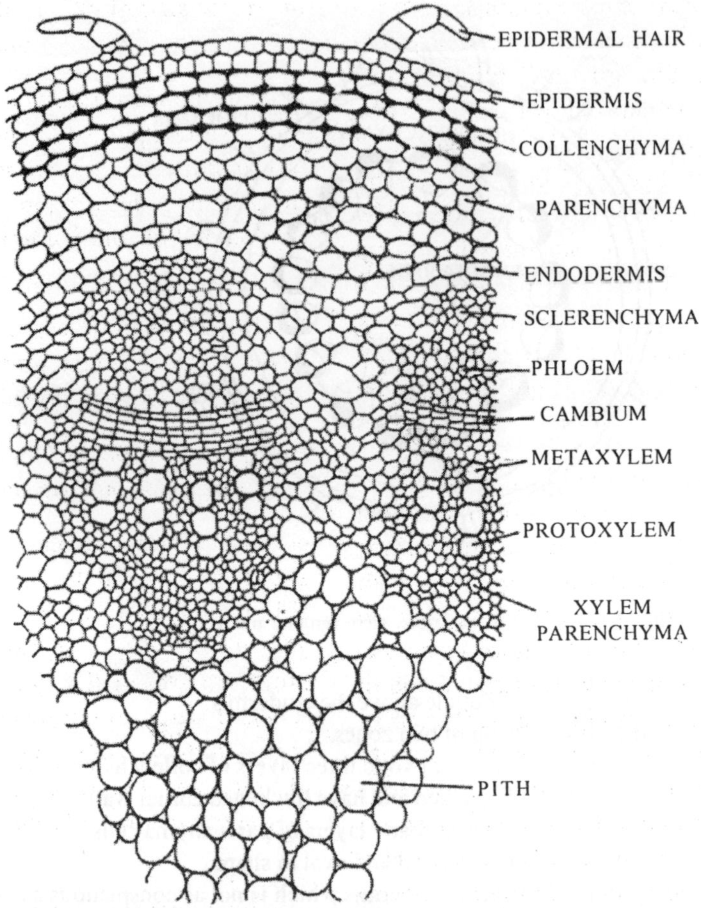

Fig. 5.5 Stems. T.S. of Xanthium stem (Detail of a sector)

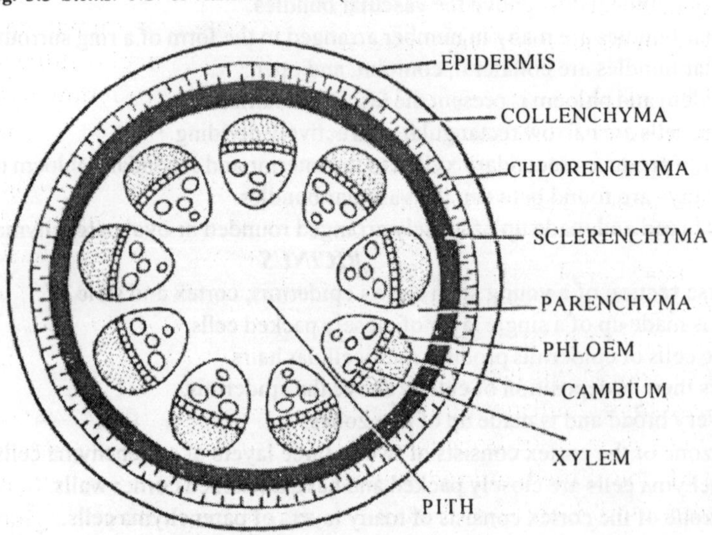

Fig. 5.6 Stems. T.S. of Ricinus stem (Diagrammatic)

10. Bounding the stele externally is the endodermis which is not as conspicuous as in the root.
11. Next to the endodermis is the pericycle which is discontinous and found in the form of patches of sclerenchyma immediately above the vascular bundles.
12. The vascular bundles are many in number arranged in the form of a ring surrounding a central pith.
13. The vascular bundles are collateral, conjoint, and open.
14. Between xylem and phloem is present the fascicular cambium.
15. The cambial cells are narrow rectangular, and actively dividing.
16. The cambium produces secondary xylem to the interior and secondary phloem to the exterior.
17. Medullary rays are found between the vascular bundles.
18. The pith is broad and made up of loosely arranged rounded or oval parenchyma cells.

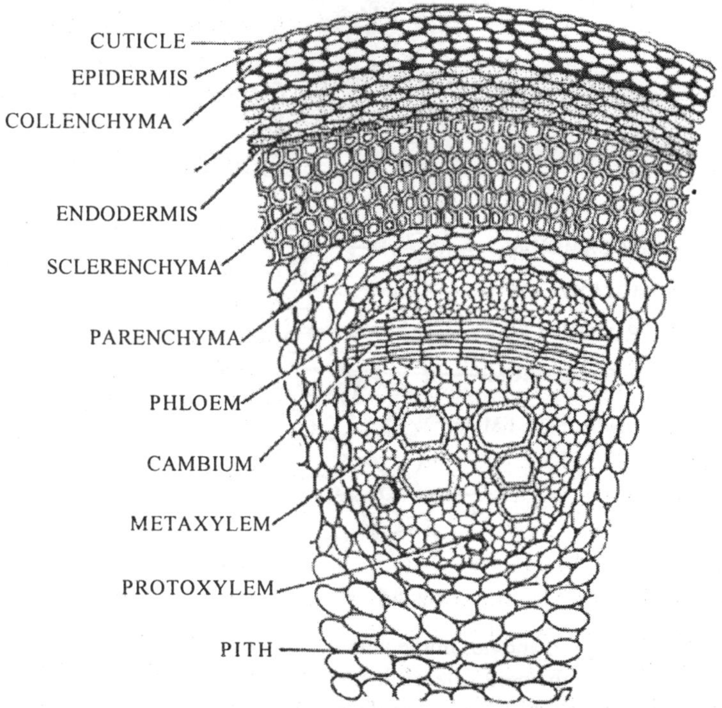

Fig. 5.7 Stems. T.S. of Ricinus stem (A portion of sector)

CUCURBITA

1. A transverse section of a young stem shows epidermis, cortex and stele.
2. Epidermis is made up of a single layer of closely packed cells.
3. Some of the cells of epidermis produce multicellular hairs.
4. Some times there is deposition of cuticle above the epidermis.
5. Cortex is very broad and is made up of two zones.
6. The outer zone of cortex consists of collenchyma cells.
7. The collenchyma is many layered near the ridges while near the forus they are either absent or very narrow.
8. The inner zone cortex consists of chlorenchyma cells consisting of chloroplasts.
9. The cells of chlorenchyma help in photo synthesis.

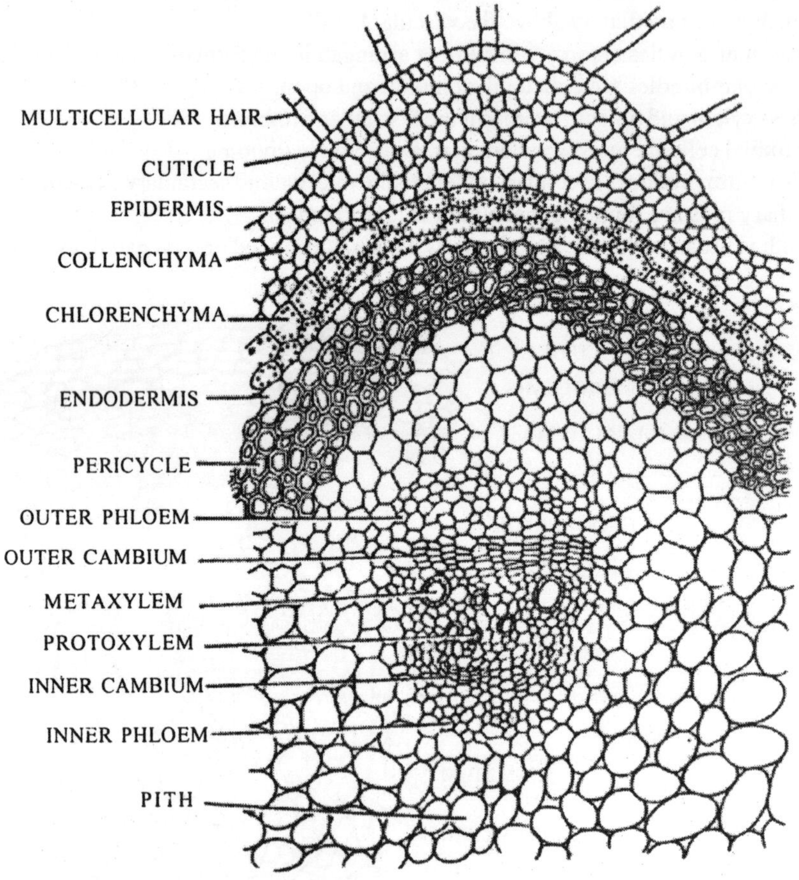

Fig. 5.8 Stems. T.S. of Cucurbita stem (Detail of a sector)

10. Endodermis is single layered and lies immediately below the chlorenchyma.
11. Next to the endodermis is the pericycle made up of a continous patch of sclerechyma.
12. The vascular bundles are embedded in a uniform mass of parenchyma tisssues often referred to as the ground tissue.
13. The ground tissue extends from the pericycle to the central pith cavity.
14. The vascular bundles are usually ten in number and arranged in two rows.
15. Outer row corresponds to the riges and inner row corresponds to the furrows.
16. The vascular bundles are conjoint, bicollateral, and open.
17. The bicollateral vascular bundles consists of outer phloem, outer xylem, outer cambium, xylem, inner cambium, and inner phloem.
18. The xylem consists of both proto and metaxylem elements.
19. There are two strips of cambia in each vascular bundle and its activity is confined to the vascular bundles.
20. Normally the outer cambial strip in the vascular bundle is the only one which shows activity.
21. Phloem occurs in two patches, the inner one towards the pith while the outer one is towards the pericycle.
22. There is a central pith cavity.

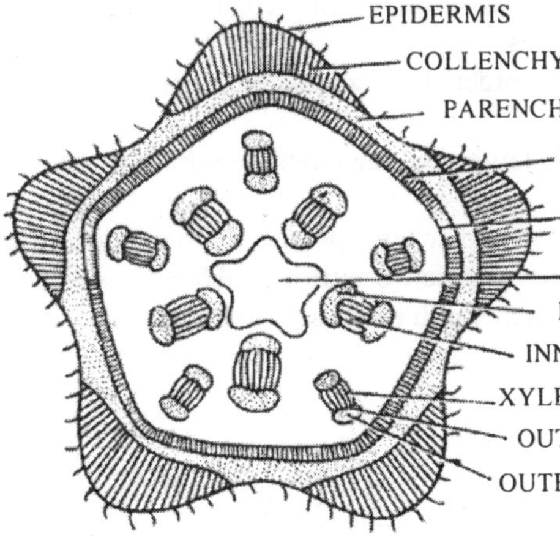

Fig. 5 9 Stems. T.S. of Cucurbita stem (Diagrammatic)

TILIA

1. The primary structure of the stem is typical to a dicotyledonous stem.
2. In the secondary structure the stem shows a rupture epidermis with cuticle deposited above it.
3. In very old stems the epidermis is completely replaced by the periderm.
4. The periderm consists of outer cork tissue, middle cork cambium, and inner secondary cortex.
5. Next to the secondary cortex is the primary cortex made up of a few layers of parenchyma cells.
6. In the secondary structure of the stem endodermis and pericycle cannot be distinguished.

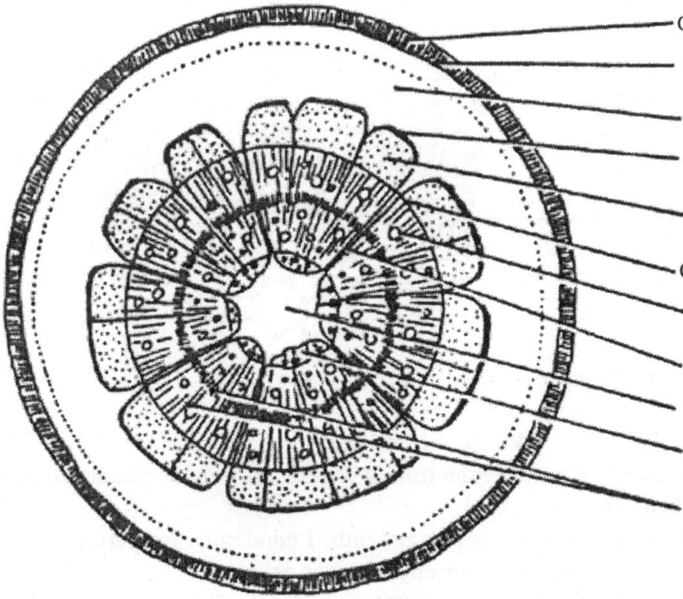

Fig. 0 Stems. T.S. of Tilia stem (Diagrammatic)

7. Secondary growth takes place by the activity of a fascicular cambium which forms a continous ring producing secondary vascular tissues.
8. Annular rings are produced in regular order. Each annual ring consists of spring wood and autumn wood which alternate.
9. The vessels of the spring wood are larger and the extent of spring wood is more in comparison with autumn wood.
10. Secondary phloem occurs in the form of patches seperated by triradiated medullary rays.
11. The medullary rays are of two types - triradiate occuring in phloem and narrow rays occuring in xylem.
12. The central region of the stem consists of pith made up of parenchyma cells.
13. Some cells of pith have mucilage.

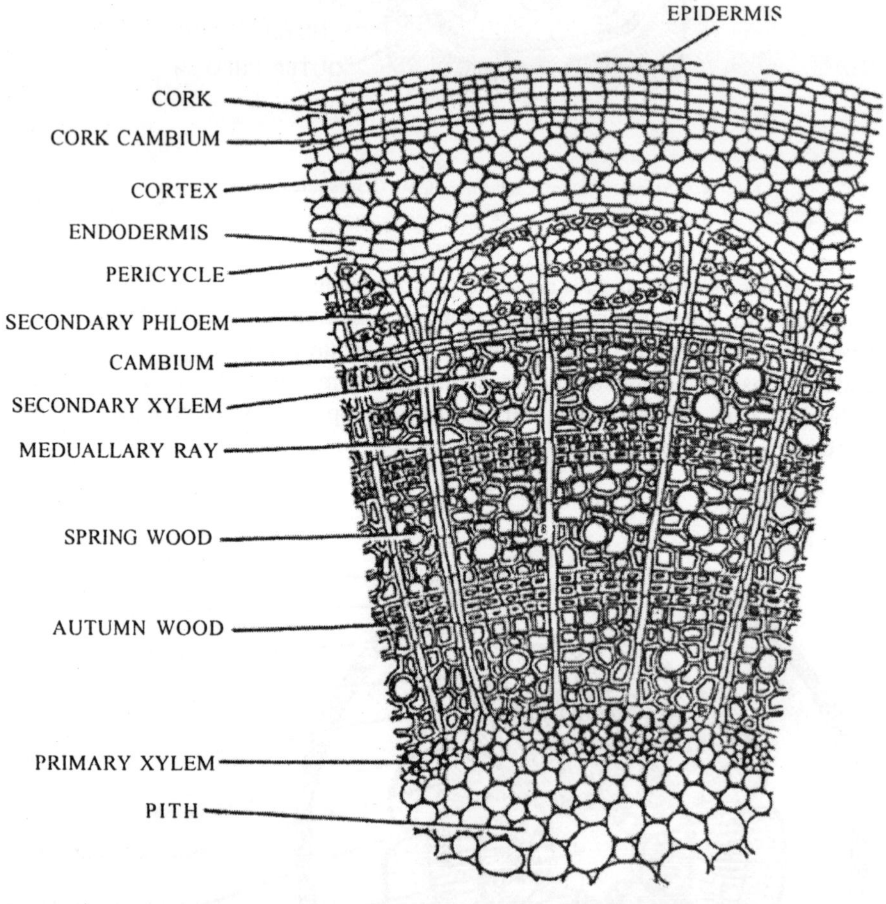

Fig. 5.11 Stems. T.S. of Tilia stem (A part cellular)

STEM OF MAIZE

1. A cross section of young stem shows an epidermis, a hypodermis and a ground tissue in which the vascular bundles are embedded .
2. There is no clear cut distinction into stele , cortex and pith. Endodermis and pericycle are absent.
3. Epidermis is made up of a single layer of compactly arranged cells.
4. It is covered with a cuticle.
5. There are no epidermal hairs.

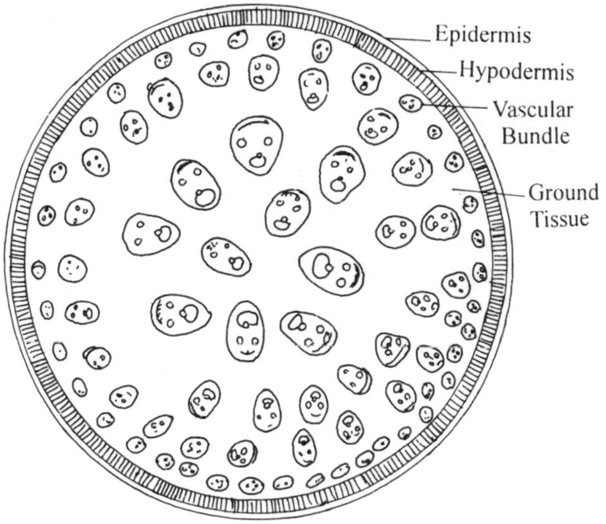

Fig. 5.12 Stems. T.S. of Maize stem (Ground plan)

6. Hypodermis is internal to the epidermis.
7. It consists of few layers of sclerenchyma cells.
8. Ground tissue is represented by loosely arranged, thin walled parenchyma cells.
9. Embedded in this tissue are the vascular bundles.
10. Vascular bundles are scattered in the ground tissue.
11. They are crowded towards the margin but smaller in size, while towards the centre they are big in size and sparsely distributed.

Structure of a vascular bundle:

1. Each bundle is oval in shape, conjoint, collateral endarch and closed (cambium is absent hence no secondary growth).
2. Surrounding the bundle there is a sclernchymatous bundle sheath.
3. The phloem is above the xylem.
4. It consists of sieve tubes and companion cells.
5. There is no phloem parenchyma.
6. The outer cells of phloem (protophloem) are crushed while the inner cells (metaphloem) are healthy.
7. Xylem is reduced; it consists of four (sometimes three) vessels (two meta and two proto) arranged in the form of 'Y'.
8. Metaxylem vessels are pitted, while protoxylem vessels are annular and spiral.
9. Sometimes in between the metaxylem vessels, small, pitted tracheas are present.
10. Below the protoxylem a lysigneous cavity is present.
11. It consists water and is known as the water containing cavity.
12. Surrounding this are present cells of wood parenchyma .

STEM OF *VANDA:*

In a cross section the following regions are seen:

Epidermis:

1. It is single layered.

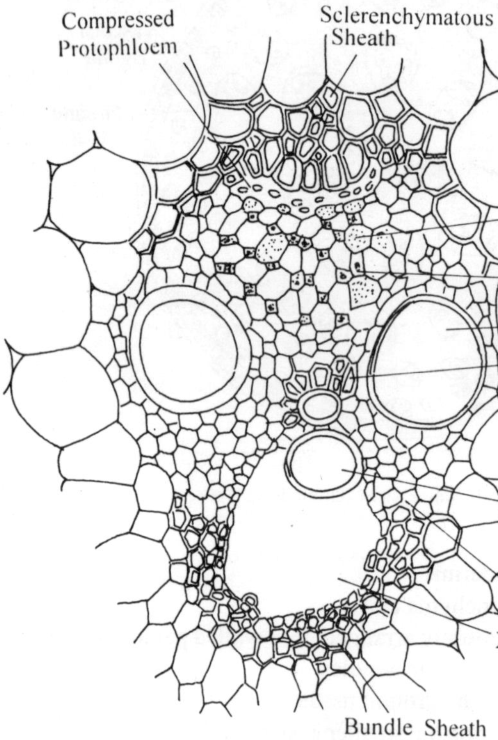

Fig. 5.13 Stems. T.S. of Maize stem (Magnified)

2. Cuticle is deposited above the epidermis.
3. There is no hypodermis.

Ground tissue:

1. Below the epidermis there are one or two layers of chlorenchyma, below which is the parenchyma consisting of thin walled, loosely arranged, oval cells.
2. This zone of the ground tissue may be called the cortex.
3. The cortex is demarcated from the rest of the central parenchymatous ground tissue by means of few layers of sclerenchyma which may be called pericycle.
4. The vascular bundles are scattered in the central ground tissue.
5. A few of the bundles at the centre are larger than the rest.
6. Vascular bundles are conjoint, collateral and closed, with endarch xylem.
7. Each vascular bundle is surrounded by a sclerenchymatous bundle sheath.
8. Structure of the phloem is same as in maize stem.
9. Xylem is reduced. There is no water containing cavity.
10. Each vascular bundle has a large vessel besides tracheids.

STEM OF *ASPARAGUS*:

In a cross section, the following regions are seen.

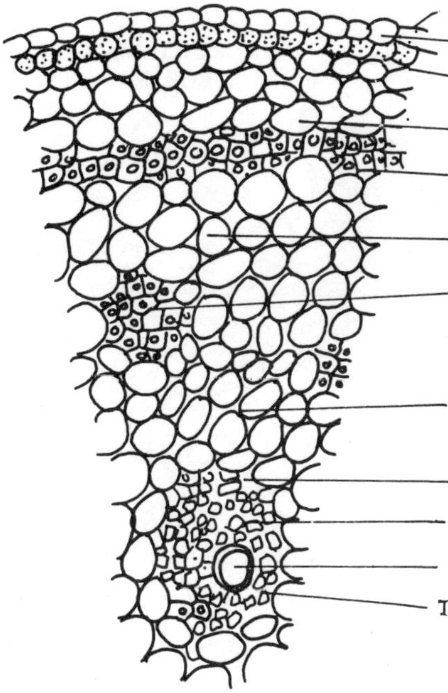

Fig. 5.14 **Stems.**
T.S. of Vanda stem (A portion enlarged).

Epidermis:

1. It is single layered and is covered by cuticle.
2. There are no epidermal hairs. There is no hypodermis.

Ground tissue:

1. It extends from the periphery to the centre of the axis and may be divided into four zones.
2. The outer zone which is immediately below the epidermis consists of a few layers of chlorenchyma with little intercellular spaces.

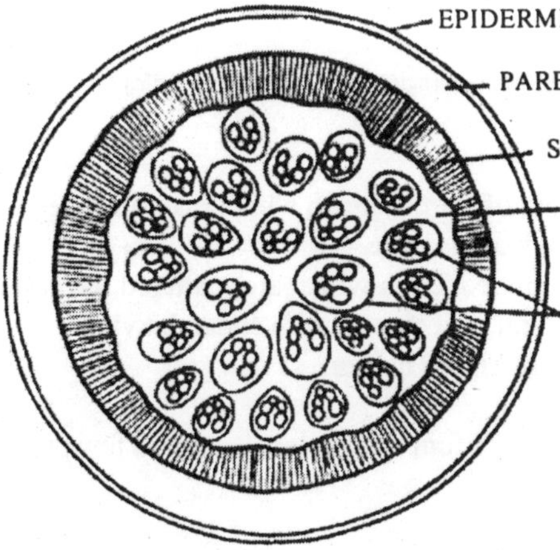

Fig. 5.15 Stems. T.S. of Asparagus stem

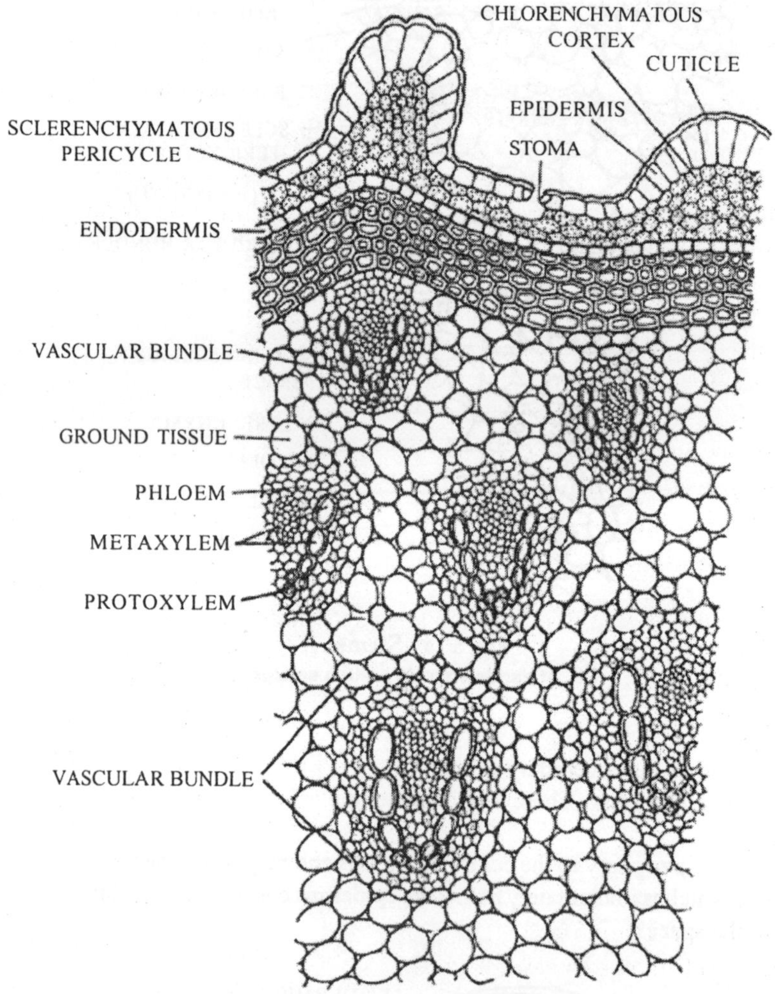

Fig. 5.16 Stems. T.S. of Asparagus stem (vascular bundle)

3. This zone may be regarded as the cortex.
4. The second zone consists of a single layer of starch sheath which may be called endodermis.
5. Next to this endodermis is the sclerenchymatous zone.
6. The innermost zone is parenchymatous embedded in which are scattered the vascular bundles.
7. The vascular bundles are conjoint, collateral, and closed, with endarch xylem.
8. Towards the periphery of the stem they are bigger in size.
9. There is no bundle sheath.
10. The xylem strand is 'V' shaped and consists of four metaxylem vessels, a few protoxylem vessels, a few tracheids and wood parenchyma.
11. There is no water containing cavity.
12. Phloem is similar in structure to that of other monocotyledons.

CANNA

1. A transverse section shows three regions namely epidermis, cortex and Stele.
2. Epidermis is the outermost layer. It is singled layered and made up of small polygonal cells.
3. The outer walls of epidermal cells are cutinized.
4. Next to the epidermis is the ground tissue.
5. Immediately below the epidermis is found the cortex.
6. The cortex consists of a few layers of parenchyma cells. These are large and polygonal.
7. Immediately below the cortex is found a single layer of chlorenchyma.
8. Next to the chlorenchyma are found patches of sclerenchyma which usually abutt the vascular bundles.
9. The entire inner region of the stem next to this consists off a uniform mass off parenchyma cells constituting the ground tissue.

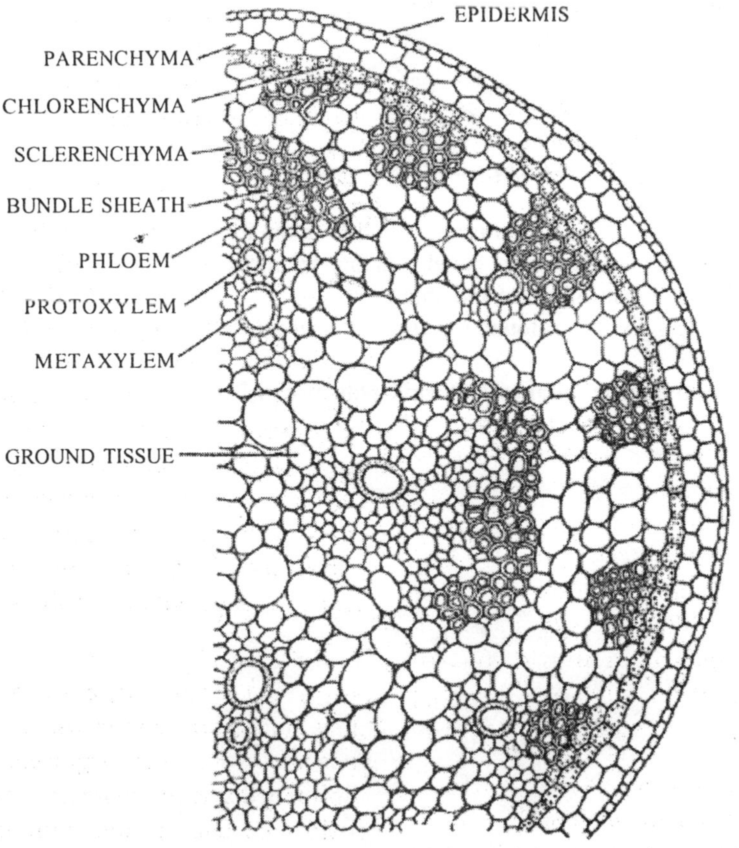

Fig. 5.17 Stems. T.S. of Canna stem (Detail of a sector)

10. The cells of the ground tissue or loosely arranged with intercellular spaces.
11. Embedded in the mass of ground tissue are found a number of vascular bundles.
12. The vascular bundles are scattered without any arrangement.
13. The vascular bundles towards the periphery are smaller in size when compared to those in the center.
14. The vascular bundles are collateral and closed.
15. There is no cambium hence there is no secondary growth.
16. Each vascular bundle is incompletely surrounded by a bundle sheath made up of sclerenchyma.

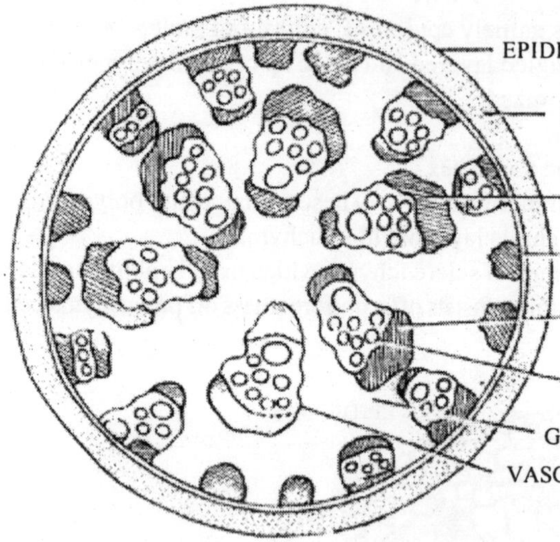

Fig. 5.18 Stems. T.S. of Canna stem (Diagrammatic)

17. The bundle sheath actually covers the vascular bundles in the form of to caps.
18. The outer cap Of sclerenchyma is more developed than the inner one.
19. Each vascular bundle consists of xylem and phloem.
20. In the vascular bundle big xylem is found towards the inner side while the phloem is found towards the outside.
21. The xylem is made up of large spiral vessels while phloem consists of seive tubes and companion cells.

CHAPTER 6

ANATOMY OF LEAVES

LEAF OF HELIANTHUS

The leaf of sunflower shows the features of a typical dicotyledonous leaf. In a transverse section it shows three well defined regions viz. epidermis, mesophyll and veins.

Epidermis:

1. There are two epidermal layers covering either surfaces of the leaf.
2. The epidermis on the upper surface is called the upper epidermis and the one on the lower surface is called the lower epidermis.
3. In some cases, covering the epidermal layers will be a layer of cuticle.
4. In the leaf of sunflower the upper epidermis is continous while the continuity of the lower epidermis is interrupted by the presence of stomata.
5. There are small pores helping in gaseous exchange and the loss of water during transpiration.
6. Each stoma is covered by two guard cells which regulate the opening and closing of the stomatal aperture.

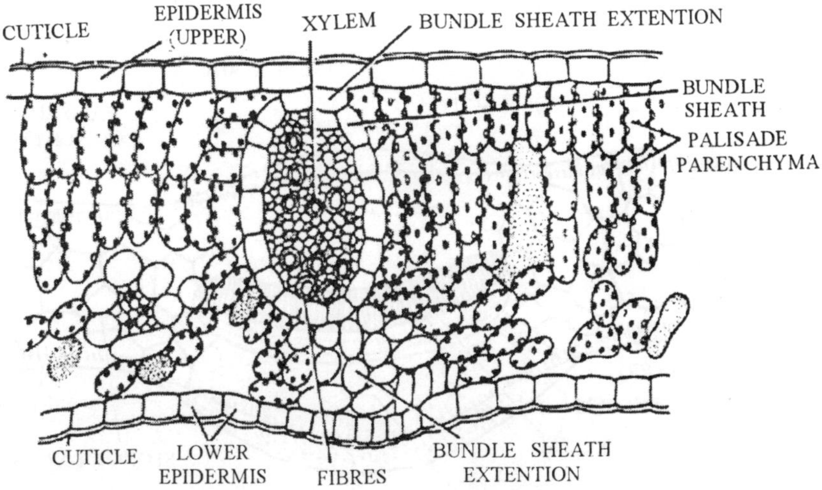

Fig. 6.1 Anatomy of Sunflower leaf (A portion enlarged)

7. The guard cells are crescent shaped with thick inner walls.
8. Internally each stoma leads into an air space.

Mesophyll:

1. In between the two epidermis layers is the mesophyll.
2. It is differentiated and has well developed palisade tissue below the upper epidermis.
3. Above the lower epidermis and below the palisade tissue is the spongy parenchyma which has loosely arranged cells with large number of air chambers.
4. Palisade tissue is chlorophyllous and it is the chief photosyntietic tissue of the plant.
5. There are a few chloroplasts in the spongy parenchyma cells also but their main function is storage.
6. There are rich in starch grains.

Vein:

1. The veins are arranged in the form of a network and in a cross section they show the vascular bundles.
2. Each vein representing a vascular bundle is surrounded by a parenchymatous envelope called the bundle sheath.
3. Within the bundle sheath are the vascular elements viz, the xylem and the phloem.
4. Unlike in stems the xylem is above and the phloem is below.

LEAF OF *DALBERGIA:*

The leaf of *Dalbergia* may be stuciied as an example of a typica dicot leaf. A cross section of the leaf shows the following features.

Epidermis:

1. There are two epidermal layers viz., upper and lower.
2. Both the layers are uniseriate and composed of barrel shaped cells with thick outer walls.
3. There is a well developed cuticle above the upper epidermis.
4. The leaves are hypostomatic (stomata found only on the lower surface).
5. The stomata internally lead to air cavities.
6. In between the two epidermal layers is the mesophyll.
7. It is differentiated and has a well developed palisade tissue below the upper epidermis.
8. Above the lower epidermis and below the palisade tissue is the spongy parenchyma which has loosely arranged cells with large number of air chambers.

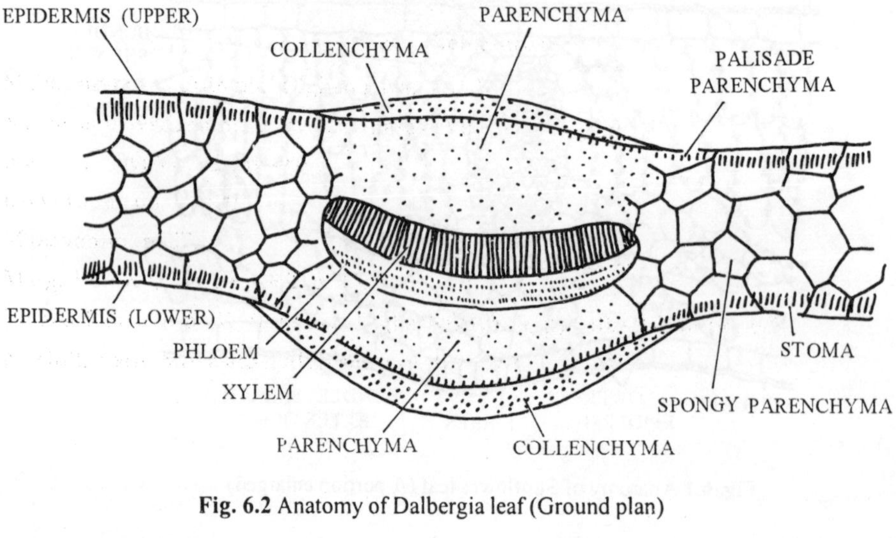

Fig. 6.2 Anatomy of Dalbergia leaf (Ground plan)

9. Palisade tissue is chlorophyllous and it is the chief photosynthetic tissue of the plant.
10. There are a few chloroplasts in the spongy parenchyma cells also but their main function is storage.
11. They are rich in starch grains.
12. Near the midrib region, internal to the epidermis is collenchyma.
13. Collenchyma cells are more towards the lower surface.
14. Immediately surrounding the vein is a bundle sheath which is parenchymatous.
15. It is very distinct in *Dalbergia* leaf.
16. The vascular bundle is *cresent* shaped and shows x/p arrangement.

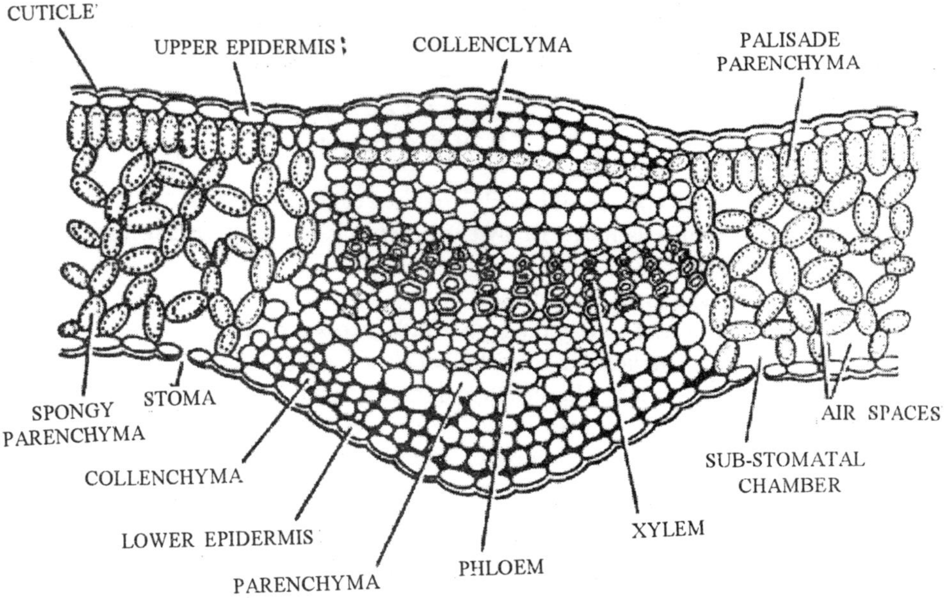

Fig. 6.3 Anatomy of Dalbergia leaf (A portion enlarged)

LEAF OF *NERIUM* (Oleander)

Nerium belongs to the family *Apocynaceae* of dicotyledons. It shows many xerophytic adaptations. A cross section of the leaf shows the following:

Epidermis:

1. The leaf is dorsiventral and shows two epidermal regions viz., upper and lower.
2. The upper epidermis is multilayered and it consists of two to three layers of barrel shaped cells arranged compactly.
3. The outermost epidermal layer is covered with a thick cuticle.
4. The structure of the lower epidermis differs from that of the upper in that it is single layered.
5. There is the deposition of a cuticle even above the lower epidermis.
6. At certain places the epidermal cells incurve forming depressions of pits.
7. The stomata are found in these depressions (hypostomatic).
8. The pits are protected with a number of *trichomes* (hairs) growing from the sides.
9. The presence of the trichomes retards the direct entry of the light and hence checks transpiration.

Mesophyll:

1. It is differentiated into palisade and spongy parenchyma.
2. Usually two or three layers of palisade tissue is present below the upper epidermis.

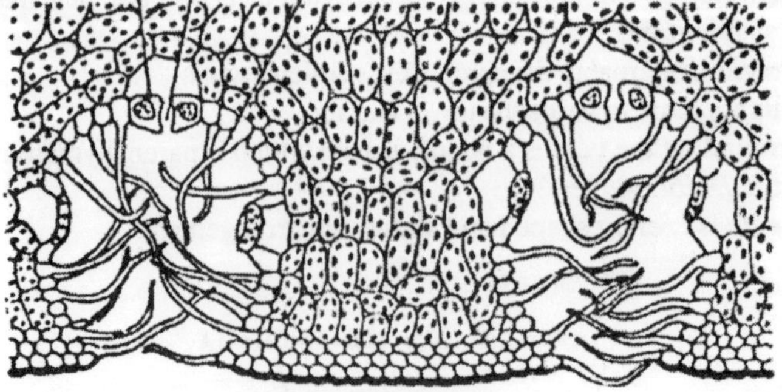

Fig. 6.4 Leaves.,
T.S. of Nerium leaf.

3. The spongy parenchyma consists of loosely arranged cells.
4. Sometimes, one layer of cells above the lower epidermis is compactly arranged resembling the palisade tissue.
5. The air cavities here and there show druses.
6. Internal to the epidermis and surrounding the midrib is collenchyma.
7. Next to collenchyma is parenchyma consisting of closely packed cells.
8. The vein is surrounded by a bundle sheath and the vasculature shows the x/p arrangement.

Xerophytic features:

1. Presence of a thick cuticle. 2. Multiple epidermis. 3. Stomata restricted to depression of the lower epidermis. 4. Presence of trichomes in the depressions to avoid direct entry of light and wind. 5. Presence of multilayered palisade tissue.

LEAF OF PADDY (*Oryza sativa*):

As an example of typical isobilateral leaf, the leaf of paddy may be studied. A cross section shows the following:

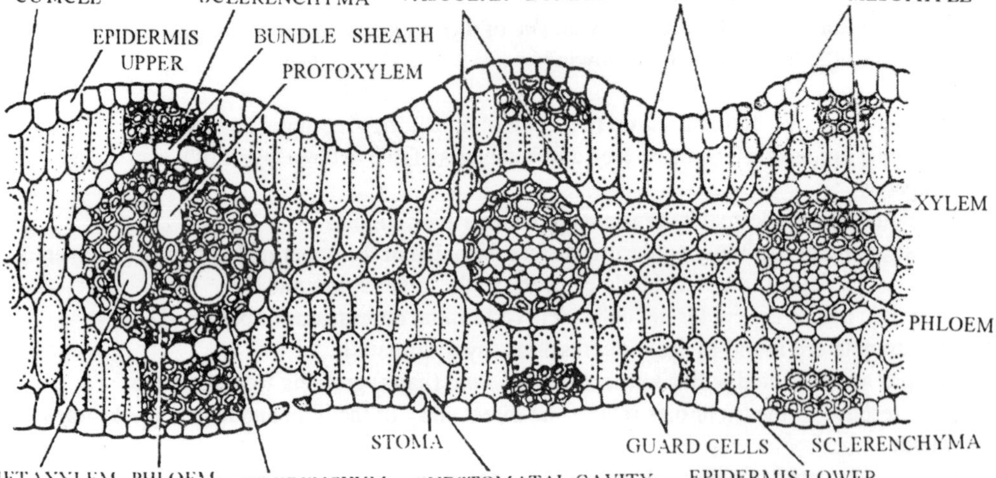

Fig. 6.5 Leaves. T.S. of Paddy leaf

Epidermis:

1. There will be two epidermal layers - upper and lower.
2. The epidermal layers are uniseriate and composed of more or less oval cells arranged compactly.
3. There is the deposition of a cuticle both one the upper epidermis as well as on the lower epidermis.
4. Besides the normal cells, the upper epidermis consists of special cells called *bulliform cells* or *motor cells*.
5. These cells are larger than the other epidermal cells and are thin walled.
6. These help in the rolling of the leaves during dry weather.
7. Leaves are *amphistomatic* (stomata are present on both the epidermal layers).
8. Each stoma internally leads into a large substomatal air chamber.

Mesophyll:

1. It is undifferentiated.
2. But the mesophyll cells below the upper epidermis may show a compact arrangement.

Vascular bundles:

1. The veins are arranged parallely and are not interconnected.
2. There are two pads of sclerenchyma supporting the vein.
3. The vein itself is surrounded by a bundle sheath made up of parenchyma cells.
4. Within the bundle sheath are present collateral vascular elements showing x/p arrangement.

LEAF OF MAIZE

A cross section shows the following.

EPIDERMIS:

1. There will be two epidermal layers upper and lower.
2. The epidermal layers are uniseriate and composed of more or less oval cells arranged compactly.
3. There is the deposition of a cuticle both on the upper epidermis as well on the lower epidermis .
4. Besides the normal cells, the upper epidermis consists of special cells called bulliform cells or motor cells.
5. These cells are larger than the other epidermal cells and are thin walled.
6. These help in the rolling of the leaves during dry weather .

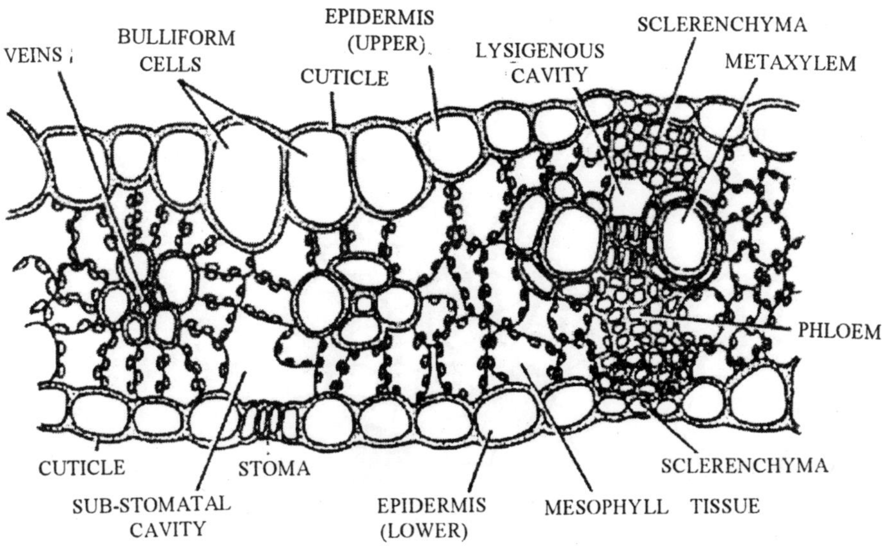

Fig. 6.6 Leaves. T.S. of Maize leaf

7. Leaves are amphistomatic (stomata are present on both the epidermal layers), Each stoma internally leads into a large substomatal air chamber.

Mesophyll:

1. It is undifferentiated.
2. But the mesophyll cells below the upper epidermis may show a compact arrangement.

Vascular bundles:

1. The veins are arranged parallely and are not interconnected.
2. There are two pads of sclerenchyma supporting the vein.
3. The vein itself is surrounded by the bundle sheath made up of parenchyma cells.
4. Within the bundle sheath are present collateral vascular elements showing X/P arrangement.

STOMATA

1. Stomata are microscopic openings present in the epidermis.
2. Each stoma has a stomatal aperture or opening surrounded by two guard cells.

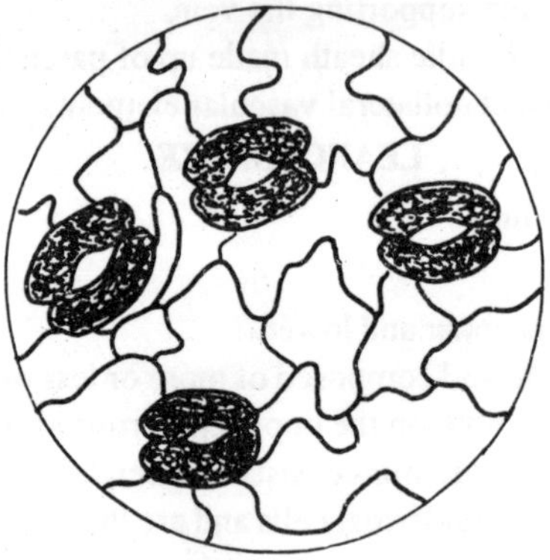

Fig. 6.7 Anomocytic stomata (Cucurbita)

3. In the leaves, stomata may be present only on the upper epidermis (*epistomatic*), only on the lower epidermis (*hypostomatic*) or on both the epidermal layers (*amphistomatic*).
4. Guard cells are generally crescent shaped in outline and have dense granular cytoplasm with chloroplasts.
5. The cell walls of guard cells are unevenly thick and this plays an important role in the opening and closing of stomata.

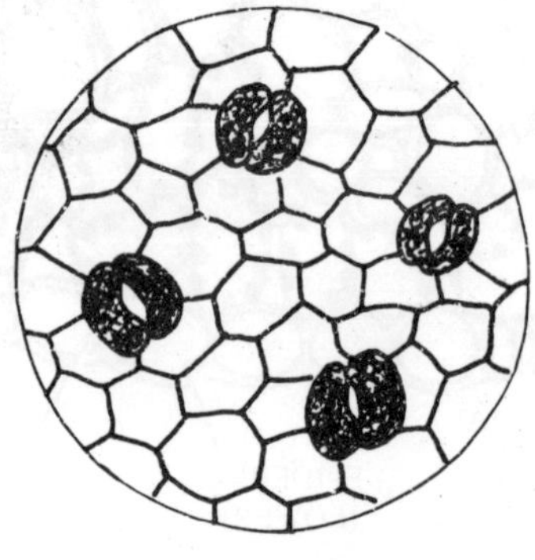

Fig. 6.8 Anomocytic stomata (Diospyros)

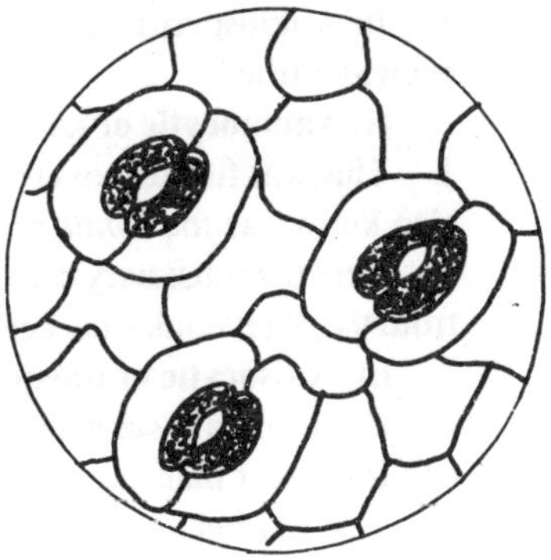

Fig. 6.9 Paracytic stomata (Ixora)

6. Usually the wall of guard cells facing the pore is thicker than the one away from the pore.
7. During day time there is increased turgor pressure in the guard cells, as a result the thin walls are pulled away so that the concavity of the front (thick) wall increases, and the aperture becomes wider.

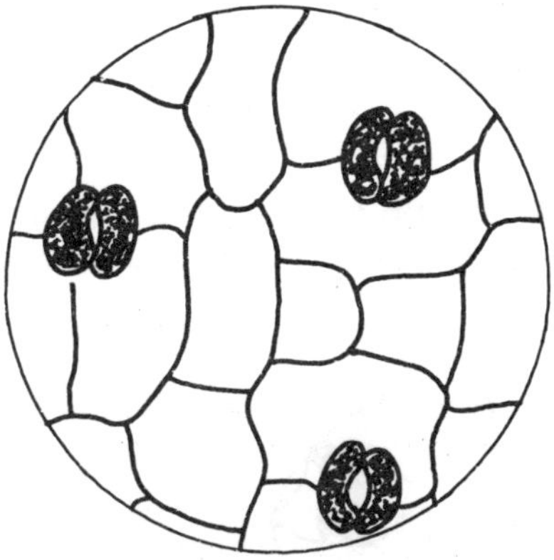

Fig. 6.10 Diacytic stomata (Dianthus)

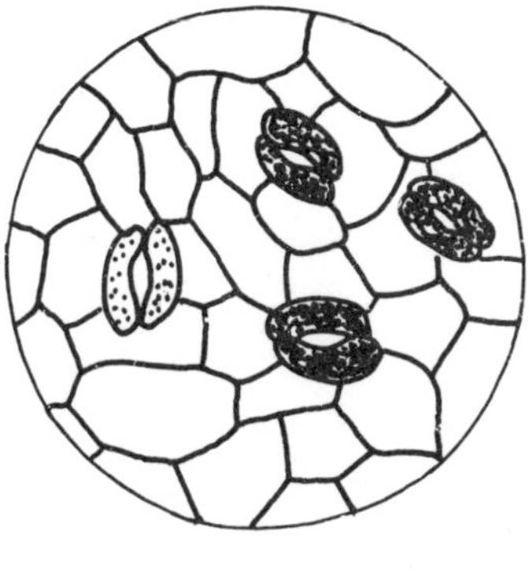

Fig. 6.11 Anisocytic stomata (Petunia)

8. Stomata are present on the aerial organs of plants like leaf, stem, sepals etc.
9. The chief distribution of stomata however is on the leaves.
10. Rarely stomata are enocuntered on petals, stamens and ovules also, but these are non functional.

TYPES OF STOMATA:

1. Classification of the types of stomata is mainly based on the nature and number of *subsidiary cells* associated with them (stomata).
2. Subsidiary cells are epidermal cells that are morphologically

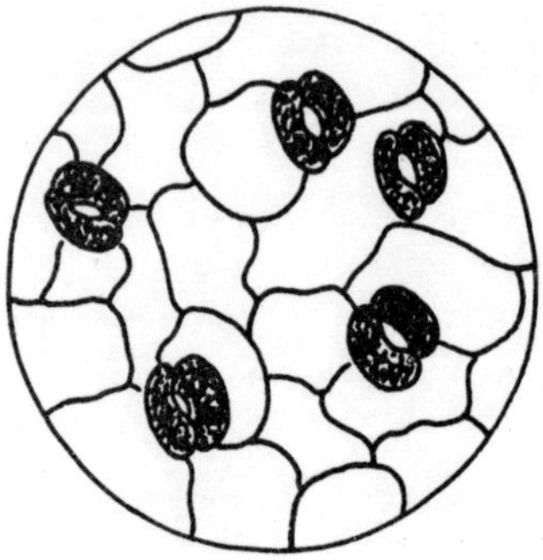

Fig. 6.12 Anisocytic stomata (Brassica)

distinct from other epidermal cells are are in close association with the stomata.

3. In Angiosperms the following main types of stomata have been identified.

(i) Anomocytic or irregular celled type:

1. This was first discovered in the family *Ranunculaceae*, hence also known as the *Ranuculaceous type*:
2. Here the subsidiary cells are absent in that they are not distinct from the other epidermal cells.

(ii) Anisocytic or unequal celled type:

1. Each stoma is surrounded by three subsidiary cells, of which one is smaller than the other two.
2. This has been reported in *Cruciferae*.

(iii) Paracytic or parallel celled type:

1. Stoma is surrounded by two equal parallely arranged subsidiary cells.
2. This is of frequent occurrence in *Rubiaceae*, hence also known as the *Rubiaceous type*.

(iv) Diacytic or cross celled type:

1. The stoma is surrounded by two subsidiary cells, which are at right angles to the guard cells unlike in the paracytic type.
2. This was first discovered in *Caryophyllaceae*, hence also known as the *Caryophyllaceous type*.

(v) Graminaceous type:

1. This is characteristic of *Graminae* (Poaceae) and *Cyperaceae*.
2. Here the subsidiary cells are dumb-bell shaped i.e., the middle portion is narrower than the ends and has a thicker wall.

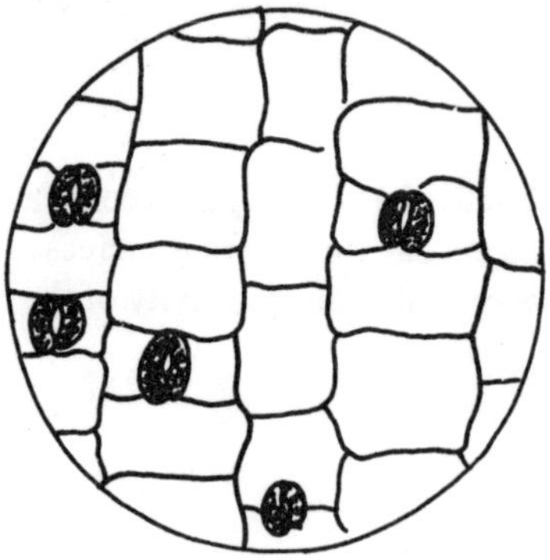

Fig. 6.13 Stomata of Rheo

Fig. 6.14 Graminaceous stomata

CHAPTER 7

ANOMALOUS STRUCTURE OF STEMS

STEM OF *ARISTOLOCHIA* (Aristolochiaceae)

1. The young stem has a typical dicotyledonous structure.
2. Pericycle is continous and is represented by sclerenchyma.
3. The vasculature consists of collateral bundles separated by wide interfascicular area.
4. During secondary growth inter and intrafascicular cambia become active and form a ring.
5. But while the intrafascicular cambium forms the vascular tissues, the segments of interfascicular cambium form only ray like parenchyma with the result the vascular bundles though increased in size, remain discrete.
6. With the increase in diameter the pith and rays get crushed.

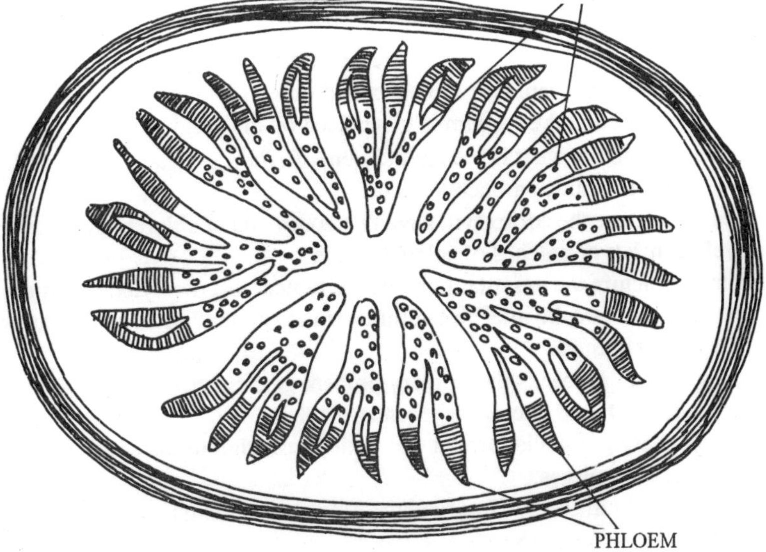

Fig. 7.1 **Anomalous secondary growth.,**
T.S. of Aristolochia stem (Diagrammatic).

7. Sometimes, portions of intrafascicular cambia also form parenchymatous rays, with the result the vascular bundles become bifurcated.
8. A stongly fluted vascular cylinder is thus formed.
9. This is an instance of the abnormal behaviour of the normal cambium.

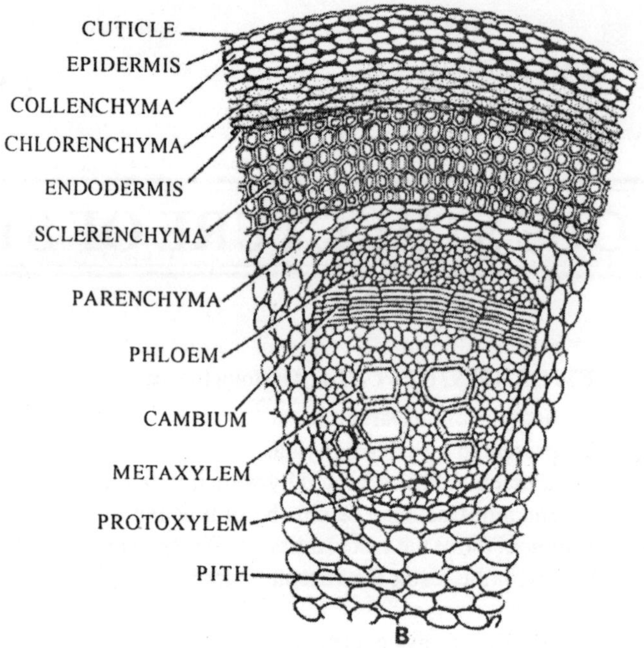

Fig. 7.2 Anatomy of stem. T.S. of young stem of Aristolochia (dicot). detail of a sector

TECOMA

1. A transverse section shows epidermis, cortex and stele in primary structure.
2. In the secondary structure the epidermis is replaced by the periderm.
3. The periderm is similar to what is seen in the stem of *Tilia*.
4. The primary cortex is mainly parenchymatous.
5. In the primary structure the cortex will have two zones - outer chlorenchyma and inner parenchyma.
6. The pericycle consists of patches of sclerenchyma.
7. In the primary structure the vascular bundles are typical to a dicotyledonus stem.
8. In the secondary structure the inter and intrafascicular cambia form a completely closed ring and cut off secondary vascular tissues.
9. As secondary growth proceeds the primary phloem gets crushed.
10. The primary xylem is found internal to secondary xylem.
11. Anomalus secondary growth is formed due to the formation of the secondary cambium internal to the primary xylem in a few places.
12. This abnormal cambium cuts off secondary xylem to the exterior and the secondary phloem to the interior.
13. In the central region is found a large pith made up of parenchyma.

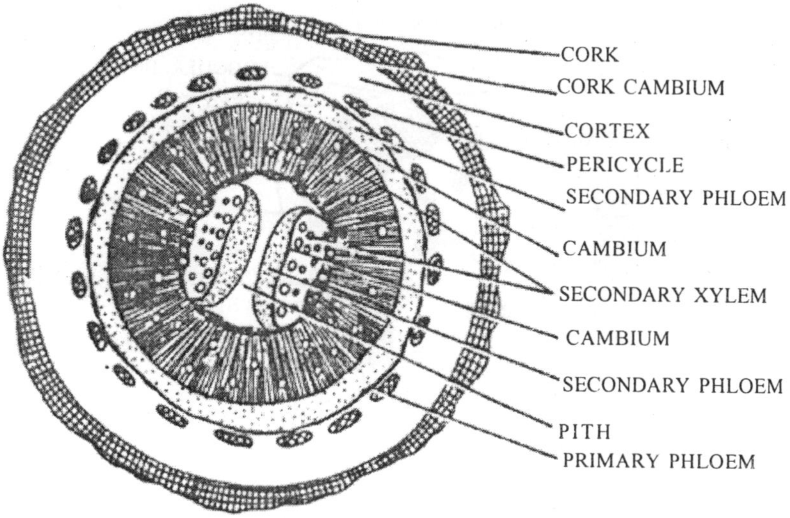

Fig. 7.3 T.S. of Tecoma stem (diagrammetic)

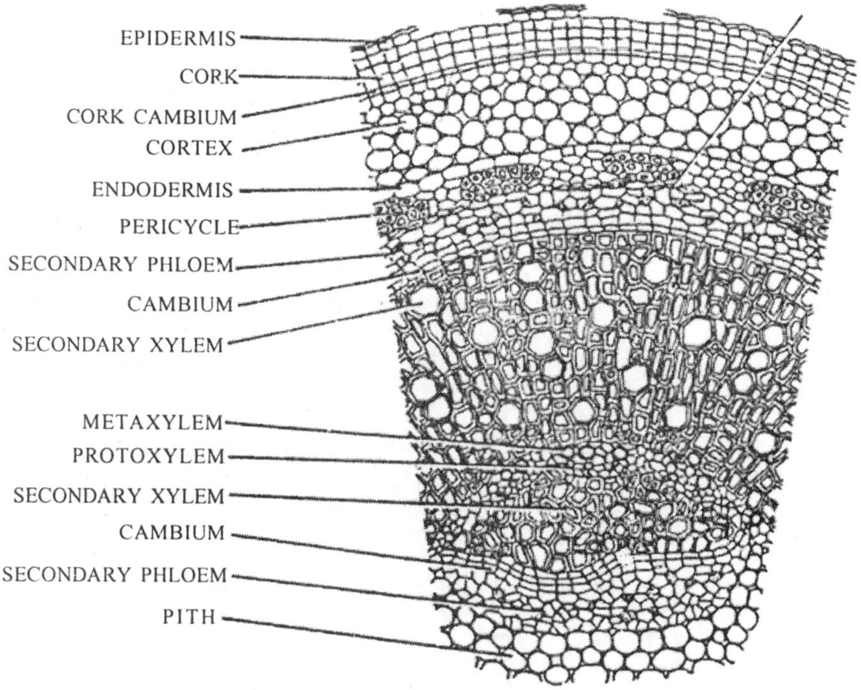

Fig. 7.4 T.S. of Tecoma stem (a part cellular)

STEM OF *AMARANTHUS* (Amaranthaceae)

The stem of *Amaranthus* shows certain anomalous structures, both in the primary structure as well as in the secondary thickening.

A cross section of the stem shows the following features.

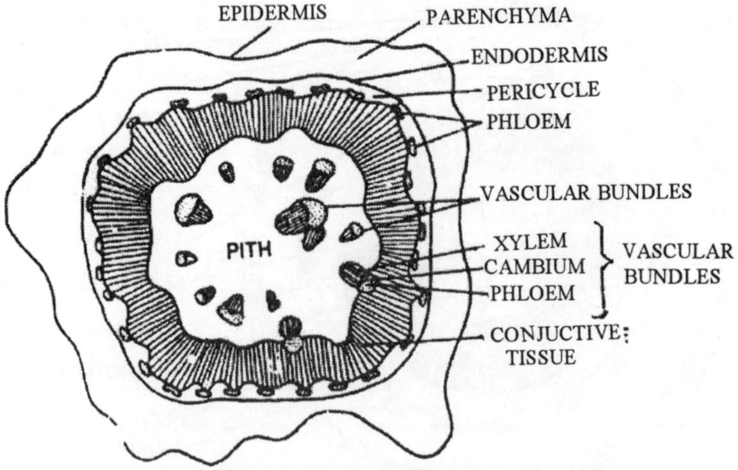

Fig. 7.5 **Anomalous secondary growth.,**
T.S. of Amaranthus stem (ground Plan).

Epidermis:

1. This is the outermost layer and is uniseriate.

Ground tissue:

1. Below the epidermis is the ground tissue made up of collenchyma, chlorenchyma and parenchyma.
2. Collenchyma is present immediately below the epidermis.

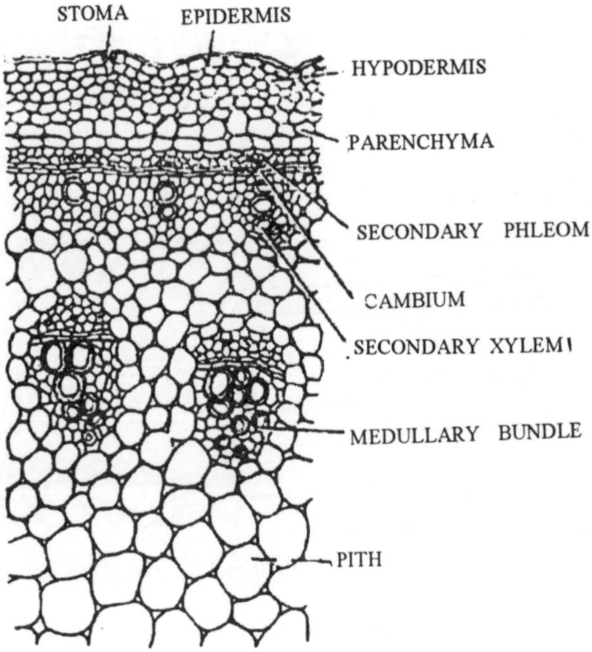

Fig. 7.6 **Anomalous secondary growth.,**
T.S. of Amaranthus stem (a portion enlarged).

3. Here and there it is interrupted by patches of chlorenchyma.
4. Below the collenchyma is a multilayered parenchymatous zone constituting the inner cortex.
5. The endodermis is not very conspicous.
6. In the primary structure the entire region inner to the endodermis is filled by loosely arranged, thin walled parenchyma cells.
7. Embedded in this, are found the scattered primary vascular bundles.
8. These are the medullary bundles.
9. There is no ring of vascular bundles in the primary structure as in a normal dicot stem.
10. The medullary bundles are conjoint, collateral, and open with endarch xylem.
11. Vascular bundles towards the periphery are smaller than those towards the centre.
12. The cambia of these bundles produce a little secondary vascular tissue (which is limited to the bundles) and very soon cease to function.
13. Secondary growth is brought about by a cambium originating outside the stele.
14. This cuts off secondary vascular bundles, which lie embedded in the conjunctive tissue around the pith region.
15. This is an example of *medullary bundles,* and *replacement of normal cambium by abnormal accessory cambium.*
16. There is anomaly both in the primary structure as well as in the secondary growth.

STEM OF *BOUGAINVILLEA* (Nyctaginaceae)

The stem in its primary structure shows an epidermis, cortex and a vascular system. The epidermis is single layered and has many multicellular hairs.

1. **Cortex** is made up of collenchyma and chlorenchyma.
2. Collenchyma is present in the form of patches.
3. Chlorenchyma is internal to collenchyma and extends between the patches of the latter.

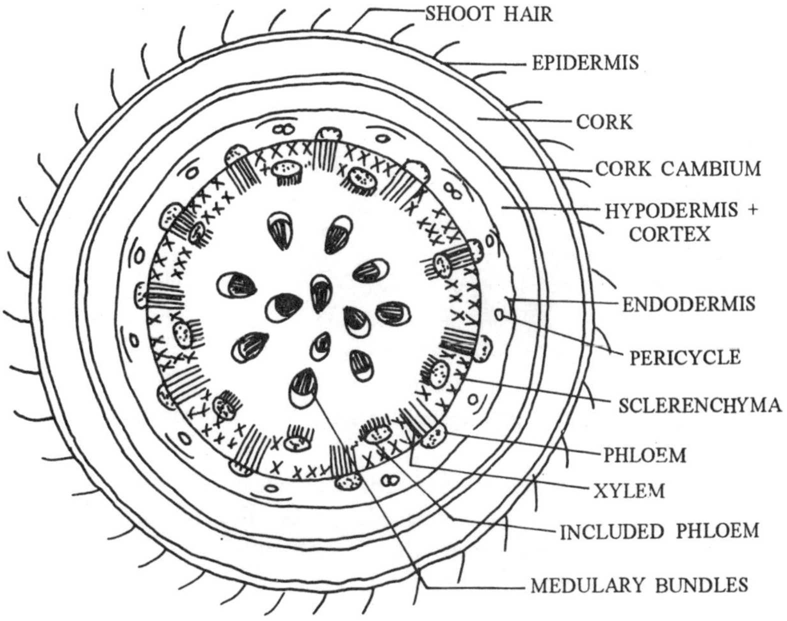

Fig. 7.7 **Anomalous secondary growth.,**
T.S. of Bougainvillea stem (ground plan).

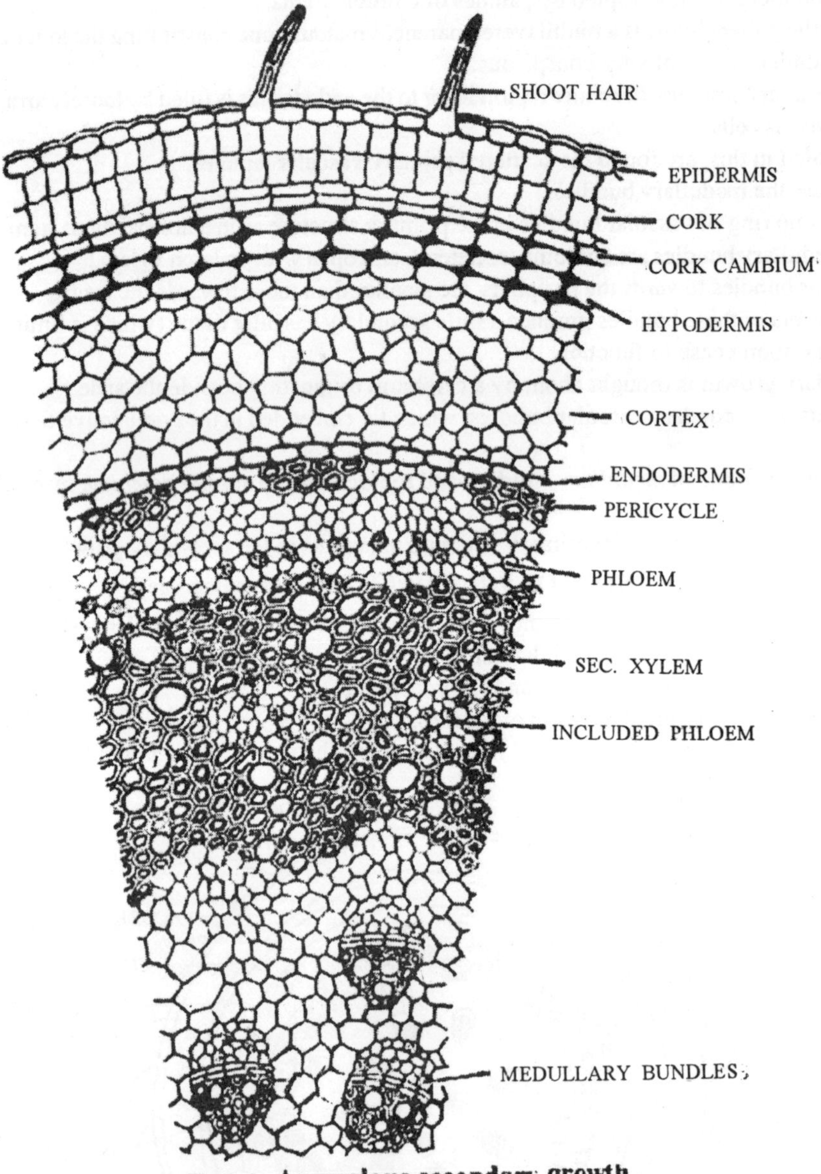

Fig.7 .8 Anomalous secondary growth.,
T.S. of Bougainvillea stem (A portion enlarged).

4. Internal to the cortex is an endodermis and a pericycle made up of one or two layers of parenchyma cells.
5. The vascular system consists of numerous medullary bundles embedded in the pith, in addition to the normal rings of conjoint, collateral, open endarch vascular bundles.
6. The medullary bundles are of two sizes.
7. There are four large bundles in the centre surrounded by many, smaller bundles.
8. Medullary bundles are similar in structure to that of normal bundles.
9. Fascicular cambium in these bundles has a limited activity.
10. During secondary growth, more than one cambial ring is formed.

11. The first cambial ring produces vasculature, only where the primary vasculature is present.
12. The interfascicular cambium produces living, thick walled conjunctive tissue.
13. After some time, this cambium ceases to function and a second cambial ring arises external to the first one.
14. The second cambial ring functions in the same way as the first. producing vasculature only in strips.
15. Successive cambial rings arise and function in the same way producing discrete secondary vascular bundles embedded in a sclerenchymatous conjunctive tissue.
16. The type of secondary growth here is same as in *Amaranthus* i.e., *medullary vascular bundles* and abnormal behaviour of *accessory cambia*.

PEPEROMIA

1. A transverse of the section shows the epidermis, cortex and the vasculature.
2. Anomalous secondary growth is seen in the stem.
3. Epidermis is made up of a single layer of compactly arranged cells.
4. There is the deposition of cuticle above the epidermis.
5. Next to the epidermis is found the cortex consisting of few layers of collenchyma cells.
6. Next to the cortex is found a uniform mass of parenchyma constituting the ground tissue.
7. Embedded in the ground tissue are found a number of scattered vascular bundles.
8. This arrangement is unusual for dicotyledons.

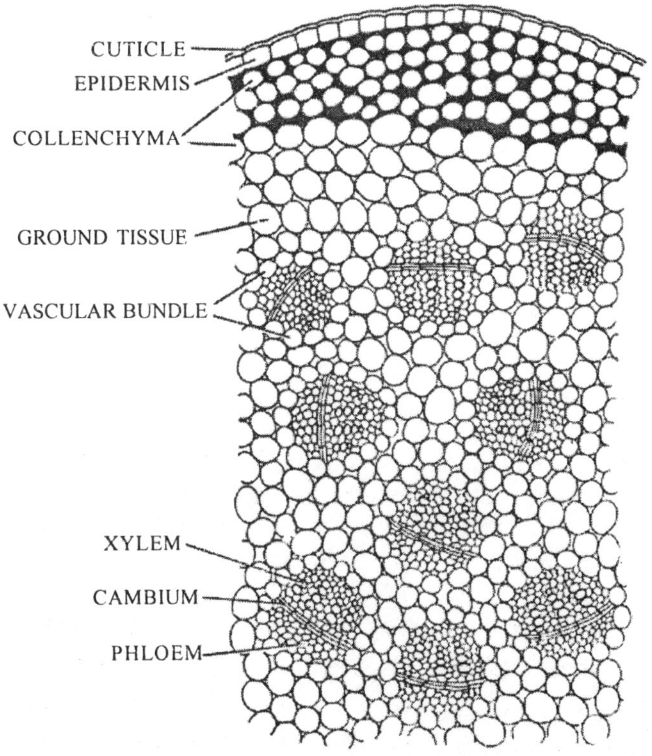

Fig. 7.9 Anatomy of anomalour stem T.S. of Peperomia (dicot stem; diagrammetic)

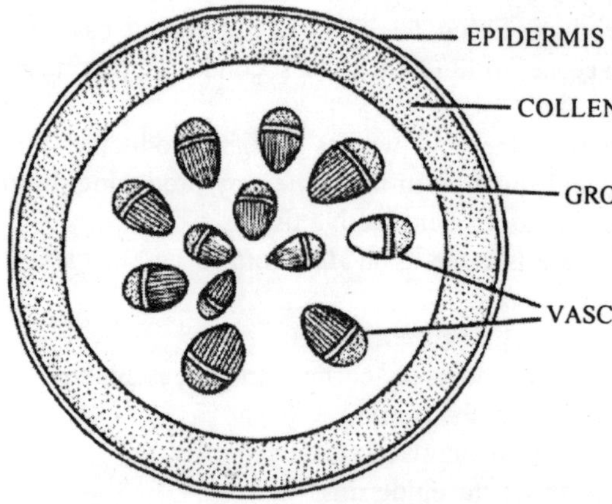

Fig. 7.10 Anatomy of anomalous stem T.S. of Peperomia (dicot stem; detail of a sector)

LEPTADEMIA

1. A transverse section of the stem show epidermis, cortex and stele.
2. Anomalous vascular structure are seen in the stem.
3. Epidermis is made up of a single layer of compactly arranged cells.
4. There is the deposition of cuticle above the epidermis.
5. Below the epidermis is found the cortex made up of chloroenchyma.
6. Sometimes below the chlorenchyma is found one or two layers of parenchyma.
7. Occasionally few stone cells may be found between parenchyma and chlorenchyma.
8. The endodermis is clearly distinguishable.
9. Next to the endodermis is found the pericycle.
10. The vasculature consists of xylem and phloem.
11. Secondary phloem is very conspicuous while the primary phloem is seen at regions opposite to the primary phloem.

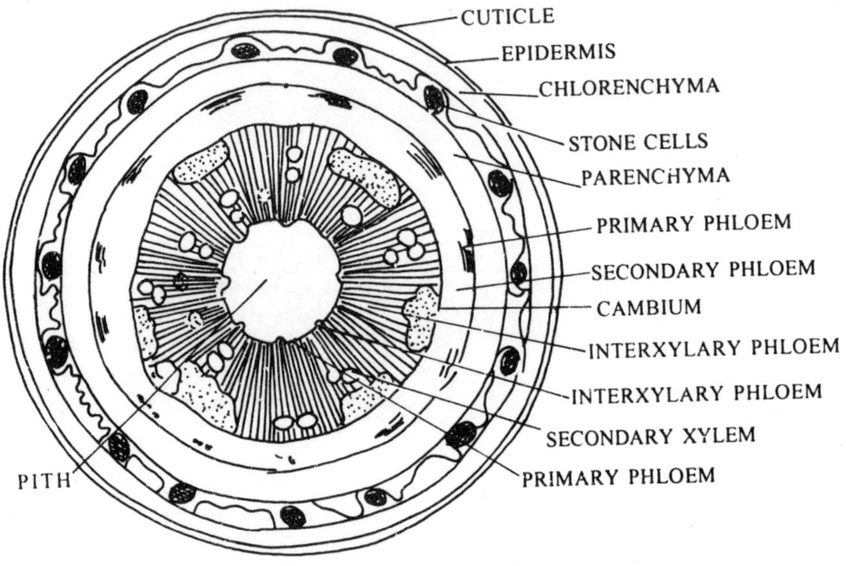

Fig. 7.11 Anatomy of anomalous stem T.S. of Leptodenia (Aselepiadaceae -dicot) stem, diagrammatic

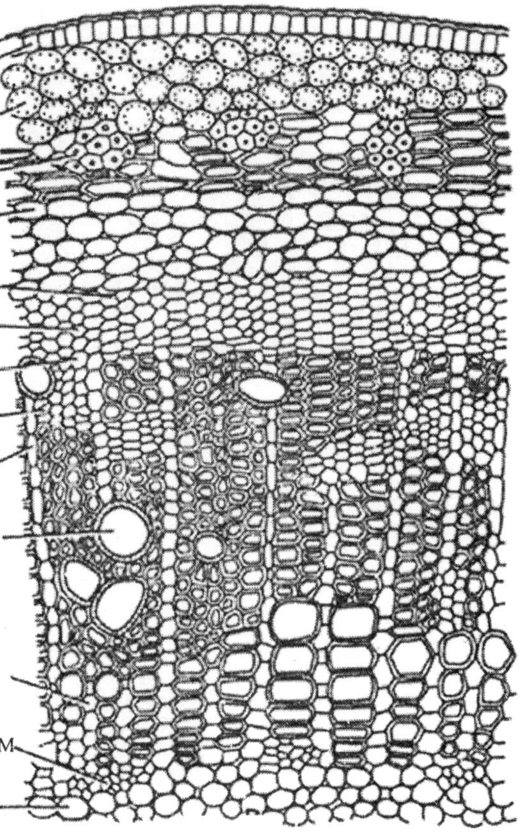

Fig. 7.12 Anatomy of anomalous stem T.S. of Leptodenia (Aselepiadaceae -dicot) stem, detail of a sector

12. Next to the secondary phloem is the cambium made up of many layers of rectuangular cells.
13. The xylem is found next to the phloem.
14. The primary xylem is found to the interior.
15. The secondary xylem consists of large vessels.
16. The secondary vasculature has unusual feature such as interxylary and intraxylary phloem.
17. The inter xylary phloem is found between the strands of the secondary xylem.
18. The intra xylary phloem is found at the border of the pith.
19. Uni or multi seritae medullary rays are found between the vascular bundles.
20. Latex in the cells are found embedded in the parenchymatous cells.

SALVADORA

1. A transverse section of the stem shows epidermis, cortex and stele.
2. Anomalous vascular structure are seen in the stem.
3. Epidermis is made up of a single layer of compactly arranged cells.
4. There is the deposition of cuticle above the epidermis.
5. Next to the epidermis is found a few layers of parenchyma.
6. Next to the parenchyma is found three to four layers of chlorenchyma.
7. In the secondary structure the epidermis is replaced by the cork tissue.

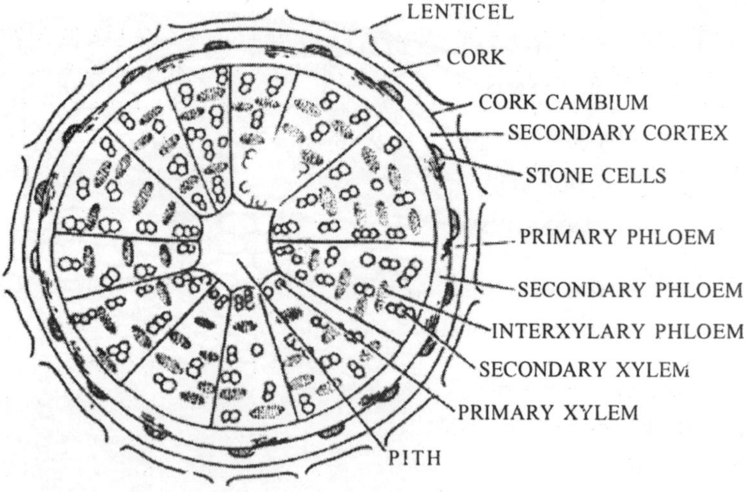

Fig. 7.13 T.S. of Salvadora stem, diagrammatic

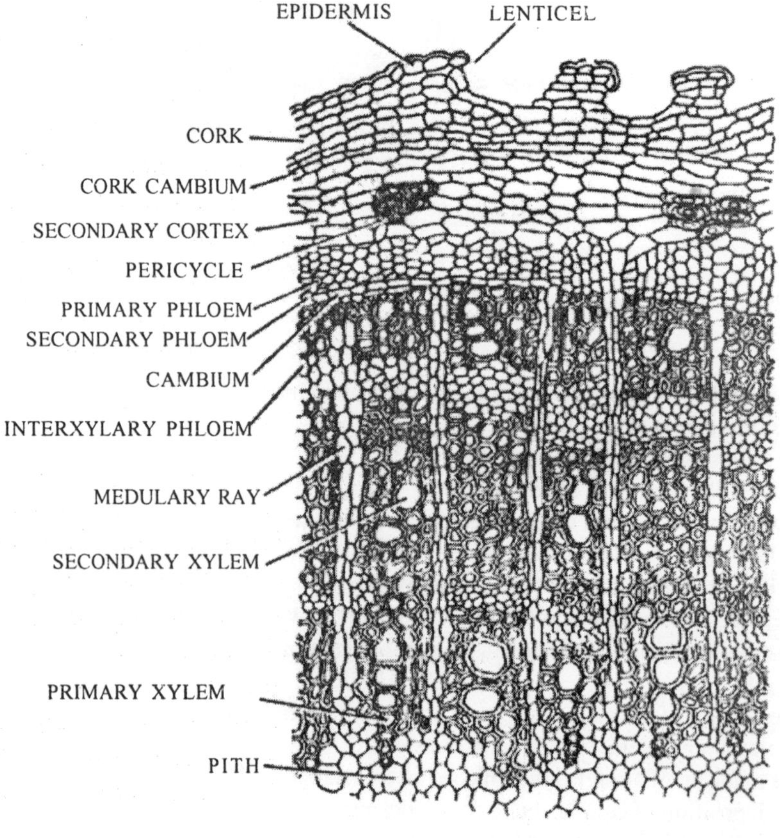

Fig. 7.14 Anatomy of anomalous stem T.S. of Salvadora stem, Detail of a sector

8. Internal to the cork is found the cork cambium and the secondary cortex.
9. Outlining the stele is found the endodermis.
10. Pericycle is found next to the endodermis and is represented by the patches of fibers oppossite the vascular bundles.
11. Internal to the pericycle is found the secondary phloem and sum amount of primary phloem.
12. The phloem bundles is seperated by uni or multiseriate medullary rays.
13. Next to the phloem is found the cambial zone.
14. The cambium is seperated into patches by the medullary rays.
15. Next to the cambium is found the secondary xylem.
16. Towards the pith is found primary xylem.
17. The secondary xylem consists of patches of phloem constituting the interxylary phloem.
18. Towards the center is found the pith made up of parenchyma.
19. The main anomalous feature in the stem is the presence of the patches of interxylary phlcem arising on the inner side of the cambium.

STRYCHNOS

1. A transverse section of the stem shows epidermis, cortex and stele.
2. Anomalous vascular structure are seen in the stem.
3. Epidermis is made up of a single layer of compactly arranged cells.
4. There is the deposition of cuticle above the epidermis.

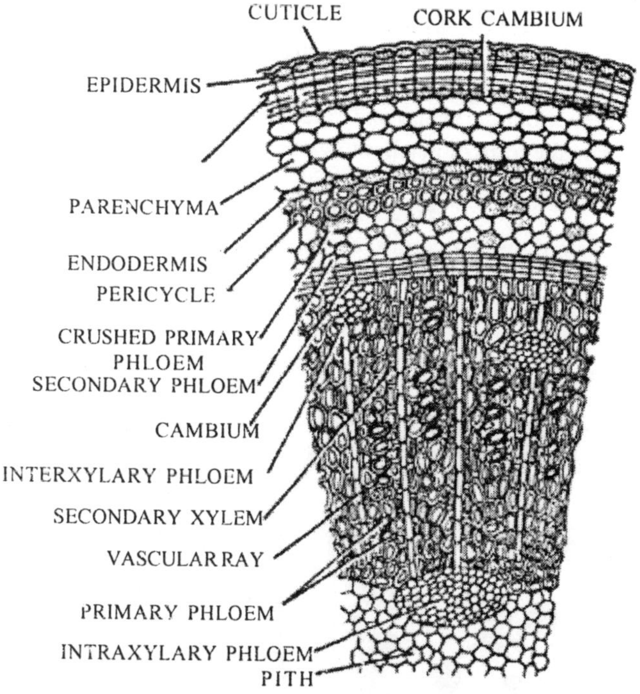

Fig. 7.15 Anatomy of anomalous stem of Strychnos (dicot), T.S. of stem, detal of sector

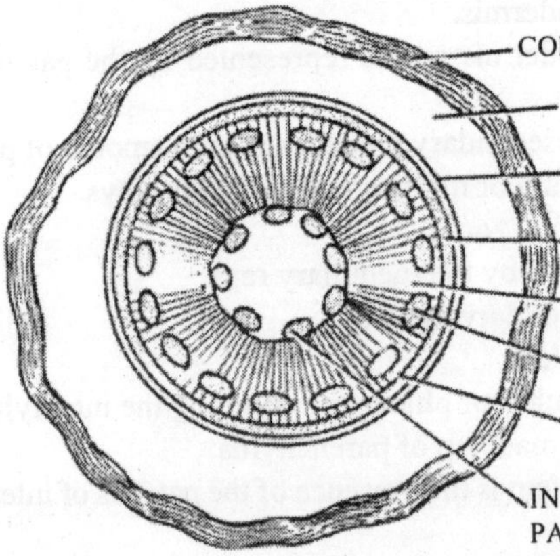

Fig. 7.16 T.S. of Strychnos stem (dicot), diagrammatic

5. Cortex consists of a few layers of chloencyma followed by parenchyma.
6. In the secondary structure the epidermis is replaced by the cork.
7. Internal to cortex is the endodermis.
8. Next to the endodermis is found a continous layer of sclerenchyma representing the pericycle.
9. The vasculature consists of conjoint, collateral and open vascular bundles in the primary structure.
10. Secondary phloem is found near the pericycle along with curshed primary phloem.
11. Uniseriate medullary rays are present connecting the pith and cortex.
12. Secondary xylem is found next to the cambium.
13. The unusual features are presence of interxylary and intraxylary phloem.
14. Intraxylary phloem is found in the form of patches towards the pith.
15. The central region consits of a wide parenchymatous pith.

MIRABILIS

1. A transverse section of the stem shows epidermis, cortex and stele.
2. Anomalous vascular structure are seen in the stem.
3. Epidermis is made up of a single layer of compactly arranged cells.

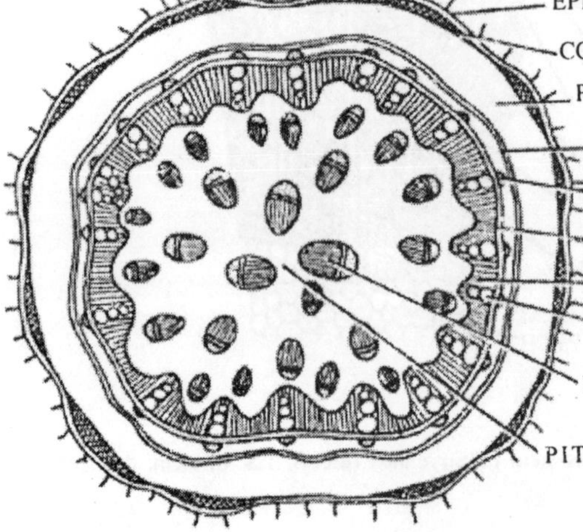

Fig. 21.17 T.S. of stem Mirabilis, Diagrammatic

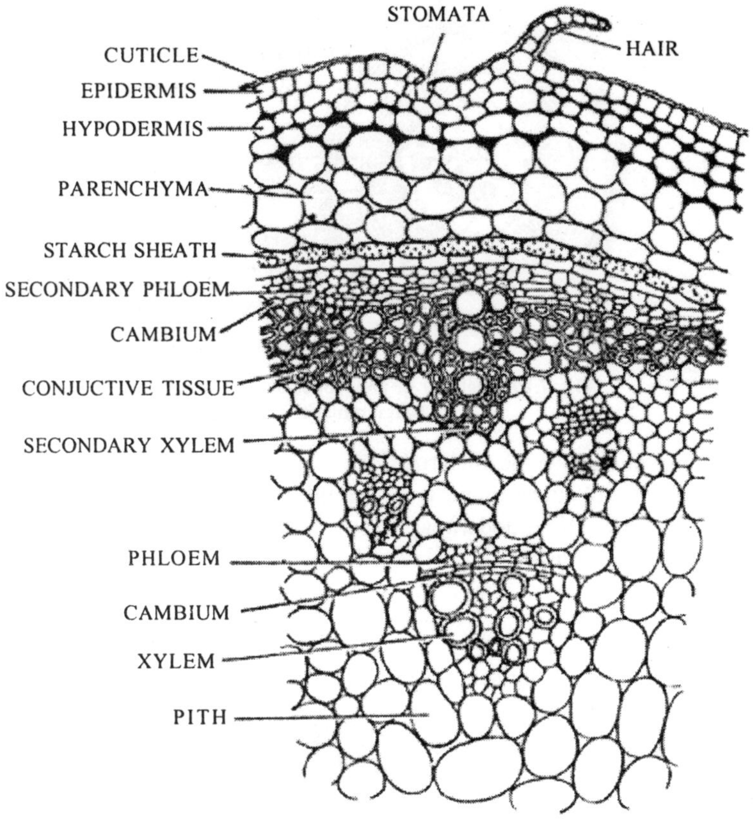

Fig. 7.18 Anatomy of anomalous stem of Mirabilis, Detail of sector of same

4. There is the deposition of cuticle above the epidermis.
5. Cortex consists of few layers of chloenchyma followed by parenchyma.
6. Internal to the cortex is the endodermis.
7. Internal to the endodermis is found pericycle made up of a single layer of parenchyma.
8. Next to the pericyle is the secondary phloem along with patches of primary phloem.
9. Next to the phloem is the cambium.
10. Internal to the cambium is the secondary xylem.
11. The central region of the stem consists of a uniform mass of parenchyma constituting the ground tissue.
12. Embedded in this ground tissue are embedded a number of vascular bundles which are reffered to as the medullary vascular cells.
13. Anomalous structures are seen in the form of a number of rings of vascular bundles occuring in the medullary region.
14. These vascular bundles are irregular arrangement.

ARGYREIA

1. A cross section of the stem is some what rectangular with four projections at the corner.
2. In the primary structure the usual epidermis, cortex and stele are seen.
3. Epidermis is single layered made up of compactly arranged cells.

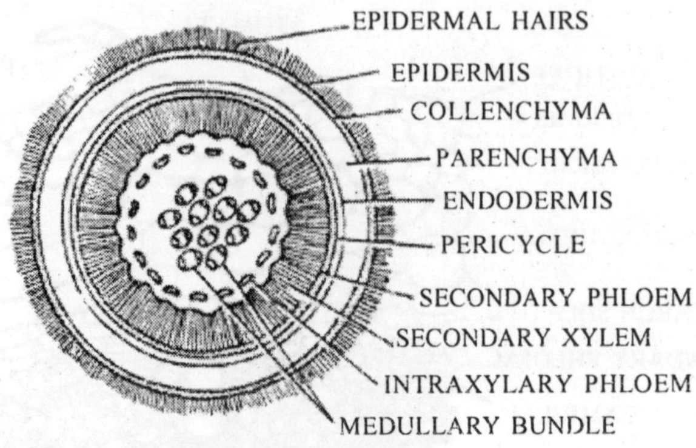

Fig. 7.19 Anatomy of anomalous stem of Argyreia Diagrammatic

4. Cuticle is deposited above the epidermis.
5. The cortex consists of outer collenchyma and inner parenchyma.
6. The cortex is followed by endodermis.
7. Next to the endodermis is the single layered parenchymatous pericycle.

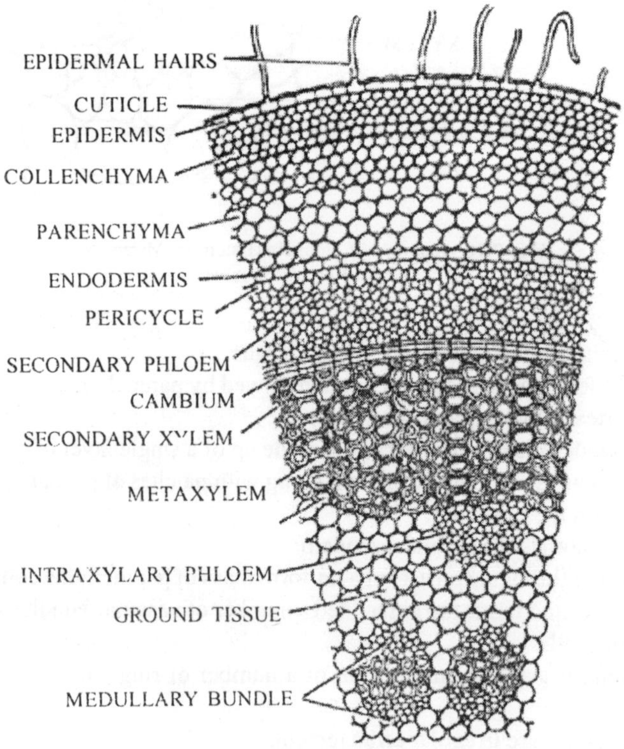

Fig. 7.20 Anatomy of anomalous stem of Argyreia, Detail of a sector of same

8. The vasculature consists of conjoint, colletral and open vascular bundles in the primary structure.
9. Next to the pericycle is the secondary vasculature consisting of secondary phloem, cambium followed by secondary xylem.
10. Central region is occupied by the pith.
11. The unusal feature is the presence of several discrete vascular bundles found embedded in the pith.
12. Another unusal feature is the presence of patches of interxylary pholem.

CHENOPODIUM

1. A cross section of the stem is some what rectangular with four projections at the corner.
2. In the primary structure the usual epidermis, cortex and stele are seen.
3. Epidermis is single layered made up of compactly arranged cells.
4. Cuticle is deposited above the epidermis.
5. The cortex is divided into two zones - outer collenchymatous and inner parenchymatous.
6. Internal to the cortex is the endodermis.
7. A parenchymatous pericylce is found internal to the endodermis.
8. The primary vascular bundles are typical to a dicotyledonous stem.
9. In the region of the pericycle develops an unusual cambium external to primary phloem.
10. It is this which bings about the abnormal secondary growth.
11. Normal seconday also occurs in the primary vascular bundles due to the activity of fasicular cambium.
12. This cambium however is active only for a short period.

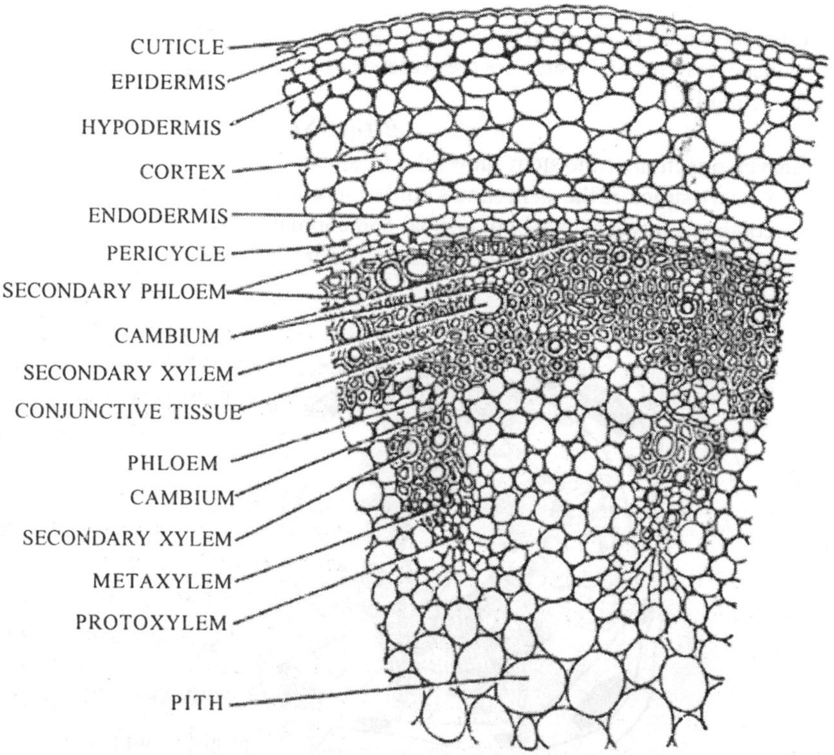

Fig. 7.21 T.S. of Chenopodium stem (a part cellular)

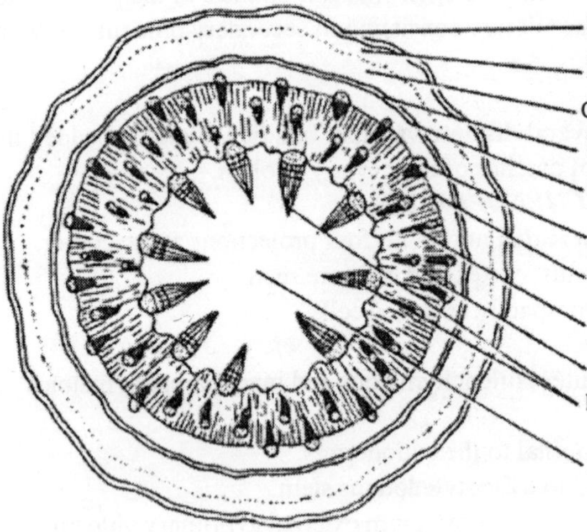

Fig. 7.22 T.S. of Chenopodium stem (diagrammetic)

13. The abnormal cambium ring produces thin walled cells towards the inner face increasing the quantity of ground tissue.
14. As a result of this the primary vascular bundles are pushed into the centre appearing like medullary vascular bundles.
15. Towards the centre of the stem is found a large parenchymatous pith.
16. The unusual features of the stem are the presence of extrastelar cambium and medullary vascular bundles (these are normal).

BIGNONIA

1. A transverse section of the stem shows the following features.
2. In the primary structure the stem shows epidermis cortex and Stele.
3. The secondary growth shows both normal and anomalous structures.
4. The outline of the section will not be exactly circular.

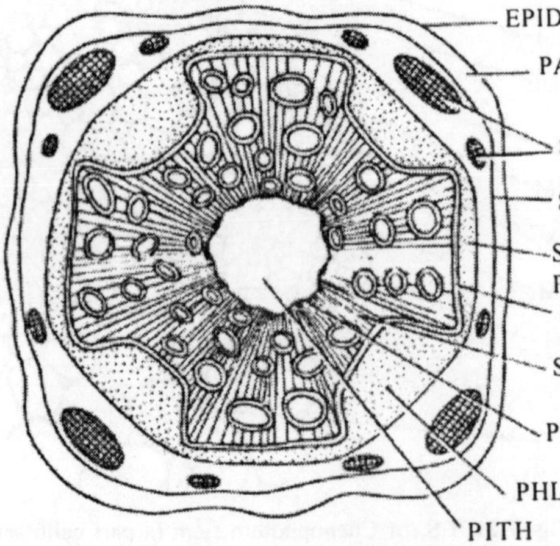

Fig. 7.23 The stem, anomalous structure. T.S. of Bignonia (dicot). detail of the structure

5. Epidermis is made up off a single layer of cells and is covered by a thick cuticle.
6. In older stems however the epidermis is replaced by the periderm.
7. The periderm is made up of Cork, cork cambium and secondary cortex.
8. The cortex is made up of the a few layers of parenchyma.
9. The endodermis is not distinct and is made up of barrel shaped cells.
10. Pericycle is represented by patches of sclerenchyma.
11. Secondary growth takes place by the activity of normal fascicular cambium.
12. The normal cambium behaves abnormally in that it cuts off more secondary phloem to the exterior and less secondary xylem towards the interior in three or four places while in other regions it cuts off more xylem towards the interior and less phloem towards the exterior.
13. As a result the secondary vascular region becomes wavey.
14. Due to this abnormal activity in many places the secondary phloem becomes 'U' shaped and projects into secondary xylem.
15. Some of the cells in the phloem become fibrous.
16. Occupying the centre region is found the pith.

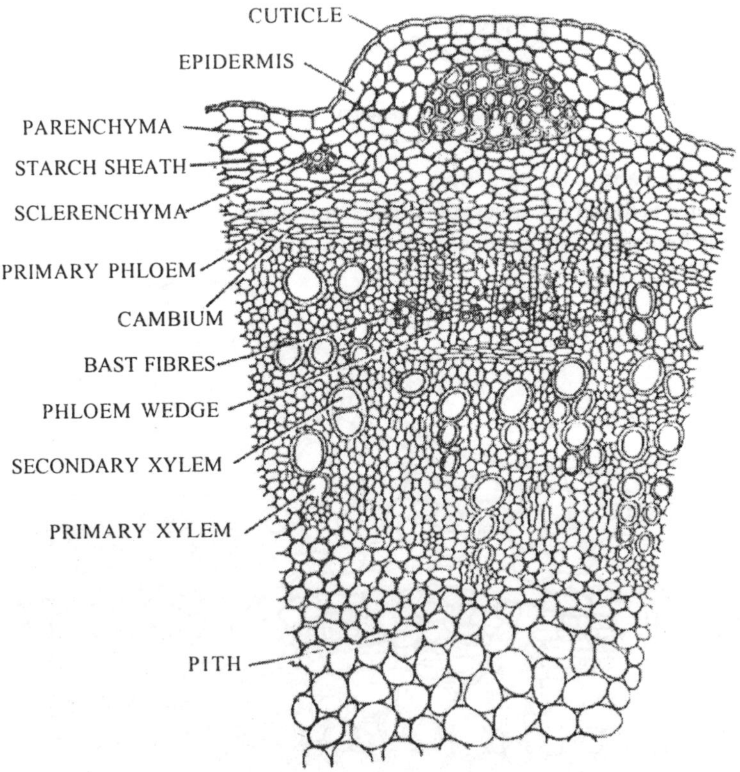

Fig. 7.24 The stem, anomalous structure. T.S. of Bignonia (dicot). diagrammetic

NYCTANTHES

1. A cross section of the stem is some what rectangular with four projections at the corner.
2. In the primary structure the usual epidermis, cortex and stele are seen.
3. Epidermis is single layered made up of compactly arranged cells.

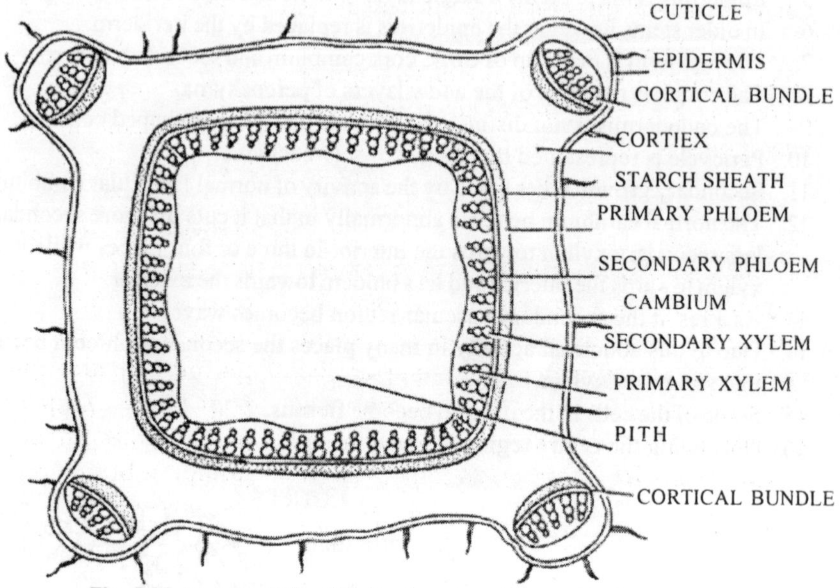

Fig. 7.25 Anatomy T.S. of Nyctanthes stem, Diagrammatic

4. Cuticle is deposited above the epidermis.
5. Next to the cortex is found few layers of collenchyma followed by a few layers of parenchyma.
6. As the stem as four ridges the exten of the parenchyma is more at these regions.

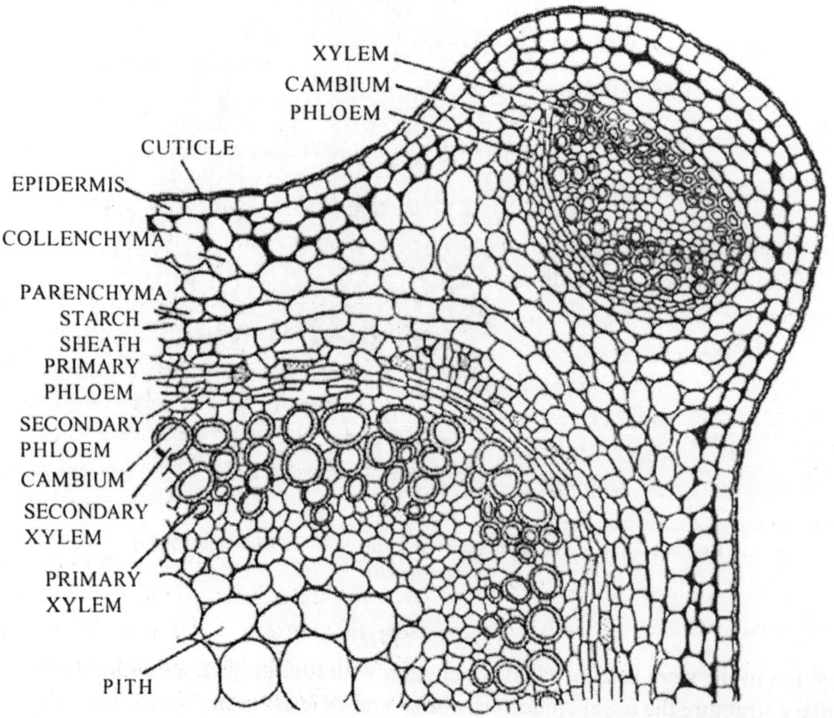

Fig. 7.26 Anatomy. T.S. of Nyctanthes stem, sector enlarged

7. Embedded in the parenchyma at the ridges are found four cortical vascular bundles.
8. These cortical vascular bundles are anomalous in origin.
9. These cortical vascular bundles are colletral and open.
10. Next to the cortex is the endodermis.
11. Endodermis is followed by pericycle and secondary vascular tissues which are normal.
12. Cambium is found between the secondary pholem and secondary xylem.
13. Occupying the secondary region is found the pith.

BOERHAVIA STEM

1. *Boerhavia diffusa* is a small semiprostrate spreading herb.
2. A transverse section of the stem shows an outermost single layered epidermis.
3. The epidermis bears a number of multicellular hairs.
4. In the primary structure, next to the epidermis internally is the hypodermis.
5. The hypodermis is made up of 2-3 layers of collenchyma which is followed by parenchymatous cortex having few chlorophyllous cells.
6. Next to the cortex is an indistinct endodermis lining externally 1-2 layered pericycle.
7. The stele consists of 3 rings of vascular bundles of which inner ring has two large centrally located medullary bundles, the middle ring is made up of 6-14 loosely arranged bundles and the outer ring has small 15-20 bundles.
8. Secondary growth is limited in the inner and middle rings.
9. As a result they slightly increase their size.
10. The inter and intrafascicular cambial strips join together to form the cambial ring.

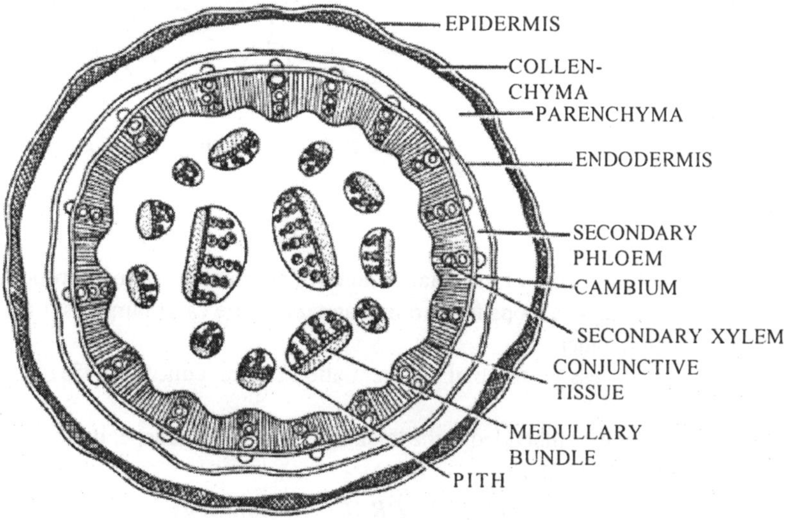

Fig. 7.27 Boerhavia diffusa. T.S. stem-diagrammetic

11. The cambial cells produce secondary xylem in the infrafascicular region and lignified conjunctive tissue in the interfascicular region on the inner side and on the outer side secondary phloem facing the secondary xylem and parenchyma opposite the conjunctive tissue.
12. After the formation of secondary tissue, the cambium ceases its activity and a new cambial ring gets differentiated by combining the secondary parenchyma cells opposite to the conjunctive tissue and the cells of pericycle outside the phloem.
13. This accessory cambium functions in a manner similar to the fascicular cambium and produces similar tissues.

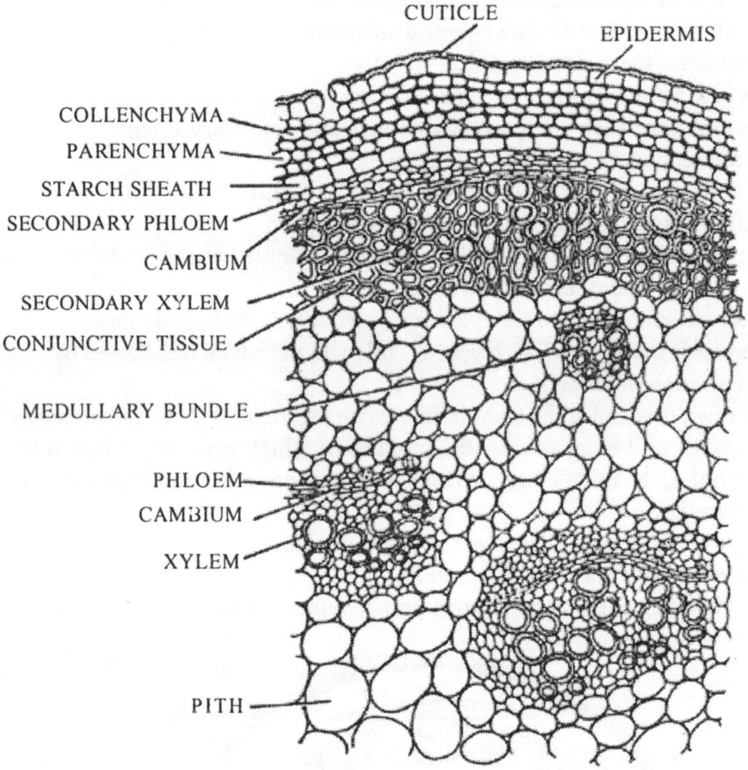

Fig.7. 28 Boerhavia diffusa. T.S. of a sector of enlarge

14. As a result of this a fourth ring of vascular bundles of secondary origin are produced.
15. Very soon this cambium also stops functioning and one more cambium gets differentiated, which also functions in a similar manner.
16. As a result of these successive cambial differentiation several concentric rings of vascular bundles are formed.
17. These vascular bundles get embedded in the thick conjunctive tissue seperated by him walled parenchyma give the appearance of growth rings.

STEM OF DRACENA (Liliaceae)

1. Generally, in monocots ther will be no secondary growth due to the lack of cambium.
2. Abnormally, secondary growth occurs in members like *Dracena, Yucca* etc.
3. The stem of *Dracena* in its primary structure consists of a normal monocotyledonous structure.
4. It has a number of conjoint, collateral and closed vascular bundles scattered in the ground tissue.
5. Secondary growth is initiated by a multilayered cambium arising outside the primary vascular bundles.
6. This abnormal cambium produces more cells towards its inner face than towards the outer face.
7. The outer derivatives develop into thin walled parenchyma whereas the inner derivatives develop partly into lignified conjunctive parenchyma and partly into secondary vascular tissues.
8. Each cambial cell divides anticlinally to form a row of four cells.

9. Further division are irregular.
10. The cells towards the periphery develop into xylem and those towards the centre develop into phloem.
11. After secondary growth, the stem of *Dracena* shows the following features.
12. Outermost layer is the epidermis.
13. Beneath the epidermis is the cork tissue.
14. Internal to this is the parenchymatous cortex which is secondary in origin.

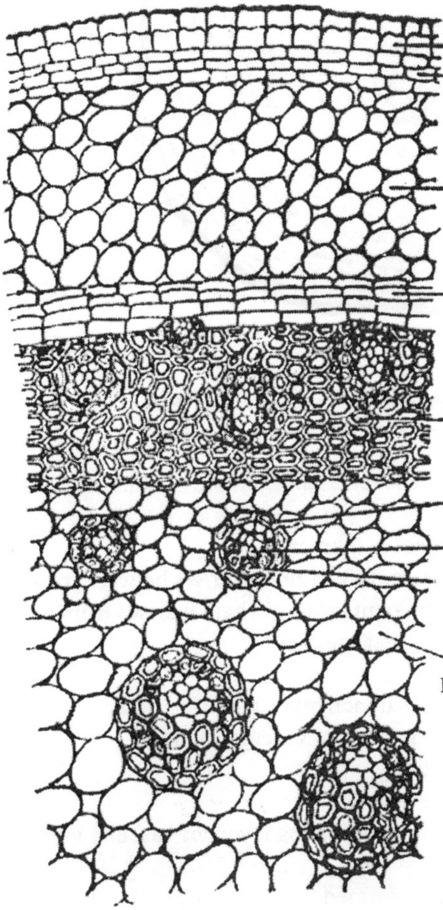

Fig. 7.29 Anomalous secondary growth..
T.S. of Dracena stem (A portion enlarged).

15. It is derived from the cambium.
16. Next to the secondary cortex is the cambium.
17. Internal to the cambium are found the secondary vascular bundles embedded in the thick walled conjunctive tissue.
18. The secondary vascular bundles are *amphivasal* (phloem surrounded by xylem on all the sides).
19. Towards the centre of the stem are found the normal primary vascular bundles.

STORAGE ROOT OF BETA *VULGARIS* (Beet root)

1. The storage roots generally represent the combination of hypocotyl and root and exhibit a great variation in their secondary structure.

Fig. 7.30 Anomalous secondary growth.,
T.S. of storage root of Beta vulgaris (ground plan).

2. The storage root of *Beta vulgaris* may be taken as an example of abnormal secondary thickening in the storage roots.
3. According to Artschwager the young root is diarach.
4. At the intiation of secondary growth, the primary cambium arises between the xylem and phloem groups, except opposite to the protoxylem and phloem groups, where it arises from the pericycle.
5. This produces the innermost ring of vascular bundles.
6. Very soon, this cambium ceases to function.
7. Second cambium arises from the phloem parenchyma outside the first ring of primary cambium.
8. The secondary cambium produces another ring of vascular bundles.
9. This also ceases to function after sometime.
10. The next cambial ring is formed in the pericycle.
11. The pericycle cells proliferate and produce alternating layers of vascular bundles and parenchyma cells of pericycle.
12. In this way a number of vascular rings are formed.
13. The vascular bundles are collateral and are separated by narrow bands of radial parenchyma, rich in reserve food.

PART TWO
EMBRYOLOGY

CHAPTER 8

FLOWER AND ORGANOGENY

Structure of a typical flower: Basically, a flower consists of three parts viz., *receptacle, sterile appendages* and *fertile appendages.*

Receptacle represents the expanded or swollen tip of the stalk(pedicel) of the flower. It is on this that the appendages of the flower are borne. The receptacle may be flat, globose or concial in shape.

The sterile appendages (accessory organs) are of two types viz., *calyx* and *corolla*. Sometimes the sterile appendages cannot be distinguished into calyx and corolla as in *Michelia*. In such cases the sterile appendages are termed *perianth lobes* or *tepals*.

Calyx is the outermost or lower most of the sterile appendages. The individual members are called sepals and they are generally green in colour. The sepals in flower may be fused or free.

Corolla is next to calyx. It is the most attractive of the floral organs. The various colours of the flowers are due to corolla. The individual members are called *Petals*. The petals in the corolla may be fused or free.

In some flowers, the sterile appendages are entirely absent. Such flowers posses only the fertile appendages and are called *achlamydeous*.

The fertile appendages are also of two types viz., *Androecium* and *Gynoecium*.

Androecium represents the collection of male reproductive structures. In arrangement, it is found next to corolla. The individual members are called *stamens*. Each stamen has three parts viz., *filament, anther* and *connective*. Filament is stalk like and it bears the anther. The portion of the filament that joins the anther is called the connective. It is in the anther (Microsporangium) that the pollen grains are formed which ultimately develop into the male gametophyte.

Gynoecium represents the collection of female reproductive structures. It is also called pistil and has three parts viz., *ovary, style* and *stigma*. Ovary is the basal swollen part of the gynoecium and is made up of *carpels* (megasporophylls). Within the carpels and attached to the plaenta are found the *ovules* (megasporangia). It is within the ovule, the embryo sac (female gametophyte) develops. Further developments i.e., fertilization, endosperm and embryo also take place in the ovule. Style is a stalk that arises from the ovary and it facilitates the growth of the pollen tube. The style terminates in a stigma which serves as a lodging for pollen grains to settle and germinate.

Organogeny: The development of floral parts is called organogeny. The floral primordium arises on the inflorescence axis (peduncle). First, bract arises as a lateral projection in the axil of which the flower originates. The floral shoot subsequently becomes flattened and forms the receptacle. The receptacle at this stage shows a group of meristematic cells with dense cytoplasm and conspicuous nucleus. The various floral organs originate from the receptacle either *spirally* or *cyclically*. In the spiral type, the recptacle shows a tendency towards indeterminate growth in a linear fashion and assumes a dome like shape. This type is regarded primitive and is seen in *Magnoliaceae*. In the cyclic type, the floral parts arise in distinct whorls. This is seen in the majority of flowering plants. An intermediate condition where the calyx and corolla are spiral, with cyclic androecium and gynoecium is seen in the *amentiferae* group.

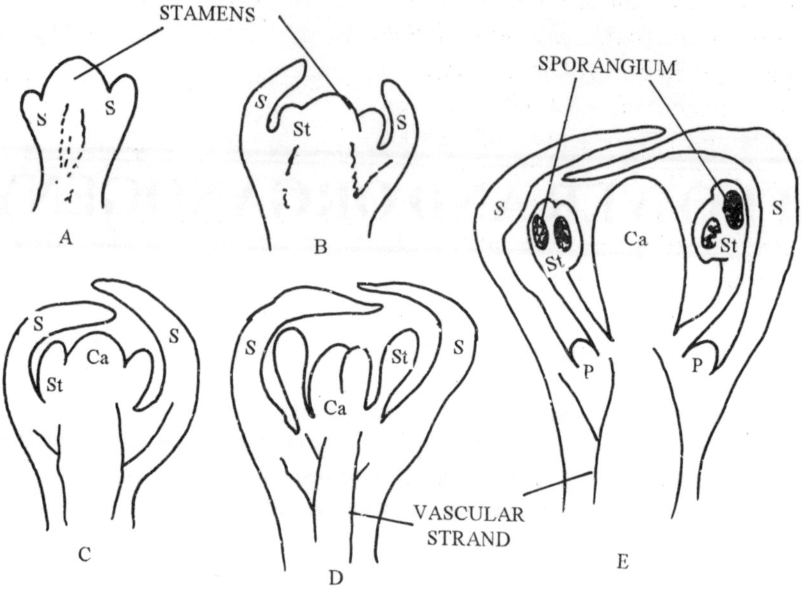

Fig. 8.1 Flower

Stages in the organogenesis of a typical flower

(S= sepal; St= stamen; P = petal ; Ca = carpel)

In the axil of the bract a conspicuous bulb like outgrowth arises and constitutes the floral primordium. By further growth, the floral primordium elongates and gives rise to two distinct regions viz., *pedicel* (stalk) and the *receptacle*. It is on the latter that the floral organs arise. In sessile flowers, the pedicel is suppressed.

The floral parts viz., K (calyx), C(corolla), A(androecium and G(gynoecium) arise in acropetalous sucession on the floral receptable. First to arise is calyx and it is followed by C.G. and A. Variations in the organogeny are met with in several members. In *Rubus* the development is K, A, G, and C while in *Rudbeckia* it is C, A, G and K.

CHAPTER 9

ANTHER AND MALE GAMETOPHYTE

Development of the microsporangium: A very young anther in a cross section shows a homogenous mass of cells bound by an epidermis. During further growth the anther assumes a four lobed shape. In each lobe a few of the hypodermal cells (sometimes one) become differentiated from the others by their prominent nucleus and dense cytoplasm. These are the *arhesporial cells*. In majority of the members the archesporium consists of several vertical rows of cells which appear as a plate of cells in a transverse section.

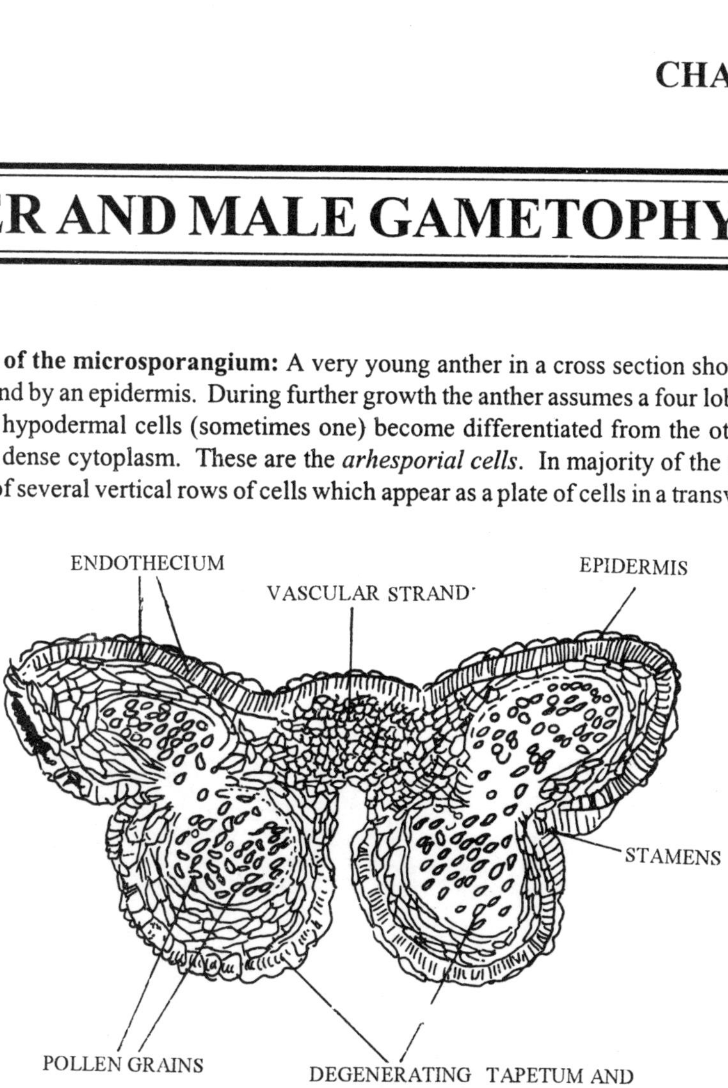

Fig. 9.1 **Microsporangium**

Transverse section of a mature anther at the time of dehiscence

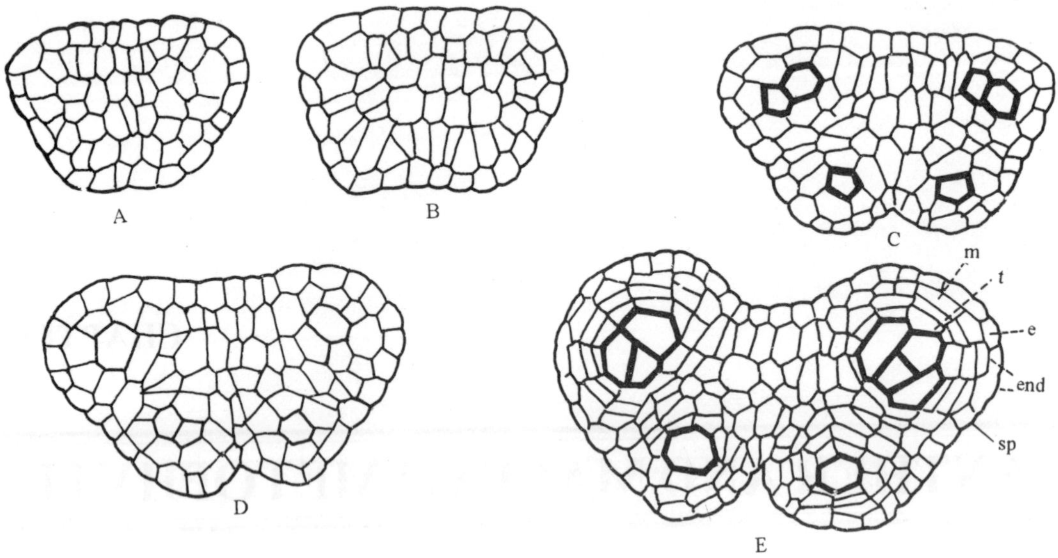

Fig.9.2 **Microsporangium**

Developmental stages of the anther (e = Show labelling on the diagram itself (redrawn from Warning, 1873)

The archesporail cells divide periclinallay cutting off *primary parietal cells* towards the epidermis and *primary sporogenous cells* towards the interior. The primary parietal layer divides to form two *secondary parietal layers*. Of these, the inner one distinguishes itself into *tapetum* while the outer divides again periclinally to form two layers of cells. Of these, the outer one forms the *endothecium* while the inner one forms the *middle layer*. The sporogenous cells develop into microspore mother cells.

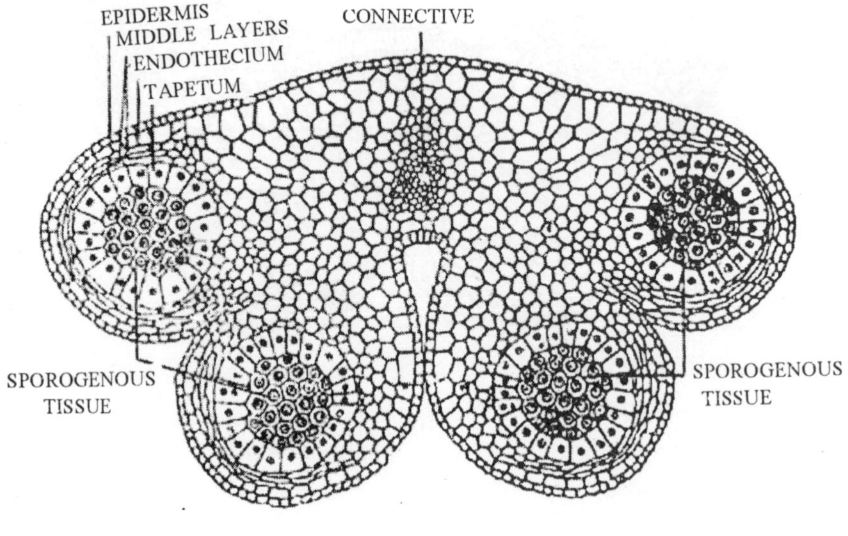

Fig. 9.3 **Microsporangium**

Transection of a fully developed tetrasporangiate anther

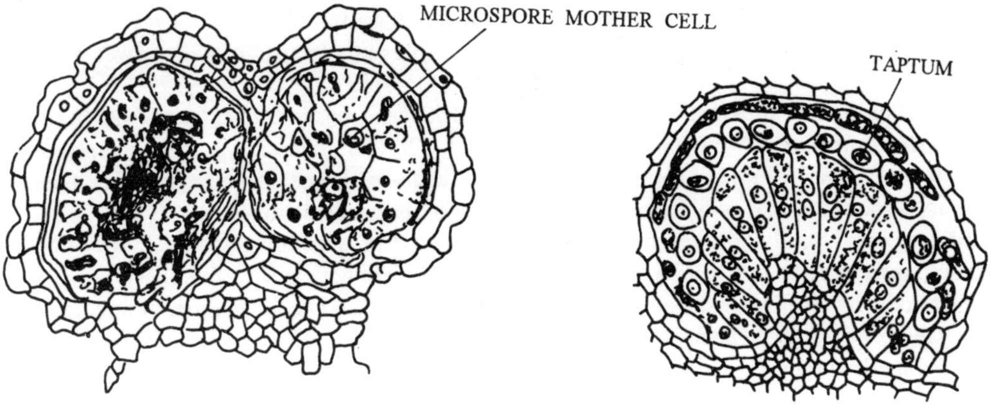

Fig. 9.4 Microsporangium. T.S. of Anther showing PMCs and tapetum

Anther Wall: A transverse section of a fully developed anther shows a group of microspore mother cells surrounded by a tapetum, a middle layer, an endothecium and an epidermis.

Epidermis: It is the outermost wall layer of the anther. During anther development, it undergoes anticlinal divisions to keep pace with the enlargement of the inner tissues. The epidermal cells become greatly stretched and flattened. In a mature anther, only the remanants of the epidermis are seen.

Endothecium: The layer of cells lying immediately below the epidermis constitute the endothecium. Endothecium attains its maximum development at the time of the anther dehiscence. Cells of this layer

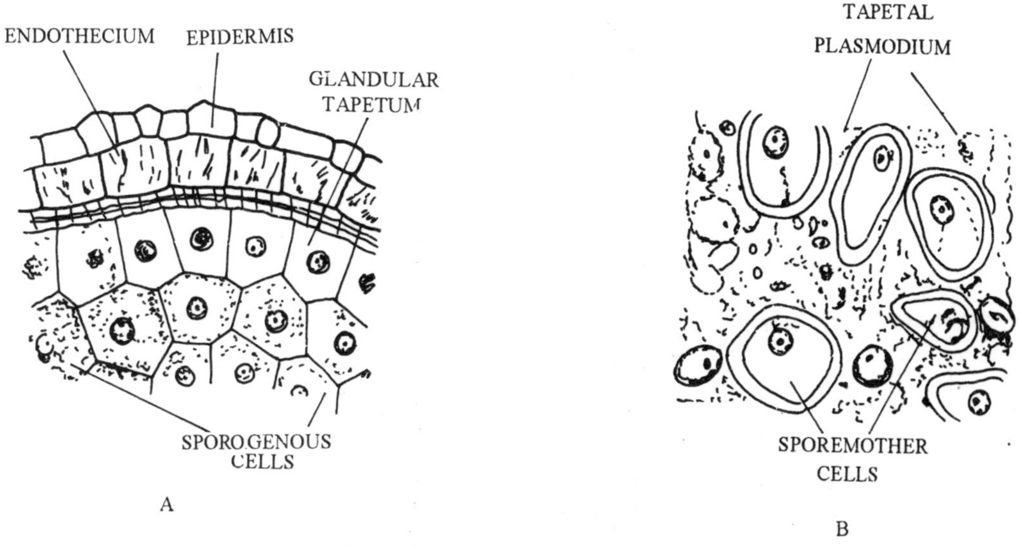

Microsporangium

Fig. 9.5 Types of Tapteum A. Glandular, B. Amoeboid

develop fibrous bands of callose which arise from the inner tangential walls (rarely from the radial walls) and run outward and upward ending near the outer wall of each cell. The outer tangential walls however remain thin.

These thickenings are apparently associated with the dehiscene mechanism of the anther. In certain members of *Hydrocharitaceae*, fibrous thickenings are absent. On the other hand in some members like *Argemone, Crinum* etc., one or more of the middle layers may acquire fibrous thickenings. In members of *Orobranchaceae*, some epidermal cells themselves may develop fibrous thickenings.

Middle layers: Usually only one middle layer is present. But, depending upon the type of anther wall development there may be one (dicotyledonous and monocotyledonous types), two (basic type) or no (reduced type), middle layers. Occassionally as in *Melastoma malabathricum*, the number of middle layers may be seven. Cells of the middle layer lack the capacity to divide anticlinally and hence the tissue cannot adjust itself to the mutiplying and expanding sporogenous cells within the sporangium. As a result, it gets crushed and cannot be seen in the mature anther. The formation of middle layers is regarded as a relic feature. In some species, on the other hand (*Ranunculus, Lillium* etc.), the middle layers persist and act as storage centres for starch and other reserves which get mobilised during later development of the wall.

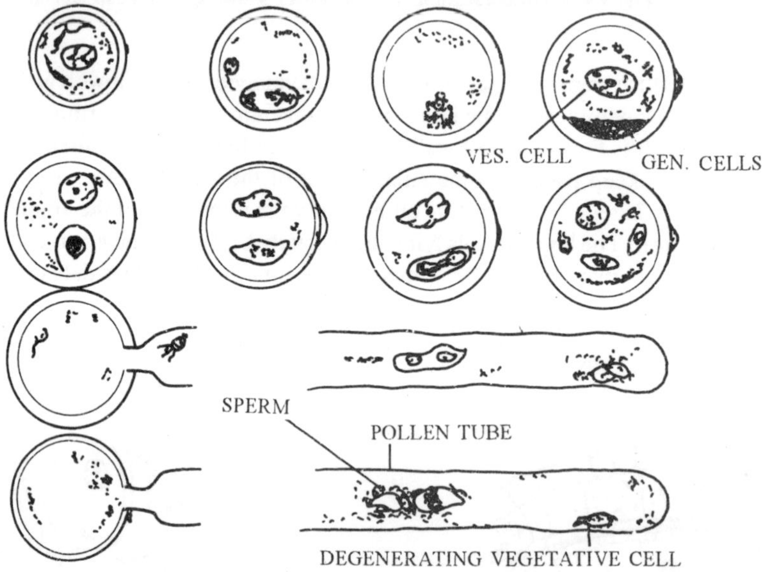

Fig. 9.6 Stages in the development of Male Gametophyte

Tapetum: This is the innermost wall layer of the anther and is of considerable physiological significance because the food material to the sporogenous cells must pass through it. Typically the tapetum is single layered with cells having dense cytoplasm and a prominent single nucleus.

The tapetum is usually derived from the secondary pareital layer, though there are reports of its derivation from the sporogenous tissue (*Antirrhinum majus*). Based on the behaviour of the cells at the time of sporogenesis, the tapetum is divided into two types viz., *glandular* and *amoeboid*.

Glandular tapetum is also known as the *secretory type* of tapetum. In this type, the cells of tapetum remain unaltered throughout the development of the microspores. Nutritional materials are contributed to the anther sac by secretions from the inner faces of the cells until their total breakdown at the time of pollen maturity. In a great majority of Angiosperms (87%) the glandular tapetum is seen.

Amoeboid tapetum is also called *invasive* or *plasmodial* type of tapetum. Here, very early during microspore development, the tapetal cells breakdown, and the protoplasts 'wander' into the anther cavity encircling the pollen mother cells. The penetration of protoplasts may take place during meiotic prophase or may be delayed until the tetrad stage. This type of tapetum is seen in *Blainvillea, Emilia, Senecio* etc.

The close association of the tapetal cytoplasm with the microspores suggests that the former must be exerting a selective influence on the transport of material into the pollen.

Sporogenous tissue: The sporogenous cells may or may not divide to increase their number. Ultimately they round off and develop into *microspore mother cells* or *pollen mother cells* (PMC).

The PMCs undergo reduction division to produce tetrads of haploid microspores. At the initiation of meiosis, the PMCs get separated from the tapetal cells. Cytokinesis or wall formation takes place at the end of meiosis and is of two types. In the *Successive type*, a dyad is formed soon after meiosis. Wall formation is by a centrifugal plate of cellulose, which starts in the centre and extends laterally. The two cells of the dyad, divide next and four spores are formed. In the *Simultaneous type* the first meiotic division is not followed by wall formation, as a result a binucleate cells is formed. Both the nuclei divide and at the end of the meiosis cytokinesis takes place by centripetal furrowing.

As stated already, after meiotic divisions microspore tetrads are formed. These are of the following types viz., *tetrahedral* (three in one plane and one in the other), *isobilateral* (all the spores in one plane), *decussate* (spores arranged crosswise), *rhomboidal* (in the form of a rhomboid), *Linear* (spores are one above the other) and 'T' *shaped* (Two spores superposed and the other two juxtaposed).

Eventually the microspores separate from the tetrad and are set free in the anther loculus.

Development of the male gametophyte: Microspores represent the first stage of the male gametophyte. After their separation from the tetrad, microspores are called pollen grains.

To start with, the pollen grain has its single nucleus situated at the centre. The first division results in the formation of two unequal cells. The larger one is the *vegetative cell* and the smaller one is the *generative cell.* The vegetative cell eventually forms the pollen tube. The generative cell is attached to the wall of the pollen grain in the beginning, but later it comes to lie freely in the cytoplasm of the vegetative cell.

CHAPTER 10

OVULE AND FEMALE GAMETOPHYTE

The ovule or megasporangium basically consists of two parts viz., *nucellus* and the *integument*. It is attached to the palcenta of the ovary by means of a short stalk called the funiculus. In a mature ovule the integument encloses the nucellus completely except for a small opening at one end called the *micropyle*. It is through this opening that the pollen tube gains entry into the ovule. The basal region of the ovule (region opposite the microphyle) is called the *chalaza*. Embedded in the nucellus is the female gametophyte or the embryo sac.

Mature ovules are classified into the following types, based principally on the position of the microphyle with reference to the funiulus.

Orthotropous: Here, the ovule is erect and the microphyle and the funiculus lie in a vertical line. This is seen in *Polygonaceae, Urticaceae, Piperaceae* etc.

Anatropous: The ovule takes a 180° turn so that the micropyle and the funiculus come to lie close to each other. This seen in almost all members of the sympetalae.

Hemianatropous: In this type, the ovule takes a 90° turn so that the microphyle is at right angles to the funiculus. This is seen in *Malpigiaceae, Primulaceae* etc.

Campylotropous: Here the ovule takes a turn a little less than 90° and is curved. This is seen in *Resedaceae, Leguminosae* etc.

Amphitropous: The curvature of the ovuie is so pronounced that it also affects the embryo sac which becomes horseshoe shaped. This is seen in *Alismaceae, Butomaceae* etc.

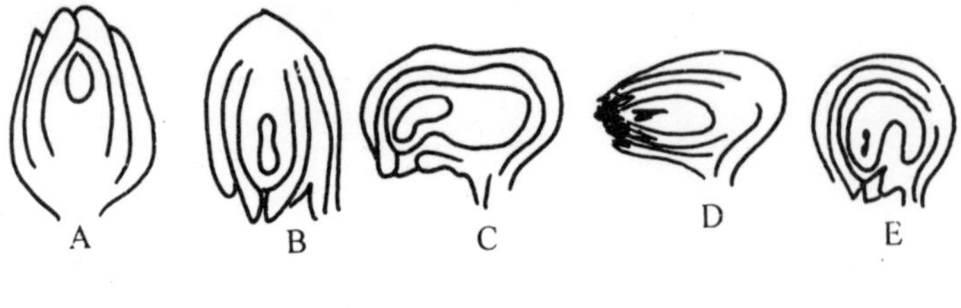

Fig. 10.1 Types of Ovules A. Orthotropous, B. Anatropous, C. Hemianatropous, D. Amphitropous, E. Campylotropous

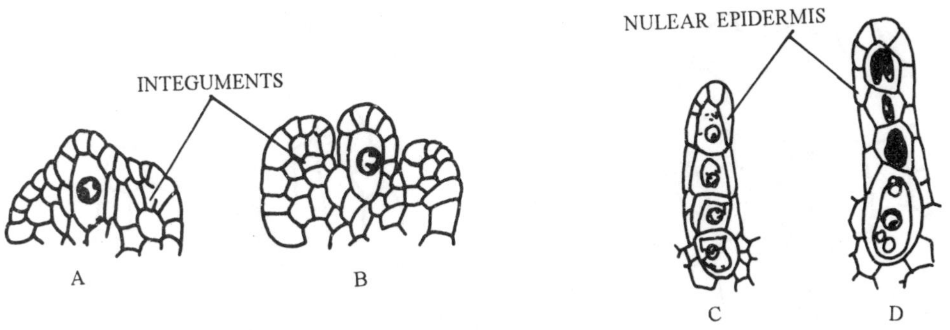

Fig. 10.2 **Female gametophyte**

Megasporogenesis in a tenuinucellate ovule (redrawn from Johri and Singh, 1959)

Circinotropous: Here the ovule first becomes anatropous. But subsequently the curvature once again lifts the micropyle to its original position so that the ovule takes a 360^0 turn. This type of ovule is seen *Plumbaginaceae* and *Cactaceae*.

Integuments: These constitute the envelopes of the ovule. In some cases as in *Loranthaceae*, the integuments are absent. Such ovules are said to be *atagmic*. Similarly ovules with one integument are called *unitegmic* and those with two, are called *bitegmic*. Rarely a third integument called *aril* is present in some members of *Ficoidaceae* (*Trianthema decandra*).

In bitegmic ovules, the micropyle may be formed only by the outer integument or only by the inner integument or by both the integuments.

Nucellus: Depending on the extent of the nucellus, the ovules are classified into two types viz., *tenuinucellate* and *crassinucellate*. In the former, the nucellus is scanty and the archesporium that arises in the hypodermal region directly functions as the megaspore mother cells (MMC), while in the latter the archesporium cuts off a few parietal cells and then functions as the MMC, so that the nucellus is bulky. A third category - *Pseudocrassinucellate* is proposed by Davis (1966), to include those ovules where the nucellar epidermis builds up the parietal tissue, while the archesporium directly functions as the MMC. But, here also the MMC becomes deep seated as in crassinucellate ovules.

MEGASPOROGENESIS

A single hypodermal cell arises in the nucellus and it differentiates itself as the *archesporial cell* by its larghe size, conspicuous nucleus and dense cytoplasm.

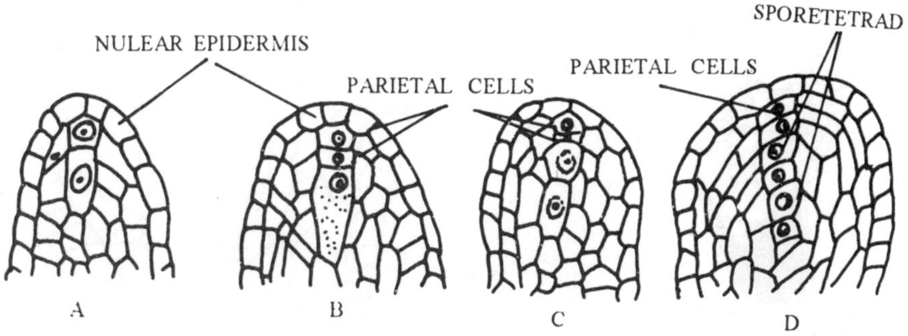

Fig. 10.3 Megasporogenesis

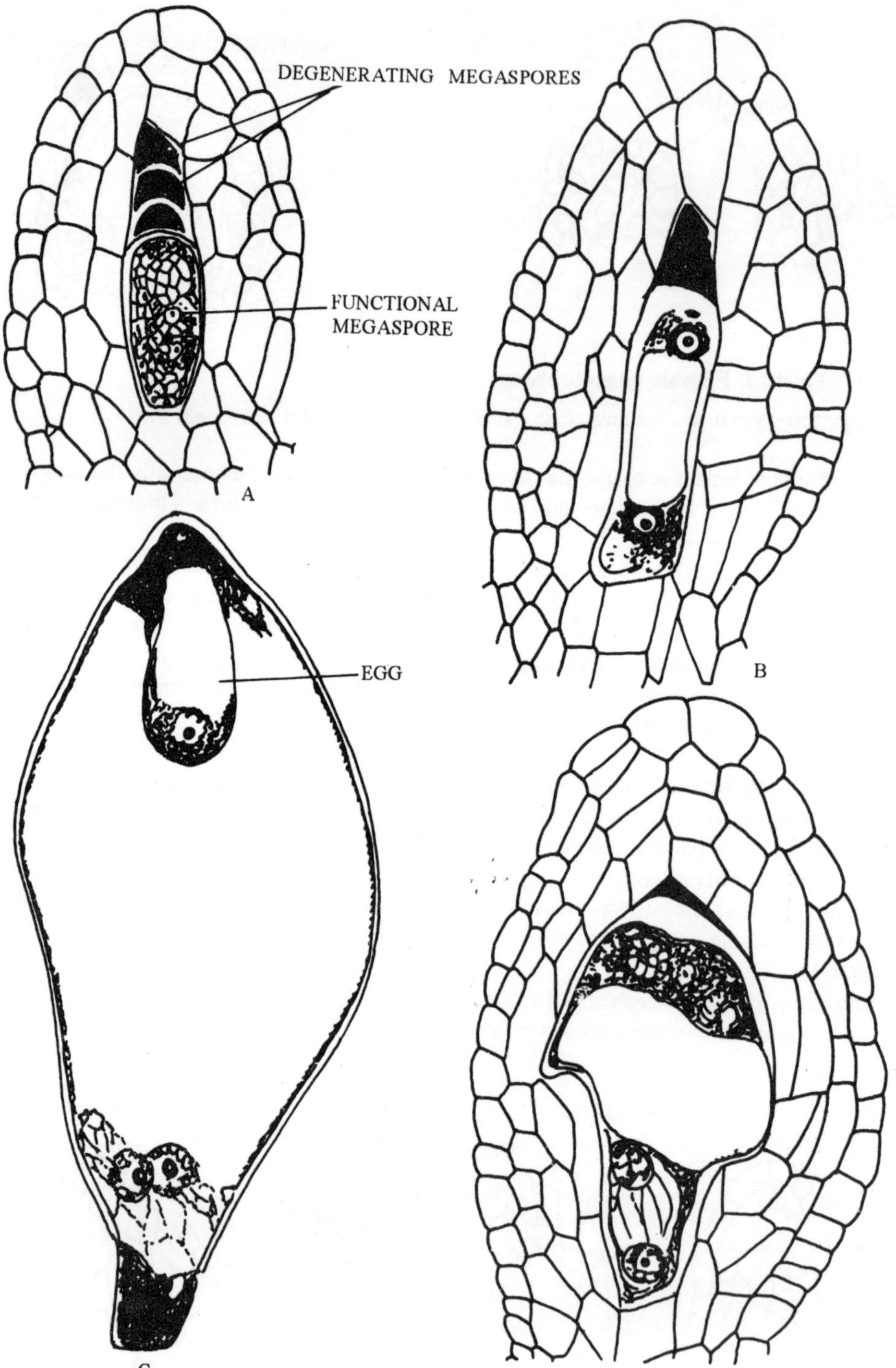

Fig. 10.4 Female gametophyte
Megasporogenesis in Hydrilla verticellata (redrawn from Maheshwari, 1950)

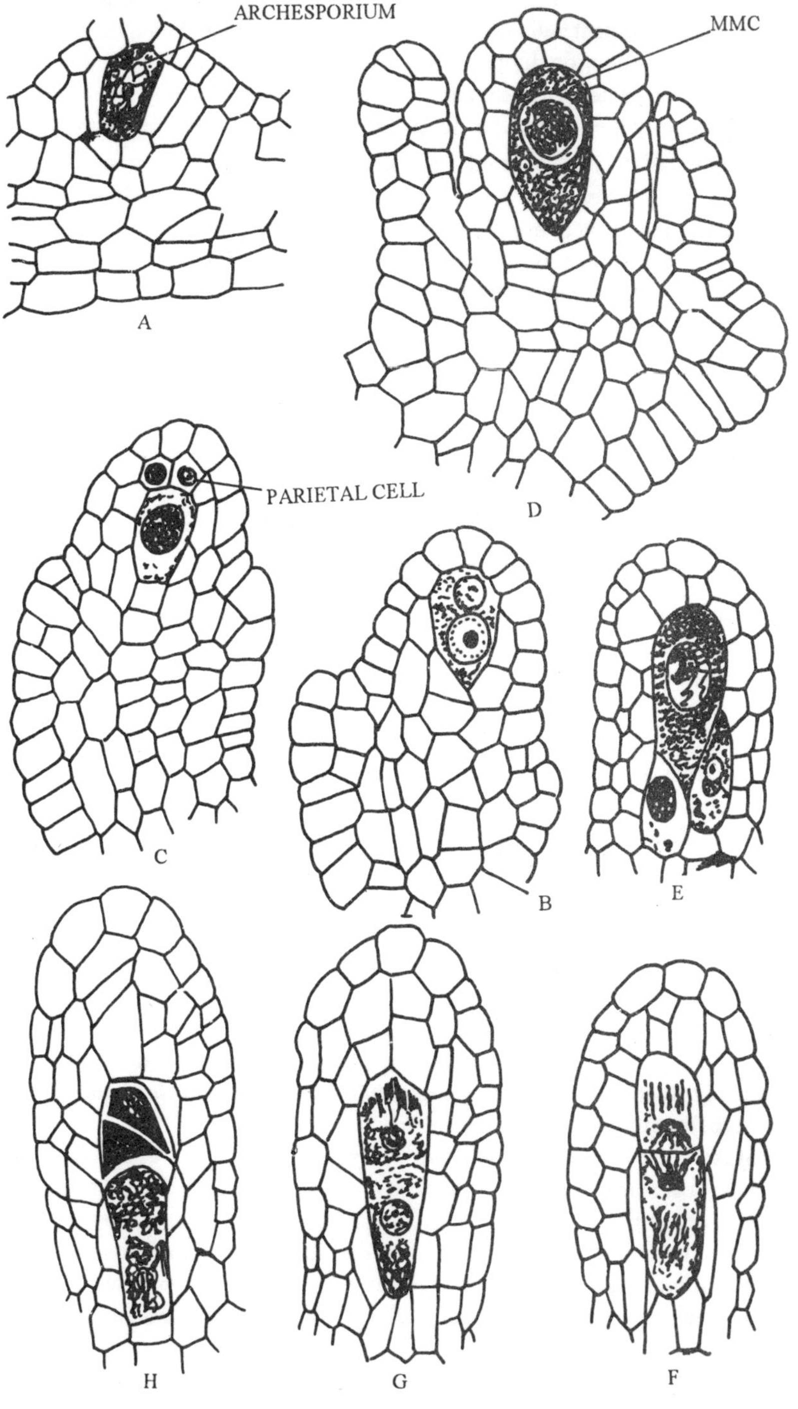

Fig. 10.5 Female gametophyte
Development of polygonum type of embryo sac in *Hydrilla verticellata* (redrawn from Maheshwari, 1950)

In tenuinucellate and pseudocrassinucellate ovules the archesporial cell directly functions as the MMC, while in crassinucellate forms it cuts off periclinally a parietal cell and then becomes the MMC.

The MMC undergoes reduction division. After the first meiotic division a dyad is formed by the laying down of a transverse wall. The second division is also transverses resulting in the formation of a linear tetrad of megaspores. Fequently the micropylar dyad divides vertically resulting in the formation of a 'T' shaped tetrad. Rarely the tetrad may be inverted T shaped (⊥), as in some Onagraceae. An isobilateral or tetrahedral tetrad of megaspores is very rare and it is reported only as an abnormality.

Of the four megaspores formed, usually the chalazal one is functional and it is called the functional megaspore. Occasionally the micropylar mehaspore may be functional. The three non functional megaspores degenerate and can be seen as black spots in a section of the ovule.

DEVELOPMENT OF THE FEMAL GAMETOPHYTE

The functional megaspore by further development forms the embryo sac (female gametophyte). Based on the number of megaspore nuclei that take part in the construction of the embryo sac, the embryo sac of Angiosperms is divided into three main types viz., *monosporic, bisporic* and *tetrasporic*. In the first one, only one spore gives rise to the embryo sac; in the second two spores and in the third all the four spores participate in the embryo sac development.

Monosporic types: The monosporic embryo sacs fall under two categories viz., **Polygonum type** and **Oenothera type**. The mature embryo sac of the former has eight nuclei while that of the later has only four nuclei.

Polygonum type: This is the commonest among flowering plants occuring in about 82% of the members. Hence it is often called the 'Normal type'. This was first discovered in *Polygonum divaricatum* (Strassburger, 1879). The MMC undergoes reduction division to form a degenerate and the lowermost (chalazal) one becomes funtional. This enlarges and its nucelus divides mitotically to form two nuclei. These two nuclei migrate to the opposite poles and a vacuole appears in between. This is the 2 - nucleate stage. The two nuclei divide again to form the four nucleate stage. All the four nuclei divide once again to form 8 nuclei (8 - nucleate stage) which are distributed at the opposite poles in two groups of four each.

After the 8-nucleate stage, there is no further division and the organisation of the embryo sac takes place. Of the four nuclei at the micropylar quartret, one migrates to the centre. Similarly from the chalazal quartret also one nucleus migrates to the centre of the embryo sac. The two nuclei fuse in the centre and form the diploid **secondary nucleus**. The three nuclei at the chalazal region form three uninucleate cells called the **antipodal cells**. The three nuclei at the micropylar region constitute the **egg apparatus**. The apparatus consists of two lateral **synergids** and a **median egg**.

In an organised polygonum type of embryo sac, 7 cells are present, namely, two synergids, one egge, a central cell consisting of the diploid secondary nucleus and three antpodal cells.

As stated already, there will be an egg apparatus, a secondary nucleus and three antipodal cell (except in *Oenothera* type) in the embryo sac.

Egg apparatus: The egg apparatus has two synergids and an egg. The synergids are elongated cells. They may be smooth or sometimes may be *hooked* or *beaked*. In some cases, the upper part of the cell is occupied by a *filiform apparatus*.

Electron microscopic studies have revealed that the filiform apparatus consists of a mass of inger like projictions of the wall into the cytoplasm. The lower region of the synergid has a large vacuole. The synergids are generally ephemeral, degenerating soon after fertilisation. But sometimes they may persist and become haustorial.

The egg is median and has a common wall with the synergids. The micropylar end of the cell has a large vacuole while the nucleus is situated at the lower end.

Central cell: This is the cell which consists of the secondary nucleus. Sometimes the two polar nuclei may not fuse for a long time and remain as free nuclei. It is from the secondary nucleus that the endosperm develops after fertilisation.

Antipodals: These are three in number and uninucleate. Sometimes they may be only two in which case one of them would be binucleate. The antiopodals generally degenerate soon after fertilisation. But frequently they may persist and become haustorial. Persistent antipodals are common in the tribe *Heliantheae* of *Asteraceae*.

CHAPTER 11

ENDOSPERM

Endosperm is the chief nutritive tissue present in the embryo sac of Angiosperms. In Gymnosperms also there is an endosperm where the female gametophyte itself provides nutrition to the developing embryo. In Angiosperms however, the embryo sac is extremely reduced and by itself is incapable of supporting the embryo. Hence the need for endosperm arises. The origin of the Angiospermous endosperm is unique. It is formed by the fusion of three nuclei, of which two are female in origin and one is male. The resultant product i.e., primary endosperm nucleus (PEN) develops into the endosperm. Generally the endosperm is triploid, but the ploidy may vary as in *Oenothera* (diploid), *Peperomia* (polyploid) etc.

There are some exceptional Angiosperms where there is no endosperm formation. These are *Orchidaceae, Podostemaceae,* and *Trapaceae.*

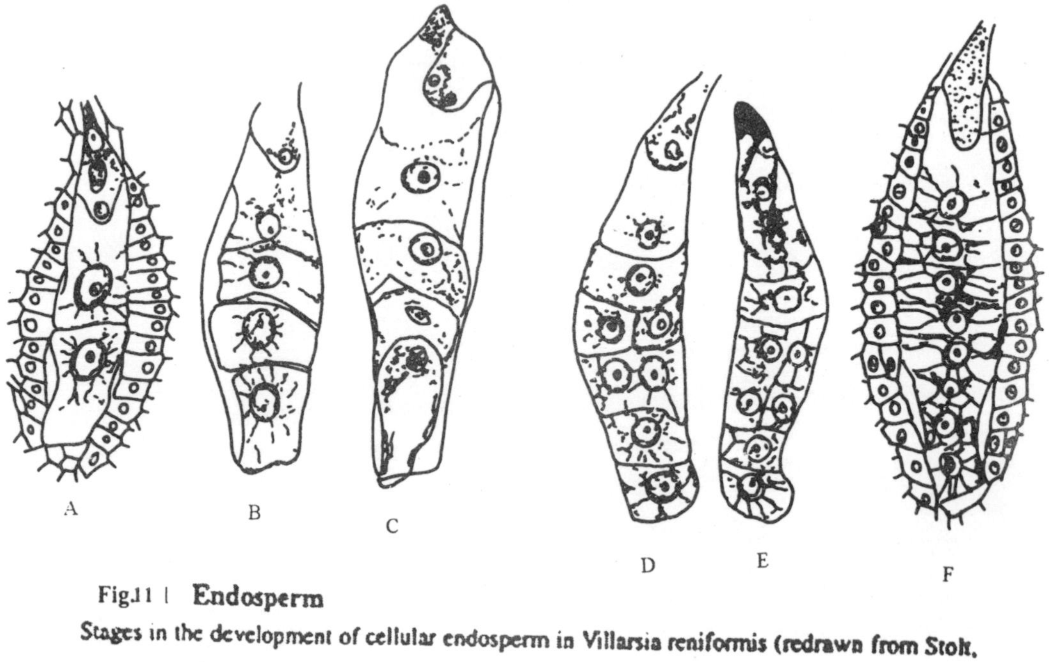

Fig.11 I **Endosperm**

Stages in the development of cellular endosperm in Villarsia reniformis (redrawn from Stolt,

The endosperm produced may be scanty or bulky. In the former, (exalbuminous) in the mature seeds there will be no endosperm as it is completely consumed by the developing embryo. In the latter (albuminous) the mature seed has plenty of endosperm and it is the nutritive tissue in the seed.

TYPES OF ENDOSPERM

Based on the behaviour of the PEN, the endosperm is divided into three main types. These are *Nuclear, ab initio Cellular* and *Helobial*. Of these, nuclear endosperm is found in 161 families, while cellular is seen in 72 families and Helobial is seen in 17 families.

Nuclear endosperm: The PEN undergoes a few free nuclear divisions, without any wall formation. As a result, lying in the cavity of the central cell, a few to several thousand nuclei are formed. The free nuclear stage may remain until it is consumed by the embryo (*Limnathes*) or it may become cellular at a later stage. In the latter case cell wall formation takes place centripetally. The entire endosperm may become cellular or a part of it may become cellular. In *Phaseolous*, the endosperm, only around the embryo becomes cellular, while the rest of it is free nuclear.

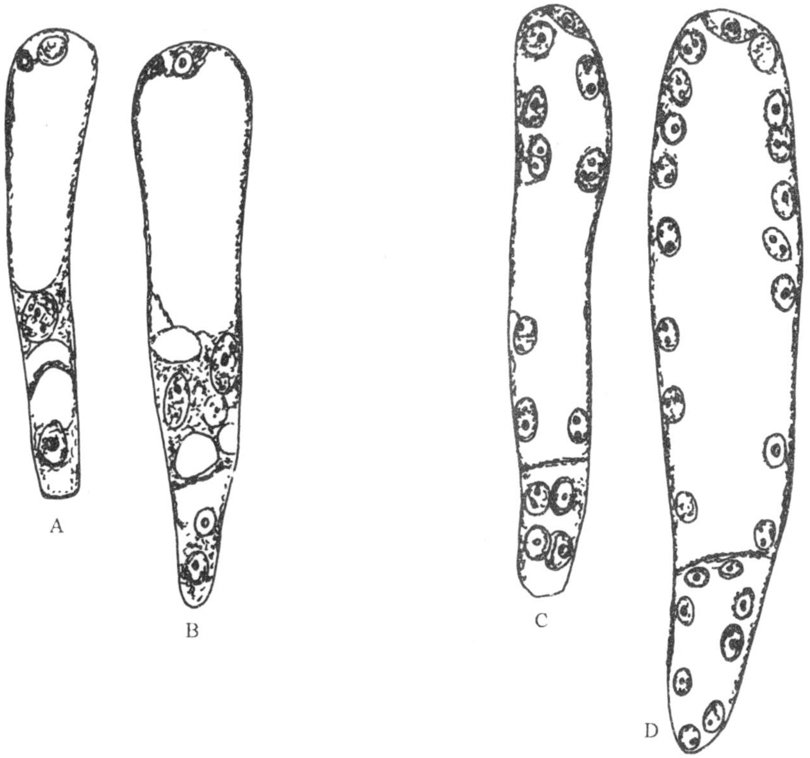

Fig. 11.2 Endosperm
Stages in the development of Helobial endosperm

ab initio Cellular: In this type, the first division of the PEN is always followed by wall formation, as a result two endosperm cells are formed. The first divisional wall may be parallel or perpendicular to the long axis of the embryo sac. Further divisions are also cellular and as a result the embryo sac gets filled by a cellular tissue.

Haustoria are very frequent in cellular endosperm also. These may arise either from the chalazal region or from the micropylar region or from both. Micropylar haustoria are seen in *Impatiens royeli* and *Hydrocera triflora* and many others. Chalazal haustoria are seen in *Magnolia obovata, Iodina rhombifolia* etc. In *Utricularia*, both chalazal and micropylar haustoria are seen.

Helobial endosperm: This is seen mostly in monocotyledons. This type of endosperm is intermediate between the nuclear and cellular types. The PEN migrates to the chalazal region of the embryo sac where it divides to form a large micropylar chamber and a small chalazal chamber. In the micropylar chamber, divisions are always free nuclear and cell wall formation, if any, starts very late. In the chalazal chamber, the nucleus remains undivided or divides only a few times. In some cases (*Phylidrum languginosum*), the chalazal chamber also becomes cellular.

CHAPTER 12

EMBRYOGENY

DICOTYLEDONOUSEMBRYO

There are about 5 types of embryo development found in dicotyledons. We will study a typical type viz., *Crucifer type*. Crucifer type of embryo is seen in the family *Cruciferae*. It is worked out in great deatail in *Capsella bursa-pastoris*. The foregoing description of embryogeny is based on the study of this plant.

The first division of the zygote is transverse and it results in the formation of a basal cell *cb* and a terminal cell *ca*. *ca* divides vertically while *cb* divides transversely to form a proembryonal tetrad which is 'T' shaped. Each of the two terminal cells (derivatives of *ca*) divide by a vertical wall at right angles to the first, resulting in a *quadrant*. The quadrant cells undergo a transverse division to form the *octant*. Of these cells, lower four give rise to the *stem tip* and *cotyledons*, while the upper four give rise to the *hypocotyl*. All these eight cells, next divide periclinally to cut off 8 outer cells which form the *dermatogen*, while the inner ones give rise to the initials of the *periblem* and *plerome*.

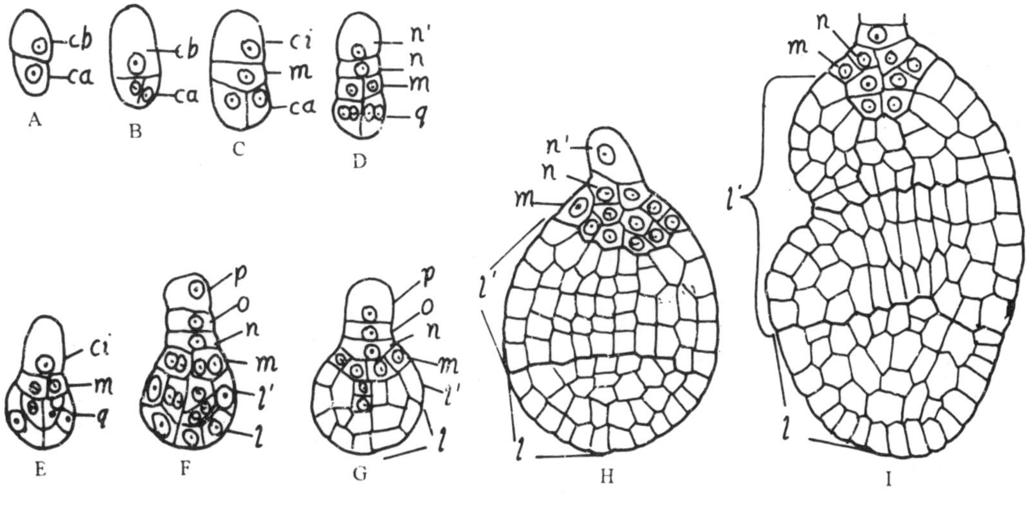

Fig. 12.1 **Embryo**

Development of embryo in Luzula forsteri (redrawn from Soueges, 1923)

Meanwhile, the two deravatives of *cb* (produced by a transverse division) namely *ci* and *m* undergo a series of transverse divisions to form a row of 6-10 *suspensor cells*. The terminal most cell of the suspensor becomes large and vesicle like and probably serves as a *haustorium*. The lowermost cell of the suspensor (*h*) functions as the *hypophysis*. The hypophysis cell divides to form two tiers of four cells each. Of these, the lower tier forms the root cortex while the upper gives rise to the root cap and the root epidermis.

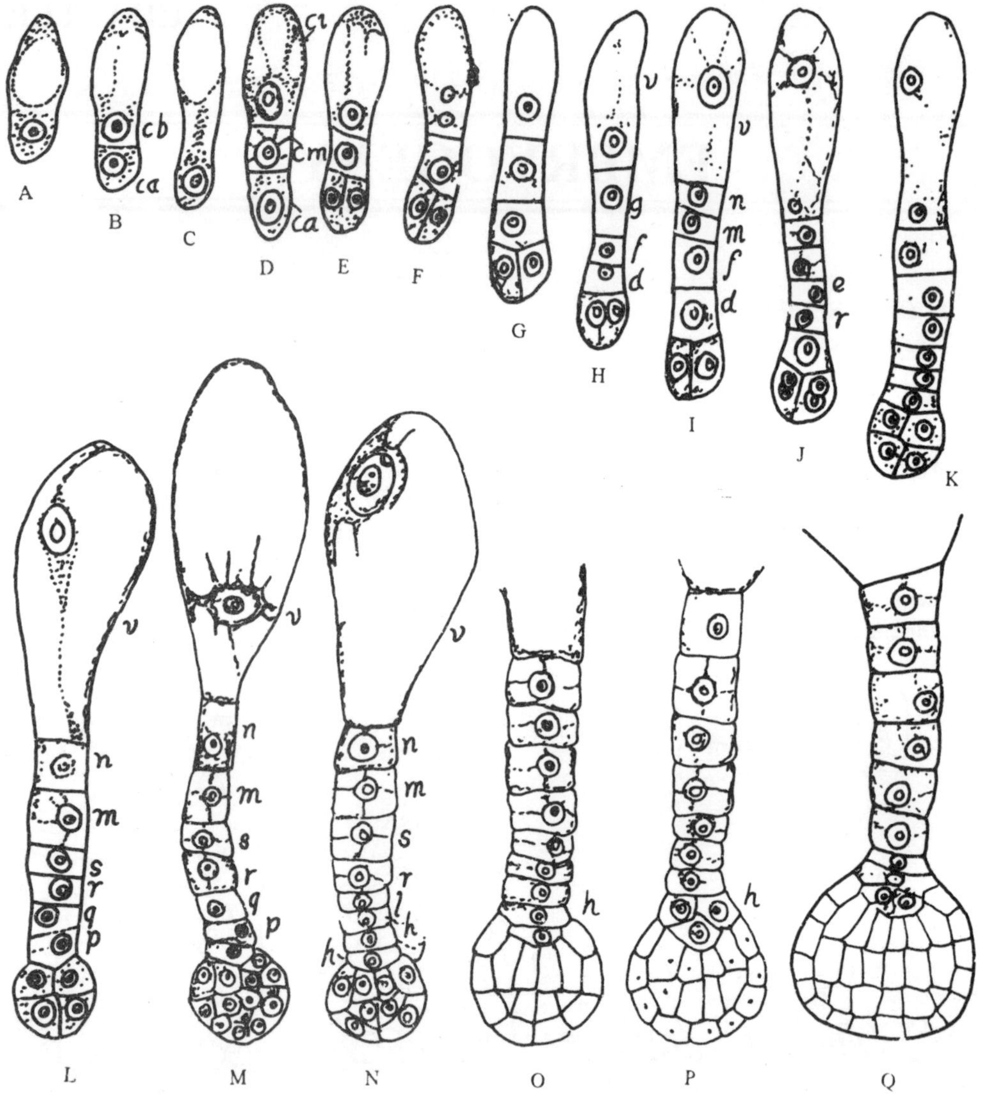

Fig. 12.2 Embryo

Early stages of embryogeny in Capsella bursa - pastoris (redrawn from Soueges, 1914, 1919)

At this stage, further growth in the embryo proper results in the formation of two cotyledons. The embryo now assumes a more or less cordate shape. The hypocotyl and cotyledons increase their size.

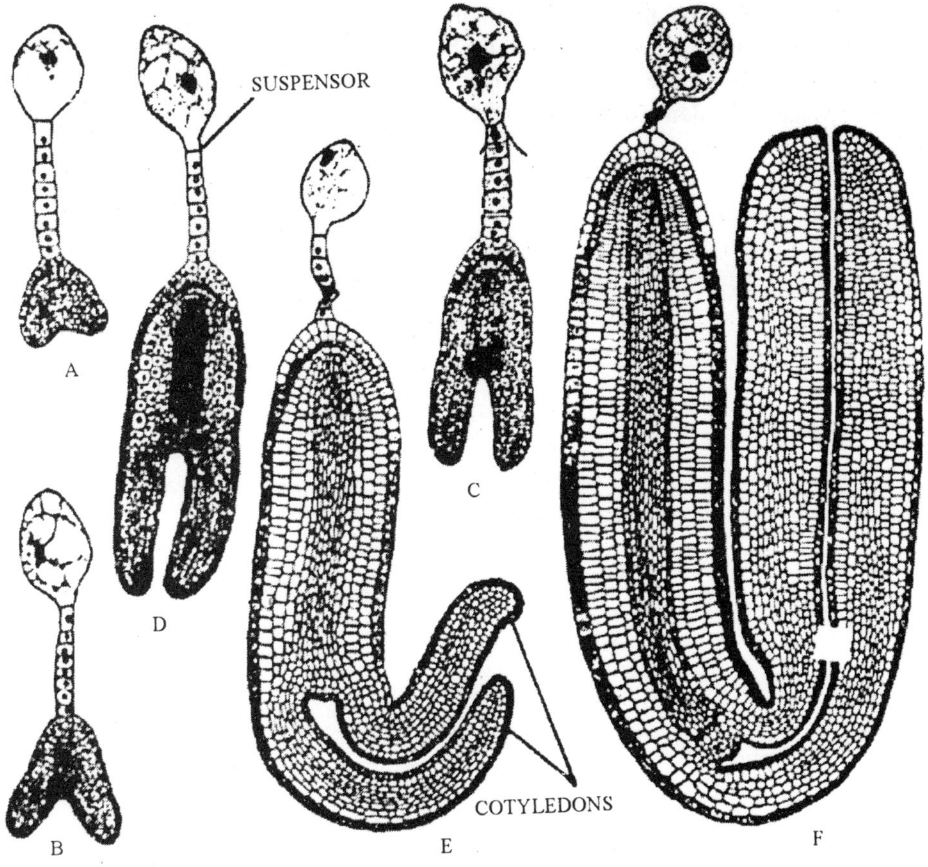

Fig. 12.3 **Embryo**

Later stages of embryogeny in Capsella bursa - pastoris (redrawn from Schaffner, 1906)

During subsequent growth the ovule becomes curved like a horse shoe and the cotyledons also are forced to assume this shape. In a mature embryo which is horse shoe shaped, the suspensor remains like a small tail.

MONOCOTYLEDONOUS EMBRYO

At stated already, there is not much of a difference in the development of embryo between monocotyledons and dictoyledons. In the mature embryo however, there is difference between the two. It is generally held that in a dicotyledonous embryo, stem tip is terminal and cotyedons are lateral, while in a monocotyledonous embryo it is *vice versa*. Some of the works of Swamy and his co-workers (Lakshmanan, 1972) however have established that even in monocotyledons, stem tip is terminal and later it is pushed towards a side due to rapid growth of the cotyledon. The developement of a typical monocotyledonous (*Luzula forsteri*) embryo proceeds as follows.

The zygote divides transversely to form a basal cell and a terminal cell. This is next followed by the formation of a T shaped proembryo. Quadrants are formed in the terminal cells by a vertical division at right angles to the previous plane of division. Periclinal divisions in the quadrant separate the dermatogen from the central cells. These cells divide transversely to form two tiers l, l'.

The tier l, gives rise to a part of the single cotyledon while the tier l' gives not only to the remaining part of the cotyledon but also to the hypocotyl and plumule.

While these developments are taking place, the basal cell divides transversely to form a row of three cells viz. m, n, and n', m and n give rise to the root and root cap while n' gives rise to a single celled suspensor. In the mature embryo the cotyledon is terminal and the stem tip is *apparently* lateral.

PART THREE
ECOLOGY

CHAPTER 13

ECOLOGY

Plants growing on the surface of the earth exhibit different characters under different environmental conditions. This is because, environment plays a profound role in the growth and development of plants. Plant also the animal life are affected by various ways. An intimate relationship exists betweeen organisms and their environment, so much so one can be regarded as theindex of the other. In short it can be safely said that plants and environment are interdependent.

Environment determines the density of plants and whether they can exist or not. The plants growing in any area fully adapt themselves to that habitat.

What is Ecology? The science of ecology also called environmental biology is extremely diversified and has transcended the limits of biology. In its broadest sense, Ecology deals with the analysis and study of inter - relationship between the organisms and their environment.

History : The term **ethology** a forerunner of ecology was first coined by a French Zoologist Isodore Geoffrey in 1859 to "study the relations of the organisms within the society and family in the aggregate and in the community".

Etymologically Ecology is derived from two Greek words *Oikos* (house) and *logos* (study), literally meaning the study of organisms in their dwelling place. The term Ecology was first used by Henry David Thoreau in 1858. According to some ecologists it was German biologist Reiter who should be credited with the coining of the word ecology. It was another German biologist Ernst Haeckel (1870) who is supposed to have used the term in the sense we understand it today. He defined ecology "By ecology we mean the body of knowledge concerning the economy of nature - the investigation of the total relations of the animals both to its inorganic and organic environment, including above all, its friendly and inimical relation with those animals and plants with which it comes directly and indirectly into contact."

Scope of Ecology : Of all the branches of science, it is ecology that comes very close to human life, because human beings also interact with the environment and are influenced by it. Present day human problems are directly or indirectly related to ecology as their solution requires an understanding of Ecology. Ecology has been contributing very much these days to socioeconomic, political and other policies of theworld. The concern to conserve the environment for their own survival has awakened many people to rise against indiscriminate destruction of natural wealth.

The changes in climate, weather, monsoon etc are all related to changes in environment. A through knowledge of ecology is very much essential to monitor all these. Ecology thus plays a very vital role in human welfare in having its influence in many fields of human endeavour..

STUDY OF VEGETATION

Vegetation study or study of the community structure is essential for understanding the dynamics of a plant community. Many of the features of a plant community such as density, species frequency etc are best expressed in terms of quantitative parameters. There are three methods of field study of the analytical characters. This are quadrat method, line transect method and point frame method. The following are some of the field experiments pertaining to the above.

1. To determine the Minimum size of the quadrat by species Area-curve method.

Requirements : Scale, four nails, A few meterres long string, hammer, graph sheet, herbarium sheet, cellotape etc.

Procedure: Hammer 3 nails to the ground in a 'L' shape and tie a piece of string. Now with another nail and a piece of string fashion a square of 10 x 10 cm. Make a note of the various species present in the area. This squarish area is called a quadrat. Increase the size of the quadrat to 20 x 20 cm^2 and note the species present. In the same way go on increasing the quadarat size to 30 x 30 Cm^2, 40 x 40 Cm^2 etc each time recording the number of species. Continue to increase the size of quadrat as long as there is increase in the number of species falling in the area. Stop increasing the size of the quadrat when there is no corresponding increase in the number of species.

Record the observations on a graph sheet. Size of the area to be marked on the X asis and the corresponding increase in the number of species on 'Y' axis. When you join the points on the sheet it would result in a curve. This is called the 'species area curve'. You will notice that initially there will be geometric increase in the number of species with the corresponding increase in the size of quadrat. Note the point where the curve starts flattening. Draw straight lines from this point to Y and X axis. The square area that will be obtained is the minimum quadrat size.

Conclusion : Species area curve is used to obtain the minimum size of quadrat while sampling the species composition of a plant community. Number of species (not individuals) in an area is a can be used as a parameter of diversity the number of quadrats. The curve obtained will indicate (where it starts to flatten) the point from which the minimum number of quadrats required can be found out.

2. To determine the minimum number of quadrats to be laid down in the field.

Requirements : Quadrat of known size, graph sheet, herbarium sheet, cellotape etc.

Procedure : Lay down about 30-50 quadrats of known size (as decided in the previous experiment) randomly at different locations in the area of study. List the various species found in each quadrat and find out the accumulation of species in each quadrat as shown in the table.

Species	1	2	3	4	5	6	7	8	9	10
A	+	+	-	-	+	-	+	-	+	-
B	-	-	+	-	+	-	-	+	-	+
C	-	+	-	+	-	+	-	-	+	+
D	+	-	-	-	-	-	+	+	+	+
E	-	-	+	+	-.-	+	+	+	+	+
F	-	-	-	-	-	+	-	-	-	-
G	+	+	+	+	+	+	+	+	+	+
H	-	-	-	-	-	-	-	-	-	+
I	+	-	-	+	-	+	+	-	+	+
J	+	+	+	-	-	-	+	-	+	-
Accumulation of total No. of species	5	4	4	4	4	5	6	4	7	7

Take a graph sheet and plot the accumulation of number of species (on the Y axis) against the numbe of quadrats. The curve obtained will indicate (where it starts to flatten) the point from which the minimum of quadrats required can be found out.

*3. **To determine the density, frequency and abundance of different species in an area***

Requirement : Quadrat of known size, graph sheet, herbarium sheet, cellotape, pencil etc.

Procedure : Lay down a known number of quadrats (as determined in experiment 2) of definite size (as determined in experiment 1) in the area of study. List the various species present in each quadrat; also count the number of individual plants in each species (in every quadrat). Tabulate the data as shown in the following table.

Name of species	No.of Quadrats land down 1 2 3 4 etc	Total No. of individuals in each species studied	No.of Quadrsts in which species species occur	Total number of quadrats	Comments
A	*2 *4 -- --	6	2	4	
B					
C					
D					

* No of individual plants occuring in that quadrat. From the data tabulated as above calculate the following.

Density : This is defined as the number of individual plants of a given species in a unit area.

$$\text{Density} = \frac{\text{Total number of individuals of the species}}{\text{Total number of quadrats studies.}}$$

$$\text{or} = \frac{\text{Toal number of quadrats in which the species occurs}}{\text{Total number of quadrats studied}} \times 100$$

Frequency : This is expressed in terms of percentage of distribution of species in a given area.

$$\text{Frequency (\%)} = \frac{\text{Total number of quadrats in which speices is present}}{\text{Total number of quatrats studied}} \times 100$$

Abundance : This refers to the total number of individuals of a species in a given area.

$$\text{Abundance} = \frac{\text{Total number of individuals in all the quadrats}}{\text{Total number of quadrats in which the species has occured.}}$$

Frequency Classes : Frequency percentages can be arranged into the following five categories based on Raunkier's (1934) classification.

Frequency Class	Frequency Percentage
A	10-20
B	21-40
C	41-60
D	61-80
E	81-100

4. To Study the vegetation by line transect method and to determine the frequency of individual speicies :

Requirements : Measuring tape, graph sheet, herbarium sheet cellotape.

Procedure : Run the measuring tape to a suitable length (1 metre) and note down the species touching the tape. This length may be called one unit of a line transect. Run several such line transects randomly in different sites in the aea of study and note down the number of species . In addition count the number of individuals of each species touching the tape in each unit of the line transect. Tabulate the data as shown in the following table.

Name of species	Units line transect 1 2 3 4 etc	Total No. of transects in which the species present	Total No. of line transects studied	Frequency Percentage	Frequency class
A	* + + + + (3) (2) (4) (1)				
B	- + + - (3) (2)				
C	- + - + (1) (2)				
D etc.					

* + indicates presence of species (not number of indivudals)
- indicates absence of species

** Value in parenthesis shows the number of individuals of particular species (A, B etc) in a given transect.

Calculate the frequency percentage for each species as in quadrat method.

Density of each species also may be calculated as in quadrat method.

5. To study the vegetation by point frame method and to determine the frequency of individual speicies.

Requirements : Point frame, graph sheet, herbarium sheet, cellotape etc.

Procedure : The point frame apparatus consists of a wooden frame (see figure) in the form of a rectangular bracket. The long horizontal arm has a number of holes to accomodate pins (nails).

Place the point frame apparatus (with 10 pins) in the field. The area occupied by the apparatus constitutes one sampling unit. Place the apparatus randomly in several places in the field and note down the number of species touching the pins. Also note down the number of individuals in each species touching the pins. Tabulate the data as in the case of transect method (substitute one unit point frame for one unit of line transect) and calculate the frequency percentage, frequency class and density.

6. To study the vegetation by Biological Spectrum Method

Requirements : Scalpel, scissors, graph sheet, spatula graph sheet, herbarium sheet, cellotape etc.

Procedure : In a given area of study examine each species as to the location of pernneating buds (or seeds) on the plant body. Note down the number of species. Now categorize the various species into the following life form types based on RaunKaier's classification as follows

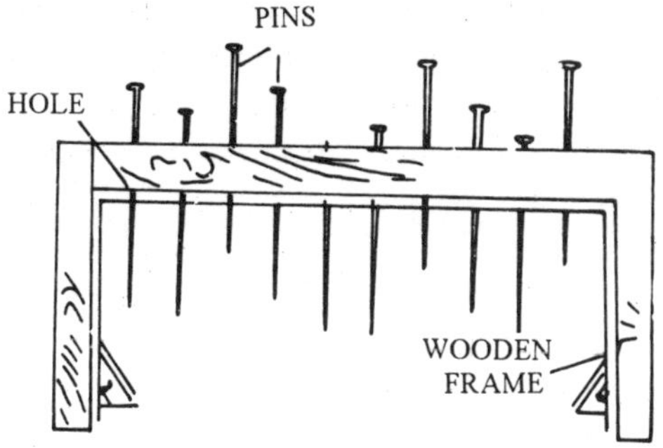

Fig 13.1 Point frame apparatus

1. Phanerophytes - Trees, shrubs with buds atleast 25 cms above ground.
2. Chamaephytes - Prostrate plants with buds on the surface or upto 25 cm above ground
3. Hemicryptophytes - Perennial plants with buds above or little below the soil surface.
4. Geophytes - Plants Rhizomatous or have under ground buds
5. Therophytes - Seasonal plants

Calculate the percentage value for each life form against the total life form in the area. With the help of a graph sheet plot the various life forms on the X axis and their percentage on the Y axis. Either a curve or histograms can be drawn on the graph sheet. This will provide the biological spectrum of the area.

Biological Spectrum may be defined as the distribution percentage of various life forms in a given area :

*7. **To determine the biomass of green plants in a given area***

Requirements : Quadrat of known size, polythene bags, spatula, hot air oven etc.

Procedure : Lay down a number of quadrats randomly in a given area. Collect the plants in the quadrats as follows :

1. Cut all the aerial parts and put them in a polythene bag.
2. Collect all the dead, fallen leaves, twigs etc and put them in another bag.
3. Dig out all the underground parts of plants clean them of soil and collect in another bag.

Weigh each portion (1, 2, 3) separately and note down the fresh weight. Now put them separately in an oven at 90°C for 24 hours and weigh them again. Note down the dry weight.

a. Fresh weight of all plant parts X_{gms}
b. Weight after (dry weight) X_{1gms}
C. Moisture content $X_{gm} - X_{1gm}$

Bio mass per quadrat (unit area) may be caluclated as follows. Suppose the quadrat is 50 x 50 cm^2 - 2500 cm^2. This means 2500 cm^2 area has X_{1gms} of dry matter. Therefore one square metre will have

$$\frac{X1 \times 100 \times 100}{2500} = X_{2gm}$$

The biomass of plant parts is X_{2gms} per square metre

7. ***To determine the vegetational cover in an area***

Requirements : Measuring tape, scissors, Vernier Calliperse, paper etc.

Procedure : Run the measuring tape for specific lengths at several places. Note down the length of plants touched by the tape.

Cut a transverse section of some plants at the ground surface. Measure the diametre of the cut end with the help of vernier scale. Tabulate the data as follows.

Name of Species	Units of tape 1 2 3 4 5 etc	Total
A B C etc.		
Total Number of Species	Length of tape for individual Plants	Total Length of all individuals

From the data above calculate the total length of tape (number of units x length of each unit). Covered by all the species and percentage of of total length of tape covered by each species. This gives the % vegetational cover of the area.

Name of species	Diametre No. of Individuals 1, 2, 3, 4 etc.	Total	Average
A B C etc			

Calculate the basal area as follows

basal area = πr^2 where

r = radius

π = 3.14

Average basal area for each species may be calculated by totalling the basal area of all species and dividing it by the number of species.

8. Study of Soil texture :

Requirement : Polythene bag, pick axe, field lens, forceps, mesh of different sizes of pores.

Procedure : Collect soil samples from various sites in polythene bags. Examine the dry sample by a field lens. Examine a moist sample also. Squeeze the soil sample under thumb and feel the texture.

Categorize the soil as follows :

1.	Particles very large	Gravel
2.	Individual grains may be seen and felt between fingers. Do not adhere when	wet Sand
3.	Individual grains are seen and felt, form a cast when moist but fall apart when squeezed	Sandy loam
4.	Dry soil mellow with a gritty feel. Moist soil forms a cast which does not break while handling.	Loam
5.	Dry, appears cloudy, soft feeling, forms a cast (both dry and wet) of plastic material when wet	Silt loam
6.	Hard when dry but forms cast which does not break while handling. Moist soil when pressed forms thin ribbon	Clay loam
7.	Very hard lumps when dry, sticks to surface when moist; forms long ribbon when squeezed	Clay

By using meshes of different pore size also soil particles can be classified.

Diameter of the Particle in mm	Classification of particle
Less than 0.002	Clay
0.002-0.02	Silt
0.02-0.2	Fine sand
0.2-2.0	Coarse sand
2.0-5.0	Fine gravel
More than 5.0	Coarse gravel

9. Study of soil temperature

Requirements : Soil thermometer (ordinary thermometer also can be used) Iron rod etc.

Procedure : (i) Dig a hold of desired depth with the help of an iron rod and directly measure the temperature by inserting an ordinary thermometer. Record the reading after about 15 minutes.

(ii) Soil thermometers which have a steel end near the mercury bulb can be directly pushed into the soil to the desired depth and temperature measured.

10. Measurement of soil pH

Requreimennts : Soil samples, water, barium sulphate, Universal indicator or pH paper.

Procedure : Collect different samples of soil in separate test tubes. To each test tube add about 5 ml of barium sulphate and 15 ml of water. Keep the solution aside untill a clear supernatant fluid is formed. Take few drops of this on a porcelain tile and add few drops of universal indicator. Note the change in colour and compare it with the colour chart for different pH values pasted on the bottle. Alternatively dip a piece of pH paper in the fluid, note the change in colour and compare it with the colour chart.

11. To determine the moisture content (%) in soil

Requirements : Soil samples, hot air oven, china balance etc.

Procedure : Place soil samples of known weight in a hot air oven at a temperature of 105-110°C for 24 hours. Weight the soil sample before keeping them in hot air oven and also weigh after heating . Calculate the moisture percentage as follows.

1. Weight of soil before heating - X_{gms}
2. Weight of soil after heating for 24 hrs X_{1gms}
3. Moisture content X_1 - X
4. % moisture content = $\frac{X-X_1}{X_1}$ X 100

12. Water holding capacity of soil

Requirements : Soil samples, water, seive, oven, dessicator, balance etc.

Procedure : Pick a number of small holes in the base of the tin box. Fill the box with air dried and seived soil. Add water so that the soil gets saturated. Keep the tin in a slanting position and hang it to a stand with the help of string. Extra water comes out of the perforations at the base. Allow it to stand for a while. When water drops stop coming out remove the soil and weigh immediately. Note down the weight. Afterwards keep the soil in a hot air oven and dry it at 105°C for 48 hrs. Cool the soil sample in a dessicator and record the weight.

Observations and Conclusion : The difference in the weight in soil samples before and after heating gives the amount of water in the soil or water holding capacity of the soil. Calculate the water holding capacity per gram of soil. Also calculate the percentage of water retained by the soil. This is generally refered to as the field capacity of the soil.

13. Calculation of carbonate content, nitrate content and base dificiency of soil samples.

Requirements : Soil samples HCL, diphenylamine, H_2SO_4 ammonium thiocyanate, H_2O_2, water, china clay, tiles, test tubes etc.

Procedure :

(a) **Carbonate :** Take a little amount of soil sample in separate test tubes. Add a few drops of Hcl to each of the tubes. Observe the rate of effervescence. Compare the rates in different soil samples. Higher rate of effervscence indicates higher carbonate content.

(b) **Nitrate content :** Prepare a soil solutions of different samples by adding 5 parts water to one part water. Take a few drops of soil solution on a porcelain plate. Add a few drops of diphenylamine and a few drops Conc H_2So_4. Blue colour formation indicates the presence of nitrates. Higher intensity of colour indicates higher nitrate content. Different soil samples may be compared based on the intensity of colour.

(c) **Base defeciency :** Take a pinch of soil sample in a tube, add saturated alcholic solution of ammonium thiocynate and shake vigourously. Allow the contents to settle. Coolect the supernatant and to it add a few drops of H_2O_2 and observe the intensity of red colouration that develops. Compare the degree of colouration in different soil samples. Higher intensity of colour indicates increased dificiency of replacable bases.

14. Test for the presence of Inorganic Salts

Requirements : Soil samples, distilled water, barium chloride solution, Hcl, H_2So_4, Nitric acid ammonium nitrate, ammonium molybdate, silver nitrate, beakers, test tubes, conical flask etc.

Procedure : Take 200 gms of soil sample in a conical flask and add 500 ml of water. Shake vigourously and keep it aside overnight to allow soluble salts to dissolve; filter the water and collect the filterate (aqueous soil extract) conduct the following tests.

(a) **Chloride :** Take 10 ml of aqueous soil extract in a test tube and add 10 ml of 0.1 N H_2So_4 to neutralize the carbonate and bicarbonate content. Now add silver nitrate to the mixture. Development of white precipitate indicates the presence of chloride.

(b) **Phosphate :** Take 10 ml of aqueous soil extract. Add a few drops of ammonium molybdate solution, Conc. nitric acid and ammonium nitrate. Development of yellow colour indicates the presence of phosphates.

(c) **Sulphate :** Take 10 ml of aqueous soil extract in a test tube and add 5 ml of Conc. HCl and boil cool and add a few ml of $BaCl_2$ solution. Development of white precipitate indicates the presence of sulphate.

15. Measurement of temperature, relative humidity and light intensity of air (atmoshphere).

Requirement : Maximum - minimum therometer, and wet thermometer (psychrometer) photometer etc.

Procedure : Temperature of the atmosphere can be measured by means of 'U' shaped thermometer in which one arm records the minimum and the other maximum temperature. The 'U' Shaped thermometer may be kept in different places like open air, shade etc and the temperature may be recorded.

Relative Humidity : This is recorded with the help of psychlometer also know as dry and wet thermometer (see fig). THe psychlometer has two thermometers - one dry and the other wet. The bulb of wet thermometer is always kept moist by dipping it in moist cotton. Evaporation of water due to normal temperature of the atmosphere causes cooling of the bulb of wet thermometer and lowers the temperature. Dry thermometer records the normal temperature. Note the difference in temperature between dry and wet thermometers. Relatiive humidity may be calculated by using a conversion table. If the relative humidity is 100% there will be no difference in temperature between the dry and wet thermometer.

16. Measurement of light intensity

Requirements : Photometer (Lux meter)

Procedure : Photometer consists of a photo electric cell and an amperemeter (microammeter). The instrument has a filter to guard against intense light. Expose the apparatus and read the value with the deflection of the needle. The photo electric cell on exposure to light converts it into electricity and the current generated is measured in the ammeter in units called lux units.

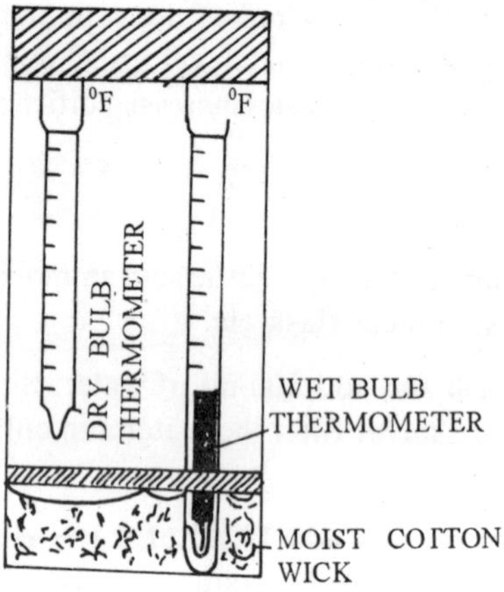

Fig 13.2 Dry and Wet Thermometer

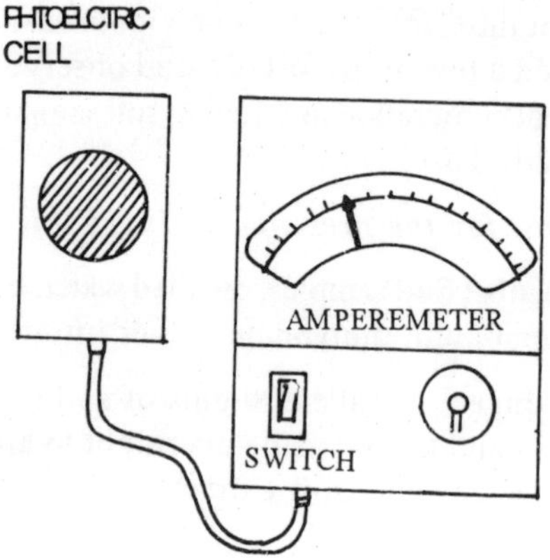

Fig 13.3 Light Meter

16. Measurement of turbidity of water :

Requirements : Secchi disc

Procedure : Secchi disc is a circular piece of metal having black and white strips. The disc has a hole in the centre through which a thread tied to a weight passes (through). The disc is lowered to a level where it will not be possible to distinguish black and white strips. If the depth at which the black and white strips lose their identity is more then the turbidity is less and vice versa. This can be compared in different ponds or tanks.

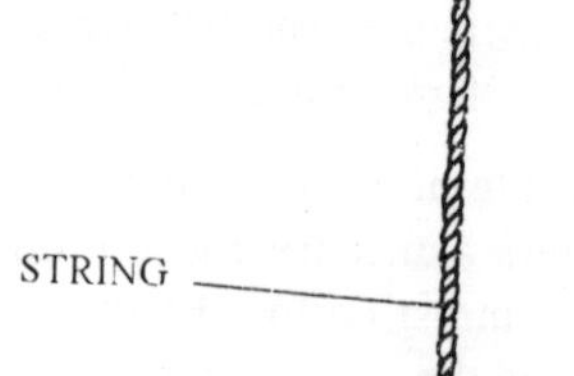

Fig 13.4 Sacchi Disc

17. Measurement of rainfall of a given area :

Requirement : Rain guage (measuring cylinder, funnel, jar, cup, oil etc).

Procedure : Keep the measuring cylinder in a jar and pour a little oil into the cylinder so that it forms approximately a 5mm thick layer. This prevents surface evaporation. Place a funnel over the measuring cylinder and fix a cap over the jar. The cap allows water only into the funnel and not into the jar. Place the apparatus in an open area and watch.

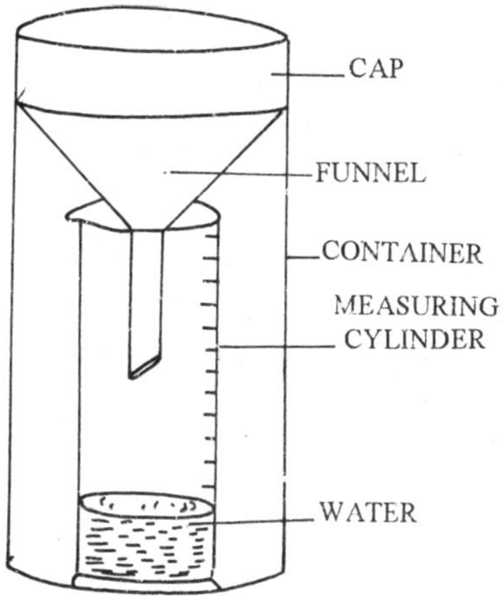

Fig 13.5 Rain gauge

Observation and Inference : Rain water percolates through the funnel and collects in the measuring cylinder. Measure the diameter of the cap and the amount of water collected. Calculate the amount of rainfall per unit area.

ECOLOGICAL GROUPS

Groups of plants belonging to one or more species that grow together in a particular area constitute an ecological community or a group.

For instance trees shrubs, climbers etc that grow in a forest form a forest community. Similarly plants that grow in a pond form a pond community. Likewise, plants that grow in desert form a desert community. Based on the environment they inhabit warming (1985) has identified eight ecological groups or communities.

HYDROPHYTES

Plants that grow in water or in soil covered with water belong to this group. Hydrophytes have a number of morphological and anatomical adaptations. These are

Morphological Adaptations :

1. **Dimorphic leaves :** Some of the plants which are partially submerged have dimorphic leaves. The underwater leaves are thin and highly dissected whereas the leaves above water are broad. This condition is known as heterophylly. Ex. *Limnophylla heterophylla, Trapa, Sagittaria sagittifolia* etc

2. **Formation of coiled pedicles and petioles :** In plants like *Vallisnaria* the pedicles and petioles are coiled so that the flowers and leaves will be always on the surface of water but do not ofter resistannce to water currents.

3. **Development of water Proofing device :** To prevent decay a waxy coating or mucilaze covering is developed on the surface to repel water.

4. **Roots :** Highly reduced, divided into thin fibrels or even absent (ultricularia). Root hairs and root caps are also absent very often.

5. **Stem :** Very thin, delicate and transparent.

6. **Guttation :** Many of the plants remove water by way of guttation.

Anatomical Adaptations

1. **Epidermis :** Cells are thin walled, contain chloroplasts

2. **Aerenchyma :** Parenchyma tissue has a large number of air cavities to store air which helps both in respiration and buoyancy (floating).

3. **Cuticle :** Absent or even when present very thin

4. **Mechanical Tissues :** These are very poorly developed. Sclerenchyma and xylem which are the principal lignified tissues are very little in extent.

5. **Stomata :** Absent in submerged leaves, but present in the upper epidermis of floating leaves

6. **Secondary growth :** Takes place very rarely.

7. **Chlorenchyma :** Very well developed.

Classification of Hydrophytes

1. **Floating hydrophytes :** Plants float on the surface of water. There are two kinds of surface hydrophytes - free floating and fixed. In the former the roots are suspended freely in water (e.g. *Eichhornia, Pistia, Salivnia* etc) Fixed floating hydrophytes have their roots fixed to the underwater soil. They occur in shallow water (e.g *Marsilea, Nymphaea* etc).

2. **Submerged hydrophytes :** They are completely under water; may be suspended freely in water (e.g. *Elodea*) or may be rooted to soil underneath the water (e.g *Vallisnaria*).

3. **Amphibious Hydrophytes :** They grow in shallow or muddy waters. Roots and parts of their stems are embedded in water while their branches and leaves are often aerial (e.g. Typha, Alisma etc). We will now study in detail morphological and anatomical adaptations in a few exapmles.

Hydrilla verticellata (Hydrocharitaceas)

Fig 13.6 Hydrilla - Habit

Morphology

1. It is a commonly occuring water plant in ponds.
2. It is a submerged hydrophyte, but rooted to the under water soil.
3. Root system is adventitious.
4. Stem is transparent, flexible and soft.
5. Leaves are small, thin and transparent. Phyllotaxy is whorsed.
6. Mucilage covers all parts of the plant body to prevent excessive absorption.

Transverse Section of stem

1. Epidermis is single layered made up of isodiametric thin walled cells.
2. Cuticle if any is very thin
3. These is a single layer of parenchymatous hypodermis
4. Cortex is made up of aerenchyma. Air chambers are irregular in size and shape and are separated. by one layered partititon of cells. Cortical cells are spherical or oval and thin walled. Air cavities are larger in size towards periphery but smaller towards the centre.

5. Internal to the aerenchymatous cortex is the endodermis surrounding the stele.

6. The stele is highly reduced. Xylem is absent and is replaced by an air cavity. Surrounding the air cavity is phloem.

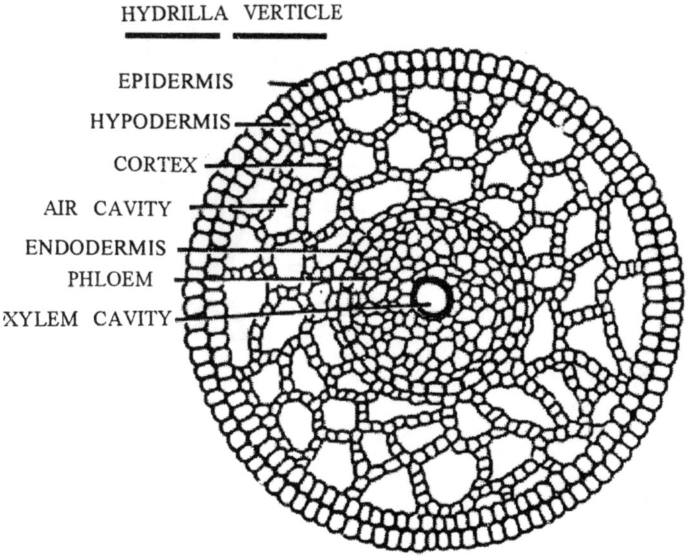

Fig 13.7 Hydrilla, T.S. of Stem

Transverse Section of Leaf

1. Leaf is dorsiventrally flattened and reduced in size.
2. The upper and lower epidermal layers are made up of a single layer of cells each. There is no cuticle but there is deposition of mucilage.
3. There are no stomata
4. Vasculature is highly reduced. There is no bundle sheath
5. Mesophyll is also highly reduced in quantity and undifferntiated
6. All cells including epidermis are chlorophyllous

Nymphaea (Nymphaeaceae)

Morphology

1. It occurs in shallow waters
2. Plant is a submerged rooted hydrophyte
3. Stem is rhizomatous growing horizontally underneath the soil.
4. Roots are adventitious.
5. Leaves are very large, and have a long flexible petiole.

6. Lamina is cordate is shape and has waxy coating as a result water does not adhere to the surface.

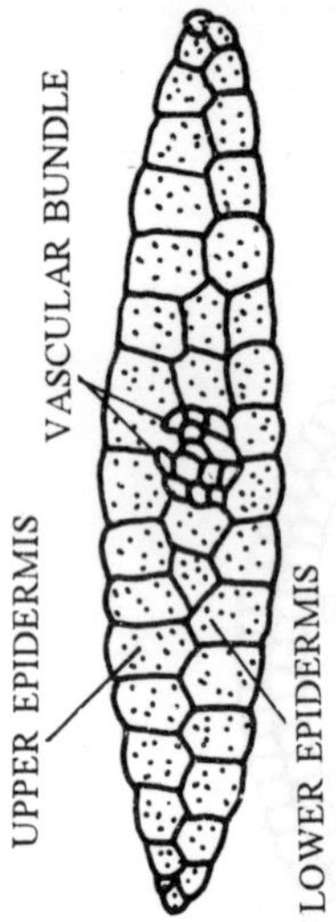

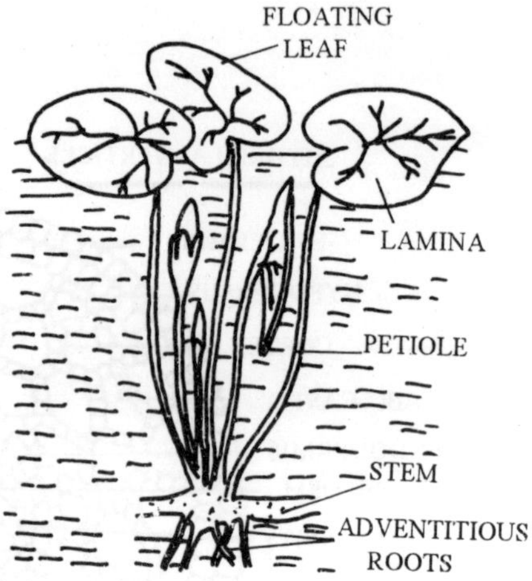

Fig 13.9. Nymphaea, Habit.

Fig 13.8. Hydrilla, T.S. of Stem

Transverse Section of Petiole

1. Outermost layer is epidermis. It is made up of thin walled barrel shaped cells. Cells are compactly arranged. There are a number of multicellular hairs arising from the epidermis. A thick coating of mucilage is found on the epidermal walls.

2. Internal to the epidermis is a few layers of collenchyma constituting the chief mechanical tissue. (Collenchyma is a living mechanical tissue whose corner walls are thickened). The collonchyma constitutes the hypodermis.

3. Next to the hypodermis is the ground tissue. Immediately next to hypodermis there are a few layers of compactly arranged cells (parenchyma) with small intercellular cavities. Next to this is the aerenchyma with large air spaces separated by uniseriate partitions.

4. Branched or stellate sclerids are found protuding into the air cavities.

5. Vascular bundles are reduced. They are scattered in the ground tissue being present mostly in the outer zone.

6. In the vascular bundle xylem is represented by a lacuna. Phloem elements are present in their normal condition.

Transverse Section of leaf (lamina)

1. The lamina is dorsiventrally flattered (bifacial)
2. Upper epidermis is single layered. waxy cuticle is deposited above the epidermis.
3. Stomata are found only in the upper epidermis (epistomatic).

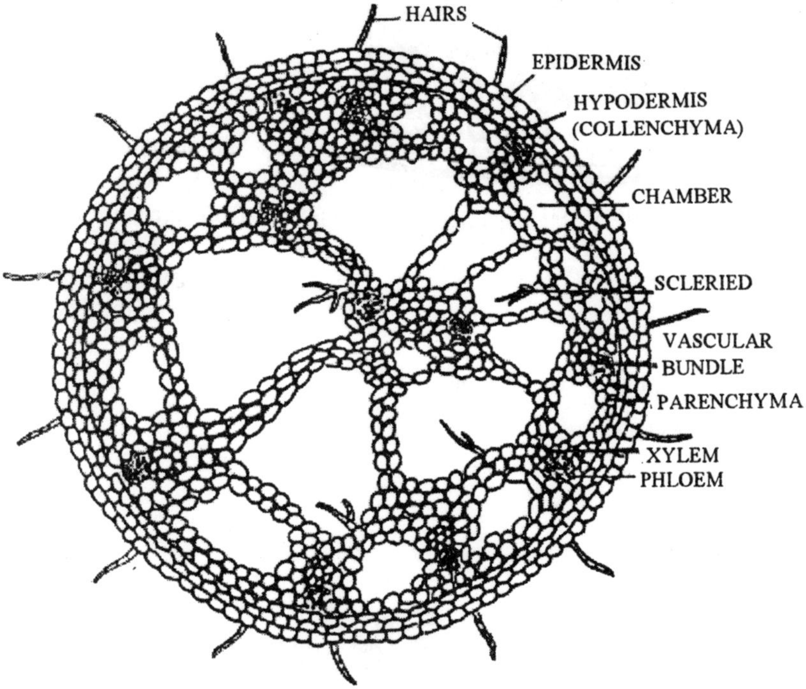

Fig 13.10 Nymphaca, T.S. of Petiole

4. Lower epidermis is also made up of a single layer of cells. There is no cuticle on lower epidermis.
5. Mesophyll is differntiated into upper palisade parenchyma and lower spongy parenchyma.
6. Palisade parenchyma is 2 layered. Cells are compact. This is the chief phtosynthetic tissue in the leaf.
7. Spongy parenchyma cells have very large air cavities between them.
8. Macro sclerids (in palisade tissue) and a sterosclerids (in spongy tissue) are present in the mesophyll.
9. Vascular bundles are poorly developed. They contain protoxylem cavities. Phloem is well developed.

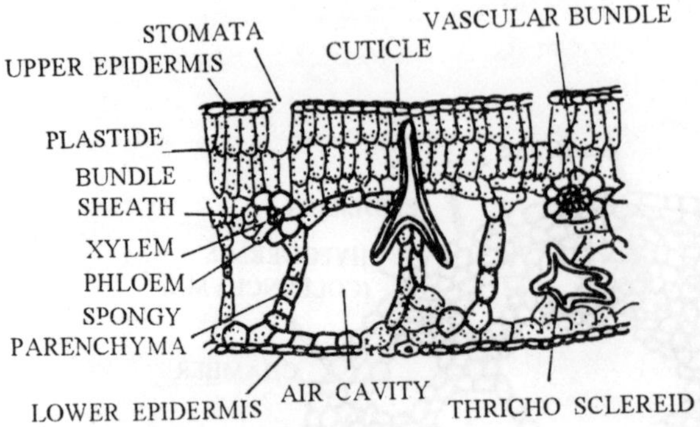

Fig 13.11 Nymphaea, T.S. of Lamina

Trapa (Trapaceae)

1. Trapa is a free floating hydrophyte grows in open tanks and lakes.
2. Leaves are dimorphic floating and submerged. Floating leaves are large. They have a swollen petiole to help in buoyancy.
3. Submerged leaves arise in pairs at each node of the stem. They are highly dissected.
4. Stem is also submerged delicate and somewhat spongy.
5. Roots are unbranched and arise from the nodes.
6. Flowers are aerial but after pollination the pedicel bends down into water and the development of the fruit is under water.

Transverse Section of Stem

1. There are three regions in a transverse section epidermis, cortex and stele.
2. Epidermis is single layered. Multi cellular hairs arise from the epidermal cells.
3. Below the epidermis is a few layers sclerenchyma constituting the hypodermis.
4. Cortex is made up of parenchyma with large air cavities.
5. sphaerarophides are found in the cortical cells
6. Internal to the cortex is the endodermis.

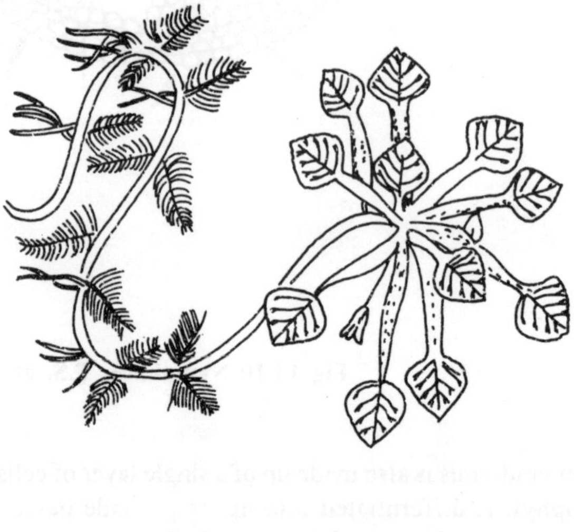

Fig 13.12 Trapa, Habit.

7. Pericycle is found internal to the endodermis
8. Vascular bundles are many in number arranged in the form of a ring surrounding a central pith.

9. Xylem is poorly developed. Phloem is found both external as well as internal to xylem.
10. Pith is wide and consists of parenchyma cells.

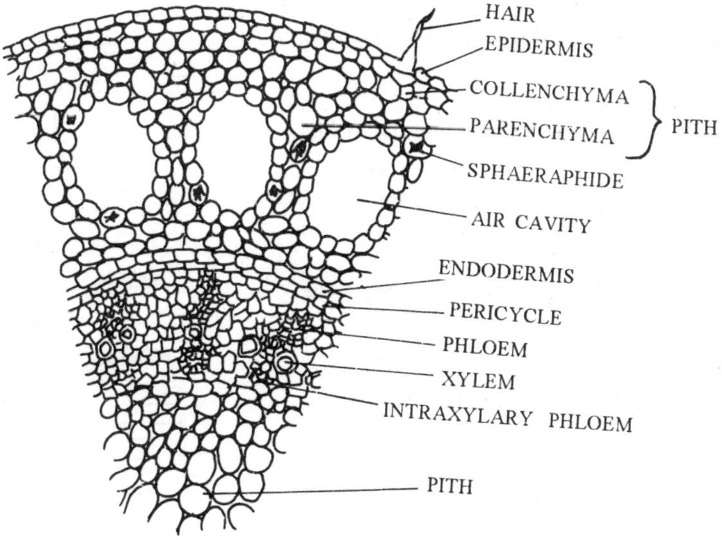

Fig 13.13 Trapa, T.S. of Stem (A part enlarged)

Transverse Section of Leaf

1. Leaf is dorsiventrally flattened.
2. Upper epidermis consists of isodiametric cells. Cuticle is deposited above the epidermis.
3. Stomata are found only in the upper epidermis (leaves are epistomatic).
4. Lower epidermis is single layered and the cells are uninterrupted.
5. Mesophyll is distinguished into palisade and spongy parenchyma
6. Palisade cells may be in one or two layers below the upper epidermis. They have a number of chloroplasts.
7. Stomata in the upper epidermis lead to substomatal cavities present between the palisade cells.

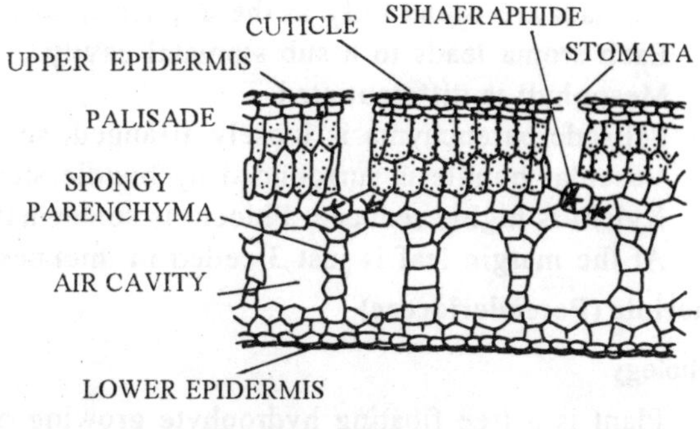

Fig 13.14 Trapa, T.S. of Leaf.

8. Cells present at the base of substomatal cavity have spaeroraphides.
9. Spongy parenchyma has large air spaces.
10. Vasculature is represented by a single vascular bundle in the midrib. Bundle sheath is parenchymatous.
11. Xylem is reduced to cavities. Phloem is well developed.

Potomogeton (Hydracharitaceae)

Morphology

1. The plant is a submerged rooted hydrophyte growing is shallow ponds and lakes.
2. Flowering branches are aerial.
3. Shoot is horizontal with vertical branches. Branches are thin delicate and long.
4. Roots are adventitious arising at the nodes
5. Leaves are semitransparent thin, thread like with a distichous phyllotaxy (arranged in two rows)
6. Mucilage covers all the underwater parts of the plant

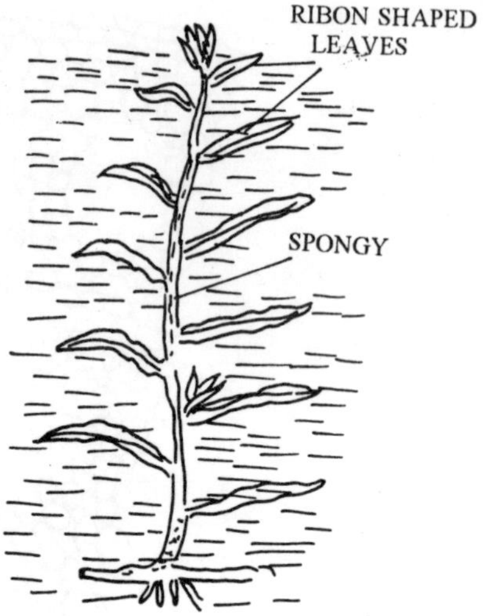

Fig 13.15 Potomogeton, Habit

Transverse Section of Stem

1. Transverse Section shows three regions - epidermis, cortex and stele.
2. Epidermis is single layered made up of barrel shaped cells. Cuticle is deposited above the epidermis.
3. Internal to the epidermis is a wide aerecnymatous cortex consisting of large air cavities separated by one celled partitions. Small patches of sclerenchyma are found in the cortex.
4. Endodermis is well developed while pericycle is indistinct
5. Xylem is reduced to a cavity while phloem is well developed.

Transverse Section of Leaf

1. Leaf is dorsiventrally flattened
2. Upper and lower epidermis are single layered arranged compactly.
3. Stomata are found only in the uppder epidermis
4. Each stoma leads to a sub stomatal cavity
5. Mesophyll is differentiated.
6. Palisade parenchyma is loosely arranged, so also spongy parenchyma.
7. Vascular bundle is surrounded by bundle sheath.
8. Xylem is highly reduced. Phloem is somewhat well developed.
9. At the margin leaf is just 3 celled in thickness.

Eichhorinia (Potenderiaceae)

Morphology

1. Plant is a free floating hydrophyte growing profusely in ponds and lakes.
2. There are a number of modifications in the root, stem and leaf to suit the aquatic habitat.
3. Stem is horizontal, cylindrical, spongy and is found just above the surface of water.
4. Stem breakes up into a number offsets which bring about vegetative propagation.
5. Roots arise in large clusters at the nodes, black in colour hairy and are called ballasting roots.

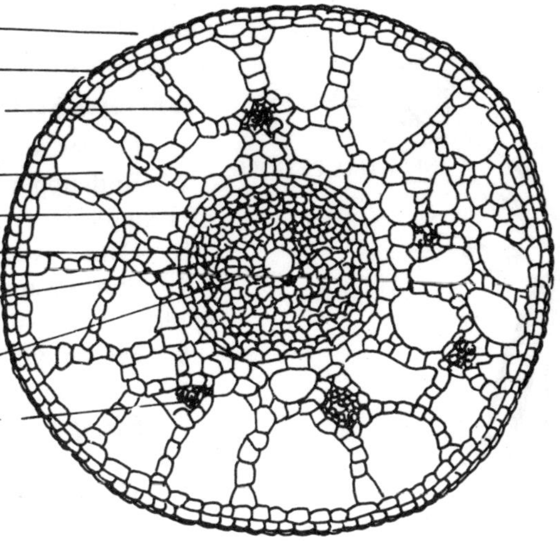

Fig 13.16 Potomogeton, T.S. of Stem

They help in balancing the plant on water.

6. Root caps are basent, instead large root packets are found.

7. Leaf has a spongy and bulbous petiole called phyllode. It helps in buoyancy.

8. Lamina is shiny due to the deposition of waxy material.

9. Pinkish flowers arise in clusters at the node

Transverse Section of Petiole

1. A T.S. shows epidermis and ground tissue in which the vascular bundles are scattered.

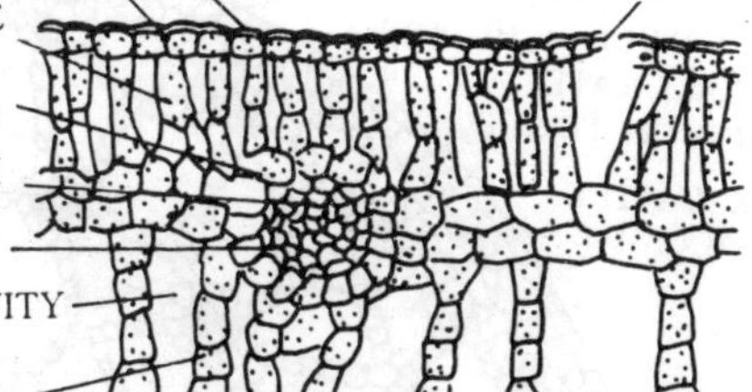

Fig 13.17 Potomogeton, T.S. of Leaf

2. Epidermis is made up of a single layer of thin walled cells. Cuticle is deposited over the epidermis.

3. Internal to the epidermis is the hypodermis made up of a few layers of compactly arranged cells. Cells are chlorophyllous.

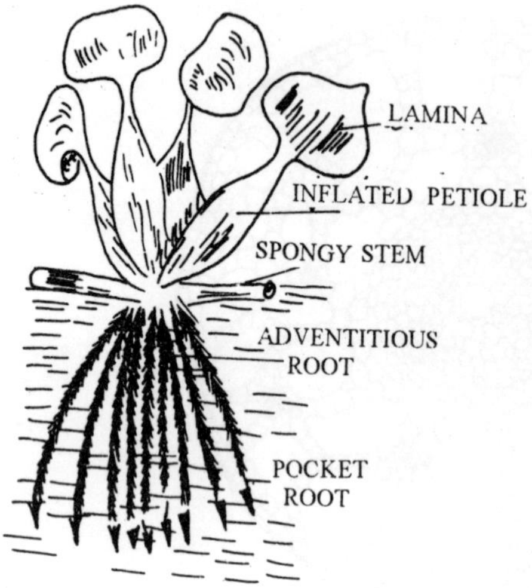

Fig 13.18 Eichhornia, Habit

4. Internal to the hypodermis is the ground tissue made up of aerenchyma with large air chambers separated by one cell thick partitions.

5. Dispersed in the ground tissue are found a number of vascular bundles. Xylem is poorly developed while phloem is well developed.

Transverse Section leaf (lamina)

1. Leaves are dorsiventally flattened
2. Both upper and lower epidermis are made up of a single layer of cells each.
3. Cuticle is deposited above the lower epidermis
4. Stomata are found in both the upper and lower epidermis (amphistomatic)
5. Mesophyll is differentiated.
6. Palisade parenchyma is one or two layered
7. Spongy parenchyma has large number of air cavities.
8. Vascular bundles have a prominent bundle sheath. Xylem however is poorly developed while phloem is well developed.

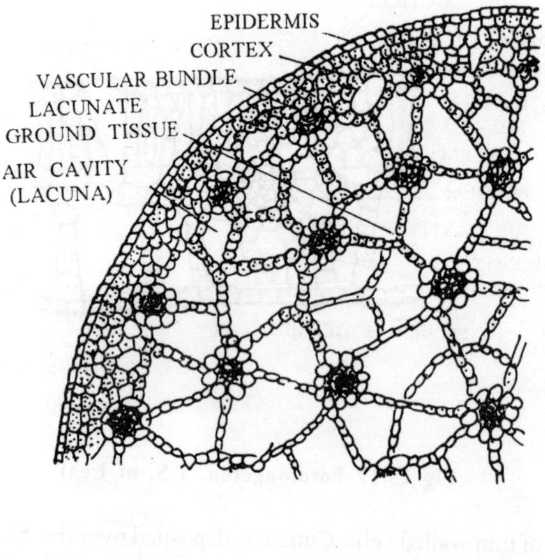

Fig 13.19 Eichhornia, T.S. of Petiole (A part enlarged)

Transverse Section of Root

1. A Transverse Section shows the three regions - epidermis, cortex and stele.
2. Epidermis is single layered made up of compactly arranged barrel shaped cells.
3. Internal to the epidermis is a few layered hypodermis followed by a broad aerenchymatous cortex.
4. Stele is surrounded by endodermis and pericycle
5. Vascular bundles are radial with exarch xylem. Xylem however is not well developed.
6. In the centre of the stele is present a parenchymatous pith.

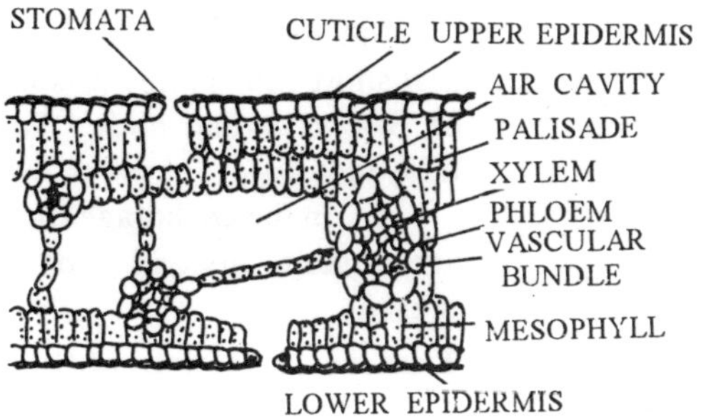

Fig 13.20 Eichhornia, T.S. of Petiole (A part enlarged)

Vallisnaria (Hydrocharitaceal)

1. Vallisnaria is a submerged rooted hydrophyte found growing in stationary or running water.
2. Roots are adventitious arising at the nodal regions of the horizontal underground stem.
3. Leaves are flat ribbon shaped and thin arising in a rosette fashion from the stem.
4. Plants are dioecious. In female plants the flowers are solitary and have a long coiled pedicle by the virtue of which the flower reaches the surface of water.
5. In the male plants flowers are in umbels which detach from the plant body and reach the surface of water.
6. Pollination is hydrophyllous (ephydrophyllous) and takes place on the surface of water.

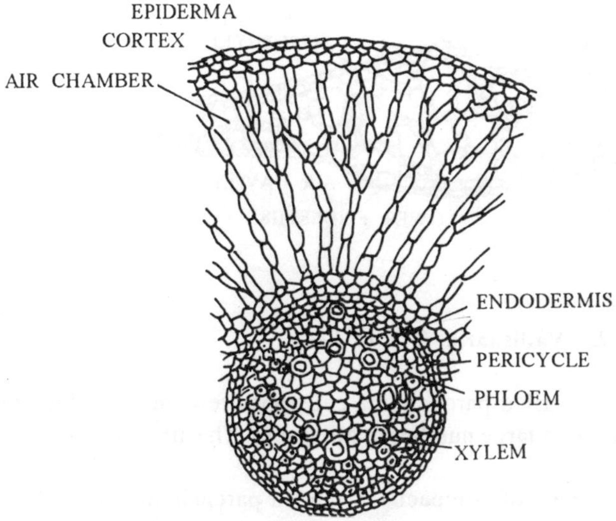

Fig 13.21 Eichhornia, T.S. of root (A part enlarged)

Transverse Section of Leaf

1. The leaf is isobilateral
2. Upper and lower epidermis are made up of a single layer of cells each. The cells have chloroplasts.
3. Cuticle and stomata are absent
4. Mesophyll is undifferentiated and consists of loosely arranged parenchyma cells with large number of air spaces.
5. Vascular bundles are reduced, Xylem is represented by a cavity while phloem is comparatively well developed.

Jussieua (Onagraceel)

Morphology

1. Plant is a free floating hydrophyte found in lake, ponds and ditches
2. Stem is horizontal growing just beneath the surface of water. It is distinguished into nodes and internodes. Stem is soft and spongy.

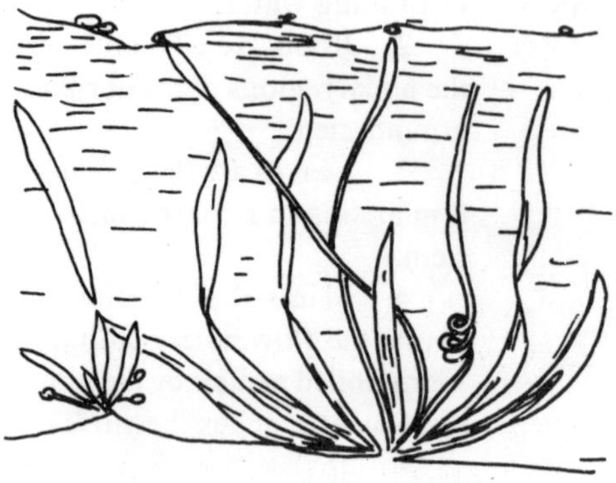

Fig 13.22 Vallisnaria, Habit

3. There are two types of roots - floating roots and ordinary roots. Ordinary roots are thin, fibrous and grow into water. Floating roots are white in colour spongy, soft short and stumpy arising in clusters of 3-4 at every node. They help in buoyancy.

4. In the anchorage roots, root hairs are absent, instead root packets are present.

Transverse Section of Stem

1. Transverse Section shows epidermis, cortex and stele.

2. Epidermis is made up of thin walled compactly arranged, radially elongated cells. Cuticle is absent

3. Next to the epidermis is the cortex divided into three zones. - outer cortex, middle cortex and inner cortex.

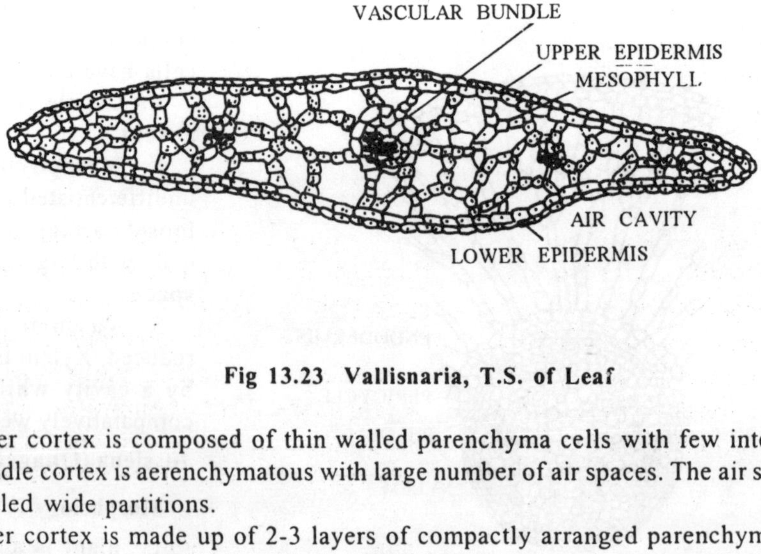

Fig 13.23 Vallisnaria, T.S. of Leaf

4. Outer cortex is composed of thin walled parenchyma cells with few intercellular spaces.

5. Middle cortex is aerenchymatous with large number of air spaces. The air spaces are separated by 3 - 4 celled wide partitions.

6. Inner cortex is made up of 2-3 layers of compactly arranged parenchyma cells.

7. Stele is surrounded by an indistinct endodermis.

8. Vascular bundles are many in number arranged in the form of a ring surrounding a central pith.

9. Vascular bundles are collateral and conjoint with poorly developed xylem but comparatively well developed phloem.

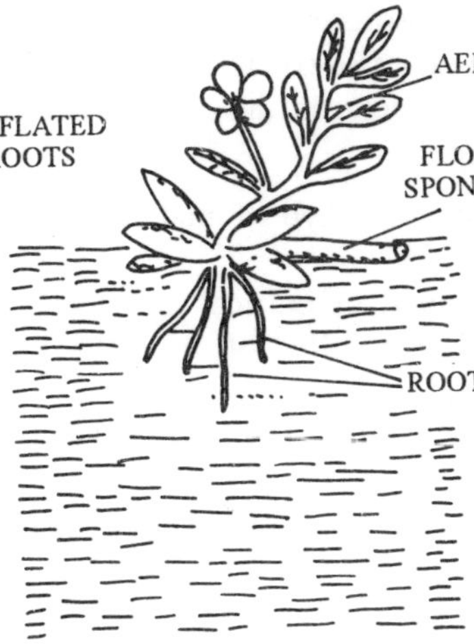

Fig 13.24 Jussieua, Habit

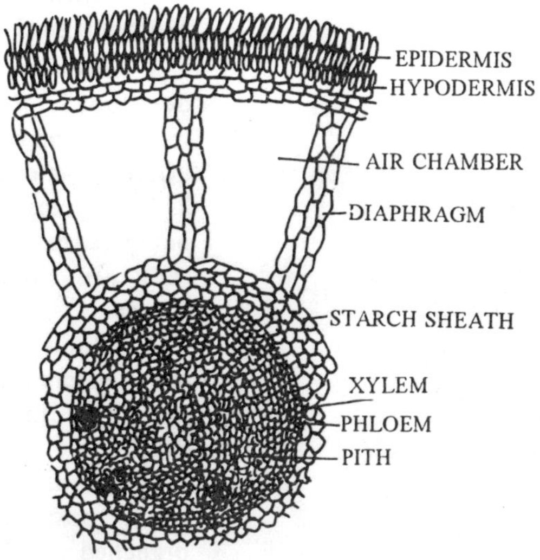

Fig 13.25 Jussiena, T.S. of Stem

Ceratophyllum

Morphology

1. Plants are submerged but rootless hydrophytes. They are feely suspended in water.
2. Stem is long, slender and delicate. Branches arise at the nodes.
3. Leaves are dichotomously dissected and arise in whorls at the nodal region. Margins of the leaves are serrate.
4. All the plant parts are covered with mucilage.
5. Mechanical breakage of the stem results in vegetative propagation.

Transverse Section of Stem

1. Transverse Section shows outer epidermis, middle cortex and the central stele.
2. Epidermis is made up of a single layer of thin walled compactly arranged cells. Cells are chlorophyllous.
3. Cortex is wide, chlorophyllous aerenchymatous with large number of air spaces.
4. Stele is surrounded by endodermis followed by a layer of pericycle.
5. Xylem is represented by an air cavity. Phloem is comparatively well developed.

Transverse Section of Leaf

1. The outline of cross section is almost circule.
2. Epidermis is single layered made up of compactly arranged thin walled chlorophyllous cells. Stomata are absent.
3. Mesophyll is undifferentiated composed of compactly arranged parenchyma cells. However 2-3 large air chammbers are present in the mesophyll.
4. Vasculare bundle is hihgly reduced and has poorly differentiated conducting tissue.

Nelumbo (Lotus)

1. Plants are rooted hydrophytes with floating leaves.
2. Leaves have a wide lamina with a long cylindrical soft air filled petiole.

3. Petiole and surface of lamina is coated with wax to prevent excessive absorption of water.
4. Some leaves may also rise above the surface of water.
5. Stomata are present on the upper epidermis.

Pistia (Araceae)

1. Free floating hydrophytes found on water surface in ponds, lakes etc.
2. Stem is horizontal, cylindrical and greenish in colour
3. Branches of stem called offsets help in vegetative propagation.
4. At each node on the stem a rosette of leaves arise in a spiral
5. Roots are adventitious arising in groups from the lower surface of the stem at the nodal regions.
6. Roots do not have root caps, instead they have root packets.

Xerophytes

Plants that grow in places where the availability of water is less are called xerophytes. Some xerophytes are also called desert plants. Xerophytes develop several morphological and anatomical adaptations to economise water.

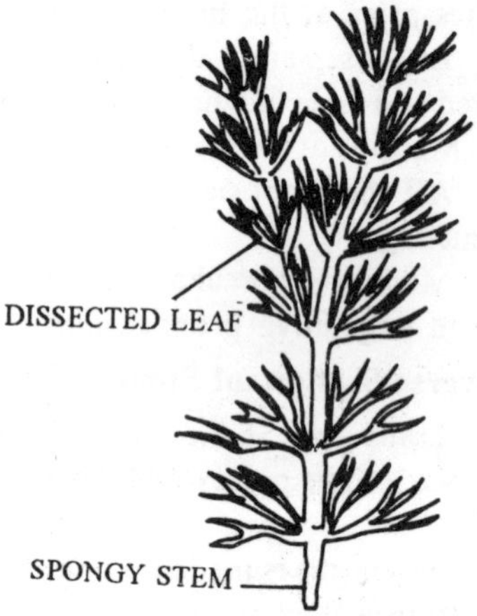

Fig 13.26 Ceratophyllum, Habit

XEROPHYTES

Morphological adaptations

1. Root system is well developed. It is both intensive and extensive for efficient water absorption.
2. Roots may be shallow but spread extensively when they are called surface feeders.
3. Roots are said to be deep feeders when they are not spreading but grow deeply into the soil.
4. Stems are stunted, thick and extensively branched.
5. In some cases stem gets modified into thick fleshy green structures called phylloclades or cladodes.

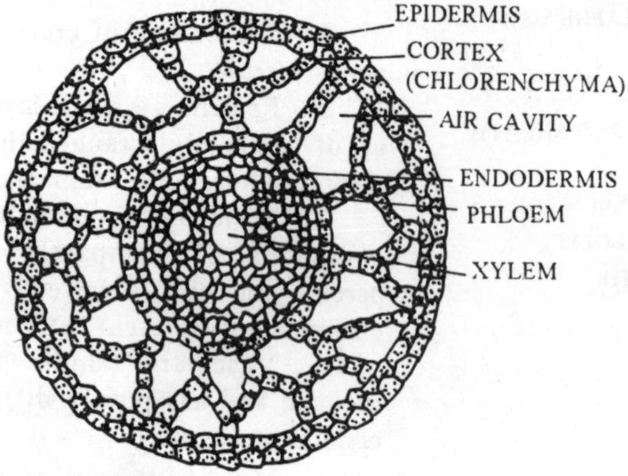

Fig 13.27 Ceratophyllum T.S. of leaf

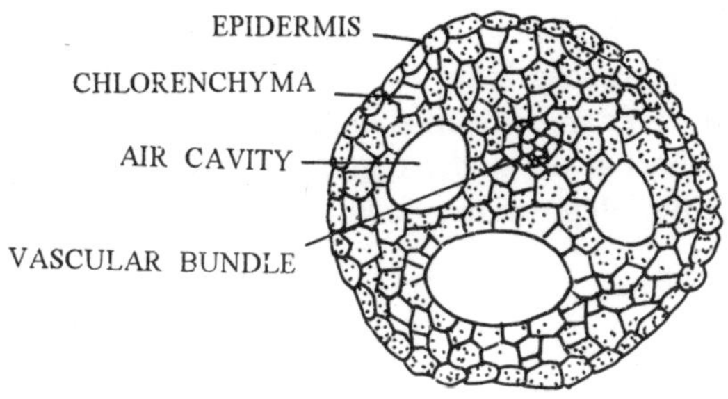

Fig 13.28 Ceratophyllum T.S. of Leaf

6. Stems may be underground in some cases or become hard and woody with thick barks.

7. Leaves are generally reduced, modified or totally eliminated.

8. Leaves may be modified into scales or spines which helps in checking transpiration as well as protects the plants from grazing animals.

9. In some plants leaves remain in a vertical posture exposing very little surface to direct sun light.

10. Some of the plants have decidous leaves which fall off during summer.

11. Leaves when present have a waxy or shiny surface to reflect right.

12. Leaves may be leathery or may have hairy outgrowths on the surface.

13. Under xeric conditions leaves may roll up and open only on the availability of water. (e.g. *Selaginella rupestris*).

14. Xerophytes produce large number of perenniating buds on the stem to help in vegetative propagation.

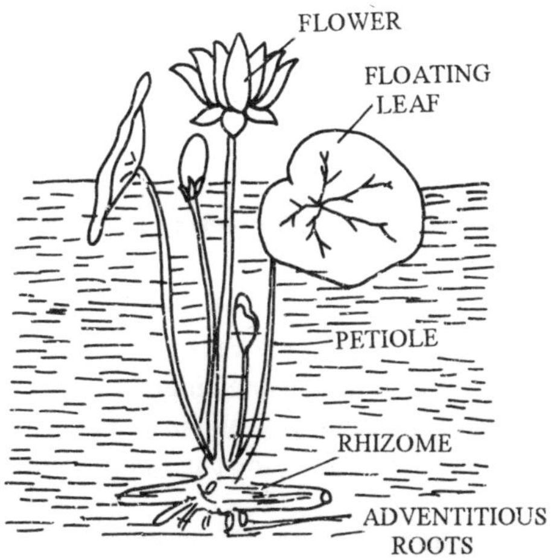

Fig 13.29 Nelumbo, Habit

Anatomical Adaptations

1. Epidermis is well developed. Sometimes multiple epidermal layers are found.
2. Cuticle is very well developed.
3. Waxy or silica coating is often found on the aerial parts.
4. Hypodermal sclerencyma is often found beneath the epidermis in the leaves.
5. Stomata are reduced in number and usually leaves are hypostomatic
6. In some plants stomata are scotoactive (i.e. open only during night).
7. Stomata are sunken, often covered by hairy outgrowths.
8. Many stomata are often found together in a sunken cavity (as in *Nerium* leaf).
9. Palisade parenchyma is very well developed; often it is many layered.
10. Air spaces are less in spongy parenchyma.
11. In succulent stems water is mixed with mucilage to prevent evaporation.
12. Mechanical tissues such as sclerenchyma, xylem etc are very well developed.
13. In some grasses motor cells are bulliform cells are present to help in the rolling up of the leaves.

We shall now study few examples of xerophytes

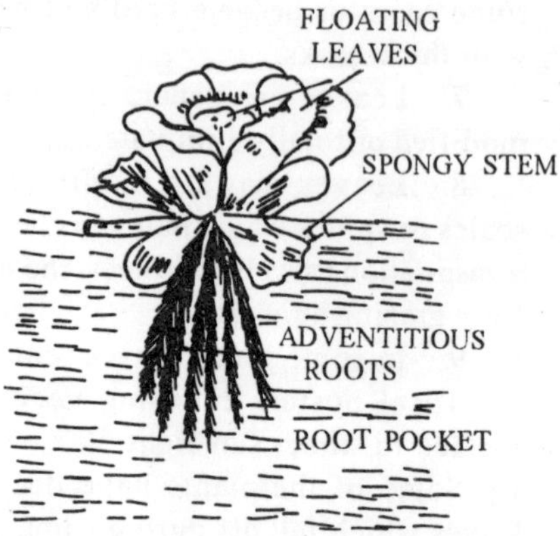

Fig 13.30 Pista, Habit

Transverse Section of Stem

1. Outline of the transverse section will be vaxy because of ridges and furrows.
2. Outermost layer is epidermis made up of a single layer of cells-cuticle is deposited above the epidermis.

Casuarina equisetifolia

Morphology

1. The plant is a tall tree.
2. Branches are green, jointed, look like leaves and many internodes long.
3. Leaves are scale like, highly reduced and present in a whorl at the nodal region.
4. Silica deposition is found on the young branches.

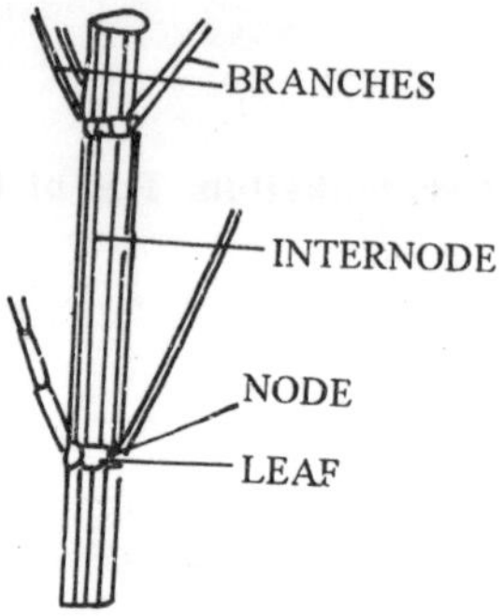

Fig 13.31 Casuarina, a branch

3. Stomata are sunken, restricted in distribution to the furrows.
4. Epidermal cells in the furrow have long multicellular hairs filling the cavity.
5. Next to the epidermis are found patches of sclerenchyma beneath the ridges.
6. Internal to the hypodermis the cortex is divided into ourter chlorenchyma and inner parenchyma
7. Chlorenchyma cells are arranged in many palisade layers constituting the chief photosynthetic tissue.
8. Embedded in the parenchymatous cortex are found a number of vascular bundles (one each below the ridge). These are called cortical vascular bundles and are collateral, conjoint, open with endarch xylem.
9. Next to the cortex is found the endodermis. Pericyle is represented by patches of sclerenchyma found between the endodermis and the stelar vascular bundles.
10. Vascular bundles are many arranged in the form of a ring surrounding a central pith. They are similar to the cortical vascular bundles in strcuture. Many phloem fibres are present in the vascular bundles.

Calotropis gigantea

Morphology

1. The plant body is a large herb or shrub.
2. All aerial parts are covered with white waxy material which acts as insulation against heat

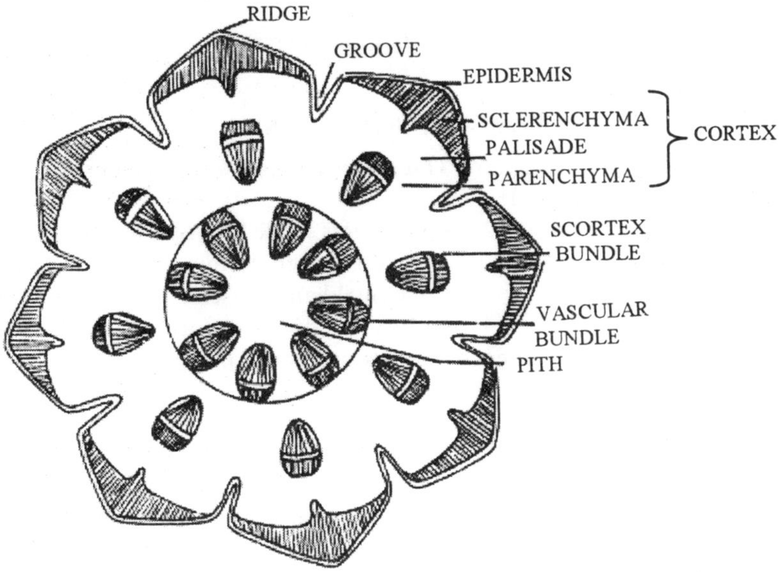

Fig 13.32 Casuarina, T.S. of Stem (Ground Plan)

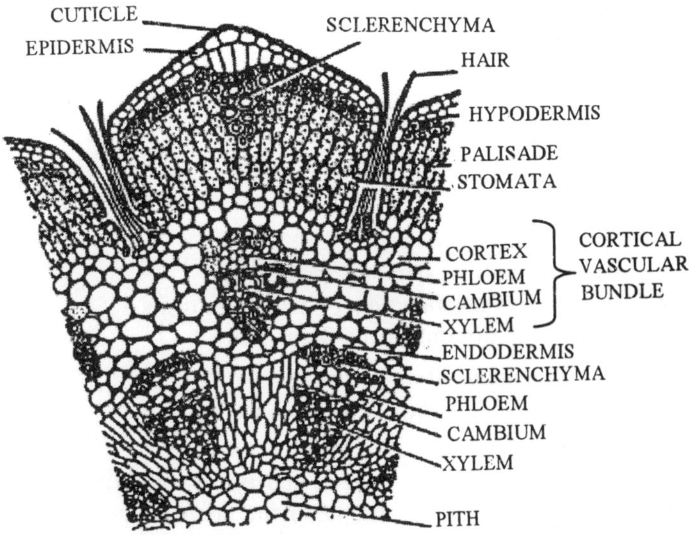

Fig 13.33 Casuarina, T.S. of Stem (A part enlarged)

3. Thick milky latex is present in the plant.
4. Leaves opposite and decussate. Inflorescence is a number.

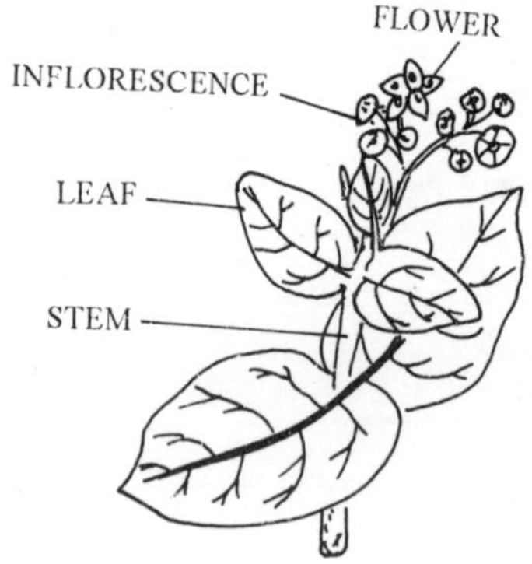

Transverse Section of Stem

1. Transverse Section shows epidermis, cortex and stele.
2. Epidermis is single layered. Thick cuticle and waxy deposition covers the epidermis.
3. Epidermal hairs present.
4. Cortex is divided into two zones - outer collenchymatous and inner parenchymatous.
5. Endodermis surrounds the stele. Pericycle is represented by patches of sclerenchyma.
6. Vascular bundles are many in number arranged in the form of a ring surrounding a central parenchymatous pith. They are bicollateral, conjoint, open with endarch xylem.
7. There are two groups of phloem in the vascular bundle (external and internal).
8. Numerous laticiferous vessels are seen in the pith.

Fig 13.34 Calotropis, Habit

Transverse Section of Leaf.

1. Leaf is dorsiventrally flattened.
2. Upper and lower epidermal layers are covered with waxy material.
3. Leaf is amphistomatic (stomata are found on both the surface).
4. Mesophyll is differentiated. Palisade cells are branched and loosely arranged.
5. Spongy parenchyma cells are also loosely arranged.
6. Vasculare bundles are covered with a prominent bundle sheath.
7. Laticiferous vessels are found in the mid rib regions.

Euphorbia tirucalli

Morphology

1. The plant body is a shrub.
2. Leaves are small in size and deciduous.
3. Stem performs the function of photosynthesis.
4. Milky latex is present in the plant body.

Transverse Section of Stem

1. A transverse section is nearly circular and shows the three regions - epidermis, cortex and stele.
2. Epidermis is single layered and is covered with thick cuticle.
3. Sunken stomata are found in the stem as the stem is the chief photosynthetic organ.
4. Cortex consists of a mixture of chlorenchyma and water storing parenchyma.
5. Latex cells are also present in the cortex.
6. Endodermis is not distinct, so also the pericycle.

7. Vascualr, bundles are arranged in a ring. They are collateral, conjoint and open with endarch xylem.

8. Pith is parenchymatous. Laticifierous cells present in pith also.

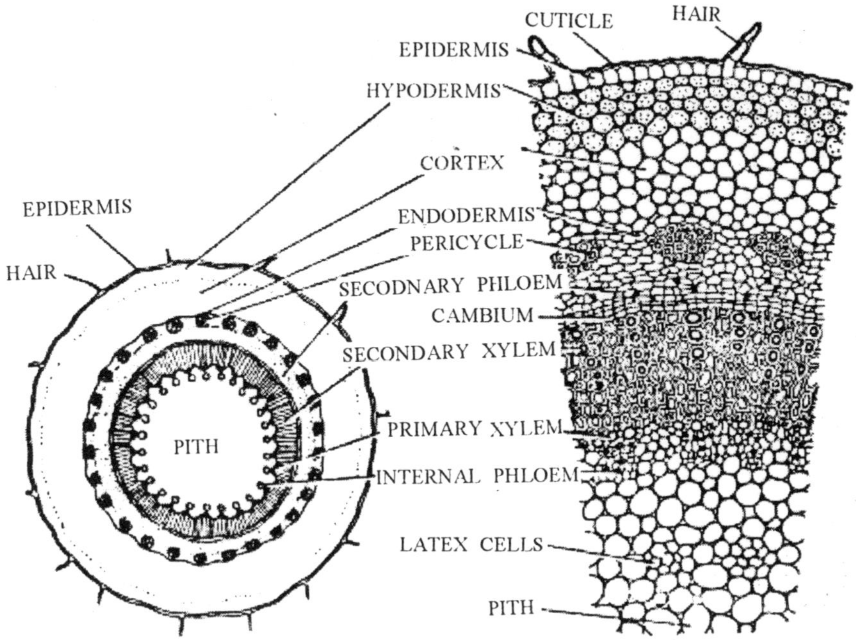

Fig 13.35 Calotropis, T.S. of Stem
A. Ground Plan
B. A part enlarged

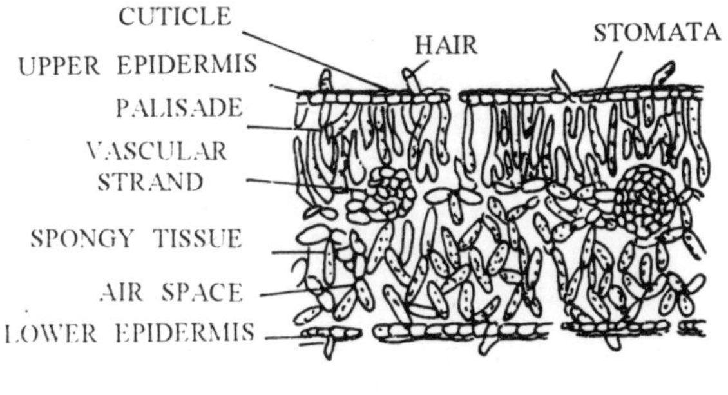

Fig 13.36 Catotropis, T.S. of Leaf

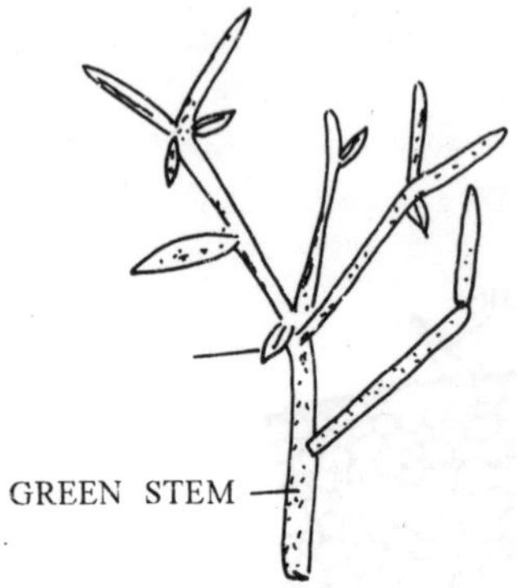

Fig 13.37 Euphorbia tirucalli, Habit

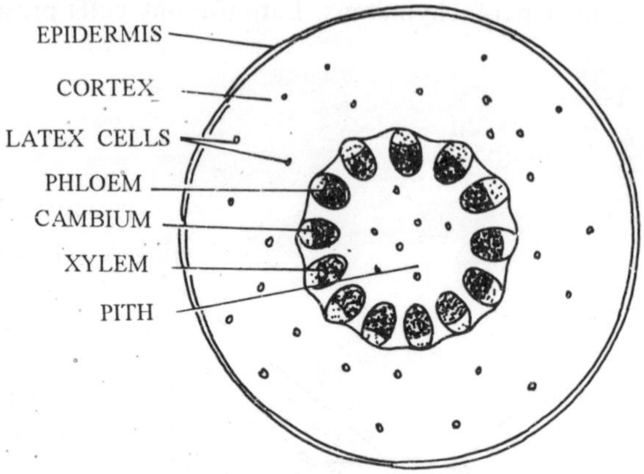

Fig 13.38 Euphorbia tirucalli, T.S. of Stem (Ground Plan)

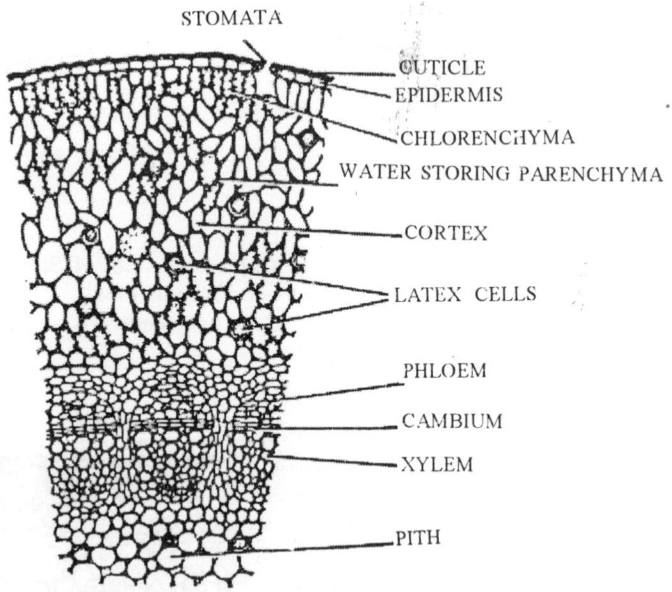

Fig 13.39 Euphorbia tirucalli, T.S. of Stem (A part enlarged)

Asparagus

Morphology

1. Plant body is a herb. Stem is cylindrical.
2. At the nodal region leaves and branches arise.
3. Leaves are reduced to scales.
4. In the axil of the scale leaves green flattened branches of one internodal length arise in clusters. These look like leaves. In reality they are modified branches know as cladodes.
5. As the leaves are reduced, cladodes take up the function of photosynthesis.
6. Roots are fasciculated, they are swollen into white tuberous structures. They store food material.

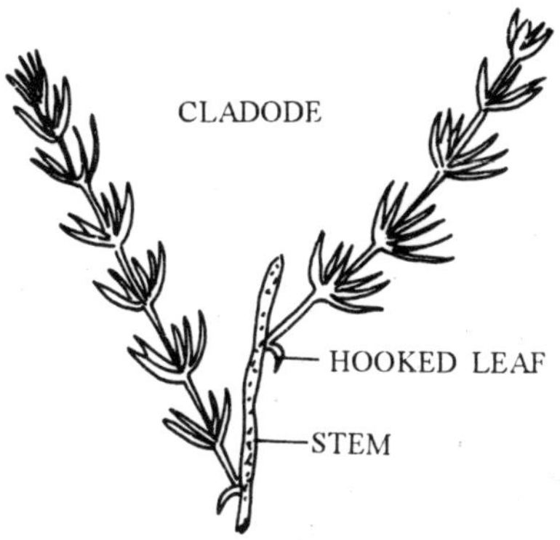

Fig 13.40 Asparagus, Habit

Transverse Section of Stem

1. A transverse section has a wavy outline due to the presence of ridges.
2. Outermost layer is epidermis which is single layered. A cuticle is deposited outside the epidermis.
3. Next to the epidermis is found a few layers of chlorenchyma constituting the hypodermis
4. Internal to the hypodermis is the endodermis.
5. Pericycle is represented by a few layers of sclerenchyma.
6. Next to the pericycle is the ground tissue made upf of parenchyma cells with intercellular spaces.
7. Vascular bundles are embedded in this ground tissue. The bundles towards the central region are larger than those towards the periphery.
8. Vascular bundles lack a bundle sheath. They are conjoint, collateral and closed with endarch xylem. Xylem is 'V' shaped.

Transverse Section of Cladode

1. The cladode is bifacial (dorsiventrally flattened).
2. Upper and lower epidermis each are single layered.
3. Towards the margins the two epidermal layers meet as the mesophyll becomes thin
4. Stomata are few in number
5. Mesophyll is differentiated, but the palisade is found all around the cladode internal to both upper and lower epidermis.
6. Internal to the palisade tissue is spongy parenchyma with many intercellular spaces.
7. There are two vascular bundles in the centre surrounded by thin walled closely packed parenchyma cells.

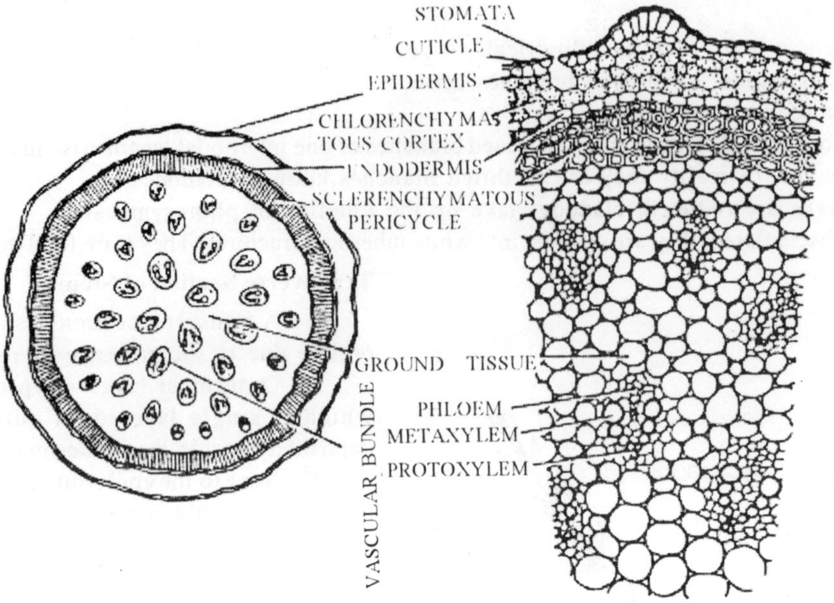

Fig 13.41 Asparagus, T.S. of stem
A. Ground Plan
B. A part enlarged

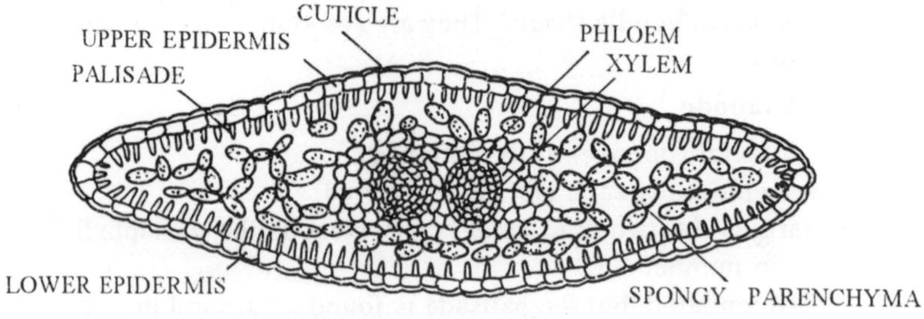

Fig 13.42 Asparagus, T.S. of Cladode

CYNODON

Morphology :

1. It is commonly called doob grass.
2. The plant usually grows in open fields, lawns etc.
3. The stem is a horizontal runner from which aerial shoots are prcduced.
4. Leaves are ribbon shaped, silica coated and have a long sheathing leaf base.

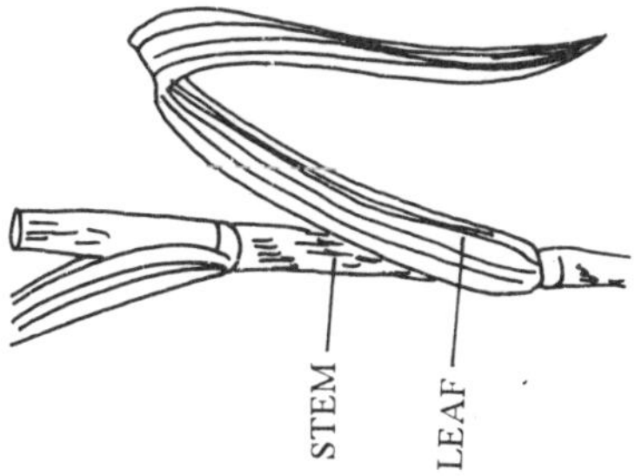

Fig 13.43 Cynodon, Habit

Transverse Section of Stem

1. Outline of of transverse section is circular.
2. Epidermis is single layered with a deposition of cuticle above.
3. Next to the epidermis internally is found a few layers of chlorenchyma.
4. Next to the chlorenchyma is a band of sclrenchyma.
5. The entire region inner to sclrenchyma is filled by a mass of parenchyma cells constituting the ground tissue.
6. A number of vascular bundles are scattered in the ground tissue.
7. Each vascular bundle is surrounded by a sclrenchymatous bundle sheath.
8. Vascular bundles are conjoint, collateral, and closed with endarch xylem. Xylem is 'V' shaped. A cavity is formed by the dissolution of protoxylem elements. Phloem is situated above the xylem.

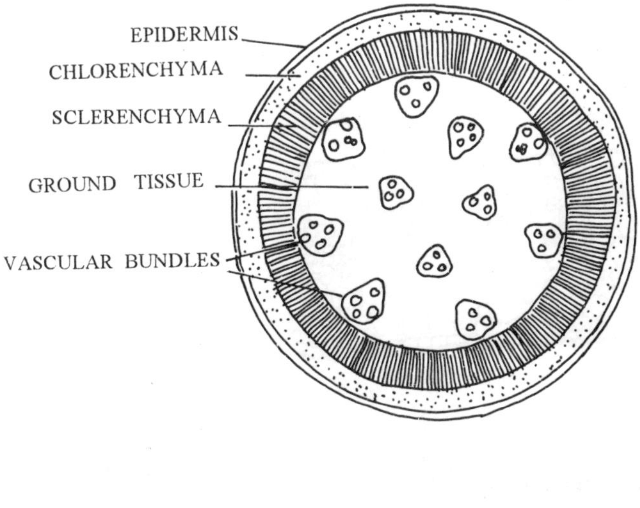

Fig 13.44 Cynodon, T.S. of Stem Ground plan

Transverse Section of Leaf

1. Leaf is isobilateral.
2. Upper and lower epidermal cells are single layered and surrounded by cuticle.
3. Stomata are found in the upper epidermis.
4. There are certain special cells in the upper epidermis called motor cells or bulliform cells. These cells help the leaf to roll up during dry conditions
5. Mesophyll is undifferentiated Chlorenchyma cells with little or no intercellular speces constitute the mesophyll tissue.

6. A number of vascular bundles are arranged paralleyly (venataion is parallel). Xylem and phloem are well developed. Occasionally pads of sclerotic cells abutt the vascular bundles both above and below.

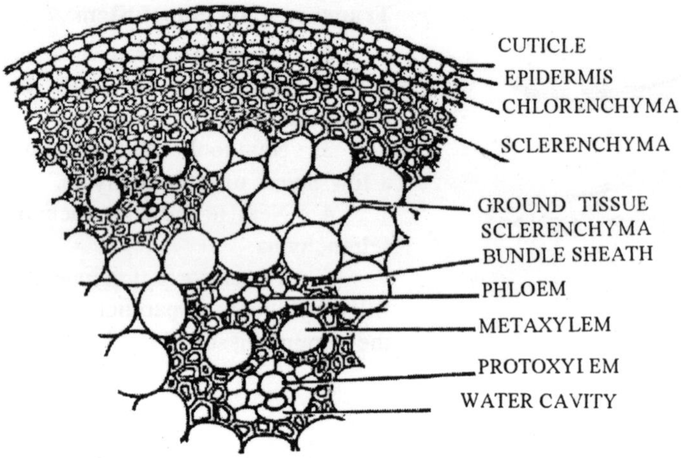

Fig 13.45 Cynodon, T.S. of Stem
(A part enlarged to show the details of vascular bundle)

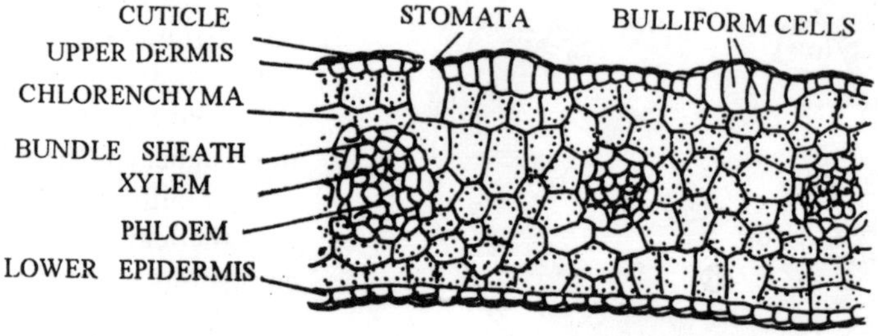

Fig 13.46 Cynodon, T.S. of Leaf

MUEHLENBECKIA

Morphology

1. Plant is a xerophytic shrub.
2. The entire stem is modified into a green flattened ribbion like structure called phylloclade.
3. Leaves are absent normally.
4. Occasionally small leaves arise from the nodal regions of the phylloclade
5. Stem takes up the function of photosynthesis
6. Flowers are borne in clusters at the nodal region.

Fig 13.47 Machlenbeckia, Habit

Transverse Section of Phylloclade

1. The phylloclade is dorsiventally flattened.
2. Upper and lower epidermis are made up of a single layers of cell each. Cuticle is found above the epidermis.
3. Internal to the epidermis all round there are patches of sclrenchyma.
4. In between the sclerotic patches chlorenchyma cells are found.
5. Endodermis is irregualr. Cells have starch grains. Pericycle is internal to the endodermis.
6. Vascular bundles are many in number; they are collateral open and endarch.
7. There is a parenchymatous pith in the centre.

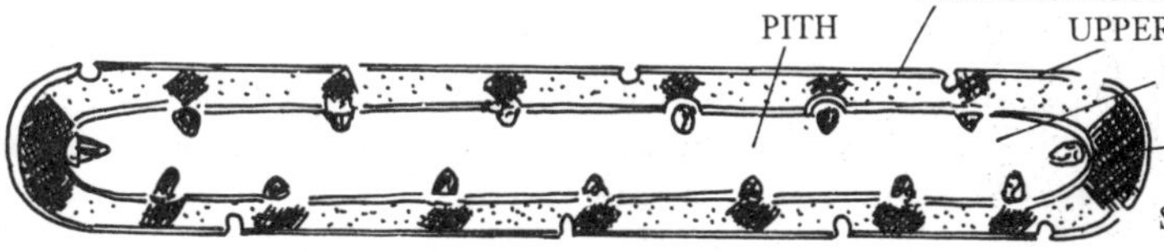

Fig 13.48 Muehlenbeckia, T.S. of Phylloclade (Ground Plan)

RUSCUS

Morphology

1. The plant is a tall herb or shrub.
2. Stem is cylindrical hard and green.
3. Branches are modified into the green flattened leaf like structures called plylloclades. Phyllocladeus are two internodes long. At the middle of the phylloclade (node) a small inflorescence arises proving the stem nature of the phylloclade.
4. Leaves are reduced to scales.

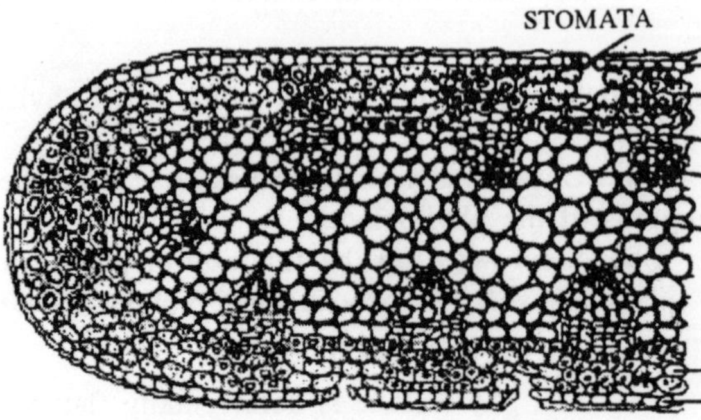

Fig 13.49 Muehlenbeckia, T.S. of Phylloclade (A part enlarged)

Transverse Section of Phylloclade

1. Epidermis is the outermost layer. It is covered by a deposition of the cuticle.
2. Stomata are present both on the upper and lower epidermis.
3. Below the upper epidermis is found a few layers of chblorenchyma to help in photosynthesis.
4. The remainder of the phylloclade has parenchyma tissue.
5. The central region consists of the vascular bundles. The vascular bundles are amphivasal (phloem surrounded by xylem) and group of parenchyma cells are found in the centre constituting the pith region. All the vascular bundles are surrounded by a common bundle sheath.
6. Additional vascular bundles are found in the 'wing' region of the phylloclade. They are also amphivasal.

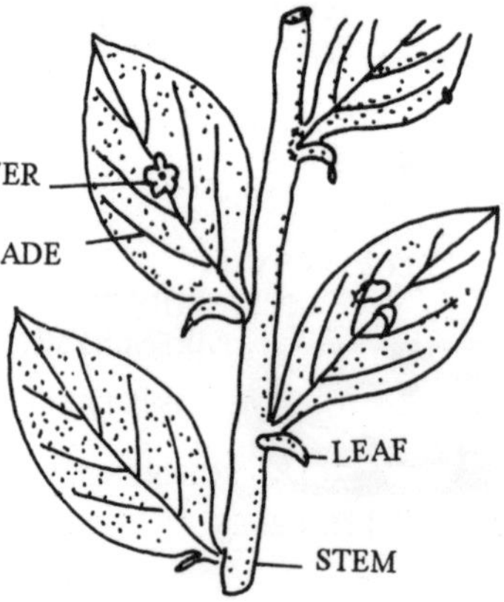

Fig 13.50 Ruscus, Habit

Opuntia

1. The entire plant body (stem) of *Opuntia* is modified into a phylloclade.
2. Leaves are absent in the adult plant. Young plants have small caducous leaves.
3. The phylloclade is flat, green, jointed, branched and succulent. It performs the function of photosynthesis.
4. Internally the phylloclade has lots of mucilage
5. On the phylloclades spines are found which represent the modifications of leaves on the axillary branches.
6. Large, red coloured attractive flowers are borne at the nodes.

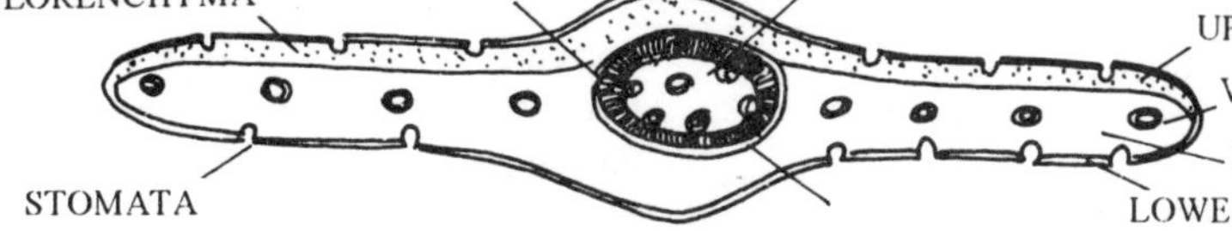

Fig 13.51 Ruscus, T.S. of Phylloclade (Ground Plan)

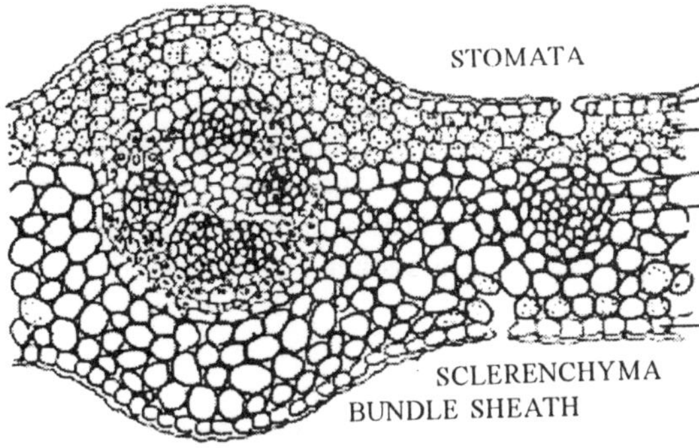

Fig 13.52 Ruscus T.S. of Phyloclade (A part enlarged)

HALOPHYTES

Plants which grow and flourish in saline soils are called halophytes. In saline or marshy soil, salts of sodium chloride, magnesium sulphate and magnesium chloride are abundant.

Rhizophora

Morphology

1. The plant body of *Rhizophora* is a small tree growing in the backwaters of sea.
2. Stem is anchored to the loose soil with the help of stilt roots.
3. In addition to the stilt roots certain special kind of roots grow apogeotropically and come above the water level. These are called pneumatophores or breathing roots. The pneumatophores have small pores called air pores which absorb air and provide the roots below as the soil has poor aeration due to water logging.
4. The seeds find it difficult to germinate in saline water due to osmotic imbalance. In order to overcome this the seeds germinate *in situ* i.e., they develop into small seedlings while still attached to the parent plant. This phenomenon is called Vivipary.
5. The seedlings get detached from the plant and fall straight into water and get dispersed due to water current.
6. Halophytes exhibit many xerophytic features such as, bushy growth, thick leathery leaves, hairy outgrowths, sunken stomata etc. This is due to physiological dryness.

PARASITES.

Parasitic plants are achlorophyllous. They depend on green plants for their nutrition. Parasities are of two types - partial and total. Partial parasites depend on the host only for some of their requirements while total parasites are wholly dependent on the host plant. A few examples are discussed below.

Viscum album

1. This is a small herb a partial parasite growing on the branches of trees.
2. The stem is jointed. Leaves arise at the nodal region.
3. The stem produces haustoria which penetrate the host stem and absorb nutrition from the stem.
4. Seeds are sticky and when birds leave it on twigs (after eating the fruit), they (seeds) germinate there with the radicle coming out as a sucker
5. The sucker after penetrating the host tissue gives out branches which form an internal absorbing system.

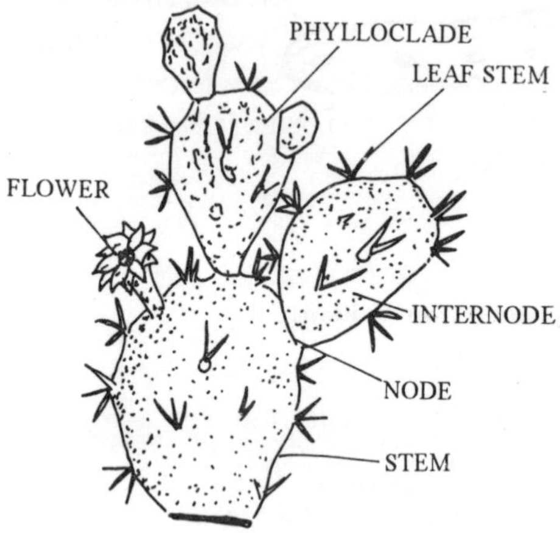

Fig 13.53 Opuntia, Habit

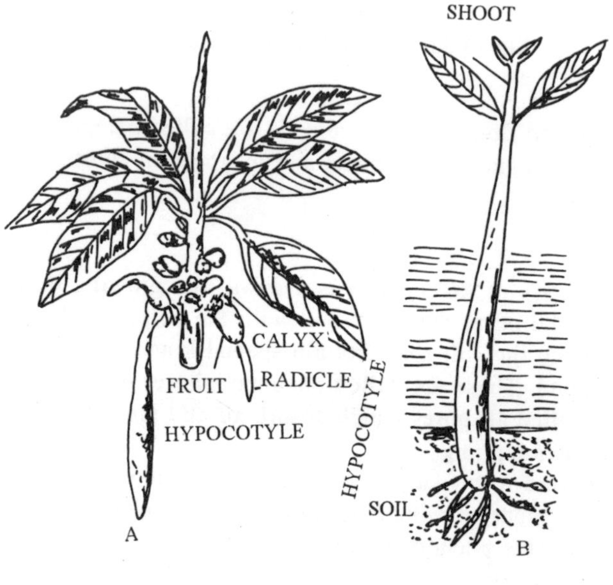

Fig 13.54 Rhizophora A. A branch showing Vivipary B. A seedling Establishing in the soil

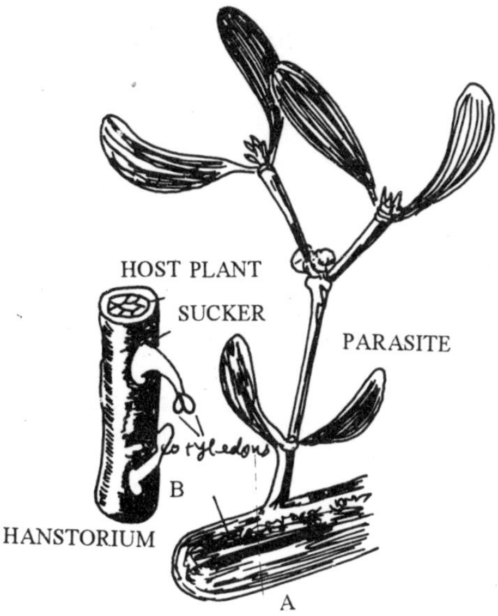

Fig 13.55 Viscum, A. Habit
B. A Stages of Germination on host stem.

Cassytha filiformis

1. The plant is a partial stem parasite.
2. The plant body consists of wixy stem that twines round the host stem.
3. Leaves and roots are absent.
4. Haustoria produced at several points from the stem penetrate the host stem and absorb nutrition.
5. The stem is greenish in colour and hence it can carry out some amount of photosynthesis.

Fig 13.56 Cassytha filformis
A partial Stem parasite (Habit)

Santalum album

1. This is a famouxs oil yielding tree.
2. Leaves are green in colour.
3. The plant is a partial root parasite.
4. Some of the lateral roots establish organic contact with the roots of other plants and absorb nutrition.

Cuscuta reflexa

1. The plant is a total stem parasite growing on host plants such as *Tecoma*.
2. The plant body consists of a yellow coloured stem that twines round the host stem.
3. Leaves and roots are absent.
4. Haustorial roots produced by the parasite penetrate the host stem reach the vasuclar bundles and absorb nutrition.

Orobanche

1. The plant is a total root parasite.
2. It has no leaves and grows on the roots of host plants.
3. The seed germinates only on coming into contact with the root of a host plant and forms an underground tuber.
4. The tuber gives rise to an aerial scape producing flowers which are birghtly coloured.

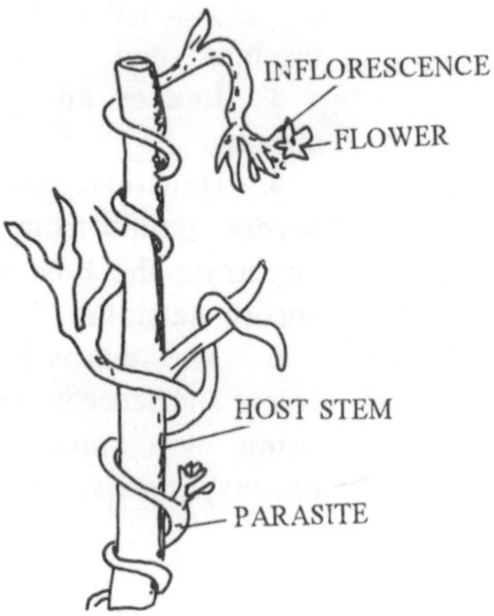

Fig 13.57 Cuscuta - A total stem Parasite, Habit

Epiphytes

Epiphytes grow on other plants only for shelter and sunlight. They are also called aerophytes. A number of plants belonging to families orchidaceae, Araceae, Cactaceae are epiphytic. *Vanda* one of the common orchid is discussed here.

Vanda roxburghii

1. The plant is a small herb growing on the branches of trees.
2. There are two types of roots - clinging roots and aerial roots (epiphytic roots).
3. Clinging roots anchor the epiphytes to the branch of the host plant.
4. Aerial roots hang down in air and absorb atmospheric moisture with the help of a tissue called velamen. This is the only source of water for the epiphyte as it does not absorb any water or nutrition from the host.
5. The epiphytic roots are green in colour and can carry out photosynthesis. Hence they are also called assimilatory roots.
6. Epiphytes develop many xeromorphic features as they get very little water. These are stunted growth, leathery leaves, thick cuticle, sunken stomata etc.

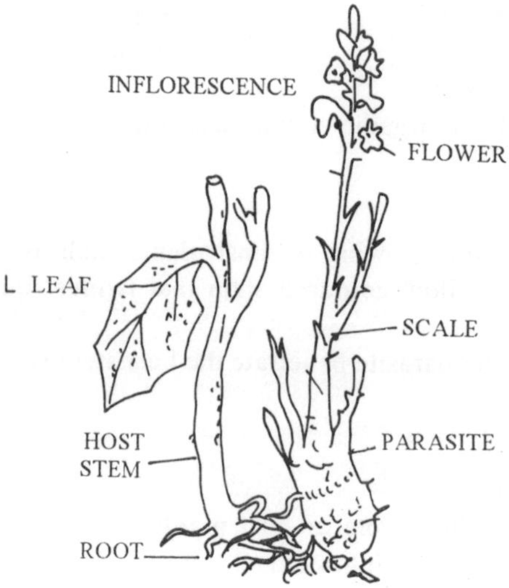

Fig 13.58 Orobanche, A total root parasite, Habit

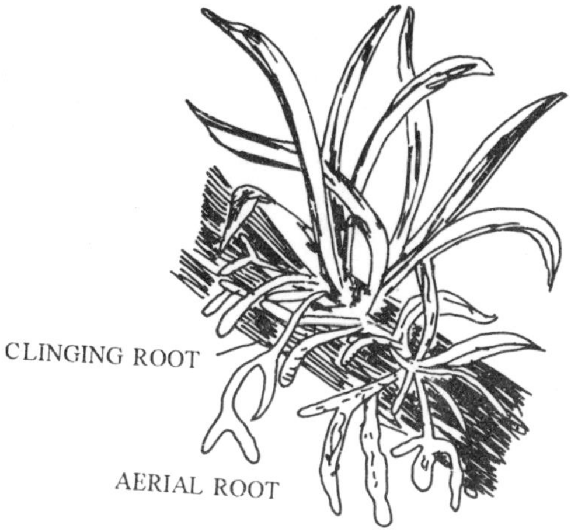

Fig 13.59 Vanda, an epiphytic Orchid

PART FOUR
PLANT PHYSIOLOGY

CHAPTER 14

PLANT PHYSIOLOGY

Green plants in general and higher vascular plants in particular provide sustenance for all forms of life including human. Plants play a vital role in the survival of man on land. Human life as we know it perhaps would have been impossible but for plants. Plants provide us the three basic necessities of life namely food, clothing and shelter. In order to benefit from plant life, a thorough knowledge of their growth, development and function is necessary. For, it is this knowledge that can help us to make attempts towards improvement of plant life. The branch of botany that deals with various functional aspects of plants is called **Plant Physiology.**

Plant physiology may be said to be the study of vital phenomena in plants. It is concerned with the processes and functions, and also responses of plants to their surroundings. The result of plant's responses to the environment namely growth, development and reproduction are also studied in plant physiology.

The process studied in plant physiology are photosynthesis, respiration, translocation etc. Each process has a continuing sequence of events and plant physiology explains and describes each one of these events. Plant physiology also attempts to explain the functions of plants at various levels. Functions of cell organelles, cells, tissues and organs are explained.

To summarize, the following are the main tasks of plant physiology.

1. Describe and explain the process
2. Describe and explain the functions at various levels in the plant organization.
3. Explain the function of chemical constituents present in the body of plants.
4. Explain the effects of the environment of process and functions in the plant system.
5. Emphasize the interrelationship and interdependence of various processes and functions, so as to understand the life processes going on in theplant.

In order to realise the above objectives, plant physiologists derive help from physics, chemistry, mathematics, statistics etc. Indeed, the role of these physical sciences in greater in plant physiology than in other branch of biology. Modern developments in these branches of physical sciences have greatly helped in unravelling the secrets of many an important physiological process in plants.

Various tools and techniques are employed, both physical and biological, in the study of plant physiology. Microscope, chromatography, Autoradiography, Electrophoresis, Tracer techniques, Spectrophotometry and a whole range of biochemical techniques form a part physiology is a dynamic discipline which continuously interacts with other disciplines and constantly updates itself in tune with advancements in other disciplines of science. Developments in the field of biotechnology, genetic engineering etc., have also helped the study of various physiology processes *in vitro.*

PLANT CELL

1. Protoplasmic Streaming

Materials : Leaves of *Elodea*, staminal hairs of *Tradescantia*, slides, cover slips, water, compound microscope.

Procedure : Collect fresh leaves or dissect out the staminal hairs. Place them on the slide and cover with the help of cover slip adding a drop of water. Observe under high power of the microscope, specially the chloroplasts in *Elodea* leaf.

Observation : The chloroplasts appear to move in a circulatory fashion around the vacuole.

In the case of *Elodea* there is a large central vacuole and the chlorplasts move in a clockwise or anticolock wise direction. This is known as Rotation. In the staminal hairs of *Tradescantia* there are many vacuoles and the contests move around each vacuole independently and it is known as Circulation.

Inference : The movement of the chloroplasts seen under the microscope is actually due to the circulation or streaming of protoplasm. The movement probably occurs due to the sol-gel transformation of protoplasm. Microtubules of cytoplasm may also contribute to the movement.

2. Importance of Living Membrane

Materials: Beet root, cork borer, water, ice, spirit lamp, beakers, test tubes, thermometer, alcohol, water bath, test tube stand.

Procedure : Cut small cylinders of (approximately one c.m length) beet root with the help of a cork borer. Place the cylinders in a beaker and wash throughly under running water (to remove the pigments that ooze out) till the exudation of colour stops.

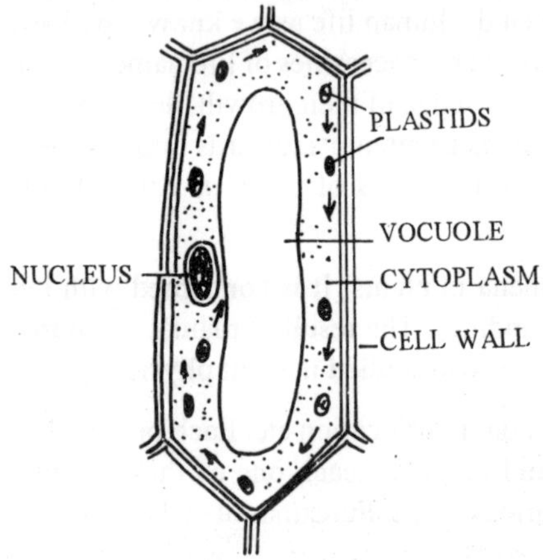

Fig 14.1 Streaming of Protoplasm in Hydrilla (Leaf cell)

Fill seven test tubes with about 20ml of water. to each test tube add 2 cylinders and keep the test tubes in water bath separately at different temperatures (25°, 30°, 40°, 50°, 60°, 75° and boiling temperature) for about 5 minutes. Also place two cylinders in ice cold water (20 ml) contained in a test tube for 5 minutes. Repeat the procedure with absolute alcohol and acetone.

Observation : Pigments from the cylinders of beet root start leaching out and the intesity of colour increases with increase in temperature with maximum being in boiling water. Increased pigment leaching is noticed in organic solvents like alcohol and acetone.

Inference : Under Normal temperature, membranes do not allow pigments to leach out, with increased temperature however permeability increases and in boiling water the membrane retentivity is lost resulting in maximum leaching. Alcohol and acetone treatments also result in membrane losing retentivity.

SOLUTIONS, SUSPENSIONS AND COLLOIDS

3. Preparation of 5% solution of sucrose (w/w)

Materials : Sucrose, water, measuring cylinder, Balance etc.

Procedure : Weigh 5 gms of analytical grade sucrose and put it into a 100 ml volumetric flask. Add little water and stir to dissolve sucrose. Now make up the volume to 100 ml.

Conclusion : A true solution of sucrose is obtained by this procedure as the solute completely dissolves in water. This is a percentage solution weight by weight (w/w) i.e 5 gms sucrose in 95 gms of water

4. Preparation of 5% solution of Sucrose (w1)

Materials : Same as in experiment No. 3.

Procedure : Weight 5 gms of analytical grade sucrose and dissolve it in 100 ml of water. Stir until sucrose completely dissolved.

Conclusion : This is a percentage solution weight by volume (w/v) i.e., 5 gms of sucrose in 100 ml of water.

5. Preparation of 80% alcohol (v/v)

Materials : Absolute alcohol, distilled (water, measuring cylinder, beaker etc.)

Procedure : Take 80 ml of absolute alcohol and to it add 20 ml of distilled water. The volume will now be 100 ml.

Conclusion : This is a percentage solution of (liquid in liquid) volume by volume. The solution is 100 ml of 80% alcohol.

6. Preparation of 1 Molar solution of Glucose

Materials : Sucrose, distilled water, balance, measuring cylinder etc.

Procedure : Weigh 180.16 grams of glucose and dissolve it in some quantity of water. Then add water to make up the final volume to 1 litre

Conclusion : 1 molar solution of glucose will have 1 mole of glucose in one litre of water. What is a mole? Gram molecular weight of a substance dissolved in one litre of water will give one mole for every litre. What is gram molecular weight? Weight in grams of a substance equivalent to its atomic weight. In the case of glucose the atomic weight is $C_6 H_{12} O_6 = 12 \times 6 + 12.6 + 16.6 = 180.16x$ i.e., 180.16 gms of glucose in 1 gram molecular weight. Hence the above solution is 1 molar (1m). To prepare a 2 molar solution dissolve 360.32 gms of glucose in distilled water and make up the volume to one litre.

7. Preparation of 1 molal solution of glucose

Materials : Same as in experiment no - 6

Procedure : Dissolve a gram molecular weight of glucose (180.16 gms) in one litre of water.

Conclusion : This is slightly different from a molar solution in that the final volume of the solution will be more than one litre. Here one litre of a solution has slightly less than 1 mole of glucose per litre of the solution.

8. Preparation of 1 normal solution of NaoH

Materials : NaoH, Distilled water, balance, measuring cylinder etc.

Procedure : Weigh 40 gms of NaoH and dissolve it in some quantity of water. Make up the final volume to one litre.

Conclusion : In a normal solution a gram equivalent of a substance is dissolved in water to finally make one litre of the solution. What is gram equivalent weight? Gram molecular weight divided by the hydrogen equivalent. In NaoH the total molecular weight is 23 + 17 = 40 gms. Since there is only one hydrogen its value is 1. Hence 40 gms divided by 1 is 40. The gram equivalent weight of NaoH is 40. In this case gram molecular weight and gram equivalent weight are same as there is only one hydrogn.

9. Preparation of 1 normal H_2So_4

Materials : Commercial, H_2So_4, Water, Measuring cylinder etc.

Procedure : Dissolve 49 gms or 28 ml H_2So_4 in one litre of water.

Conclusion : 49 gms is the gram equivalent weight of H_2So_4. Since it cannot be weighed it can be converted to volume by the formula V=M/D. Specific gravity of H_2So_4 is 1.84. Hence 49/1.8 = 26 ml. Since H_2So_4 is 8% pure about 28ml of H2SO4 is required to prepare 1 litre of 1N solution of H_2So_4.

10. Preparation of a Suspension

Materials : Barium Chloride Solution, test tubes, dilute H_2SO_4, Test tube stand etc.

Procedure : Pour a little quantity of $BaCl_2$ solution in a test tube. Add a few ml of sulphuric acid. Shake the mixture thoroughly.

Observation : A white precipitate of barium sulphate settles down.

Conclusion: In a suspension, particles remain dispersed in the medium. They do not dissolve. Suspensions are generally unstable and gradually the particles settle down and the pure solvent can be separated.

11. Preparation of an Emulsion

Materials : Groundnut oil or mustard oil, water, test tube etc.

Procedure : Take little amount of oil (10ml) in a test tube. Add 20 ml of water and shake vigorously.

Observation : Oil and water are imiscible. Water is polar while oil is non polar, further the density of oil being less than that of water, it (oil) floats on the surface of water. . Thus two layers (oil and water) are formed in this emulsion.

12. Demonstration of Brownian Movement

Materials : Microscope, latex of *Calotropis*, or *Nerium* India ink, slide, cover slip etc.

Procedure : Place a drop of the latex on the slide and cover it with cover slip. Observe it under the high power objective of the microscope

Dilute India ink abou t10 times and follow the same procedure as for latex.

Observation : Small particles in the liquid get illuminated and move to and fro in the microscopic field.

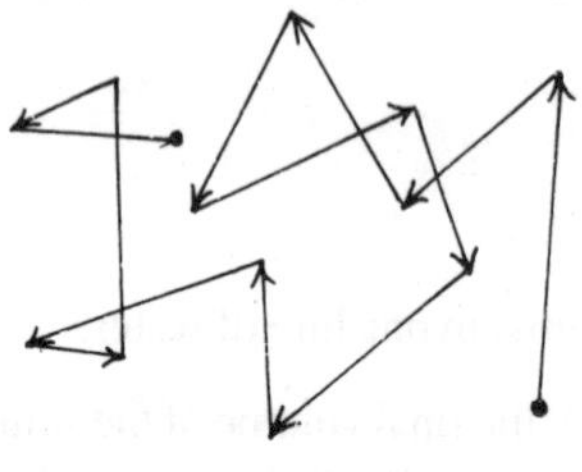

Fig 14.2 Demonstration of Brownian Movement

Conclusion : The particles are in constant motion due to their inherent kinetic energy in a colloidal system. Uneven bombardment of the particles by the dispersion medium results in this movement. This movement was first discovered by Brown, hence it is called Brownian movement.

13. Demonstration of Tyndall effect

Materials : 1% glucose solution, fruit juice, beaker, pencil, torch, water.

Procedure : Take two beakers and pour glucose solution in one and fruit juice in the other. Keep the beakers in darkness. Direct the torch towards each beaker and switch on the light so that the beam of light passes through the solution or fruit juice. Observe the behaviour of the light beam in both the cases.

Observation : The beam of light will be normal in the case of beaker having glucose solution., while in the case of fruit juice the light beam gets scattered.

Conclusion : Fruit juice being a colloideal solution the light beam gets scattered as the particles are in constant motion. This phenomenon of scattering of light while passing through a colloidal solution is known as Tyndall effect.

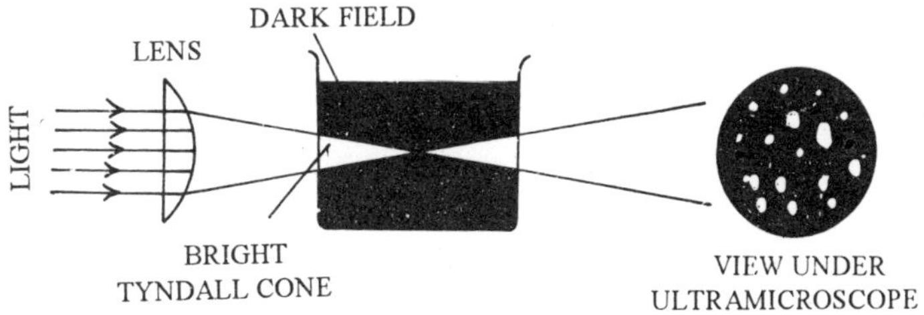

Fig 14.3 Demonstration of Tyndall effect

14. Separation of Starch and Salt by dialysis

Materials : Collodion bag, starch, Nacl, water, test tube, spirit lamp, silver nitrate, iodine, thread, stand, glass rod, beakers etc.

Procedure : Prepare 2 % solution of Nacl and a colloidal solution of starch (Add 1 gm of starch to 50 ml of boiling water with constant strirring). Mix the two solutions in equal volumes. Fill the mixture in a dialysis bag (collodion bag) and immerse it as shown in the figure in a beaker containing distilled water. After about 1 or 1 1/2 hours, test the water of the beaker with silver nitrate and Iodine Solutions separately.

Observation : On treated with silver nitrate, the water turns milky showing the presence of salt. It does not turn blue idicating starch has not diffused out of the dialysis bag.

Conclusion : The collodion membrane is differentially permeable, wnile it allows the salts to go out, it does not allow starch to go out.

Movement of Substances

15. Demonstration of Diffusion :

Materials : Beaker, water, $CuSo_4$, crystals.

Procedure : Place few crystals of copper sulphate in a beaker containing water and observe.

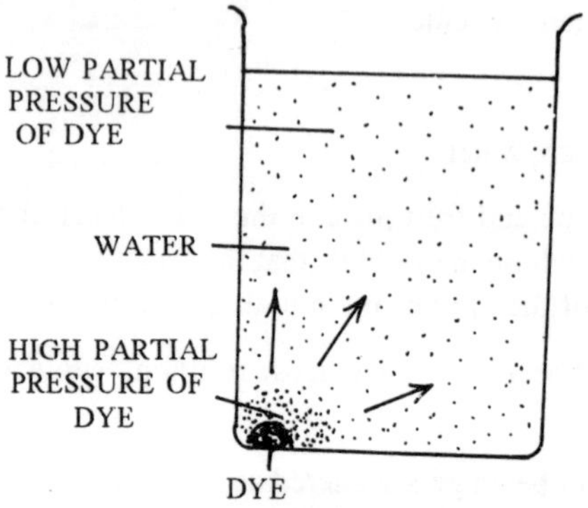

Fig 14.4 Demonstration of Diffusion

Observation : The crystals dissolve slowly in water. The particles may be seen moving away rapidly from the crystal to other parts in water. After the crystals completely dissolve this unidirectional movement stops and there will be to and fro movement.

Conclusion : The above experiment demonstrates the diffusion of solid particles. Diffusion is defined as the particles of solid, liquied organs distributing themselves evenly throughout the available place.

16. Demonstration of Imbibition

Materials : Dry seeds, water, pertridish.

Procedure : Soak the seeds in water overnight. Observe the ssize of the seeds. Check the termperature of water.

Observation : The seeds become swollen. After some time the seed coat breaks open. There will be slight rise in termperature of water.

Conclusions : Seeds swell due to imbition of water. The cotyledons swell at a much faster rate than seed coat, the seed coat breaks open exposing the cotyledons. This helps in germination. Imbibition involves both *absorption* as well as *adsorption*. Due to adsorption water molecules are held over the surface of the seed. As they cannot move they lose their energy in the form of heat into water; hence the rise in temperature of the water.

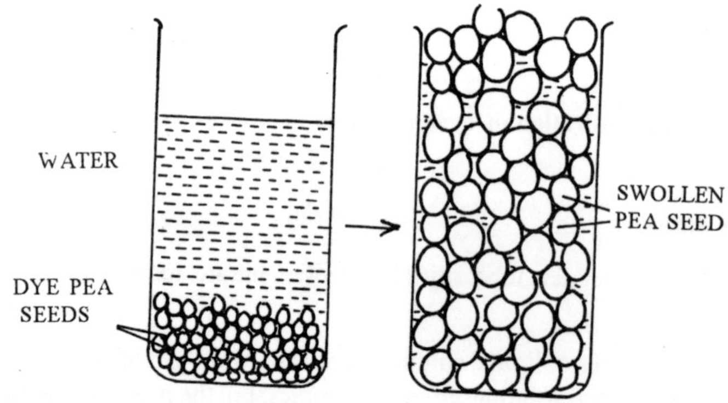

Fig 14.5 Demonstration of Imbition

16. Demonstration of Imbition Pressure

Materials : Imbibitional pressure instrument (disatometer), dry bean seeds, water and weights.

Procedure : Keep the apparatus as shown in the figure. The apparatus consists of a cylindrical jar fitted with a metallic piston. The metallic piston has a flat pan on its top to which weights can be attached. To the upperr end of the piston is atached a pointer (indicator) which glides above and below on a graduated arc scale fitted to a stand.

Keep some dry bean seeds in the apparatus and add water into it. Press the piston over the seeds by attaching weights on the pan. Note down the initial reading on the scale. Leave the experiment for about two hours. Note the final reading on the scale.

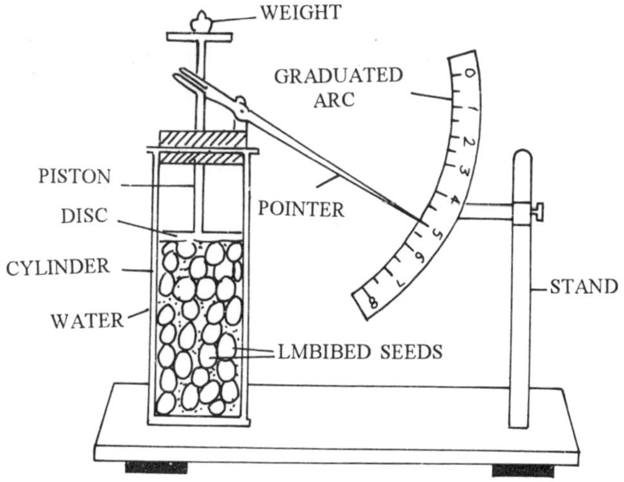

Fig 14.6 Apparatus for measurement of Imbibition Pressure

Observation : As the seeds soak in water they become swollen and push the piston above. As a result the indicator moves down on the scale certain degrees.

Conclusion : Due to imbibition the seeds swell and as they are confined in a closed container they develop a pressure and as a result the piston moves up. If we put enough weight on the pan so that the piston does not move up, that weight or force equals the imbibitional pressure developed by the seeds.

17. Physical Demonstration of Osmosis (Thistle Funnel Experiment)

Materials : Egg membrane, animal bladder or parchment paper, 5% sugar solution, thistle funnel, beaker, water, stand etc.

Procedure : Take a beaker half filled with water. Take a thistle funnel and tie its mouth with egg membrane parchment paper or animal bladder. Fix the thistle funnel to a stand inversley so that the funnel is immersed in water. Fill the funnel with 5% sugar solution. Note the level by making a mark on the stem of the thistle funnel. Allow the experiment to stand for about 30 minutes. Then note the final level of sugar solution. Taste the water in beaker

Observation : The level of sugar solution in the thistle funnel has gone up. Water in the beaker does not taste sweet.

Conclusion : The egg membrane is selectively permeable only to water and not to sugar. Hence, as the solvent concentration is higher in beaker than in the thistle funnel. Water migrates into the thistle funnel and the level of sugar solution rises up. This phenomenon is known as osmosis which may be defined as Migration of solvent molecules from the region of *their* higher concentration to the region of *their* lower concentration across a selectively permeable membrane.

18. Physiological Demonstration of Osmosis (Potato Osmoscope)

Materials : Potato tubers, 5% sugar solution, water, petridish, pins.

Procedure : Select a large size raw potato and carefully peel off the skin. Cut into two pieces horizontally.

Scoop out potato tissues from the centre and fashion a small cup as shown in the figure. Pour a little of sugar solution into the potato cup. Stick a pin inside the cup to note down the level of sugar solution. Keep the potato cup (it is now called a potato osmoscope) in the petridish containing water. Leave the apparatus aside for about 30 minutes. At the end of this period note the final level of sugar solution in the potato osmoscope and stick a pin to mark the level.

Observation : It is observed that the level of the sugar solution has gone up inside the potato cup due to entry of water from petridish.

Conclusion : Potato tissues act as a selectively permeable barrier and allow only water but not solute particles to pass through. As a result water which is in higher concentration in the petridish migrates to the potatocup. This migration is due to osmosis. (For definition of osmosis see the previous experiment).

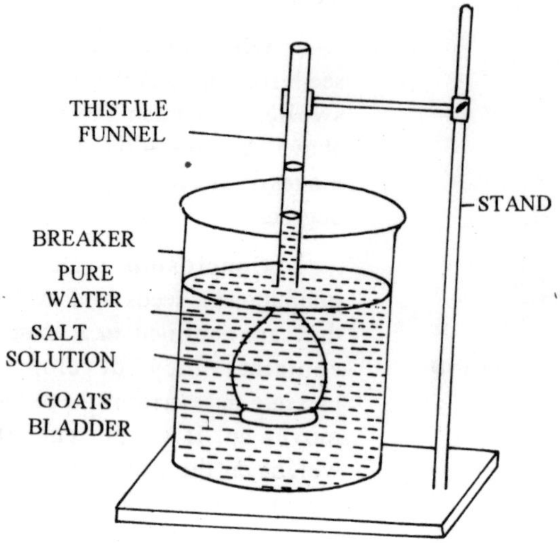

Fig 14.7 Physical Demonstration of Osmosis (Thistle funnel experimcnt)

This expeiment is known as physiological demonstration of osmosis as it is seen in living tissues, while the thistle funnel experiment in physical demonstration because it is seen in membrane removed out of the living tissues.

19. Prepartions of a Semi permeable Membrane (Cullodion bag)

Materials : Collodion solution (Cellulose nitrate dissolved in alcohol and ether), water, test tube glass rod, toluol etc.

Procedure : Pour a little collodion solution into a test tube. Keep the test tube in a slanting position and revolve it slowly so that the solution forms a coating on the inner walls of the test tube. Decant the extra solution. Allow it to settle. Then give a second coating of the solution in the same manner. Aftr some time the solution dries up into a membrane which can be carefully peeled off from the interior of the test tube.

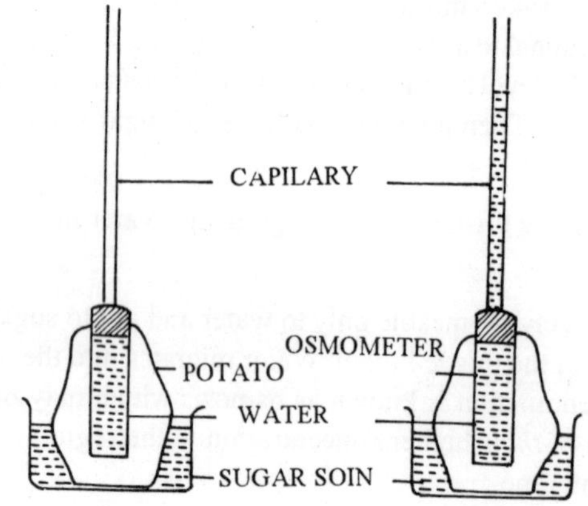

Fig 14.8 Demonstration (Physiological) of Osmosis using Potato Osmoscope

Collodion sac prepared as above can be preserved in water to which a few drops of toluol have been added. This membrane has the property of semipermeability

20. Demonstration of plasmolysis

Materials : Onions or Tradescantia leaf, safety blade, sugar solutions of different concerntrations, cover slips, slides, water, microscope etc.

Procedure : Onion peel can be obtained by tearing the onion leaves obliques when a thin transparent layer comes out. Semilarly leaf epidermal peels also can be obtained. Set up a series of slides and place a drop or two of water, 0.25 m, 0.5 m, 0.75 m and 1.0 m sugar solution on the slides as shown below.

Slide :	1	2	3	4	5
	Water	0.25m	0.5m	0.75m	1.0m

(Sugar Solution)

on each slide place a piece of the onion peel and put the cover slip. Observe the shape of the cell, cytoplasm etc and draw a sketch. Leave the preparation aside for 15 minutes. Observe again.

Observation : In slide No. 1 (water) the cells retain their original size and shape and in slide 2 (0.25m) there is slight shrinking of the cytoplasm. A small space will be left between the cell wall and plasma membrane. In slide No. 3 (0.5m) the cytoplasmic shrinkage is considerable and the plasma membrane becomes separate from the cell wall. In slides 4 and 5, cytoplasm completely loses contact with the cell wall and shrinks to its minimum size. The cytoplasm will be either concave or convex in shape.

Conclusion : In ordinary water, the cell retains its shape and size or slight increate might be there due to entry of water. Water is thus hypotonic in comparison with cell sap. In slides 2, 3, 4 and 5 the solution will be increasingly hypertonic as the shrinkage in cytoplasm becomes more due to exit of water. This exit of water from a living cell when placed in a hypertonic solution is known as plasmolysis.

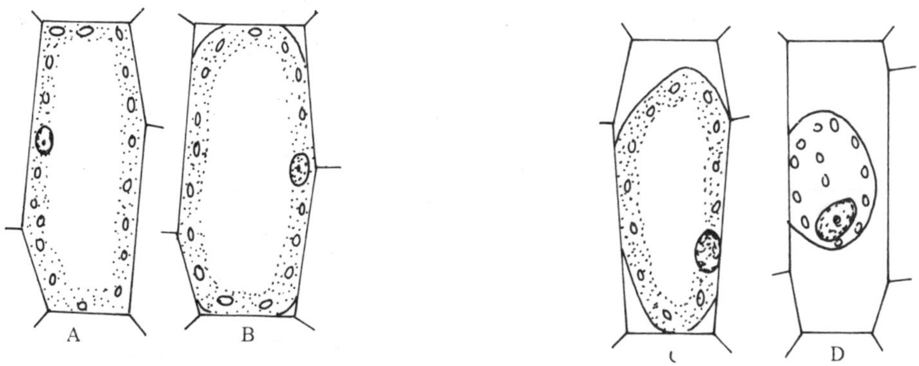

Fig 14.9 Demonstration of Plasmolysis

21. Determination of Osmotic Concentration of Cell sap by Plasmolytic Method

Materials : 1m glucose solution, onion peel, cavity slides, microscope etc.

Procedure : Prepare the following series of dilutions of glucose solutions from 1m glucose solutions.

0.1m, 0.2m, 0.3m, 0.4m, 0.5m, 0.6, 0.7m, 0.8m, 0.9 and 1 m

Take a few drops of the above concentration and place a piece of the onion peel. Observe the shape and size of cells. Leave them aside for 30 minutes and observe.

Observation : Note down as to the increase, decrease or no change of the cell size in different concerntrations of the solution.

Conclusion : The concentration of the solution where there in neither increase or decrease of the onion peel cells equals the concentration of the cell sap. This is the isotonic concentration. Wherever there is increase, it is hypotonic and wherever there is decrease it is hypertonic. For instance if 0.2m is the isotonic concentration osmotic pressure concentration) of cell sap can be calculated as follows.

OP = 22.4 atm x Molarity of external solution x Absolute temperature + Room Temperature (Approximately $24^{0}C/273$

OP = $22.4 \times 0.2 \times 273 + 24/273$

OP = 4.56

22. Measurement of Diffusion Pressure deficit (Water potential)

Materials : Potato tubers, cork borer, measuring cylinder 1m glucose solution, test tubes, balance, filter paper, scale, test tube stand.

Procedure : Prepare a stock solution of 1m glucose in distilled water (dissolve 180.16 gms in water to make the volume to 1000 ml) and prepare serial dilutions from it as follows.

0.1m, 0.2m, 0.25m, 0.3m, 0.35m, 0.4m and 0.5m. Take 15 ml of each of these dilutions in separate test tubes and keep them in a stand. Take 15 ml of distilled water in another test tube.

Peel the skin of potato and with the help of a cork borer scoop out cylindrical pieces of potato. Cut them into 24 pieces of equal length. Wash them in water and dry with the help of filter paper. Separate the potato cylinders into 8 groups of 3 pieces each. Weigh each group separately and note down the weight. Now put each group into separate test tubes containing glucose solution of different concentrations mentioned. Also place one group into distilled water kept in another test tube. There will be seven dilutions of glucose solution and another will be distilled water. Keep the tubes for about 45 minutes. At the end of this period remove groups (3 pieces each) of potato cylinders from the test tubes, wipe them with filter paper and weigh them separately. Note down the loss, increase or no change in weight in each group. Tabulate the observations as follows

No.	Contents	Initial Weight of potato cylinders	Final Weight of potato cylinder	Difference + or -
1	0.1m	4.5 gms	4.9	0.4+
2	0.2m	4.5 gms	4.7	0.2+
3	0.25m	4.5 gms	4.5	0.0
4	0.3m	4.5 gms	4.4	0.1-
5	0.35m	4.5 gms	4.3	0.2-
6	0.4m	4.5 gms	4.2	0.3-
7	0.5m	4.5 gms	4.1	0.4-
8	Distilled water	4.5 gms	5.0	0.5+

(The weights given above are arbitarary and serve only to illustrrate the example)

Observations : From the above it will be noticed that maximum increase in weight is noticed (0.5 gms) in the case of water, while there is slight increase in 0.1 m and 0.2 m, whereas in molarities above 0.3 there is decrease. In 0.25m however there is neither decrease nor increase. This concentration equals the concentration of the cell sap.

Conclusions : In concentrations below 0.25m there is entry of water into potato cylinders hence their weight increases. In concentrations above 0.2m there is exit of water from the potato cylinders hence their weight decreases. At 0.25m there is neither entry nor exit of water hence no change in weight. This concentration may be used to calculate the osmotic potential (OP) as follows.

OP = CRT

(Where C= molar concentration of the solution, R = gas constant equivalent to 0.082 and T=absolute temperature i.e. 273 + room tmperature Room temperature may be approximately 24°C)

Hence OP = 0.25 x 0.082 x 277

(This value will be a negative value as OP is always calculated as lack of so much pressure to accomodate certain quantity of water to bring its value to zero. When maximum water is present the pressure is zero).

23. Comparison of concentrations of two solutions in relation to cell sap by turrgidity method

Materials : Fresh stem of any herbaceous plant such as *Synedrella*, *Amaranthus*, *Gynura* etc, Salt solution 1(weak) and 2 (strong), watch glasses.

Procedure : Cut the stem (internode) of the herbaceous plant into a cylinders of about 4 cms. Split into two halves logitudinally. Drop one half into salt solution 1 and the other into solution 2 taken in watch glasses. 2 leave them aside for about 30 minutes. Observe the changes in the stem.

Observation : The piece immersed in a weak solution (solution1) will bend outwards i.e. towards the epidermis and the piece put in strong solution (solution 2) will bend inwards i.e. towards the pith.

Conclusion : In a hypotonic solution (solution 1) the epidermal cells will not absorb water readily due to the presence of cuticle. The pith cells however absorb water (due to osmosis) readily and increase their size. As a result due to the streching of the pith cells the stem bends outwards. In a hypertonic solution (solution 2) the pith cells lose water much faster than the epidermal cells. As a result they (pith cells) shrink resulting in the inward curvature of the stem.

24. Development of turgor pressure due to osmosis

Materials : Collodion sac, 10% sugar solution, water, beaker, thread, stand etc.

Procedure : Prepare a collodion sac as explained in experiment No. 19. Fill the sac with 10% sugar solution, twist the open end bend it back and tie it securely with the help of a thread. Immerse the sac in water contained in the beaker after securing it to a stand. Observe the changing of shape of the sac.

Observation : It is noticed that the collodion sac starts bulging due to the entry of water from the beaker.

Conclusion : The collodion membrane is selectively permeable. It allows water only but not solutes. Since the solvent concentration in the beaker is more, water enters into the collodion sac due to osmosis. As water enters, the sugar solution presses against the inner wall of the collodion sac. This pressure is called the turgor pressure. One to this pressure the collodion sac bulges.

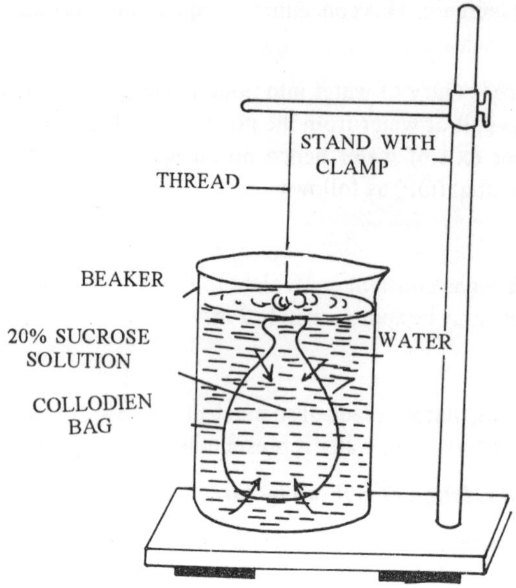

Fig 14.10 Demonstration of Turgor Pressure

25. Demonstration of endosmosis and exosmosis Using boiled Eggs

Materials : Wide mouth bottles, 10% salt solution, water, boiled eggs

Procedure : Take two boiled eggs and carefully remove the shell. Let the egg be intact without any damage. Select two wide mouth bottles (A and B) where the diametre of the mouth of the bottle should be such that, the deshelled egg can be squeezed through (the mouth) with some difficulty. Let bottle A have distilled water and bottle B have 10% salt solution. Squeeze one egg each into these bottles. Allow them to be immersed in the liquids for 30 minutes. At the end of this period try to remove the eggs from the bottles by inverting these bottles.

Observation : It will be very difficult to take out the egg from bottle A without breaking it as it would have swollen. From bottle B however, the egg comes out easily as it would have shrunken when compared with the original size.

Conclusion : The liquid (water) in bottle A is hypotonic when compared with the sap in the egg. The egg is surrounded by a selectively permeable membrane and it acts as an osmotic system and draws water into it. As a result the egg swells. This is called Endosmosis as water enters into a confined system (egg). In bottle B the liquied is a hypertonic solution whose concentration is more than that inside the egg. It is a weak solvent. Hence water goes out of the egg and into the bottle. As a result the egg shrinks in size and can easily come out of the bottle. This is called exosmosis as water comes out of a confined system (egg).

Absorption of Water

26. Absorption of water by roots

Materials: A potometer, beaker, balsam plant, rubber tube, bent glass tube, measuring cylinder, knife etc.

Procedure : Select a potometer which has a reservoir fitted to a glass jar and a graduated capillary tube attached at another end (see figure). Pull out a healthy balsam plant from the soil with intact roots. Wash the root system under running water to remove all adhering soil particles. Insert the root system into the jar filled with water through a split cork. Now cut off the shoot system a little above the cork and connect the stump to a bent glass tube with the help of a rubber tube. Let the free end of the tube be inserted into a measuring cylinder. Carefully introduce an air bubble into the graduated tube and dip it into a beaker filled with water. Leave the apparatus to stand for some time. Observe the exudation of water into the measuring cylinder and absorption of water through the graduated tube.

Observation : The air bubble moves from one end of the graduated tube to another end as the water is absorbed by the root system. By opening the reservoir the air bubble can be pushed back and the movement of the air bubble can be monitored. The quantity of water absorbed by the plant can be calculated by measuring the distance travelled by the airbubble. The liquied that comes out of the stump gets collected in the measuring cylinder is due to root pressure as a result of water absorption. It may be noted that the amount of water collected in the measuring cylinder is approximatelly equivalent to the amount of water absorbed.

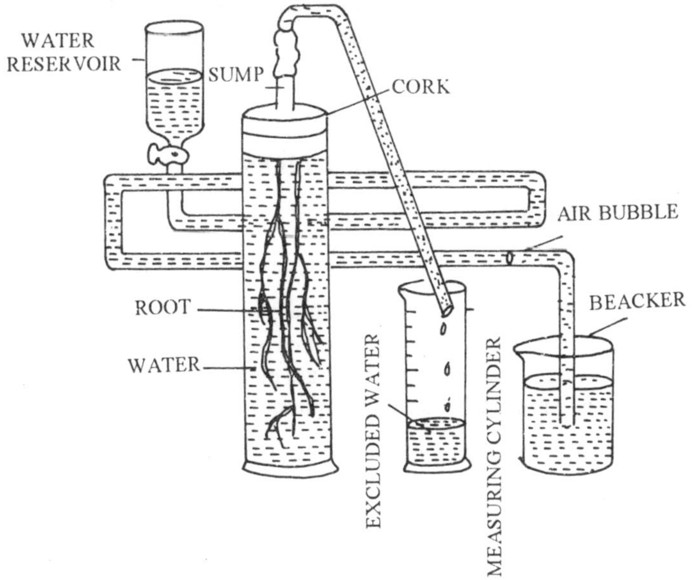

Fig 14.11 Demonstration of Water Obsorption

Conclusion : This experiment demonstrates the absorption of water through roots and its movement into the stem.

ASCENT OF SAP

27. Root Pressure

Materials : A healthy potted plant, rubber tube, blade or knife, mercury manometer, water, stand etc.

Procedure : Take a healthy potted plant. Keep it in sunlight. Water the plant well. After some time cut off the shoot a few inches above the soil level and attach a mercury manometer to it with the help of a rubber tube. Fill the tube below the bulb of manometer with water and above the bulb with mercury. Note the initial level of mercury. Keep the apparatus for a while. Note the level of mercury again.

Observation : There is increase in the level of mercury in the manometer.

Conclusion : Water being continuosly absorbed by the roots enters the cortext without much resistence. As the water reaches endodermis it does not have an easy passage and it has to get collected. But with the continued absorption of water and the unyielding endodermis a hydrostatic pressure is built up and as a result water is forced into the vasculature and it rises up to a certain level in the stem. This results in a rise in the level of mercury. This phenomenon of water rise in the stem due to the hydrostatic pressure built up by roots is known as root pressure.

28. Demonstration of Path of Ascent of Sap

Materials : Balsam plant eosin, water, beaker, slide, cover slip, microscope.

Procedure : Cut out the root portion of the balsam plant and keep it in a beaker containing water coloured with eosin. Leave it aside for about one hour. Observe the plant after that period. Cut a cross section of the stem and observe it under the microscope.

Observation : It will be seen that in the transparent stem of the balsam plant coloured lines or streaks appear indicating the path of upward movement of water. Transverse section of the stem observed under the microscope shows that only xylem vessels are coloured indicating that eosin coloured water moves up through them.

Conclusion : This experiment shows that ascent of sap in the stem is through the xylem elements

TRANSPIRATION

29. Demonstration of transpiration by bell Jar Method

Materials : A healthy potted plant, glass plate, vaseline oil, cloth or polythene bag, thread etc.

Procedure : Select a healthy potted plant and keep it on a glass plate. Cover the top of the pot with a polythene bag or oil cloth leaving a hole for the plant. Tie the cloth securely to the pot. Cover the pot with a bell jar and apply vaseline to the edges of the bell on the glass plate to make it air tight. Now keep the entire set up in sun light. Observe the apparatus after about 30 minutes.

Observation : Small drops of water appear on the inner side of the bell jar.

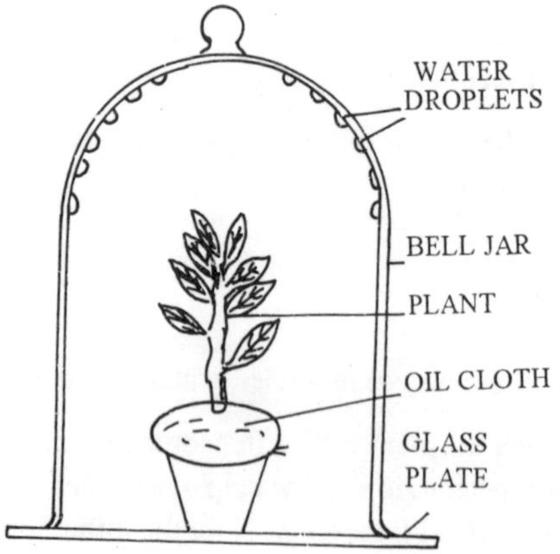

Fig 14.12 Demonstration of Transpiration (Bell jar method)

Conclusion : This experiment demonstrates transpiration. As the water vapouurs come out of the plant they get condensed on the walls of the bell jar and form water drops.

30. Demonstration of unequal transpiration from the leaf surfaces by Cobalt chloride method

Materials : A healthy potted plant with dorsivental leaves, filter paper, cobalt chloride solution (5%) slides, thread, desicator etc.

Procedure : Prepare a slightly acidic solution of cobalt chloride (5% solution). Dip pieces of filter paper in the $CoCl_2$ solution, squeeze the excess solution from the filter paper and dry in a desicator. Select an intact leaf in a healthy potted plant. With the help of a cotton wad clean both surfaces of the leaf. Take two pieces of cobalt chloride paper and place one piece on either surface of the leaf. Secure the $CoCl_2$ paper to the leaf by placing dry glass slides on them and tying them with the help of thread. Keep the entire set up in bright light and watch the change in colour of the filter papers on either side of the leaf.

Observation : Within few minutes the $cocl_2$ paper attached to lower epidermis which is blue in colour turns to pink whereas the $cocl_2$ paper attached to the upper epidermis turns pink only later.

Conclusion : Dry $Cocl_2$ paper will be blue in colour. They turn pink only in the presence of moisture. The paper on lower epidermis turns pink first because of more moisture output. Similarly the paper on upper epidermis turns pink later because of low moisture output. Moisture ouput in the form of vapours in due to transpiration. This experiment shows the rate of transpiration to be more in lower epidermis than in upper epidermis. Quantity of water vapours lost from an area due to transpiration is directly proportional to the number of stomata in that area. This experiment also demonstrates that the lower epidermis has more number of stomata per unit area than the upper epidermis.

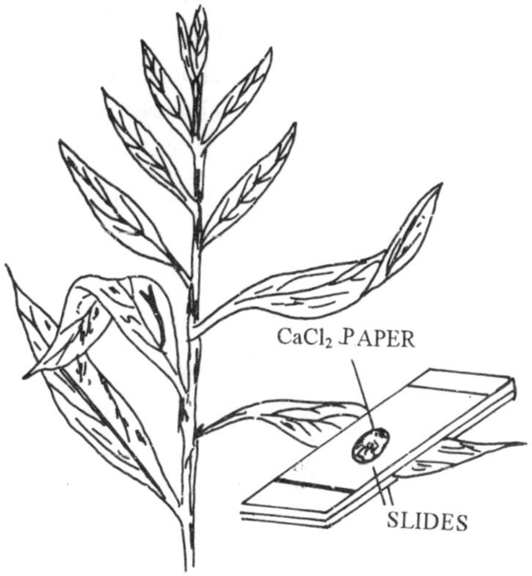

Fig 14.13 Demonstration of Unequal Transpiration (Cobalt Chloride method)

31. Comparison of rates of transpiration between upper and lower epidermis of leaf by bell jar Method

Materials : A healthy potted plant, two small bell jars, test tubes, thread, stand, mercury manometer, cork, desicator, grease, anhydrous $CaCl_2$, balance

Procedure : Select a healthy leaf in a well watered plant and place the open end of the bell jars on the upper and lower epidermis as shown in the figure. Fix the two bell jars to a stand. Now with the help of a split cork insert a mercury manometer to the other end of each bell jar. This is to disconnect the air supply between the exterior and interior. Carefully weigh about 6 grams of $CaCl_2$ kept in a desicator. Divide into two equal portions and eqaulise the weight of both the test tubes. Keep one test tube in the lower bell jar and hang another with the help of a thread in the upper bell jar. Keep the apparatus in sunlight for about 1 hour. At the end of this period remove the test tubes with $CaCl_2$ from both the bell jars and weigh them separately. Note down the increase or decrease in weight.

Observation : It is noticed that the weight of $CaCl_2$ tube in the lower bell jar has increased more than that of the tube in the upper bell jar.

Conclusion : The rate of transpiration is more in the lower epidermis than in the upper epidermis. As a result $CaCl_2$ kept in the lower bell jar absorbs more moisture and weighs more than the $CaCl_2$ tube kept in the upper bell jar. Higher rate of transpiration in the lower epidermis is due to more number of stomata per unit area than in the upper epidermis.

32. Unequal Transpiration in the lower and Upper epidermis - 4 leaves experiment

Materials : Four fresh dicotyledonous leaves, vaseline, thread, stand etc.

Precedure : Select four healthy dicotyledonous leaves of approximately same age from a single plant. Apply vaseline to the cut end of the petiole and hang them on to a thread streched between two stands by securely tying them. Smear vaseline selectively to the surfaces of the leaf as follows -

(a) Leaf 1 : Vaseline on both surfaces
(b) Leaf 2 : Vaseline on only upper surface
(c) Leaf 3 : Vaseline on only lower surface
(d) Leaf 4 : No vaseline (control)

Let the apparatus stand for about 2 hours

Observation : Leaf 4 wilts first followed by leaf 2 and 3, while leaf 1 does not wilt and remains fresh.

Conclusions : Leaf 4 wilts first because of unchecked transpiration from both upper and lower epidermis while leaf wilts next because of unchecked transpiration from lower epidermis. Leaf 3 takes a little longer because transpiration from the lower surface which is always high is checked because of smearing vaseline. Leaf 1 remains fresh for a longer time because transpiration from both upper and lower epidermis is checked. Since he leaf 2 wilts earlier than leaf 3 it is clear that the rate of transpiration is more in lower epidermis than in upper epidermis.

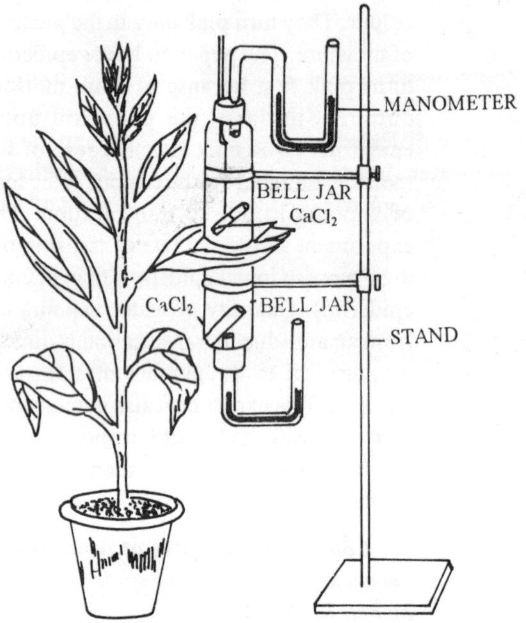

Fig 14.14 Comparison of rate of Transpiration between upper and lower surface of leaf

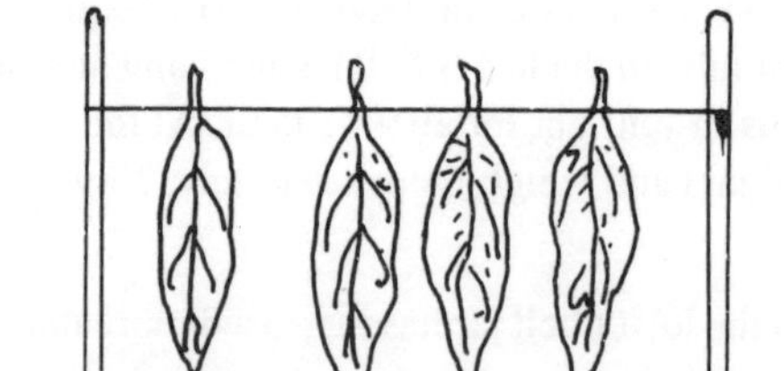

Fig 14.15 Four leaves Experiment

33. Comparison of rate of absorption and rate of Tranpiration

Materials : Wide mouthed bottle with a graduated side tube, oil, split cork, a healthy plant, balance.

Procedure : Select a wide mouthed bottle with a graduated side tube and fill it with water. Put a drop of oil on to the water in the side tube to check evaporation. Now pull out a healthy plant from soil with roots. Wash it thoroughly and fix it to the mouth of the bottle with the help of a split cork so that the root system is submerged in water. Weigh the entire apparatus and note down the weight. Similarly note down the level of water in the side tube.

Keep the apparatus in sunlight for about one hour. At the end of this period weigh the apparatus and also note the level of water in the side tube.

Observation : There would be reduction in the weight of the aparatus and at the same time the level of water would have come down in the graduated side tube.

Conclusions : Lowering of water level in the side tube is due to absorption of water by the plant while reduction in weight is due to loss of water by way of transpiration. The difference in the weight of the apparatus is equal to the amount of water absorbed by the plant. This shows that the amount of water absorbed is approximately equal to the amount of water transpired.

Fig 14.16 Apparatus to compare transpiration and Absorption

34. Development of suction force due to transpiration

Materials : An open ended glass jar with a side tube, a healthy plant, stand, beaker, mercury, water, glass tube etc.

Procedure : Take a glass jar with a side tube. Fill in water and close all the three ends with the help of corks. To the lower end of the jar fix a capillary glass tube and dip the other end of the tube in a beaker containing mercury. Initially the capillary tube is also filled with water. Fix a plant with roots into the side tube with the help of a split cork. Make the apparatus air tight and keep it in sun light. Observe after about 45 minutes.

Observation : It is observed that the level of mercury rises in the capillary tube.

Conclusion : As water evaporates due to transpiration a suction force develops and water is made to enter the plant. This results in the rise in the level of mercury in the capillary tube. Suction force can be calculated as follows.

Sf = $\pi r^2 \times 13.6$ where

sf = suction force

13.6 is the relative density of mercury and value of π is 22/7.

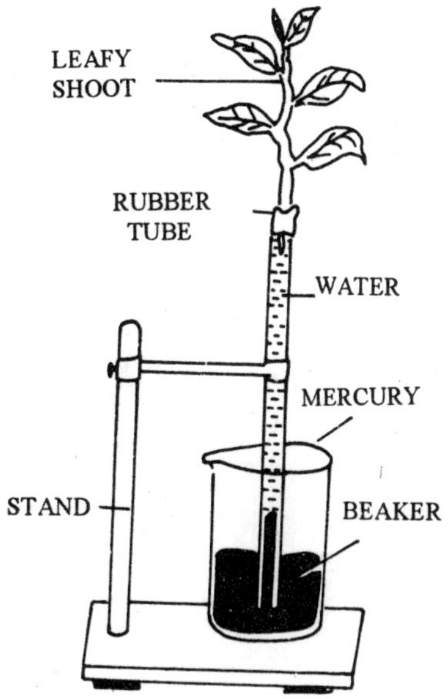

Fig 14.17 Development of suction force due to transpiration

35. Measurement of rate of transpiration by Ganong's potometer

Materials : Ganongs potometer, beaker, plant, eosin water.

Procedure : The apparatus consists of a horizontal graduated tube which is bent in opposite directions at either ends. One bent end is wide and the other end is narrow. A reservoir is fixed to the horizontal tube near the wider end. The reservoir has a stop cock to regulate the water flow. The apparatus is filled with water by opening the reservoir. A twig or a small plant is fixed to the wider arm through a split cork. The othe bent end of the horizontal tube is dipped into a beaker containing water coloured with eosin. An air bubble is introduced into the graduated tube at the narrow end (arm). As transpiration takes place, the air bubble will move from one end of the graduated tube to the other. After it has reached the reservoir, it (reservoir) is opened to push the air bubble back and then its movement can be watched again.

Observation : Record the time that is needed for the air bubble to move from one end to the other; make many observations to obtain the average of time taken to travel the distance.

Conclusion : The Ganong's potometer measures the rate of transpiration indirectly by measuring the rate of absorption. The rate of transpiration is approximately equal to the rate of absorption. By using the following formula the amount of water transpired may be caluclated.

$V = \pi \underline{r}^2 l$

Where V = value of water, $\pi = 22/7$, $\underline{r}$ = radius of graduated tube, 1 = distance travelled by the air bubble.

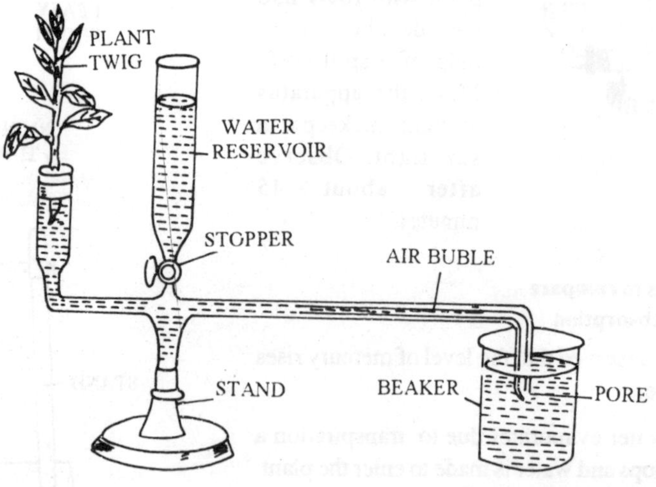

Fig 14.18 Ganong's Potometer

36. Measurement of Transpiration by Farmers' Potometer

Materials : Farmer's potometer, beaker, water, vaseline, plant

Procedure : The Farmer's potometer consists of a wide mouthed bottle. A cork with three holes is fitted on to the mouth of the bottle. One whole is fixed with a thrice bent tube with a scale fixed on to it horizontally, the second hole has a reservoir with a stopper fitted onto it and through the third hole fix a twig under water.

Fill the apparatus with water, open the reservoir so that the tube gets filled with water. Make the apparatus air tight by applying vaseline. Dip the free bent end of the horizontal tube into a beaker containing water. Introduce an air bubble into the tube. Keep the apparatus in sunlight for some time.

Observation and *Conclusion* : Same as in experiment No. 35.

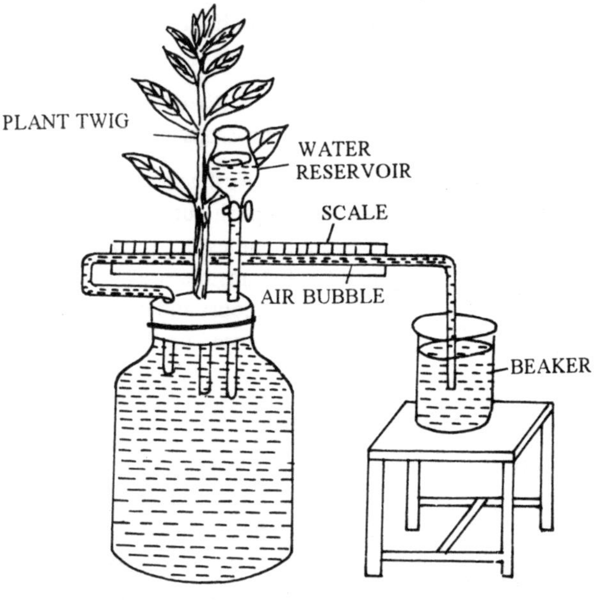

Fig 14.19 Farmer's Patometer

37. Demonstration of Continuity of air spaces in the leaf

Materials : Wide mouthed bottle, cork with two holes, aspirator, bent glass tube, vaseline, a leaf with a long petiole such as *Colocasia*, lotus etc.

Procedure : Take a wide mouthed bottle which is half filled with water. Close the mouth with a two holed cork. Introduce a leaf into the bottle through one hole so that the petiole is dipped in water. Through the other hole fix a bent tube so that the tip of the tube shoudl be few centimetres above the water level. Connect the other end of the tube to an aspirator. Apply vaseline to all connecting points so that the apparatus is air tight. Start the aspirator and observe the cut end of the petiole.

Observations : Air bubbles start coming out of the petiole as long as the aspirator is on.

Conclusion : As the aspirator sucks air the suction is transmitted through the petiols along the intercellular spaces and up to stomata. Stomata suck air and it travels down the petiole and released into water. This is possible only when the air space is continious within the plant.

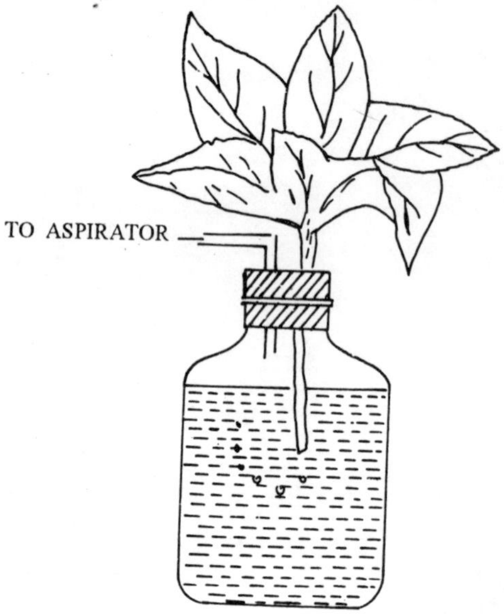

Fig 14.20 Demonstration of Continuity of Air Spaces in Plant

38. Demonstration of Guttation

Materials : *Nasturtium* plant in a pot, glass plate, bell jar, plastic sheet, vaseline, thread.

Procedure : Keep the potted planted (*Nasturtium* on a glass; plate cover the pot with a plastic sheet with a hole to allow the plant to be exposed. Invert the bell jar on the pot. Let the mouth of the bell jar at the narrow end be covered with a single holed cork. Fix a bent tube through this whole so that one end is in the bell jar while the other end is fixed to an aspirator. Make the apparatus air tight by applying vaseline. Keep the apparatus in sunlight and switch on the aspirator and observe.

Observation : As the air inside is sucked by the aspirator water drops appear on the margins of leaves.

Conclusion : As the air is sucked, hydathodes present in the leaves push out water due to the sucking force. Water here comes out physically and it is known as guttation. In transpiration however water comes out in the form of vapours and it does not settle on the leaves, but gets condensed on the bell jar.

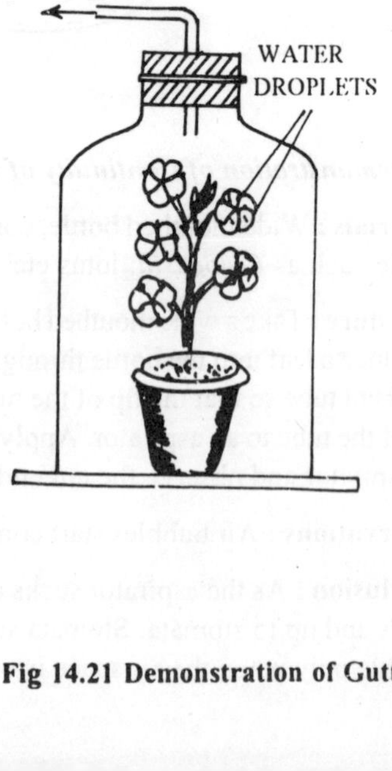

Fig 14.21 Demonstration of Guttation

39. Measurement of Stomatal Opening by Darwin's Potometer

Materials : Darwin's prometer, A healthy potted plant, water, clips, stopwatch, beaker, adhesive etc.

Procedure : Darwin's prometer has a 'T' shaped tube. Attach a rubber tubing to both the arms of 'T'. Close one of the rubber tubings with the help of a clip while to another attach a small cup. Fix the cup to the lower surface of a leaf of a potted plant with an adhesive. Smear vaseline to the edges of the cup to make it air tight. Dip the stem of 'T' in a beaker containing water open the clip and suck water through it. Water column rises in the stem of 'T' tube. Note the level. Separate the leaf from the plant and after sometime note the level of water in 'T'.

Observation : When the leaf is attached to the plant the level of water in 'T' goes down gradually while the level remains unaffected if the leaf is detached from the plant.

Conclusions : When air is sucked through the rubber tubing its pressure is lowered in the potometer. In the intact leaf stomata are open and transpire, with the result water vapours enter the cup and push down the water column. When the leaf is detached stomata remain closed, transpiration does not occur and water level in the 'T' tube remains same.

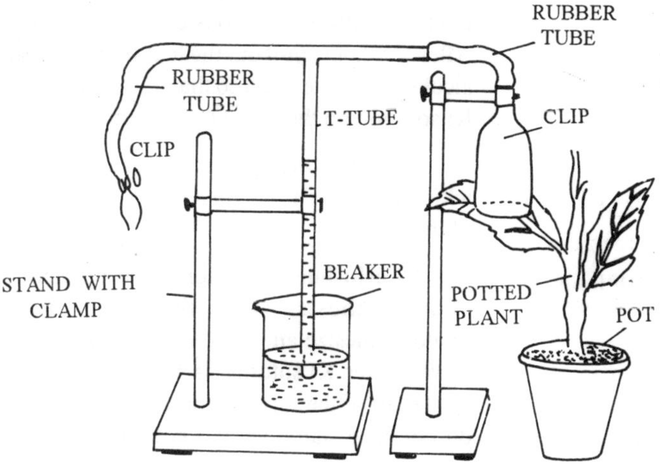

Fig 14.22 Demonstration of Stomatal opening by Darwin's potometer method

PHOTOSYNTHESIS

40. Evolution of O_2 during Photosynthesis

Materials : Beaker, watch glass, funnel, test tube, *Elodea* or *Hydrilla* plant, Pyrogallol.

Procedure : Take a glass beaker half filled with water and put some pieces of *Hydrilla* or *Elodea* plant into it. Invert a glass funnel over it. Fill a test tube with water and carefully invert it over the stem of the funnel so that the mouth of the test tube is under water. Keep the apparatus in sunlight and observe

Observation : Small air bubbles appear from the cut ends of braches of the plants and they enter the test tube. Gradually water in the test tube gets downwardly displaced and it (test tube) becomes filled with the gas. The gas can be tested by one of the following methods.

(a) cover the mouth of the test tube and carefully take out the test tube. Introduce a burning splinter into the test tube. It glows brightly.

(b) Dissolve some amount of pyrogallol in water. The gas is dissolved in pyrogallol and the level of the water in the test tube rises again.

Conclusion : Both the tests performed above will show that the gass is oxygen. The plants kept in the beaker photosynthesize in the presence of sunlight and release oxygen which comes out of the plant in the form of air bubble.

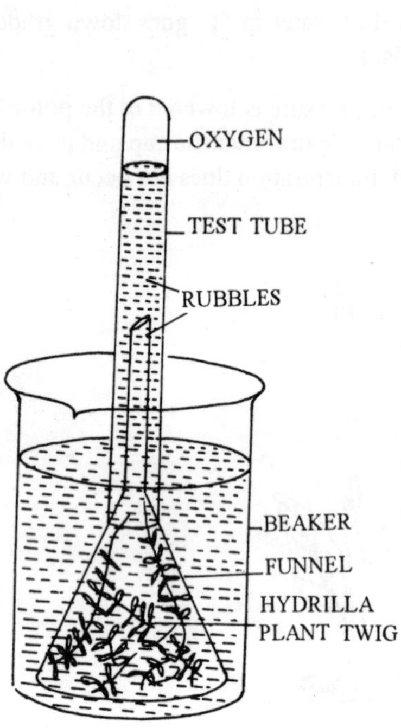

Fig 14.23 Evolution of O_2 during photosynthesis

41. Chlorophyll is necessary for photosynthresis

Materials: Plant with variegated leaves such as *Codeum*, *Coleus*, *Manihot* etc, butter paper, pencil 70% alcohol, potassium iodide solution, water bath

Procedure: Select a potted plant with variegated leaves and keep it in darkness for 72 hours so that it gets destarched. Bring out the plant and place it in bright sunlight. Select a leaf with variesated colours and placing a butter paper on the leaf trace the outline of various colours with the help of a pencil. After few hours pull out the leaf from the plant, boil it in 70% alcohol in a water bath so that the leaf gets decolourized. Pour KI solution on the leaf. Observe the change in colour.

Observation: Blue colour appears only in those areas which were green. This can be checked in the outline diagram of the leaf.

Conclusion: Chlorophyll pigment is necessary for photosynthesis. Only the green areas in the leaf photosynthesize and starch is present in those regions. That is why when KI solution is poured they turn blue (indication of starch). Other areas (non green) do not photosynthesize, do not have starch and do not turn blue on contact with KI solution.

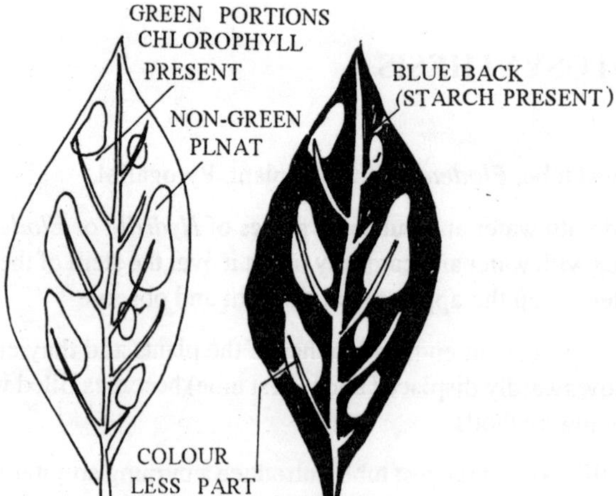

Fig 14.24 Chlorophyll is necessary for Photosynthesis

42. Light is necessary for Photosynthesis (Ganong's light screen experiment)

Materials: A potted plant, Ganong's light screen, beaker, 7% alcohol, KI solution, test tube, water bath.

Procedure: Keep a potted plant in darkness for about 72 hours so that it is starch free.

Fix a ganong's light screen to one of the leaves and place the pot in bright sunlight. After few hours pluck the leaf from the plant, remove the light screen and decolourize it by boiling (in a test tube having 70% alcohol) it in a water bath. Take out the decolourized leav and pour KI solution.

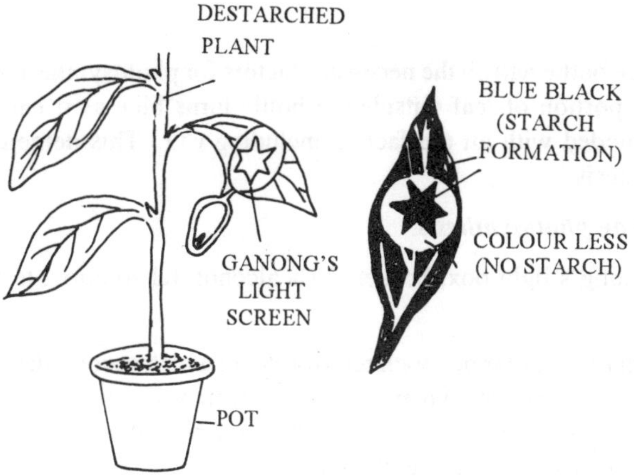

Observation : The area of leaf covered by light screen remains bleached while the other areas turn blue on contact with KI solution.

Conclusion : The area covered by the light screen does not get light hence starch is not manufactured in that area. As a result it does not turn blue on contact with KI solution. Other areas get light so they manufacture starch and turn blue on contact with KI solution.

Fig 14.25 Light is necessary for photosynthesis

43. CO_2 is necessary for photosynthesis (Mohl's Half leaf experiment)

Materials : Wide mouthed bottle, split cork, KOH, Healthy plant.

Procedure : Keep a potted plant in darkness for about 72 hours to remove starch. Select a leaf from such a plant and without removing it insert half a portion into a bottle through a split cork. Before inserting the leaf pour some quantity of KoH liquid into the bottle. Make the apparatus air tight by smearing vaseline to the edges of the cork. Keep the apparatus in sunlight for about 2 hours. At the end of the period remove the leaf from the bottle detach it from the plant and boil it is 70% alcohol to decolourize it. Pour Iodine solution and observe the change in colour.

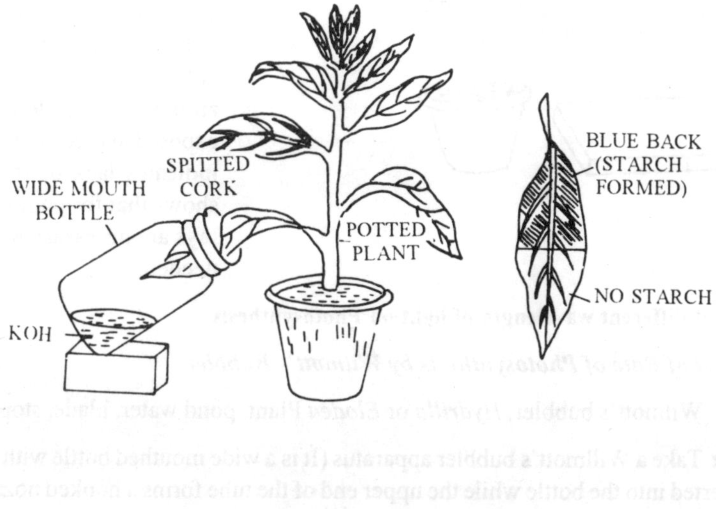

Fig 14.26 Mohl's half leaf experiment

Observation : The part of the leaf inside the bottle will remain bleached whereas the portion outside turns blue.

Conclusion : The portion of leaf inside bottle had all the necessary factors for photosynthesis except CO_2 because it was absorbed by KoH. The portion of leaf outside the bottle turns blue as it has starch manufactured by photosynthesis. It was provided with all the factors including CO_2. This demonstrates clearing that CO_2 is necessary for Photosynthesis.

44. Effect of different wave lengths of light on photosynthesis

Materials : A healthy potted plant, Ganong's light box, beaker, 70% alcohol, KI solution, test tube, water bath.

Procedure : Ganong's light box is a small rectangular box with removable lid which accomodates glass plates of different colours. Select a leaf from a healthy potted plant which has been destarched (by keeping it in darkness for 72 hours) and insert it into Ganong's light box. Let the box have three colour plates viz. red (660 μm), blue (455 μm) and green (530 μm). Keep the apparatus in sunlight with this arrangement the leaf will be exposed to three wave lengths of light viz red, blue and green of the visible spectrum. After about 2-3 hours remove the leaf from light box detach it from the plant, decolourize it by boiling it in 70% alcohol in a water bath. Pour Iodine solution over the leaf and observe the change in colour.

Observation : Three portions of the leaf were exposed to three different rays of light. The segment of leaf exposed to red light turns deep blue, the part of leaf exposed to blue rays turns light blue and the part of leaf exposed to green light remians almost unchanged

Conclusion : Part of the leaf exposed to red rays turning deep blue indicates concentration of starch, light blue colour in the region exposed to blue rays is an indication of less starch and the region exposed to green light remaining unchanged indicates lack of starch. This experiment shows that for photosynthesis red and blue rays are necessary while green rays are not needed.

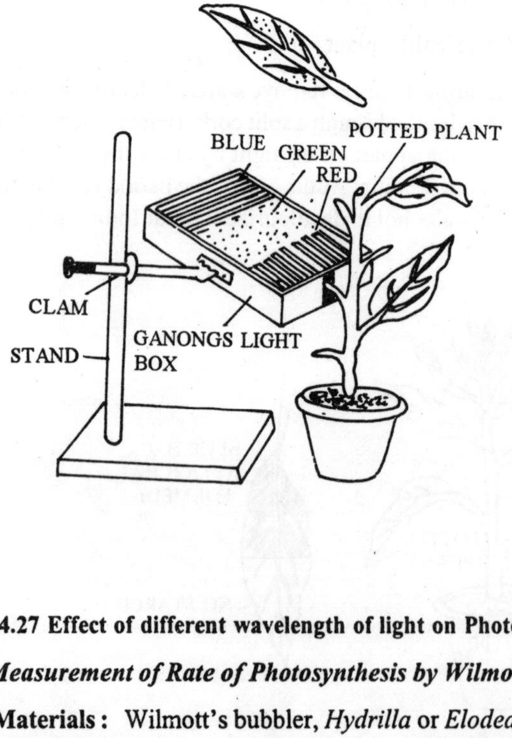

Fig 14.27 Effect of different wavelength of light on Photosynthesis

45. Measurement of Rate of Photosynthesis by Wilmott's Bubbler

Materials : Wilmott's bubbler, *Hydrilla* or *Elodea* Plant, pond water, blade, stop watch etc.

Procedure : Take a Willmott's bubbler apparatus (It is a wide mouthed bottle with a single holed cork. A glass tube is inserted into the bottle while the upper end of the tube forms a hooked nozzle inside a broad tube called reservoir).

Fill the willmotts bubler with pond water. Select a branch of *Elodea* or *Hydrilla*, cut the shoot under water and place the cut end inside the tube fitted in the cork (see figure). Dip the other end of the plant in the bottle and close the stopper. Now pour some water into the reservoir above the bottle so that the nozzle is under water. Apply vaseline to the cork and keep the apparatus in sunlight and observe.

Observation : Small air bubbles escape from the cut end of the stem travel through the glass tube and finally come out through the nozzle. These bubbles which are uniform in size can be counted per unit time by using a stop watch

Conclusion : Release of air bubbles (O_2) from the cut end of the plant indicates the rate of photosynthesis. The number of bubbles per unit time can be calculated to measure the rate of photsynthesis. This is only an arbitary measurement.

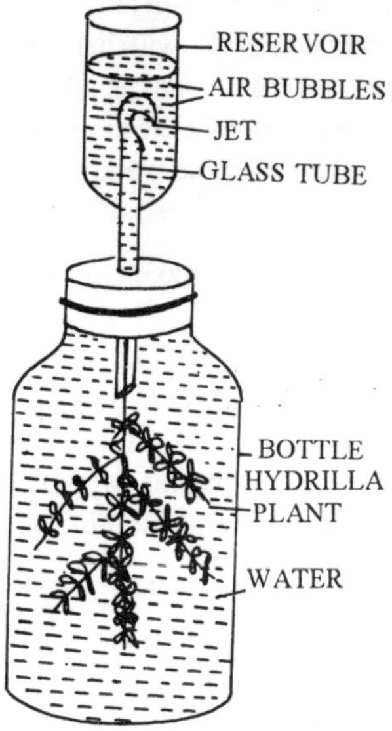

Fig 14.28 Wilmotl's bubble experiment

46. Demonstration of presence of starch in chloroplasts

Materials : Filaments of *Spirogyra* or leaves of *Elodea*, Chloral hydrate Iodine solution, slide, cover glass, microscope etc.

Procedure : Mount a filament of *Spirogyra* a green alga under the microscope. Note down the shape of chloroplasts in the cells. Stain the cells with chloral hydrate Iodine solution and observe under the microscope.

Observation: The chloroplasts in *Spirogyra* filament are spiral in shape. After staining the entire spiral band turns blue black. The chloroplasts turn blue black due to the presence of starch.

47. Study of photosynthesis by gass analysis (Ganong's Photosynthometer)

Materials: Photosynthometer, leaves, test tube, caustic potash, carbon dioxide generator.

Procedure: Ganong's photosynthometer consists of three parts - (1) a glass bulb A(2) a graduated tube C and (3) a connecting link B provided with a stop cock. B fits into the neck of A at the lower end and at the upper end the lower end of tube C fits into it. The total volume of the whole apparatus is approximately 103 ml. There are two holes a and b in A and B respectively so that they are at the same level when A fits into B.

Place a few fresh green leaves in the bulb of 'A' together with a little quantity of water. Now invert the graduated tube 'C' close its stop cock and fill it with 5 ml of water. Fix 'C' on to 'B' and close the stoplock of latter and fill water into lower half portion of 'B' and carefully covering it by hand dip into a beaker with water. Clamp the tube so that water in the beaker is on level with hole b. Open the stop cock of B as well as of 'C' and connect 'C' to a carbon dioxide generator. Stop the Co_2 entry into tube 'C' as soon as the water in tube 'C' falls to the level of water in beaker. Tube 'C' now has 5 ml of Co_2 (by displacing 5 ml of water). Close both the stop cocks ('C' and 'B\) and fit the whole thing ('B' and 'C') on to the bulb 'A'. Now the openings a and b (in 'A' and 'B' parts) are on the same level. this will equalise the pressure inside the bulb A to the atmospheric level. Twist 'B' slightly now so that a and b move apart and the connection becomes airtight. Place the apparatus in sunlight and open the stopcock of 'B' to allow CO_2 into bulb A (with leaves). Allow the apparatus to stand for 6 - 8 hours. At the end of this period, remove 'B' and 'C' together from 'A' ande dip them in water. Separate 'B' from 'C' underwater and open the stop cock of 'C' so that water enters into it up to zero mark. Fill a test tube with caustic potash and connect it to 'C' near the stop lock with the help of a rubber tube provided with a pinch cock. Lift the graduated tube ('C') from water by closing the lower end with hand and open the pinch cock as well as the stop cock to allow potash fluid to flow into 'C'. Shake thoroughly and drain the potash back to the tube and close the pinch cock as well as the stop cock. Now, still closing the lower end of 'C' with hand dip it into water, disconnect the test tube with caustic potash. Place the tube (in water) in such a way that the zero mark is on level with water. Take out the hand and water will rise into the tube up to a mark.

Observation and Conclusion : The amount of water which rises up in the tube is equal to the amount of CO_2 used up for photosynthesis. Originally 5 ml of CO_2 is present. Now if water rises up to 2 ml, it means 2 ml of CO_2 is used up for photosynthesis and 3 ml is left unused. (It is possible that this may not give an accurate amount of CO_2 used in photosynthesis, because CO_2 is slightly soluble in water).

48. Extraction of Chlorophyll Pigments By Solvent wash method

Materials : Fresh leaves (spinach, tecoma etc), or 3 grams of dry leaf powder, petroleum ether, Acetone, methyl alcohol,methanolic KOH, Separating funnel, stand, beaker, pestle, mortar etc.

Procedure : Take 5 gms of fresh leaves of spinach or *Tecoma.* Wash them thoroughly. Wipe off moisture with the help of a filter paper. Place them in a pestle and mortar and grind with 50 ml of ice cold acetone. Collect the liquid and filter. Discard the residue and collect the filterate. The filterate (acetone extract) consists of chlorophyll a, chlorphyll b, *carotene* and *xanthophyll.*

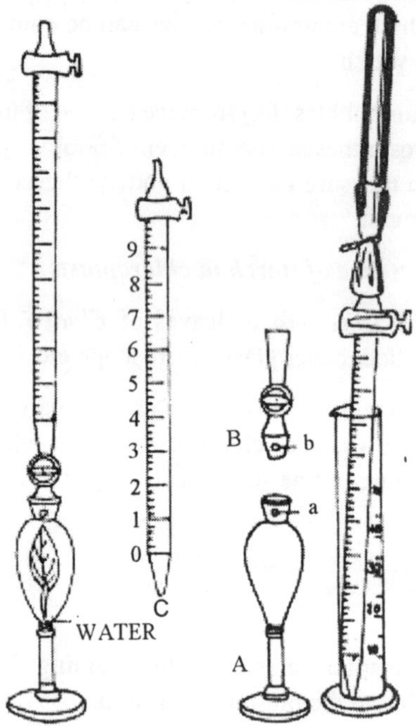

Fig 14.29Ganong's Photosynthometer

Instead of fresh leaves, leaf powder may be used. Collect fresh leaves and dry them in an oven. Powder the dry leaves. Weigh 3 gms and soak it in 40 ml of 80% acetone containing a pinch of CaCo3. Stir the solution and filter. The filterate acetone extract consists of all the pigaments mentioned above.

Add 60 ml of petroleum ether to 40 ml of acetone extract in a separateing funnel. Gently rotate the funnel. Slowly add 75 ml of distilled water along the sides of the funnel to avoid emulsification. Rotate the funnel again gently and fix it to a stand. Gradually two layers of the liquid are formed. Of these the lower is acetone - water layer. This may be discarded. The upper layer (petroleum ether layer) consists of all the pigments. Add 50 ml of distilled water gently to the petroleum ether layer. Again two layers are formed. Discard the lower layer. The upper layer may be washed again with 50 ml of distilled water and discarding the lower layer.

Add to the petroleum ether layer 50 ml of 92 % methyl alcohol. Gently rotate the separating funnel. Allow it to stand. Tow layers are formed viz. the lower methyl alcohol layer and the upper petroleum ether layer. Draw the two layers into separate beakers. Lower layer consists of chlorophyll b and xanthophyll The upper layer consists of chlorophyll a and carotene. Further procedure for two layers is as follows -

1. **Upper Petroleum ether layer :** Take 30 ml of this layer and to it add 15 ml of freshly prepared 30% methyl alcoholic potassium hydroxide solution in a separating funnel. Rotate gently. Two layers are formed. Add 30 ml water and shake. The upper petroleum ether layer consists of carotene while the lower methanolic KoH layer consists of chlorophyll a. Collect these two layers in separate tubes.

2. **Lower Methyl Alcohol Layer :** Take 50 ml of methyl alcohol layer in a separating funnel and add 50 ml of diethyl ehter. Mix them thoroughly by rotating the funnel. Two layers are formed. Discard the lower methyl alcohol layer. The upper layer has thepigments. Add 15 ml of methanolic KOH to the upper layer. Shake gently for about 10 minutes. Add 30 ml of distilled water and shake again. Two layers are formed. The lower methanolic KoH fraction consists of chlorophyll b, while the upper diethyl either layer consists of xanthophyll. Collect them separately.

Observation and Conclusion : The four pigments separated can be characterised by checking their peak absorption spectra. The following is a flow chart for the process given above.

Leaves of Spinach, Tecoma etc
(5 gms)
↓ Grind in 50 ml of ice cold acetone and filter
Leaf extract
↓ Add 60 ml of petroleum ether and rotate gently
Separating funnel
↓
Wash gently with 75 ml of distilled water
↓
Two layers of the liquid
↓

Upper petroleum ether layer | Lower acetone - water layer discard

Upper petroleum ether layer
↓
Wash with distilled water three times and discard lower layer every time
↓
Add 50 ml of 92% methyl alcohol
↓ Roate gently
Two layers
↓

Upper petroleumether layer
↓
Add 15 ml of methanolic KoH and rotate gently
↓
Shake for 10 minutes
↓
Add 30 ml of distilled water and shake
↓
Two layers
↓
Petroleum ether layer **(Carotene)** | Methanolic KOH layer (Chlorophyll a)

Lower methyl alcohol layer. Collect in alcohol funnel
↓
Add 50 ml of diethyl ehter
↓
↓
Wash with 30 ml of water
↓
Two layers
↓
Upper ether layer ↓ | Lower methyl alcohol layer Discard

Add 15 ml of methanolic KoH
(30% and Shake
↓
Add 30 ml water and shake
↓
Two layers
↓

Upper diethyll ether layer (**xanthophyll**)	Lower Methanolic Koh Layer (**Chlorophyll b**)

49. Separation of leaf pigments by circular paper Chromotography

Materials : Two large petridishes, Whatman No. 1 filter paper, acetone, chlorophyll extract.

Procedure : Take a circular piece of whatman No. 1 filter paper and apply the chlorophyll extract to the centre of the paper in the form of a spot. Be careful not to allow the spot to spread. Use a hair dryer if necessary to dry the spot. Cut the filter paper from the periphery to the centre in the form of a small strip to form a 'wick'. Now place the filter paper over a small petridish (diameter must be smaller than that of the filter paper) containing about 30 ml of 20% acetone. Bend the wick in such a way that it dips in the acetone. Cover a large petridsh over the paper to prevent evaporation. Keep it for about 30 minutes. Watch the progress of the solvent. Remove the filter paper before the solvent reaches the outer periphery (of the paper). Dry it and observev the rings that appear. Measure the distance of the rings from the centre.

Observation and Conclusion : Four rings of coloured streaks appear. Of these the one closest to the centre of the filter paper is chlorophyll b, followed by chlorophyll a, xanthophyll and carotene. Rf values for each ring may be calculated as follows.

$$Rf = \frac{\text{Distance travelled by ring of pigment}}{\text{Distance travelled by the solvent}}$$

Rf value will be always less than 1 as distance travelled by the solvent is more than the distance travelled by the compound.

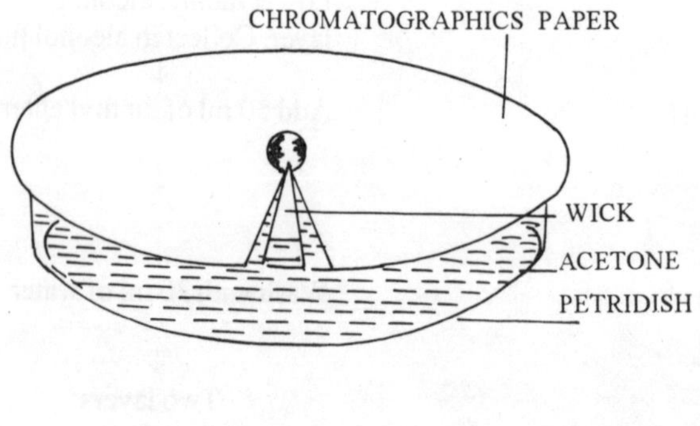

Fig 14.30 Circular paper chromatography for separation of leaf pigments

50. Separation of leaf pigments by Ascescending paper Chromatography

Materials : Cylindrical or rectangular jar with a tight fitting lid, chlorophyll extract, whatman No.1. Filter paper, petroleum ether, acetone, hook or thread, micro pipette, pencil etc.

Procedure : Prepare the chlorophyll extract as explained in experiment 48. Cut out a long strip of whatman No. 1 filter paper and draw across a pencil line about 2-3 centimeters from the end. With the help of a micropipetle draw a little quantity of concentrated chlorophyll extract and spot it on the centre of the pencil line. Allow the spot to dry (use a hair dryer if necessary). Apply some more quantity of the extract to the same region. Be careful not to allow the spot to expand on the paper.

Now select a jar of suitable size (longer than the filter paper strip) with a tight fitting split cork. Pour 50 ml of the solvent (petroleum ether : acetone -= 100 : 12) into the jar. Fix the upper end of the filter pape (i.e away from the spot region) to the split cork and insert the paper into the jar so that the spot end dips into the solvent below. The paper should be dipped in such a way that the spot is above the level of the solvent. Apply vaseline to the edges of the cork. Allow it to settle for about 30 - 45 minutes. Keep observing.

Observation : The solvent migrates up along the paper strip and carries the spot along with it. Gradually 4 different coloured spots apper at different distances from the original spot. Remove the paper from the jar before the solvent reaches the upper edge of the pape and dry it.

Conclusion : The four spots that appear are (starting from a distance closest to the original spot) . Chlorphyll b, chlorophyll a, xathophyll and carotene. Calculate the Rf values as given in experiment 49.

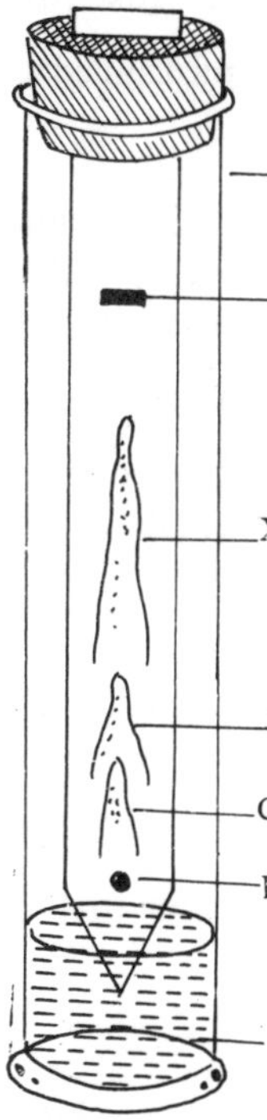

Fig 14.31 Ascending paper chromatography for separation of leaf pigments

51. Demonstration of Hill's reaction in isolated Chloroplasts (Dye reduction method)

Materials : Spinach leaves, Sodium phosphate buffer (Nacl solution 0.25 ml, 0.1m potassium dihydrogen phosphate), 0.1%, 2-6 dichlorophenol indophenol (DCPIP), Ice bath, water bath, pestle and mortar, cheese cloth, test tubes, aluminium foil etc.

Procedure : There are two major steps (1) preparation of chloroplast suspension and (2) reduction of the dye DCPIP.

(1) Preparation of Chloroplast Suspension : Prepare an ice bath and cool all the reagents, chemicals and glass ware used in the experiment. Weigh about 10 grams of fresh spinach leaves, wash them thoroughly and cut them into pieces. Remove, petiole, midrib and other hard materials and collect only the lamina portion. Macerate this tissue in pestle and mortar kept in an ice bucket. Alternatively the tissue can be homogenised in a waring blender with the operations conducted in a cold room.

Add 20 ml of sodium phosphate buffer at pH 6.5 (mix 0.24 m Nacl and 0.1m KH_2 Po_4 in the ratio of 25 : 10 ml). Filter the suspension with two layers of cheese cloth. Discard the residue and collect the filterate. Spin the filterate in a cooling centrifuge at 3000 rpm for 15 minutes. Discard the super natant and suspend the pellet in buffer and spin again. Discard the supernantant and resuspend the pellet in buffer. Stir it and transfer it to a tube kept in ice bath. This is the suspension of isolated chlorplasts.

(2) Reduction of Dye : Take two test tubes and add ingredients as follows :

A. 5.0 ml buffer + 0.6 ml distilled water + 2.0 ml of chloroplast suspension

B. Same as above.

Cover tube 'A' with aluminium foil or black paper so that no light can reach the chloroplast suspension. Let tube B be uncovered. To both 'A' and 'B' add a few drops of DCPIP. Keep both the tubes in water bath at room temperature and subject them to strong illumination.

After some time observe the contents of both tubes.

Observation : It is seen that the blue colour of the dye remains unchanged in the tube covered with black paper while in the tube exposed to light the dye would be decolourized.

Conclusions : During light reaction of photosynthesis water splits into H and OH. While OH recombines to form water, H is accepted by a hydrogen acceptor (NADP) which gets reduced. In the present experiment DCPIP is used instead of NADP. Chloroplasts receive light and water splits to form H and OH. DCPIP accepts H, gets reduced and becomes colourless

DCPIP + $2H_2O$ DCPIP (reduced + O_2

This experiment demonstrates that isolated chloroplasts can carry out photosynthesis even in a cell free medium.

RESPIRATION

52. Demonstration of Anaerobic respiration

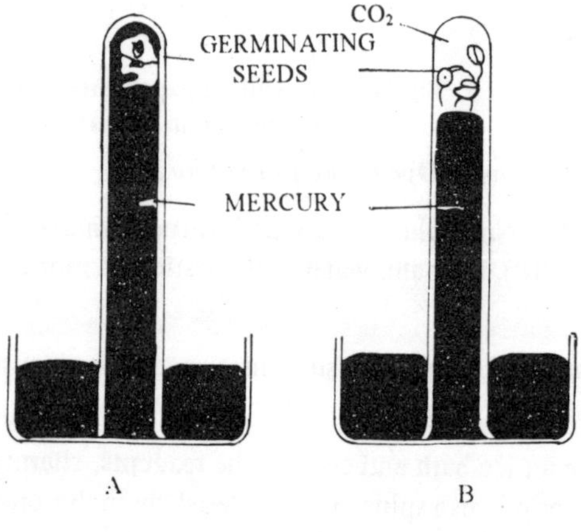

Fig 14.32 Demonstration of Anaerobic Respiration

Materials : Petridish, test tube, germinating seeds, mercury, KOH crystals, forceps etc.

Procedure : Fill a test tube with mercury and invert it over a petridish also containing mercury. Now lift the test tube lightly and introduce few germinating seeds (gram, pea etc with testa removed) with the help of a forceps. The seeds will rise to the top of the test tube. Keep the test tube back in its position and allow the apparatus to stand for about 45 minutes.

Observation : A gas starts building up at the test tube and it pushes the mercury down. Now introduce a piece of caustic potash. This absorbs the gas and the level of mercury goes up and fills the test tube.

Conclusion : Germinating seeds have a high rate of respiration. As they are kept in mercury they do not get oxygen and hence respire anaerobically. Since introduction of KOH absorbs the released gas it (gas) must be CO_2. This experiment demonstrates that CO_2 is released during anaerobic respiration

53. Demonstration of Fermentation (Kuhne's fermentation flask)

Materials : Kuhne's flask, yeast extract or toddy, 5% glucose solution, KOH pellets

Procedure : Kuhne's flask consists of a slightly slanted glass tube with a side bulb attached to it. It has a basal stand. The standed tube has no opening at the top while the side tube with a bulb has an opening which can be closed with a stopper.

Fill the Kuhne's flask with 5% sugar solution in such a way that the slanted tube is full but the bulb is only filled up to half its capacity. Add yeast extract or few drops of toddy into the glucose solution. Close the opening of the bulb with a stopper or with cotton. Let the apparatus stand for some time.

Observation and Conclusion : Small bubbles of gas start rising in the slanted tube and push the level of the solution below with the result a gas filled space is created at the top. If this is allowed to continue the glucose solution fills the side bulb and it may overflow if there is no stopper. Introduce a piece of KOH into the tube through the bulb so that it reaches the gas filled top, the gas is absorbed and the glucose solution once again moves to the top of the tube. The gas must be CO_2 as it is absorbed by KOH. This shows that CO_2 is released during fermentation. This is different from anaerobic respiration because here the respiratory substrate is outside the cell (glucose solution). Alcohol is formed during fermentation.

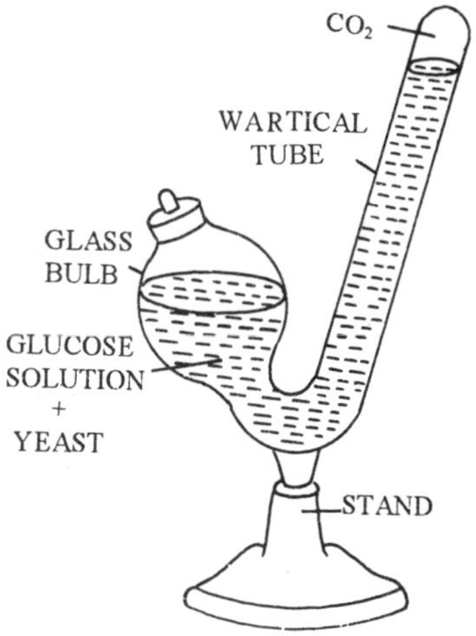

Fig 14.33 Kuhne's fermentation flask

54. Release of Co_2 during aerobic respiration (conical flask method)

Materials : Conical flask, A twice bent tube, test tube, thread, germinating seeds, KOH solution, vesline beaker, saline water.

Procedure : Take a conical flask and place some germinating seeds in it. Close the mouth with a single holed cork. Introduce a twice bent tube through the hole. Take a small quantity of KOH solution in a test tube and hang it inside the flask with the help of a thread. The thread will be held tight with the help of the cork. Let the free end of the bent tube be dipped in a beaker containing saline water. Apply vaseline to the cork to make it air tight. Let the apparatus stand for some time

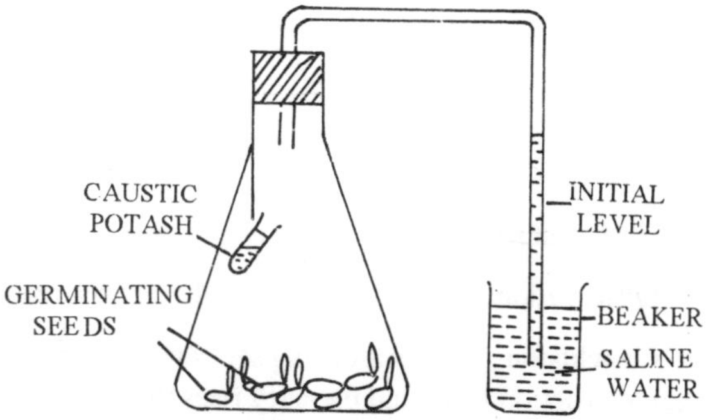

Fig 14.34 Release of CO_2 during aerobic respiration

Observation : Water level in the tube rises after some time.

Conclusion : As the germinating seeds respire to release Co_2 which is absorbed by KoH. This creates a kind of suction pressure which is transfered along the tube into the beaker containing saline water making it to rise.

55. Evolution of heat during respiration

Materials : Two thermos flasks, germinating seeds, mercuric chloride, rubber cork etc.

Procedure : Take two thermos flasks. Keep germinating seeds in one and mercuric chloride treated (killed) seeds in the other. With the help of single holed corks introduce a thermometer into each of these flasks and close the mouth (of the termo flask). Note the inital temperature in both the flasks. After 45 minutes note the temperature again. Record your observations.

Observation : In the thermos flask with killed seeds there will be neither increase or decrease in termperature as recorded by the thermometer. In the other flask having germinating seeds there will be an increase in temperature as recorded by the thermometer.

Conclusion : During respiration there is release of energy, while most of it used for the synthesis of ATP, some amount escapes as heat and this results in increased temperature in the flask with germinating seeds. In the other flask where the seeds are killed, there is no increase in temp as there is no respiration.

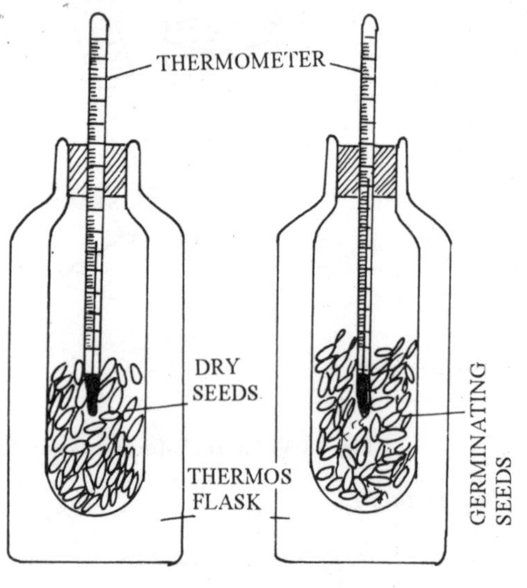

Fig 14.35 Evolution of heat during respiration

56. Demonstration of use of O_2 during Respiration

Materials : Conical flask, beaker, bent glass tube, trough, mercury, germinating seeds, caustic potash stand etc.

Procedure : Keep some germinating seeds in a conical flask. Also keep a small vial containing KOH inside the flask. Now close the mouth of the conical flask with the help of a cork having a single hole. Through this hole introduce a twice bent glass tube. Let the the free end of the tube be inserted in a beaker containing mercuryh. Allow the experiment to stand for some time.

Observation : Mercury kept in the beaker rises in the glass tube up to 15 cm height and then stops.

Conclusion : Germinating seeds respire and release Co_2 which is immediately absorbed by KOH. This creates a suction force and mercury rises up in the glass tube. After reaching a height of 15 cms rise of level of mercury stops indicating the stoppage of respiration and consequently the production of CO_2.15 cms height of mercury in the tube is approximately one fifth of normal height at which atmospheric pressure maintains it. This means that the seeds must have utilized 1/5th of air. The share of O_2 in atmospheric gases is also 1/5th. It may therefore be assumed that seeds have absorbed O_2 (for respiration) which accounts for 1/5th of air.

The rise in level of mercury is equivalent to the amount of CO_2 produced. This means the production of Co_2 during respiration is about 1/5th of the volume of atmosphere. During respiration the amount of O_2 utilized is equivalent to the amount of Co_2 released ($C_6H_{12}O_6 + 6O_2 + 6H_2o$ $6Co_2 + 2H_2o$). Hence this indicates that 1/5th of volume of air i.e, the entire O_2 content is used up. As there is no further amount of O_2 left in the conical flask respiration stops and the rise in the level of mercury also stops.

57. Demonstration of evolution of CO_2 during aerobic respiration by gas flow method

Materials : Potted healthy plant, bell jary black cloth, wide mouth bottles, 'U' tube, soda lime, bent tubes, lime water, baryta water, aspirator, vaseline etc.

Procedure : Take a bell jar and invert it over a healthy potted plant. Close the top of the bell jar with a two holed cork. Connect two bent tubes to these holes. To one tube connect a bottle containing baryta water. This bottle also should be fitted with a two holed cork. Insert a bent tube into this bottle with the other end of the tube being connected to an aspirator. Now the other bent tube from the bell jar should be connected serially to two bottles having lime water and then onto a 'U' tube containing soda lime. Smear vaseline to all the connections and cover the bell jar with a black cloth.

Set up the apparatus in such a way that air necessary for respiration first enters the 'U' tube followed by two bottles containing lime water and then into bell jar. As a result the air that enters the bell jar will be free of CO_2 as it gets absorbed by lime water in bottles. As the bell jar is covered by black cloth there will be no photosynthesis and hence no chance of CO_2 released during respiration to be utilized for photosynthesis. The air now enters the bottle containing baxyta water and then to the aspiration.

Observation and conclusion : Baryta water turns milky showing that the air from bell jar (with the plant) has CO_2. The air entering the bell jar is Co_2 free while the air leaving bell jar has CO_2. Hence the CO_2 must have been released by the plant during aerobic respiration.

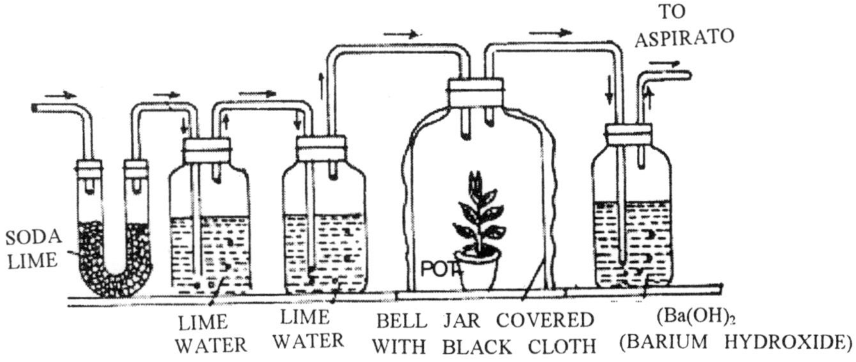

Fig 14.36 Evolution of CO_2 during aerobic respiration (Gas flor method)

58. Evolution of CO_2 during an aerobic respiration

Materials : Same as experiment 57 and pyrogallol.

Procedure : Set up the experiment in the same manner as in 57, except instead of two bottles having lime water, replace lime water, in one bottle with pyrogallol. As the air enters thsese bottles, one (lime water) absorbs CO_2 and the other (pyrogallol) absorbs O_2 so that the air entering bell jar has neither CO_2 nor O_2. Still the air that leaves bell jar has CO_2 indicating respiration in the absence of O_2. This demonstrates anaerobic respiration.

59. Measurement of Repiratory quotient (RQ) of different respiratory substances using Ganong's respirometer

Materials : Ganong's respirometer. Water, saline water, stand, cotton, different germinating seeds such as wheat, rice (carbohydrate), castor, mustard (fat), pea (protein), succulent plant such as *Opuntia* or *Bryophyllum*.

Procedure : Ganong's respirometer basically consits of a vertical graduated tube connected a stopper. The neck of the bulb as well as the stopper have a small hole each so that they are in line at a particular level. The other end of the graduated tube is attached to a levelling tube with the help of rubber hole. Take some quantity of saline solution and pour it into the levelling tube so that both tubes get filled about one half their capacity. Now open the stopper and keep some wet cotton in the bulb together with some germinating seeds (any one mentioned above - in materials). Place the stopper and rotate it slightly in the mouth of the bulb so that the holes in them are in line. The air pressure inside the bulb now equals that of outside. Adjust the levelling tube (lowering or raising) in such a way that the level of saline water in the graduated tube reaches zero level Rotate the stopper now to cut off contact with external air. Let the apparatus stand for some time.

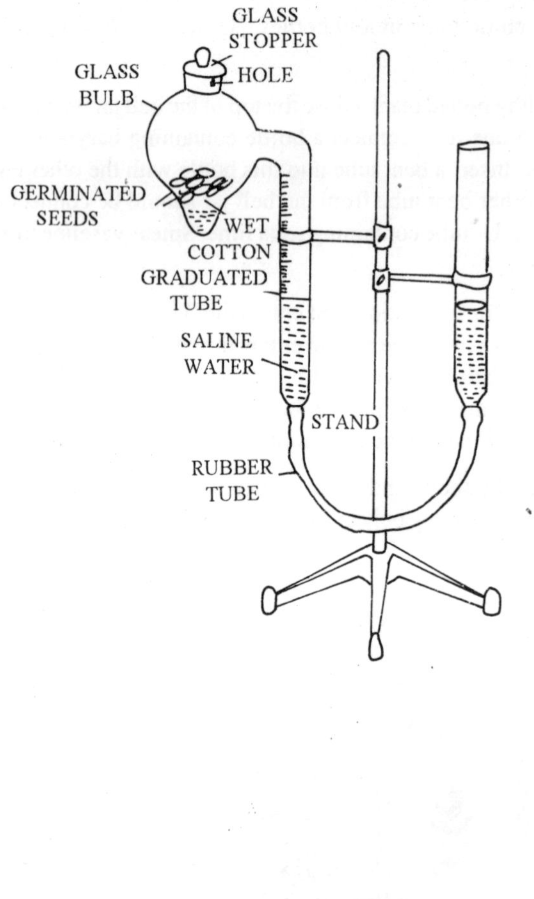

Fig 14.37 Ganong's respirometer

Observation : The Level of saline solution in the graduated tube remains stationary

Conclusion : No change in the level of saline solution indicates that the amount of O_2 used is equivalent to the amount of CO_2 released. There will be no change in volume of air in the bulb. Hence it does not press down nor suck the saline solution from the graduated tube. Thus the RQ is equivalent to 1 or unity. RQ is calcualted as follows.

$$RQ = \frac{\text{Amount of } Co_2 \text{ evolved}}{\text{Amount of } O_2 \text{ used}}$$

In case of **Carbohydrates** (Wheat, Rice,) such as sugars vol of CO_2 evolved is equivalent to CO_2 used.

$$C_6HO_6 + 6O_2 + 6H_2O \rightarrow 6CO_2 + 12H_2O$$

$$RQ = \frac{\text{Amount of } CO_2 \text{ evolved}}{\text{Amount of } O_2 \text{ used}} = \frac{6}{6} = 1 \text{ or unity}$$

When **fats** are respiratory substrates (castor, mustard etc) the level of saline water in the graduated tube will rise. Fats have less O_2 and hence require more O_2 for oxidation. O_2 used in respiration is more than CO_2 evolved and RQ will be less than unity. When palmition is the respiratory subbstrate -

$$2C_{51}H_{98}O_6 + 145O_2 \longrightarrow 102CO_2 + 98\,H_2O$$

$$RQ = \frac{102}{145} = 0.7$$

When the respiratory substrate is a protein (pea, gram etc) the level of saline solution will rise because proteins also require more O_2 in respiration than the amount of CO_2 evolved. As a result volume of air in the bulb gets decreased and the saline water rises being sucked by the air. (Usually RQ for proteins will be 0.8 - 0.9)

In case of **organic acids** being respiratory substrates, more amount of Cos will be released than O2 utilized. As the volume of air increases in the bulb it presses down the saline solution and its level will go down. Here RQ will be more than unity.

$$C_4H_6O_5 + 3O_2 \longrightarrow 4CO_2 + 3H_2O$$

$$RQ = 4/3 = 1.33$$

Caluclation of actual RQ in an experiment using Ganong's respirometer

1. Note the initial level of saline solution in the graduaated tube (l_1)
2. Final level (l_2)
3. The difference between the two $l_1 - l_2 = V_1$ (The value of V_1 may be negative or positive)

Insert a piece of KOH into the graduated tube. As the total CO_2 is absorbed the level will rise further. This rise may be called l_3.

The difference between l_2 and l_3 $(l_3 - l_2) = V_2$

V_2 denotes the actual volume of CO_2 evovled during respiration. While $V_2 + V_1$ gives the volume of CO_2 used during respiration.

60. Comparison of Respiration and Photosynthesis

Materials : Two long necked flasks, petridishes, stands, mercury, KOH pellets, green leaves, petals etc.

Procedure : Take two flasks A and B. Place few fresh moist green leaves in flask 'A' and moist petals in 'B'. Invert the flasks over troughs of mercury so that the neck is immersed in mercury. Fix the flasks to separate stands with the help of clamps. Keep the apparatus under diffuse light for about 10 - 12 hours. At the end of this period introduce KOH pellets into both the flasks and observe.

Observation : There is very little rise in the level of mercury in flask 'A' while in flask 'B' there is a considerable rise. In flask 'A' green leaves photosynthesize and utilize CO_2 rleased during respiration. Hence there is little if any free CO_2 left in the flask. In flask 'B' where petals are kept the entire CO_2 is in the flask when KoH is introduced it obsorbs CO_2 and hence there is rise in the level of mercury.

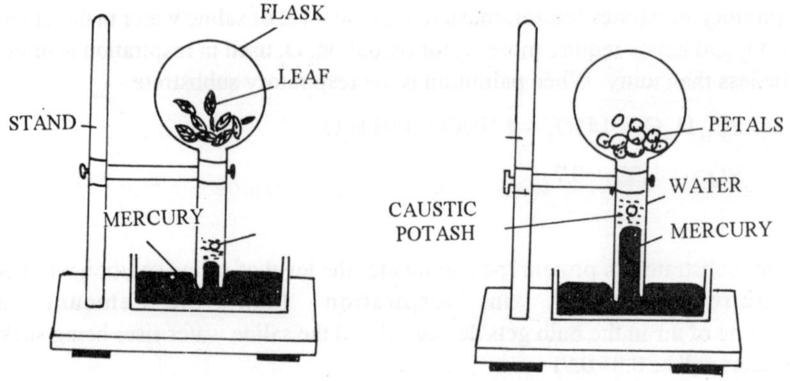

Fig 14.38 Comparison of Respiration and Photosynthesis

ENZYMES

61. Test of Oxidase

Materials : Fresh apple, tray,

Procedure : Cut thin slices of apple and place them on a tray. The pieces turn brown after some time. This shows that the phenolic compounds get oxidised in the presence of atmospheric oxygen due to the action of enzyme oxidase.

The enzyme oxidase catalyses the oxidation of the substrate by removing hydrogen from them and transfering it to molecular oxygen as follows.

$$AH_2 + O \quad A + H_2o$$

62. Test for Peroxidase

Materials : Potato tuber, Buchner's funnel, beaker, mortar, Buchner's funnel, muslin cloth, test tubes, 1% guaiacum (Benzidine) solution, $3H_2o_2$.

Procedure : Crush a fresh potato in a mortar. Filter the slurry through two layers of muslin cloth in a Buchner's funnel. Tkae the filterate (3 ml) in a test tube and add 0.5 ml of 1% Benzidine (dissolve 2 gms of guaiacum in 100 ml of absolute alcohol). Immediately add to the test tube a few drops of 3% H_2O_2 solution. Note down the change in colour of the mixture.

Observation : The colour rapidly changes to blue

Conclusion : Phenolic compounds get oxidized rapidly in the presence of H_2O_2. This shows the presence of peroxidase in the potato tubers. If boiled potato (the enzyme will be denatured) is taken blue colour will not appear. The reaction is as follows :

$$AH_2 + H_2o_2 \quad A + 2H_2o$$

63. Test for amylase

Materials : Fresh potato, starch solution, germinating barley grains, citrate - phosphate buffer (pH 5.6), Iodine solution, test tube, spirit lamp, water etc.

Procedure : This involves two steps (i) preparation of enzyme extract and (ii) assay for enzyme activity.

Preparation of enzyme extract : Mash germinating barlye grains in cold citrate phosphate buffer (pH 5.6) and filter the extract through two layers of chees cloth. The filterate serves as the enzyme source. Keep it in ice bath.

Enzyme assay : Take two clean test tubes A and B. Add 2ml of enzyme extract to each of the test tubes. Boil the enzyme extract in A in a water bath. Cool it to room temperature. Now to bath tubes add freshly prepared potato starch solution (1 ml) and shake gently. Keep the tubes aside for 10 minutes. Test the contents of A and B separately at periodic intervals with Iodine solution for the presence of starch.

Observation and Conclusion : In tube A the contents always turn blue indicating the presence of starch. In tube 'B' the blue colour will be intense in the beginning but later it turns lighter and finally no colour formation will be there. This indicates the gradual digestion of starch due to the activity of amylase. In tube 'A' the enzyme will not be active as it gets denatured due to boiling.

NITROGEN METABOLISM

64. Demonstration of nitrates in green plants :

Materials : A healthy potted plant, slide, cover slip calcium nitrate solution (1.5%), diphenylamine (1%) , microscope.

Procedure : Select a healthy potted plant. Add to the pot 100 ml of 1.5 % calcium nitrate solution. Keep the plant in darkness for 72 hours and then transfer it to bright light. Cut transverse section of lamina and petiole of the leaf and mount them in few drops of diphenylamine sulphate solution (1% diphenylamine dissolved in 50% H_2SO_4). Cover them with cover glass and observe under the microscope.

Observation : A deep blue colour appears

Conclusion : The colour indicates the presence of nitrates.

65. Experiment to demonstrate that atmospheric N_2 is not directly absorbed by green plants :

Materials : Glass troughts (one big and one small), bell jar, dilute sulpuric acid, petridish, funnel tube, sterilized N_2 free sand distilled water.

Description of the apparatus : The apparatus consists of a large glass trough (A) which is half filled with dilute sulphuric acid. A bell jar (B) is inverted inside the trough and placed on three corks so that it is raised above the floor of the trough. A small trough (C) is placed inside the bell jar on a large petridish so that it is not in contact with H_2SO_4 in the large trough. Trough 'C' is filled with water and a small glass pot is kept inside the trough (C). The inverted bell jar has a small inlet tube in the side which is connected on the inside to trough 'C' and on the outside to a funnel. This facilitates addition of water from time to time. The top of the bell jar is closed with a cork having an inlet tube which allows CO_2 inside the bell jar.

Procedure : Collect some legume seeds and estimate their total N_2 content by chemical analysis collect the ash also and estimate the N_2 content. Take some quantity of sterilized N_2 free sand and add it to the small glass pot inside trough 'C' together with some quantity of ash. Sow some legume seeds (the seeds from the same plant on which chemical analysis has been done) in the pot and allow them to germinate. Let them grow into small plants. After a definite period of time (15-12 days) pull out the plant along with roots. Estimate the total N_2 present in the whole plant. Compare it with the amount of N_2 already estimated in the seeds.

Observation and Conclusion : Comparison of the N_2 content shows that the plant does not have more N_2 than the seeds. This shows that the plant has not absorbed gaseous N_2 from atmosphere. Higher plants can only absorb N_2 in the form of nitrates or nitrates. In this experiment the sand from which roots absorb nutrition is N_2 free. The only available form of N_2 for plants is gaseous N_2. Dilute H_2SO_4 absorbs traces of ammonia if any in the bell jar.

GROWTH AND HORMONES

66. Measurement of Growth by arc auxanometer

Materials : Arc auxanometer, weight, thread, a potted plant.

The apparatus : The arc auxanometer consists of a tripod stand and a pulley connected to a pointer which moves on an arc scale. A thread is made to pass over the pulley tightly. One end of the pulley is tied to a weight.

Procedure : Take a healthy potted plant and tie its tip to thread which is held over the pulley in such a way that the thread does not slacken. Keep it aside and observe.

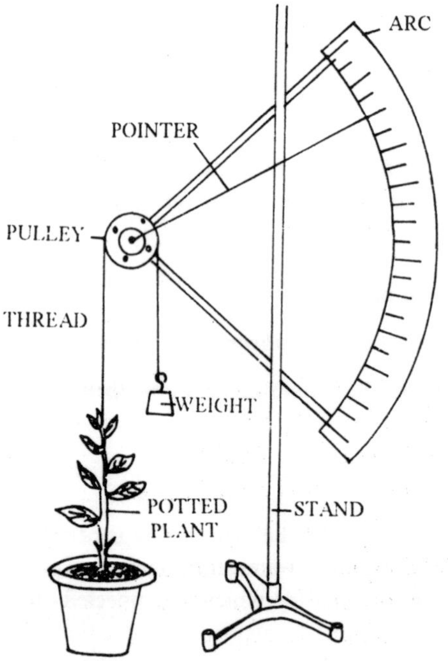

Fig 14.39 Arc Auxamometer

Observation and Conclusion : As the stemp top grows, it will push the thread up and the tight thread will turn the pulley. The pulley in turn will move the pointer over the graduated arc scale. The extent of movement from the initial to the final in a given period of time can be measured. The distance moved by the pointer does not give the actual linear growth, because the arc is divided into 90° (one quarter of a circle i.e., 360 / 0.25) and not linear measurements. Further the measurement on the graduated arc is magnified because of the length of the pointer. The actual growth is calculated as follows-

Inital position of pointer = A

Final position of pointer = B

Actual distance travelled by the pointer during a given period of time = B - A

Acutal growth = B - A x Radius of pulley

in length length of pointer (from the centre of the pulley

For example if the radius of pulley is 10 cms and the length of the pointer is 50 cms the magnification is 10 times. If the pointer has moved 5 cms in 20 hours the actual growth is as follows.

<u>5 x 10</u> = 1 cm in 20 hours

50 Therefore Growth per hour = 0.5 mm

67. Measurement of growth by Pfeffer's Auxanometer :

Materials : Pfeffer's auxanometer, potted plant, thread etc.

The apparatus : Pfeffer's auxanometer consists of a vertical stand and a self rotating drum over which a smoked paper is glued. The drum rotates on its axis through a clock work mechanism at the rate of one turn per every hour. Attached to the stand on a shaft are two pulleys one small and the other bitg. The smaller pulley is attached to the central axis of the larger pulley. A thread passes over the small pulley, one end of which is tied to a weight. Another thread also passes over the large pulley with weights tied to both ends. A pointer is attached to one end of this thread slightly above the weight. This pointer touches the smoked paper wrapped around the rotating drum.

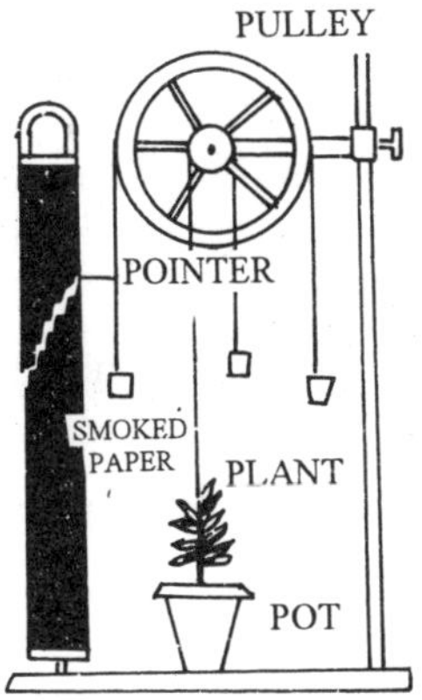

Procedure : Take a healthy potted plant and tie its tip to the free end of the thread coming over the small pulley. Observe.

Observation and Conclusion : As the stem tip grows the small pulley moves together with large pulley. This results in the pointer continuously marking a spiral line on the rotating drum's smoked paper.

The perpendicular distance between two adjoining spirals is the hourly growth of the plant. Actual growth rate may be calculated taking into consideration the radii of both the pulleys and the rate of rotation of drum.

Fig 14.40 Pfeffer's Auxanometer

68. Measurement of growth by space markers

Materials : Indeliable ink (India ink), germinating seeds

Procedure : Select few germinating seeds where radicle has emerged out into the tap root. Using a thin nib mark graduations at regular intervals a little behind the root tip. Observe.

Observation and Conclusion : It is observed that the distance between the graduations have increased. This increased distance is the actual linear growth of the root.

69. Effect of Auxins (Indole acetic acid - IAA) on plant growth

Materials : Oat grains, Auxin solution (10 mg/l of IAA), 10% sucrose solution, water, circular filter paper, graph sheet etc.

Procedure : Soak some oat (*Avena sativa*) grains in water and place them on a wet filter paper in a petridish in dark. After about 2 days when coleoptiles are 2 - 3 cms long, remove their tips and discard them. Cut sections of about 5 mm long from the decapitated coleuptile tips. Collect about 15 sections, and keep them in three separate petridishes (each having five sections) as follows :

Petridish	A	Water
Petridish	B	Sucrose solution
Petridish	C	Sucrose solution + 1 ml auxin solution.

Keep the petridishes back in darkness and observe.

Observation : It is observed that *Avena* coleoptile pieces kept in petridish C have grown much longer than those in A and B. The shortest length is found in A.

Conclusion : Coleoptile elongation is minimum in 'A' (water) as it has no nutrition. While slight growth is found in B because of sucrose solution. In 'C' however these is maximum elongation as auxins promote coleoptile growth.

Note : Different concentrations of IAA may be tried to find out their effect on growth

70. Effect of Gibberellins on plant Growth

Materials : Seeds of pea, tomato, lettuce etc, petridish, circular filter paper, gibberellic acid solution (100 mg/litre), measuring cylinder etc.

Procedure : Germinate seeds in dark on wet filter paper placed in a petridish. When the seeds have germinated for about 48 hours, transfer them to petridishes - (a) filter paper with water and (b) filter paper with gibberellic acid solution. Keep the petridishes in light for 72 hours and observe.

Observation : It is seen that hypocotyls in the petridish with gibberellic acid solution become much more elongated than the one with only water.

Conclusion : Gibberellin has an effect of elongation of cell and hence the hypocotyls become longer in length.

71. Effect of Cytokinin on expansion of Cotyledons

Materials : Seeds of cucumber, pumpkin or ash gourd, petridish, filter paper, 10 mg kinetin solution, water, graph paper etc.

Procedure : Germinate the seeds on wet filter paper placed on a petridish in dark. Select germlings with about 2 - 3 mm of hypocotyl and remove the seed coat. Separate the cotyledons and float them on petridishes - (a) containing only water and (b) containing kinetin solution. The abaxial surface of the cotyledons should be in contact with kinetin solution. Keep them aside for 72 hours and observe.

Observation : It is seen that cotyledons kept in kinetin solution have expanded much more than those kept in water. This can be tested by placing them on graph paper.

Conclusion : Kinetin a growth hormone is known to induce rapid cell division thus bringing about expansion of the cotyledons.

PLANT MOVEMENTS

72. Geotropism (Horizontal Klinostat)

Materials : Klinostat, a healthy plant

The appratus : Horizontal klinostat essentially consists of a clockwork mechanism. Instead of the hands at the region of the dial there is an axis to which is attached a small metallic tumbler. The clock is fitted to a stand. When the clock is working the tumbler slowly rotates. It remains stationary when the clock is not working.

Procedure : A small potted plant is kept in the tumbler as shown in the figure. Observe what happens to the plant under the following conditions

(a) when the tumbler is rotating (clock working)
(b) When the thumbler is not rotating (clock not working)

Observation : When the tumbler is stationary i.e., clock not working, after few days the shoot bends up while the root bends down (downward curvature of the root may be observed by removing the soil). When the tumbler is rotating, the plant continues to grow horizontally i.e, parallel to ground. There is no upward curvature in the stem and downward curvature in the root.

Conclusion : When the clock is not working auxins accumulate in the lower surface (ground facing side) of stem and root. Auxin accumulation has opposite effects on stem and root. In stem auxin accumulation increases cell division while in root it decreases cell division. As a result of this, shoot bends up because there is more growth on the lower surface than the surface away from ground. In root the auxin accumulated lower surface has less growth than the surface away from gravity hence it bends down. The unequal accumulation is due to gravity.

When the tumbler is rotating there will not be gravitational effecti on any part of the plant as auxins can not accumulate in any one surface. As the plant is also rotating there is equal distribution of auxins. Hence the plant (shoot or root) does not bend but grows horizontally.

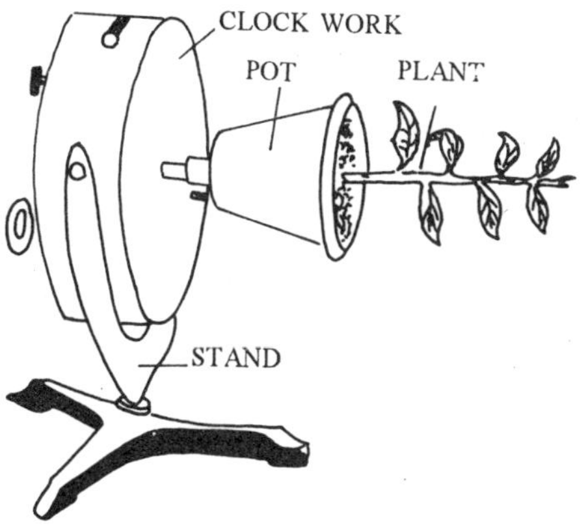

Fig 14.41 Horizontae Klinostat

73. Effect of Geotropism (Vertical Klinostat)

Materials : Vertical klinostat, germinated seeds with hypocotyl emerging, cork, pin, etc.

The apparatus : A vertical klinostat is essentially like a horizontal klinostat except that the face of the clock work mechanism will be directed up. In addition, a disk is attached to the rotating axis instead of tumbler. When the clock is operating the disc rotates. Corks are fitted at regular intervals near the periphery of the disc.

Procedure : Attach a few germinated seeds to the cork with the help of pins. Observe the behaviour of seedlings under the following conditions :-

(a) When the disc is rotating slowly
(b) When the disc is not rotating
(c) When the disc is rotating fast

Observation : When the disc is rotating slowly i.e 2.5 turns per hour the roots will diverge and shoots converge at an angle of 45°. The bending is due to the fact that centrifugal force (rotating force) and gravitational force act together and bring about the slanting growth.

When the disc is not rotating only geotropic force will act on the plant. Shoot will grow up and root down.

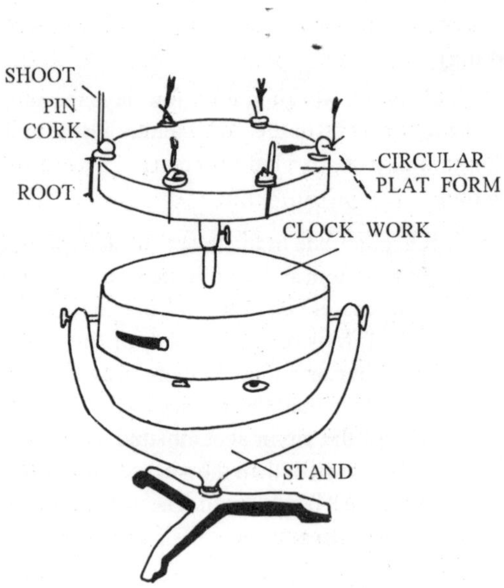

Fig 14.42 Vertical Klinostat

When the disc is rotating fast the geotropic force is nullified by the centrifugal force. As result the plant grows horizontally the root away from the disc and the shoot towards the centre of the disc.

74. Phototropism (Heliotropism)

Mateials : Phototrophic chamber (box), A potted plant.

A phototrophic chamber is a small wooden box painted black inside. It has a hinged door at the top (to open and close) and a small window on the side.

Procedure : Keep a potted plant in the phototrophic chamber close the lid. Keep the box in such a way that the window of the box is directed towards the source of light. Let the box stay in place for about 2 - 3 days.

Observation : It is observed that the shoot bends towards the window of the box, comes out and then once again bends up and grows vertically.

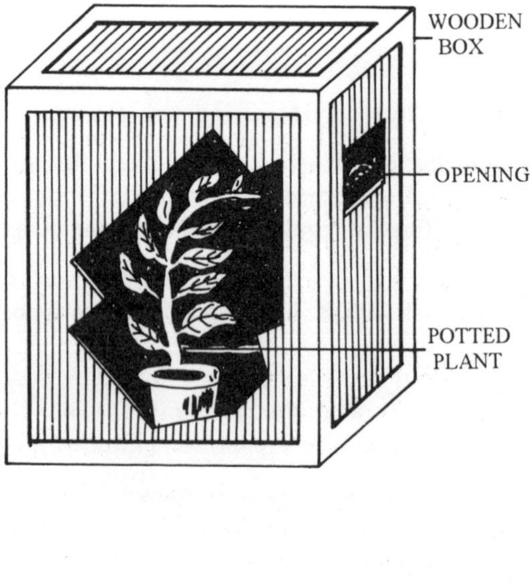

Fig 14.43 Phototropic Chamber

Conclusion : The plant gets light only through the side window. Unidirectional light falling on the side of the stem destroys the auxins on that part of the stem. Auxins accumulate in the shaded region. As a result there is more growth in the shaded region than the lighted region and the shoot bends and comes out of the window of the box. When the shoot comes out of the box the light strikes the upper surface of the horizontal stem making auxins to concentrate on the lower surface. Hence the stem bends up then onwards as the light is not undirectional on the apex it grows up. This is known as phototropism i.e., growth towards light. While shoots are positively phototropic roots are negatively phototropic i.e. they grow away from light (towards gravity)

GERMINATION

75. Viability (Germinability) test for seeds :

Materials : Barley seeds, water, petridish, filter paper, 0.1 solution of tetrazolium chloride.

Procedure : Soak barley seeds in water for about 2-3 hours. Cut the seed exactly into 2 parts slicing through the embryo. Keep the halves in petridishes containing a solution of tetrazolium chloride. The seeds should be completely immersed in the solution. Keep the petridishes in dark for about 45 minutes and observe.

Observation : Some of the seeds have stained embryos while others are colourless.

Conclusion : Viable seeds respire and can change the colourless tetrazolium compound to a coloured (reduced) compound formazon. Formazon is red in colour and thus the embryos are stained. In non viable seeds there is no respiration and hence the tetrazolium salt remains colourless. Percentage of germination of seeds can be calculated as follows :-

Number of seeds used for experiment = A
Number of seeds stained = B
Percentage germination B/A x 100
Microchemical Tests

CARBOHYDRATES

76. Test for Cellulose

Materials : Paper or Cotton fibres, Iodine solution, water 75% sulphuric acid, slides, cover slips, microscope etc.

Procedure : Place a few cotton fibres or very thin shreds of paper in a drop of water. Observe the cell wall under the microscope. Add a few drops of Iodine and a few drops of H_2SO_4.

Observation : The fibres turn yellowish brown first and then turn blue.

Conclusion : Blue colour indicates the presence of cellulose

77. Test for Cutin

Materials : *Ficus* or *Nerium* leaf, microscope, slides, coverslips, watch glass, water, KOH solution etc.

Procedure : Cut cross sections of *Nerium* leaf, mount the sections on a slide, stain them with KOH solution. Put coverglass and observe the colour of the outermost deposition on the epidermis.

Observation : The deposition takes a yellow colour.

Conclusion : Yellow colour indicates the presence of cutin.

78. Test for Hemicellulose

Materials : Soyabean seeds, Iodine, water, slide, cover glass, glycerine etc.

Procedure : Cut a thin section of the seed, place it in water and stain it with iodine solution for few minutes. Observe under the microscope.

Observation : Sections particularly the walls take a blue colouration.

Conclusion : Blue colouration indicates the presence of hemicellulose.

79. Test for Suberin

Materials : Natural cork, sudan IV (alcoholic), 50% ethyl alcohol, slide, cover glass, water etc.

Procedure : Cut thin sections of cork. Stain them with sudan IV for about 15 minutes. Wash excess stain with 50% ethyl alcohol. Transfer the sections to water for about 5 minutes and finally mount in glycerine. Put a cover glass over the section and observe under the microscope.

Observation and conclusion : All the suberized portions take red stain, indicating the presence of suberin in the cell wall.

80. Test for Lignin

Materials : Wood shavings, sections of woody material, 1% alcoholic phloroglucin, 25% Hcl, 1% neutral aqueous $KmNO_4$, Amonium hydroxide or sodium bicarbonate, slide, cover glass, water etc.

Procedure : (1) Treat the material (section or wood shavings), with phloroglucin on a slide and put a coverglass. Flood the section with HCL along the edges of the cover glass. Observe.

(2) Treat the material with potassium permanaganate ($KmNO_4$) solution for about 15-20 minutes. Wash the material with 25% Hcl. Follow it up with several washes of water. Add a few drops of either NH_4OH or sodium bicarbonate solution.

Observation and Conclusion : In the first method lignified walls take up a red-violet colour. In the second method a deep red colour develops.

81. Test for Mucilage :

Materials : Seed coat of linseed, 10% $CuSO_4$ solution, 10% KOH solution, water, slide, cover glass etc.

Procedure : Cut thin sections of the seed coat and soak them in 10% $CuSO_4$ solution for about 20 minutes. Wash them in water and transfer to 10% KOH solution. Observe the sections under microscope after mounting them in glycerine.

Observation and conclusion : Cells having mucilage are stained bright blue.

82. Test for Latex

Materials : Latex of *Calotropis*, *Asilepias*, *Euphobia* etc sucrose, Conc H_2SO_4, test tube, water etc.

Procedure : Extract the latex in alcohol (mix alcohol and latex thoroughly; centrifuge and take the supernatant) and add an equal amount of sucrose and sulphuric acid.

Observation and Conclusion : The colour turns pinkish purple. This indicates the presence of latex.

83. Test for Glucose (Fehling's test)

Materials : Glucose solution, Test tube, spirit lamp, freshly prepared solution A and B.

Procedure : Mix about 1.5 ml each of Fehyling's solution A and B in a test tube and add 3ml of glucose solution. Boil on spirit lamp. Cool it and observe.

Observation and Conclusion : A brick red pricipitate develops indicating the presence of glucose. Fehlings solution has cupric sulphate, sodium hydroxide, and potassium sodium tartarate. When mixed with glucose and boiled, the aldehyde groups of glucose are oxidized while the cupric salt gets reduced to cuprous oxide which forms a brick red pricipitate insoluble in water.

$$C_6H_{12}O_6 + 2Cu\,(oH)_2\ C_6H_{12}O_2 + Cu_2o + H_2o$$

84. Test for Glucose (Bendecit's test)

Materials : Glucose solution, Bendict's solution, spirit lamp, test tube etc.

Procedure : Add 1 ml of glucose solution to 3 ml of Benedict's solution taken in a test tube; warm the mixture on the flame of spirit lamp untill a red brown colour appears.

Observation and conclusion : The colour changes from blue to green and finally to rust brown indicating the presence of reducing sugars.

85. Test for Sucrose :

Materials : Sucrose solution, Fehling's solution A and B, NaOH solution, HCl litmus paper, spirit lamp

Proceudre : Take 3 ml of sucrose solution in a test tube and few drops of conc. HCl and boil on the flame of spirit lamp, cool. Add a little NaOH to neutrialize the mixture. Test the pH with litmus paper. Mix Fehlings solution A and B. Add 3 ml of this mixture to the neutralized mixture of sucrose solution. Boil again till a brick red precipitate appears.

Observation and Conclusion : Sucrose is a non reducing sugar, but on acid hydrolysis it gets broken up into glucose and fructose which are reducing sugars. These on treatement with Fehling's solution give a brick red precipitate.

86. Test for Starch

Materials : Starch, supspension, test tube, Iodine solution, spirit lamp.

Procedure : Take 5 ml of starch suspension in a test tube and add equal volume of idoine solution. watch the change in colour.

Observation and Conclusion : Starch gives a blue colour on reacting with iodine. The colour disappears on boiling but reappears on cooling. Starch molecules form a complex with iodine solution under normal temperature.

PROTEINS

87. Biuret Test

Materials : Gram flour, test tube, 40% NaOH, 1% CuSo4 solution.

Procedure : Mix some quantity of gram flour and prepare a solution. Take 5 ml of this solution in a test tube and add 2 ml of NaOH and one drop of $Cuso_4$ solution. Mix the contents by stirring and note the colour change.

Observation and Conclusion : A pale purple or pink colour appears indicating the presence of proteins. Biuret reagent reacts with the peptide bond of proteins and produces the colour.

88. Xanthoproteic Test

Materials : Gram flour, con. HNO_3, conc NaOH solution, spirit lamp etc.

Procedure : Mix a little gramflour with water and prepare a solution. Take about 3 ml of this solution in a test tube and add equal volume of conc. HNO_2. A thick white precipitate appears. Boil the solution, the precipitate turns canary yellow. Cool the test tube with cold water and add NaOH to neutralize and then to make it alkaline. Observe the colour.

Observation and Conclusion : Yellow colour turns to orange showing the presence of proteins. Aromatic aminoacids present in the protein when heated with HNO_3 form nitroderivative of aromatic aminoacids. These nitroderivatives on treatment with NaOH form salts which are orange coloured.

FATS

89. Test for Fat

Materials : Seeds of almond, sunflower etc., water, osmic acid, sudan III (0.5% in 70% alcohol), glycerine etc.

Procedure : Cut thin sections of the material; stain them with sudan III for about 10 minutes. Wash the sections with water. Observe the colour under microscope.

Observation and Conclusion : Materials turn red on microscopic observation. It is seen that the oil droplets have taken the red stain. In another method a few drops of osmic acid is added to the material taken in a test tube. Formation of black colour indicates the presence of fat.

90. Test for Fat (oil)

Materials : Olive oil, water, petroleum ether, sudan III (0.5% dye in 70% alcohol)

Procedure : Take a few ml of the oil in a test tube and add a few drops of sudan III. Shake well. The oil turns red. Pour half the contents into another tube. Add water to one tube and petroleum ether to another. Shake both the tubes gently and observe.

Observation and Conclusion : The red coloured oil does not mix with water hence it floats forming a red coloured ring over water. However when petroleum ether is added, oil dissolves in it and a homogenous red solution is formed.

91. Test for presence of fatty acids by saponification :

Materials : Butter, 10% alcoholic KOH, test tube, litmus paper, spirit lamp, water, NaOH solution.

Procedure : Melt the butter and take 1 ml of the oil in a test tube and add 2 ml of alcoholic KOH. Boil the mixture for about 2 minutes and add 5 ml of distilled water. Boil again for 5-6 minutes. Cool and add a few drops of conc HCl to make the mixture slighly acidic. Test the pH with litmus paper. A layer of fatty acids is formed at the top. Transfer it to another test tube and add a dilute solution of NaOH and heat.

Observation : The solution turns soapy

Conclusion : The soapy solution consists of fatty acids. By the addition of NaCl solution, the soapy solution precipitates indicating the presence of the fatty acids.

92. Test to Determine the saponification value of Fat

Materials : Butter, petroleum ether, 0.5 N alcoholic KOH, 0.5 N HCl, buretle, relex condenser, conical flask etc.

Procedure : Take one gram of butter in a conical flask (250 ml). Add 3 ml of fat solvent (petroleum ether) and shake to dissolve butter. Add 25 ml of KOH solution and attach a reflex condeser to the flask and heat the contents in a water bath for 30 minutes. Cool the mixture to the room temperature. Titrate the mixture using a burette against 0.5 NHCl. Note the reading; conduct a blank experiment (without butter) and note the reading.Find out the difference in values.

Observation and conclusion : The difference in titrated value between blank (without fat) and experiment (with fat) will be the number of mg of 0.5 N KOH required for the saponification of 1 gm of butter. This value is called the saponfication value for one gm of fat.

NUCLEIC ACIDS

(Microchemical test)

93. Test for RNA and DNA :

Materials : Onion roots, azure B solution, butanol (0.25 mg/ml in citrate buffer at pH 4), watch glass, microscope.

Procedure : With the help of a razor blade cut thin long sections of root tips of onion. Transfer them to a watch glass and stain them using Azure B solution for about 2 hrs. Wash the sections in distilled water and transfer them to another watch glass. Flood the sections with butanol for about 30 minutes. Mount the sections on a slide, put cover glass and observe under microscope.

Observation and Conclusion : Certain portions of the cells turn greenish blue and certain others turn deep blue. Greenish blue colour shows the presence of DNA and deep blue colour indicates RNA.

PART FIVE
CYTOLOGY

CHAPTER 15

CYTOLOGY

TECHNIQUES OF CYTOLOGICAL STUDY

Cytological study in practicals involves the study of chromosomes which are visible during cell division. Both mitosis and meiosis are studied with the help of microscopic observation of cells during division. Young flower buds are chosen for meiotic study while freshly emerging root tips are used for mitotic study.

In flower buds anthers are selected as they have microspore mother cells which undergo reduction division (meiosis) to produce pollen grains ultimately. Ovule also can be chosen as a material for meiotic study as the megaspore mother cell (MMC) will be undergoing reduction division to produce megaspores. But it is not a suitable material as in the entire ovule there will be only a single MMC undergoing meiosis. In the case of anther however there will be a large number of microspore mother cells dividing meiotically.

Root tips are selected for the purpose of mitotic study as the meristematic cells there will be dividing mitotically. In some cases shoot tips (apical buds) also serve as an ideal material for mitotic study.

There are two methods for the microscopic study of flower buds and root tips. These are the *Smear* and the *Squash techniques.*

Smear : This involves the spreading of a single layer of cells for the purpose of study. Cells are selected when they are actively dividing. One of the prerequisite for such a study would be fixing the material at suitable stage. This is done by fixing material at different periods of time and then selecting one which is most suitable from the point of view of division stages. Microsporocytes of *Tradescantia, Rheo, Chlorophytum* etc and root tips of onion *Ficus* etc are killed (fixed) in carnoy's or other suitable fluid at an appropriate time.

Perfectly clean slides are very essential for smears. For this purpose they are immersed for a long time in a mixture of sulphuric acid and potassium bichromate. They are then throughly washed in running water and dried in a hot air oven or wiped clean with lint free cloth. Fresh young anthers are dissected out of the flower buds and placed in the centre of the slide. Using a sharp razor blade one end of the anther is cut and by gently applying the pressure microsporocytes are sqeezed out. The debris like anther wall and other cellular material is removed.

Staining : Few drops of acetocarmine stain is added on the microsporocytes and a cover glass is put on the material. A pressure is applied gently on the cover glass in order to spread the smear of the microsporocytes.

The slide is gently warmed over the spirit lamp. Excess stain may be removed with the help of blotting paper. The edges of the cover slip is sealed with melted wax in order to prevent evaporation. The slide is now ready for the observation of meiotic stages.

Squash Technique : This method is adopted for root top mitotic studies. An onion bulb is placed on a coupling jar filled with water. The lower stock portion of the bulb should be immersed in water. This willl facililate the quick emergence of roots. After the roots are about 1 cm long the tips should be cut.

Pretreatment : Before squashing, the root tips are to be pretreated with either (a) calchicine or (b) .8 hydroxy quinoline

(a) Prepare a 0.5% or 1% aqueous solution of commercially available colchicine. Take about 10ml (depending on the quantity of the material to be squashed) or more of colchicine solution in a watch glass and drop the root tips into it. Let the root tips remain in this solution for about 30 minutes at room temperature.

(b) Prepare an aqueous 0.29 solution of 8 hydroxy quinoline and treat the rot tips in this solution for 2½ to 3 hours at 12-14°C temperature.

Staining : After pretreatment roots should be washed in distilled water and treated with acetic alcohol (1:3) solution for 15-30 minutes. Acetic alcohol is a fixative and the roots may be preserved in this fluid for a long time.

Take few root tips in a watch glass and add 2% aceto orcein in HCL in the ratio of 9 drops of stain +1 drop of HCL for every two root tips. At this stage the watch glass should be warmed over the flame of spirit lamp until the fumes of acetic acid escape. Care should be taken not to boil the mixture. Cool the material for about 15-20 minutes. Now take a root tip on a clean side, place a drop or two of acetoorcein stain put a cover glass and gently press with the back of a pencil or any flat object. Reamove excess stain by blotting paper. Seal the edges by metled wax to prevent evaporation. The slide is now ready for observation.

Permanent Mounting : The methods mentioned above are for temporary microscopic observation. They are to be made permanent as follows if they have to be stored for a long time. Slides that are to be made permanent must at least be 24 hours old after squashing. Select such slides and pass them through terrtiary butyl alcohol and acetic acid series taken in a coupling jar as follows :

(a) 1:1 acetic acid : tertiary butyl alcohol
(b) 1:3 acetic acid : tertiary butyl alcohol
(c) only tertiary butyl alcohol.

Scrape and remove all the wax used for sealing the cover glass. Then keep the slides in (a) for 5-10 minutes. Within this period the cover glass slips off the slide. Do not discard the cover glass. Material will be adhering to both the slide and coverglass. Hence both of them should be processed. Transfer both the slide and coverglass separately to (b). In order to facilitate handling the delicate cover glas a tooth pick may be split a little at the tip and the cover glass edge inserted on to it. Leave both the slide and cover glass in (b) for 10 minutes and then pass them on to (c) for 5 minutes. The slide and cover glass are now ready for mounting.

Mounting : As both the slide and cover glass has the material, both of them should be used in mounting. But the cover glass should not be put on the same slide. As both these have the material there will be over lapping and the stages will not be clear. Hence both the slide and coverglass having the material should be mounted separately with another cover glass and another slide. Select a clean slide and put a drop of euparol. Now carfully place the coverglass on euparol. The mounting medium spreads to the edges of the coverglass. Extreme care should be taken to see that the side of the coverglass having material must face the slide and come into contact with euparol. Now take the original slide with the material, place a drop of euparol and gently slide a fresh clean cover glass. With this procedure, a single temporary slide can be made into two permanent slides. Dry the slides in a hot air oven till the cover glass does not move on the surface of the slide. Label them properly. Permanent slides are now ready for microscopic observation.

Micrometry (Microscopic Measurements)

This is a method employed to make measurements of microscopic objects like bacteria fungi, yeast cells, fibres, pollen grains, chromosomes etc. Two objects are necessary for microscopic measurements – an ocular micrometer and a stage micrometer. Ocular micrometer is a circular glass disc which has fine graduations engraved on one surface. The fine graduations can be seen only with the help of a microscope. The graduations engraved on the ocular micrometer are arbitrary though they are uniformly placed. Hence they have to be calibrated with the stage micrometer. A stage micrometer is essentially a glass slide with graduations. But these are definitely placed. Each line is 0.01 µm (10 micrometer) apart.

What is Calibration? An ocular micrometer by itself cannot be used for measurement because the lines are arbilary. Whereas in a stage micrometer the lines are 10 micrometers apart. In which case why use an ocular micrometer. One can use a stage micrometer placed on the stage for microscopic measurements. But if the stage of the microscope already has a stage micrometer it will not be possible to keep any other microscopic object (on a slide) for the purpose of observation. Hence the lines of ocular must be made to coincide with the stage micrometer and the distance between the lines of ocular known in terms of micrometers (µ) This is known as calibration.

After calibration there is no need to use the stage micrometer (provided one is using the same microscope and same magnification lenses).

How Calibration is done? The ocular micrometer disc is to be placed between the lenses of the eye piece. In order to do this remove the eye piece tube from the microscope, carefully unscrew the top lens, place the micrometer disc on a metal diaphragm below and screw the lens back. Replace the eyepiece tube on the microscope. In some microscopes the micrometer disc can be simply inserted at the bottom of the eye piece.

To calibrate the ocular micrometer for a given objective (45, 10, 60, or 100 x) it is necessary to super impose the scales of ocular and stage micrometers. After placing the ocular micrometer disc in the eye piece, place a stage micrometer on the stage and focus by using the coarse and fine adjustment knobs. Both the scales will be in sharp focus. Now carefully adjust the slide micrometer in such a way that the lines of both coincide as shown in the figure. Here seven ocular divisions match up with one stage micrometer division of 0.01 µm (10mm). In other words the value of 7 ocular divisions in 0.07/7 = 0.00143 mm. Since there are 1000 mm in 1 millimeter, each division in the ocular is 1.43 µm apart. After this value is known, the stage micrometer may be replaced with a slide of material to be measured. Calibration has to be done for each objective separately because the magnifications are different. Calibrate the ocular micrometer with the stage micrometer with 10 x 45 x 60 x and 100 x (oil immersion). A few examples of calibrating calculations are given below.

Calculations :

(a) 4 divisions of stage micrometer line up with 20 divisions of the ocular micrometer. What is the distance between each ocular division (OD) OD = 4/20 x 0.01 = 0.002 or 2 µm.

(b) 50 divisions of ocular are equal to 75 divisions of stage micrometer. What is the distance between each OD. OD = 75/50 x 0.01 = 0.014 or 15 µm.

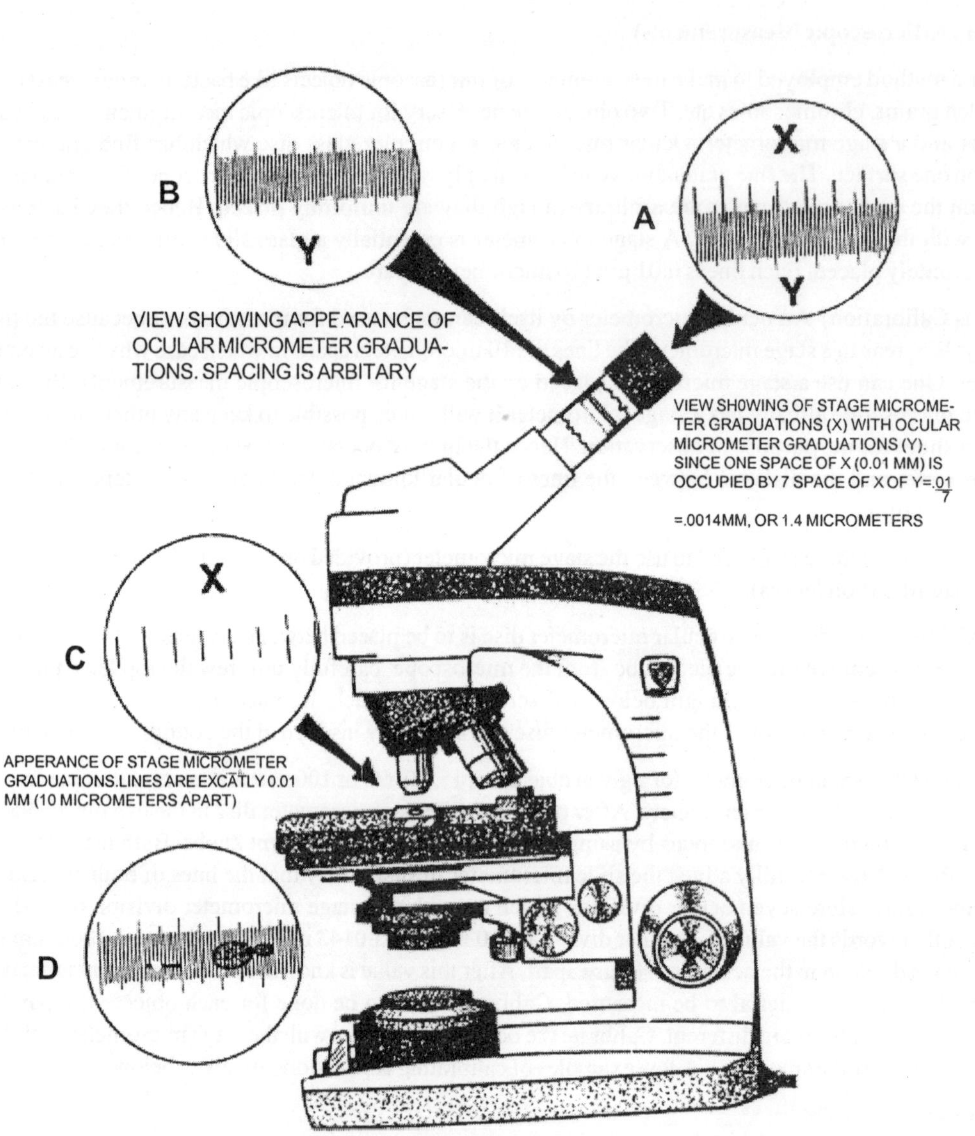

Fig 15.1 Calibrations of ocular micrometer

MEIOSIS

Meiosis or reduction division occurs in higher plants and animals at the time of gametogenesis. Meiosis consists of the following stages. Meiosis I (First meiotic division) and meiosis II (Second meiotic division). Between the two is interkinesis.

Meiosis I has the following stages

Interphase

Prophase I

(a) Proleptolene (Proleptonema)
(b) Leptotene (Leptonenna)
(c) Zygotene (Zygonema)
(d) Pachytene (Pachynema)
(e) Diplotene (Diplonema)
(f) Diakinesis
2. Prometaphase
3. Metaphase I
4. Anaphase I
5. Telophase I.

Meiosis II has the following stages.

1. Prosphase II
2. Metaphase II
3. Anaphase II
4. Telophase II

The characteristic cytological features for the identification of these stages are given below.

Interphase

1. Nucleus appears to be resting, but it is highly active as it prepares the nucleus for division.
2. Synthesis of RNA and DNA will be maximum
3. Nuclear membrane intact, nucleolus can be seen clearly.

PROPHASE

Proleptotene

1. Nucleus enlarges in size
2. Chromosomes are thin and hazy (except – for sex chromosomes)
3. Chromosomes appear double stranded.

Leptotene

1. Chromosomes are distinct, long and less coiled.
2. Chromosomes appear as threads with regularly placed beads or granules called chromomeres
3. The homologous chromosomes are associated at their tips where they are attached to the nuclear membrane
4. Leptotene chromosomes may be uniformly distributed or localized into boquets.
5. In some cases chromosomes may form tangled mass known as synzetic knot.
6. The centriole duplicates and each daughter centriole migrates to the opposite poles of the cell (in animal cells)

Zygotene

1. Chromosomes are shorter and thicker
2. Lengthwise pairing of the homologus chromosomes begins as they are attracted towards each other
3. Of the two homologues one is paternal and the other maternal. These are called bivalents.
4. Each homologus chromosome has two threads i.e. chromatids. In each bivalent there will be four chromatids.

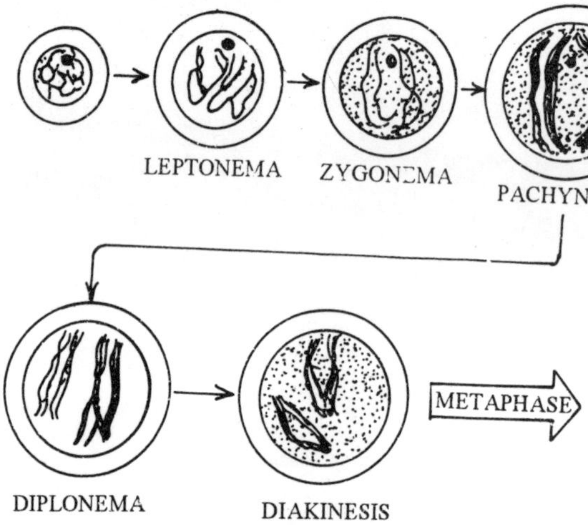

Fig 15.2 Stages in Prophase I of meiosis

Pachytene

1. Chromosomes of a bivalent are associated in tetrad (four chromatid) condition. Each chromatod pair is united by a centromere.
2. Due to longitudinal contraction pachytene chromosomes appear as thick threads.
3. The two bivalents twist around each other as relational coils (the coil of one twining with the other in opposite direction.
4. The chromatids also twist (coiling) resulting in tension.
5. Crossing over or chiasma takes place at this stage between non sister chromatids (opposing chromatids)
6. Two chromatids will be always out of perview of cross over.
7. Crossing over results in breakage and reunion of segments of non sister chromatids. After a break the same segments do not reunite. The chromatid which breaks at one point will unite with the segment of opposing nonsister chromated.
8. Crossing over may occur at 1, 2, or 3 points depending on the length of the bivalents.
9. Nucleolous remains unroganised. It may increase in size. It is found to be associated with certain chromosomes. These are called nucleolar organizing chromosomes.

(Note : Only details which can be seen under a light microscope are given. For example synaptonemal complex is organised at this stage. But this can be seen only under an electron microscope).

Diplotene

1. Homologous chromosomes start separating (repulsion), but are held together at one or more points called Chiasmata
2. Chiasmata are clearer in diplotene than in pachytene because of increased condensation of chromosomes.
3. Chiasma may be formed at the ends (terminal chiasma) or in the middle (interstitial chiasma) of the chromosomes.
4. Terminalization or uncoupling of chromosomes begins due to repulsion between homologous chromosomes.

Diakinesis

1. This is the final stage of prophase I.
2. Bivalent chromosomes are very thick and lightly contracted.
3. Bivalents are distributed more towards the periphery of the nucleus.
4. Terminal chiasma may be retained maintaining contact between homologous chromosomes.
5. Nucleolous detaches from the chromosome and finally disappears.

Prometaphase I

1. Nuclear membrane disappears.
2. Centrioles are joined by spindles which develop from tubules (in animal cells)
3. Chromosomes reach their maximum contraction

Metaphase I

1. Spindle becomes clearly visible and becomes more pronounced
2. Bivalent chromosomes arrange themselves on the equatorial plate.
3. Centromeres of the bivalents lie towards the poles while the arms lie towards the equator
4. Centromere is attached to the spindle fibres. These are called chromosomal fibres.

Anaphase I

1. Each homologous chromosome of each bivalent moves towards the opposite pole.
2. Homologues do not separate in all instances immediately, while shorter chromosomes can be seen separated, longer chromosomes will still be in contact. (they take a longer time to separate).
3. Chromosomes assume 'V' shape as they are pulled towards the opposite poles.
4. The centromere does not divide.

Telophase I

1. Chromosomes reach respective poles, spindles persist for some time
2. Chromosomes undergo despiralization, becoming indistinct
3. Nucleolous reappears, so also the nuclear membrane. In some cases nucleolous may not reappear at this stage.
4. Cell wall is formed in plants in the form of a plate. In animals cells cell membrane is formed by furrowing.
5. In some of the plant cells cytokinesis will not take place at this stage.

Interkinesis

1. The daughter nuclei or daughter cells undergo a resting period
2. It is of a very short duration

Meiosis II

Second meiotic division is essentially a mitotic division in which both the chromosomes and cell divide.

Prophase II

1. Centrioles divide into two and two pairs of centrioles are formed. They migrate to the opposite poles (in animal cells)

2. The microtubules arrange themselves at right angles to the previous plane.
3. The arms of the chromosome remain separate and there is no relational coiling.
4. Nucleolous as well as nuclear envelop disappear with the formation of achromatic figure.

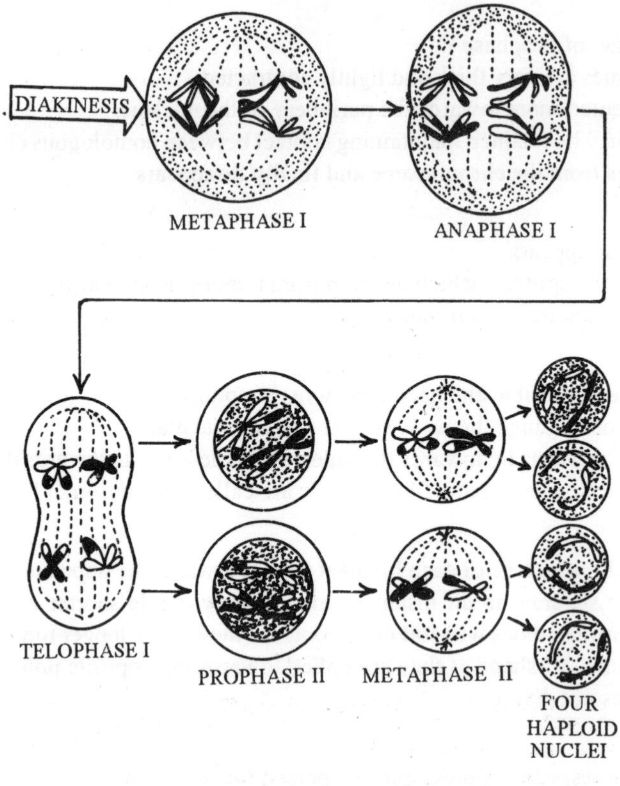

Fig 15.3 Stages in meiosis (after prophase)

Metaphase II

1. It is of very short duration
2. Chromosomes with their chromatids arrange themselves on the equatorial plane.
3. The centromeres lie on the equator, while the arms are directed towards the poles.

4. The centromere divides, thus separating the two chromatids which eventually become separate chromosomes
5. Centromere is attached to the chromosomal spindle fibres in the achromatic figure.

Anaphase II

1. The daughter chromosomes start their migration towards the opposite poles.
2. Chromosomes are not very thick and short like in analphas I, but are like the anaphase daughter chromosome of mitosis.

Telophase II

1. The chromosomes having reached the opposite poles uncoil and become less distinct.
2. Nuclear membrane and nucleolous reappear resulting in the information of a daughter nucleus.
3. The nucleus gets into the interphase state.
4. The nuclei are haploid.

Cytokinesis

1. Wall formation begins either as a furrow or as a plate and two daughter cells are formed.
2. At the end of meiosis four daughter cells are formed.
3. In some cells cell wall would not have been formed after meiosis I as a result four nuclei are formed. Cell walls are now laid down resulting in the formation of four cells.

MITOSIS

Mitosis occurs in the somatic cells of multicellular organisms and it is the chief means of multiplication of cells contributing to the increase in size oft the individual.

For the convenience of study, mitosis may be divided into two stages – karyokinesis and cytokinesis. Karyokinesis is again divided into the following phases – a) Interphase, b) prophase c) metaphase d) anaphase and e) telophase. The following are the characteristics of these stages.

Interphase

1. No morphological changes are visible in the nucleus.
2. It was originally thought to be a resting phase between two nuclear divisions.
3. Recent studies have indicated that interphase is the most active as thesynthesis of RNA and DNA takes place in order to prepare the cell for division
4. The interphase nucleus is large and conspicuous, so also the nucleolus. (Details of G, S and S_2 phase are not given as they are theoretical observations)

Prophase

1. Cells become spheroidal with increase in its viscosity and refractivity (in animal cells)
2. Chromonema starts condensing and ultimately towards the end of prophase thick threads called chromosomes are formed. Chromosomes contact up to $1/25^{th}$ of their original length. Specific number of chromosomes are formed.
3. Each chromosome longitudinally splits into two identical chromatids connected by a centromere
4. The two chromatids are coiled round each other and cannot be easily separated
5. Nuclear matrix begins to accumulate on the chromatids and the chromosomes assume a hazy outline.
6. Centrioles which have undergone duplication during interphase and situated above the nucleus start moving in opposite directions (in animal cells)
7. Astral rays and spindle fibres (amphiaster) appear. But amphiaster is not visible at this stage as it does not take stain (in animal cells)
8. Nucleolous decreases in size and finally disappears.
9. Towards the end of prophase, nuclear membrane disappears.

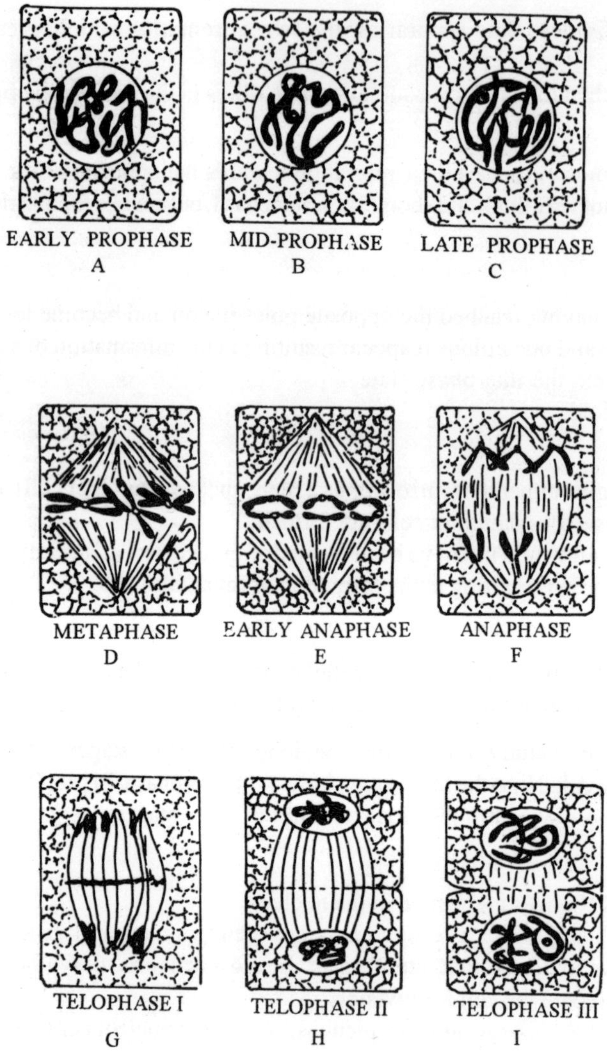

Fig 15.4 Stages of mitosis in a plant cell

Prometaphase

1. Nuclear envelop totally gets disentegrated

Metaphase

1. Duration is short
2. In early metaphase highly condensed and shortened chromosomes arrange themselves radially on the equatorial plane. It is called the metaphase plate or the equatorial plate
3. Chromosomes lie at the equator with the centrioles pointing towards the poles attached to the spindle fibres (in animal cells)
4. In plant cells chromosomes are irregularly scattered on the equatorial plate. Smaller chromosomes lie at

the centre, larger chromosomes towards the peripherey.

5. Chromatids become separated from each other
6. Centromere divides so that the chromatids get separated each having its own centromere.

Anaphase

1. After the division of the centromere, the original number of chromosomes gets doubled
2. Exactly half of them will begin to migrate to one pole and the other half to the other pole.
3. Spindle fibres are distinct at this stage. They are of three kinds (a) Chromosomal spindle fibres attached to the chromosomes, (b) Spindle fibres attached to the asters known as continuous spindle fibres and (c) spindle fibres between departing chromosomes called interzonal spindle fibres (in animal cells)
4. In plants all the spindle fibres are said to be anastral as they are formed without centrioles or asters.
5. During the polar migration chromosomes appear as rod shaped 'L' shaped or V shaped 'L' shaped chromosomes are called heterobrachial (unequal arms) while V shaped chromosomes are called homobrachial (in equal arms).
6. The some instances some chromosomes do not move and remain at the equatorial plate only. They are called laggards and are without centromere.
7. Usually sex chromosomes migrate slowly (generally found in animal cells)
8. At the end of anaphase clusters of chromosomes appear at each pole of the cell.

Telophase

1. Chromosomes begin to uncoil into their strands and become distinct
2. Nucleolus begins to appear as the nucleonemal strands become thicker
3. Spindles disappear
4. Asters lose their fibres and are reconverted to centrioles (in animal cells)
5. Nuclear membrane reappears forming a regular nucleus.

Cytokinesis

1. In plants cell plate is formed between two groups of chromosomes
2. Cell plate is formed at the centre and growing on either side joins the cell wall.
3. In animal cells cytokinesis is by furrowing.